T0074428

Atmospheric Evolution on Inhabited and Lifeless Worlds

As the search for Earth-like exoplanets gathers pace, in order to understand them, we need comprehensive theories for how planetary atmospheres form and evolve. Written by two well-known planetary scientists, this text explains the physical and chemical principles of atmospheric evolution and planetary atmospheres, in the context of how atmospheric composition and climate determine a planet's habitability. The authors survey our current understanding of the atmospheric evolution and climate on Earth, on other rocky planets within our Solar System, and on planets far beyond. Incorporating a rigorous mathematical treatment, they cover the concepts and equations governing a range of topics, including atmospheric chemistry, thermodynamics, radiative transfer, and atmospheric dynamics, and provide an integrated view of planetary atmospheres and their evolution. This interdisciplinary text is an invaluable onestop resource for graduate-level students and researchers working across the fields of atmospheric science, geochemistry, planetary science, astrobiology, and astronomy.

David C. Catling is a Professor in Earth and Space Sciences at the University of Washington, Seattle, who studies planetary surfaces, atmospheres, and habitability. He actively participates in the research of NASA's Astrobiology Institute and is the author of *Astrobiology: A Very Short Introduction* (2013). He has taught courses in planetary atmospheres, planetary geology, astrobiology, and global environmental change at undergraduate and graduate levels. He was also an investigator for NASA's Phoenix Mars Lander, which successfully operated in the arctic of Mars during 2008.

James F. Kasting is an Evan Pugh Professor of Geosciences at The Pennsylvania State University, and an acknowledged expert on atmospheric and climate evolution. He is the author of the popular book, *How to Find a Habitable Planet* (2010) and coauthor of the introductory textbook, *The Earth System* (3rd edn, 2009). Dr. Kasting has received numerous awards, including that of Fellow of the American Geophysical Union, the Geochemical Society, the International Society for the Study of the Origin of Life (ISSOL), the American Academy for the Advancement of Science, and the American Academy of Sciences. He received the Oparin Medal from ISSOL in 2008, and the Stanley Miller Medal from the National Academy of Sciences in 2016.

Atmospheric Evolution on Inhabited and Lifeless Worlds

David C. Catling
University of Washington

James F. Kasting
Pennsylvania State University

CAMBRIDGE
UNIVERSITY PRESS

CAMBRIDGE
UNIVERSITY PRESS

University Printing House, Cambridge CB2 8BS, United Kingdom

One Liberty Plaza, 20th Floor, New York, NY 10006, USA

477 Williamstown Road, Port Melbourne, VIC 3207, Australia

4843/24, 2nd Floor, Ansari Road, Daryaganj, Delhi – 110002, India

79 Anson Road, #06–04/06, Singapore 079906

Cambridge University Press is part of the University of Cambridge.

It furthers the University's mission by disseminating knowledge in the pursuit of
education, learning, and research at the highest international levels of excellence.

www.cambridge.org
Information on this title: www.cambridge.org/9780521844123

© David C. Catling and James F. Kasting 2017

First published 2017

Printed in the United Kingdom by TJ International Ltd. Padstow Cornwall

A catalogue record for this publication is available from the British Library.

Library of Congress Cataloging-in-Publication Data
Names: Catling, David (David Charles) | Kasting, James F.
Title: Atmospheric evolution on inhabited and lifeless worlds / David C. Catling,
University of Washington, James F. Kasting, Pennsylvania State University.
Description: Cambridge : Cambridge University Press, 2017. | Includes bibliographical
references and index.
Identifiers: LCCN 2016030819 | ISBN 9780521844123 (hardback : alk. paper)
Subjects: LCSH: Atmosphere. | Atmosphere, Upper. | Planets–Atmospheres. |
Geochemistry.
Classification: LCC QC861.3 .C38 2017 | DDC 551.5–dc23 LC record available at
https://lccn.loc.gov/2016030819

ISBN 978-0-521-84412-3 Hardback

Contents

Preface

For millennia, philosophers have offered mere opinions about whether life exists beyond the Earth, but amazingly it will soon be possible to replace such conjectures with data. In other words, we stand on the brink of solving the question of "are we alone?" The answer may come from the spectrum of an exoplanet, from rocks on Mars, or from some unexpected source. In any case, we now know that our galaxy contains billions of exoplanets, some of which may be inhabited. This problem focuses our attention on what makes a planet habitable. When we look at Earth, Mars and Venus, we see that an atmosphere – through composition and climate – plays a critical role in distinguishing lifeless from inhabited planets. Thus, the topic of *atmospheric evolution* is essential. We need to know where atmospheres come from, how atmospheres remain stable, how the mixture of gases in an atmosphere changes over billions of years from the origin of a planet to its current or future state, and whether an atmosphere can provide a climate conducive to life.

We have been working on various aspects of atmospheric evolution for two (DCC) and four decades (JFK), respectively. Both of us have been fortunate in receiving continual support from NASA's Exobiology Program and the NASA Astrobiology Institute. We have previously published reviews on relevant topics, such as the physics and chemistry of atmospheres (Catling, 2015) (which we expand in Chapters 1–5), the origin and habitability of the Earth (Kasting and Catling, 2003) (which we bring up-to-date in Chapters 6–11), oxygenation of Earth's atmosphere (Catling, 2014; Kasting, 2013) (which is newly reviewed in Chapter 10), astrobiology (Catling, 2013) (a theme throughout this book), and searching for habitable planets (Kasting, 2010) (which is the topic of Chapter 15). Our reviews have been at different levels, and, for the researcher, useful information has remained very scattered across the scientific literature. Such

dispersal arises because aspects of astronomy, geology, geochemistry, and atmospheric science all contribute to the formation and evolution of atmospheres. On a planet with life, we must add biology also. Consequently, we wrote this book to gather our knowledge of planetary atmospheres into a framework of atmospheric evolution and habitability. Our intended reader is any interested researcher or graduate student.

In this book, we are concerned with inhabited planets such as Earth, as well as lifeless ones. But the Earth itself changed from being lifeless to inhabited. We also consider other inhabited planets in the context of possible life on early Mars or potential life on exoplanets that might be remotely detectable. As mentioned above, whether a planet is inhabited or lifeless is a key motivation to study atmospheric evolution. All of these considerations led us to the title of this book.

The book has three parts. In Part I, we concisely describe principles of atmospheres (structure, radiation, chemistry, and motions) that are needed to appreciate atmospheric evolution and habitability. Part II describes origins of atmospheres and the evolution of Earth's atmosphere and climate. Finally, Part III turns to other worlds, including Mars, Venus, outer planet satellites, and exoplanets.

Because atmospheric evolution ranges over a vast swath of disciplines that stretches the limits of our expertise, colleagues have kindly provided essential help. In alphabetical order, we thank the following people for commenting on individual chapters: Dorian Abbot, Don Brownlee, John Chambers, Nick Cowan, Colin Goldblatt, David Grinspoon, Paul Hoffman, Dick Holland, Edwin Kite, Conway Leovy, Ralph Lorenz, Vikki Meadows, Tyler Robinson, Adam Showman, Jon Toner, and Steve Warren. We also thank Beth Tully for her patience in helping us with the diagrams. We owe special thanks to a couple of departed friends on our list who are greatly missed. These colleagues always generously shared their ideas and unfailingly provided eager encouragement. They were Conway Leovy (1933–2011) and Heinrich ("Dick") Holland (1927–2012). We would also like to acknowledge James C.G. Walker, who was a mentor to JFK, and who in 1977 wrote a previous book titled *Evolution of the Atmosphere* that has guided thinking on this topic for about the past 40 years. Finally, we thank Vince Higgs at CUP for enormous patience in waiting for us to complete this book.

PART I
Principles of Planetary Atmospheres

1 The Structure of Planetary Atmospheres

Interactions between the atmosphere, ocean, solid planet, biosphere, and space are responsible for the evolution of the Earth's atmosphere. Other planets have simpler global systems that merely consist of an atmosphere surrounding a rocky surface without a liquid ocean, or, in the case of gas giants, atmospheres that thicken inwards towards higher-density layers and cores. In the first five chapters, we describe general physical and chemical principles that apply to all or most atmospheres. There are many excellent introductory texts on atmospheric physics and chemistry, e.g., Walker (1977), Hartmann (1994), Houghton (2002), Hobbs (2000), Wallace and Hobbs (2006), and Andrews (2010). These books focus on Earth's atmosphere. Our approach here is to discuss the fundamentals of atmospheres using examples from a variety of planetary atmospheres. Thus, Chapters 1–5 are an introduction to the concepts of planetary atmospheres, often with a view to how the basics apply to issues of atmospheric evolution. We refer back to these fundamentals in subsequent chapters.

Some other books also discuss planetary atmospheres. Ingersoll (2013) is an excellent readable primer. Chamberlain and Hunten (1987) has a particular emphasis on upper atmospheres, Yung and Demore (1999) focus on photochemistry, Pierrehumbert (2010) has broad insightful coverage with an emphasis on radiation, while Sanchez-Lavega (2010) is a valuable general reference text for planetary atmospheres.

The key uniqueness of this book is that we emphasize atmospheric evolution, which, needless to say, we consider the most interesting aspect of planetary atmospheres!

1.1 Vertical Structure of Atmospheres

1.1.1 Atmospheric Temperature Structure: An Overview

A planet's atmosphere, through its composition and greenhouse effect, controls climate, which is weather that has been averaged over some period of time such as a month, season, year, or longer. The mean surface temperature is a key parameter of importance for habitability. If the surface temperature is not suitable for liquid water, life as we know it will probably not exist there, except possibly hidden in the subsurface.

The same principle applies to extreme places on the Earth. For example, the coldest place on Earth's surface is Vostok, Antarctica, with temperatures ranging from –14 °C (259 K) to a record extreme of –89.2 °C (184 K) (Turner et al., 2009). Consequently, Vostok is frozen and lifeless. In part, the coldness of Vostok arises from its height of 3450 m above sea level. How the air temperature varies with altitude is one of the most basic features of planetary atmospheres, the *atmospheric structure*, which is our starting point for this book.

The Earth's atmosphere provides the nomenclature for describing vertical regions of planetary atmospheres. Terminology for Earth's atmospheric layers was developed in the early 1900s (Lindemann and Dobson, 1923; Martyn and Pulley, 1936). Because of this historical origin, the same terminology only approximately applies to other planetary atmospheres.

Earth's atmosphere consists of five layers of increasing altitude (Fig. 1.1). Going upwards, new layers begin where there is a temperature inflection.

(1)   The *troposphere* goes from the surface to the *tropopause*, which lies at a height of 11 km (0.2 bar) in the US Standard Atmosphere but varies in altitude from ~8 km at the poles to ~17 km in the tropics. The global average tropopause pressure is 0.16 bar (Sausen and Santer, 2003), varying 0.1 to 0.3 bar from tropics to pole (Hoinka, 1998). *Tropos* is ancient Greek for the 'turning' of convection and *pause* is Greek for "stop."

(2)   The *stratosphere* (from Greek *stratus* for "layered") goes from the tropopause to the *stratopause* at ~50 km altitude, where the air pressure is about 100 Pa (1 mbar).

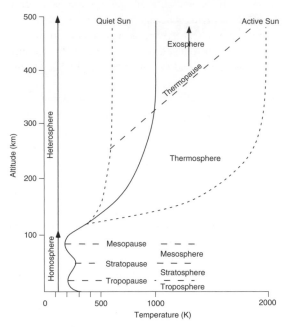

Figure 1.1 The nomenclature for vertical regions of the Earth's atmosphere, shown schematically.

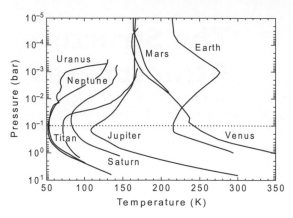

Figure 1.2 Thermal structure of the atmospheres of various planets of the Solar System. The dashed line at 0.1 allows you to see the feature of a common tropopause near ~0.1 bar for the thick atmospheres, despite the differences in atmospheric composition. See Robinson and Catling (2014) for sources of data.

(3) The *mesosphere* extends from the stratopause to the *mesopause* at ~85 km altitude, where the air pressure is about 1–0.1 Pa (0.01–0.001 mbar)

(4) The *thermosphere* goes from the mesopause to the *thermopause* at about ~250–500 km depending on varying ultraviolet from the Sun. Above the thermopause, the atmosphere becomes isothermal.

(5) The *exosphere* lies above the thermopause and joins interplanetary space. Unlike the other layers that are defined by the temperature profile, the exosphere is where collisions between molecules are so infrequent that they can usually be neglected. The *exobase* is the bottom of the exosphere and nearly coincides with the thermopause.

The terminology developed for Earth's lower atmosphere depends upon the presence of the ozone layer in the stratosphere. Ozone absorbs ultraviolet (UV) sunlight, which causes temperature to increase with height above the troposphere and defines the stratosphere. Other planets, such as Mars, do not have UV absorbers to induce a stratosphere, so the geocentric nomenclature for atmospheric layers breaks down (Fig. 1.2). However, we generally find analogs in other planetary atmospheres for a troposphere, mesosphere, thermosphere, and exosphere. On Titan and the giant planets, absorbers of shortwave sunlight (UV, visible, and near-infrared) produce stratospheres.

Convection is the key process in tropospheres. On Earth, radiation heats the surface, so air is warmed near the ground and lifted upwards by buoyancy. Consequently, air parcels convect to places of lower pressure where they expand and cool. The net effect of lofted parcels that cool, and sinking ones that warm, is to maintain an annual average temperature decrease from Earth's warm surface to the cold upper troposphere of about 6 K km^{-1} when globally averaged (see Sec. 1.1.3).

Earth's troposphere contains ~80% of the mass of the atmosphere and this mass, along with the composition of the air, renders the troposphere fairly opaque to thermal-infrared (IR) radiation emanating from the planet's surface. In general, somewhat below the tropopause, atmospheres become semi-transparent to thermal-IR radiation. Consequently, transfer of energy by radiation in the upper troposphere replaces convection as the means of upward energy transfer at a *radiative–convective* boundary. Efficient emission of thermal-IR radiation to space accounts for the temperature minimum at the tropopauses of many planetary atmospheres, which occurs above the radiative–convective boundary for atmospheres with stratospheres.

For planets with thick atmospheres, the tropopause temperature minimum occurs where the air has thinned to roughly ~0.1 bar pressure (Fig. 1.2). Remarkably, this rule applies to Earth, Titan, Jupiter, Saturn, Uranus, and Neptune, and the mid-to-high latitudes of Venus (Tellmann *et al.*, 2009), despite vast differences in atmospheric composition. This commonality occurs because the broadband opaqueness to thermal-IR in upper troposphere is pressure-dependent with similar scaling – varying approximately with the square of the pressure – despite the differences in atmospheric composition. Since all these atmospheres have strong and roughly similar opaqueness (within the same order of magnitude) to thermal-IR at moderately deep pressures, such as 1 bar, the weak square-root dependence of pressure on the upwardly decreasing opacity ensures that the tropopause minima are all near 0.1 bar, within a factor of 2 or so (Robinson and Catling, 2014). Thus, a "0.1 bar tropopause minimum rule" could apply to exoplanet atmospheres, provided that they have temperature minima at the base of stratospheres. Of course, such a rule can't apply to atmospheres without stratospheric absorbers and tropopause temperature minima.

In Earth's stratosphere, ozone (O_3) warms the air by absorbing solar UV over a wavelength range of about 200–300 nm. Stratospheric warming also occurs on the giant planets and Titan because aerosols (fine, suspended particles) and methane absorb in the UV, visible, and near-IR. Warming creates an *inversion* – a vertical region of the atmosphere where temperature increases with height (Fig. 1.2). Cooler air is inhibited from rising into warmer air, so the air becomes stratified and mixes slowly in the vertical. Mars and Venus, which have atmospheres containing 96% CO_2, do not have strong shortwave absorbers above their tropospheres and so lack inversions in the global average. However, an inversion is not required to prevent convection. We will see later that convection effectively ceases when the rate at which temperature decreases with altitude becomes smaller than a critical value (Sec. 1.1.3). For example, the atmosphere on the early Earth, which lacked oxygen and an ozone layer, would still have had relatively slow mixing above a troposphere, with a stratosphere of more stable air.

The terrestrial stratosphere is very dry. Atmospheric motions control the entry of tropospheric water vapor into the stratosphere, and only ascending air in the tropics can break through into the stably stratified stratosphere. The tropical tropopause is the coldest part of the troposphere and freezes out most of the water that passes through it. This tropical tropopause "*cold trap*" limits stratospheric water vapor to about 3–4 *parts per million by volume* (ppmv), meaning that water makes up 3–4 molecules out of every million air molecules.

Earth's mesosphere has the lowest air temperature. Here ozone concentrations have declined while CO_2 still radiates to space efficiently. The mesopause forms the lower boundary of the *thermosphere* and generally occurs at or a little below the ~ 0.1 Pa pressure level in all planetary atmospheres in the Solar System. Mesopauses are extremely cold because the atmospheric density becomes sufficiently low around this altitude ($<10^{14}$ molecules cm^{-3}) that energy-exchanging collisions between molecules become less frequent than emission of photons.

Consequently, a molecule excited by absorption of a photon or a rare collision loses internal energy by radiation rather than kinetic energy through collisions. The altitude where this happens is the level of *radiative relaxation* (Curtis and Goody, 1956; López-Puertas and Taylor, 2001).

Above the mesopause, Earth's thermosphere is extremely thin (e.g., $<10^{12}$ molecules cm^{-3} above 120 km, compared to $\sim 10^{19}$ cm^{-3} at the surface) and responds rapidly to variations in solar radiation shorter than ~ 100 nm in wavelength – the extreme ultraviolet (EUV). Daily temperature changes are very large here. Similar EUV absorption occurs high up in all planetary atmospheres, where EUV, x-rays, and chemical reactions break up molecules into atoms. The mean free path (which is inversely proportional to density) becomes so big that the combination of a continuous flux of ionizing radiation and infrequent collision between ions and electrons produces an *ionosphere* – on Earth, a layer of electrons and ions from the mesosphere into the exosphere.

The rate of heating of the thermosphere is about one millionth that of Earth's surface but the average temperature is ~ 1000 K at 250 km altitude versus 288 K at the Earth's surface (Fig. 1.1). Temperature measures the average kinetic energy or velocity of atoms or molecules and in the thermosphere the high energy of UV photons absorbed is converted to kinetic energy. Atoms and ions cannot vibrate or rotate and thus are not good radiators in the infrared, unlike the molecules of the lower atmosphere. So kinetic energy generated by absorption of UV or collisions with hot (i.e. fast-moving) electrons cannot be transferred to molecular vibrations. Consequently, the high thermospheric temperature reflects the inefficiency of removing heat. Energy must be removed somehow, however, and this is achieved by conduction when atoms and ions infrequently collide and transfer heat downwards from the hot thermosphere to the cold mesopause.

Atmospheric composition makes a large difference to the temperature of a thermosphere. Because the atmospheres of Venus and Mars have much more CO_2 (96%) than Earth's atmosphere (0.04%), radiative cooling of their upper atmosphere is more pronounced. Despite heating from EUV, the upper atmospheres are relatively cold: ~ 270 K for Mars and ~ 250 K for Venus. To explain Venus' cold thermosphere despite its location much closer to the Sun, we must also consider the effect of photochemical reactions, which are described in Ch. 3.

At the exobase, collisions are so infrequent that energetic atoms can escape into space. Today, only H and He escape from Earth significantly. The other molecules and atoms are sufficiently massive that they are bound by

Earth's gravity and only a negligible fraction of them ever attain escape velocity. However, fast escape from past, present and future atmospheres is extremely important in understanding the evolution of planetary atmospheres in our own Solar System and those of exoplanets (Ch. 5).

In Fig. 1.1, we also divide the Earth's atmosphere into two basic compositional layers, a *homosphere* below ~ 100 km altitude, where the bulk constituents of the atmosphere are homogeneously mixed, and a *heterosphere* above this level. This nomenclature is based on vertical mixing rather than on temperature. The air in the heterosphere is sufficiently thin that molecular diffusion becomes important compared to mixing by turbulence or large-scale winds. In this diffusive regime, molecules separate in altitude according to mass. Gravity, of course, exerts the same acceleration on heavy objects, or atoms, as it does on light ones, as proved by Galileo. But pressure increases downward in the gravitational field, and lighter atoms tend to diffuse toward lower pressure, so the net effect is that in the heterosphere the abundance of heavy species decreases more rapidly with altitude than light ones.

As a result of diffusive separation, atomic oxygen becomes the most abundant constituent between ~ 250–1000 km altitude, helium becomes the most abundant species between ~ 1000–2500 km, and atomic hydrogen dominates above that. The tendency for species to separate diffusively according to mass exists at all altitudes, but in the homosphere the timescale for diffusive separation is longer than the timescale of turbulent mixing, so the latter dominates. For example, near-surface air has a diffusive separation timescale ~ 10^5 years, compared to mixing by weather in a few days.

The diffusive timescale decreases exponentially with altitude as the mean free path increases, and the *homopause* is the transition from turbulent to diffusive mixing, i.e., on Earth at ~ 100 km height,

diffusive separation time = mixing time

Consequently, above the homopause, where diffusive separation is faster than mixing, concentrations of O, H_2, and He increase relative to CO_2, N_2, O_2, and Ar, and the mean molar mass of air decreases.

Diffusive separation occurs in all planetary atmospheres at low pressure. It can be measured by monitoring the distribution of some relatively heavy gas and finding where it drops off relative to lighter gases. Typically, diffusive separation begins to occur within a few orders of magnitude of 0.01 Pa (or 0.1 μbar), as shown in Table 1.1, noting that characteristic motions of each atmosphere are influential in controlling the homopause level (Leovy, 1982b).

1.1.2 Atmospheric Composition and Mass

The vertical structure of atmospheric pressure and density determine where physical and chemical processes take place by affecting the vertical distribution of thermal radiation, which emanates from the planet's surface, and incoming sunlight through wavelength absorption and scattering. The three-dimensional distribution of temperature determines the distribution of pressure, which drives an atmosphere's circulation or wind system. To connect distributions of pressure, temperature, and density, we use two equations: (1) an *equation of state* that relates the three variables, and (2) a *hydrostatic equation* that links pressure to the mass per unit area of overlying atmosphere. The latter equation is, by definition, always approximate, as no atmosphere is ever truly without motion in the vertical direction. But it is an excellent approximation in the lower atmospheres of all known planets and in the upper atmospheres of most of them. We now describe these two fundamental equations.

Table 1.1 Homopause levels. (Sources: Atreya *et al.* (1991), p. 145; Atreya *et al.* (1999).)

Planet	Altitude (km)	Pressure (μbar)	Number density (molecules cm^{-3})
Venus	130–135	0.02	7.5×10^{11}
Earth	~100	0.3	10^{13}
Mars	~130	0.002	10^{10}
Jupiter	~385 above 1 bar level	1	1.4×10^{13}
Saturn	~1140 above 1 bar level	0.005	1.2×10^{11}
Titan	800–850	~0.0006	2.7×10^{10}
Uranus	~354–390 above the 1 bar level	~20–40	$1–2 \times 10^{15}$
Neptune	~586–610 above the 1 bar level	~0.02	10^{13}

1.1.2.1 *The Ideal Gas Equation*

The equation of state applicable to common atmospheric situations is the ideal gas law,

$$p = nkT = (n\overline{m})\left(\frac{k}{\overline{m}}\right)T = \rho\overline{R}T \tag{1.1}$$

Here k is Boltzmann's constant ($k = 1.381 \times 10^{-23}$ J K^{-1}), $\overline{m}$ is the mean mass of molecules in the atmosphere, n is the number of molecules per unit volume or *number density*, $\rho = n\overline{m}$ is the mass density, and $\overline{R} = k/\overline{m}$ is the specific gas constant. The universal gas constant R (8.314 J K^{-1} mol^{-1}) is related to the *specific gas constant* by $\overline{R} = R/\overline{M} = k/\overline{m}$, where we have introduced $\overline{M}$, the mean molar mass of the atmosphere (in kg mol^{-1}). In turn, $\overline{M}$ is related to the mean molecular mass, $\overline{m}$, of individual molecules by $\overline{M} = N_A\overline{m}$, where

N_A is Avogadro's number, 6.022×10^{23} molecules mol^{-1}. The ideal gas law can be written in a variety of forms given in Box 1.1. Deep in the atmospheres of gas giant planets, the ideal gas law will fail because gas molecules are so close together that they begin to attract or repel each other so the gas is no longer "ideal." A modified equation of state is then needed.

The mean molar mass is defined as follows. Planetary atmospheres consist of mixtures of gases, which obey Dalton's Law of partial pressures. Dalton's Law says that each gas exerts a partial pressure p_i proportional to its number density n_i,

$$p_i = n_i kT \tag{1.2}$$

The total pressure of the gas is made up of the sum of partial pressures, $p = \sum p_i$, so it follows that in eq. (1.1)

Box 1.1 Different Forms of the Ideal Gas Law

(1) One basic form of the gas law is:

$$pV = n_m RT \tag{B1}$$

Here, n_m = number of moles in volume V at pressure p and temperature T. $R = 8.314$ J K^{-1} mol^{-1} = **universal gas constant**. Equation (B1) incorporates **Boyle's Law** (pV = const.), **Charles' Law** (V/T = const.), and **Avogadro's hypothesis** ($n_m = pV/RT$, i.e., gas samples at the same p and T contain same number of moles n_m).

(2) Rather than moles, we can use the total number of molecules, N. The number of molecules per mole is Avogadro's number, N_A, so $N = n_m N_A$. The equation of state is then

$$pV = \frac{N}{N_A}RT, \quad \text{or} \quad pV = NkT \tag{B2}$$

Here we have introduced $k = R/N_A$ = Boltzmann's constant = 1.381×10^{-23} J K^{-1}.

(3) A third way of expressing the ideal gas law is to use molecules per unit volume:

$$p = nkT \tag{B3}$$

where $n = N/V$ is the number density of molecules . (In SI units, n is in molecules m^{-3}, but frequently molecules cm^{-3} is used in atmospheric science, which requires other consistent units.)

(4) A fourth expression uses mass density rather than number density. Density = mass/volume = $\rho = M/V = N\overline{m}/V$, where $\overline{m}$ = mean molecular mass and N = number of molecules in volume V. Hence

$$p = \frac{\rho kT}{\overline{m}} \tag{B4}$$

(5) A fifth expression (commonly used in studies of planetary atmospheres) can be derived from eq. (B4), by substituting for $k = R/N_A$ and noting that $N_A\,\overline{m}$ = molar mass, $\overline{M}$. This gives:

$$p = \frac{\rho RT}{\overline{m}N_A} \Rightarrow p = \frac{\rho RT}{\overline{M}} \Rightarrow p = \rho\overline{R}T \tag{B5}$$

Note that $\overline{M}$, the *mean molar mass*, is sometimes called the *mean molecular weight*. Frequently, $\overline{R}$ is described as the *gas constant*, and should not be confused with the *universal gas constant*, R [J K^{-1} mol^{-1}]. The *specific gas constant* $\overline{R}$ is per unit mass [J K^{-1} kg^{-1}] and depends on the gas composition, unlike R.

$$\rho = \sum_i m_i n_i, n = \sum_i n_i, \overline{m} = \frac{\sum_i m_i n_i}{n} \qquad (1.3)$$

where m_i is the mass of the ith molecule. In Earth's lower atmosphere, the mean molar mass, $\overline{M}$, of dry air of 0.02897 kg mol^{-1} and $\overline{R} = R/\overline{M} = 287$ J K^{-1} kg^{-1}.

To show how we use the equations, given Earth's global mean temperature of 288 K and sea-level pressure of 1.013×10^5 Pa, eq. (1.1) gives the air density as 1.2 kg m^{-3} to which we can compare the air density on Mars. If the Martian atmosphere were pure CO_2, we would expect $\overline{M} = 0.044$ kg mol^{-1}, the molar mass of CO_2. But the atmosphere is 96% CO_2 by volume, with

N_2 and Ar making up most of the remainder (Table 1.2), and since N_2 and Ar are lighter molecules than CO_2, $\overline{M} = 43.5$ mol^{-1}, which gives $\overline{R} = R/\overline{M} = 191$ J kg^{-1} K^{-1}. The global average barometric pressure and temperature on the surface of Mars are ~ 610 Pa and ~ 215 K, respectively, so eq. (1.1) gives the average air density on the Martian surface as 0.015 kg m^{-3}, approximately 1% of that on Earth.

1.1.2.2 The Composition of Planetary Atmospheres

We express the amount of a gas in a planetary atmosphere as a *mixing ratio*, f_i, which usually means *volume mixing ratio* (Table 1.2). This is the mole fraction of a gas in a

Table 1.2 Properties of Venus, Earth, and Mars and their atmospheric compositions. Note the large disparity in the atmospheric pressures of the planets when interpreting their atmospheric composition. For example, 3.5% N_2 in the atmosphere of Venus represents 3.3 bars of N_2 (considerably more than the Earth), indicating a planet that is either more volatile-rich than the Earth or (more likely) much more efficiently outgassed. Abbreviation: ppm = parts per million by volume. (Sources: Krasnopolsky and Lefevre (2013); Lodders and Fegley (1998); Pollack (1991); Franz *et al.* (2015).)

Parameter	Venus		Earth		Mars	
Mean surface pressure (bar)	95.6 at a modal radius		1.0 at sea level		0.006	
Mean surface temperature (K)	740		288		215	
Mass relative to Earth (5.97×10^{24} kg)	0.815		1.0		$0.107 \approx 1/9$	
Mean radius relative to Earth (6371 km).	0.950		1.0		$0.532 \approx 1/2$	
Key gases in atmosphere (by volume)	CO_2	$96.5 \pm 0.8\%$	N_2*	78.084%	CO_2	$95.7 \pm 1.6\%$
	N_2	$3.5 \pm 0.8\%$	O_2*	20.946%	N_2	$2.03 \pm 0.03\%$
	SO_2¶	• 150 ± 30 ppm (22–42 km)	H_2O	0.1 ppm–4% (varies)	Ar	$2.07 \pm 0.02\%$
		• 25 ± 150 ppm (12–22 km)	Ar	9340 ppm	O_2	$0.173 \pm 0.006\%$
	Ar	70 ± 25 ppm	CO_2*§	• ~ 280 ppm (pre-industrial)	CO	749 ± 3 ppm
	$^{36+38}$Ar	75 ± 35 ppm		• 400 ppm (year 2015)	H_2O	$\sim 0.03\%$ (varies)
	H_2O¶	30–70 ppm (0–5 km)			He	10 ppm
	CO¶	• 45 ± 10 ppm (cloud top)	Ne	18.18 ppm	H_2	15 ± 5 ppm
		• 17 ± 1 ppm (12 km)	^{4}He	5.24 ppm	Ne	2.5 ppm
	He	12 (+24/-8) ppm	CH_4*	1.8 ppm	Kr	0.3 ppm
	Ne	7 ± 3 ppm	Kr	1.14 ppm	O_3	0–80 ppb
	H_2¶	2.5 ± 1 ppm (50–60 km)	H_2*	0.55 ppm	H_2O_2	0–40 ppb
			N_2O*	~ 320 ppb	SO_2	< 0.3 ppb†
	HCl	0.4 ppm (70 km)	CO*	125 ppb		
	^{84}Kr	50 ± 25 ppb				

* Under varying degrees of biological influence.
† Indicates lack of volcanic outgassing.
§ CO_2 is currently increasing by ~ 2 ppm/yr due to fossil fuel use by humans.
¶ Altitude-dependent.

mixture of gases. By Avogadro's hypothesis, the mole fraction is simply the ratio of the number of molecules N_i of a gas to the total number of molecules N. To summarize:

$$f_i = \frac{N_i}{N} = \text{number fraction} = \frac{p_i}{p} = \text{partial pressure} = \frac{V_i}{V} = \text{volume fraction} = \frac{n_i}{n} \tag{1.4}$$

The relations in eq. (1.4) follow from the ideal gas law (eq. (B3)) and Dalton's Law. Here, V_i is the volume occupied by molecules of the gas in total sample volume, V.

We can also define a *mass mixing ratio*, μ_i, which is ratio of the mass of a particular gas to the total mass of the gas mixture sample, or

$$\mu_i = \frac{N_i m_i}{N \overline{m}} = \frac{m_i}{\overline{m}} \frac{p_i}{p} = \frac{m_i}{\overline{m}} f_i \tag{1.5}$$

Here, we have used eq. (1.4) to relate the mass mixing ratio to the volume mixing ratio.

An awkward exception to the above relationships is the way that meteorologists describe water vapor in the terrestrial atmosphere. The volume mixing ratio of H_2O is usually given in meteorology as the number of H_2O molecules expressed as a ratio with respect to the total number of air molecules in a sample *excluding* H_2O. Similarly, the mass mixing ratio of H_2O is given as the mass of H_2O with respect to the mass of air in a sample *excluding* H_2O. Fortunately, the differences are minor between these definitions and those in eqs. (1.4) and (1.5) because the H_2O amount is generally small. However, we can imagine atmospheres, for example, the *runaway greenhouse* atmosphere of early Venus, where this is not true. So, it is important to be clear about what we mean by "H_2O mixing ratio" in such cases.

What are the types of chemical composition of planetary atmospheres? Sometimes, atmospheres of Solar System bodies are grouped as follows: (i) N_2 atmospheres (Earth, Titan, Triton, and Pluto), (ii) CO_2 atmospheres (Venus and Mars), (iii) H_2-rich giant planets, and (iv) extremely tenuous atmospheres. However, a chemically meaningful categorization for relatively thick atmospheres falls into only two broad categories: *reducing* and *oxidizing*. This classification indicates the overall chemical character of an atmosphere and the form in which elements are likely to exist, e.g., CH_4 versus CO_2 for carbon. With this scheme, Solar System atmospheres group as follows (where the values in parentheses are the surface atmospheric pressures in units of bar (1 bar = 10^5 Pa) from Lodders and Fegley (1998)).
(i) *Reducing atmospheres*: Jupiter, Saturn, Uranus, Neptune, and Titan (1.5).

(ii) *Oxidizing atmospheres*: Earth (1.0), Mars (6×10^{-3}), and Venus (95.6 at the modal radius).
Reducing atmospheres are relatively rich in reducing gases, which are typically hydrogen-bearing gases, such as methane (CH_4), hydrogen itself (H_2), and possibly ammonia (NH_3). An atmosphere with CO would also be reducing and *hot Jupiter* exoplanets (Jupiter-like gas giants that orbit within 0.5 Astronomical Units of their parent stars) can have carbon in the form of CO rather than CH_4 in their upper atmospheres (Moses *et al.*, 2011; Sharp and Burrows, 2007). Billions of years ago, the atmospheres of the very early Earth (Ch. 10) and early Mars (Ch. 12) may also have been reducing, and so these atmospheres have probably switched places in our above classification scheme as they evolved and became oxidizing.

Oxidizing atmospheres are poor in hydrogen-bearing reducing gases and include O_2-rich Earth's and Mars' atmosphere, which is CO_2-rich with $\sim 0.1\%$ O_2 and only 15 ± 5 ppmv H_2 (Krasnopolsky and Feldman, 2001). Venus, as often, is a complex case. Its upper atmosphere is oxidizing, and it even has a very tenuous ozone (O_3) layer at 90–105 km altitude on the nightside (Montmessin *et al.*, 2011). However, in the hot dense Venusian atmosphere below ~ 20 km altitude, sulfur can exist in various redox states (Fegley, 2014; Krasnopolsky and Pollack, 1994) and so the air there could legitimately be described as weakly reducing. Unlike Earth and Mars, O_2 is absent on Venus with <0.3 ppmv abundance.

One useful aspect of classifying atmospheres into reducing versus oxidizing is that the redox character indicates the possible chemistry of aerosols, which are fine particles in atmospheres. In particular, oxidizing atmospheres with sulfur gases tend to produce aerosols of sulfate at altitude (McGouldrick *et al.*, 2011). On Venus, we find clouds of sulfuric acid (H_2SO_4) and there is an analogous but very thin sulfate haze in the Earth's stratosphere at ~ 20–25 km altitude called the *Junge layer* (Crutzen, 1976; Robock, 2000). When early Mars was volcanically active, its atmosphere should also have had sulfate aerosols (Settle, 1979; Smith *et al.*, 2014; Tian *et al.*, 2010). The reducing atmospheres in the Solar System generate hydrocarbon aerosols, which we find on the giant planets (Irwin, 2009; Moses *et al.*, 2004; West *et al.*, 2004; West *et al.*, 2009; West *et al.*, 1991) and Titan (Ch. 14) (Krasnopolsky, 2010; Lavvas *et al.*, 2008a, b; Wilson and Atreya, 2004). Hydrocarbon aerosols were also probably present in the anoxic, weakly reducing atmosphere of the early

Earth (Pavlov *et al.*, 2001; Wolf and Toon, 2010; Zerkle *et al.*, 2012) (see Ch. 9).

Some Solar System bodies have very thin atmospheres or even *exospheres*, which are gases so rarefied that they are effectively collisionless. Tenuous atmospheres can be grouped as follows, with surface pressure given in bars in parentheses.

(i) *N₂-rich atmospheres above surfaces covered in N_2 ice* include Triton ($\sim(14$–$20)\times10^{-6}$), the largest moon of Neptune, and Pluto ($\sim15\times10^{-6}$). These tenuous N_2-rich atmospheres (with minor CH_4) on Triton and Pluto are essentially at the vapor equilibrium with N_2 (and CH_4) ice at prevailing surface temperatures.

(ii) *O_2-rich exospheres and thin atmospheres above moons covered in water ice* include Jupiter's moons Europa ($\sim10^{-12}$–10^{-13} O_2) (Smyth and Marconi, 2006), Ganymede ($\sim10^{-12}$ O_2) (Hall *et al.*, 1998), and Callisto ($\sim10^{-9}$ bar of O_2 with some CO_2) (Carlson, 1999; Cunningham *et al.*, 2015; Kliore *et al.*, 2002; Liang *et al.*, 2005), and Saturn's moons Rhea ($\sim10^{-12}$ O_2 and CO_2) (Teolis *et al.*, 2010) and Dione ($\sim10^{-12}$ O_2) (Tokar *et al.*, 2012). The O_2 comes from the decomposition of surface water ice by radiolysis and sputtering (Johnson *et al.*, 2009). *Radiolysis* is the chemical alteration of ices by charged particles, and *sputtering* is when charged particles collide and eject atoms or molecules. The faster escape of hydrogen compared to oxygen leaves oxygen lingering behind. Charged particles come from *magnetospheres*,

a mass-mixing ratio that varies from about 0.1 ppmv in Antarctica to about 4% in wet tropical regions. Ozone is an example of a species affected by chemical reactions, so that ozone concentrations vary with altitude.

The specific gas constant $\overline{R}$ varies if there are changes in gas concentrations. Generally, this variation is only important in atmospheres in which some species undergoes phase changes. Because of condensable gases, such as water vapor in the atmosphere of Earth, methane in the atmosphere of Titan, or ammonia in Jupiter's atmosphere, some atmospheric literature calculations define a *virtual temperature*. This fictitious temperature has the purpose of allowing the ideal gas law to be satisfied using the gas constant for the dry gas alone. The virtual temperature T_v satisfies

$$T_v = \frac{\overline{R}}{\overline{R}_d}T = \frac{T}{[1-(e/p)(1-\varepsilon)]}, \text{ where } \varepsilon \equiv m_c/\overline{m}_d \quad (1.6)$$

Here, m_c is the molecular mass of the condensable gas, $\overline{m}_d$ is the mean molecular mass for the dry gas, $\overline{R}_d$ is the gas constant for the dry gas, e is the vapor pressure of the condensable species, and p is total pressure (e.g., see Wallace and Hobbs (2006), p. 67). If we define the *specific gas constant for the condensable species* as $R_c = R/M_c$, we can easily derive eq. (1.6). Using Dalton's Law of partial pressures $p = p_d + e$, where p_d is the partial pressure of dry air, the ideal gas law gives the total density as:

$$\rho = \frac{(p-e)}{\overline{R}_d T} + \frac{e}{R_c T} = \left(\frac{p}{\overline{R}_d T} - \frac{e}{\overline{R}_d T}\right) + \frac{e}{R_c T} == \frac{p}{\overline{R}_d T}\left[1 - \left(\frac{e}{p}\right)(1-\varepsilon)\right]$$

which are envelopes of ions around planets controlled by the planetary magnetic fields. The higher density of O_2 on Callisto suggests that its atmosphere is collisional rather than an exosphere (Cunningham *et al.*, 2015).

(iii) *Volcanogenic atmospheres* include SO_2 on Io ($\sim10^{-7}$–10^{-9}), a moon of Jupiter, and a water vapor–CO_2 exosphere on Enceladus, a moon of Saturn. See Ch. 14 for details.

(iv) *Exospheres above rocky surfaces* include Mercury ($\sim10^{-15}$) and the Moon (3×10^{-15}). Both have ballistic exospheres with atoms hopping across rocky surfaces.

Of the above thin atmospheres, those of Pluto, Triton, Io, and perhaps Callisto are sufficiently dense to generate characteristic patterns of atmospheric motions or *atmospheric dynamics*.

Atmospheric compositions can vary because of phase changes or chemical reactions. A species affected by phase changes is water vapor in Earth's atmosphere, with

where $\overline{R}_d/R_c = m_c/\overline{m}_d = \varepsilon$ because $\overline{R}_d = R/\overline{M}_d = k/\overline{m}_d$ and $\overline{R}_c = R/M_c = k/m_c$. The last equality in the equation above may be rearranged and written as

$$p = \overline{R}_d \rho T_v \quad (1.7)$$

where T_v is given by eq. (1.6). Consequently, the gas constant in the ideal gas form of eq. (1.7) remains the same in all cases, provided we replace the temperature with the virtual temperature T_v.

Because water vapor has a molecular mass of 18 atomic mass units (a.m.u.), which is lighter than the mean molecular mass of the dry gas in Earth's atmosphere of 28.97 a.m.u., the density of moist air is *less* than the density of dry air at the same pressure and temperature. Similarly, "moist" air on Titan with condensable methane is less dense than "dry" Titan air. In contrast, on Jupiter, the condensable gas ammonia is heavier than the mean molecular mass of the dry gas (a mixture of H_2 and He), so the density of "moist" air is greater than that of the "dry" gas. Consequently, virtual temperature due to H_2O or CH_4, respectively, is slightly *greater* than actual temperature for Earth and Titan, while virtual temperature accounting for variable NH_3 concentration is slightly *less* than actual temperature in the atmosphere of Jupiter.

1.1.2.3   The Hydrostatic Equation and Barometric Pressure

A key property of atmospheres is pressure, which decreases exponentially with height. Pressure declines with altitude because when air settles under gravity, high pressure lower down in the atmosphere pushes upwards against the weight of overlying air. At higher altitudes there is less weight of air above, so pressure is lower. If the weight of overlying air were not balanced in this way, the atmosphere would contract or expand until it reached equilibrium. The balance of forces involved is *hydrostatic equilibrium*, an approximation that on Earth is generally valid over a horizontal scale larger than a few kilometers.

Considering a column of air of unit cross-sectional area and height Δz allows us to derive the equation for hydrostatic equilibrium (Fig. 1.3). The decrease in the pressure Δp over a height increment of Δz arises from the force caused by the weight of air at the base of the column of height Δz. This column of air has volume 1 m^2 $\times \Delta z$, so its mass is simply $\rho \Delta z$, where ρ is the air density. Pressure is defined as force over area, so

$$\Delta p = \frac{\text{weight}}{\text{area}} = \frac{-(\text{mass of column}) \times g}{1} = -(\rho \Delta z)g$$

where g is gravitational acceleration. In atmospheric science, pressure is given in various units: bar, millibar, atmospheres, or Pascals (the SI unit). The conversion factors are 1 bar = 10^5 Pa = 1000 mbar, and 1 atm = terrestrial mean sea level pressure = 1.01325 bar = 1013.25 mbar = 1.01325 $\times$ 10^5 Pa. Rewritten in calculus form, the above equation is

$$\frac{\partial p}{\partial z} = -g(z)\rho(z), \quad p(z) = \int_z^\infty g(z)\rho(z)dz \tag{1.8}$$

In eq. (1.8), acceleration due to gravity depends on height z, i.e., $g = g(z)$, which could be approximated as $g(z) \approx g_0[r_0/(r_0 + z)]^2$, where r_0 is a reference radius, usually a mean solid surface on a rocky planet or a cloud-top radius on a gas giant, z is the geometric height above that radius, and $g_0 = GM/r_0^2$. This approximation is good for most atmospheric applications on terrestrial planets.

The acceleration in the local vertical also depends on latitude because of planetary rotation. Let us define Ω, the *planetary angular rotation rate* or angular frequency. On Mars, for example, $\Omega = 2\pi/(\text{sidereal rotational period}) = 2\pi/(88\ 642.663\ \text{s}) = 7.088 \times 10^{-5}\ \text{s}^{-1}$, and on Jupiter, $\Omega = 2\pi/(35\ 727.3\ \text{s}) = 1.76 \times 10^{-5}\ \text{s}^{-1}$. At any point on the planet, there will be an outward centrifugal acceleration given by $\Omega^2 r_a$, where r_a is the radial distance from the axis of rotation (Fig. 1.4). At the equator, r_a is just the radius of the planet R_p, whereas at the poles r_a is zero. If r is the radial distance from the center of the planet, the centrifugal acceleration at latitude ϕ is $\Omega^2 r\cos\phi$. If we turn this vector through angle ϕ, we get the upward component of this acceleration in the local vertical, $\Omega^2 r\cos^2\phi$. Hence, a first-order correction to g is to subtract this component. Thus, the corrected gravitational acceleration can be expressed as

$$g(r,\phi) \approx \frac{GM}{r^2}\left[1 - \frac{\Omega^2 r^3 \cos^2\phi}{GM} + \cdots \right] \tag{1.9}$$

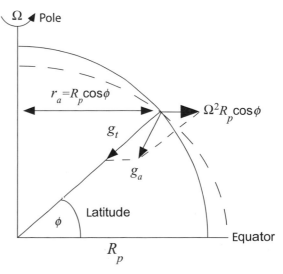

Figure 1.4 The effect of rotation at angular velocity Ω of a spherical planet of radius R_p on the gravity vector at latitude ϕ. In practice, the planet will adjust to an oblate spheroid (dashed); g_t = true gravitational acceleration, g_a = apparent gravitational acceleration = the resultant of the true gravity vector and outward centrifugal acceleration (exaggerated).

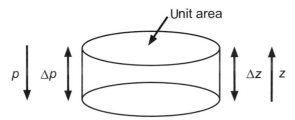

Figure 1.3 Diagram illustrating the pressure decrement Δp across an air column of height increment Δz.

where M is the mass of the planet below radius r and $G = 6.674 \times 10^{-11}$ N m^2 kg^{-2} is the universal gravitational constant. The correction term in eq. (1.9) accounts for the centrifugal acceleration at latitude ϕ. For the terrestrial planets, this term produces correction to the sphericity of gravitational potential surfaces of about a few parts per thousand. Higher-order corrections also appear because of the departure from spherical symmetry of the mass distribution caused by rotation or asymmetric topography (such as on Mars). The correction to g due to rotation reaches as much as 7% on giant planets. The planetary shape of Jupiter, for example, is distinctly deformed from spherical.

Figure 1.4 shows how the addition of gravitational and centrifugal acceleration vectors should produce a gravitational acceleration vector offset from the local vertical on a spherical planet. However, in reality, all planets, even rocky ones, are fluid over geological time-scales and bulge at their equators. The bulge adjusts for the centrifugal acceleration of a planet's rotation so that the local g-vector becomes perpendicular to the local surface of the planet. With an oblate planet, there is no apparent g offset from the local vertical but the *magnitude* of the gravitational vector must still be corrected.

1.1.2.4 Atmospheric Scale Height

The *scale height* of an atmosphere is the vertical distance over which pressure or density drops by a factor of $1/e = 1/(2.7183) \sim 1/3$. The *pressure scale height* will generally differ slightly from the *density scale height* because they are connected via the ideal gas law through temperature, which varies with height. The pressure scale height equals the density scale height only in an isothermal atmosphere. The hydrostatic equation and the ideal gas equation combine to give us an expression for the pressure scale height, H. From eq. (B4), we can express the air density as $\rho = \overline{m}p/kT$, which can be substituted in eq. (1.8) to produce

The integral form of the hydrostatic equation (eq. (1.11)) can be simplified further if we assume that the atmosphere is isothermal, well mixed, and that the variation of gravity with height is small. For most planetary atmospheres, these assumptions are not as bad as they seem. For example, the Earth's atmosphere is well mixed by turbulence to uniform composition of the major gases below the ~ 100 km altitude homopause (Sec. 1.1.1), and over the same vertical distance the temperature lies within 50 K, or 20%, of 250 K. Assuming an isothermal atmosphere at temperature $\overline{T}$, we obtain,

$$p = p_s \exp\left(-\frac{z}{H}\right), \text{where } H = \frac{k\overline{T}}{\overline{m}g} = \frac{R\overline{T}}{g} \qquad (1.12)$$

We can also think of the scale height in terms of a Boltzmann energy distribution. In an isothermal atmosphere the average molecular thermal energy is $\sim k\overline{T}$. If all this energy were used to lift an average molecule to a height H, the gravitational potential energy would be $\overline{m}gH$. However, for a gas at fixed temperature, molecules have a Maxwell–Boltzmann distribution of energies and some will be more energetic. But not many molecules have much more energy than $k\overline{T}$, so the atmosphere thins out above a scale height. Thus the height in $\overline{m}gH \sim k\overline{T}$ defines the top of the first energy level in an exponential Boltzmann distribution.

For most planetary atmospheres, the pressure scale height lies between 5–20 km because the ratio $\overline{T}/\overline{m}g$ is roughly similar for both terrestrial and giant planets, given that small $\overline{T}$ (and large g for Jupiter) compensates for small $\overline{m}$ in hydrogen-rich gas giants. On Earth, if we take a near-surface temperature of 288 K, the pressure scale height is ~ 8.4 km, whereas for a typical mid-tropospheric temperature of 250 K, it is 7.3 km. On Titan, the near-surface temperature $\overline{T} \sim 90$ K, $g = 1.34$ m s^{-2}, and $\overline{M} = 28.6$ g mol^{-1}, so the pressure scale height is $H = (8.314$ J K^{-1} mol$^{-1} \times 90$ K$)/(0.0286$ kg mol$^{-1} \times 1.34$ m s$^{-2}) \sim 20$ km. This shows how an atmosphere on a small body can be extended vertically because of its small gravity.

$$\frac{\partial p}{\partial z} = -\left(\frac{\overline{m}(z)g(z)}{kT(z)}\right)p \quad \Rightarrow \frac{\partial p}{p} = -\left(\frac{\partial z}{H}\right), \text{where } H = \frac{kT(z)}{\overline{m}(z)g(z)} = \frac{\overline{R}(z)T(z)}{g(z)} \qquad (1.10)$$

Integration of the above equation from the surface ($z = 0$) to an arbitrary altitude z gives

$$p = p_s \exp\left(-\int_0^z \left(\frac{dz}{H}\right)\right) \qquad (1.11)$$

where p_s is the surface pressure.

1.1.2.5 Geopotential Height and Geometric Height

To avoid dealing with the variation of g with height, a transformation is sometimes used that is analogous to the use of virtual temperature that we encountered earlier. Geopotential height Z, is the height of a given point in the atmosphere in units proportional to the potential energy of a unit mass at this height relative to some reference altitude. The potential energy per unit mass at height z is the integral of potential energy to that height, which is the *geopotential*, given by

$$\Phi(z) = \int_0^z g\,dz \quad \text{and} \quad d\Phi = g\,dz \quad [\text{m}^{-2}\text{s}^{-1} \text{ or J kg}^{-1}]$$

(1.13)

The *geopotential height*, defined by

$$Z = \frac{\Phi(z)}{g_0} = \frac{1}{g_0}\int_0^z g(z,\phi)\,dz \quad [\text{m}], \quad \text{where } g_0 = \frac{GM}{r_0^2}$$

(1.14)

can be used as the dependent variable in place of geometric height, where ϕ is latitude and r_0 is a reference radius. Since $z << r_0$, it follows that $g_0 \approx g(z,\phi)$ and $Z \approx z$. For example, Z differs only 1.55% from z at 100 km on Earth. Using the hydrostatic equation in basic form, $dp = -g\rho\partial z = -\rho d\Phi$, we have, with the ideal gas law:

$$dp = -\left(\frac{p}{RT}\right)d\Phi = -\left(\frac{p}{RT}\right)g_0 dZ \Rightarrow \frac{dp}{p} = -\frac{1}{H}dZ$$

Consequently, the expression for the pressure decrease with geopotential height is

$$p(Z) = p_0 \exp\left(-\int_0^Z \frac{dZ}{H}\right)$$

(1.15)

where the scale height is

$$H(Z) = \frac{\overline{R}T(Z)}{g_0} = \frac{\overline{R}_d T_v(Z)}{g_0}$$

(1.16)

The difference in geopotential height between two pressure surfaces is called *thickness*:

$$\Delta Z = Z(p) - Z(p_0) = \int_p^{p_0} H(p)\frac{dp}{p} = \overline{H}\ln\left(\frac{p_0}{p}\right) \propto \overline{T}$$

(1.17)

where $\overline{H}$ is the log-pressure average scale height between p and p_0; it is proportional to $\overline{T}$, the log-pressure mean temperature.

Equation (1.17) is known as the *hyposometric equation* and has a practical use that one can construct a distribution of heights of isobaric surfaces given the temperature distribution (measured with remote sensing, for example) and the height of one isobaric surface that is measured or assumed. This approach directly relates the temperature distribution to its influence on pressure variations, which, in turn, cause winds in an atmosphere. Equation (1.17) indicates how air expands vertically with increasing temperature. Thus lateral gradients of

temperature cause lateral gradients of pressure to develop at altitude. For example, the 500 mbar surface in Earth's mid-troposphere (~5–6 km altitude) tends to slope downwards from warm tropics to cold high latitudes. Greater height of a pressure surface is associated with greater pressure below because pressure increases with depth, so there is generally higher pressure towards the tropics. This situation causes strong eastward winds in northern midlatitudes because when account is taken of Earth's rotation, winds tend to flow such that low pressure is to the left of the flow in the northern hemisphere (Sec. 4.1.2).

In hydrostatic equilibrium, p is a unique function of z for a given distribution of $T/\overline{m}$ from eq. (1.11). So pressure or any function of pressure can be used an alternative vertical coordinate to geometric height z. For example, some dynamical studies use a "log-pressure" vertical coordinate. This approximates height and is defined by

$$z^* = -H_{\text{ref}}\ln\left(\frac{p}{p_{\text{ref}}}\right), \text{ where } H_{\text{ref}} = \frac{kT_{\text{ref}}}{gm_{\text{ref}}}$$

(1.18)

Here H_{ref} is a reference scale height, p_{ref} is a reference pressure, and T_{ref} and m_{ref} are some global mean reference values.

1.1.2.6 The Relationship Between Surface Pressure and Mass of an Atmosphere

When considering atmospheric evolution, or comparing different planetary atmospheres, it is useful to examine how much material is in an atmosphere. The surface pressure, p_s, integrates the mass, M_{col}, of a column of air per unit area because, by definition, p_s = weight/(unit area) = $M_{\text{col}}g$. So the mass of a column of air per unit area is given by

$$M_{\text{col}} = \frac{p_s}{g}$$

(1.19)

On Earth, for example, the mass of a column of air above sea level is ~(101 325 Pa)/(9.8 m s^{-2}) ~ 10^4 kg m^{-2}, which compares to 164 kg m^{-2} (610 Pa/3.72 m s^{-2}) above an average location on the surface of Mars.

Highly accurate calculation of the mass of an atmosphere is more involved. This is because eq. (1.19) is a flat Cartesian approximation to a planet that in reality is an oblate spheroid. The surface pressure is actually slightly less than the weight per unit area because curved geometry allows the weight of the atmosphere to be supported by lateral pressure forces as well as the pressure forces at the bottom of the atmosphere (Bannon *et al.*, 1997; Sanudo *et al.*, 1997; Trenberth and Guillemot, 1994).

1.1.2.7 Column Abundance

The column abundance of a gas, which is the number of molecules per unit area above a surface, is commonly used to describe gases that absorb radiation, such as methane or ozone. For a surface at altitude z, the column abundance $N_{col}(z)$, is given by

$$N_{col}(z) = \int_z^\infty n(z)dz \qquad (1.20)$$

where $n(z)$ is the number of molecules of the gas per unit volume at altitude z. The column mass is $N_{col}(z)\overline{m}$, so the pressure at altitude z, is $p(z) = N_{col}(z)\overline{m}g$. Thus we can derive an expression for the column abundance in terms of scale height H as follows:

$$N_{col}(z) = \frac{p(z)}{\overline{m}g} = \frac{n(z)kT(z)}{\overline{m}g} \Rightarrow N_{col}(z) = n(z)H(z)$$

$$(1.21)$$

For the last expression on the right-hand side, we recognized the scale height $H(z) = kT(z)/\overline{m}g$.

Column abundance is usually expressed in units of molecules per cm^2, but the planetary science literature sometimes uses the unit "cm-atm" or "cm-amagat".[1] This is the number of centimeters of a column produced if the gas were taken to STP (Standard Temperature and Pressure) conditions. The relationship with $N_{col}(z)$ is a normalization, as follows, by a reference number density:

$$\mathcal{N} \text{ cm-atm} = \frac{N_{col}(z)}{n_0}, \text{ where } n_0 = \frac{p_{STP}}{kT_{STP}} = 2.687 \times 10^{19} \text{ molecules cm}^{-3} \qquad (1.22)$$

where $N_{col}(z)$ is in units of molecules cm^{-2}. Here n_0 is the *Loschmidt number*, the molecular number density in cm^{-3} of an ideal gas at STP conditions of $p_{STP} = 1$ atm = 101 325 Pa and $T_{STP} = 273.15$ K.

As an example of the use of eq. (1.22), observations of peak ozone levels on Mars by NASA's *Mariner 9* orbiter spacecraft are reported as a column abundance of ~60 μm-atm (Barth, 1974). Application of eq. (1.22) with $\mathcal{N} = 60 \times 10^{-4}$ cm-atm gives $N_{col} = 1.6 \times 10^{17}$ molecules O_3 cm^{-2}. This can be compared to a typical ozone column abundance on Earth of ~8×10^{18} molecules O_3 cm^{-2}, some ~50 times greater. The large difference with Earth indicates that Mars does not have an ozone layer shield for shortwave ultraviolet radiation,

even when its ozone column abundance is maximal. Consequently, the surface of Mars is sterilized by short-wave ultraviolet (Ch. 12).

A specific unit of column abundance also exists for ozone. This is the Dobson Unit (DU), named after Gordon Dobson, the Oxford professor who built the first spectrometer to measure Earth's ozone layer in the 1920s. The column abundance in Dobson Units is related to $\mathcal{N}$ and $N_{col}(z)$ by the following expression:

$$\text{Dobson Units} = 1000 \, \mathcal{N} \text{cm-atm } O_3 = \frac{1000 N_{col}(z)}{n_0}$$

$$(1.23)$$

One DU refers to a layer of ozone that would be 0.001 cm (or 0.01 mm) thick at STP. Earth's ozone column abundance varies with latitude and time, but is typically about 300 DU, equivalent to a layer of 3 mm thickness at STP, or 0.3 cm-atm.

Yet another measure of column abundance is used for atmospheric water vapor, which is the depth of water that would result if all the water were condensed out from an atmospheric column onto the surface. This is expressed as *precipitable centimeters* (pr cm) on Earth and as *precipitable microns* (pr μm) on Mars. On Earth, the typical water column abundance is 3 pr cm. In contrast, the Martian atmosphere contains an average 10 pr μm of water. Precipitable water can be readily converted into a columnar mass. The mass of the water per unit area is (precipitable depth) × (the density of water). Thus 10 pr μm on Mars is equivalent to $(10 \times 10^{-6}$ m) × $(1000$ kg m$^{-3}) = 0.01$ kg H_2O m^{-2}.

1.1.3 Convection and Stability

Pressure and density must generally decrease with increasing height because of hydrostatics but the way that temperature varies with altitude is less certain *a priori*. Consequently, *atmosphere structure* is usually shorthand for how the temperature changes with height, as mentioned in Sec. 1.1.1. Air temperature can decrease or increase with height depending on the circumstances (Sec. 1.1.1.1), but the dominant physical processes ultimately responsible are radiation and convection. These processes may also interact with other atmospheric motions, e.g., the generation of atmospheric waves can modulate the atmospheric structure (e.g., Sec 4.4.2; Sec. 4.6).

---

[1] When number density is normalized by the Loschmidt number (i.e., n/n_0) the result is in units of amagat (or "amg"), named after French physicist Emile Amagat (1841–1915).

Convection occurs when air parcels become buoyant, meaning less dense than their surroundings. Several processes can drive buoyancy: radiative heating, contact with a warm planetary surface, release of latent heat when a substance condenses (e.g., water vapor in Earth's atmosphere, see Sec. 1.1.3.5), and heat from the interior of a planet on Jupiter, Saturn, or Neptune. Buoyant air parcels rise and displace adjacent air, causing mixing. Upward-moving air parcels rise into regions of lower ambient air pressure, where they expand and cool. Conversely, sinking parcels contract and warm. In convective equilibrium, we expect atmospheric temperature to decrease with height, as indeed is generally observed above the surfaces of Earth and Mars (during the daytime), Venus and Titan, as well as in the tropospheres of giant planets (Fig. 1.2).

The rate that temperature T, decreases with altitude z, is the *lapse rate*. This is denoted by $\Gamma = -(dT/dz)$ and commonly expressed in units of K km^{-1}. The negative sign convention generally gives a positive number for Γ in tropospheres. For example, on Earth, the global average lapse rate is ~ 6 K km^{-1} in the troposphere.

Let us consider a parcel that ascends or descends such that its pressure adjusts to its surroundings. In an idealized case where the temperature of the air parcel warms by contraction or cools by expansion with no exchange of energy with its surroundings either by conduction or radiation, the temperature change is said to be *adiabatic* (from the Greek for "impassable" (to heat)). In this case, the temperature change with altitude is the *adiabatic lapse rate*, $\Gamma_a = -(dT/dz)_a$.

Convection occurs when the ambient lapse rate exceeds the adiabatic lapse rate, $\Gamma > \Gamma_a$. This means that a rising adiabatic parcel will be at a warmer temperature than ambient surroundings after accounting for adiabatic cooling, and so will continue to rise (Fig. 1.5). The result of such an *unstable* atmosphere would be extremely thorough mixing so that the ambient temperature profile would adjust to an adiabatic one. Thus, we think of Γ_a as a limiting lapse rate, so that in nature we expect to find *stable* atmospheres with $\Gamma \leq \Gamma_a$. On average this is observed in planetary atmospheres (Table 1.3).

However, we should be wary that *superadiabatic* conditions can occur, as follows: (i) in Earth's tropics Γ can exceed the *moist adiabatic lapse rate* (see later, Sec. 1.1.3.6), which gives rise to large thunderstorms; (ii) within a few meters of planetary surfaces in daytime, Γ can exceed Γ_a by a factor of hundreds or thousands, especially in deserts. For example, on Mars the atmosphere cannot keep up with the daytime warming of the surface and so "warm towers" or "thermal vortices" develop that are the convective equivalent of thunderstorms without the clouds or rain. Often such vortices pick up dust and become *dust devils* (reviewed by Balme and Greeley, 2006).

In the cloudless terrestrial atmosphere, the ambient lapse rate is often superadiabatic near the ground and nearly adiabatic at higher levels. Consequently, gases released at the surface travel upwards and are dispersed. However, sometimes an *inversion* arises, meaning that the lapse rate is negative and temperature increases with height. In this case, the atmosphere is highly stable. In urban areas, we commonly experience pollution trapped near the ground when there is a temperature inversion.

1.1.3.1 The Dry Adiabatic Lapse Rate and Observed Lapse Rate

By considering an adiabatic parcel of air, we can derive a formula for the adiabatic lapse rate. We start with the first law of thermodynamics, which is conservation of energy, i.e.,

heat input = (change in internal energy) + (work done)

$$dq = du + dw \qquad (1.24)$$

Here dq, du, and dw are energy changes per unit mass, which we indicate in lower case. Now, the definition of specific heat at constant volume is $c_v = (du/dT)_v$, where the v subscript indicates constant volume, from which

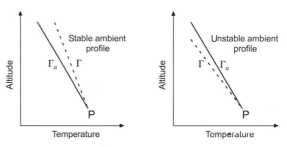

Figure 1.5 Atmospheric profiles showing the adiabatic lapse rate (Γ_a) as a solid line and the ambient lapse rate ($\Gamma = -dT/dz$) as a dashed line. The left-hand profile is stable because $\Gamma < \Gamma_a$, i.e. the decrease of temperature with altitude is smaller than in the adiabatic case. A parcel displaced adiabatically from point P will be colder and denser than its surroundings and so will tend to return to P. In the right-hand graph when $\Gamma > \Gamma_a$, the atmosphere is unstable to convective mixing. A parcel will continue rising from point P. However, the adiabatic assumption will break down, so the parcel will lose its heat to the air aloft such that the ambient profile will move towards adiabatic equilibrium.

Table 1.3 Thermodynamic properties and lapse rates on various planets.

Body	Main gases	Molar mass, M g mol^{-1}	$\bar{R}$ J kg^{-1} K^{-1}	g m s^{-2}	c_p J kg^{-1} K^{-1}	$\Gamma_a = g/c_p$ troposphere K km^{-1}	Γ observed in troposphere K km^{-1}
Venus	CO_2, N_2	43.45	189	8.901	930	9.5	~8.0
Earth	N_2, O_2	28.97	287	9.81	1004	9.8	~6.5
Mars	CO_2, N_2	43.5	191	3.72	850	4.4	~2.5
Jupiter	H_2, He	2.22	3745	24.25	10 988	2.1–2.45*	1.9
Saturn	H_2, He	2.14	3892	10.0	10 658	0.7–1.1	0.85
Titan	N_2, CH_4	28.67	290	1.36	1044	1.3	1.0–1.4
Uranus	H_2, He	2.3	3615	8.80	8643	0.7–1.1	0.75
Neptune	H_2, He	2.3	3615	11.1	8643	0.85–1.34	0.95
HD209458b	H_2	2.0	4160	18.5	14 300	1.3	~0.2 (model)**

* On giant planets, g varies from equator to pole.
** Menou and Rauscher (2009).

we get $du = c_v\, dT$. Also a small increment of work dw is given by "force $\times$ incremental distance," which is equivalent to "pressure $\times$ incremental volume change." If we define specific volume (volume per unit mass) as $\alpha = 1/\text{density} = 1/\rho$, we have $dw = p\, d\alpha$. (Note that some books use "pdV" but then V would have to be in units of m^3/kg and in Box 1.1 we defined V as volume (units of m^3), so we don't use V here). Substituting for du and dw, we have

$$\underbrace{dq}_{\substack{\text{heat added/kg} \\ \text{(zero for adiabatic)}}} = \underbrace{c_v dT}_{\substack{\text{change in internal heat/kg}}} + \underbrace{p d\alpha}_{\substack{\text{work done/kg}}} \tag{1.25}$$

Here, dq is a small heat input per unit mass (e.g., from contact with a warm surface, or zero if adiabatic), dT is a small rise in temperature, and $d\alpha$ is a small increase in parcel volume per kg. From the ideal gas law $p = \rho \bar{R} T$, it follows that $p\alpha = \bar{R}T$. By applying the product rule $[d(xy) = x\, dy + y\, dx]$ to T/p, we get

$$d\alpha = (\bar{R}/p)dT - (\bar{R}T/p^2)dp$$

Substituting for $d\alpha$ in the conservation of energy equation (1.25), we have

$(\bar{R}T/p)dp$ with $-gdz$ by using the hydrostatic equation (eq. (1.8)), i.e., $dp = -g\rho dz$.

Now consider an adiabatic parcel of air that moves up or down. Adiabatic means no exchange of heat with the surroundings. Thus, with $dq = 0$, the last expression in eq. (1.26) gives us the change in air parcel temperature dT, as the parcel moves up or down distance dz and expands or contracts adiabatically. When we rearrange this equation, we get the definition of the *dry adiabatic lapse rate*:

$$\Gamma_a = -\left(\frac{dT}{dz}\right) = \frac{g}{c_p} \tag{1.27}$$

This little equation provides the temperature change with altitude of a dry parcel of air moving up and down through an atmosphere in hydrostatic equilibrium.

Given the specific heat capacities of the atmospheres of Venus, Earth, and Mars as 930, 1004, and 850 J K^{-1} kg^{-1}, and g values of 8.87, 9.81, and 3.72 m s^{-2}, respectively, we can readily calculate from eq. (1.27) that Γ_a (K km^{-1}) = 9.8 (Earth), 9.5 (Venus), 4.4 (Mars). However, typical observed lapse rates in the troposphere of these planets (Table 1.3) are Γ (K km^{-1}) $\sim$ 6 (Earth), $\sim$8.0 (Venus), $\sim$2.5 (Mars). On average, the atmospheres do not cool as much as expected with height. The explanations are as follows. In Earth's lower

$$dq = c_v dT + p\left[\frac{\bar{R}}{p}dT - \frac{\bar{R}T}{p^2}dp\right] = (c_v + \bar{R})dT - \left(\frac{\bar{R}T}{p}\right)dp \Rightarrow dq = c_p dT + gdz \tag{1.26}$$

In the final expression, we've used $c_p = c_v + \bar{R}$ from kinetic theory, which connects specific heat at constant pressure (c_p) to that at constant volume. Also, we replaced

atmosphere, water vapor condenses, which is an exothermic (heat releasing) process. On Mars, dust absorbs visible sunlight and warms the atmosphere, while on

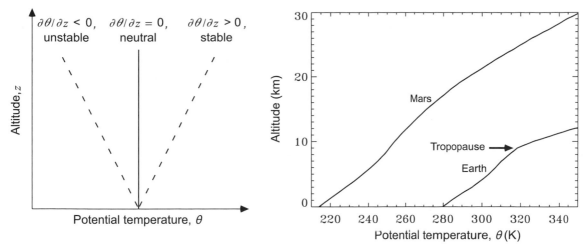

Figure 1.6 Left: The relationship between profiles of potential temperature θ, and stability. Right: Averaged stable potential temperature profiles in midlatitudes of Mars and Earth. Note the marked increase in static stability in the Earth's stratosphere.

Venus sulfur gases exothermically condense into H_2SO_4 (sulfuric acid) particles.

1.1.3.2 Potential Temperature and Generalized Lapse Rates in Tropospheres

Potential temperature is an alternative thermodynamic variable to absolute temperature, and its slope with height gives direct insight into whether an atmosphere is stable or unstable to convection. This variable is easily derived by setting $dq = 0$ in eq. (1.26) and substituting $(\overline{R}T/p)dp$ for gdz from the hydrostatic equation. This gives:

$$\frac{dT}{T} = \left(\frac{\overline{R}}{c_p}\right)\frac{dp}{p} \tag{1.28}$$

This equation indicates that fractional changes in temperature and pressure are in direct proportion for adiabatic processes, with a constant of proportionality given by

$$\frac{\partial \ln T}{\partial \ln P} = \frac{\overline{R}}{c_p} \equiv \kappa \tag{1.29}$$

Integrating eq. (1.29) between a reference pressure p_{ref} and arbitrary pressure p, we get,

$$[\ln T']_\theta^T = \kappa[\ln p']_{p_{ref}}^p \Rightarrow \ln\left(\frac{T}{\theta}\right) = \kappa \ln\left(\frac{p}{p_{ref}}\right) = \ln\left(\frac{p}{p_{ref}}\right)^\kappa$$

Thus,

$$T = \theta\left(\frac{p}{p_{ref}}\right)^\kappa \tag{1.30}$$

and

$$\theta = T\left(\frac{p_{ref}}{p}\right)^\kappa \tag{1.31}$$

Equation (1.31) is called *Poisson's (adiabatic state) Equation* and defines the variable θ, which is *potential temperature*. Mathematically, θ is a constant of integration equivalent to the actual temperature at a reference pressure p_{ref}. Physically, θ is the temperature a parcel of air would have if it were taken to pressure p_{ref} dry adiabatically. Consequently, θ can be regarded as an alternative thermodynamic variable to temperature so that eq. (1.30) is an alternative equation of state.

On Earth, usually $p_{ref} = 10^5$ Pa and $\kappa = 287$ J K^{-1} kg^{-1} /1004 J K^{-1} kg^{-1} = 0.286. Potential temperature θ is generally greater than actual temperature when the pressure is less than 1000 mbar, and vice versa. The importance of θ is that it is constant with altitude ($\partial\theta/\partial z = 0$) when the lapse rate is adiabatic. Deviations from this constancy indicate conditions of convective stability ($\partial\theta/\partial z > 0$) and instability ($\partial\theta/\partial z < 0$) indicated in Fig. 1.6.

If we invert eq. (1.31), it gives the dry adiabatic temperature variation in the convective part of a troposphere as:

$$T = T_{ref}\left(\frac{p}{p_{ref}}\right)^{\overline{R}/c_p} \Rightarrow T = T_{ref}\left(\frac{p}{p_{ref}}\right)^{(\gamma-1)/\gamma} \tag{1.32}$$

Here, T_{ref} is a reference temperature at a reference pressure p_{ref}, while γ is the ratio of specific heats c_p/c_v. The exponent in eq. (1.32) is from the relationship of specific heats:

$$\bar{R} = c_p - c_v \Rightarrow \frac{\bar{R}}{c_p} = 1 - \frac{1}{\gamma} = \frac{(\gamma - 1)}{\gamma} \tag{1.33}$$

Kinetic theory allows us to relate the ratio of specific heats γ to the degrees of freedom, N_{dof}, for the primary atmospheric constituent(s), where

$$\gamma = 1 + \frac{2}{N_{dof}} \tag{1.34}$$

Often atmospheres are dominated by linear diatomic gases, such as H_2 (Jupiter, Saturn, Uranus, and Neptune), N_2 (Titan), or an N_2–O_2 mixture (Earth). These molecules have three translational and two rotational degrees of

causing oscillations that eventually will be frictionally damped. Atmospheres have a characteristic angular frequency of oscillation, called the *buoyancy frequency* or *Brunt–Väisälä* frequency N_B, which we derive below.

Consider a parcel of air of volume V and temperature T that is displaced by height δz from height z at environmental temperature T to a place where the environmental temperature is T_e (Fig. 1.7). The parcel cools adiabatically to a new temperature $T_p = T - \Gamma_a \delta z$, while the environment cools *less* at the ambient lapse rate as $T_e = T - \Gamma \delta z$. The difference in density between the parcel (ρ_p) and air (ρ_e) drives a buoyancy force, which we can write in the form of Newton's second law:

$$\underbrace{\left(\cancel{V}\rho_p\right)}_{\text{mass}} \underbrace{\frac{d^2(\delta z)}{dt^2}}_{\text{acceleration}} = \underbrace{-g\left(\rho_p - \rho_e\right)\cancel{V}}_{\text{buoyancy force}} \Rightarrow \frac{d^2(\delta z)}{dt^2} = g\left(\frac{\rho_e}{\rho_p} - 1\right) = g\left(\frac{T_e}{T_p} - 1\right) \tag{1.36}$$

Substituting for temperatures T_p and T_e, we get:

$$\frac{d^2(\delta z)}{dt^2} = g\left(\frac{T - \Gamma_a \delta z}{T - \Gamma \delta z} - 1\right) = g\left(\frac{\Gamma \delta z - \Gamma_a \delta z}{T - \Gamma \delta z}\right) \approx g\left(\frac{\Gamma - \Gamma_a}{T - \cancel{\Gamma \delta z}}\right)\delta z \Rightarrow \ddot{\delta z} + \frac{g}{T}(\Gamma_a - \Gamma)\delta z = 0$$

freedom, so that $N_{dof} = 5$ and $\gamma = 7/5 = 1.4$ for these worlds. In the case of the CO_2-dominated atmospheres of Venus and Mars, there are three translational, two rotational, and ~ 0.3 vibrational degrees of freedom, so $N_{dof} = 6.3$ and $\gamma = 1.3$ (Bent, 1965). Consequently, the dry adiabatic temperature T (eq. (1.32)) varies with altitude expressed as pressure as $p^{0.3}$ and $p^{0.2}$ for diatomic and CO_2 dominated atmospheres, respectively. Such a relationship for dry adiabats is general and will also apply to dry tropospheres of exoplanets.

In reality, we observe that all the tropospheres we know (in the Solar System) do not follow a dry adiabat because condensation of volatiles during convection releases latent heat and lowers the lapse rate below that of the dry adiabatic lapse rate. The convective part of a tropospheric temperature structure can be modified from eq. (1.32) to

$$T = T_{ref}\left(\frac{p}{p_{ref}}\right)^{\alpha_c(\gamma - 1)/\gamma} \tag{1.35}$$

For planets in the Solar System, α_c is an empirical factor, typically around 0.6–0.9, which represents the average ratio of the true lapse rate in the planet's convective region to the dry adiabatic lapse rate.

1.1.3.3 Static Stability of Atmospheres

In stable conditions, a vertically displaced air parcel will tend to return to its previous height and overshoot,

This is an equation of motion (compare "$\ddot{x} + \omega^2 x = 0$" in elementary physics textbooks), which we may write with an angular frequency N_B, as

$$\frac{d^2(\delta z)}{dt^2} + N_B^2(\delta z) = 0 \tag{1.37}$$

where

$$N_B^2 \equiv \frac{g}{T}(\Gamma_a - \Gamma) \equiv \frac{g}{T}\left(\frac{dT}{dz} + \frac{g}{c_p}\right) \tag{1.38}$$

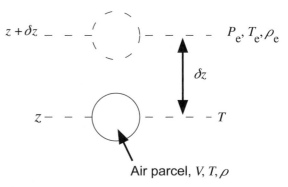

Air parcel, V, T, ρ

Figure 1.7 Schematic diagram indicating the conditions for a parcel of air of volume V, temperature T, and density ρ that is displaced adiabatically over a vertical distance δz into an environment with air of pressure P_e, temperature T_e, and density ρ_e.

In a statically stable atmosphere, $\Gamma < \Gamma_a$; thus $N_B^2 > 0$ and solutions to eq. (1.37) are simple harmonic motion. The displaced parcel oscillates up and down at a natural angular frequency N_B. For Earth's lower troposphere, $dT/dz \sim -6$ K km^{-1} and $T \sim 270$ K so $N_B^2 = (9.81$ m s^{-2}/ 270 K)(-6×10^{-3} K m^{-1} + 9.81 m s^{-2}/1004 J K^{-1} kg^{-1}), which gives $N_B = 0.0117$ s^{-1}, equivalent to a period of oscillation of $2\pi/N_B \approx 9$ minutes.

In a statically unstable atmosphere, $\Gamma > \Gamma_a$ and $N_B^2 < 0$. This means that N_B is imaginary and leads to exponential solutions of eq. (1.37). A displaced parcel would continue to rise with increasing speed. The resulting convection would drive the lapse rate down towards Γ_a by carrying warmer parcels upwards and cooler parcels downward.

The Brunt–Väisälä frequency can also be expressed in terms of potential temperature. First, we take logs of eq. (1.31) to get

$$\ln \theta = \ln T + \kappa(\ln p_{ref}) - \kappa(\ln p)$$

If we differentiate w.r.t. z, remembering that $d(\ln x) = (1/x)dx$, we have

$$\frac{d(\ln \theta)}{dz} = \frac{d(\ln T)}{dz} + \kappa\frac{d(\ln p_{ref})}{dz} - \kappa\frac{d(\ln p)}{dz}$$

$$\frac{1}{\theta}\frac{d\theta}{dz} = \frac{1}{T}\frac{dT}{dz} - \frac{\kappa}{p}\frac{dp}{dz} = \frac{1}{T}\frac{dT}{dz} + \frac{\kappa\rho g}{p}$$

In the very last step we have used the hydrostatic equation (1.8) to substitute for dp/dz. Using the definition of $\kappa = \overline{R}/c_p$ and the ideal gas law (B5), $p = \rho\overline{R}T$, we have

$$\frac{1}{\theta}\frac{d\theta}{dz} = \frac{1}{T}\frac{dT}{dz} + \frac{g}{c_pT} \Rightarrow \underbrace{\frac{g}{\theta}\frac{d\theta}{dz}}_{N_B^2} = \frac{g}{T}(-\Gamma + \Gamma_a) \qquad (1.39)$$

Comparing eq. (1.38), we see that the square of the Brunt–Väisälä frequency is given by

$$N_B^2 = \frac{g}{\theta}\frac{d\theta}{dz} \qquad (1.40)$$

Stable atmosphere solutions are defined by positive N_B^2 and positive $d\theta/dz$. This means that that a displaced parcel will experience a restoring force and return to its previous position, albeit with oscillations. When $d\theta/dz = 0$, then $N_B = 0$, and we have neutral stability, where the atmosphere is just stable and the lapse rate equals the adiabatic lapse rate. If $d\theta/dz < 0$, the atmosphere is unstable to free convection (Fig. 1.6).

1.1.3.4 The Planetary Boundary Layer (PBL)

On planets with rocky surfaces, the *Planetary Boundary Layer* (PBL) is a near-surface region where turbulence generated by thermal convection produces an air–ground temperature differential and rapid variation with height of temperature and wind. PBL fluxes of heat, mass and momentum drive the circulation of an atmosphere.

To illustrate the structure of the PBL, consider Mars, shown schematically in Fig. 1.8. Just above the surface, air is so strongly restrained by friction with the ground and by its own intermolecular friction (viscosity), that its motion is nearly nonexistent. This is the *laminar boundary layer* (point A on Fig. 1.8) where momentum, heat and volatiles are transported mainly by molecular diffusion. It has depth ~ 1 cm on Mars compared to a few millimeters on Earth. The next region (marked B) is the turbulent or *surface boundary layer* (or *Prandtl layer*), which on Earth extends up to altitude 5–50 m. On Mars, it reaches a height of a few to a few hundred meters. This is the most turbulent

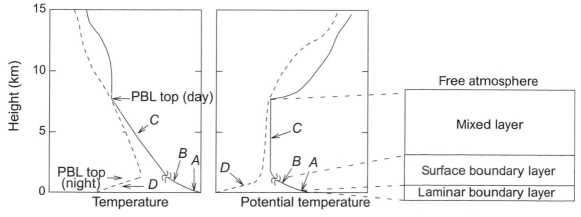

Figure 1.8 Schematic depiction of the planetary boundary layer (PBL) on Mars, indicating day and night conditions. Heights of the top of the laminar boundary layer (A) and surface layer (B) are exaggerated. (Following a schematic by Leovy (1982a).)

part of the atmosphere. Flow is in planes parallel to the surface and strongly shears (i.e. changes direction) with height. After sunset, radiational cooling may lead to a temperature inversion (D). Further up is a much thicker convective layer (C) called the *mixed layer* (or *Ekman layer*) that extends up to the top of the well-mixed (constant θ) zone. In this layer, away from the equator, the local vertical component of the angular momentum of a rotating planet can become important in affecting the direction of flow. Above C, the PBL ends and an increase of potential temperature θ (i.e., static stability) greatly reduces turbulent mixing. The region above the PBL is the *free atmosphere* or *geostrophic level*.

The depth of the PBL varies. On Earth, the PBL typically extends to 0.5–1 km, but can be tens of meters or ~2–4 km in certain circumstances. In early morning on Mars, the PBL may be less than 100 m deep but on a clear, summer afternoon typical thermal convection produces a PBL 5–10 km thick (based on afternoon cumulus cloud shadows, probe entry measurements, dust devil shadows, and atmospheric models). During Martian nighttime, a *nocturnal jet* (10–20 m/s) at a height of ~100 m may form above the inversion layer because the temperature inversion decouples the winds from the surface friction. After sunrise, convection in the PBL destroys the jet.

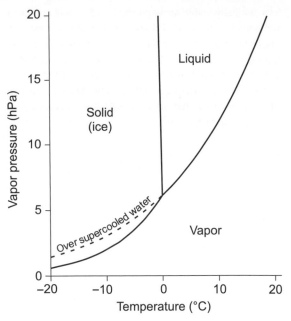

Figure 1.9 The saturation vapor pressure (SVP) for water. The line at the right shows the boundary between liquid and vapor. The dashed line shows the SVP over supercooled water, whereas the solid line beneath it is for SVP over ice. The higher SVP over super-cooled water compared to ice causes more rapid evaporation.

1.1.3.5 Condensable Species, Entropy, and the Clausius–Clapeyron Equation

In various planetary atmospheres, there are condensable species that affect atmospheric structure through their change of state from vapor to liquid, or vapor to solid: H_2O on Earth, CH_4 and HCN on Titan, H_2O and CO_2 on Mars, H_2SO_4 (SO_3 + H_2O) on Venus, H_2O, NH_3 on Jupiter and Saturn, and N_2 on Triton. Species can also form rain (e.g., water on Earth or CH_4 on Titan), snow (e.g., CO_2 in Mars' wintertime polar atmosphere), and clouds (all of the aforementioned species). Condensables play a critical role in atmospheric evolution, such as the *runaway greenhouse effect*, which we meet later.

The key concept in understanding the behavior of a condensable is saturation vapor pressure. At any given temperature, the *saturation vapor pressure* (SVP) is the vapor pressure exerted in equilibrium with condensed liquid or solid phase. If we plot the SVP versus temperature, a curve is defined that represents the transition between vapor and liquid phases (or vapor and solid phases) (Fig. 1.9). Once a condensable gas vapor reaches SVP, it must condense into liquid or solid. The *Clausius–Clapeyron equation* defines the curve of Fig. 1.9, or more generally, SVP as a function of temperature for a

condensable gas or a change in the melting point of a solid with pressure, such as CO_2 on Mars. On Earth, we are accustomed to thinking of the Clausius–Clapeyron equation as defining the saturation vapor pressure of water, but an analogous equation can be written for any condensable gas in any atmosphere.

Before quantifying how condensable species affect lapse rates, we must have a slight diversion and discuss *entropy*. Rudolf Clausius (1822–1888) coined the word *entropy*, from the Greek τροπή, meaning *transformation*. The basic motivation for his definition of entropy is that the integral of the quantity dQ/T (= heat absorbed (or lost)/ temperature) along a reversible path between two states A and B is independent of path. This means that you can define a fundamental thermodynamic function S that depends only on the initial and final states of a reversible process. If S_A and S_B are the values of this function in the states A and B, S is defined as

$$S_B - S_A = \int_A^B \frac{dQ}{T} \quad \text{or} \quad dS = \frac{dQ}{T} \quad (1.41)$$

If we define a reference state "O," the state S can be defined for any state X as:

$$S_X = S_O + \int_O^X \frac{dQ}{T} \quad (1.42)$$

The third law of thermodynamics give us an entropy reference value S_O as follows: at the absolute zero of temperature (0 K) the entropy of every chemically homogeneous solid or liquid body is zero. In the real world, the transformation from state O to state X involves irreversible processes along a path, as opposed to a reversible path. In idealized thermodynamics, it is often assumed that every irreversible transformation that occurs in nature can also be achieved through a reversible process for which eq. (1.42) is valid.

Although the above seems abstract, entropy is a measurable physical quantity with units of J K^{-1}. For example, when one mole of water ice turns into liquid water the entropy change is 22 J K^{-1} mol^{-1} (= 6008 J mol^{-1}/273.15 K), where we have used a melting temperature of 273.15 K, the latent heat of fusion of water 6008 J mol^{-1}, and eq. (1.41) in the form $\Delta S = \Delta Q / T$.

We can derive the Clausius–Clapeyron equation from basic thermodynamics. We will see that the key concept in understanding where the Clausius–Clapeyron equation comes from is that two phases are in *equilibrium* along the SVP versus T line, so the Gibbs free energy change of the phase transition is zero everywhere along the line. The Gibbs free energy is that available to do useful work and at equilibrium the change of Gibbs free energy is zero.

In order to change unit mass of a condensable from liquid (or ice) to vapor at constant temperature and pressure, a certain amount of energy, called the *latent heat* or, more properly, *enthalpy l_c*, is absorbed. Enthalpy per unit mass has units J kg^{-1}. If the change is from liquid to vapor, l_c is the *enthalpy (latent heat) of vaporization*; from ice to liquid it is the *enthalpy (latent heat) of fusion*, and from ice to vapor it is the *enthalpy (latent heat) of sublimation*. The first law of thermodynamics, eq. (1.24), gives

$$l_c = \int_{q_1}^{q_2} dq = \int_{u_1}^{u_2} du + p \int_{\alpha_1}^{\alpha_2} d\alpha = u_2 - u_1 + p(\alpha_2 - \alpha_1)$$

$$= \text{(change in internal energy in making vapor)} + \text{(work done in making vapor)}$$

(1.43)

Here u is the internal energy per unit mass, $p = e_s$ is the equilibrium or saturation vapor pressure, and α is the specific volume. Subscript 1 refers to the condensed phase and subscript 2 to the vapor phase. The change of phase takes place at constant temperature, so from the definition of entropy (eq. (1.41)), it follows that $l_c = T\delta s = T(s_2 - s_1)$, which can be substituted for l_c on the left-hand side of (1.43) to give

$$T(s_2 - s_1) = u_2 - u_1 + p(\alpha_2 - \alpha_1)$$
$$\Rightarrow u_1 + p\alpha_1 - Ts_1 = u_2 + p\alpha_2 - Ts_2$$

(1.44)

The combination $u + p\alpha - Ts$ is called the Gibbs function, or G, which, from eq. (1.44), is constant during an isothermal change of phase. (Elsewhere, G is often written as $G = H - Ts$ where H is enthalpy or $H = u + p\alpha$, per unit mass).

Now if the phase change takes place at a higher temperature, $T + dT$, and a correspondingly higher equilibrium vapor pressure, $e_s + de_s$, we can write the Gibbs function as $G + dG$. To obtain dG we take the derivative of $u + p\alpha - Ts$, as follows:

$$
\begin{aligned}
dG &= du + p\,d\alpha + \alpha\,dp - T\,ds - s\,dT \quad ;\text{sub. for } dq = du + p\,d\alpha \\
&= \quad dq \quad + \alpha\,dp - T\,ds - s\,dT \quad ;\text{sub. for } dq = T\,ds \\
&= \alpha\,dp - s\,dT
\end{aligned}
$$

(1.45)

Since G remains constant during the isothermal change of phase, $G_1 + dG_1 = G_2 + dG_2$ and $dG_1 = dG_2$. So, from eq. (1.45)

$$\alpha_1 dp - s_1 dT = \alpha_2 dp - s_2 dT$$
$$\Rightarrow dp(\alpha_2 - \alpha_1) = dT(s_2 - s_1)$$

(1.46)

Thus, replacing p above with e_s (the saturation vapor pressure), the slope of the Clausius–Clapeyron curve is defined by

$$\frac{de_s}{dT} = \frac{\delta s}{\delta \alpha} = \frac{\text{entropy gained per kg from condensed to vapor}}{\text{volume increase per kg from condensed to vapor}}$$

(1.47)

Since enthalpy $l_c = T\delta s$, it follows that we can substitute for δs in eq. (1.47) as follows:

$$\frac{de_s}{dT} = \frac{l_c}{T\delta\alpha} = \frac{l_c}{T\left(\alpha_{\text{vapor}} - \alpha_{\text{condensed}}\right)} \approx \frac{l_c}{T\alpha_{\text{vapor}}}$$

The specific volume of vapor is much larger than the specific volume of the condensed phase ($\alpha_{\text{vapor}} \gg \alpha_{\text{condensed}}$), so we can ignore the latter. By substituting $\alpha_{\text{vapor}} = R_c T / e_s$, where R_c is the specific gas constant for the condensable species vapor, we obtain

$$\frac{de_s}{dT} = \frac{l_c e_s}{R_c T^2}, \quad \text{or} \quad \frac{d(\ln e_s)}{dT} = \frac{l_c}{R_c T^2}$$

(1.48)

This equation is a differential form of the *Clausius–Clapeyron equation*. We can also obtain an integral form by integrating from a reference temperature T_0 to temperature T, to give saturation vapor pressure as a function of temperature:

$$e_s(T) = e_s(T_0) \exp\left(\int_{T_0}^{T} \frac{l_c}{R_c} \frac{dT}{T^2}\right) \approx e_s(T_0) \exp\left[\frac{l_c}{R_c}\left(\frac{1}{T_0} - \frac{1}{T}\right)\right]$$

$$(1.49)$$

The second expression in eq. (1.49) is approximate because l_c changes slightly with temperature.

What are some properties of the Clausius–Clapeyron equation? Applied to water vapor over liquid water, we take $T_0 = 273.15$ K and $e_s(T_0) = 611$ Pa; $R_c = 461$ J K^{-1} kg^{-1} is the gas constant and $l_c = 2.5 \times 10^6$ J kg^{-1} is the specific enthalpy of vaporization for water. We find that SVP roughly doubles for every 10 K temperature rise. The exponential form of eq. (1.49) indicates that SVP is very sensitive to temperature. Noting that $R_c = k/m_c$, and given that exponents are additive, we deduce from eq. (1.49) that $e_s \propto \exp(-m_c l_c/kT)$. This is a Boltzmann equation, which gives us the insight that the numerator in the exponent is the energy required to free a water molecule from its neighbors while the denominator is the average molecular energy.

For some bodies, the Clausius–Clapeyron equation gives direct insight into the entire atmosphere. On Mars, CO_2 condenses on the cold poles at a temperature ~ 148 K, which corresponds to a CO_2 SVP of ~ 600 Pa. This pressure is the typical surface air pressure on Mars. Consequently, Mars' CO_2 ice caps may be buffered at a temperature near 148 K (Leighton and Murray, 1966). On Pluto and Triton, the atmospheric pressure is just the SVP of N_2 over N_2 ice at temperatures of 38 K and ~ 40 K, respectively (Brown and Ziegler, 1979; Lellouch et al., 2011b; Zalucha et al., 2011). Perhaps on some hot exoplanets, rock-vapor atmospheres are similarly in equilibrium with molten surfaces.

To consider the effect of condensable species on atmospheric temperatures, we define the volume mixing ratio, f_c, of the condensable species according to eq. (1.4),

$$f_c = \frac{n_c}{n} = \frac{e_c}{p}$$

where the ns are number densities, e_c is the partial pressure of the condensable species, and p is the total pressure. Similarly, we define a mass mixing ratio by eq. (1.5) as:

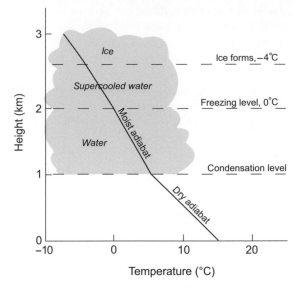

Figure 1.10 A rising air parcel and the lifting condensation level, which forms the base of cloud at 1 km altitude. The cloud is shown schematically as the shaded region.

Now consider a parcel that rises adiabatically from the surface, which is at pressure p_{ref} and temperature T_{ref}. In the adiabatic parcel, potential temperature θ starts out as T_{ref} and remains constant at T_{ref}, while the actual temperature will fall via eq. (1.30) as $T = T_{ref}(p/p_{ref})^\kappa$, or equivalently, $p = p_{ref}(T/T_{ref})^{1/\kappa}$. Substituting for p in eq. (1.50), where e_{parcel} is the partial pressure of the condensable, gives

$$e_{parcel}(T) = \frac{p\mu_{cm}}{\varepsilon} \Rightarrow e_{parcel}(T) = \frac{\mu_{cm}p_{ref}}{\varepsilon}\left(\frac{T}{T_{ref}}\right)^{1/\kappa}$$

$$(1.52)$$

The parcel behaves as dry air until it reaches the SVP temperature, when $e_{parcel} = e_s$, at an altitude called the *lifting condensation level*. After this point, cloud particles form from the condensable and the parcel is warmed. Instead of the temperature dropping at the dry adiabatic lapse rate, the temperature drops less sharply with altitude at a *saturated adiabatic lapse rate* on the SVP curve (Fig. 1.10).

$$\mu_{cm} = \varepsilon f_c = \varepsilon\frac{e_c}{p}, \quad \text{where } \varepsilon = \frac{m_c}{\overline{m}} = \frac{\text{molecular mass of condensable}}{\text{mean molecular mass of air}} \approx \frac{m_c}{m_d} \qquad (1.50)$$

For water vapor in Earth's atmosphere, $\varepsilon = 0.622$. We can also define saturation mixing ratios as

$$f_{cs} = \frac{e_s(T)}{p}, \qquad \mu_{cms} = \varepsilon\frac{e_s(T)}{p} \qquad (1.51)$$

SVP also gives us general insight into the formation of clouds (Fig. 1.11). On average, clouds will tend to form where the mean temperature–pressure conditions reach the SVP for particular consensable species (Sanchez-Lavega et al., 2004).

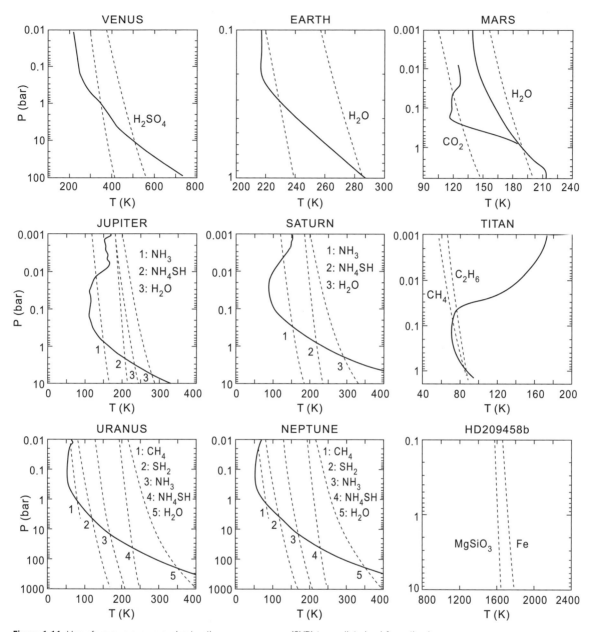

Figure 1.11 Use of vapor pressure and saturation vapor pressure (SVP) to predict cloud formation in Solar System atmospheres and hot Jupiter HD209458b. The solid line is the typical vertical profile of pressure (p) versus temperature (T). For Mars, an annual average p–T profile is shown as well as a colder profile. Dashed lines are the SVP curves for various condensables, assuming a fixed volume mixing ratio of the condensable. It is assumed that particles condense when the partial pressure (e) reaches the saturation vapor pressure (e_s), i.e., where $e = f_c p(T) \geq e_s(T)$, where f_c is the volume mixing ratio of the condensable. Thus, the base of clouds are marked where a dashed line crosses the solid p–T profile. Below this level, the phase is vapor because the ambient air is too hot. Mixing ratios are assumed to calculate the dashed curves: Venus 2 ppm and 2000 ppm H_2SO_4 (where liquid clouds form); Earth 250 ppm and 1.5% H_2O, giving ice and liquid clouds, respectively; Mars 0.95% CO_2 and 300 ppm H_2O giving ice clouds; Jupiter 200 ppm NH_3, 36 ppm NH_4SH, 50 ppm H_2O, 0.17% H_2O giving solid phase clouds except for the higher water level; Saturn 200 ppm NH_3, 36 ppm NH_4SH, 0.17% H_2O, giving solid particles; Titan 5% CH_4, 10 ppm C_2H_6, giving solid particles; Uranus and Neptune 2% CH_4, 37 ppm SH_2, 200 ppm NH_3, 36 ppm NH_4SH, 0.17% H_2O, giving solid particles; HD209458b 75 ppm $MgSiO_3$ and 68 ppm Fe, giving solid particles. (From Sanchez-Lavega et al. (2004) Reproduced with permission. Copyright 2004, American Association of Physics Teachers.)

1.1.3.6 *The Saturated Adiabatic Lapse Rate (SALR)*

The saturated (or "moist") adiabatic lapse rate occurs when chilled moist air is held at saturation by warming associated with continual condensation of excess vapor. Textbooks and papers commonly talk of "release of latent heat," which has been criticized as obscuring the physics (Bohren and Albrecht, 1998, p. 290). Condensation is a warming process because as water vapor molecules approach within a few molecular diameters of liquid water they are attracted to it and increase their kinetic energy.

Recall the equation used in deriving the adiabatic lapse rate in Sec. 1.1.3.1 for a unit mass air parcel,

$$dq = c_p dT + gdz$$

For present purposes, we consider the specific heat capacity c_p, to be that for an air–water vapor mixture. If mass $\delta\mu_{cms}$ condenses then $dq = -l_c\delta\mu_{cms}$, from the definition of the enthalpy ("latent heat") of phase change. Hence, substitution for dq in the above equation gives,

$$l_c\delta\mu_{cms} + c_p dT + gdz = 0 \tag{1.53}$$

At saturation, the condensable species' mass mixing ratio is equal to the saturation mass mixing ratio given by eq. (1.51), $\mu_{cms} = \varepsilon e_s/p$. Taking logs and differentiating (1.51),

$$\frac{\delta\mu_{cms}}{\mu_{cms}} = \frac{\delta e_s}{e_s} + \frac{\delta\varepsilon}{\varepsilon} - \frac{\delta p}{p} \tag{1.54}$$

Now, $\delta e_s = (de_s/dT)\delta T$ and from eq. (1.48), $(1/e_s)(de_s/dT) = l_c/(R_cT^2)$. Also $\delta p/p = -g\delta z/(\overline{R}T)$ from the hydrostatic equation (1.10). Putting all of these in eq. (1.54) gives

$$\frac{\delta\mu_{cms}}{\mu_{cms}} = \frac{l_c\delta T}{R_cT^2} + \frac{g\delta z}{\overline{R}T} \tag{1.55}$$

We can substitute $\delta\mu_{cms}$ from (1.55) into eq. (1.53), as follows,

$$c_p\delta T + gdz + l_c\mu_{cms}\left(\frac{l_c\delta T}{R_cT^2} + \frac{g\delta z}{\overline{R}T}\right) = 0$$

$$\Rightarrow \delta T c_p\left(1 + \frac{l_c^2\mu_{cms}}{c_pR_cT^2}\right) + g\delta z\left(1 + \frac{l_c\mu_{cms}}{\overline{R}T}\right) = 0$$

If we now divide through by δz and rearrange we get

This equation does not allow for the precipitation, suspension or further phase change of the condensate and so is only approximate. However, the mass of the condensate is usually small compared to the gas phase. This lapse rate is also called the *pseudo-adiabatic lapse rate* if condensates are assumed to fall out of the parcel. Static stability is controlled by Γ_{sa} under saturated conditions, i.e., saturated air is stable if $\Gamma < \Gamma_{sa}$. Indeed, in Earth's tropics, data indicate that moist convection controls static stability up to ~ 0.4 bar (Zelinka and Hartmann, 2011). On Earth, in warm humid air near the ground, $\Gamma_{sa} \approx 4$ K km^{-1}, whereas $\Gamma_{sa} \approx 6$–7 K km^{-1} in the middle troposphere, and Γ_{sa} is close to dry adiabatic near the top of the troposphere where the air is cold and dry.

The assumption that the condensable mass mixing ratio is small enough that we can ignore the condensate is not true for CO_2 on Mars or for H_2O in a runaway greenhouse atmosphere. Also the neglect of precipitation is violated by rain on Earth. Additional limits are as follows. (1) That phase change is an equilibrium process, which should not be true for the partitioning of spin isomers of H_2 in giant planet atmospheres. Hydrogen has *ortho*- and *para*-hydrogen forms where the spins of the two atomic nuclei in the molecule are parallel (*ortho*) or antiparallel (*para*). Exothermic conversion of *ortho*- to *para*-hydrogen increases the temperature of an air parcel and is analogous to latent heat release from condensation. (2) That the latent heat of condensation is approximately constant, which is true for most compounds only over a narrow temperature range. More general formulations of the moist adiabatic lapse rate for major constituent condensation are given by Ingersoll (1969) and Kasting (1988, 1991).

1.2 Condensable Species on Terrestrial-Type Planets

Below we discuss CO_2–H_2O atmospheres. The more exotic atmosphere of Titan, the largest moon of Saturn, has methane as a condensable. Methane rainfall and clouds on Titan are discussed in Sec. 14.4.

1.2.1 Pure Water Atmospheres

We can gain some insight into the behavior of condensable volatiles by thinking about idealized atmospheres.

$$\Gamma_{sa} \equiv \text{saturated adiabatic lapse rate (SALR)} = -\left(\frac{dT}{dz}\right)_{sa} = \frac{g}{c_p}\frac{\left(1 + \dfrac{l_c\mu_{cms}}{\overline{R}T}\right)}{\left(1 + \dfrac{l_c^2\mu_{cms}}{c_pR_cT^2}\right)} \tag{1.56}$$

For example, consider a "water world," with water as the sole volatile. We imagine an atmosphere consisting of water vapor as the only contributor to the surface pressure, which, for the sake of argument, we take as 10^4 Pa (100 mbar). One might imagine a gas giant exoplanet that has migrated inward toward its parent star so that an ice-covered satellite orbiting the gas giant now resides where sunlight melts its surface. In Fig. 1.12, suppose a parcel of pure water vapor starts at point A with $T = 340$ K and $p = 10^4$ Pa. The parcel is cooled and follows the path A to B, given that the pressure, which is just the weight of atmosphere above, remains constant. At point B, however, the parcel is at the saturation vapor pressure and so the pure water vapor atmosphere will condense into droplets and precipitate. With further cooling, the parcel would follow the follow the line of stability between the liquid and gas phases, until the triple point is reached, whereupon solid precipitates. Further cooling beyond the freezing point would result in equilibrium between gas and solid phases, at least locally. The pressure does not drop this way in Earth's polar atmosphere in wintertime, where the pressure–temperature combination (10^5 Pa and $T < 0$ °C) lies within the solid phase part of the H_2O phase diagram, because Earth's atmosphere is not pure water, so even if virtually all H_2O is removed, N_2, O_2, Ar, and the other gases maintain a pressure far above the saturation vapor curve. (Note that the freezing point of water is weakly dependent on pressure: 273.16 K at the triple point pressure of water and 273.15 K ($\equiv$ 0 °C) at 10^5 Pa.)

1.2.2 Atmospheres with Multiple Condensable Species

Let us now consider another hypothetical planetary atmosphere with a mixture of CO_2, Ar, and H_2O gases. We take 10^3 Pa = 10 mbar as the (Mars-like) total pressure on this planet, which is the sum of the partial pressure of all gases. Now imagine that we cool a parcel of air that starts at an initial temperature of 340 K (Fig. 1.12). Water condenses at the point where the partial pressure contributed by the water vapor reaches the saturation vapor pressure. Subsequently, the partial pressure contributed by water vapor is fixed in equilibrium with liquid water and then ice at colder temperatures. The water partial pressure is small so the total pressure becomes $\approx P_{CO2} + P_{Ar}$. With further cooling, eventually the CO_2 solid–gas phase boundary is reached and CO_2 will begin to precipitate. Locally, the CO_2 partial pressure will fall to a level supported by the solid–gas equilibrium at the local surface temperature. The total pressure will now be P_{Ar}

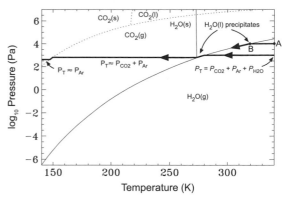

Figure 1.12 Phase diagram of H_2O (solid lines) and CO_2 (dashed). The path from A to B and then along the saturation vapor pressure curve is for cooling a pure H_2O atmosphere starting at a total pressure of 10^4 Pa and temperature of 340 K. The lower path is for a hypothetical Ar–CO_2–H_2O atmosphere starting at the same temperature and a total pressure P_T of 10^3 Pa. As the mixture is cooled, liquid water condenses when the vapor pressure becomes equal to the saturation vapor pressure. At cooler temperatures, CO_2 ice condenses until the only volatile purely in the gas phase is Ar. (Following a concept from Mutch et al. (1976).)

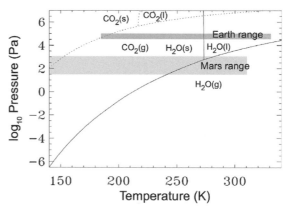

Figure 1.13 Phase boundaries for pure H_2O (solid lines) and CO_2 (dashed). Shaded areas show the range of *total* surface pressure and temperature on Earth and Mars. Earth's surface falls below 190 K in Antarctica but CO_2 does not condense because the CO_2 partial pressure of ~40 Pa is far below the ~10^5 Pa CO_2 saturation pressure at such temperatures. In contrast, the Martian atmosphere consists of 96% CO_2, so CO_2 condenses at the Martian poles. (From Jacob (1999). Reproduced with permission. Copyright 1999, Princeton University Press.)

plus the solid–gas vapor equilibrium of CO_2 at the prevailing surface temperature.

We can apply these ideas to planets. Figure 1.13 shows a plot of the saturation vapor pressure for water and CO_2. Boxes indicate the range of temperatures and pressures on the surface of Earth and Mars. The "Mars range" box shows that generally the equilibrium for water is between ice and vapor, while CO_2 condenses at the poles, represented by the top left-hand corner of the box. After N_2, argon is the second most abundant non-condensable gas on Mars. Analogous to Fig. 1.12, enriched atmospheric Ar has been detected during southern winter as CO_2 frost forms (Sprague *et al.*, 2004; Sprague *et al.*, 2007). This was done using a Gamma-Ray-Spectrometer (GRS) on *Mars Odyssey*. Gamma rays are emitted from [41]Ar decay, which is created from neutron capture by atmospheric [40]Ar.

1.2.3 Water in the Present-Day Martian Atmosphere

Figure 1.14 shows an expanded pure water phase diagram and limits of pressure and temperature on Mars. The spatially and seasonally averaged surface atmospheric pressure on Mars is $\sim$600 Pa (6 mbar), but in low-lying regions such as the Hellas basin, the atmospheric pressure exceeds 1200 Pa. The *boiling point* of a liquid is the temperature at which the saturation vapor pressure equals the total external pressure regardless of the source of that external pressure. For example at 283 K, the saturation vapor pressure of water is $\sim$1200 Pa (Fig. 1.14). At this temperature, liquid water will be stable against boiling if the total pressure against it is 1200 Pa or more, even if some gas other than H_2O, such as CO_2 or N_2, supplies the external pressure. Thus, in principle, liquid water is thermodynamically possible on Mars today in very low-lying locations.

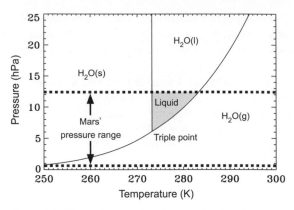

Figure 1.14 Pure phase water diagram showing the range of surface pressures on Mars from the top of Olympus Mons (<0.5 hPa) to the depths of Hellas basin (>12 hPa). Liquid water is stable on Mars against boiling in the shaded zone but not against rapid evaporation.

However, such liquid water would not be stable over time against evaporation, which occurs when the actual vapor pressure in the air is less than the vapor pressure of the liquid water. This is generally true on both Earth and Mars, i.e., liquid water evaporates. However, evaporation rates for Mars are very high (Ingersoll, 1970) so liquid water is unlikely to persist (Hecht, 2002). One caveat is that water that is very salty can have a lower vapor pressure and would be more stable against evaporation on Mars. In addition, certain salts can depress the freezing point far below 0 °C (see Table 12.9).

2 Energy and Radiation in Planetary Atmospheres

2.1 Energy Sources and Fluxes on Planets

2.1.1 Planetary Energy Sources

All planetary atmospheres are affected by the supply of free energy (i.e., energy that can do work and be dissipated), which includes the following possibilities: (1) radioactive decay, (2) accretional energy, (3) energy released by internal differentiation of the planet, such as core formation, (4) tidal energy, and (5) light from a parent star. A combination of energy sources (1)–(4) tends to make planetary interiors hotter than planetary surfaces, so that heat is transferred outwards to a planet's exterior. But this heat flux at the planets' exterior is usually much smaller than absorbed stellar insolation, except for gas giants such as Jupiter (where internal heat is nearly comparable) or extra-solar bodies with great tidal heating. Despite the general difference in the magnitude of fluxes, stellar radiation does not induce the high temperatures found in planetary interiors because internal heat only escapes slowly through solid material whereas stellar energy is balanced by infrared radiation that is usually readily emitted back to space. Exceptions are steam atmospheres, which are highly opaque to infrared, and strongly irradiated exoplanets.

Radioactive energy from planetary interiors is produced by the decay of radiogenic elements, of which the main three on Earth are uranium (U), thorium (Th) and potassium (K). By considering conservation of energy, the change in internal heat must equal the heat input from radioactive decay minus the heat loss Q_{hf} from the surface of the planet. On Earth, the global average for Q_{hf} is 0.087 W m^{-2}, whereas on the Moon, Q_{hf} is about 0.015 W m^{-2} (Schubert *et al.*, 2001).

Geothermal activity and outgassing should scale with heat flow Q_{hf}. However, the scaling with Q_{hf} may be different for each planet, depending on its tectonic regime and the thickness of its lithosphere, which is the outer rigid layer within which heat is transferred by conduction rather than convection. For the Earth, volcanic outgassing is thought to scale roughly with Q_{hf} to a power between 1 and 2. The upper value comes from the idea that outgassing depends on the rate of seafloor area creation, which theory suggests should scale as Q_{hf}^2 (Schubert *et al.*, 2001, p. 597). Since Q_{hf} was certainly greater in the past, then the rate of volcanism and outgassing would have been correspondingly greater, with consequences for atmospheric composition.

Accretional energy comes from the conversion of kinetic energy during giant impacts when planets were formed by the coalescence of smaller bodies. Accretional heat was enough to melt terrestrial planets (Newsom and Sims, 1991). When a terrestrial planet melts, its molten exterior becomes a magma ocean, and this overlies an iron-rich core that is at least partly molten. The surface of this magma ocean should chill and solidify because of radiation of heat to space. However, the skin would be continuously ruptured by the vigorous convection of hot material underneath.

Differentiation, such as core formation. During terrestrial planetary formation, iron sinks to a planet's core because iron is dense, and this causes heating as gravitational energy is converted to thermal kinetic energy. We consider this source of internal energy in more detail in Ch. 6 when we discuss the origin of the Earth and planets.

Tidal heating. Just as the Earth's oceans experience tides due to the pull of the Moon and the Sun, so the solid Earth itself is distorted by these forces and produces tides with amplitude up to 10 cm. This squeeze and release generates a small amount of heat that is often neglected for the Earth. However, tidal energy dissipated in the oceans due to lunar and solar tides is ~10% of the surface heat flow. In Earth's long-term evolution, tidal dissipation from ocean tides is important. Because of oceanic tidal friction, the Earth's rotation rate is slowing down (see Sec. 6.8.2).

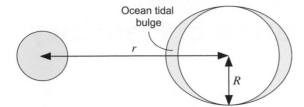

Figure 2.1 The Moon and Earth separated by distance r, illustrating tidal forces. Both the Earth and Moon orbit a common center of mass. In a co-rotating frame the Moon pulls the ocean on the Earth's near side more than the centrifugal force of orbital motion about the center of mass pulls the ocean away. On the far side of the Earth, the centrifugal force of orbital motion about the center of mass pulls the ocean outward relative to the lunar gravitational force. The net effect produces tidal bulges on both near and far sides.

For satellites of the outer planets, tidal heating can dominate the heat flow from the interior. Tidal heat makes Jupiter's moon, Io, the most volcanically active body in the Solar System, while Europa's tidal heating provides an energy source of potential importance for possible life in the Europa's probable subsurface ocean.

For bodies separated by radial distance, r, from center to center of each body (Fig. 2.1), the tidal amplitude, which arises from a differential force, is proportional to $1/r^3$. The gravitational force (F) produced by a body of mass M on a 1 kg test mass is given by Newton's law of gravity, $F = GM/r^2$, where G is the universal gravitational constant. The dependence of tidal force on distance r is given by differentiation:

$$\delta F = \frac{dF}{dr}\delta r = \frac{d}{dr}\left(\frac{GM}{r^2}\right)\delta r \propto \frac{M}{r^3} \qquad (2.1)$$

Because the size of the tidal force varies as $1/r^3$, a small change in separation distance r has a large effect on the size of the tidal bulge.

The amount of energy dissipated by tides can be big. For exoplanets with very eccentric orbits close to their parent stars, tidal heating can be large enough to significantly affect climate and drive atmospheric escape (Barnes *et al.*, 2013). Tidal energy dissipation $\dot{E}_{tide}$ (in Watts) for a synchronously orbiting satellite is a much stronger function of separation distance than the tidal force and depends on $1/r^6$ (Ferraz-Mello *et al.*, 2008; Goldreich and Soter, 1966; Murray and Dermott, 2001, p. 173)

$$\dot{E}_{tide} = \frac{\omega E_{max}}{Q_{factor}} \propto \frac{GM^2 r_{sat}^5 e^2}{Q_{factor} r^6} \qquad (2.2)$$

Here, ω is the angular frequency of flexing (often denoted in the literature as the mean motion "n," which is 2π divided by the orbital period), E_{max} is the peak potential energy stored in the tidal distortion, e is the eccentricity of

an orbit, r_{sat} is the radius of a satellite about the large body of mass M, and Q_{factor} is the tidal dissipation parameter (or "quality factor") of the satellite. The Q_{factor} parameter describes the inverse ratio of the energy lost per cycle to the maximum potential energy stored in the tidal distortion:

$$Q_{factor} = \frac{2\pi E_{max}}{\Delta E} \qquad (2.3)$$

The Q_{factor} parameter increases with lower loss and is the number of cycles for the energy to decrease by $1/e$. For example, the Q_{factor} for Io is 36 and for Jupiter it is $\sim 3 \times 10^4$; for dry bodies the Q_{factor} tends to be around 50–100 (Segatz *et al.*, 1988).

Stellar radiation is the fifth energy source, which we discuss at length below because light from the host star strongly influences a planet's climate.

2.1.2 Radiation From the Sun and Other Stars
2.1.2.1 Spectral Types
Our Sun is a star. Stars differ in their luminosity and color, and the latter indicates the temperature of their outer atmospheres. Stellar temperatures and spectra are important when we consider planetary habitability in other solar systems in Ch. 15. From hot, blue-white to relatively cold infrared, the letters OBAFGKM[1] designate stellar spectral types. Cooler R- and N-type stars (today referred to as carbon, or C-type, stars) and S-type stars are additional stellar classes. But the C- and S-type stars are invariably giant or supergiant stars and differ substantially from the others in having peculiar heavy-metal abundances.

Three spectral types, L stars (very low mass red dwarfs and brown dwarfs) and T and Y stars (brown dwarfs) have been added since the 1990s. Sky surveys found that L and T stars were abundant and that even cooler Y stars exist (Burgasser *et al.*, 2002; Cushing *et al.*, 2011; Kirkpatrick *et al.*, 1999). These stars are comparatively cool (<2500 K) and emit radiation mostly at infrared wavelengths. L stars have deep H_2O absorption bands, but no CH_4 absorption except for late L stars; T dwarfs have H_2O and CH_4 features, while Y stars (~300 K to 500 K) have H_2O, CH_4 and NH_3 bands (Helling and Casewell, 2014). Overall, the complete sequence of spectral types is now OBAFGKMLTY.

Amongst the spectral types are some famous stars visible to the naked eye. Sirius, the brightest star in our sky is A-type, Polaris (the pole star) is an F star, Aldebaran is a K star, and Betelgeuse is an M supergiant.

[1] Generations of students have remembered stellar spectral types with the mnemonic: "Oh Be A Fine Girl/Guy Kiss Me". The addition of LTY dwarfs changes the ending to "My Lips Tonight, Yahoo!".

But we cannot see the most common type of star with the naked eye, namely M dwarfs, which are red dwarfs.

Types O–M are subdivided into ten subcategories. For example, the hottest B star is a B0, followed by B1, B2, and so on; B9 is followed by A0. Our Sun is a G2-type star. For historical reasons, the hotter stars are called early types and the cooler stars late types, so that "B" is earlier than "F," for example. Table 2.1 shows how an *effective temperature* characterizes each spectral type (see Box 2.1). This temperature applies to a star's

Table 2.1 Effective temperature (T_{eff}) as a function of the spectral type for main sequence stars.

Spectral type	O5	B0	A0	F0	G0	K0	M0	L0
T_{eff} (K)	40000	25000	11000	7600	6000	5100	3600	2200

Box 2.1 Blackbody Spectra and "Effective Temperature"

Every object gives out a spectrum of radiation by virtue of its finite temperature. We can see this in a coal fire: the hottest parts are blue and emit a wavelength that is shorter than that coming from the parts that glow red or even cooler parts that radiate in the infrared.

A blackbody spectrum is an idealized spectrum that depends only on the temperature of the source. The temperature uniquely determines the intensity of emitted radiation as a function of wavelength, including the wavelength at which the intensity is a maximum. Moreover, when the radiation output is summed over all wavelengths, the total flux in Watts per square meter depends only on temperature.

Historically, the concept of blackbody radiation derived from considering the radiation field within a closed cavity that perfectly absorbs and emits radiation at all wavelengths. At equilibrium, production and loss of radiation balance and the intensity of the radiation field is uniform. Max Planck (1858–1947) found the spectrum to be uniquely related to the cavity wall temperature, T, by a mathematical expression, the *Planck function*, denoted by the symbol, $B_\lambda(T)$ where the λ subscript indicates that B_λ is specified at a particular wavelength, λ.

The graph below shows blackbody spectra at different temperatures from planets (300 K) to the Sun (~6000 K). A higher temperature shortens the wavelength at which the flux is maximal (where curves intersect the dotted line) and increases the total flux, which is the area under each curve. Dashed lines enclose visible wavelengths.

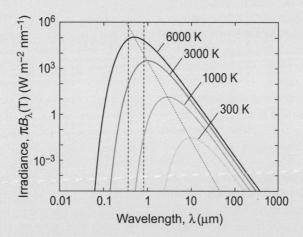

As a first approximation, celestial bodies such as the Earth or Sun are often considered as blackbodies, i.e., objects that give out blackbody spectra of radiation characterized by a single radiating temperature. This temperature is 5780 K for the Sun and 255 K for a mean radiating temperature of the Earth's troposphere. In reality, the spectral output of the Sun and Earth are complicated by species that absorb and emit radiation in their atmospheres. But the total flux output can still be characterized by an equivalent blackbody temperature, which we call the *effective temperature*.

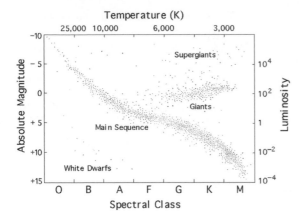

Figure 2.2 Hertzsprung–Russell diagram for nearby stars, with the stellar luminosity plotted relative to a solar luminosity on the right-hand vertical axis of 1. The absolute magnitude is a scale proportional to $\log_{10}$(luminosity), where one step corresponds to a change in brightness of factor $100^{1/5} \approx 2.512$. (Source: NASA.)

photosphere, which is the outer region of the star's atmosphere where most of the emitted electromagnetic energy originates. Although a photosphere is gaseous, the photosphere is also called a star's "surface."

A plot of stellar luminosity against effective temperature is the *Hertzsprung–Russell* (H-R) *diagram*, where temperature is always plotted "backwards" from high to low on the horizontal axis so that it runs in the same sense as the series of spectral types. Figure 2.2 shows an H-R diagram. Nearby stars are plotted because distances to nearby stars can be readily measured by parallax, which gives their absolute luminosities. Stars lie mainly on a diagonal band from upper left to lower right called the *main sequence*. These stars shine because of hydrogen fusion in their interiors, which requires a minimum stellar mass of 75–80 Jupiter masses (M_J). Brown dwarfs (LTY types) are below this mass limit and cannot fuse ^1H but those heavier than $13M_J$ can fuse deuterium, i.e., ^2H.

Main sequence stars are called *dwarfs* to distinguish them from larger, intrinsically brighter *giants* or *supergiants* that lie above and to the right of the main sequence. The Sun is a G dwarf, and the majority of stars are even smaller dwarfs and farther to the right in the main sequence. Main sequence stars are said to be *ordinary dwarfs* to distinguish them from *white dwarfs*, which are faint hot objects located below and to the left of the main sequence. The white dwarfs are stellar remnants of mass less than 1.4 solar masses. Thus, there are G giants, G ordinary dwarfs, and even G white dwarfs all with similar effective temperature. However, the giants are much more luminous than the dwarfs because of greater stellar radii.

Main sequence stars with approximately solar composition have an empirical *mass–luminosity relation* between stellar mass M_* and luminosity L_*, which Salaris and Cassisi (2005, p. 139) give as

$$L_* \propto M_*^{4.5} \quad (\sim 0.5 M_\odot < M_* < 2 M_\odot)$$
$$L_* \propto M_*^{3.6} \quad (2 M_\odot < M_* < 20 M_\odot) \tag{2.4}$$

where $M_\odot$ is the solar mass. The Sun is made up of about 73% hydrogen by mass with most of the remainder helium, and only ~2% of heavier elements.

2.1.2.2 *The Solar Constant: Solar Radiation at the Top of Planetary Atmospheres*

The Sun's properties and location are important in the way they affect planetary atmospheres of the Solar System. Some key solar parameters are follows:

$$\left.\begin{array}{l} \text{mass} = M_\odot = 1.99 \times 10^{30} \text{ kg} \\ \text{luminosity} = L_\odot = 3.83 \times 10^{26} \text{ W} \\ \text{radius} = R_\odot = 6.96 \times 10^8 \text{ m} \end{array}\right\} \tag{2.5}$$

At one Astronomical Unit (AU), the mean distance of the Earth from the Sun, the Sun's energy spreads over the surface area of a sphere, $4\pi r^2$, with radius $r = 1$ AU = 1.495 98 $\times$ 10^{11} m. Consequently, the *solar constant*, which is the solar flux at Earth's mean orbital distance, is $L_\odot/(4\pi r^2) = (3.83 \times 10^{26} \text{ W})/(4\pi \times (1.495 98 \times 10^{11} \text{ m})^2)$ = 1.36 $\times$ 10^3 W m^{-2}. More precisely, satellites orbiting the Earth measure the solar constant as 1360.8±0.5 W m^{-2} (Kopp and Lean, 2011), consistent with ground-based inferences (Chapman *et al.*, 2012). This value is somewhat less than an older number of 1366 W m^{-2} (Lean and Rind, 1998) that was affected by instrumental issues of stray light.

The solar "constant" also changes over a variety of timescales. Because of the eccentricity of Earth's orbit, solar flux varies annually from 1316 W m^{-2} to 1407 W m^{-2} at the extremes. Over decades, the solar constant cycles up and down by ~0.1%, a small amount. The cycle is correlated with the number of sunspots, where the average length of time between sunspot maxima is about 11 years. *Sunspots* are relatively dark regions on the Sun's surface. Because sunspots are cooler (~4000 K) than their surroundings (~6000 K), one might expect the solar luminosity to decrease with the number of sunspots. However, the opposite occurs (Fig. 2.3) because bright regions known as *plages* (pronounced "plah'jes") surround sunspots and bright, hot patches called *faculae* accompany sunspots.

On timescales of millennia, solar variability (the sunspot number) is deduced from isotope proxies. The flux of cosmic rays is modulated by solar activity and carbon-14

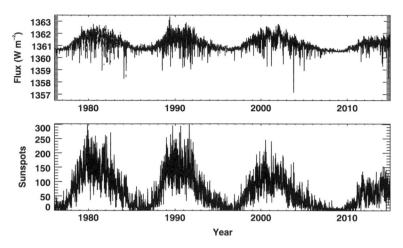

Figure 2.3 The correlation of daily solar irradiance at 1 AU and sunspot number for the years 1976–2014. Flux data are from PMOD (Physikalisch-Metorologisches Observatorium Davos), Switzerland, using measurements from the ESA/NASA SoHO (Solar and Heliospheric Observatory) mission. Sunspot data are from SILSO (Sunspot Index and Long-term Solar Observations), Royal Observatory of Belgium.

and beryllium-10 form when these rays hit the atmosphere (Beer, 2000). Solar activity is inferred from ^{14}C in tree rings and ^{10}Be in ice cores over the past 11 000 years (Solanki et al., 2004), but the effect on the solar constant and Earth's climate is modest (Solanki and Krivova, 2003). However, over tens of thousands of years, the solar flux at the top of a planet's atmosphere is affected by changes in orbital parameters of a planet, according to Milankovitch theory. This is important for the climate of Earth (Ch. 11) and Mars (Ch. 12). On yet longer timescales of millions and billions of years, the solar luminosity (and luminosity of other main sequence stars) gradually increases, as described in Ch. 11.

2.1.2.3 The Solar Spectrum

The spectrum of sunlight consists of a continuum from gamma rays to radio waves with a peak flux around 0.5 μm and a superimposed line structure. In the visible and infrared, there are thousands of absorption lines, known as *Fraunhofer lines*, characteristic of chemical elements in the photosphere. For ultraviolet wavelengths shorter than 185 nm, lines tend to be seen in emission. With a solar constant of 1361 W m^{-2}, a blackbody spectrum at an effective temperature of 5780 K corrected for the distance of a planet provides a good approximation to the visible and infrared portions of the solar spectrum. For ultraviolet wavelengths of 210–300 nm, the effective temperature of the Sun tends to be lower, ~5000 K. However, a spike in flux occurs at 121 nm, well above the Sun's nominal blackbody flux, which is the *Lyman-α* wavelength, corresponding to the transition between ground and first excited state of hydrogen atoms. Lyman-α radiation is important for providing energy to the upper atmospheres of planets and for driving

chemistry, e.g., breaking up methane molecules in the stratospheres of giant planets, Titan, or the early Earth.

2.2 Planetary Energy Balance and the Greenhouse Effect

2.2.1 Orbits and Planetary Motion

Variations in *solar insolation*, which is the flux of energy per unit area (W m^{-2}), affect the radiation balance that sets planetary temperatures. Insolation changes with orbital position. Figure 2.4 shows a general planetary orbit, with a semi-major axis a, and semi-minor axis b. From Fig. 2.4, it can be seen that the *orbital eccentricity e* is defined by the ratio of the distance $a \times e$ from the center of the ellipse to either focus divided by the semi-major axis, a. In terms of major and minor axes of the orbital ellipse, eccentricity is given by

$$e^2 = 1 - \left(\frac{b}{a}\right)^2 \tag{2.6}$$

For circular orbits, $e = 0$. By Kepler's first law, planetary orbits are elliptical, which means that $b/a < 1$ so that $0 < e < 1$.

The *true anomaly f_t* is the angle between perihelion and planetary position (Fig. 2.4). This angle sets the heliocentric (planet-to-Sun) distance $r_\odot$, as follows

$$r_\odot = \frac{a(1 - e^2)}{1 + e \cos f_t}$$
$$\left. \begin{array}{l} r_{\odot,perihelion} = a(1 - e) \\ r_{\odot,aphelion} = a(1 + e) \end{array} \right\} \tag{2.7}$$

The flux at a planet varies inversely with the square of the heliocentric distance, $r_\odot$.

The orbital eccentricity value of different Solar System planets varies widely, and larger values have consequences for climates. Pluto's large eccentricity $e = 0.25$ sets this Kuiper Belt Object apart from the other planets, and Pluto's orbital position drastically affects its atmosphere. Mars, with $e = 0.093$, also experiences climatically significant solar flux variation between perihelion and aphelion. In contrast, the effect of Earth's eccentricity of $e = 0.017$ on our current climate is modest. Earth's eccentricity varies over time, however, from 0 to 0.06, so eccentricity has had a much stronger effect on climate in the past than it has today.

2.2.2 Time-Averaged Incident Solar Flux

The flux of sunlight on a planet of radius R can be calculated by considering Fig. 2.5(a). In this diagram, an elemental ring of planetary surface is defined by angle ϕ, which goes from 0 to $\pi/2$ radians. The element of surface has area $2\pi(R \sin \phi)Rd\phi$, and the component of sunlight normal to the surface is $S_\odot \cos \phi$, where $S_\odot$ is the solar flux at the top of the atmosphere. Consequently, the power integrated over a hemisphere is

$$\text{total power [Watts]} = 2\pi R^2 S_\odot \int_0^{\pi/2} \cos \phi \sin \phi \, d\phi \quad (2.8)$$

We can substitute $x = \sin\phi$ and $dx = \cos\phi \, d\phi$, so that eq. (2.8) can be rewritten as:

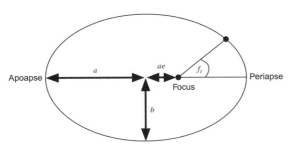

Figure 2.4 Basic elements of an elliptical orbit, showing semi-major axis a, semi-minor axis b, and true anomaly f_t. The eccentricity of the orbit is e.

$$\text{total power} = 2\pi R^2 S_\odot \int_0^1 x dx = 2\pi R^2 S_\odot \left[\frac{x^2}{2}\right]_0^1 = \pi R^2 S_\odot$$

$$(2.9)$$

This power is incident on the sunlit hemisphere. A hemisphere has area $2\pi R^2$. Hence the area-averaged incident flux (in W m^{-2}) is $\pi R^2 S_\odot / 2\pi R^2 = S_\odot/2$. But the hemisphere is illuminated only half the day, so

$$\text{area-and time-averaged incident flux } \left[\text{W m}^{-2}\right] = \frac{S_\odot}{4}$$

$$(2.10)$$

In essence, we divide $S_\odot$ by 2 to account for glancing angles on a spherical planet and we divide by 2 again because only one side of the rotating planet is in daylight.

Another way to understand eq. (2.10) recognizes that the factor of 1/4 represents the ratio of disk to sphere areas $(\pi R^2/4\pi R^2)$. The instantaneous solar flux intercepted by a planet given by eq. (2.9) is equivalent to that on a projected disk of area πR^2 (proved by eq. (2.9)) compared with a total planetary area of $4\pi R^2$ (Fig 2.5(b)). For the Earth, the globally averaged insolation at the top of the atmosphere given by eq. (2.10) is $(1360.8\pm0.5$ W m^{-2})/4 $= 340.2\pm0.1$ W m^{-2}, where $S_\odot$ is from Kopp and Lean (2011).

2.2.3 Albedo

Solar System planets reflect sunlight by different amounts. Consequently, the energy flux given by eq. (2.10) is not entirely absorbed by a planet's surface or atmosphere. The fraction of incident power that is reflected is the *albedo*. There are different ways of expressing the albedo. The *monochromatic albedo* is the fraction of incident power that gets reflected or scattered back to space at a given frequency of light:

$$A_v = \frac{(\text{reflected or scattered power at frequency } v)}{(\text{incident radiation power at frequency } v)}$$

$$(2.11)$$

(a)   (b)

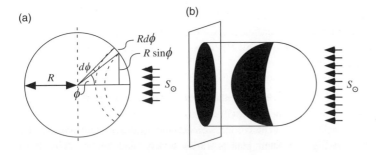

Figure 2.5 (a) An elemental ring of surface defined by angle ϕ on a planet of radius R. The incident solar flux is $S_\odot$. (b) The equivalent area of intercept for the solar flux is a projected disk of area πR^2 compared with a total sphere area of $4\pi R^2$.

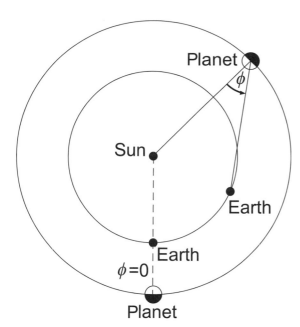

Figure 2.6 The phase angle ϕ for determining geometric albedo is the angle between lines connecting the observer, planet and Sun. *Quadrature* is when the angle subtended at the Earth by the directions to the Sun and a superior planet is 90°.

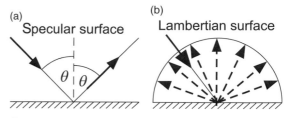

Figure 2.7 Different types of reflection. (a) Pure *specular reflection*, where the angle of incidence equals the angle of reflection, e.g., a mirror. (b) A *Lambertian* surface reflects radiation evenly in all directions. This surface is to be *diffusely* reflecting.

opposition, meaning when the Sun, Earth and planet form a line (Fig. 2.6).

Geometric albedo is related to the Bond albedo by the expression:

$$A_b = qA_g \qquad (2.14)$$

where q is a *phase integral* which reflects the variation of the intensity of radiation over the phase angle ϕ. Karttunen (2007), pp. 149–151, and Seager (2010), pp. 33–39, provide a step-by-step derivation of the phase integral. We note that if the sphere is perfectly reflecting (Bond albedo A_b is unity) and Lambertian (isotropic scatterer) the phase integral q is 1.5 and the geometric albedo is 2/3. Thus 1/3 of the radiation is scattered out of the line-of-sight. In general, even if the Bond albedo is not unity, the geometric albedo is 2/3 of the Bond albedo if the planet is a spherical Lambertian scatterer. Bond and geometric albedos are equal when $q = 1$, which means that reflection from the planet behaves like a Lambertian disk of the same diameter.

Finally, in photometry, reflectance observations of planetary surfaces and atmospheres are often given as a *radiance factor*, I/F, said as "I over F." The radiance factor is the reflected intensity (radiance) I at a given wavelength and viewing angle for an incident solar flux density (irradiance) defined as "πF" in this case (see Sec. 2.4.1.1 for the

If we integrate this over all frequencies, we get the *Bond albedo* named after astronomer George Bond (1825–1865) and synonymous with *planetary albedo*,

$$A_b = \frac{(\text{total reflected or scattered radiation power})}{(\text{incident radiation power})}$$
$$(2.12)$$

For climate calculations, we use the Bond albedo.

In astronomy, a *geometric albedo* is often used. For Solar System planets viewed from the Earth, the phase angle ϕ, as shown in Fig. 2.6, is the angle between incident sunlight and radiation received on Earth. When $\phi = 0$, sunlight is observed in pure backscatter. The geometric albedo is defined by

$$A_g = \frac{F(\phi = 0)}{F_{\text{Lambert-disk}}} = \frac{(\text{reflected flux at zero phase angle})}{(\text{flux reflected from a Lambertian disk of the same cross-section})} \qquad (2.13)$$

Here, $F_{\text{Lambert-disk}}$ is the flux reflected by a disk with a Lambertian surface of the same cross-section as the planet at the same distance from the Sun. A Lambertian surface is one that reflects all incident radiation isotropically, as illustrated in Fig. 2.7, such that its brightness is the same in all directions of view. For example, a wall painted in matt white is very roughly Lambertian. Thus, the geometric albedo is the fraction of incident light reflected in the direction of the observer of an outer planet measured at

definition of "flux density"). A perfectly reflecting Lambertian surface viewed at normal incidence has $I/F = 1$. Hapke (2012) reviews I/F and similar quantities.

2.2.4 Planetary Equilibrium Temperature

The Bond albedo enables us to calculate the effective temperature (i.e., blackbody equivalent temperature) that a planet attains in equilibrium with sunlight. If the Bond albedo is A_b, then using eq. (2.10), the time-averaged solar flux absorbed by a planet is $(1 - A_b)S_p/4$, where S_p is the solar flux at the orbital position of the planet. The Stefan–Boltzmann Law (derived later in Sec. 2.4.1.5) gives the emitted flux of radiation from a blackbody at absolute temperature T as σT^4, where $\sigma = 5.6697 \times 10^{-8}\,\mathrm{W\,m^{-2}\,K^{-4}}$ is the Stefan–Boltzmann constant. A planet is a good absorber of infrared radiation and therefore a good emitter, so a planet can be approximated by a blackbody at infrared wavelengths. If we assume that a planet's internal heat can be neglected, an equilibrium temperature, T_{eq}, is achieved from a balance of incoming and outgoing radiation, as follows:

absorbed solar radiation flux = outgoing infrared radiation flux

$$(1 - A_b)\frac{S_p}{4} = \sigma T_{eq}{}^4$$

(2.15)

We can apply eq. (2.15) to the inner planets. Earth's Bond albedo is 0.3, so the left-hand side of eq. (2.15) is $\sim 238\,\mathrm{W\,m^{-2}} = (1 - 0.3) \times (1361\,\mathrm{W\,m^{-2}}/4)$. At equilibrium, this energy flux must be emitted back to space by radiation from the Earth. Solving eq. (2.15) gives an equilibrium temperature, $T_{eq} = 255$ K. This temperature is an effective blackbody temperature, defined in Box 2.1, and is clearly not equal to Earth's actual surface temperature. Table 2.2 shows actual mean surface emission temperatures of Earth, Venus, and Mars compared to equilibrium temperatures calculated from eq. (2.15). The difference between the surface temperature and equilibrium temperature in Table 2.2 arises because of the *greenhouse effect* caused by the presence of an atmosphere around each planet.

2.2.5 The Greenhouse Effect

On planets with atmospheres, the atmosphere warms up and provides a source of thermal infrared radiation to the surface in addition to the shortwave solar flux. Thus, the global average surface temperature of the Earth, T_s, is

about 288 K rather than the blackbody emission temperature of 255 K (Table 2.2). The *greenhouse effect* can be explained as follows. The atmosphere warms up because it absorbs infrared radiation from the Earth below. Because the atmosphere is warm it radiates, and some of this radiation travels back down to the Earth. Consequently, the surface of the Earth is hotter than it would be in the absence of an atmosphere because it receives energy from two sources: sunlight and the heated atmosphere.

On a planet with a dense atmosphere, the difference between T_{eq} and T_s is a convenient way to express the magnitude of the greenhouse effect

$$\Delta T_g \equiv T_s - T_{eq}$$

(2.16)

For the Earth, $\Delta T_g = 288 - 255 = 33$ K.

We now qualify the above considerations. On an airless body, such as the Moon, there is no greenhouse effect, so $\Delta T_g = 0$ K. The Moon's surface receives solar radiation only, so we might expect its mean surface temperature to be the equilibrium temperature, according to eq. (2.15). However, the arithmetic mean global temperature is lower. A slowly rotating airless body does not have an atmosphere or ocean to transport heat efficiently. Because surface conduction is inefficient, the lateral transfer of heat is minimal and each local surface rapidly reaches thermal equilibrium with incoming solar radiation in relative isolation. Although global energy balance must still apply, a global average temperature is not a good measure of radiation balance because local emission has a T^4 scaling by the Stefan–Boltzmann Law, so the actual temperature representative of mean surface emission is biased towards warmer areas. For airless bodies, one must

Table 2.2 A comparison of equilibrium temperatures calculated using eq. (2.15) and the mean global surface temperatures of inner planets. (1 AU = 1.496×10^{11} m.)

Planet	Bond albedo (dimensionless)	Semi-major axis (AU)	Equilibrium temperature (K)	Mean global surface temperature of emission (K)	Greenhouse effect (K)
Venus	0.76[a]	0.7233	229	737[d]	508
Earth	0.3[b]	1.00	255	288	33
Mars	0.25[c]	1.5236	210	215[e]	5

[a] Moroz *et al.* (1985). [b] Palle *et al.* (2003). [c] Pleskot and Miner (1981). [d] At the mean radius of Venus (Fegley, 2014). [e] For Mars, this value is computed from the global and annual mean value of T_s^4, where T_s is the surface temperature; the temperature 215 K is fourth root of this value via eq. (2.17) (Haberle, 2013). See Sec. 2.2.5 for why a simple arithmetic global mean is less than this value for Mars.

use a time-dependent, spatially resolved model (Pierre-humbert, 2010 p. 152; Vasavada *et al.*, 2012; Vasavada *et al.*, 1999) and the average emission temperature must be calculated with a T^4 weighting.

Although Mars is not an airless body, its atmosphere is so thin that heat transfer is quite inefficient too. The arithmetic global mean surface temperature of Mars is estimated to be 202 K (Haberle, 2013), whereas the equilibrium temperature is 210 K (Table 2.2). This apparent paradox is resolved by realizing that one needs to compute the fourth power of the surface temperature T_s^4 at all locations on the planet and average this quantity globally and annually to obtain $\overline{T_s^4}$. Then, the global mean effective *surface emission temperature* T_{se} and corresponding greenhouse effect are defined as follows:

$$T_{se} = \left[\overline{T_s^4}\right]^{1/4}, \qquad \Delta T_g \equiv T_{se} - T_{eq} \qquad (2.17)$$

For Mars, T_{se} is 215 K so that the greenhouse effect is ~5 K by eq. (2.17).

For the Earth two factors allow the global mean surface temperature to be an accurate reflection of the global mean emission temperature. First, there is efficient lateral heat transfer by the atmosphere and oceans. Second, a feedback effect of water vapor in the atmosphere causes the emission of radiation from the surface to have a net linear scaling with surface temperature, which we explain in Sec. 2.3.1.

For Earth, there is debate about whether a greenhouse effect should be expressed in terms of radiative fluxes rather than a global mean temperature enhancement, ΔT_g in eq. (2.16) (Schmidt *et al.*, 2010; Zeng, 2010). In our view, using ΔT_g is no problem if we remember that it is directly related to fluxes of IR radiation. When we say 33 K warming compared to an Earth without an atmosphere, we mean, "all other things being equal in a hypothetical scenario," i.e., if you imagined the same albedo and a mean isothermal surface. Of course, an imaginary Earth with no atmosphere and ocean would not have the same albedo and, even if it did, hypothetically, the mean surface temperature would differ from 255 K because of the nonlinear T^4 scaling of local thermal emission on an airless world. We use the convention of ΔT_g because it is rooted in two observables: it indirectly quantifies the roughly 150 W m^{-2} difference in flux between ~390 W m^{-2} from a measurable 288 K mean global surface (emission) temperature and ~240 W m^{-2} at the top of the atmosphere from a globally equivalent blackbody temperature of 255 K that can be measured by satellites.

In Earth's modern atmosphere, the two most important greenhouse gases are H_2O and CO_2. In clear skies,

H_2O is responsible for ~2/3 of the 33 K warming, although it acts in a different manner than CO_2 because it is near its condensation temperature in the troposphere. Effectively, H_2O is a slave to CO_2 and other greenhouse gases: if CO_2 levels increase and warm the Earth, then warmer surface waters produce a larger vapor pressure of H_2O, which amplifies the greenhouse effect (Lacis *et al.*, 2013; Lacis *et al.*, 2010). In Earth's current climate, doubling CO_2 itself leads to ~1.2 K mean global temperature rise. However, various feedbacks in the climate system can cause the actual temperature to eventually increase 2.0–4.5 K. Based on observations, the positive feedback due to water vapor is ~2 W m^{-2} K^{-1} and 0.3 W m^{-2} K^{-1} in the troposphere (Dessler *et al.*, 2008) and stratosphere, respectively (Dessler *et al.*, 2013). More will be said about water vapor feedback in Sec 2.3.

After water vapor, CO_2 accounts for most of the remaining 1/3 of the greenhouse effect in a clear sky atmosphere (Kiehl and Trenberth, 1997; Stephens and Tjemkes, 1993). Lesser contributions, around 2–3 K total, come from CH_4, O_3, N_2O, and various anthropogenic chlorofluorocarbons (CFCs). These trace constituents are powerful greenhouse gases on a molecule per molecule basis in Earth's present atmosphere. For example, in the current atmosphere, each CH_4 molecule increases radiative forcing 34 times more than a CO_2 molecule over a century (Stocker *et al.*, 2013). This is partly because CH_4, N_2O, O_3, and CFCs absorb within the ~8–12 μm wavelength region, an otherwise mostly transparent *window* through which most of the Earth's thermal-IR radiation energy fluxes to space. But it is also partly because these last four trace gases are all much less abundant in Earth's atmosphere than is CO_2. When one compares equal amounts of CH_4 and CO_2, CO_2 is by far the better greenhouse gas because the shorter wavelength wing of its strong absorption band at 15 μm is closer to the Wien peak of the outgoing thermal-IR radiation (see Sect. 2.4.1.5) than is the 7.6-μm CH_4 band (Kiehl and Dickinson, 1987).

Equations (2.15) and (2.16) show that the mean surface temperature, T_s, of a rocky planet with an atmosphere depends on three factors:

(a)   solar flux, which is set by astronomical geometry and solar physics,

(b)   Bond albedo A_b, which is set by clouds, aerosols, and the nature of the surface, and

(c)   the greenhouse effect, which is set by the composition of the atmosphere.

For the evolution of climate, any changes in T_s (e.g., low-latitude glaciation) can only be understood by appealing to changes in one or more of these factors.

The most problematic factor is the way clouds affect both planetary albedo A_b and the greenhouse effect. Clouds cause most of the albedo on some planets, such as Earth,[2] and all of the albedo on planets such as Venus. (However, if you could magically take the clouds away from Venus, the planet would still have a high albedo of ~0.6 because dense CO_2 is good at scattering shortwave photons). On Earth, clouds contribute about 0.15 of the 0.3 Bond albedo, which cools the Earth. However, icy cirrus clouds also contribute to greenhouse warming because they are largely transparent to visible radiation but opaque to thermal IR. When clouds are taken into account, it is found that the global contributions to the greenhouse effect are ~50% from water vapor, ~25% from clouds, ~20% from CO_2, and the rest from other gases (Schmidt *et al.*, 2010). Here the percentages refer to fractions of the total longwave (LW) greenhouse forcing of 155 W m^{-2}: the global annual average upwelling LW flux ($\sigma T^4 = 390$ W m^{-2}, with $T = 288$ K) minus the top-of-atmosphere mean longwave flux, currently estimated as 235 W m^{-2} (Kiehl and Trenberth, 1997). However, when we look at the climate as a whole with satellite observations, while clouds enhance the greenhouse effect by ~30 W m^{-2}, they reduce Earth's absorbed shortwave radiation by −47 W m^{-2} (Loeb *et al.*, 2009). The net effect is that clouds *cool* the Earth by −17 W m^{-2}, which is large (Loeb *et al.*, 2009; Ramanathan and Inamdar, 2006).

Clouds can be observed and parameterized in Earth's present atmosphere, but their exact properties are difficult to predict for atmospheres other than our own (Marley *et al.*, 2013). For this reason, climate calculations for early Earth or for other Earth-like planets are subject to considerable uncertainty.

Clouds can also be influenced by biology. Clouds consist of water vapor that condenses on cloud condensation nuclei (CCN), which are submicroscopic particles. In Earth's atmosphere, a large contribution to CCN comes from sulfate salts, which can be affected by biology in the following ways. Some marine algae creates a gas, dimethlyl sulfide (CH_3SCH_3), or "DMS," that is subsequently oxidized and hydrated to form CCN (Charlson *et al.*, 1987). In the troposphere, where there is biological ammonia, most of the sulfate aerosols are mixtures of ammonium sulfate ((NH_4)$_2SO_4$) and bisulfate ((NH_4)HSO_4). For example, if sulfuric acid is derived from the oxidation of a sulfur-bearing gas, ammonia can neutralize the sulfuric acid derived as follows,

$$2NH_3 + H_2SO_4 \rightarrow (NH_4)_2SO_4 \qquad (2.18)$$

There is a correlation between marine stratus clouds and phytoplankton chlorophyll (Falkowski *et al.*, 1992), but whether this system has positive, negative or zero feedback on climatic temperature is currently unknown. However, other feedbacks, which we discuss below in Sec. 2.3, are more constrained.

In the stratosphere, sulfate aerosols are sourced from volcanic sulfur (SO_2 or H_2S that is subsequently oxidized) as well as biological gases, mainly carbonyl sulfide (OCS). Oxidation and hydration causes sulfuric acid (H_2SO_4) aerosols to form in the stratosphere, where their scattering of solar radiation contributes to planetary albedo.

2.2.6 Giant Planets, Internal Heat, and Equilibrium Temperature

As we discussed in Sec. 2.2.4, planets absorb radiation from the Sun and lose energy by radiating to space in equal amounts at equilibrium, but this idea assumes that we can ignore internal heat sources. For terrestrial planets, the internal heat flux is small. For example, the Earth's average heat flow is only ~90 mW m^{-2} (Schubert *et al.*, 2001) compared to the globally averaged net solar flux of 238 W m^{-2}. But in the Solar System Jupiter, Saturn and Neptune radiate significantly more energy than they absorb. For example, Jupiter radiates ~13.6 W m^{-2} while it absorbs only ~8.2 W m^{-2} from the Sun. An internal energy flux of 5.44 ± 0.43 W m^{-2} makes up the difference (Hanel *et al.*, 1981; Pearl and Conrath, 1991).

Internal heat on giant planets comes from accretion and differentiation. Gravitational energy is released in these processes. Saturn has higher internal heat than expected from its size and is believed to have been in state of internal differentiation for billions of years, in which immiscible helium "rains out" of a deep interior layer of metallic hydrogen towards Saturn's core, with release of gravitational energy (Stevenson and Salpeter, 1977b). Support comes from a relatively low He/H ratio of 0.6 in Saturn's atmosphere relative to solar compared to 0.81 on Jupiter. Because Saturn is smaller and cooler in its interior than Jupiter, the phase separation of helium has been going for far longer than inside Jupiter, where it began more recently. The fact that Uranus has little internal heat flow, 42 ± 47 mW m^{-2} (Pearl *et al.*, 1990), unlike the other giant planets is puzzling and may be due to different internal structure.

---

[2] Terrestrial clouds cover $68 \pm 0.3\%$ of the surface for >0.1 in optical depth clouds or ~$56 \pm 0.3\%$ for >2 in optical depth. Also, there is 10%–15% more cloud cover over oceans than land (Stubenrauch *et al.*, 2013).

For giant planets with significant internal heat fluxes, the *effective temperature* T_{eff} (of an equivalent blackbody) is greater than the *equilibrium temperature*, T_{eq} where outgoing IR balances incoming sunlight. Consequently, there is an internal heat flux, F_i, given by

$$F_i = \sigma\left(T_{eff}^4 - T_{eq}^4\right) \tag{2.19}$$

The total emitted flux is modified to include this internal heat,

outgoing radiation flux = absorbed solar flux + internal heat flux

$$\sigma T_{eff}^4 = (1 - A_b)\frac{S_p}{4} + F_i \tag{2.20}$$

2.3 Climate Feedbacks in the "Earth System"

Feedbacks in the Earth system can amplify the mean temperature of the planet in positive feedback or stabilize it in negative feedback. There are many potential feedbacks for climate, of which four important ones are:

(1) positive feedback from atmospheric water vapor;
(2) positive feedback from ice–albedo;
(3) negative feedback from outgoing radiation on short timescales; and
(4) negative feedback from the carbonate–silicate cycle on geological timescales.

Water vapor acts as a positive feedback on the climate system because it is near its condensation temperature. As the climate cools, the saturation vapor pressure drops. If the relative humidity (the ratio of vapor pressure to saturation vapor pressure) remains constant, then the water vapor concentration in the atmosphere will decrease proportionately. Less water vapor results in a smaller greenhouse effect, which in turn results in further cooling. Just the opposite happens if the climate warms: atmospheric H_2O increases, thereby increasing the greenhouse effect and amplifying the initial warming. This particular feedback loop is very important for Earth's climate. It essentially doubles the effect of climatic perturbations, such as changes in solar flux or in atmospheric CO_2 (Stevens and Bony, 2013).

The second feedback, the *ice–albedo feedback*, describes the interaction between surface temperature and the fraction of Earth's surface covered by snow and ice. An increase in surface temperature causes a decrease

in snow and ice cover, which decreases the albedo and so causes further temperature increase in *positive* feedback. Conversely, a decrease in surface temperature increases the albedo due to snow and ice cover, which amplifies the decline in surface temperature. The snow-and-ice albedo feedback loop played a major role in the advances and retreats of the *Pleistocene* (~2.6 Ma–10 ka) ice sheets (Lorius *et al.*, 1990) and is an important positive feedback in anthropogenic climate change.

The climate system must also contain negative feedbacks, or else it would be unstable. The most basic negative feedback is the interaction between surface temperature T_s, and the outgoing infrared flux, F_{IR}. As T_s increases, F_{IR} increases. Earth cools itself by emitting infrared radiation; thus, as F_{IR} increases, T_s decreases. This creates a negative feedback loop. This feedback loop is the reason that Earth's climate is stable on short timescales. On long timescales ~1 m.y., we will see in Ch. 11 that an important climate feedback involves the interaction between atmospheric CO_2 and surface temperature, called the carbonate–silicate cycle (Berner, 2004; Walker *et al.*, 1981). In this feedback, if the climate warms, the weathering rate on land increases and removes carbon dioxide, which acts in negative feedback on surface temperature.

2.3.1 Climate Sensitivity

An illustration of the way the Earth's climate works in response to feedback comes from considering a simple energy balance, following Budyko (1969). Imagine the Earth as a blackbody with a surface temperature in Kelvin of $T = T_0 + \Delta T$. If T_0 is a reference temperature, such as water's freezing point (273.15 K), then we can write $\Delta T = T_s$, the surface temperature in °C. Thus, the emitted thermal-IR flux by the Stefan–Boltzmann Law (Sec. 2.4.1.5) is

$$F_{IR} = \sigma T^4 = (T_0 + T_s)^4$$

We can expand the right-hand side of this equation as follows:

$$\sigma(T_0 + T_s)^4 = \sigma(T_0 + T_s)^2(T_0 + T_s)^2 = \sigma\left(T_0^2 + 2T_sT_0 + T_s^2\right)\left(T_0^2 + 2T_sT_0 + T_s^2\right)$$
$$= \left(\sigma T_0^4\right) + \left(4\sigma T_0^3 T_s\right) + \cdots$$

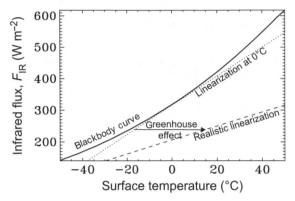

Figure 2.8 Outgoing infrared radiation at the top of the terrestrial atmosphere F_{IR}, as a function of global mean surface temperature. The solid line is the blackbody case. A linearization to this curve about 0 °C gives a surface temperature of –16.6 °C for an outgoing flux of 239 W m^{-2}. A more realistic linear model has a mean surface temperature of 15 °C and a significantly different slope or climate sensitivity. The arrow at F_{IR} = 239 W m^{-2} shows the magnitude of the greenhouse effect, which causes the temperature offset between the two linearizations.

By taking only the first two terms, we can linearize the emitted IR flux as

$$F_{IR} = \underbrace{a}_{\sigma T_0^4} + \underbrace{bT_s}_{4\sigma T_0^3 T_s}; \qquad T_s \text{ in } °C \qquad (2.21)$$

If the Earth behaved truly as an airless blackbody, then this linearization would give $a = \sigma(273.15)^4 = 315.659$ W m^{-2}

and $b = 4\sigma(273.15)^3 = 4.623$ W m^{-2} °C^{-1}. Given an incoming or outgoing radiation of ~239 W m^{-2}, the value of T_s would be –16.6 °C, close to 255 K for the blackbody curve.

For Earth's actual climate, the values of a and b are empirically closer to $a = 206$ W m^{-2} and $b = 2.2$ W m^{-2} °C^{-1}. Inserting these numbers in eq. (2.21), we derive:

$$239 \text{ W m}^{-2} = 206 \text{ W m}^{-2} + (2.2)(T_s) \qquad (2.22)$$

From eq. (2.22), we calculate $T_s = 15$ °C. Equation (2.21) is plotted in Fig. 2.8. Because the value of b is smaller in the real climate than for a blackbody, the surface temperature, T_s, is more sensitive to changes in F_{IR} (or the solar constant S_0, since F_{IR} is proportional to S_0). Thus, the surface temperature increases more for a given change in F_{IR} than it would in the blackbody case. The main reason for the higher sensitivity is water vapor feedback. If the

climate warms, the water vapor increases and prevents IR radiation from escaping.

Climate sensitivity λ, is defined as the ratio of the change in global mean surface temperature ΔT_s to the change in climate forcing ΔQ (where Q has units of W m^{-2}).

$$\Delta T_s = \lambda \Delta Q \qquad (2.23)$$

The change in forcing could be a change in the solar constant or greenhouse effect. In the case of the latter, an accurate value of λ is important for predicting the anthropogenic influence on climate change.

We can evaluate the climate sensitivity to changes in the solar constant for a blackbody approximation and compare our simple but more realistic linearized model of Earth's climate system represented by eq. (2.21). For the blackbody case, let us ignore any temperature dependence of albedo and assume that the effective temperature and surface temperature are linearly related. Radiation balance gives us

$$F_{IR} = \sigma T^4 = \frac{(1 - A_b)}{4} S_\odot$$

Then

$$\frac{d(F_{IR})}{dS_\odot} = \frac{(1 - A_b)}{4} = \left(\frac{d(F_{IR})}{dT}\right)\left(\frac{dT}{dS_\odot}\right) \qquad (2.24)$$

Hence,

$$\lambda = \frac{dT}{dS_\odot} = \frac{(1 - A_b)}{4}\left(\frac{1}{d(F_{IR})/dT}\right) = \frac{(1 - A_b)}{4(4\sigma T^3)} = \frac{T}{4}\left(\frac{(1 - A_b)}{4\sigma T^4}\right) = \frac{T}{4S_\odot} \qquad (2.25)$$

Consequently, λ = (255 K)/(4 × 1361 W m^{-2}) = 0.047 K (W m^{-2})$^{-1}$. This means that a 1% change in the solar constant $S_\odot$ of 0.01 × 1361 = 13.61 W m^{-2} would produce a 13.61 × 0.047 = 0.64 K change in surface temperature. However, this Stefan–Boltzmann feedback is too small and unrealistic. If we consider our linearized model, we have

$$F_{IR} = a + bT = \frac{(1 - A_b)}{4} S_\odot \qquad (2.26)$$

If we now apply eq. (2.24), we obtain

$$\lambda = \frac{dT}{dS_\odot} = \left(\frac{d(F_{IR})/dS_\odot}{d(F_{IR})/dT}\right) = \frac{[(1 - A_b)/4]}{b} \qquad (2.27)$$

Consequently, in this case the climate sensitivity is almost double that before, λ = (1– 0.3)/(4 × 2.2 W m^{-2} K^{-1}) = 0.080 K (W m^{-2})$^{-1}$. In this case, a 1% change in the solar constant would produce a change in surface temperature of about 1.1 K. The almost factor of 2 difference between this and the Stefan–Boltzmann case comes about because we have included an empirical water vapor positive feedback, which doubles climate sensitivity.

We can also estimate how much temperature is raised by an increase in the greenhouse effect. Radiative model calculations suggest that a doubling of CO_2 from 300 ppmv to 600 ppmv would cause an extra 4 W m^{-2}

$$\tau_e = \left(\frac{p_{emission}c_p(\Delta T_{emission})}{g}\right) \Big/ \left(\frac{(1-A_b)S_p}{4}\right) = \frac{4p_{emission}c_p(\Delta T_{emission})}{(1-A_b)S_p g} \quad (2.29)$$

in the greenhouse effect and warm the surface. From eq. (2.26), $dT/d(F_{IR}) = 1/b = 0.45$ K (W m^{-2})$^{-1}$, so that the surface temperature would be expected to change by 4 W m^{-2} × 0.45 K (W m^{-2})$^{-1}$ ~2 K, assuming no other positive or negative feedbacks. This simple calculation produces an answer that is near the lower end of the range predicted by sophisticated, three-dimensional simulations of global warming. The IPCC (Intergovernmental Panel on Climate Change) suggests a temperature increase of (1.5–4.5) K for doubled CO_2, based on calculations by a suite of different climate models (Stocker et al., 2013). Additional positive feedbacks from decreased sea ice and changes in cloud cover generally suggest ~3 K warming for doubled CO_2 (Hansen et al., 2008).

2.3.2 The Emission Level and Radiative Time Constants

For planetary thermal equilibrium, it is useful to think of a level at some height, $z_{emission}$, in the atmosphere where the temperature T equals the effective temperature T_{eff}. We call this level the *emission level*, i.e.,

$$T_{surface} = T_{eff} + \Gamma z_{emission} \quad (2.28)$$

where Γ is the lapse rate. Given that the Earth's troposphere has a typical mean lapse rate $\Gamma = 6$ K km^{-1}, $T_{surface}$ = 288 K, and T_{eff} = 255 K, the emission level is at a height of about 5 km on Earth.

We can think of energy absorbed from the Sun as being balanced by emission at the emission level at temperature T_{eff}. The column mass at this level, M_{ce}, is given by eq. (1.19) as $M_{ce} = p_{emission}/g$, where $p_{emission}$ is the pressure at the emission level. If a temperature change $\Delta T_{emission}$ is forced by the absorption of solar radiation, the heat change due to heat capacity of the column of atmosphere is given by

$$\text{heat change per m}^2 = M_{ce}c_p(\Delta T) = \frac{p_{emission}c_p(\Delta T_{emission})}{g}$$

where c_p is the specific heat capacity and g is gravitational acceleration. For sunlight,

$$\text{power absorbed per m}^2 = \frac{(1-A_b)S_p}{4}$$

Consequently, we can define a time constant as follows:

If $\tau_{day} = 1$ day, we obtain a typical diurnal temperature variation at the emission level associated with average solar input of

$$\Delta T_{emission} = \frac{(1-A_b)S_p g \tau_{day}}{4p_{emission}c_p} \quad (2.30)$$

Typical values of $\Delta T_{emission}$ are 2 K (Venus), 2 K (Earth), 80 K (Mars), and 0.001 K (Jupiter). Diurnal radiative effects are clearly important on Mars. In contrast, terrestrial radiative processes act slowly and can be largely ignored in short-term atmospheric developments. In the long-term, however, they dominate sources and sinks of energy. Also from eq. (2.30), we can find the pressure level where $\Delta T_e = T_{eff}$ as follows: 700 mbar (Venus), 8 mbar (Earth), 2 mbar (Mars), and 0.05 mbar (Jupiter).

The radiative time constant is sometimes defined differently from above. For example, if the pressure in eq. (2.29) is taken as surface pressure, and we consider heating a full column of air to temperature T_{eff}, then using eq. (2.15) to substitute for $\sigma T_{eff}^4 = S(1-A_b)/4$, we get

$$\tau_{rad1} = \frac{M_c c_p}{\sigma T_{eff}^3} = \frac{p_s c_p}{g\sigma T_{eff}^3} \quad (2.31)$$

where M_c is the total columnar mass. Equation (2.31) applies to planets with surfaces.

2.4 Principles of Radiation in Planetary Atmospheres

The way that solar radiation interacts with atmospheres and surfaces of planets is essential for understanding planetary habitability, given that liquid water is needed for Earth-like life. The solar energy absorbed is balanced by the transmission of thermal infrared radiation through a planet's atmosphere and back to space. The extent to which thermal infrared radiation is absorbed depends upon the greenhouse gases in the atmosphere, and these gases link the evolution of atmospheric composition to climate. In order to appreciate how climate depends upon atmospheric composition, we need to understand *radiative transfer*, the physics of how electromagnetic radiation passes through media. For our purposes, the medium is an atmosphere of matter that scatters, absorbs, and emits radiation. However, before getting into details, we start with some basic definitions.

2.4.1 Basic Definitions and Functions in Radiative Transfer

2.4.1.1 Solid Angles and Spherical Polar Coordinates

The study of radiation often involves considering the amount of energy in a solid angle. A *solid angle*, expressed in units of steradians (sr), is defined as:

$$\text{solid angle}, \Omega = \frac{\text{area } \sigma_A \text{ of spherical surface}}{r^2} \quad [\text{sr}]$$
$$(2.32)$$

Thus, for a complete sphere, $\Omega = 4\pi r^2 / r^2 = 4\pi$ steradians. *Spherical polar coordinates* are also often used in radiative transfer where a *zenith angle* θ, and an *azimuthal angle* ϕ, are defined according to Fig. 2.9(a). This figure also shows an elemental area, $d\sigma_A = (rd\theta)(r\sin\theta d\phi)$, from which we define a differential solid angle, $d\Omega$ as

$$d\Omega = \sin\theta d\theta d\phi \quad (2.33)$$

2.4.1.2 Radiometric Quantities

In radiative transfer, it is important to understand the distinction between radiance, irradiance, and net flux. Briefly, these are defined as follows.

(a) *Radiance = intensity, I*, is the power per unit frequency interval crossing unit area in unit solid angle in a given direction at some point in space. It has units of W m^{-2} Hz^{-1} sr^{-1} (or it can be per unit wavelength, e.g., W m^{-2} nm^{-1} sr^{-1}). Blackbody radiation is usually expressed as intensity denoted by B.

(b) *Flux density = irradiance, F*, is radiance integrated over a solid angle, usually a hemisphere, i.e., total power per unit frequency crossing unit area (a scalar quantity). In a spectral region, it is specified in units of W m^{-2} Hz^{-1} (or per unit wavelength, e.g. W m^{-2} nm^{-1}). Strictly speaking, "flux" only refers to the power in units of Watts, but that definition is almost never used. In most literature, "flux" is identical to "flux density." If we integrate over the whole spectrum, the result is a total flux density, expressed in W m^{-2}. A mnemonic for the difference between *radiance* and *irradiance* is that the "i" in the latter helps you to remember that irradiation involves integration of radiance over a solid angle.

(c) *Net flux ($\underline{F}$)* takes into account energy transfer in all directions, and $\underline{F}$ refers to a unit area in a specific direction, i.e., $\underline{F}$ is a vector. $F^+ - F^-$ is the component of $\underline{F}$ in that direction. Net flux is specified in W m^{-2} for a given frequency interval.

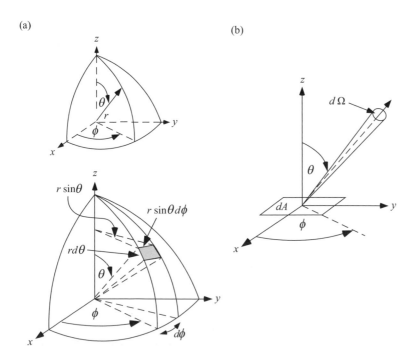

(a)

(b)

Figure 2.9 (a) Spherical coordinates: zenith angle θ, azimuthal angle ϕ, and radius r. (b) Illustration of the angles that define the radiance fluxing through an element of area dA through elemental solid angle $d\Omega$ in a direction defined by zenith angle θ and azimuthal angle ϕ.

2.4.1.3 Heating Rate

The heating rate in a planetary atmosphere is the net power dissipated (or removed from) the medium per unit volume. It is often expressed in units of K day^{-1}. The heating rate depends on the divergence of $\underline{F}$, which is the gradient of $\underline{F}$ as reflects inflow and outflow. Thus the heating rate due to the difference in upward and downward fluxes in the vertical (z) direction is given by

$$\frac{dT}{dt}\rho_{air}c_p = -\mathrm{div}\underline{F}, \qquad \text{which, in the } z \text{ direction, } = -\frac{d}{dz}(F^+ - F^-) \tag{2.34}$$

where F^+ is in the same direction as z. (For understanding divergence, recall that for a vector $\underline{v} = v_1\mathbf{i} + v_2\mathbf{j} + v_3\mathbf{k}$, we define $\mathrm{div}\,\underline{v} - \partial v_1/\partial x + \partial v_2/\partial y + \partial v_3/\partial z = \nabla \cdot \underline{v}$, and that divergence measures outflow minus inflow). Alternatively, sometimes arrows are used to indicate the direction of fluxes, in which case we can rewrite eq. (2.34) as

$$\frac{dT}{dt}\rho_{air}c_p = -\left(\frac{dF\uparrow}{dz} - \frac{dF\downarrow}{dz}\right) = \frac{d}{dz}(F\downarrow - F\uparrow) \tag{2.35}$$

where $F\uparrow$ is the upward flux.

2.4.1.4 Hemisphere Radiation and the Relationship Between Flux and Intensity

The geometry for a pencil beam of radiation passing through an elemental area, dA, is shown in Fig. 2.9(b). The radiant energy dE_v in time interval dt and frequency range v to $v + dv$ crossing elemental area dA is given by

$$dE_v = I_v\cos\theta\,dA\,d\Omega\,dv\,dt \tag{2.36}$$

where $dA\cos\theta$ is the effective area of intercept. From earlier, an element of solid angle is $d\Omega = \sin\theta\,d\theta\,d\phi$, which we can substitute. The *monochromatic flux density*, i.e., the flux density at a particular frequency, is defined by the normal of the radiance I_v integrated over the entire upper hemisphere,

$$F_v{}^+ = \frac{E_v}{At} = \int_0^{\phi=2\pi}\int_0^{\theta=\pi/2} I_v\cos\theta\sin\theta\,d\theta\,d\phi - 2\pi\int_0^{x=1} I_v x\,dx \tag{2.37}$$

In the last step, we have substituted $x = \sin\theta$ and $dx = \cos\theta\,d\theta$ and assumed I_v has no azimuthal dependence. Note that we can also write flux density for the hemisphere on the opposite side as:

$$F_v^- = \int_0^{\phi=2\pi} d\phi \int_\pi^{\theta=\pi/2} I_v\cos\theta\sin\theta\,d\theta \tag{2.38}$$

Considering the upper hemisphere, if the irradiance is isotropic (i.e., $I_v(\theta) = I_v$), we can integrate the expression in eq. (2.37) as follows:

$$F_v^+ = 2\pi I_v\left[\frac{x^2}{2}\right]_0^1 = \pi I_v \quad [\text{unit}: \text{ W m}^{-2}\text{Hz}^{-1}] \tag{2.39}$$

Note that integration over a hemisphere is why we frequently see a "pi" written in front of radiances. Note also

that the net flux transported through an area will be $F_v - F_v^+ - F_v^-$.

Incidentally, the factor of pi is a perennial source of confusion for generations of students of radiative transfer because of a peculiar convention in a foundational textbook by Chandrasekhar. He chose to absorb the factor of pi into his definition of net flux as "πF_{Av}" instead of using F_v (Chandrasekhar, 1960, p. 3). Here, we use an "A'" subscript in F_{Av} to denote Chandrasekhar's "astrophysical radiative flux". In Chandrasekhar's notation, eq. (2.37) would be written $F_{Av}^+ = 2\int_0^{x=1} I_v x\,dx$, for comparison.

The *total flux density* or *irradiance* (which has units of W m^{-2}) is obtained by integrating over all frequencies v, i.e.,

$$F = \int_0^\infty F_v dv \tag{2.40}$$

2.4.1.5 Blackbodies: the Planck Function, the Stefan–Boltzmann Law, and Wien's Law

The Planck function is an intensity that derives from Planck's Law for the energy of a photon,

$$E = hv \tag{2.41}$$

where $h = 6.626 \times 10^{-34}$ J s. It describes the blackbody radiation spectrum (Box 1.2), as a function of temperature. There are three forms of the Planck function that are commonly encountered. First, when the spectrum is in frequency units, we have

$$B_v dv = \frac{2hv^3}{c^2}\frac{dv}{(e^{hv/kT} - 1)}, \quad \text{where } B_v \text{ is in W m}^{-2}\text{ Hz}^{-1}\text{ sr}^{-1} \tag{2.42}$$

Second, using $c = v\lambda$, => $dv = -(c/\lambda^2)d\lambda$, we can rewrite eq. (2.42) in terms of a wavelength spectrum

$$B_\lambda d\lambda = \frac{2hc^2}{\lambda^5} \frac{d\lambda}{(e^{hc/\lambda kT} - 1)}, \quad \text{where } B_\lambda \text{ is in W m}^{-2} \text{ m}^{-1} \text{ sr}^{-1}$$

$$(2.43)$$

Third, we can rewrite the function in terms of wavenumber units (usually cm^{-1}), which are used in spectroscopy. Since $\tilde{v} = 1/\lambda$, $d\tilde{v} = -(1/\lambda^2)d\lambda$, so

$$B_{\tilde{v}} d\tilde{v} = 2hc^2 \frac{\tilde{v}^3 d\tilde{v}}{(e^{hc\tilde{v}/kT} - 1)}, \quad \text{where } B_{\tilde{v}} \text{ is in W m}^{-2} \left(\text{cm}^{-1}\right)^{-1} \text{sr}^{-1} \qquad (2.44)$$

There are two key inferences from the Planck function: the Stefan–Boltzmann Law and Wien's Law.

The Stefan–Boltzmann Law results from eqs. (2.39) and (2.40) applied to the Planck function (2.42). We can use the substitution $x = hv/kT$ and the standard integral

$$\int_0^\infty \frac{x^3 dx}{e^x - 1} = \frac{\pi^4}{15}$$

Thus,

$$F = \int_0^\infty \pi B_v(T) dv = \frac{2\pi h}{c^2} \int_0^\infty \frac{v^3 dv}{(e^{hv/kT} - 1)} = \frac{2\pi k^4 T^4}{c^2 h^3} \int_0^\infty \frac{x^3 dx}{(e^x - 1)} = \left(\frac{2\pi^5 k^4}{15c^2 h^3}\right) T^4 \qquad (2.45)$$

The group of constants in front of T^4 on the right-hand side comprise the Stefan–Boltzmann constant, σ (= 5.67 $\times$ 10^{-8} W m^{-2} K^{-4}), as follows,

$$F = \sigma T^4 = \left(\frac{2\pi^5 k^4}{15c^2 h^3}\right) T^4 \qquad (2.46)$$

Wien's Law gives the spectral location of the maximum emission of the Planck function,

$$\frac{\partial B_\lambda}{\partial \lambda} = 0 \quad \Rightarrow \lambda_{max} T = 2.897 \times 10^{-3} \text{ m K} \qquad (2.47)$$

For the Sun, characterized by an effective temperature of 5780 K, Wien's Law indicates that the peak emission occurs at a wavelength of $\lambda_{max} = (2.897 \times 10^{-3} \text{ m K})/(5780 \text{ K}) = 0.5 \mu\text{m}$, in the visible. One nuance of Wien's Law is that when we work in wavenumbers, the maximum occurs at a wavenumber corresponding to a wavelength λ_{max} given by $\lambda_{max} T = 5.100 \times 10^{-3}$ m K, which is not the same as in the case of the wavelength curve. Thus, strictly speaking, the position of the Planck maximum has no meaning unless we say which spectral units we are using.

2.4.1.6 *Kirchoff's Law*

Media often differ from perfect blackbodies. One defines an absorptivity, a_v, which is the fraction of energy at frequency v per unit frequency falling on a body that is absorbed. In the case of a blackbody, it perfectly absorbs and $a_v = 1$ for all frequencies. Generally, $a_v < 1$ for all frequencies and some of the radiation is reflected or transmitted rather than absorbed.[3] When we assume that a_v is some fixed value less than 1 for all frequencies, we define what is known as a *gray body*. For example, a *gray atmosphere* is an approximation using a broadband value of a_v less than 1 for radiative calculations.

In thermodynamic equilibrium, with a steady temperature of the medium, a body emits as much as it absorbs. We can introduce the emissivity, which is the ratio of the spectral radiance at a particular frequency or wavelength to that of a blackbody, i.e., $\varepsilon_v =$ (spectral radiance at frequency v)/ $[B_v(T)]$. Kirchoff recognized a curious fact: in thermodynamic equilibrium, the absorptivity at a particular frequency depends on the temperature of the substance. Kirchoff's Law says that

$$(\text{emissivity})_v = (\text{absorptivity})_v, \quad \text{or} \quad \varepsilon_v = a_v \qquad (2.48)$$

Thus, a fractional absorber at a particular frequency and given temperature is also a fractional emitter at the same frequency in thermodynamic equilibrium. Strictly, eq. (2.48) is valid only for a specific viewing direction, unless ε_v (and so a_v) is independent of direction.

2.4.2 Radiative Transfer in the Visible and Ultraviolet

We now have enough basic definitions to consider the passage of radiation through an atmosphere and we start with shortwave (ultraviolet (UV) and visible) because it has simpler physics than that in the thermal infrared (IR). In the IR, the complication is that a planetary atmosphere itself, by virtue of its finite temperature, generates radiation at IR wavelengths in addition to the incident

---

[3] Absorptivity (and hence emissivity) can exceed unity in the case of small particles that have a larger interaction cross-section with radiation than their geometric cross-section. The rule that absorptivity < 1 applies strictly only to extensive objects.

sunlight. In the visible and UV parts of the spectrum, we can neglect radiative emission at temperatures typical of planetary surfaces and atmospheres, which greatly simplifies calculations.

We can easily justify our neglect of shortwave emission by using eqs. (2.43) and (2.39). At the short wavelength end of the planetary infrared spectrum in the near-infrared at 1 μm, the monochromatic flux radiance falling on a horizontal plane just above the Earth's surface from radiation from the Earth is πB_λ (288 K, 1 μm) = 7.1×10^{-8} W m^{-2} m^{-1}. In contrast, the monochromatic flux radiance at 1 μm from the Sun is

$$(1 - A_b)\left(\frac{R_\odot}{r_{AU}}\right)^2 \pi B_\lambda (5780 \text{ K, 1 μm}) \tag{2.49}$$

which computes to 2.76×10^8 W m^{-2} m^{-1}. Here, A_b is the Bond albedo of the Earth (≈ 0.3), $R_\odot$ is the radius of the Sun (6.96×10^8 m), and r_{AU} is the Earth–Sun distance of 1 AU or 1.49598×10^{11} m. Because the visible flux is $\sim 10^{15}$ greater, we can neglect the planetary thermal radiation at this wavelength; solar radiation clearly dominates, as seen in Fig. 2.10.

Given that solar and thermal-IR radiation fields have little overlap, we often treat them independently. In this approximation, the atmosphere is assumed to be transparent to solar radiation but largely opaque to thermal radiation from the Earth. This was the approach we used in estimating a planetary equilibrium temperature earlier in Sec. 2.2.4. However, planetary atmospheres are not completely transparent to solar radiation: they absorb in the UV and near-IR parts of the solar spectrum. Since relatively little energy is contained in the UV and near-IR parts of the visible spectrum, the energy balance calculations are not affected much.

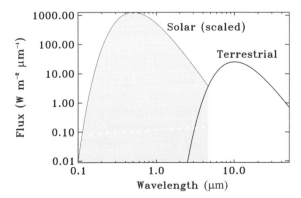

Figure 2.10 Blackbody curves of the Sun and Earth, where the solar flux density is that at 1 AU distance scaled by a factor of $1 - A_b$, where A_b is Earth's Bond albedo. The curves cross near ~4 μm. Solar and thermal spectra peak at ~0.5 and 10 μm, respectively.

The structure and color of atmospheres are strongly affected by shortwave absorption and scattering by gases. In the terrestrial atmosphere, H_2O vapor absorbs near-IR solar radiation (1–4 μm region) and contributes 1 K per day heating to the mid-troposphere at ~5 km altitude. Stratospheric ozone absorbs UV (0.3–0.2 μm), and all gases absorb EUV (i.e., <100 nm) in the thermosphere. Ozone also weakly absorbs in the visible spectrum in the Chappuis bands continuum (~450–850 nm), with peak absorption in the yellow-orange at ~600 nm (which is why pure ozone gas is a pale blue color).

The blue color of Earth's sky comes from Rayleigh scattering of the solar spectrum by N_2 and O_2 molecules along with the response of the human eye to color (Bohren and Fraser, 1985; Hoeppe, 2007; Smith, 2005). At very low Sun angles during twilight and dusk, transmitted light from the Earth's horizon should be yellow, which is the color remaining after Rayleigh scattering out of the light; but absorption of yellow by the ozone layer ensures a blue hue. Edward Hulburt first realized that the long pathlength through ozone dominates over Raleigh scattering under these circumstances (Hulburt, 1953). Yellow-orange absorption in the transmission of starlight through exoplanet atmospheres is also a possible way to detect ozone (e.g., Yan et al., 2015). Such a measurement is of interest because the ozone on Earth is derived from biogenic O_2.

The color of the Earth when viewed as a single pixel from Saturn's orbit or beyond has been described as a Pale Blue Dot, which arises from the scattering and absorption properties of the Earth (Sagan, 1994). Rayleigh scattering of sunlight back to space from the atmosphere primarily causes the blue color, not the ocean (Crow et al., 2011; Krissansen-Totton et al., 2016b). Clouds, which reflect fairly uniformly in the visible, cause the paleness. In detail, ozone absorption in the Chappuis bands sits at the flat bottom of a U-shape in the 400–1000 nm reflectance spectrum, reducing the reflectivity by 7%–8% (Robinson et al., 2014a). An upward slope in reflectance occurs from 600 nm to 900 nm because of vegetated land surfaces (Arnold et al., 2002; Fujii et al., 2010; Seager et al., 2005; Tinetti et al., 2006).

In the giant planet atmospheres, parts of the solar visible and near-IR spectrum are absorbed. In particular, Jupiter, Saturn, Uranus, Neptune, and Titan all have strong CH_4 bands in the visible and near-IR (543 nm, 619 nm, 725 nm, 844 nm, 865 nm, 888 nm) (Karkoschka, 1994, 1998). The light we see coming from Neptune is sunlight scattered by its uppermost atmosphere. CH_4 absorbs in the orange-red part of the spectrum so that scattered sunlight is depleted in red light, and Neptune appears bluish-green. The color of Uranus is attributable to the same process.

2.4.2.1 The Beer–Lambert–Bouguer "Extinction Law" and the Meaning of Opacity

The *Beer–Lambert–Bouguer Law* describes how a beam of light is exponentially attenuated with distance in an optically homogeneous medium. Pierre Bouguer (pronounced "boo-gair") first described the law in 1729, followed by Johann Lambert in 1760 and August Beer in 1852. Despite Bouguer being first, the law is commonly called *Beer's Law* or *Lambert's Law*. For simplicity, we call it *the Extinction Law*.

When a beam of light passes through a substance, it is found experimentally that the loss of intensity depends linearly on both the incident intensity I_v, and the amount of optically active matter (i.e., absorbing or scattering material) along the beam direction, proportional to an elemental path, ds:

$$dI_v \propto I_v ds \qquad \Rightarrow dI_v = k'_v I_v ds \qquad (2.50)$$

The constant of proportionality is the *extinction coefficient k'_v*. In general, extinction of radiant energy is caused by absorption and scattering. Bouguer's insight was that the exact process doesn't matter for the formulation of the law. The extinction coefficient is the sum of an absorption coefficient and a scattering coefficient, and generally depends on temperature and pressure because these affect the absorption of gases.

The extinction coefficient can be defined in three ways:

(i) the **extinction coefficient** or **volume extinction coefficient**, in terms of pathlength:

$$k'_v = \frac{dI_v}{I_v ds} \qquad [\mathrm{m^{-1}}] \qquad (2.51)$$

(ii) the **mass extinction coefficient**, in terms of mass path, $dM = \rho_a ds$, where ρ_a is the mass density of the absorber or scatterer [kg m^{-3}]:

$$k_v = \frac{dI_v}{I_v \rho_a ds} \qquad [\mathrm{m^2\,kg^{-1}}] \qquad (2.52)$$

(iii) the **extinction cross-section**, in terms of column number density, $dN_c = n_a ds$, where n_a is the number density [m^{-3}] of the optically active gas:

$$k''_v = \frac{dI_v}{I_v n ds} \qquad [\mathrm{m^2}] \qquad (2.53)$$

In fact, for particles, a dimensionless extinction efficiency can be defined in terms of k_v', as follows

$$Q_e = \frac{k''_v}{A_{\mathrm{xs}}} \qquad (2.54)$$

where A_{xs} is the geometric cross-section, such as πr^2 for a sphere of radius r. For terrestrial cloud droplets, $Q_e \sim 2$ in the visible part of the spectrum, for example.

If we integrate eq. (2.50), we get the usual form of the Extinction Law:

$$I_v = I_{v0} e^{-\tau_v} \Rightarrow \text{transmissivity } T_v = e^{-\tau_v} \qquad (2.55)$$

The law indicates that intensity decays exponentially along the path. Here I_{v0} is the intensity at distance $s = 0$ and τ_v is the (extinction) *optical path* or *optical thickness* or *opacity* along the beam direction, which is a dimensionless quantity defined as:

$$\tau_v = \int_0^s k'_v ds' = \int_0^s k_v \rho_a ds' = \int_0^s k''_v n_a ds' \qquad (2.56)$$

Sometimes, particularly in astronomy, the extinction coefficient k_v is called "opacity," with potential for confusion. Optical thickness τ_v is a measure of the strength and number of *optically active* particles (meaning those that scatter and/or absorb) along a beam. Equation (2.56) also shows that factors of ρ_a or n_a relate all three extinction coefficients.

2.4.2.2 Direct Beam Solar Flux and the Meaning of "Optical Depth"

The Extinction Law can be formulated for the case of solar irradiance, i.e., flux, noting that the above is for radiance. Consider a *plane-parallel* atmosphere, which is an approximation that ignores the spherical nature of a planet and assumes that atmospheric properties are only a function of altitude, z. We consider a direct beam at zenith angle θ, and ignore diffuse radiation scattered in from different angles.

In Fig. 2.11, both flux density F and distance ds are measured positive downwards. The flux is attenuated according to the pathlength, ds. Trigonometry relates the pathlength to altitude through $-dz = ds \cos\theta$, so that $ds = -dz/(\cos\theta)$. For an absorber density, ρ_a, and mass absorption coefficient, k_{abs}, over the visible range, the decrease in flux over pathlength ds is given by eq. (2.52) as:

$$dF = -k_{\mathrm{abs}} \rho_a F ds \qquad (2.57)$$

By substituting for ds in terms of dz, we obtain

$$\cos\theta \frac{dF}{F} = k_{\mathrm{abs}} \rho_a dz \qquad (2.58)$$

We define *optical depth, τ, as the optical path (or optical thickness) measured vertically downwards from the top of the atmosphere* to some altitude,

$$\tau = \int_z^\infty k_{abs}\rho_a dz' = \int_{\tau=\tau}^{\tau=0} -d\tau' \qquad (2.59)$$

Note that we need to be careful with the sign in eq. (2.59). Thus, (2.58) becomes:

$$\cos\theta \frac{dF}{F} = -d\tau \qquad (2.60)$$

which integrates to

$$F = F_\infty e^{-\tau/\cos\theta} \qquad (2.61)$$

Here, F_∞ is the downward flux density at the top of the atmosphere where $z = \infty$. In (2.61), the flux density decreases exponentially along the slant path (inclined at angle θ to the vertical) with a total optical thickness given by $\tau/\cos\theta$.

2.4.2.3 Direct and Diffuse Solar Fluxes

The optical depth of atmospheres can often be relatively high. For example, on Mars the broadband visible optical depth in global dust storms rises to ~5 (Colburn *et al.*,

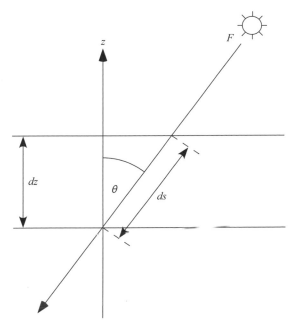

Figure 2.11 The slanted extinction path *ds*, of a direct solar beam through a slab of air in a plane-parallel atmosphere. The Sun is at a zenith angle of θ.

1989; Kahn *et al.*, 1992), which means that direct beam sunlight is attenuated by a factor of $e^{-5} \approx 1/148$. However, this apparently large factor bears little relationship to the sunlight reaching the ground because much of the visible light is scattered by the dust and reaches the ground as diffuse light (Haberle *et al.*, 1993a).

In reality, the total flux at the ground is the sum of two components (Monteith, 2012):

$$\text{total solar flux at ground} = (\text{vertical component of direct flux}) + (\text{diffuse flux}) \qquad (2.62)$$

The *direct flux* is the flux normal to the surface of the beam from the solar disk (plus a small component scattered forward), i.e., eq. (2.61) gives F and then $F\cos\theta$ gives the vertical component of F. The *diffuse flux* is all other scattered radiation from the sky and from clouds, either scattered or transmitted. (The sky component includes a bright white glow around the Sun, known as the *aureole*, which is caused by forward scattering of sunlight by aerosol particles.) Another term for the total flux at the ground is the *global flux*, which is generally considered as the solar radiation (0.3–3 μm) received from a solid angle of 2π steradians on a horizontal surface.

On Earth during daylight, the direct beam rarely exceeds 75% of the solar constant, i.e. ~1030 W m^{-2}, with the remaining 25% attributable to scattering and absorption in roughly equal proportions. In a cloudless sky, the diffuse flux is maximally ~200 W m^{-2} at noon and no more than ~10%–15% of the total flux. If the sky is obscured by dense cloud, the diffuse flux is the total flux at the ground.

The calculation of the diffuse flux is more complicated than the direct flux. Typically, *aerosols* (defined as suspended solid or liquid particles) are involved in the scattering and absorption, so we must know their properties.

2.4.2.4 Extinction, Absorption, and Scattering

The extinction coefficient mentioned previously is the sum of an absorption coefficient and a scattering coefficient, i.e.,

$$k_\nu = a_\nu + s_\nu \qquad (2.63)$$

Both a_ν and s_ν typically depend on temperature, so that k_ν is determined from lab measurements. Absorption or scattering can be due to gases or aerosols. The ratio of s_ν to extinction is known as the *single scattering albedo*:

$$\tilde{\omega} = \frac{s_\nu}{a_\nu + s_\nu} \qquad (2.64)$$

The value of $\tilde{\omega}$ is <0.5 for strongly absorbing substances and 1 for purely scattering.

Absorption is when a molecule or atom absorbs a photon, producing an excited state. The lifetime of excitation varies from $\sim10^{-8}$ s in the shortwave to ~0.1–10 s in the infrared, and at sufficiently high pressures, molecular collisions generally occur before the excited atom or molecule is able to undergo *radiative decay* and emit a photon in a random direction. Thus, the energy gets transferred to translation, rotation, or vibration that readily interchange energy. Excess vibrational or rotational energy gets converted into kinetic energy by collisions amongst the molecules and is released as heat. The process of changing the photon energy into heat is *thermalization* or *quenching*. For short wavelength photons, such as EUV (i.e., <120 nm) in the upper atmosphere, absorption can also cause *photo-ionization* or *photo-dissociation*. Aerosols can also absorb photons directly and warm the atmosphere during daytime, e.g., dust on Mars (Pollack *et al.*, 1979) or black carbon on Earth (Ramanathan and Carmichael, 2008).

Scattering is when the direct beam of radiation is attenuated because light is redirected out of the original direction of propagation as a result of an interaction with aerosols or when radiative decay of an excited molecule or atom emits a photon in a random direction. Gas molecules scatter solar radiation (*Rayleigh scattering* – see below). Aerosol particles (dust, smoke or cloud droplets) with dimensions comparable to the wavelength of radiation scatter radiation according to *Mie scattering*. Mie theory solves Maxwell's equations for the electromagnetic field around a dielectric particle. Although scattering of photons can result in loss of radiation from the beam, photons can also be scattered into the beam. This is the problem of *multiple scattering*, which occurs in atmospheres loaded with aerosols.

An insight into various forms of scattering can be gleaned from the classical physics of a molecule acting as a simple harmonic oscillator. We consider an electron bound to a nucleus with a "spring constant," which undergoes oscillations at a single natural frequency when irradiated. The electric field of the incoming radiation drives the electron up and down and the electron radiates because it is accelerated. In this case, we use the mathematics of a forced oscillator. The following expression for the resulting total scattering cross-section is derived in *The Feynman Lectures on Physics*, pp. 32–36 (Feynman *et al.*, 1963), where it is discussed in the context of light scattering. For simplicity, the expression omits a damping term in the denominator:

$$\sigma_s = \frac{8\pi r_e^2}{3}\left[\frac{v^4}{(v^2 - v_0^2)^2}\right] \tag{2.65}$$

Here v_0 is the resonant frequency and r_e is the *classical electron radius*, which is derived by equating the electrostatic energy of an electron with its relativistic rest energy, giving $r_e = (q_e^2/4\pi\varepsilon_0)/(m_e c^2) = 2.82 \times 10^{-15}$ m, where q_e is the electron charge, ε_0 is the permittivity of free space, m_e is the mass of an electron, and c is the speed of light.

How does eq. (2.65) apply? If we know the electric field E, the average of the square of this field times $\varepsilon_0 c$ gives the flux (in W m^{-2}) through a surface normal to the direction in which the radiation is going, i.e.

$$F = \varepsilon_0 c \langle E^2 \rangle \tag{2.66}$$

Molecules or atoms in a gas are randomly located so that the power radiated by the total number of molecules in any direction is just the sum of the power scattered by each molecule. So we can multiply the radiated power by the number of free gas molecules. But if N molecules are joined together in a lump such as a tiny aerosol, which is much smaller than the wavelength of radiation, then the electric field of the aggregated molecules will move in phase in response to incident radiation. The amplitude of the scattered radiation will be the sum of the N radiated electric fields in phase, so that the power will be N^2 times the scattered power of a single molecule by eq. (2.66).

A dimensionless *size parameter* is used to help to diagnose the type of scattering. This parameter is defined in terms of r_s, the scatterer radius, and λ the wavelength of radiation, as follows:

$$x = \frac{2\pi r_s}{\lambda} \tag{2.67}$$

With the above reasoning, the scattering cross-section of eq. (2.65) and size parameter implies four cases.

(1) *Thomson scattering* happens when free electrons are produced by photoionization at very high altitudes in a planetary atmosphere. These electrons do not have resonant frequencies or quantized levels, so $v_0 = 0$ and the total cross-section is independent of frequency, i.e., $\sigma_s = 8\pi r_e^2/3$ = 6.65×10^{-25} cm^2 for an electron.

(2) *Rayleigh scattering* occurs when $v_0 \gg v$, and $\sigma_s \propto v^4 \propto \lambda^{-4}$. In this case, the size of the scattering particle is much smaller than the wavelength of radiation, i.e., $x \ll 1$. For example, gas molecules Rayleigh scatter visible light. As the size r_s, of scatterers increases, the number of molecules they contain scales as r_s^3, so the scattering power of the particles and effective cross-section grows in proportion to r_s^6 (see Fig. 2.12). Rayleigh scattering, of

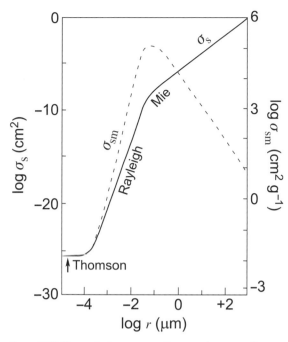

Figure 2.12 The scattering cross-section σ_s and cross-section per unit mass σ_{sm}, as a function of the effective radius r, of a scattering particle, for 0.6 μm radiation (orange visible light). Four regimes of scattering occur: Thomson, Rayleigh, Mie, and geometric. Radius r ranges from the size of an electron to 1 mm. (Redrawn, based on a similar diagram by Lewis (2004), p.185.)

2.4.3 Radiative Transfer in the Thermal Infrared

In calculating the transmission of purely visible and UV radiation passing through an atmosphere, we saw that we did not need to consider the thermal emission in the atmosphere because thermal emission at planetary temperatures has negligible radiative output in the shortwave. For example, with a visible–UV spectrometer on the surface of Earth or Mars looking upwards, we could calculate the column abundance of a species from measured atmospheric absorption across a particular band without having to know the thermal structure of the atmosphere.

In the thermal-IR, we need to consider the attenuation of IR radiation through a path *and* the temperature-dependent emission in the IR at points along the path. Consequently, radiative transfer in the IR is inherently more complicated than in the visible and UV and forms the subject of a large literature.

2.4.3.1 *Spectral Visualization of the Greenhouse Effect*

Satellite measurements of the emission spectrum of the Earth show that the Earth–atmosphere system is not a blackbody in reality (Fig. 2.13). In comparing the level of a blackbody spectrum corresponding to the N. African surface temperature in Fig. 2.13 of ~320 K, one sees "holes" or dips in the spectrum due to gas absorption. The base of the dips corresponds to the effective temperature of the emission. Photons from the bottom of the dips emanate not from the surface but from higher up in the atmosphere where it is colder. In particular, the 20 μm water band has emission at the temperature of the lower troposphere. The bottom of the absorption dip for carbon dioxide emits at the temperature near the tropopause, shown schematically in Fig. 2.13(b).

In Fig. 2.13(a), the total area of the emitted spectrum would be the total emitted power per unit surface area. If we had a globally averaged version of Fig. 2.13(a), the total emitted power would correspond to that of a blackbody at 255 K, which we calculated in Sec. 2.2.4 from energy balance. Gaseous absorption (the area of the "holes" in the real spectrum) obviously decreases the total area and the total power emitted. So in order to maintain equilibrium between power received from the Sun and power emitted, the surface temperature (which defines the blackbody curve where there is an atmospheric window, e.g., 8–12 μm) must increase well above 255 K. Thus, Fig. 2.13(a) or any similar spectrum obtained from orbit can be interpreted as a visualization of Earth's greenhouse effect.

course, is responsible for the blue sky of the Earth because atmospheric gas molecules scatter blue photons more than red in all directions.

(3)   *Mie scattering* arises when the size of a scatterer becomes comparable to the wavelength in the size parameter range $0.1 \leq x \leq 50$. In this case, the scattering particles are no longer all in phase because they are too far apart compared to the wavelength of incident radiation. Then, the situation is no longer so simple because of complex relationships between scattered intensity, scattering angle, and the resultant superposition of electromagnetic radiation. Fine dust, smoke, or cloud droplets produce *Mie scattering*, often with forward scattering. In models, a parameter used to account for the directionality of scattering is the asymmetry factor, g_s. This parameter indicates whether the scattering direction is forward ($0 < g_s \leq 1$), backward ($-1 \leq g_s < 0$) or isotropic ($g_s = 0$) (e.g., see Petty (2006)).

(4)   *Geometric scattering* happens when the scatterer size is much greater than the wavelength of incident radiation, $x > 50$. Then, the scattering cross-section grows in proportion to the geometric area, πr_s^2, as shown in the solid line of Fig. 2.12.

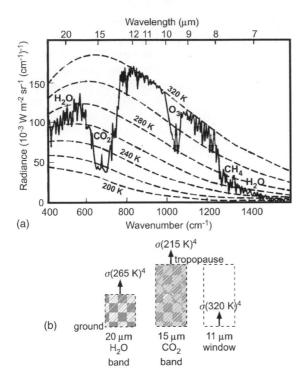

(a)

(b)

Figure 2.13 (a) The spectral emission function for noon over a vegetated region of the Niger Valley in N. Africa. Dashed lines show blackbody curves for particular temperatures. (Adapted from Hanel *et al.* (1972).) (b) A schematic showing how to interpret the meaning of parts of the curve in (a). The arrows indicate from where blackbody fluxes originate, according to the Stefan–Boltzmann Law. (Part (b) follows a concept from Jacob (1999), p. 132.)

2.4.3.2 *General Equation of Radiative Transfer*

We now derive the general equation of radiative transfer, which describes how radiation passes though a medium in any coordinate system. If a beam of monochromatic radiance I_v [W m^{-2} Hz^{-1} sr^{-1}] passes through an elemental path ds, the change of intensity of the beam will be as follows:

intensity change = emission − extinction

$$dI_v = dI_v(\text{emitted}) - dI_v(\text{extinguished})$$

(2.68)

Using our previous expression (2.52) for the mass extinction coefficient (k_v [m^2 kg^{-1}]), we can express the extinction component using the Extinction Law, so that

$$dI_v(s) = dI_v(\text{emitted}) - k_v\rho_a I_v(s)ds$$

(2.69)

where ρ_a is the density of the absorbing and/or scattering gas. The increase in intensity due to emission and multiple scattering is defined as:

$$dI_v(\text{emitted}) = j_v\rho_a ds = k_v J_v(s)\rho_a ds$$

(2.70)

where we define a *source function* J_v[W m^{-2} Hz^{-1} sr^{-1}] such that

$$J_v = j_v/k_v$$

(2.71)

where j_v is the *source function coefficient* (also called the *emission coefficient*) due to scattering and thermal excitation. It follows that (2.69) can be rearranged as follows:

$$dI_v(s) = (k_v\rho_a ds)J_v - (k_v\rho_a ds)I_v \Rightarrow \frac{dI_v}{k_v\rho_a ds} = J_v - I_v$$

(2.72)

This is the *general radiative transfer equation* without any particular coordinate system imposed and without any assumptions about the form of the source function.

2.4.3.3 *Schwarzschild's Equation: For Blackbody Emission With No Scattering*

Schwarzschild's equation is when we assume (a) that the gas is in local thermodynamic equilibrium (LTE) and (b) we consider a non-scattering medium. It is named after astrophysicist Karl Schwarzschild who first considered such a solution to the radiative transfer equation for the Sun's atmosphere in 1914. LTE means that the source function defined by (2.71) is given by the Planck function, i.e.,

$$J_v = B_v(T)$$

(2.73)

Hence, the equation of radiative transfer can be rewritten as

$$\frac{dI_v}{k_v\rho_a ds} = B_v(T) - I_v \quad \text{Schwarzschild's equation} \quad (2.74)$$

Because we are neglecting scattering, k_v is now the mass *absorption* coefficient rather than the mass extinction coefficient.

We shall consider two solutions to eq. (2.74) (Schwarzschild's equation): (1) a general form for the solution; (2) the case of a plane parallel atmosphere.

2.4.3.4 *A General Solution to Schwarzschild's Equation*

We obtain a general solution to Schwarzschild's equation by considering a path for radiation without a specific coordinate system and integrating the equation. We define a monochromatic optical path between points s and s_1 (Fig. 2.14), as

$$\tau_v = \int_s^{s_1} k_v(s')\rho_a(s')ds'$$

(2.75)

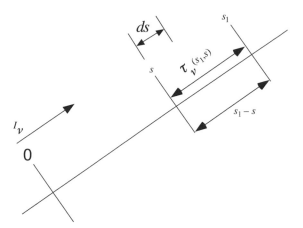

Figure 2.14 An optical path between points s and s_1 for light of frequency v and radiance I_v. The optical thickness between s and s_1 is $\iota_v(s_1, s)$.

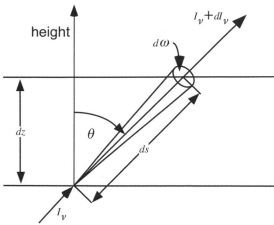

Figure 2.15 A beam of monochromatic radiance I_v passes through a medium at angle θ from the vertical. An incremental distance ds is traversed and the beam has a solid angle $d\omega$. The vertical incremental distance is $dz = ds\cos\theta$.

using a primed dummy variable s' because s is the start of the path and a limit of integration. We specify our direction such that optical depth decreases as s increases (to be analogous to altitude), so that the elemental optical depth between s and s_1 is

$$d\tau_v = -k_v(s')\rho_a(s')ds' \tag{2.76}$$

Then eqn. (2.74) becomes

$$\frac{dI_v(s)}{d\tau_v(s_1, s)} = I_v(s) - B_v[T(s)] \tag{2.77}$$

We can multiply this equation by an integrating factor, $e^{-\tau_v(s_1,s)}$ and integrate between 0 and s_1, as follows:

2.4.3.5 Plane-Parallel Solutions to the Radiative Transfer Equation: Upward and Downward Intensities

We now consider radiative transfer in a plane-parallel atmosphere. We define our coordinate system as a beam of radiance I_v passing upward through a layer of depth dz at an angle to the vertical of θ. Then we get $dz = ds\cos\theta$ (Fig. 2.15). Returning to generality by using the source function, J_v, we have from eq. (2.74),

$$\cos\theta \frac{dI_v}{k_v\rho_a dz} = J_v - I_v \tag{2.80}$$

$$dI_v(s)e^{-\tau_v(s_1,s)} + I_v(s)\left(-d\tau_v(s_1,s)e^{-\tau_v(s_1,s)}\right) = -B_v[T(s)]e^{-\tau_v(s_1,s)}d\tau_v(s_1,s)$$

$$\int_0^{s_1} d\left[I_v(s)e^{-\tau_v(s_1,s)}\right] = -\int_0^{s_1} B_v[T(s)]e^{-\tau_v(s_1,s)}d\tau_v(s_1,s) \tag{2.78}$$

which has solution (substituting in the integral for $d\tau_v$ from eq. (2.76)),

$$I_v(s_1) = I_v(0)e^{-\tau_v(s_1,0)} + \int_0^{s_1} B_v[T(s)]e^{-\tau_v(s_1,s)}k_v\rho_a ds \tag{2.79}$$

$$I_v(s_1) = I_v(0)e^{-k_v\rho_a s_1} + \int_0^{s_1} B_v[T(s)]e^{-k_v\rho_a(s_1-s)}k_v\rho_a ds \quad \text{for constant } k_v \text{ and } \rho_a$$

Here the first term is the absorption attenuation due to the absorber from 0 to s_1, in the form of the Extinction Law. The second term is the emission contribution along the path 0 to s_1. Note that to integrate (2.79), we need to know T, ρ_a, and k_v along the path.

Let us re-introduce the optical depth measured downwards from the top of the atmosphere (eq. (2.59)),

$$\tau = \int_z^\infty k_v\rho_a dz' = \int_{\tau=\tau}^{\tau=0} -d\tau' \tag{2.81}$$

Top of atmosphere ———————— $\tau = 0$

$\tau = \tau$

$\tau' - \tau$

$- \ - \ - \ - \ - \ - \ - \ - \ \tau = \tau'$

Bottom of atmosphere ———————— $\tau = \tau_*$

Figure 2.16 Optical depth coordinate system for a solution to the radiative transfer equation.

consider the (a) upward intensity and then (b) the downward intensity at any level. After that, we simplify the expressions to derive (c) intensities at the top of atmosphere, which a satellite would measure, and at the ground.

(a) Upward Intensity at any Level τ

Upward intensities range from vertically upwards to horizontal, equivalent to $\mu > 0$ where the zenith angle θ ranges from 0 to 90°. We want to integrate from the level where the optical depth is τ to the bottom of the atmosphere where $\tau = \tau_*$, as in Fig. 2.16. Consequently, given that τ is a limit of integration, we adopt the use of the dummy variable $\tau' - \tau$, according to scheme of Fig. 2.16. To obtain the upward intensity at any level where the optical depth is τ, we multiply (2.82) by an integrating factor $e^{-(\tau' - \tau)/\mu}$. Here we have dropped the v subscript on τ, and assume it is implicit. So we have the following:

$$e^{-(\tau'-\tau)/\mu}dI_v(\tau') + I_v(\tau')\left(-\frac{d\tau'}{\mu}e^{-(\tau'-\tau)/\mu}\right) = -J_v(\tau')e^{-(\tau'-\tau)/\mu}\frac{d\tau'}{\mu}$$

$$\Rightarrow d\left[I_v(\tau')e^{-(\tau'-\tau)/\mu}\right] = -J_v(\tau')e^{-(\tau'-\tau)/\mu}\frac{d\tau'}{\mu} \quad \Rightarrow d\left[I_v(\tau')e^{-(\tau'-\tau)/\mu}\right]\Big|_\tau^{\tau_*} = -\int_\tau^{\tau_*}J_v(\tau')e^{-(\tau'-\tau)/\mu}\frac{d\tau'}{\mu}$$

$$I_v(\tau)\uparrow = I_v(\tau_*)e^{-(\tau_*-\tau)/\mu} + \int_\tau^{\tau_*}J_v(\tau')e^{-(\tau'-\tau)/\mu}\frac{d\tau'}{\mu} \quad (1 \geq \mu > 0) \tag{2.83}$$

We also define $\mu \equiv \cos\theta$, the cosine of the zenith angle. Then, we can rewrite (2.80) as

$$\mu\frac{dI_v}{d\tau_v} = I_v - J_v \tag{2.82}$$

This is the basic equation for the problem of radiative transfer in plane-parallel atmospheres. Here $J_v[W \ m^{-2} \ Hz^{-1} \ sr^{-1}]$ and $I_v[W \ m^{-2} \ Hz^{-1} \ sr^{-1}]$ are functions of μ and τ, and in the most general form, they are also functions of azimuth angle.

Here, the terms on the right-hand side are as follows.

First term: the absorption attenuation due to the absorbing medium of the radiation along the path from the ground where $\tau = \tau_*$ to the $\tau = \tau$ level. Here $I_v(\tau_*)$ represents the upward intensity at the ground.

Second term: upwards emission contribution along the path from the ground to the $\tau = \tau$ level. Here $J_v(\tau')$ corresponds to emission in the upward direction.

We could work in terms of altitude, but if we transform eq. (2.83) in terms of z, it looks more complicated, as follows:

$$I_v(z)\uparrow = I_v(z=0)e^{-\frac{1}{\mu}\int_0^z k_v(z')\rho_a(z')dz'} + \int_0^z k_v(z')\rho_a(z')J_v(z')e^{-\frac{1}{\mu}\int_{z'}^z k_v\rho_a dz}\frac{dz'}{\mu} \tag{2.84}$$

We can solve (2.82) using the same procedure that we used to solve (2.77). We take the atmosphere as being bounded at the top and bottom (Fig. 2.16). We

(b) Downward Intensity at any Level τ

To obtain the solution for the downward intensity ($\mu < 0$) at any level where the optical depth is τ, we replace μ

by $-\mu$, multiply (2.82) by an integrating factor $e^{\tau/\mu}$ and integrate from the top of the atmosphere where $\tau = 0$ to the $\tau = \tau$ level. This yields

$$I_v(\tau) \downarrow = I_v(\tau = 0)e^{-\tau/\mu} + \int_0^\tau J_v(\tau')e^{-(\tau-\tau')/\mu}\frac{d\tau'}{\mu} \quad (-1 \le \mu < 0) \tag{2.85}$$

Here, the terms on the right-hand side are as follows. First term: the absorption attenuation due to the absorbing medium of the radiation along the path from the top of the atmosphere to the $\tau = \tau$ level. Here, $I_v(\tau = 0)$ is the downward intensity at the top of the atmosphere. Second term: the emission contribution along the path from the top of the atmosphere to the $\tau = \tau$ level. Here, $J_v(\tau')$ corresponds to emission in the downward direction.

(c) Intensities Seen from Space or from the Surface
Frequently, one measures the emergent intensity at the top of the atmosphere with a satellite or the emergent intensity at the bottom of an atmosphere with a surface-based instrument. Emerging intensity at the top of the atmosphere is derived from (2.83) by setting $\tau = 0$ in the integral limit, i.e.,

$$I_v(\tau = 0) \uparrow = I_v(\tau_*)e^{-\tau_*/\mu} + \int_0^{\tau_*} J_v(\tau')e^{-\tau'/\mu}\frac{d\tau'}{\mu} \tag{2.86}$$

Emerging intensity at the bottom of the atmosphere is derived from (2.85) if we set $\tau = \tau_*$, i.e.,

$$I_v(\tau_*) \downarrow = I_v(\tau=0)e^{-\tau_*/\mu} + \int_0^{\tau_*} J_v(\tau')e^{-(\tau_*-\tau')/\mu}\frac{d\tau'}{\mu} \tag{2.87}$$

Recall from previous discussion (Sec 2.3.1) that we can derive a hemispherical flux from the intensity at a particular frequency by integrating over a hemisphere. Then we derive a total flux density [W m^{-2}] by integrating over all frequencies.

2.4.4 Level of Emission and the Meaning of "Optically Thick" and "Optically Thin"
Below we show that absorbed radiation and emitted radiation come from a region in the atmosphere near the region of unit optical depth for a particular wavelength or frequency.

Commonly, we talk of the atmosphere below the emission (or absorption) level of photons as *optically thick* ($\tau_v \gg 1$) and the atmosphere above the emission (or absorption) level as *optically thin* ($\tau_v \ll 1$). For example, IR photons from a planet's atmosphere "escape" to space from the level where the optical depth drops to unity or below. This is a very useful concept in understanding the remotely measured spectrum of radiation emitted from planetary atmospheres. We applied this idea in discussing Fig. 2.13.

In the shortwave, incoming solar UV photons are absorbed where the level reaches unity optical depth for a particular UV wavelength. Hence, in the UV, this level can be thought of as where action occurs, such as heating or ionization. We consider shortwave absorption and infrared emission levels, in turn.

2.4.4.1 The Level of Shortwave Absorption
At what level do some atmospheric gases, such as ozone in the Earth's atmosphere or methane in giant planet atmospheres, absorb shortwave sunlight and heat the atmosphere? The vertical distribution of any monochromatic component of the solar flux absorbed in a planetary atmosphere can be calculated as follows. The monochromatic flux crossing a surface normal to the solar beam is

$$F_{v\odot} = I_{v\odot}\delta\Omega \tag{2.88}$$

where $I_{v\odot}$ is the monochromatic solar radiance in the small solid angle subtended by the Sun at the planet, $\delta\Omega$. Within the planetary atmosphere at height z, the direct flux can be deduced from the Extinction Law, which accounts for absorption by overlying atmospheric layers,

$$F_v = \mu F_{v\odot}e^{-\tau_v/\mu} \tag{2.89}$$

where $\mu \equiv \cos\theta$ is the cosine of the zenith angle. The contribution to volume heating rate Q is given by $Q = -(dF_{net}/dz)$ where F_{net} is the net flux, that is, the upward minus the downward flux [see eqns. (2.34) and (2.35)]. In our case the solar flux is downward, so

$$Q = \frac{dF_v}{dz} = \frac{dF_v}{d\tau}\frac{d\tau_v}{dz} = \left(-\frac{1}{\mu}\mu F_{v\odot}e^{-\tau_v/\mu}\right)(-k_v\rho_v) = (k_v\rho_v F_{v\odot})e^{-\tau_v/\mu} \tag{2.90}$$

where we have used our definition of optical depth $d\tau_v/dz = -k_v\rho_v$. Note how function (2.90) will change with height.

- At the top of the atmosphere, $\tau_v(z)$ is small and the exponential factor remains near 1, i.e., $e^{-\tau_v/\mu} \to e^0$. Consequently, the overall function increases downward as the absorption term $k_v\rho_v$ in the curved brackets increases.
- Eventually, the increase in optical depth becomes significant, so the term $\exp(-\tau_v/\mu)$ begins to dominate over the downward increase of $k_v\rho_v$, and the volume heating rate begins to decrease exponentially downwards.

Consequently, the volume heat rate, Q, has a single maximum in the planetary atmosphere.

To find the maximum of Q, we differentiate eq. (2.90) w.r.t. z and set $dQ/dz = 0$. We use the product rule of calculus, where the first term $(k_v\rho_v)$ multiplies the second term $F_{v\odot}e^{-\tau_v/\mu}$. Hence,

$$\frac{dQ}{dz} = 0 = (k_v\rho_v)\left(F_{v\odot}e^{-\tau_v/\mu}\frac{(k_v\rho_v)}{\mu}\right) + F_{v\odot}e^{-\tau_v/\mu}\frac{d(k_v\rho_v)}{dz}$$

so

$$\frac{(k_v\rho_v)^2}{\mu} = -\frac{d(k_v\rho_v)}{dz} \qquad (2.91)$$

In this differentiation, we have used $d\tau_v/dz = -k_v\rho_v$ from our definition of optical depth. Using this definition yet again, we note that $d(k_v\rho_v)/dz = -d^2\tau_v/dz^2$ and $(k_v\rho_v)^2 = (d\tau_v/dz)^2$. So eq. (2.91) can be rewritten (swapping left and right sides) as

$$\frac{d^2\tau_v}{dz^2} = \sec\theta\left(\frac{d\tau_v}{dz}\right)^2 \qquad (2.92)$$

The density of absorbing gases usually drops approximately exponentially with height, and the absorption coefficient is usually either constant or a function of pressure that can be expressed as an approximate power law relationship. (The pressure dependence is because of *pressure* or *collision-induced broadening* of absorption lines that we meet later in Sections 2.5.6 and 2.5.7.2.) Optical depth also usually varies approximately exponentially with height, so that

$$\tau_v = \tau_{v0}\exp\left(-\frac{z}{H_\tau}\right) \qquad (2.93)$$

where H_τ is a scale height for optical depth. Consequently,

Substituting these approximations into eq. (2.92), we find the following:

$$\frac{1}{H_\tau^2}\tau_v = \sec\theta\left(\frac{1}{H_\tau^2}\tau_v^2\right)$$

$$\Rightarrow \tau_v\sec\theta = 1, \qquad \tau_v = 1 \text{ when } \theta = 0 \qquad (2.94)$$

This important result tells us the following.

At any wavelength, the maximum volume heating rate, or the maximum of any related physical effect such as ionization or dissociation, occurs at the level where the optical depth is unity for normal incidence radiation, or at a level that is higher by a factor of $\sec\theta$ for solar radiation incident at any other zenith angle θ.

The level of absorption forms a layer with a characteristic width. From eqs. (2.90) and (2.93), we can deduce that the volume absorption rate decreases by a factor of e within a height range H_τ above the maximum, and decreases even more rapidly below the maximum. So the monochromatic absorption peaks near or above $\tau_v = 1$ and has a characteristic width $\sim H_\tau$, the scale height for optical depth, which is comparable to or smaller than a density scale height. Such a vertical structure is called a *Chapman layer*. These atmospheric layers where action is occurring (heating, ionization, photodissociation) correspond to specific wavelengths of solar radiation.

2.4.4.2 The Level of Emission of Infrared

A matter of practical utility is the measurement by spacecraft of IR radiation emitted from a planetary atmosphere or surface. Consequently, we consider monochromatic radiation emitted at zenith angle θ according with the Planck function and detected by a spacecraft outside the atmosphere at $z = \infty$. To simplify matters, we neglect the radiance emitted from the lower boundary of the atmosphere, which is a good approximation for atmospheres with large optical depth to the surface. Then, in our expression for upward intensity (eq. (2.84)) we neglect the first term, which is due to surface radiation, and keep the second. We also adjust the integral limits to bound the whole atmosphere and assume blackbody emission, i.e., $J_v = B_v$. Hence,

$$I_v(\infty)\uparrow \approx \sec\theta\int_0^\infty k_v\rho_a B_v \exp\left[-\sec\theta\int_z^\infty k_v\rho_a dz'\right]dz$$

$$= \sec\theta\int_0^\infty B_v W_v dz \qquad (2.95)$$

where

$$W_v(z) \equiv k_v(z)\rho_a(z)\exp\left[-\sec\theta\int_z^\infty k_v(z')\rho_a(z')dz'\right] \qquad (2.96)$$

is a "weighting function." Thus according to eq. (2.95), the outgoing monochromatic radiance is a weighted vertical integral of the Planck function, with a weight factor given by eq. (2.96).

In fact, the form of the weighting function means that radiation of a certain frequency is emitted from a distinct layer. Function W_v has the same form as the factor multiplying the monochromatic solar flux in eq. (2.90) for the volume absorption rate, Q. Consequently, except for a possible additional lower boundary term, the radiance measured by the satellite is equal to the secant of the zenith angle times the weighted integral of the Planck function, where the weighting factor has all the properties of the vertical distribution of incoming absorbed monochromatic solar radiation. In particular, the emitted monochromatic radiance comes from a layer with characteristic thickness of order of a scale height in optical depth or smaller centered at the level where

$$\tau_v \sec \theta = 1, \qquad \tau_v = 1 \text{ when } \theta = 0$$

This is the same condition as eq. (2.94), which shows some unity between infrared and shortwave radiation.

2.4.5 Radiative and Radiative–Convective Equilibrium

The average vertical temperature profile and climate of a planet depends, principally, on radiation and convection. The temperature profile calculated according to radiative equilibrium in an atmosphere like the Earth's is found to be superadiabatic near the surface, which is unstable to vertical motion. Hence convection ensues and establishes a stable adiabatic, subadiabatic, or moist adiabatic vertical temperature profile. This lapse rate will persist up to some level, the *radiative–convective boundary*, above which radiative equilibrium tends to dominate the average vertical profile of temperature. To show that atmospheres behave like this, we consider some simple equilibrium models.

2.4.5.1. *Atmospheric "Skin Temperature"*

To discuss atmospheric thermal structure, we start with *atmospheric skin temperature*, which is the asymptotic temperature at high altitudes of an upper atmosphere that is optically thin in the thermal-IR and transparent to shortwave radiation.

To derive an equation for skin temperature, we consider a two-layer model of an atmosphere on a planet with an effective temperature T_{eff}. A flux of IR radiation given by σT_{eff}^4 (eq. (2.46)) goes upwards from an emission level in the lower atmosphere (where IR photons "escape") into

an overlying slab of optically thin gas at temperature T_{skin} (Fig. 2.17). We assume that the slab is a gray absorber with an absorptivity equal to the emissivity, ε, by Kirchoff's Law (eq. (2.48)). In equilibrium, the outgoing IR flux absorbed by the slab must equal the IR flux absorbed, as follows:

$$2\varepsilon\sigma T_{skin}^4 = \varepsilon\sigma T_{eff}^4 \quad \Rightarrow T_{skin} = \frac{T_{eff}}{2^{1/4}} \qquad (2.97)$$

Thus, the skin temperature in an optically thin, gray upper atmosphere is colder than the effective temperature by a factor of $2^{-1/4}$ or ~0.84. For the Earth with $T_{eff} = 255$ K, we calculate $T_{skin} = 214$ K.

In planetary atmospheres that absorb little shortwave radiation, the skin temperature is the asymptotic

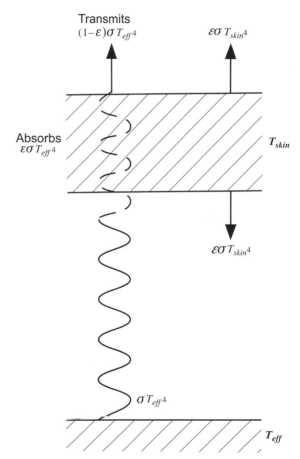

Figure 2.17 A two-layer atmospheric model to understand the concept of atmospheric *skin temperature*. The lower atmosphere is represented by layer that is optically thick in the infrared and emits at the effective temperature, T_{eff}. The upper atmosphere is represented by a layer that is optically thin in the infrared and has a temperature T_{skin} and emissivity ε, which is equal to absorptivity by Kirchoff's Law. By considering conservation of energy in the upper layer, we determine that $T_{skin} = 2^{-1/4} T_{eff}$ (see text).

temperature of a nearly isothermal stratosphere. For real atmospheres with stratospheric inversions, the skin temperature provides an estimate of the tropopause minimum temperature. For example, the Earth's observed global mean tropopause temperature is 208 K (Han *et al.*, 2011), which differs from T_{skin} by only 3%.

2.4.5.2 *Illustrative Radiative–Convective Thought Experiments*

To understand radiative–convective equilibrium, it is useful to consider some thought experiments. In Fig. 2.18(a), we imagine a planet that is rotating rapidly so that we can apply time- and spatial-average energy balance, in which the planet attains an isothermal mean temperature in equilibrium with sunlight. The surface temperature will attain an effective temperature, T_{eff}. We imagine an envelope of gas around this world that is a poor greenhouse gas, e.g., a thin layer of pure N_2. We also assume no convection, only radiation, and that the air is optically thin in the thermal-IR. The air reaches the skin temperature, while the surface radiates through this air to space (Fig. 2.18(a)). The attenuation of IR radiation is small and the greenhouse effect is negligible.

We then take one step closer to reality by imagining air with finite heat capacity that convects because it is in contact with a warm surface. In particular, the

discontinuous T gradient in Fig. 2.18(a) is unstable to convection. The result is the temperature profile in Fig. 2.18(b).

Finally, we imagine air that contains greenhouse gases and is optically thick to thermal-IR near the surface. Then the emission level will be at the altitude where the mass of air between the emission level and space becomes optically thin to thermal-IR (Fig. 2.18(c)). The emission level is no longer at the surface pressure but at a lower pressure, $p_{emission}$. This *emission level* is located where a column of air from space down to $p_{emission}$ has an optical depth of about unity. If there is a dimensionless mass mixing ratio of the absorber, μ_a, which is constant with height, and if the absorber has a broadband (gray), pressure-independent mass absorption coefficient k_a (m²/kg), the optical depth $\tau_{emission}$ at the emission level is

$$\tau_{emission} = k_a\mu_a\left(\frac{p_{emission}}{g}\right) \approx 1 \qquad (2.98)$$

Here, we have combined eq. (2.59) and eq. (1.19), where g is gravitational acceleration. We can think of the region below pressure level $p_{emission}$ as sliced up into blackbody slabs each of unity optical depth (Goody and Walker, 1972, p. 56). Each slab absorbs thermal-IR radiation, and IR photons only escape to space from the top slab.

If there is an even greater mixing ratio of greenhouse gas, μ_a, the IR emission level moves up in altitude to a

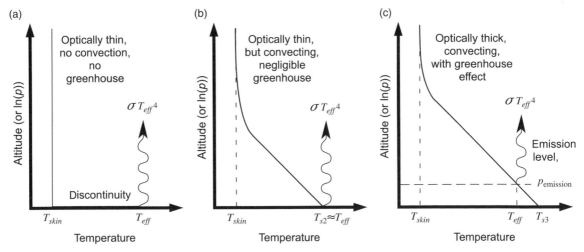

Figure 2.18 (a) Thought experiment with an infrared-transparent atmosphere and no convection. The surface reaches a temperature of the effective temperature, T_{eff}, while the optically thin atmosphere attains the skin temperature, T_{skin}. There is a temperature discontinuity at the surface. (b) Thought experiment with negligible greenhouse gases but convection. A convective profile sets in. The surface temperature T_{s2} is T_{eff}. (c) Thought experiment with greenhouse gases and convection. The near-surface atmosphere is optically thick, so that emission at the effective temperature takes place at a higher level where the pressure is $p_{emission}$. The surface temperature T_{s3} exceeds T_{eff} because of a greenhouse effect.

lower pressure and the surface temperature increases. The skin temperature remains the same.

Modeling the climate of any planet involves more realism for increasing accuracy. The complexity includes actual gas mixtures, p- and T- and wavelength-dependent

$$\frac{dF^+}{d\tau^*} = F^+ - \pi B \qquad -\frac{dF^-}{d\tau^*} = F^- - \pi B, \quad \text{where } \pi B(\tau^*) = \sigma T^4(\tau^*) \qquad (2.99\text{a,b})$$

absorption and scattering, clouds and their feedbacks, three dimensions, atmospheric dynamics and heat transport, ocean dynamics and heat transport, ice sheet dynamics, and feedbacks involving biogeochemical cycles. To generally understand planetary climates, we need to consider these issues for all possible atmospheres. This is part of what makes discoveries of exoplanets with atmospheres a very grand challenge.

2.4.5.3 Radiative Equilibrium in a Gray Atmosphere

To be more quantitative, we can use a simple radiative equilibrium model. Several assumptions make the problem amenable to analytical solution.

(1)   We consider a gray atmosphere, which means that the extinction coefficient is independent of frequency or, equivalently, wavelength.

(2)   We assume an atmosphere transparent to solar radiation, i.e. visible and other shortwave radiation.

(3)   We adopt a so-called *two-stream* approximation, in which we treat the upward (F^+) and downward (F^-) fluxes independently.[4]

(4)   We also use the *diffuse approximation* in which we replace radiances I and B in Schwarzschild's radiative transfer equation (i.e., eq. (2.82) of the form $dI/d\tau = I - B$) with irradiance, i.e. flux, provided that we scale the optical depth τ by an appropriate factor as $\tau^* = D\tau$, where D is the *diffusivity factor*. Basically, D takes account of the slant paths and allows us to work with fluxes instead of radiances. An approximation justified by comparison to numerical results is $\tau^* = 1.66\tau$ (Armstrong, 1968; Rodgers and Walshaw, 1966), where τ is the actual optical depth and τ^* is the scaled optical depth with $D = 1.66$. In other treatments, D is taken as 3/2 (Goody and Yung, 1989; Weaver and Ramanathan, 1995). We take

the scaled optical depth τ^* as our vertical coordinate, which increases downward from the top of the atmosphere ($\tau^* = 0$) to a value at the ground of $\tau^* = \tau_0$. With the above assumptions, Schwarzschild's equation (2.74) becomes

Here, upward and downward fluxes and the spectrally integrated Planck function B depend on τ^*. Signs in eq. (2.99) are such that (by convention) the IR optical depth decreases along the upward path. The flux density is πB by eq. (2.39). Thus, if $T(\iota^*)$ is known, upward (F^+) and downward (F^-) radiative fluxes can be calculated.

We define the net radiative flux as:

$$F_{\text{net}} = F^+ - F^- \qquad (2.100)$$

In radiative equilibrium, the heating rate must be zero, i.e., the heating rate per unit volume $Q = -dF_{\text{net}}/dz = 0$ (see eq. (2.34)). Hence,

$$F_{\text{net}} = F^+ - F^- = \text{constant} = F_0 \qquad (2.101)$$

The value of F_0 can be found by considering the top of the atmosphere where the longwave flux into the top of the atmosphere, $F^-(0)$, is negligible. Thus, the net upward longwave flux is $F^+(0)$, which is the outgoing thermal radiation. This must balance the incoming net absorbed shortwave radiation, $F_{\text{net,a}}^{\odot}$ (positive downward):

$$F_{\text{net,a}}^{\odot}(\tau^* = 0 \text{ or } \tau = 0) = (1 - A_b)\frac{S_{\odot}}{4} \qquad (2.102)$$

where A_b is the planet's Bond albedo and $S_{\odot}$ is the solar flux incident at the top of the atmosphere. The value of $F_{\text{net,a}}^{\odot} = F_0$ is ~239 W m^{-2} for the modern Earth.

We next apply eq. (2.101) to (2.99). First, we subtract (2.99b) from (2.99a),

$$\frac{d(F^+ + F^-)}{d\tau^*} = F^+ - F^- = F_{\text{net}} \quad \Rightarrow \quad \frac{d(F^+ + F^-)}{d\tau^*} = F_{\text{net,a}}^{\odot} \qquad (2.103)$$

and then sum (2.99a,b):

$$\frac{d(F_{\text{net}})}{d\tau^*} = \frac{dF_{\text{net,a}}^{\odot}}{d\tau^*} = F^+ + F^- - 2\pi B = 0 \qquad (2.104)$$

Hence $F^+ + F^- = 2\pi B$, which can be substituted into (2.103) to give

$$2\frac{d(\pi B)}{d\tau^*} = F^+ - F^- = F_{\text{net,a}}^{\odot} \qquad (2.105)$$

---

[4] There are many other forms of two-stream approximation. A common one is Eddington's approximation in which the intensity field is represented by a Legendre polynomial series, truncated at the second term. This approximation treats the upward and downward fluxes as independent.

We can integrate this equation. For a limit on the integral, we note that at the top of the atmosphere where $\tau^* = 0$, the downward longwave flux $F^-(0) \approx$ zero, so eq. (2.104) gives $\pi B(\tau^* = 0) = F^+(0)/2$, which equals $F_{net}^\odot/2$ by eqs. (2.101) and (2.102). Hence

$$\int_{F_0/2}^{\pi B} d(\pi B) = \frac{F_{net,a}^\odot}{2} \int_0^{\tau^*} d\tau^* \Rightarrow \pi B = \frac{1}{2}F_{net,a}^\odot \tau^* + \frac{1}{2}F_{net,a}^\odot \Rightarrow \pi B = \frac{1}{2}F_{net,a}^\odot(\tau^* + 1)$$

$$\Rightarrow \boxed{\sigma T^4(\tau) = \frac{F_{net,a}^\odot}{2}(1 + D\tau)} , \qquad T(\tau) = \left[\frac{F_{net,a}^\odot}{2\sigma}(1 + D\tau)\right]^{1/4} \tag{2.106}$$

The boxed expression in eq. (2.106) is a common result of simple, purely radiative equilibrium models given in textbooks (e.g., Goody and Yung (1989), p. 392; Andrews (2010), p. 85) and reviews (Weaver and Ramanathan, 1995).

We can also derive upward and downward fluxes, including the upward flux at the surface. Since $F^+ + F^- = 2\pi B$ (eq. (2.104)) then eq. (2.106) can also be written as

$$F^+ + F^- = F_{net,a}^\odot(1 + \tau^*) \tag{2.107}$$

Adding (2.107) and (2.101), gives a solution for the upward radiative flux,

$$F^+ = \frac{1}{2}F_{net,a}^\odot(2 + \tau^*) \Rightarrow F_{ground}^+ = \frac{1}{2}F_{net,a}^\odot(2 + \tau_0^*) = F_{net,a}^\odot\left(1 + \frac{\tau_0^*}{2}\right) \tag{2.108}$$

where F_{ground}^+ is the flux from the ground in the lowest layer at scaled optical depth τ_0^*. We see a greenhouse effect because F_{ground}^+ is greater than the flux to space ($F_{net,a}^\odot$). Substitution of eq. (2.108) into eq. (2.107) gives the downward radiative flux:

$$F^- = \frac{1}{2}F_{net,a}^\odot \tau^* \tag{2.109}$$

The flux F_{ground}^+ is σT_{ground}^4, where T_{ground} is surface temperature:

$$\pi B(T_{ground}) = \sigma T_{ground}^4 = F_{net,a}^\odot\left(1 + \frac{\tau_0^*}{2}\right) \tag{2.110}$$

If we apply eq. (2.110) to the Earth with $F_{net,a}^\odot$ ~239 W m^{-2} and T_{ground} ~288 K, the optical depth at the ground is $\tau_0 = \tau_0^*/1.66 = 0.8$. This value is consistent with other purely radiative models of Earth's total gray IR optical depth (Stephens and Tjemkes, 1993). However, a gray radiative–convective calculation gives a broadband IR optical depth ~2 for modern Earth (Robinson and Catling, 2012, 2014). The value of 0.8 is smaller in the purely radiative case of eq. (2.110) because we are using an observed value of ground temperature that, in fact, is

cooler than a purely radiative model would predict because of heat transfer by tropospheric convection in the real atmosphere. If we apply eq. (2.110) to Venus, with $F_{net,a}^\odot$ ~163 W m^{-2} (lower than for Earth because of Venus' high albedo of 0.76) and $T_{ground} = 735$ K, the value of $\tau_0 = \tau_0^*/$

1.66 = 121. This big optical depth indicates an extreme greenhouse effect at the bottom of Venus' thick atmosphere, but again a more realistic broadband optical depth would be even larger.

A characteristic of purely radiative equilibrium models is a temperature discontinuity between the ground and the atmosphere immediately above. In the lowest layer of the atmosphere, we have scaled optical depth τ_0^*, so that eq. (2.106) gives

$$\pi B(T_{a\text{-ground}}) = \sigma T_{a\text{-ground}}^4 = \frac{1}{2}F_{net,a}^\odot(1 + \tau_0^*) \tag{2.111}$$

Thus,

$$\pi B(T_{ground}) - \pi B(T_{a\text{-ground}}) = \frac{1}{2}F_{net,a}^\odot$$

or, in terms of temperature,

$$2\sigma\left(T_{ground}^4 - T_{a\text{-ground}}^4\right) = F_{net,a}^\odot \tag{2.112}$$

As we described in Sec. 2.4.5.2, a temperature discontinuity is unphysical. Convection will ensure and dominate the temperature profile in the lower atmosphere.

Fluxes from the simple radiative model are shown in Fig. 2.19(a). The net thermal IR flux at the top of the atmosphere is $F_{net,a}^\odot$ upwards, showing equilibrium with absorbed incoming sunlight. The fact that there is a linear dependence of πB on scaled optical depth $\tau^* = 1.66\tau$, means that there is an exponential increase of πB as we go down through the atmosphere because of increasing absorber mass. Recall eq. (2.57), $d\tau = -k_{abs}\rho_a dz$, where k_{abs} and ρ_a are the absorption coefficient and absorber mass density. This relationship implies an isothermal stratosphere because the straight line of πB vs. τ converts into a curved line on a height vs. temperature graph.

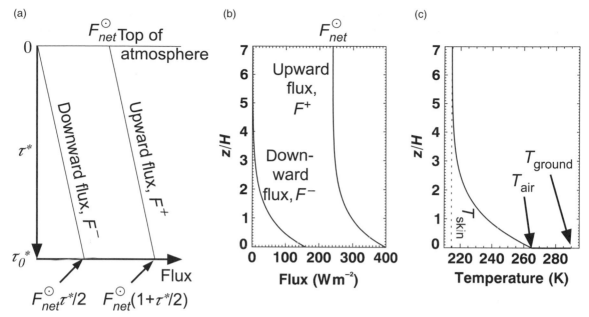

Figure 2.19 Fluxes and temperatures for the purely radiative model described in the text. (a) Downward irradiance F^- and upward irradiance F^+ as a function of scaled optical depth τ^*. (b) The irradiance plotted as a function of altitude z in units of scale height H. A total optical depth τ_0 of 0.8 is assumed, equivalent to a total scaled optical depth $\tau_0^* = D\tau_0 = 1.328$ for diffusivity parameter $D = 1.66$. The net absorbed solar flux is taken as $F_{net}^\odot = 239$ W m^{-2}. (c) The corresponding temperature profile. For an effective temperature $T_{eff} = 255$ K, the skin temperature of the upper atmosphere is $T_{skin} = 214.4$ K, the near-surface air temperature is $T_{air} = 264.9$ K, and the ground temperature is $T_{ground} = 289.6$ K.

We can plot fluxes and temperature in terms of altitude if we make the hydrostatic assumption and assume a value for the surface optical depth. We assume that k_{abs} is constant and that ρ_a varies with altitude z according to the hydrostatic equation as $\rho_a(z) = \rho_a(0)\exp(-z/H_a)$. Then the scaled optical depth will vary as $\tau^*(z) = \tau_0^* \exp(-z/H_a)$. Substituting for $\tau^*(z)$, the equations for the upward (2.108) and downward (2.109) radiative fluxes as a function of altitude are:

$$F^+(z) = \frac{1}{2}F_{net,a}^\odot \left(2 + \tau_0^* e^{-(z/H_a)}\right), \qquad F^-(z) = \frac{1}{2}F_{net,a}^\odot \tau_0^* e^{-(z/H_a)} \qquad (2.113)$$

Similarly, the temperature profile and ground temperature from eqs. (2.106) and (2.110) are

$$T(z) = \left[\frac{F_{net,a}^\odot}{2\sigma}\left(1 + \tau_0^* e^{-(z/H_a)}\right)\right]^{1/4} = T_{skin}\left(1 + \tau_0^* e^{-(z/H_a)}\right)^{1/4},$$

$$T_{ground} = \left[\frac{F_{net,a}^\odot}{2\sigma}\left(2 + \tau_0^*\right)\right]^{1/4} = T_{skin}\left(2 + \tau_0^*\right)^{1/4}$$

$$(2.114a,b)$$

In eq. (2.114), we used the fact that $\left(F_{net,a}^\odot/2\sigma\right)^{0.25}$ is equivalent to $\left(\sigma T_{eff}^4/2\sigma\right)^{0.25} = T_{eff}/2^{0.25}$, the skin temperature,

T_{skin}, given earlier in eq. (2.97). In eq. (2.114a), at the top of the atmosphere $\tau^* = 0$, so the temperature asymptotes to T_{skin}. Also, the effective temperature occurs at an emission level where $T(z) = \left(F_{net,a}^\odot/\sigma\right)^{0.25} = T_{eff}$, which is when $\tau^* = 1$. Radiative fluxes and temperature as a function of altitude are shown in Fig. 2.19, assuming a gray optical depth $\tau_0 = 0.8$, so that the scaled optical depth is $\tau_0^* = D\tau_0$ ~1.3. Given a net absorbed solar flux of 239 W m^{-2}, these values give a ~290 K ground temperature.

Further realism can be introduced by non-gray analytic radiative models (Parmentier and Guillot, 2014; Parmentier et al., 2015), analytic radiative–convective models (Robinson and Catling, 2012), and finally numerical radiative-convective models, which were pioneered by Manabe and Strickler (1964). The basic difference between radiative and radiative-convective profiles can be seen in the conceptual diagram of Fig. 2.20(a). Figure 2.20(b) shows results from a numerical radiative–convective model for the Earth, which unlike Fig. 2.20(a) incorporates a stratospheric inversion due to ozone

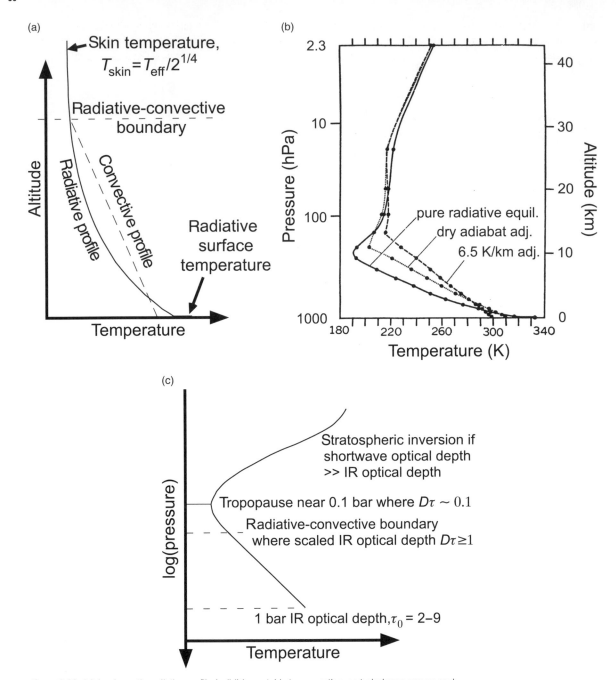

Figure 2.20 (a) A schematic radiative profile (solid) is unstable to convection, so turbulence ensues and an adiabatic, convective profile (dashed) is established. This intersects the radiative profile at a radiative–convective boundary and is close to the real atmospheric profile. In the upper stratosphere, the isothermal skin temperature, T_{skin} is a simple function of the effective temperature, T_{eff} of the planet. (b) Results from a radiative model for the Earth that includes the effects of stratospheric ozone causing an inversion above a tropopause minimum. (From Manabe and Strickler (1964). Reproduced with permission. Copyright 1964, American Meteorological Society.) (c) Schematic diagram following Robinson and Catling (2014), showing the typical thermal structure of dense atmospheres in the Solar System (Earth, Titan, and the giant planets) that have stratospheric inversions. Such atmospheres are optically thick in the IR at 1 bar, but higher up is a radiative–convective boundary where the scaled IR optical depth $D\tau$ is about unity (or greater for tropospheres that absorb significantly in the shortwave). Further up in height, a tropopause temperature minimum occurs at an optically thin IR optical depth near 0.1 bar pressure.

absorption. Appendix A describes a one-dimensional (1-D) numerical radiative–convective code that we have used in many studies of planetary atmospheres, including the early Earth.

2.4.5.4 Radiative–Convective Equilibrium and the Condition for Stratospheric Inversions

We can do better than our simple radiative equilibrium model if we join a gray radiative equilibrium model for a stratosphere and upper troposphere to a convective profile below, following treatments given by Catling (2015) and Robinson and Catling (2012, 2014). The key equation to solve is derived from differentiating eq. (2.104) w.r.t. τ,

$$\frac{d^2 F_{net}}{d\tau^2} = D \left(\underbrace{\frac{d(F^+ + F^-)}{d\tau}}_{DF_{net}} - 2\pi \frac{dB}{d\tau} \right)$$

$$\Rightarrow \frac{d^2 F_{net}}{d\tau^2} - D^2 F_{net} = -2\pi D \frac{dB}{d\tau} \qquad (2.115)$$

Here, we have used eq. (2.100) for F_{net}. The Stefan–Boltzmann Law gives the profiles of blackbody radiance $B(\tau)$ and temperature $T(\tau)$ from eq. (2.115) if $F_{net}(\tau)$ is known and if we use a top-of-atmosphere boundary condition on the net flux from eq. (2.102).

To consider the profile of net flux, $F_{net}(\tau)$, we allow the atmosphere to absorb some of the incoming shortwave flux. Shortwave absorption can produce

attenuation is a reasonable approximation for stellar radiation, so the profile of net absorbed stellar flux $F_{net}^{\odot}$ can be described as follows (Robinson and Catling, 2012):

$$F_{net}^{\odot}(\tau) = \underbrace{F_{tropo}^{\odot} e^{-k_{tropo}\tau}}_{\text{tropospheric (and surface) solar absorption}} + \underbrace{F_{strato}^{\odot} e^{-k_{strato}\tau}}_{\text{stratospheric solar absorption}} \qquad (2.116)$$

Here, $F^{\odot}_{strato}$ and $F^{\odot}_{tropo}$ are the top-of-atmosphere net absorbed stellar fluxes in the stratosphere and troposphere, respectively, where $F^{\odot}_{tropo}$ also includes surface absorption for planets with surfaces. Dimensionless parameters $k_{strato} = \tau_{sws}/\tau$ and $k_{tropo} = \tau_{swt}/\tau$ control the attenuation of solar flux in two channels, where τ_{sws} and τ_{swt} are short-wave optical depths and τ is the gray thermal-IR optical depth.

A generalized energy balance at any atmospheric level is between the net thermal IR flux and the absorbed stellar flux plus any internal energy flux, F_i from a planet's interior (which is important on giant planets or certain tidally heated rocky exoplanets):

$$F_{net}(\tau) = F_{net}^{\odot}(\tau) + F_i \qquad (2.117)$$

By combining eq. (2.116) with (2.117) and inserting into eq. (2.115), we obtain

$$\frac{d\sigma T^4}{d\tau} = \frac{1}{2D}\left[\left(D^2 - k_{tropo}^2\right)F_{tropo}^{\odot}e^{-k_{tropo}\tau} + \left(D^2 - k_{strato}^2\right)F_{strato}^{\odot}e^{-k_{strato}\tau} + D^2 F_i \right] \qquad (2.118)$$

which integrates to:

$$\sigma T^4(\tau) = \sigma T^4(0) + \frac{1}{2D}\left[\frac{D^2 - k_{tropo}^2}{k_{tropo}}\left(1 - e^{-k_{tropo}\tau}\right)F_{tropo}^{\odot} + \frac{D^2 - k_{strato}^2}{k_{strato}}\left(1 - e^{-k_{strato}\tau}\right)F_{strato}^{\odot} + D^2 F_i \tau \right] \qquad (2.119)$$

The flux at the top of the atmosphere in this equation ($\sigma T^4(0)$) can be eliminated to derive a general expression in three steps. First, we use the definition of F_{net} (eq. (2.100)) to rewrite eq. (2.104) as

$$\frac{dF_{net}}{d\tau} = D\left(F_{net} + 2F^- - 2\sigma T^4\right) \qquad (2.120)$$

Second, we solve for σT^4 using eqs. (2.116) and (2.117):

$$\sigma T^4(\tau) = F^-(\tau) + \frac{1}{2}\left[\frac{D + k_{tropo}}{D} F_{tropo}^{\odot} e^{-k_{tropo}\tau} + \frac{D + k_{strato}}{D} F_{strato}^{\odot} e^{-k_{strato}\tau} + F_i \right] \qquad (2.121)$$

stratospheric temperature inversions, while absorption in troposphere can stabilize upper troposphere against convection, which is notable on Titan. Exponential

Third, we obtain $\sigma T^4(0)$ by inserting $\tau = 0$ in eq. (2.121) and noting that $F^-(0)$ is zero. Inserting $\sigma T^4(0)$ into eq. (2.119) gives a final expression for the temperature profile in the radiative part of a gray atmosphere that lies above a radiative–convective boundary:

$$\sigma T^4(\tau) = \frac{F^\odot_{strato}}{2}\left[1 + \frac{D}{k_{strato}} + \left(\frac{k_{strato}}{D} - \frac{D}{k_{strato}}\right)e^{-k_{strato}\tau}\right] + \frac{F^\odot_{tropo}}{2}\left[1 + \frac{D}{k_{tropo}} + \left(\frac{k_{tropo}}{D} - \frac{D}{k_{tropo}}\right)e^{-k_{tropo}\tau}\right] + \frac{F_i}{2}(1 + D\tau)$$

$$(2.122)$$

If there is no internal heat flux ($F_i = 0$) and no solar attenuation ($k_{strato} = k_{tropo} \to 0$), eq. (2.122) reduces to simple radiative equilibrium, eq. (2.106), using L'Hôpital's rule.

Equation (2.122) allows us to derive the condition under which an atmosphere has a pronounced tropopause temperature minimum and thus a stratospheric temperature inversion. We remove terms in tropospheric shortwave attenuation by taking $k_{tropo} \ll 1$, which is true for most real atmospheres. Then, setting the derivative w.r.t. τ of the resulting expression to zero, the optical depth at the tropopause minimum τ_{tp}, is (Robinson and Catling, 2014):

$$\tau_{tp} = \underbrace{\frac{1}{k_{strato}}}_{\sim 1/100} \ln\left[\underbrace{\frac{F^\odot_{strato}}{F^\odot_{tropo} + F_i}}_{\sim 1/10}\underbrace{\left(\frac{k_{strato}^2}{D^2} - 1\right)}_{\sim 10^3}\right]}_{\sim 5} \quad (2.123)$$

The numbers under parts of this equation are typical values in Solar System atmospheres, which yield a small gray IR optical depth at the tropopause temperature minimum of $\tau_{tp} \sim 0.05$. The tropopause minimum is at a much lower pressure than the radiative–convective boundary, which is where the optical depth of ~ 1 or greater, in the case of atmospheres with considerable tropospheric shortwave absorption such as Titan's. Furthermore, eq. (2.123) is only valid if

$$k_{strato}^2 > D^2\left[1 + \frac{(F^\odot_{tropo} + F_i)}{F^\odot_{strato}}\right] \quad (2.124)$$

Equation (2.124) is the general condition for a temperature minimum and stratospheric inversion to exist. Remembering that $k_{strato} = \tau_{sws}/\tau$, the inequality indicates that the shortwave optical depth τ_{sws} in a stratosphere must be large compared to the thermal-IR optical depth, τ. In Solar System atmospheres with well-developed tropopause temperature minima, typical fluxes in eq. (2.124) give $k_{strato} \sim 10^2$.

The inequality of eq. (2.124) indicates the radiative properties that cause stratospheric temperature inversions

in some atmospheres but not others. Earth's stratospheric ozone ensures that $k_{strato} \sim 90$, so shortwave heating is large relative to radiative cooling from CO_2, water vapor, and ozone. Consequently, Earth has a stratospheric inversion. On the giant planets and Titan, k_{strato} is $\sim 10^2$ because stratospheric aerosols absorb shortwave radiation strongly and methane absorbs near-IR solar radiation. Thermal-IR emission cools outer planet stratospheres inefficiently through a collision-induced absorption continuum of H_2–H_2 and H_2–He on the giant planets, and the bands of acetylene (C_2H_2, 13.7 µm), ethane (C_2H_6, 12.1 µm) and methane (7.6 µm) (Yelle *et al.*, 2001; Zhang *et al.*, 2013) on the giant planets and Titan. In contrast, CO_2 in the stratospheres of Mars and Venus has weak shortwave absorption compared to strong IR cooling from CO_2 emission. Consequently, the global mean temperature profiles of Mars and Venus lack stratospheric inversions.

To join a convective temperature profile to the radiative equilibrium profile described by eq. (2.122) as $T(\tau)$, we must convert optical depth to pressure or vice versa. Temperature as a function of pressure in the convective part of the troposphere is given by eq. (1.35). Combining the definition of differential optical IR depth $d\tau = -k_a\rho_a dz$ (eq. (2.59), where k_a is a gray mass absorption coefficient, ρ_a is the absorber mass density, and dz is the differential altitude) with hydrostatic equilibrium $dp/dz = -g\rho$ (where g is gravitational acceleration and ρ is atmospheric density), we get $d\tau \propto k_a dp$ if the absorber is well-mixed. Pressure-broadening and collision-induced absorption (see Sec. 2.5.7) cause a dependence where $k_a \propto p$ below middle stratospheres. Thus, integration of $d\tau \propto k_a dp \propto pdp$ gives $\tau \propto p^{n_p}$ where $n_p = 2$. More generally, a gray IR optical depth is often approximated by a power law relationship with pressure,

$$\tau = \tau_0\left(\frac{p}{p_{ref}}\right)^{n_p} \quad (2.125)$$

where τ_0 is the optical depth at reference pressure p_{ref}, such as 1 bar. As noted, $n_p = 2$ in a troposphere and lower stratosphere applies for a well-mixed IR absorber. The terrestrial atmosphere is more complex in detail: the 8–12 µm IR window, $n_p = 2$ scaling for the 15 µm CO_2 band, and $n_p = 4$–5 scaling for water bands at 6.3 µm and beyond ~20 μm. However, calculations show that $n_p = 2$ is a good overall gray approximation for the tropospheres and lower stratospheres of Earth, Titan, and the giant planets (Robinson and Catling, 2012, 2014).

A radiative–convective temperature profile is obtained using the scaling relationship between pressure and optical depth (eq. (2.125)). The τ–p scaling is inserted into the equation for the T–p profile in the convective part of the troposphere (eq. (1.35)) to give a temperature profile in terms of optical depth rather than pressure. In turn, an analytic expression for the convective upwelling flux can be derived, which gives an analytic solution to radiative–convective equilibrium (Robinson and Catling, 2012). The temperature in an atmosphere follows a convective profile up to a radiative–convective boundary. Above the boundary, the radiative temperature profile of eq. (2.122) applies. The treatment can be extended to include scattering (Heng *et al.*, 2014).

The physics just described allows us to understand why dense atmospheres in the Solar System with stratospheric inversions tend to have global mean tropopause temperature minima within a factor of ~2 of 0.1 bar pressure (Fig. 1.1, Fig. 2.20(c)). If an atmosphere satisfies the temperature minimum condition (eq. (2.124)) then the IR optical depth at the minimum is close to $\tau_{tp} \sim 0.05$. The dense atmospheres of Earth, Venus, the giant planets, and Titan are all optically thick at depth, with an average IR optical depth at 1 bar within a few units of $\tau_0 \sim 5$, despite the great diversity of compositions. From eq. (2.125), the tropopause pressure is $p_{tp} \approx (0.05/5)^{0.5} = 0.1$ bar. This commonality is surprising because gravity ought to affect optical depth (c.f. eq. (2.98)). However, the radiative properties of gases and the shared τ–p scaling caused by common broadening processes turn out to dominate. Also, the different greenhouse effects of the aforementioned bodies, represented by a gray optical depth τ_0 at 1 bar, ranging ~2–10, have a weak influence on tropopause pressure because the latter decreases only as the inverse square root of τ_0, i.e., $p_{tp} \propto \tau_0^{-1/2}$ from eq. (2.125).

Finally, we note a word of caution: the total gray optical depth is not a real physical property of atmospheres but merely a fitting parameter for a gray model to real observations, such as surface temperature and absorbed solar flux. Broadband optical depths should be viewed as semi-quantitative measures that give us greater understanding of planetary climates. Thus, it is usually never worthwhile laboriously computing a gray optical depth τ_0 from wavelength-dependent optical depths. Instead, τ_0 should be fit.

2.4.5.5 Runaway Greenhouse: a Limit on Outgoing Longwave Radiation

The dependence of surface temperature in radiative equilibrium (e.g., eq. (2.110)) on the optical depth can lead to an unstable climate. For wet, Earth-like planets, the optical depth depends on the concentration of water vapor, which is the major greenhouse gas. The connection to water vapor is behind the possibility of a *runaway greenhouse effect*, which is an upper limit on outgoing infrared flux F^+ at the top of the atmosphere that can be developed by a planet with liquid water on its surface.

To understand the runaway greenhouse concept, consider a saturated atmosphere on a warmer Earth. The surface pressure will be proportional to the partial pressure of water, p_{H2O}, which is an exponential function of temperature. In turn, the infrared optical depth, τ_{IR}, is proportional to the column mass or surface pressure via eq. (1.19). Consequently, τ_{IR} is proportional to an exponential function of temperature via the Clausius–Clapeyron equation (1.49). If the solar flux is large enough to produce a sufficiently warm surface, τ_{IR} becomes so high that the atmosphere becomes essentially completely opaque to outgoing thermal-IR radiation across all wavelengths. Earth's 8–12 μm window closes. The atmosphere then radiates to space at a level where $\tau_{IR} \sim 1$, analogous to a star's photosphere except in the thermal-IR. Essentially no thermal-IR fluxes to space directly from the surface. It turns out that if the atmosphere is close to saturation with water vapor, the outgoing flux at the top of the atmosphere F^+ reaches a limit defined by the properties of water and the gravity of the planet, which is $F^+_{limit} \sim 300$ W m^{-2} for Earth. If the solar flux exceeds F^+_{limit} then the excess incoming energy that cannot be radiated away is used to evaporate the oceans inexorably. Once the oceans are all evaporated, the surface melts. Energy balance is achieved when a hot upper atmosphere radiates to space through near-IR windows, which are wavelengths where water vapor is relatively transparent. This *runaway greenhouse effect* is the most widely accepted explanation of the climate history of Venus (see Ch. 13).

2.5 Absorption and Emission of Radiation by Atmospheric Gases

In Sec. 2.4, we described how to calculate the transmission of UV, visible, and IR radiation through atmospheres, given appropriate absorption and scattering coefficients, but we didn't describe *why* certain gases absorb or scatter at particular wavelengths. We now consider this issue, and how absorption and emission depend upon pressure and temperature. In doing so, we address the question of why gases such as N_2 and O_2 in the Earth's atmosphere are not greenhouse gases, whereas others are, such as H_2O, CO_2, O_3, and CH_4. We also examine how on giant planets, H_2O and CH_4 also absorb in the IR, but there is a predominance of IR absorbers of hydrogen-containing gases such as H_2 itself, NH_3 and C_nH_n.

2.5.1 Overview of Absorption Lines

To understand how gases absorb, we start with the concept that electromagnetic radiation is quantized into elementary particles, *photons*, each with an energy given by Planck's Law, eq. (2.41). A gas molecule can absorb a photon if the energy of the photon is close to a quantized energy jump within the molecule that is needed for vibration, an increase in rotational speed, or the redistribution of charge when an electron leaps to a higher orbit. Consequently, we write the overall internal energy of a gas molecule as:

$$E_{total} = E_{electronic} + E_{vibrational} + E_{rotational} + E_{translational}$$
(2.126)

Here, $E_{electronic}$ is the energy associated with electron orbits, and the other terms refer to vibrational energy, rotational kinetic energy, and translational kinetic energy, respectively.

Collisions between molecules, which depend on $E_{translational}$ and therefore temperature, redistribute energy between the various forms. As discussed in Sec. 2.4.2.4, if collisions are sufficiently rapid compared to the radiative decay time for the molecules, energy exchanges between the forms in eq. (2.126) and excess vibrational or rotational energy is converted into heat. The heating establishes *Local Thermodynamic Equilibrium* (LTE) where a local volume of the atmosphere emits and absorbs like a blackbody with a characteristic brightness temperature. This is generally a good assumption apart from at high altitude (e.g., Earth's mesosphere), where LTE breaks down because of infrequent collisions.

In general, electronic transitions require more photon energy than vibrational ones, which in turn demand more energy than changes in rotational speed. For example, a single quantum of energy between two allowed rotational frequencies of a molecule tends to be relatively small and so would correspond by Planck's Law (eq. (2.41)) to the energy of a low frequency photon in the IR or microwave part of the spectrum. The correspondence between wavelengths and types of transition is as follows.

Dominant transition	Wavelength range	Band
Electronic	< 1 μm	X-ray to UV to visible
Vibration	1–20 μm	Near-IR to far-IR
Rotation	>20 μm	Far-IR to microwave[*]

[*] Except H_2 in warm, dense atmospheres, which can have rotational transitions at shorter wavelengths.

The changes in states of molecules that give rise to absorption lines often involve combinations of transitions, such as a simultaneous low-energy rotational transition and a higher energy vibrational or electronic transition. This produces considerable structure in the absorption line spectrum of atmospheres. For quantum mechanical details, the reader is referred to Ch. 13 of McQuarrie and Simon (1997).

When we examine the spectrum of radiation from a planetary atmosphere, we see discrete lines, groups of neighboring lines called *bands*, and smoother varying absorption called the *continuum* (Fig. 2.21), all of which we consider below. Spectral intervals where atmospheric absorption is very weak are termed *windows*.

Various windows exist in Earth's atmospheric spectrum can affect the climate or are useful for astronomical observations. The most well-known window is at 12.5–8 μm where much thermal infrared radiation fluxes to space. Astronomers using ground-based telescopes observe through atmospheric windows to characterize celestial objects with infrared photometry (Bessell, 2005). Their J, H, K, L, and M pass bands are at 1.2, 1.6, 2.2, 3.5, and 4.8 μm, respectively, while N and Q bands are at longer wavelengths of 10.5 and 21 μm, corresponding to relatively transparent atmospheric transmission.

2.5.2 Electric and Magnetic Dipole Moments

The ability of a gas molecule to absorb radiation depends on the electric or magnetic field distribution across the molecule. Whether gas molecules have a *dipole moment* determines if they undergo rotational or vibrational transitions, which are important in the greenhouse effect and spectroscopy of planetary atmospheres. A dipole moment comes in two kinds: electric and magnetic. A molecule with a distribution of positive and negative charges $\pm Q_c$ separated by distance l has an electric dipole moment, $p_d = Q_c l$ in units of coulomb·meter. For example, an H_2O vapor molecule has a slight positive charge on its hydrogen atoms and slight negative charge on the oxygen atom, and possesses a permanent electric dipole moment

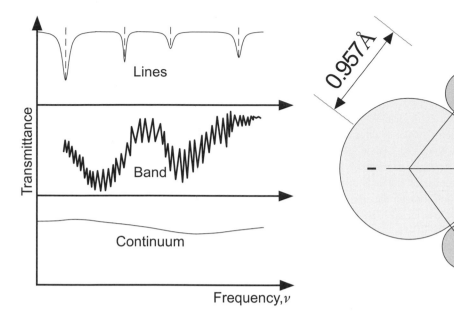

Figure 2.21 Schematic diagram of absorption lines, a band, and a continuum. Note: lines occur at a smaller frequency scale and make up the bands.

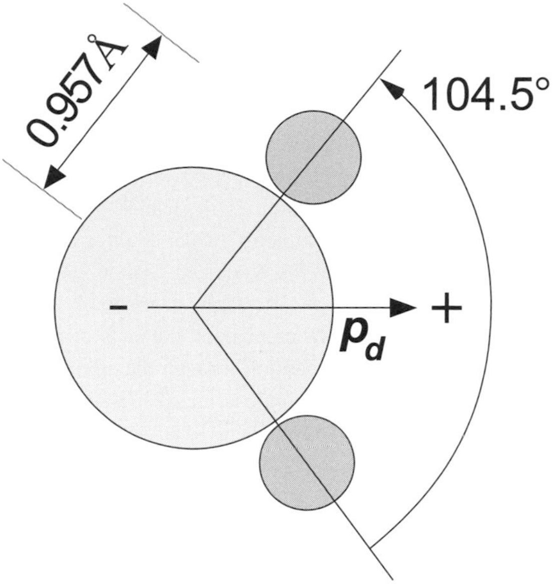

Figure 2.22 The electric dipole moment vector $\boldsymbol{p_d}$ of a water vapor molecule, H_2O.

of 6.1×10^{-30} C·m (Fig. 2.22). An electric field vector $\boldsymbol{E}$ will exert a torque on an electric dipole, given by $\boldsymbol{p_d} \times \boldsymbol{E}$, which tends to align the molecule's dipole with the field. Thus, electromagnetic radiation can induce rotation in molecules with electric dipoles.

Molecular oxygen O_2 possesses a magnetic dipole moment. O_2 does not possess an electric dipole moment because its atoms are identical with no separation of charge, so the electric field in an electromagnetic wave does not interact with an isolated molecule. Homoatomic molecules such as O_2, N_2, and H_2 *can* develop transient induced electric dipoles through collisions that distort their electron clouds, allowing them to interact weakly with the electric field of passing electromagnetic waves, but we consider those later in Sec. 2.5.6. The magnetic field in an electromagnetic wave can torque a molecule with a magnetic dipole moment into rotation. An analogy is the needle of a compass, which possesses a magnetic dipole moment and so points northwards when torqued by the Earth's magnetic field. The magnetic dipole moment is defined by $\mu = I_c A$ in units of A.m^2, where I_c is current and A is the vector area of a loop. A magnetic field vector $\boldsymbol{B}$ will exert a torque on a magnetic dipole, given by $\mu \times \boldsymbol{B}$, which is directly analogous to the electric dipole case.

Why do some molecules have magnetic dipoles? Electrons orbiting an atom classically define a current in a loop. But individual electrons pair up with opposite spins, causing no net magnetic moment for a molecule

as a whole. Molecular oxygen O_2, however, has two unpaired electrons. The resulting magnetic dipole allows O_2 to have rotational absorption bands in the microwave at ~60 and 118 GHz. These absorptions are used in satellite observations of temperatures in Earth's atmosphere (Janssen, 1993).

Other common gases may or may not have dipoles. For example, N_2 has neither an electric nor magnetic dipole and so has no pure rotational spectrum, except through collisions (Sec. 2.5.6). Although CH_4 and CO_2 also have no electric or significant magnetic dipoles, their bonds bend, which breaks molecular symmetry and allows a temporary electric dipole with associated rotation–vibrational transitions. We now examine these transitions in more detail.

2.5.3 Rotational Transitions

Rotational transitions occur because molecules spin like tumbling acrobats with a frequency that is the number of revolutions per second. A rotational transition occurs when a photon of appropriate energy is absorbed and the rotational frequency of a molecule is boosted in a quantized way, similar to how a household fan only spins at certain speeds. We describe rotational motion with equations that are analogous to linear motion (Table 2.3).

How quickly a molecule spins depends on its moment of inertia I. This quantity is just the sum of the products of elemental mass m_i and the square of the radial distance r_i from the axis of rotation:

$$I = \sum_i m_i r_i^2 \qquad (2.127)$$

In general, we can define three principal moments of inertia about three axes through the center of mass of a molecule that are aligned along axes of rotational or reflectional symmetry. In spectroscopy, these are denoted I_A, I_B and I_C, going from smallest to largest. But if a molecule has symmetry, two or all three of these moments of inertia may be equivalent (Table 2.4). For example, a pancake has one large moment of inertia about its radius and two equivalent moments of inertia about orthogonal axes passing through the edge of the pancake. In the case of a single atom, such as neon (Ne), $I_A = I_B = I_C \approx 0$, and so Ne has no rotational transitions. A linear molecule, such as carbon dioxide (O=C=O) has

top molecules with two equal and one different moment of inertia, and (3) *asymmetric top* molecules with three different moments of inertia.

The physics of rotational spectra are most easily understood by considering a diatomic linear molecule behaving as a *linear rotor*. The molecule has only $I_B = I_C$, which we can call one moment of inertia I. The total angular momentum L, associated with the rotation is quantized according to quantum mechanics by

$$L = \frac{h}{2\pi}\sqrt{J(J+1)} \qquad (2.128)$$

where $J = 0, 1, 2, \ldots$ is the *rotational quantum number*. Hence, using the expressions in Table 2.4, the kinetic energy of rotation for quantum number J is:

$$E_J = \frac{1}{2}I\omega^2 = \frac{L^2}{2I} = \frac{h^2}{8\pi^2 I}(J(J+1)) = hB(J(J+1)), \qquad \text{where } B = \frac{h}{8\pi^2 I}$$

an axis that passes through the nuclei, where $I_A \approx 0$, while the moments of inertia around the other perpendicular axes are equal and non-zero, $I_B = I_C > 0$. It is an example of a *linear rotor*. For nonlinear polyatomic molecules, three categories occur: (1) *spherical top* molecules with all moments of inertia equal, (2) *symmetric*

Quantity B is called the *rotational constant*. (Sometimes the rotational constant is given as $h/(8\pi^2 Ic)$ when working in wavenumbers, i.e. units of cm^{-1}). Only rotational transitions between adjacent quantum states J and $J + 1$ are allowed, so the energy change in a transition is

$$\Delta E = hB[(J+1)(J+2) - J(J+1)] = hB\left[J^2 + 3J + 2 - J^2 - J\right] = 2hB(J+1)$$

Table 2.3 Equations of linear and rotational motion. Here m is mass, F is force, a is acceleration, v is velocity, T is torque, ω is angular velocity, and I is moment of inertia.

	Linear	Rotational or Angular
Force and acceleration	Force, $F = ma$	Torque, $T = I\, d\omega/dt$
Kinetic energy	½ mv^2	½ $I\omega^2$
Momentum	mv	$I\omega$

Consequently, the frequency of a photon for a rotational transition in a linear rotor is

$$v = \frac{\Delta E}{h} = 2B(J+1) \qquad (2.129)$$

The pure rotational absorption spectrum is therefore a series of lines uniformly separated by frequency differences of $\Delta v = 2B$. However, in reality, as a molecule rotates faster with increasing J, centrifugal force causes the bond to stretch slightly and the frequency spacing to

Table 2.4 Rotational symmetry and moments of inertia in common molecules present in atmospheres of giant and terrestrial planets.

Molecular type	Moment of inertia	Example gases
Monoatomic	$I_A = I_B = I_C \approx 0$	Ar, He, Ne
Linear (or linear rotor)	$I_A \approx 0$, $I_B = I_C > 0$	O_2, N_2, CO_2, CO, N_2O, C_2H_2
Spherical top (or spherical rotor)	$I_A = I_B = I_C > 0$	CH_4, GeH_4
Symmetric top (or symmetric rotor)	$I_A \neq 0 < I_B = I_C$	C_2H_6, C_2H_4
	Or $I_C > I_A = I_B$	NH_3, PH_3, CH_3Cl
Asymmetric top	$I_A \neq I_B \neq I_C$	H_2O, O_3, H_2S, SO_2, C_3H_8

change. A higher-order "centrifugal distortion" correction to eq. (2.129) can account for this behavior.

Our treatment above only considers a diatomic molecule but similar expressions can be derived for more complicated molecules. These have n moments of inertia I_n each associated with J_n rotational quantum numbers (McQuarrie and Simon, 1997). The absorption of a photon can also cause simultaneous changes rotational and vibrational quantum numbers, which we now consider.

2.5.4 Vibrational Transitions

A molecule vibrates when the constituent atoms of molecules move toward and away from each other at some frequency. Because molecular bonds behave somewhat like springs, a diatomic molecule acts like a simple harmonic oscillator with a resonant frequency v_0 determined by the masses of the atoms and a "spring constant." In quantum mechanics, however, the actual oscillation frequency is quantized according to

$$v = \left(v + \frac{1}{2}\right)v_0$$

where "v" (the letter vee) is a *vibrational quantum number*, $v = 0, 1, 2, \ldots$. Consequently, the allowed energy levels are separated by an integer times the energy at the classical resonant frequency:

$$E_V = hv_0\left(v + \frac{1}{2}\right)$$

The lowest vibrational state (v = 0) has a finite energy of E= $hv_0/2$ rather than zero. This *zero point energy* is consistent with Heisenberg's uncertainty principle because if the oscillating particle were stationary the uncertainty on its position would be zero, giving it infinitely uncertain momentum, which cannot be so. The transition from the lowest energy state to the next highest, or vice versa, is known as the *fundamental*. This typically produces the strongest line for statistical reasons: at typical planetary temperatures and pressures near STP, most of the molecules of a gas are in the ground state, Usually, the quantum mechanical rule is that only transitions of $\Delta v = \pm 1$ are allowed. However, large displacements of atoms in molecules in so-called anharmonic oscillations give rise to cases where $\Delta v = \pm 2, \pm 3$, etc., which produce *overtone* bands that are less intense than the fundamental absorption.

Because vibrational and rotational transitions often occur together, the combined transition frequency can be slightly less or greater than the pure vibrational transitional frequency, depending on whether the rotational quantum number J goes up or down. Consequently, *ro-vibrational* spectra consist of a series of closely spaced lines fanning out either side from the pure vibrational line frequency. The low frequency side corresponding to changes $\Delta J = -1$ is called the P branch, while the higher frequency side with changes $\Delta J = +1$ is called the R branch. The central frequency line with $\Delta J = 0$ is called the Q branch, but it may not exist in certain cases (e.g., most diatomic molecules) where quantum mechanics requires $\Delta J \neq 0$ (Fig. 2.23). By convention, lines in the P and R branches are labeled according to the initial value of the rotational quantum number J, i.e., R(0), R(1),..., and P(0), P(1),....

In another case, superposed $\Delta J = -2$ or $+2$ transitions occur, which are called *electric quadrupole transitions*. Instead of the PQR branch structure, there are branches called O ($\Delta J = -2$) and S ($\Delta J = +2$). Homonuclear diatomic molecules can engage in these weak transitions, despite their lack of a permanent electric dipole. Giant planet atmospheres have electric quadrupole absorption lines of H_2 in the far IR, for example.

Molecules with more than two atoms have a large variety of vibrations, and the theory for their vibrations concerns the superposition of normal modes. These

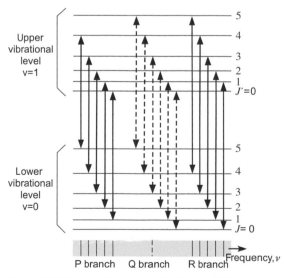

Figure 2.23 The meaning of the P, Q, and R branches in vibrational–rotational transitions, showing a vibrational transition Δv and superposed rotational transitions, ΔJ. The P branch corresponds to $\Delta v =1$ with rotational transitions $\Delta J = -1$. The Q branch corresponds to $\Delta v =1$ and $\Delta J = 0$ and the R branch corresponds to $\Delta v =1$ and $\Delta J = +1$. The lower shaded panel shows the appearance of the lines in the spectrum schematically, noting that the Q branch is offset from P and R branches for clarity in order to show the Q-branch ΔJ transitions.

multiple modes require more vibrational quantum numbers to describe all the energy levels than in a diatomic molecule, but the number of modes is given by some simple rules. A linear molecule consisting of N atoms has $3N - 5$ normal modes (or degrees of freedom), whereas a nonlinear molecule of N atoms has $3N - 6$ normal modes. Examples of normal modes of vibration of some common gas molecules are shown in Fig. 2.24. In order for the absorption of a photon to excite a mode, the dipole moment must vary during the normal mode motion. When this occurs for IR photons, the mode is said to be *infrared active*. Otherwise, the mode is *infrared inactive*.

The vibrational absorption bands of CO_2 are of great importance for the atmospheres and climates of Venus, Earth, and Mars. CO_2 has $3 \times 3 - 5 = 4$ normal vibration modes shown in Fig. 2.24. There is no change in the dipole moment during the symmetric stretch ν_1, so this mode is infrared inactive. Two bending modes are equivalent and are treated as a single mode denoted ν_2. In fact, the 15 μm rotation–vibration band is centered on the ν_2 fundamental. This is actually a PQR branch structure but individual lines are indistinct in the spectra of Venus, Earth, or Mars because of pressure broadening (see below). The center of the 15 μm band is remarkably absorbing. On Earth at 1 bar, air with a typical concentration of CO_2 absorbs ~95% of 15 μm radiation over a pathlength of only 1 m!

Methane is also an important absorber on Earth, Titan, and the giant planets (Fig. 2.25). It has $3(4) - 6 = 9$ normal modes, but 5 are equivalent by symmetry, leaving only four unique modes, ν_1 to ν_4. The vibration–rotation band at 7.6 μm, centered on the ν_4 fundamental is important for the Earth's climate (see Ch. 11). At 7.6 μm, methane is evident in emission in spectra of the giant planets and Titan, indicating the presence of a warm CH_4-containing layer above a cooler layer – a stratosphere – on all these planets (Fig. 2.25).

Hot bands. We have already mentioned that at pressures and temperatures near STP most molecules in a gas tend be in the ground state. However, at high temperatures molecules can exist in excited vibrational states because the high kinetic energy of collisions excites vibrations. "Hot bands" refer to transitions where the vibrational state does not involve the ground state. Hot bands are also important in gases at low pressure, where collisions are infrequent and there is a departure from Local Thermodynamic Equilibrium, or non-LTE, which allows higher vibrational bands to become densely populated, as reviewed by López-Puertas and Taylor (2001).

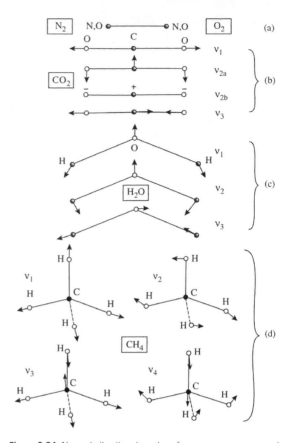

Figure 2.24 Normal vibrational modes of some common gas molecules. The arrows indicate a half-cycle direction of motion, which is followed by movement in the opposite direction to complete the full cycle. Signs +, – indicate motion out of and into the page. (a) Homonuclear molecules N_2 and O_2 have a single stretching mode of vibration. (b) CO_2 has a symmetric stretch (ν_1), which is optically inactive, and an asymmetric stretch ν_3 with a fundamental at 4.3 μm. The bending mode ν_2, with a fundamental at 15 μm, has two equivalent (or *degenerate*) modes 2a and 2b through a 90° rotation. N_2O, which is "NNO" structurally (not shown), behaves similarly to CO_2 with ν_1 to ν_3 modes at fundamentals of 7.8, 17.0, and 4.5 μm, respectively. (c) H_2O has a bending mode ν_2, with a fundamental at 6.3 μm. Water vapor's symmetric stretch ν_1 (fundamental at 2.74 μm) and asymmetric stretch ν_3 (fundamental at 2.66 μm) contribute to near infrared absorption. O_3 (not shown) behaves similarly to H_2O with ν_1 to ν_3 modes at fundamentals of 9.01, 14.3, and 9.6 μm, respectively. (d) CH_4 has four distinct bending modes, ν_1 to ν_4, but only ν_3 and ν_4 are infrared active. Modes ν_3 (with a fundamental at 3.3 μm) and ν_4 (with a fundamental at 7.6 μm) are important bands in spectra of planetary atmospheres (Modified from Thomas and Stamnes (1999).)

2.5.5 Electronic Transitions

Electronic transitions concern the energy levels of electron orbits around the nucleus of an atom. There are larger energy differences between these levels than in vibrational transitions, so electronic transitions of outer

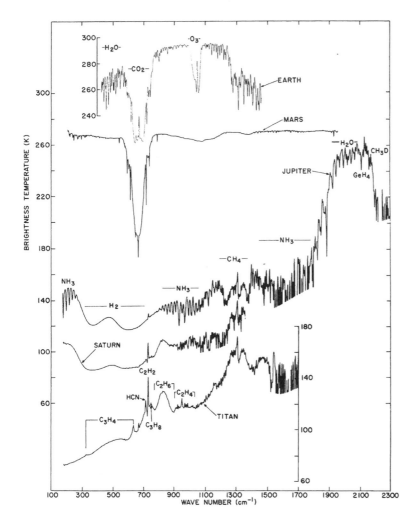

Figure 2.25 Spectra of the Earth, Mars, Jupiter, Saturn, and Titan from the Infrared Interferometer Spectrometer (IRIS) instruments on Nimbus 4, Mariner 9, and Voyager 1 and 2 spacecraft. (From Hanel (1981), Courtesy of NASA.) Carbon dioxide at 15 μm (667 cm^{-1}) is a common feature of the atmospheres of Earth, Mars, and Venus (not shown), but only the Earth has prominent ozone absorption at 9.6 μm (1043 cm^{-1}) and abundant water bands. The giant planets, including Uranus and Neptune, which are not shown, have absorption bands due to hydrogen, ammonia and methane. Titan has a rich spectrum showing hydrocarbons and nitriles. Note the emission of stratospheric acetylene, C_2H_2, at 13.7 μm (729 cm^{-1}).

electrons are associated with absorption and emission at shorter wavelengths, from the near-IR to the UV. For inner electrons, which are tightly bound to the nucleus, absorption and emission occur in the x-rays. Superposed upon the electronic energy levels are the smaller vibrational transitional levels, which in turn have superposed rotational energy levels, as discussed earlier. Consequently, each electronic transition appears as a series of closely spaced lines. All molecules, including those lacking a permanent dipole, such as N_2, have electronic spectra because a dipole moment change always accompanies a change in charge distribution.

2.5.6   Collision-Induced Absorption: Giant Planets, Titan, Early Earth, and Venus

Homonuclear molecules such as H_2 or N_2 lack a permanent electric dipole moment but they still interact with electromagnetic radiation. In fact, absorption by H_2 dominates the far IR (10–1000 μm) spectrum of the giant planets (Fig. 2.25) and H_2 is a key greenhouse gas on Titan (Sec. 14.4.2). We have already met electric quadrupole transitions that allow H_2 to absorb. Of more importance are collisions at high pressure that induce a temporary electric dipole, allowing radiative transitions amongst vibrational and rotational states that would never occur in the isolated molecule (Trafton, 1966, 1998). This process is called *collision-induced absorption* (CIA). It is also called pressure-induced absorption, but this term has the potential for confusion with pressure broadening, so we avoid it. Although the absorption is weak, broad absorption features occur when the gas is abundant and there is a long pathlength. Minor CIA of N_2 and O_2 occurs in the Earth's present atmosphere with radiative effect of little importance (Farmer and Houghton, 1966), but in other atmospheres CIA dominates absorption.

CIA is important in reducing atmospheres of the present and past. CIA results from H_2–H_2 and H_2–He collisions on the giant planets and H_2–N_2, N_2–N_2 and N_2–CH_4 collisions on Titan (McKay *et al.*, 1989). H_2–N_2 and H_2–CO_2 collisions may have been important in warming prebiotic Earth (Wordsworth and Pierrehumbert, 2013) and early Mars (Ramirez *et al.*, 2014a), respectively, as first suggested by Sagan (1977). CIA involving H_2 is particularly effective for greenhouse warming because the moment of inertia of this molecule is very small, causing its rotational energy levels to be widely spaced (see eq. (2.129)). Consequently, at Earth-like temperatures H_2 absorbs across the thermal-IR spectrum.

Dense CO_2 atmospheres are another case where CIA is important. On Venus, CIA from CO_2–CO_2 enables transitions that contribute significant infrared opacity to the atmosphere (Moskalenko *et al.*, 1979). Models of the early Earth that use high partial pressures of CO_2 to offset a fainter Sun ~4 billion years ago must also consider CIA (Kasting and Ackerman, 1986; Wordsworth *et al.*, 2010).

2.5.7 Line Shapes and Broadening

Broadening processes are essential effects in the radiative transfer of planetary atmospheres. If there were not processes in atmospheres that broadened the absorption lines, there would be no greenhouse effect, the Earth would be uninhabitable, and you would not be reading this book. Absorption lines at precise frequencies are infinitesimal delta functions but in order for energy to be absorbed there must be absorption over a wavelength range. The three line-broadening processes are as follows.

- *Natural broadening*: Heisenberg's uncertainty principle expressed in terms of energy (E) and time (t), states that $\Delta E\,\Delta t \geq \hbar/2$ where $\hbar = h/2\pi$. Excited states have a natural lifetime τ_D with respect to decay or τ_C between collisions, so that energy levels must be spread out in energy by at least $\Delta E \sim (\hbar/2\tau_D)$ or $(\hbar/2\tau_C)$. Using $\Delta E = h\Delta v$, this translates to a spread of frequency of $\Delta v \sim 1/4\pi\tau_D$ or $\sim 1/4\pi\tau_C$. However, apart from in the UV, this broadening is miniscule compared to the other two processes described below so we will not discuss it further.
- *Doppler broadening:* Gas molecules move randomly relative to the source of radiation. Such motion produces absorption and emission at frequencies that are Doppler-shifted compared with the rest position of a spectral line.
- *Pressure (or collisional) broadening:* Collisions between molecules provide or remove energy during radiative transitions so that emission and absorption occur over a broader wavelength range.

A useful generalization is that pressure broadening of IR absorption prevails in tropospheres and lower stratospheres, whereas Doppler broadening dominates at higher altitudes. However, the transition from pressure to Doppler broadening occurs higher up for longer wavelength lines because broadening is wavelength-dependent.

Doppler-broadened and pressure-broadened lines have different shapes, as shown in Fig. 2.26. In the pressure-broadened case, the profile is called a *Lorentz line shape*. In general, we describe a line shape by

$$k_v = Sf(v - v_0) \qquad (2.130)$$

where k_v is the absorption coefficient, S is the *line strength*, and $f(v - v_0)$ is the *line shape factor* (or *function*) at frequency v about a center frequency of the line, v_0. The line shape is normalized to unit area, so that the line strength is related to the absorption coefficient as follows:

$$S = \int_0^\infty k_v dv \qquad \int_0^\infty f(v - v_0)dv = 1 \qquad (2.131)$$

The absorption coefficient can be expressed in different ways (Sec. 2.4.2.1), so that the units of S need to be consistent with that choice. For example, if k_v is the volume absorption coefficient in units of cm^{-1} then S is in $cm^{-1}\,s^{-1}$ when working in frequency units (Hz), or S is in cm^{-2} when working in wavenumber units (cm^{-1}).

2.5.7.1 *Doppler-Broadened Line Shape*

The shape of Doppler-broadened lines comes from the physics of the Doppler effect. When a source emitting radiation moves toward an observer at a relative speed V_x, which is the component of the velocity along the line of sight to the observer, the frequency v of a detected photon

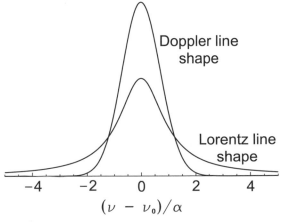

Figure 2.26 Lorentz and Doppler line shapes for the same line widths α and strengths.

is higher than the frequency v_0 emitted in the moving frame. The difference is given by the Doppler shift formula:

$$v = v_0 \left(\frac{1 + V_x/c}{1 - V_x/c} \right)^{1/2} \approx \frac{v_0}{1 - V_x/c} \approx v_0 \left(1 + \frac{V_x}{c} \right) \tag{2.132}$$

The approximations in eq. (2.132) arise from assuming $V_x \ll c$, the speed of light. For a Doppler shift in recession, we would swap all minus signs for plus signs and vice versa. Rearranging the last expression in eq. (2.132) gives

$$\Delta v = v - v_0 = v_0 \frac{V_x}{c} \tag{2.133}$$

The one-dimensional version of the Maxwell–Boltzmann distribution of gas molecule velocities gives the probability that a molecule has a speed between V_x and $V_x + dV_x$ as

$$P(V_x)dV_x = \left(\frac{m}{2\pi kT} \right)^{1/2} \exp \left(-\frac{mV_x^2}{2kT} \right) dV_x \tag{2.134}$$

where m is the molecular mass, k is Boltzmann's constant, and T is temperature. From eq. (2.133), we see that $V_x = c\Delta v/v_0$ and $dV_x = (c/v_0)d(\Delta v)$, which we can substitute in eq. (2.134) to give the probability that the detected photon is at frequency v shifted by an amount Δv from frequency v_0:

$$\text{Prob}(v - v_0 = \Delta v)d(\Delta v) = \text{Prob}\left(V_x = c\frac{\Delta v}{v_0} \right)dV_x$$

$$= \left(\frac{m}{2\pi kT} \right)^{1/2} \exp \left[-\frac{mc^2}{2kT} \left(\frac{\Delta v}{v_0} \right)^2 \right] \frac{c}{v_0} d(\Delta v) \tag{2.135}$$

If we substitute

$$\alpha_D = \frac{v_0}{c} \sqrt{\frac{2kT}{m}} \tag{2.136}$$

in eq. (2.135), we get the Doppler line profile:

$$f_D = \frac{1}{\alpha_D \sqrt{\pi}} \exp \left(-\left[\frac{v - v_0}{\alpha_D} \right]^2 \right) \tag{2.137}$$

We describe the width of the line by the *half-width at half-maximum*, $\alpha_{1/2}$, which is the value of $v - v_0$ at which the absorption falls by half. Solving $f_D(\alpha_{1/2})/f_D(0) = \frac{1}{2}$ gives:

$$\alpha_{1/2} = \alpha_D \sqrt{\ln 2} \tag{2.138}$$

Doppler broadening depends on temperature as $T^{1/2}$ but not on pressure. This makes intuitive sense: the higher the

temperature, the larger the line width because molecules acquire a wider range of speeds.

2.5.7.2 Pressure-Broadened Line Shape

Unlike Doppler broadening, pressure broadening has no exact theory. The Lorentz line shape approximation is given by

$$f_L = \frac{\alpha_L}{\pi \left[(v - v_0)^2 + \alpha_L^2 \right]} \tag{2.139}$$

where α_L is the half-width at half-maximum, which is related to the time between collisions by $\alpha_L = (2\pi\tau_C)^{-1}$. This equation can be derived by taking the Fourier transform of a burst of light that decays exponentially with time constant τ_C, e.g., see Zdunkowski *et al.* (2007), pp. 205–209. The Lorentz half-width is proportional to the pressure and inversely proportional to temperature to some empirical exponent N_L (where $N_L = \frac{1}{2}$ to 1 depending on the gas),

$$\alpha_L \propto \frac{p}{T^{N_L}} \quad \text{or} \quad \alpha_L = \alpha_{\text{ref}} \left(\frac{p}{p_{\text{STP}}} \right) \left(\frac{T_{\text{STP}}}{T} \right)^{N_L} \tag{2.140}$$

In this equation, T_{STP} and p_{STP} and are standard temperature and pressure values 273.15 K and 1013 mbar, respectively, and α_{ref} is empirical. If we divide eq. (2.136) by eq. (2.140), and put a typical exponent $N_L = \frac{1}{2}$ in the latter, the result is the ratio of Doppler to pressure broadening:

$$\frac{\alpha_D}{\alpha_L} = \frac{v_0 p_{\text{STP}} T}{\alpha_{\text{ref}} cP} \left(\frac{2k}{mT_{\text{STP}}} \right)^{1/2} \simeq (5 \times 10^{-11} \text{ Pa Hz}^{-1}) \left(\frac{v_0}{P} \right) \tag{2.141}$$

For the last approximation, we have used terrestrial values of $m = 30$ a.m.u. and a mid-tropospheric temperature $T \sim$ 225 K. Values of α_{ref} range from 0.1 to 1 cm^{-1} in wavenumbers and we have used $\alpha_{\text{ref}} = 0.07$ cm^{-1} (close to the value for CO_2), which in frequency units is $\alpha_{\text{ref}} = 0.07$ cm^{-1} $\times$ c $\times$ 100 cm/m = 2.1×10^9 Hz. Thus, for the prominent 15 μm or 2×10^{13} Hz band of CO_2, Doppler and pressure broadened line widths become equal at ~10 hPa, which is in the stratosphere at ~31 km altitude. For the 2.5 mm or 118 GHz microwave line for O_2, eq. (2.141) indicates equal line widths at ~6 Pa, which is ~68 km altitude (Elachi and Van Zyl, 2006). See Fig. 2.27.

Before leaving line shapes, we note that the Lorentz and Doppler line shapes can be convolved into a *Voigt profile* line shape, which is a useful representation when both broadening processes contribute. The Voigt profile equation is more complicated and requires numerical solution.

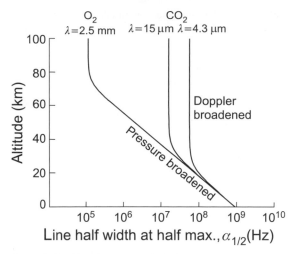

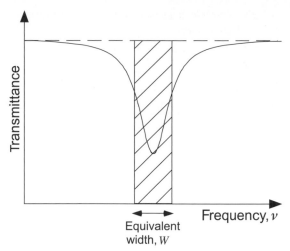

Figure 2.27 Although wavelength-dependent, in the thermal-infrared *pressure broadening* generally dominates in tropospheres, whereas *Doppler broadening* dominates above middle stratospheres. This graph shows the broadening of CO_2 infrared lines and an O_2 microwave line as a function of altitude on Earth. (After a similar diagram by Elachi and van Zyl (2006), p. 453, originally by J. Waters of JPL.)

Figure 2.28 Equivalent width, W, is the width of a fully absorbing rectangle with the same area as the dip in the signal caused by the absorption line.

2.5.8 Continuum Absorption

In addition to spectral lines and bands, the continuum is smoothly varying absorption over a broad range of wavelengths. In the infrared window of the Earth's atmosphere from 800 to 1200 cm^{-1} (12.5 to 8 μm), weak continuum absorption is mainly due to water vapor (Shine *et al.*, 2012). However, the absorption coefficient is proportional to the *square* of the water vapor density, which is unusual. Two ideas are proposed for this continuum absorption. The first is that the continuum comes from the distant wings of very many lines in the far IR. The second idea is that continuum absorption arises from clusters of H_2O molecules, including dimers ($H_2O \cdot H_2O$), trimers, and polymers.

In contrast, in the shortwave, continuum absorption arises from well-known processes, including *photoionization*, when photons are energetic enough to create ions, and *photodissociation* when photons break or dissociate molecules into atoms. Of course, photodissociation is very important for atmospheric chemistry, as we describe in Sec. 3.1.

2.5.9 Band Transmission and Weak and Strong Absorption

Rather than deal with the varying depth of broadened lines, it is sometimes convenient to have a fundamental

integrated measure of the extent to which a spectral line can reduce radiation. Such a quantity is the *equivalent width*, which is the width of a completely absorbed rectangle having the height of the continuum and the same absorption area as the line (Fig. 2.28). In frequency units, the equivalent width is defined as

$$W = \int_{\Delta v}(1 - e^{-\tau_v})dv = \int_{\Delta v}(1 - \mathcal{T}_v)dv = \Delta v(1 - \overline{\mathcal{T}}_{band}) = \Delta v \mathcal{A}_{band}$$

(2.142)

where τ_v is the optical path at frequency v (eq. (2.56)), $\mathcal{T}_v$ is the transmission, and Δv is the spectral interval. The other quantities are the average *band transmission* and *band absorption* over the spectral band, which are defined as follows:

$$\overline{\mathcal{T}}_{band} = \frac{1}{\Delta v}\int_{\Delta v} \mathcal{T}_v dv \qquad \mathcal{A}_{band} = 1 - \overline{\mathcal{T}}_{band} \qquad (2.143)$$

If the mass absorption coefficient is $k_{m,v}$, we define a *mass path* u_a as a function of absorber density, ρ_a, over an arbitrary pathlength l, from point s_1 to s_2,

$$u_a = \int_{s_1}^{s_2} \rho_a(s)ds, \qquad = \rho_a l \text{ for constant } \rho_a \qquad (2.144)$$

Then, the transmission is $\mathcal{T}_v = \exp(-k_{m,v}u_a)$. There are two limiting cases that help us understand how increases of certain gases will affect greenhouse warming in atmospheres or how to use measured absorption in spectral lines to derive gas abundances.

2.5.9.1 The Weak Line Limit (for an Optically Thin Gas)

If absorption is weak, the transmission and equivalent width are

$$\mathcal{T}_v = \exp\left(-k_{v,m}u_a\right) \approx 1 - (k_{v,m}u_a) \qquad W = \int_{\Delta v}(1 - \mathcal{T}_v)dv = \int_{\Delta v}k_{v,m}u_a\,dv = Su_a \qquad (2.145)$$

Here, we have used the definition of the line strength S (eq. (2.131)) in the last step. This linear law between W and u_a is called the *weak approximation*. It tells us that if the gas is optically thin (i.e., $k_{m,v}u_a \ll 1$), doubling the amount of absorbing gas or pathlength will double the absorption, which is shown as the linear section of Fig. 2.29(b).

2.5.9.2 Strong Line Limit (for an Optically Thick Gas)

The strong line limit applies when there is complete absorption in the line center, shown on the right-hand side of Fig. 2.29(a). When this happens, we say that the line center has *saturated*. In this case, adding extra absorbing gas can only increase absorption in the wings of the line, which has a nonlinear relationship between absorber amount and absorption. We can derive this relationship using the Lorentz profile (eq. (2.139)). The equivalent width, using eqs. (2.130), (2.139), and (2.142), will be given by

We also absorbed the minus sign in substituting for dx in eq. (2.147), given that the limits on the integral in eq. (2.147) should otherwise reverse in changing from x to β. The result that $W \approx 2(S\alpha_L u_a)^{1/2}$ is known as the *square root law*.

A log–log plot of the equivalent width W versus the mass path of the absorber u_a, is called the *curve of growth* (Fig. 2.29(b)), in which we can see the progression from the "weak" linear relationship to the "strong" square root relationship.

What does the curve of growth mean? If a trace greenhouse gas, such as CCl_2F_2 in the Earth's atmosphere, doubles in abundance, its band-averaged absorption will double along a path. On the other hand, a gas such as CO_2, which is saturated in strongest bands such 15 μm, needs to quadruple in abundance in order to double its band-averaged absorption. In fact, the radiative forcing change due to perturbations in the abundances of these particular gases in the current Earth's climate can each be represented by simplified analytical expressions that reflect the line broadening. There is a linear relationship of forcing to CCl_2F_2 concentration and a logarithmic one for CO_2 concentration (e.g., Sec 6.3.5 of Ramaswamy *et al.* (2001), noting that hot climates with much higher CO_2 levels deviate from purely logarithmic (Caballero and Huber, 2013)).

$$W = \int_{v-v_0=-\infty}^{\infty}[1 - \exp\left(-Sf_L u_a\right)]d(v - v_0) = \int_{-\infty}^{\infty}\left[1 - \exp\left(\frac{-S\alpha_L u_a}{\pi\left[(v-v_0)^2 + \alpha_L^2\right]}\right)\right]d(v - v_0) \qquad (2.146)$$

When the line center is completely absorbing, we need only consider the wings of the line. Consequently, $(v - v_0)^2 \gg \alpha_L^2$, so α_L^2 can be omitted from the denominator. Noting that the line is symmetrical, we can use substitutions to compute the integral as follows:

Curves of growth are also useful in deriving the abundance of gases from planetary spectra. For example, one of the authors (DCC) when an undergraduate, measured reflected sunlight from Jupiter's cloud-tops using a small optical telescope, grating spectrometer, and

$$W \approx 2\int_0^{\infty}\left[1 - \exp\left(\frac{-S\alpha_L u_a}{\pi x^2}\right)\right]dx = \sqrt{\frac{S\alpha_L u_a}{\pi}}\int_{\beta=0}^{\infty}\frac{[1 - \exp(-\beta)]}{\beta^{3/2}}d\beta \qquad (2.147)$$

Using standard integral $\displaystyle\int_0^{\infty}\frac{[1 - \exp(-\beta)]}{\beta^{3/2}}d\beta = 2\sqrt{\pi} \quad \Rightarrow W \approx 2\sqrt{S\alpha_L u_a} \qquad (2.148)$

To solve the integral in eq. (2.147), we made the following substitutions in eq. (2.147),

$$x = v - v_0, \ \ \beta = \frac{S\alpha_L u_a}{\pi x^2}, \ \ \text{so} \ \ \frac{d\beta}{dx} = -2\frac{S\alpha_L u_a}{\pi x^3} \ \ \text{and} \ \ dx = \frac{-1}{2}\sqrt{\frac{S\alpha_L u_a}{\pi}}\frac{d\beta}{\beta^{3/2}}$$

(a)

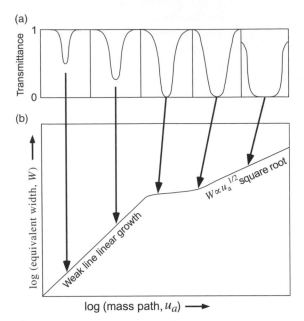

(b)

Figure 2.29 How the *curve of growth* of equivalent width W, is related to changes in line shape. (a) The schematic change in the line shape as the absorber density and/or pathlength increases. (b) The related change in the log–log plot of equivalent width versus the mass path u_a, i.e., the product of absorber density and pathlength.

charged-couple device detector, which revealed absorptions due to methane vibration–rotation bands at wavelengths centered at 543 nm and 619 nm. From the measured absorption, one can calculate equivalent widths. Then one can use laboratory data of the curve of growth for methane in the linear weak limit to calculate the column abundance of methane above Jupiter's cloud-tops.

2.6 Calculating Atmospheric Absorption in Climate Calculations

Above, we have considered single lines, but for calculating the climate of a planet, we have to deal with thousands to millions of lines with a great range of line widths rather than isolated lines, and often these lines overlap.

Consequently, sophisticated methods have been devised to quantify atmospheric absorption as a function of wavelength.

Line-by-line calculations add up the effects of all the contributions of known individual lines. This method is as accurate as possible (and line lists are continually updated to provide even higher accuracy), but computers take a very long time to calculate broadband fluxes at every level in an atmosphere, which proves impractical for repeated climate calculations. Consequently, two other approaches have been developed to reduce the computational time by orders of magnitude.

• *Band models* rely on a simple analytical form for the positions and strengths of lines, which can be solved much faster than line-by-line calculations. Lines are assigned mean line strengths and are spaced randomly or in some other arrangement, which allows analytical calculation over some spectral interval.

• The *correlated k-distribution method* recasts the very complicated variation of absorption coefficients k_v over a spectral interval Δv into a smooth, monotonic, cumulative probability function of the fractional occurrence of k_v within Δv. This approach is a statistical technique that speeds up calculations. Each spectral interval Δv is sufficiently narrow that the Planck function within the interval is treated as a function of temperature only, without frequency dependence. Because the cumulative probability function is smooth and monotonic, the k_v values can be summed with relatively large bins, which allows for fast computation of transmissivity. See Appendix A for a description of this method in a one-dimensional (1-D) radiative–convective climate model.

Validation against line-by-line calculations shows that band models and k-distribution methods can produce realistic results if applied correctly. Petty (2006) gives a very readable introduction to such methods, so we will not describe these models further. More detailed treatments are given in lengthier books devoted to atmospheric radiation by Liou (2002), Thomas and Stamnes (1999), Goody and Yung (1989) and Zdunkowski (2007). Pierrehumbert (2010) also includes detailed discussion.

3 Essentials of Chemistry of Planetary Atmospheres

Atmospheric composition is a fundamental concern of atmospheric and climate evolution. Consequently, in this chapter, we outline some general principles that apply to the chemistry of atmospheres. We also provide a brief overview of the chemistry of Earth's atmosphere, and some common aspects of the chemistry of other planetary atmospheres. More detailed chemistry of particular planets is discussed in later chapters.

3.1 General Principles

Two chemical approaches in understanding atmospheric evolution are equilibrium chemistry and kinetics. *Chemical equilibrium* is when the concentrations of chemicals in a chemical reaction or set of reactions have reached a condition of no further change because forward and reverse reaction rates are equal. A chemical equilibrium approach can help determine the atmospheric composition at the surface of Venus or in a "Hot Jupiter" exoplanet where high temperature and pressure help to establish conditions that are close to equilibrium. However, thermodynamics does not tell us about the speed of reactions, which is often crucial in determining whether reactions are important. Furthermore, atmospheric chemistry is also strongly influenced by UV radiation from the Sun, or from the parent star for an exoplanet. This radiation is emitted from the star's hot photosphere or from even hotter regions higher up in the stellar atmosphere; hence, its effective "temperature" is much higher than the planet's temperature. Thus, the composition of *all* planetary atmospheres is more accurately determined by the rates of chemical and photochemical reaction between all components.

The science dealing with rates of reaction is *kinetics*. In particular, the chemical composition of the atmosphere of planets such as Earth is very different from a thermodynamic equilibrium mixture. The competition between the rates of hundreds of reactions in a planetary atmosphere actually sets the atmospheric composition. Moreover, on geologic timescales we have to consider rates of reaction between gases and surface minerals on rocky planets. One way to think about the difference between the two approaches of equilibrium and kinetics is that chemical equilibrium tells you what "should happen" if everything reacted to the lowest energy state whereas kinetics tells you what actually "does happen."

3.1.1 Essentials of Thermodynamic Chemical Equilibrium

Consider a chemical reaction between reactants B and C, which makes products G and H,

$$b\text{B} + c\text{C} = g\text{G} + h\text{H} \tag{3.1}$$

where b, c, g, and h are *stoichiometric coefficients*. We define the equilibrium constant for this reaction as

$$K_{eq} = \frac{\text{activity product of products}}{\text{activity product of reactants}} = \frac{a_\text{G}{}^g a_\text{H}{}^h}{a_\text{B}{}^b a_\text{C}{}^c} \tag{3.2}$$

where a_B, a_C, a_G and a_H are the *activity* of each chemical species at equilibrium. Activity is an effective concentration that takes into account non-ideal behavior associated with the attraction and repulsion of molecules in real gases or aqueous solutions. Thus, activity (rather than concentration) determines how a species *acts* with respect to reactivity. For example, if we put acid "HA" into NaCl (sodium chloride) solution, A^- anions will be surrounded by Na^+ cations and so A^- will be shielded and less "active" than it would be on the basis of ideal assumptions. Activity is defined as a *dimensionless* quantity by normalizing non-ideal concentrations to a standard unity concentration. We can apply this concept to gases and aqueous solutions.

Gases. For gases, we express concentrations in terms of *fugacity*, which has units of atm or bar. The fugacity f_i

of gas i is the partial pressure p_i multiplied by a dimensionless *fugacity coefficient* ϕ_i, which is a correction factor for non-ideal conditions depending on the pressure–temperature regime and molecular nature of the gas. The activity of a gas species i is then defined as the fugacity of i divided by a unity standard pressure:

$$a_i = \frac{f_i}{p_{\text{standard}}} = \frac{\phi_i p_i}{p_{\text{standard}}} \tag{3.3}$$

Unfortunately, there are two conventions for the standard pressure: it may be defined as either 1 atm (101 325 Pa) or 1 bar (10^5 Pa) depending on the database. So the convention should be stated, although the difference is often negligible for many calculations.

In the case where gases are nearly ideal, the fugacity coefficient $\phi_i \approx 1$ and the fugacity is simply equal to the partial pressure, i.e., $f_i \approx p_i$. Then using partial pressures reasonably approximates the equilibrium constant for a reaction between gases:

$$K_{eq} \approx \frac{p_G{}^g p_H{}^h}{p_B{}^b p_C{}^c} \tag{3.4}$$

Aqueous solutions. The equilibrium constant formula of eq. (3.2) also applies to aqueous solutions. In aqueous chemistry, the activities express non-ideal aqueous concentrations. For example, very salty solutions deviate from ideal conditions because the ions are prone to interact and shield each other, as mentioned above. In aqueous solutions, the *activity* a_i of species i is usually defined in terms of the *molality* m_i (moles per kg) multiplied by a dimensionless *activity coefficient* γ_i that accounts for interactions between ions, as follows:

$$a_i = \frac{\gamma_i m_i}{m_{\text{standard}}}, \quad \text{where } m_{\text{standard}} = 1 \text{ mol kg}^{-1} \tag{3.5}$$

For dilute solutions, $\gamma_i \approx 1$, so the activity is numerically equal to the molal concentration of each species, i.e., $a_i \approx m_i$, where the division by a 1 molal standard concentration is often assumed and not written out explicitly. In the near-ideal case, replacing the activities in eq. (3.2) with molal concentrations gives an approximation for the equilibrium constant. In applying this equation, we set activities of pure substances such as solids or pure liquids to unity because they cannot affect the concentrations of other species.

Fortunately, it is relatively easy to calculate the equilibrium constant K_{eq} if appropriate thermodynamic data

are known for chemical products and reactants. The necessary thermodynamic data are Gibbs free energies of formation, which are tabulated in databases for standard conditions of 1 atm (sometimes 1 bar) and 298.15 K. *Gibbs free energy* is energy that is available to do useful work. This is different from energy that has lost its usefulness by being dispersed amongst molecules, such as energy dissipated by friction into the random motion of molecules. At constant pressure P and temperature T, denoted by subscripts,

$$G_{T,P} = H - TS \tag{3.6}$$

where quantities are often expressed per mole, with $G_{T,P} =$ Gibb's free energy [kJ mol^{-1}], $H =$ enthalpy ("heat content") [kJ mol^{-1}] and $S =$ entropy [kJ mol^{-1}·K^{-1}].

Equilibrium is a minimum Gibbs free energy state so a system moves toward equilibrium by releasing free energy (through a change, ΔG) and doing work. For changes of Gibbs free energy at constant pressure and temperature,

$$\Delta G_{T,P} = \Delta H - T\Delta S \tag{3.7}$$

where ΔH is the heat absorbed (negative ΔH is exothermic and positive ΔH is endothermic), and ΔS is the change in entropy. Equation (3.7) is fundamental in chemistry. If $\Delta G_{T,P}$ is positive in a system at a given pressure P and temperature T, a reaction cannot proceed. If $\Delta G_{T,P}$ is negative, a reaction should proceed to equilibrium, with the caveat that it may not, if rates of reaction are too slow, and kinetics trumps thermodynamics. With the rule that $\Delta G_{T,P}$ needs to be negative, we can see from eq. (3.7) that reactions are favored by *release of heat* (ΔH is negative) and an *increase of entropy* of reaction products versus reactants (ΔS is positive).

Under standard conditions, the Gibb's free energy change for a reaction is readily calculated via:

$$\Delta G_{\text{reaction}}^0 = \sum G_{\text{products}}^0 - \sum G_{\text{reactants}}^0 \tag{3.8}$$

where superscript "0" indicates the standard state. The summations on the right hand side of eq. (3.8) are for Gibbs free energies of formation (ΔG_f^0) that can be found in tables (see below for an example). Also, by convention, elements in their standard states are assigned $\Delta G_f^0 = 0$, where the standard state is when the pure element is in its most stable form at 25 °C and 1 atm. The equilibrium constant for reaction (3.1) is related to $\Delta G_{\text{reaction}}^0$ by

$$K_{eq,298K} = \frac{\text{activity product of products}}{\text{activity product of reactants}} = \frac{a_G{}^g a_H{}^h}{a_B{}^b a_C{}^c} = \exp\left(\frac{-\Delta G_{\text{reaction}}^0}{RT}\right) \tag{3.9}$$

where T is absolute temperature and R (=8.314 J mol^{-1} K^{-1}) is the universal gas constant.

We can calculate the equilibrium constant under standard conditions, given $\Delta G^0_{reaction}$ from eq. (3.8). This allows us to calculate unknown activities of species in a mixture at equilibrium from the equilibrium constant, eq. (3.9). Note, as a shortcut, eq. (3.9) can be written as:

$$\log_{10}K_{eq} = -\Delta G^0_{reaction}/5.708 \text{ for kJ mol}^{-1}, \quad \text{or } \log_{10}K_{eq} = -\Delta G^0_{reaction}/1.364 \text{ for kcal mol}^{-1} \tag{3.10}$$

It is important to realize that you cannot calculate K_{eq} from eq. (3.9) by changing T from 298.15 K when $\Delta G^0_{reaction}$ comes from tables and is referenced to 298.15 K. In eq. (3.9), K_{eq} and $\Delta G^0_{reaction}$ are for the same temperature. To assess how K_{eq} changes with temperature or pressure, we would need to go back to our definition of $\Delta G_{T,P}$ (eq. (3.7)) and determine how enthalpy and entropy are affected by temperature or pressure. For this topic, we refer the reader to detailed discussions in a variety of thermodynamics textbooks, e.g., Chapters 4 and 8 of Nordstrom and Munoz (1994), Ch. 9 of Anderson (2005), or Ch. 13 of Anderson and Crearer (1993).

To make the above concepts clearer, consider the combustion of methane to produce carbon dioxide and water:

$$CH_4 + 2O_2 \rightleftharpoons CO_2 + 2H_2O \tag{3.11}$$

The equilibrium constant for this reaction under standard conditions (1 bar, 25 °C) can be determined from the Gibb's free energy of formation values, ΔG^0_f, in kJ/mol of CH$_4$ = –50.72, CO$_2$ = –394.359, and H$_2$O = –237.129. By definition, ΔG^0_f values are for the formation of the compounds from their standard state elements (e.g., for water, H$_2$(g) + ½O$_2$(g) = H$_2$O(l)). Also by definition, elements in their standard states have ΔG^0_f = 0, which applies to O$_2$ in eq. (3.11). So using eq. (3.8), we have

$$\Delta G^0_{reaction} = \sum G^0_{products} - \sum G^0_{reactants} = [(2 \times -237.129) - 394.359] - (-50.72)] = -817.9 \text{ kJ mol}^{-1}$$

The negative $\Delta G^0_{reaction}$ means that the reaction is possible. To make further progress, we need to calculate the equilibrium constant. Consequently, using eq. (3.10), $\log_{10}K_{eq}$ = $\Delta G^0_{reaction}/5.708$ = (–(–817.9)/5.708) = +143.3.

The equilibrium constant for eq. (3.11) allows us to answer the question of how much methane should exist in equilibrium in a 1 bar atmosphere at 25 °C containing fixed levels of 0.21 bar of O$_2$ and ~380 ppmv CO$_2$ above liquid water. Using eq. (3.4), and recognizing that the activity of water is unity, the calculated partial pressure of methane would be astonishingly small:

$$p_{CH4} \sim \frac{(p_{CO2})(a_{H2O})}{(p_{O2})^2 K_{eq}} = \frac{380 \times 10^{-6} \times 1}{(0.21)^2 \times 10^{143.29}} = 10^{-145} \text{ bar}$$

Since the Earth's atmosphere consists of ~10^{44} molecules, a partial pressure of 10^{-145} bar is equivalent to less than a single methane molecule in the entire atmosphere! However, 1.8 ppmv CH$_4$ coexists with 0.21 bar O$_2$ in the current atmosphere, which unequivocally demonstrates that gases in Earth's atmosphere are in strong chemical disequilibrium, and demands a considerable flux of CH$_4$ into the atmosphere to sustain the methane concentration. About 90% of the methane flux is biological, while the rest is from geologic sources (Kirschke *et al.*, 2013), so biology sustains the disequilibrium methane abundance in Earth's atmosphere.

In reality, the combustion reaction of eq. (3.11) does not occur directly in the atmosphere but only indirectly via many photochemical reactions ultimately initiated by solar photons. Consequently, the composition of terrestrial air is *not* determined by equilibria but by kinetics, a topic considered below.

Before leaving the topic of chemical equilibrium, however, we note that thermodynamics contains many subtleties and that we have omitted some important concepts such as chemical potential, redox parameters Eh and pe, and Gibbs free energy minimization. The interested reader should consult Anderson (2005) for a lucid introduction and Anderson and Crearer (1993) or Fegley and Osborne (2013) for in-depth detail. Also Van Zeggeren and Storey (1970) is a classic introduction on the subject of computing chemical equilibrium using methods such as Gibbs energy minimization.

3.1.2 Chemical Kinetics of Atmospheric Gases

We can write a general reaction of species B and C to make species G and H as follows:

$$bB + cC \rightarrow gG + hH \tag{3.12}$$

The reaction rate is the number of reactions per unit volume and unit time,

$$R_r = k_r n_B{}^m n_C{}^n \tag{3.13}$$

where n_B and n_C are number densities, often in units of molecules cm^{-3}. (Square brackets, i.e., [B] is a common alternative notation for number density.) Here, the exponents are determined experimentally and give the *order of the reaction*, m+n. A newcomer should not confuse this reaction rate expression with the expression for the equilibrium constant, eq. (3.2), where stoichiometric

$$\tau_i = \frac{M_i}{F_{\text{out}}} = \frac{M_i}{F_{\text{in}}} = \frac{\text{steady state amount of constituent } i \text{ in a reservoir}}{\text{rate of removal (or input) of constituent } i \text{ from a reservoir}} \tag{3.20}$$

coefficients are used as exponents. The value k_r is the *rate coefficient* or *rate constant*, which has units depending on the order of the reaction and the concentration units. If $m = 1$ and $n = 0$, the reaction is first order and k_r has units of s^{-1}. If $m = n = 1$, the reaction is second order and k_r has units of cm^3 molecule^{-1} s^{-1}. And so on.

The rates of consumption or appearance of species in eq. (3.12) are related as follows:

$$\frac{d[\text{B}]}{dt} = -b \times \text{rate} = -bR_r, \quad \frac{d[\text{G}]}{dt} = gR_r, \text{etc., so}$$

$$\text{reaction rate}, R_r = -\frac{1}{b}\frac{d[\text{B}]}{dt} = -\frac{1}{c}\frac{d[\text{C}]}{dt} = \frac{1}{g}\frac{d[\text{G}]}{dt} = \frac{1}{h}\frac{d[\text{H}]}{dt} \tag{3.14}$$

Consider an example of using eq. (3.13). The oxidation of methane in the Earth's atmosphere is initiated with an attack by a hydroxyl (OH) species:

$$\text{OH} + \text{CH}_4 \rightarrow \text{CH}_3 + \text{H}_2\text{O} \tag{3.15}$$

This reaction is the rate-limiting step in the destruction of methane. In the modern atmosphere, typical number densities of CH_4 and OH are $n_{OH} = 8 \times 10^5$ molecules cm^{-3} and $n_{CH4} = 4.3 \times 10^{13}$ molecules cm^{-3}. Given an average tropospheric rate constant $k_{rCH4} = 3.6 \times 10^{-15}$ cm^3 molecule^{-1} s^{-1} for reaction (3.15), we can estimate the *chemical lifetime* of methane, which is the average time a molecule spends in the atmosphere before being destroyed. The units of the rate constant tell us that the reaction is first order, so

$$\tau_X = \frac{1}{\sum_i k_{ri} n_{Yi}} \tag{3.19}$$

More generally, the *residence time* of a species i is defined in geochemistry as

In the above example for methane gas, our reservoir was the atmosphere, but we can also define a residence time for other types of species in other reservoirs. The reservoir in eq. (3.20) could be defined as any part of the planet's system, such as the atmosphere, the ocean, the crust or the mantle. For example, the residence time of sulfate (SO_4^{2-}(aq)) in the modern ocean is $\sim 10^7$ years.

3.1.3 The Importance of Free Radicals

Reaction (3.15) is typical of atmospheric reactions because it involves a *free radical species*, which in this case is the hydroxyl radical, OH. In fact, the only reactions that proceed at appreciable rates in the atmosphere, with few exceptions, are those involving at least one radical species. A *free radical* is a species that has an unpaired electron in its outer shell. Consequently, free radicals react vigorously to attain a more stable bonding state. In general, a species with an odd number of electrons will be a radical, such as atomic chlorine, Cl. An exception is ground state atomic oxygen (O), which is called a *biradical* because it has eight electrons of which two are unpaired. Some textbooks indicate radicals by writing a dot after them, e.g., HO· or Cl·.

Radicals have high free energies because they ultimately originate from an input of free energy from solar photons when molecules are broken apart:

$$\tau_{\text{CH4}} = \frac{n_{\text{CH4}}}{\text{destruction rate}} = \frac{\cancel{n_{\text{CH4}}}}{k_{r\text{CH4}} n_{\text{OH}} \cancel{n_{\text{CH4}}}} = \frac{1}{\left(3.6 \times 10^{-15} \text{cm}^{-3} \text{ molec s}^{-1}\right)\left(8 \times 10^5 \text{cm}^{-3}\right)} \sim 12 \text{ years} \tag{3.16}$$

This example demonstrates that for a reaction of the form,

$$\text{X} + \text{Y} \rightarrow \text{products}, \quad \text{rate constant}, k_r \tag{3.17}$$

the *chemical lifetime* of reactant X is

$$\tau_X = \frac{1}{k_r n_Y} \tag{3.18}$$

If there are many reactions of X of the type of (3.17) with many species Y_i, the lifetime is

nonradical + $h\nu \rightarrow$ radical + radical
(e.g., $Cl_2 + h\nu(\lambda < 450\,\text{nm}) \rightarrow Cl + Cl$) $\tag{3.21}$

Subsequent reactions with nonradical molecules must propagate the free energy to other types of radicals because it is not possible to pair up all the electrons involved, i.e.,

radical + nonradical $\rightarrow$ radical + nonradical
(e.g., $Cl + CH_4 \rightarrow CH_3 + HCl$) $\tag{3.22}$

The example above is an *abstraction* reaction where an atom is pulled away from a nonradical species. In this case, it is *hydrogen abstraction* from the methane. Free radicals are lost in reactions with other free radicals. The loss is either through *disproportionation* when two non-radicals are produced,

$$\text{radical} + \text{radical} \rightarrow \text{nonradical} + \text{nonradical} \\ (\text{e.g., OH} + \text{HO}_2 \rightarrow \text{H}_2\text{O} + \text{O}_2) \tag{3.23}$$

or *recombination* when a single molecule is formed,

$$\text{radical} + \text{radical} + \text{M} \rightarrow \text{nonradical} + \text{M} \\ (\text{e.g., NO}_2 + \text{OH} + \text{M} \rightarrow \text{HNO}_3 + \text{M}) \tag{3.24}$$

Thus, the chain of reactions initiated by sunlight and propagated by radicals can eventually terminate through reactions such as (3.23) and (3.24). In reaction (3.24), the "M" species is any molecule in the atmosphere. "M" serves the purpose of removing excess energy because otherwise the nonradical species formed could split apart. In fact, reaction (3.24) is an example of a three-body or *termolecular* reaction, which we now consider.

3.1.4 Three-Body (Termolecular) Reactions

If two gases, A and B, combine to produce a single product AB, chemical energy is released when the combined product AB is formed and AB will split apart unless a third body is able to dispose of the energy. When the

pressure, the reaction rate coefficient can be expressed differently and have no dependence on [M]. We can understand this by decomposing eq. (3.25) into its component reactions:

$$\text{A} + \text{B} \rightarrow \text{AB}^* \qquad (k_{\text{AB}^*}) \tag{3.27}$$

$$\text{AB}^* \rightarrow \text{A} + \text{B} \qquad (k_{\text{A+B}}) \tag{3.28}$$

$$\text{AB}^* + \text{M} \rightarrow \text{AB} + \text{M}^* \qquad (k_{\text{AB}}) \tag{3.29}$$

Here, the asterisk indicates an excited state with excess energy and the *k*s in parentheses are rate coefficients. The rate of formation of AB is determined by eq. (3.29), i.e.,

$$\frac{d[\text{AB}]}{dt} = k_{\text{AB}}[\text{AB}^*][\text{M}] \tag{3.30}$$

We can assume that the excited complex AB* reacts as soon as it is produced. Consequently, in steady state, we can equate production and loss of AB* as follows:

$$\underbrace{k_{\text{AB}^*}[\text{A}][\text{B}]}_{\text{gain of AB}^*} = \underbrace{k_{\text{A+B}}[\text{AB}^*] + k_{\text{AB}}[\text{AB}^*][\text{M}]}_{\text{losses of AB}^*} \tag{3.31}$$

We can rearrange this equation to make [AB*] subject and then eliminate [*AB**] from eq. (3.30), to give,

$$\frac{d[\text{AB}]}{dt} = -\frac{d[\text{A}]}{dt} = -\frac{d[\text{B}]}{dt} = \frac{k_{\text{AB}^*}k_{\text{AB}}[\text{A}][\text{B}][\text{M}]}{k_{\text{A+B}} + k_{\text{AB}}[\text{M}]} \tag{3.32}$$

This equation has a *low-pressure limit* with rate constant, k_0 [cm^6 molecule^{-2} s^{-1}]

$$\frac{d[\text{AB}]}{dt} = -\frac{d[\text{A}]}{dt} = -\frac{d[\text{B}]}{dt} = \frac{k_{\text{AB}^*}k_{\text{AB}}[\text{A}][\text{B}][\text{M}]}{k_{\text{A+B}}} = k_0[\text{A}][\text{B}][\text{M}] \quad \text{when } [\text{M}] << \frac{k_{\text{A+B}}}{k_{\text{AB}}} \tag{3.33}$$

and *high-pressure limit* with rate constant, k_∞ [cm^3 molecule^{-1} s^{-1}]

$$\frac{d[\text{AB}]}{dt} = -\frac{d[\text{A}]}{dt} = -\frac{d[\text{B}]}{dt} = k_{\text{AB}^*}[\text{A}][\text{B}] = k_\infty[\text{A}][\text{B}] \quad \text{when } [\text{M}] >> \frac{k_{\text{A+B}}}{k_{\text{AB}}} \tag{3.34}$$

third body is present in the collision, the result is a three-body or *termolecular* reaction:

$$\text{A} + \text{B} + \text{M} \rightarrow \text{AB} + \text{M} \tag{3.25}$$

where "M" represents an atmospheric molecule that has an identity of no concern. Usually "M" is the most abundant species, such as N_2 in the Earth's atmosphere. The reaction rate of eq. (3.25) is proportional to the product of the concentrations of the species that must collide simultaneously,

$$\text{Reaction rate, } R_r \left(\text{cm}^{-3}\text{s}^{-1}\right) = k_r[\text{A}][\text{B}][\text{M}] \tag{3.26}$$

Here k_r is the three-body reaction rate coefficient, with units of cm^6 molecule^{-2} s^{-1}. However, in the case of high

3.1.5 Temperature Dependence of Reaction Rates

A reaction rate "constant" k_r generally depends on thermal energy kT, through an Arrhenius equation as

$$k_r = A \exp\left(-E_a/kT\right) \tag{3.35}$$

where E_a is an *activation energy* and A is a *pre-exponential factor*. The latter reflects the collision rate of reactants and the molecular orientations that must be satisfied for a favorable reaction. Temperature dependence of gas phase reactions is determined experimentally, but a few general intuitive rules apply.

– Reactions between radicals and molecules are common and generally have positive activation energy E_a so that k_r increases with temperature.

– Reactions between two radicals are fast and generally have negative activation energy E_a so that k_r decreases with temperature.

– Termolecular reactions are faster at lower temperatures when the molecules approach each other at slower speeds.

– Reactions between two nonradical molecules are usually very slow unless they are able to react on particle surfaces.

– The reactions that occur at Earth-like temperatures are exothermic or very slightly exothermic.

As we have already mentioned, the initiation of chemical reaction chains requires the production of radicals, which is ultimately driven by sunlight. So we next consider photolysis rates.

3.1.6 Photolysis

In the photolysis of molecule A, its number density n_A [molecules cm^{-3}] diminishes with a photodissociation or photolysis rate as follows:

$$A + h\nu \rightarrow \text{products}, \quad \text{photolysis rate} = \frac{\partial n_A}{\partial t} = -j_A n_A$$

(3.36)

Here, j_A is the *photolysis rate coefficient* or *photodissociation coefficient* [s^{-1}]. This rate depends upon several factors as follows.

• The incident flux of solar photons from all directions on a volume of air called the *actinic flux* [photons cm^{-1} s^{-1} nm^{-1} or photons cm^{-1} s^{-1} over a band]. *Actinic* means capable of causing chemical reactions. The actinic flux is different from irradiance that we met earlier because it measures the number of photons crossing a surface unit area coming from any direction as opposed to the perpendicular photon energy flow.

• Molecular properties of species A:
  • *absorption cross-section* σ [cm^2 molecule^{-1}], which varies with wavelength and has a size range 0–10^{-16} cm^2, with 10^{-18}–10^{-17} cm^2 being big;
  • *quantum yield q* (0 to 1), which is the probability that absorption of a photon will cause photolysis of molecule A. Absorption of a photon may lead to some other process such as radiation or breaking of a different bond, so q gives the fraction that cause the photolysis of interest.

Equation (3.36) is a first-order differential equation, so if nothing else happened, the characteristic lifetime of molecule A, would be an exponential decrease according to exp($-j_A$t) with an *e*-folding time of $1/j_A$.

We can consider how coefficient j_A is calculated in a plane-parallel atmosphere. The actinic flux $I_\lambda d\lambda$ is the number of photons crossing horizontal unit area (1 cm^2) from any direction in a spectral interval of wavelength λ to $\lambda + d\lambda$. So the number of molecules of species i photolyzed per second per unit volume in spectral interval λ to $\lambda + d\lambda$ is:

$$r_i = [q_{i\lambda}\sigma_{i\lambda}I_\lambda(z)d\lambda]n_i(z) = j_{i\lambda}n_i(z) \quad \text{photolysis rate cm}^{-3}\,\text{s}^{-1}$$

(3.37)

Here, $q_{i\lambda}$ is the quantum yield and $\sigma_{i\lambda}$ is the absorption cross-section for species i (usually in cm^2 molecule^{-1}). Integration over wavelength gives the photolysis rate constant j_i:

$$j_i = \int_\lambda q_{i\lambda}\sigma_{i\lambda}I_\lambda d\lambda \quad \text{s}^{-1}$$

(3.38)

This can easily be implemented in a computer program where $d\lambda = \Delta\lambda$ if we have laboratory data for the variation of $\sigma_{i\lambda}$ over λ at intervals of $\Delta\lambda$. If we ignore all scattering and reflection, and assume an overhead sun, then the actinic flux at a given wavelength is $I_\lambda = F_\lambda/(hc/\lambda)$, where we divide the solar flux F_λ by photon energy. More often, scattering needs to be included, and so more complicated radiative transfer methods are needed to calculate the actinic flux. In this case, the actinic flux is simply the *mean intensity* (in energy units) divided by the photon energy. Appendix B contains a description of numerical 1-D photochemical model.

3.2 Surface Deposition

Another aspect of atmospheric chemistry that is important is the rate of deposition of gases to the ground, which is divided up into two types: dry and wet.

Dry deposition is the loss to the surface in the absence of precipitation of any kind, although the surface itself could be wet or dry. For example, in the driest desert on Earth, the Atacama in Chile, atmospheric nitrate formed in reaction (3.24) dry deposits onto the ground, and over time this process has created a substantial deposit of nitrate in this desert (Catling *et al.*, 2010; Michalski *et al.*, 2004). On Mars, atmospheric sulfate formed from past volcanic emissions similarly should have accumulated on the surface, so dry deposition can alter the geochemistry of a planet's surface (Settle, 1979; Smith *et al.*, 2014).

Wet deposition is when a gas is incorporated into rain, snow, fogs, or clouds and deposited to the surface.

We express the vertical deposition flux F_i of species i to a surface as

$$F_i = v_d n_i$$

(3.39)

where v_d is the deposition velocity and n_i is the concentration of the constituent at some reference altitude.

The value of v_d depends on the type of surface (e.g., whether bare soil or water) and is obtained from field and laboratory measurements, noting that some theories exist to attempt to predict v_d (Seinfeld and Pandis, 2006). For deposition of long-lived species over oceans, v_d can be related to the so-called *piston velocity* of the species within the surface ocean. We shall save that discussion for Sec. 9.7.1, where it plays an important role in simulating atmosphere–ocean gas exchange on the early Earth.

3.3   Earth's Stratospheric and Tropospheric Chemistry

The unique richness of Earth's atmosphere in O_2 is ultimately responsible for the character of much of the atmospheric chemistry that occurs in both the stratosphere and troposphere, which we now consider. In addition, the presence of substantial amounts of water vapor is a primary contributor to chemistry in the troposphere. We start with the stratosphere because the stratosphere is chemically simpler and less diverse in the number of chemical species than the troposphere. Our description of Earth's atmospheric chemistry, particularly the tropospheric chemistry, is an extremely brief overview. A large number of books deal with the subject in more detail, ranging from introductory texts (e.g., Graedel and Crutzen (1993), Hobbs (2000), Jacob (1999)) to monographs (Brasseur and Solomon (2005), Warneck (2000), Wayne (2000), Finlayson-Pitts (2000)).

3.3.1   Earth's Stratospheric Chemistry

The key feature of Earth's stratosphere is the ozone (O_3) layer, which shields Earth's surface from biologically harmful solar UV radiation. UV radiation is that between x-rays (≤ 4 nm) and blue visible light (400 nm). Radiation below about 200 nm is strongly absorbed by CO_2, O_2, and other gases. So even the thin CO_2 atmosphere of Mars prevents UV below 200 nm from reaching the surface of

from the 200–300 nm radiation because of stratospheric ozone, which ultimately derives from O_2.

The first theoretical attempt to explain the presence of the ozone layer was made by Sydney Chapman in 1930, using a set of reactions with species containing oxygen atoms only, namely O, O_2, and O_3. His basic theory provided an over-estimate of the peak ozone concentration in the stratosphere because it ignored catalytic cycles that destroy ozone, which were discovered later. We now know that catalysts include species such as nitric oxide (NO), chlorine radicals (Cl), and hydroxyl radicals (OH).

The *Chapman scheme* for creating and destroying ozone can be written as follows, where the k and j values in parentheses are rate constants and photolysis rate constants, respectively:

$$O_2 + h v (\lambda < 242 \text{ nm}) \rightarrow O + O \qquad (j_2) \qquad (3.40)$$

$$O_2 + O + M \rightarrow O_3 + M \qquad (j_2) \qquad (3.41)$$

$$O_3 + h v \rightarrow O_2 + O(^3P) \qquad (j_2) \qquad (3.42)$$

$$O_3 + O \rightarrow 2 O_2 \qquad (k_3) \qquad (3.43)$$

To make ozone, an O atom from O_2 photolysis (eq. (3.40)) combines with another O_2 (eq. (3.41)). The bond enthalpy of O_2 (498 kJ mol^{-1}) corresponds to the energy of a 242 nm photon. Consequently, absorption of radiation below 242 nm dissociates O_2 into O atoms. These O atoms are in the ground state, denoted $O(^3P)$. (An explanation of so-called *term symbols*, such as $O(^3P)$ and $O(^1D)$, requires a discussion of quantum chemistry, which is given in Appendix C). In reaction (3.41), where O atoms react with oxygen molecules to form ozone, "M" denotes any air molecule (usually the most abundant ones N_2 or O_2), which acquires the excess energy liberated by the reaction and dissipates the energy through collisions with other air molecules.

Ozone is destroyed by photolysis (eq. (3.42)). The bonds in O_3 are weaker than in O_2, so ozone can absorb photons at wavelengths <1140 nm (Chappuis band) to make ground state O. However, photons with wavelengths below 310 nm in the Hartley band can efficiently generate an O atom in its first electronically excited state $O(^1D)$, as follows·

$$O_3 + h v (\lambda < 310 \text{ nm}) \rightarrow O_2 + O(^1D)$$
$$O(^1D) + M \rightarrow O(^3P) + M \text{ ("quenching" of } O(^1D) \text{ to ground state)} \qquad (3.44)$$
$$O_3 + h v \rightarrow O_2 + O(^3P) \qquad (j_3)$$

Mars, but 200–300 nm radiation gets through. These relatively short UV wavelengths of 200–300 nm are particularly destructive for biology, partly because DNA absorbs strongly in this region. On Earth, we are protected

The $O(^1D)$ is rapidly stabilized to ground state oxygen $O(^3P)$, by collision with other gas molecules represented by "M". (Alternatively, it can react with other species such as H_2O, as described in Sec. 3.3.2.) The energy absorbed in these reactions heats the stratosphere. However, eq. (3.42) is not a net sink for O_3 because the O atom may go on to generate more O_3. The net loss of O_3 is eq. (3.43), in which both O_3 and O are converted into stable O_2 molecules.

To determine the ozone equilibrium distribution, we can consider a rate equation, which expresses the production and loss of ozone:

$$\underset{\text{ozone production}}{\frac{\partial [O_3]}{\partial t} = k_2[O][O_2][M]} \underset{\text{ozone loss}}{- j_3[O_3] - k_3[O][O_3]} \qquad (3.45)$$

Similarly, for the production and loss of atomic oxygen we get

$$\frac{\partial [O]}{\partial t} = \underset{\text{production of O from photolysis}}{2j_2[O_2] + j_3[O_3]} \underset{\text{loss of O from reactions with } O_2 \text{ or } O_3}{- k_2[O][O_2][M] - k_3[O][O_3]}$$
$$(3.46)$$

We can also add eqs. (3.45) and (3.46) to give:

$$\frac{\partial [O_3]}{\partial t} + \frac{\partial [O]}{\partial t} = 2j_2[O_2] - 2k_3[O][O_3] \qquad (3.47)$$

In steady state, we set the time derivatives to zero. From eq. (3.45) we get:

$$\underset{\text{ozone production}}{k_2[O][O_2][M]} = \underset{\text{ozone losses}}{j_3[O_3] + k_3[O][O_3]} \qquad (3.48)$$

In the stratosphere, $j_3 \gg k_3[O]$, so we can neglect the second term on the right-hand side of eq. (3.48), and derive the equilibrium concentration of atomic oxygen, as:

$$[O] = \frac{j_3[O_3]}{k_2[O_2][M]} \qquad (3.49)$$

Substitution for [O] into eq. (3.47) (with time derivatives set to zero) yields:

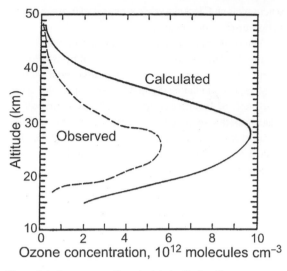

Figure 3.1 An ozone profile calculated with the Chapman reactions at the equator overestimates the ozone compared with observations over Panama at 9° N on November 13, 1970. The reason is that natural catalysts that destroy ozone are omitted from the oxygen-only Chapman reactions. (Adapted from Seinfeld and Pandis (1998). Reproduced with permission. Copyright 1998, John Wiley and Sons.)

(3.43) destroys it, while reactions (3.41) and (3.42) simply convert one form of odd oxygen into another. That is why reaction (3.43) is the key destruction process for ozone in the Chapman mechanism.

The Chapman scheme ignores catalytic cycles that destroy ozone and also dynamical transport of ozone and other relevant species. In Fig. 3.1, we see a Chapman profile calculated for equatorial conditions. Until the mid 1960s, it was believed that the Chapman reactions principally governed ozone destruction and production.

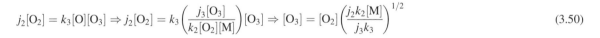

$$j_2[O_2] = k_3[O][O_3] \Rightarrow j_2[O_2] = k_3 \left(\frac{j_3[O_3]}{k_2[O_2][M]} \right)[O_3] \Rightarrow [O_3] = [O_2] \left(\frac{j_2 k_2[M]}{j_3 k_3} \right)^{1/2} \qquad (3.50)$$

This is the Chapman expression for the ozone concentration, which can be evaluated as a function of height if we know the vertical profile of the terms on the right-hand side. At high altitudes, [M] and $[O_2]$ are small, so $[O_3]$ will be small. Deep in the atmosphere, j_2 (the photolysis rate constant for O_2 dissociation) will be tiny so that $[O_3]$ will be small again. Thus, we expect a vertical profile where $[O_3]$ grows to a maximum and then diminishes, as observed (Fig. 3.1).

Another way to think about the Chapman mechanism is in terms of *odd oxygen*, denoted $O_x \equiv O + O_3$. "Even oxygen," then, is just O_2, although that term is rarely used. Reaction (3.40) creates odd oxygen, reaction

However, observations showed that they overestimated ozone, as depicted in Fig. 3.1. In the 1970s, various catalytic cycles of ozone destruction were identified, which were added to the Chapman scheme. Catalytic cycles that destroy ozone follow this basic scheme:

$$X + O_3 \rightarrow XO + O_2 \qquad (3.51)$$
$$XO + O \rightarrow X + O_2 \qquad (3.52)$$
$$\text{Net}: \quad \overline{O_3 + O \rightarrow 2O_2}$$

where X is a free radical, which can be Cl, Br, H, OH, or NO. The net result of the catalytic cycle is to remove odd oxygen species rapidly and effectively speed up the last reaction in the Chapman scheme (eq. (3.43)), which is the net destruction of ozone.

In the 1980s, it became apparent that additional mechanisms could destroy ozone in the cold stratospheres of the polar regions. In Antarctica, it was discovered that an *ozone hole* occurs annually in the austral spring (Farman *et al.*, 1985). In this case, the culprits are chlorine free radicals released from anthropogenic CFCs (chlorofluorocarbons) when CFCs break up under UV light in the stratosphere (Russell *et al.*, 1996). The climatic conditions in the Antarctic also play a role. *Polar stratospheric clouds* (PSCs) form during the dark, cold polar winter in each hemisphere. The atmospheric chemistry that we have discussed so far involves *homogeneous reactions*, which are reactions between species that are all in the gas phase. In Antarctica, *heterogeneous reactions*, which are those that happen on solid surfaces, occur on PSC particles (Solomon, 1999). Such reactions convert unreactive forms of chlorine, such as chlorine nitrate and HCl, into Cl_2:

$$\underbrace{ClONO_2}_{\text{chlorine nitrate}} + HCl \rightarrow Cl_2 + \underbrace{HNO_3}_{\text{nitric acid}} \qquad (3.53)$$

This reaction also means that *odd nitrogen* ($NO_x = NO + NO_2$, pronounced "nox") compounds are taken up through the chlorine nitrate into unreactive nitric acid (HNO_3) on the PSC particles. It is odd nitrogen that can convert reactive chlorine into unreactive chlorine nitrate via:

$$ClO + NO_2 + M \rightarrow ClONO_2 + M \qquad (3.54)$$

Once the Sun comes up in September above Antarctica, Cl_2 is released from PSC particles and photolyzes into Cl free radicals, which results in a springtime ozone hole.

Dynamics are also important. Above the South Pole, the wintertime stratospheric circulation is a vortex of eastward winds, which isolates the air from lower latitudes. This prevents the influx of new odd nitrogen from lower latitudes during winter and also new ozone once the hole forms, given that the vortex tends not to break up until late Spring.

Although ozone is depleted, a similar extent of depletion does not occur in the Arctic spring. Atmospheric gravity waves from meridional mountain ranges (Sec. 4.4) disrupt the development of a coherent Arctic vortex, unlike the case of the Antarctic continent, which is surrounded by ocean water. Odd nitrogen is thus present in high enough concentrations in the Arctic to tie up much of the chlorine in unreactive chlorine nitrate.

3.3.2 Earth's Tropospheric Chemistry

Biology is the single most important control of the bulk chemistry of the troposphere.

All major gases (except Ar) are mediated by biological cycles, which notably include O_2, N_2, and CO_2. In addition, important trace gases have biological sources (Fowler *et al.*, 2009), including the following:

– dimethyl sulfide (CH_3SCH_3), or DMS, from ocean phytoplankton (Liss *et al.*, 1997),
– carbonyl sulfide COS from microbes, which makes its way to the stratosphere and is the dominant source of the *Junge layer* (~20–25 km altitude) of high albedo stratospheric sulfate aerosols during periods of volcanic quiescence (Crutzen, 1976; Montzka *et al.*, 2007),
– methane (CH_4) (Wuebbles and Hayhoe, 2002) and nitrous oxide (N_2O) (Cicerone, 1989) from microbes, both of which are greenhouse gases.

In many cases, biogenic fluxes of trace gases exceed inorganic sources. For example, today's flux of DMS of 1.1 Tmol S yr^{-1} is a larger sulfur source to the atmosphere than volcanic emissions of 0.3–1 Tmol S yr^{-1} (Warneck, 2000, p. 611). Biogenic methane also comprises ~90% of total methane emissions, with geological sources accounting for the remainder (Etiope and Klusman, 2002). And even much of the geologic source comes from the decayed remnants of organisms, as discussed in Sec. 7.5.2.

A basic question concerns the fate of trace gases, such as methane or DMS, which have continuous sources of billions of tonnes per year. What prevents them accumulating? Where do they go? The answer ultimately lies in the presence of substantial water vapor in the Earth's atmosphere (unlike Mars or Venus), which allows the hydroxyl radical (OH) to be generated in sufficient quantities to strongly control the abundances of trace gases (Prinn, 2014). A newcomer might naively think that O_2, which is vastly more abundant than OH, would be an important oxidant, but direct gas phase reactions with a nonradical species such as O_2 are unimportant because they are far too slow. Instead, OH is the dominant oxidizing radical in Earth's tropospheric chemistry, and has been dubbed the "*detergent of the atmosphere*" by Nobel Prizewinning atmospheric chemist Paul Crutzen (Crutzen and Zimmermann, 1991; Riedel and Lassey, 2008).

The hydroxyl radical, OH, cleans the air of hydrocarbons gases such as CH_4, converts NO_x trace gases into nitric acid, and helps oxidize sulfur-containing gases from biology and volcanoes into sulfates. In the modern atmosphere, the OH radical derives from water vapor in a chemical pathway that was first deduced by Hiriam Levy in the 1970s. Ozone absorption ≤340 nm (in the Huggins bands) generates $O(^1D)$, most of which is quenched (like in eq. (3.44)). But a few percent reacts with H_2O as follows (Levy, 1971):

$$H_2O + O(^1D) \rightarrow OH + OH \tag{3.55}$$

The hydroxyl radical is then available to attack various species, including the following:

– CO, which has a lifetime ~3 months, and ends up oxidized to CO_2,

– CH_4, which has a lifetime ~10 years and ends up oxidized to CO_2 and H_2O,

– H_2S, COS, and DMS, which are oxidized to make SO_2. In turn, SO_2 is oxidized to make sulfate (SO_4^{2-}) aerosols. In the troposphere, where there is sufficient ammonia, most of the aerosols end up as ammonium sulfate (($NH_4)_2SO_4$) and bisulfate (($NH_4)HSO_4$). In the stratosphere, SO_2 is oxidized to H_2SO_4.

The global diurnally averaged concentration of OH in the lower troposphere is ~10^6 radicals cm^{-3} (Prinn *et al.*, 1995), which is tiny compared to background air ~10^{19} molecules cm^{-3}. Because of its reactivity, the lifetime of OH is very short, ~1 s. Only a few tropospheric gases can escape oxidation by OH, notably anthropogenic chloro-fluorocarbons and nitrous oxide, N_2O, which instead are decomposed by shortwave UV when they are mixed up to the stratosphere.

Ozone is another major player in tropospheric chemistry largely because it helps generate OH, as we have mentioned. The troposphere contains about ~10% of the ozone column abundance and in remote areas, ozone is present at concentrations of ~30–40 ppbv. Tropospheric ozone derives from two major classes of precursors: Oxides of nitrogen, NO_x = NO + NO_2, and reactive hydrocarbon compounds from biology and anthropogenic activity. Most of the direct emission of NO_x to the atmosphere is in the form of NO from lightning or pollution. NO_2 is produced when the NO is oxidized, but the overall net reaction in the troposphere is mediated by non-methane reactive hydrocarbons (NMHCs) (e.g., C_2–C_5 compounds, including butane (C_4H_{10}), ethylene (C_2H_4), isoprene (C_5H_8), etc., which are produced by vegetation and marine plankton), as follows:

$$NMHCs + NO + h\nu \rightarrow NO_2 + products \; [\text{a net reaction of many}] \tag{3.56}$$

Subsequent photolysis of NO_2 can generate O atoms,

$$NO_2 + h\nu \; (\lambda \leq 410 \text{ nm}) \rightarrow NO + O \tag{3.57}$$

Almost all the O atoms combine with O_2 to make O_3 via eq. (3.41). For completeness, we also note that in the remote troposphere, the oxidation of carbon monoxide and methane by OH in the presence of NO produces O_3 as a side-product.

It is important to realize from this discussion that a major difference between Earth's stratosphere and troposphere is that whereas NO_x *destroys* ozone in the stratosphere, NO_x *generates* ozone in the troposphere (Crutzen, 1979). The distinction arises because the reactive hydrocarbons that are present in the troposphere are absent at the high altitudes of the stratosphere, so that eq. (3.56) does not apply. In addition, in stratospheric chemistry, O atoms are important because they are generated by shortwave UV photolysis but in the troposphere, O atoms are much less abundant. Consequently, eq. (3.52), which causes NO to catalyze O_3 destruction in the stratosphere, does not apply in the troposphere.

The importance of tropospheric ozone is its role in generating OH, when the ozone is photolyzed followed by eq. (3.55) or when hydroperoxy radicals destroy ozone:

$$HO_2 + O_3 \rightarrow OH + O_2 \tag{3.58}$$

Without OH, the Earth's atmosphere would be completely different. The trace gases that OH scavenges include greenhouse gases (e.g., CH_4 and other hydrocarbons) and substances that would be poisonous to higher forms of life if they were not removed (e.g., H_2S). The generation of sufficient OH is ultimately tied to two basic aspects of Earth's atmospheric composition: abundant O_2 (and hence ozone) and water vapor.

3.4 CO_2 Stability on Venus and Mars

In contrast to the Earth, the most striking aspect of the neighboring atmospheres on Venus and Mars is the predominance of CO_2 at levels of 96.5% and 96%, respectively. Later, when we consider atmospheric evolution, it will be reasonable to conclude that similar CO_2-dominated atmospheres probably exist on some rocky planets in extrasolar systems. The chemistry of the atmospheres of Mars and Venus are described in more detail in Ch. 12 and Ch. 13, respectively, but here we describe one common aspect. Early studies of CO_2 in both atmospheres ran into the so-called *CO_2 stability problem*. Comparatively small CO and O_2 abundances did not support initial predictions that CO_2 should be photolyzed and destroyed. On Mars, there is ~10^4 times less CO_2 than on Venus, so that CO_2 on Mars should be converted to CO and O_2 on a much more rapid timescale than on Venus. The basic photodissociation process for CO_2 is

$$CO_2 + h\nu (< 205 \text{ nm}) \rightarrow CO + O \tag{3.59}$$

followed by

$$O + O + M \rightarrow M + O_2 \tag{3.60}$$

where M is any molecule. For quantum mechanical reasons, the direct recombination of O with CO is forbidden (because it requires the spin flip of an electron). In the absence of recombination, all Martian CO_2 would convert to CO and O_2 in ~3500 years, but clearly this is not so. Recombination must be effective on Venus also because Venusian CO is a relatively minor constituent, ~20 ppm at 20 km altitude and 45 ppm above the clouds, while O_2 is undetectably small.

McElroy and Donahue (1972) and Parkinson and Hunten (1972) solved the problem for Mars. Essentially, OH radicals (derived from water dissociation or hydrogen peroxide photolysis) catalyze the recombination of CO and O in cycles such as

$$OH + CO \quad \rightarrow CO_2 + H$$
$$H + O_2 + M \rightarrow HO_2 + M$$
$$\underline{HO_2 + O \quad \rightarrow OH + O_2}$$
$$Net: \quad CO + O \quad \rightarrow CO_2$$

Collectively, H, OH, and HO_2 are called *odd hydrogen* (HO_x).

On Venus, and on the modern Earth, Cl plays a similar role to odd hydrogen species on Mars. On Earth, Cl and ClO are lumped together to form *odd chlorine* (ClO_x). Together, they catalyze ozone destruction via the cycle described by reactions (3.51) and (3.52). But Prinn (1971) recognized the importance of ClO_x chemistry on Venus long before it role in ozone destruction was recognized on Earth. Cl radicals are produced from the photodissociation of HCl on Venus, and these enable the following catalytic cycle (Pernice *et al.*, 2004; Yung and DeMore, 1982):

$$Cl + CO + M \quad \rightarrow ClCO + M$$
$$ClCO + O_2 + M \quad \rightarrow ClCO_3 + M$$
$$ClCO_3 + X \quad \rightarrow Cl + CO_2 + XO$$
$$XO + O \quad \rightarrow X + O_2$$
$$Net: \quad CO + O \quad \rightarrow CO_2$$

where X = O, H, Cl, SO, or SO_2.

Consequently, once again, we see the importance of radical-driven catalytic cycles. In this case, such catalytic cycles maintain the very large abundances of CO_2 of 96% and 96.5% in the atmospheres of Mars and Venus, respectively (Table 1.2).

3.5 CO_2 and Cold Thermospheres of Venus and Mars

In Ch. 1, we noted how the thermosphere temperatures of the atmospheres of Earth, Venus and Mars are strikingly different, so that the exobase temperatures are ~1000 K, ~250 K, and ~270 K. It is perhaps puzzling that Venus'

exobase should be colder than Mars, despite the proximity of Venus to the Sun.

We can understand the disparate exobase temperatures in terms of radiation and chemistry. In Earth's atmosphere, the major constituents are O_2 and N_2, which have no electric dipole moment so that all rotational and vibrational transitions are forbidden (ignoring collision induced absorption). Also, atomic species formed by photolysis in the thermosphere do not radiate effectively. Collisions can excite neutral atomic oxygen, O, to levels $^3P_0$ and $^3P_1$ just above the ground state level $^3P_2$. De-excitation causes emission at 63 and 147 μm, but only the 63 μm is significant, and only between ~100 to ~140 km. However, this IR radiative cooling is small compared to EUV heating (Banks and Kockarts, 1973, pp. 22–25).

The coolants in Earth's thermosphere, CO_2 and NO, are relatively ineffective because of their low concentrations. In contrast, on Mars and Venus, the atmospheres are almost pure CO_2, which makes the upper atmospheres cold through efficient radiative cooling. However, Venus' atmosphere is cooled even further because atomic oxygen is more abundant by a factor of ~10 in the upper atmosphere of Venus than Mars and facilitates cooling via vibrationally excited CO_2, i.e.,

$$O + CO_2 \rightarrow CO_2^* + O$$

The excited CO_2 decays by emission of photons with 15 μm wavelength. The enhanced density of O depends the solar flux, which causes more photodissociation of CO_2 on Venus, and a scale height that is smaller for the same temperature by a factor proportional to gravitational acceleration (see eq. (1.10)). Thus the factor of $10 \approx 4.5 \times$ (8.80 m s^{-2})/(3.72 m/s^{-2}), where 4.5 (=$(1.52/0.72)^2$) is the solar flux at Venus relative to Mars (Krasnopolsky, 2011).

3.6 Methane and Hydrocarbons on Outer Planets and Titan

In contrast to the oxidized species found in the atmospheres of Earth, Mars, and Venus, hydrogen-bearing species dominate the redox of the atmospheres of the giant planets and Titan. In particular, the photochemistry of CH_4 and its photolysis products are important for the thermal structure and chemistry of the stratospheres on the giant planets (Irwin, 2009; Moses *et al.*, 2004; West *et al.*, 2004; West *et al.*, 2009; West *et al.*, 1991) and Titan (Krasnopolsky, 2010; Lavvas *et al.*, 2008a, b; Wilson and Atreya, 2004). But there is a critical difference between the giant planets and Titan, namely, the ability of hydrogen to escape from Titan but not from the giant planets. The result is that there is a net, irreversible destruction of

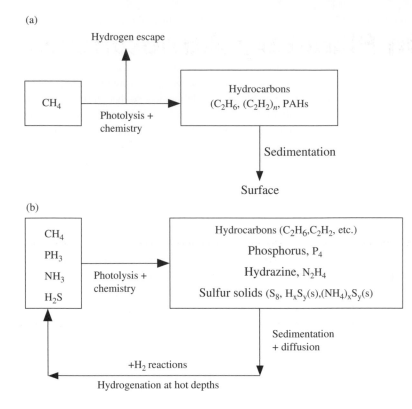

Figure 3.2 (a) An open cycle where hydrogen escapes on Titan irreversibly produces large carbon-number hydrocarbons. (b) The closed cycles of hydrogen-bearing species on giant planets. Hydrocarbons, phosphorus, sulfur, and hydrazine condensables can be generated from methane, phosphine, hydrogen sulfide, and ammonia. But deep cycling reconstitutes the light gases.

methane on Titan (Yung *et al.*, 1984), whereas on the giant planets, methane and other hydrogenated species, such as ammonia or phosphine, are regenerated.

The C–H bonds in methane are broken by solar UV photons with wavelength < 145 nm, especially by those at the Lyman-α wavelength (121.6 nm) where there is a large flux. Many complex photochemical reactions occur, but overall, net reactions can be summarized as:

$$2CH_4 + \text{photons} \rightarrow \underbrace{C_2H_6 + H_2}_{\text{ethane+hydrogen}} \tag{3.61}$$

$$2nCH_4 + \text{photons} \rightarrow \underbrace{(C_2H_2)_n + 3nH_2}_{\text{polyacetylene+hydrogen}} \tag{3.62}$$

$$6nCH_4 + \text{photons} \rightarrow \underbrace{(C_6H_6)_n + 9nH_2}_{\text{polyaromatic hydrocarbons+hydrogen}} \tag{3.63}$$

The last schematic reaction is important on Titan, where benzene serves as a precursor to *polyaromatic hydrocarbons* (PAHs), which are compounds based on fused benzene rings. This chemistry could also have been important on early Earth, where it probably led to the formation of organic haze prior to the rise of atmospheric O_2 (Pavlov *et al.*, 2001; Wolf and Toon, 2010; Zerkle *et al.*, 2012) (see Sec. 11.6.1).

Atmospheric hydrogen can escape to space on Titan and so only accumulates to a small level where its rate of escape to space balances its net rate of production from photochemistry. As a result, complex products of methane photolysis (ethane, acetylene, ethylene, and larger hydrocarbons, as well as organic molecules that incorporate nitrogen) are irreversibly generated and fall out of Titan's atmosphere to the surface because they are liquid or solid (Atreya *et al.*, 2009) (Fig. 3.2).

On the giant planets, hydrogen cannot escape and hydrogen is the major constituent. Consequently, hydrocarbons that are produced from methane photolysis on giant planets can eventually mix back down into the deep, hotter ($\sim$1000 K) atmospheric regions where they react with abundant H_2 to reform methane. Other hydrogenated species, such as ammonia (NH_3) and phosphine (PH_3) behave similarly. Ammonia and phosphine are photolyzed in giant planet atmospheres but the resulting products are mixed down to depths where hydrogenation reactions cyclically reconstitute ammonia or phosphine (e.g., Lewis, 2004).

Fundamentally, whether hydrogen can escape from an atmosphere or not has a profound effect on the overall evolution of the planet's chemistry (Fig. 3.2). The issue of hydrogen escape is explored in more detail in Ch. 5.

4 Motions in Planetary Atmospheres

Atmospheric motions are an inevitable response to a planet's rotation, gravity, differential stellar heating, and other forces. *Dynamics* are the relationships between the heating, the forces, and the winds that they drive.

Dynamics are important for climate for many reasons and so affect a planet's climate evolution and habitability. Winds transport heat, e.g., from warm tropics to cold poles. Furthermore, winds can affect atmospheric chemistry. Air movement carries condensable species responsible for clouds, such as water on Earth and Mars, ammonia on Jupiter, or methane on Titan. The albedo of clouds is so important to the radiation balance that the presence of clouds and their net global cooling effect makes the Earth habitable (e.g., Goldblatt *et al.*, 2013). Dynamics also strongly affects the distribution of non-condensable trace gases such as ozone in Earth's stratosphere. Furthermore, winds can modify a surface, such as long-term erosion or deposition of solid material on Mars.

Of course, winds are highly variable in time, changing hourly and daily, and in this chapter we concentrate on winds averaged over days or months and over large areas, often of continental scale for Earth. Generally, winds have a preferred direction that either persists throughout a year or season. Such average winds are the *general circulation* of a planetary atmosphere.

We note that winds can drive secondary circulations, in particular, ocean currents on Earth, which strongly affect climate by transporting heat and mass. However, to limit the scope of this chapter, we restrict ourselves to atmospheric dynamics. For ocean dynamics, the reader is referred to textbooks by Marshall and Plumb (2008) and Williams and Follows (2011).

This chapter was originally going to be contributed as a guest chapter by Conway Leovy (1933–2011). His rough draft, which has been expanded and developed, provided essential guidance and insight.

The role of atmospheric dynamics in the long-term evolution of atmospheres and planetary habitability remains relatively unexplored. There are a few examples of the influence of dynamics on the Earth's Precambrian climate and the habitability of exoplanets that have been identified, which we mention as we go along. But this chapter is merely a point of departure to introduce some key concepts of the dynamics of planetary atmospheres. We anticipate much more research in this area in the future.

4.1 Introductory Concepts

4.1.1 Forces, Apparent Forces, and the Equation of Motion

At the heart of fluid dynamics are forms of Newton's second law expressed per unit mass. The second law gives acceleration in a particular direction as:

$$a_i = \frac{\sum F}{m} \tag{4.1}$$

Here, $\sum F$ is a sum of forces, m is the mass, and a_i is the acceleration of a parcel of air in an *inertial* coordinate system, which is that fixed in space outside the planet–atmosphere system.

In dynamics, various forces F cause air to flow, including *real forces*, such as gravity, and *apparent forces* that arise because we conventionally use a *non-inertial* frame of reference that co-rotates with the planet and is therefore accelerated. An everyday example of real and apparent forces occurs in an upwardly accelerating elevator, where scales would measure your weight (a force) in excess of your true weight. Upward acceleration of the frame of reference results in an apparent force in the *opposite* direction of a real acceleration that, if neglected, gives an incorrect weight in applying Newton's second law. The real forces in atmospheric dynamics are pressure gradients, friction and gravity, while the apparent forces that arise from choosing a co-rotating frame of reference are centrifugal and Coriolis forces.

4.1.1.1 The Pressure Gradient Force

In Sec 1.1.2.3, we discussed hydrostatic balance, where gravitational acceleration g balances a pressure (p) gradient force per unit mass over vertical distance z of $-1/\rho \, (\partial p/\partial z)$, where ρ is air density. In this chapter, the large-scale motions we consider will always be close to hydrostatic equilibrium, so that the vertical equation of motion simplifies to the hydrostatic equation (eq. (1.8)). Later, in Sec. 5.10, we will consider a special non-hydrostatic case of hydrodynamic escape.

The pressure gradient forces F_{PG} per unit mass in horizontal directions are analogous. Defining local x- and y-axes pointing in the local horizontal eastward and horizontal northward directions, we have:

$$\frac{F_{x,\,PG}}{m} = -\frac{1}{\rho}\frac{\partial p}{\partial x}, \quad \frac{F_{y,\,PG}}{m} = -\frac{1}{\rho}\frac{\partial p}{\partial y}, \quad \text{or in 3-D,} \quad \frac{\mathbf{F}_{PG}}{m}$$
$$= -\frac{1}{\rho}\nabla p \qquad (4.2)$$

Here, the minus sign indicates that the force acts from high to low pressure. As an example, a pressure gradient of 2.5 hPa over 100 km horizontal distance in terrestrial midlatitudes exerts a force per unit mass of $(-1/(1.2 \text{ kg m}^{-3}))$ $\times (250 \text{ Pa}/10^5 \text{ m}) = -2.1 \times 10^{-3} \text{ m s}^{-2}$, and would produce gale-force winds of $\sim 20 \text{ m s}^{-1}$.

4.1.1.2 Centrifugal and Coriolis Forces

When dealing with planetary atmospheres, our coordinate system is fixed with respect to the planet's surface. A planet rotates with an angular rate of magnitude $\Omega = 2\pi/\tau_d$, where τ_d is the rotation period (or sidereal day). For Earth, $\Omega = 2\pi/(86\,164 \text{ s}) = 7.29 \times 10^{-5} \text{ rad s}^{-1}$. The apparent forces per unit mass that arise for an observer in the rotating frame are as follows.

(1) A *Coriolis force*, which is perpendicular to both the velocity vector $\mathbf{v}$ in the rotating frame and to the rotation vector $\boldsymbol{\Omega}$.

(2) A *centrifugal force*, which has magnitude $\Omega^2 R$, where R is the distance from the point with position vector $\mathbf{r}$ to the rotation axis, directed perpendicularly away from the rotation axis. Usually, this force is merely regarded as a correction to gravity g, which we have already discussed in Sec. 1.1.2.3.

A pedagogical illustration (from McIlveen (1992, p. 187)) of how apparent forces arise is a toy train with an imaginary toy passenger going around a circular track at train speed V on a turntable that has angular rotation Ω and

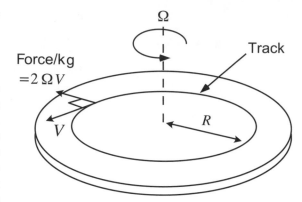

Figure 4.1 Circular motion on a rotating turntable illustrates the origin of the Coriolis force. Accelerations lost from measurements made in the co-rotating frame create equal and opposite apparent accelerations (see text).

tangential speed $V_t = \Omega R$. The train speed is $V_{total} = V + V_t$, as viewed externally (Fig. 4.1). If the train is stationary ($V = 0$), the passenger on the train measures no centripetal acceleration. The difference of his measurement and the actual radially inward centripetal acceleration of the track is $0 - V_t^2/R = -\Omega^2 R$. The net result is that the passenger experiences a *force* per unit mass in the direction *opposite* to the real inward centripetal acceleration, which is a centrifugal force of magnitude $\Omega^2 R$ acting radially outwards. If the train moves at speed V, the centripetal acceleration is $V_{total}^2/R = (V + \Omega R)^2/R$ as viewed from the exterior, but the co-rotating passenger sees only V^2/R. The difference in accelerations is now $V^2/R - (V^2/R + 2\Omega V + \Omega^2 R) = -\Omega^2 R - 2\Omega V$. The passenger not only experiences a centrifugal force $(-\Omega^2 R)$ but also a Coriolis *force* of magnitude $2\Omega V$, which acts to the right of velocity vector, as shown in Fig. 4.1.

A planet is a rotating sphere, not a disk. For a sphere, the Coriolis acceleration is $2\Omega V$ in magnitude only at the poles, where the planet's angular rotation vector $\boldsymbol{\Omega}$, is vertical. Elsewhere, the Coriolis acceleration depends on the latitude because the vector $\boldsymbol{\Omega}$ is no longer perpendicular to the local surface. To see how the scaling operates, we have to consider a spherical coordinate system.

4.1.1.3 The Spherical Coordinate System

Atmospheric dynamics is commonly described with local Cartesian coordinates on a spherical planet (Fig. 4.2(a)). The x-, y- and z-axes point in the horizontal eastward, horizontal northward, and vertical directions, respectively, while the winds in these directions are the *zonal wind u* in the west-to-east or "zonal" direction, the *meridional wind v* along a meridian in the south-to-north direction, and the

(a)

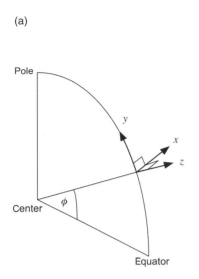

(b)

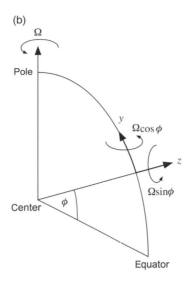

Figure 4.2 (a) Local Cartesian system at a point on a planet at latitude ϕ. The x distance is along a line of latitude, the y distance is along a meridian, and z is altitude. (b) Components of the planetary rotation about local vertical and meridional axes. Here, Ω is the angular rotation rate of the planet's spin.

vertical wind w upwards. The $z = 0$ level is sea level on Earth. On other planets, a mean pressure level or a geopotential surface serves as a reference surface.

Here and later, we also use the word *prograde* to mean in the same direction as planetary rotation. Although we use prograde interchangeably with "eastward" for the Earth, prograde is more general and applies to any spin direction. *Retrograde* is the opposite direction to planetary rotation. Specifying "prograde" or "retrograde" for winds is useful because there are two conventions for which pole of a planet is the North Pole. NASA often defines the North Pole of a planet with a right-hand rule: If curled fingers represent the direction of rotation, the thumb points north. Using this definition, Venus has an obliquity of 177.4° and is upside down and rotating in a conventional counter-clockwise direction when viewed from above its North Pole. Similarly, Uranus is tilted at 97.9° degrees with respect to its orbital axis. In contrast, the International Astronomical Union (IAU) has the convention that the North Pole of a planet points upwards from the plane defined by the counter-clockwise orbital motion of the planet. With this definition, Venus and Uranus have obliquities of 2.6° and 82.1°, respectively, to their orbits, and both planets spin "backwards" (to the west) when viewed from above the orbital plane.

In our x–y–z coordinate system, we define a wind vector (m s^{-1}) as:

$$\mathbf{v} = u\mathbf{i} + v\mathbf{j} + w\mathbf{k} \qquad \text{3-D wind vector} \qquad (4.3)$$

where $\mathbf{i}$, $\mathbf{j}$, and $\mathbf{k}$ are orthogonal unit vectors. Because the vertical wind is generally much smaller than horizontal winds, often we neglect w and work with a horizontal wind vector:

$$\mathbf{v}_h = u\mathbf{i} + v\mathbf{j} \qquad \text{2-D horizontal wind vector} \qquad (4.4)$$

At any latitude ϕ, a planet's angular velocity vector $\mathbf{\Omega}$ (rad s^{-1}) can be resolved into two orthogonal components along the local y-axis and vertical, as shown in Fig. 4.2(b):

$$\mathbf{\Omega} = \mathbf{j}(\Omega\cos\phi) + \mathbf{k}(\Omega\sin\phi) \qquad (4.5)$$

Given $\mathbf{\Omega}$, we can calculate the Coriolis force per unit mass with vector algebra as:

$$\frac{-\mathbf{F}_{\text{Coriolis}}}{m} = -2\mathbf{\Omega} \times \mathbf{v} = -2\Omega \begin{vmatrix} \mathbf{i} & \mathbf{j} & \mathbf{k} \\ 0 & \cos\phi & \sin\phi \\ u & v & w \end{vmatrix}$$

$$= \mathbf{i}(-2\Omega(w\cos\phi - v\sin\phi)) - \mathbf{j}(2\Omega u\sin\phi)$$
$$+ \mathbf{k}(2\Omega u\cos\phi) \qquad (4.6)$$

(For a readable derivation of the vector form of forces in a rotating frame, see Taylor (2005).) Here the negative sign is because the Coriolis acceleration would be positive in the inertial frame, so in the co-rotating frame it is subtracted, analogous to outward centrifugal versus inward centripetal force in circular motion. If we consider a southward wind v in the northern hemisphere (i.e., positive latitude ϕ, and no zonal or vertical wind so $u = w = 0$), then the Coriolis acceleration vector is $\mathbf{i}(2\Omega v\sin\phi)$. With negative v, the acceleration is in the negative x-direction (westward), so the southward moving parcel of air experiences acceleration to the right of motion. This is the origin of terrestrial trade winds. Air traveling towards the equator in the northern tropics is deflected westward, so that the resultant winds blow from the northeast to the southwest (Sec. 4.2.2).

We can simplify eq. (4.6) by neglecting terms in w because vertical wind speeds are usually much smaller than horizontal ones. Also the vertical component term in **k** of the Coriolis acceleration is often much smaller than the vertical acceleration of gravity. Then, the Coriolis force per unit mass is

$$= \mathbf{i}(-2\Omega(w\cos\phi - v\sin\phi)) - \mathbf{j}(2\Omega u\sin\phi) + \mathbf{k}(2\Omega u\cos\phi) \approx \mathbf{i}(2\Omega v\sin\phi) + \mathbf{j}(-2\Omega u\sin\phi)$$
$$= \mathbf{i}(fv) + \mathbf{j}(-fu)$$

$$(4.7)$$

Here, we have introduced a new variable,

the Coriolis parameter, $f = 2\Omega\sin\phi$ $\qquad$ (4.8)

Given a typical zonal wind speed of $u = 10$ m s^{-1} on Earth, the magnitude of the Coriolis force per unit mass in mid-latitudes (e.g., $\phi = 45°$) is $fu = 10^{-3}$ m s^{-2}, which is similar in magnitude to the pressure gradient acceleration we estimated in Sec. 4.1.1.1. This is no coincidence. The pressure gradient acceleration often approximately balances the Coriolis acceleration on rapidly rotating planets, such as Earth (Sec. 4.1.2.1).

4.1.1.4 The Equations of Motion and the Role of Advection

Considering all the forces gives a vector form of Newton's second law for fluid flow, specifying the acceleration of a unit mass of air in a frame co-rotating with the planet:

planet below (defined more precisely in Sec. 4.5). The deep atmospheres in the Solar System all exhibit superrotation or strong differential rotation, and the dynamical effects of external tidal torques remains to be fully investigated.

If the centripetal acceleration is taken to the right-hand side of eq. (4.9), by convention it is treated as centrifugal force per kg, which we can lump in with gravity (because both are the gradient of a potential) to form an effective gravity g that we measure. An effective gravity applies because planets deform and become oblate (Sec. 1.1.2.3). Thus, neglecting tidal torques, we can rewrite eq. (4.9) as:

$$\frac{d\mathbf{v}}{dt} = \underbrace{-\frac{1}{\rho}\nabla p}_{\text{pressure gradient force/kg}} - \underbrace{(2\mathbf{\Omega}\times\mathbf{v})}_{\text{Coriolis force/kg}} - g\mathbf{k} + \mathbf{F}_{\text{visc}}$$

$$(4.10)$$

Here, $\mathbf{F}_{\text{visc}}$ is the frictional or viscous force per unit mass. Eq. (4.10) is the conventional basic equation of dynamics called the *momentum equation*.

For practical use, we expand the above equation into three components of acceleration in the x, y, and z directions of eastward, northward and upward. We will not go

$$\underbrace{\frac{d\mathbf{v}}{dt}}_{\text{acceleration in co-rotating frame}} + \underbrace{(2\mathbf{\Omega}\times\mathbf{v})}_{\text{Coriolis acceleration}} + \underbrace{[\mathbf{\Omega}\times(\mathbf{\Omega}\times\mathbf{r})]}_{\text{centripetal acceleration}} = \underbrace{\frac{\mathbf{F}_g}{m}}_{\text{force of gravity}} + \underbrace{\frac{\mathbf{F}}{m}}_{\text{all other forces}} \qquad (4.9)$$

Here, the last term includes pressure gradient forces, frictional or viscous forces, and external tidal forces (from the Sun or parent star, from a moon, or from a planet being orbited if we're considering the atmosphere of a large moon). Viscous forces are usually small and neglected, but they necessitate a constant resupply of angular momentum to the equatorial jets of Jupiter, Saturn, Venus, and Titan (Hide, 1969) as well as the eastward phase of a stratospheric circulation on Earth known as the quasi-biennial oscillation. External torques are also small and conventionally neglected. However, some authors have suggested that these forces give rise to torques that balance the viscous torques in deep atmospheres. For example, solar gravitational torques acting on an atmospheric mass distribution that is non-uniform in longitude because of solar heating has been proposed as the cause of a *superrotation* in the upper atmosphere of Venus (Gold and Soter, 1971). Superrotation is where the equatorial atmosphere rotates faster than the solid

through the derivation because the dedicated reader can find it in various textbooks, e.g., a vector form in Appendix B of Andrews (2010), a geometric derivation in Ch. 2 of Holton and Hakim (2013), and another derivation in Ch. 4 of Jacobson (2005). Suffice it to say that scaling theory allows one to drop small terms so that large-scale motions simplify to the following equations when we neglect vertical motions and friction:

Vertical (eq.(1.8)) : $\qquad \dfrac{\partial p}{\partial z} = -g\rho$ $\qquad$ (4.11)

Zonal : $\dfrac{du}{dt} - \underbrace{\left(f + \dfrac{u\tan\phi}{a}\right)}_{\text{Coriolis term \ Centrifugal term}} v = -\dfrac{1}{\rho}\dfrac{\partial p}{\partial x}$ $\quad$ (4.12)

Meridional : $\dfrac{dv}{dt} + \underbrace{\left(f + \dfrac{u\tan\phi}{a}\right)}_{\text{Coriolis term \ Centrifugal term}} u = -\dfrac{1}{\rho}\dfrac{\partial p}{\partial y}$

$$(4.13)$$

Here, a is the planetary radius at the origin of the x, y, z surface coordinates and f is the Coriolis parameter (eq. (4.8)). Terms fv and $-fu$ are the x and y components of the Coriolis acceleration discussed earlier. The Coriolis force is perpendicular to the wind vector, to the right in Earth's northern hemisphere and to the left of the wind vector in Earth's southern hemisphere. So-called "metric terms" $(utan\phi/a)v$ and $-(utan\phi/a)u$ are the x and y components

Another key concept in fluid dynamics is advection. The *substantial*, *total*, *material* or *Stokes derivative* of the form dX/dt (written DX/Dt in some textbooks) represents the rate of change of some quantity X with respect to time *following* a fluid parcel, unlike $\partial X/\partial t$, the change with respect to time at a *fixed point*. The relationship between these derivatives of any flow-field quantity that is a function of x, y, z and t is:

$$\underbrace{\frac{d(\ )}{dt}}_{\text{substantial derivative}} \equiv \underbrace{\frac{\partial(\ )}{\partial t}}_{\text{local derivative}} + \underbrace{u\frac{\partial(\ )}{\partial x} + v\frac{\partial(\ )}{\partial y} + w\frac{\partial(\ )}{\partial z}}_{\text{advective terms}} = \frac{\partial(\ )}{\partial t} + \mathbf{v}\cdot\nabla(\ ) \quad (4.14)$$

of *cyclostrophic* acceleration, which is the acceleration arising from the excess or deficit of centrifugal acceleration due to motion relative to the planetary rotation. Equations (4.12) and (4.13) are often called *the momentum equations*.

Here, $\mathbf{v}$ is the 3-D velocity vector (eq. (4.3)), and we have used $\mathbf{a}\cdot\mathbf{b} = a_1b_1 + a_2b_2 + a_3b_3$, along with the gradient operator, $\nabla = \partial/\partial x + \partial/\partial y + \partial/\partial z$. Box 4.1 illustrates a physical interpretation of advective terms. When the quantity in the dot product is $\mathbf{v}$, i.e., $\mathbf{v}\cdot\nabla\mathbf{v}$, the nonlinearity makes atmospheric dynamics "interesting" for weather forecasting, or difficult, depending on your disposition. However, in uniform flow where the velocity spatial derivatives are zero, equations can become linear.

Box 4.1 Understanding Advective Terms

(1) Suppose we have a temperature field $T(x, y, z, t)$ that is constant in time ($\partial T/\partial t = 0$) but spatially varying, as shown in contours below. As an observer moves from place to place at velocity $\mathbf{v}$ the rate of change of temperature T is

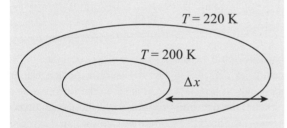

$$\frac{dT}{dt} = \mathbf{v}\cdot\nabla T$$

$$\text{e.g.,} \quad \frac{\Delta T}{\Delta t} = \left(\frac{\Delta x}{\Delta t}\right)\left(\frac{\Delta T}{\Delta x}\right) = \frac{20\ \text{K}}{\Delta t}$$

$T = 200\ \text{K}, t = t_1 \qquad\qquad T = 200\ \text{K}, t = t_2$

(2) Now suppose contours change position with a velocity $\mathbf{v}$ but not value. An observer traveling at the same velocity sees no change in T with time, so $dT/dt = 0$ because $\partial T/\partial t$ is balanced by the *advection* of temperature, $\mathbf{v}\cdot\nabla T$.

4.1.2 Characteristic Force Balance Regimes in Atmospheres

To provide physical insight, the x–y–z equations of motion, (4.11)–(4.13), can be simplified further when different forces dominate and, in particular, they can then provide an intuitive understanding of the effect of planetary rotation on atmospheric circulation. We consider cases that apply according to whether a planet rotates rapidly or slowly. We then look at how a balance between heating and cooling leads to large-scale ascent and descent, which provides our starting point for discussing how heat is transported in large circulation cells on Earth, Mars, Venus, Titan, and Earth-like exoplanets (Sec. 4.2).

4.1.2.1 *Geostrophic Balance: Fast Rotating Planets*
If a planet rotates rapidly enough, the Coriolis terms in f in eqs. (4.12) and (4.13) balance the pressure gradient in so-called *geostrophic balance*, at least in the extratropics. This balance is an excellent approximation when two interrelated conditions hold,

$$\frac{d(u, v)}{dt} \ll f, \quad \frac{u\tan\phi}{a} \ll f \quad (4.15)$$

so that we can eliminate the d/dt and metric terms in eqs. (4.12) and (4.13). The first condition applies when a dimensionless number, the *Rossby number*, is $\ll 1$:

Rossby number, $Ro = \dfrac{\text{relative acceleration}}{\text{Coriolis acceleration}} = \dfrac{U^2/L}{fU} = \dfrac{U}{fL}$

(4.16)

Here, U and L are characteristic horizontal velocity and length scales of the motions. Horizontal acceleration in eq. (4.15) has scale $d(u, v/dt) \sim U/t \sim U^2/L$. Since the maximum possible length scale is planetary radius a, the second condition in eq. (4.15) should be valid if the first holds, which we find on rapidly rotating planets: Earth, Mars, and the giant planets of the Solar System.

We can consider the Rossby number on various planets. On Earth, $U \approx 10$ m s^{-1} and $L \approx 1000$ km = 10^6 m in mid-latitudes, while $f = 10^{-4}$ s^{-1} at 45° N from eq. (4.8). Hence $Ro \approx 0.1$, and winds should be nearly geostrophic. Mars has a similar rotation rate and Rossby number to Earth, so geostrophy applies to Mars too. Jupiter has peaks winds ~100 m s^{-1} at the boundaries of its bright zones and dark belts and a characteristic wind speed somewhat slower as $U \approx 50$ m s^{-1}, $L \approx 10\,000$ km = 10^7 m, and $f \approx 1.8 \times 10^{-4}$ s^{-1}. Thus, $Ro \approx 0.03$, so geostrophy again is reasonable. Saturn, with faster winds, has $Ro \approx 0.1$.

When both of the conditions in eq. (4.15) are satisfied, the horizontal flow described by eqns. (4.12) and (4.13) simplifies to geostrophic flow, as follows:

geopotential Φ (the gravitational potential energy per unit mass defined in eq. (1.13)) on a constant pressure surface is indicated by the "p" subscript. The second form is more compact because the pressure gradient force per unit mass in 3-D is simply $\nabla\Phi$, i.e., the geopotential variation along a constant pressure surface for a system in hydrostatic equilibrium.

The geostrophic wind has four properties.
(1)   It flows tangential to the pressure gradient, i.e., parallel to isobars. Equivalently, it's parallel to geopotential contours on a constant pressure surface.
(2)   In any small latitude range, speed is proportional to the spacing of those contours.
(3)   The low-pressure region is to the left of the wind vector in the northern hemisphere on Earth, giving rise to a "Low to the Left" mnemonic known as the *Buys-Ballot law* after an early Dutch meteorologist. The converse is true in the opposite hemisphere.
(4)   The wind speed implied by a given geopotential contour spacing (or isobar spacing, approximately, in altitude coordinates) is inversely proportional to sine of latitude, i.e., increases at lower latitudes as $f \to 0$.
As a consequence of the first and third properties, winds blow along the isobars of terrestrial low-pressure-centered *cyclonic* systems in the sense shown in Fig. 4.3. The

$$u_g = -\frac{1}{f\rho}\left(\frac{\partial p}{\partial y}\right)_z, \quad v_g = \frac{1}{f\rho}\left(\frac{\partial p}{\partial x}\right)_z \quad ; \text{ or } \quad u_g = -\frac{1}{f}\left(\frac{\partial \Phi}{\partial y}\right)_p, \quad v_g = \frac{1}{f}\left(\frac{\partial \Phi}{\partial x}\right)_p \qquad (4.17)$$

Here, the "g" subscript indicates geostrophic winds (i.e., the approximation assumes $Ro = 0$). Eq. (4.17) gives equivalent relations in two commonly used coordinate systems. In altitude coordinates on the left, the "z" subscript indicates that the partial derivative is evaluated along a surface of constant altitude z. In the other system,

opposite sense occurs around *anticyclonic* systems. In reality, as depicted in Fig. 4.3, surface friction causes the winds to flow at a slight angle to the isobars near a planet's surface. This creates convergence and divergence, i.e., inward and outward airflow, on low- and high-pressure areas, respectively.

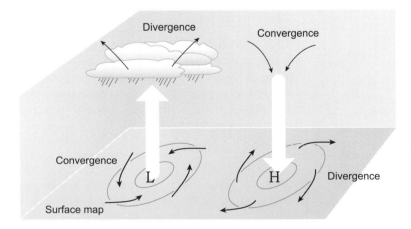

Figure 4.3 Winds in cyclones (low-pressure systems) and anticyclones (high-pressure systems) in the northern hemisphere of the Earth. Divergent and convergent airflow near the surface and aloft are also shown.

It may seem odd that purely geostrophic air flows at right angles to the pressure gradient instead of down the gradient. However, air moving because of a pressure gradient will veer sideways in response to the Coriolis force on a rotating planet and *in time* reach the state of geostrophy. In eq. (4.16) for the Rossby number, L/U is a timescale τ for a flow feature to move across its own diameter. We can estimate the timescale by writing the Rossby number as $Ro = 1/f\tau = 1/[(\Omega(2\sin\phi)\tau] = \tau_d/[(2\pi)(2\sin\phi)\tau]$, where τ_d is a sidereal day. At $\phi = 45°$ latitude, $Ro \approx \tau_d/9\tau$, so if $Ro < 0.1$, which is required for geostrophy, τ must exceed a sidereal day. Thus, geostrophy is a balance developed by large-scale flow that takes longer than a sidereal day to cross a characteristic vortex dimension. The flow is therefore inevitably affected by rotation, which operates over that timescale.

4.1.2.2 "Thermal Wind" (or "Thermal Windshear") Balance

If there is a temperature gradient in the horizontal direction, a column of air (if isothermal) will shrink or grow relative to neighboring columns, setting up different horizontal pressure gradients at altitude and producing changing wind vectors with height. The *thermal wind* (or *thermal windshear*) equation relates *wind shear* – defined as the change of wind speed and direction with height – to horizontal temperature gradients. Differentiation of the geostrophic wind equation and substitution of the hydrostatic equation gives the thermal wind equation. For the meridional wind v_g, the algebra is as follows:

$$fv_g = \frac{1}{\rho}\left(\frac{\partial p}{\partial x}\right) = \frac{\overline{R}T}{P}\left(\frac{\partial p}{\partial x}\right),$$

differentiate w.r.t. z and use the hydrostatic and ideal gas eqns:

Here, we've retained possible variability of $\overline{R}$ in the log-pressure coordinate form on the right. These thermal wind equations show that winds as a function of latitude and height (or log pressure) can be deduced from the temperature distribution in a planetary atmosphere. Temperatures can be measured by remote sensing of emitted thermal radiation. Of course, we need a boundary condition to integrate eq. (4.18), which can be winds measured at a level such as cloud top on a giant planet or surface of a rocky planet. For the latter, we could alternatively use the distribution of surface pressures measured with barometers. For rocky planets where we use known winds at some level, often an assumption is made that the surface wind is negligible with the price of some inaccuracy.

The great insight is that the thermal wind equation links geostrophic flow to atmospheric thermodynamics. For example, in a common situation where the distribution of incoming solar radiation sets up a temperature decrease from equator to pole, the equation predicts that prograde zonal winds increase with altitude. We will find that it is also insightful to think about thermal windshear in a "backwards" way: If we detect prograde zonal winds increasing with altitude on a rapidly rotating planet (e.g., through cloud motion), there must be a corresponding meridional temperature gradient.

4.1.2.3 Cyclostrophic Balance

On Earth, we sometimes encounter *cyclostrophic balance* when a strong local pressure gradient force is balanced by centrifugal force. For example, when spinning air in a tornado moves with tangential speed V around a radius R, this balance is as follows:

$$\underbrace{\frac{V^2}{R}}_{\text{centrifugal acceleration}} = \underbrace{\frac{1}{\rho}\frac{\partial P}{\partial R}}_{\text{pressure gradient acceleration}} \qquad (4.19)$$

$$f\frac{\partial v_g}{\partial z} = \overline{R}T\frac{\partial^2}{\partial z\partial x}(\ln p) \approx \overline{R}T\frac{\partial}{\partial x}\left(\frac{1}{p}\left(\frac{\partial p}{\partial z}\right)\right) = \overline{R}T\left(\frac{g}{\overline{R}T^2}\frac{\partial T}{\partial x}\right) \quad \text{with} \quad \frac{\partial p}{\partial z} = -g\rho = -\frac{gP}{\overline{R}T}$$

Here, we ignored vertical variations in T and $\overline{R}$, and so inserted an approximation sign. Manipulation of the zonal geostrophic wind equation is analogous, giving the results:

A tornado may have $V \sim 50$ m s^{-1}, $R \sim 100$ m, and a correspondingly strong pressure gradient. In such systems, planetary rotation is not "felt" by the much faster circulation.

$$f\frac{\partial u_g}{\partial z} \approx -\frac{g}{T}\frac{\partial T}{\partial y}, \quad f\frac{\partial v_g}{\partial z} \approx \frac{g}{T}\frac{\partial T}{\partial x} \quad ; \text{or} \quad f\frac{\partial u_g}{\partial(\ln p)} = \left(\frac{\partial(\overline{R}T)}{\partial y}\right)_p, \quad f\frac{\partial v_g}{\partial(\ln p)} = -\left(\frac{\partial(\overline{R}T)}{\partial x}\right)_p \qquad (4.18)$$

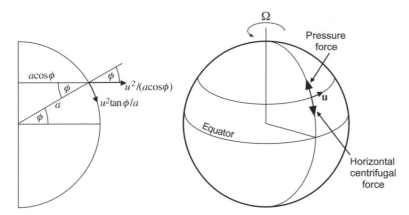

Figure 4.4 The geometry of cyclostrophic balance at latitude ϕ on a slowly rotating planet, e.g. Venus. The zonal wind speed is u, the planetary radius is a, and Ω is the planet's angular rotation rate.

For planets or moons as a whole, we find cases such as Venus and Titan where there is little planetary rotation, so $\Omega \approx 0$, f is small, and geostrophic balance does not apply. On Venus, a scale analysis at midlatitudes gives characteristic horizontal scale $L \approx 1000$ km $= 10^6$ m, characteristic wind speed $U \approx 10$ m s^{-1}, and $f = 2\Omega \sin \phi = 2(2.98\times 10^{-7}$ s$^{-1})(\sin 45°) = 4.2\times10^{-7}$ s^{-1} at 45° N, so that the Rossby number, $Ro \approx 10$ from eq. (4.16). However, if the atmosphere itself rotates sufficiently rapidly over some altitude range, cyclostrophic balance prevails (Leovy, 1973). At latitude ϕ, the centrifugal acceleration due to the zonal wind is $u^2/(a \cos \phi)$, because $a\cos \phi$ is the distance to the rotation axis (Fig. 4.4). This has a local component, parallel to the planet's surface, of $(u^2\tan \phi)/a$, giving

$$\frac{u^2 \tan \phi}{a} = -\frac{1}{\rho}\frac{\partial p}{\partial y}; \quad \text{or} \quad \frac{u^2 \tan \phi}{a} = -\frac{\partial \Phi}{\partial y} \quad (4.20)$$

We can also derive eq. (4.20) from eq. (4.13) by neglecting the Coriolis and substantial derivative terms. Physically, we have a picture where the centrifugal force balances an equator-to-pole pressure gradient (Fig. 4.4).

In evaluating the integral, we assumed an isothermal scale height $H = \overline{R}T/g$(eq. 1.12) and used $dz' = -H(dp/p)$ (eq. 1.10) and ignored vertical variations in $\overline{R}$.

Data show that the cyclostrophic approximation works for slowly rotating planets. Equations (4.20) and (4.21) are good approximations in the middle atmospheres of Venus and Titan given the observed increase of wind speed with altitude on Venus from *Pioneer Venus* probes (Seiff *et al.*, 1980) and on Titan from *Huygens* probe measurements (Bird *et al.*, 2005).

We can apply the theory to slowly rotating exoplanets. Consider the hot Jupiter HD209458b, which has a 3.5-day period and radius 1.35 R_J where R_J = Jupiter's radius of 69 911 km. Though not purely cyclostrophic, let's assume this approximation. We have some of the inputs to eq. (4.21) if we take $\overline{R} = 4160$ J K^{-1} kg^{-1} (Table 1.3), tan $\phi = 1$ at 45° latitude, and a day–night contrast $\Delta T \sim 300$ K operating over a meridional scale of $\Delta y = 10^5$ km. We cannot solve eq. (4.21) without knowing u^2 at a reference level but we can assess the magnitude of Δu^2. Between pressure levels of 0.01 to 1 bar, we obtain:

$$\Delta u^2 = \left[(4160 \times 300 \times (1.35 \times 69\,911\,\text{km}) \times \ln(1/0.01))/(10^5\text{km})\right] = 5.4 \times 10^6 \text{m}^2\text{s}^{-2}$$

We can also derive a *cyclostrophic thermal wind equation* that tells us how the zonal wind changes with altitude in response to a meridional temperature gradient. We use the same method as before: differentiation of the wind equation (4.20) w.r.t. z and substitution of ideal gas and hydrostatic equations,

The peak winds on Jupiter are ~ 100 m s^{-1}, so a Δu^2 value of more than two orders larger on HD209458b indicates stronger thermal wind shear. If we assume a small zonal velocity at 1 bar then $\Delta u \sim 2$ km s^{-1}. Other scale analysis (eqs. (49) and (50) of Showman *et al.* (2010)) gives the same Δu result. Thus, fast zonal winds are expected on

$$\frac{u^2 \tan \phi}{a} = -\frac{\overline{R}T}{p}\frac{\partial p}{\partial y} \Rightarrow \frac{\tan \phi}{a}\frac{\partial (u^2)}{\partial z} \approx -\overline{R}T\frac{\partial}{\partial y}\left(\frac{1}{p}\frac{\partial p}{\partial z}\right) = -\frac{\overline{R}\overline{T}g}{\overline{R}\,T^2}\frac{\partial T}{\partial y}$$

$$u^2(z, \phi) = u^2(z_0, \phi) - \frac{\overline{R}a}{\tan \phi}\int_{z_0}^{z}\frac{1}{H}\frac{\partial T}{\partial y}\,dz' \approx u^2(z_0, \phi) - \frac{\overline{R}a}{\tan \phi}\frac{\Delta T}{\Delta y}\ln\left(\frac{p_0}{p}\right) \quad (4.21)$$

hot Jupiters, which is consistent with hotspots that are offset from a planets' substellar points by fast winds (Agol et al., 2010; Knutson et al., 2007) and detection of fast winds from Doppler-shifted absorption lines (Showman et al., 2013a; Snellen et al., 2010). In the future, similar dynamics may be detected on slowly rotating Earth-like exoplanets.

4.1.2.4 Thermodynamic Balance: Large-Scale Ascent and Subsidence of Air

In Sec. 2.4.5, we assumed that that atmospheric structures are described well by 1-D radiative–convective equilibrium, but in fact large-scale motions disturb the temperature field through horizontal advection and also compression and expansion associated with ascent and subsidence. We can examine the latter using the associated thermodynamics.

The first law of thermodynamics expressed in terms of energy changes per unit mass, dq, was given in

---

$$\text{substitute } \frac{Q}{c_p} \approx -\alpha_N \left(T - T_{eq}\right), \qquad \text{so that } \quad w(\Gamma_a - \Gamma) = -\alpha_N \left(T - T_{eq}\right) \qquad (4.26)$$

---

eq. (1.27) as $dq = c_p\, dT - \alpha\, dp$, where $\alpha = 1/\text{density} = 1/\rho$ and c_p is the specific heat capacity. Dividing by δt and c_p, and taking the limit $\delta t \to 0$, we obtain:

$$\frac{dT}{dt} = \frac{\alpha}{c_p}\left(\frac{dp}{dt}\right) + \frac{Q}{c_p} \qquad (4.22)$$

Here, $Q = dq/dt$ [J kg^{-1} s^{-1}] is the *diabatic heating* per unit mass caused by radiation, convection, or molecular conduction. We expand dT/dt, according to eq. (4.14):

$$\frac{dT}{dt} = \left(\frac{\partial T}{\partial t} + u\frac{\partial T}{\partial x} + v\frac{\partial T}{\partial y}\right) + w\frac{\partial T}{\partial z} = \frac{dT'}{dt} + w\frac{\partial T_0}{\partial z} \qquad (4.23)$$

In the last equality, T' represents the horizontal and time variable component of temperature – shown in the brackets in the first equality, and T_0 is the horizontal and time average temperature, which is assumed a function of altitude only. Thus, using eq. (4.23) and the relation that $dp/dt \approx -g\rho w = -gw/\alpha$, where w is the vertical velocity, we can re-write eq. (4.22) as

$$\underbrace{\frac{dT'}{dt}}_{\text{time variable component}} + \underbrace{w\left(\frac{\partial T_0}{\partial z} + \frac{g}{c_p}\right)}_{\text{work done}} = \underbrace{\frac{Q}{c_p}}_{\text{heat gain}} \qquad (4.24)$$

In atmospheric science, this form of the first law of thermodynamics is the *thermodynamic energy equation*, which describes how energy input or loss causes a rate of temperature change [K s^{-1} or K day^{-1}] for a parcel of air

that moves up or down. The term dT'/dt is often smaller than the other two terms and its omission gives

$$\underbrace{w\left(\frac{\partial T_0}{\partial z} + \frac{g}{c_p}\right)}_{\text{static stability}} = \underbrace{\frac{Q}{c_p}}_{\text{heat gain}} , \quad \text{or } w(\Gamma_a - \Gamma) = \frac{Q}{c_p} \qquad (4.25)$$

Here, the term in brackets on the left-hand side is *static stability*, i.e., the difference between the adiabatic lapse rate $\Gamma_a = g/c_p$ (e.g., $\Gamma_a = 9.8$ K km^{-1} for Earth) and the observed lapse rate $\Gamma = -(dT/dz)$ (e.g., $\Gamma \approx 6$ K km^{-1} in the global mean for Earth), which we met previously in eq. (1.35). Equation (4.25) expresses a balance between compressional heating (or expansional cooling) on the left-hand side and diabatic heating (or cooling) on the right.

To understand thermodynamic balance, it helps to replace the diabatic heating term with the *Newtonian cooling approximation*, in which temperature relaxes back to an equilibrium reference temperature field described by T_{eq}:

Here, α_N is a Newtonian cooling coefficient [s^{-1}], where $\alpha_N = 1/\tau_{rad}$, where τ_{rad} is the radiative time constant (Sec. 2.3.2). The quantity T_{eq} is the temperature of radiative–convective equilibrium when radiation and convection are the only major heat sources and sinks. Equation (4.26) simply says that that heating or cooling due to vertical motion is balanced by radiative relaxation of the temperature back to an equilibrium temperature.

The above theory suggests that the observed atmospheric temperature field T can be used with theoretical expectations of T_{eq} to infer large-scale vertical velocities. Note that static stability $(\Gamma_a - \Gamma)$ is generally positive, e.g., typically a few K km^{-1} for the atmospheres of the terrestrial planets (Sec. 1.1.3.1). Thus, if we measure $T < T_{eq}$, then eq. (4.26) requires that w is positive and large-scale ascent is taking place to produce a temperature colder than equilibrium because of expansion and cooling. Where $T > T_{eq}$, it follows from eq. (4.26) that w is negative and large-scale subsidence is producing compressional heating and a temperature that exceeds that of radiative–convective equilibrium. In fact, such regions can be seen in zonal mean meridional temperature cross-sections of real planetary atmospheres, such as Mars' atmosphere at equinox shown in Fig. 4.5. At pressures below ~0.5 mbar, the atmospheric temperature *increases* with latitude from the equator to high latitudes. The atmosphere cannot be in radiative equilibrium, which

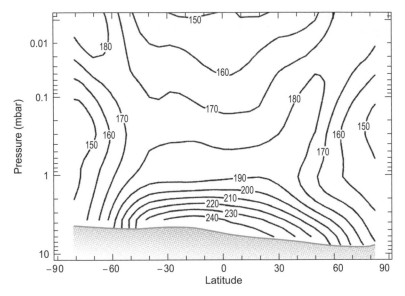

Figure 4.5 Zonal mean meridional atmospheric temperatures at northern hemisphere autumnal equinox on Mars derived from measurements by the Thermal Emission Spectrometer (TES) instrument on NASA's *Mars Global Surveyor* spacecraft. The solid line at the lower boundary is the zonal mean atmospheric pressure at the surface. (Adapted from Smith *et al.* (2001). Reproduced with permission. Copyright 2001, John Wiley and Sons.)

implies dynamical effects. Large-scale meridional circulation in both hemispheres has ascent at low latitudes that causes cooling aloft and descent at mid to high latitudes that, at altitude, produces enough compressional heating to completely reverse one's expectation of a warm-to-cold equator-to-pole temperature gradient.

4.2 The Zonal-Mean Meridional Circulation and Thermally Driven Jet Streams

4.2.1 The Two Types of Jet Stream: Thermally Driven and Eddy Driven

One of the main features of atmospheric circulation on all planets is the jet stream structure – part of the arrangement of west–east wind systems over altitude and latitude. Because of planetary rotation, prograde winds tend to dominate (west-to-east for Earth), and usually they reach a maximum strength near the tropopause. Here, the term *jet stream* describes any wind maximum, whether it occurs in the prograde or retrograde direction.

Earth's atmosphere illustrates two types of jet stream. Seasonally and zonally averaged zonal winds are shown in Fig. 4.6(a). Below Earth's ~10^2 hPa tropopause, there are *subtropical tropospheric jet streams* at 30–40° latitude with a single such jet on average in each hemisphere. Of course, inhabitants of mid-latitudes know that a *polar front jet stream* is commonly present. But unlike the subtropical jet, the polar front jet stream is narrow, only partly continuous, and very mobile, so that it does not show up in the annual-average latitude-height section of

Fig. 4.6(a). The subtropical jet arises from the temperature gradient between equator and higher latitudes combined with planetary rotation. This thermal gradient produces a large overturning of the atmosphere – the "Hadley circulation" discussed below. In contrast, the polar front jet stream is driven by eddies, i.e., the turbulent wind currents around low- and high-pressure disturbances in mid to high latitudes. We think of these two types of jets as *thermally driven* and *eddy driven*, respectively (Lee and Kim, 2003; Showman *et al.*, 2013b). They exemplify two types of jet expected in planetary atmospheres that we discuss below (Sec. 4.2.2 and Sec. 4.3).

On different planets, jet streams may be single or multiple in each hemisphere (Fig. 4.7); and they can also be strong or weak. In addition, jets can meander in latitude, such as on Earth and probably on Mars and Venus also, or can appear relatively straight. Terrestrial jets wander from *trough* regions of equatorward displacement to *ridge* regions of poleward shift. Meanders are often associated with transient eddies or waves. We describe such waves with a *zonal wavenumber*, which is the number of full wavelengths around an entire circle of latitude (Fig. 4.8). Zonal wavenumbers vary greatly from one planet to another because of different dynamics and a planet's size. The typical wavenumber of the meanders on Earth is ~6, on Venus it is 1 or 2, while on Mars, *Viking Lander* and orbital data suggest that it is 2–4. On Jupiter, the wavenumber can be >40. Jets appear straight in low spatial resolution images of Jupiter or Saturn, but much turbulence is evident at high spatial resolution.

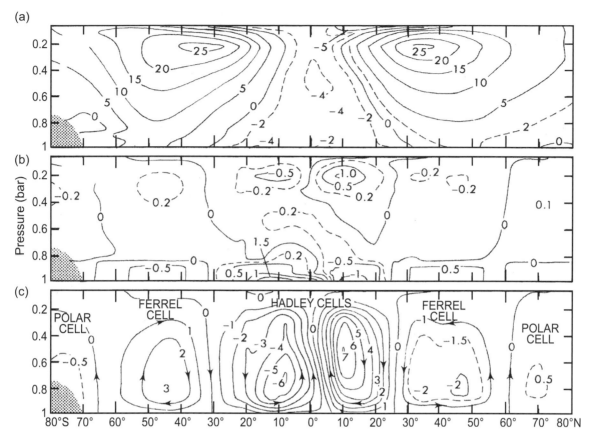

Figure 4.6 (a) Earth's annual mean cross-section of the zonal wind (m s^{-1}). *Westerlies* (from the west, going eastward) are shown as solid contours, *easterlies* (from the east, going westward) are dashed contours. Below the tropopause (~0.1 bar) are *subtropical jet streams* at ~30–40° in each hemisphere. (b) The annual average meridional wind (m s^{-1}). Solid lines are northward, dashed lines are southward. (c) Inferred mass flow streamlines (10^{10} kg s^{-1}) indicate Hadley cells, mid-latitude Ferrel cells, and weak polar cells in each hemisphere in the annual average. (Adapted from Peixoto and Oort (1984). Reproduced with permission. Copyright 1984, American Physical Society.)

4.2.2 The Hadley Circulation and Subtropical Jets

George Hadley (1685–1786) attempted to explain the winds in the Earth's tropics with a low latitude circulation (Hadley, 1735). The circulation that bears his name is often depicted as symmetric about the equator (Fig. 4.9 (a)), which is reasonable for the annual mean (Fig. 4.6) but actually atypical most of the time on planets with seasons, as discussed below. The same basic principles apply to a circulation that is not zonally symmetric provided zonal variations are small compared with meridional variations.

In a conceptual picture of the Hadley circulation, warm air rises near the equator (on Earth, primarily in convective cores of tall cloud systems) and spreads poleward from a level of maximum ascent near the tropopause. Air raised from near the surface at the equator has a

greater angular momentum than that at higher latitudes by virtue of frictional coupling to planet's rotation. However, the flow aloft is less affected by friction. So, atmospheric parcels aloft tend to have eastward angular momentum relative to the surface as they move poleward, forming eastward winds and a subtropical jet stream. At upper levels, the poleward-moving air cools and sinks. On Earth, this occurs in the subtropics, where the resulting high-pressure zones of dry air create deserts such as the Sahara, Patagonian desert, and American southwest. The cycle is completed by return flow in low latitudes at low levels, on Earth in a boundary layer 1–2 km deep. In the return flow, conservation of angular momentum leads to the development of surface easterly (westward) winds. These equatorward, westward flows are the *trade winds*.

Hadley's original idea involved rising motion near the equator and symmetry about the equator, but Hadley circulations on Earth and Mars are often highly asymmetric about the equator and also not uniform in longitude. Latitudinal asymmetry occurs because the rising branch is usually shifted well into one hemisphere and the descending branch shifted well into the other (Fig. 4.9(b)).

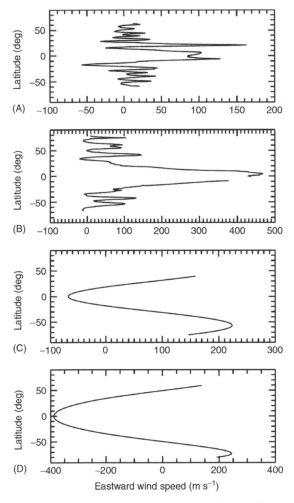

Figure 4.7 Jet streams can be single or multiple on different planets. Zonal-mean zonal wind from cloud tracking for (A) Jupiter, (B) Saturn, (C) Uranus, and (D) Neptune show peaks that correspond to jets. Jupiter and Saturn have 20 east or west jets, while Uranus and Neptune appear to have three broad jets. (Source: Showman *et al.* (2010).)

On Earth, longitudinal asymmetry also results from continent–ocean contrasts, topography, and motion localized to small-scale convective cells that produce heavy rainfall in the *intertropical convergence zone* (ITCZ) where low-level trade winds converge. Earth's ITCZ is located close to, but not on the equator (Fig. 4.10). Descent also preferentially occurs in large-scale high-pressure areas of the subtopical Atlantic, Pacific, and Indian Oceans. Large-scale propagating waves also play a role in shaping the jet (Held and Phillips, 1990; Lee and Kim, 2003).

4.2.3 Symmetric Hadley Circulation Theory

Hadley did not attempt to answer two basic questions.

(1) What controls the most poleward latitude of the sinking motion?

(2) What sets the maximum zonal wind speed in the subtropical jet?

Two centuries later, Held and Hou (1980) provided a quantitative model of a zonally symmetric Hadley circulation based on three principles: (1) the tendency for parcels moving meridionally to conserve angular momentum, (2) geostrophic or cyclostrophic balance, and (3) a balance between energy gain by heating of the system and radiative loss to space. Although their model is a great conceptual model, it doesn't describe what happens in the real Hadley cells of the Earth, as discussed below. Nonetheless, it illustrated key concepts and moved the field forward. Below we follow its discussion similar to that in reviews by Showman *et al.* (2013b) and Schneider (2006).

In the Held–Hou model, the upper branch of the Hadley circulation conserves angular momentum per kg by assumption. If there is no relative zonal wind at the equator (latitude $\phi = 0$), the angular momentum per kg there is Ωa^2. If angular momentum per kg is conserved in transport of air parcels from the equator to latitude ϕ, then:

$$\Omega a^2 = (\Omega a \cos \phi + u) a \cos \phi \qquad (4.27)$$

The terms on the right can be understood by examining Fig. 4.11. A point on the planet's surface at latitude ϕ

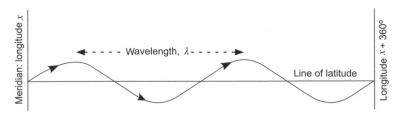

Figure 4.8 The *zonal wavenumber* is the number of complete wavelengths around a circle of latitude. In this case, the zonal wavenumber is 2.

(a)

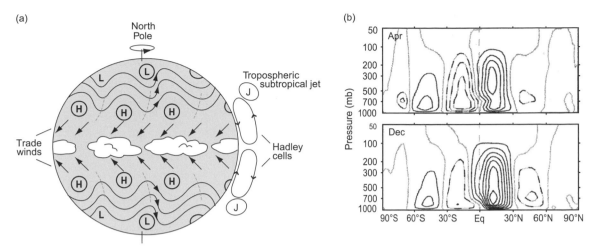

(b)

Figure 4.9 (a) Idealized symmetric Hadley cells (Redrawn from Wallace and Hobbs (2006).) (b) A seasonally asymmetric Hadley circulation is revealed by reanalysis of observations in zonally averaged mass flow streamlines. Compare monthly averages for April (top), which is near the equinox, and December (bottom), which is near the southern summer solstice. Contours are 2×10^{10} kg s^{-1}. (Source: Dima and Wallace (2003). Reproduced with permission. Copyright 2003, American Meteorological Society.)

(a)

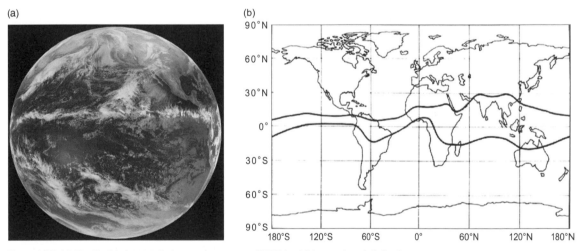

(b)

Figure 4.10 (a) The Earth's intertropical convergence zone (ITCZ) is visible as a band of cloud across the eastern Pacific from west to east (courtesy: NOAA/NASA GOES project). (b) The latitude of the ITCZ changes seasonally and the annual average latitude of the ITCZ is north of the equator.

moves with tangential velocity $\Omega a \cos \phi$, to which we add the local zonal wind speed. However, the radius to the planetary rotation axis is $a \cos \phi$, so that the angular momentum per unit mass is given by the expression on the right of eq. (4.27). Solving for the zonal wind and using the identity $\sin^2\phi \equiv 1 - \cos^2\phi$, we have:

$$u \cos \phi = \Omega a \left(1 - \cos^2\phi\right) \quad \Rightarrow u = (\Omega a)\frac{\sin^2\phi}{\cos\phi} \quad (4.28)$$

We remind ourselves that this equation applies for when the ITCZ is at the equator. For example, suppose that in the upper branch of Earth's Hadley circulation air is carried from the equator, where $u = 0$ and $\Omega a = (7.3 \times 10^{-5}$ rad s$^{-1})(6378 \times 10^3$ m$) \approx 465$ m s^{-1}, to latitude 30°. Equation (4.28) gives the zonal wind speed at 30° as $u = 134$ m s^{-1}.

The zonal wind speed in the upper branch of Earth's Hadley circulation seasonally reaches ~30–70 m s^{-1} in the core of the subtropical jet and is not as extreme as predicted by eq. (4.28) because friction and pressure forces reduce the angular momentum as a parcel travels. In detail, eddies associated with waves propagating from

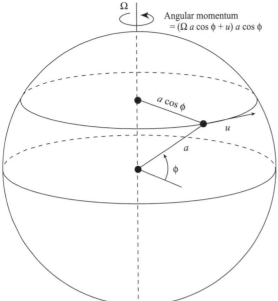

Figure 4.11 Diagram illustrating the zonal angular momentum at latitude ϕ, which can be compared to that at the equator to help understand the Hadley circulation. The zonal wind speed is u, the planetary radius is a, and the planet's angular rotation rate is Ω.

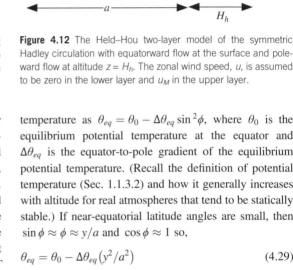

Figure 4.12 The Held–Hou two-layer model of the symmetric Hadley circulation with equatorward flow at the surface and poleward flow at altitude $z = H_h$. The zonal wind speed, u, is assumed to be zero in the lower layer and u_M in the upper layer.

the midlatitudes tend to provide westward angular momentum that decelerates the upper branch of the Hadley cell (Bordoni and Schneider, 2008, 2010; Randel and Held, 1991; Schneider and Bordoni, 2008). In any case, surface wind speeds must be far lower because of friction. Consequently, there must be a big increase of eastward wind speed with height near the jet. According to the thermal wind equation for the zonal wind (4.18), there needs to be a strong meridional temperature gradient $\partial T/\partial y$ at the jet latitude, with poleward temperature T decreasing rapidly. The gradient in the radiative–convective equilibrium temperature T_{eq} should vary slowly and smoothly with latitude, but the actual temperature gradient is concentrated in the vicinity of the jet stream. On the equatorial side of the subtropical jet, $T < T_{eq}$ implies large-scale ascent (Sec. 4.1.2.4)). Meanwhile on the polar side, $T > T_{eq}$ implies large-scale descent.

temperature as $\theta_{eq} = \theta_0 - \Delta\theta_{eq}\sin^2\phi$, where θ_0 is the equilibrium potential temperature at the equator and $\Delta\theta_{eq}$ is the equator-to-pole gradient of the equilibrium potential temperature. (Recall the definition of potential temperature (Sec. 1.1.3.2) and how it generally increases with altitude for real atmospheres that tend to be statically stable.) If near-equatorial latitude angles are small, then $\sin\phi \approx \phi \approx y/a$ and $\cos\phi \approx 1$ so,

$$\theta_{eq} = \theta_0 - \Delta\theta_{eq}\left(y^2/a^2\right) \tag{4.29}$$

Also, using $\sin\phi \approx y/a$ and $\cos\phi \approx 1$, eq. (4.28) simplifies to $u = \Omega y^2/a$.

The thermal wind equation can be used to predict the variation of temperature with latitude that we would expect if angular momentum conservation sets the zonal wind. In terms of potential temperature, the thermal wind is:

$$f\frac{\partial u}{\partial z} \approx -\frac{g}{\theta_0}\frac{\partial\theta}{\partial y} \Rightarrow \left(\frac{2\Omega y}{a}\right)\frac{u_M}{H_h} = -\frac{g}{\theta_0}\frac{\partial\theta}{\partial y} \Rightarrow \frac{\partial\theta}{\partial y} = \left(-\frac{\theta_0}{g}\right)\left(\frac{2\Omega y}{a}\right)\frac{\Omega y^2}{aH_h} \tag{4.30}$$

Held and Hou's two-layer axisymmetric model for estimating the size of the Hadley circulation and its strength is shown in Fig. 4.12, and ignores variation of density with altitude. They assumed a latitudinal distribution of radiative–convective equilibrium potential

using $f = 2\Omega\sin\phi = 2\Omega y/a$ for the Coriolis parameter and assuming a height H_h to the upper branch of the Hadley cell as shown in Fig. 4.12. Integration of eq. (4.30) in θ and y then gives the variation of temperature with latitude as:

$$\theta = \theta_{M0} - \left(\frac{\Omega^2 \theta_0}{2ga^2 H_h}\right) y^4 \tag{4.31}$$

where θ_{M0} is the potential temperature at the equator (to be determined) and the "M" subscript reminds us that this temperature field derives from momentum conservation.

Lastly, two assumed constraints provide a solution. First, heating and cooling across the Hadley circulation are assumed to balance. With Newtonian cooling,

$$\int_0^Y \frac{d\theta}{dt} dy = 0, \quad \Rightarrow \int_0^Y -\alpha(\theta - \theta_{eq}) dy = 0, \quad \Rightarrow \int_0^Y \theta dy = \int_0^Y \theta_{eq} dy \tag{4.32}$$

where Y is the maximum meridional extent of the Hadley cell. (Note that $Y = a\phi_H$, where ϕ_H (radians) is the high latitude boundary of the cell.) The second constraint is that the temperature (4.31) is continuous with latitude at the subtropical edge of the Hadley cell and assumed equal to the radiative equilibrium temperature (4.29). These constraints produce two equations for the two unknowns, Y and θ_{M0}:

$$\theta_{M0} Y - \left(\frac{\Omega^2 \theta_0}{10ga^2 H_h}\right) Y^5 = \theta_0 Y - \left(\frac{\Delta\theta_{eq}}{3a^2}\right) Y^3; \quad \theta_{M0} - \left(\frac{\Omega^2 \theta_0}{2ga^2 H_h}\right) Y^4 = \theta_0 - \left(\frac{\Delta\theta_{eq}}{a^2}\right) Y^2 \tag{4.33}$$

These equations can be solved for the meridional extent of the Hadley cell, Y:

$$Y = \sqrt{\frac{5gH_h}{3\Omega^2}\left(\frac{\Delta\theta_{eq}}{\theta_0}\right)}, \quad \text{or } \phi_H = \sqrt{\frac{5gH_h}{3\Omega^2 a^2}\left(\frac{\Delta\theta_{eq}}{\theta_0}\right)} \tag{4.34}$$

With annual-mean values for the Earth of $\theta_0 \approx 260$ K, $\Delta\theta_{eq} \approx 70$ K and $H_h = 15$ km, and with $a = 6380$ km, $g = 9.8$ m s^{-2}, and $\Omega = 7.2 \times 10^{-5}$ rad s^{-1}, we get $Y = 3570$ km or $\phi_H \approx 30°$, which is remarkably close to reality considering the simplicity of the model.

Equation (4.34) also implies that the latitudinal extent of the Hadley circulation should contract on a planet that rotates faster, i.e., with larger angular velocity Ω. Essentially, rapid rotation confines the Hadley circulation to low latitudes with the subtropical jet at the high latitude boundary. Earth-like GCM (general circulation model) simulations that vary planetary rotation rate support this inference (Del Genio and Suozzo, 1987; Kaspi and Showman, 2015; Navarra and Boccaletti, 2002; Showman et al., 2013b). Conversely, as a planet's rotation rate slows, the Hadley circulation gets wider, heat is transported farther, and the equator-to-pole temperature contrast diminishes. If the Rossby number Ro exceeds ~10,

the planet can be called an "all tropics" world in terms of its dynamics (Showman et al., 2013b).

A limitation of the Held and Hou model is its neglect of turbulent eddies. In detail, the model predicts the strength of Earth's Hadley circulation as ~10^{10} kg s^{-1}, which is an order of magnitude too small (cf. Fig. 4.6(c)); also, the exact dependence of Hadley cell width on rotation rate and other parameters is not supported in detail by GCMs. So alternative theories have been suggested. The leading idea is that midlatitude eddies confine the Hadley circulation (e.g., Bordoni and Schneider, 2010). Midlatitude eddies tend to generate eastward flow in the upper troposphere, so the application of the Coriolis force would imply an equatorward component of the flow in opposition to the poleward upper branch of Hadley cells. When the temperature changes rapidly in the horizontal and the wind changes strongly with height, the atmosphere is said to have a *baroclinic* structure. For example, lines of constant temperature tilt steeply downward in the midlatitudes of Mars' troposphere even at equinox (Fig. 4.5), and similar sloping isotherms occur in Earth's midlatitude troposphere, particularly in the winter hemisphere. In contrast, isotherms are relatively flat in the tropics. Midlatitude baroclinicity gives rise to *baroclinic instability*, which is when flow is driven by the extraction of potential energy from a latitudinal temperature gradient, giving rise to large-scale eddies.

Hadley cells perhaps extend to the latitude where the troposphere becomes baroclinically unstable (Held, 2000). Effectively, energy transported from the tropics by the Hadley circulation is passed on to baroclinic eddies at the subtropics, which transport energy to the poles. With a two-layer model of baroclinic instability and the latitudinal temperature gradient implied by angular momentum conservation (eq. (4.31)), Held (2000) derives a different latitudinal extent as follows:

$$\phi_H = \left(\frac{gH}{\Omega^2 a^2}\left(\frac{\Delta\theta_{vert}}{\theta_0}\right)\right)^{1/4} \tag{4.35}$$

Here, $\Delta\theta_{vert}$ is the potential temperature difference going vertically from the surface to the top of the Hadley cell. Equation (4.35) has a much weaker power law dependence on rotation rate and other variables compared to eq. (4.34) and provides a better fit to some GCM simulations (Frierson *et al.*, 2007; Lu *et al.*, 2009; Schneider, 2006). However, as mentioned above, winds in the real atmosphere do not obey simple angular-momentum conservation, so the model is simplistic. In fact, a theory that fully and clearly explains the strength and extent of Hadley cells remains elusive (Levine and Schneider, 2011).

The region of subsidence at subtropical latitudes on the Earth corresponds to descent on the equatorward flank of baroclinic waves, which forms part of the *Ferrel Cell*, a weak meridional circulation over midlatitudes (Fig. 4.6 (c)). The ascent of the Ferrel Cell occurs on the poleward flank of baroclinic waves and is driven by the mechanical energy of midlatitude storms. One should bear in mind that the Ferrel Cell is a temporal average and at any point in time, the circulation is dominated by very sporadic airflow associated with midlatitude storms and fronts.

4.2.4 Asymmetric Hadley Circulations on Earth and Mars, and Monsoons

As mentioned previously, the actual Hadley circulation on Earth is not symmetric about the equator but usually has the rising branch displaced toward one hemisphere or the other (Cook, 2004; Dima and Wallace, 2003; Lindzen and Hou, 1988). In general, a shift can be due to the tilt of the planet's rotation axis, which ensures that the sub-solar latitude is away from the equator for all but a very short part of the year. Or it can be due to other factors, as we next consider.

On average, Earth's Hadley circulation is stronger in the southern hemisphere because of the hemispheric difference in the distribution of oceans and continents. Earth's ITCZ lies at ~7° N in the annual average and is stronger at the solstices than near the equinoxes (Fig. 4.10). Because the ITCZ is offset from the equator, surface flow crosses the equator toward the opposite hemisphere. As it does so, it can curve from the Coriolis force from westward to eastward, and become a monsoon circulation. A *monsoon* is where there is a seasonal reversal of winds associated with latitudinal movement of the ITCZ. Monsoons are also effectively seasonal intensifications of the Hadley circulation in longitudinal bands because of land–sea temperature contrasts or a mountain range barrier. Such circulations are important in the continents of America, Africa, and Asia where summer heating shifts the rising branch of the Hadley circulation far from the equator.

Monsoon-like circulations also occur on Mars. Mars' orbit has relatively high eccentricity with more incoming solar radiation near the southern summer solstice when the Hadley circulation is stronger than during southern winter solstice (Fig. 4.13).

Absorption of solar radiation by dust is an important component of atmospheric heating on Mars that affects

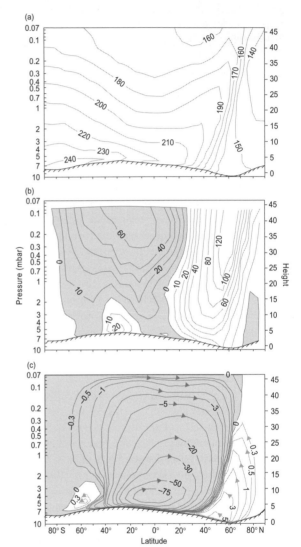

Figure 4.13 Illustration of the Martian Hadley circulation at northern winter solstice, as calculated by the NASA Ames Mars general circulation model. The zonal mean temperature distribution (top panel) shows a temperature gradient maximum at 50–60° N in the winter hemisphere where there is a subsiding branch of the Hadley circulation. (b) Zonal mean winds, where shading corresponds to westward winds. Note the eastward jet stream at 60° N, 30 km altitude. Note also monsoon-like eastward winds near the surface in the southern subtropics. (c) Mass flow streamlines (10^8 kg s^{-1}) for the overturning thermally driven circulation. Clockwise flow is in the shaded plane. Note the large latitudinal extent of the zonally asymmetric Hadley circulation. (Source: Leovy (2001). Reprinted by permission from Macmillan Publishers Ltd: (C. Leovy, 412, 245–249) © 2001.)

the monsoon circulation. Local dust storms occur throughout the year, but during southern summer near perihelion, dust raised into the atmosphere can spread globally and greatly increase atmospheric heating on a planet-wide scale. In some years, planet-wide dust storms are associated with a big increase in the intensity of the latitudinally asymmetric Hadley circulation (Lewis et al., 2007). Winds generated by the intensifying Hadley circulation contribute to widespread lifting of dust. The monsoon circulation becomes intense and dust is raised into the atmosphere throughout the monsoon latitude band, where it intensifies solar heating. The resulting positive feedback helps initiate and maintain these planet-wide storms. Not only does the dust spread to high latitudes (~55° N) in the expanding Hadley circulation, it is also lifted up to ~60 km altitude. Such dust storms mix constituents between the lower and middle atmosphere and presumably influence atmospheric chemistry.

4.2.5 Hadley Circulations on Venus and Titan

Much less is known about Hadley circulations on Venus and Titan. However, their study might tell us about circulations expected on slowly rotating exoplanets, which could be important for habitability, in some cases. Venus has a solar constant of 2614 W m^{-2}, but only a paltry ~30 W m^{-2} is absorbed in the lower atmosphere and ~20 W m^{-2} at the surface (Titov et al., 2007). Meager heating in a dense atmospheric layer can drive only a very weak Hadley circulation. Because the constraining effect of rotation is small, this circulation probably extends to high latitude (cf. eqs. (4.34) and (4.35)). Surface wind markings measured by the Magellan spacecraft radar are spatially variable with an unknown time of emplacement, but a general equatorward direction at higher latitudes and an east–west direction nearer the equator suggests a possible Hadley circulation over a broad latitude range (Greeley et al., 1997). Emitted infrared radiation as a function of latitude has a much flatter shape on Venus than on Earth, which is evidence for a smaller equator-to-pole temperature gradient and poleward transport of heat (Taylor et al., 1983).

A second Hadley-like circulation appears to exist near Venus' cloud top level (~60–70 km) where ~100 W m^{-2} is absorbed, which is most of the non-reflected, incoming solar radiation (Baker and Leovy, 1987). A strong temperature gradient and a weak maximum in the zonal wind speed relative to the background flow near 55° latitude suggest that this circulation extends to about this latitude in both hemispheres. Poleward of this latitude, the bottom of an approximately isothermal stratosphere drops in altitude and a tropopause minimum in the temperature profile becomes evident, unlike the absence of a temperature minimum in the equatorial regions (Tellmann et al., 2009).

The situation on Titan is less clear. As on Venus, the heating per unit mass of the lower atmosphere is small so any thermally driven circulation must be weak. Evidence for prevailing winds in the lowest part of the Titan atmosphere comes from dunes in dark areas in low latitudes (Jaumann et al., 2009). The dunes indicate eastward flow – exactly opposite to the direction expected for the trade winds of a Hadley circulation. Numerical simulations suggest that the dunes may form during brief periods when equatorial zonal winds reverse and are relatively strong (Tokano, 2009, 2010). In particular, occasional methane storms in low latitudes around equinox (because of strong updrafts then) may couple surface winds to fast superrotating eastward winds aloft (Charnay et al., 2015).

4.2.6 Mean Meridional Circulation and Planetary Habitability

The zonal-mean meridional circulation discussed above is one of the most obvious areas where dynamics affects planetary climate. On Earth, the Hadley circulation creates dry, cloud-free conditions in its subtropical descending branch (Sec. 4.2.2). The runaway greenhouse effect sets the inner edge of the habitable zone around a star for Earth-like planets (Sec. 2.4.5.4 and Sec. 13.4). In 3-D models examining the runaway greenhouse, dry areas associated with the descending Hadley cell branch prove important for increasing the solar flux threshold at which an Earth-like planet goes into a runaway state (Leconte et al., 2013). Another example concerning the dry descending branch of the Hadley circulation is a 3-D model of Snowball Earth (Sec. 11.10). This branch creates a latitudinal zone of bare ice rather than snow-covered ice, which lowers the albedo in that region, enabling a belt of water to persist in the tropics (Abbot et al., 2011).

The ascending branch of Hadley-like circulation also has potentially important climatic effects. On tidally locked Earth-like exoplanets that have the same face towards their parent star, strong dayside convection can create a high albedo circular patch of sub-stellar water clouds that allows the inner edge of the habitable zone to be closer to the star than suggested by 1-D climate models (Yang et al., 2013). Three-dimensional (3-D) models also show that atmospheric dynamical heat transport can warm the nightside of tidally locked Earth-like exoplanets and maintain their habitability by preventing the freeze-out of an atmosphere on the nightside (Joshi, 2003; Joshi et al., 1997).

4.3 Eddy-Driven Jet Streams and Planetary Waves

Hadley's theory said nothing about Earth's atmospheric circulation in midlatitudes, where we find polar jet streams and a different circulation regime. As mentioned earlier, jets can also be forced by systematically structured eddies superimposed on the zonal mean flow. A familiar example of such eddies are terrestrial traveling cyclones and anticyclones. Another example is the stationary wave structure imposed on the zonal mean flow by lower boundary forcing and differential heating due to the distribution of continents and oceans on Earth.

Multiple jets in the atmospheres of Jupiter and Saturn, and the high latitude jets of Uranus and Neptune shown in Fig. 4.7 are probably eddy driven. Earth's atmosphere has two jets in the southern hemisphere much of the time. In northern hemisphere winter, there are also often two jets: the subtropical jet associated with the Hadley circulation and an eddy-driven jet at higher latitude that sometimes merges with the subtropical jet. Even on Mars and Venus, the single jet in each hemisphere – though primarily thermally driven – is modified by large-scale stationary and transient eddies.

4.3.1 Vorticity

To discuss eddy-driven jets, we need to introduce some concepts of fluid dynamics concerned with rotating air. The rotation at a point in a fluid could be specified in terms of radial distance from the center of rotation and the velocity field, but it is described more concisely through a single quantity, *vorticity*, which we now define.

Relative vorticity ζ, is a measure of rotation about a vertical axis relative to the planet's surface, defined *positive if flow is counter-clockwise*, i.e., a right-hand rule: with fingers in the direction of curved flow, "thumb up" is positive. In solid body rotation, relative vorticity is equivalent to 2ω, where ω is the instantaneous angular velocity. For example, if tangential speed $V = 10$ m s^{-1} in solid body rotation about radius $r = 300$ km, the relative vorticity $\zeta = 2$ $(V/r) = 2(10$ m s$^{-1})/ (300 \times 10^3$ m$) = 6.7 \times 10^{-5}$ s^{-1}. In a

general flow field, *vorticity* at any point is the component of the curl of the wind about the local vertical. Thus, we take the **k** component of the expression in Box 4.2 to get the relative vorticity, i.e., the vorticity relative to the frame of reference of the planetary rotation:

$$\zeta = -\frac{\partial u}{\partial y} + \frac{\partial v}{\partial x} \qquad (4.36)$$

We must also consider the effect of planetary spin. A fluid blob also possesses a component of planetary angular velocity of $\Omega \sin \phi$ about the local vertical (Fig. 4.2(b)). Doubling it gives *planetary vorticity* $2\Omega\sin \phi$, which is the Coriolis parameter, f (eq. (4.8)). The *absolute vorticity* η is the sum of relative and planetary vorticity:

$$\eta = \zeta + f = \left(-\frac{\partial u}{\partial y} + \frac{\partial v}{\partial x}\right) + f \qquad (4.37)$$

Relative vorticity is expressed in atmospheres as a combination of horizontal shear (the change of velocity transverse to the flow, e.g., imagine eastward winds that weaken as you go north) and curvature (the rotation of the downstream flow direction). This explains why relative vorticity is 2ω in solid body rotation: a shear contribution of the change in tangential velocity with radius, $\partial V/\partial r = \partial(\omega r)/\partial r = \omega$, is added to a curvature contribution of ω from the change in the tangential velocity with angle, $V(\partial(\text{angle})/ \partial(\text{path})) = V(2\pi/2\pi r) = V/r = \omega$. In Earth's northern hemisphere, positive relative vorticity characterizes curvature in the counter-clockwise direction and/or sheared flow with decreasing flow speed toward the left of the flow direction. Positive relative vorticity occurs in cyclones, such as the surface low-pressure system in Fig. 4.3, and so is called *cyclonic vorticity*. Negative relative vorticity is expressed as curvature in the clockwise direction and/or as shear flow with speed increasing toward the left of the flow direction, which is referred to as *anticyclonic vorticity* (such as the surface high-pressure system on the right of Fig. 4.3). However, because an opposite relationship applies in the southern hemisphere where the sign of f is opposite, we think of *cyclonic vorticity* as spin about a vertical axis in the same direction as the vertical component of the planetary rotation and

---

Box. 4.2 Curl of a Vector Velocity Field Measures the Rotation of the Air

With a wind vector $\mathbf{v} = u\mathbf{i} + v\mathbf{j} + w\mathbf{k}$ (eq. (4.3)), the curl in Cartesian coordinates is:

$$\text{curl } \mathbf{v} = \nabla \times \mathbf{v} = \begin{vmatrix} \mathbf{i} & \mathbf{j} & \mathbf{k} \\ \frac{\partial}{dx} & \frac{\partial}{dy} & \frac{\partial}{dz} \\ u & v & w \end{vmatrix} = \mathbf{i}\left(\frac{\partial w}{\partial y} - \frac{\partial v}{\partial z}\right) + \mathbf{j}\left(\frac{\partial u}{\partial z} - \frac{\partial w}{\partial x}\right) + \mathbf{k}\left(\frac{\partial v}{\partial x} - \frac{\partial u}{\partial y}\right)$$

anticyclonic vorticity as the opposite. (To add complication for the uninitiated, dynamicists also talk about *curvature* of a flow, such as the flow of a wave circling round a planet. When a flux line is concave up, like a bowl, curvature is defined to be positive. Negative curvature applies to convex up, like a hill.)

The change of vorticity in time tells us how atmospheric circulations emerge and can be expressed in a *vorticity equation*. To derive it, we neglect the small term $u \tan\phi/a$ in the equations of motion, (4.12) and (4.13), add momentum forcing terms F_x and F_y to the left sides of (4.12) and (4.13), expand the substantial derivatives using (4.14), operate on the equations with $-\partial/\partial y$ and $+\partial/\partial x$, respectively, and add. This is somewhat tricky, so we show the full expansion and explain how it is simplified:

shear. This twisting usually results from small-scale motions. The second term with brackets results from (baroclinic) horizontal shears of pressure and density.

To simplify, we neglect the vertical velocity and baroclinic terms. Also, if vertical velocity is negligible ($w = 0$), there is no horizontal divergence, i.e., no expansion or narrowing of columns of air resulting from vertical shrinking or stretching, so the divergence term ($\eta\Delta$) vanishes. Then eq. (4.38) becomes very simple:

$$\frac{d\eta}{dt} = S_\eta \qquad (4.39)$$

Here, in purely horizontal flow, the substantial derivative is the horizontal form, $d/dt = \partial/\partial t + u\,\partial/\partial x + v\,\partial/\partial y$. Equation (4.39) shows that if there is no forcing by friction or

column of $\partial/\partial x$ terms   column of $\partial/\partial y$ terms

$$\underbrace{\frac{\partial}{\partial t}\left(\frac{\partial v}{\partial x} - \frac{\partial u}{\partial y}\right)}_{\zeta} + u\frac{\partial}{\partial x}\left(\frac{\partial v}{\partial x}\right) + \boxed{\frac{\partial u}{\partial x}\frac{\partial v}{\partial x}} - u\frac{\partial}{\partial y}\left(\frac{\partial u}{\partial x}\right) - \boxed{\frac{\partial u}{\partial y}\frac{\partial u}{\partial x}} +$$

$$v\frac{\partial}{\partial y}\left(\frac{\partial v}{\partial x}\right) + \boxed{\frac{\partial v}{\partial x}\frac{\partial v}{\partial x}} - v\frac{\partial}{\partial y}\left(\frac{\partial u}{\partial y}\right) - \boxed{\frac{\partial v}{\partial y}\frac{\partial u}{\partial x}} +$$

$$w\frac{\partial}{\partial z}\left(\frac{\partial v}{\partial x}\right) + \frac{\partial w}{\partial x}\frac{\partial v}{\partial z} - w\frac{\partial}{\partial z}\left(\frac{\partial u}{\partial y}\right) - \frac{\partial w}{\partial y}\frac{\partial u}{\partial z} +$$

$$\boxed{+ f\frac{\partial u}{\partial x}} + u\frac{\partial f}{\partial x} + \boxed{f\frac{\partial v}{\partial y}} + v\frac{\partial f}{\partial y} =$$

$$\frac{-1}{\rho}\left(\frac{\partial^2 p}{\partial x\partial y}\right) + \frac{1}{\rho}\left(\frac{\partial^2 p}{\partial x\partial y}\right) + \frac{1}{\rho^2}\left(\frac{\partial p}{\partial y}\frac{\partial \rho}{\partial x}\right) - \frac{1}{\rho^2}\left(\frac{\partial p}{\partial x}\frac{\partial \rho}{\partial y}\right) + \frac{\partial F_y}{\partial x} - \frac{\partial F_x}{\partial x}$$

The double-underlined terms sum to the $d\zeta/dt$; also, the six terms in boxes sum to $(f + \zeta)(\partial u/\partial y + \partial v/\partial y)$. Using $\eta = f + \zeta$ (eq. (4.37)), and noting that $v\,df/dy = df/dt$, the above simplifies to:

stirring and $S_\eta = 0$, the absolute vorticity does not change in time and so is said to be conserved in horizontal flow. Because absolute vorticity (η) is the sum of relative (ζ) and planetary vorticity (f), it means that as air shifts

$$\frac{d\eta}{dt} + \eta\Delta - \left(\frac{\partial w}{\partial y}\frac{\partial u}{\partial z} - \frac{\partial w}{\partial x}\frac{\partial v}{\partial y}\right) + \frac{1}{\rho^2}\left(\frac{\partial \rho}{\partial x}\frac{\partial P}{\partial y} - \frac{\partial P}{\partial x}\frac{\partial \rho}{\partial y}\right) = -\frac{\partial F_x}{\partial y} + \frac{\partial F_y}{\partial x} = S_\eta \qquad (4.38)$$

Here, $\Delta = \partial u/\partial x + \partial v/\partial y$ is the *divergence* of the horizontal wind, which is a tendency of outward flow from a point if positive or inward flow (*convergence*) if negative. Terms in F_x and F_y define a vorticity forcing term S_η, which can arise from friction or from stirring by eddies. The first bracketed term in eq. (4.38) corresponds to the inter-conversion between vorticity about axes, i.e., "twisting" of horizontal vorticity because of vertical wind

latitude and f changes, the air must acquire relative vorticity and start spinning, which proves to be a useful concept, as we shall see.

4.3.2 Jet Forcing by Stirring or Friction

The vorticity equation can help us see how jets are forced by eddies and how this can account for multiple jets that switch from eastward to westward on the giant planets at different latitudes and no doubt also in atmospheres of some large rapidly rotating exoplanets. Jet formation from eddies is also reviewed by Showman *et al.* (2013b) and in Ch. 12 of Vallis (2006).

To understand vorticity forcing by eddies and its relationship to jets, we can consider a case where there is a uniform zonal mean flow of speed $\bar{u}$ and an eddy fluctuation deviation from the average that is primed, e.g., $u = \bar{u} + u'$. We also assume a small meridional perturbation $v = v'$. Here, the zonal mean value of the fluctuation is zero, $\overline{u'} = 0$. These primed values arise from transient eddies that are associated with rapidly developing and decaying weather disturbances of the kind commonly experienced in terrestrial midlatitudes. The average of the product of fluctuation components, e.g., $\overline{u'v'}$, is called the *covariance* and is non-zero when u' and v' are correlated.

Using the above theory, a term such as $u(\partial\eta/\partial x)$ within the vorticity equation (4.39) can be written:

$$u\frac{\partial\eta}{\partial x} = (\bar{u} + u')\frac{\partial(\bar{\eta} + \eta')}{\partial x} = \bar{u}\frac{\partial\eta'}{\partial x} + u'\frac{\partial\eta'}{\partial x} \approx \bar{u}\frac{\partial\eta'}{\partial x} \quad (4.40)$$

Here, we use the fact that $\bar{\eta}$ has no dependence on t or x. With such "perturbation theory," we linearize the time-dependent vorticity equation (4.39) by expanding the substantial derivative and using eq. (4.40):

$$\left(\frac{\partial}{\partial t} + (\bar{u} + u')\frac{\partial}{\partial x} + v'\frac{\partial}{\partial y}\right)(\bar{\eta} + \eta') = S_\eta' \quad \Rightarrow \quad \left(\frac{\partial}{\partial t} + \bar{u}\frac{\partial}{\partial x}\right)\eta' + v'\frac{\partial}{\partial y}\bar{\eta} = S_\eta' \quad (4.41)$$

To obtain an equation for the vorticity variance (also called *enstrophy*) in terms of the covariances between meridional velocity, vorticity, and vorticity forcing, we multiply eq. (4.41) by η' and take the zonal average (deleting terms with a zonal mean of zero):

How does eq. (4.42) apply? Assume that eddies are generated at and cross latitude "1," ϕ_1, as shown in Fig. 4.14. If the meridional gradient of absolute vorticity $\partial\bar{\eta}/\partial y$, exceeds 0 everywhere (because f increases with latitude and $\partial f/\partial y$ is positive (Sec. 4.3.3)), eq. (4.42) shows that, in the time-mean horizontal flow, eddy stirring is associated with a poleward flux of vorticity across latitude circle ϕ_1 by the eddies represented by $\overline{v'\eta'}$ at ϕ_1. This increases vorticity in the area of the planet poleward of latitude ϕ_1.

We next see how the increase of vorticity above latitude ϕ_1 drives a prograde jet at ϕ_1. Whereas vorticity measures rotation of a fluid at a point, *circulation* [unit m^2 s^{-1}] quantifies rotation over an area. The vorticity within an area is related to the circulation around a closed curve enclosing that area by *Stokes' theorem* (e.g., see Fleagle and Businger (1980), p. 407, for a derivation):

$$\Gamma_C = \oint V_s \, ds = \iint \eta dA \quad (4.43)$$

Here, V_s is the component of velocity along the arc ds (positive when in the sense of planetary rotation), dA is an element of area, and Γ_C is the circulation. Applying Stokes' theorem to the polar cap area bounded at its southern extent by latitude ϕ_1, we have:

$$2\pi(a\cos\phi_1)\bar{u} = \int_{\lambda=0}^{2\pi}\int_{\phi=\phi_1}^{\pi/2} \eta a^2 \cos\phi \, d\phi \, d\lambda = 2\pi[\eta]a^2 \int_{\phi=\phi_1}^{\pi/2} \cos\phi \, d\phi \quad (4.44)$$

where λ is longitudinal angle and a is planetary radius. On the left, we evaluated the line integral by averaging around latitude ϕ_1 of circumference $2\pi(a\cos\phi_1)$, and on the right, we inserted an element of area in latitude–longitude coordinates, $a^2\cos\phi \, d\phi \, d\lambda$ (see Fig. 2.9(a)), and wrote the average vorticity over the polar cap region as $[\eta]$. Equation (4.44) shows that a poleward vorticity

$$\overline{\eta'\frac{\partial\eta'}{\partial t}} + \overline{\bar{u}\eta'\frac{\partial\eta'}{\partial x}} + \overline{v'\eta'\frac{\partial\bar{\eta}}{\partial y}} = \overline{S_\eta'\eta'} \Rightarrow \frac{1}{2}\frac{\partial\overline{(\eta')^2}}{\partial t} + \frac{\bar{u}}{2}\frac{\partial\overline{(\eta')^2}}{\partial x} + \overline{v'\eta'}\frac{\partial\bar{\eta}}{\partial y} = \overline{S_\eta'\eta'}$$

$$\Rightarrow \underbrace{\overline{v'\eta'}}_{\substack{\text{poleward vorticity} \\ \text{transport by eddies}}} = \underbrace{-\frac{\partial\overline{(\eta')^2}/\partial t}{2\partial\bar{\eta}/\partial y}}_{\text{vorticity variance term}} + \underbrace{\frac{\overline{S_\eta'\eta'}}{\partial\bar{\eta}/\partial y}}_{\text{eddy forcing term}} \quad (4.42)$$

flux by eddies that increases the average vorticity $[\eta]$ in the cap region poleward of ϕ_1 also increases the mean zonal wind $\bar{u}$ at latitude ϕ_1 (Fig. 4.14).

Now suppose that stirring generates eddies along ϕ_1, forcing a prograde jet at that latitude, and eddies propagate poleward to a second latitude ϕ_2 where there is dissipation and hence negative covariance between eddy vorticity forcing and eddy vorticity itself, i.e., $\overline{S_\eta' \eta'}$ is negative. According to eq. (4.42), along latitude ϕ_2, the eddy vorticity flux $\overline{v'\eta'}$ will be equatorward (negative), which means that a retrograde jet will be forced. Thus, if eddies can propagate north and south, both prograde and retrograde jets can be generated. This argument provides an explanation for how multiple jets can arise from eddies on giant planets or in atmospheres on rapidly rotating planets in general.

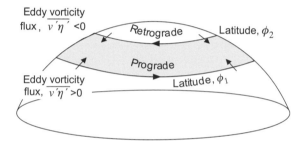

Figure 4.14 Illustration of the connection between vorticity flux across latitude circles and zonal flow acceleration. Vorticity flux due to eddy stirring $\overline{v'\eta'}$ is northward across latitude ϕ_1 and southward across latitude ϕ_2 because of dissipation (small arrows). This flux accelerates zonal flow in the prograde direction along ϕ_1 and in the retrograde direction along ϕ_2 (bigger arrowheads). The shaded zone is the zone around which circulation is increasing according to Stokes' theorem (see text).

Another view of eddy forcing of jets in horizontal flows follows from considering zonal angular momentum, which turns out to be mathematically equivalent to considering vorticity transport. The poleward flux of zonal angular momentum per unit mass is $\overline{u'v'}$ [(m^2 s^{-1})/s], which is the covariance between zonal and meridional components of the eddy wind velocity. Positive $\overline{u'v'}$ corresponds to a northward flux of zonal momentum and negative $\overline{u'v'}$ corresponds to a southward flux. Eddy wind component covariance corresponds to a tendency for systematic tilt of eddies. Where $\overline{u'v'} > 0$ (positive v' and u'), vector addition of u' and v' implies that eddies have a southwest to northeast tilt. Conversely, $\overline{u'v'} < 0$ produces a northwest to southeast tilt (Fig. 4.15). Idealized GCM simulation even show those tilts as precipitation patterns (Frierson *et al.*, 2006). On the Earth, warm subtropical air is rotating faster than cold polar air and tilted waves associated with eddies in northern midlatitudes will transfer eastward momentum and heat northwards, while cold air is carried southwards (Fig. 4.15(b)).

The equivalence of a northward flux of eddy vorticity $\overline{v'\eta'}$ with the convergence of zonal mean momentum flux (expressed through its meridional gradient, $-\partial\left(\overline{u'v'}\right)/\partial y$) means that zonal angular momentum flows into the source region for eddies from both north and south and feeds an eastward jet there. This flux of zonal momentum to a region of already high zonal momentum is said to be "up-gradient." The equivalence with vorticity transport can be shown by using the fact that non-divergent flow can be fully described by a *stream function*, ψ, which is a scalar field for a fluid flow with contour lines that are streamlines of the flow. An advantage of working with a stream function rather than a velocity field is that we need

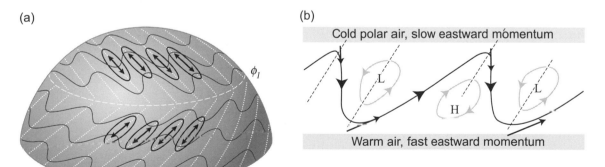

Figure 4.15 (a) Streamlines of an eastward zonal flow with superimposed wave (heavy solid lines with arrowheads). Momentum flux converges at latitude ϕ_1 where it drives an eastward jet. Thin ovals with thin slanting double-headed arrows represent the eddy perturbations north and south of ϕ_1. Such a repeating pattern could drive prograde and retrograde jets on giant planets. (b) Titled eddies in the midlatitudes of the Earth transport eastward angular momentum from the subtropics to the poles, along with heat.

only find one function rather than two (u and v) because the stream function ψ is defined as:

$$u = -\frac{\partial \psi}{\partial y}, \quad v = \frac{\partial \psi}{\partial x} \qquad (4.45)$$

Relative vorticity, defined by eq. (4.36), can be written in terms of ψ, as follows,

$$\zeta = -\frac{\partial u}{\partial y} + \frac{\partial v}{\partial x} \Rightarrow \zeta = \frac{\partial^2 \psi}{\partial x^2} + \frac{\partial^2 \psi}{\partial y^2} = \nabla^2 \psi \qquad (4.46)$$

where ∇ is the horizontal Laplace operator. By introducing perturbation forms of eq. (4.45), i.e., $u' = -d\psi'/\partial y$ and $v' = d\psi'/\partial x$ into eq. (4.46), we get ζ'. Thus, the poleward vorticity flux in terms of ψ' is as follows (eliminating the x derivative of zonal mean quantities that are zero):

$$\overline{v'\eta'} = \overline{v'\zeta'} = \overline{\frac{\partial \psi'}{\partial x}\left(\frac{\partial^2 \psi'}{\partial x^2} + \frac{\partial^2 \psi'}{\partial y^2}\right)} = \frac{\partial}{\partial y}\overline{\left(\frac{\partial \psi'}{\partial x}\frac{\partial \psi'}{\partial y}\right)} \Rightarrow \overline{v'\eta'} = -\frac{\partial(\overline{u'v'})}{\partial y} \qquad (4.47)$$

Equation (4.47) says that meridional vorticity transport (left) and the meridional flux of zonal mean angular momentum (right) by eddies are equivalent; they are just alternative ways ways to think about eddy forcing of jets.

4.3.3 Planetary Waves

Jets can meander back and forth in latitude because of waves in the circulation, including some comparable in size to the planet itself. Oscillations and propagating waves can occur in any medium in which perturbations from equilibrium are resisted by a restoring effect. In planetary atmospheres, the meridional gradient of absolute vorticity (arising from planetary rotation) provides a restoring effect that produces waves. To understand how such waves arise,

In differentiating f, we have used $\delta y = a\delta\phi$, from spherical geometry. The meridional planetary vorticity gradient β, is normally much larger in magnitude than the second term, the vorticity gradient due to the mean zonal flow $-\partial^2 \bar{u}/\partial y^2$, so that β dominates the expression for $\partial\eta/\partial y$. For this reason, we were safe to assume earlier that $\partial\eta/\partial y > 0$ in eq. (4.42). The restoring effect due to the planetary vorticity gradient is called the *beta effect*.

The beta effect can be appreciated by considering a gas parcel blowing from west to east with zero relative vorticity that drifts across latitude ϕ_1 because of some perturbation, such as flow over mountains. As the parcel moves poleward, planetary vorticity f increases but absolute vorticity ($\eta = \zeta + f$) is conserved so relative vorticity ζ must decrease. This decrease is expressed in part by an increase in anticyclonic curvature (clockwise in Earth's northern hemisphere), so the parcel trajectory curves equatorward and crosses latitude ϕ_1 with zero relative vorticity. Equatorward of ϕ_1, relative vorticity increases, cyclonic curvature develops and the trajectory returns again toward latitude ϕ_1. In this way, in the absence of dissipation or stirring, the gas parcel sinusoidally oscillates about ϕ_1 as it moves downstream. This is the mechanism of *Rossby waves*, which are members of a larger class of planetary waves influenced by the restoring effect of the planetary vorticity gradient (Fig. 4.16). Typical meanders in Earth's midlatitude jet stream with wavenumbers of 4 to 6 are guided by Rossby waves.

We can derive a wave equation for Rossby waves in horizontal flow by considering the vorticity equation linearized with a perturbation about the mean zonal flow $\bar{u}$, eq. (4.41). Neglecting the vorticity forcing term S_η, we have:

$$\left(\frac{\partial}{\partial t} + \bar{u}\frac{\partial}{\partial x}\right)\eta' + v'\frac{\partial\overline{\eta}}{\partial y} = 0 \quad \Rightarrow \quad \frac{\partial(\nabla^2 \psi')}{\partial t} + \bar{u}\frac{\partial(\nabla^2 \psi')}{\partial x} + \beta\frac{\partial(\psi')}{\partial x} = 0 \qquad (4.49)$$

where we have substituted for the following:

$$\eta' = \frac{\partial^2 \psi'}{\partial x^2} + \frac{\partial^2 \psi'}{\partial y^2} = \nabla^2 \psi' \quad \text{and} \quad \frac{\partial\overline{\eta}}{\partial y} = \beta \quad \text{and} \quad v' = \frac{\partial\psi'}{\partial x} \qquad (4.50)$$

we assume a zonal mean zonal wind $\bar{u}$ and differentiate absolute vorticity η (eq. (4.37)) w.r.t. y:

For simplicity, we assume that the approximately spherical geometry of the planet can be neglected so that we can

$$\frac{\partial\eta}{\partial y} = \beta - \frac{\partial^2 \bar{u}}{\partial y^2}, \quad \text{where } \beta \equiv \frac{\partial f}{\partial y} = \frac{\partial(2\Omega \sin\phi)}{\partial y} = \frac{1}{a}\frac{df}{d\phi} = \frac{2\Omega \cos\phi}{a} \qquad (4.48)$$

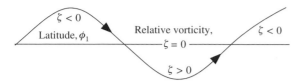

Figure 4.16 Illustration of the Rossby wave mechanism. A fluid parcel in the northern hemisphere moves through the wave pattern from west to east. Starting from the left, as the parcel crosses its equilibrium latitude ϕ_1, the relative vorticity $\zeta = 0$. North of ϕ_1, conservation of absolute vorticity requires $\zeta < 0$ and so the trajectory curves southward anticyclonically. South of ϕ_1, the trajectory curves northward cyclonically and $\zeta > 0$.

work with horizontal flow in a local Cartesian coordinate system in order to examine mechanisms and gain intuitive understanding.

With the additional assumption that the zonal mean zonal wind $\bar{u}$ is constant in the vicinity of latitude ϕ_1, we guess wave-like solution to (4.49), sinusoidal in x and y:

$$\psi'(x, y, t) = \Psi e^{i(\omega t - kx - ny)} = \Psi e^{i\xi} = \Psi(\cos\xi + i\sin\xi)$$
(4.51)

Here, the expression includes a wave amplitude factor Ψ, an angular frequency ω, and zonal and meridional wavenumbers $k = 2\pi/\lambda_x$ and $n = 2\pi/\lambda_y$, where λ_x and λ_y are zonal and meridional wavelengths, respectively. (We remind ourselves that, in physical space, ψ', is identified with the real part of the complex function). Substituting eq. (4.51) in eq. (4.49), allows us to derive a *dispersion relation*, meaning how the eastward phase speed c, of the Rossby wave varies with wavelength:

$$(-\cancel{i}\omega + \cancel{i}\bar{u}k)(k^2 + n^2) - \cancel{i}k\beta = 0 \Rightarrow \omega = k\bar{u} - \frac{k\beta}{(k^2 + n^2)}$$

$$c = \frac{\omega}{k} = \bar{u} - \frac{\beta}{(k^2 + n^2)}$$
(4.52)

Equation (4.52) gives the phase speed relative to the planet's surface. If we take $n = 0$, we see that Rossby waves propagate westward relative to the zonal wind at phase speeds that increase with increasing wavelength or smaller wavenumber, given $\lambda_x = 2\pi/k$. Stationary Rossby waves when $c = 0$ occur only in eastward zonal flows since $\beta^2/(k^2 + n^2)$ is positive. If we set $n = 0$ and $c = 0$, eq. (4.52) gives $k^2 - \beta/\bar{u}$ and we see that stationary waves have a west–east wavelength of $\lambda_x = 2\pi(\bar{u}/\beta)^{0.5}$. Taking $\bar{u} = 20$ m s^{-1} in Earth's middle troposphere at 45° latitude where $\beta \sim 1.6 \times 10^{-11}$ m^{-1} s^{-1}, the stationary wavelength is ~7000 km. Four of these waves would girdle the Earth. Even longer Rossby waves would propagate westward relative to the Earth's surface.

Rossby waves can steer and affect the amplitude of smaller, faster-moving baroclinic waves (see below) such

that low-pressure cyclones tend to form in the troughs of the waves, and high-pressure anticyclones tend to form in the crests of the waves. To see Earth's Rossby waves in data, the faster-moving baroclinic waves need to be smoothed out by averaging over several days. Then, the height of 500 mbar surfaces, for example, reveal Rossby waves, usually moving slowly eastward a few degrees longitude per day. Similar Rossby waves have been observed on Mars in midlatitudes in spring and winter with wavenumber 1–2 (Hollingsworth and Barnes, 1996; Wilson *et al.*, 2002).

If we relax the constraint of constant $\bar{u}$, the stream function amplitude will vary with meridional distance. An appropriate form of $\Psi(y)$ is then

$$\frac{\partial^2 \Psi}{\partial y^2} + n^2(y)\Psi = 0$$
(4.53)

where substitution of eq. (4.53) and eq. (4.51) into eq. (4.49) gives n as

$$n^2(y) = \frac{\beta}{(\bar{u} - c_x)} - k^2$$
(4.54)

where c_x is the eastward phase speed of the wave. Equation (4.53) has the form of a 1-D wave equation that is common in acoustics and optics so we can use concepts from those fields to discuss what it means.

The factor $n^2(y)$ is the square of the 1-D *refractive index* – analogous to the refractive index in optics. The analogy includes the concepts of *wave activity* – a wave property that depends on the square of the amplitude of the wave that is approximately conserved in propagation through a variable medium (e.g., Vallis, 2006, pp. 299–302), *wave packets* or *groups*, which are bounded disturbances consisting of closely spaced Fourier components, and *group velocity* ($d\omega/dk$), which is an approximate velocity of wave packet propagation. Waves can propagate in the y direction when $n^2(y) > 0$ but not when $n^2(y) < 0$, so latitudes at which $n^2(y)$ changes sign are of special importance.

Under two circumstances, $n^2(y)$ changes sign. In the first, $n^2(y) = 0$ when $[\beta/(\bar{u} - c_x)] - k^2 = 0$ in eq (4.54), and the eastward phase speed c_x, of the wave matches that of the freely propagating Rossby wave discussed previously. These are *turning point latitudes*. When a wave with specified frequency and zonal wave number ω and k propagating in the $+y$ direction (northward) encounters a turning point at $y = y_{\text{turn}}$, it is refracted back toward its origin. The wave cannot propagate beyond y_{turn}. Pairs of turning points can form a waveguide such that waves generated between the turning points are trapped. For example, prograde jets may have turning points for waves with the appropriate values of ω and k on both sides of the jet core. Waves generated by stirring or other processes near the jet core with ω and k that have turning points on the edges of the jet will be trapped in the waveguide. Only waves with $c_x < \bar{u}_{min} - \beta/k^2$ can escape the wave-guide, where $\bar{u}_{min}$ is the minimum mean zonal flow speed on either side of the eastward jet. Usually these will be waves with small values of k (long wave-lengths) and rapid westward phase speed. All other waves are blocked by the mean zonal flow.

The second condition under which $n^2(y)$ changes sign is $(\bar{u} - c_x) = 0$, which occurs at *critical latitudes*, denoted y_{crit}. As a critical latitude is approached through regions with allowed wave propagation, $n^2(y)$ tends to infinity through positive values by inspection of eq. (4.54). On the opposite (non-propagating) side of y_{crit}, $n^2(y)$ approaches infinity at y_{crit} through negative values. Oscillation behavior near y_{crit} is very different at critical latitudes than at turning point latitudes. As y_{crit} is approached through positive values, energy of waves with the values of ω and k appropriate to that value of y_{crit} is absorbed. Wave oscillation rate increases to infinity, but wave amplitude diminishes to zero. No propagation of waves past y_{crit} is allowed.

This can be illustrated by considering waves in eastward flow on the poleward side of a prograde jet (Fig. 4.17). Assume they are generated at latitude y_{gen}. Outside of the generation region, planetary wave packets, consisting of disturbances with a narrow range of ω and k propagate poleward. But as these packets encounter critical latitudes for the specified values of ω

and k in the packets, wave energy will be absorbed. In this way, wave packets generated in the vicinity of a particular latitude, y_{gen}, where the resulting eddy vorticity generation drives eastward mean zonal flow, can propagate poleward or equatorward to a distant latitude y_{crit}. At latitude y_{crit}, the waves are absorbed and the resulting eddy vorticity dissipation drives westward acceleration. This combination of generation, propagation, and dissipation can increase jets going in both directions and strengthen the zonal mean wind shear between jets.

Near critical latitudes, meridional wavelengths become short and the parcel trajectories begin to stretch so that parcels from distant origins are brought close together. This allows small scale eddies to efficiently mix properties and dissipate variations. Rossby waves are reversible distortions of contours of absolute vorticity (Fig. 4.16), but at a critical latitude the strong distortion of these contours is irreversible (Fig. 4.18). The process bears a rough analogy to ocean waves propagating shoreward from deep water. As waves approach a beach, the waves overturn, break, and dissipate. Because of this analogy, the atmospheric process of dissipation near critical latitudes is also referred to as *wave breaking*. Critical latitudes for planetary waves are more complex than waves on the beach because under certain conditions critical latitudes can reflect certain discrete Fourier solutions of the wave equation called *normal modes*. Nevertheless their most important role is absorption of planetary waves (Randel and Held, 1991).

4.3.4 Effects of Vertical Variation

In contrast to our consideration of waves in Sec. 4.3.3, real atmospheric flows are obviously not strictly horizontal. Vertical variations of wind and corresponding horizontal variations of temperature are of great importance. Strictly horizontal flows with no temperature variations on isobars are termed *barotropic*, while more general flows with vertical wind variations, non-zero vertical velocity, and horizontal temperature gradients are termed *baroclinic*, as we met earlier.

For baroclinic flow, the concept of vorticity remains useful but must be generalized to include vertical variations of wind and temperature. An analogous quantity, *potential vorticity* is conserved for a fluid element in the absence of forcing, i.e., mixing, friction, or heat transfer. Potential vorticity allows for the generation of absolute vorticity through vertical stretching or shrinking of fluid columns. Mass conservation means that vertical stretching of a column corresponds to horizontal narrowing.

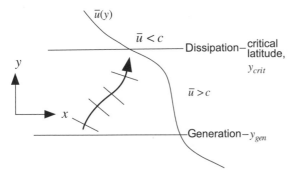

Figure 4.17 A schematic diagram of mean zonal wind $\bar{u}$ as a function of meridional distance y showing regions of planetary wave generation and dissipation. Waves of phase speed c are generated at latitude y_{gen} where $\bar{u} > c$ and propagate to the critical latitude y_{crit} where $\bar{u} = c$. Wave energy is absorbed near y_{crit}. The wave packet trajectories are depicted by the curved heavy line, and wave phase surfaces are shown as the thin lines crossing the trajectories. Eastward flow is accelerated near latitude y_{gen} and decelerated near latitude y_{crit}.

(a)

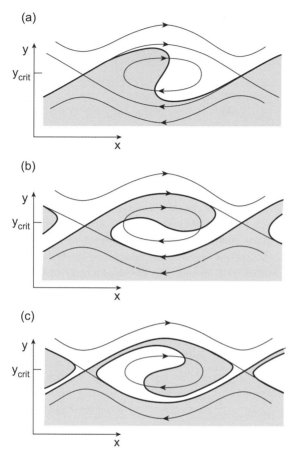

(b)

(c)

Figure 4.18 Depiction of three phases (a)–(c) in the evolution of a breaking planetary wave in zonal (x) and meridional (y) space. The critical latitude is y_{crit} and the y scale is highly exaggerated. Thin lines indicate streamlines and the center, where streamlines are closed, is known as a "Kelvin's cats eye." The thick line is a contour of absolute vorticity, where the shaded region has less absolute vorticity than this contour value. As the wave breaks, vorticity contours curl up tighter and tighter. (Redrawn from Andrews *et al.* (1987), p. 256.)

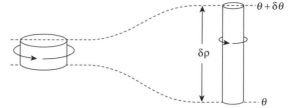

Figure 4.19 Illustration of Ertel's potential vorticity. A fluid column moves isentropically, i.e., within contours of potential temperature as shown. As it stretches vertically, it shrinks horizontally, increasing its spin rate.

that links the concept of vorticity to thermodynamic constraints is called *Ertel's potential vorticity*, Q_P,

$$Q_P = \frac{\eta}{\rho}\frac{\partial \theta}{\partial z} \qquad (4.56)$$

where θ is potential temperature and $\partial \theta / \partial z = -(g\rho)(\partial \theta / \partial p)$ measures static stability, as discussed in Sec. 1.1.3.2. Here, η is evaluated on surfaces of constant θ. Quantity Q_P is often expressed in potential vorticity units, PVUs, of 10^{-6} K kg^{-1} m^2 s^{-1}. In adiabatic flow, an increase in static stability $\partial \theta / \partial z$ corresponds to vertical shrinking and horizontal expansion of fluid columns so that, if Q_P is conserved, the absolute vorticity η must decrease according to eq. (4.56). Correspondingly, decreasing static stability in adiabatic flow corresponds to vertical stretching, horizontal shrinking, and increasing η. Equation (4.56) shows that potential vorticity involves potential temperature variations and so Q_P can be generated and dissipated by diabatic heating as well as by stirring and friction.

A dramatic example of generation of vorticity by heating is a hurricane. Such a cyclonic storm with very low central pressure draws air near the surface inward toward a wall cloud surrounding the core, or "eye," of the storm. In the wall cloud, air rapidly ascends and because tropical storms are very moist at low levels, condensation, precipitation, and rapid release of latent heat energy take place. As a result, atmospheric columns in the eye wall stretch rapidly in the vertical, increasing or maintaining both the high rate of spin (vorticity) and the low level inflow to the eye wall cloud.

Ertel's potential vorticity is not easy to work with analytically. So an alternative is *quasi-geostrophic (QG) potential vorticity* q_g, an approximate form valid when the Rossby number (Ro) is small and flows are approximately geostrophic. QG equations are widely used in dynamics because they facilitate study of large-scale, relatively low-frequency motions characteristic of atmospheric circulations. Derivation of an equation for QG potential vorticity

Horizontal shrinking of columns of spinning fluid increases the spin rate, just as the spin rate of an ice skater increases as she draws her arms in toward her body (Fig. 4.19). Spin is virtually always present and cyclonic because of the dominant cyclonic planetary vorticity.

In a constant density fluid, the potential vorticity P_V [m^{-1} s^{-1}] is

$$P_V = \frac{\eta}{h} = \frac{(f + \zeta)}{h} \qquad (4.55)$$

where h is the incompressible (volume-conserving) height of a fluid column (e.g., Wallace and Hobbs, 2006, p. 289). When density ρ varies with height, an illuminating quantity requires detailed work. The form given here is derived in Satoh (2004), pp. 59–63 and pp. 113–114. See also Andrews *et al.* (1987) pp. 118–122. We skip straight to the result:

$$q_g + q_g' = \eta + \left(\frac{f_0}{\rho_0}\right)\frac{\partial}{\partial z}\left(\frac{\rho_0\theta'}{(\partial\theta_0/\partial z)}\right) = (f_0 + \beta y) + \zeta' + \left(\frac{f_0}{\rho_0}\right)\frac{\partial}{\partial z}\left(\frac{\rho_0\theta'}{(\partial\theta_0/\partial z)}\right) \qquad (4.57)$$

$$\Rightarrow q_g + q_g' = \underbrace{(f_0 + \beta y)}_{\text{basic state}} + \underbrace{\nabla^2\psi' + \left(\frac{1}{\rho_0}\right)\frac{\partial}{\partial z}\left(\rho_0\frac{f_0^2}{N^2}\frac{\partial\psi'}{\partial z}\right)}_{\text{perturbation}} \qquad (4.58)$$

Here, f_0 represents the value of the Coriolis parameter f at a reference latitude ($f = f_0 + \beta y$), N is the Brunt–Väisälä frequency defined in Sec. 1.1.3.3, ρ_0 and θ_0 represent height-dependent zonal and time-mean values of density and potential temperature ($\rho_0 = \rho_s\exp(-z/H)$, where H is density scale height and ρ_s a reference density), and θ' represents horizontal and time-dependent departures from θ_o. In eq. (4.58), the QG potential vorticity is written in terms of a QG perturbation stream function to obtain an equation in a single variable. Like the stream function for horizontal flow, the QG stream function satisfies eqs. (4.45) and (4.46) for geostrophic zonal and meridional winds, i.e., $u_g = -\partial\psi'/\partial y = (-1/f_0\rho_0)(\partial p'/\partial y)$ and $v_g = \partial\psi'/\partial x$. The vertical variation of ψ' ($=p'/(f_0\rho_0)$) is found by differentiating it w.r.t. z:

$$\frac{\partial\psi'}{\partial z} = \frac{1}{f_0}\frac{\partial}{\partial z}\left(\frac{p'}{\rho_0}\right) = \frac{g}{f_0}\left(\frac{\rho'}{\rho_0}\right) \Rightarrow \frac{\partial\psi'}{\partial z} = \left(\frac{N^2}{f_0}\right)\left(\frac{\theta'}{(\partial\theta_0/\partial z)}\right) \qquad (4.59)$$

Here, we used $N^2 = (g/\theta_0)(\partial\theta_0/\partial z)$(eq. 1.40) and a perturbed hydrostatic equation $\partial p'/\partial z = -\rho'g$. Equation (4.59) explains the last term in brackets in eq. (4.57) vs. eq. (4.58).

It turns out that that there's a "free lunch": absolute vorticity η can be formally replaced by QG potential vorticity q_g. So, after linearization about the zonal mean flow $\bar{u}(y,z)$, the vorticity equation (4.41) can be replaced by its exact QG analog:

$$\left(\frac{\partial}{\partial t} + \bar{u}\frac{\partial}{\partial x}\right)q_g' + v'\frac{\partial\overline{q_g}}{\partial y} = S_q' \qquad (4.60)$$

Then, after multiplying by the basic state density ρ_0, eq. (4.42) is similarly replaced by

$$\rho_0\overline{v'q_g'} = -\frac{\partial\rho_0\overline{(q_g')^2}/\partial t}{2\,\partial\overline{q_g}/\partial y} + \frac{\rho_0\overline{S_q'q_g'}}{\partial\overline{q_g}/\partial y} \qquad (4.61)$$

$\underbrace{\phantom{\rho_0\overline{v'q_g'}}}_{\substack{\text{poleward potential}\\\text{vorticity transport}\\\text{by eddies}}} \quad \underbrace{\phantom{\frac{\partial\rho_0}{2}}}_{\substack{\text{potential vorticity}\\\text{variance term}}} \quad \underbrace{\phantom{\frac{\rho_0 S}{\partial}}}_{\text{eddy forcing term}}$

Here, the zonal mean QG potential vorticity gradient is

$$\frac{\partial q_g}{\partial y} = \beta - \frac{\partial^2\bar{u}}{\partial y^2} + \left(\frac{1}{\rho_0}\right)\frac{\partial}{\partial z}\left[\left(\rho_0\frac{f_0^2}{N^2}\right)\frac{\partial\bar{u}}{\partial z}\right] \qquad (4.62)$$

One key physical change is that the eddy vorticity forcing term S_q' in eq. (4.61) now includes contributions from eddy potential vorticity generation by heating.

In addition, the relationship between zonal momentum flux and absolute vorticity fluxes (eq. (4.47)) is replaced by its analog for QG potential vorticity,

$$\overline{v'q_g'} = -\frac{\partial\overline{(u'v')}}{\partial y} + \left(\frac{f_0}{\rho_0}\right)\frac{\partial}{\partial z}\left(\frac{\rho_0\overline{v'\theta'}}{(\partial\theta_0/\partial z)}\right) \qquad (4.63)$$

We see that the meridional potential vorticity flux has a contribution from the meridional flux of zonal mean momentum $\overline{u'v'}$, as with absolute vorticity, but there is now also a contribution from the zonal mean meridional heat flux represented by $\overline{v'\theta'}$.

Analogous to the way we looked for wave solutions of (4.49), it is possible to look for wave solutions of eq. (4.60), setting $S_q' = 0$, as follows,

$$\psi' = e^{(z/2H)}\Psi(y,z)e^{i(\omega t - kx)} \qquad (4.64)$$

where H is the density scale height, ω an angular frequency, and k is the zonal wavenumber. Mak (2011), pp. 166–172, provides a detailed discussion of such wave solutions to eq. (4.60). We note that the wave propagates in the x-direction with amplitude that depends on height. The factor $\exp(z/2H)$ anticipates that the amplitude of the wave perturbation should grow as altitude increases with the inverse square root of the background density ρ_0. In the absence of dissipation, the wave kinetic energy density $E \sim \rho_0\psi'^2$ is independent of height (cf. ½(mass) × (velocity)^2). So, if this energy E remains constant with altitude, the amplitude of the perturbation grows as $\psi' \sim \rho_0^{-0.5}$. Since density ρ_0 varies with altitude in the barometric law as $\sim\exp(-z/H)$, the amplitude variation goes as $(\exp(-z/H))^{-0.5}$, which is $\exp(z/2H)$. Such a relationship has consequences for atmospheric chemistry because eddy diffusion (at least to certain altitudes) is often parameterized in photochemical models of planetary atmospheres proportional to $\rho_0^{-0.5}$. The same $\rho_0^{-0.5}$ relationship also applies to the vertical amplitude growth of buoyancy (gravity) waves (Sec. 4.4.2).

The 1-D wave equation for the stream function complex amplitude in barotropic flow (eq. (4.53)) with a 1-D refractive index can also be replaced by a 2-D analog for the amplitude of the quasi-geostrophic stream function Ψ (y, z) that incorporates a 2-D squared refractive index. The result is analogous to the results with horizontal flow. But instead of turning latitudes and critical latitudes, there are *turning surfaces* where refraction occurs for specified ω, k where $(\bar{u} - c)$ is large, and *critical surfaces* where $(\bar{u} - c)$ approaches zero, the wave oscillation rate is large, and waves are absorbed (Fig. 4.20). Andrews *et al.* (1987), pp. 183–186 provides more discussion on this topic.

Unfortunately, the analogy between absolute vorticity and potential vorticity expressed in eqs. (4.42) and (4.61) is imperfect (there is no totally free lunch) because absolute vorticity and potential vorticity are not the same quantities and their fluxes correspond to different responses of the zonal mean flow. In the QG-limit, acceleration of the zonal-mean zonal wind depends on the poleward potential vorticity flux (see Ch. 7 of Vallis (2006) for discussion). But the 3-D case is more complicated than the 2-D case because potential vorticity is proportional both to absolute vorticity and static stability (eq. (4.56)), so its poleward flux corresponds to some combination of increase of the mean absolute vorticity and the mean static stability of the background flow in the region poleward of the latitude of the flux. Nevertheless, the general tendency is for prograde jets to be forced in regions in which eddies are generated and retrograde jets to be forced in regions in which eddies are dissipated (Held, 1975; Thompson, 1971).

4.3.5   Planetary Wave Instability

When the restoring effect of the potential vorticity gradient vanishes, gas parcels do not undergo stable wave-like motions in the meridional plane, but can drift away from their initial latitude and altitude. Such motions are said to be unstable. *Instability* of planetary waves can occur where $\partial \bar{q}_g / \partial y$ as expressed in eq. (4.61) changes sign. Sign changes of $\partial \bar{q}_g / \partial y$ (given by eq. (4.62)) occur with some combination of large positive horizontal curvature of the zonal mean flow $\partial^2 \bar{u} / \partial y^2$ or large negative curvature or large positive vertical shear of the mean zonal wind, $(1/\rho_0) \partial / \partial z (\rho_0 B \partial \bar{u} / \partial z)$, where $B = f_0^2 / N^2$. Through the thermal wind (eq. (4.18)), $\partial \bar{u} / \partial z$ is linked to the horizontal gradient of zonal mean temperature. When $\partial^2 \bar{u} / \partial y^2$ is the dominant factor in eq. (4.62), the growing unstable eddies derive their energy from the kinetic

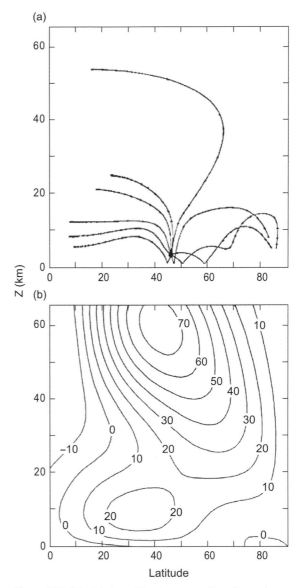

Figure 4.20 Calculated meridional cross-section of rays (wave packet trajectories) for stationary Rossby waves, i.e, where phase speed c, relative to the surface is c = 0. The diagram depicts Earth's wintertime middle atmosphere. Crosses mark daily time intervals. (a) Rays refract away from the zonal mean ($\bar{u}$) wind maxima shown in (b) and toward zonal mean wind minima (the $\bar{u} = 0$ contours) where $\bar{u} - c$ approaches zero. (b) Zonal mean wind distribution. Because zonal wind is the major factor in refractive index, most rays are refracted toward weak zonal mean winds and away from strong zonal mean winds. (Redrawn from Andrews *et al.* (1987), p. 186.)

energy of the zonal mean flow and the instability is referred to as *barotropic instability* or *Rayleigh-Kuo instability*. When $(1/\rho_0) \partial / \partial z (\rho_0 B \partial \bar{u} / \partial z)$ is the dominant factor, unstable eddies draw their energy from the mean meridional gradient of temperature and the instability is referred to as *baroclinic instability*, as mentioned earlier.

The effect of planetary wave instabilities on the mean zonal flow depends on the statistical properties of the life cycles of unstable eddies as they grow, propagate, and dissipate. Eddy growth corresponds to vorticity that generates stirring and prograde acceleration of zonal mean flow. Propagation carries eddy wave packets to altitudes and latitudes separated from the altitude and latitude of origin, while eddy dissipation corresponds to vorticity dissipation and retrograde zonal mean flow acceleration.

Barotropic instability accelerates prograde flow in the latitude–altitude region in which the instability is centered. Large positive values of $\partial^2 \bar{u}/\partial y^2$ often occur in the negative shear zones (i.e., where the zonal wind decreases to the right of the flow) on the flanks of prograde jets where the vorticity due to the shear is anticyclonic. Barotropic instability tends to reduce the positive curvature in shear zones, noting that if a jet stream follows the trough of the wave, there is positive (cyclonic) vorticity due to curvature on both the north and south side of the jet. As unstable eddies develop, they propagate to latitudes where they decay and accelerate retrograde zonal mean flow. The scale of the most unstable eddies, which will be close to the scale of the dominant planetary waves produced by the instability, depends on the strength of the zonal mean wind and the β effect, corresponding to the condition $\partial^2 \bar{u}/\partial y^2 > \beta$. There is a meridional length scale associated with this instability, called the *Rhines radius*, $r_{\text{Rh}} \sim (2U/\beta)^{1/2}$, where U is the typical speed of the jet (Rhines, 1975). In fact, the number of bands on the face of a gas giant is roughly given by a semi-circle circumference divided by the Rhines radius, $\sim (\pi a)/r_{\text{Rh}}$ (Cho and Polvani, 1996; Menou *et al.*, 2003). This scaling for banding applies if the flow is highly turbulent and forced at small scales by baroclinic instabilities or convection but can be violated on bodies such as hot Jupiters where flow is forced at more global scales.

For baroclinic flows in the case where vertical curvature of the mean zonal wind (i.e., $\partial^2 \bar{u}/\partial z^2$) and vertical variations of the factor $B = f_0^2/N^2$ are small, the baroclinic term in eq. (4.62) reduces to:

wind eq. (4.18) indicates that this vertical shear corresponds to a horizontal gradient of ~5 °C over 20° latitude, which is much less than the temperature gradient expected from solar heating. Consequently, the production of turbulent eddies from baroclinic instability is inevitable in the midlatitudes of Earth-like planets.

Baroclinic instability (e.g., Pierrehumbert and Swanson, 1995) is fostered by the constraint of the lower boundary where the vertical velocity vanishes, and it is most commonly found near a lower solid boundary of an atmosphere in regions of strong vertical shear of the mean zonal wind. The horizontal scale of the most unstable eddies in this instability is proportional to the *Rossby radius of deformation*,

$$r_{\text{Ro}} = \frac{NH}{f_0} \qquad (4.66)$$

where H is the scale height, N is the Brunt–Väisälä frequency, and f_0 is a reference Coriolis parameter. Equation (4.66) indicates that eddies decrease in size with increasing planetary rotation rate, which should be relevant for interpreting exoplanet atmospheres.

We can estimate Rossby radii. In Earth's midlatitude troposphere, $r_{\text{Ro}} \sim 700$ km, using $f_0 \sim 10^{-4}$ s^{-1}, $H \sim 7$ km, and $N \sim 10^{-2}$ s^{-1} (see Sec. 1.1.3.3). With a typical mean zonal wind of $\bar{u} \sim 10$ m s^{-1}, the time for a disturbance to grow is ~700 km/10 m s^{-1} ~ 1 day, which agrees with observations of weather systems. Since $r_{\text{Ro}} \sim 10^3$ km, only one eddy-driven jet can fit within Earth's baroclinic midlatitudes. On Jupiter, $r_{\text{Ro}} \sim [(0.003 \text{ to } 0.02) \times 27] / [2 \times (2\pi / 35730) \times \sin(45°)] \approx 325$–2200 km, which is far smaller than the planetary size and allows for multiple jets. Here, we have used Jupiter's rotation period of 35 730 s, a scale height of $H \sim 27$ km at the 1 bar level, and a range of N from ~0.003 s^{-1} measured by the *Galileo* probe in an unusual cloudless spot (Magalhaes *et al.*, 2002) to ~0.02 s^{-1}, which might be more typical (Shetty and Marcus, 2010).

The Rossby radius of deformation has the following interpretation. Suppose that an eddy develops and generates vorticity by vertical stretching and corresponding horizontal divergence that spreads the eddy laterally. Initially, when the eddy is small, it is constrained primarily

$$\left(\frac{1}{\rho_0}\right) \frac{\partial}{\partial z} \left[(\rho_0 B) \frac{\partial \bar{u}}{\partial z} \right] \approx \left(\frac{B}{\rho_0}\right) \left(-\left(\frac{\rho_0}{H}\right) \frac{\partial \bar{u}}{\partial z} + \rho_0 \frac{\partial^2 \bar{u}}{\partial z^2} \right) \propto -\left(\frac{1}{H}\right) \frac{\partial \bar{u}}{\partial z} \qquad (4.65)$$

Here, we have used the barometric law $\rho_0 = \rho_s \exp(-z/H)$ and differentiated w.r.t. z using the product rule. Thus, instability occurs with sufficiently large positive vertical shear of the mean wind, $\partial \bar{u}/\partial z$, e.g., when $\partial \bar{u}/\partial z$ exceeds ~1 m s^{-1} km^{-1} in terrestrial midlatitudes. The thermal

by the static stability factor N, but as it continues to grow, it "feels" the constraining effect of the planetary vorticity f, which limits further lateral extent as the eddy begins to spin. The distance from the origin of the eddy at which planetary vorticity limits the spread is $\sim r_{\text{Ro}}$.

4.3.6 Eddy-Driven Jets on the Outer Planets: Shallow Layer Atmospheres

The jet structure on Jupiter and Saturn with strong eastward equatorial jets and alternating eastward and westward jets at higher latitudes (mapped from cloud motions, Fig. 4.7) is a puzzle that historically has provoked two classes of explanations. The first is based on winds in shallow layers overlying horizontally uniform, convective interiors; the second concerns winds that are surface manifestations of atmospheric dynamics that extend deeply (Del Genio et al., 2009; Dowling, 1995; Ingersoll et al., 2004; Vasavada and Showman, 2005).

The giant planet atmospheres are difficult to model because of their large horizontal extent, great depth, uncertainty about properties and processes in the deep atmospheres, and relatively small Rossby radius. Explicit models must use horizontal scales well below the Rossby radius, which requires huge computing power. Consequently, simpler models are often used. One of these is a barotropic *shallow layer atmosphere* in which the layer depth D_L, satisfies $D_L \ll r_{\text{Ro}}$ and $D_L \ll r_{\text{Rh}}$. According to one view, the winds are shallow and they are weak below a "weather layer" extending to a depth where the pressure is ~10 bar (Ingersoll and Cuzzi, 1969).

A shallow barotropic atmospheric layer can be stirred by forcing from below. We have discussed the role of stirring by baroclinically unstable eddies in accelerating prograde and retrograde jets but another stirring process may be applicable to the outer planets. Forcing can be imposed by convective elements driven by convective instability in the deep atmosphere impinging on the statically stable upper layer and spreading laterally. The spread will continue until its predominantly divergent motion is converted to rotational motion through the influence of planetary rotation (the term $\eta\Delta$ in eq. (4.38)). This will occur at a radial distance from a convective core of ~r_{Ro}.

There is direct evidence that convection impinges from below on Jupiter's statically stable layer where we observe the winds. Patches of rapidly growing and evolving cloud are observed. Some patches correspond to lightning flashes, indicative of convective energy release (Little et al., 1999), which is consistent with deep, moist (water) convection. Terrestrial lightning requires both liquid water droplets and frozen particles to undergo collisions that generate charge, followed by the gravitational separation of unlike charges on small and large particles, which creates an electric field. By analogy, Jupiter's lightning suggests that convection extends below the liquid water condensation level at ~5 bars. This is not to say that water vapor is the only condensable gas

that can release latent heat energy on Jupiter. Ammonia and sulfur-containing gases condense too. Molecular hydrogen also comes in *ortho* and *para* states, corresponding to parallel and antiparallel spins of the hydrogen nuclei, respectively, and latent heat release can arise from the conversion between these two states. But water vapor has much larger possible latent heat release and so potential to generate eddies and jets (Lian and Showman, 2010). It follows that the most intense Jovian convective storms are likely to be ~90 km deep, in contrast to Earth where the depth of convection does not exceed ~10 km. There is also evidence for deep convection in the form of rapidly evolving small-scale cloud systems and lightning on Saturn (Dyudina et al., 2010).

If this type of convection is responsible for driving the large-scale jets on Jupiter and Saturn, it must migrate around the planet to cause features spread over longitude. In a possible scenario, slow convection throughout the deep interior of the planet reaches the layer in which water condensation can take place. Convection imposes strong static stability upon its surroundings because of stabilizing downward motion outside of the convective clouds. Thus, most of the atmosphere will present a stable barrier near the base of the condensation layer. Over time, the build up of energy from below will break the barrier allowing localized deep convection to occur. Gradually changing conditions near the barrier or in the atmosphere above might allow regions of convection to migrate.

4.3.7 Eddy-Driven Jets on the Outer Planets: Deep Atmospheres

In contrast to shallow layer models, another category of simulations for giant planets are *deep models*. The huge amount of energy from planetary accretion causes heat to be lost from within giant planets and the high interior opacities lead to convection and thus temperature and density profiles that should be very nearly adiabatic (Guillot et al., 2004). Atmospheres are not shallow and circulation systems must extend deep into the quasi-adiabatic interiors. Since an adiabatic atmosphere cannot support significant horizontal density variations, deep atmospheric regions must also be barotropic so jet stream structures in these atmospheres can extend to great depth. Direct evidence for strong, deep winds of 180 m s^{-1} at 20 bar and 420 K comes from the *Galileo* probe that descended into a Jupiter at an entry point of 6.5° N (Atkinson et al., 1997).

In a deep atmosphere, convection currents combined with the effects of planetary rotation organizes the internal winds into geostrophic flow on cylindrical

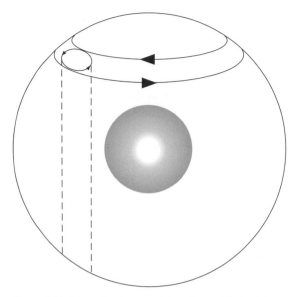

Figure 4.21 Schematic diagram of Busse's model of winds on Jupiter. Models and lab experiments show that the combination of convection and rapid rotation could organize fluids into differentially rotating cylinders that are concentric with the planet's rotation axis. The fluid would move in columns (dashed). The surface manifestation of the rotating columns (small arrows) would be a series of zonal winds that reverse in direction (large arrows).

surfaces parallel to the planetary rotation axis, extending throughout the whole molecular hydrogen layer (Busse, 1976, 1983). The jet stream structure in the upper troposphere will be the upper manifestation of these cylindrical structures (Fig. 4.21). Prograde and retrograde upper tropospheric jets will occur where the cylindrical flow that intercepts the upper troposphere is prograde and retrograde respectively. Cylindrical structures can penetrate through the entire planet at low latitudes where they do not intercept a solid or liquid deep interior. At higher latitudes, they can penetrate only to the solid or liquid interior. The flow in these models is driven by the convective heat flux from the deep interior.

Some models of this interior circulation produce realistic looking results for Jupiter's upper tropospheric jet stream structure, e.g., Heimpel *et al.* (2005), Heimpel and Aurnou (2007). But one fundamental difficulty with these models is that they require larger energy input rates than are available from the Jupiter's combined internal and solar heat fluxes. So they cannot be considered a realistic solution to the problem of explaining the jet stream structures of Jupiter and Saturn. Moreover, the Jupiter-like jet patterns depend on the existence of an impermeable lower boundary condition in these models. So a second basic problem is that giant planets lack the internal boundary that is crucial in these models for getting a Jupiter-like flow.

4.3.8 A Shallow Atmosphere Model Coupled to the Deep Interior of Outer Planets

A third class of models includes shallow atmospheres driven by baroclinic instability, which produce realistic looking simulations of the upper tropospheric jet stream structure (Williams, 1979, 2003); however, the relationships of the upper tropospheric circulation to the circulation in the deep troposphere and the global energy budget of the Jovian atmosphere are not addressed.

Detailed modeling of the complete system including the upper tropospheric circulation and deep atmosphere circulation is still precluded, but Schneider and Liu (2009) have used a hybrid model that may incorporate important effects of both the upper and lower troposphere on the jet stream structure. Their model couples a shallow upper troposphere to the deep atmosphere via large scale overturning circulations generated by eddy forcing in the upper troposphere. They hypothesize that deep meridional circulations transfer momentum downward from the upper troposphere to the partially ionized deep atmosphere at $\sim 10^5$ bar where circulation is damped by *ohmic dissipation*, i.e., when electrical energy turns into heat as electric currents flow through a resistance. Such magnetohydrodynamic (MHD) drag affects the zonal flow only off the equator. The shallow atmosphere eddy-driven jets modeled in the upper troposphere are presumed to extend downward to the dissipation region but the model goes down to only ~ 3 bar, so an artificial drag is imposed to represent the effect of the MHD drag in the outer atmosphere.

Results from an extended run of the Schneider and Liu model that uses the appropriate solar and internal energy input fluxes show stable jet structure resembling Jupiter's at both high and low latitudes (Fig. 4.22). In this model, middle and high latitude jets are forced by baroclinically unstable eddies whose instability arises from the meridional gradient of incoming solar radiation and temperature. Near the equator, strong prograde jet structure is produced by convection associated with the planetary internal heat flux. Convection arising from a uniformly distributed internal heat flux is not strong enough to drive jets at high latitudes, and low-latitude temperature gradients and baroclinic instability are not strong enough to drive the equatorial jet. Both solar heating and internal heat flux are required to produce the low and high latitude jet structure. This hybrid model shows promise for Saturn as well as Jupiter; however, its depth to only 3 bar and dry convective adjustment does not deal with latent heating, which is likely critical for generating horizontal temperature contrasts. Neglect of vertical momentum transport also means that realistic convective plumes are not captured.

(a)

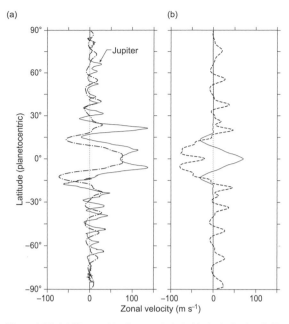

(b)

Figure 4.22 (a) Observed Jupiter zonal winds (dash-dot line) at 0.65 bar compared with zonal winds from the model of Schneider and Liu (2009) (solid line) including both convective forcing and forcing by baroclinically unstable waves. (b) Calculation of Schneider and Liu without baroclinic instability (dashed line) and without internal convection (solid line). (From Schneider and Liu (2009) Reproduced with permission. Copyright 2009, American Meteorological Society.)

4.3.9 Ice Giants: Uranus and Neptune

Given the multiple jet structures and strong low latitude prograde winds found on Jupiter and Saturn, the upper tropospheric winds measured from cloud drifts on the ice giants, Uranus and Neptune, are surprising (Fig. 4.7). On both planets, there is a broad band of strong retrograde winds at low latitudes and a high latitude prograde jet. In fact, the retrograde jet on Neptune reaches ~600 m s^{-1}, the fastest in Solar System atmospheres.

The reasons for the marked differences in circulation between the ice giants and gas giants are unclear. Why is there only one prograde jet at high latitudes on the two outer planets? Why are low latitudes filled with a broad smooth band of retrograde zonal winds? One observation is that heat flux from the interior is much weaker on the ice giants than gas giants (Sec. 2.2.6). There are also physical differences at depth. Water changes into ions (OH$^-$ and H$_3$O$^+$) and ammonia changes into NH$_4^+$ at much shallower depths (>0.1 Mbar) on the ice giants than the >3 Mbar pressure required for the sea of hydrogen ions and electrons inside the gas giants. Thus, the neutral molecular atmospheres of Jupiter, Saturn, Uranus, and Neptune extend to different depths: 0.8, 0.5, 0.8, and 0.85 planetary radii or depths of 15 000 km, 30 000 km, 5500 km, and 3500 km, respectively.

One possibility for the dynamical disparities is that the upward heat fluxes in the upper tropospheres of the two ice giants are too weak to drive convection or generate baroclinic instabilities almost everywhere. Convective forcing may only occur where the static stability in the upper layer is sufficiently weak. On Neptune that could happen at high latitudes where the annual average incoming solar radiation flux is weak, but this idea leaves unanswered the question of why Uranus, whose rotation axis lies close to the ecliptic (98° obliquity), would exhibit zonal winds so closely resembling Neptune's (30° obliquity). Nevertheless, if vertical heat fluxes were confined to high latitudes for whatever reason, prograde jets forced by eddies would be confined to high latitudes as well. Across the tropics and subtropics, retrograde winds could be maintained by persistent vorticity mixing. This picture of the mechanism of the Uranus and Neptune circulations is consistent with the general conclusions of Schneider and Liu (2009) for Jupiter.

4.4 Buoyancy Waves and Thermal Tides

There is another class of waves besides planetary waves that influences the structure of planetary atmospheres. Such waves are central for understanding motions and temperatures in the terrestrial mesosphere of the Earth, for example. In their simplest form, these waves are confined to the vertical and one horizontal coordinate. Buoyancy acts as a restoring force that causes a vertically displaced gas parcel to oscillate about its equilibrium position in a statically stable atmosphere. Such waves are often called *gravity waves*, but because buoyancy is the restoring force, we refer to them as *buoyancy waves*. If the wave is vertically *propagating*, it is said to be *internal*, and if it decays with height and is "trapped" and *evanescent*, it is called *external*. Nappo (2012) proposes using the descriptive terms "propagating" and "evanescent" and refraining from the obscurity of "internal" and "external." We follow this suggestion.

Like planetary waves, buoyancy waves can transport momentum from one region of the atmosphere to another and accelerate or decelerate the atmospheric mean flow far from the point of wave origin. Even though the mechanisms and geometries are different, some key concepts developed for planetary waves apply to the mechanisms by which buoyancy waves affect the mean zonal flow.

4.4.1 Mechanism and Properties of Buoyancy Waves

In Sec. 1.1.3.3, we saw that a gas parcel oscillating in the vertical and subject to the restoring force of buoyancy in a statically stable atmosphere will oscillate at angular frequency $\omega = N$ where N, the buoyancy frequency, is given by eqs. (1.39) and (1.41) as

$$N^2 = \frac{g}{T}\left(\frac{dT}{dz}+\frac{g}{c_p}\right) = \frac{g}{\theta}\frac{d\theta}{dz} \qquad (4.67)$$

Here, θ is potential temperature. Purely vertical oscillations are idealized and in real atmospheres, the simplest approximation for buoyancy waves is an oscillation at some angle to the vertical in an x–z plane. For such waves, Fig. 4.23 shows what is meant by the terminology of a *wave front* of constant phase, which is perpendicular to a *wave vector* for a wave that propagates.

Consider a buoyancy wave oscillation tipped at an angle β with respect to the vertical in the x–z plane, as shown in Fig. 4.24(a). We imagine a parcel of air of unit mass displaced vertical distance δz from A to B up a slope distance of $\delta s = \delta z/\cos\beta$. The parcel will oscillate about point A as part of a propagating transverse wave with a wave vector κ as shown in the inset of Fig. 4.24(a). The oscillation's angular frequency, ω, is the projection of N on the slope, $\omega = N\cos\beta$. Earlier, we saw that the downward buoyancy force was $N^2\delta z$ for a purely vertically

displaced parcel (Sec 1.1.3.3). This force has a component *up* slope A–B of $-(N^2\delta z)\cos\beta$. The acceleration up the slope can also be written $d^2(\delta s)/dt^2 = d^2(\delta z/\cos\beta)/dt^2$. Hence, Newton's second law is:

$$\frac{d^2}{dt^2}\left(\frac{\delta z}{\cos\beta}\right)+N^2\delta z\cos\beta = 0 \;\Rightarrow\; \frac{d^2(\delta z)}{dt^2}+\left(N^2\cos^2\beta\right)\delta z = 0$$
$$(4.68)$$

In eq. (4.68), we recognize simple harmonic motion of the standard form "$\ddot{X}+\omega^2 X = 0$". Consequently, the parcel will oscillate at an angular frequency,

$$\omega = N\cos\beta = \frac{Nk}{\left(k^2+n^2\right)^{1/2}} \qquad (4.69)$$

This equation relates angular frequency to the wave structure, i.e., it is a dispersion relation. In eq. (4.69), we used the wave vector geometry of the Fig. 4.24(a) inset where $\cos\beta = k/|\kappa|$ and $|\kappa| = \left(k^2+n^2\right)^{1/2}$. According to eq. (4.69), as the oscillation tips into the vertical ($\beta = 0°$, $\cos\beta = 1$) the frequency approaches N, fluid particles oscillate vertically, and waves propagate horizontally. If tipped towards the horizontal ($\beta = 90°$, $\cos\beta = 0$), the angular frequency becomes very low and approaches

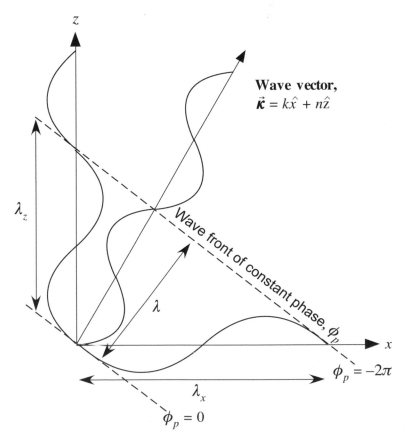

Wave vector,
$\vec{\kappa} = k\hat{x} + n\hat{z}$

Wave front of constant phase, ϕ_p

λ_z

λ

λ_x

$\phi_p = -2\pi$

$\phi_p = 0$

Figure 4.23 Wave vector κ and associated wave fronts and wavenumbers k and n for a 2-D wave of wavelength λ in an $x\,z$ plane. Negative values of phase angle ϕ_p indicate where the wave front passed a stationary observer at (0,0) earlier than the subsequent wave fronts. (Adapted from Nappo (2013), p.16.)

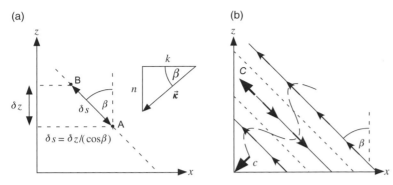

Figure 4.24 (a) A fluid parcel is displaced from point A to B along a slope inclined at angle β from the local vertical. The vertical displacement is δz, while the slope distance is δs. The resulting oscillation has a wave vector $\vec{\kappa}$ with horizontal wavenumber k and vertical wavenumber n, shown in the inset. (b) An x–z plane representation of buoyancy wave phase surfaces. Propagating waves are tipped at the angle β to the local vertical. Gas parcel trajectories relative to the background flow go parallel the phase surfaces. The directions of phase velocity c and group velocity C are indicated by the heavy arrows. Curiously, the vertical components of c and C are in opposite directions: i.e., the phase velocity component is downward when the group velocity component is upward.

zero. This is equivalent to a situation of very strong stratification, i.e., small ω/N. Towards this limit, the fluid particles oscillate almost horizontally and waves propagate almost vertically.

We can derive from eq. (4.69) expressions for the vertical components of the phase velocity c_z and group velocity C_z. The latter is important because the flux of wave energy is in the direction of group velocity:

$$c_z = \frac{\omega}{n} = \frac{Nk}{n\left(k^2 + n^2\right)^{1/2}}, \quad C_z = \frac{\partial \omega}{\partial n} = -\frac{Nnk}{\left(k^2 + n^2\right)^{3/2}}$$
(4.70)

Equation (4.70) shows that C_z and vertical propagation of wave activity vanishes when either n or k is zero, i.e., when the oscillation tips into the vertical or horizontal. It also indicates the peculiar property that the vertical component of the group velocity C_z is opposite in sign to the vertical component of phase velocity c_z. Buoyancy waves propagating wave energy upward will be propagating phase downward, as shown in Fig. 4.24(b).

It is possible to derive a wave equation for buoyancy waves called the *Taylor–Goldstein* equation, analogous to eq. (4.53) for planetary waves using the linearized equations of motion, continuity, and thermodynamic energy in the x-z plane, including a zonal mean wind. The method is to assume small variations from basic state variables of density, vertical velocity, pressure and potential temperature. The interested reader can find the algebra in many textbooks (e.g., Holton, 2004, pp. 196–201); a comprehensive treatment is given in the book by Nappo (2012) devoted to buoyancy waves. Suffice it to say that

perturbations in the state variables follow a waveform with an amplitude that varies as $\sim e^{\left(z/2H_\rho\right)} e^{i(kx-\omega t)}$, where the factor $\exp(z/2H_\rho)$ accounts for the vertical variation of atmospheric density, where H_ρ is the density scale height.

An approximate dispersion relation derived from detailed analysis is the same as eq. (4.69) except that the zonal mean wind Doppler shifts the angular frequency:

$$\omega_r = (\omega - k\bar{u}) \approx \frac{Nk}{\left(k^2 + n^2\right)^{1/2}}$$
(4.71)

$$\Rightarrow n^2(z) \approx \frac{k^2 N^2(z)}{\omega_r^2(z)} - k^2$$
(4.72)

Here, ω_r is the relative angular frequency in coordinates moving with the background zonal flow, $\bar{u}$.

Equation (4.72), like its analog for planetary waves, tells us about the propagation of buoyancy wave packets. Variation of $n^2(z)$ determines the wave behavior. Heights called *turning point levels* are where $n^2(z)$ approaches zero when $\omega_r^2/k^2 = (c - \bar{u})^2$ becomes large and/or when N^2 (representing static stability) becomes small. *Critical levels* are where the local oscillation rate approaches infinity when $\omega_r^2 = (c - \bar{u})^2$ approaches zero. At a critical level where $c = \bar{u}$, the mean flow absorbs the buoyancy wave and the wave cannot propagate any higher.

4.4.2 Wave Generation, Breaking, and Impact on the Zonal Mean Flow

The two main mechanisms for generating buoyancy waves are flow over topography and interaction of large-scale flow with convective cells. In Fig. 4.25, an

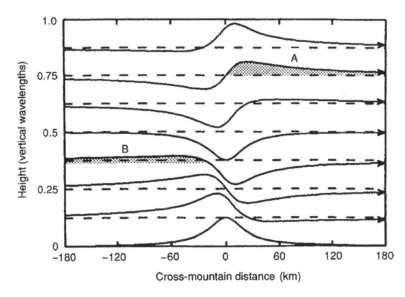

Figure 4.25 A vertically propagating buoyancy wave produced by flow over an isolated ridge line. Phase tilts westward (upstream or to the left) with height if you consider a line going through troughs. Wave perturbation pressure is a maximum on the west-facing slope and minimum on the east-facing slope. Shaded areas A and B are where streamlines are displaced upward from the dashed equilibrium lines are places where clouds may form. (Source: Holton and Hakim (2013), p. 297.)

eastward wind blows across a north–south oriented ridge projecting into the page, producing a *mountain wave*, which is a buoyancy wave generated when stable air passes over a mountain. On the west-facing slopes, pressure, and upward vertical (w') and westward horizontal velocity (u') perturbations are positive while on the eastward slopes, pressure, upward vertical and westward horizontal velocity perturbations are negative. The wave phase tilts backwards with altitude relative to the mean flow, indicating upward flux of westward momentum through the correlation $\overline{u'w'}$ of the wave. The westward momentum extracted from the solid surface acts as a drag on the surface eastward wind. Pressure pushes eastward on the hill and thereby on the solid planet, which is essentially unaffected, but the hill pushes back on the atmosphere, which is affected. The upward flux of westward wave momentum affects the atmosphere in the altitude range over which the wave is dissipated.

Topographically forced waves have phase speeds at or close to $c = 0$, i.e., they're stationary. Non-zero frequencies and horizontal phase speeds occur only because of transience: the initiation, change, or termination of the flow. Similarly, waves produced by interaction with convection generally have horizontal phase speeds close to zero because convective cells tend to move slowly compared to the background flow at the upper levels of the convective cells where buoyancy waves are generated. We therefore focus on the behavior of solutions in which c is close to zero and $\omega_r^2 / k^2 = \bar{u}^2$.

Near turning point levels, buoyancy waves will be refracted toward their level of origin. Thus, turning points can trap buoyancy waves in a waveguide. Such waves can

be easily identified because they persist as wave trains far downstream from the source, usually a topographic obstacle to the flow. Such *lee waves* are often rendered visible by a series of parallel clouds trains on Earth and Mars (Fig. 4.26). Trapped buoyancy waves are typically found beneath jet streams because the rapid increase of background flow $\bar{u}$ with height produces the conditions for trapping.

Buoyancy wave packets of relatively small initial amplitude that escape being trapped in the troposphere propagate upward toward the stratosphere and mesosphere, and there they significantly affect the atmospheric circulation. In Earth's lower mesosphere, there are westward and eastward jets around 50–70 km altitude in the summer and winter hemisphere, respectively (Fig. 4.27). Why do winds diminish and reverse in their direction above 70 km? The answer involves buoyancy waves.

In the winter hemisphere, vertically propagating buoyancy waves with eastward phase speeds will be absorbed at critical levels where $c = \bar{u}$. Small amplitude waves whose phase speed is westward or near zero (the speed of the underlying surface) can penetrate through the tropospheric zonal winds and will break in the lower mesosphere (above 85 km altitude) where their amplitudes increase because of density decrease. These effects are illustrated schematically in Fig. 4.28. The wave breaking causes the winter hemisphere eastward zonal wind speed to decrease with height in the mesosphere and this decrease further enhances wave breaking. Conversely, in the summer hemisphere, tropospheric winds are usually eastward (westerly), but relatively weak so

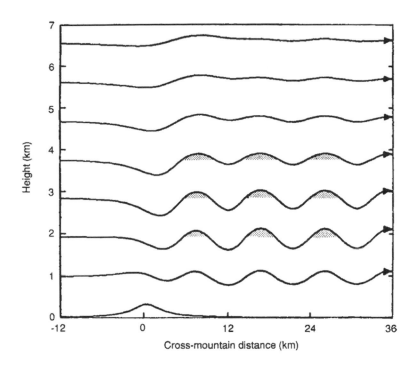

Figure 4.26 Topographically forced buoyancy wave trapped by upward increase in wind and/or upward decrease in static stability. Shaded regions are where so-called lee wave clouds may form. (Source: Holton and Hakim (2013), p. 299.)

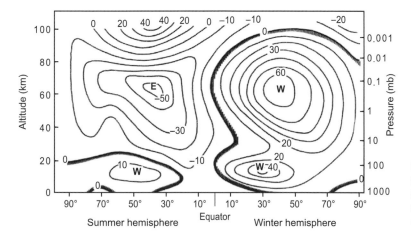

Figure 4.27 Earth's zonal mean winds for winter and summer. E = easterly (westward) wind. W = westerly (eastward) wind. (Redrawn from Andrews et al. (1987), p. 8.)

buoyancy waves generated by flow over topography can propagate through the troposphere. But they encounter a critical layer in the stratosphere and break, modifying the stratospheric temperature and wind structure. Waves whose phase speeds are outside of the range of the background zonal winds penetrate to the mesosphere. In practice, in the summer hemisphere, these waves are those with moderate eastward phase speeds that tend to be produced on the eastward phase speed tail of the convectively generated wave spectrum. Since their initial amplitudes are very low, they break and dump eastward

momentum in the upper mesosphere. Consequently, westward winds in Earth's summer mesosphere peak in a jet the lower mesosphere and decrease with height in the upper mesosphere.

The net effect of upward propagating buoyancy waves on the stratosphere and mesosphere is described as *closing off* all jets: the tropospheric jet, and the eastward and westward jets of the winter and summer mesosphere. The buoyancy waves act as a powerful drag on the background zonal flow and make the upper atmosphere "feel" the rotation speed of the underlying planet.

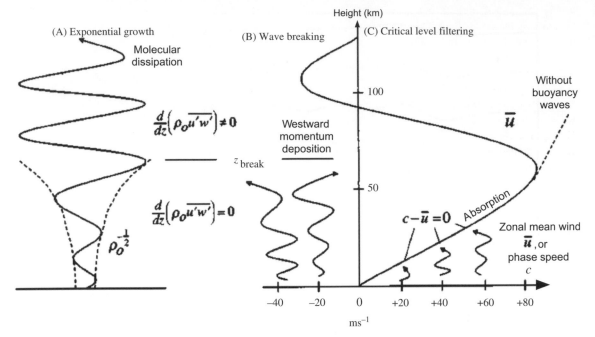

Figure 4.28 A schematic diagram illustrating the effects of buoyancy waves on the Earth's upper atmosphere. (A) Without dissipation, the amplitude of vertically propagating buoyancy waves grows in height as $\rho_0^{-1/2}$. (B) Waves with westward phase speeds carry westward momentum upwards until they break in the upper atmosphere because their amplitudes grow large. This transfers westward momentum to the zonal mean zonal wind, which "closes" a mesospheric jet and reserves the wind speed. (C) Buoyancy waves with eastward phase speeds cannot penetrate beyond critical levels where $c = \bar{u}$. The effect of (B) and (C) is to close an eastward (westerly) mesospheric jet such as that shown in the winter hemisphere of Fig. 4.27. (Adapted from Forbes (2002). Reproduced with permission. Copyright 2002, John Wiley and Sons.)

Buoyancy waves affect the global circulation of the mesosphere. The vertical gradient of the upward flux of horizontal momentum $(\rho_0 \overline{u'w'})$ is a force on the atmosphere that is approximately balanced in the steady-state zonal momentum equation by meridional flow:

$$\underset{\text{Coriolis acceleration}}{-f\bar{v}} \approx \underset{\text{force per kg}}{-\frac{1}{\rho_0}\frac{d}{dz}\left(\rho_0\overline{u'w'}\right)} = F_x \qquad (4.73)$$

Here, $\bar{v}$ is the zonal mean meridional wind. The air flows southward, i.e., $\bar{v}$ is negative, both in the northern hemisphere summer where $F_x > 0$ and $f > 0$ and in the southern hemisphere winter where $F_x < 0$ and $f < 0$. Because of this net summer-to-winter meridional flow, to conserve mass, air ascends at the summer pole and subsides on the winter pole. Because ascent causes adiabatic cooling and subsidence produces warming (Sec. 4.1.2.4), Earth's mesopause at ~90 km altitude is 130–160 K in the summer mid to high latitudes compared to 200–220 K at winter mid to high latitudes. Such a gradient is the opposite of that expected from purely radiative equilibrium and shows the importance of waves.

4.4.3 Atmospheric Tides
4.4.3.1 *Principles of Thermal Tides*

All externally heated rotating planetary atmospheres are subject to gravitational and thermal tides. Atmospheric gravitational tides are generated in the same way as ocean and solid planet tides and they exert torques on underlying solid or liquid layers. Acting over long periods of time, tidal torques can profoundly affect the rotation of a planet. Earth's rotation rate has slowed substantially over the past few billion years as a consequence of the lunar gravitational tide acting on the ocean (Sec. 6.8.2) with potential consequences for atmospheric dynamics and climate. The same side of Venus faces the Earth every time the Earth and Venus are aligned with the Sun in so-called inferior conjunction. It is possible that this Earth-locked rotation of Venus has resulted from the action of the gravitational tides of the Sun and Earth producing surface torque that ultimately slowed the solid planet Venus to its present resonance state (Dobrovolskis, 1980; Dobrovolskis and Ingersoll, 1980; Gold and Soter, 1969; Ingersoll and Dobrovolskis, 1978; Yoder, 1997). Here, we initially examine how thermal tides modify atmospheric structure on relatively short timescales, and then discuss the issue of effects of thermal tides on rotation rate and climate in Sec. 4.4.3.2.

Thermal tides are global oscillations in the atmospheric pressure and wind fields caused by the heating and cooling associated with the apparent motion of the Sun over the surface of a planet. They are very large-scale waves with periods that are some fraction of a solar day, i.e., the noon-to-noon interval. The Fourier component with a solar daylength is the *diurnal tide*, while that with half a daylength is the *semi-diurnal* tide. Tides are classified as *migrating* or *non-migrating*. Migrating tides migrate with the Sun's apparent motion and so have constant phase with respect to solar heating as it moves around the planet once per solar day. Non-migrating tides are generated by interactions between migrating tides and non-uniform topography, such as the planetary scale topographic variations of Earth and Mars. Migrating tidal propagation can only be retrograde because the planet's surface moves underneath in a prograde sense away from the subsolar point of heating. However, non-migrating tides can propagate in either prograde or retrograde longitudinal directions. Because migrating tides are the dominant tides on Earth and Venus, we focus on them.

The theory of thermal tides involves considering buoyancy waves in a spherical geometry that includes planetary rotation. When one develops the linearized wave perturbation equations in this way, a key result is that the global-scale waves can only propagate vertically if their angular frequency ω is greater than the Coriolis parameter, i.e., $\omega > f$ (Lindzen, 1971). Since $f = 2\Omega \sin \phi$, where ϕ is latitude, it follows that the diurnal tide with $\omega = \Omega$ cannot propagate vertically beyond $30°$ latitude, because at that latitude $f = 2\Omega \sin(30°) = \Omega$. At latitudes outside $\pm 30°$, the diurnal tide acts like a Rossby wave and it does not propagate vertically away from its heat sources. In contrast, semi-diurnal tides can vertically propagate at all latitudes because $\omega > f$ for any ϕ.

In a more complete treatment, one considers the daily solar heating variation Q' as a function of latitude, longitude, altitude and time, i.e., $Q'(\phi, \lambda, z, t)$, relative to zonal mean heating. Q' is decomposed into its Fourier components. In turn, these components are expanded as a series of *Hough functions* that depend only on latitude and describe horizontal structure, along with a heating function that depends on altitude (Chapman and Lindzen, 1970). Following Forbes (2002), the magnitude of the atmospheric tidal fields X (where X could be velocity, pressure or density, for example) in the horizontal structure of each tidal mode can be simplified to the following expression:

$$X(\phi, t) = A \cos (\omega t + s\lambda - \varphi) \qquad (4.74)$$

Here, s is the zonal wavenumber and A and φ are latitude-dependent amplitudes and phases. The dominant components are the $s = 1$ component, the diurnal migrating tide, and the $s = 2$ component, the semi-diurnal migrating tide. We can specify the time as a local time seen by an observer on the surface, $t_{LT} = t + \lambda/\Omega$. Substituting for t and setting $s = 1$ or 2 corresponding to $\omega = \Omega$ or 2Ω, respectively, we obtain:

$$X_{\text{diurnal}}(\phi, t) = A_1 \cos (\Omega t_{LT} - \varphi_1) \qquad (4.75)$$

$$X_{\text{semi-diurnal}}(\phi, t) = A_2 \cos (2\Omega t_{LT} - \varphi_2) \qquad (4.76)$$

We see that both diurnal (4.75) and semi-diurnal (4.76) tidal fields are independent of longitude (at least on a uniform sphere). However, the diurnal and semi-diurnal tides have quite different properties over latitude and height. On both Earth and Mars, the semi-diurnal tide behaves as a global buoyancy wave from pole to pole with very large vertical wavelength, typically ~100 km. Its amplitude increases with height, and because it has large vertical wavelength, the response reflects little interference between heating in different layers of the atmosphere. The diurnal tide acts as a buoyancy wave only between latitudes $\pm 30°$. Within this latitude range, the diurnal migrating tide propagates vertically and its amplitude increases with height above the heat source. Because of its relatively short vertical wavelength (~30 km on Earth and Mars), there is significant interference between heating at different levels, and its amplitude is both smaller and less stable relative to its heat sources than the semi-diurnal tide.

4.4.3.2 Observations and Implications of Atmospheric Tides for Planetary Climates

Migrating and non-migrating tides have been observed on Earth (Fig. 4.29) and Mars, and migrating tides have been observed on Venus (Migliorini *et al.*, 2012; Pechmann and Ingersoll, 1984; Peralta *et al.*, 2012; Schofield and Taylor, 1983) (Fig. 4.30). Venus is an interesting case. With its 243 day rotation period and solar day of 117 Earth days, Venus hardly rotates with respect to the Sun, so why should it have thermal tides? Even though the equatorial surface rotates at a speed of only 4 m s^{-1}, the middle atmosphere near the cloud top level of ~65–70 km altitude rotates at ~110 m s^{-1} with respect to both the planet and the Sun. The thermal tides in the middle atmospheric reflect this atmospheric rotation, *not* the rotation of the solid planet. The air rotates past the sub-solar point and migrating waves are established that

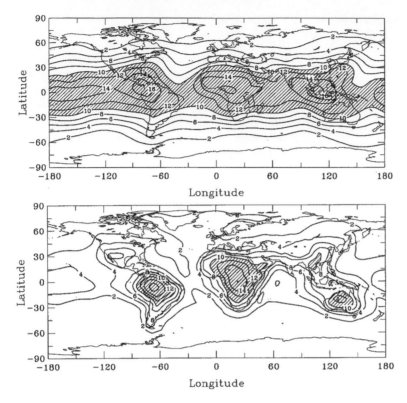

Figure 4.29 The annual average amplitudes of semi-diurnal (top) and diurnal surface pressures on a latitude–longitude map in units of 0.1 mbar = 10 Pa. Each contour is separated by 20 Pa. Note the maxima in low latitudes and longitudinal variability of the diurnal tide due to differences in heating and non-uniform land mass distribution. (Source: Dai and Wang (1999). Reproduced with permission. Copyright 1999, American Meterological Society.)

are stationary with respect to the Sun but move opposite to the direction of the mean flow (Gierasch *et al.*, 1997). Titan's middle atmosphere also rotates at a significant rate, so thermal tides are expected in the middle and upper atmosphere of Titan.

The principal forcing factors for thermal tides depend on the properties of a planet's atmosphere. On Earth, forcing is by stratospheric heating due to absorption of solar radiation by ozone between ~20–60 km, heating in the troposphere by absorption of solar radiation by water vapor, and heating in the tropical troposphere by the diurnal cycle of deep convection. On Mars, the main tidal forcing is the daily cycle of radiative and convective heating of the lowest few km under clear conditions and heating of a deep layer due to absorption of solar radiation by dust when the atmosphere is moderately or very dusty. On Venus, the main forcing for the thermal tides in the middle atmosphere is heating by absorption of solar radiation in the cloud layer between 45 and 70 km (Fig. 4.30). Absorption of solar radiation by methane and other hydrocarbons in the middle atmosphere of Titan should be capable of generating significant thermal tides despite the large distance of Titan from the Sun.

Thermal tides have significant influence on the atmosphere of Mars. Feedback between tidal surface winds and

lifting of dust high in Mars' atmosphere may contribute to the generation of planet encircling dust storms (Leovy *et al.*, 1973). Large amplitude tides may break and also contribute to vertical mixing in the Martian middle atmosphere.

Atmospheric thermal tides are characterized by an asymmetric mass distribution with respect to the substellar point (caused by thermal inertia), which can affect the rotation rate of a planet that's close to its host star (Cunha *et al.*, 2015; Laskar and Correia, 2004; Leconte *et al.*, 2015). The gravitational pull from a star on the asymmetric planetary atmosphere accelerates or decelerates rotation in the atmosphere, and the atmospheric angular momentum can be transferred to the solid planet by frictional coupling (Fig. 4.31). Such an effect probably accounts for the slow retrograde rotation of Venus (Correia and Laskar, 2001, 2003; Correia *et al.*, 2003; Dobrovolskis and Ingersoll, 1980; Ingersoll and Dobrovolskis, 1978) but the same physics could also drive rocky planets away from tidal synchronicity, even for planets with 1 bar atmospheres (Leconte *et al.*, 2015). Since synchronous rotation is assumed for some models of planetary habitability (e.g., Yang *et al.*, 2013) then atmospheric tides potentially could influence habitability.

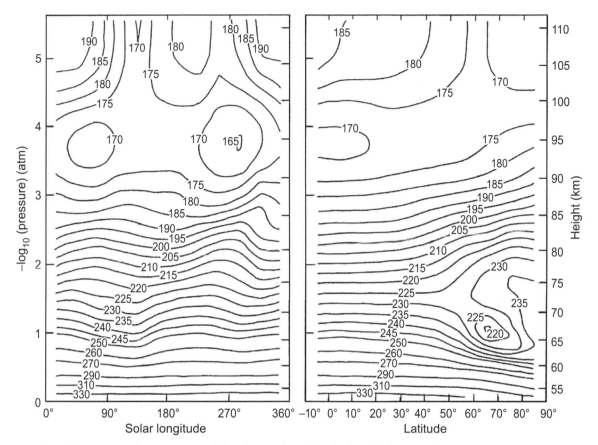

Figure 4.30 Temperature (K) cross-sections of Venus' atmosphere above 55 km altitude averaged between 0–30° N. The field is shown in Sun-following coordinates where zero longitude is noon (left) and in zonal mean cross-section (right). From above the cloud-tops (70 km) to 100 km, a clear semi-diurnal variation is revealed. The phase in the solar longitude plot suggests a Sun-following eastward-propagating oscillation forced at lower altitudes. (Redrawn from Gierasch *et al.* (1997), p. 463.)

4.5 Superrotation

Superrotation is prograde atmospheric rotation at equatorial latitudes. (More specifically, superrotation is a region where the angular momentum per unit mass of the atmosphere about the rotation axis is a local maximum in the latitude–height cross-section and exceeds the surface angular momentum at the equator.) Because the angular momentum of a planet's surface is maximal at the equator, zonal mean torques cannot drive equatorial superrotation (Hide, 1969, 1970), so their generation and maintenance has been something of puzzle. We have already encountered superrotation on Jupiter and Saturn. Superrotation on slowly rotating planets such as Venus and Titan where the equatorial atmosphere rotates much faster than the solid planet is particularly perplexing.

A variety of data indicate superrotation on Venus. Cloud tracking shows superrotating winds (Belton *et al.*,

1991; Peralta *et al.*, 2007). The temperature distribution for Venus in the right panel of Fig. 4.30 can also be used to infer fast winds in the middle atmosphere. A meridional temperature gradient from warm equator to ~50° latitude occurs at 60–80 km altitude. The zonal wind in cyclostrophic balance (eq. (4.21)) with this meridional temperature distribution increases with height at low latitudes to ~110 m s^{-1} at 65–70 km altitude, which is 2.5 times faster than terrestrial jet streams and 60 times faster than the solid planet below. The wind is sometimes called the "four-day rotation."

Evidence also exists for Titan's superrotation. Prior to the *Cassini–Huygens* mission, *Voyager* infrared data (Flasar *et al.*, 1981) and stellar occultation data (Hubbard *et al.*, 1993) implied superrotating zonal winds. Tracking of the *Huygens* probe indicated speeds ~100 m s^{-1} at 130 km height (Bird *et al.*, 2005) while Earth-based

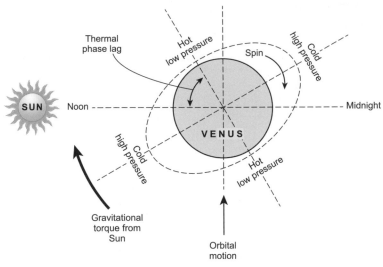

Figure 4.31 Schematic diagram illustrating the principle of how atmospheric tides can affect the long-term rotation rate of a planet. Venus is our example, but the same principle should apply to exoplanets with sufficiently thick atmospheres. Thermal tides redistribute mass between warm and cold parts of the atmosphere. There is a low pressure (i.e., low mass) where the atmosphere is warm and high pressure where the atmosphere is cold. Thermal inertia means that the hottest time is not local noon but later in the afternoon. By symmetry, a maximum of cold, high atmospheric mass occurs before local noon. Because the atmospheric mass wave of the semi-diurnal tide (indicated with the dashed-line oval) has a maximum before noon, the Sun's gravity exerts a torque on this atmospheric mass. When the torque is transmitted via friction to the solid planet below, it accelerates the rotation in the prograde sense shown. Solid body tides raised on Venus by the Sun produce a solid bulge, which the spin of Venus tends to move ahead of the line connecting the Sun and Venus. The Sun, acting on this solid body bulge, tends to cause a torque in the opposite (retrograde) direction to that produced by the atmospheric tidal torque. The solar torque due to solid body tides, on its own, would tend to cause Sun-synchronous rotation. But the solar torque due to thermal atmospheric tides can impart net rotation and prevent synchronous rotation. (Adapted from a diagram by Schubert (1983), p. 720. Reproduced with permission from University of Arizona Press. Copyright 1983, The Arizona Board of Regents.)

spectroscopic measurements suggest ~200 m s^{-1} (with large error bar) at 200 km (Kostiuk *et al.*, 2006). Infrared measurements from *Cassini* in 2004 (soon after northern winter solstice in October 2002) indicate a jet near 0.1 mbar with a core of 190 m s^{-1} spanning 30–60° N (Achterberg *et al.*, 2008).

Net torque is required to maintain the angular momentum of superrotation against frictional loss to the surface. A general proposal is that eddies could provide the necessary torque (e.g., Gierasch, 1975), although gravitational torque acting on the atmosphere is an alternative (Gold and Soter, 1971). Eddies may act primarily in the horizontal plane as planetary or Rossby waves, or they can act largely in the vertical plane as buoyancy waves. In either case, they must provide a net convergence of angular momentum on the equator at the jet level (Del Genio and Zhou, 1996; Mitchell and Vallis, 2010; Showman and Polvani, 2011). In the case of Jupiter and Saturn, the required equatorward momentum fluxes may be primarily due to convectively driven large-scale eddies

acting in the horizontal (Lian and Showman, 2010; Schneider and Liu, 2009).

The mechanism on Venus is unclear. Some models produce strong superrotation due to horizontal eddy torques while others generate little or no superrotation (e.g., Del Genio and Zhou, 1996; Hollingsworth *et al.*, 2007; Lebonnois *et al.*, 2010; Yamamoto and Takahashi, 2009). None of the models is able to simulate the entire Venus atmosphere at the scale of buoyancy waves and convection, and numerical studies generally have to assume a background superrotation at the cloud base. Superrotation also reduces the effective gravity causing a equatorial bulge, which should require a quasi-hydrostatic treatment (Tokano, 2013).

While some models may suggest that horizontal eddics can drive Venus' superrotation, the coincidence between the level of the equatorial wind maximum and the level of maximum solar heating implies that thermal tides play a role at least in determining the location and magnitude of the superrotation maximum (Gierasch *et al.*,

1997; Leovy, 1987; Pechmann and Ingersoll, 1984). A further indication of the role of thermal tides is the direct observation of the semi-diurnal tide in the atmosphere above the 65–70 km level of the superrotation maximum (Fig. 4.30, left panel). Upward propagation of wave activity related to this tidal component is associated with downward flux of prograde momentum toward the cloud top level. Below that level, observations do not reveal the presence of a tide, but it is likely that the semi-diurnal tide transports prograde angular momentum upward as well as downward toward the level of maximum solar heating near the cloud tops. It is possible that both horizontal and vertical eddies contribute with horizontal eddies forcing the background superrotation ($\sim$40 m s^{-1}) below the jet and thermal tides shaping the jet level winds.

Some models have succeeded in producing superrotation on Titan. However, the mechanism can be unclear (Crespin et al., 2008; Rannou et al., 2004), or the superrotation occurs only under special conditions (Lebonnois et al., 2012; Newman et al., 2011), or the magnitude of the superrotating zonal winds is too low (Friedson et al., 2009; Lora et al., 2015). The coincidence between the levels of maximum solar heating and maximum superrotation suggests that thermal tides again may play a role in determining the location and magnitude of the superrotation peak.

Finally, because two slowly rotating Solar System bodies exhibit superrotation in the middle atmosphere, we pose the following conjecture: Is atmospheric superrotation an inherent property of slowly rotating exoplanets subject to zonal mean and tidal heating? The idea may be testable through Doppler-shifted lines. Also, in extreme cases, superrotation may be detectable as prograde displacement of the temperature field in infrared spectra, like those predicted originally for hot Jupiters (Showman and Guillot, 2002). Shifted hot spots are detected on various hot Jupiters: HD 189733b (Knutson et al., 2007; Showman et al., 2013a), HD 209458b (Zellem et al., 2014), and WASP-43b (Stevenson et al., 2014).

4.6 Transport by Eddy-Driven Circulations

4.6.1 The Brewer–Dobson Circulation and Mesospheric Circulation

The generation or modification of jet stream structure changes the thermal structure according to the thermal wind equation (4.18), and so generates temperature fields that are not in radiative-convective equilibrium. These disequilibrium temperature fields are balanced by compressional heating and expansional cooling due to large-scale subsidence and rising motion (eqs. (4.25) and (4.26)) along with horizontal advection.

In the lower stratosphere of Earth where the subtropical jet stream decreases in amplitude with height as a result of wave driving, temperature decreases from high latitudes to the equator so that there is a temperature minimum, or *cold trap*, at the tropopause on the equator (Fig. 4.32(a)). Since this minimum corresponds to temperatures cooler than radiative-convective equilibrium, in the zonal mean, air rises in the tropical lower stratosphere with compensating stratospheric subsidence on the poleward side of the jet (Fig. 4.32(b)). These rising and sinking motions are not zonally uniform and, particularly in the sinking region, take place in large part in eddies. Nevertheless, the net effect is the *Brewer–Dobson Circulation*: a mean meridional overturning circulation that carries air through the tropical tropopause, poleward in the lower stratosphere at subtropical and middle latitudes, and downward at mid and high latitudes (Brewer, 1949; Dobson, 1956). This circulation maintains a distribution of stratospheric ozone in which the highest ozone concentrations are found at high latitudes despite the fact that ozone is produced photochemically in the tropical stratosphere where sunlight is most intense. Because water vapor or ice particles transported upward in the Brewer–Dobson Circulation must pass through the cold trap, the Brewer–Dobson Circulation also accounts for Earth's very dry stratosphere with only 3–4 ppmv H_2O.

Earlier, we encountered Earth's mesospheric meridional circulation, which is associated with a meridional temperature gradient that is far from radiative equilibrium (Sec. 4.4.2). Similar to the Brewer–Dobson Circulation, wave breaking and eddy dissipation drive this circulation and close off mesospheric jets above the level of the jet maxima.

Temperature distributions on Venus, Mars, and Titan also indicate eddy driven meridional circulations in their middle atmospheres. In Venus' middle atmosphere, temperatures are warmer near the poles than the equator, indicative of subsidence at high latitudes and ascending motions at low latitude (Fig. 4.30, right panel between 70 and 100 km altitude and $\sim$50° latitude and the pole). Such a circulation would rapidly mix constituents vertically through the mesosphere. Since thermal tides are strong on Venus, it is likely that breaking of the tides is primarily responsible for driving this circulation. Similar warm areas near the pole and relatively cool areas at low latitudes are found in high-resolution vertical temperature profiles of the Martian atmosphere (McCleese et al., 2010). This circulation system may be driven by a

(a)

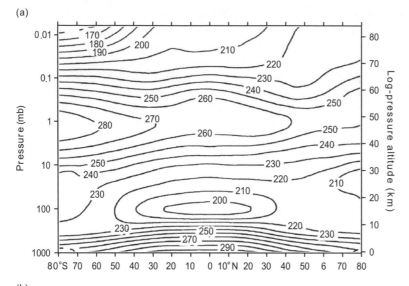

(b)

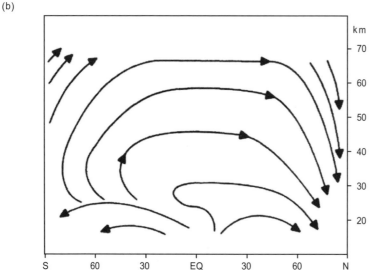

Figure 4.32 (a) January temperature distribution in Earth's atmosphere. Temperature increases from equator to pole in the lower stratosphere and from summer pole to winter pole in the upper mesosphere. (Redrawn from Andrews *et al.* (1987), p. 222.) (b) Schematic streamlines of the circulation at the southern summer solstice deduced from eq. (4.25) and the equation of mass continuity. In the stratosphere, we see poleward motion of air, the Brewer–Dobson circulation, driven by planetary waves. Higher up is another circulation: air rises into the summer mesosphere and descends on the winter pole, which is driven mainly by buoyancy waves (Redrawn from Andrews *et al.* (1987), p. 304.)

combination of thermal tides and buoyancy waves. The mesosphere of Titan also has warmer temperatures near the poles than near the equator and the resulting circulation is responsible for downward transport of photochemically produced hydrocarbons in the polar regions (Fig. 4.33). Since forcing of buoyancy waves is likely to be weak, these circulations may also be driven largely by the thermal tides.

4.6.2 Implications of Large-Scale Overturning Circulations for Atmospheric Evolution

The transport of trace species through the middle and upper atmosphere and their ultimate escape to space can be strongly influenced by the dynamics of the middle

atmosphere. For example, upward transport of water vapor and the subsequent escape of hydrogen on Earth are very sensitive to the temperature of the "cold trap" at the equatorial tropopause (Fig. 5.9). The temperature of this cold trap is controlled by the Brewer–Dobson Circulation, which ascends and cools near the equator. But where does the Brewer–Dobson Circulation come from? It is forced by planetary and buoyancy wave activity, as described above. Dynamically modulated cold trapping in the atmospheres of Venus and Mars also influences the upward flux of condensable trace species. Furthermore, the pressure level at the homopause influences the abundance of trace species at the escape level. This pressure level is determined by turbulence arising from the breaking of buoyancy waves and thermal tides (Leovy, 1982b),

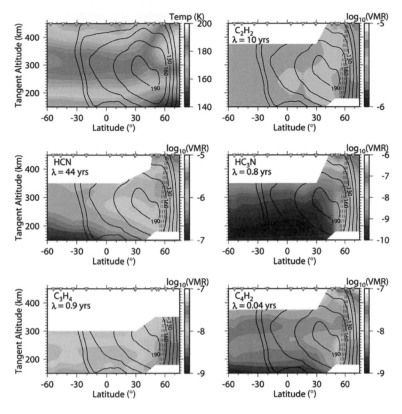

Figure 4.33 Titan zonal mean temperatures (K), cyclostrophic zonal mean winds (m s^{-1}), and trace constituent concentrations (as $\log_{10}$ volume mixing ratio (VMR)) above 150 km altitude. The photochemical lifetimes at 300 km are denoted by λ. The zonal mean winds exhibit strong equatorial superrotation as well as seasonal effects, and the constituents exhibit strong downward transport in the region of the temperature maximum near the winter pole. Blue dashed lines denote a dynamical mixing barrier (a region of steepest horizontal potential vorticity gradient) that confines a polar vortex. (Source: Teanby *et al.* (2008). Reproduced with permission. Copyright 2008, John Wiley and Sons.) (A black and white version of this figure will appear in some formats. For the color version, please refer to the plate section.)

so the dynamics of these waves can potentially affect the escape of trace gases through their influence.

One-dimensional (1-D) models of the composition of middle atmospheres generally employ an eddy diffusion scheme to account for vertical mixing (Appendix B), but the mechanism of vertical exchange is not well understood, so these schemes are usually *tuned* to give overall agreement with observations. Vertical transport in planetary middle atmospheres takes place largely through large-scale wave driven overturning circulations (Sec. 4.6.1). The role of these circulations has yet to be incorporated in a sophisticated way into 1-D photochemical models. Understanding their general behavior is important for attempting to model exoplanet photochemistry, including the formation of hazes that may be common features in some exoplanet atmospheres.

Large-scale wave-driven overturning circulations may even control the bulk composition of planetary atmospheres. For example, CO_2 on Mars is maintained against photodissociation by vertical transport that rapidly brings CO and O down from the mesosphere to the lower atmosphere where recombination to CO_2 is catalyzed (Sec. 3.4). The eddy diffusion approach requires an unreasonably large eddy diffusion coefficient to account for

this transport, but transport by middle atmospheric circulations and episodic planetwide dust storms (which are likely associated with atmospheric tides (Sec. 4.4.3.2) can plausibly do the job.

4.7 Atmospheric Dynamics and Habitability: Future Prospects

We close this chapter by noting that atmospheric dynamics can affect atmospheric evolution and the habitability of planets in many ways. The various possibilities are yet to be fully explored and earlier we only identified a few cases. In general, the meridional transport system and the formation of jets (whether thermally forced or forced by planetary waves) influences cloud distribution, the surface climate, the transport of trace species through the atmosphere, and the movement of sand and dust near the surface, and hence even the geology in some cases. An example of geological influence is on Mars, where surface winds and meridional transport determine regions of long-term dust deposition and erosion (Haberle *et al.*, 2003). In this chapter, we have pointed out how the meridional circulation (Sec. 4.2.6), atmospheric tides (Sec. 4.4.3.2), and wave-driven overturning circulations in the middle atmosphere can be important for atmospheric evolution or planetary habitability (Sec. 4.6.2). There are likely many more links between atmospheric dynamics and atmospheric evolution and habitability.

Ultimately, we would like to be able to use our dynamical understanding to infer properties of the general circulation and climate of exoplanets (Showman *et al.*, 2010; Showman *et al.*, 2013b). The principles outlined above are a starting point for predicting the distribution and character of jet streams, overturning circulation, and vertically and horizontally propagating waves, all of which may influence the habitability of a planet and the atmospheric evolution.

5 Escape of Atmospheres to Space

So far, our discussion of atmospheric evolution has concentrated on atmosphere and climate fundamentals. Climate constrains possible life and, as we will see later in this book, the way that climate is thought to have evolved can explain many environmental differences between Earth, Venus, and Mars. Climate is closely tied to the composition of a planet's atmosphere, which determines the greenhouse effect. Consequently, to understand how climate has changed over time, we must consider how atmospheric composition has evolved. In turn, we must examine how atmospheric gases can be lost.

Gases are lost at an atmosphere's upper and lower boundaries: the planet's surface and interplanetary space. In this chapter, we consider the latter. Studies of the Solar System have shown that some bodies are vulnerable to atmospheric escape (Hunten, 1990). Indeed, many smaller objects, e.g., most moons and essentially all asteroids, are airless because of escape, making the theory of atmospheric escape crucial for explaining differences in surface volatiles. Escape processes can help us understand the lack of atmospheres on the Moon and Mercury, the barren nature of the Galilean satellites versus Titan (Griffith and Zahnle, 1995; Gross, 1974; Zahnle *et al.*, 1992), why the atmosphere of Mars is thin (Brain and Jakosky, 1998; Melosh and Vickery, 1989; Zahnle, 1993b), the red color of the Martian surface (Hartman and McKay, 1995; Hunten, 1979c), the lack of oceans on Venus (Kasting and Pollack, 1983) (see Ch. 13), and possibly the oxidizing nature of the Earth's atmosphere and surface (Catling *et al.*, 2001) (See Ch. 10).

We can group various types of atmospheric escape into three categories following Catling and Zahnle (2009). (i) *Thermal escape* is when irradiation from a parent star (or, less commonly, a very high heat flux from a planet or moon interior) heats an atmosphere, causing atmospheric molecules to escape to space. Two end-member approximations of thermal escape are appropriate under different circumstances: *Jeans' escape*, where individual molecules evaporate into a collisionless exosphere, and *hydrodynamic escape*, which is a bulk outflow with a velocity driven by atmospheric heating that induces an upward pressure gradient force (e.g., Johnson *et al.*, 2013d; Walker, 1982). (ii) *Suprathermal* (or *nonthermal*) *escape* is where individual atoms or molecules are boosted to escape velocity because of chemical reactions or ionic interactions. Finally, (iii) *impact erosion* is where atmospheric gases are expelled *en masse* as a result of large body impacts, such as the cumulative effect of asteroids hits. Of these three types, nonthermal escape is generally slow because if it were fast the molecules would collide and the escape would be in the thermal category. Theory suggests that the two mechanisms that can most efficiently cause substantial atmospheric loss are hydrodynamic escape driven by stellar irradiation (Lammer *et al.*, 2008; Sekiya *et al.*, 1981; Sekiya *et al.*, 1980a; Watson *et al.*, 1981; Zahnle *et al.*, 1990; Zahnle and Kasting, 1986) and impact erosion (Griffith and Zahnle, 1995; Melosh and Vickery, 1989; Walker, 1986; Zahnle *et al.*, 1992). In addition, hydrodynamic escape from early hydrogen-rich atmospheres on the terrestrial planets is relevant for observations of noble gases and their isotopes, as discussed in Ch. 6, because such escape can drag along heavier gases.

In this chapter, we focus particularly on the escape of hydrogen, for two reasons. First, hydrogen is the lightest gas and consequently the most prone to escape. Second, later in the book, we will see that substantial loss of hydrogen can affect the redox chemistry of a planet's atmosphere and surface, changing the chemical character of a planet. Rocky planets, as a whole, become more oxidized when hydrogen escapes to space. This oxidation occurs irrespective of whether the hydrogen is transported through the atmosphere as H_2, H_2O, CH_4, HCN, NH_3, or some other H-bearing compound. Oxidation occurs because the hydrogen atom that escapes ultimately derives from some oxidized form of hydrogen such as water (H_2O), water of hydration in silicate rocks (–OH), or hydrocarbons (–CH). It was in these compounds that hydrogen was originally incorporated into planets like the Earth. Consequently, when hydrogen escapes, matter somewhere on a planet's surface or subsurface is irreversibly oxidized.

Oxidation is most obvious if we consider hydrogen that escapes after atmospheric water vapor undergoes photolysis. Consider water vapor photolysis and escape in the upper atmosphere of the Earth. In this case, the oxygen left behind can oxidize the Earth's surface so that any further oxygen produced (by photolysis and hydrogen escape) is less likely to be taken up by the crust and more liable to remain in the atmosphere. However, today's abiotic production rate of oxygen is $\sim 10^2$ times smaller than the rate of O_2 production from photosynthesis and, hence, plays a negligible role in the atmospheric oxygen budget. It is nonetheless important to understand such abiotic oxygen, both because of its possible effect on very early life on this planet and because of its future significance in interpreting spectra that may be obtained from exoplanets.

The effect of the escape of hydrogen in oxidizing surfaces is also widely considered to be responsible for the oxidized states of Venus and Mars, as illustrated by the red color of the Martian surface (Hartman and McKay, 1995; Hunten, 1979c). Ancient hydrogen escape has also been proposed as a means of oxidizing the Earth's atmosphere, crust, and mantle (Catling et al., 2001; Kasting et al., 1993a; Zahnle et al., 2013) (see Ch. 10).

5.1 Historical Background to Atmospheric Escape

The idea of the escape of gases from the Earth's atmosphere is as old as kinetic theory and has an unusual history. A Scottish amateur scientist, John Waterston (1811–1883), first developed a theory of gases in which the mean kinetic energy of each species was proportional to temperature, and he also introduced the notion of atmospheric escape (Haldane, 1928, pp. 209–210). However, the Royal Society rejected Waterston's paper describing kinetic theory in 1845, and it remained unknown until Lord Rayleigh rediscovered the manuscript in 1891. By then, Waterston's ideas had been overtaken by the work of Clausius, Maxwell, and Boltzmann, while Waterston disappeared in 1883, presumed to have drowned near Edinburgh.

Later, the Irish physicist George Stoney (who gave us the term *electron*) understood that a few gas particles in the high-velocity tail of a Maxwell–Boltzmann distribution of velocities would have sufficient energy to escape from a planet's upper atmosphere even if an average particle did not (Stoney, 1898, 1900a, b, c, 1904). This process is nowadays called *Jeans' escape* after Sir James Jeans, who described its physics in *The Dynamical Theory of Gases* (1954, first edition 1904). At that time, in the early twentieth century, balloon soundings in Earth's lower atmosphere were extrapolated to the entire upper atmosphere, which was assumed to be isothermal at ~220 K. The hot, 1000–2000 K thermosphere was unknown. Consequently, Jeans incorrectly calculated an exceedingly low escape rate of hydrogen.

Later, the Space Age provided data from rocket soundings. As a result, in the next major treatment of atmospheric escape, Spitzer (1952) corrected Jeans' earlier mistake by using more realistic thermospheric temperatures. From the 1950s to the present day, data have become directly available on the number density of hydrogen and the temperature in the upper atmosphere. Measurements include satellite drag through the thermosphere, in situ mass spectrometer measurements, and images of the *geocorona*, which is a glow at the Lyman-α wavelength (121 nm) caused by resonant scattering of solar ultraviolet (UV) by a cloud of atomic hydrogen that surrounds the Earth. UV images taken by spacecraft show the hydrogen atoms. Atoms are on ballistic trajectories back to Earth, escaping, or in orbit (Fig. 5.1).

For astrobiology, we note that about half of the hydrogen atoms seen in Fig. 5.1 derive from decomposition of methane (CH_4), ~90% of which enters the atmosphere from the biosphere. Most of the other half of the H atoms originates from the photodissociation of water vapor. In Fig 5.1, we catch a glimpse of some of the 93 000 tonnes of hydrogen that escape each year (or 3 kg/s) from the Earth.

In the past 60 years, planetary exploration and astronomy have widened our perspective of both atmospheric escape and of *aeronomy*, the study of processes in the rarefied atmosphere from the stratosphere to interplanetary space. Space science led to the recognition of suprathermal escape, hydrodynamic escape, and impact erosion, as discussed in various reviews (e.g., Ahrens, 1993; Chamberlain, 1963; Hunten, 1990, 2002; Hunten and Donahue, 1976; Hunten et al., 1989; Johnson et al., 2008c; Lammer, 2013; Shizgal and Arkos, 1996; Strobel, 2002; Tinsley, 1974; Walker, 1977). Recently, the discovery of exoplanets has made atmospheric escape a fundamental consideration in understanding exoplanetary

Table 5.1 Mechanisms for the escape of atmospheric gases and ions.

Impact erosion (different approximations)	Thermal escape (different end-member approximations)	Suprathermal (or nonthermal) escape (different mechanisms)
Walker "cookie cutter" Ahrens' "bomb analogy" Melosh "tangent plane"	Jeans' escape Hydrodynamic escape	Photochemical escape Charge exchange Ion pickup Sputtering The polar wind Bulk removal

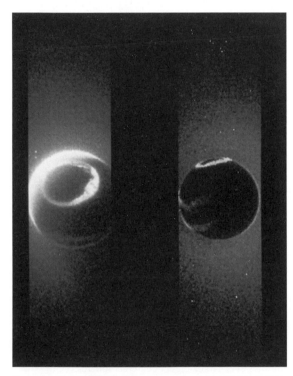

Figure 5.1 Earth imaged in the vacuum ultraviolet (VUV) by NASA's *Dynamics Explorer 1* (Rairden *et al.*, 1986). *Left panel*: View with the spacecraft at 16 500 km altitude above 67° N latitude at 2017 UT on October 14, 1981. Glow beyond the limb of the planet (red false color) is due to Lyman-α (121 nm) solar radiation resonantly scattered by Earth's extended hydrogen atmosphere or *geocorona*. Energetic hydrogen atoms in the geocorona are escaping to space. Features on the Earth's disk (dayglow from the sunlit atmosphere, a northern auroral oval, and equatorial airglow) are due to the emission of atomic oxygen at 130.4 and 135.6 nm and emission in the Lyman–Birge–Hopfield band of N_2 (140–170 nm). Isolated points of light are background stars that are bright in the VUV. *Right panel*: A view of Earth's dark hemisphere at 0222 UT on February 16, 1982, with the Sun behind Earth. Spacecraft altitude and latitude are 19,700 km and 13° N, respectively. Equatorial airglow straddles the magnetic equator in the pre-midnight sector. (Image credit: NASA.) (A black and white version of this figure will appear in some formats. For the color version, please refer to the plate section.)

atmospheres and whether they vanish, persist, or change composition (Koskinen *et al.*, 2014; Lammer *et al.*, 2003b; Luger and Barnes, 2015; Luger *et al.*, 2015; Owen and Jackson, 2012).

5.2 Overview of Atmospheric Escape Mechanisms

Table 5.1 summarizes the three principal categories of escape of atmospheric gases. Below, we give a brief overview of each type of escape. Then, the rest of the chapter examines the physics describing common ways that atmospheric gases escape, with particular emphasis on the two forms of thermal escape.

5.2.1 Thermal Escape Overview

Thermal escape is when heating of an atmosphere allows molecules to escape. In basic models, the theory assumes neutral species with a Maxwellian velocity distribution, which occurs when collisions between molecules are frequent. The "Jeans' escape" and "hydrodynamic escape" end-member approximations to thermal escape apply under different circumstances of atmospheric heating that we summarize below and discuss in further detail in Sec. 5.10.1.

Jeans' escape is when a relatively small number of high-energy molecules in the tail of the thermal distribution of velocities of molecules have sufficient kinetic energy to escape into a nearly collisionless exosphere from the collisional atmosphere below (see Sec. 5.6 for the physics). This process is important for the loss of hydrogen, a low-mass species that more easily attains escape speed at a given temperature. As such, Jeans' escape was likely influential in the atmospheric evolution of all the early terrestrial planets. Jeans' escape currently accounts for a non-negligible fraction of hydrogen escaping from Earth, Mars, and Titan, but it is negligible for

Venus because of a cold upper atmosphere combined with relatively high gravity.

Hydrodynamic escape occurs when heating in the collisional region of an atmosphere causes an upward pressure gradient force that drives a bulk, radial outflow (see Sec. 5.10 for the physics). Under such collisional circumstances, the pressure force can remain active up to very high altitudes with the result that the whole upper atmosphere expands as a fluid into space and gases attain escape velocity.

Hydrogen-rich atmospheres on relatively low-gravity rocky planets or very hot hydrogen-rich atmospheres on bigger planets are susceptible to hydrodynamic escape, which can drag along heavier gases in a way that is moderately mass fractionating (see Sec. 5.11). Shklovskii (1951) and Öpik (1963) first discussed the concept that heavy gases might be dragged along by a large hydrogen escape flux from primitive atmospheres but, compared with other escape mechanisms, hydrodynamic escape only received limited attention prior to the 1980s (Gross, 1972; McGovern, 1973; Ziering and Hu, 1967; Ziering et al., 1968). The lack of attention is probably because hydrodynamic escape was not thought to be active on any planet-sized body in the Solar System.

However, increasing evidence from within our Solar System and beyond suggests that warm hydrogen-rich atmospheres are prone to undergo hydrodynamic escape. The depletion of some light isotopes of noble gases in the atmospheres of Earth, Venus, and Mars, suggests that hydrodynamic escape may have operated very early in Solar System history if the very earliest atmospheres on these planets had been composed of a significant fraction of hydrogen in any chemical form. Beyond our Solar System, the gas giant HD 209458b, which orbits a Sun-like star at 0.05 AU, has hot H atoms beyond its Roche lobe, containing O, C^+, and Si^{2+}, presumably dragged there by hydrodynamic hydrogen flow (Linsky et al., 2010; Vidal-Madjar et al., 2003; Vidal-Madjar et al., 2004; Vidal-Madjar et al., 2008). Near-UV spectra suggest that hot Jupiter WASP-12B also has metals in its Roche lobe (Fossati et al., 2013; Fossati et al., 2010; Haswell et al., 2012).

Hydrodynamic escape and Jeans' escape are both approximations to thermal escape in different ways. Hydrodynamic escape is approximate because it neglects the fact that at very high altitudes there will eventually be few collisions and so a fluid description of the flow becomes invalid. Classical Jeans' escape neglects distortions of the particle velocities away from a Maxwellian distribution because of mass motion.

Specific conditions indicate when it is appropriate to apply the approximations of classical Jeans escape or purely fluid flow for hydrodynamic escape. The Jeans case applies when the atmosphere is essentially hydrostatic and when stellar heating of a thermosphere results in production and loss of electrons and ions, which conduct heat down to the lower thermosphere and mesopause where heat is efficiently radiated away (see Sec 1.1.1). Such atmospheres have roughly isothermal upper thermospheres. Jeans evaporation of atoms or molecules occurs from a static atmosphere into an essentially collisionless exosphere.

A hydrodynamic situation can occur when the heating of an upper atmosphere is strong enough to drive a bulk outflow. The bulk upward flow can attain the speed of sound in the collisional domain at an altitude called the *sonic level*. The speed of sound u_{sound}, can be compared with the root mean square speed of the molecules from kinetic theory u_{rms}, as follows,

$$u_{sound} = \left(\frac{\gamma p}{\rho}\right)^{1/2} \quad u_{rms} = \left(\frac{3p}{\rho}\right)^{1/2} \tag{5.1}$$

where p is pressure, ρ is density, and γ is the ratio of specific heats. Because $\gamma \sim 1.4$ for linear diatomic gases (e.g., H_2), we can see from comparing u_{sound} and u_{rms} that gas traveling at the speed of sound moves at a velocity similar to the mean thermal velocity of molecules, which is responsible in kinetic theory for providing the pressure of a gas (i.e., $p = (1/3)\rho u_{rms}^2$). In such a fast-moving fluid, a pressure gradient drives an upward bulk flow and the velocity increases above the sonic level to supersonic and then escape velocity. Because the density of the atmosphere decreases with altitude, and matter must be conserved, the flow velocity in such a case increases with altitude in order to maintain a constant mass flux [kg s^{-1}] through ever-larger planet-centered spheres. Under these circumstances, the fluid equations of hydrodynamic theory are reliable approximation, as noted by Walker (1977 pp. 149–151; 1982). The vertical profiles of density and velocity are relatively unaffected when the transition to the nearly collisionless domain occurs above the level from sonic to supersonic flow (Holzer et al., 1971). Figure 5.2 shows a schematic diagram of these two end-member cases of thermal escape: Jeans' escape and transonic hydrodynamic escape.

In hydrodynamic escape, the temperature profile depends on the balance of adiabatic cooling from the expansion of the atmosphere and absorption of stellar radiation. If adiabatic cooling dominates, atmospheric temperature can decline with increasing altitude. However, temperature can also increase with altitude if

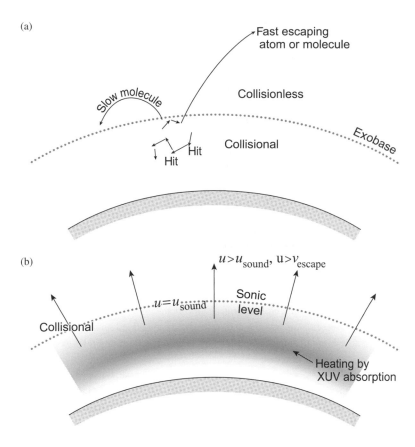

Figure 5.2 Schematic diagram of thermal escape end-members. (a) *Jeans' escape* is escape of molecules or atoms with an upward-directed component of velocity bigger than the escape velocity from the collisional part of the atmosphere into a virtually collisionless exosphere. (b) *Transonic hydrodynamic escape* is where the upper atmosphere has sufficient heating to produce a radial outward velocity u, at the speed of sound (u_{sound}) in the collisional part of the atmosphere at the radius labeled *sonic level*. Heating is typically by soft x-rays and extreme ultraviolet radiation (EUV). The atmosphere flows out to the vacuum of space driven by a pressure gradient (with the boundary condition set at the sonic level) and the supersonic flow reaches escape velocity, v_{escape}. At very high levels, the atmosphere will have very few collisions and the fluid assumption of hydrodynamic escape breaks down. However, under the transonic circumstances depicted, the fluid equations extrapolated to infinity provide a reasonable approximation to the density, temperature, and flow profiles.

absorption of stellar radiation is strong (e.g. Sekiya *et al.*, 1981; Sekiya *et al.*, 1980a).

When heating is smaller, or if there exists a sufficient backpressure at the top of the atmosphere, the outgoing flow may remain *subsonic* at all levels. An example of the latter might be escape in the direction towards the parent star, where eventually the escaping, and partly ionized, gas encounters the bow shock formed by interaction with a strong stellar wind. Perhaps more importantly, in an atmosphere that is weakly heated, or in which hydrogen is not sufficiently abundant, the atmosphere expands, but it is invalid to apply purely fluid hydrodynamic equations because the atmosphere becomes increasingly collisionless above some level. The application of hydrodynamic equations is a reasonable approximation if the mean free path remains smaller than the density scale height below the sonic level. But in the case when expanding gas becomes collisionless without reaching the speed of sound, it still exerts pressure, but that pressure cannot be calculated in the normal statistical way, which assumes a Maxwellian velocity distribution. Sophisticated models can be used to treat the transition from collisional to

rarefied domains in order to calculate a realistic temperature density, temperature and flow structure. Such methods are based on the *Boltzmann equation*, which in its most general form is an equation for the time evolution of the velocity distribution function of species in a gas mixture as a result of external forces and collisions (see Ch. 3 of Schunk and Nagy, 2009).

Using the Boltzmann equation for a single-component atmosphere, Merryfield and Shizgal (1994) found that escape can be fractionally (~30%) greater than Jeans' escape due to streaming of particles from the heavy, denser regions below. Another such model of gas flow is "direct simulation Monte Carlo" (DSMC), in which a large set of particles is followed subject to collisions, heating and gravity (Bird, 1994; Volkov *et al.*, 2011a). DSMC models show that purely hydrodynamic models that were once applied to N_2 escape from Pluto (where a sonic level probably does not occur in the collisional domain) produce an erroneous atmospheric structure of temperature and density (Erwin *et al.*, 2013; Johnson *et al.*, 2013d). Another key prediction of DMSC models, at least for single component atmospheres, is a sharp transition from "Jeans-like" escape to rapid, transonic hydrodynamic escape as the thermal energy of upper atmosphere gas molecules is increased relative to their gravitational binding energy (Volkov and Johnson, 2013; Volkov *et al.*, 2013; Volkov *et al.*, 2011b).

The physics of hydrodynamic escape of planetary atmospheres is analogous to the solar wind – the fully ionized, electrically neutral plasma that is a supersonic expansion of the solar corona as a result of a pressure difference between the corona and the interstellar medium. (See Ch. 4 of Kivelson (1995) for an introduction to the solar wind.) Essentially, the solar corona – the plasma that we see around the Sun in a total solar eclipse – is so hot that the Sun's gravity cannot hold on to it and it flows out as the solar wind. Consequently, hydrodynamic escape of a planetary atmosphere is sometimes called a *planetary wind*. However, there are differences. The key one is that the fluid description of solar wind in the vacuum of space works because of quasi-collisional effects caused by electromagnetic interactions between charged particles, whereas neutral species generally dominate planetary atmospheres and cannot be assumed to behave like that.

Finally, we note that some authors use the term *blow-off* interchangeably with "hydrodynamic escape" (e.g., Volkov *et al.* (2013); Hunten (1973)). Unfortunately, there are different definitions about the meaning of the term blowoff. Chamberlain and Hunten (1987, p. 377) state that blowoff is "when an escaping light gas is able to carry heavier constituents along with it." Lammer (2013, p.30) describes blowoff as when "the whole exosphere evaporates" because "the mean thermal energy...of gases at the exobase level exceeds their gravitational energy." Because of such difference in definitions, in the rest of the chapter, we avoid the term blowoff.

5.2.2 Suprathermal (or Nonthermal) Escape, in Brief

Suprathermal escape refers to loss processes that affect either neutral species or ions that attain a velocity significantly greater than that corresponding to the background neutral temperature. Consequently, suprathermal escape is also called *nonthermal escape* (Hunten, 2002), Most suprathermal processes involve ions, which may themselves have a Maxwellian velocity distribution but with a temperature exceeding that of the neutral population. Various types of suprathermal escape are as follows.

Photochemical escape occurs when atoms resulting from various photochemical reactions attain sufficient energy to escape to space. Such escape can happen when a neutral species is photoionized by solar EUV radiation and recombines with an electron to form a fast neutral atom. Photochemical escape is important for the loss of C, O, and N from Mars (Sec 12.2.4).

Charge exchange is where a fast ion can impart its charge to a neutral atom through collision, and become a fast neutral atom with escape velocity (Sec 5.7, below). In today's terrestrial atmosphere, charge exchange is usually the dominant mechanism for hydrogen escape, although it is exceeded by Jeans' escape at solar maximum.

Ion pickup occurs when atmospheric ions are exposed to an electric field from the magnetized solar wind. Atmospheric particles are ionized either by solar UV radiation (photoionization) or by charge exchange. Acceleration of ions due to the electric field can cause some ions to reach escape velocity, whereas others head into the atmosphere where they may cause sputtering, as we describe next.

Sputtering occurs when ions that have been picked up by the magnetic field embedded in the solar wind impact a planetary atmosphere and undergo charge exchange. Charge exchange neutralizes the ions, which can impart their large energies to surrounding particles by collision. Upward-directed energetic particles can then escape. This process may have been important on early Mars, after it lost its magnetic field and was no longer shielded from the solar wind.

The polar wind is a stream of hydrogen ions (protons) that flows upward near the poles where Earth's magnetic field lines are more or less vertical. These field lines do not necessarily reconnect, or they do so only sporadically, and so these hydrogen ions are eventually swept away by the solar wind.

Bulk removal is caused by instabilities at the solar wind–atmosphere interface that can strip away large portions of ionized atmosphere (Perez-de-Tejada, 1987) or cause ion outflow (Hartle and Grebowsky, 1990) from planets, such as Mars and Venus, that lack a protective magnetic field. This process is currently poorly understood.

5.2.3 Impact Erosion, in Brief

Impact erosion occurs when the hot vapor plume or high-speed ejecta associated with a large asteroid or comet impact imparts sufficient kinetic energy to atmospheric molecules for them to escape en masse (Sec. 5.12). The impactor is vaporized along with part of the target body. This erosion process affects smaller target bodies more strongly than larger ones and could have been important for removing virtually the entire early Martian atmosphere (Sec. 12.3.3). Impact erosion may also explain why Titan has a thick atmosphere whereas the Galilean moons of Jupiter, which are subject to more energetic impacts, remain largely barren (Zahnle *et al.*, 1992).

5.2.4 The Upper Limit of Diffusion-Limited Escape, in Brief

Diffusion-limited escape of hydrogen is an *upper limit* on the escape rate set when the escape rate of hydrogen is not controlled by processes at high altitude but is regulated by the rate that hydrogen can diffuse up from the lower atmosphere (see Sec. 5.8.3, 5.8.4 and 5.8.9 for details). For example, in Earth's current atmosphere, hydrogen does not simply escape via Jeans' escape determined by the temperature of the exobase. Instead, both Jeans' escape and suprathermal processes remove hydrogen rapidly from the exobase, and the rate-limiting step is the relatively slow upward diffusion of hydrogen through the layer of background air between the homopause and exobase. Basically, the hydrogen escape rate is limited both by the supply of hydrogen from below and upward diffusion. Diffusion-limited escape is a remarkably successful theory that appears to apply to hydrogen escape from the current atmospheres of Earth, Venus, Mars, and Titan, as well as can be determined from the available data (Sec. 5.8.4 and Sec. 5.9).

Diffusion-limited escape can also apply as an *upper limit* to hydrodynamic escape. In this case, the upper limit on the rate of escape is set by diffusion of hydrogen through a layer of background air of heavier species beneath the level where hydrogen is accelerated radially outwards because of heating caused by the absorption of shortwave light from the parent star.

5.3 Breakdown of the Barometric Law

We begin our more detailed description of atmospheric escape by showing that the barometric law, which describes the vertical pressure distribution in the lower parts of Earth's atmosphere, must break down at some altitude above the surface. Our discussion follows that of Walker (1977), pp. 147–151.

In Ch. 1, we showed that (averaged over horizontal distances of several km), atmospheric pressure varies with altitude z according to the *hydrostatic equation* (eq. (1.12)):

$$p(z) = p_{surf} \exp\left(-z/H_a\right) \qquad (1.12)$$

Here p_{surf} is the surface pressure and $H_a = kT/(mg)$ is the atmospheric scale height, where k is Boltzmann's constant, T is temperature, m is mean molecular mass, and g is gravitational acceleration.

We can write eq. (1.12) in differential form as follows:

$$\frac{1}{p} dp = -\frac{mg}{kT} dz = -\frac{1}{H_a} dz \qquad (5.2)$$

To extend this relation high up in the atmosphere, the variation of g with altitude must be considered according to $g = GM/r^2$, where G is the universal gravitational constant $(6.672 \times 10^{-11}$ N m^2 kg^{-2}), M is the mass of the planet, and r is the radial distance from a planet's center. Then, eq. (5.2) may be rewritten as

$$\frac{1}{p} dp = -\frac{GMm}{r^2 kT} dr \qquad (5.3)$$

Integrating from the surface at radial distance r_{surf} up to radial distance r yields

$$p(r) = p_{surf} \exp\left[\frac{GMm}{kT}\left(\frac{1}{r} - \frac{1}{r_{surf}}\right)\right] \qquad (5.4)$$

Now, consider what happens as $r \to \infty$. In this case, we obtain

$$p_\infty = p_{surf} \exp\left(-\frac{GMm}{kT r_{surf}}\right) \qquad (5.5)$$

If this result were valid, it would imply that the pressure at infinity is small but finite; hence, the atmosphere would have infinite mass. For example, using $m = 29$ atomic mass units (the mean molecular mass in the Earth's lower atmosphere), $M = 5.97 \times 10^{24}$ kg, a temperature typical for the thermosphere of $T = 1000$ K, and $r_{surf} = 6371 \times 10^3$ m, eq. (5.5) predicts that $p_\infty/p_{surf} \sim 2 \times 10^{-95}$. If we take $m = 1$ a.m.u. instead, reflecting the fact that the uppermost atmosphere is composed mostly of atomic hydrogen, we get $p_\infty/p_{surf} \sim 5 \times 10^{-4}$. Neither result is physically realistic, but they demonstrate that the atmospheric pressure would be significant at high altitudes were the barometric law to apply in this way.

In fact, the actual pressure at the upper boundary of the Earth's atmosphere depends on location. It is highest in the sunward direction where the solar wind impinges on the magnetosphere, i.e., the region where the Earth's magnetic field dominates. At the subsolar *magnetopause*, which is at a distance of ~10 Earth radii, the ram pressure of the solar wind is ~3 nPa and balances plasma pressure within the magnetosphere. By comparison, the calculated pressures at infinity for the two cases above are ~10^{-90} Pa and 50 Pa, respectively. For the pure atomic H case, the solar wind would not be able to supply the necessary backpressure even in the sunward direction. In the anti-sunward direction, the effective backpressure should be essentially zero. The barometric law obviously cannot apply at great distances from the Earth.

How can the breakdown of the barometric law be resolved? Two different possibilities exist: one that applies to the present terrestrial atmosphere and one that may have applied to the primitive atmospheres of the terrestrial planets.

For the present atmosphere, the key is that the atmosphere becomes virtually collisionless at some altitude. Once the atmosphere becomes collisionless, the atmosphere is no longer in a hydrodynamic regime of "continuum flow" but "free molecular flow." Equations (5.3)–(5.5) no longer apply once the velocity distribution in the collisionless region deviates from a Maxwellian distribution. The *exobase* or *critical level* is the altitude above which the virtually collisionless region occurs.

A different situation may have existed for Earth's primitive atmosphere shortly after it formed at 4.5 Ga and possibly for hundreds of millions of years afterwards. This very early atmosphere is thought to have been hydrogen-rich (see Sec. 6.5.2). In this case, absorption of EUV and x-rays in the upper atmosphere should have driven a bulk outflow from the upper atmosphere, which was not hydrostatic. Upwards-flowing hydrogen would have been pushed along by a pressure differential under hydrodynamic escape. The atmosphere literally should have expanded into the vacuum of space. We discuss the physics of *hydrodynamic escape* in detail in Sec. 5.10.

The fractional decrease of the upward flux of particles, Φ, from distance $r' = r$ to $r' = \infty$ due to collisions is given by

$$\frac{\Phi_\infty}{\Phi_r} = \exp\left[-\int_r^\infty n(r')\sigma_c dr'\right] \qquad (5.6)$$

Here, σ_c is the collision cross-section of a molecule and $n(r')$ is the number density of *all* molecules at distance r'. The decrease of upward-directed particles caused by collisions is analogous to the Beer–Lambert–Bouguer Law in radiative transfer (Sec. 2.4.2.1). Thus, the term in the exponential is analogous to optical depth, except that the photon absorption cross-section has been replaced by a molecular collision cross-section and we deal with molecules instead of photons.

If the probability of escaping above the exobase is e^{-1} (i.e., $\Phi_\infty/\Phi_r = e^{-1}$) then the term inside the square brackets of eq. (5.6) must equal minus one for $r = r_{exob}$, i.e.,

$$\int_{r_{exob}}^\infty n(r')\sigma_c dr' = 1 \qquad (5.7)$$

From eq. (1.21), the above integral is related to the atmospheric molecular column density, $N(r)$, overlying the exobase at radius r_{exob} by

$$N(r) = \int_{r_{exob}}^\infty n(r')dr' \cong n(r_{exob})H_a(r_{exob}) \qquad (5.8)$$

Thus, at the exobase, the number density n and scale height H_a, respectively, are:

$$n(r_{exob}) \equiv \frac{1}{\sigma_c H_a(r_{exob})}, \quad H_a(r_{exob}) \equiv \frac{1}{\sigma_c n(r_{exob})} = \text{m.f.p.} \quad \text{for air of density } n(r_{exob}) \qquad (5.9)$$

5.4 The Exobase or "Critical Level"

The *exosphere* is the uppermost layer of an atmosphere that is essentially collisionless. This means that the mean free path is so long that collisions can largely be neglected. We denote the height of the bottom of the exosphere, i.e., the exobase, as the radius r_{exob}, above a planet's center. The *exobase* is defined as the height where a proportion e^{-1} (~1/3) of fast, upward-directed particles experience no collisions and hence escape. Equivalently, the exobase can be defined as the altitude at which the mean free path of a molecule (in the horizontal direction) is equal to the local scale height. (The reason for stating "horizontal direction" is explained below.)

In the expression for the local scale height in eq. (5.9), the right-hand side is approximately the definition of mean free path ℓ_{mfp}, for air with number density $n(r_{exob})$, i.e., ℓ_{mfp} along a horizontal path at an altitude at radius r_{exob}. This explains the exobase definition given earlier. Strictly, for molecules with a Maxwellian velocity distribution, the relationship is $\ell_{mfp}(r) = 1/(\sqrt{2}n\sigma_c)$, but we ignore the $\sqrt{2}$ factor.

Another common way of quantifying the importance of collisions is the *Knudson number*, K_n, which is defined as follows:

$$K_n(r) = \frac{\ell_{mfp}(r)}{H_a(r)} = \frac{\text{mean free path}}{\text{local scale height}} \qquad (5.10)$$

The Knudson number grows with altitude and the exobase occurs where $K_n = 1$.

Following Jeans (1954), we can combine eqs. (5.8) and (5.9) to write the condition for the exobase as

$$\sigma_c N(r_{exob}) = 1 \qquad (5.11)$$

The thoughtful reader will reflect that the exobase where particles can escape is analogous in its definition to the level of unity optical depth where photons can escape (Sec. 2.4.4), merely by swapping photons for molecules, as noted above.

The number density at Earth's exobase can be estimated from eq. (5.9), taking $\sigma_c \sim 3 \times 10^{-15}$ cm^2. Gravity will be smaller at high altitude, i.e., at 500 km altitude, $r_{exob} = 6870$ km and $g = 8.44$ m s^{-2}. Thus, $H_a = RT/Mg =$ (8.314 J mol^{-1} K^{-1})(1500 K)/[(0.016 kg mol^{-1}) (8.44 m s^{-2})] $\cong$ 92 km, or roughly 100 km (= 10^7 cm), where we use a molar mass appropriate for atomic oxygen, the dominant constituent of the thermosphere. Thus, $n_{exob} \cong 3.6 \times 10^7$ cm^{-3}, which is roughly similar to the value at the exobase on most planets and satellites. For this exobase number density, the corresponding altitude on Earth varies from solar minimum to solar maximum, but is typically around 450–500 km.

Of course, the exobase is an idealized concept. We have assumed that the transition from continuum flow to free molecular flow is sharp, but in reality it is gradual. However, we will show in Sec. 5.5.1 that defining the exobase in this way does not significantly affect calculations. It is also important to recognize that the dominant species at today's terrestrial exobase is atomic oxygen, which does not escape. The O atoms provide a static background from which hydrogen atoms evaporate off into space. The process of evaporation is described in Sec 5.6 below.

5.5 Escape Velocity

In Earth's present atmosphere, hydrogen atoms are lost to space by reaching escape velocity at the exobase. The escape velocity v_e, is attained when the kinetic energy of an atom of mass m equals its gravitational potential energy:

$$\frac{1}{2}mv_e^2 = \frac{GMm}{r} \qquad (5.12)$$

or

$$v_e = \sqrt{\frac{2GM}{r}} \qquad (5.13)$$

Here, M is the mass of the planet, G the universal gravitational constant, and r the radius from the center of the planet.

Atoms that have speeds in excess of v_e and whose velocities are directed upwards have a $1/e$ (i.e., 37%) probability of escaping at the exobase by avoiding collisions above it. For Earth, eq. (5.13) gives $v_e \sim 10.8$ km s^{-1} at the exobase. Escape velocities for other planets are 5.0 km s^{-1} for Mars, 10.4 km s^{-1} for Venus, and 60.2 km s^{-1} for Jupiter.

Two salient points arise from eq. (5.13). First, for planets of roughly similar mean density, we find $v_e \propto \sqrt{R_p^3/R_p} \propto R_p$, where R_p is the planetary radius. Thus, Mars' escape velocity is roughly half that of Earth or Venus because Mars has about half the diameter of these planets. But the escape energy is ~1/4 as high for Mars as for Earth or Venus because the energy is $\propto v_e^2$. Second, although eq. (5.13) shows that the escape velocity is independent of the mass of the escaping molecule, lighter molecules more easily attain escape velocity for a given kinetic temperature than heavier molecules. This is a consequence of the *equipartition theorem* in kinetic theory for an ideal gas with purely translational energy. This theorem states that molecules of different masses have the same average kinetic energy, given that the mixture of gases has a well-defined temperature.

The mean thermal velocity of a hydrogen atom at Earth's exobase can easily be calculated. If we assume that atoms at the top of the atmosphere are in thermal equilibrium, the mean thermal energy of particles is (3/2) kT, so that the kinetic energy is:

$$\frac{1}{2}mv^2 = \frac{3}{2}kT \qquad (5.14)$$

Taking a thermosphere temperature of ~1000 K, the root mean square speed is $v = (3kT/m)^{1/2} = [3 \times (1.38 \times 10^{-23}$ J K$^{-1}) \times (1000$ K)/(1.67 \times 10^{-27}$ kg)]$^{1/2} \approx 5$ km s^{-1}. This speed is lower than the escape speed of 10.8 km s^{-1} at the exobase. Thus, even the lightest atom, hydrogen, does not have sufficient mean energy to escape from Earth. Instead, it is energetic atoms in the tail of the Maxwell–Boltzmann distribution of velocities that escape rather than those with typical speed, as illustrated schematically in Fig. 5.3.

Given that only the energetic atoms escape, integration over the Maxwell–Boltzmann velocity distribution is needed to calculate the escape flux of a particular gas, as discussed below in Sec. 5.6. A rule-of-thumb approach found in elementary textbooks is that if the average speed of a molecule or atom exceeds one-sixth of the escape speed, then escape is generally possible for that species. The average speed obviously depends on the temperature at the exobase. Consequently, Fig 5.4 illustrates the

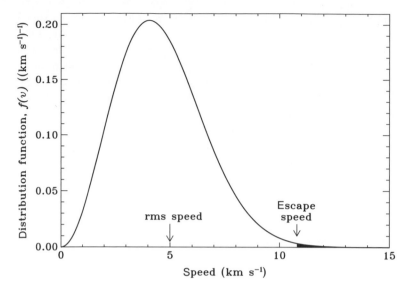

Figure 5.3 The Maxwell–Boltzmann speed distribution function $f(v)$, for hydrogen atoms at 1000 K. The escape speed at Earth's exobase and the root mean square (rms) speed of the atoms are indicated. Only those atoms in the energetic tail of the speed distribution with speeds exceeding the escape speed (shaded) are able to undergo *Jeans' thermal escape*. In contrast, atoms with the most probable speed at the peak of the distribution, or with the slightly higher rms speed, are unable to escape.

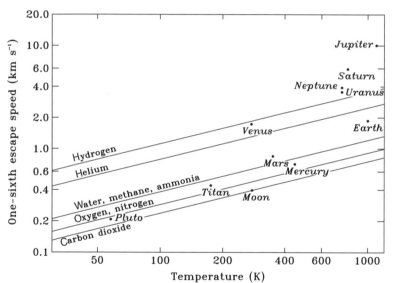

Figure 5.4 A schematic graph showing gas species that are prone to undergo thermal escape from planets in the Solar System on the basis of the "rule of thumb" approach discussed in the text. For bodies with substantial atmospheres, the temperature corresponds to the average exobase temperature. For the Moon and Mercury, the temperature is the mean surface temperature. The sloping lines correspond to the root mean square speed of the various gas molecules at the given temperature.

stability of gas species against Jeans' escape for planets in the Solar System using exobase temperatures. This diagram is a rough summary, but its general inferences are valid: Nothing escapes from the giant planets, whereas the smaller, rocky planets lose light gases.

In viewing Fig. 5.4, one should remember that *suprathermal* mechanisms that are not based on thermal equilibrium distributions of molecules allow light gases to escape. For example, O, C, and N can escape slowly from Mars via photochemical or ionic reactions; Fig 5.4 does not include such effects, which we discuss in Sec. 12.2.4.

5.6 Jeans' Thermal Escape of Hydrogen

5.6.1 Concept and Mathematical Derivation

Our purpose in what follows is to derive a general expression for the Jeans' thermal escape flux of hydrogen and other gases. Molecules in a collisional gas have a Maxwell–Boltzmann speed distribution (Box 5.1) and we have to integrate over the "tail" of the distribution above the escape speed to obtain the escape flux.

The number of molecules with speeds between v and $v+dv$ is given by

Box 5.1 The Maxwell–Boltzmann Velocity Distribution

Derivation of the Maxwell–Boltzmann distribution starts with the translational kinetic energy of each molecule, $E = \frac{1}{2}mv^2$. Then the probability that a molecule has a particular speed is proportional to the Boltzmann factor, i.e.,

$$\text{probability of particular speed} \propto \exp\left(-E/kT\right) = \exp\left(-mv^2/2kT\right)$$

For the distribution of velocities, imagine a sphere representing velocity parameter space, centered on v_x, v_y, v_z axes. Within a shell of thickness Δv at radius v, the volume is proportional to the square of the radius times the thickness, i.e. $v^2\Delta v$. So the number of speed states between v and $v+\Delta v$ is $\propto v^2\Delta v$. Thus, the probability of finding a given molecule in the speed range v to $v+\Delta v$ follows the proportionality

$$\begin{pmatrix} \text{probability of finding a} \\ \text{given molecule in speed} \\ \text{range } v \text{ to } v + dv \end{pmatrix} \propto \begin{pmatrix} \text{no. of microstates} \\ \text{in the speed range} \end{pmatrix} \times \begin{pmatrix} \text{probability of finding} \\ \text{the molecule in a given} \\ \text{microstate in the speed range} \end{pmatrix}$$

$$f(v)dv \propto v^2 \exp\left(-\frac{mv^2}{2kT}\right)dv$$

A proportionality constant is then selected to satisfy the condition $\int_0^\infty f(v)dv = 1$, i.e. that all molecules must be in some state. This results in eq. (5.15).

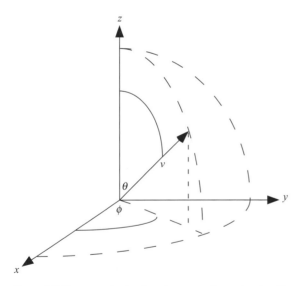

Figure 5.5 The geometry for Jeans' escape of a molecule with velocity v, in spherical polar coordinates, where θ is zenith angle and ϕ is azimuthal angle.

$$nf(v)dv = \frac{4n}{\sqrt{\pi}}\left(\frac{m}{2kT}\right)^{3/2} v^2 \exp\left(-\frac{mv^2}{2kT}\right)dv \qquad (5.15)$$

Here, n represents the total number density of the constituent being considered.

Now we make the further assumption that the velocity distribution is isotropic, so that the same number of molecules travels in every direction. In spherical polar coordinates, with azimuth angle ϕ and polar angle θ, an element of solid angle $d\Omega$, is given by the following (see Fig. 5.5):

$$d\Omega = \sin\theta d\theta d\phi \qquad (5.16)$$

Integrating around azimuth angle ϕ, gives $d\Omega = 2\pi \sin\theta d\theta$. Thus, the fraction of upward traveling molecules is $2\pi \sin\theta d\theta/4\pi$. In turn, the number of molecules with velocities between v and $v+dv$ travelling at an angle between θ and $\theta+d\theta$ from the vertical is given by

$$\frac{nf(v)dv2\pi \sin\theta d\theta}{4\pi} = \frac{1}{2}nf(v)dv \sin\theta d\theta \qquad (5.17)$$

The vertical flux of molecules, Φ, is found by multiplying by the vertical component of velocity, $v\cos\theta$, and integrating over the upwards hemisphere from $\theta = 0$ to $\pi/2$:

$$\text{vertical flux with speed } v = \Phi dv = \int_0^{\pi/2} \frac{1}{2}nf(v)dv\, v\cos\theta \sin\theta d\theta$$
$$(5.18)$$

Noting that $d(\sin^2\theta)/d\theta = 2\sin\theta\cos\theta$, we can evaluate the integral in eq. (5.18) using

$$\int_0^{\pi/2} \cos\theta \sin\theta d\theta = \frac{1}{2}\left[\sin^2\theta\right]_0^{\pi/2} = \frac{1}{2} \qquad (5.19)$$

Thus, the vertical flux of molecules with velocity v is given by

$$\Phi dv = \frac{1}{4}nf(v)vdv \qquad (5.20)$$

The total escape flux (with SI units of particles m^{-2} s^{-1}) is found by setting $n = n_{exob}$, the number density at the exobase, substituting for $nf(v)dv$ from eq. (5.15), and integrating over all velocities exceeding the escape velocity, v_e

$$\Phi_{esc} = \int_{v_e}^{\infty} \Phi dv = \frac{n_{exob}}{\sqrt{\pi}} \left(\frac{m}{2kT}\right)^{3/2} \int_{v_e}^{\infty} v^3 \exp\left(-\frac{mv^2}{2kT}\right) dv \tag{5.21}$$

To evaluate the integral, we substitute $x = v^2$, $dx = 2vdv$, and integrate by parts, giving,

$$\text{escape flux} = \Phi_{esc} = \frac{n_{exob}}{\sqrt{\pi}} \left(\frac{m}{2kT}\right)^{3/2} \left(\frac{kT}{m}\right) \left(v_e^2 + \frac{2kT}{m}\right) \exp\left(-\frac{mv_e^2}{2kT}\right) \tag{5.22}$$

One can simplify this expression by using eq. (5.13) to substitute for the square of the escape velocity at level r_{exob},

$$v_e^2 = \frac{2GM}{r_{exob}} \tag{5.23}$$

Also, let us define, v_s, which is the most probable speed in the Maxwell–Boltzmann speed distribution function

$$v_s \equiv \left(\frac{2kT}{m}\right)^{1/2} \tag{5.24}$$

We also introduce the (Jeans) escape parameter, λ_J, which is the ratio of gravitational potential energy GMm/r to thermal energy $\sim kT$ (Chamberlain, 1963),

$$\lambda_J \equiv \frac{GMm}{kTr}; \text{ at exobase, } \lambda_{Jexo} \equiv \frac{GMm}{kTr_{exob}} \equiv \frac{r_{exob}}{H_{exob}} \equiv \frac{v_e^2}{v_s^2} \tag{5.25}$$

Then, eq. (5.22) becomes

$$\text{escape flux} = \Phi_{esc} = \frac{n_{exob}}{2\sqrt{\pi}} \frac{1}{v_s} \left(v_e^2 + v_s^2\right) \exp\left(-\frac{v_e^2}{v_s^2}\right)$$

or

$$\Phi_{esc} = \frac{1}{2\sqrt{\pi}} n_{exob} v_s (1 + \lambda_{Jexo}) e^{-\lambda_{Jexo}} \tag{5.26}$$

This is a convenient expression by which one may evaluate the Jeans escape flux.

Equation (5.26) for the Jeans escape rate can differ from the actual rate of escape because of evaporative cooling of the exobase and distortions in the velocity distribution function. Several papers have suggested that the expression *overestimates* the number of high-velocity molecules at the exobase that escape because escape depletes the high-speed molecules and can cool the background gas. The overestimate is ~20%–30% when hydrogen escapes from a two- or multi-component gas model (Brinkmann, 1971; Pierrard, 2003; Shizgal and Blackmore, 1986) for Earth or Titan (Tucker et al., 2013). Figure 5.6 shows a variety of estimates of the correction factor for the Earth in the range ~0.6-0.8. Fortunately, we can often ignore such discrepancies in atmospheric evolution studies given that other uncertainties dominate, such as the bulk composition of ancient atmospheres.

Volkov et al. (2011a; 2011b), who considered the angular distribution of the velocity distribution at the exobase for a single-component gas, found that Jeans escape can *underestimate* the actual thermal escape in that case. The ratio of actual escape to the Jeans escape rate at the exobase in their model was ~1.4–1.7 for an atmosphere with Jeans' parameter ranging from 6 to 15. This arose because some molecules gained energy to escape from

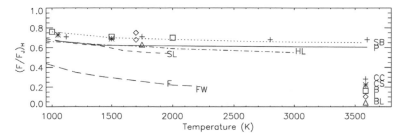

Figure 5.6 The ratio of the non-Maxwellian escape flux of hydrogen, F, to the Maxwellian Jeans' escape flux, F_J, versus the temperature at the Earth's exobase (from Pierrard, 2003). The solid line (P) shows the results of Pierrard (2003). Results are compared with those obtained with Monte Carlo simulations represented by different symbols: CC (Chamberlain and Campbell, 1967); L (Lew, 1967); B (Brinkmann, 1970); CS (Chamberlain and Smith, 1971); BL (Barakat and Lemaire, 1990). The other lines correspond to analytic solutions: HL (Hays and Liu, 1965); F (Fahr, 1976); FW (Fahr and Weidner, 1977); SL (Shizgal and Lindenfeld, 1980), SB (Shizgal and Blackmore, 1986).

collisions above the exobase. See Sec. 5.10.1 for further discussion of such enhanced Jeans-like escape.

The value of the escape parameter, λ_{Jexo}, is important in eq. (5.26). When the gravitational binding energy (GMm/r_{exob}) is much larger than kT, λ_{Jexo} is large, causing a small exponential factor $e^{-\lambda_{Jexo}}$ in eq. (5.26) and a small escape rate. In contrast, if the gravitational binding energy is much smaller than kT and λ_{Jexo} is small, the exponential factor in eq. (5.26) approaches unity, gas expands into the vacuum of space, and an atmosphere is unstable such as on the Moon.

One criticism of the above derivation is that the definition of the exobase is somewhat arbitrary. Why should it be the height where $\sigma N(r) = 1$? After all, only H atoms whose velocities are directed precisely upwards will have a $1/e$ chance of escaping. Those headed off at some other angle will have a somewhat lower probability of doing so. Fortunately, it can be shown that the escape flux is only weakly sensitive to the exact value of r_{exob} (c.f., Walker (1977)). Suppose, for example, that we picked some other altitude r_{exob}' as the location of the exobase. Let the number density at that altitude be n_{exob}'. Then, using the analog to eq. (5.4), but picking the reference point to be r_{exob} rather than r_{surf}, and taking T to be constant, allows us to write

$$n_{exob}' = n_{exob} \exp\left[\frac{GMm}{kT}\left(\frac{1}{r_{exob}'} - \frac{1}{r_{exob}}\right)\right] \tag{5.27}$$

Using the definition of λ_{Jexo} above allows us to rewrite this as

$$n_{exob}' e^{-\lambda_{Jexo}'} = n_{exob} e^{-\lambda_{Jexo}} \tag{5.28}$$

The Jeans escape flux evaluated at distance r_{exob}' would be related to the flux at distance r_{exob} by

$$\frac{\Phi'}{\Phi} = \frac{n_{exob}'\left(1 + \lambda_{Jexo}'\right)e^{-\lambda_{Jexo}'}}{n_{exob}\left(1 + \lambda_{Jexo}\right)e^{-\lambda_{Jexo}}} = \frac{1 + \lambda_{Jexo}'}{1 + \lambda_{Jexo}} \approx 1 \tag{5.29}$$

Because λ_J is only a slowly varying function of r, the Jeans escape flux is relatively insensitive to the exact altitude at which it is evaluated.

5.6.2 Effusion Velocity

To provide some physical insight into the implications of eq. (5.26), let us calculate the effective *effusion velocity* (m s^{-1}) for hydrogen escaping from Earth's atmosphere, which is the average rate at which hydrogen atoms or molecules are drifting upwards. Using eq. (5.26), we can write

$$\text{effusion velocity}, v_J = \frac{\Phi_J}{n_{exob}} = \frac{1}{2\sqrt{\pi}}v_s(1 + \lambda_{Jexo})e^{-\lambda_{Jexo}} \tag{5.30}$$

In Earth's upper atmosphere, the temperature varies from ~1000 K at solar minimum to 1500 K or more at solar maximum. As a result of these hot temperatures, molecular hydrogen is broken down efficiently into atomic hydrogen by the reaction

$$H_2 + O \rightarrow H + OH \tag{5.31}$$

Thus, the dominant form of hydrogen at the exobase on Earth is H, rather than H_2. This is not true on Mars or Titan, where the upper atmosphere is much colder. There, H_2 and H are both important hydrogen-bearing constituents at high altitudes.

If we calculate the numbers for atomic H at Earth's exobase, we find $\lambda_c = 7$ and $v_J = 0.87$ m s^{-1} for $T_\infty = 1000$ K, and $\lambda_{Jexo} = 4.7$ and $v_J = 110$ m s^{-1} for $T_\infty = 1500$ K. This tells us two things. First, it shows why elements heavier than He do not escape from Earth's atmosphere. The next lightest gas-forming element, carbon (C), has a mass number of 12 and, thus, $\lambda_{Jexo} > 50$. For example, at 1500 K, λ_{Jexo} for C is 12 times the value for atomic hydrogen, giving $12 \times 4.7 = 56.4$. Because λ_{Jexo} appears as a negative exponential in eq. (5.26), this effectively precludes thermal loss of C or any heavier element from Earth's atmosphere. Second, it shows that effusion velocities for atomic H vary widely from solar minimum to solar maximum. As we will see in the next section, Jeans' escape is the dominant escape mechanism for hydrogen at solar maximum, but is outweighed by suprathermal hydrogen loss processes at solar minimum. In fact, overall suprathermal (or nonthermal) processes dominate the time-averaged loss rate.

5.7 Suprathermal (Nonthermal) Escape of Hydrogen

Thermal loss is only one of several possible mechanisms by which gases can escape from atmospheres. Various *suprathermal loss processes* (also called *nonthermal*) dominate Earth's current hydrogen escape. Suprathermal molecules or atoms are particles whose velocities exceed the expected values from the Maxwellian distribution because they acquire kinetic energy in ways other than purely thermal collisions. What suprathermal processes have in common is that a boost from a chemical reaction or electrical or magnetic acceleration imparts escape velocity to single particles.

The two most important suprathermal hydrogen loss mechanisms for Earth are as follows.

(a) H–H$^+$ charge exchange. In this process, neutral H atoms in the upper atmosphere exchange charge with fast-moving ("hot") H$^+$ ions in Earth's plasmasphere

(a) (b) (c)

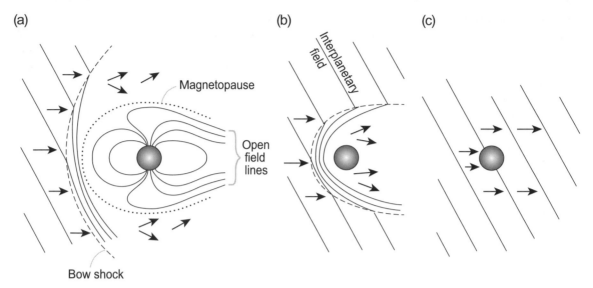

Figure 5.7 (a) Interaction of the solar wind with a planet that has a significant magnetic field, e.g. Earth (and similar also for Jupiter, Mercury, Saturn, and Uranus). Inside the magnetopause, the magnetic field is dominated by that of the planet and solar wind particles that are thermalized at the bow shock (where the wind is brought to rest) flow around the magnetopause and have little influence on the atmospheric evolution. Open field lines, however, allow ions to escape from the poles if they exceed the escape velocity. (b) Interaction of the solar wind with a planet without a significant magnetic field, e.g. present-day Mars or Venus. A bow shock is formed much closer to the planet due to an induced field in the ionosphere. Thermalized solar wind particles can interact directly with the atmosphere. (c) The solar wind collides directly with a body that has low electrical conductivity, no ionosphere/atmosphere, and no magnetic field, e.g. the Moon. (Adapted from Fig. 3.8 in Lewis and Prinn (1984).)

$$H + H^+(hot) \rightarrow H^+ + H(hot) \qquad (5.32)$$

The *plasmasphere* is the region, including the ionosphere (60–3000 km) and magnetosphere, where large numbers of ionized species are present. The ion temperature in Earth's plasmasphere is ~ 5000–20 000 K. It is much hotter than the neutral temperature because of heating by Coulomb collisions, i.e. acceleration due to ion charge attraction and repulsion. When these hot ions exchange charge and become neutral hydrogen atoms, they retain their original high velocities, which can be approximated by a Maxwellian distribution at a much higher temperature. Hence, the fraction of these hydrogen atoms that exceed escape velocity is far higher than for the neutral background population. According to model calculations by Yung *et al.* (1989), about 75% of Earth's hydrogen escapes suprathermally when averaged over time. Charge exchange reactions account for ~50% of the escape. At solar maximum, however, the majority of the hydrogen follows the Jeans escape path.

(b) The polar wind. Near the magnetic poles of the Earth, the magnetic field lines are open over about 1/40 of the Earth's surface, that is, they do not connect to field lines that re-enter the Earth's surface (see Fig. 5.7(a)). Thus, hydrogen ions that are accelerated upwards along these field lines can escape. A well-known mechanism exists to produce such upward acceleration. When solar extreme ultraviolet radiation and x-rays ionize neutral atoms and molecules in the upper atmosphere, the lighter (and more mobile) electrons congregate higher up than the heavier, less mobile ions. This produces a *charge separation electric field* that has the effect of accelerating positively charged ions in the upward direction. The dominant ion in the F region (150–800 km) of the ionosphere, where most of atomic ion production occurs, is O^+. These oxygen ions are too heavy to escape. Rather, they remain fixed and sustain the electric field, while hydrogen atoms that become ionized in Earth's polar regions are accelerated out to space. According to Yung *et al.* (1989), about 15% of the hydrogen that escapes from Earth does so by this process. Details of the polar wind process are reviewed by Schunk (1988) and Ganguli (1996).

For present-day Earth, the loss mechanisms for hydrogen in order of contribution are 60%–90% for charge exchange, 10%–40% for Jeans' escape, and ~10%–15% for the polar wind (Liu and Donahue, 1974; Maher and Tinsley, 1977; Yung et al., 1989). In fact, hydrogen escape occurs efficiently by a variety of mechanisms once hydrogen reaches the exobase. As we shall see below, upward diffusion through the static background thermosphere below the exobase limits the rate of hydrogen escape on Earth today.

It is worth noting that solar wind particles have a more direct interaction with the atmospheres on planets that do not have global dipole magnetic fields, such as Mars or Venus (Fig. 5.7(b)). In such cases, the interaction is primarily with an ionosphere rather than a magnetosphere. For small planets, such as Mars, this can lead to substantial atmospheric loss over time (Sec. 12.2.4). Of course, on bodies without atmospheres and little electrical conductivity, e.g., the Moon, the solar wind impinges directly on the surface (Fig. 5.7(c)).

5.8 Upwards Diffusion and the "Diffusion-Limited Escape" Concept

Under a variety of circumstances (covering bodies as diverse as Venus, Earth, Mars, and Titan), the flux of hydrogen escaping to space is not constrained by the rate of removal of hydrogen at the exobase but is limited by the slower, upwards supply of hydrogen through the atmosphere below the exobase. This so-called *limiting flux* will be shown to be linearly proportional to the hydrogen mixing ratio at the homopause. To derive this result, we must consider upward diffusion of hydrogen through an atmosphere. Such diffusion can take place either by molecular diffusion or by eddy diffusion, which we consider in turn.

5.8.1 Molecular Diffusion

To understand how hydrogen diffuses upward through an atmosphere, we start by considering a binary mixture of two gases, denoted 1 and 2. The relative diffusion velocity of gas 1 with respect to gas 2 in its most general 3-D form is given by Banks and Kockarts (1973 Part B, p. 33 ff.) as:

Terms on the right-hand side of this equation account for diffusion because of gradients in concentration, mass and temperature, respectively, as indicated. The last right-hand side term is diffusion caused by differential forces. Here,

n_1, n_2 = number densities of gases 1 and 2 (in m^{-3} or cm^{-3})

$n = n_1 + n_2$

m_1, m_2 = molecular masses of gases 1 and 2

$m = (n_1 m_1 + n_2 m_2)/(n_1 + n_2)$

k = Boltzmann's constant (1.38×10^{-23} J K^{-1})

T = temperature (K)

$\vec{F}_1, \vec{F}_2$ = accelerations acting on particles of gas 1 and 2 from external forces, e.g. gravity or an electric field

D_{12} = binary diffusion coefficient

α_T = thermal diffusivity ($\cong -0.25$ for H or H_2 in terrestrial air, ~0 for gases of comparable molecular mass)

∇ = the gradient operator, or $\vec{i} \frac{\partial}{\partial x} + \vec{j} \frac{\partial}{\partial y} + \vec{k} \frac{\partial}{\partial z}$ in Cartesian coordinates.

The binary diffusion coefficient, D_{12}, can be written as

$$D_{12} = \frac{b_{12}}{n} \qquad (5.34)$$

where b_{12} is a binary diffusion parameter that is found empirically to vary as $b = AT^s$ where A and s are constants for particular binary gases mixtures. Diffusion parameters for H and H_2 in terrestrial air at 300 K are given in the Table 5.2.

The vector equation (5.33) describes a 3-D distribution of particles. To apply this to a plane-parallel planetary atmosphere, we simplify to just the vertical component. For neutral gases in a gravitational field, the external force term is zero because the gravitational acceleration, g, is independent of a particle's mass, i.e.,

Table 5.2 Binary diffusion parameters for H and H_2 in terrestrial air. (Data from Banks and Kockarts (1973 Part B, pp.40–41).)

Gas 1	Gas 2	A	s	b_{12} ($cm^{-1}s^{-1}$)
H_2	Air	2.7×10^{17}	0.75	1.95×10^{19}
H	Air	4.8×10^{17}	0.70	2.60×10^{19}

$$\vec{v} = \vec{v}_1 - \vec{v}_2$$

$$= -D_{12} \left[\underbrace{\frac{n^2}{n_1 n_2} \nabla \left(\frac{n_1}{n} \right)}_{\text{conc. gradient term}} + \underbrace{\frac{m_2 - m_1}{m} \nabla (\ln p)}_{\text{mass gradient term}} + \underbrace{\alpha_T \nabla (\ln T)}_{\text{temp. gradient term}} - \underbrace{\frac{m_1 m_2}{mkT} \left(\vec{F}_1 - \vec{F}_2 \right)}_{\text{force gradient term}} \right] \qquad (5.33)$$

$\vec{F}_1 = \vec{F}_2 = g$. This would not be true for an ionized gas in the presence of an electric field where the opposite forces acting on positively charged ions and on electrons give rise to a phenomenon referred to as *ambipolar diffusion*.

With the above simplifications, let the vertical component of velocity be denoted by w. Then, eq. (5.33) becomes

$$w_1 - w_2 = -D_{12}\left[\frac{n^2}{n_1 n_2}\frac{d(n_1/n)}{dz} + \frac{m_2 - m_1}{m}\frac{1}{p}\frac{dp}{dz} + \frac{\alpha_T}{T}\frac{dT}{dz}\right]$$

$$(5.35)$$

The differential in the first term in the square brackets can be expanded as follows,

$$\frac{d}{dz}\left(\frac{n_1}{n}\right) = \frac{1}{n}\frac{dn_1}{dz} - \frac{n_1}{n^2}\frac{dn}{dz}$$

so that the first term in the brackets becomes

$$\frac{n^2}{n_1 n_2}\frac{d(n_1/n)}{dz} = \frac{n}{n_1 n_2}\frac{dn_1}{dz} - \frac{1}{n_2}\frac{dn_2}{dz} \qquad (5.36)$$

We apply these relations to a light gas moving through a stationary, heavier background gas, so species 1 = H or H_2, and species 2 = air. Then, by assumption, $w_2 = 0$ for the static air. Let us further assume that species 1, which we will henceforth denote by the subscript 'i', is a minor constituent. That is,

$$n_1 \equiv n_i << n_2$$
$$n \equiv n_1 + n_2 \cong n_2 \qquad (5.37)$$
$$m \cong m_2$$

Then, by using eq. (5.36) and relationships (5.37), we can rewrite eq. (5.35) as

$$w_i = -D_i\left[\frac{1}{n_i}\frac{dn_i}{dz} - \frac{1}{n}\frac{dn}{dz} + \left(1 - \frac{m_i}{m}\right)\frac{1}{p}\frac{dp}{dz} + \frac{\alpha_T}{T}\frac{dT}{dz}\right]$$

$$(5.38)$$

The ideal gas law, $p = nkT$ gives $n = p/kT$, so $\ln n = \ln p - \ln T - \ln k$. Taking d/dz of this last expression, gives

$$\frac{1}{n}\frac{dn}{dz} = \frac{1}{p}\frac{dp}{dz} - \frac{1}{T}\frac{dT}{dz} \qquad (5.39)$$

Substituting $(1/n)(dn/dz)$ from eq. (5.39) into eq. (5.38) yields

$$w_i = -D_i\left[\frac{1}{n_i}\frac{dn_i}{dz} - \frac{m_i}{m}\frac{1}{p}\frac{dp}{dz} + \left(\frac{1 + \alpha_T}{T}\right)\frac{dT}{dz}\right] \qquad (5.40)$$

where we have used the fact that two of the terms containing dp/dz cancel. Finally, we use the barometric law (eq. (1.10)) to write

$$\frac{1}{p}\frac{dp}{dz} = -\frac{1}{H_a} = -\frac{mg}{kT} \qquad (5.41)$$

Consequently, the second term in the square brackets of eq. (5.40) can be written,

$$\frac{m_i}{m}\frac{1}{p}\frac{dp}{dz} = \frac{m_i g}{kT} \equiv \frac{1}{H_i}$$

where H_i represents the scale height of species "i". Then, eq. (5.40) becomes

$$w_i = -D_i\left[\frac{1}{n_i}\frac{dn_i}{dz} + \frac{1}{H_i} + \left(\frac{1 + \alpha_T}{T}\right)\frac{dT}{dz}\right] \qquad (5.42)$$

The *flux* associated with molecular diffusion (also called "Fickian diffusion"), which we will denote by using a superscript "*mol*", is then

$$\Phi_i^{mol} = n_i w_i = -D_i n_i\left[\frac{1}{n_i}\frac{dn_i}{dz} + \frac{1}{H_i} + \left(\frac{1 + \alpha_T}{T}\right)\frac{dT}{dz}\right]$$

$$(5.43)$$

Here the first term ($D_i\, dn_i/dz$) on the right-hand side is the familiar form of *Fick's First Law*, which expresses how the flux of molecules of species i across unit area in unit time is proportional to the concentration gradient of that species, dn_i/dz.

5.8.2 Eddy Diffusion

Thus far, we have described the individual motions of gas molecules of trace species "i" with respect to molecules of a static background atmosphere. In the lower atmosphere, though, most of the mass transport occurs not by diffusion of individual molecules but, rather, by turbulent, macroscopic eddies or by advection of air parcels, again of macroscopic scale. For convenience, aeronomers lump all such transport into the single process of *eddy diffusion*. Its magnitude is parameterized by an *eddy diffusion coefficient, K*. Eddy diffusion, by its very nature, acts so as to reduce gradients in relative species concentrations. Thus, if we define the *volume mixing ratio* of species i as $f_i \equiv n_i/n$, then we can write the flux due to eddy diffusion as

$$\Phi_i^{eddy} = -Kn\frac{df_i}{dz} \qquad (5.44)$$

The total flux due to both molecular and eddy diffusion is their sum,

$$\Phi_i = \Phi_i^{mol} + \Phi_i^{eddy} \qquad (5.45)$$

The value of the eddy diffusion coefficient, K, is not precisely defined, unlike the molecular diffusion coefficient,

(a)

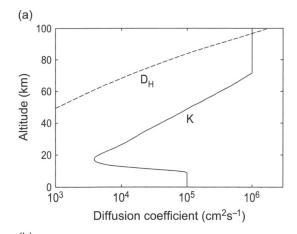

(b)

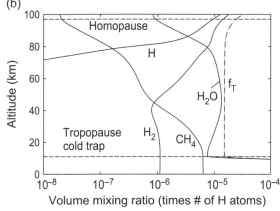

Figure 5.8 (a) Vertical profiles of the eddy diffusion coefficient, K, and the molecular diffusion coefficient for atomic hydrogen, D_H. (b) Vertical mixing ratio profiles of major hydrogen-bearing species, weighted by the number of H atoms in each species, e.g., methane at ground level is 1.8 ppmv $\times$ 4 = 7.2 ppmv H. The curve labeled f_T represents the total hydrogen mixing ratio defined in the text. The value of f_T begins to increase near the homopause because of diffusive separation of species that favors H and H_2.

D_i. No theory yet gives us the exact magnitude of K; rather, K is determined empirically using tracer studies (Hunten, 1975; Massie and Hunten, 1981; Gutowsky, 1976; National Research Council, 1979). A typical example is the eddy diffusion profile shown in Fig. 5.8(a).

Theory provides some guidance on how K should vary in certain altitude regimes. For example, in the upper stratosphere and mesosphere, most of the mass transport is caused by turbulent eddies generated by the breaking of upwards-propagating gravity waves (Garcia and Solomon, 1985; Leovy, 1964). The amount of energy dissipated by such waves is predicted to vary as the inverse square root of density (see Sec. 4.3.4 and 4.4.1), and so in this region it is assumed that $K \propto n^{-1/2}$. Because the

spatial scale of the eddies is relatively small, parameterizing mass transport as "diffusion" is a good approximation in this region. By contrast, in the lower stratosphere and troposphere, much of the mass transport occurs as a result of large-scale advection, so the eddy diffusion approximation is less well justified.

At some altitude (~100 km on Earth), D_i becomes greater than K. This altitude is the *homopause* (as defined in Sec. 1.1.1). Although K increases as $n^{-1/2}$ in the mesosphere, the molecular diffusion coefficient, D_i, is proportional to $1/n$ (eq. (5.35)) and, hence, increases more rapidly with height than K. The region below the homopause, where eddy diffusion dominates, is the *homosphere* where air is mixed, or homogenized, by the processes that we have just discussed. Species that lack strong chemical sources or sinks, e.g., N_2, have constant mixing ratios in the homosphere. This can be easily demonstrated by setting $\Phi_i^{eddy} = 0$ in eq. (5.44). Then, $df_i/dz = 0$ as well, which implies that f_i is constant with altitude.

The region above the homopause, where molecular diffusion dominates, is the *heterosphere*. Here, mixing ratios of lighter species increase with altitude. This can be easily demonstrated by considering a non-reactive species such as N_2 for which the vertical flux is essentially zero. Then, eq. (5.43) says that

$$\frac{1}{n_i}\frac{dn_i}{dz} = -\frac{1}{H_i} - \left(\frac{1+\alpha_T}{T}\right)\frac{dT}{dz} \qquad (5.46)$$

If we neglect thermal diffusion (i.e., set $\alpha_T = 0$), this implies that the partial pressure, $p_i = n_i kT$, of each individual species varies with its own scale height, i.e.

$$p_i = p_0 \exp\left(-\frac{z-z_0}{H_i}\right) \qquad (5.47)$$

where p_0 is the partial pressure at some reference height z_0. Partial pressure $p_i(z)$ is the analog of the barometric law (eq. (1.12)) for the total atmospheric pressure, $p(z)$.

5.8.3 Diffusion-Limited Escape of Hydrogen

We now derive a very useful expression for the maximum upward flux of an escaping gas, such as hydrogen. As shown below, the escape of hydrogen from Earth's atmosphere is limited by the rate at which it can diffuse upwards through the homopause. Physically, one may think of the static background atmosphere, N_2 and O_2 in Earth's case, as providing a frictional resistance that retards the upward flow of hydrogen. The significance of this phenomenon was first pointed out by Donald Hunten (1973) in a paper that was motivated by his work on the escape of hydrogen from Saturn's moon, Titan. It was also well described by Walker (1977) and provided the basis for his pioneering work on predicting Earth's prebiotic O_2 concentrations (Walker, 1978). The limiting flux concept has proven extremely valuable in understanding the behavior of atmospheres on rocky planets because, to the degree that data are available, diffusion-limited escape explains the H escape rate from Earth, Venus, Mars, and Titan (Sec. 5.9).

We begin by recasting the expression for the molecular diffusion flux (eq. (5.43)) in a form similar to that of the eddy diffusion flux (eq. (5.44)). First we note that

$$\frac{df_i}{dz} = \frac{d}{dz}\left(\frac{n_i}{n}\right) = \frac{1}{n}\frac{dn_i}{dz} - \frac{n_i}{n^2}\frac{dn}{dz}$$

Hence,

$$n\frac{df_i}{dz} = \frac{dn_i}{dz} - \frac{n_i}{n}\frac{dn}{dz} = n_i\left(\frac{1}{n_i}\frac{dn_i}{dz} + \frac{1}{H_a} + \frac{1}{T}\frac{dT}{dz}\right) \quad (5.48)$$

where we have used eq. (5.39) and hydrostatic eq. (5.41) to eliminate dn/dz. By comparing eq. (5.48) with the molecular diffusion flux (eq. (5.43)), we can write

$$\Phi_i^{mol} = -D_i n\frac{df_i}{dz} + D_i n_i\left[\frac{1}{H_a} - \frac{1}{H_i} - \frac{\alpha_T}{T}\frac{dT}{dz}\right] \quad (5.49)$$

If we now combine eq. (5.49) with eddy diffusion (eq. (5.44)) and total flux (eq. (5.45)) equations, we can write the total flux of species "i" as

$$\Phi_i = \underbrace{-(K + D_i)n\frac{df_i}{dz}}_{\text{counter-gradient flux term}} + \underbrace{D_i n_i\left(\frac{1}{H_a} - \frac{1}{H_i} - \frac{\alpha_T}{T}\frac{dT}{dz}\right)}_{\text{limiting flux term}}$$

$$(5.50)$$

The first term on the right-hand side of eq. (5.50) is called the *counter-gradient flux* and is denoted by Φ_c

$$\Phi_c = -(K + D_i)n\frac{df_i}{dz} \quad (5.51)$$

The magnitude of this term is proportional to the gradient in species mixing ratio, df_i/dz. It turns out that this term is unable to sustain any net upwards diffusion of hydrogen. Physically, the reason is because f_i must decrease with altitude in order to drive upwards diffusion via this term. But if f_i decreases with height, then n_i will decrease even more rapidly. The upward flux is equal to $n_i w_i$, where w_i is a velocity. Rapidly decreasing n_i would require rapidly increasing w_i, which is not physically possible. (See Walker, 1977, for a mathematical derivation of this result.) Thus, a maximum must occur in Φ_i (eq. (5.50)) when $df_i/dz = 0$ and $\Phi_c = 0$.

The second term on the right-hand side of eq. (5.50) is the *limiting flux* or the *diffusion-limited flux* and is denoted by Φ_l, as follows:

$$\Phi_l = D_i n_i\left(\frac{1}{H_a} - \frac{1}{H_i} - \frac{\alpha_T}{T}\frac{dT}{dz}\right)$$

$$\approx D_i n_i\left(\frac{(m_a - m_i)g}{kT} - \frac{\alpha_T}{T}\frac{dT}{dz}\right) \quad (5.52)$$

As may be ascertained from the form of this equation, this term is entirely due to the difference in molecular weight between the escaping gas (presumed to be hydrogen) and the background atmosphere and to the thermal diffusivity of hydrogen. This expression is typically applied either at the homopause or in the lower stratosphere.

We can simplify the limiting flux equation as indicated by the strike-throughs in eq. (5.52). Temperature gradients are small, so $dT/dz \sim 0$ and the thermal diffusion term is generally neglected. Furthermore, for a light gas (H or H_2) diffusing through air (an N_2–O_2 mixture for Earth, or CO_2 for Mars and Venus), $H_i \gg H_a$, so eq. (5.52) simplifies to

$$\boxed{\text{diffusion limited flux, } \Phi_l \cong \frac{D_i n_i}{H_a} = \frac{b_i f_i}{H_a} \propto f_i} \quad (5.53)$$

where we have used eq. (5.34), which relates D_i to the binary diffusion parameter b_i.

It should be remembered that eq. (5.53) was derived for a minor constituent. It can, however, be applied to a major constituent as well (c.f. Walker, 1977) if one replaces the term f_i with $f_i/(1+f_i)$. This form should be used in cases where f_i exceeds a few percent.

5.8.4 Application of Diffusion-Limited Hydrogen Escape to Earth's Atmosphere

To apply the limiting flux equation to the Earth, we begin by evaluating some of the parameters. At the Earth's homopause, the temperature is ~208 K, $b_H \cong 2.73\times 10^{19}$ cm^{-1}s^{-1} and $b_{H2} \cong 1.46\times10^{19}$ cm^{-1}s^{-1}. H_2 is several times more abundant that H at the homopause, 5.2×10^7 cm^{-3} vs. 1.8×10^7 cm^{-3} (Liu and Donahue, 1974). Rather than calculating the flux of each species separately, let us combine their mixing ratios and use a weighted average value for b_i of 1.8×10^{19} cm^{-1}s^{-1}. At the homopause, the scale height, H_a, is ~6.36 km, so $b_i/H_a \cong 1.8\times10^{19}$ cm^{-1}s^{-1}/6.36$\times10^5$ cm = 2.8×10^{13} cm^{-2} s^{-1}.

We now define the *total hydrogen mixing ratio*, f_T(H), as the sum of the mixing ratios of hydrogen in all of its chemical forms, weighted by the number of hydrogen atoms each species contains. Thus,

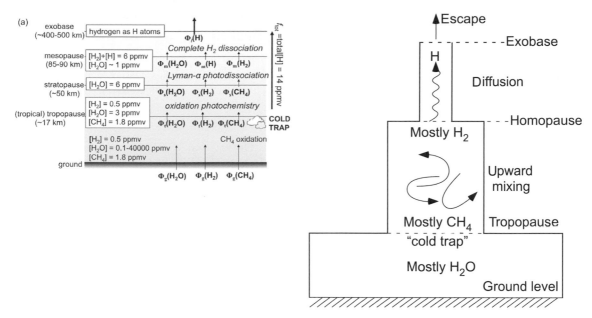

Figure 5.9 (a) A schematic diagram of hydrogen-bearing species in Earth's atmosphere showing processes responsible for the profiles shown in Fig. 5.8(b). Hydrogen is input from the ground in fluxes (Φ) of water, hydrogen and methane. At the tropical tropopause, a cold trap limits the flux of hydrogen from water vapor into the stratosphere. The total H mixing ratio f_T, just below the homopause sets the diffusion-limited escape flux of hydrogen to space (Φ$_l$), which is reflected by f_T in the lower stratosphere above the tropical tropopause cold trap. (b) A schematic diagram showing the two principal "bottle-necks" for hydrogen escape: the tropical tropopause cold trap and diffusion above the homopause.

$$f_T(H) = f_H + 2f_{H_2} + 2f_{H_2O} + 4f_{CH_4} + \cdots \qquad (5.54)$$

These terms, along with $f_T(H)$ itself, are shown in Fig. 5.8 (b). At the Earth's homopause, almost all hydrogen exists as either H or H_2, so only the first two terms are important. We could evaluate $f_T(H)$ there by using rocket measurements of the concentrations of these two species. However, there is an easier way to evaluate $f_T(H)$ that is also much more useful in studying atmospheres in general, which we now examine.

To find a simple way to estimate the total hydrogen mixing ratio, $f_T(H)$, we start with fundamentals by writing the *continuity equation* for species i as

$$\frac{d\Phi_i}{dz} = P_i - L_i \qquad (5.55)$$

Here, P_i and L_i represent chemical production and loss rates per unit volume. For a species with no chemistry, $d\Phi_i/dz = 0$, so Φ_i is constant with altitude. But we know that $\Phi_i = \Phi_l$ at the homopause. Consequently, the total flux at all altitudes below and down to the lower stratosphere must be equal to the diffusion-limited flux. Equation (5.50) therefore implies that $df_i/dz = 0$, that is, the total hydrogen mixing ratio should be constant with altitude. This result is not exact because the species scale height, H_i, and thermal

diffusivity, α_T, change as one hydrogen-containing compound is transformed into another. However, these changes can be compensated by gradients in f_i that are so small as to be negligible. Thus, despite all the complexity of atmospheric photochemistry, we arrive at the simple result:

The rate of escape of hydrogen is proportional to the concentration of hydrogen compounds in the lower stratosphere.

We can see how this works by looking at a schematic diagram showing the abundance of hydrogen in all its forms in the Earth's atmosphere as a function of height shown in Fig. 5.8(b) and schematically in Fig. 5.9(a). The total hydrogen mixing ratio f_T is essentially preserved above the tropopause, as verified by satellite measurements of hydrogen-bearing species (Harries *et al.*, 1996).

We can write the diffusion limited escape flux as a simple linear equation. Using eq. (5.53), we can represent the escape flux as

$$\Phi_l \cong (\text{constant}) f_T(H) \qquad (5.56)$$

We deduced a constant of 2.8×10^{13} cm^{-2} s^{-1} from our previous calculation. However, a more detailed calculation that takes account of other terms in eq. (5.52) suggests that the constant is 2.5×10^{13} cm^{-2} s^{-1} (Hunten and Strobel 1974), i.e.,

$$\Phi_l \cong \left(2.5 \times 10^{13}\right) f_T(H) \left[\text{atoms cm}^{-2}\text{s}^{-1}\right] \qquad (5.57)$$

Thus, provided we know $f_T(H)$ we can easily calculate the diffusion-limited escape flux. Here, $f_T(H)$ is simply the total hydrogen mixing ratio, in all its forms, above the cold trap at the tropopause, which can be readily measured from balloon-borne instruments in Earth's present atmosphere or computed theoretically for other atmospheres.

Diffusion-limited escape is an *upper limit*. The hydrogen cannot escape any faster than the rate of upwards diffusion through lower levels of the atmosphere by eddy and molecular means. Thus, conceptually, the Earth has two "bottlenecks" that limit the upward supply of hydrogen (Fig. 5.9(b)): the cold-trap of the tropical tropopause where water freezes out and the region above the homopause where diffusion is rate limiting.

In Earth's present atmosphere, the concentrations of hydrogen-bearing gases in the lower stratosphere are 1.8 ppmv CH_4, ~3 ppmv H_2O, and 0.55 ppmv H_2. From these measurements, we can calculate the diffusion-limited escape rate for hydrogen from the Earth. First we calculate $f_T(H)$, which is 14×10^{-6} (= $[2(0.55) + 4(1.8) + 2(3)] \times 10^{-6}$).

Then the escape rate from eq. (4.51) is

$$\Phi_l \cong \left(2.5 \times 10^{13}\right) f_T(H) = \left(2.5 \times 10^{13}\,\text{cm}^{-2}\text{s}^{-1}\right)\left(14 \times 10^{-6}\right) = 3.5 \times 10^8 \text{ atoms cm}^{-2}\text{s}^{-1}.$$

Given that the area of the Earth is 5.1×10^{18} cm^2, the area-integrated escape rate is $(3.5 \times 10^8$ atoms cm^{-2} s^{-1}) $\times$ $(5.1 \times 10^{18}$ cm^2) = 1.8×10^{27} atoms s^{-1} = 5.6×10^{34} atoms yr^{-1} = 9.3×10^{10} moles H atoms yr^{-1} = 93 000 tonnes of H yr^{-1}.

Measurements of exospheric temperatures and hydrogen densities prove that the limiting flux concept works well (Bertaux, 1975). Essentially, the total flux consists of the sum of suprathermal and Jeans' escape fluxes. The sum remains constant at about the limiting flux value at the homopause. As the Jeans flux varies due to changing exospheric temperature with the solar cycle, the suprathermal flux component changes to make up the balance.

The limiting flux concept does not work for terrestrial helium. Helium escape is rate-limited by the removal process at the top of the atmosphere to a flux ~100 times lower than diffusion-limited escape. Instead helium escapes efficiently as an ion along the open magnetic field lines at high latitudes (Axford, 1968; Johnson and Axford, 1969), with the result that the lifetime of helium in Earth's atmosphere is ~10^6 years.

5.9 Diffusion-Limited Hydrogen Escape Applied to Mars, Titan, and Venus

Now that we have discussed the theory for hydrogen escape and the limiting flux concept, it is interesting to apply it to other planetary atmospheres.

5.9.1 Mars

On Mars, simulated homopause temperatures vary as a function of season and range from ~150 K to ~200 K (e.g., Bougher *et al.*, 2000). Let us take $T \sim$ 180 K, corresponding to the dayside homopause at ~135 km altitude. Here, the most abundant hydrogen species is H_2 (Nair *et al.*, 1994). The binary diffusion coefficient of H_2 in CO_2 is approximated a function of temperature (Marrero and Mason, 1972),

$$b(H_2 \text{ in } CO_2) = \frac{3.1 \times 10^{-6}}{k} T^{0.75} e^{-11.7/T} \left[\text{cm}^{-1}\text{s}^{-1}\right]$$

$$(5.58)$$

A temperature of 180 K yields $b = 1.0 \times 10^{19}$ cm^{-1} s^{-1} (for H_2 in CO_2), and a scale height H_a = 9.3 km. Consequently, the limiting flux from eq. (5.53) is

$$\Phi_l \cong \left(1.1 \times 10^{13}\right) f_T(H) \qquad (5.59)$$

Spectroscopy suggests that the H_2 mixing ratio on Mars is ~15±5 ppmv in the lower atmosphere (Krasnopolsky and Feldman, 2001). Insertion of $f_T(H) = (30 \pm 10) \times 10^{-6}$ in eq. (5.59) gives a diffusion-limited escape rate of $(3.3 \pm 1.1) \times 10^8$ H atoms cm^{-2} s^{-1}.

Note that the low gravity on Mars leads, counter-intuitively, to a *smaller* diffusion-limited escape flux, according to eq. (5.53). This is because the atmospheric scale height, $H_a = kT/mg$, appears in the denominator of the expression, making $\Phi_l \propto g$. Essentially, the rate of diffusion of the light escaping gas relative to the heavier background gas is enhanced when the planet's gravitational attraction is stronger. This scaling will break down for very large planets when gravity-dependent removal processes at the exobase become inefficient and the limiting flux no longer applies to hydrogen.

Hydrogen escape fluxes cannot be observed directly but are inferred from the vertical profile of hydrogen. *Mariner* 6 and 7 Lyman-α observations imply an escape flux of $(1-2) \times 10^8$ H atoms cm^{-2} s^{-1} if a

Maxwellian velocity distribution is assumed (Anderson, 1974; Anderson and Hord, 1971). *Mars Express* data have been interpreted to indicate that atomic hydrogen in Mars' exosphere has two populations: a suprathermal one (~1000 K) with a low number density and a colder one with a greater number density. The suprathermal population may arise from ionic reactions or charge exchange reactions with the solar wind (Galli *et al.*, 2006a; Galli *et al.*, 2006b). Such data imply a Jeans escape rate of $4.3^{+5.6}_{-2.5} \times 10^8$ H atoms cm^{-2} s^{-1} (Zahnle *et al.*, 2008). Lyman-α and -β data from *Rosetta* were acquired during a gravity assist swing-by of Mars (while en route to comet 67P/Churyumov–Gerasimenko) and used to infer a somewhat lower escape rate (no error bar) of ~0.8×10^8 H atoms cm^{-2} s^{-1} (Feldman *et al.*, 2011).

Overall, the limiting flux concept appears to work within uncertainty in explaining the current hydrogen escape rate for Mars. Since diffusion-limited escape depletes the typical abundance of water vapor in the atmosphere (~5×10^{19} cm^{-2} H atoms) in ~5 kyr, atmospheric water vapor must be replenished. One nuance is that *Mars Express* and *Hubble Space Telescope* data suggest that the H escape rate, although consistent with limiting flux, can decline when there is a decrease in the source of water vapor during large dust storms (Chaffin *et al.*, 2014; Clarke *et al.*, 2014).

5.9.2 Titan

On Titan, the homopause altitude is constrained by data from the *Cassini* spacecraft to be ~800–900 km (Strobel *et al.*, 2009) while the exobase is around 1400–1500 km altitude (see Sec. 12.4.3). Photochemical models suggest that most hydrogen escapes as H$_2$, with ~30% as H (Yung *et al.*, 1984).

The *Huygens* probe measured hydrogen-bearing species in Titan's atmosphere, with CH$_4$ ~1.48±0.09% in the stratosphere and H$_2$ ~1010±160 ppmv (Niemann *et al.*, 2010). At the homopause, the temperature is T~150 K, which gives a scale height $H_a = RT/Mg = $ [(8,314 J mol^{-1} K^{-1})(150 K)]/[0.0286 kg mol^{-1}) (0.733 m s^{-2})] = 59.5 km. The binary diffusion parameter for H$_2$ in N$_2$ at 150 K is $b = 1.1 \times 10^{19}$ cm^{-1} s^{-1}. Hence the limiting flux obtained from eq. (5.53) is

$$\Phi_l \cong \frac{b_i}{H_a} f_T = \frac{1.1 \times 10^{19}}{5.95 \times 10^6} f_T = 1.8 \times 10^{12} f_T \qquad (5.60)$$

Since methane is the dominant hydrogen-bearing species we must take into account the overall chemistry of methane photolysis and the fate of its products. Methane

can be destroyed through different photolysis paths with differing yields of hydrogen:

$$2CH_4 \rightarrow C_2H_2 + 3H_2 \text{ (ionospheric)} \qquad (5.61)$$

$$2CH_4 \rightarrow C_2H_4 + 2H_2 \text{ (direct)} \qquad (5.62)$$

$$2CH_4 \rightarrow C_2H_6 + H_2 \text{ (catalytic)} \qquad (5.63)$$

Hydrogen atoms contained in the hydrocarbon products are lost as rainout to the surface whereas the H$_2$ is subject to escape. If we assume that every two CH$_4$ molecules produce an H$_2$ molecule (eq. (5.63)), then the total mixing ratio of H atoms will be f_T = (0.0148/2)2 + 2(0.001) = 0.0168, giving a limiting flux of 3×10^{10} atoms cm^{-2} s^{-1}. More detailed model calculations suggest an escape flux of 2.0×10^{10} atoms cm^{-2} s^{-1} (Table 1 in Lebonnois *et al.* (2003)), which is similar to the column-integrated destruction rate of CH$_4$ ~1.5×10^{10} cm^{-2} s^{-1} due to photolysis (Yung *et al.*, 1984).

The limiting flux for H escape from Titan is supported by observations. Utilizing *Cassini* ion-neutral mass spectrometer (INMS) data, Bell *et al.* (2010a; 2010b) find that hydrogen escapes at close to the Jeans rate, while others have suggested that the rate is sometimes enhanced with energy input by Saturn's magnetospheric particles (Cui *et al.*, 2011). The inferred escape rate is $(2.0–2.1) \times 10^{10}$ H atoms cm^{-2} s^{-1} (Bell *et al.*, 2010a; Cui *et al.*, 2008), which is essentially the diffusion-limited rate evaluated at the homopause (Bell *et al.*, 2014; Strobel, 2012).

5.9.3 Venus

We can calculate a diffusion-limited H escape flux for Venus of ~3×10^7 H atoms cm^{-2} s^{-1}, based on spectroscopic evidence for a total atomic hydrogen abundance of a few ppmv at the homopause. This total hydrogen mixing ratio comes from summing HCl ~0.5 ppmv (Bézard *et al.*, 1990) and H$_2$O ~ 1 ppmv (Fink *et al.*, 1972) above the cloud tops. This diffusion-limited escape flux agrees reasonably well with estimates of the globally averaged escape flux based on *Pioneer Venus* measurements of upper atmospheric composition and temperature, which in units of ~10^7 atoms cm^{-2} s^{-1} are 1.7 (Hodges and Tinsley, 1981), 0.2 (McElroy *et al.*, 1982) and 2.7 (Kumar *et al.*, 1983).

On Venus, Jeans' escape of hydrogen is negligible because of the relatively high, Earth-like gravity and low exospheric temperature. The upper atmosphere of Venus is cold, with a ~300±25 K dayside exobase temperature, because of strong radiative cooling via decay of vibrationally excited CO$_2$ (see Sec. 3.5). At 275 K, the thermal

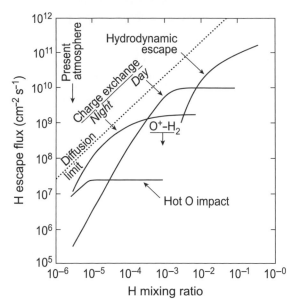

Figure 5.10 Calculated hydrogen escape fluxes from Venus. From left to right: increasing H concentrations to ~1% eventually permit hydrodynamic escape. At lower concentrations of H, charge exchange processes are the dominant source of hot H atoms. However, in all cases, the rate of escape is limited by diffusion through the atmosphere from lower layers (dashed line). (From Kumar *et al.* (1983).)

escape rate is only ~10^4 atoms cm^{-2} s^{-1}. On Venus, escape of suprathermal hydrogen completely dominates.

Mariner 5 and *10, Venera 9–12* and *Venus Express* experiments all found evidence for a suprathermal population of hydrogen in Venus' exosphere with a temperature of ~1000 ± 500 K and exobase density of ~10^3 cm^{-3}, compared to a cold, background hydrogen density of ~10^5 cm^{-3} at the exobase and temperature 300 ± 25 K (dayside) or 150 K (nightside) (Bertaux *et al.*, 1978; Bertaux *et al.*, 1982; Chaufray *et al.*, 2012; Kumar *et al.*, 1983). Suprathermal escape mechanisms that are considered to be important for Venus include (Hodges, 1999; Lammer *et al.*, 2006): (1) charge exchange between neutral hydrogen atoms and hot H$^+$ or O$^+$ ions, (2) charge exchange with hot oxygen atoms produced by CO$_2$ photolysis or from electron impact dissociation of CO$_2$, and (3) dissociative recombination of H-containing ions. Protons (H$^+$), produced by photoionization or solar wind interaction, can also escape, as measured at solar minimum at a rate of 3×10^6 cm^{-2} s^{-1} by an ion spectrometer on *Venus Express* (Barabash *et al.*, 2007).

Figure 5.10 is a summary plot of the various processes responsible for H escape from Venus, including on early Venus when hydrodynamic escape likely operated. The dashed line shows the diffusion-limited flux,

which represents the maximum escape rate from the sum of these processes.

5.10 Hydrodynamic Escape

5.10.1 Conditions for Hydrodynamic Escape

In Sec. 5.3, we saw how the barometric law breaks down at high altitude and how a gravitationally bound atmosphere at a finite temperature must lose mass to the vacuum of space. We noted the two end-member cases of thermal escape processes associated with this breakdown, namely, Jeans' escape and hydrodynamic escape. Jeans' escape is the more accurate description when dealing with escape of a light gas, e.g., hydrogen, from a heavier and static background atmosphere. Hydrodynamic escape is a better approximation when the background atmosphere itself is escaping and driven by a pressure gradient force between the dense atmosphere below and vacuum above, or, equivalently, when the upper atmosphere is hydrogen-dominated and sufficiently heated. An important aspect of hydrodynamic hydrogen escape for atmospheric evolution is that heavy molecules can achieve escape velocity through collisions with hydrogen, and be dragged upward. Thus, rapid flow of hydrogen will carry away heavy gases into space, even though such gases would be too heavy to undergo Jeans' escape.

The classical Jeans' approximation treats the transition from collisional to collisionless as a discontinuity, but this concept begins to break down once the height of the exobase becomes defined by the escaping gas, typically hydrogen, rather than by some static background gas. As we noted earlier (Sec. 5.6), in a multi-component atmosphere, the classical Jeans' formula can overestimate the escape flux of hydrogen because it fails to account for depletion of molecules in the high-energy tail of the Maxwellian velocity distribution. Also, the Jeans approximation ignores the reduction in temperature caused by the loss of fast, escaping particles. This error of 20%–30% for Earth is not large because that tail is replenished through collisions with non-escaping molecules. In the transitional region from collisional to non-collisional gas, collisions between the atoms or molecules of the minor escaping gas are unimportant.

However, once the escaping gas becomes the dominant constituent in the thermosphere, the situation is more akin to a single-component atmosphere and the net result can be an anisotropic velocity distribution rather than an isotropic Maxwellian distribution. By integrating the non-linear Boltzmann equation for an atmosphere of atomic hydrogen, Merryfield and Shizgal (1994) found that, at high altitudes, a population of escaping hydrogen atoms was enhanced by streaming of particles from warmer, denser air below. There was also upward conduction of heat due to the escape of the particles from the tail of the distribution and some particles above the exobase gained escape velocity due to occasional collisions. Consequently, in such a single component model, Jean's escape was found to *underestimate* actual escape by a factor of ~1.3 for a Jeans' parameter of ~6.5 (eq. (5.25)). Using a direct simulation Monte Carlo model (DSMC), Volkov *et al.* (2011a; 2011b) similarly found that Jeans escape in a single-component atmosphere *underestimates* actual escape at the exobase by a factor of ~1.4–1.7 for a Jeans' parameter ranging from 6 to 15.

Sophisticated models, such as DSMC, are required for precise estimates of strong "Jeans-like" escape to bridge the collisionless and hydrodynamic regimes when the Jeans parameter is small but escape is not strong enough for a purely fluid description of hydrodynamic escape to be justified. Apart from DSMC models, another method that has been used extensively to simulate the polar wind (Lemaire *et al.*, 2007; Tam *et al.*, 2007) is to make an approximation to the Boltzmann equation with a number of moments, e.g., Grad's 13-moment approximation (Cui *et al.*, 2008; Grad, 1949; Schunk, 1977; Ch. 3 of Schunk and Nagy, 2009).

The theory of hydrodynamic escape for a planetary atmosphere is best put in context by referring back to the old literature concerning the hydrodynamic nature of the solar wind. In the 1960s, Joseph Chamberlain and Eugene Parker debated the question of whether the solar wind was subsonic or transonic. Parker thought that it was transonic while Chamberlain thought that it was not. Ultimately, Parker was proved correct.

The debate happened because both subsonic and transonic solutions exist for the equations of hydrodynamic outflow of the solar wind. The wind, which is an expansion of the Sun's extremely hot corona, is fully ionized. The particles moving within it are charged and are subject to long-range electrostatic forces; hence, the wind is always in the collisional hydrodynamic regime. More precisely, the dynamics of the solar wind are described by magneto-hydrodynamics, which includes the influence of the magnetic field, but early treatments of the solar wind ignored this complication. In general, the solutions of the full equations of hydrodynamic outflow are difficult to obtain. If, however, one ignores the energy equation and assumes isothermal outflow, then the solution is analytic (Box 5.2, eq. (B5.11)). These isothermal outflow solutions are shown in Fig. 5.11.

The isothermal outflow solutions for the solar wind fall into six categories, but only three are physically

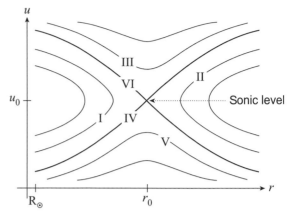

Figure 5.11 Solutions to the isothermal, time-independent, hydrodynamic escape equations (Box 5.2, eqs. (B5.11) or (B5.12)), but for an ionized gas. Six different classes of solution are shown (I to VI). Solution IV, which increases in speed and passes through the *sonic level* (also called the *critical point*) and becomes supersonic, is the solar wind solution. (Following Parker (1963).)

plausible. Two solutions pass through the *critical point* or *sonic level*, r_0, which is the radius at which the flow velocity is equal to the isothermal sound speed, $u_0 = \sqrt{kT/m}$. Variables are defined in Box 5.2. Solution IV in Fig. 5.11, which starts from low velocity near the Sun and becomes supersonic at large distances, is the transonic escape solution. This solution was eventually shown by spacecraft data to be correct for the solar wind. Solution VI has low velocity at large distances and high velocity near the Sun. It represents infall of material, called *Bondi accretion* (Bondi, 1952; Shu, 1991 pp. 77–81). It could apply, for example, to capture of gas from a surrounding solar nebula, although the isothermal assumption would need to be abandoned in this case. This solution is thus more illustrative than practical.

The other categories of solutions (I, II, III, V) fall into the four quadrants delineated on the graph by solutions IV and VI. Double-valued solutions in quadrants I and II do not connect the surface of the Sun ($r << r_0$) to locations far from the Sun ($r >> r_0$) and are physically meaningless. Quadrant III solutions have supersonic speeds at the Sun and are also unphysical. Apart from solutions IV and VI, only solutions in the bottom center quadrant (V) are potentially meaningful. These subsonic solutions have low velocity near the Sun, a peak velocity at distance r_0, and return to low velocities at large distances.

For an escaping fluid like the solar wind, which is collisional, the question of whether the flow is subsonic or transonic depends on the boundary conditions at large distances from the Sun. If the pressure of the interstellar medium is low, which it is, then the escape will be

Box 5.2 Equations of Hydrodynamic Escape and Isothermal Approximations Pertaining to Them

The equations describing hydrodynamic escape, and fluid dynamics in general, tend to be rather opaque to those who do not deal with them on a regular basis. For many of us, there is much to be learned from a simplified description of a problem because one can gain physical insight from analytic approximations. Hydrodynamic escape is amenable to an analytic simplification with certain assumptions. In general, one needs to deal with three equations: conservation of mass, momentum, and energy. In spherical geometry, these equations can be expressed as follows (using equations of Chamberlain and Hunten (1987), p. 71–73, in spherical geometry).

Conservation of mass

$$\underbrace{\frac{\partial \rho}{\partial t}}_{\text{density change}} = \underbrace{-\frac{1}{r^2}\frac{\partial}{\partial r}(r^2 \rho u)}_{\text{radial mass inflow or outflow (divergence)}} \tag{B5.1}$$

Conservation of momentum

$$\underbrace{\frac{\partial(\rho u)}{\partial t}}_{\text{momentum change}} = \underbrace{-\rho u \frac{\partial u}{\partial r}}_{\text{momentum inflow or outflow}} \underbrace{-\frac{\partial p}{\partial r}}_{\text{pressure gradient force}} \underbrace{-\rho \frac{GM}{r^2}}_{\text{gravitational force}} \tag{B5.2}$$

Conservation of energy

$$\rho c_{\mathrm{p}} \frac{\partial T}{\partial t} = \frac{1}{r^2}\frac{\partial}{\partial r}\left(r^2 \kappa \frac{\partial T}{\partial r}\right) - \frac{kTu}{m}\frac{\partial \rho}{\partial r} - \rho c_{\mathrm{v}} u \frac{\partial T}{\partial r} + q \tag{B5.3}$$

Here, t is time, r is radial distance from the planet's center, ρ is mass density, u is radial velocity, p is pressure, G is the universal gravitational constant, M is the planet's mass, c_{p} and c_{v} are the specific heats at constant pressure and volume, κ is the thermal conductivity, k is Boltzmann's constant, and q is the specific heating (less cooling) rate per unit volume. Note that if there is no radial flow then $u(r) = 0$ and eq. (B5.2) reduces to the hydrostatic equation, $\partial p/\partial r = -g\rho$.

The last of the equations above can be expressed in a number of forms. If one neglects the term in eq. (B5.3) involving $\partial T/\partial r$, the three equations are often referred to as *the Euler equations* of fluid dynamics. Techniques for solving these equations, e.g., Godunov's method, are described in Toro (1999) and LeVeque (2002). In general, numerical solution is complex, partly as a consequence of the existence of shock waves in the solutions and partly because of possible transitions from subsonic to supersonic flow. Simpler numerical techniques, specifically the Lax–Friedrichs method, have been used to find transonic solutions to these equations (Tian *et al.*, 2005), but artificial numerical diffusion in such techniques can violate mass and energy conservation and cause order of magnitude underestimation of the escape rate compared to a more accurate "constrained interpolation profile" (CIP) scheme (Kuramoto *et al.*, 2013). More elementary methods can be used to find steady-state (time-independent) solutions (Watson *et al.*, 1981).

Simplification is possible if one ignores both time dependence and the energy equation (B5.3) and assumes steady-state, isothermal expansion. Under these assumptions, eqns. (B5.1) and (B5.2) can be rewritten as

$$\frac{1}{r^2}\frac{d}{dr}(r^2 \rho u) = 0 \quad \text{mass conservation} \tag{B5.4}$$

$$\rho u \frac{du}{dr} = -\frac{dp}{dr} - \rho \frac{GM}{r^2} \quad \text{momentum conservation} \tag{B5.5}$$

Parker (1963) first presented a solution of these coupled equations, as follows. First, we eliminate dp/dr from equation (B4.5) using the ideal gas

$$p = \rho kT/m \tag{B5.6}$$

Differentiation of eq. B5.6 with respect to r gives,

$$\frac{dp}{dr} = \frac{kT}{m}\frac{d\rho}{dr} \tag{B5.7}$$

where m is molecular (or atomic) mass, and we have used an isothermal approximation by setting $dT/dr = 0$. Substituting equation (B5.7) back into (B5.5) and dividing by ρ yields

$$u\frac{du}{dr} = -\frac{kT}{m}\frac{1}{\rho}\frac{d\rho}{dr} - \frac{GM}{r^2} \tag{B5.8}$$

Now, we use mass conservation equation (B5.4). Integrating with respect to r yields

$$r^2\rho u = \text{constant} \equiv C \tag{B5.9}$$

Here the constant C is related to the escape flux. The physical meaning of eq. (B5.9) is seen if we multiply by 4π, i.e., $4\pi r^2\rho u = F$, where F is the mass flux [kg s^{-1}] through the surface area of a sphere. Now take logarithms of eq. (B5.9) (i.e., $2\ln r + \ln \rho + \ln u = \ln C$) and differentiate with respect to r to get

$$\frac{2}{r} + \frac{1}{\rho}\frac{d\rho}{dr} + \frac{1}{u}\frac{du}{dr} = 0 \Rightarrow \frac{1}{\rho}\frac{d\rho}{dr} = -\frac{1}{u}\frac{du}{dr} - \frac{2}{r} \tag{B5.10}$$

We will use eq. (B5.10) to eliminate $(1/\rho)d\rho/dr$ in equation (B5.8). For convenience, we also define

$$u_0^2 = \frac{kT}{m} \qquad r_0 = \frac{GMm}{2kT}$$

Here, u_0 is the isothermal sound speed, while r_0 is related to the Jeans escape parameter (eq. (5.25)), $r_0 = r(\lambda_J/2)$. Dividing equation (B5.8) by kT/m, making the above substitutions, and using eq. B5.10, gives

$$\frac{u}{u_0^2}\frac{du}{dr} = -\frac{1}{\rho}\frac{d\rho}{dr} - \frac{2r_0}{r^2} = \left(\frac{1}{u}\frac{du}{dr} + \frac{2}{r}\right) - \frac{2r_0}{r^2}$$

$$\Rightarrow \frac{1}{u}\frac{du}{dr}\left(1 - \frac{u^2}{u_0^2}\right) = \frac{2r_0}{r^2} - \frac{2}{r} \tag{B5.11}$$

This is a differential form of *Bernoulli's equation*, named after Daniel Bernoulli (1700–1782). In essence, the meaning of Bernoulli's equation is that absent any input or output of energy, when fluid is accelerated, the pressure drops. We will see how this applies to hydrodynamic escape shortly.

Bernoulli's equation has a wide range of mathematical solutions, some physical and some unphysical (see Fig. 5.11). As discussed further in the main text, the particular solution that is of physical interest to the hydrodynamic escape problem is the *transonic* solution that starts at low velocities near the planet and accelerates to high velocities at great distance. The distance, $r = r_0$, at which the flow goes supersonic, $u = u_0$, is termed the *critical point*. To avoid confusion with the critical level or exobase for Jeans' escape, we prefer to call it the *sonic level*. Note that both sides of equation (B5.11) vanish at the sonic level. This is what leads to the mathematical complexity of Bernoulli's equation.

Bernoulli's equation also has an integral form. Equation (B5.11) can be integrated term by term. We do this taking the limits from r to r_0 and u to u_0, which gives us the transonic solution:

$$\frac{1}{u}\frac{du}{dr} - \frac{1}{u_0^2}u\frac{du}{dr} = -\frac{2r_0}{r^2} - \frac{2}{r} \Rightarrow \int_{u_0}^{u}\frac{1}{u}du - \frac{1}{u_0^2}\int_{u_0}^{u}u\,du = \int_{r_0}^{r}\frac{2r_0}{r^2}dr - \int_{r_0}^{r}\frac{2}{r}dr$$

$$\Rightarrow \ln\left(\frac{u}{u_0}\right) - \frac{1}{2}\left[\frac{u^2}{u_0^2} - 1\right] = -2\left[\frac{r_0}{r} - 1\right] - 2\ln\left(\frac{r}{r_0}\right) \tag{B5.12}$$

$$\Rightarrow \ln\left(\frac{u}{u_0}\right) - \frac{1}{2}\left(\frac{u}{u_0}\right)^2 = -\frac{2r_0}{r} - 2\ln\frac{r}{r_0} + \frac{3}{2}$$

Consider the solution at large distances from the planet. Recall that the concept of hydrodynamic escape was motivated by the fact that the mass of a static atmosphere is infinite if the barometric law remains valid (Sec. 5.8). For the case of transonic escape, though, the solution at large distances is quite different. As $r \to \infty$, the two largest terms in equation (B5.12) give

$$\frac{1}{2}\left(\frac{u}{u_0}\right)^2 \approx 2 \ln \frac{r}{r_0}$$

or

$$u \approx 2u_0 \left(\ln \frac{r}{r_0} \right)^{\frac{1}{2}} \tag{B5.14}$$

Inserting this back into equation (B5.9) shows that the mass density decreases at large distances as

$$\rho \approx \frac{C}{r^2 u} \propto \frac{1}{r^2 \ln r} \tag{B5.15}$$

The total atmospheric mass M_{atm} is given by

$$M_{atm} = 4\pi \int_{r_0}^{\infty} \rho r^2 dr \tag{B5.16}$$

Thus, M_{atm} is bounded because, in the integral, ρ decreases faster than $1/r^2$ for this transonic solution, i.e., as $1/(r^2 \ln r)$ according to eq. (B5.15). Thus, the transonic hydrodynamic escape solution is the one of physical interest for a planet like Earth that is embedded in a tenuous interplanetary medium.

transonic, and a termination shock will be created near the boundary with the interstellar medium where the solar wind slows down. The heliosphere is the region around the Sun dominated by the solar wind and its edge with the interstellar medium is the heliopause, which is somewhat more distant than the termination shock. Voyager 1 (launched in 1977) reached the heliopause at 122 AU distance in 2012, by measuring a sudden decrease in heliospheric ions by a factor $>10^3$ and increase in the intensity of galactic cosmic ray nuclei (Krimigis *et al.*, 2013; Stone *et al.*, 2013). Theoretically, higher background pressures could lead to subsonic outflow, or even to inflow if the background pressure were high enough.

For planetary winds, the background pressure would only be important for extremely strong stellar winds. As mentioned in Sec. 5.3, the ram pressure of the current solar wind on Earth is small. A planet has its host star and the impinging stellar wind on one side as compared to nearly empty interplanetary space on the other. Possibly, hydrodynamic flow far from a planet might be bent around in the anti-stellar direction by strong stellar wind pressure. Meanwhile, escape from the opposite side of the planet, like a comet's tail, would still be possible.

To assess whether an assumption of spherically symmetric hydrodynamic flow is reasonable, we can compare the planetary wind pressure at the sonic level with the stellar wind pressure. The planetary wind pressure is the static pressure plus the dynamic pressure, $p + \rho u^2$, where p is static pressure, ρ is density, and u is the flow speed. Calculations show that generally the planetary wind pressure exceeds the present solar wind pressure by orders of magnitude (Fig. 5.12), although we must bear in mind that stronger winds are possible from young stars.

Another complication is that planetary winds are expected to be largely neutral, and hence become collisionless at some great height. But if the exobase is beyond the sonic level, then the flow should become transonic and fluid equations are a good approximation. It can be demonstrated mathematically that the flow at the sonic level is independent of anything that occurs beyond that distance, so that the sonic level is a boundary condition.

If the exobase is below the sonic level, then the pressure force will be weak, the flow will remain subsonic, and flow velocity should eventually decrease at very high altitudes. We can quantity the speed as the Mach number M, the ratio of the flow speed, u, to the speed of sound. As shown by Walker (1977 p. 149), a subsonic flow with $M \ll 1$ causes expansion of an atmosphere and an outward velocity, but the inertial term, $\rho u(du/dr)$, in the equation of motion (eq. (B5.5)) is negligible. Consequently, the density profile is unaffected by the expansion and has an exponential, barometric form, as one can deduce from eq. (B5.5) with a negligible inertial term. Such flow should not be treated with purely

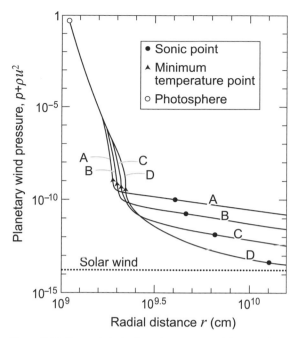

Figure 5.12 Transonic hydrodynamic escape model results from Sekiya, M. *et al.* (1980a, b) for a primordial H_2-rich atmosphere showing the pressure ($p + \rho u^2$) of the planetary wind as a function of distance. The four cases A–D have a net absorbed solar EUV flux of 1, 0.1, 0.01, and 0.001 W m^{-2}, respectively. This flux is absorbed in the outer atmosphere through ionization of H_2 where it heats the gas and drives hydrodynamic escape. The photosphere is a much lower altitude level where solar visible light is assumed to be absorbed and mostly reemitted in thermal infrared, so that it is not available to drive escape. Expansion of the atmosphere with increasing flow velocity typically causes a minimum temperature point, marked by triangle (▲). The sonic level (●) typically has a higher pressure than today's solar wind pressure, which is marked with a dashed horizontal line. (From Sekiya, M. *et al.* (1980). Reproduced with permission. Copyright 1980, The Physical Society of Japan.)

hydrodynamic fluid equations but kinetic models, as discussed above. Such models show that a subsonic regime in a single-component fluid produces an escape rate that is enhanced up to a factor of ~1.3–2 compared to the Jeans' escape rate (Merryfield and Shizgal, 1994; Volkov et al., 2011a), which can reasonably be called Jeans-like because of the small factor (Johnson et al., 2013d).

The question of subsonic flow arose in the context of old suggestions that escape of N_2 might be hydrodynamic on Pluto (Hunten and Watson, 1982; McNutt, 1989; Trafton, 1980; Trafton et al., 1997), and rather than transonic, that escape was in a so-called "slow hydrodynamic escape" regime where fluid equations might be appropriate (Krasnopolsky, 1999; Strobel, 2008a; Tian and Toon, 2005). Pluto is a low-gravity body so that even for N_2 at once presumed temperatures ~80–90 K, the Jeans parameter (eq. (5.25)) at the exobase is fairly small,

ranging ~4 to 6 depending on model type and solar UV variability (Erwin et al., 2013).

If a sonic level does not lie in the collisional regime, we would *not* expect fluid equations to be accurate and, indeed, purely fluid equations applied to Pluto produce erroneous profiles compared to DSMC calculations (Erwin et al., 2013; Johnson et al., 2013d). While the purely fluid assumption underestimates the escape rate only slightly, it produces a very inaccurate temperature–density structure compared to more complete calculations (Fig. 5.13), which is important because temperature and density profiles are observable with remote sensing, whereas escape fluxes have to be inferred. In particular, the exobase in the purely fluid model is calculated to be at a much lower altitude and far colder than in more realistic models. When Jeans escape is evaluated at such an exobase of such a model, it leads to the erroneous conclusion that the escape rate greatly exceeds Jeans' escape. In fact, Pluto's escape is in a Jeans regime from a ~70 K exobase due to cooling from HCN and C_2H_2; and escape rates are only ~$10^{23}N_2$ molecules s^{-1} (Gladstone et al., 2016). Thus, Fig. 5.13 is merely illustrative.

Subsonic hydrodynamic escape models have also been discussed for Titan (Strobel, 2008b) but, again, given that no sonic level lies below Titan's exobase, such an approach is physically problematic, as demonstrated by comparison to more sophisticated kinetic models (Bell et al., 2014; Cui et al., 2008) and DSMC models (Tucker and Johnson, 2009), including three-component ones (Tucker et al., 2013). DSMC models show that hydrogen escapes from Titan at about the diffusion-limited rate, as mentioned previously (Sec. 5.9.2), while methane escapes at a negligible Jeans rate. Consequently, Titan's escape of hydrogen is ultimately limited by condensation of methane at the tropopause cold-trap.

5.10.2 Energy-Limited Escape

Particles in the solar wind obtain the energy needed to escape from heating in the solar corona at the base of the flow. That energy is transported outwards by thermal conduction, which is efficient in a fully ionized wind. This process does not work for planetary winds, because thermal conduction is inefficient for neutral particles. Instead, planetary winds are powered by absorption of extreme ultraviolet (EUV) radiation from a host star. EUV nominally spans wavelengths from 10 nm to 100 nm although the lower bound is indistinct and extends into soft x-rays. All EUV wavelengths below 91.2 nm can be directly absorbed by atomic hydrogen because this *Lyman limit* wavelength is where a photon has enough energy to ionize the H atom,

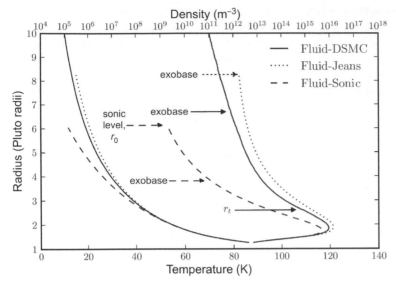

Figure 5.13 Comparison of simulations of atmospheric escape from Pluto with models using a single component, N_2, for solar-medium conditions at 32 AU. Curves on the left map to the upper axis of number density. Curves on the right map to the left vertical axis, which shows radial distance in number of Pluto radii. The radius of Pluto is 1153 km. *Solid lines* show a fluid-"direct simulation Monte Carlo" model, where fluid and DSMC methods are coupled at radius r_t, at a Knudsen number of 0.1. This model has an escape rate of 2.6×10^{27} N_2 molecules s^{-1}. *Dashed lines* show a hydrodynamic escape model, where fluid equations have been applied despite a sonic level above the exobase (Strobel, 2008a). This model has an escape rate of 2.5×10^{27} N_2 molecules s^{-1} but produces inaccurate number density and temperature profiles. *Dotted lines* show a model with an upper boundary of classical Jeans escape, which has an escape rate of 2.6×10^{27} N_2 molecules s^{-1}. (Modified with permission from Johnson *et al.* (2013d).)

temporarily creating a free proton and free electron. Indeed, most elements strongly absorb EUV because the outer electron binding energy (or, equivalently, the ionization potential) is typically smaller than EUV photon energy.

The maximum rate at which a gas can escape from an atmosphere in the Solar System can be calculated by equating the globally averaged flux of incoming solar EUV radiation S_{EUV}, with the energy carried out by the escaping particles. Assume for now that the escaping gas is atomic hydrogen. Each hydrogen atom has gravitational energy GMm/r, where G is the universal gravitational constant, m is the hydrogen atom mass, and r is the radius of the relevant escape level. Consequently, the energy-limited flux of hydrogen, Φ_{el}, can be approximated by the following expression (Watson *et al.*, 1981)

$$\Phi_{el} = \frac{S_{EUV}}{(GMm/r)} \qquad (5.64)$$

To evaluate expression (5.64), we need an estimate for the solar EUV flux in the past. Despite being fainter in

the visible (see Sec. 11.1), data from other stars suggest that the young Sun was significantly brighter at UV and EUV wavelengths (Claire *et al.*, 2012; Ribas *et al.*, 2005; Zahnle and Walker, 1982). The reason is that the Sun, like other stars, should have rotated more rapidly in its youth, before being slowed by torques exerted by its magnetic field as it interacted with the escaping solar wind. Faster rotation increases the strength of the solar magnetic dynamo. This increase causes increased flare activity and heats up the layers of the Sun's atmosphere, the chromosphere and corona. The Sun's photosphere is where most light comes from, but short-wavelength radiation is emitted from the hot chromosphere and corona.

Ribas *et al.* (2005) report short wavelength fluxes from 0.1 nm to 118 nm for the Sun and other young, solar-type stars based on data from the Far UV Spectroscopic Explorer (FUSE) satellite. Modifying their eq. (1) to include only wavelengths below 92 nm gives

$$S(< 92 \text{ nm}) = 23.3\tau_{age}^{-1.23} \text{ erg cm}^{-2}\text{s}^{-1} = 23.3 \times 10^{-3}\tau_{age}^{-1.23} \text{ W m}^{-2} \qquad (5.65)$$

This yields $S = 3.6$ erg cm^{-2} s^{-1} (=3.6 mW m^{-2}) at a current solar age, τ_{age}, of 4.56 billion years. Estimating the EUV flux that is available to drive escape requires division of this number by 4 to account for the ratio of the planet's surface area to its cross section. The flux should also be scaled upward to account for the larger effective cross-section of the atmosphere compared with the planet and downward to account for inefficiency of EUV heating. Not all EUV can drive escape because some absorbed energy is lost by radiation to space. Following Watson *et al.* (1981), we'll assume a heating efficiency of 0.15 and a geometric enhancement factor of 2; thus, S_{EUV}= (3.6 × 0.15 × 2)/ 4 = 0.3 erg cm^{-2} s^{-1}.

Now, let's use these numbers to calculate the *energy-limited escape rate* of hydrogen from the modern Earth. Plugging in values, and setting r equal to the radius of the Earth, yields $\Phi_{el} \cong 3 \times 10^{11}$ H atoms cm^{-2}s^{-1}. This is a very large number, nearly 1000 times greater than the diffusion-limited escape rate for hydrogen calculated in Sec. 5.8.4 (3.5×10^8 cm^{-2} s^{-1}). It says nothing about how fast hydrogen escapes today, as the modern Earth, with its hydrogen-poor atmosphere, is *not* in the energy-limited escape regime. It shows instead that hydrogen escape could conceivably have been a very important process early in Earth's history if the atmosphere was more hydrogen-rich.

For illustrative purposes, let's calculate the amount of time that it would take for the hydrogen in Earth's oceans to escape at this rate. The oceans would be ~3 km deep if they were spread evenly over the globe, which is equivalent to a column mass of 3×10^5 g cm^{-2}. Only 2/18 of this is hydrogen, so the hydrogen column mass is 3.3×10^4 g cm^{-2}, and its column density is ~2×10^{28} H atoms cm^{-2}. The lifetime of this hydrogen is thus 2×10^{28} H atoms cm^{-2}/3×10^{11} H atoms cm^{-2} s^{-1} $\approx 6.7 \times 10^{16}$ s or ~2 billion years. Equivalently, over two oceans' worth of hydrogen could have been lost over the Earth's history if hydrogen escaped at the energy-limited rate. Indeed, the actual number is more than five times higher than this if one accounts for the high early EUV flux predicted by eq. (5.65) by integrating that equation. Clearly, hydrogen escape has the potential to alter the water inventory of the Earth or other Earth-like planets, given the right conditions.

One caveat is that escape of hot hydrogen can reach a limit less than the energy limit called *radiation-recombination-limited escape*. With a high EUV flux, a hydrogen-rich upper atmosphere, e.g., on a hot Jupiter, can thermostat to a temperature ~10^4 K because the energy input is balanced by radiative recombination and Lyman-α cooling rather than adiabatic cooling of the expanding gas through "*pdV*" work (Murray-Clay *et al.*, 2009). Consequently, a radiation-recombination-limited escape rate is less than the energy-limited escape limit, and is found to vary as $\sqrt{S_{EUV}}$ rather than linearly as in eq. (5.64) (Murray-Clay *et al.*, 2009; Owen and Jackson, 2012).

5.10.3 Density-Limited Hydrodynamic Escape

The energy-limited escape flux predicted by eq. (5.64) is also not likely to be achieved when availability of hydrogen is limiting, particularly in a multi-component atmosphere on a rocky world. If more hydrogen is available at the base of the expansion, then a greater percentage of the absorbed EUV energy can be utilized to drive escape. When hydrogen is scarce, much of this EUV energy is either absorbed and radiated to space or conducted downwards through the expanding thermosphere. Higher hydrogen densities also increase the spatial extent of the atmosphere, thereby increasing the total absorbed energy that powers escape.

The importance of hydrogen density is illustrated by the calculations of Watson *et al.* (1981), shown in Fig. 5.14. Transonic solutions to the hydrodynamic escape equations are labeled A to E in Fig. 5.14 and correspond to a progressive increase of the number density at the lower boundary. The solutions were found using the *shooting method*, whereby one integrates both outward and inward from the sonic level, attempting to match the boundary conditions on each side. Figure 5.14(b) shows how the number density increases from case A to case E. Case E has an enormous number density at the bottom of the model, 120 km, and is not physically plausible. The case E escape flux equals, or even slightly exceeds, the energy-limited flux because energy limit is not absolute given that the effective cross-section of the atmosphere increases with increasing number density. The large escape flux in case E causes strong adiabatic cooling within the flow, resulting in a deep temperature minimum near 2000 km altitude (Fig. 5.14(a)). This also is physically unreasonable, but it demonstrates that this solution is indeed near the energy-limited escape rate. The low-hydrogen cases, A and B, are more physically realistic and have escape rates that are 20%–50% of the energy-limited escape rate.

A similar dependence of the escape rate on hydrogen density has been demonstrated for the process of water loss from Venus during a runaway greenhouse (Kasting and Pollack, 1983). Their calculations are like those shown by the transonic curve IV in Fig. 5.11. In Kasting

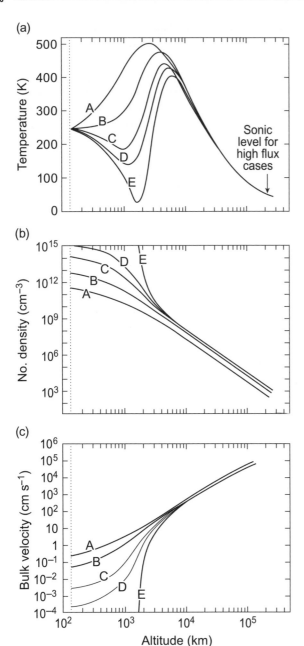

(a)

(b)

(c)

Figure 5.14 Temperature (a), number density (b), and upward velocity (c) versus altitude for a transonically escaping pure atomic H terrestrial atmosphere. (After Watson *et al.* (1981).). The sonic level is shown by the arrow in panel (a) at 2×10^5 km or ~30 planet radii. An EUV heating efficiency of 0.15 was assumed in the calculations. The escape fluxes (H atoms cm^{-2} s^{-1}) for the five cases, normalized to the Earth's surface, are: (A) 5.7×10^{10}, (B) 2.2×10^{11}, (C) 2.9×10^{11}, (D) 3.1×10^{11}, and (E) 3.5×10^{11}. The escape flux for case E is close to the energy-limited escape flux.

and Pollack, the factor that was varied was the H$_2$O mixing ratio at the lower boundary. Chemistry included within the model converted this to atomic

H within the expanding upper atmosphere. The solutions generated in this model were subsonic, but the authors argued that the escape rates were close to the transonic escape rate because the peak Mach number in the flow (the velocity divided by the sound velocity) exceeded ~0.7.

5.10.4 Maximum Molecular Mass Carried Away in Hydrodynamic Escape

Jeans' escape is an exponential function of molecular mass (Sec. 5.6), and consequently is important only for hydrogen on Earth, but this is not necessarily true in hydrodynamic escape because heavier species can be dragged along.

Consider an atmosphere composed largely of H$_2$ that flows out into space. (This may be more realistic for the early Earth than the pure atomic H atmosphere modeled by Watson *et al.* (1981).) The escaping hydrogen will drag along well-mixed minor gases. If there were no diffusion, there would be no separation by mass and the mixing ratios of the minor constituents would remain constant. Diffusion, however, allows the heavier gases to flow downwards under the action of gravity and this, in principle, means that heavier species will be retained on the planet to a degree depending on their mass. Whether the discrimination in mass is important or not depends on the relative magnitudes of the hydrogen out-flow and the diffusion velocity of the minor constituent (Hunten, 1979a; Hunten *et al.*, 1987; Sekiya *et al.*, 1980a, b).

In a full treatment of outflow in a hydrodynamically escaping atmosphere, an equation of motion is used that includes acceleration terms (Zahnle and Kasting, 1986). These terms have an important influence on the escape flux of the light, major gas constituent. However, their effect is small for the escape of heavy, minor gases. This is because diffusion processes occurring between the homopause and the sonic level where the outflow velocity is subsonic determine the flux of a heavy gas molecule. At these relatively low altitudes, the ambient density closely follows the barometric equation and acceleration terms in the momentum equation are negligible. Once the fluxes of heavy constituents are established they must obey the continuity equation at higher altitudes.

Consider the diffusion of a heavy, minor constituent relative to the ambient light gas. Our treatment follows Walker (1982) and Hunten *et al.* (1987). See also Chamberlain and Hunten (1987), Ch. 7. We denote the masses, fluxes, vertical velocities relative to the planet, and number densities of the light gas 1 and heavy gas 2 by

m_1 and m_2, F_1 and F_2, w_1 and w_2, and n_1 and n_2, respectively. In an isothermal atmosphere, the relative velocities are determined by diffusion. Following eq. (5.35) and its manipulation in Sec. 5.8.1, and neglecting the minor terms containing thermal diffusivity, the relative velocities can be written as

$$w_1 - w_2 = \frac{n_1 w_1}{n_1} - \frac{n_2 w_2}{n_2} = \frac{F_1}{n_1} - \frac{F_2}{n_2} = -\frac{b}{n_2}\left[\frac{1}{n_1}\frac{dn_1}{dr} + \frac{m_1 g}{kT}\right]$$
(5.66)

$$w_2 - w_1 = \frac{F_2}{n_2} - \frac{F_1}{n_1} = -\frac{b}{n_1}\left[\frac{1}{n_2}\frac{dn_2}{dr} + \frac{m_2 g}{kT}\right]$$
(5.67)

In these equations, r indicates the distance from the center of the planet. Also we have used the relationship $F = nw$ between vertical flux F, number density n, and vertical velocity w, and we express the diffusion coefficient as $D = b/n$. Each of these equations can be rearranged to put the number density gradient on the left-hand side:

$$\frac{dn_1}{dr} = -\frac{m_1 g}{kT}n_1 + \frac{1}{b}(n_1 F_2 - n_2 F_1)$$
(5.68)

$$\frac{dn_2}{dr} = -\frac{m_2 g}{kT}n_2 + \frac{1}{b}(n_2 F_1 - n_1 F_2)$$
(5.69)

If we add eqs. (5.68) and (5.69), the last terms on the right-hand side cancel, and we get

$$\frac{d}{dr}(n_1 + n_2) = -(n_1 m_1 + n_2 m_2)\frac{g}{kT}$$
(5.70)

This equation gives the variation of the total number density ($n = n_1 + n_2$) with altitude, and is a differential form of the barometric law. We define the mole fraction, or mixing ratio, of heavy gas 2 as

$$X_2 = \frac{n_2}{n_1 + n_2} = \frac{n_2}{n}$$
(5.71)

The logarithm of eq. (5.71) is $\ln X_2 = \ln n_2 - \ln n$, which we can differentiate with respect to radial distance r, to give

$$\frac{1}{X_2}\frac{dX_2}{dr} = \frac{1}{n_2}\frac{dn_2}{dr} - \frac{1}{n}\frac{dn}{dr}$$
(5.72)

If we now substitute from eq. (5.69) for the first term on the right-hand side of eq. (5.72) and from eq. (5.70) for the second term, we get

We note that $1 - X_2 = X_1$, or $-(m_2 - X_2 m_2) = -X_1 m_2$, which we can apply to collected terms in $m_2 g/kT$. Thus, eq. (5.73) rearranges to

$$\frac{1}{X_2}\frac{dX_2}{dr} = \frac{1}{b}\left(F_1 - \frac{X_1}{X_2}F_2\right) - (m_2 - m_1)\frac{X_1 g}{kT}$$
(5.74)

If heavy gas 2 is carried along efficiently by light gas 1, then we can assume that the mole fraction X_2 will be constant with height, so the left-hand side will be zero, giving

$$(m_2 - m_1)g = \frac{kT}{b}\left(\frac{F_1}{X_1} - \frac{F_2}{X_2}\right)$$
$$\Rightarrow (m_2 - m_1)g = \frac{nkT}{b}(w_1 - w_2)$$
(5.75)

This equation expresses a balance of forces. A molecule of gas 2 will be subject to a downward gravitational force of $m_2 g$ and an upward buoyancy force of $m_1 g$, resulting in a net downward force of $(m_2 - m_1)g$, given by the left-hand side of eq. (5.75). This net downward force will be balanced by an upward viscous drag, proportional to the difference in velocities of the molecules, $w_1 - w_2$, given by the right-hand side of eq. (5.75). This is illustrated in Fig. 5.15.

Equation (5.75) can be interpreted in terms of fluxes of the two gases. Consider a mass of gas 2 sufficiently heavy that it is not dragged along out into space by the lighter gas 1. Putting $F_2 = 0$ in eq. (5.75), we obtain the required mass:

$$m_{crossover} = m_1 + \frac{kTF_1}{bgX_1}$$
(5.76)

This mass is called the *crossover mass* and is the smallest mass for which the flux of constituent 2 is zero. It can be interpreted thus.

• If $m_2 > m_{crossover}$ then the buoyancy force is not enough to compensate for the gravitational force and viscous drag acting on molecules of gas 2, and gas 2 will not be lifted out of the atmosphere. The mole fraction X_2 will decrease with altitude with a scale height that is the diffusive equilibrium value augmented by an amount depending on the flux of gas 1. Meanwhile, the mole fraction X_1 will approach a value of 1 at high altitude.

• If $m_2 = m_{crossover}$ then the drag force is just sufficient to balance the net downward force on molecules of gas 2, but gas 2 will not be lifted out of the atmosphere.

$$\frac{1}{X_2}\frac{dX_2}{dr} = \frac{1}{n_2}\left(-\frac{m_2 g}{kT}n_2 + \frac{1}{b}(n_2 F_1 - n_1 F_2)\right) - \frac{1}{n}\left(-(n_1 m_1 + n_2 m_2)\frac{g}{kT}\right)$$
$$= -\frac{m_2 g}{kT} + \frac{1}{b}\left(F_1 - \frac{X_1}{X_2}F_2\right) + (X_1 m_1 + X_2 m_2)\frac{g}{kT}$$
(5.73)

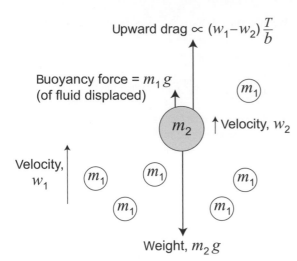

Figure 5.15 The forces acting on a heavy gas molecule of mass m_2 moving at an upward velocity w_2 immersed in a light gas flowing upwards in the diffusively separated upper atmosphere. The light gas has molecules of mass m_1 that move at an upward velocity of w_1. Absolute temperature is T and $b = Dn$ where D is the diffusion coefficient and n is total number density. If the initial total upward force on molecule 2 is greater than the downward force of its weight minus buoyancy, molecule 2 will be accelerated until the upward and downward forces come into balance.

- If $m_2 < m_{crossover}$ then constituent 2 will be carried along by the flux of light gas and swept out to space. Mole fractions X_1 and X_2 will be independent of altitude.

We can derive an expression for the flux of heavy gas 2. Rearranging eq. (5.75), we get

$$F_2 = \frac{X_2}{X_1} F_1 \left(1 - (m_2 - m_1) \frac{bgX_1}{kTF_1} \right) \qquad (5.77)$$

If we then note that $bgX_1/kTF_1 = 1/(m_{crossover} - m_1)$ from eq. (5.76), we can substitute in eq. (5.77) to give

$$F_2 = \frac{X_2}{X_1} F_1 \left(1 - \frac{m_2 - m_1}{m_{crossover} - m_1} \right)$$
$$= \frac{X_2}{X_1} F_1 \left(\frac{m_{crossover} - m_2}{m_{crossover} - m_1} \right) \qquad (5.78)$$

The important point from eq. (5.78) is that if heavy constituent 2 is swept out of the atmosphere into space, then its flux, F_2, will vary linearly with its mole fraction X_2 and molecular mass m_2. Note that the crossover mass is different for different gases, however, because the binary diffusion coefficient, b, in eq. (5.76) varies with species. The linear dependence on molecular mass contrasts with the exponential dependence on molecular mass in Jeans' escape described earlier in Sec. 5.6.2.

In the concept of crossover mass, there is a maximum molecular mass than can be carried away. Gases with masses exceeding $m_{crossover}$ are not affected. Actually, numerical results that incorporate nonlinear terms show a very slow loss of gases with molecular masses exceeding $m_{crossover}$ (Zahnle and Kasting, 1986) but the same results also show that eq. (5.78) is a good approximation for the escape of a trace constituent provided that

$$\left(\frac{m_{crossover} - m_2}{m_{crossover} - m_1} \right) > \frac{m_1}{m_2} \qquad (5.79)$$

For noble gases escaping in hydrogen, eq. (5.77) is a good approximation for Ne, Ar, Kr, and Xe, but is not so good for He.

The crossover mass can be calculated as a function of the hydrodynamic escape flux. Consider the energy-limited escape flux that was calculated for atomic hydrogen on Earth in Sec. 5.10.2. Let's divide that flux by 2 to convert it to an H_2 flux, yielding $F_1 = 1.5 \times 10^{11}$ H_2 molecule cm^{-2} s^{-1}. Substituting this value of F_1 into the equation for the crossover mass, eq. (5.76), and dividing by the mass of an H atom, m_H, to convert to atomic mass units yields

$$M_{crossover} - M_1 = \frac{kTF_1}{bgm_HX_1} \qquad (5.80)$$

Here, $M_{crossover}$ is the molecular mass of the heavy species in a.m.u. and M_1 (= 2 a.m.u.) is the molecular mass of H_2. Assume $T = 400$ K (from Fig. 5.14) and binary diffusion parameter $b = 2 \times 10^{19}$ cm^{-1} s^{-1} from Table 5.2. Taking $g = 980$ cm s^{-2} and $X_1 = 1$ yields $M_{crossover} - M_1 = 0.25$. In other words, escape of H_2 at the energy-limited rate from the modern Earth would be incapable of dragging along any heavier gas, including He.

Now consider gases that might have been dragged away earlier in Earth's history when the solar EUV flux was higher. Assume that the escape flux was energy-limited, i.e., eq. (5.65). H_2 escape fluxes and corresponding crossover masses are listed in Table 5.3. Evidently, gases as heavy as neon (molecular mass 20 or 22), N_2 or CO, might have escaped during the first 100–200 million years of Earth's history.

This same analysis, based on eq. (5.80) can be applied to other planets if one scales the EUV flux by orbital distance and adjusts for the planet's gravity. For Mars, interestingly, these two factors almost cancel: the solar flux is lower by a factor of 2.3, whereas gravity is lower by a factor of 2.6. Hence, the predicted crossover masses for Mars are nearly the same as for Earth.

Table 5.3 Energy-limited escape fluxes of H_2 and the corresponding maximum mass of a molecule (the crossover mass) that can escape by being dragged along by the hydrogen to space at different times in Earth's history.

Time after Earth's formation (billions of years)	EUV enhancement compared to today	Energy-limited escape rate of H_2 (cm^{-2} s^{-1})	Crossover mass, $M_{crossover}$ (a.m.u.)
0.1	110	1.7×10^{13}	30
0.2	47	7.0×10^{12}	14
0.5	15	2.3×10^{12}	5.8
1.0	6.5	9.8×10^{11}	3.6
2.0	2.8	4.2×10^{11}	2.7
4.56	1	1.5×10^{11}	2.25

5.11 Mass Fractionation by Hydrodynamic Escape

5.11.1 Fractionation Theory

We now consider how the quantity of heavy gases will be fractionated by mass when hydrodynamic escape is integrated over time because then we can see if predictions are consistent with data, such as those of noble gas isotopes. In the previous section, we saw how the flux of a heavy constituent varies linearly with its mass in eq. (5.78). Let us assume that the inventory of light gas is denoted by N_1 and that the inventory of heavy gas is denoted by N_2. If we substitute N_2/N_1 for X_2/X_1 in eq. (5.78), we get

$$\frac{F_1}{F_2} = \left(\frac{m_{crossover} - m_1}{m_{crossover} - m_2}\right)\frac{N_1}{N_2} \tag{5.81}$$

Because escape fluxes are proportional to their reservoirs, the evolution of the reservoirs can be treated as a *Rayleigh fractionation* process, which can be thought of as analogous to a distillation process where the heavy isotope concentration depends increasingly on the depletion of the light component.

The fundamental equation for Rayleigh fractionation is as follows, where dN is an infinitesimal number of particles removed per unit time,

$$\frac{dN_1}{dN_2} = \text{(fractionation factor)} \times \frac{N_1}{N_2} = (1+y)\frac{N_1}{N_2} \tag{5.82}$$

Here, we write the fractionation factor as $1+y$, which is usually slightly larger than unity, so that y is a very small value. Defining the fractionation factor this way gives the degree to which one gas escapes relative to another. Other symbols are sometimes employed for this "$1+y$" factor in the literature, such as R (Yung *et al.*, 1988) or x (Zahnle and Kasting, 1986). The present notation is

convenient for dealing with cases where the mass difference between species 1 and species 2 is small compared to their total mass, as is the case for most noble gas isotopes. Bearing in mind that fluxes are $F_1 = dN_1/dt$ and $F_2 = dN_2/dt$, comparison of eq. (5.82) with eq. (5.81) shows that

$$(1+y) = \left(\frac{m_{crossover} - m_1}{m_{crossover} - m_2}\right) \tag{5.83}$$

We proceed from the basic Rayleigh fractionation equation, eq. (5.82), by integrating. We assume initial inventories indicated by a superscript of 0, as follows:

$$\int_{N_1^0}^{N_1} \frac{dN_1}{N_1} = (1+y)\int_{N_2^0}^{N_2} \frac{dN_2}{N_2} \Rightarrow \ln\left(\frac{N_1}{N_1^0}\right) = \ln\left(\frac{N_2}{N_2^0}\right)^{(1+y)}$$

$$\Rightarrow \left(\frac{N_2}{N_2^0}\right) = \left(\frac{N_1}{N_1^0}\right)^{1/(1+y)} \Rightarrow \left(\frac{N_2}{N_2^0}\right) = \left(\frac{N_1}{N_1^0}\right)^{\frac{(m_{crossover} - m_2)}{(m_{crossover} - m_1)}}$$
$$\tag{5.84}$$

Figure 5.16 shows a plot of eq. (5.84) for an example crossover mass, of 100 a.m.u. The graph shows how the depletion of the heavier gas (N/N_0) increases with the depletion of the lighter gas, where the lines of increasing slope indicate the latter. The depletion of the heavy gas also depends on its particular mass, m_2, shown on the horizontal axis. However, eq. (5.84) is a simplification because we are assuming that y, and by implication the crossover mass, is constant in time. In reality, as the solar EUV flux decreases over time, the crossover mass decreases and also the hydrogen escape flux (Table 5.3). This would cause the slope of lines in Fig. 5.16 to decrease with time because heavier gases would cease to evolve while lighter gases would continue to change. This would cause curved lines in Fig. 5.16, concave downwards.

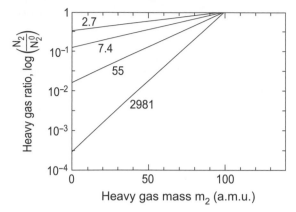

Figure 5.16 The evolution of the inventory of heavy gas 2 relative to its initial inventory as a function of molecular mass, m_2. For this graph, a constant crossover mass of 100 is assumed and a constant hydrogen escape flux. The numbers on the diagonal lines indicate indicate values of N_0^1/N_1, which shows the depletion of the lighter gas inventory. Thus the lines at the bottom of the plot correspond to later times in an evolutionary history. (Adapted from Hunten *et al.* (1987).)

5.11.2 Applications of Mass Fractionation in Hydrodynamic Escape: Noble Gas Isotopes

In Ch. 6, we discuss how elemental and isotopic abundances of gases provide constraints on the origin and evolution of the atmospheres of the rocky planets. In particular, the noble gases provide good tracers of atmospheric evolution for three reasons. First, noble gases tend to reside in the atmosphere because of their (near) chemical inertness. Second, the fractionation of different isotopes of non-radiogenic noble gases tells us about atmospheric evolution because lighter isotopes are lost preferentially to a degree that depends on ancient atmospheric conditions. Third, radiogenic noble gases – those derived from radioactive decay of other elements – act as chronometers of planetary evolution. Of course, all of these inferential principles are tempered by the fact that the patterns of abundance and fractionation in noble gases are complex and presently not fully understood.

Nonetheless, hydrodynamic escape could account for the isotopic fractionation of some of the noble gases found in planetary atmospheres. On the Earth, hydrodynamic escape could explain the isotopic differences between the $^{20}Ne/^{22}Ne$ ratio of ~9.8 in the atmosphere versus that in the Earth's upper mantle, ~10–13 (Pepin, 1991; Sasaki and Nakazawa, 1988; Zahnle *et al.*, 1990). Hydrodynamic escape allows the lighter neon isotope to

escape preferentially. Hydrodynamic escape has also been invoked to explain the Martian $^{36}Ar/^{38}Ar$ ratio (Bogard, 1997), which is isotopically heavy (4.2±0.1 (Atreya *et al.*, 2013)) compared with a terrestrial ratio of 5.32 and the average carbonaceous chondrite value of ~5.3 (Pepin, 1989). The fractionation of xenon on Earth (Hunten *et al.*, 1987; Pepin, 1991, 2000; Pepin and Porcelli, 2006; Sasaki and Nakazawa, 1988) and Mars (Pepin, 1991) has also been attributed to hydrodynamic escape. These applications of hydrodynamic escape are discussed below.

Terrestrial neon. The Earth's mantle is enriched in light neon isotopes relative to the atmosphere, and escape could have made the atmosphere isotopically heavy. Both $^{20}Ne/^{22}Ne$ and $^{21}Ne/^{22}Ne$ ratios are higher in the mantle. $^{20}Ne/^{22}Ne$ in mid-ocean ridge basalts ranges from near the atmospheric value (9.8) up to 13, while $^{21}Ne/^{22}Ne$ ranges from near air (0.029) to 0.07 (Farley and Neroda, 1998). The $^{20}Ne/^{22}Ne$ ratio of the material from which Earth accreted was probably ~12.5–13.6 (Farley and Poreda, 1993).

Why is atmospheric $^{20}Ne/^{22}Ne$ smaller than in the mantle? Either the atmosphere was partially derived from an external isotopically light neon source during late bombardment or the atmosphere has been modified by escape. In the latter case, models show that hydrodynamic escape can drag off ^{20}Ne in preference to ^{22}Ne, and reduce the solar $^{20}Ne/^{22}Ne$ ratio to the observed atmospheric value (Hunten *et al.*, 1987; Zahnle *et al.*, 1990). A hydrogen-rich upper atmosphere is required for this to have occurred. Also the escape must have happened early in Earth's history when the solar EUV flux was much higher than it is today (Claire *et al.*, 2012; Ribas *et al.*, 2005; Walter and Barry, 1991; Zahnle and Walker, 1982). Both conditions would have been met in an impact-produced steam atmosphere that occurred continuously during the main accretion period, and intermittently thereafter (Matsui and Abe, 1986a, b; Zahnle *et al.*, 1988).

Zahnle *et al.* (1990) showed that fractionation of neon would have occurred in a steam atmosphere as a byproduct of hydrodynamic hydrogen escape regulated at the diffusion-limit through an atmosphere of a major background constituent, such as CO_2, N_2, or CO. Unlike the $^{20}Ne/^{22}Ne$ ratio, there is no clear distinction between mantle and atmospheric $^{36}Ar/^{38}Ar$, which suggests that Earth's argon was unaffected by the hydrodynamic escape. The reason neon can escape while argon cannot is that neon is less massive than the likely background gases in the atmosphere (CO_2, N_2, or CO). Figure 5.17 shows that it would have taken only ~10 m.y. to produce

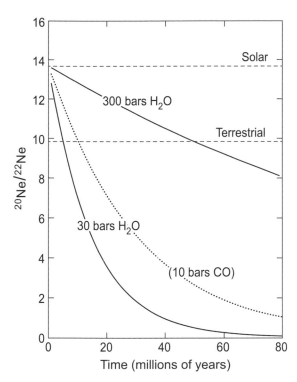

Figure 5.17 Neon isotope fractionation produced by hydro-dynamic escape of hydrogen in various steam atmospheres during accretion of the Earth. Three cases are shown. The calculated timescale for neon fractionation is consistent with the expected lifetime of impact-induced steam atmospheres on early Earth. (From Fig. 15 of Zahnle et al. (1990).)

the observed neon fractionation with a 30 bar steam atmosphere and 10 bars of CO. Atmospheres with less CO take shorter times, while thicker atmospheres take longer.

Martian argon. From analysis of trapped pockets of Martian air in the impact glass of the EET79001, Wiens *et al.* (1986) deduced a $^{36}Ar/^{38}Ar$ value of 4.1±0.2 within the uncertainty of 3.6 ± 0.44 obtained by Swindle *et al.* (1986). The ratio in the Martian atmosphere has been measured by the *Curiosity Rover* as 4.2±0.1 (Atreya *et al.*, 2013). This ratio is considerably less than 5.305±0.008 in the Earth's atmosphere (Lee *et al.*, 2006) or 5.50±0.01 in the solar wind derived from samples collected by the *Genesis* mission (Pepin *et al.*, 2012; Vogel *et al.*, 2011). Thus, light argon isotopes have been preferentially lost from Martian air relative to heavy argon.

Atreya *et al.* (2013) argue that solar wind sputtering since the end of heavy bombardment accounts for the $^{36}Ar/^{38}Ar$ ratio, but in our opinion it is possible that ancient hydrodynamic escape was the main fractionation

mechanism. In sputtering, the solar wind picks up and accelerates ions, and a fraction of the energetic ions or neutrals impacts the exobase, causing Ar escape (Hutchins and Jakosky, 1996; Hutchins *et al.*, 1997; Jakosky *et al.*, 1994; Jakosky and Phillips, 2001). Sputtering is fractionating because argon isotopes are diffusively separated above the homopause. Models estimate that 75%–99% of ^{36}Ar is lost (Hutchins *et al.*, 1997). But if this much Ar is lost, then it needs to be replenished from volcanism to be consistent with the Ar/Kr ratio on Mars because Kr is not subject to sputtering. Moreover, neon, which is even more prone to sputtering than Ar, needs even more replenishment. However, the estimated volcanic outgassing on Mars is too small by one or two orders of magnitude to do the job (Hutchins and Jakosky, 1996).

Early hydrodynamic escape provides an alternative for the fractionation of Martian argon (Pepin, 1991; Zahnle, 1993a; Zahnle *et al.*, 1990). Of course, if argon escapes and fractionates, neon must also. Martian atmospheric $^{20}Ne/^{22}Ne$ appears to be ~10, somewhat similar to the terrestrial atmospheric ratio, though some data are consistent with lower values for Mars (Bogard *et al.*, 2001; Bogard and Garrison, 1998). If the original Martian ratios of $^{36}Ar/^{38}Ar$ and $^{20}Ne/^{22}Ne$ were 5.35 and 13.7, respectively, then diffusion-limited hydrodynamic escape results in a $^{20}Ne/^{22}Ne$ ratio no greater than 9.5 ± 1.3, consistent with observation (Fig. 5.18). The presence of abundant CO_2 or a hydrogen escape flux sufficient to drag away neon but not argon would result in a yet lower $^{20}Ne/^{22}Ne$ ratio.

Terrestrial xenon. Interpretation of xenon is complicated because xenon has nine stable isotopes, several of which have been affected by the decay of extinct radionuclides. Also, xenon, with atomic weight 131.3, should be less depleted and less fractionated than krypton with atomic weight 83.8. But the opposite is observed. Krypton is depleted in the terrestrial atmosphere by a factor of 3.3×10^4 relative to solar composition while xenon is depleted by a factor of 4.8 $\times 10^4$. Nonradiogenic xenon isotopes are also much more strongly fractionated compared to krypton isotopes. The unexpected paucity of xenon is known as the *missing xenon paradox* (Ojima and Podosek, 2002; Pepin, 1991; Tolstikhin and O'Nions, 1994).

Vigorous hydrodynamic escape could produce the observed fractionation pattern in xenon (Hunten *et al.*, 1987; Pepin, 1991, 2006; Sasaki and Nakazawa, 1988) but additional circumstances must have led to no correspondingly large fractionation in krypton isotopes, which are less massive. There are three possible solutions. First,

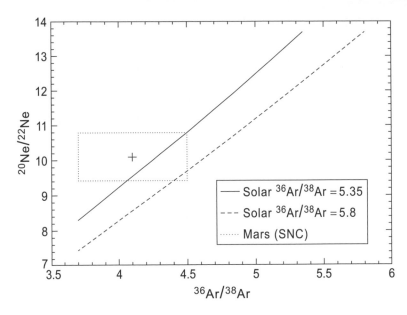

Figure 5.18 The ^{20}Ne/^{22}Ne ratio produced in fractionating argon by hydrodynamic escape from Mars, assuming an initial ^{20}Ne/^{22}Ne of 13.7. (From Fig. 7 of Zahnle (1993).)

Pepin (1991) suggested that xenon fractionation is a significant remnant of an early escaped atmosphere, whereas neon, argon and krypton were later outgassed from the mantle while xenon was not because it was incorporated into the core. Pepin (1991) assumed that xenon behaves as a siderophile at high pressure to justify why it partitions into the core. Second, xenon's low solubility in silicate melts could be used to argue that it was the most strongly partitioned into the earliest atmosphere. Third, xenon may have escaped as an ion during hydrodynamic escape of hydrogen ions along the open magnetic field lines at high latitudes (Zahnle, 2000). Ions interact strongly with each other and consequently cross-sections for ion–ion interactions are large. In contrast to xenon, krypton is extremely difficult to ionize and would not be subject to such ionic escape.

A very intriguing report is that the pattern of xenon's nine stable isotopes was lighter in the Archean than today, based on analysis of fluid inclusions in Archean barites and quartz (Hebrard and Marty, 2014; Pujol et al., 2011). This relationship would require xenon to escape to space during the Archean and mass fractionate, long after the early period of very high solar EUV. If xenon escaped as an ion, a plausible explanation is that it was dragged by rapid hydrogen escape in a *polar wind* (Sec. 5.7). Such hydrogen escape would be expected from an anoxic Archean atmosphere relatively rich in H_2 and CH_4 (e.g., Catling et al., 2001). However, whether the xenon data provide evidence for such theories is uncertain because the ancient xenon might be a mixture of modern air with an unfractionated mantle component (Pepin, 2013).

Martian xenon. Like Earth, Mars also has *missing xenon*, except more severely. The nonradiogenic isotopes are ~80 times less abundant and have a fractionation pattern generally similar to the Earth's. On the other hand, the ^{129}Xe derived from the decay of ^{129}I (half-life 17 m.y.) is about one third that of Earth. The high ratio of radiogenic to nonradiogenic xenon implies that escape took place very early before ^{129}I had undergone several half lives. Fractionation of the nonradiogenic isotopes on Mars can also plausibly be explained by mass fractionation during hydrodynamic escape of hydrogen (Pepin, 1991). However, explaining why krypton is not also strongly fractionated must be considered, as for the Earth. The same kind of explanations for the Earth can be proffered for Mars.

We have not discussed Venus above because Venus is anomalous in its noble gas abundances. On a gram per gram of planet basis, Venus is remarkably well endowed with nonradiogenic argon (^{36,38}Ar) and neon. It has ~60 times more ^{36}Ar than Earth, for example. It is plausible that Venus stochastically accreted a large (>600 km) comet from the outer Solar System, where temperatures would have been cold enough for argon to condense (Owen and Bar-Nun, 1995). The chance of such a single event happening is about 25% (Zahnle, 1998). This probability is large enough for plausibility and on the other hand small enough that Earth need not have suffered a similar fate.

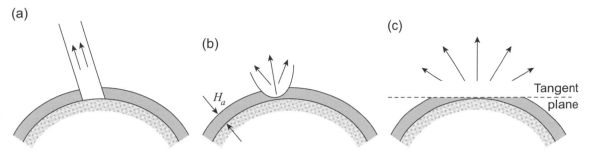

Figure 5.19 Various semi-analytical approximations to atmospheric impact erosion. (a) In the *cookie-cutter* approximation, the mass of gas that escapes is similar to that intercepted by the impactor multiplied by a factor close to unity (Walker, 1986). (b) In the *massless point explosion approximation*, an accelerating shock wave drives off the atmosphere of scale height (Ahrens, 1993). (c) In the *tangent plane model* of atmospheric impact erosion, a sufficiently massive impactor removes the entire atmosphere above a plane tangent to the planet. The impactor is required to have an impact velocity above a threshold of twice the escape velocity of the target (Melosh and Vickery, 1989).

5.12 Impact Erosion of Planetary Atmospheres

A large impact on a planet is very unlike an impact in everyday experience, such as throwing a stone into mud, because extraterrestrial impactors possess enormous kinetic energy and they vaporize in a process akin to a thermonuclear explosion. Consequently, a sufficiently large and energetic impactor can heat atmospheric gases to escape velocity while very high-speed ejecta can accelerate atmospheric gases to escape speed.

Early in a planet's history, there are many large impacts, as witnessed by the craters on Mars, the Moon, Mercury, and other bodies, so impact erosion could have been an important process for early loss of atmospheres on vulnerable bodies. Whether a body is subject to atmospheric impact erosion depends primarily on an object's escape velocity and whether it experiences a high impact velocity regime by virtue of its orbital position in a planetary system.

The velocity of an impactor depends on the escape velocity, v_e, which sets a minimum impact velocity by energy conservation for an object falling in from infinity, and a median encounter velocity, v_{enc}, which depends on the type and origin of the impactor:

$$v_{impact}^2 = v_e^2 + v_{enc}^2 \tag{5.85}$$

As a rough guideline, if the Keplerian orbital velocity of a planet around the Sun is v_{orb}, asteroids and Kuiper Belt comets tend to hit planets with moderate median encounter velocity, $v_{enc} \approx 0.5v_{orb}$, because these intruders are prograde and in the ecliptic plane (Bottke *et al.*, 1995; Zahnle *et al.*, 1992). In contrast, Oort Cloud comets have typical $v_{enc} \approx 1.7v_{orb}$. Thus, in general, planets close to

their host stars that move at higher orbital velocities compared to those farther away tend to suffer energetic impacts (Lissauer, 2007). Of course, such planets also receive higher stellar irradiation and are more prone to thermal escape too, so their atmospheres are doubly vulnerable to escape.

Four concepts have been used to estimate the atmospheric erosion effect of impact. Walker (1986) suggested that an atmosphere between an impactor and its target is heated by multiple shocks with the net effect that an atmospheric cross-section comparable to the impactor is lost: a fraction $\sim r_{imp}^2/R_p^2$, where r_{imp} is the radius of the impactor and R_p is the planet's radius (Fig. 5.19(a)). This is a so-called *cookie-cutter approximation* to impact erosion. Second, others have suggested that impacts behave like massless point explosions (Ahrens, 1993). The shock wave from the explosion accelerates through the escape velocity as it propagates up into exponentially thinner gas (Fig. 5.19(b)). The fraction of atmosphere to escape in this approximation scales as $\sim 10H_a^2/R_p^2$, where H_a is the atmospheric scale height.

A third and popular model for estimating cumulative impact erosion is that of Melosh and Vickery (1989) who deduced that escape driven by high-speed impact ejecta makes erosion more efficient and that a sufficiently large and energetic impact can erode all of the atmosphere above a plane tangent to the planet (Fig. 5.19(c)). This model is called the *tangent plane approximation*. A much larger amount of atmosphere escapes when the momentum of the ejecta is not much impeded by the inertia of the intervening atmosphere, and the ejecta move faster than the escape speed. Tangent plane erosion requires impact velocities that Melosh and Vickery (1989) estimated as $>2v_e$, cautioning that this threshold was uncertain. The mass of gas to escape relative to the total atmospheric mass in a qualifying impact is $\sim H_a/2R_p$. Many models of impact erosion for Mars and icy satellites have used tangent-plane erosion. They assume that loss of atmospheric mass, $m_{tangent}$, above a plane tangent to the planet occurs if $v_{impact}/v_e > 2$ and the mass of the impactor, $m_{impactor}$, exceeds a critical mass that was originally proposed to be $m_{tangent}$ (Melosh and Vickery, 1989) but later revised to a multiple of $m_{tangent}$ (Pham *et al.*, 2011; Vickery and Melosh, 1990).

A fourth form of impact erosion could occur when an impactor is so big and fast that a shock wave propagates through a planet and erupts at the surface, particularly the antipode (Chen and Ahrens, 1997; Genda and Abe, 2003, 2005). The expelled solid surface can carry along much atmosphere. For example, on a planet with a deep, thick gas envelope, air above a critical isobar might be accelerated to the escape velocity, given that shock waves accelerate as they move into thinner gas. Currently, research on this mechanism is limited. However, simulations of late stage terrestrial planet accretion suggest that roughly half of the collisions between planets strip off the outer mantle of the larger planet while obliterating the smaller planet (Agnor and Asphaug, 2004), so the effect on atmospheres ought to be important.

The rate of net change of volatiles $\dot{M}_{atm}$ for a planet will be the difference between the mass rate delivered ($\dot{M}_{deliv}$) and eroded ($\dot{M}_{erode}$), i.e.,

$$\dot{M}_{atm} = \dot{M}_{deliv} - \dot{M}_{erode} \qquad (5.86)$$

The delivery rate of volatiles $\dot{M}_{deliv}$, depends on the volatile mass fraction in impactors, which is typically ~0.01–0.1, and how much impactor mass escapes relative to the mass of the impactor. The erosion rate $\dot{M}_{erode}$ depends on the fraction of mass that escapes that is atmospheric gas or volatiles from impactor or target. Both $\dot{M}_{deliv}$ and $\dot{M}_{erode}$ have to be estimated from a double integral of the impactor velocity distribution over all velocities and the mass density of the flux of impactors over all masses. In turn, $\dot{M}_{atm}$ must be integrated over time. The literature gives details for such models using the tangent plane approximation (Manning *et al.*, 2006b; Pham *et al.*, 2011; Schlichting *et al.*, 2015; Zahnle *et al.*, 1992; Zahnle, 1993b).

In the Solar System, the mass density of the flux of impactors can be derived from the derivative with respect to mass of the time-dependent cumulative number flux of impactors, which varies as $\sim m^{-b}$. Parameter b is the spectral slope of the cumulative mass distribution, often estimated as ~0.5–0.8. It is likely that b is a natural outcome

of collisional cascades (Dohnanyi, 1972), and so such distributions may also apply to atmospheric impact erosion for exoplanets.

The most sophisticated impact erosion models are 3-D numerical models. Hydrocodes are numerical models that deal with shock physics and solve the mass, momentum and energy conservation equations as a function of time a grid (e.g., Barr and Canup, 2010; Barr and Citron, 2011; Kraus *et al.*, 2011; Pierazzo *et al.*, 2008; Senft and Stewart, 2007, 2008, 2011). Another numerical approach, *smoothed particle hydrodynamics* (SPH), models bodies as a large number of discrete of discrete, often spherically symmetric particles that are sometimes fuzzy (i.e., with spatial kernels), whose individual dynamics and compositional identities are followed in time.

Such numerical models have been applied to impact erosion for terrestrial planets (Maindl *et al.*, 2015; Shuvalov, 2009; Shuvalov *et al.*, 2014) and Titan (Artemieva and Lunine, 2005; Korycansky and Zahnle, 2011).

5.13 Summary of the Fundamental Nature of Atmospheric Escape

In this chapter, we have discussed various mechanisms for the escape of gases from planetary atmospheres, noting how thermal escape, suprathermal (nonthermal) escape, and impact erosion are three basic categories.

Today in the Solar System, no gases attain escape velocity from the gas giants, but the rocky planets cannot hold on to light gases such as hydrogen. On Venus, Earth, Mars, and Titan, the escape of hydrogen from current atmospheres is described well by the diffusion-limited flux. In this limit, the escape rate is set by two factors: the amount of hydrogen in all its chemical forms at the homopause, and the diffusion of hydrogen above that level to the exobase, the bottom of the exosphere. For Earth, the total mixing ratio of hydrogen in all its forms in the lower stratosphere above the "cold trap" at the tropopause can be used to calculate the diffusion-limited flux because this mixing ratio is similar to that at the homopause. Jeans' escape is often a minor component of the time-average escape flux from Earth's exobase. Suprathermal escape mechanisms dominate on Earth in the time-average and contribute essentially the entire hydrogen flux from the cold upper atmosphere of Venus.

The early terrestrial planets may have had more hydrogen-rich atmospheres after they formed for $\sim 10^7$–10^8 years. Also on Earth, hydrogen-rich steam atmospheres would have formed intermittently because of large, ocean-vaporizing impacts during heavy bombardment. Under such circumstances, the absorption of higher ultraviolet radiation flux from the youthful Sun likely drove a bulk hydrodynamic outflow of hydrogen-rich upper atmospheres.

In hydrodynamic escape, heavy atoms can be dragged along when collisions with hydrogen push the heavy atoms upward faster than gravity pulls them downward. This can lead to loss of heavy gases and mass fractionation of different isotopes of noble gases. The isotopic patterns of noble gases on Earth and Mars are consistent with fractionation by early hydrodynamic hydrogen escape. It is possible that major gases, such as carbon dioxide and nitrogen, were also lost from Mars during hydrodynamic escape.

Looking ahead to later chapters, we will see that hydrogen escape has had an important influence on the chemical evolution of the atmospheres and surfaces of Venus and Mars. In Ch. 10, we will see that hydrogen escape may also have affected the evolution of the oxidation state of the Earth's atmosphere and, as a consequence, biological evolution.

Finally, impact erosion can be effective early in a planetary system's history, particularly for small bodies. In our own Solar System, apart from Mercury, Mars was the planet most prone to impact erosion. Moons around Jupiter were also vulnerable. For exoplanets, small bodies close to parent stars will be vulnerable because their large orbital velocities imply bigger impact speeds. In conclusion, escape processes are fundamental for understanding the existence and evolution of planetary atmospheres.

PART II
Evolution of the Earth's Atmosphere

6 Formation of Earth's Atmosphere and Oceans

In the previous chapters, we provided essential chemistry and physics of planetary atmospheres needed for the rest of the book. Now, we turn to the evolution of Earth's atmosphere – a topic that will occupy most of the following six chapters. Earth is, of course, the best-studied planet, and it is also the one of greatest intrinsic interest because it harbors life, including us. One of the great goals of planetary science, which we will discuss in Ch. 15, is to determine whether truly Earth-like planets exist around other stars and if they're inhabited. To pursue that investigation, we need to be well informed about how Earth's atmosphere evolved and what kept our own planet habitable. Here, we start at the very beginning of atmospheric evolution on Earth: the origin of the atmosphere.

6.1 Planetary Formation

6.1.1 Formation of Stars and Protoplanetary Disks

In 1755, Immanuel Kant (1724–1804) qualitatively proposed that the Solar System formed from gravitational collapse of a cloud of diffuse matter, and in 1796, Pierre-Simon Laplace (1749–1827) provided a rough scientific outline for this theory. Today, it is generally accepted that both stars and planets form from the collapse of interstellar clouds of gas and dust. In the case of the Solar System, the central parts of the cloud collapsed to form the Sun, and the remainder of the material was spun out by rotation into a flattened disk, called the *solar nebula* (Boss and Ciesla, 2014). What was originally an amorphous cloud flattened into a disk because matter contracting within the plane of rotation was resisted by gas pressure and centrifugal force (experienced within the co-rotating frame of reference), whereas matter contracting from above either pole of the initial cloud did so more easily, opposed only by gas pressure. Similar gaseous

nebulae around other stars are called *protoplanetary disks*. Such disks evolve in a few million years into *debris disks*, which consist of solid debris without the gas.

The nebula theory is supported by the detection of circumstellar nebulae and debris disks around young stars. For example, the Atacama Large Millimeter/submillimeter Array (ALMA) has revealed a pattern of dark and bright concentric rings at ~1–3 mm wavelengths in a protoplanetary disk surrounding the star HL Tauri with a spatial resolution of a few AU (ALMA-Partnership *et al.*, 2015) (Fig. 6.1(a)). This 1.3 solar mass star is ~450 light years away and ≤1–2 m.y. old. The dark rings are perhaps regions where planet formation is taking place. Whether gaps have been cleared by planets or are places where smaller solid grains are coagulating is unresolved at the time of writing. One suggestion is that some dark rings correspond to condensation of ices such as water (D1 in Fig. 6.1(a)) and ammonia hydrates (D2) (Blake and Bergin, 2015).

Figure 6.1(b) shows a Hubble Space Telescope visible wavelength picture of a debris disk 63 light years away around the star β-Pictoris. The visible part of the disk extends to over 100 AU from the star – well beyond the ~30 AU orbit of Neptune in our own Solar System. The disk is warped due to perturbation from a large planet. Figure 6.1(c) shows an image obtained from the Very Large Telescope (VLT) in Chile operated by the European Southern Observatory (Lagrange *et al.*, 2010) using *adaptive optics* (a technique described in Sec. 15.2.1). The small bright dot to the upper left of the (dark) star is the planet, β-Pictoris-b, which has about 9 Jupiter masses, an orbital radius of 8-15 AU, and an effective temperature of 1500 ± 300 K (Bonnefoy *et al.*, 2011). β-Pictoris is a bright, bluish main sequence star of spectral type A5V, which is about 1.75 times the Sun's mass and ~8–20 million years in age. Consequently, this system is not a perfect analog for our own Solar System's

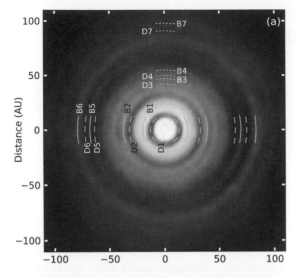

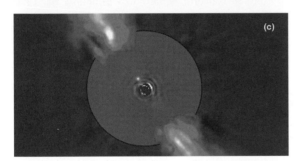

Figure 6.1 (a) A pattern of bright concentric rings (labeled B1, B2, etc.) separated by dark rings (labeled D1, D2, etc.) around the star HL Tauri, imaged by the *Atacama Large Millimeter/submillimeter Array* (ALMA) at 1 mm wavelength. (Source: ALMA-Partnership (2015).) (b) The disk of Beta Pictoris seen in visible light by the *Hubble Space Telescope*. The central star is blocked out in the photo and a faint secondary disk, inclined at 4°, is seen in scattered light. (Courtesy of NASA, ESA.) (c) Near infrared photograph of Beta Pictoris taken by the *Very Large Telescope* (VLT) in Chile. The star is again blocked out. The white dot to the upper left of the star is an 8-Jupiter-mass planet aligned with the disk at 8 AU from the star. A separate disk image from ESO's 3.6 m telescope has been grafted onto the central VLT image in this photo. (Courtesy of ESO/ A.-M. Lagrange *et al.*) (A black and white version of this figure will appear in some formats. For the color version, please refer to the plate section.)

past; nevertheless, it provides direct evidence for planet formation in a circumstellar disk. Something similar happened around our own Sun, albeit on a somewhat smaller scale.

6.1.2 The Planetesimal Hypothesis

The exact steps in planet formation are still a matter of research. Most astronomers think that planet formation is initiated by *accretion* of solid materials that condenses from a disk. The term accretion refers to the process by which orbiting particles collide with each other, eventually forming *planetesimals*. Planetesimals are conventionally considered to be objects 0.1–10 km across but there are several competing models for planetesimal formation and some recent models form 100–1000 km planetesimals directly from centimeter-size pebbles or meter-scale boulders in the nebula in a single event. In the traditional model, a ~10 km planetesimal has enough gravity to perturb the motion of other planetesimals and accrete mass from smaller ones. So bodies become fewer in number over time. In regions where growth of a few bodies outpaces the others, *runway accretion* leads a *planetary embryo* or *protoplanet* of diameter 10^3–10^4 km, and eventually to planets.

The physics for the process of planetesimals accreting into planets originated with the astronomer Viktor Safronov (1917–1999) (Safronov, 1972). Rocky, *terrestrial planets* such as Venus and Earth are formed almost entirely from lumps of such solid material. *Gas giants* such as Jupiter and Saturn are thought to have solid cores that formed by accretion. Once these cores grew larger than about 10–15 Earth masses, though, they were able to capture more and more gaseous hydrogen and helium from the surrounding solar nebula in positive feedback (Inaba and Ikoma, 2003). The largest gas giant, Jupiter, grew to over 300 Earth masses and although it has a composition enriched in elements heavier than helium compared to the Sun (Guillot, 1999), most of the material, which is H or He, must have been captured gravitationally from the nebula. This process is the *core accretion model* of giant planet formation (e.g., Pollack *et al.*, 1996).

Some astronomers have argued that gas giant planets possibly formed by gravitational collapse of the disk itself (e.g., Boss, 2005, 2006, 2008, 2012; review by Helled *et al.*, 2014). In the cool outer regions of the disk, numerical simulations show that the gas can clump into Jupiter-sized objects within a few orbital periods. Whether this *disk instability* (*DI*) (or *gravitational instability, GI*) model is viable or not might be resolved if a spacecraft determines whether Jupiter really does have a core of rock and ice through detailed study of its gravitational field. But such analysis is challenging even for NASA's *Juno* orbiter mission because the core is only a few percent of the total mass (Helled *et al.*, 2011). Also, the result may not yield a definitive answer because cores can form even in the GI model when grains entrained in the gas sediment out under gravity (Helled *et al.*, 2008).

Currently, not many astronomers favor the DI mechanism for forming planets partly because stars that have giant planets possess high metallicity, i.e., a relatively high abundance of elements heavier than H and He, which makes sense if rocky cores are important for giant planet formation (Fischer and Valenti, 2005; Johnson *et al.*, 2010). Planets of less than four Earth radii form over a wide range of host star metallicity (Buchhave *et al.*, 2012) but metallicity differences are still linked to the occurrence of rocky planets and gas dwarfs (Buchhave *et al.*, 2014; Wang and Fischer, 2015). Overall, evidence favors the *core accretion* model for forming both terrestrial and giant planets, but considerable uncertainty still exists and the core accretion and DI models may not be mutually exclusive or may each have roles under different circumstances.

The details of the planetary accretion process are not completely understood although various steps have been identified (e.g., reviews by Chambers (2014), Johansen *et al.* (2014), Lunine *et al.* (2011), Pfalzner *et al.* (2015), Raymond *et al.* (2014)). Initially, solid particles that condensed out of the solar nebula would have been gravitationally attracted to the nebular midplane, where they would have collided with each other and clumped together to form larger and larger particles. This is followed by four stages of growth: (1) planetesimal formation, (2) runaway growth, (3) oligarchic growth, and (4) late state accretion. Figure 6.2 shows the timescales

associated with the size of objects at each stage. Stages (1)–(3) make planetary embryos in what we call the "traditional model." A recent model, which we discuss below, suggests that the accretion of *pebbles* (centimeter-sized objects) can rapidly make embryos while gas is still in the disk, however (Jansson and Johansen, 2014; Lambrechts and Johansen, 2012).

The first phase of getting from centimeter-sized objects up to kilometer-sized planetesimals has generally been thought a theoretical challenge. Gas pressure slows the orbital motion of gas molecules more than large dust particles. So the fast-moving clumps experience a headwind, leading to orbital decay. Thus, gas drag can cause bodies to fall into the Sun. This process is fastest for meter-size objects, so the problem is called the *meter-size catastrophe* or *barrier* (Weidenschilling, 1977). Various mechanisms to overcome this difficulty have been suggested, including gravitational instability and clumping of bodies between turbulent eddies (Cuzzi *et al.*, 2008).

In fact, the physics of the so-called catastrophe may instead be a solution to planet formation. If a radial pressure bump arises in a nebula (e.g., from turbulence) pebbles should drift radially into the bump from both inner and outer sides (Jansson and Johansen, 2014; Whipple, 1972). If the pressure p varies with orbital radius r with gradient dp/dr, then on the inner side of a bump, where $dp/dr > 1$, gas has super-Keplerian velocity and particles are forced by the gas to move outward. In contrast, on the outer side where $dp/dr < 1$, the gas is sub-Keplerian and particles are dragged inward. Pebbles pile up at the bump and may induce core accretion (Chatterjee and Tan, 2014). Due to gas drag, capture of pebbles can rapidly form gas giant cores (Levison *et al.*, 2015).

In the traditional model, once planetesimals reached beyond a kilometer in size, the remaining three steps of the accretion process are reasonably well understood. Two key factors in the growth of larger bodies are called *gravitational focusing* and *dynamical friction*. The collision cross-section of a given body is enhanced beyond the geometric cross-section by a gravitational focusing factor, F_g,

$$F_g = 1 + \frac{v_{esc}^2}{v_{rel}^2} \tag{6.1}$$

where v_{esc} is the body's escape velocity (proportional to a body's size for objects of the same density) and v_{rel} is the relative velocity of nearby accreting bodies. Most encounters do not lead to collisions, but gravitational tugs change the orbits of planetesimals. *Dynamical friction* is the statistical process by which large bodies involved in many encounters tend to acquire circular, co-planar orbits, while

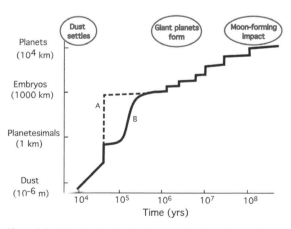

Figure 6.2 A sketch showing the characteristic timescales and sizes of objects in the formation of objects in the Solar System. Paths A is for a model where centimeter to meter-size objects clump quickly into planetary embryos. Path B represents the standard picture of *runaway growth* up to embryos. Paths A and B join around the *oligarchic growth* phase. (From Raymond *et al.* (2010). Reproduced with permission. Copyright 2010, John Wiley and Sons.)

small bodies are perturbed into eccentric, inclined orbits. Because the orbits of larger planetesimals remain nearly circular, they tend to pass each other slowly so that v_{rel} is small and F_g is large, enhancing the likelihood of collision. So, this growth stage is called *runaway growth*. During this phase the largest planetesimal in each orbital zone consumes most nearby planetesimals. These large planetesimals, though, still represent only a small fraction of the total mass. Runaway growth ends once the mass of the largest bodies becomes gravitationally important, probably when they reach the range of 10^{-5}–10^{-3} Earth masses.

When each region of the disk contains a single planetary embryo, along with numerous small planetesimals, the third growth stage, called *oligarchic growth*, begins. Runaway growth slows and larger embryos stir up the velocities of nearby planetesimals more than smaller ones, so that smaller embryos catch up in their growth. During oligarchic growth, the embryo "feeding zones" are about 10 Hill radii in width. The Hill radius defines a sphere within which a body's gravity is more influential for the motion of another body than is the Sun's gravity. Hence, a Hill radius is defined as a function of the ratio of the mass of the body, M, to the mass of the sun, $M_\odot$:

$$r_{Hill} = a \left(\frac{M}{3M_\odot} \right)^{1/3} \tag{6.2}$$

where a is the semi-major axis of the orbit around the Sun. As an aside, the Hill radius is equivalent to the distance of the L_1 Lagrange point, which lies along a line between the Sun and a body in its orbit. At the L_1 point for the Earth, for example, the gravitational pull of the Earth is just enough that a body at L_1 feels less effective gravity from the Sun and orbits in 1 year with the same

angular velocity as the Earth. Inserting masses into eq. (6.2), the Hill radius of the present Earth is ~1% of an AU; for Jupiter, r_{Hill} ~0.3 AU.

Oligarchic growth ends when planetary embryos contain about half of the solid mass in a particular region, while the other half resides in planetesimals (Kenyon and Bromley, 2006). The result is the formation in ~10^5 years of Moon-to-Mars-sized embryos at 1 AU and in ~10^6 years of protoplanetary cores of 1–10 Earth masses beyond 4 AU. These cores then sweep up nebula gas and become giant planets within a few million years. When many planetesimals are lost, dynamical friction lessens, so embryos excite the eccentricities and inclinations of other embryos, as shown in Fig. 6.3. With less gravitational focusing, the rates of collisions become more infrequent. So, a prolonged, ~10^8 year phase of *late-stage accretion* ensues for the terrestrial planets, which involves giant impacts with bodies the size of the Moon or Mars. We'll return to this point below, as it has important consequences for the formation of Earth's ocean and atmosphere. Thus, during late-stage accretion, embryos coalesce into inner planets through embryo–embryo collisions. The Earth, for example, probably formed from the collisional accretion of tens of Moon to Mars-sized bodies.

The traditional accretion theory described above presumes that the disk of dust and gas was largely gone during the latter stages of accretion in the terrestrial planet zone. This scenario is self-consistent, as the timescale for the dissipation of such disks, a few million years (Alexander *et al.*, 2006; Haisch *et al.*, 2001; Hartmann *et al.*, 2005; Russell *et al.*, 2006), is shorter than the time scale for the final assembly of rocky planets, which is ~30–100 million years (Fig. 6.2).

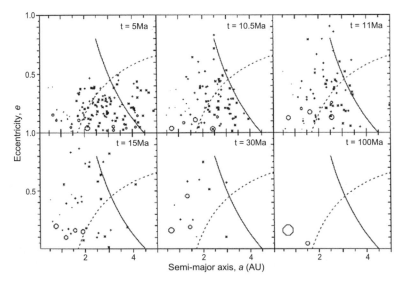

Eccentricity, *e*

Semi-major axis, *a* (AU)

Figure 6.3 Snapshots of the accretion process taken at various intervals, according to the model of Morbidelli *et al.* (2000). (Reproduced with permission. Copyright 2000, John Wiley and Sons.) This particular model was initialized with 5.5 Earth masses of planetary embryos distributed between 0.7 and 4 AU, along with 100 asteroids of negligible mass. Asteroids with initial semi-major axes *a*, of 2–2.5 AU are shown as crosses, whereas those with initial semi-major axes beyond 2.5 AU are denoted by asterisks. The solid and dashed curves represent the boundaries of the present-day asteroid belt, with aphelion ($a(1 + e)$) and perihelion ($a(1 - e)$) distances corresponding to 4.5 AU and 1.7 AU, respectively, where *e* is eccentricity.

Jupiter had to form much faster than the inner planets in order to capture large amounts of gas from the nebula. A factor that favored accretion is that Jupiter should have formed beyond the *ice line* (or *snow line*) in the nebula, where water ice could condense. Oxygen is the third most abundant element in the Sun, ~0.05% by number, and so condensation of H_2O ice would have provided relatively large amounts of solid material, thereby allowing the accretion process to proceed quickly at Jupiter's orbital distance. Also, Jupiter's greater distance from the Sun allowed a wider feeding zone for a proto-Jupiter, following eq. (6.2). Water ice was also available farther out in the nebula, but the orbital times were longer, and so Saturn, Uranus, and Neptune accreted less solar nebula material. Indeed, the latter two planets are commonly termed *ice giants*, as opposed to *gas giants*, as they are both strongly depleted in H and He compared to the Sun.

6.1.3 Planetary Migration: When Did the Gas and Dust Disappear?

An alternative line of thought about accretion models, sometimes called the *Hayashi school*, was developed in Japan. Chushiro Hayashi and those who followed him assumed that the *terrestrial* planets grew to large sizes in the presence of significant dust and gas (Hayashi *et al.*, 1979; Hayashi *et al.*, 1985). In this model, accretion proceeds faster with dust and gas present. When applied to our own Solar System, the model implies that Earth's primordial atmosphere should have contained gas of solar composition. As we discuss below, data from noble gases suggest that the present atmosphere was not derived directly from a gas of solar composition. But it is possible that a single large impact event (e.g., the Moon-forming impact) that occurred late during the accretion process could have removed an earlier solar-composition atmosphere. So, the Hayashi model cannot be easily dismissed on these grounds.

Many models for Earth's final assembly have concentrated on the gas-free accretion scenario. The bulk of the atmosphere and oceans must then have formed from solid materials that condensed out of the solar nebula and were present in the planetesimals from which Earth formed. Observations of exoplanets, though, show that not all planetary systems form in the same way. About 0.5%–1% of Sun-like stars have *hot Jupiters* – giant planets orbiting very close (<0.5 AU, typically 0.04−0.05 AU) to their parent stars (Howard, 2013). It should be impossible to form giant planets at very small distances because of the high temperature of the gas, tidal disruption, and Keplerian shear where material closer to

star orbits faster than material farther away. Hence, such planets must have formed farther out and then *migrated* in to closer orbital distances (reviewed by Chambers (2009)). Such migration is possible only in the presence of substantial gas and dust in the disk. So, the accretion process proposed by the Hayashi school may well apply to other planetary systems. In our own Solar System, the contribution from gas-assisted accretion versus planetesimal accretion may vary in the different formation regimes of terrestrial planets, asteroids, or objects beyond the ice line (e.g., Johansen *et al.*, 2015).

Planet migration also occurs in some recent models of the Solar System. In the *Grand Tack* model, Jupiter migrates inward to 1.5 AU during the first 0.6 m.y., then back outward to ~5 AU once Saturn forms (Hansen, 2009; O'Brien *et al.*, 2014; Walsh *et al.*, 2011). Such migration can stunt the growth of Mars by truncating the distribution of solids beyond 1 AU. Producing a small Mars has proved challenging for other planet formation models.

In general, we should keep an open mind about how our Solar System actually formed. We will hopefully learn much more about this process over the next few decades from observing what has happened around other stars.

6.2 Volatile Delivery to the Terrestrial Planets

In this book, we are primarily concerned with how planetary atmospheres form and evolve. A key issue, then, is how did the Earth obtain its *volatiles*? Volatile compounds, to an atmospheric scientist, are those that have relatively low melting or boiling points, so that they are present as liquids or gases in a planet's hydrosphere or atmosphere. Key volatiles for the Earth include H_2O, carbon, nitrogen, and sulfur. These (plus phosphorus) are also the so-called "SPONCH" elements from which life is made.

6.2.1 The Equilibrium Condensation Model

Astronomers and planetary scientists have been concerned with the question of volatile delivery ever since they started thinking about how planets might be built. One early thinker on this topic was John Lewis (of MIT and then University of Arizona). Lewis developed the *equilibrium condensation model* for planetary formation (e.g., Lewis and Prinn, 1984)). In this model, Lewis assumed that a nebula of gas and dust that had the same overall composition as the Sun surrounded the growing protosun. Lewis also assumed that the solar nebula contained the same amount of mass as the Sun – an estimate that may be too high by at least a factor of 10 – but we can ignore this aspect of his model, because it has only a weak effect on his predictions. (A nebula that includes more than about 0.1 solar masses would have rapid transport of most of the mass inwards onto the central star, together with radial expansion of the remaining disk and is unlikely to evolve to the present Solar System.) The basic idea is that initially the solar nebula is hot for a variety of reasons: high density prevents radiation from the protosun escaping, friction within the nebula generates heat, and gravitational potential energy has been converted to kinetic energy during infall of material. After reaching a peak temperature, the nebula cools. The order in which different materials should condense from such a nebula as it cools is shown in Table 6.1. In summary, the condensation sequence begins with highly refractory metals and oxides, and this is followed by Ni–Fe metal, silicates, sulfides, hydrated minerals, and finally, ices.

The *refractory* metals and oxides near the top of Table 6.1 are compounds that vaporize only at extremely high temperatures, >1500 K. The species making up the lower part of the list below silicates are volatile compounds with relatively low vaporization temperatures. Missing from the table are H_2 and He, which condense only at extremely low temperatures, and hence were probably always present in the solar nebula as gases.

Lewis proposed that the list of Table 6.1 explains much of what we observe about the composition of planets in our own Solar System. The terrestrial planets formed closer to the Sun, from silicates and iron–nickel alloy, which are the compounds that could condense in the warm inner parts of the solar nebula. In Lewis's model, the nebula became cool enough somewhat interior to 5 AU to allow H_2O to condense out as water ice; hence this distance is called the *ice line* (or *snow line*), as mentioned previously (Fig. 6.4). The presence of Jupiter at 5.2 AU from the Sun is thus nicely explained in Lewis's model. In the current Solar System, water-rich (~10 wt%) asteroids occur beyond ~2.7 AU (Gradie and Tedesco, 1982).

The Lewis model also purports to explain why Earth has some water (~0.1 wt%) whereas Venus does not. According to Fig. 6.4, Earth forms just outside the region where hydrated silicates should have condensed out of the nebula (represented in Fig. 6.4 by tremolite $(Ca_2Mg_5Si_8O_{22}(OH)_2)$ and serpentine $((Mg, Fe)_3Si_2O_5(OH)_4)$). Venus, by contrast, is well inside of this boundary. With a little tweaking of the nebula temperature profile, Earth would have received water from this mechanism, whereas Venus would not. So, for many years, Lewis and his colleagues argued for a dry origin for Venus.

Table 6.1 Materials that condense from the nebula as it cools. The shading of rows indicates chemical groups of substances. Note that the exact equilibrium condensation temperature depends on assumptions about total pressure and nebula composition. Here a pressure of 10^{-4} bar is assumed (Lodders, 2003).

Temp/ K	Substance	General groups (Comments)
~1800	Highly refractory metals, W, Os, Ir, Re	*Refractory metals*
1677	Corundum, Al_2O_3	*Refractory oxides*
1593	Perovskite, $CaTiO_3$	
1397	Spinel, $MgAl_2O_4$	
1360	Nickel–iron metal, Ni, Fe	*Ni–Fe (core-forming metals)*
1347	Pyroxene, $CaMgSi_2O_6$ (diopside)	*Silicates (rock-forming minerals)*
1354	Olivine, Mg_2SiO_4 (forsterite), or Fe_2SiO_4 (fayalite)	(60% of Earth's crust)
<1000	Alkali feldspars, $(Na, K)AlSi_3O_8$	
~700	Troilite, FeS	*Sulfides*. (Chalcophile (sulfur-loving) elements also include Zn and Pb).
550–330	Minerals with –OH or H_2O in their formulae	*Hydrated minerals*

Ionic substances above ▲, Molecular substances below ▼

~180	Water ice, H_2O	*Ices* (Caveat: The form in which C or N condenses
~120–130	Ammonia ice, $NH_3 \cdot H_2O$	depends upon the availability of water and
40–78	Methane ice, $CH_4 \cdot 7H_2O$ or CH_4 ice	kinetics. If there is not enough water, they will
50–60	Nitrogen ice, $N_2 \cdot 6H_2O$ and $N_2 \cdot 7H_2O$	not condense as clathrates, e.g., graphite could condense at higher temperature (Lodders, 2003).)

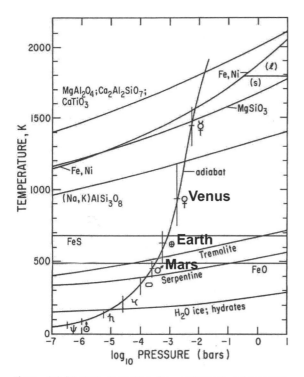

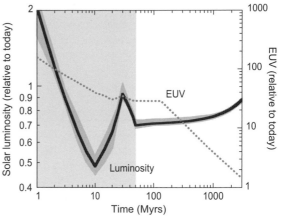

Figure 6.5 Changes in luminosity for the Sun's first 3 billion years. A time of zero corresponds to the age of the Solar System of 4.5673 Ga. The shaded region is the pre-main sequence time. The shaded path shows the probable solar luminosity, with its spread giving the uncertainty. (Adapted from Zahnle *et al.* (2007). Reproduced with permission of Springer. Copyright 2007, Springer Science + Business Media B.V.)

Figure 6.4 Diagram illustrating the equilibrium condensation model of planetary formation. The curved line running through the middle represents an adiabat extending radially along the midplane of a $1M_{Sun}$ solar nebula. The other curves represent boundaries at which various minerals would condense, assuming solar composition for the nebula. Standard astronomical symbols mark the planets. (From Lewis and Prinn (1984), p. 61.)

The equilibrium condensation model for planetary formation is no longer considered viable, for a number of reasons. For one, radial mixing of planetesimals during the latter stages of accretion means that Earth and the other terrestrial planets are composed of material that originally condensed over a wide range of orbital distances. This provides other mechanisms by which Earth may have obtained its water, as discussed further below. Of equal importance is the fact that hydrated silicates are now considered kinetically difficult to form (Prinn and Fegley, 1989). Chemical reactions between gaseous and solid materials proceed extremely slowly at these relatively low temperatures. Hydrated silicates are indeed found in meteorites, but they are now thought to have formed by alteration of silicate minerals by liquid water within meteorite parent bodies (Bunch and Chang, 1980). Thus, many of the detailed predictions of the equilibrium condensation model are no longer accepted. However, the simple prediction of why our Solar System contains rocky planets on the inside and gas or ice giants farther out is still an important success.

One further nuance of the Lewis model concerns the early history of solar luminosity (Zahnle *et al.*, 2007). The Sun has steadily brightened since it settled down onto the *main sequence* (Sec. 2.1.2) at ~4.52 Ga (see Ch.11 for climatic implications of this brightening). By convention, the Solar System formation clock starts ticking at the age of the oldest objects, which are calcium-aluminum inclusions (CAIs) in chondritic meteorites, dated at 4.5673 ± 0.0002 Ga (Amelin *et al.*, 2010; Connelly *et al.*, 2016). During the ~50 m.y. of pre-main-sequence time from then until 4.52 Ga, solar luminosity changed considerably as the Sun contracted and then went through nuclear fusion ignition (Fig. 6.5). This period of time happens to correspond to planet formation, and so would have affected volatiles. For example, at 1 AU, water was in the form of vapor at 2 m.y. but was in the ice phase at 10 m.y. when solar luminosity went through a minimum. These issues remain to be fully explored in planet formation models. Where the snow line was during the nebula phase also depends upon how rapidly the nebula cooled, which depends upon assumed opacities (Lesniak and Desch, 2011; Mulders *et al.*, 2015).

6.2.2 Modern Accretion Models

With the development of faster computers, the simulations of accretion have become more detailed. One illustrative simulation is shown in Fig. 6.6. This particular calculation extended from the Sun to 5 AU. That was enough to include the four innermost planets in our own

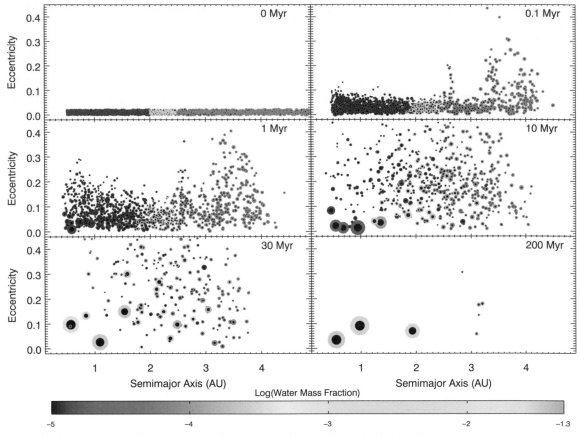

Figure 6.6 Snapshots of a particular rocky planet accretion simulation for the region inside 5 AU around the Sun. The horizontal axis is the planet's semi-major axis, i.e., its mean distance from the Sun. The dots represent large planetesimals, some of which will grow into planetary embryos and planets. The position of the dot on the vertical axis indicates the planetesimal's eccentricity. The size of the dot indicates the mass of the planetesimal or planet, and its color shows the fraction of its total mass that is made up of water. The simulation was terminated after 200 million years. (From Raymond *et al.* (2006).) (A black and white version of this figure will appear in some formats. For the color version, please refer to the plate section.)

Solar System, but not the four giant planets. The calculation started from a swarm of 1886 planetesimals of various sizes, the average mass being about half the mass of the Moon. These planetesimals were initially assumed to be orbiting at various distances from the Sun, randomly picked between 0.4 and 5 AU (top left panel). The initial eccentricities and inclinations of the orbits were assumed to be zero. (Orbital eccentricity is defined in Sec. 2.2.1. The inclination is the angle of the planet's orbital plane with respect to the average, or *invariant,* plane of the system.) A Jupiter-mass planet, not shown, was assumed to be on a circular orbit just outside the calculation, at 5.5 AU. The colors of the dots represent the water content of the planetesimals, with blue showing water-rich bodies and red representing dry ones. The planetesimals change from red to blue going from a few tenths of an AU out to

5 AU. Water-rich planetesimals containing 5% water by mass are present beyond 2.5 AU.

Several interesting phenomena occur in such simulations, only a few of which will be mentioned here. Within a few hundred thousand years following the start of the simulation shown, the planetary embryos began to drift both inwards and outwards from their initial positions, and they were excited to higher eccentricities and inclinations. All of this happened because of the way they perturb each other gravitationally. Most importantly, water-rich planetesimals from beyond 2.5 AU were scattered inward towards the inner parts of the planetary system. Some of these ended up being incorporated into planets that remain close to the Sun. In this particular simulation, a 2-Earth-mass planet formed at 0.98 AU, very close to Earth's actual orbital distance. Besides being considerably larger than the real Earth, this planet was also much more water-rich. The fraction of Earth's total mass that is water (including water in Earth's mantle) is estimated around 0.1 wt% (see Sec. 6.3), which would make it yellowish-green in this figure. By contrast, the blue planet in the figure has a water mass fraction close to 10^{-2}. Such a planet, if it existed, would have oceans that were 30–40 km deep, as compared to only about 3–4 km on Earth.

The simulation shown in Fig. 6.4 also produced a 1.5-Earth-mass planet at 0.55 AU and a 1-Earth-mass planet at 1.9 AU. Both of these planets also have lots of water. The innermost one, though, is well inside the inner edge of the liquid water *habitable zone*, as we will see later on in Ch. 15. If such a planet formed in a real planetary system, it would probably lose its water by the runaway greenhouse mechanism described in Ch. 13. The outermost planet is close to, or beyond, the outer edge of the habitable zone, and so any water on that planet's surface may be frozen, depending upon the strength of the greenhouse effect of its atmosphere and the planet's albedo.

The key point from this simulation and others like it is that the planets formed within the inner parts of planetary systems can be either much wetter or much drier than Earth. That is because some planets just happen to incorporate large, water-rich planetesimals from outside 2.5 AU, whereas other planets do not. Once again, the results are stochastic. If one does many simulations, though, and counts the terrestrial planets that are formed, one finds more that are water-rich than water-poor (Raymond *et al.*, 2004). Hence, if these simulations realistically represent planetary formation, there should be lots of rocky planets with at least as much water as Earth.

6.2.3 D/H Ratios and Their Implications for Water Sources

Another potential source of water and other volatiles is comets. Comets are small bodies with diameters ranging from kilometers to tens of kilometers that are composed of roughly equal mixtures of ice and rock. They originate from beyond Neptune's orbit (~30 AU). The comets that we observe today come from two regions: the *Oort Cloud* or the Kuiper Belt (see Fig. 6.7). The Oort Cloud is a spherical shell of 10^{12}–10^{13} comets surrounding the Solar System that extends outward to roughly 100 000 AU, or approximately 1.6 light years, beyond which the Sun's gravitational influence can be overcome by other stars (Levison and Dones, 2014). The Oort Cloud is the source of most *long-period comets*, many of which are observed only once. The orbits of these comets are randomly distributed in space and are nearly parabolic, indicating that their source region must be spherical and extremely distant. The *Kuiper Belt* is a donut-shaped reservoir of comets that lies within the plane of the Solar System,

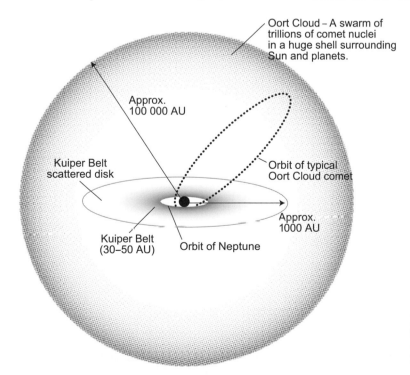

Oort Cloud – A swarm of trillions of comet nuclei in a huge shell surrounding Sun and planets.

Approx. 100 000 AU

Kuiper Belt scattered disk

Orbit of typical Oort Cloud comet

Approx. 1000 AU

Kuiper Belt (30–50 AU)

Orbit of Neptune

Figure 6.7 Diagram illustrating the relationship of the Oort Cloud and the Kuiper Belt to the Solar System. The orbit of a typical Oort Cloud comet is shown.

but beyond the orbit of Neptune, between 30–50 AU with scattered objects out to ~1000 AU. It is estimated to contain over 70 000 objects with size exceeding 100 km and is the source of most *short-period comets*, many of which are in low-inclination orbits that are prograde (i.e., in the same direction that the Sun rotates and that the planets orbit).

Because they contain large amounts of water ice, comets could, in principle, have been the source of Earth's water. This idea was popular amongst planetary scientists for many years (see, e.g., Chyba, 1987), especially as it might also explain the abundance pattern of noble gases in Earth's atmosphere (Owen *et al.*, 1992). We'll return to this issue in Sec. 6.4. Comets also contain other volatile materials, including complex organic carbon compounds, and so some scientists (Chyba, 1990; Oro, 1961) have suggested that they might have contributed directly to the origin of life. The main problem with this latter idea is that the proportion of interesting organic molecules that survive impact with the Earth is very small (Pasek and Lauretta, 2008; Pierazzo and Chyba, 1999).

Over the past two decades, the idea that comets supplied most of Earth's water has fallen out of favor. One reason is that the deuterium to hydrogen ratio became known for different Oort Cloud comets: Halley,

Hyakutake, Hale-Bopp, 2002T7, and Tuttle. The D/H ratio of terrestrial seawater has a value of $(1.558 \pm 0.001) \times 10^{-4}$. The D/H ratio in the five aforementioned Oort Cloud comets average about twice that value (Hartogh *et al.*, 2011b; Robert, 2001). Consequently, Oort Cloud comets could not have accounted for most of the Earth's water. As we will discuss in more detail in Ch. 13, the D/H ratio in a planet's atmosphere can increase with time if the planet loses hydrogen faster than it loses deuterium, but it cannot go back in the other direction. However, comet 103P/Hartley 2 has a D/H ratio of $(1.6 \pm 0.24) \times 10^{-4}$, similar to Earth's oceans (Hartogh *et al.*, 2011b). This object is a *Jupiter-family comet* (JFC), which means that it is a short-period comet in the ecliptic plane sourced from the Kuiper Belt. However, its nitrogen isotope ratio does not support the idea that JFCs contributed significantly to Earth's water. The Earth has a $^{15}N/^{14}N$ ratio of 3.678×10^{-3} whereas the ratio measured in HCN and CN in 103P/Hartley 2 (and all other comets) is ~1.8 times higher. By contrast, chondritic meteorites (see below, Sec. 6.3), which come from the asteroid belt, have an average D/H ratio that is close to that of Earth's oceans. For example, carbonaceous chondrites have D/H $= (1.4 \pm 0.1) \times 10^{-4}$ (Fig. 6.8) (Marty and Yokochi, 2006), and their $^{15}N/^{14}N$ ratio is also comparable to the

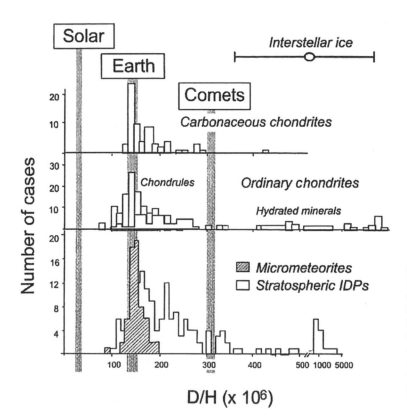

Figure 6.8 The deuterium/hydrogen (D/H) ratio in different reservoirs of the solar system. The vertical gray lines shows the Sun, Earth, and Oort Cloud comets. From top to bottom, the histograms show values for carbonaceous chondrites, ordinary chondrites, and micrometeorites and Interplanetary Dust Particles (IDPs). (From Marty and Yokuchi (2006). Reproduced with permission. Copyright 2006, The Mineralogical Society of America.)

bulk Earth's. Furthermore, measurements on 67P/Chur-yumov–Gerasimenko (a JFC) by the *Rosetta* spacecraft show a D/H = $(5.3 \pm 0.7) \times 10^{-4}$, which is ~3 times higher than on Earth (Altwegg *et al.*, 2015). So, overall the data are consistent with the idea that planetesimals from the asteroid belt region were the major source of Earth's water. In fact, isotopic and mass balance suggests that comets could have contributed no more than ~10% of the Earth's water (Dauphas *et al.*, 2000).

A second strike against comets as the main source of Earth's water is that whereas abundant asteroids are expected to be scattered into the inner solar system, only about 1 in 3 million comets hits the Earth after Jupiter forms (Levison *et al.*, 2001). Thus, an unfeasible number of comets would need to have been scattered for the Earth to have accreted its ocean solely from such bodies (Zahnle, 1998).

Finally, we note that asteroids and comets lie on a continuum from warm, rocky planetesimals with no ice in the inner solar system to cold, ice-rich planetesimals far from the Sun. The asteroid–comet division is therefore one of taxonomic convenience rather than absolute distinction.

6.3 Meteorites: Clues to the Early Solar System

In considering the formation of the Solar System, we assumed that the nebula had the same composition as

the Sun. Direct evidence supports this idea. The composition of the Sun is known, at least partially, through spectroscopy, while *chondritic* meteorites (defined below) have preserved the composition of the nebula because they are leftovers of planet formation. The chondrite and solar compositions match, excluding gas-forming elements lost from the meteorites and lithium consumed on the Sun during nuclear reactions (Fig. 6.9).

Traditionally, meteorites were divided into three categories: (1) *irons*, which are predominantly iron and thought to be pieces of metallic asteroid cores, (2) *stony*, which are mostly silicates, and (3) *stony-irons* which are a mixture of silicates and iron that presumably sampled both asteroid mantles and cores. However, more recently, meteorites have been classified into *chondrites*, *achondrites* and *primitive achondrites*, as shown in Fig. 6.10, because this classification better reflects their origins (Weisberg *et al.*, 2006).

Meteorite classification depends on composition. Chondrites are often defined as containing *chondrules*, which are globules of silicate minerals, up to a few millimeters in size, interpreted as rounded particles of rapidly cooled silicate melt formed by condensation or re-melting of dust in the solar nebula. However, some chondrites do not contain chondrules. So, a more general definition of a chondrite is a meteorite with nearly solar-like composition, excluding the gas-forming elements. Achondrites do not contain chondrules and are pieces of

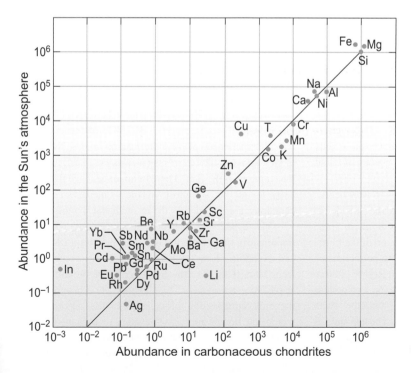

Figure 6.9 The close match between the elemental abundance determined from spectroscopic measurements of the solar photosphere and the CI carbonaceous meteorites. Data are normalized such that $Si = 10^6$. Lithium (Li) is more abundant in CI carbonaceous chondrites than the Sun because this element is used in solar nuclear fusion. (Adapted from McBride and Gilmour (2004), p. 37.)

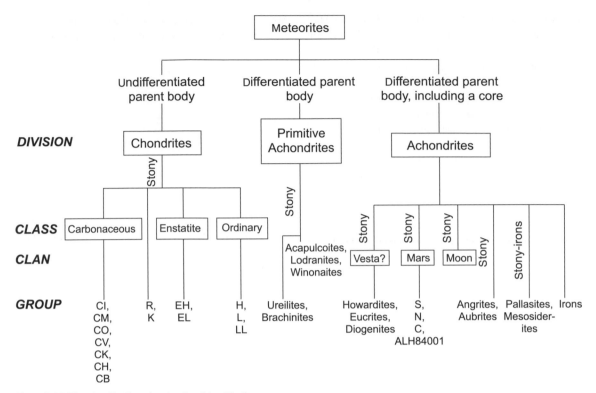

Figure 6.10 The classification of meteorites (simplified).

igneous rock that crystallized from magmas or lavas of parent bodies that had differentiated into a core, mantle, and crust. They include meteorites from Mars, the Moon, the asteroid Vesta, and other unknown differentiated asteroids. The third division shown on Fig. 6.10, the *primitive achondrites*, contains meteorites from asteroids that were heated to the point of melting but not of large-scale differentiation when a metallic core formed.

Below the three divisions of Fig. 6.10, the nomenclature can get complicated and include classes, clans, groups, and subgroups. For simplicity, we show only some key terminology. Three important classes of chondrites are as follows.

- *Ordinary chondrites*, which contain 5–15 wt% Fe–Ni, and make up 97% of all chondrites in the worldwide meteorite collection. Some geochemists think of these chondrites as similar to the bulk composition of the Earth, albeit imperfectly because the Earth has no perfect match to any chondrite or achondrite (Drake and Righter, 2002).
- *Carbonaceous chondrites*, contain organic compounds with CI and CM groups at 2–3.5 wt% C, and others with less, CK ~0.1 wt% C and CH ~0.8 wt% C. The CH and CB groups are also relatively rich in metals at 40–50 wt% Fe compared to 18.2 wt% Fe in CI.

- *Enstatite chondrites*, so named because they contain enstatite ($MgSiO_3$, a pyroxene mineral) as the dominant silicate mineral. They also contain an average of $20\pm$ 9 wt% metal and are notable for being the only chondrite class that has a wide variety of elements with isotope compositions identical to the Earth (Javoy *et al.*, 2010). Amongst these, carbonaceous chondrites appear least processed, which means that while parent body processes, such as aqueous alteration, may have destroyed minerals, the elemental composition has remained intact. They haven't experienced element segregation because the ratios of their non-volatiles elements (such as Fe, Si, Mg, Al, Ca) match those in the Sun. The CI chondrites have this similarity for all but the most volatile elements, so that CI chondrites are regarded as the most chemically primitive.

One aspect of meteorite composition, which is important in discussing the origin of the Moon (Sec. 6.8) and explaining how we know that certain meteorites are from Mars (Ch. 12), concerns the relative proportions of stable oxygen isotopes, ^{16}O, ^{17}O, and ^{18}O. These differ in the bulk composition of celestial bodies according to where the bodies formed. One idea to account for this trend is that photodissociation of nebula gases (particularly CO) fractionated the isotopes of oxygen in a way that depended upon distance from the Sun (Clayton, 2002; Lyons and Young, 2005; Thiemens, 2006; Thiemens and Heidenreich, 1983). Subsequent condensation into solids caused the variation of the oxygen isotopes with heliocentric distance to be preserved.

However, exactly how the O isotopes fractionated may be more complicated. At least some of the models have CO photodissociation taking place far from the inner Solar System, with the liberated oxygen later transported inwards in the form of water ice or some other O-bearing species. In these models, the isotopic differences between Earth, Mars, and the various meteorite types must have arisen at a later stage sometime during planet formation. There also does not seem to be a clear trend with heliocentric distance: ordinary and carbonaceous chondrites lie in opposite directions in O isotope space compared with Earth. Nonetheless, the oxygen isotopes in meteorites serve as a geochemical fingerprint for their provenance. For example, if the O isotope ratios in the bulk silicate of a meteorite match those from the clan of meteorites known to come from Mars (because they contain inclusions of Martian air), they serve as acceptable proof that the meteorite is also Martian.

The volatile content of meteorites is important because such volatiles ended up composing the Earth's atmosphere, ocean and life (e.g., Fegley and Schaefer, 2010). Water in the oceans and crust is ~0.03% of the Earth's mass. Estimates for how much water is inside the Earth vary but a few oceans worth in the mantle is typical, e.g., Lecuyer *et al.* (1998) estimate 0.3–3 oceans, equivalent to an extra 0.008–0.08 wt%. More recent estimates for Earth's bulk water are higher, 0.1%–0.3 wt% (Marty, 2012). We can compare ordinary chondrites, which contain ~0.1 wt% C, ~0.03 wt% N, and ~0.3 wt% water. The CI meteorites in Fig 6.10 are the most volatile-rich of the carbonaceous chondrites and contain an average of 3.5 wt% C, 0.3 wt% N, and up to ~10 wt% water (Fegley and Schaefer, 2010; Kerridge, 1985). It is possible that many planetesimals that went on to form Earth had already melted and differentiated (e.g., Kruijer *et al.*, 2014), and were not primitive objects like chondrites, and so had lost most of their volatiles. But the volatile abundance of the Earth can be explained if only a small proportion of our planet was accreted from volatile-rich chondritic material contained in either planetesimals or embryos scattered into the inner solar system. Indeed, planet formation models predict such scattering and show that embryos can contribute most of the water (Fig. 6.6). Volatiles inside embryos and planetesimals are expected to vaporize into gases during accretion in the process of *impact degassing*, as we describe below in Sec. 6.5.

6.4   The Implications of the Abundances of Noble Gases and Other Elements

Aside from the accretion of solids, the gravity of planetary embryos larger than Mars would have been enough to capture some nebula gas, mostly H_2 and He (Inaba and Ikoma, 2003). But today, the atmospheres of Venus, Earth, and Mars do not resemble solar composition at all, and they could not have been directly derived from a solar gas, as we discuss in more detail below. For now, and to illustrate the problem, suppose that the atmosphere of a terrestrial planet had started out with a solar composition and that hydrogen and helium escaped to space. In this case, primordial methane and ammonia would have been oxidized to CO_2 and N_2 with the loss of hydrogen, and the result would be an atmosphere of roughly 60% CO_2, 20% Ne, and 10% N_2, with a remainder of minor gases. This composition is not observed in any terrestrial planet atmosphere, with neon being strikingly underabundant (e.g., 18 ppmv in the Earth's atmosphere). Consequently, the present atmospheres on the rocky planets are described as *secondary atmospheres*. This distinguishes them from *primary atmospheres* of solar composition captured from the nebula, such as the atmosphere of Jupiter, or atmospheres directly descended from solar composition in the way that we just described.

6.4.1   Atmophiles, Geochemical Volatiles, and Refractory Elements

In thinking about the origin of atmospheres, it is useful to divide up the chemical elements into volatile and refractory elements. *Refractory* elements are those that tend to stay in solid compounds with very high melting and boiling points, e.g., Fe. We can sub-categorize the volatile elements into *geochemical volatiles*, such as K, Zn, or Cl, which are volatile at moderately high temperatures, and *atmophiles*, which are elements that tend to form liquids or atmospheric gases at typical planetary temperatures. Examples of liquids are Earth's oceans and liquid methane in Titan's polar lakes. The most important atmophiles are C, H, O (at least in H_2O or CO_2), N, and the noble gases.

If we examine the average composition of Venus, Earth, and Mars, we can compare the abundance of some elements to the solar abundance (Fig. 6.11). We then see that all three inner planets are depleted in geochemical volatiles compared to the Sun, although Mars is less depleted, presumably because Mars formed farther out in a cooler part of the solar nebula where geochemical volatiles, such as chlorine, were more abundant (Wänke

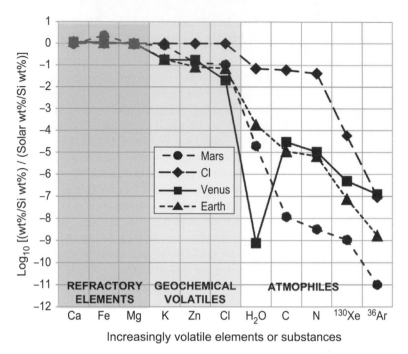

Figure 6.11 The relative abundance of some refractory, geochemical volatile, and atmophile elements and substances in the average compositions of CI carbonaceous chondrites, Venus, Earth, and Mars compared to solar abundance. Following convention, both solar and planetary compositions are normalized to the Si mass fraction. Data sources: solar abundance from Anders and Grevasse (1989), noble gas data from Pepin (1991), CI meteorite C and N abundance from Fegley and Schaefer (2010), elemental model compositions of Mars and Venus from Lodders and Fegley (1998), and elemental model composition of Earth from McDonough (2003).

and Dreibus, 1994). While we see that all atmophiles are depleted many orders of magnitude compared to solar abundances, the relative abundance of atmophiles differs greatly for Venus, Earth, and Mars.

Atmophile contrasts between Venus, Earth, and Mars provide clues to the divergent fates of their atmospheres. For example, the lack of water on Venus (only ~40 ppm in its atmosphere) is explained by its loss through a runaway greenhouse effect, which we describe in Ch. 13. In Ch. 12, we also discuss how the general lack of atmophiles on Mars can be explained mostly by their escape to space. This, of course, is a consequence of Mars' small mass (~1/9 of an Earth mass) and low gravity. For all the terrestrial planets, the depletion of atmophile elements relative to solar abundance suggests that direct capture of atmospheres of solar composition from the solar nebula cannot explain the composition of present atmospheres on the terrestrial planets. Patterns of noble gas abundance provide additional evidence for how terrestrial planet atmospheres formed (e.g., Moreira (2013), Halliday (2013)), as we now discuss.

6.4.2 Noble Gases

The noble gases, He, Ne, Ar, Kr, and Xe are unreactive and so are the strongest atmophiles. Their condensation temperatures are so low that they should have remained as atomic gases throughout the solar nebula, at least within the orbit of Neptune. On planets, primordial noble gases

ought to reside in atmospheres rather than solids, except to the degree that they have been trapped in planetary interiors either because they were dissolved from primary atmospheres in melted rock during planetary accretion or were part of primary accretion material that was implanted by the solar wind.

The Nobel prize-winning chemist Francis Aston first described the evidence for loss of a primary atmosphere when he quantified the extreme scarcity of the noble gases compared to the other elements on the Earth (Aston, 1924). Modern data, similar to those of Aston, are shown in Fig. 6.12. We consider only nonradiogenic noble gases, i.e., gases that are *not* formed from the decay of parent radioactive elements. In Fig. 6.12, the amount of neon relative to nitrogen is especially notable. Nitrogen and neon should have been roughly similar in abundance in the solar nebula, but on Earth the Ne/N ratio is a tiny $\sim 10^{-5}$. Clearly, the noble gases contribute a miniscule amount to the mass of the Earth compared to other elements. Consequently, the Earth either did not take up the gaseous component of the solar nebula or it lost it, otherwise Ne/N would be about unity. So, the Earth must have accreted its present atmophiles from solids. Studies of meteorites suggest that C and N were largely brought in by organic material (e.g., Bergin *et al.* (2015)), while water was delivered as ice or in hydrous minerals (Sec. 6.3, above).

In a similar manner, we can examine the abundances of noble gas isotopes on Venus and Mars from data obtained by the *Viking* landers for Mars and the *Pioneer*

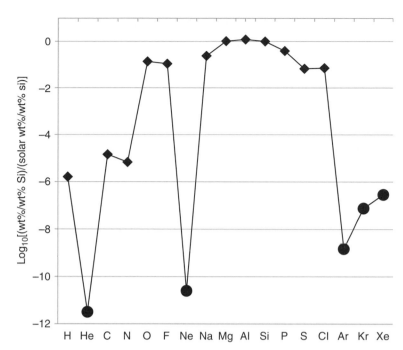

Figure 6.12 Relative abundances of various elements on Earth compared to the Sun. Abundances are normalized to the abundance of Si on the Earth and Sun, respectively. Data show the ratio of Earth/Sun normalized abundances. Noble gases (filled circles) are very depleted compared to other elements (filled diamonds). Data: solar abundance from Anders and Grevasse (1989), noble gases from Pepin (1991), and elemental composition for Earth from Lodders and Fegley (1998).

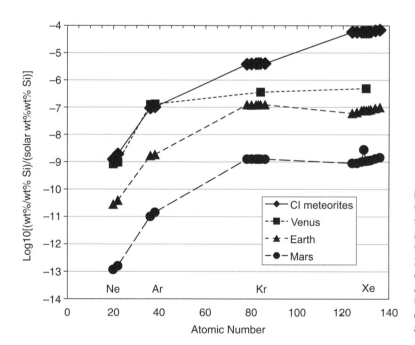

Figure 6.13 Abundances of noble gas isotopes in Cl carbonaceous chondrites and terrestrial planets relative to their solar abundance. Abundances are normalized to Si, as indicated on the vertical axis. Xenon isotopes on Earth and Mars are clearly mass-fractionated. Mars also has an unusually high relative abundance of the ^{129}Xe, formed from the radioactive decay of ^{129}I. Data from Pepin (1991) and references therein.

Venus and *Venera* spacecraft for Venus (Fig. 6.13). We exclude helium because this element escapes to space easily from rocky planets. In Fig. 6.13, the huge depletion of noble gases on Mars and Venus compared to solar abundance implies that these planets accreted volatiles as solid compounds, just like the Earth did. Another common aspect between the three terrestrial planets in Fig. 6.13 is that the noble gases are generally increasingly enriched with greater atomic mass, which suggests that lighter gases have escaped or that heavier gases were brought in more easily. In fact, stable isotope ratios in the terrestrial atmospheres, which generally are expressed as the heavy/light isotopes for a particular element, tend to be larger than their solar equivalents. This pattern can often be explained by preferential escape of the lighter isotope to space, as we discussed in Ch. 5.

While the abundance patterns of noble gases for Venus, Earth, and Mars all have large depletions and broadly similar trends compared to solar abundances, distinctions between them provide clues to the origin of each atmosphere. For example, argon and neon abundances are unusual for Venus. Nonradiogenic argon (^{36}Ar and ^{38}Ar) on Venus exceeds the chondritic abundance (Fig. 6.13), whereas neon approaches it. To put it another way, argon and neon are 70 and 20 times more abundant on Venus than on Earth, respectively. Why? One possibility is that Venus, by chance, accreted a big (>600 km) Kuiper Belt comet, where cold temperatures of <30 K and <17 K allowed argon and perhaps neon to condense as an ices, respectively (Owen and Bar-Nun, 1995). The probability of such a cometary accretion is fair, about 1 in 4 (Zahnle, 1998). A second explanation relies on the idea that during late accretion, early Venus experienced a *runaway greenhouse effect* (defined in Sec. 13.4) during which all its water was vaporized into a thick atmosphere. Under such circumstances, numerical simulations show that giant impacts would fail to remove components of a Venusian primary atmosphere, whereas on the Earth, explosive conversion of a liquid ocean to steam by giant impacts would transmit the shock to the atmosphere, causing atmospheric escape to take away noble gases previously captured from the nebula (Genda and Abe, 2005).

Radiogenic argon, ^{40}Ar, provides clues to Venus' atmospheric and interior evolution. Argon-40 is produced from the radioactive decay of ^{40}K with a half-life of 1.25 b.y., and so should gradually accumulate in an atmosphere if outgassing is efficient. Indeed, this is the reason why argon is the third most abundant gas in the Earth's atmosphere. However, Venus' atmosphere has only ~1/4 of the ^{40}Ar found in the Earth's atmosphere. A possible explanation is that Venus is not efficiently outgassed, so its volcanism must have been relatively quiescent for most of its history, whereas Earth's volcanism was more or less continuous as a consequence of plate tectonics. It turns out that tectonics and volcanism underpin the long-term climate stability and volatile recycling needed for a biosphere (Ch. 11), so if life ever existed on an early, clement Venus, we can surmise that it would have been geologically short-lived.

The isotopes of xenon and neon on Earth and Mars are also fractionated, as shown by the slope of the closely packed data points in Fig. 6.13. These slopes are possibly informative about atmospheric evolution. They suggest an episode of early atmospheric escape driven by the extreme ultraviolet (EUV) radiation from the early Sun,

which was hundreds of times greater than today (e.g., Claire *et al.*, 2012), as shown in Fig. 6.5. We discussed these issues further in Sec. 5.11.2. For now, we will just consider how noble gas and geologic data imply that much of the atmosphere and hydrosphere on the Earth was formed relatively quickly.

6.4.3 Early Degassing

We have discussed how volatiles were accreted as part of solids, so that the Earth's atmosphere and ocean must have been made from volatiles released from the Earth's interior, but there has been historical debate about how quickly this happened. The process by which gases are released from solids to form the atmosphere and the ocean is called *degassing* or *outgassing*. An old view promoted by the geologist William Rubey was that volatiles degassed onto an almost airless body that had lost its primary atmosphere so that the ocean and secondary atmosphere built up gradually (Rubey, 1951, 1955). This idea is now seen as incorrect because volatiles would have been rapidly degassed to the surface of the Earth during accretion, as we describe below in Sec. 6.5. Indeed, various lines of evidence support early degassing. That said, Rubey was right about other things, as we shall discuss in Ch. 7.

Unequivocal geological evidence shows that oceans were present in the early Archean. The data come from the Isua area in West Greenland, which preserves a ~3.8 Ga belt of layered sedimentary and volcanic rocks (e.g., Nutman, 2006). Although the rocks have been heated up to several hundred degrees Celsius and compressed to pressures of thousands of bar by metamorphism, they provide evidence for material originally deposited under water. First, they include *pillow basalts* (Komiya *et al.*, 1999), which were formed when submarine volcanic lava erupted and squeezed out underwater like toothpaste to form bulbous, pillow-like structures. Second, some sediments were laid down on the seafloor or in shallow water according to their sedimentology and geochemistry (Moorbath *et al.*, 1973; Nutman *et al.*, 1997). There is also evidence of marine life because graphite grains in the sedimentary rocks are enriched in ^{12}C, which is consistent with their derivation from microbes that tend to concentrate ^{12}C (Ohtomo *et al.*, 2014; Rosing, 1999). Overall, the Isua rocks confirm that the Earth had acquired oceans, an atmosphere, and probably life when it was only 700 million years old.

More subtle geochemical evidence suggests that oceans were present even earlier at ~4.3 Ga. Fragments of 4.5–4.0 Ga crust persist as small (<0.5 mm) *zircons*, which are grains of zirconium silicate ($ZrSiO_4$) (reviewed by Harrison, 2009). Zirconium silicate is so durable that it remains even after its parent rock has vanished through erosion. Particularly old zircons, >4.0 Ga, are found embedded in a fossilized gravel bar called the Jack Hills conglomerate in western Australia. These zircons contain inclusions of quartz, which imply their production from silica-rich igneous rocks such as granites before 4.0 Ga. Thus, the zircons are possible evidence for continental rocks. But they also provide evidence for oceans. The zircons are enriched in ^{18}O. Typically, surface rocks acquire an enrichment of ^{18}O when surface waters weather them to produce clay minerals. It is well known from more modern rocks that this ^{18}O-enrichment can be passed on to igneous rocks if the clay-rich surface rocks are buried and melted. So, these ancient zircons imply that liquid water existed on the surface of the Hadean Earth and that continental rocks not only existed but were being recycled through weathering and transport (Cavosie et al., 2005; Mojzsis et al., 2001; Trail et al., 2007; Wilde et al., 2001).

Noble gases also suggest early degassing (Allegre et al., 1987). The present fluxes of primordial noble gases are far too slow to account for their amount in the atmosphere today (Holland, 1984; Tajika, 1998). For example, the current outgassing flux of ^{36}Ar is a factor of ~4400 too slow to account for the amount of ^{36}Ar in Earth's atmosphere if it had been added at that rate over geologic time. Consequently, most of the ^{36}Ar was put into the atmosphere during an earlier time at a high outgassing rate. Indeed, the reason for such a small flux of ^{36}Ar today is that the primordial noble gas isotopes (such as ^{3}He, ^{20}Ne, ^{36}Ar, and ^{130}Xe) are severely depleted in mantle-derived rocks compared to atmospheric values, consistent with the idea of early degassing. By contrast, the isotopes such as ^{40}Ar that are derived from radioactive decay are enriched inside the solid Earth. For example, the Earth's atmospheric value of $^{40}Ar/^{36}Ar$ is 298.6, which is roughly a hundred times smaller than the upper mantle ratio of $^{40}Ar/^{36}Ar$ of 32 000±4000 and 27 times smaller than values from mantle plumes, such as ~8000 for Hawaii (Trieloff et al., 2003). This suggests that ^{36}Ar was outgassed early, while ^{40}Ar accumulated over time in the mantle from decay of ^{40}K, which has a 1.25 b.y. half-life. In fact, the $^{40}Ar/^{36}Ar$ ratio of 143±24 in 3.5 Ga quartz inclusions suggests that considerable K was extracted from the mantle into crust in the Archean (Pujol et al., 2013), consistent with other indications of early crust (e.g., Iizuka et al., 2015). However, not all the ^{40}Ar produced from decay has outgassed and about half remains in the Earth's interior (Marty, 2012; Turner, 1989), depending upon an assumed inventory of ^{40}K.

In the atmosphere, some noble gases that are daughter products of comparatively short-lived radionuclides are notably missing, which suggests that they were released early and escaped to space. The best example is ^{129}Xe, which has the radioactive parent ^{129}I with a half-life of 15.7 m.y. Let us use a standard notation and denote the xenon-129 derived from radioactive decay of ^{129}I as $^{129}Xe*$ to distinguish it from solar xenon. During the first 110 m.y. of Earth history, which corresponds to seven half-lives of ^{129}I, 99% of the $^{129}Xe*$ would have been produced. The amount of $^{129}Xe*$ trapped in the minerals of chondritic meteorites tells us that the ratio of original ^{129}I relative to the stable ^{127}I isotope was ~10^{-4}. An estimate for the amount of ^{127}I in the bulk silicate Earth[1] is 11 ppb by mass (Kargel and Lewis, 1993), so that after full decay of all ^{129}I in the Earth, we should expect 3×10^{13} moles of $^{129}Xe*$ to have been produced. But the Earth's atmosphere contains only 4.2×10^{12} moles of ^{129}Xe and only ~7% of this, or 2.9×10^{11} moles, is estimated to be from the decay of ^{129}I, while the rest is solar and non-radiogenic (Pepin, 2000). Hence ~99% of radiogenic $^{129}Xe*$ – the amount produced during the first 110 m.y. – is missing. Thus, $^{129}Xe*$ appears to have outgassed early and to have been lost during or shortly after 110 m.y. It is unlikely that $^{129}Xe*$ was gradually lost later in Earth history because xenon is the heaviest (non-anthropogenic) gas in the atmosphere and is currently unable to escape. However, the process of hydrodynamic escape (see Sec. 5.10), which is thought to have been driven by the much higher extreme ultraviolet (EUV) output from the young Sun, might account for such xenon loss.

One other scenario that is sometimes invoked by geochemists to explain an apparent decoupling between atmospheric noble gases and those in the mantle is called a late veneer. The idea is that atmospheric noble gases were probably lost during the Moon-forming impact, and that they may have been replenished by material that was accreted later, particularly cometary material that arrived during the late heavy bombardment (see Sec. 6.7). Stable krypton and xenon isotopes ($^{82,84,86}Kr$ and $^{124,126,130}Xe$) measured in well gases that are thought to represent upper mantle composition lie on a mixing line between isotopically heavy carbonaceous chondrite material and air, and are distinct from solar values (Holland et al., 2009). This suggests that the mantle derived its noble gases from an accreted component similar to

---

[1] The term bulk silicate Earth means "mantle + crust + hydrosphere" and is synonymous with "primitive mantle," which is the theoretical reservoir that differentiated into a crust, depleted mantle, and hydrosphere.

carbonaceous chondrites, whereas the atmosphere comes from a different source in which noble gases were depleted in the light isotopes. According to its proponents, a *late veneer* of cometary material could account both for isotopic components of the noble gases and a nearly solar Kr/Xe ratio in the atmosphere, as observed (Dauphas, 2003; Owen *et al.*, 1992).

6.5 Impact Degassing, Co-accretion of Atmospheres, and Ingassing

6.5.1 Laboratory Evidence for Impact Degassing

An effective mechanism for producing early degassing was the high-energy impacts that occurred during accretion. Planet formation models suggest that at least some of the bodies accreted by the Earth would have been rich in volatiles (Sec. 6.2.2). How much of this material would have been released during impact?

In the 1980s, Tom Ahrens and colleagues, as well as some Russian scientists, performed experiments in which they fired high-speed projectiles into mineral targets (Gerasimov and Mukhin, 1984; Lange and Ahrens, 1982, 1986). They discovered that the heat of impact shocks liberated volatiles, including water, from carbonaceous chondrites (Tyburczy *et al.*, 1986). Complete devolatilization was achieved at shock pressures of 20–40 GPa. Shock pressures can be related to the velocity of planetesimal impacts during accretion by assuming reasonable material properties for the planetesimals. According to theory, degassing becomes important for impact velocities exceeding 5 km s^{-1}, which occur for (Mars-sized) planetesimals, ~1/10 of an Earth mass. On that basis, once the radius of the Earth reached about a third of its present value, a dense, steam atmosphere should have formed (Matsui and Abe, 1986a). Later studies showed that other gases would have also been released with a composition that depends mainly on the iron content of the accreting planetesimals (Hashimoto *et al.*, 2007; Schaefer and Fegley, 2010; Zahnle *et al.*, 2010), as we discuss below.

6.5.2 Formation of Steam and Reducing Atmospheres During Accretion

The steam atmosphere formed from impact degassing would have provided insulation for the heat flux deposited by impacts on the growing planet, along with a substantial greenhouse effect. Calculations by Matsui and Abe (1986b) showed that such a "thermal blanket" would have caused the entire surface of the planet to melt during the late accretion, creating a *magma ocean*. Actually, irrespective of the atmospheric thermal blanket, a magma ocean of hundreds of km depth is an inevitable

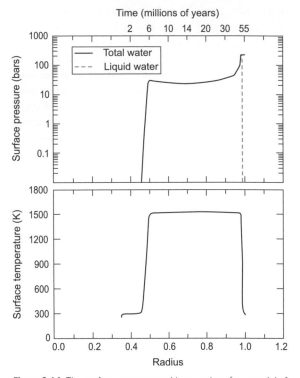

Figure 6.14 The surface pressure and temperature for a model of impact degassing during accretion of the Earth. In this model, degassing exceeds escape to space when the growing planet reaches roughly 0.5 of an Earth radius. At that point, the atmosphere is opaque to the thermal infrared in a "runaway greenhouse" state and the surface melts. The steam atmosphere eventually collapses when the planet nearly reaches the current Earth radius. (Adapted from Zahnle *et al.* (1988) Reproduced with permission from Elsevier. Copyright 1988.)

consequence of giant impacts during late accretion (Tonks and Melosh, 1993). The surface pressure of the atmosphere at this time was controlled by the solubility of H_2O in the melt and should have been ~100 bar.

Figure 6.14 shows the steam atmosphere model of Zahnle *et al.* (1988), which was similar to that of Matsui and Abe, except with more elaborate H_2O absorption coefficients. Related calculations including more detailed mineralogy and gaseous CO_2 have also been published (Elkins-Tanton, 2008; Kuramoto and Matsui, 1996). Zahnle *et al.* explicitly kept track of the amount of water trapped inside the growing planet, along with exchange of water between the atmosphere and surface and escape of hydrogen to space. Because the hydrogen comes from water initially, such escape of hydrogen oxidizes the mantle. According to this calculation (Fig. 6.14, top panel), the atmospheric pressure increased from ~30 bar to 240 bar right near the end of accretion, as the magma ocean solidified. This water then condensed out to form the ocean. The amount of water in Earth's current ocean,

1.4×10^{21} kg, is equivalent to a surface pressure of 270 bar; hence, this model ends up with ~90% of the observed amount of surface water. About 140 bar of water was directly emplaced in the atmosphere by impacts; so subsequent outgassing provides roughly 100 bar of water, or about 40% of the modern ocean.

During impact degassing, other gases besides H_2O should have been released (Schaefer and Fegley, 2010; Zahnle *et al.*, 2010). If the infalling material had the composition of CI carbonaceous chondrites, equilibrium calculations at high pressure and temperature suggest that the gas composition would have been relatively oxidized, with H_2O and CO_2 as the predominant constituents and minor H_2 (Fig. 6.15 (b)). On the other hand, if accreting planetesimals were similar to ordinary chondrites, H_2-rich atmospheres with CO and H_2O as secondary components are expected, because of the presence of iron as a strong reducing agent (Fig. 6.15 (a)). So, the atmosphere during accretion could have been much more reduced than modern volcanic gases, whose relatively oxidized composition we discuss in Ch. 7.

6.5.3 Ingassing

Evidence suggests that the Earth has also experienced an opposite process from degassing, which is *ingassing*

(or *regassing*). Ingassing likely happened during accretion and has certainly occurred over geologic time from the subduction of volatile elements captured in sedimentary minerals.

If the proto-Earth grew larger than Mars in the presence of the nebula, it should have captured a *primary atmosphere* of nebular gas (Inaba and Ikoma, 2003), which would have partially dissolved in the molten surface produced by impacts (Fig. 6.16(a)). Subsequently, noble gases could have been trapped inside the Earth once molten material cooled. Evidence for solar noble gas might reside in ^3He and neon (Harper and Jacobsen, 1996; Jacobsen *et al.*, 2008), which we consider in turn.

Helium has radioactive and primordial components. The two forms are: (1) ^4He, which is present inside the Earth mainly as a consequence of radioactive decay in the mantle and crust of ^{235}U, ^{238}U, and ^{232}Th, and (2) ^3He, which is a primordial remnant and thus relevant for this discussion. Helium diffuses easily, and ^3He comes out of the mantle at a rate of 3.7 kg yr^{-1} (Schubert *et al.*, 2001, pp. 574–577). The inferred abundance of ^3He in the deep Earth is similar to that in meteorites, so sometimes geochemists refer to the ^3He as part of an *undegassed mantle reservoir* (Graham, 2002; Porcelli and Elliott, 2008). What this actually means is that ^3He must have been

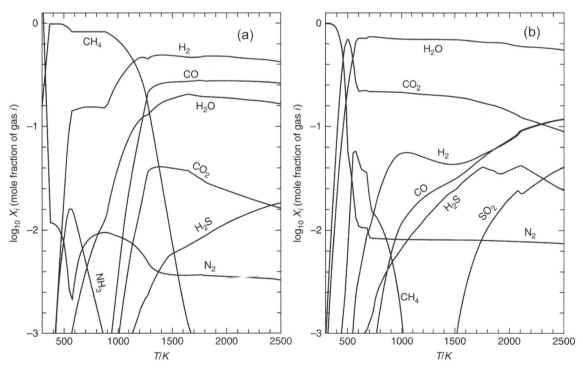

Figure 6.15 (a) Strongly reducing atmospheres of accretion, calculated assuming gas compositions in equilibrium with CI chondrite composition at 100 bars. (b) Weakly reducing atmospheres of accretion, calculated assuming gas compositions in equilibrium with ordinary H-type chondrites at 100 bar. (From Zahnle *et al.* (2010) Reproduced with permission from Cold Spring Harbor Laboratory Press. Copyright 2010.)

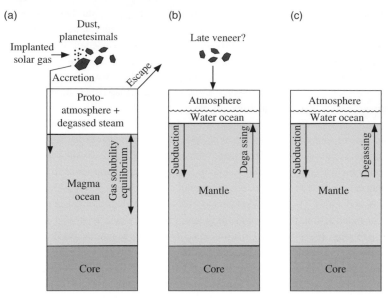

Figure 6.16 Conceptual models of early to late atmospheres, following Jacobsen *et al.* (2008)). (a) When the proto-Earth planetary embryo was larger than Mars, it could have captured a primitive atmosphere of solar composition. A magma ocean produced by impacts of accretion should have allowed a small fraction of the primitive atmosphere to dissolve in equilibrium, in the process of *ingassing*. Dust and planetesimals would have added a chondritic component to the growing Earth, including a possible solar component from implantation by the young solar wind. The hydrogen-rich primitive atmosphere was either eroded by the Moon-forming impact or escaped given the high flux of extreme UV from the young Sun. (b) The magma ocean froze into the mantle, possibly trapping a dissolved solar noble gas component of ^{3}He and Ne. Collapse of the steam atmosphere from the Moon-forming impact and outgassing produced an ocean and atmosphere during the Hadean. Small imbalances between outgassing and subduction of sediments were likely important for returning carbon, nitrogen and sulfur atmophiles to the mantle. Possibly a "late veneer" of volatiles was added during Late Heavy Bombardment. (c) Subsequently, the atmosphere was maintained over geologic time by a balance between outgassing and ingassing.

added by a mechanism such as equilibrium dissolution of a primary atmosphere in a magma ocean because impact degassing ought to ensure that ^{3}He could not have been accreted directly in solids (unless impact degassing is less than 100% efficient).

Evidence that neon may have dissolved in a magma ocean from a primary atmosphere in a similar way to helium comes from the inferred ratio of ^{22}Ne/^{20}Ne ~12.5 in the mantle, which is similar to ^{22}Ne/^{20}Ne = 12.5±0.2 in gas-rich meteorites (Porcelli and Wasserburg, 1995). However, any primary atmosphere was probably doomed. Vigorous escape to space of a hydrogen-rich primitive atmosphere is expected to remove it, and a *secondary atmosphere* formed from impact degassing would eventually dominate (Fig. 6.16(b)).

Once the Earth had a crust and liquid ocean (and after the Moon-forming impact, discussed in the next section), the ocean and atmosphere were maintained by a balance between ingassing of geologically transported volatiles and outgassing from volcanoes and metamorphism (Fig. 6.16(c)). But some volatiles must have suffered an imbalance, perhaps during the Hadean eon represented by Fig. 6.16(b). Carbon, nitrogen, and sulfur have apparently partly returned to the mantle after early degassing. The ratio of carbon to ^{36}Ar fluxes coming out of the mantle today is C/^{36}Ar ~8 ×10^9, whereas the ratio of the total amount of carbon to ^{36}Ar in the crust, atmosphere and ocean is only 1.6×10^6. At current rates, the surface reservoir of carbon would accumulate in 5 b.y., whereas the ^{36}Ar would take an unfeasible 22 000 b.y. Clearly, the ^{36}Ar degassed early, while it seems likely that carbon returned to the mantle after similar early degassing (Walker, 1990). The fate of most of the CO_2 degassed into the atmosphere after the Moon-forming impact was probably to be incorporated into oceanic crust and subducted within ~10^8 yrs (Zahnle *et al.*, 2007; Zahnle and Sleep, 2002). Others have suggested that organic carbon has been subducted into the mantle over geologic time (Hayes and Waldbauer, 2006).

Evidence suggests that significant amounts of nitrogen and sulfur have also been incorporated into sediments and subducted into the mantle. Measurements of mantle N correlate with ^{40}Ar, the daughter product of radioactive ^{40}K, which indicates that mantle N comes from subducted rocks in which NH_4 substituted for K^+ in minerals such as clays or micas (Goldblatt et al., 2009; Marty and Dauphas, 2003). We discuss nitrogen further in Ch. 11. Tallying up the nitrogen inventories of the modern Earth suggests that almost ~60% of the nitrogen is in the mantle (Table 11.1). Sulfur isotopes and mass balance provide evidence that sulfur returned to the mantle as subducted sulfides. This process may have been efficient during the middle Proterozoic when rivers supplied the ocean with sulfate produced by oxidative weathering of continental sulfides, but the low-oxygen seafloor was conducive to the bacterial formation of sulfide, which could then be subducted (Canfield, 2004), as suggested earlier by Catling et al. (2002).

6.6 Moon Formation and its Implications for Earth's Volatile History

6.6.1 The Giant Impact Hypothesis

In the 1970s, various lines of evidence led to two independent suggestions that the Moon formed from the debris of a collision of a Mars-sized body (called *Theia*) with the proto-Earth (Hartmann and Davis, 1975; Jöns, 1985). One important finding from lunar samples brought back by the Apollo astronauts was that the Moon has a bulk composition similar to that of the Earth's upper mantle when we look at the major elements such as Si and Mg (Wänke, 2001). Also, the stable isotopes of oxygen ^{16}O, ^{17}O, and ^{18}O occur in the same relative proportions in the Earth and the Moon to a very high precision of several parts in a million (Herwartz et al., 2014). The oxygen isotope ratios in bulk silicates of various celestial bodies vary depending on where the bodies were formed in the Solar System (Sec. 6.3 above). Thus, the oxygen isotopes suggest that the Moon and Earth have a common origin and that the Moon is not a captured object. Alternatively, turbulent mixing in the molten disk aftermath of the giant impact equilibrated the oxygen isotopes (Pahlevan and Stevenson, 2007). Tungsten and silicon isotope ratios are also nearly identical (Dauphas et al., 2014).

By the late 1980s, the theory that the Moon formed as a result of a large collision with a Mars-sized object had gained acceptance (Stevenson, 1987). The theory, in brief, is that about 50–100 m.y. after the formation of the Solar System, Theia coalesced with the proto-Earth (both of which had already differentiated), causing much or all of the Earth to melt (Kleine and Rudge, 2011). The likely age of this impact is ~4.47 Ga (Bottke et al., 2015). Under such conditions, the density of molten iron caused it to separate into the Earth's core. However, the impact ejected considerable debris into orbit around the Earth from Theia material and the Earth's mantle. This debris accreted to form the Moon.

The giant impact theory explains several geochemical observations. The Moon's crust is depleted in *siderophile* elements by a factor of 0.2–0.003 compared with the Earth's crust and mantle. Siderophile elements, such as nickel (Ni), iridium (Ir), molybdenum (Mo), and germanium (Ge), are those that tend to be scavenged by metallic iron during melting. The impact model accounts for these data because there should be a low abundance of siderophile elements in the silicate mantle debris that gave rise to the Moon. Also, the Moon is poor in geochemical volatiles such as K and Rb by a factor of ~0.5 times terrestrial abundances (Taylor, 2001, p. 384). If the Moon formed from vaporized debris, then geochemical volatiles could have escaped.

The giant impact theory also explains geophysical observations. The Earth is comparatively high in its mean density in the trend of decreasing densities of the planets as a function of distance from the Sun, while the Moon is much less dense than the Earth (3.344 g cm^{-3} vs. 5.515 g cm^{-3}). These phenomena make sense if the Earth acquired iron from Theia's core when the two bodies coalesced, leaving the Moon with a small, inner iron core only 240 km in radius beneath a ~90 km thick fluid shell (Weber et al., 2011). This lunar core has an estimated <6 wt% light alloying components such as sulfur, as compared to up to ~10 wt% in the Earth's core (Birch, 1964; Poirier, 1994), again indicating the Moon's depletion of geochemical volatiles. The angular momentum of the Earth–Moon system is anomalously high compared to other planet–moon systems such that if all the lunar mass and momentum were put into the Earth it would rotate every 4 hours. Calculations show that this high angular momentum problem can be solved if Theia struck the proto-Earth obliquely at a speed comparable to the Earth's escape velocity of 11.2 km/s. So, overall the giant impact origin of the Moon succeeds in explaining a variety of geochemical and geophysical observations.

6.6.2 The Post-Impact Atmosphere and Loss of Volatiles

What was the effect of the Moon-forming impact on Earth's early atmosphere? If we assign a time of zero to the impact itself, essentially four environmental stages

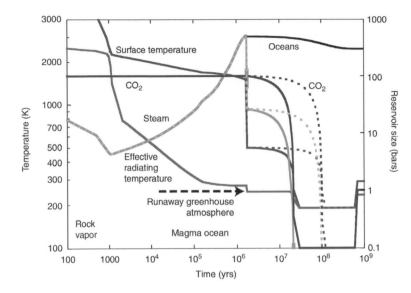

Figure 6.17 A self-consistent model of the environmental consequences of the Moon-forming impact. See text for a description of the various phases. (From Zahnle *et al.* (2010) Reproduced with permission from Cold Spring Harbor Laboratory Press. Copyright 2010.)

should have followed (Zahnle *et al.*, 2007). These are illustrated in Fig. 6.17.

(1) From 0 to 10^3 years, when the atmosphere consists of a cloud of ~2500 K vaporized rock. The most abundant atmospheric gases are SiO and O_2, but geochemical volatiles such as Na and Cl are also present in the atmosphere.

(2) From 10^3 to ~2 m.y., when a deep magma ocean exists and volatiles such as water and CO_2 are partitioned between the ~2000 K melt and the atmosphere, depending on solubility. Whether the atmosphere is strongly reducing (mostly H_2 and CO) or weakly reducing (mostly H_2O and CO_2 with minor reducing gases) depends on how quickly the molten iron acquired from Theia sinks into the Earth's core. Generally, post impact atmospheres in equilibrium with crust or bulk silicate Earth consist mostly of H_2O and CO_2 (Lupu *et al.*, 2014). Radiative cooling is controlled by steam in the atmosphere, and the atmosphere remains largely opaque to thermal infrared emanating from the surface, which maintains the high surface temperature. But eventually, the mantle starts to solidify upwards from some depth.

(3) From 2 m.y. to ~10^8 yr: A solid surface has formed, an ocean condenses (which is salty from the Na and Cl that were once in the post-impact atmosphere), and the atmosphere consists of 100–200 bar of CO_2. The thick CO_2 atmosphere is gradually removed during this period by reacting with the crust.

(4) From ~10^8 to ~10^9 yr: During this time, the thickness of the Hadean atmosphere is unknown. Whether the Hadean climate was clement or cold depends principally on the amount of greenhouse gases because the young Sun was ~30% fainter than the modern Sun. Some have

argued for a generally cold and frozen Hadean Earth because more vigorous tectonics at that time would have removed CO_2 via the incorporation of CO_2 into the basaltic seafloor and subsequent subduction (Sleep and Zahnle, 2001). Others suggest that much more CO_2 would have been present, up to 10 bar (Walker, 1985), and that the surface could have been quite hot, 80–90 °C (Kasting and Ackerman, 1986). The nature of Hadean climate and atmosphere remains unresolved.

Earlier, we mentioned the model of Genda and Abe (2005) for the loss of volatiles during giant impacts. Assuming that an ocean existed on the Earth before Theia hit it, their calculations suggest that the Moon-forming impact would have removed little water (Abe, 2011). However, the ocean would have been vaporized explosively, accelerating a significant fraction of the overlying atmosphere to greater than the escape velocity. Hence, most of the preexisting atmosphere would have been blown away while the ocean would have survived. The atmosphere that emerged afterwards was thus mostly from degassed volatiles of the magma ocean formed during the impact event. Consequently, the Moon-forming impact and earlier impacts during accretion are likely to have removed any primary atmosphere by promoting its escape to space.

6.7 "Late Heavy Bombardment": Causes and Consequences

After the Moon-forming impact, large bodies left over from planetary accretion were still present within the Solar System and would have caused significant impacts during the Hadean. Evidence for large impacts comes from the cratering record on the Moon and Mars. Earth would have received ~20 times the impactor flux of the Moon, given Earth's larger gravity and area, so the big lunar craters imply that about a hundred ~100-km-diameter bodies and thousands of 10-km bodies pelted the early Earth. The timing of these impact events is debated (reviewed by Hartmann *et al.*, 2000), but arguments have been made that impacts were concentrated during the interval 4.1–3.8 Ga, which is called the *Late (Heavy) Bombardment* (LHB) or *lunar cataclysm* (Tera *et al.*, 1974). The alternative point of view, illustrated in Fig. 6.18, is that there was a smooth, exponential decline in the cratering rate back to the origin of the Moon, and that the LHB was just the tail of this decline or a statistical spike in the tail. However, very early high impact fluxes should have created huge impact basins on the surfaces of the Moon and Mars that are unseen, along with a contamination of the lunar crust with siderophile elements (such as iridium) contained in asteroids, which is also not observed. Overall, the terrestrial planets seem to have unusually large basins that formed relatively late: Imbrium and Orientale on the Moon; Hellas, Isidis, and Argyre on Mars; Rembrandt and Caloris on Mercury.

Nonetheless, evidence for the duration and magnitude of the LHB is debated (see review by Fassett and Minton (2013)). The motivation for a narrow, 100–200 m.y. LHB was a clustering of basin ages near 3.9 Ga derived from radiometric dating of lunar samples: Nectaris (3.92), Crisium (3.91), Serenitatis (3.89), and Imbrium (3.85) (Ryder *et al.*, 2000). However, *Apollo* sample radiometric

clocks appear to have been contaminated by material from the very large (~1160 km) Imbrium impact (Haskin, 1998; Mercer *et al.*, 2015). Moreover, data from lunar zircons and breccias suggest older impact events at 4.3–4.1 Ga (Norman *et al.*, 2010; Norman and Nemchin, 2014). On the other hand, lunar meteorites have few ages >4.0 Ga (Cohen *et al.*, 2000). Also, meteorites thought to be from the ~530 km asteroid Vesta have a spread of shock ages ranging 3.4-4.1 Ga but few from 4.4–4.1 Ga (Bogard, 1995; Marchi *et al.*, 2013). It has also been argued that the LHB is supported by a ~7 times enhanced abundance of iridium (~150 ppt) in 3.8 Ga Isua sedimentary rocks compared to modern crust (~20 ppt) (Jorgensen *et al.*, 2009). However, contamination of Isua sedimentary rocks with basaltic debris containing ~200 ppt Ir makes this result doubtful. Overall, the data suggest a more protracted LHB rather than a narrow interval near 3.9 Ga. In addition, impact spherules and models suggest a gradual decline of impacts in the Archean (Johnson and Melosh, 2012; Lowe *et al.*, 2014). For example, there were probably ~70 impactors from 3.7–1.7 Ga comparable to the Chicxulub impactor (Bottke *et al.*, 2012).

Various hypotheses have been put forth to account for the LHB, all of which involve some source of planetesimals that became unstable about 600 m.y. after the Solar System formed. The most popular hypothesis has been the *Nice model*, which originated in Nice, France, and which relies on planetary migration (Gomes *et al.*, 2005). If giant planets encounter planetesimals and eject

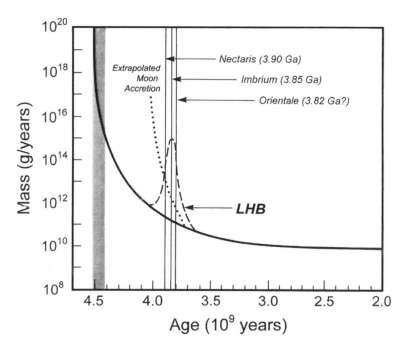

Figure 6.18 A schematic diagram showing basic data that motivates the idea of a Late Heavy Bombardment (LHB) pulse of impacts. Ages of large lunar impact basins have been interpreted to cluster near 3.9 Ga, as shown. Backwards extrapolation of an accretion curve that includes such large impactors (dotted line) would imply accretion of Moon-sized objects (~10^{26} g) at 4.1 Ga, which is unlikely; hence, the idea of a late pulse of bombardment. The solid line shows a backwards extrapolation of the current impactor flux to the origin of the Solar System. The vertical gray bar indicates the time of the formation of the Moon. (From Koeberl (2006). Reproduced with permission. Copyright 2006, Geological Society of America.)

them either in or out of the Solar System, conservation of angular momentum requires that the giant planets move slightly. The Nice model shows that for reasonable initial conditions, an evolution of orbits can cause Saturn and Jupiter to go through a 2:1 *mean motion resonance*, in which Saturn orbits the Sun once for every two Jupiter orbits. This resonance, which can occur in the models around 4.0–3.9 Ga, given appropriate initial conditions, creates a regular gravitational nudge that makes the orbits of Saturn and Jupiter more eccentric. In turn, the orbits of the other two outer planets are affected. Neptune and Uranus move outwards and themselves acquire eccentric orbits. Neptune can initially be inside the orbit of Uranus

asteroids (Chambers, 2007), and another in which Uranus and Neptune form between Jupiter and Saturn and then get scattered outwards, disturbing icy planetesimals (Thommes *et al.*, 2002).

In any case, the largest impactors of the LHB would have had a severe effect on the Earth's atmosphere, oceans and any life at the time. Essentially, the environmental consequences were similar to the Moon-forming impact, except without a magma ocean, shorter in duration, and much smaller in scale (Fig. 6.19).

Impactors larger than a certain mass, $m_{impactor}$, would have vaporized all the water of mass m_{ocean} in the ocean into steam, according to the following energy balance:

$$\text{kinetic energy} = (\text{heat to vaporize ocean}) + (\text{heat to reach critical temperature}, 647\text{ K})$$
$$\tfrac{1}{2}m_{impactor}v^2 = m_{ocean}L + m_{ocean}c_p\Delta T$$

(6.3)

and then overtake it in the migration outwards, and both planets, particularly the farthest one, scatter small icy bodies, some of which enter the inner Solar System. Meanwhile, changes in Jupiter's orbit cause resonances that increase the orbital eccentricities of asteroids, flinging some of them into the inner Solar System and expelling others entirely. Such havoc could explain several features of the Solar System: the low mass of the population of objects beyond Neptune, the low mass in the asteroid belt (~1/25 of the Moon's mass), and the LHB.

Other studies suggest that the Nice model would produce terrestrial planet eccentricies and inclinations that are larger than those observed, so that the model cannot account for the LHB (Kaib and Chambers, 2015). Other hypotheses for the LHB have been proposed, including one in which a small planet (called "Planet V") in the asteroid belt is ejected by Jupiter and produces a rain of

where $L = 2.5 \times 10^6$ J kg^{-1} (at 273 K) is the latent heat of vaporization of water, and $c_p = 1900$ J kg^{-1} K^{-1} is the specific heat of water vapor. Taking the temperature change $\Delta T = 647$ K $- 273$ K $= 374$ K, a typical asteroid collision speed of $v = 14$ km s^{-1}, $m_{ocean} = 1.4 \times 10^{21}$ kg, and assuming that ~¼ of the impact energy is spent evaporating water while the rest enters the solid Earth or radiates to space, we get 2×10^{20} kg for the required impactor mass. For 3000 kg m^{-3} density, the impactor is a ~500 km diameter object, similar to the asteroid Vesta (Sleep and Zahnle, 1998; Sleep *et al.*, 1989; Zahnle and Sleep, 1997). Statistically, 0–4 impactors larger than 1000 km capable of global sterilization should have hit the Earth between the Moon-forming impact and ~3.8 Ga, while 3–7 ocean-vaporizing impactors larger than 500 km should have hit (Marchi *et al.*, 2014). Each ocean vaporizing impact could potentially have sterilized the

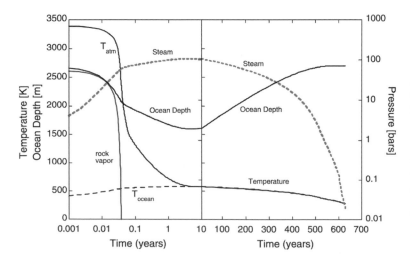

Figure 6.19 The environmental consequences on the early Earth of an impactor that released 10^{27} J, comparable to that that caused the 2100 km-wide Hellas basin on Mars. Ocean depth, ocean temperature, and atmospheric temperature are shown as a function of time, along with the pressure of rock vapor and steam. (From Nisbet *et al.* (2007a). Reproduced with permission of Springer. Copyright 2007, Springer Science + Business Media, Inc.)

Earth, although microbes in the deep continental subsurface (if it existed) might have been able to survive (Sleep and Zahnle, 1998).

Large but non-sterilizing impacts would boil off the top tens or hundreds of meters of the ocean (Segura *et al.*, 2013). Layers of marine silica-rich sinter at ~3.3 and 3.2 Ga could be evidence of partial ocean evaporation from such impacts (Lowe and Byerly, 2015).

The environmental consequences shown in Fig. 6.19 are for an impact comparable to that which caused the Hellas basin on Mars (2100 km wide, 8 km deep) or the lunar South-Pole Aitken crater (2500 km diameter, 13 km deep). The post-impact stages include a month with a rock vapor atmosphere that heats and evaporates seawater, making a steam atmosphere that then lasts for ~10^2 years before radiative cooling to space causes condensation. The significances of such an event would be to devastate photosynthetic life. However, hydrothermally hosted life in the deep ocean could perhaps survive. Consequently, such impacts may explain why a comparison of genomes of organisms on Earth today suggests that the common ancestor of all life was a hyperthermophilic ("heat-loving") microbe (Gogarten-Boekels *et al.*, 1995; Sleep *et al.*, 1989). Such microbes would have survived the LHB. Alternatively, warm hydrothermal vents might just have been good places for life to originate (Martin *et al.*, 2008).

6.8 The Early Atmosphere: the Effect of Planetary Differentiation and Rotation Rate

6.8.1 Core Formation and its Effect on Atmospheric Chemistry

The timing of differentiation of the Earth into a core, mantle, and crust had consequences for the composition of the early atmosphere. If the source region for volcanic gases was highly reducing then the gases released would have been reducing also, i.e., rich in hydrogen or hydrogen-bearing gases such as methane or ammonia. Consequently, when the proto-Earth still had metallic iron in its mantle, gases released into the atmosphere would have been very reducing, as illustrated in Fig. 6.15(a) and Fig. 6.20(a). In simple outline, once the molten iron sank to the core, the gases would have been only weakly reducing, i.e., predominantly H_2O instead of H_2, mainly CO_2 instead of CO or CH_4, and N_2 rather than NH_3, as shown in Fig. 6.15(b) and Fig. 6.20(b). (The actual story of the evolution of mantle oxidation state is more complex, as discussed in Ch. 7.)

Theory and geochemical data suggest that the core of the proto-Earth formed during accretion. Then, after the

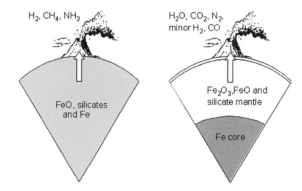

Figure 6.20 Effect of core formation on volatiles. *Left:* Before molten iron sank from the mantle into the core, the chemical effect was to produce gases that were highly reducing. *Right:* Once the iron separated into the core, the gases would have become only weakly reducing.

Moon-forming impact, further accreted iron settled to the core quickly, so that a highly reducing atmosphere may have persisted only tens of millions of years after this impact (but see Sec. 7.3). We note that further impacts of metal-containing asteroids (i.e., the composition of ordinary or enstatite meteorites) could have generated temporary reducing atmospheres later in the Hadean, of course.

Isotopes provide clues about the rate of the Earth's accretion and differentiation into a core and mantle (Kleine and Rudge, 2011). The most useful isotopes are radiogenic parent–daughter pairs where one element is a *siderophile*, meaning it has a tendency to separate out with molten iron, and the other is not, making it a *lithophile*. In this case, during core formation, the extent to which a radioactive lithophile remains in the mantle and the daughter product goes into the core depends on their relative affinities for iron. The now-extinct radionuclide ^{182}Hf (hafnium) decays to ^{182}W (tungsten) with a 9 m.y. half-life. The tungsten isotope ratio of ^{182}W relative to nonradiogenic ^{180}W in Earth's mantle and crust is slightly higher than the carbonaceous chondrite value that represents primordial Solar System material. Consequently, a fraction of the Earth's core must have formed during the 9 m.y. lifetime of the hafnium parent, because radioactive hafnium is such a strong lithophile that it remained in the mantle, which allowed the mantle, but not the core, to accumulate excess ^{182}W. Core formation also decreased the mantle's lead/uranium ratio, Pb/U, because Pb is a siderophile whereas U is not, so Pb was lost to the core. Measurements of the ratio of ^{206}Pb (produced by the decay of ^{238}U with a half-life of 4468 m.y.) and ^{207}Pb (produced by the decay of ^{235}U with a half-life of 703 m.y.) relative to nonradiogenic ^{204}Pb

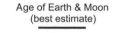

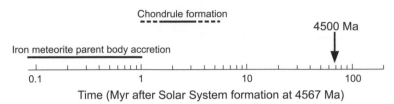

Figure 6.21 Timescales after the formation of the Solar System, which occurred when calcium–aluminum inclusions (CAIs) formed in chondrite meteorites at 4.567 Ga. (Data from Kleine *et al.* (2011).)

suggest a segregation of lead and uranium ~50–150 m.y. after Solar System formation (Wood and Halliday, 2010).

Understanding the implications of both Hf–W and U–Pb isotope systems in detail is model-dependent. Results are determined by assumptions about whether accretion declined exponentially with time or by some other functional form, and whether the core formed in equilibrium or disequilibrium between metal and silicate. Despite these uncertainties, the data suggest that the Earth grew to 95% of its current size between 30 m.y. and 120 m.y. after Solar System formation, and that the core formed within the same period. These data are consistent with the ages of the oldest rocks on the Moon, 4.46 ± 0.04 Ga (Norman *et al.*, 2003). A summary of timescales for the events in the early Solar System is shown in Fig. 6.21.

6.8.2 Day Length, the Lunar Orbit, and the Early Steam Atmosphere

The angular momentum of the Earth–Moon system has affected atmospheric evolution, while at the same time it is possible that the early atmosphere influenced the evolution of the angular momentum system, as we describe below. Because of tides raised in the ocean by the Moon and the friction that this exerts, the Earth's rotation rate is slowing down (Goldreich, 1966). Geology provides direct evidence. Rhythmites are stacked laminae of sandstone, siltstone and mudstone, which display periodic variations in thickness because of tidal, lunar and annual cycles (Coughenour *et al.*, 2009). Their study indicates that there were about 400 ± 7 sidereal days (rotation periods) per year at ~620 Ma, consistent with a 21.9 ± 0.4 hour day (Williams, 1997). Extrapolation (although not straightforward for reasons described below) would suggest that, at ~3.5 Ga, the Earth had a ~15–16 hour day. As described in Sec. 4.2, a more rapidly rotating planet would have affected the Earth's climate system through its influence on atmospheric dynamics.

The overall angular momentum of the Earth–Moon system is conserved (if we ignore angular momentum exchange with the orbit of Earth around the Sun), so as the Earth's rotation rate has declined, the angular momentum has increased in the Moon's orbit through a greater Earth–Moon distance. Essentially, the Moon has receded from the Earth as the Earth's rotational angular momentum has gradually been transferred to the Moon's orbital angular momentum.

The current lunar recession rate is 3.82 ± 0.07 cm per year, which has been measured using the round-trip Earth–Moon time for laser light shined on reflectors placed on the Moon's surface by Apollo astronauts (Dickey *et al.*, 1994). This rate of recession reflects a high tidal dissipation rate on the modern Earth. The dissipation rate must have been ~3 times lower on average during the past because otherwise, running the system backwards, the Moon would have spiraled into the Earth at ~1.4 Ga (Walker *et al.*, 1983). Tidal dissipation occurs mostly when ocean waves encounter continents or shallow shelves, and hence depends critically on global geography. Unusual phenomena such as the very high tidal amplitude in the Bay of Fundy in the northwest Atlantic Ocean can contribute significantly to the dissipation rate. Fortunately, evidence from geology supports the idea that the tidal dissipation rate was slower in the past. In addition to rhythmites, there are a variety of other data. It is possible to count daily growth bands between prominent bands, interpreted as annual, in corals, bivalves and brachiopods, in order to determine the day length in the Phanerozoic (reviewed by Williams, 2000). Cyclic deposition in the Weeli Wolli Formation, Australia, which is a *banded iron formation* sedimentary rock (see Sec. 10.3.2), has also been used to estimate a day length of 17.1 ± 1.1 hours at 2.45 Ga.

The effect of the early atmosphere may help solve a second conundrum of the lunar orbital evolution (Zahnle *et al.*, 2007). The Moon's orbit is currently inclined by ~5° with respect to Earth's equatorial plane (the ecliptic).

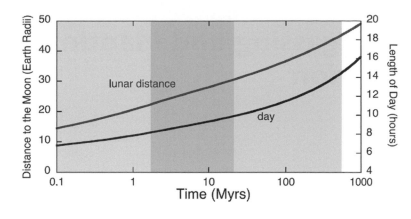

Figure 6.22 The Earth–Moon distance (left vertical axis) and length of the Earth's day (right vertical axis) versus time after the Moon-forming impact. The first shaded period 0.1–2 Myr corresponds to the presence of a steam atmosphere after the impact. The second shaded period corresponds to a thick CO_2 atmosphere that wanes with time. The third shaded period corresponds to the rest of the Hadean. The current distance to the Moon is ~60 Earth radii. (Adapted from Zahnle *et al.* (2007). Reproduced with permission of Springer. Copyright 2007, Springer Science + Business Media, B.V.)

The giant impact hypothesis predicts that the Moon should have formed within the equatorial plane, so some process must have increased its inclination. According to Touma and Wisdom (1998), this can be accomplished if the Earth–Moon system passed through two resonances that occurred soon after Moon formation. For these resonances to change the Moon's inclination significantly, though, the Moon must have been receding much more slowly than standard theory would predict.

The problem of slower lunar recession after the Moon-forming impact can be resolved by considering the atmospheric effect of the impact on Earth's surface and interior (Zahnle *et al.*, 2007). The impact would likely have created a molten Earth with a hot core. Tidal energy cannot be dissipated efficiently in a molten planet, which would greatly limit the rate of outward migration of the Moon. Also, the thermal blanketing of a steam atmosphere formed after the Moon-forming impact would help maintain the magma ocean by restricting the outgoing thermal radiation to space. As the mantle begins to solidify at depth, the tidal energy dissipation acting in this subsurface region would slow the freezing. But this tidal energy could not be lost from the Earth as a whole any faster than the *runaway greenhouse* threshold set by a steam atmosphere opaque to the infrared, which is ~300 W m^{-2} derived from the properties of water vapor (see Sec. 13.3.2 for an analytical derivation of this *runaway greenhouse* flux limit). The tidal heating and steam atmosphere would be coupled: the magma ocean would be maintained, while the leakage of tidal heat from the mantle would help sustain the thermal blanketing effect of the atmosphere. Eventually, slow migration of the Moon away from the Earth allowed the mantle to freeze. At this point, about 2 m.y. after the Moon forming impact, the steam greenhouse collapsed. Because of the limit on the outgoing infrared radiation set by the steam atmosphere, the evolution of the lunar orbit is slowed by a factor of 100–1000 times compared to models that ignore the effect of atmospheres and assume that tides are raised on a solid body (Zahnle *et al.*, 2007). Although 2 m.y. seems short, it is important on a logarithmic scale because this is the time when the Moon's recession should be expected to be fastest. A possible history of the Earth's day length and lunar distance is shown in Fig. 6.22.

7 Volcanic Outgassing and Mantle Redox Evolution

In the previous chapter we discussed how Earth's atmosphere formed. Here, we begin the discussion of how Earth's atmosphere has evolved over time. Volcanic outgassing is one of the most important processes influencing atmospheric evolution because it affects both atmospheric composition and redox state, that is, the balance between reduced and oxidized gases, e.g., H_2 and O_2. The redox state of volcanic gases is, in turn, influenced by the redox state of Earth's mantle – a topic that will require us to delve into some aspects of petrology.

To motivate our discussion of volcanic outgassing, we briefly review what previous authors have said about the composition of the prebiotic atmosphere. This is a crucial part of understanding the difference between inhabited and uninhabited worlds. Our own, more detailed, analysis will be saved for Ch. 9, but the discussion here will show why volcanic outgassing rates and mantle redox state are important. Furthermore, some of these older ideas about the primitive atmosphere are so deeply entrenched in popular and scientific thinking that it is difficult to discuss any of these concepts without understanding how they originated.

7.1 Historical Context: Strongly and Weakly Reduced Atmospheres

Past theories of the prebiotic atmosphere have been heavily influenced by the perceived requirement that it provide a suitable environment for the origin of life. As we discuss further in Ch. 9, this constraint may be more apparent than real, because there are other pathways to make complex organic compounds besides atmospheric chemistry. However, serious thinking about the earliest atmosphere began with the origin of life (e.g., see Chang *et al.* (1983)), so this is where we start.

A modern scientific theory for the origin of life was first developed by the Russian biochemist Alexander Oparin and published in Russian (Oparin, 1924) and later, several times, in English (Oparin, 1938, 1957, 1968). Similar but less detailed views were published several years later by the British geneticist and evolutionist John Haldane (Haldane, 1929). The *Oparin–Haldane hypothesis*, as it later came to be called, suggested that life was formed from complex organic compounds that were synthesized by photochemical or thermochemical processes (e.g., lightning) in a highly reduced atmosphere containing methane, ammonia, and other compounds rich in hydrogen. The process of assembling non-living compounds into organic precursors to biological entities was termed *chemical evolution*. Oparin's model for the early atmosphere superceded that of the American paleontologist Henry Osborn (1917), cited by Oparin (1938). Osborn had proposed that the early atmosphere consisted primarily of CO_2 and H_2O. The pendulum eventually swung back in favor of Osborn's idea, but it took nearly a century for it to do so.

Oparin's theory of the prebiotic atmosphere received a tremendous boost in the early 1950s when the American geochemist Harold Urey and his graduate student Stanley Miller performed a famous experiment in which they synthesized possible prebiotic compounds from plausible early atmospheric gases (Miller, 1953, 1955). A diagram of Miller's experimental apparatus is shown in Fig. 7.1.

Urey was aware that methane and ammonia had been discovered in the atmospheres of Jupiter and Saturn, and, through a chain of logic, argued for their presence in Earth's earliest atmosphere. He reasoned that these gases were present on the giant planets because their gravitational attraction was strong enough to prevent hydrogen from escaping. Deep convective mixing also plays a role, as it allows carbon, nitrogen, and hydrogen to be mixed downward to regions where temperatures and pressures are high enough for thermodynamic equilibrium. At high pressures in the presence of large amounts of H_2, the

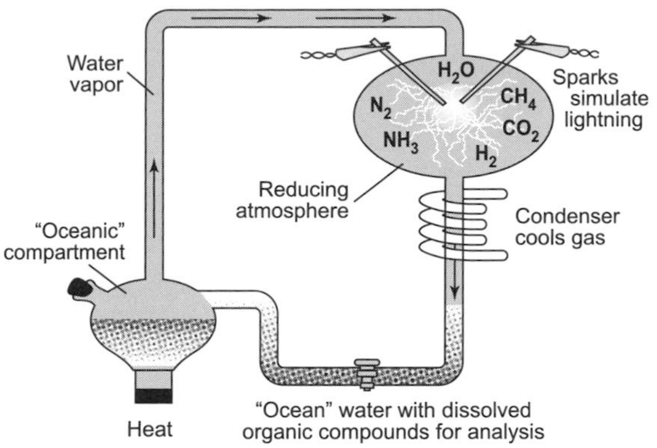

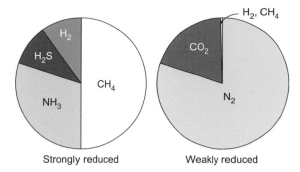

Figure 7.1 Diagram illustrating the Miller–Urey experiment.

Figure 7.2 Pie chart showing representative examples of strongly and weakly reduced atmospheres.

stable form of carbon is methane, CH_4, and that of nitrogen is ammonia, NH_3. Urey further reasoned, perhaps not quite correctly, that if one went back to a time before hydrogen had had a chance to escape, Earth's primitive atmosphere would have resembled those of the giant planets. The logic here is suspect because, as we saw in Ch. 5, the timescale for hydrogen loss is short enough that Earth's atmospheric hydrogen has probably always been more or less in steady state, with production of hydrogen from volcanoes and impacts balanced by escape of hydrogen to space and reduction of oxidized volatiles to reduced ones (e.g., $CO_2 \rightarrow CH_2O$). One should bear in mind, however, that Urey was working well before astronomers had deduced the timescales for solar nebula evolution and planetary formation. If the fully formed Earth had been embedded in a dense solar nebula from which it captured large amounts of H_2, then Urey's picture of the initial conditions for life might be reasonable. We argued in Sec. 6.4.2, though, that any primordial captured atmosphere was probably dissipated during the accretion process itself and replaced with one derived from impact degassing.

Urey's ideas about early atmospheric composition became firmly ingrained in scientific thought when Stanley Miller performed his historic experiment, which is recounted in detail by Miller's student Jeffrey Bada (Wills and Bada, 2000). Miller set up a pair of electrodes within a flask containing CH_4, NH_3, and H_2O. Hydrogen sulfide (H_2S) was also included in some experiments. Over periods of hours to days, the electrodes provided spark discharges to the mixture, simulating lightning in the early atmosphere. Then, Miller analyzed reddish-brown material that accumulated in the water and on the walls of the flask. This residue contained all sorts of complex hydrocarbons, including amino acids, the building blocks of proteins. Both the scientific community and

the general public were encouraged that we were well on the way to understanding the origin of life. Carried along by this sense of optimism, the CH_4/NH_3-rich gas mixture used in the experiment became a common model of primitive atmosphere composition.

At about this same time, unnoticed by nearly everyone except his fellow geochemists, William Rubey of the US Geological Survey was developing his own ideas of early atmospheric composition (Rubey, 1951, 1955). Rubey tabulated data on the gases emanating from modern volcanoes. These gases are dominated by CO_2 and H_2O, rather than CH_4 and NH_3. Indeed, NH_3 concentrations are too low to be detected in gases released from surface volcanism (Holland, 1984), and CH_4/CO_2 ratios are $\ll 10^{-6}$ (Giggenbach, 1997; Ryan et al., 2006; Taran and Giggenbach, 2003). Nitrogen in volcanic emissions is probably in the form of N_2, but this is difficult to measure because of mixing with the atmosphere before the gases can be collected.

Rubey hypothesized that the Earth was initially devoid of a gaseous envelope and that the atmosphere formed by volcanic outgassing; hence, it should have consisted predominantly of CO_2 and N_2, rather than CH_4 and NH_3. The oxidation state of such an atmosphere is described as *weakly reduced*, as it would have contained small quantities of hydrogen while being virtually devoid of O_2 (see Fig. 7.2). CO_2 and N_2 can be said to have a neutral oxidation state, as we will explain in Ch. 8. Of course, Rubey's model has conceptual difficulties as well, as we now think that Earth was enveloped in an impact-degassed atmosphere throughout the latter part of its formation (see Sec. 6.5 and further discussion below).

Other authors developed various permutations of the Miller–Urey and Rubey scenarios. Holland et al. (1962) described a three-stage model of atmospheric evolution. Stage 1 was a highly reduced atmosphere similar to that considered by Urey and Miller, which would have lasted until Earth's core formed. During this time the upper mantle would still have contained metallic iron, and the volcanic gases that were released from it would have been highly reduced (i.e., dominated by CH_4, NH_3, and H_2) as shown in Fig. 6.20. Stage 2 was a weakly reduced atmosphere similar to that proposed by Rubey, while Stage 3 referred to the modern O_2-rich atmosphere. In Holland's model, as in other contemporary models of planetary accretion, the Earth formed from the accretion of large numbers of small planetesimals over a long period of time, and hence had a relatively cool interior. We saw in Ch. 6, though, that this model is no longer accepted. Rather, evidence suggests that the Earth formed over a time scale of 10^7–10^8 years by a series of collisions with increasingly large impactors. The biggest impacts melted large portions of the mantle and allowed the core to form concurrently with the Earth itself (Stevenson, 1983) (Sec. 6.8). Hence, Holland's Stage 1, highly reducing atmosphere may have been present only during the main accretion period. After that time, both volcanic gases and the atmosphere they formed should have been more oxidized than he assumed.

That said, the issue of how the composition of Earth's upper mantle changed over time, and how this may have affected the oxidation state of volcanic gases, is still not entirely resolved. As we discuss further below, the mantle contains some ferric (oxidized) iron (Fe^{3+}), whereas meteorites contain mostly ferrous (Fe^{2+}) and metallic iron (Fe^0), both of which are more reduced. So, iron within Earth's mantle must have been oxidized either during or following accretion. Evidence from Ce anomalies in zircons indicates that the mantle was already oxidized by 4.4 Ga (Trail *et al.*, 2011), suggesting that the oxidation occurred during the process of planetary accretion, although the Hadean and early Archean crust may have been more reducing than later (Yang *et al.*, 2014b). (See further discussion in Sec. 7.3.) Life likely originated after 4.4 Ga, and so the atmosphere at the time of life's origin was perhaps less reduced than Miller and Urey had envisioned.

7.2 Volcanic Outgassing and Metamorphic Degassing of Major Volatile Species

We now turn to the discussion of outgassing itself. Let's start by considering how and where volcanic gases are emitted. As we will see, the term *volcanic outgassing* is really a catchall phrase that has been used to describe a variety of physical mechanisms by which gases are injected into the atmosphere. The discussion that follows is based on work by several different authors, including Holland (1978, 1984, 2002, 2009), Sleep and Zahnle

(2001), Zahnle and Sleep (2002) Catling *et al.* (2001), Sleep (2005), Canfield (2005), Canfield *et al.* (2006), and Hayes and Waldbauer (2006).

We should point out that while there is general agreement regarding the observed volcanic outgassing rate of CO_2, published outgassing rates of H_2 and SO_2 differ by factors of up to 10. Some of the disagreement results from differences in accounting procedures, but some of it results from the difficulty of making measurements, along with the fact that different outgassing venues exhibit different gas compositions. That is why it is important to understand the mechanisms by which volcanic gases are released, as well as the techniques by which they are measured (e.g., see review by Oppenheimer *et al.* (2014)).

7.2.1 Mechanisms of Volcanic Outgassing

We consider volcanism to be any process during which rocks are melted to form magmas from which gases may be evolved. We will see below that gases may be evolved from the solid Earth without having it melt. However, the gases released from magmas are special because they are hot enough to be close to thermodynamic equilibrium. We will henceforth reserve the term "volcanic" to describe such gases and will use other terms to describe gases given off at lower temperatures.

Volcanic gases are emitted in several different tectonic settings (see Fig. 7.3). *Arc volcanism* occurs in places such as the Cascade mountain range in the northwestern United States or the Aleutian Islands off the coast of Alaska, where an oceanic plate is subducted beneath a neighboring continental plate. The subducted seafloor is often layered with sediments containing carbonates, organic carbon, and abundant trapped water. These volatiles are heated as they are transported down to hotter regions of the Earth, and they can be explosively vaporized in some situations. The violent eruption of Mount St. Helens in Washington State in 1980 is a good example. If the sediments that are subducted are rich in sulfate minerals, then the released gases may be rich in sulfur dioxide, SO_2. In Plinian eruptions (those characterized by high columns of gas and ash), some SO_2 may be injected directly into the stratosphere where it is oxidized to SO_3, which then hydrates and condenses to form sulfate (H_2SO_4) aerosols. The eruption of Mount Pinatubo in the Philippines in 1991 is the best quantified example of such a process (Robock, 2000). Stratospheric aerosols are not germane to the present topic but are an important driver of short-term climate variability.

Point volcanism, or *hotspot volcanism*, occurs in places such as Hawaii, where a hotspot, probably caused

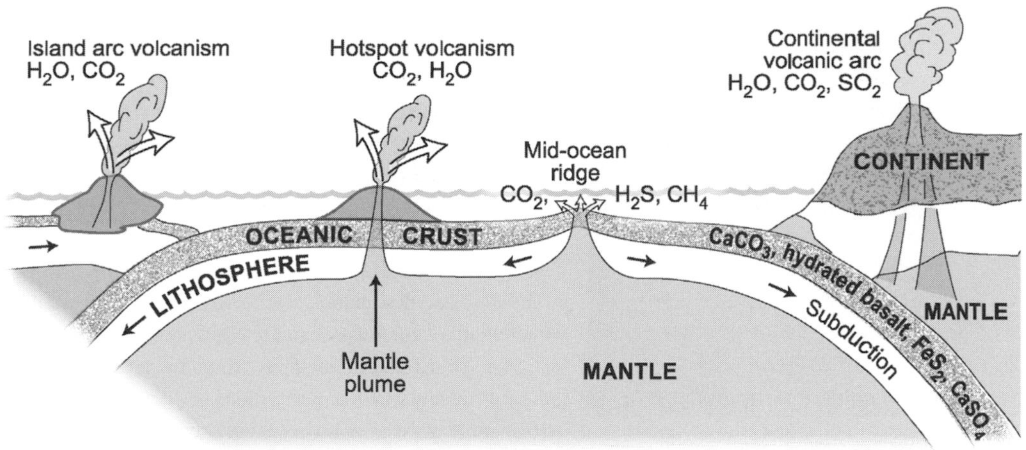

Figure 7.3 Diagram illustrating the different types of volcanism described in the text. Mid-ocean ridge, hotspot (point), and arc volcanism are all important sources of volatiles to Earth's atmosphere.

by an underlying mantle plume, is found beneath either an oceanic or continental plate. The area around Yellowstone Park in northwestern Wyoming is an example of a continental hotspot. The magmas produced by point volcanism are less volatile-rich than those formed in arc volcanism. They are also more *mafic* (rich in magnesium and iron minerals), whereas magmas produced by arc volcanism are more *felsic* (rich in silica, SiO_2). The word "mafic" is a blend of "magnesium" and "ferric" – the latter from the Latin word for iron, ferrum. Mafic rocks contain olivine (M_2SiO_4) and/or pyroxene ($M_2Si_2O_6$) minerals with roughly equal amounts of plagioclase feldspar (($Ca,Na)Si_2O_6$). Here, the "M" can represent any of several different metals, including Mg and Fe for olivine, and significant Ca and minor Al, Ti, and Na for pyroxene.

The distinction between igneous rocks is related to the types of volcanism and the volatiles given off. The seafloor itself is composed of basalt, which is a mafic rock. The gases evolved from basaltic volcanoes, such as those in Hawaii, are usually released gradually, not explosively. Hence, unlike the gases given off by arc volcanism, they are emitted only into the troposphere. We emphasize this point because some authors have suggested that volcanic gases associated with major flood basalts, e.g., the Siberian traps, have been a major driver of climate change and mass extinctions during the Phanerozoic Eon (Saunders and Reichow, 2009). Some CO_2-induced warming might indeed result from such extended volcanic events, but the sharp negative surface temperature excursions caused by stratospheric aerosols should not be associated with this process. Tropospheric volcanic aerosols rain out quickly and do not create measurable climate excursions.

Ultramafic igneous rocks are more magnesium-rich than mafic ones and contain at least 90% olivine and pyroxene. *Peridotite* is an ultramafic rock that is a mixture of olivine and pyroxene minerals found in the mantle and *dunite* is an end-member peridotite of mainly olivine, which can be forced up from mantle depths to the surface in tectonic collisions (e.g., Twin Sisters Mountain in Washington State, is made of dunite). Below in Sec. 7.5, we note that volcanic gas compositions would have been affected if the oceanic and continental crust had been more ultramafic in the past.

In addition to arc volcanoes and hotspots, the largest amount of volcanism in terms of volumetric rock production occurs at the *mid-ocean ridges* where new seafloor is being created at a rate of ~20 km^3 yr^{-1}. The mid-ocean ridges are permeated by hydrothermal circulation systems that cool the fresh, hot rock and spawn *hydrothermal vents* in places where warm or hot water flows into the overlying ocean.

Hydrothermal vents come in two main types. (i) Axial hydrothermal vents are hot (350–400 °C), acidic (pH = 4 to 5), and are caused by seawater flowing through fractures in seafloor that is heated by magma. These vents contain large quantities of dissolved ferrous iron and sulfide, which react with each other to form insoluble iron sulfides when they mix with the cold surrounding seawater. This has earned them the name *black smokers* (Fig. 7.4). The fluids released by black smokers contain gases derived from the mantle, as demonstrated by the presence of dissolved ^{3}He. (Note that ^{3}He, unlike ^{4}He, is not generated by radioactive decay, and hence must be primordial.) (ii) Off-axis vents are cooler (~30–90 °C), alkaline vents, which produce fluids with pH of 9–11. The water is heated not by magma but by

Figure 7.4 The top of a black smoker chimney on the south-eastern edge of the iguanas vent field imaged on the Galapagos 2005 expedition. (NOAA image ID: expl1233, courtesy of UCSB, University of S. Carolina, NOAA, WHOI.) (A black and white version of this figure will appear in some formats. For the color version, please refer to the plate section.)

chemical reactions of water with ultramafic rocks. The most studied off-axis vent field is the *Lost City hydrothermal field* on the flank of the mid-Atlantic ridge (Kelley *et al.*, 2001, 2005), but other ultramafic vents have also been found in the mid-Atlantic (Konn *et al.*, 2015). At the time of writing, only one other vent field, the Pescadero Basin in the Gulf of California, is known to have carbonate chimneys like those at Lost City. The fluids emanating from the Lost City vents are rich in methane that may originate from rock-water reactions. This methane is not truly "volcanic" in the sense defined above, because it is not released from magma; rather, it illustrates another mechanism by which the solid Earth can release gases into the atmosphere (or ocean, in this case). Similar ultramafic vents may be important beyond Earth, including at the bottom of a large sea on Saturn's moon, Enceladus (Hsu *et al.*, 2015).

Some authors also distinguish "cold" *hydrothermal seeps*, which are <30–60 °C. These are places on the seafloor where hydrocarbon-rich fluids emerge. They are often associated with deep-sea brine pools caused by dissolution of salt deposits and the release of sedimentary hydrocarbons.

7.2.2 Outgassing and Metamorphic Degassing of CO_2

Jarrard (2003) has published estimates for the CO_2 fluxes associated with arc volcanism, $(2.5\pm0.7)\times10^{12}$ mol yr^{-1}, and ridge volcanism, $(0.7–3)\times10^{12}$ mol yr^{-1}. Other estimates of median fluxes of 1.9 $\times10^{12}$ mol yr^{-1} and 2.2 $\times10^{12}$ mol yr^{-1} for arcs and ridges, respectively

(Fischer, 2008), lie within Jarrard's error bars. The CO_2 release rate from hotspot (or plume) volcanism is difficult to measure, but Jarrard suggests that it is comparable to that from arc volcanism. One particular hotspot volcano that has been studied extensively is Sicily's Mt. Etna. Helium and carbon isotope data suggests that most of its CO_2 is mantle-derived (Allard, 1997) because CO_2 solubility is high in its carbon-rich alkaline (low in silica) magma. Consequently, Etna's CO_2 flux alone is thought to be $\sim1\times10^{12}$ mol yr^{-1} (Brantley and Koepenick, 1995; Kerrick, 2001), or about one-eighth of the total global release rate of CO_2.

The total volcanic CO_2 flux from the sum of the aforementioned contributions is about $(8.5\pm2)\times10^{12}$ mol yr^{-1}, of which $(6.5\pm1.6)\times10^{12}$ mol yr^{-1} is subaerial (Table 7.1). This global flux is consistent with the average value of literature estimates that range (3 to 10) $\times10^{12}$ mol yr^{-1}, as tabulated by Gerlach (2011).

We have a somewhat better estimate of the total outgassed (volcanic + metamorphic) CO_2 flux than we do of the individual components. This is especially true for subaerial volcanism, where one can equate the volcanic outgassing rate with the consumption of CO_2 by silicate weathering followed by carbonate deposition. The silicate weathering rate, which dominates CO_2 loss today, can be estimated from measured concentrations of bicarbonate ions and various cations in river water, along with the known total delivery rate of river water to the oceans. That flux consumes about 11.5×10^{12} mol CO_2 yr^{-1} (Berner *et al.*, 1983). Two moles of CO_2 are required for each mole of (Ca–Mg) silicate that is consumed, following the stoichiometry

$$CaSiO_3 + 2CO_2 + H_2O \rightarrow Ca^{++} + 2HCO_3^- + SiO_2 \quad (7.1)$$

But one gets half of this CO_2 back when carbonate is deposited in sediments because the precipitation reaction may be written as

$$Ca^{++} + 2HCO_3^- \rightarrow CaCO_3 + CO_2 + H_2O \quad (7.2)$$

Hence, the required subaerial CO_2 outgassing flux needed to balance the carbonate-silicate cycle is just half the silicate weathering rate, or $\sim5.7\times10^{12}$ mol CO_2 yr^{-1}, which is similar to the estimate shown in Table 7.1.

The relatively large uncertainties in all of these fluxes leave room for an additional CO_2 production mechanism. The additional CO_2 source, which could be as large as 2×10^{12} mol yr^{-1}, is thought to arise from *carbonate metamorphism* occurring in rift valleys as well as mountain belts such as the Himalayas in Tibet or the Apennines in Italy (Kerrick, 2001; Kerrick *et al.*, 1995). There, carbonate rocks are squeezed and heated (but not melted), and they release large amounts of CO_2 in the process.

Table 7.1 Outgassing rates of major (neutral) species.

Process	CO_2	H_2O	$SO_2{}^c$	$N_2{}^b$
Arc volcanism	2.5 ± 0.7^a	80 ± 20^b	0.8 ± 0.4^b	0.0375 ± 0.027
Hotspot volcanism	2 ± 1^a	15 ± 7^b	1.0 ± 0.5^b	0.03 ± 0.02
Metamorphism	2 ± 1^d	?	?	0.015 ± 0.015
TOTAL subaerial	6.5 ± 1.6	95 ± 20	1.8 ± 0.6	0.083 ± 0.037
Ridge volcanism	2 ± 1^a	~0	~0	0.0038 ± 0.0015
TOTAL global	8.5 ± 2	95 ± 20	1.8 ± 0.6	0.09 ± 0.04

Fluxes in units of 10^{12} mol yr^{-1}. Question marks indicate "uncertain."

[a] Jarrard (2003).
[b] See text.
[c] SO_2 is treated as a neutral species in our discussions of reduced atmospheres and as a reduced gas in our discussions of oxidized atmospheres. See Chapters 8 and 10.
[d] Kerrick et al. (1995), cited in Berner (2004, his Table 4.1) has at least ~1. Lee et al. (2016) suggest 1.6 ± 0.8 from the E. African rift alone. We assume 2 ± 1.

Reduced gases, primarily CH_4 and H_2, may be released, as well (Etiope et al., 2008; Morner and Etiope, 2002; Svensen and Jamtveit, 2010). This *metamorphic* CH_4 and H_2 play an important role in some models of atmospheric evolution (e.g., Catling et al., 2001). We will return to this question in Ch. 10.

Recently, Burton et al. (2013) have published an estimate for the subaerial CO_2 outgassing flux of 540 Mt CO_2 yr^{-1}, or 12.3×10^{12} mol yr^{-1}. This is higher than our estimate by a factor of ~1.9. The reason may be because they calculate an average assuming a linear distribution of outgassing amongst volcanoes, but the real distribution appears to be biased towards the highest emitter in a power law (Pedone et al., 2014). With a power law distribution, the average outgassing flux is a factor of ~2–3 lower. Burton et al. also estimate that an additional 300 Mt CO_2 yr^{-1} (6.6×10^{12} mol yr^{-1}) is produced by low temperature metamorphism, giving their total subaerial CO_2 flux of ~19×10^{12} mol yr^{-1}. According to eq. (7.1), this would produce a bicarbonate flux in rivers of twice this value, or 38×10^{12} mol yr^{-1}, which is almost four times the amount attributed to silicate weathering. We suspect that riverine bicarbonate fluxes are measured more reliably than are volcanic outgassing rates, so we will stick with our lower CO_2 outgassing estimates here, while acknowledging that higher estimates exist.

Although it is tangential to the discussion here, atmospheric CO_2 is also consumed by weathering of carbonate minerals on land – mostly calcite and dolomite. The reaction for calcite weathering is just the reverse of reaction (7.2). This does not affect the atmospheric CO_2 balance over long timescales because the CO_2 is returned to the atmosphere–ocean system when carbonate is precipitated. But is important if one attempts to calculate the carbon isotopic mass balance for the ocean, as we do in the next chapter (Sec. 8.5). The total amount of bicarbonate released by this process is ~23.6×10^{12} mol yr^{-1}, or about twice the amount derived from silicate weathering (Berner et al., 1983). So, this process is a major input to the oceanic bicarbonate reservoir.

7.2.3 Subaerial Outgassing of H_2O, SO_2, H_2S, and N_2

The global volcanic flux of H_2O can be estimated by comparing H_2O and CO_2 concentrations in subaerial volcanic gases and then scaling to the global flux of CO_2 (Holland, 2002) (Submarine volcanic outgassing of H_2O can be neglected because mid-ocean ridge basalts are exceedingly dry and because we are not concerned here with changes to ocean volume.) Alternatively, the H_2O flux from arc volcanism can be estimated by looking at the amount of water in subducted sediments and crust and assuming that they are all returned by outgassing (Jarrard, 2003). A comparison of these two methods gives some feeling for the uncertainties. According to Holland (2002), the ratio of H_2O to CO_2 in average volcanic gases is ~30. If we multiply the total CO_2 flux from arc and point volcanism by this number, the flux of H_2O is then $30 \times$ (4.5×10^{12} mol yr^{-1}) $\cong 140 \times 10^{12}$ mol yr^{-1}. This number is almost certainly too high, however, because the gases released from hotspot volcanoes do not contain reprocessed, hydrated marine sediments and are therefore considerably drier than those released from arc volcanoes. The H_2O/CO_2 ratio in Hawaiian (hotspot) volcanoes is about 7 or 8 (Holland, 1978, p. 289; Walker, 1977, his Table 5.5).

If we multiply the hotspot CO_2 flux (2×10^{12} mol yr^{-1}) by this number, we get 15×10^{12} mol H_2O yr^{-1}. The arc volcanic H_2O flux is reduced to $\sim80\times10^{12}$ mol yr^{-1}, giving a total of 95×10^{12} mol H_2O yr^{-1}. By comparison, Jarrard (2003) gives 102×10^{12} mol yr^{-1} as the H_2O flux from arc volcanism. Given the large uncertainties in the measurements, the agreement between these numbers is surprisingly good.

It is difficult to measure N_2 in volcanic gases because N_2 is the bulk gas in air. With care, though, contamination can be excluded (Giggenbach and Matsuo, 1991), and measured ratios of N_2/CO_2 or N_2/SO_2 from volcanic/metamorphic fumaroles can be used along with better-constrained CO_2 or SO_2 fluxes to estimate N_2 fluxes. Using an N/C ratio of 0.03 ± 0.02 from data in Berner (2006b) and previous CO_2 fluxes, we estimate N fluxes of 0.075 ± 0.054, 0.06 ± 0.04, 0.03 ± 0.03, and 0.06 ± 0.04 from arcs, hotspots and metamorphism, respectively, in units of 10^{12} mol N yr^{-1}. Fischer (2008) gives a median flux of 0.07×10^{12} mol N yr^{-1} for arcs, which agrees well. Marty and Zimmerman (1999) estimate a ridge flux of $3.8\pm1.5\times10^{9}$ mol N_2 yr^{-1}. Dividing the first set of numbers by 2 to account for N_2 vs. N fluxes, these fluxes are listed in Table 7.1.

We can take the total global volcanic N_2 flux from Table 7.1, $(9\pm4)\times10^{10}$ mol yr^{-1}, and see how long it would take to replenish the atmosphere in N_2. In addition to this volcanic flux, weathering provides a nitrogen source. Most of the N released from oxidative weathering of organic material on the continents undergoes rapid biological denitrification and is released as either N_2O or N_2. The N_2O is then photochemically converted to N_2 within the atmosphere. If we take $(7.5\pm1.7)\times10^{12}$ mol yr^{-1} for the weathering rate of organic carbon (Holland, 2002), multiply by an N/C ratio of 0.02 ± 0.003 in continental rocks (Berner, 2006b), and then divide by 2 for stoichiometry, we get an N_2 flux of $(7.5\pm1.7)\times10^{10}$ mol yr^{-1}. That is slightly smaller than the global volcanic N_2 flux listed in Table 7.1. Adding the volcanic and weathering N_2 fluxes yields a total N_2 production flux of $(1.7\pm0.4)\times10^{11}$ mol yr^{-1}. The mass of the atmosphere is $\sim5.2\times10^{18}$ kg, or 1.8×10^{20} moles (assuming a mean molar mass of 28.96 g). The N_2 volume mixing ratio is 0.78, so the total N_2 amount is $\sim1.4\times10^{20}$ moles. Dividing this number by the total N_2 source flux gives a replenishment time of ~0.85 billion years. This represents the *geologic residence time* of atmospheric N_2, i.e., the lifetime of N_2 against long-term geologic processes.

The *biological residence time* of atmospheric N_2 should be considerably shorter than the geologic one. The total rate of biological nitrogen fixation today is estimated to be 413×10^{12} g N yr^{-1} (Fowler *et al.*, 2013), or 1.5×10^{13} mol N_2 yr^{-1}, about half of which is anthropogenic. So, the lifetime of N_2 against natural biological cycling is only 1.4×10^{20} moles/7×10^{12} mol yr^{-1} = 20 million years. However, this form of recycling should not affect atmospheric N_2 amounts because the inventory of fixed nitrogen stored within living biomass is relatively small.

To balance the nitrogen cycle, the burial rate of nitrogen in marine sediments should be approximately equal to this value. If we take 1×10^{13} mol yr^{-1} for the burial rate of organic carbon (Holland, 1978), multiply by the N/C ratio of 0.04 ± 0.02 in marine sediments (Berner, 2006b), and again divide by 2, we get an N_2 removal flux of $(2\pm1)\times10^{11}$ mol yr^{-1}, which is in good agreement with the production flux given above (although the uncertainty is even larger than shown because the organic carbon burial rate is also uncertain by a factor of ~2). So, while it appears that the geological nitrogen cycle is in reasonably good balance, it is also clear that atmospheric N_2 could change appreciably on billion-year timescales, as suggested by some authors (Goldblatt *et al.*, 2009).

Let us now switch our attention to sulfur gases, keeping the focus on subaerial volcanism. (Submarine volcanism will be discussed later in the chapter.) Direct satellite measurements of the SO_2 flux from arc volcanism yield values of $(2.3–3.1)\times10^{11}$ mol yr^{-1} (Halmer *et al.*, 2002). The accompanying H_2S flux has been estimated to be about a factor of 10 lower (Berresheim and Jaeschke, 1983), although some assessments have a huge range, $(0.044–1.1)\times10^{12}$ mol H_2S yr^{-1} (Halmer *et al.*, 2002). SO_2 predominates in high temperature (~1200 $^\circ$C) surface emissions because thermodynamic equilibrium favors the formation of small molecules, like H_2. Hence, the equilibrium reaction

$$SO_2 + 3H_2 \leftrightarrow H_2S + 2H_2O \qquad (7.3)$$

is pushed to the left. But these measurements capture only the SO_2 injected by explosive volcanism. More SO_2 is evidently released in a relatively passive manner. Holland (2002) finds that the ratio of total sulfur to H_2O in arc volcanism is ~0.01 and that of carbon to H_2O is ~0.03 (see his Fig. 6). Combining these numbers suggests that the ratio of sulfur to carbon is ~0.3, which agrees with data tabulated by Symonds *et al.* (1994). If we multiply our earlier estimate of the flux of CO_2 from arc volcanism by this number, then the rate of SO_2 release should be $\sim8\times10^{11}$ mol yr^{-1}.

Outgassing of SO_2 from hotspot volcanism is difficult to estimate accurately because the number of available measurements is small. Hotspot volcanoes appear to be sulfur-rich, however. The observed SO_2/CO_2 ratio in Hawaiian volcanic gases is ~0.5 (Walker, 1977, his Table 5.5), which agrees with the estimate from Holland (2002) (see his Fig. 6). Multiplying our estimated hotspot CO_2 outgassing rate by this ratio yields a large SO_2 source (~1×10^{12} mol yr^{-1}).

We are interested ultimately in the effect of volcanic outgassing on atmospheric oxidation state. Hence, we need to know the flux of reduced gases that can react with O_2. H_2O does not do so, of course, but it is accompanied by H_2, which does. Likewise, CO_2 is accompanied by CO (for surface volcanism) or CH_4 (for submarine volcanism), both of which can react with O_2. Both SO_2 and H_2S also react with O_2, yielding sulfuric acid or sulfate in an oxidized environment. Thus, we treat SO_2 as a reduced gas in our discussion of oxidized atmospheres (Ch. 10). In a reduced atmosphere, however, SO_2 can either be oxidized to sulfate or reduced to elemental sulfur or H_2S (see, e.g., Pavlov and Kasting, 2002). So, we treat SO_2 as a neutral species in our discussions of reduced, or weakly reduced, atmospheres (see Ch. 8).

Because surface volcanic gases are released directly from hot magmas, we can estimate the outgassing rates of reduced species from thermodynamic equilibria. To do so, however, we need to understand the oxidation state of magmas generated in the upper mantle. We include a brief discussion of this complicated topic in the next section, as it is critical to understanding volcanic gas composition on Earth, as well as how that composition might differ on other planets (e.g., Gaillard and Scaillet, 2009).

7.3 Oxidation State of the Mantle

Earlier in this chapter, we described the three-stage atmospheric evolution model of Holland *et al.* (1962). Highly reduced volcanic gases were emitted from the Stage 1 atmosphere, when metallic iron was present in the upper mantle. More oxidized volcanic gases were emitted later after the metallic iron had migrated to the core. As noted earlier, Stage 1 may have existed only during the accretion period itself, and thus may not be relevant to conditions surrounding the origin of life. But it is still useful to study this model, as it suggests an evolutionary path for atmospheric composition starting from the very earliest times. It may also be useful in analyzing the evolution of Mars' atmosphere, as the smaller mass of that planet may have led to a more reduced upper mantle, as inferred from meteorite studies (Bridges and Warren, 2006; Wadhwa, 2008).

7.3.1 Oxidation State of the Present Upper Mantle

The actual composition of Earth's upper mantle is complex. The mantle is composed primarily of various magnesium silicate minerals, such as olivine and pyroxene (see Sec. 7.2.1). Several of the elements present in the mantle can change their valence state, depending on the local *fugacity* of O_2, abbreviated as f_{O_2} (Frost, 1991). For non-petrologists, the term "fugacity" represents the effective partial pressure of a gas in thermodynamic equilibrium with a particular mineral assemblage. But oxygen itself is not a gas species in the rock. Instead, the dissociation of H_2O, CO_2, and SO_2 or the reduction of Fe^{3+} can supply oxidizing power, for which f_{O_2} is a theoretical equivalent. The most abundant elements in the upper mantle that can respond to changes in f_{O_2} are Fe, C, and S.

In the laboratory, petrologists construct various artificial mineral assemblages, which they use to calibrate oxygen fugacities. These assemblages generally involve iron minerals. Iron has three valence states that occur naturally: Fe^0 (elemental, or metallic, iron), Fe^{2+} (ferrous iron), and Fe^{3+} (ferric iron). The superscript represents the valence state, which in this case represents the deficit of electrons relative to protons. When combined with oxygen, the latter two ions form wüstite (FeO) and hematite (Fe_2O_3), respectively. Recall that oxygen is always assigned a valence of –2 (except for elemental oxygen (0) or superoxides (–1/2)), because it requires two electrons to complete its outer shell. There is also an intermediate iron oxide mineral, magnetite (Fe_3O_4), which contains two Fe^{3+} ions and one Fe^{2+} ion. Magnetite happens to be particularly stable thermodynamically, so it is formed over a wide range of f_{O2} values.

In the mantle, ferrous iron is typically a constituent of silicate minerals; hence, petrologists sometimes add quartz, SiO_2, to these iron oxide minerals in order to use them as mantle surrogates. In the presence of quartz, and at relatively low f_{O2} and low pressure, the stable iron phases are Fe^0, Fe_2SiO_4 (fayalite), and Fe_3O_4. The iron in fayalite is ferrous, as in FeO. (Think of fayalite as two FeOs plus SiO_2.) Suppose now that one begins with a mixture of iron and silica at very low f_{O2} and then gradually adds oxygen. At low f_{O2}, the silicates are typically ignored, so one deals only with Fe and FeO. Initially, all of the iron is present as metallic iron, Fe^0. As oxygen is added, some of the metallic iron is converted to FeO. Assuming that the system is kept well mixed, the oxygen fugacity of the mixture is defined by the equilibrium reaction

$$Fe + \frac{1}{2}O_2 \xrightarrow{K_4} FeO \qquad (7.4)$$

The equilibrium constant for this (reversible) reaction, K_4, depends on pressure, P, and absolute temperature, T. For conditions relevant to surface volcanism, we take $P = 5$ atm and $T = 1200$ °C, or 1473.15 K (Holland, 1984). The oxygen fugacity for this reaction can be determined from a parameterization (Mueller and Saxena, 1977)

$$\log f_{O_2} = -\frac{A}{T} + B + \frac{C(P-1)}{T} \qquad (7.5)$$

The values of the coefficients A, B, and C are given in Table 7.2. (The third term is negligible for pressures relevant to Earth's surface.) For these data, we obtain $f_{O_2} = 10^{-11.9}$. As oxygen is added to the system, f_{O_2} remains constant at this particular value until all of the initial Fe^0 is converted to FeO; hence, the mineral assemblage acts as an oxygen *buffer*. This particular oxygen buffer is the iron–wüstite (IW) buffer. One can think of it as the boundary between the stability fields of Fe^0 and Fe^{2+}.

If one now continues to add oxygen to the system, some of the FeO (which combines spontaneously and without oxygen consumption with SiO_2 to form Fe_2SiO_4) is converted to Fe_3O_4. The reaction here may be written as

$$3\,Fe_2SiO_4 + O_2 \leftrightarrow 2\,Fe_3O_4 + 3\,SiO_2 \qquad (7.6)$$

Referring again to the empirical coefficients in Table 7.2, we obtain $f_{O_2} = 10^{-8.5}$. The oxygen fugacity remains at this level until all of the fayalite is converted to magnetite plus quartz. This buffer is termed the quartz–fayalite–magnetite buffer, often written as QFM or FMQ.

The relatively simple behavior described above applies to a pure system involving only a single silicate mineral, fayalite. The actual mantle contains a variety of iron-bearing minerals, including olivine, pyroxene, and spinel. Instead of jumping abruptly from the IW buffer to the QFM buffer as oxygen is added, these form a *solid solution* in which the oxygen fugacity varies smoothly with added O_2. Petrologists use Mössbauer spectroscopy to directly measure the Fe^{+3}/Fe^{2+} ratio in these minerals (Luth and Canil, 1993; Mattioli and Wood, 1986; Wood and Virgo, 1989). One can then develop *oxybarometers* from reactions such as (Wood and Virgo, 1989)

$$6\,Fe_2SiO_4 + O_2 \leftrightarrow 3\,Fe_2Si_2O_6 + 2\,Fe_3O_4 \qquad (7.7)$$

Investigators using this technique find laterally varying f_{O2} values that are related to different tectonic environments (Woodland and Koch, 2003). Reported f_{O2} values in the subcontinental lithosphere range from QFM $-$ 1 to QFM $+$ 2. Values from the oceanic mantle lithosphere are much lower, QFM $-$ 1 to QFM $-$ 3. Furthermore, there is an apparent decrease in f_{O2} with depth by as much as 3 log units between 80 and 150 km (*ibid.*). Evidently, the oxidation state of the upper mantle is heterogeneous and is not controlled by any simple redox buffer. That said, the shallow continental and upper lithospheric oceanic upper mantle, from which volcanic gases are released, is generally considered to be near QFM (Kasting *et al.*, 1993a). So, we shall follow Holland (1978, 1984) and assume that this is indeed the case.

7.3.2 How the Mantle Became Oxidized

The word "oxidized" means different things to different types of geochemists. To those who study planetary accretion and core formation, the mantle as a whole is oxidized because it contains more FeO than would be expected on the basis of metal–silicate equilibrium. Reviews of this topic are provided by Rubie *et al.* (2015b) and Rubie and Jacobson (2015). The extent to which the iron core equilibrated with the silicate mantle during core formation can be constrained by measuring the abundance of siderophile elements in mantle xenoliths. (A mantle *xenolith* is a fragment of the upper mantle that becomes encased in an igneous rock. As explained in Sec. 6.8, a *siderophile* is an element that preferentially partitions into molten iron, as compared to molten silica.) Studies show that the abundance of strongly and moderately siderophile elements can be approximately matched if equilibration occurs at high pressures and temperatures, for example at the base of a 1000-km deep magma ocean formed by the Moon-forming impact. At low pressures, e.g., 1 bar, the observed abundance of siderophile elements in the mantle is much too high. These data led Wänke (1981) to propose a two-stage model for Earth's

Table 7.2 Constants for calculating the f_{O2} of solid buffers.

Buffer	Abbreviation	A	B	C	Note
SiO_2–Fe_2SiO_4–Fe_3O_4	QFM	25 738	9.00	0.092	1
Fe–FeO	IW	27 215	6.57 0	0.0552	2

[1] Wones and Gilbert (1969).
[2] Eugster and Wones (1962).

accretion, in which the first 60% of the planet formed from highly reduced material originating from within 1 AU of the Sun, and the last 40% consisted of more oxidized material from farther away. Such *heterogeneous accretion* models are still in line with the types of dynamical accretion models discussed in Sec. 6.2. Indeed, the best fit to the siderophile abundance data come from models in which core formation proceeds by way of a series of multiple giant impacts, just as the dynamical models predict should occur (Rubie *et al.*, 2011; Rubie *et al.*, 2015a).

In order to make this process work quantitatively, though, the mantle needs to become more oxidized during the accretion process. According to Rubie and Jacobson (2015), the most likely reaction involves uptake of Si by the core, leaving FeO behind in the mantle:

$$\underset{core}{2Fe} + \underset{mantle}{SiO_2} \rightarrow \underset{mantle}{2FeO} + \underset{core}{Si} \tag{7.8}$$

When this reaction is combined with heterogeneous accretion in which the first 60%–70% of the incoming material is reduced, and the last 30%–40% is oxidized, one can create a model that successfully explains both the FeO content of the mantle and the observed siderophile abundances. So, one may think of this as an elaboration on Wänke's original proposal.

While the process just described may explain the abundance of FeO in the mantle, it does not account for the relatively high oxidation state of the present upper mantle. At the end of accretion, the heterogeneous accretion model of Rubie *et al.* (2011) predicts a mantle f_{O2} of ~IW – 2 (see their Fig. 4c). As mentioned earlier, the oxidation state of the present upper mantle is closer to QFM, or IW+4.

Kasting *et al.* (1993a) estimated that the upper mantle above 700 km contains about 8 wt% atomic Fe, 2% of which is ferric. A similar estimate is given by Kuramoto and Matsui (1996) and McCammon (2005) albeit budgeting as the oxide, Fe_2O_3. The estimated mass of material above 700 km is 1×10^{24} kg, so this corresponds to ~3×10^{22} mol Fe^{+3}. Oxidation of 4 moles of Fe^{+2} to Fe^{+3} requires 1 mole of O_2,

$$4FeO + O_2 \rightarrow 2Fe_2O_3 \tag{7.9}$$

Hence, the total O_2 that must have been added to the upper mantle to bring it up from the IW buffer is about 7.5×10^{21} moles. Even more O_2 would be needed if the mantle started at IW– 2. By comparison, the mass of the ocean is 1.4×10^{21} kg, or 7.8×10^{22} mol, so the total amount of O_2 in the ocean (as H_2O) is about 4×10^{22} mol. This O_2 is only available, of course, if H_2 is lost.

The upper mantle thus contains the equivalent of about 20% of an ocean of oxygen, as Fe^{+3}.

How this extra oxygen (or oxidizing power) got into the mantle has been a topic of debate for decades. Holland (1984) suggested that it came from H_2O that was originally contained in the mantle, which then outgassed H_2 and left oxygen behind. The basic idea here is sound, but the oxidation process is complicated because the accreting Earth was a dynamic environment. Much of the oxidation may have occurred during the time that a steam atmosphere existed and Earth's surface was molten from its blanketing effect (see Sec. 6.5.2). In the Zahnle *et al.* (1988) calculations depicted in Fig. 6.14, about 100 bar of water, or 40% of the present ocean, was outgassed near the end of the main accretion period as the magma ocean solidified. This water then condensed out to form the ocean. The outgassed H_2O in this model contains twice as much oxygen as is needed to oxidize the upper mantle; however, at least half of it would have to have been outgassed as H_2 in order to do so. As the mantle became more oxidized, the H_2/H_2O ratio in volcanic gases would have decreased (see next section), so the amount of oxygen left behind may have been smaller than needed. A calculation by Hamano *et al.* (2013) suggests that at least half of the ferric iron in the upper mantle could have been produced by this mechanism. The results from the models of Zahnle *et al.* and Hamano *et al.* depend on several poorly known parameters, most importantly the water content of the accreting material. The calculations shown in Fig. 6.14 assume 0.1 wt% water – a value typical of ordinary chondrites. But, as discussed in Ch. 6 and above, the material that initially condensed at Earth's orbit could have been much drier than this. Water would have been supplied primarily by water-rich planetesimals originating from farther out in the solar nebula. The process would have been stochastic, and so it is difficult to estimate how much water would actually have been delivered. Metallic iron-rich planetesimals could also have hit the Earth late during accretion, resetting the mantle oxygen fugacity to a lower value. So, whether this process would have brought Earth's mantle its present oxidation state by the end of accretion is uncertain.

The upper mantle could also have become oxidized following accretion. Kasting *et al.* (1993a) proposed that O_2 was added slowly by subduction of hydrated oceanic crust, and carbonates, followed by outgassing of H_2 and CO. In their model, the upper mantle did not become fully oxidized until around 2.3 Ga, and this affected the timing of the rise of atmospheric O_2. This hypothesis requires the recycling and outgassing of a large amount of H_2O, which may be difficult if water is largely released during subduction. Furthermore, data on the redox state of ancient rocks may not support it. Petrologists measuring the Cr (Delano, 2001) and V (Canil, 1997, 2002; Li and Lee, 2004) content of Archean rocks have argued that the average upper mantle oxygen fugacity has not changed by more than ~0.5 $\log_{10}$ units since at least 3.5 Ga, and probably 3.8 Ga. Also, an analysis of Ce anomalies in zircons has suggested that this statement may hold true back to as early as 4.35 Ga (Trail et al., 2011). However, other redox-sensitive data allow for 0.5–1.0 $\log_{10}$ unit change during the Archean, so mantle redox evolution continues to be debated (Aulbach and Stagno, 2016; Nicklas et al., 2017).

Analysis of the oxidation state of undegassed mantle-derived melt samples has shown that subduction causes magmas in island arcs to become more oxidized (Kelley and Cottrell, 2009). The oxidation process is believed to occur from chemical components associated with water-rich fluids, such as Fe^{3+} dissolved in a silicate-rich fluid, rather than from water itself. Recently, Bell and Anbar (2013) have proposed that the subcontinental lithosphere, including parts of the upper mantle, may have been oxidized in this manner. If so, then gases emanating from arc volcanism could have become more oxidized with time, even if large parts of the upper mantle were unaffected.

Another mantle oxidation mechanism that operates on intermediate time scales has been proposed by Frost et al. (2004), Wade and Wood (2005), and Wood et al. (2006). (See also Frost and McCammon (2008).) The model is illustrated in Fig. 7.5. At the high pressures encountered near the base of a deep magma ocean, ferrous iron should have disproportionated to ferric plus metallic iron

$$3\,Fe^{+2} \rightarrow 2\,Fe^{+3} + Fe \qquad (7.10)$$

Some of the metallic iron – it must be at least 10% for the mechanism to work – segregated out and went into the core, perhaps aided by continued downward migration of molten iron from large impacts. The rest of the metallic iron, along with virtually all of the ferric iron, remained behind in the mantle. Paradoxically, because of the high prevailing pressure, the lower mantle remained highly reduced even though the ferric iron content was quite high (as much as 2 wt%, or 1/4 of total iron) (McCammon, 2005). The upper mantle was then oxidized as some of this ferric iron was lifted upward by convection. So, this mechanism requires no O_2 addition (and thus no water) whatsoever. The timescale for upper mantle oxidation depends on the convective overturn time, and it most certainly requires whole mantle convection. A strong point of this model is that it accounts for the apparently heterogeneous oxidation state of the upper mantle, pointed out previously, because mixing may have been incomplete. This theory might allow some portions of the upper mantle to remain relatively reduced for as much as several hundred million years, if convective mixing was slow.

An additional strong point of this hypothesis, as pointed out by its proponents, is that it can explain a reduced redox state of Mars' mantle. Oxygen fugacities inferred from Martian meteorites (with a variety of geochemical techniques) range from slightly below the iron–wüstite (IW) buffer to the QFM buffer, with an average that is reducing compared to Earth (Bridges and Warren, 2006; Stanley et al., 2011; Wadhwa, 2001, 2008) (see Table 12.8). Mars, being smaller, never achieved such high pressures in its interior, and so its mantle remained closer to the initial IW state.

7.4 Release of Reduced Gases From Subaerial Volcanism

With some uncertainty that arises because of the complexity of Earth's real mantle, we can estimate volcanic outgassing rates of H_2 and reduced carbon species. Recall from Sec. 7.2 that the subaerial fluxes of CO_2 and H_2O from arc and hotspot volcanism combined are roughly 4.5×10^{12} mol yr^{-1} and 9.5×10^{13} mol yr^{-1}, respectively. Following Holland (1984), we assume a temperature of 1200 °C and a pressure of 5 atm. (In arc volcanism, the gases are released explosively at pressures exceeding atmospheric pressure.) We further assume, as did Holland, that the magmas from which these gases are released have f_{O2} values near QFM.

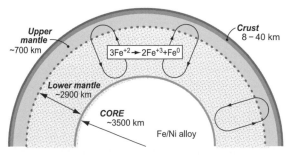

Figure 7.5 Diagram illustrating the mantle oxidation mechanism of Wade and Wood (2005) and Wood et al. (2006). Ferrous iron, Fe^{2+}, disproportionates into ferric iron, Fe^{3+}, and metallic iron, Fe^0, near the base of a deep magma ocean. Metallic iron sinks to join the core, while ferric iron is carried up into the upper mantle by convection.

We can write the equilibrium reactions as

$$H_2O(g) \xleftrightarrow{K_{11}} H_2 + \frac{1}{2}O_2 \qquad (7.11)$$

$$CO_2 \xleftrightarrow{K_{12}} CO + \frac{1}{2}O_2 \qquad (7.12)$$

Using data from Table 7.2, along with thermodynamic free energies from the US National Institute of Standards and Technology (NIST) online database, we get $K_{11} = 1.31 \times 10^{-6}$ bar$^{0.5}$ and $K_{12} = 3.24 \times 10^{-6}$ bar$^{0.5}$. But

$$K_{11} = \frac{f_{H_2} \cdot f_{O_2}^{0.5}}{f_{H_2O}} \qquad (7.13)$$

Thus, the ratio of H_2 to H_2O is given by

$$\frac{f_{H_2}}{f_{H_2O}} = \frac{K_{11}}{f_{O_2}^{0.5}} = 0.023 \qquad (7.14)$$

Here, we have taken $f_{O_2} = 10^{-8.5}$ atm, the value for the QFM buffer at 1200 °C and 5 atm pressure. Similarly,

$$\frac{f_{CO}}{f_{CO_2}} = \frac{K_{12}}{f_{O_2}^{0.5}} = 0.056 \qquad (7.15)$$

If we now multiply the ratios given in (7.14) and (7.15) by the outgassed fluxes of H_2O and CO_2, respectively, we obtain fluxes of 2.2×10^{12} mol yr^{-1} for H_2 and 2.5×10^{11} mol yr^{-1} for CO.

These numbers are uncertain for a variety of reasons in addition to the uncertainties in the fluxes of H_2O and CO_2. Average equilibration temperatures, for example, may be lower than assumed here; this would result in lower amounts of both H_2 and CO. Hayes and Waldbauer (2006) assume an H_2/H_2O ratio of 0.01, based on data from Giggenbach (1996). Subducted slabs might also have higher f_{O_2} values than does the average upper mantle, which could also reduce the amount of H_2 and CO released. Certainly, evidence indicates that arc magmas are more oxidized than basalts generated at mid-ocean ridges (Fig. 7.6), although there is debate about whether this results from oxidants in the subducting slab (e.g., H_2O, Fe^{3+}, S^{6+}, CO_2) (Evans et al., 2012; Kelley and Cottrell, 2009; Lee et al., 2005) or late stage differentiation (Lee et al., 2005; Lee et al., 2010) perhaps associated with degassing (Burgisser and Scaillet, 2007; Holloway, 2004).

If we use Hayes and Waldbauer's estimated H_2/H_2O ratio in place of the value from eq. (7.14), and reduce CO emissions by the same amount, the release rates of H_2 and CO become 1×10^{12} mol yr^{-1} and 1×10^{11} mol yr^{-1}, respectively. Our revised number for H_2 outgassing agrees well with the subaerial H_2 outgassing rate of 0.9×10^{12} mol yr^{-1} estimated by Holland (2009), so we shall use these lower numbers in the analyses presented in the next two chapters.

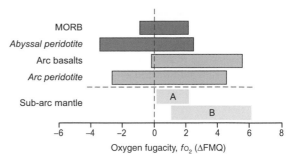

Figure 7.6 Oxygen fugacity estimates for a variety of tectonic settings, following Stamper et al. (2014) and references therein. (Reproduced with permission of Springer. Copyright 2014, Springer-Verlag Berlin Heidelberg 2014.) For the sub-arc upper mantle, the range labeled "A" comes from Lee et al. (2005; 2010) and Mallmann and O'Neill (2009); range "B" is that proposed by Evans et al. (2012).

Other authors have argued that the H_2 flux from subaerial volcanoes is significantly lower than the values just derived. For example, Canfield et al. (2006) estimate a volcanic H_2 flux of $(1.8–5.0) \times 10^{11}$ mol yr^{-1}. They derive this range of values by comparing the observed amount of H_2 in volcanic gases with that of SO_2 and then multiplying by the volcanic flux SO_2 flux measured from satellites, $(2.3–3.1) \times 10^{11}$ mol yr^{-1} (Sec. 7.2.3). But we have already pointed out that this latter number may be too small because of passive SO_2 degassing, and thus the H_2 flux of Canfield et al. would be too small.

Before leaving this section, it is instructive to estimate the flux of methane from surface high-temperature volcanism. We have mentioned that methane is not observed, and we can justify this observation with theory. If thermodynamic equilibrium pertains, then the methane abundance can be calculated from the equilibrium reaction

$$CO_2 + 2H_2O \xleftrightarrow{K_{16}} CH_4 + 2O_2 \qquad (7.16)$$

Taking thermodynamic data from Table 7.2, we obtain

$$K_{16} = \frac{f_{CH_4} \cdot f_{O_2}^2}{f_{CO_2} \cdot f_{H_2O}^2} = 4.31 \times 10^{-29} \qquad (7.17)$$

This can be rearranged to yield

$$\frac{f_{CH_4}}{f_{CO_2}} = K_{16}\left(\frac{f_{H_2O}}{f_{O_2}}\right)^2 = 9.4 \times 10^{-11} \qquad (7.18)$$

Hence, as mentioned in Sec. 7.1, methane is not predicted to be emitted from surface volcanoes, a result that was shown previously by Holland (1978, 1984). Temperatures are too high and pressures too low for this five-atom molecule to be stable. Where methane is observed, it is attributed either to serpentinization reactions (see next section) or to the thermal decomposition of organic matter, rather than to magmatic processes (Taran and Giggenbach, 2003).

The same type of analysis can be used to show that ammonia, NH_3, is not expected to be emitted from surface volcanism at modern mantle oxidation states (see Holland, 1984). NH_3 is also too large a molecule to be stable at high temperatures. Even at the IW buffer ($f_{O_2} \sim 10^{-12}$ atm), the predicted ratios of f_{CH_4}/f_{CO_2} and f_{NH_3}/f_{N_2} are only of order 10^{-3}. That is why Rubey, who assumed that the atmosphere formed from surface volcanism, was led to a different conclusion from that of Oparin or Miller and Urey.

7.5 Reduced Gases Released From Submarine Volcanism and Hydrothermal Systems

7.5.1 H_2S and H_2

As pointed out earlier, significant amounts of volcanism also occur at the mid-ocean ridges. Indeed, mid-ocean ridge volcanism dominates rock production and heat flow. Approximately 90% of the geothermal heat flow makes its way to the surface via the mid-ocean ridges, although it is given off over a wider area as the seafloor cools.

Mid-ocean ridge outgassing is a significant source of volatiles. The outgassing of CO_2 has already been discussed but submarine outgassing is also an important source of reduced species. Measurements of the composition of vent fluids have been made from submersibles (such as Woods Hole Oceanographic Institution's *Alvin*) for several decades. These fluids can be very different in composition, depending on where one is looking. The main reductants in hot, axial vent fluids are H_2S, H_2, and dissolved ferrous iron, Fe^{2+}. The H_2S concentration is highly variable and ranges from 3 to 80 mmol kg^{-1} at different vent sites (Von Damm, 1990, 1995). Based on these measurements, Holland (2002, his Table 1) estimated average dissolved H_2S concentrations of 7 mmol kg^{-1} in fluids emanating from the hot, axial hydrothermal vents. This is accompanied by ~2 mmol kg^{-1} of dissolved H_2 and 20 mmol kg^{-1} of dissolved CO_2. To convert these concentrations to fluxes, one needs to multiply by the total flux of water coming out of the axial vents, ~5×10^{13} kg yr^{-1} (German and Seyfried, 2014; Mottl and Wheat, 1994; Wolery and Sleep, 1976). (The water flux is calculated from the so-called "heat flow anomaly," i.e., the difference between the total heat released at the ridges from cooling of magma and the measured heat flow in seafloor near the ridges.) This yields submarine H_2S and H_2 fluxes of 3.5×10^{11} mol yr^{-1} and 1×10^{11} mol yr^{-1},

respectively. We return to the H_2 flux below, because the value calculated here might be too small.

We should point out that calling this submarine H_2S "volcanic" is something of a misnomer, according to our terminology, as the sulfur is not being released directly from magma. Isotopic studies suggest that much of it is leached from the surrounding basalts (Alt, 1995), as its $\delta^{34}S$ values are typically closer to that of mantle sulfides (0‰) than to seawater sulfate (21‰). Hence, the H_2S is termed "magmatic" because it comes from rocks which themselves came from magma.

Sulfate is also reduced to sulfide during passage through modern hydrothermal vents. In this way, some H_2S is derived from the reduction of seawater sulfate provided the reduction process occurs at temperatures below 300 °C (Arthur, 2000). Most sulfate does not participate in this reaction, however, as it is precipitated as anhydrite, $CaSO_4$, at temperatures above 150 °C. The anhydrite is thought to mostly redissolve prior to subduction. The sulfate reduction process does not release significant amounts of reduced gases because about 80% of the reduced sulfate forms pyrite, FeS_2 rather than H_2S (Ono *et al.*, 2007). It was not an important process on the early Earth because dissolved sulfate concentrations were probably quite low, but it would have become a significant oxygen sink later on once sulfate concentrations rose. The rate at which this reaction occurs today can be estimated by examining rates of seafloor oxidation. According to Lecuyer and Ricard (1999), some $(0.5-1.9) \times 10^{12}$ mol O_2 equivalent is consumed by oxidizing Fe^{2+} in basalt to Fe^{3+} in magnetite (Fe_3O_4) during aqueous oxidation of seafloor. The main oxidant is sulfate, rather than dissolved O_2, because its abundance, 28 mmol kg^{-1}, is far higher than that of O_2, ~0.2 mmol kg^{-1}. We can write the oxidation reaction as

$$22\,FeO + 2\,H_2SO_4 \rightarrow FeS_2 + 7\,Fe_3O_4 + 2\,H_2O \quad (7.19)$$

The upper bound on the estimated reaction rate would require completely oxidizing ~1 km (out of 6 km total depth) of fresh oceanic crust, which is more than is observed (Sleep, 2005). Following Sleep, we take 1×10^{12} mol O_2 equivalent yr^{-1} as an average value for oxygen consumption during hydrothermal alteration of seafloor.

As indicated above, the actual submarine H_2 flux may be higher than 1×10^{11} mol yr^{-1}. According to Sleep (2005), the rate of serpentinization of seafloor is ~2×10^{11} mol O_2 equivalent per year, about one-fifth of the total seafloor oxidation rate. If most of this results in H_2 production, as seems likely, the submarine H_2 flux should be twice this value, or 4×10^{11} mol yr^{-1}. This may be compared with independent estimates of 1.9×10^{11} mol yr^{-1}

(Keir, 2010) and $(0.48–6)\times10^{11}$ mol yr^{-1} (Cannat *et al.*, 2010). We will use 4×10^{11} mol H$_2$ yr^{-1} here to remain consistent with Sleep's estimate of seafloor oxidation.

Before leaving this topic, we should point out that the composition of mid-ocean ridge vent fluids on the Archean Earth could have been quite different from today. Higher mantle temperatures may have led to higher degrees of partial melting at the mid-ocean ridges. Keller and Schoene (2012) suggest that the percentage of fractional melting has decreased from 35% during the Archean to ~10% today. If so, then the Archean seafloor should have been more ultramafic, and hence more prone to serpentinization (Kasting, 2013; Kasting and Canfield, 2012). (Serpentinization is described in more detail in the following section.) This could have resulted in significantly greater H$_2$ concentrations in vent fluids.

Another aspect of the Archean Earth is that widespread formation of banded iron formations requires that the deep ocean contained large quantities of dissolved ferrous iron (Holland, 1973b). This, in turn, requires that ferrous iron was more abundant than sulfide in hydrothermal vent fluids, which it should have been if oceanic sulfate concentrations were low (Kump and Seyfried, 2005; Walker and Brimblecombe, 1985). Lower vent pressures caused by shallower mid-ocean ridges (Isley, 1995; Kasting *et al.*, 2006) may have contributed to the dominance of iron over sulfide (Kump and Seyfried, 2005). (Shallower ridges are predicted by Archean plate tectonic models with thick oceanic crust, which are discussed in Sec. 7.6.) The ferrous iron could have become a source of H$_2$ to the atmosphere if it was upwelled and oxidized by solar UV radiation (Braterman *et al.*, 1983), although some lab experiments show photo-oxidation to be negligible compared to other processes (Konhauser *et al.*, 2007a). These postulated changes in the composition of hydrothermal vent fluids play a role in some theories for the rise of atmospheric oxygen (see Ch. 10).

7.5.2 CH$_4$

Some CH$_4$ is also present within the fluids released at hot, axial vents. Welhan (1988) reported very low CH$_4$ concentrations, 0.1–0.2 μmol, in fluids emanating from the East Pacific Rise (EPR) along with CH$_4$/CO$_2$ ratios of 1%–2%. If one accepts Holland's (2002) estimate of 20 mmol kg^{-1} for dissolved CO$_2$, this yields higher CH$_4$ concentrations of (0.2–0.4) mmol kg^{-1}. These numbers are roughly consistent with those predicted from a thermodynamic equilibrium analysis at conditions applying deep within the vent systems (Kasting and Brown, 1998; Welhan, 1988). We'll use 0.3 mmol kg^{-1}, giving us

an axial CH$_4$ flux of 1.5×10^{10} mol yr^{-1} – an insignificant amount.

Much higher dissolved CH$_4$ concentrations, 1–2 mmol kg^{-1}, have been found in cooler (50–60 °C) vent fluids emanating from the off-axis Lost City vent field on the Mid-Atlantic ridge (Kelley *et al.*, 2005). Since that initial discovery, several other off-axis vent fields have been discovered along this spreading ridge (Konn *et al.*, 2015). Reactions of ultramafic rock (peridotite) (defined in Sec. 7.2.1) with water deep within the oceanic crust may generate this methane (McDermott *et al.*, 2015). When exposed to water, such ultramafic rocks are altered to form serpentine minerals – in the process of *serpentinization* – and these minerals tend to exclude iron. The iron then enters a stable phase, which happens to be magnetite, Fe$_3$O$_4$. As noted earlier, this mineral contains one Fe^{+2} ion and two Fe^{+3} ions. Because iron is being oxidized, something else must be reduced. In some cases, iron and nickel are reduced to metal alloys. However, laboratory experiments (Berndt *et al.*, 1996) have shown that hydrogen is also produced from oxidation of water. This reaction can be written as:

$$3\,FeO + H_2O \rightarrow Fe_3O_4 + H_2 \qquad (7.20)$$

When dissolved CO$_2$ is present in the water, CH$_4$ can be produced, as well (*ibid.*). One has to be careful about assuming that all CH$_4$ in vent fluids is produced by this process, as methane could also be generated biologically from CO$_2$ and H$_2$ by methanogens living within the vent systems. However, isotopic studies of the fluids at Lost City suggest that the methane observed there is abiotic (Proskurowski *et al.*, 2008).

Although CH$_4$ can be produced by serpentinization, the submarine flux of CH$_4$ is evidently much smaller than that of H$_2$. The ratio of CH$_4$:H$_2$ in ultramafic vent fields is about 1:15 (Cannat *et al.*, 2010; Keir, 2010). This leads to an estimated CH$_4$ flux of $(2–2.5)\times10^{10}$ mol yr^{-1} (*ibid.*). Multiplying Sleep's estimated H$_2$ flux by 1/15 yields a value of 3×10^{10} mol yr^{-1}, and we adopt this value for Table 7.3. This value compares with a present global biological methane flux of ~535 Tg CH$_4$ yr^{-1}, or 3.3×10^{13} mol yr^{-1} (Houghton *et al.*, 1995). Thus, the abiotic methane flux from seafloor hydrothermal systems is smaller than the biological flux by a factor of more than 1000.

The source of abiogenic methane from continental hydrothermal systems has been estimated as up to 5 Mton yr^{-1}, or about 3×10^{11} mol yr^{-1} (Fiebig *et al.*, 2009). Again, this is small compared to the biological methane flux, but large enough to impact the global redox budget. If low temperature (200–800 °C) metamorphism

Table 7.3 Present outgassing rates of reduced species.

Surface volcanism	Flux[*]
H_2	1.0
CO	0.1
H_2S	0.03
CH_4 (hydrothermal)	0.3
Submarine volcanism	
H_2	0.4
H_2S	0.35
CH_4 (axial)	0.01
CH_4 (off-axis)	0.03

[*] Fluxes in units of 10^{12} mol yr^{-1} (1×10^{12} mol yr^{-1} = 3.74×10^9 $cm^{-2}s^{-1}$). All values are uncertain by at least a factor of 2. Derivations of these numbers are given in the text.

happened on the early Earth in crust that was more reducing, calculations show that the CH_4/CO_2 ratio in emissions increases by ~3 orders of magnitude if the crustal oxygen fugacity in highly reduced areas is decreased from QFM–1 to QFM–3, assuming thermodynamic equilibrium (Claire *et al.*, 2006). Methane is also produced thermogenically (i.e., from thermal breakdown of old organic matter by metamorphism) at rates of ~20–40 Mton or 1.25–2.5×10^{12} mol yr^{-1} (Etiope *et al.*, 2009). This flux is important for the modern methane budget, but it could have been smaller during the Archean, depending on the amount of organic matter available, and should have been virtually absent on the prebiotic Earth.

One last caveat concerning outgassing fluxes of methane should be added. If the early seafloor and continental crust were both more ultramafic during the Archean, as postulated above, then the abiotic source of methane could have been much larger than it is today. This might be useful in some models of life's origin, as CH_4 is needed to form the key prebiotic molecule, hydrogen cyanide (HCN) photochemically (Zahnle, 1986). But it also suggests that methane by itself would be a dubious biosignature, if identified in the atmosphere of an exoplanet. We should bear this in mind when we discuss atmospheric biosignatures in Ch. 15.

7.6 Past Rates of Volcanic Outgassing

Thus far, we have assumed that the total volcanic outgassing rate is equal to its modern value. We are interested in the early Earth, though, and so we should consider how this rate might have changed in the distant past. Our

analysis will be speculative, as there is an ongoing debate amongst geophysicists as to how geothermal heat flow and volcanism have changed over time. One needs to understand how plate tectonics has evolved in order to answer these questions, and researchers are far from reaching agreement on this topic. Below, we briefly discuss several different points of view.

On one point there is general agreement: Earth's interior must have been hotter in the past, partly because of increased radiogenic heat production (primarily from U, K, and Th) and partly because of greater leftover heat from accretion and core formation. Geophysicists include these factors in so-called *thermal evolution* models, in which they attempt to model the evolution of mantle temperature and heat flow, typically using highly parameterized models of mantle convection. Conventional thermal evolution models predict that geothermal heat flow should have been higher in the past as a consequence of the Earth's hotter interior. For example, Davies (1980) predicted that heat flow could have been 2–5 times higher than today at 3.0 Ga and 4–8 times higher at 3.5 Ga. Sleep and Zahnle (2001) adopted a somewhat lower heat flow for the early Earth, 2–3 times present, but they argued that both oceanic crust production and volcanic outgassing should scale as the square of heat flow; hence, volcanic outgassing rates during the Archean could have been 4–9 times higher than today.

Other geophysicists have reached quite different conclusions, however, partly because there is some (surprising) evidence for *low* surface heat flow early in Earth's history. For example, Hopkins *et al.* (2008) analyzed Jack Hills zircons of Hadean age and concluded that heat flow in the region of their formation was only 75 mW m^{-2}. This is slightly lower than today's global average of 90 mW m^{-2}, and at least 3–5 times lower than conventional thermal history models would predict for that time period. The authors suggest that this could be because the Jack Hills area sat above a subduction zone, where heat flow was abnormally low. Subducting oceanic lithosphere creates thicker crust that lowers the mean geothermal gradient for the same temperature difference between surface and Mohorovičić discontinuity, following Fourier's Law. Of course, advected heat causes arc volcanism above subduction zones also. But the result from the zircons could also mean that conventional thermal history models, which assume that plate tectonics was operating more or less as it does today, are incorrect.

Other geophysical modelers have formulated detailed hypotheses for how plate tectonics may have changed over time. Many of them agree that higher mantle temperatures should have led to an increased depth of partial melting at the ridges and, hence, to increased oceanic crustal thickness (Bickle, 1986; Burke *et al.*, 1976; Davies, 2002; Moores, 1986, 1993, 2002; Sleep, 2007; Sleep and Windley, 1982). (See Davies (2006) for a refutation of this hypothesis, and Korenaga (2008b) for a criticism of that refutation.) That increased thickness, combined with possible dehydration of the lower part of this crust from the high temperatures, makes the oceanic lithosphere stiffer and more buoyant, and thus harder to subduct, thereby slowing down the operation of plate tectonics. The result could be that geothermal heat flow in the distant past was comparable to or lower than today (Korenaga, 2006, 2007, 2008a, b). Korenaga reconciles this with modern heat flow measurements by adopting a low convective Urey ratio (ratio of mantle heat production/mantle heat flow) of ~0.2, compared to a typically assumed value ~0.4 (Butler and Peltier, 2002). He argues that this allows his model to avoid the "thermal catastrophe" (complete surface melting) that can occur when conventional models are integrated backwards in time. (Heat production from radioactive decay increases at earlier times because of higher concentrations of U, K, and Th. If the additional heat released cannot be removed by convection, then the whole upper mantle melts.) If Korenaga is correct, then volcanic outgassing rates in the distant past could have been more or less the same as at present.

Another hypothesis is the "heat-pipe" model suggested by Moore and Webb (2013). They draw an analogy with Jupiter's moon, Io, which has an extremely high estimated surface heat flow of >2 W m^{-2} caused by tidal heating. Io is famous for being the most volcanically active body in the Solar System, as evidenced by the photographs taken of 150 active volcanoes by *Voyager*, *Galileo*, *Cassini*, and *New Horizons* spacecraft. Io supports 20-km-high mountains, however, implying that its crust is relatively thick. It does so by losing most of its internal heat through volcanism, i.e., by direct magma generation at the surface. The remainder of the relatively thick, cool crust sinks slowly and eventually melts at its base. This heat flow mechanism might account for the "vertical tectonics" that has been suggested by other authors (e.g., Robin and Bailey, 2009) to explain formation of continental crust on the Archean Earth. The implications for early Earth's volatile budget have not been rigorously explored; however, one would imagine that high rates of volcanism would lead to efficient delivery of volatiles from the mantle to the surface.

We conclude, as we warned initially, that thermal evolution modeling of the Earth is an ongoing effort

and that no results should be taken for granted. Note, though, that even if Korenaga is correct and surface heat flow was low, this does not imply that the net production rate of reduced gases was necessarily the same as today. As pointed out in the previous section, higher degrees of partial melting at the ridges should have led to greater amounts of ultramafic oceanic crust, and this in turn could have led to higher production rates of H_2 and CH_4 from serpentinization. Likewise, production of H_2 in continental hydrothermal systems might also have been enhanced. The widespread occurrence of komatiites (an ultramafic type of rock) in Archean greenstone belts supports the idea that rock types were different at that time. Indeed, the boundary between the Archean and Proterozoic Eons, although set by committee as exactly 2.5 Ga, is characterized by differences in rock types. This is also very close to the time (2.45 Ga) when atmospheric O_2 increased to appreciable concentrations for the first time. It is thus plausible to hypothesize that changes in abiotic H_2 and/or CH_4 production may have contributed to this change.

7.7 Summary

Ideas about the nature of the prebiotic atmosphere have changed dramatically over the past century. Opinion has swung back and forth as to whether the early atmosphere was dominated by CO_2 and N_2 or whether it contained significant quantities of highly reduced gases such as CH_4 and NH_3. In part, changing concepts of how the redox state of Earth's interior evolved have fueled the debate over the composition of the early atmosphere. Although some data indicate that the average redox state of the upper mantle has remained relatively constant for at least the past 3.5 billion years, we have argued here that the upper mantle could have been more reduced (but still devoid of metallic iron) during the first few hundred million years of Earth's history if upward transport of ferric iron from the deep mantle was delayed and if other mantle oxidation mechanisms (e.g., subduction of hydrated crust, followed by outgassing of H_2) had not yet had sufficient time to operate. When combined with ongoing bombardment by reduced planetesimals, this leaves open the possibility that the atmosphere was still relatively reduced at the time when life originated, provided that this event occurred prior to ~3.8 Ga.

Once the upper mantle reached its present oxidation state, the redox state of magmatic volcanic gases should have been similar to that observed today. But the elemental composition could have changed if, for example, arc volcanism became relatively more important compared to point volcanism over time. Changes in volcanic gas composition (in particular, the relative fluxes of C- and S-bearing gases) play an important role in some models of the rise of O_2, to be discussed in Ch. 10. Also important for O_2 evolution is that the release rate of reduced gases generated by serpentinization reactions could have remained higher than today if ultramafic rocks were more prevalent on the seafloor or in continental environments. Thus, changes in the production rate of reduced gases may have influenced the redox evolution of the atmosphere–ocean system.

8 Atmospheric and Global Redox Balance

We devoted considerable effort in the previous chapter to understanding volcanic outgassing rates and the oxidation state of volcanic gases. The reason for doing so is that outgassing of reduced gases directly affects the *atmospheric redox budget*. The term "redox budget" is used to describe the balance between chemical reductants (electron donors) and chemical oxidants (electron acceptors). Earth's present atmosphere is oxidizing because its second most abundant gas, O_2, is a powerful electron acceptor. The early atmosphere was likely to have been reducing because O_2 fluxes were outweighed by fluxes of electron donor gases, such as H_2, CH_4, and CO. On the early Earth, the atmospheric redox budget was controlled by loss of hydrogen to space at the top and by volcanic outgassing and rainout or surface deposition of reduced and oxidized gases at the Earth's surface, so this budget includes exchange of oxidized and reduced gases between the atmosphere and the ocean. These are quantities that one can keep track of with an atmospheric photochemical model. The first part of this chapter is concerned with how to do this atmospheric budgeting. It is left as an exercise for the reader to verify that the timescale over which the early atmospheric redox budget must balance is of order 30 000 years.[1]

A second redox budget that must also be balanced over somewhat longer timescales is that of the combined atmosphere–ocean system. We shall refer to this tally sheet as the *global redox budget*. By definition, exchange of gases between the atmosphere and the ocean does not count in this budget. The global redox budget is more difficult to compute, as it involves exchange of oxidants and reductants with the crust and the mantle through processes that are not easy to observe, although much of the discussion in the preceding chapter is relevant. This budget is crucial to understanding the rise of atmospheric O_2, so we shall return to it again in Ch. 10. It also controls the O_2 content of the modern atmosphere on time scales of millions of years (see Sec. 10.2.4), and so various authors (Berner, 2004, 2006a; Berner and Canfield, 1989; Catling, 2014; Claire *et al.*, 2006; Holland, 1978, 1984, 2002; Kasting, 2013; Kasting and Canfield, 2012) have attempted to quantify it. Our focus here, and in Ch. 9, will be to try to identify the differences between the global redox budget on the early Earth and today. If we truly understood these differences, we would know precisely why atmospheric O_2 concentrations rose when they did.

The definitions made above are not the only way that one might think of redox budgets. Some authors (e.g., Hayes and Waldbauer, 2006) have focused on exchange of reducing power between Earth's crust and its mantle. The continents are more oxidized than the mantle from which they were generated, and so they represent a large, long-term sink for O_2. We can include this O_2 sink in our global redox budget without worrying about whether there are corresponding redox changes in the mantle. Such changes in the mantle are not required, as much of the oxygen that has gone into the continental crust probably came from the biotically mediated breakdown of water in the atmosphere–ocean system, followed by escape of hydrogen to space (Catling *et al.*, 2001; Zahnle *et al.*, 2013). The planet as a whole is progressively oxidized by this process, which may also have played a role in oxidizing the upper mantle (see Sec. 7.3.2). However, an inferred apparent constancy of average upper mantle redox state since the rock record began perhaps implies that we need

---

[1] If escape to space was the dominant loss process for H_2 in the early atmosphere, then the lifetime against escape is the appropriate timescale over which the atmospheric redox budget must balance. To calculate this lifetime, find the atmospheric column mass from eq. (1.19). Convert to molecules cm^{-2}, then multiply by the H_2 volume mixing ratio, $f(H_2)$, to get the number of H_2 molecules in a vertical column. Divide by the diffusion-limited escape rate, eq. (5.57), assuming that H_2 is the dominant H-bearing constituent, to get the lifetime. The answer should be independent of $f(H_2)$.

not explicitly keep track of how much oxygen has gone into the mantle or how much hydrogen has come out of it.

8.1 Principles of Redox Balance

To begin, we note that reduction and oxidation processes must balance: when one compound is oxidized, another must be reduced to conserve electrons. Geochemists generally keep track of redox balance in terms of O_2 equivalents. For example, photosynthesis by green plants produces O_2 by the following reaction

$$CO_2 + H_2O \rightarrow CH_2O + O_2 \qquad (8.1)$$

Here, "CH_2O" (literally carbohydrate) is shorthand for more complex forms of organic carbon. Carbon is reduced in this process, i.e., it gains electrons, while two oxygen atoms give up electrons to form O_2. When one mole of CH_2O is buried in sediments, one mole of O_2 is produced. By compiling the various sources and sinks for O_2, one can estimate a global O_2 budget for the modern Earth. (The modern atmospheric O_2 budget is largely irrelevant, as O_2 is freely exchanged between the atmosphere and ocean.) We will construct such a budget in Ch. 10 when we discuss controls on atmospheric O_2.

On the early Earth, as we will demonstrate, O_2 was a trace atmospheric constituent. Instead of keeping track of O_2, it makes more sense to keep track of the reduced species, H_2 or H_2 equivalents. In terms of redox balance, H_2 is related to O_2 by the reaction

$$2H_2 + O_2 \leftrightarrow 2\,H_2O \qquad (8.2)$$

Thus, 2 moles of H_2 can react with 1 mole of O_2. Water is freely available at Earth's surface, so it is convenient to define water as having a "neutral" oxidation state. We will generalize the concept of neutral oxidation states below. For the early Earth, it is useful to construct both an atmospheric H_2 budget and a global H_2 budget. The atmospheric H_2 budget can be used as a modeling tool: it provides a check on the chemical scheme and numerical accuracy of photochemical models, and it can be used to analyze biologically driven gas exchange processes occurring at the atmosphere–ocean interface, as discussed in the following chapter. The global H_2 budget determines the concentration of H_2 and other reduced gases in the early atmosphere, along with the eventual transition to an O_2-rich atmosphere.

8.2 H_2 Budget of the Prebiotic Atmosphere: Approximate Solution

We can use the data on volcanic outgassing rates provided in Ch. 7, along with the information on hydrogen escape to space from Ch. 5, to estimate the H_2 concentration in the

(a) The atmospheric redox budget

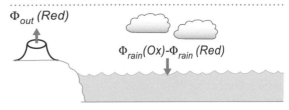

$$\Phi_{esc}(H_2) \cong 2.5 \times 10^{13}\, f_T(H_2)$$

Homopause (~100 km)

$$f_T(H_2) = f(H_2) + 0.5\, f(H) + f(H_2O) + 2\, f(CH_4) + \cdots$$

Tropopause (~15 km)

$\Phi_{out}(Red)$

$\Phi_{rain}(Ox)$-$\Phi_{rain}(Red)$

(b) The global redox budget

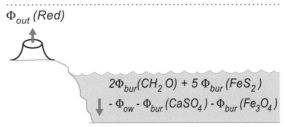

$$\Phi_{esc}(H_2) \cong 2.5 \times 10^{13}\, f_T(H_2)$$

Homopause (~100 km)

$$f_T(H_2) = f(H_2) + 0.5\, f(H) + f(H_2O) + 2\, f(CH_4) + \cdots$$

Tropopause (~15 km)

$\Phi_{out}(Red)$

$$2\Phi_{bur}(CH_2O) + 5\,\Phi_{bur}(FeS_2)$$
$$- \Phi_{ow} - \Phi_{bur}(CaSO_4) - \Phi_{bur}(Fe_3O_4)$$

Figure 8.1 Diagrams illustrating (a) the atmospheric hydrogen budget, and (b) the global redox budget. Volcanic outgassing of hydrogen, $\Phi_{out}(H_2)$, is balanced by escape to space at the diffusion limit, $\Phi_{esc}(H_2)$. The latter depends on the mixing ratio of hydrogen in all its forms at the homopause. For the prebiotic Earth, most of the hydrogen should have been in the form of H_2.

early atmosphere. We'll focus here on the prebiotic atmosphere, although the same methodology can still be applied after life had arisen, but prior to the origin of oxygenic photosynthesis. The idea, which was proposed originally by Walker (1977, 1978), is to balance volcanic outgassing of hydrogen with escape of hydrogen to space. These are two of the terms illustrated in the redox budgets shown schematically in Fig. 8.1. The other terms, and the differences between Fig. 8.1(a) and Fig. 8.1(b), will be discussed below. Walker considered only the flux of H_2 from surface volcanism, so let's use that flux, 1×10^{12} mol yr^{-1} (see Table 7.3), as the total input of hydrogen. According to reaction (8.2), each mole of H_2 consumes 0.5 moles of O_2. So, the potential consumption rate of O_2 by reduced volcanic gases is about 5×10^{11} mol yr^{-1}.

To determine the abundance of O_2 in the prebiotic atmosphere, we can compare the O_2 consumption rate with the abiotic production rate of O_2. (We'll rephrase the question in terms of H_2 balance momentarily.) Let's consider how O_2 can be produced in the absence of photosynthesis. In an atmosphere containing CO_2 and H_2O, numerous photochemical pathways form O_2. The fastest of these is the sequence

$$CO_2 + hv \rightarrow CO + O \qquad (8.3)$$

$$H_2O + hv \rightarrow H + OH \qquad (8.4)$$

$$O + OH \rightarrow O_2 + H \qquad (8.5)$$

These reactions, and others like them, form O_2 in photochemical models of the primitive atmosphere, but they

are not *net* sources of O_2. If the oxygen atoms come from CO_2, then CO is left behind. Eventually, this CO will react with O_2 to reform CO_2. Similarly, if the oxygen atoms come from H_2O, this will leave behind H_2. If that H_2 remains in the atmosphere, then it will eventually recombine with O_2, and so the net O_2 production should be zero. If the hydrogen escapes to space, however, then 0.5 moles of O_2 are produced for every mole of H_2 that escapes. So, the abiotic production rate of O_2 should be equal to one-half the escape flux of H_2 that originates from water vapor photolysis. (Note the nuance, however, that the escape of hydrogen to space originating from any hydrogen-bearing compound, including gases without O atoms such as CH_4 or H_2S, will oxidize the Earth as a whole. Since O atoms are not liberated from these gases, their photolysis does not create O_2 molecules directly.)

As discussed in Ch. 5, calculating the actual escape flux of hydrogen is complicated because the escape rate can be limited either by diffusion through the homopause or by energy considerations at the exobase. However, an upper limit on the escape flux is available because hydrogen cannot escape faster than the diffusion limit (eq. (5.57))

$$\Phi_l(H) \cong 2.5 \times 10^{13} f_T(H) \text{ atoms cm}^{-2}\text{s}^{-1} \qquad (8.6)$$

where, from eq. (5.54), the total hydrogen atom volume mixing ratio $f_T(H)$ in all its forms is

$$f_T(H) = f_H + 2f_{H_2} + 2f_{H_2O} + 4f_{CH_4} + \cdots \qquad (8.7)$$

The escaping hydrogen from H_2O is given by the third term on the right of eq. (8.7). Hence, we require knowledge of the H_2O volume mixing ratio in the stratosphere, f_{H_2O}. Today, that value is ~3 ppmv in the lower stratosphere, below the level where methane oxidation occurs. It is limited by condensation at the cold, equatorial tropopause (Sections 1.1.1 and 4.6.1; Fig. 5.9). This region is often referred to as the atmospheric *cold trap* because it limits the amount of water vapor in the stratosphere, just as a cold trap in a gas flow line limits the concentration of water vapor or other volatile gases passing through it.

When budgeting hydrogen it is convenient to express the escape flux in terms of H_2 molecules rather than H atoms. Equation (8.6) retains the same form if Φ_l and f_T are both expressed in units of H_2 rather than H. The magnitude of $f_T(H_2)$ is just one-half that of $f_T(H)$. Plugging 3 ppmv of H_2O into eq. (8.7) and using this in eq. (8.6), gives the escape rate of H_2 molecules from H_2O:

$$\Phi_l(H_2) \cong (2.5 \times 10^{13}) \cdot (3 \times 10^{-6}) = 7.5 \times 10^7 \, H_2 \text{ molecules cm}^{-2}\text{s}^{-1} \qquad (8.8)$$

The production rate of O_2 is half the escape rate of H_2, or 3.75×10^7 O_2 molecules cm^{-2} s^{-1}. Following atmospheric scientists' convention, we shall henceforth suppress the word "molecules" and express escape fluxes in units of "cm^{-2} s^{-1}". Converting this number back into geochemists' units gives an O_2 production rate of 1.0×10^{10} mol O_2 yr^{-1}. By comparison, the potential consumption rate of O_2 by reaction with volcanic hydrogen estimated above was 5×10^{11} mol yr^{-1}. If these numbers are correct, the outgassed flux of hydrogen today is ~50 times larger than that needed to consume all of the abiotically produced O_2. As we will see, the total hydrogen outgassing rate should be larger than the number used here; hence, the ratio between potential O_2 consumption and abiotic O_2 production is probably even larger than just calculated.

The bulk of the outgassed hydrogen could not actually have reacted with O_2, of course, because there was not enough O_2 around for it to do so. Thus, the remainder of the outgassed H_2 must have either escaped to space or have been converted to reduced organic compounds and buried in sediments. We will argue in the next chapter that the abiotic formation rate of such compounds, largely formaldehyde (H_2CO), was probably much less than the hydrogen escape rate, and can thus be neglected to first order.[2] If we further assume that hydrogen was escaping

---

[2] Earlier predictions that a "primordial oil slick" of more complex organic compounds might have formed on the prebiotic Earth (Lasaga *et al.*, 1971) depend on CH_4 being a major bulk constituent. As we saw in Ch. 7, this possibility is often considered unlikely.

to space at the diffusion-limited rate, then we can estimate the atmospheric H_2 mixing ratio by balancing the H_2 escape rate, $\Phi_l(H_2)$, with the H_2 outgassing rate, $\Phi_{out}(H_2)$. To do so, first convert the H_2 outgassing rate into photochemists' units. The conversion factor is: 1 mol yr^{-1} = 3.74×10^{-3} cm^{-2} s^{-1}. Thus, $\Phi_{out}(H_2) = 1\times10^{12}$ mol yr^{-1} $\cong 4\times10^9$ cm^{-2} s^{-1}. Then, equating outgassing with escape (in units of cm^{-2} s^{-1}) gives

$$\Phi_{out}(H_2) \cong \Phi_{esc}(H_2)$$
$$4 \times 10^9 \cong 2.5 \times 10^{13} f(H_2) \qquad (8.9)$$

or

$$f(H_2) \cong \frac{4 \times 10^9}{2.5 \times 10^{13}} = 1.6 \times 10^{-4} \qquad (8.10)$$

The value calculated, 1.6×10^{-4}, or 160 ppmv, is essentially the same as that in Walker (1978 his Table 1) at the same H_2 outgassing rate. By comparison, the H_2 mixing ratio in the present atmosphere is 0.55 ppm, so our estimated prebiotic H_2 concentration is ~300 times higher than today's value.

We have made several assumptions along the way, however, that may have caused our H_2 estimate to be too low. First, the H_2 outgassing rate on the early Earth could have been significantly higher than today, although it is not easy to calculate by how much (see Sec. 7.6). Furthermore, we have neglected outgassing of other reduced gases, such as CH_4 and CO. Of these, CH_4 is potentially the most important. According to Table 7.3, the combined flux of CH_4 from continental and mid-ocean ridge hydrothermal systems is ~4×10^{11} mol yr^{-1}. As discussed further in the next section, each mole of outgassed CH_4 is equivalent to 4 moles of H_2. We can see this from the reaction

$$CH_4 + 2H_2O \rightarrow CO_2 + 4H_2 \qquad (8.11)$$

CO_2, like H_2O, is another "neutral" oxidation state compound. (We will explain this statement in the next section.) Hence, the effective input of hydrogen from CH_4 is 4· (4×10^{11} mol yr^{-1}) = 1.6×10^{12} mol yr^{-1}, which is higher than the estimated outgassing rate of H_2 itself. So, if this CH_4 was converted to H_2, the atmospheric H_2 mixing ratio should increase to ~4×10^{-4}, or 400 ppmv. If the actual escape rate was slower than the diffusion limit, atmospheric H_2 concentrations could have been even higher.

A factor that is slightly more complex to analyze is our assumption that the stratospheric water vapor mixing ratio was the same as today, ~3 ppmv. Here, by "stratosphere," we mean the quasi-isothermal region above a troposphere in a radiative equilibrium profile (e.g., Fig. 2.20(a)), rather than a true stratosphere associated with a temperature inversion, which on the modern Earth

is caused by the ozone layer. As noted earlier, stratospheric water vapor is limited by the efficiency of the cold, high, equatorial cold trap. How might this have changed in the past? The answer depends on the nature of the early atmosphere and on the amount of heating provided by the young Sun. As we shall see in Sec. 11.1, the Sun at 4.0 Ga was about 25% less luminous than today. This should have tended to make the primitive stratosphere colder, and therefore dryer. If atmospheric CO_2 concentrations were also higher at that time, as seems likely, then the stratosphere should have been even colder because the additional CO_2 would have radiated energy efficiently to space (e.g., Kasting and Ackerman, 1986). As is true for most greenhouse gases, an increase in CO_2 concentrations warms the surface but cools the stratosphere. For all of these reasons, then, the atmospheric H_2 mixing ratio calculated in eq. (8.10) should represent a lower bound on the actual prebiotic H_2 mixing ratio, and any corresponding O_2 concentration should represent an upper bound on O_2. In the next chapter, we'll show how O_2 can be calculated in such an atmosphere using a photochemical model.

8.3 Rigorous Treatment of Atmospheric Redox Balance

The analysis above is sufficient to demonstrate that O_2 should not have been a major constituent of the prebiotic atmosphere and to determine the approximate concentration of atmospheric H_2. As mentioned above, credit for this insight should go to Walker (1977). The reason why H_2 dominated over O_2 is that the input of reductants to the atmosphere exceeded the input of oxidants. In performing this analysis, though, we neglected some processes that could conceivably have influenced the atmospheric redox state. We have already mentioned one of these, namely, outgassing of reduced gases other than H_2. We looked at CH_4 above, but we need to generalize this concept to include other reduced species. Rainout of photochemically produced oxidants and reductants could also have been important, provided that these compounds were somehow removed at the Earth's surface. Below, we outline a general formalism for treating all such atmospheric redox processes.

Consider the volcanic gas CO. If one puts CO and water vapor together in a flask at room temperature, nothing happens. But if one heats the flask up to >200 °C and provides a suitable metal-oxide catalyst, they react according the *water–gas shift reaction*

$$CO + H_2O \rightarrow CO_2 + H_2 \qquad (8.12)$$

In the atmosphere, this reaction does not occur directly. However, the same net reaction can be catalyzed by the photochemical reaction sequence

$$H_2O + hv \rightarrow H + OH \qquad (8.13)$$

$$CO + OH \rightarrow CO_2 + H \qquad (8.14)$$

$$\underline{H + H + M \rightarrow H_2 + M} \qquad (8.15)$$

$$Net: \quad CO + H_2O \rightarrow CO_2 + H_2 \qquad (8.16)$$

(The "M" in reaction (8.15) represents a third molecule necessary to carry off the excess energy of the collision.) This same net reaction can occur by other pathways as well. In terms of its overall effect, it doesn't matter what the exact kinetic pathway might be. If carbon enters the atmosphere as CO and then leaves the atmosphere as CO_2, then H_2 must be left behind.

Other reduced volcanic gases, H_2S for example, can also affect the atmospheric redox budget. Suppose that H_2S is oxidized by atmospheric photochemical reactions and that it leaves the atmosphere as sulfur dioxide, SO_2 dissolved in rainwater. The oxidant (the hydroxyl radical, OH) again derives from H_2O. So, we can write the net oxidation reaction as

$$H_2S + 2H_2O \rightarrow SO_2 + 3H_2 \qquad (8.17)$$

Again, H_2 is produced, but this time there are three molecules of H_2 for every molecule of H_2S that is oxidized in this manner.

Both of these reactions produced H_2 from other reduced volcanic gases. But photochemical processes can produce H_2, as well. One example is the production of hydrogen peroxide, H_2O_2, from photolysis of water vapor

$$H_2O + hv \rightarrow H + OH \quad (\times 2) \qquad (8.18)$$

$$OH + OH + M \rightarrow H_2O_2 + M \qquad (8.19)$$

Then add

$$\underline{H + H + M \rightarrow H_2 + M} \qquad (8.20)$$

$$Net: \quad 2H_2O \rightarrow H_2O_2 + H_2 \qquad (8.21)$$

Hydrogen peroxide is soluble, and so it is removed from the atmosphere by rainout and direct surface deposition. So, rainout of H_2O_2 constitutes a source for H_2 in the atmospheric hydrogen budget. If the H_2O_2 is consumed by reaction with ferrous iron or other reduced compounds, then it is also a source for H_2 in the global redox budget. Reduced compounds were probably abundant at Earth's surface prior to the rise of O_2; hence, even if volcanoes had ceased to operate for some time on the early Earth, atmospheric H_2 should have remained relatively abundant

and O_2 should have remained scarce. Photochemical simulations verify this prediction (e.g., Kasting et al., 1984b; Segura et al., 2007).

H_2 can also be consumed by photochemical reactions followed by rainout of soluble reduced species. For example, consider the production of formaldehyde, H_2CO, by the following mechanism (Pinto et al., 1980)

$$CO_2 + hv \rightarrow CO + O \qquad (8.22)$$

$$H_2O + hv \rightarrow H + OH \qquad (8.23)$$

$$H + CO + M \rightarrow HCO + M \quad (\times 2) \qquad (8.24)$$

$$HCO + HCO \rightarrow H_2CO + CO \qquad (8.25)$$

Then, add

$$O + OH \rightarrow O_2 + H \qquad (8.26)$$

$$H + O_2 + M \rightarrow HO_2 + M \qquad (8.27)$$

$$H + HO_2 \rightarrow 2OH \qquad (8.28)$$

$$\underline{H_2 + OH \rightarrow H_2O + H \, (\times 2)} \qquad (8.29)$$

$$Net: \quad CO_2 + 2H_2 \rightarrow H_2CO + H_2O \qquad (8.30)$$

We had to work hard to balance this cycle because the (ground state) O atoms produced from CO_2 photolysis (8.22) do not react directly with H_2 at low temperature; instead, they must be converted to O_2, then HO_2 and OH in order to do so. The high flux of solar UV photons in the absence of an ozone screen allows this to happen readily on the early Earth. In practice, one does not need to understand the details of the chemistry if one is doing a photochemical model calculation, because the model calculates these catalytic cycles for you.

The net effect of the above cycle (eq. (8.30)) is that each molecule of formaldehyde (H_2CO) that is produced consumes two molecules of H_2. Formaldehyde is soluble

Table 8.1 Redox effect of various species on the atmospheric hydrogen budget.*

Reductants	H_2 equivalents
CO	1
H_2S	3
CH_4	4
NH_3	1.5
H_2CO	2
S_8	16
Oxidants	H_2 equivalents
H_2O_2	-1
H_2SO_4	-1

* The reference oxidation state for sulfur in this table is SO_2.

and removed from the atmosphere by rainout and by direct surface deposition into the oceans. This constitutes a sink for H_2 in the atmospheric hydrogen budget.

Whether rainout of formaldehyde is a sink for H_2 in the *global* redox budget is less clear. H_2CO could have been photolyzed into H_2 and CO in the surface ocean and then re-emitted to the atmosphere. In that case, its net effect on global redox would be zero. However, it could also have polymerized via the *formose reaction* (when formaldehyde polymerizes in a hot alkaline solution) to form a tarry substance that was then buried in sediments (e.g., see Benner, 2009, p. 140). This, of course, would have drawn down atmospheric H_2 according to eq. (8.30). We will demonstrate in the next chapter that the amount of hydrogen used to produce formaldehyde was probably a small fraction of the amount lost to space, and so we will henceforth ignore it in the prebiotic global redox budget.

One might argue that, since H_2 itself is soluble, rainout of H_2 should also be counted as a sink for atmospheric hydrogen. This would be true if H_2 was being consumed within the ocean, for example by reducing CO_2 to organic matter. On an abiotic Earth, though, H_2 was probably not consumed in this manner, and thus the ocean should have achieved saturation with respect to atmospheric H_2. If so, then any rainout of H_2 should have been balanced by a return flux of H_2 from the ocean to the atmosphere, and the net loss of H_2 would be zero. Once life had evolved, H_2 should have been consumed by organisms living within the ocean, for example by methanogens, and so then one must keep track of fluxes of H_2 and other gases into and out of the ocean.

If we return now to the atmospheric redox budget, it is easy to see the general principle: If an outgassed species is oxidized during its passage through the atmosphere, then H_2 is generated; if an outgassed species is reduced during its time in the atmosphere, then H_2 is consumed. The amount of H_2 produced or consumed can be calculated from the stoichiometry of the relevant chemical reactions.

To apply the redox balance concept to an entire suite of different gases, one must define reference oxidation states for the various volatile elements. All researchers who have computed redox budgets for the Earth have done this implicitly because they require such reference points to do the calculation. The choice of reference oxidation states is rarely discussed, though, and this is a point of confusion. For certain elements we choose different reference oxidation states for reduced and oxidized atmospheres because it simplifies the budget calculation. Reference oxidation states are defined for convenience, not for any more fundamental reason.

For some elements, the choice of reference oxidation state is obvious. In the case of hydrogen, nearly all of it at Earth's surface is present in the form of H_2O. Because H_2O also enters and leaves the atmosphere freely through evaporation and precipitation, it would be hopeless to keep track of those fluxes in a photochemical model. If we define H_2O as the reference oxidation state for hydrogen, then this problem goes away. Hydrogen enters into the redox budget only if it enters or leaves the atmosphere either in reduced form, as H_2, or in oxidized form, as OH, HO_2, or H_2O_2.

Convenient reference oxidation states for carbon and nitrogen are also easy to determine. As shown earlier in the chapter, carbon is outgassed primarily as CO_2, and it leaves the atmosphere–ocean system primarily as carbonate. In both of these compounds, carbon has the identical oxidation state of +4. So, we can take CO_2 as the reference state for carbon. Nitrogen is present in the atmosphere almost exclusively as N_2, so N_2 is an obvious choice for its reference oxidation state.

Sulfur is the other major volatile element to consider. Here, the choice of reference state is not so obvious. Indeed, it is convenient to define the reference oxidation state differently for the modern oxidized atmosphere and for an early reduced atmosphere. Most of the sulfur outgassed from surface volcanoes today, and probably in the past as well, is in the form of sulfur dioxide, SO_2. In the present atmosphere, virtually all SO_2 is oxidized to H_2SO_4 (or sulfate ion) either in the atmosphere or in the surface ocean. Hence, from a modern atmospheric standpoint, SO_2 acts as a reduced volcanic gas, and it is treated as such in most analyses of the present global redox budget (e.g., Holland, 2002). We follow the same convention in Ch. 10 by defining H_2SO_4, or sulfate, as the reference oxidation state for sulfur for the modern atmosphere. Doing so has the added advantage that we do not need to keep track of the loss of sulfate from the oceans by precipitation of anhydrite ($CaSO_4$) within off-axis mid-ocean ridge vent systems. But if that sulfate is reduced to pyrite, FeS_2, which is deposited in sediments, then this process constitutes a loss of O_2 from the atmosphere-ocean system.

In a low-O_2 environment, the behavior of sulfur gases is quite different. Outgassed SO_2 can either be oxidized to H_2SO_4 or reduced to H_2S or S_8 within the atmosphere itself; hence, sulfur gases play a swing role in both the atmospheric redox budget and the global redox budget, as emphasized by Holland (2002, 2009). Thus, in photochemical models that include sulfur gases (Pavlov and Kasting, 2002; Pavlov et al., 2001; Zahnle et al., 2006), SO_2 is typically treated as the neutral sulfur species.

Rainout of H_2SO_4 is then analogous to rainout of H_2O_2, i.e., it represents removal of a photochemical oxidant and is thus a source of atmospheric H_2 by way of the net reaction

$$SO_2 + 2H_2O \rightarrow H_2SO_4 + H_2 \qquad (8.31)$$

Conversely, rainout of elemental sulfur, S_8, is a sink for atmospheric H_2 according to

$$8SO_2 + 16H_2 \rightarrow S_8 + 16H_2O \qquad (8.32)$$

Consider now a weakly reduced atmosphere that has various reduced volcanic gases being added to it. Label the outgassing flux of species i as $\Phi_{out}(i)$. For the sake of specificity, we will include H_2, CO, CH_4, and H_2S. Other gases could be added to this list if they were thought to be quantitatively important. (Remember that SO_2 is being treated as neutral.) Then, the total outgassing rate of reduced gases, in units of H_2 equivalents, is given by

$$\Phi_{out}(Red) = \Phi_{out}(H_2) + \Phi_{out}(CO) + 4\Phi_{out}(CH_4) + 3\Phi_{out}(H_2S) \qquad (8.33)$$

The stoichiometric coefficients in eq. (8.33) are obtained by writing reactions similar to (8.12) and (8.17) in which H_2O is used as the oxidant and H_2 is used as the reductant. Outgassing rates for the volcanic gases listed in eq. (8.33) are given in Table 8.2, along with their weighted contribution to the atmospheric hydrogen budget.

Table 8.2 Outgassing rates of reduced species and their effect on the atmospheric H_2 budget for a weakly reduced atmosphere.

	Flux*	Weighted flux**
Surface volcanism		
H_2	1.0	1.0
CO	0.1	0.1
H_2S	0.03	0.1
CH_4 (hydrothermal)	0.3	1.2
Subtotal		2.4
Submarine volcanism		
H_2	0.4	0.4
H_2S	0.35	1
CH_4 (axial)	0.01	0.04
CH_4 (off-axis)	0.03	0.12
Subtotal		1.5

* Fluxes in units of 10^{12} mol yr^{-1} (where 10^{12} mol yr^{-1} = 3.74×10^9 molecules cm^{-2}s^{-1}). All values are uncertain by at least a factor of 2.
** Fluxes in 10^{12} mol H_2 equivalents yr^{-1} when weighted by the stoichiometric coefficients from Table 8.1. The neutral sulfur species is taken as SO_2.

The early atmosphere was also losing oxidized and reduced species through rainout and surface deposition into the ocean, as discussed above. (Mathematical parameterizations for these processes are described in Appendix B.) Label the combined loss from these processes as $\Phi_{rain}(Ox)$ for photochemically produced oxidants and $\Phi_{rain}(Red)$ for photochemically produced reductants. Then, a rigorous expression of atmospheric redox balance, replacing eq. (8.9) above, is

$$\Phi_{out}(Red) + \Phi_{rain}(Ox) = \Phi_{esc}(H_2) + \Phi_{rain}(Red) \qquad (8.34)$$

The rainout terms (which include both rainout and surface deposition) consist of sums of the loss rates of reduced or oxidized species, again weighted by their appropriate stoichiometric coefficients. The atmospheric redox budget is illustrated schematically in Fig. 8.1(a).

As mentioned earlier, the atmospheric redox budget expressed in eq. (8.34) can be used to test the accuracy of one's photochemical model. If the transport equations are written in conservation form (see Appendix B), the model should automatically balance the atmospheric redox budget. Equation (8.34) can then be used to check the accuracy of the chemical scheme. If the individual chemical reactions are not balanced, or if a species is produced but not consumed, then eq. (8.34) will be unbalanced as well. Likewise, if some processes included within the model are not redox balanced (e.g., lightning production of nitrogen oxides), this will also appear as an imbalance. Hence, eq. (8.34) is a useful tool for checking the accuracy of photochemical calculations in which redox balance is important, as it is for virtually all calculations involving the early Earth.

8.4 Global Redox Budget of the Early Earth

Let's now return to the issue of the global redox budget of the early Earth. The global redox budget is defined for the combined atmosphere-ocean system. It is illustrated schematically in Fig. 8.1(b). Because the atmosphere and ocean are considered together in this budget, the rainout terms in eq. (8.34) are excluded. But we must now consider the burial of reduced and oxidized compounds in sediments, along with oxidation of the continents and seafloor. Today, both reduced carbon and reduced sulfur are important components of marine sediments. Sulfur cycling was probably slow on the anoxic early Earth, so the main reductant in sediments would likely have been organic carbon, CH_2O. We'll show in the next chapter how organic carbon could have been produced even in the absence of life. In terms of the global redox budget, burial of one mole of organic carbon is equivalent to loss of two moles of H_2:

$$CO_2 + 2H_2 \rightarrow CH_2O + H_2O \tag{8.35}$$

Iron also affects redox balance at the Earth's surface, so we will need to define a reference oxidation state for iron. Most of the iron in the mantle, and thus in fresh basalts, is ferrous iron, and this is how it should have entered the ocean back in the distant past. In a reduced environment, much of this iron exits the ocean in the same oxidation state, as ferrous silicates. We can avoid having to keep track of this process by defining the reference state for iron to be FeO, or Fe^{++}. Some of this ferrous iron would have been oxidized to ferric iron by various processes occurring at the Earth's surface, usually exiting the system as magnetite, Fe_3O_4. Much of this magnetite was part of banded iron formations, or BIFs, which will be discussed in Ch. 10. BIFs also include iron in other oxidation states, e.g., as hematite, Fe_2O_3, but we will ignore that complication here. Some ferric iron should also have been produced by *serpentinization* of seafloor and of continental rocks, as discussed in Sec. 7.5.2. Burial of one mole of magnetite produces one mole of H_2:

$$3FeO + H_2O \rightarrow Fe_3O_4 + H_2 \tag{8.36}$$

So, when budgeting redox on the early Earth, we will count deposition of ferric iron as a hydrogen source.

Sulfur outgassing could have affected the global redox budget, as well. From inspection of the last column in Table 8.2, it is clear that submarine H_2S is an important hydrogen source (oxygen sink). However, its net effect on global redox is smaller than suggested by Table 8.2 because much of the sulfur is removed as pyrite in marine sediments. We can see this by writing the reaction

$$2H_2S + FeO \rightarrow FeS_2 + H_2 + H_2O \tag{8.37}$$

Thus, the net effect of outgassing one mole of H_2S, even today, is to produce only 0.5 moles H_2 or, equivalently, to consume only 0.25 moles O_2. In the early oceans, virtually all of the vented H_2S may have reacted immediately with dissolved Fe^{+2} to form pyrite. This is consistent with deep-water, *volcanic massive sulfide*-related exhalative sedimentary rocks, which consist mainly of sulfides in the Archean (Slack and Cannon, 2009) (although many of these deposits formed at high temperatures well below the seafloor).

Outgassing of sulfur dioxide could also have had a significant effect on both the atmospheric and global redox state of the early Earth, even though we have classified it as neutral. The reason is that sulfur can leave the atmosphere in a variety of redox states, including sulfate, S_8, and H_2S. All three compounds would have been cycled biologically within the Archean biosphere (e.g., Halevy *et al.*, 2010). Removal of sulfur from the combined atmosphere–ocean system was probably mostly in the form of pyrite, as evidenced by the absence of marine sulfate evaporites during this time (Holland, 2006), although some minor amount of marine sulfate was removed in bedded barites prior to 3.2 Ga (Huston and Logan, 2004). If we assume the stoichiometry for complete removal of S in pyrite, then each mole of SO_2 outgassed would have consumed 2.5 moles of H_2

$$2SO_2 + 5H_2 + FeO \rightarrow FeS_2 + 5H_2O \tag{8.38}$$

which means that SO_2 acts as an oxidant. If SO_2 was outgassed at the modern rate, $\sim 1.8 \times 10^{12}$ mol yr^{-1} (Table 7.1), then the consumption of H_2 should have been 2.5 times this value, or 4.5×10^{12} mol yr^{-1}. But this is greater than the total modern production rate of H_2 listed in Table 8.2. So, to explain the absence of evaporite sulfates in the Archean, either the SO_2 outgassing rate must have been lower or the net H_2 outgassing rate must have been higher than estimated here, or both.

A lower SO_2 outgassing rate in the Archean is easy to justify. Following Oppenheimer *et al.* (2014) (and references therein), the average $\delta^{34}S$ value of SO_2 released by arc volcanoes is $\sim 5\permil$. Seawater sulfate has a $\delta^{34}S$ value of $+21\permil$, and mantle S has $\delta^{34}S = 0\permil$. From mass balance, we can therefore deduce that about $\sim 1/4$ of the SO_2 originating from arc volcanism today comes from recycled seawater sulfate. Thus, a low abundance of sulfate in the Archean oceans might indeed have resulted in proportionally lower volcanic release rates of sulfur gases. Indeed, evidence from multiple sulfur isotopes, to be discussed in Ch. 10, supports this hypothesis (Ono *et al.*, 2003).

On the modern Earth, some outgassed SO_2 is oxidized to sulfate and buried as gypsum. We can represent this schematically in the redox budget by using sulfuric acid as a surrogate for the mineral gypsum ($CaSO_4 \cdot 2H_2O$) because sulfur has the same oxidation state in both compounds

$$SO_2 + 2H_2O \rightarrow H_2SO_4 + H_2 \tag{8.39}$$

Burial of one mole of gypsum releases one mole of H_2, so SO_2 acts as a reductant in this case.

To make our global redox budget complete, we will include one other loss process for oxygen (or, equivalently, one other production process for hydrogen): oxidative weathering of the continents and seafloor, denoted by Φ_{OW}. Today, oxidative weathering of the continents is the major loss process for O_2 over geologic timescales; oxidative weathering of the seafloor, mainly by dissolved sulfate, is also a sink for O_2 (see Sec. 10.2.3). Both gaseous O_2 and sulfate would have been scarce on the early Earth, and so oxidative weathering would likely have been negligible at that time. But it would immediately have become important as soon as atmospheric O_2 concentrations rose.

If one puts all of these processes together, the global redox budget of the early Earth can be written in terms of H_2 equivalents as

$$\Phi_{out}(\text{Red}) + \Phi_{OW} + \Phi_{burial}(\text{CaSO}_4) + \Phi_{burial}(\text{Fe}_3\text{O}_4)$$
$$= \Phi_{esc}(\text{H}_2) + 2\,\Phi_{burial}(\text{CH}_2\text{O}) + 5\,\Phi_{burial}(\text{FeS}_2)$$
$$(8.40)$$

Here, the source terms for hydrogen (sinks for O_2) are on the left, and the sinks for hydrogen (sources for O_2) are on the right. The second and fourth terms on the left both represent oxidation of the crust, but we list them separately because oxidative weathering requires the presence of free O_2 and/or sulfate, whereas formation of magnetite does not. The precise form of this equation reflects our choice of H_2 as the unit of redox currency for the early Earth, as well as the choice of SO_2 and FeO as reference oxidation states for sulfur and iron. It also makes the stoichiometry of iron burial explicit by assuming that magnetite and pyrite were the predominate minerals that were formed.

In Ch. 10, we present an alternative form of the global redox budget equation for the modern Earth, eq. (10.6). This equation is time dependent, changes units from H_2 to O_2, and uses different S and Fe reference states (sulfate and ferric iron, respectively), as well as using different terminology for the various individual fluxes. These changes result partly from different preferences in the literature for how to most clearly express redox balance. But the underlying principle is the same in both cases: in steady state, the flux of available reductants into the atmosphere is equal to the flux out.

With this more complete global redox budget in hand, let us return to the question of how much hydrogen should have been present in the Archean atmosphere. It is conservative to use modern outgassing rates of reduced gases in a model of the Archean Earth. In reality, both the total input and the proportion of reduced gases were probably larger than they are today, in part because significant quantities of H_2 could also have been generated by BIF deposition and by serpentinization of ultramafic crust. The mixing ratio of atmospheric H_2 that could have been generated under such circumstances is difficult to estimate, but it could well have been considerably higher

than the 160–400 ppmv estimated in Sec. 8.2. The answer depends also on whether hydrogen would have continued to escape at the diffusion-limited rate, and this in turn depends on the balance between radiative cooling in the upper atmosphere and strong heating caused by absorption of EUV radiation from the active young Sun, as well as suprathermal loss processes, as discussed in Ch. 5. These are still active areas of research.

8.5 Organic Carbon Burial and the Carbon Isotope Record

Carbon isotopes provide another powerful tool for examining the global redox budget. They have been used extensively to estimate how atmospheric O_2 concentrations may have varied over the Phanerozoic Eon (Berner, 2004, 2006a; Berner and Canfield, 1989). They can also be used to analyze global redox balance during the Proterozoic and Archean Eons, although one must be careful in doing so for reasons discussed below. Carbon isotopes can thus be useful in discriminating between different theories for the rise of atmospheric O_2 to be discussed here and in Ch. 10.

Once life had originated on Earth, a new mechanism became available for removing hydrogen, or its redox equivalent, from the atmosphere–ocean system, namely, creation and burial of organic carbon. As mentioned earlier, some organic carbon may have been generated in the prebiotic world, but the amount would probably have been small. In the next chapter we will describe different processes by which primitive organisms might have produced organic matter. For this discussion, though, we focus on the process by which most organic matter has been produced throughout the latter half of Earth's history, namely, *oxygenic photosynthesis*, which is represented by eq. (8.1). Along with other authors, we presume that oxygenic photosynthesis has dominated biological productivity for at least the past 2.5 billion years, and this makes it somewhat easier to analyze the carbon isotope record during that time.

When organisms fix CO_2 to make organic carbon, they typically fractionate the carbon isotopes because ^{12}C, the most abundant stable isotope, is fixed more readily than is ^{13}C. (A third isotope, ^{14}C, is radioactively unstable with a half-life of 5730 years, and so cannot survive in rocks older than a few tens of thousands of years.) This allows us to use the carbon isotope record to keep track of organic carbon burial, and thus its influence on the global redox budget. The amount that carbon isotopes are fractionated depends on the particular mechanism by which CO_2 is reduced. Organisms that make their metabolic living through oxygenic photosynthesis fix carbon by way of the reductive pentose phosphate cycle, or Calvin cycle (Schidlowski et al., 1983). This process discriminates against ^{13}C by approximately 25‰–35‰ relative to the CO_2 from which it starts. (The symbol "‰" indicates parts per thousand, or "permil," short for "per mille," where mille is Latin for thousand.)

The relative change in the $^{13}C/^{12}C$ ratio is typically reported in delta notation, $\delta^{13}C$, defined as

$$\delta^{13}C \equiv \left[\frac{\left(^{13}C/^{12}C\right)_{sample} - \left(^{13}C/^{12}C\right)_{standard}}{\left(^{13}C/^{12}C\right)_{standard}} \right] \times 1000$$
(8.41)

The standard $^{13}C:^{12}C$ ratio (0.0112372) is from the Pee Dee belemnite – an extinct cephalopod fossil (superficially squid-like) made of marine carbonate from the Pee Dee Formation in South Carolina, and so the $\delta^{13}C$ is on the so-called "PDB" scale. Typical marine carbonates are close to PDB in $^{13}C:^{12}C$ ratio and so have $\delta^{13}C \approx 0$‰. The reductive acetyl-CoA pathway used by *methanogens* and *acetogens* (organisms that produce methane and acetate, respectively) also produces large fractionations, while other carbon fixation mechanisms tend to produce less (House et al., 2003b). However, the amount of fractionation also depends on the growth conditions, and so the interpretation of $\delta^{13}C$ values is not straightforward.

Today, CO_2 emitted from volcanoes has a $\delta^{13}C$ value of about −5‰ PDB, which is an input value, $\delta^{13}C_{in}$, for

that will be also be described in Chapters 10 and 11. Rivers carry nearly all of this CO_2 as well, so the CO_2 input from weathering of all types is called the *riverine flux*. The mean $\delta^{13}C$ value of riverine carbon is also about −5‰ (Holser et al., 1988), suggesting that the mean isotopic composition of the continents has adjusted itself over time to be equal to the mantle value (or, more correctly, so that ^{12}C and ^{13}C are released during weathering at their mantle ratios).

Now, consider mass balance. The atmosphere and oceans act as a single, combined system on time scales longer than about 1000 years (the turnover time for the deep ocean). Over timescales $>10^5$ years (the residence time of carbon in the system), the total amount of carbon entering the system from riverine plus volcanic input must be approximately balanced by removal of carbon by burial of carbonates and organic carbon in sediments. (Small imbalances in these fluxes result in net increases or decreases in total carbon in the system.) The same must be true separately for each isotope of carbon, ^{12}C and ^{13}C. Assume for the time being that the difference in carbon isotopic composition of carbonates and organic carbon, $\Delta_B = \delta^{13}C_{carb} - \delta^{13}C_{org}$, is 25‰. All carbon is buried either as carbonate or organic carbon, so the fractions of each type, f_{carb} and f_{org} must sum to unity

$$f_{carb} + f_{org} = 1$$
(8.42)

Considering mass balance, and using equation (8.42), it follows that

$$\delta^{13}C_{in} = f_{org}\delta^{13}C_{org} + f_{carb}\delta^{13}C_{carb} = f_{org}\delta^{13}C_{org} + \left(1 - f_{org}\right)\delta^{13}C_{carb} = \delta^{13}C_{carb} - f_{org}\Delta_B$$
(8.43)

the atmosphere–ocean system. The number −5‰ is the upper mantle value measured in xenoliths (inclusions in igneous rocks which have survived melting) (Mattey, 1987), while most mantle-derived diamonds have $\delta^{13}C$ of -5 ± 1‰ (Shirey et al., 2013). Modern marine carbonate sediments have $\delta^{13}C$ values close to zero. Organic carbon in sediments has values that are much more scattered, but that typically average around −25‰. The $\delta^{13}C_{in}$ value appears to have been constant over geologic time because peridotitic xenoliths have $\delta^{13}C$ of −5‰ independent of age (Pearson et al., 2003), as too do mantle-derived basalts and carbonatites (Mattey, 1987). This CO_2 is eventually consumed by weathering of silicate rocks on the continents, creating bicarbonate ions (HCO_3^-) that flow into the oceans via rivers (see Sec. 11.4.1.).

CO_2 also enters the atmosphere–ocean system from weathering of organic carbon and carbonates, processes

Rearranging, the fraction of outgassed CO_2 removed as organic carbon, f_{org}, is given by

$$f_{org} = \frac{\delta^{13}C_{carb} - \delta^{13}C_{in}}{\Delta_B} = \frac{\delta^{13}C_{carb} + 5}{25}$$
(8.44)

For the modern system, $\delta^{13}C_{carb} \cong 0$‰; hence, $f_{org} \cong 0.2$. This means that roughly 20% of riverine CO_2 is buried as organic carbon, while the other 80% is buried as carbonates. Alternatively, the $C_{org}:C_{carb}$ ratio in marine sediments is about 1:4. These ratios are reflected in balanced models of the long-term carbon cycle, e.g., Berner et al. (2004; 1983). According to these authors, the burial flux of calcite, $CaCO_3$, is 18.4×10^{12} mol yr^{-1} (which is similar to estimates by Holland (1978, p. 165) and estimates summarized by Francois and Walker (1992)), while that of organic carbon is $\sim 5 \times 10^{12}$ mol yr^{-1}, yielding $C_{org}:C_{carb}$ very close to 1:4. Other authors (e.g. Holland, 2002) estimate an organic carbon burial flux that is twice as high as Berner's value, which would require a carbon input flux higher than most current estimates (such as that by Burton *et al.* (2013)) to be consistent with the carbon isotope record,

Now, let's further assume, following Holland (1984, 2002), that the continents are in steady state with respect to carbon, that is, they are neither gaining nor losing carbonate or organic carbon or, if they are, they always vary in a 4:1 ratio. Then, to preserve isotopic mass balance in the atmosphere–ocean system, 20% of volcanically outgassed CO_2 would need to be buried as organic carbon. According to Table 7.1, the total CO_2 outgassing rate is $(8.5 \pm 2) \times 10^{12}$ mol yr^{-1}, so 1.7×10^{12} mol yr^{-1} of this CO_2 would need to be reduced to organic carbon. Because the continents have now been left out of the equation, the required reducing power can only come from volcanic gases. If we express this power in terms of H_2 equivalents, the stoichiometry is given by reaction (8.35), which shows that 2 moles of H_2 are required to reduce 1 mole of CO_2 to organic carbon. Is there enough hydrogen available to do this? According to Table 8.2, the total H_2 outgassing rate, including surface and submarine processes, is 3.9×10^{12} mol yr^{-1}. Equation (8.35) says that the amount of organic carbon generated should be half this value, or $\sim 2 \times 10^{12}$ mol yr^{-1}. The estimated value of f_{org} is thus $\sim$(the CH_2O flux)/(the volcanic CO_2 flux) = 2×10^{12} mol yr^{-1}/(8.5×10^{12} mol yr^{-1}) $\cong 0.24$. This is a little higher than the value of 0.2 predicted from our crude isotopic mass balance, but equal to the value of 0.23–0.24 derived from more detailed analyses (Hayes and Waldbauer, 2006; Krissansen-Totton *et al.*, 2015). Thus, the volcanic outgassing and organic carbon burial rates assumed here are consistent with constraints from carbon isotopes.

In principle, this type of analysis can provide constraints on past global redox balance and, by implication, on past concentrations of atmospheric O_2. Isotope geochemists have measured $\delta^{13}C$ values in thousands of different carbonate rocks of all ages. A recent compilation is shown in Fig. 8.2. The figure also shows $\delta^{13}C$ values of organic matter from kerogenous sediments; *kerogen* is dispersed, highly degraded organic matter in rocks.

Focusing on the carbonates, it can be seen that the $\delta^{13}C$ values exhibit a wide spread, with large positive excursions at around 2–2.4 Ga and 0.6–0.8 Ga. The earlier of these excursions, called the *Lomagundi Event* (see Sec. 10.4.1) has $\delta^{13}C$ values of over +10‰. Plugging +10‰ into equation (8.44) yields an organic carbon burial fraction, $f_{org} \cong 0.6$, which is three times higher than today. A simple interpretation would be that this represents an increased burial flux of organic carbon, which may have caused the rise of atmospheric O_2. When one looks at this in more detail, however, the change in burial rate appears to be as much a consequence of the O_2 rise as a cause; in other words, they are part of a coupled system. But we'll save the rest of this story for Ch. 10.

For now, let us simply note that the average $\delta^{13}C$ value of carbonates has remained roughly constant at a value of $\sim$0‰. Organic carbon exhibits a much wider range of isotopic values, but on average its $\delta^{13}C$ value is about -30‰. (There is a dramatic negative excursion at 2.8–2.6 Ga with $\delta^{13}C$ values near -60‰, but we shall save that story for Sec. 10.4.1.) Taken at face value, this implies that the fraction of outgassed CO_2 buried as organic carbon has remained close to 20‰ throughout geologic time. Different authors have interpreted this observation differently. Schidlowski *et al.* (1983) called it "a continuous isotopic signal of biological (viz., autotrophic) activity

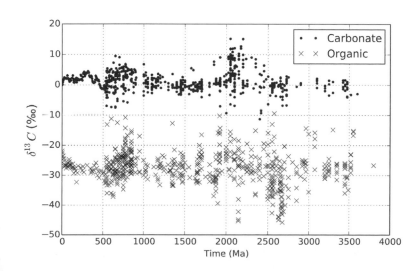

Figure 8.2 The carbon isotopic record of marine carbonates (dots) and organic carbon (crosses) where each point represents the average per formation per author. Samples that are lacustrine, heavily metamorphosed, or from banded iron formations have been excluded. From Krissansen-Totton *et al.* (2015); see references therein for data sources. (Reproduced with permission. Copyright 2015, American Journal of Science.)

beginning at least 2.5 (if not 3.8) Ga ago." That statement is probably secure. But the data are also consistent with stronger interpretations. For example, some authors (Holland, 1984, 2002, 2009; Kump and Barley, 2007; Kump *et al.*, 2001) have taken this as evidence that the relative burial rates of organic and carbonate carbon have remained fixed at a 1:4 ratio most of the time, apart from brief excursions. Still others (Nisbet and Sleep, 2001; Schidlowski, 1988) have gone even further and suggested that oxygenic photosynthesis has been operating since 3.5 Ga, or even 3.8 Ga, as that could explain why the organic carbon burial rate and the average isotope fractionation, Δ_B, both appear to have remained near modern values since that time. This is a much stronger claim that would be exceedingly important if it were true. But geochemists do not agree about these stronger interpretations.

When one analyzes the carbon isotope record in more detail, it appears that f_{org} has not remained constant throughout geologic time. This is true even with a conventional carbon cycle model where carbon input to the atmosphere–ocean system is balanced by carbon burial in sedimentary organic carbon and carbonates. For example, Hayes and Waldbauer (2006) suggest that f_{org} was 0.14–0.16 throughout much of Earth's history, then rose unsteadily to ~0.24 during the past 900 million years. A more sophisticated statistical analysis indicates that f_{org} had a mean value of 0.15 ± 0.02 in the Archean (3.8–2.5 Ga) compared to a mean of 0.23 ± 0.02 during the Phanerozoic (since 0.54 Ga) (Krissansen-Totton *et al.*, 2015) (Fig. 8.3). Canfield (2005) also calculated lower f_{org} values ~0.1 in the Archean.

(a)

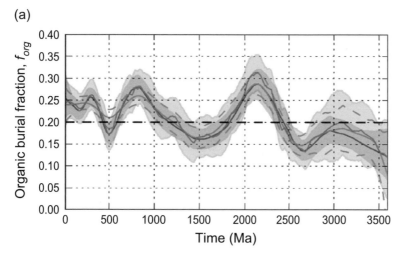

(b)

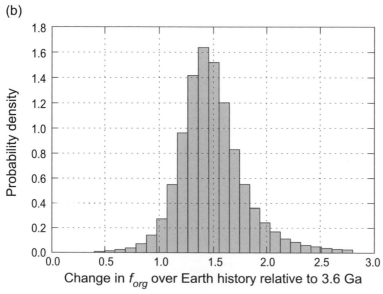

Figure 8.3 (a) Time series of the organic carbon burial fraction, f_{org}, based on the isotopic data shown in Fig. 8.2. The curves show different statistical analyses. *Blue*: locally weighted scatterplot smoothing (LOWESS); *red*: kernel regression; *green*: Kalman smoothing. Shading or dashed lines of corresponding color show 95% confidence intervals. All statistical techniques indicate the same general pattern. (b) The probability distribution of the change in organic carbon burial fraction, f_{org}, from 3.6 Ga to present based on a generalized least squares regression bootstrap analysis. The most probable change is a factor of ~1.5 increase and the 95% confidence range is an increase of ~1.2 to ~2.0. However, the increase in f_{org} relative to 3.6 Ga is mainly in the past ~400 Ma with no statistical difference between average f_{org} in the Archean (3.8–2.5 Ga) versus "boring billion" (1.8–0.8 Ga). From Krissansen-Totton *et al.* (2015). (A black and white version of this figure will appear in some formats. For the color version, please refer to the plate section.)

Analysis also shows that the difference in carbon isotopic composition of carbonates and organic carbon, Δ_B, did not stay constant (Des Marais *et al.*, 1992; Hayes and Waldbauer, 2006; Krissansen-Totton *et al.*, 2015). The mean value of Δ_B was $31 \pm 3‰$ in the late Archean (2.8–2.5 Ga), which can be compared with $26 \pm 3‰$ in the early Archean (3.8–2.8 Ga), and $29 \pm 1‰$ in the Phanerozoic(Krissansen-Totton *et al.*, 2015). The increase in the late Archean suggests redox change prior to the Great Oxidation Event. Newly available oxidants, such as sulfate, would allow methane oxidation by methanotrophic microbes to contribute isotopically light carbon to sedimentary organics because methanotrophy causes strong concentration of ^{12}C (Hayes, 1994; Schoell and Wellmer, 1981).

In any case, if the organic carbon/carbonate carbon burial ratio, f_{org}, has remained roughly near 1:4, this is unlikely to have been by accident. Even if f_{org} has varied, some process must have kept it within bounds. At least two different mechanisms for regulating this ratio have been suggested. The first mechanism involves phosphorus limitation (Junge *et al.*, 1975; Schidlowski, 1988). Living marine organic matter has an average C:P ratio of ~106:1. This is part of the *Redfield ratio* of C, N, and P elements within organic material in the deep ocean (Redfield, 1958). Empirically, C:N:P $\cong$ 106:16:1, according to Redfield's analysis. The P:C ratio in rocks undergoing weathering might be just high enough to create organic matter out of one-fifth of the available carbon. This hypothesis suffers from the fact that terrestrial organic matter has C:P ratios of 250–1000, and some of it gets buried in sediments (Kump *et al.*, 2010). Also, the C:P ratio in sediments depends on the dissolved O_2 content of the overlying seawater because P is remobilized (redissolved) much more efficiently if the O_2 content is low (Van Cappellen and Ingall, 1996). So, one might expect to see a secular decrease in organic carbon burial (i.e., a decrease in the $\delta^{13}C$ values of carbonates) around 2.4 Ga when the surface ocean became oxygenated (Ch.10). But no such change is observed. Indeed, as pointed out above, if there is a change it goes the other way. But the issue is actually more complex, as the reason that P is remobilized in modern anoxic waters is because they are typically rich in sulfide. The sulfide ties up binding sites on ferric oxyhydroxides in sediments, making them unable to bind P (Bjerrum and Canfield, 2002). Hence, these latter authors argued that P was actually removed very efficiently in the sulfide-free Archean oceans. But this hypothesis itself was challenged, because the high abundance of dissolved silica in Archean ocean water may have tied up those same sites, again leading to efficient

P recycling (Konhauser *et al.*, 2007b). So, the hypothesis that P availability controls the organic carbon burial fraction remains murky, at best.

The second idea relies on the Earth system being nearly closed and time invariant with respect to redox changes (Holland, 1973a, 1984). If the available reducing power from volcanic gases had remained constant during the Earth's history (or, more precisely, if it scaled linearly with outgassed CO_2), and if all of it was used to reduce CO_2 to CH_2O, then f_{org} should have remained constant with time. But if hydrogen escape to space was an important term in the global redox budget during the Archean, as postulated in Sec. 8.2, then less hydrogen should have been available to reduce CO_2, and the CH_2O/carbonate burial ratio should have been lower than today – just the opposite of what phosphorus limitation would predict, and in agreement with the lower Archean f_{org} calculated by Canfield (2005), Hayes and Waldbauer (2006), and Krissansen-Totton *et al.* (2015). So, if these authors are correct about f_{org}, then perhaps global redox balance is indeed responsible for controlling the $\delta^{13}C$ record. Furthermore, because average f_{org} remains low until the Phanerozoic (Hayes and Waldbauer, 2006; Krissansen-Totton *et al.*, 2015), this may also suggest that hydrogen escape continued to be an important term in the global redox budget during the Proterozoic, even after O_2 levels had risen (Catling *et al.*, 2002). We'll return to this thought in Ch. 11 when we discuss greenhouse gases and climate.

8.6 Redox Indicators for Changes in Atmospheric Oxidation State

As we will discuss further in Ch. 10, one of the major unresolved questions about atmospheric evolution is what controlled the timing of the rise of atmospheric O_2. A whole host of different hypotheses have been advanced, none of which seems entirely convincing. In preparation for this discussion, we will use this section to define two different "redox state parameters" that have been used to frame this debate.

8.6.1 Holland's *f*-Value Analysis

Depending on how strongly one believes in the constancy of the carbon isotope record and its interpretation, one can decide whether or not to use it as a constraint on global redox evolution. Holland (2002, 2009) included a 4:1 inorganic/organic carbon burial ratio explicitly in his model for atmospheric O_2 increase. We will describe this model in some detail because it provides a good illustration of how redox balance works, and it may also turn out to be broadly correct if the organic carbon burial fraction is fixed by non-redox-related factors. Holland incorporates the fixed burial ratio into his definition of the quantity "f" (not to be confused with f_{org}), which represents the fraction of outgassed sulfur that is removed from the atmosphere–ocean system as pyrite, FeS_2. The derivation of f, which involves stoichiometry similar to that discussed in this chapter, requires some effort and is given by Holland (2002). Following Kasting *et al.* (2012), we have modified f to include S_2 as one of the released volcanic gases. The result is

$$f = \frac{m(H_2) + 0.6\, m(CO) - 0.4\, m(CO_2) + 3\, m(H_2S) + 4\, m(S_2)}{3.5[m(SO_2) + m(H_2S) + 2\, m(S_2)]} + \frac{1}{3.5} \tag{8.45}$$

Here, $m(i)$ represents the initial molar concentration (or volume mixing ratio) of species "i" within a volcanic gas mixture. The value of f is defined such that volcanic gases fall into four different groups.

Group 1: ($f > 1$) Contains sufficient hydrogen to reduce 20% of outgassed CO_2 to organic matter and all of the outgassed sulfur to pyrite and still have some leftover H_2. Creates a reducing atmosphere.

Group 2: ($f = 1$) Contains exactly the right amount of H_2 to join Group 1, but with no leftover hydrogen.

Group 3: ($0 < f < 1$) Contains enough H_2 to reduce 20% of CO_2 to organic matter, but not enough to reduce all sulfur to pyrite.

Group 4: ($f < 0$) Contains insufficient H_2 to reduce 20% of CO_2 to organic matter.

According to this analysis, an average f value >1 (Group 1) is needed in order to maintain a reducing atmosphere. If the average f value of volcanic gases is <1, then the atmosphere should become oxidizing.

Figure 8.4 (from Holland, 2002) shows tabulated f values for gases collected from different volcanoes. Most of the values fall between 0 and 1, putting them into Group 3 of Holland's classification. As just explained, this is consistent with the fact that today's atmosphere is highly oxidized. Holland then calculates that a decrease in mantle oxygen fugacity by 0.76 $\log_{10}$ units back in the Archean would have been sufficient to make $f > 1$, and thus to keep the atmosphere in a reduced state. This is less than the 2 $\log_{10}$ unit decrease in mantle f_{O_2} postulated by Kump *et al.* (2001) for the Archean, but it may still be too much of a change to be consistent with oxybarometers based on Cr and V (see Sec. 7.3.2).

Other factors besides mantle redox state can cause changes in Holland's f factor, as well. The form of eq. (8.45) suggests that it should be sensitive to the ratios of total carbon-bearing (ΣC) and sulfur-bearing (ΣS) species in volcanic gases to that of H_2O. This dependence is shown explicitly in Fig. 8.5(a). Low $\Sigma C/H_2O$ ratios promote high f values because relatively less CO_2 must be reduced for a given H_2 input. (The H_2/H_2O ratio in Holland's model is fixed at the QFM oxygen fugacity buffer, as described in Sec. 7.4.) Low $\Sigma S/H_2O$ ratios also promote high f values, provided that the $\Sigma C/H_2O$ ratio is below the critical threshold where 20% of the CO_2 can be reduced by H_2.

Holland points out that lower dissolved sulfate concentrations in the Archean oceans should have led to low $\Sigma S/H_2O$ ratios and correspondingly high f values in recycled, arc volcanic gases, thereby promoting a reduced atmosphere. Different types of volcanoes also have different average $\Sigma C/H_2O$ and $\Sigma S/H_2O$ ratios, as shown in Fig. 8.5(b). In general, convergent plate volcanoes (arc volcanoes) have lower ratios than do divergent plate volcanoes (mid-ocean ridges) and hotspot volcanoes. So, a switch in the dominant mode of volcanism from mid-ocean ridges and hotspots to arc volcanoes could also

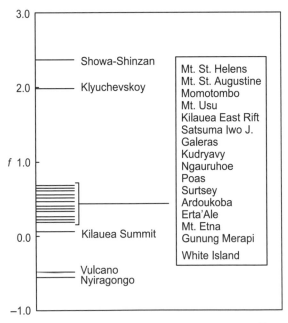

Figure 8.4 Measured "f" values for modern volcanic gases. (From Holland (2002). Reproduced with permission from Elsevier. Copyright 2002.)

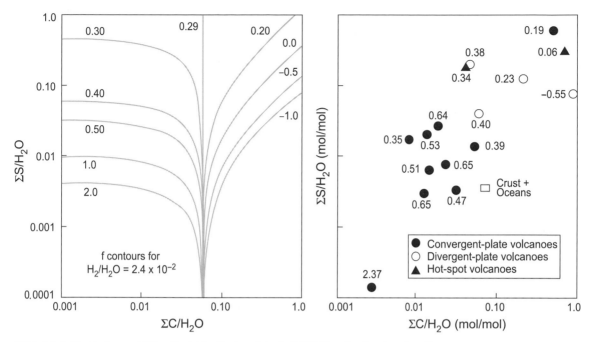

Figure 8.5 (a) Dependence of "f" on the total sulfur-to-water ratio ($\Sigma S/H_2O$) and carbon-to-water ratio ($\Sigma C/H_2O$) in volcanic gases. (b) The $\Sigma S/H_2O$ and $\Sigma C/H_2O$ ratios for different modern volcanoes. The numbers next to the symbols are the "f" value. (From Holland (2002). Reproduced with permission from Elsevier. Copyright 2002.)

have caused an increase in global redox state, according to this model.

Holland (2009) developed a numerical model based on this f-value analysis with an explicit dependence on mode of outgassing. Mantle degassing of H_2O and CO_2 decreased as $e^{-\lambda t}$ in his model, where λ is an inverse time constant representing the cooling rate of the interior. The rate of H_2 supply from the mid-ocean ridges and hotspots declined with it. Outgassing of CO_2 and SO_2 from arc volcanism first increased as the sedimentary reservoirs of carbon and sulfur grew, and then decreased as the Earth cooled off. Key to the success of this model is the assumption that the H_2O content of gases produced by arc volcanism remained roughly constant over time because the rate of hydration of seafloor did not change. Hence, the $\Sigma C/H_2O$ and $\Sigma S/H_2O$ ratios increased with time, leading to lower f values, and eventually to a switch in atmospheric oxidation state (i.e., the rise of O_2) when average f values fell below unity. The timing of this switch depends on the assumed value of λ and on other model parameters (see Fig. 8.6). With the right choice of parameters, the increase in atmospheric O_2 happens right near 2.4 Ga, in agreement with observations.

Holland's model for the rise of atmospheric O_2 is one that a geochemist should love, as it is based entirely on

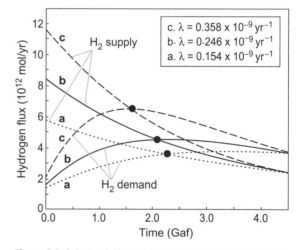

Figure 8.6 Calculated H_2 supply and demand rates for three different values of the cooling time constant, λ, from the model of Holland (2009). (Reproduced with permission from Elsevier. Copyright 2009.) The points at which the supply and demand curves intersect mark the rise of atmospheric O_2. Time is in billions of years starting from the Earth's formation.

global redox balance. The model must be at least partly correct, as the requirement of global redox balance is a fact, not a hypothesis. Importantly, the model predicts that global redox balance depends on both the oxidation state of volcanic gases and on their elemental composition. Although the model contains numerous assumptions and parameters, some of these are not in dispute. For example, many different authors have emphasized the idea that sulfur was less mobile, and therefore less available, in the Archean (e.g., Canfield *et al.*, 2006; Ono *et al.*, 2003; Walker and Brimblecombe, 1985). Let us not forget, though, that the assumption of a 4:1 carbonate:organic carbon ratio in sediments is hardwired into Holland's model – so much so that biology is never mentioned in either of Holland's papers, even though it is universally agreed that O_2 levels would never have risen without the influence of cyanobacteria. Thus, we clearly need to understand what controls the carbonate:organic carbon burial ratio if we hope to have a convincing theory for why atmospheric O_2 levels rose.

8.6.2 The Catling and Claire K_{OXY} Parameter

Catling and Claire (2005) and Claire *et al.* (2006) have formulated a different type of redox state parameter that is better suited for use with the flux-based analysis presented earlier. Their parameter, which they call K_{OXY}, represents the ratio of oxygen sources to oxygen sinks, *not including oxidative weathering or hydrogen escape*. The idea is that oxidative weathering only occurs when O_2 is in excess, whereas significant amounts of hydrogen escape only occur when H_2 is in excess. The notation used by Claire *et al.* was different from that used here. (See Kasting (2013) for a comparison and Sec. 10.7.1, eq. (10.17) for the Claire *et al.* formulation.) Here, we define a similar K_{OXY} parameter based on global redox balance, eq. (8.40). The result is

$$K_{OXY} = \frac{2\Phi_{burial}(CH_2O) + 5\Phi_{burial}(FeS_2)}{\Phi_{out}(Red) + \Phi_{burial}(Fe_3O_4)} \qquad (8.46)$$

The terms in the numerator represent oxygen sources (hydrogen sinks), whereas the terms in the denominator represent oxygen sinks (hydrogen sources). Note that the Φ_{OW} and $\Phi_{esc}(H_2)$ terms from eq. (8.40) do not appear

here, for the reason just given. Burial of gypsum, $\Phi_{burial}(CaSO_4)$, also does not appear because, like oxidative weathering, it requires an excess of oxygen. Equation (8.46) also differs from Claire *et al.* formulation by the addition of the anaerobic iron-oxidation term, $\Phi_{burial}(Fe_3O_4)$, in the denominator. Recall that this term represents processes, such as serpentinization or BIF deposition, that oxidize iron without the need for oxidants such as O_2 or sulfate.

The form of eq. (8.46) illustrates the advantage of choosing SO_2, rather than sulfate, as the reference redox state for sulfur. If one takes sulfate as the reference state, then SO_2 is included in the flux of reduced gases. This is potentially misleading, because SO_2 is an *oxidant* in the reduced Archean environment, whereas it is sometimes a *reductant* in today's environment (when removed as gypsum).

The point of defining K_{OXY} is that it can be used to analyze different mechanisms for causing atmospheric O_2 to rise (see Kasting (2013), for a detailed discussion). Values of $K_{OXY} < 1$ lead to a reducing atmosphere; values of $K_{OXY} > 1$ lead to an oxidizing atmosphere. According to Claire *et al.* (2006), $K_{OXY} \approx 6$ today, while in Sec. 10.7.1, we calculate a revised value of $K_{OXY} \approx 3$ using a more comprehensive budget of O_2 sinks. In any case, the atmosphere is shifted well over to the oxidized side. Thus, one needs to either boost the O_2 source or reduce the O_2 sinks by a considerable amount in order to switch over to a reduced atmosphere back in the Archean. We will use this parameter to perform our own analysis of the rise of O_2 when we get to Ch. 10.

In summary, both the atmospheric and global redox budgets must be considered when trying to model the composition of the early atmosphere or when trying to understand the reasons why that composition changes over time. Formulating those redox budgets quantitatively is relatively easy; evaluating the terms in them, particularly in the global redox budget, is much tougher. In the next two chapters, we'll show how these budgets have been used to try to simulate atmospheric evolution.

9 The Prebiotic and Early Postbiotic Atmosphere

In the last chapter we estimated the H_2 mixing ratio in a weakly reduced atmosphere on the early Earth by balancing the total outgassing rate of H_2 against loss of H_2 to space. The answer that we obtained was about 4×10^{-4}, or 400 ppm, although it could have been significantly higher if H_2 outgassing rates were greater than today or if hydrogen escape to space was inefficient. Here, we use this number to estimate the concentrations of O_2 and other species. To do so, we will make use of one-dimensional (1-D) photochemical model results.

As discussed in Appendix B, a 1-D photochemical model is a numerical computer code in which one assumes a plane-parallel atmosphere, puts the Sun at some angle to the vertical, and then calculates the photodissociation rates of different gases, along with the chemical interactions between the various trace species that are produced. Vertical transport of long-lived gases is parameterized using the concept of eddy and molecular diffusion. Production of trace species by lightning and removal of soluble gases by rainout and surface deposition are also included in such models.

9.1 N$_2$ and CO$_2$ Concentrations in the Primitive Atmosphere

To perform photochemical simulations, we need to estimate the surface pressure of the primitive atmosphere and the concentrations of the major atmospheric gases, N_2 and CO_2. N_2, of course, is the major component of Earth's present atmosphere. Its mixing ratio today is about 0.78 by volume, so its partial pressure is just under 0.8 bar. As we'll discuss further below, N_2 is relatively unreactive, so its geologic cycle is slow (see Sec. 7.2.3). Roughly one-fourth of Earth's nitrogen, 4×10^{18} kg N, resides in the atmosphere as N_2. About half that amount is present in continental rocks and marine sediments, and more than twice that amount is thought to be present in the mantle

(as shown in an inventory in Table 11.1) (Goldblatt *et al.*, 2009; Johnson and Goldblatt, 2015). Whether nitrogen was distributed in this same manner on the early Earth is unsettled. Goldblatt *et al.* argue on theoretical grounds that more N_2 should have been partitioned into the atmosphere during the Archean, in which case the surface pressure could have been higher by a factor of 2 or more. However, data show that the N_2 partial pressure, pN_2, was possibly lower than at present in the late Archean. The data include $N_2/^{36}Ar$ ratios in ancient fluid inclusions (Marty *et al.*, 2013), raindrop imprints (Som *et al.*, 2012), and bubbles in lava flows (Som *et al.*, 2016). We discuss these data and their implications for climate in Sec. 11.8.

The effect on atmospheric photochemistry of different pN_2 levels should be small. Rayleigh scattering caused by this additional N_2 might reduce the fluxes of short wavelength UV radiation in the lower atmosphere; also, three-body chemical reaction rates would be speeded with higher pN_2. These effects, however, are probably less important than other uncertainties in the model, such as the volcanic outgassing rates discussed in the previous two chapters. Consequently, the calculations here will assume that the N_2 partial pressure has remained approximately constant throughout Earth history.

The other major gas in a weakly reduced primitive atmosphere would likely have been CO_2. The abundance of CO_2 is determined largely by the carbonate-silicate cycle and, to a lesser extent, by the organic carbon cycle. These cycles will be described in some detail in Secs. 10.2.2 and 11.4.1, respectively. A key point is that CO_2 is a greenhouse gas and that higher concentrations of greenhouse gases were needed to compensate for lower solar luminosity early in Earth history. Higher CO_2 concentrations were probably part of this story (Kasting, 1987; Walker *et al.*, 1981), although we should acknowledge at the outset that there is an ongoing debate as to how high they might have been (e.g., Rosing *et al.*, 2010). Most of the models that we discuss below assume relatively high CO_2 concentrations, which we will express in terms of PAL (Present Atmospheric Level) of CO_2. For the purpose of our discussion we'll define 1 PAL as 3×10^{-4} bar, or 300 ppmv. This is closer to the preindustrial CO_2 concentration, 280 ppmv, than the value in 2015 of ~400 ppmv. We'll also show some calculations with much lower CO_2 levels, though, as there may be other ways to solve the faint young Sun problem (e.g., other greenhouse gases, such as methane, or changes in Earth's albedo).

To provide context for the following discussion and for Chs. 10 and 11, Fig. 9.1 gives an overview of Earth history. It highlights some key events in the coupled evolution of the atmosphere and biosphere, including the first signs of life in the Archean Eon, major glacial episodes, and the flip in redox state of the atmosphere around 2.4 Ga.

9.2 Prebiotic O_2 Concentrations

The most important trace atmospheric constituent in a weakly reduced prebiotic atmosphere is O_2. Even in a reduced early atmosphere, O_2 would have been produced in small quantities by several different processes, as described below. O_2 is important for several reasons. First, it acts as a poison to prebiotic chemistry; hence,

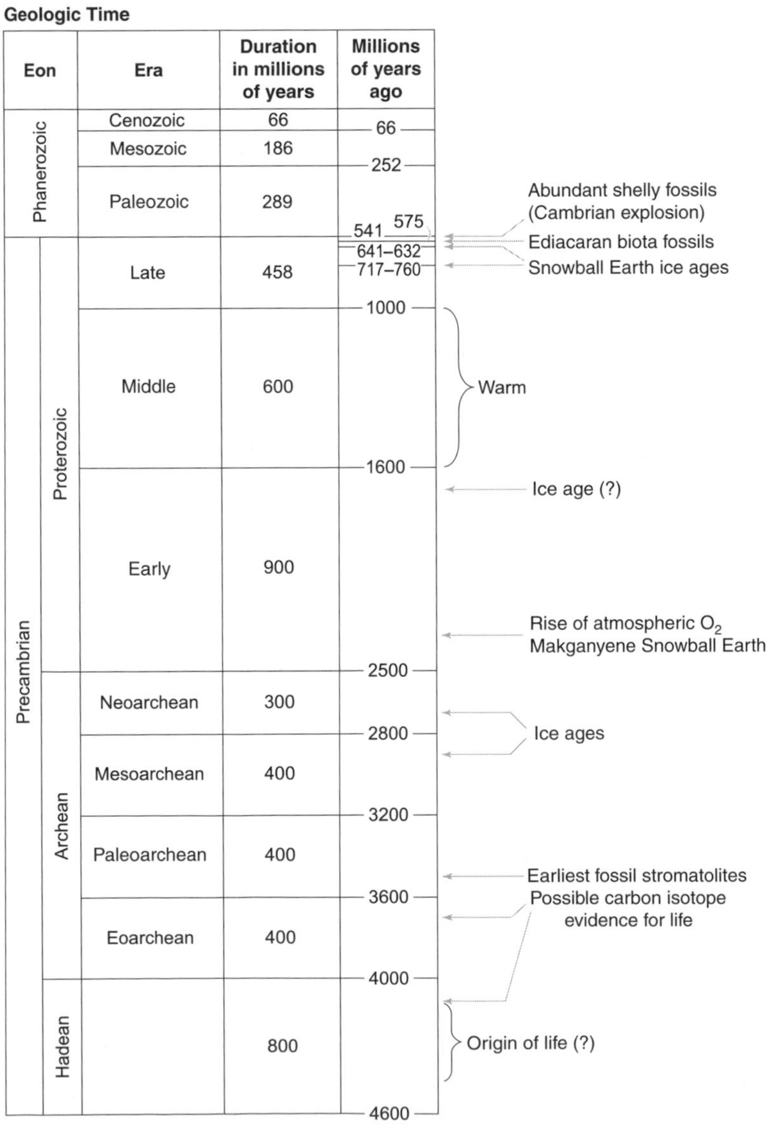

Geologic Time

Eon	Era	Duration in millions of years	Millions of years ago
Phanerozoic	Cenozoic	66	66
Phanerozoic	Mesozoic	186	252
Phanerozoic	Paleozoic	289	541 575
Proterozoic	Late	458	641–632 717–760 1000
Proterozoic	Middle	600	1600
Proterozoic	Early	900	2500
Archean	Neoarchean	300	2800
Archean	Mesoarchean	400	3200
Archean	Paleoarchean	400	3600
Archean	Eoarchean	400	4000
Hadean		800	4600

Abundant shelly fossils (Cambrian explosion)
Ediacaran biota fossils
Snowball Earth ice ages
Warm
Ice age (?)
Rise of atmospheric O_2
Makganyene Snowball Earth
Ice ages
Earliest fossil stromatolites
Possible carbon isotope evidence for life
Origin of life (?)

Figure 9.1 Geologic time scale showing major evolutionary events and Precambrian ice ages.

nearly all endogenous theories of life's origin require that O_2 was scarce on the early Earth. Second, because it is produced by photosynthesis, O_2 is a possible biosignature molecule on exoplanets, as we'll discuss in Ch. 15. But, to use it as such, one first needs to understand how much O_2 can be produced abiotically.

Many authors have addressed this problem. The earliest paper by Harteck and Jensen (1948) calculated the amount of oxygen produced from the photodissociation of water and associated hydrogen escape. They estimated that the modern amount of atmospheric oxygen could have been regenerated 50 times. However, they overestimated the effect because they incorrectly assumed that the concentration of stratospheric water vapor was set by the *highest* temperature of the tropopause rather than the lowest temperature of the tropical tropopause, which Gordon Dobson soon discovered (Dobson, 1956). Also, as Bernal (1951, p. 69) pointed out, the abiotic level of O_2 would be very low because of potentially large sinks on oxygen.

Later, Berkner and Marshall (1965; 1964, 1966, 1967) also assumed, correctly, that O_2 would have been produced in the early atmosphere by photolysis of H_2O, followed by escape of hydrogen to space. They then assumed, incorrectly, that O_2 should have accumulated until it shielded tropospheric H_2O from photolysis. This allowed them to estimate that prebiotic O_2 concentrations should have been of the order of 10^{-4} to 10^{-3} PAL. (1 PAL of O_2 is equivalent to 21% by volume within a 1-bar atmosphere.) We have already seen in the previous chapter, however, that the H_2 content of the early atmosphere would have been determined by the global redox budget. The H_2 content of the atmosphere determines the O_2 content through photochemistry, as discussed further below; hence, Berkner and Marshall's original hypothesis cannot be correct.

The next author was Brinkmann (1969), who also assumed that O_2 would have been generated from H_2O photolysis. However, he assumed, incorrectly, that precisely one-tenth of the H atoms produced by photolysis would have escaped to space. This caused his model to predict inordinately large prebiotic O_2 levels, of the order of 0.27 PAL. We know, in reality, that the hydrogen escape rate cannot exceed the diffusion limit (Ch. 5); hence, Brinkmann's result must be wrong, as well. Below, we discuss how prebiotic O_2 concentrations are calculated in more modern atmospheric models.

9.2.1 Dependence of O₂ on CO₂

The concentration of O_2 in the prebiotic atmosphere can be calculated theoretically just as the O_3 concentration is calculated in models of the modern atmosphere. The results depend on the abundances of other atmospheric gases, especially CO_2 and H_2. As mentioned in Ch. 8, CO_2 is readily dissociated by photons with wavelength <204 nm and provides a source of atomic oxygen via the reaction

$$CO_2 + hv \rightarrow CO + O \tag{9.1}$$

Similarly, H_2O can be photolyzed at ultraviolet wavelengths <240 nm

$$H_2O + hv \, (\lambda < 240 \text{ nm}) \rightarrow H + OH \tag{9.2}$$

In some literature, the photolysis cross-section of H_2O is considered negligible at UV wavelengths longer than 200 nm (Brasseur and Solomon, 2005, p. 311), but actually the optical depth can be significant because H_2O is relatively abundant. Certainly, the energy threshold for breaking the H–OH bond corresponds to a wavelength of 240 nm. In our models, e.g., Kasting and Walker (1981) and all subsequent Kasting group papers, the H_2O photolysis cross-section extends to 240 nm using a log-linear extrapolation. Extrapolating in this manner allows H_2O to be photolyzed all the way down to the surface in low-O_2 atmospheres, and this, in turn, helps keep surface O_2 low by providing H and OH radicals that can catalyze its reaction with H_2 and CO.

Between 175–190 nm the quantum yield in reaction (9.2) is unity for production of H and OH. Then O_2 can be produced by either

$$O + OH \rightarrow O_2 + H \tag{9.3}$$

or

$$O + O + M \rightarrow O_2 + M \tag{9.4}$$

Reaction (9.4) is important in the stratosphere, where H_2O (and hence OH) is scarce. In model calculations of the early Earth, it results in the creation of an "oxygen layer," crudely analogous to the ozone layer on modern Earth (Fig. 9.2).

The more CO_2 is present, the more O_2 is produced but the O_2 is consumed by reaction with H_2 at lower altitudes. Figure 9.2 shows calculated O_2 mixing ratio profiles for three different atmospheric CO_2 levels, ranging from 1 to 1000 PAL. Figure 9.3 shows the mixing ratios of major long-lived species for the 1000-PAL case. Even at low CO_2 levels, significant concentrations of O_2 are predicted to be present in the upper stratosphere near 60 km altitude where O_2 is being produced. The O_2 from this region flows downward into the troposphere where it is consumed by reaction with H_2, catalyzed by the by-products of water vapor photolysis. For example, one such catalytic cycle is

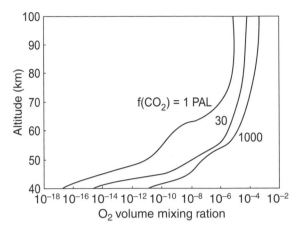

Figure 9.2 Calculated O_2 mixing ratio profiles in a prebiotic atmosphere for an H_2 input of 4×10^9 molec $cm^{-2}s^{-1}$, or 1.07×10^{12} mol yr^{-1} and for different mixing ratios of atmospheric CO_2, denoted here as $f(CO_2)$. A 1-bar atmosphere is assumed. H_2 is well mixed and satisfies the atmospheric redox balance eq. (8.34). H_2 mixing ratios for the three cases are: 1 PAL–1×10^{-4}; 30 PAL–7×10^{-5}, 1000 PAL—2.5×10^{-4}. "PAL" means "times the Present Atmospheric Level." (From Kasting *et al.* (1984b). Reproduced with permission of Springer. Copyright 1984, D. Reidel Publishing Company.)

$$H + O_2 + M \rightarrow HO_2 + M \qquad (9.5)$$

$$H + HO_2 \rightarrow 2\,OH \qquad (9.6)$$

$$H_2 + OH \rightarrow H_2O + H \quad (\times 2) \qquad (9.7)$$

$$Net : O_2 + 2\,H_2 \rightarrow 2\,H_2O \qquad (9.8)$$

Alternatively, atomic hydrogen may react with CO, forming the formyl radical HCO, and then HCO can react with O_2

$$H + CO + M \rightarrow HCO + M \qquad (9.9)$$

$$HCO + O_2 \rightarrow HO_2 + CO \qquad (9.10)$$

As a consequence of reactions such as these, the predicted O_2 mixing ratio declines rapidly with decreasing altitude, reaching values of 10^{-17} to 10^{-11} near the surface. The formyl radical can also react with itself to produce formaldehyde, H_2CO

$$HCO + HCO \rightarrow H_2CO + CO \qquad (9.11)$$

This last reaction may have played a role in prebiotic synthesis of organic compounds, an issue to which we will return later in the chapter, in Sec. 9.3.1

Actually, surface O_2 concentrations would be even lower than shown in Fig. 9.2 were it not for an additional O_2 production process that has been included in the photochemical model. Lightning in a weakly reduced atmosphere can generate O_2 via the high-temperature equilibrium reactions

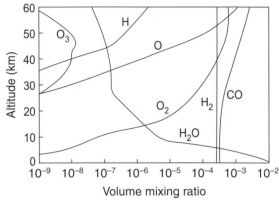

Figure 9.3 Vertical profiles of mixing ratios of various species for the 1000 PAL CO_2 case in Fig. 9.2. (From Kasting *et al.* (1984b). Reproduced with permission of Springer. Copyright 1984, D. Reidel Publishing Company.)

$$2\,CO_2 \leftrightarrow 2\,CO + O_2 \qquad (9.12)$$

$$2\,H_2O \leftrightarrow 2\,H_2 + O_2 \qquad (9.13)$$

A similar high-temperature process results in formation of NO in today's atmosphere via

$$N_2 + O_2 \leftrightarrow 2\,NO \qquad (9.14)$$

The modern global-average rate of NO production by lightning is about 2.9×10^9 Tg N yr^{-1}, or roughly 8×10^8 $cm^{-2}s^{-1}$ (Borucki and Chameides, 1984; Martin *et al.*, 2002). The estimated uncertainty is large, (0.8-8)$\times$ 10^9 Tg N yr^{-1}, so we will round this number up to 1×10^9 cm^{-2} s^{-1}. If one corrects for the relative abundances of N_2, O_2, and CO_2, one can scale this lightning production rate of NO and O_2 to the prebiotic atmosphere by assuming that these gases equilibrate at an effective "freeze-out temperature" of $\sim$3000–3500 K (Chameides and Walker, 1981; Kasting, 1985; Kasting *et al.*, 1985). The calculated production rates of NO in the high-CO_2 atmosphere shown in Fig. 9.3 is about 5×10^8 $cm^{-2}s^{-1}$, or half that in the present atmosphere. The production rates of O_2 are 2–3 times higher, reflecting the calculated ratio of $O_2 : NO$ at the freeze-out temperature.

The calculations shown in Figs. 9.2 and 9.3 can be used to illustrate the principle of atmospheric redox balance expressed by eq. (8.34) in the previous chapter. In this particular model, which did not contain sulfur gases, the principal photochemically produced oxidant was H_2O_2, and the principal photochemically produced reductant was H_2CO. Tropospheric number density profiles for these two species are shown in Fig. 9.4. As the assumed CO_2 concentration is increased, both H_2O_2 and H_2CO become more abundant; hence their rainout rate and their contribution to the atmospheric H_2 budget increase

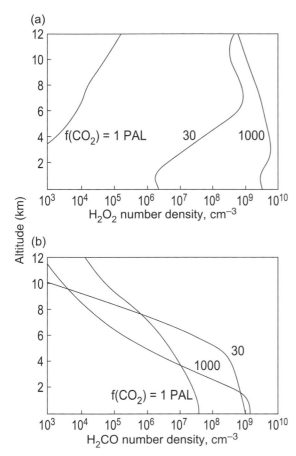

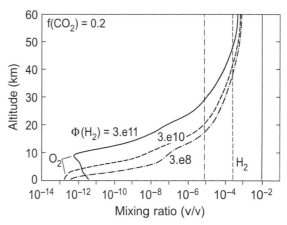

Figure 9.5 Calculated O_2 and H_2 vertical profiles for three different total H_2 outgassing rates, $\Phi(H_2)$, listed on the figure in units of $cm^{-2}s^{-1}$. A surface pressure of 1 bar and an atmospheric CO_2 mixing ratio (denoted fCO_2) of 0.2 have been assumed.

Figure 9.4 Tropospheric number density profiles of H_2O_2 (top panel) and H_2CO (bottom panel) for the three cases shown in Fig. 9.2. (From Kasting *et al.* (1984b). Reproduced with permission of Springer. Copyright 1984, D. Reidel Publishing Company.)

correspondingly. Consequently, the atmospheric H_2 concentration changes in each calculation, even though the assumed surface influx of H_2 is the same in all three cases: 4×10^9 $cm^{-2}s^{-1}$, or 1.07×10^{12} mol yr^{-1} (about one-fourth of the modern total H_2 outgassing rate estimated in the previous chapter). The calculated H_2 mixing ratios are in the range $(0.7-2.5) \times 10^{-4}$. These numbers are lower than predicted by eq. (8.10) because a higher proportionality factor between Φ_l and f_T (4×10^{13} instead of 2.5×10^{13}) was used in this particular calculation. In reality, the appropriate constant depends on the relative proportion of H and H_2 at the homopause level. H, being lighter, has the higher diffusion coefficient, as shown in Table 5.2.

9.2.2 Dependence of O₂ on H₂

The concentration of atmospheric O_2 should also depend on the abundance of H_2, as it is the photochemically

catalyzed reaction with H_2, which is the major sink for O_2. Figure 9.5 shows vertical O_2 and H_2 profiles for low, medium, and high H_2 outgassing rates using an updated version of a photochemical model similar to that used by Segura *et al.* (2007). The atmospheric CO_2 mixing ratio was set to 0.2 in these calculations, and the total H_2 volcanic outgassing rate spans a range around the estimated present-day value of $\sim 4 \times 10^{12}$ mol yr^{-1}. In Fig. 9.5, H_2 is well mixed vertically and ranges in concentration from 8×10^{-6} to 0.01. O_2 generally decreases with increasing H_2, although this trend reverses near the surface at very high H_2, probably because the H_2 helps to destroy the radical species that consume O_2.

Figure 9.6 shows the calculated terms in the atmospheric redox budget (eq. (8.34)) over the entire range of H_2 outgassing rates. Both the H_2 mixing ratio (top panel) and the hydrogen escape rate to space, $\Phi_{esc}(H_2)$ (solid curve, bottom panel), increase linearly at high outgassing rates. That's because nearly all of the outgassed H_2 escapes to space at the diffusion-limited rate. But, at low outgassing rates, both the H_2 mixing ratio and the escape rate level are nearly independent of the outgassing rate. In this regime, the redox balance is mostly between rainout of oxidants, $\Phi_{rain}(Ox)$ and rainout of reductants, $\Phi_{rain}(Red)$. The soluble oxidants in this model are dominated by oxides of nitrogen (NO, NO_2, and HNO) while the major soluble reductant is formaldehyde, H_2CO. A detailed list of photochemical oxidants and reductants can be found in Segura *et al.* (2007). The rainout rate of H_2CO is shown explicitly on the diagram: it is equal to roughly one-half $\Phi_{rain}(Red)$ because the weighting coefficient of formaldehyde in the hydrogen budget is 2. Note that, at high outgassing rates, the dominant sink for H_2

(a)

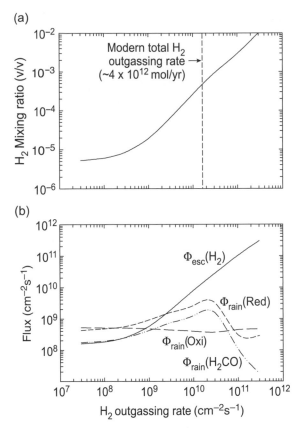

(b)

Figure 9.6 (a) H_2 mixing ratio vs. total H_2 outgassing rate. (b) Factors contributing to the global redox budget, including rainout fluxes of oxidants, $\Phi_{rain}(Ox)$, and reductants, $\Phi_{rain}(Red)$. The removal rate of formaldehyde (H_2CO) is also shown as $\Phi_{rain}(H_2CO)$. The hydrogen escape flux to space is $\Phi_{esc}(H_2)$. The rainout terms become important at low H_2 outgassing rates. These calculations include the cases shown in Fig. 9.5.

is escape to space, not rainout of formaldehyde. This supports our assertion from the previous chapter that most of the outgassed hydrogen on the prebiotic Earth was lost to space, and not recycled into organic material.

9.2.3 Effect of Higher UV Fluxes on O_2 and O_3

The above calculations demonstrate how the atmospheric redox system should have worked on the early Earth, but they ignore a factor that could have had important effects on atmospheric photochemistry. As discussed in Sec. 5.10.2, despite being fainter overall, the young Sun was brighter at UV and EUV wavelengths (e.g., see eq. (5.65)).

In the 1980s, Canuto *et al.* (1982; 1983) suggested that the high UV flux from the young Sun could have created significantly higher concentrations of O_2 and O_3 in the Earth's early atmosphere. They simulated an early Earth-like planet orbiting an extremely active, young

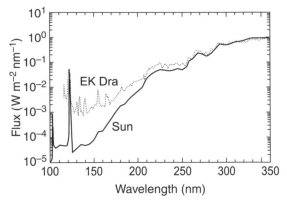

Figure 9.7 Present UV flux from the Sun and from a young, solar-analog star, EK Dra. (From Segura *et al.* (2007). Reproduced with permission from ESO. Copyright 2007.)

T-Tauri star. T-Tauri stars are now thought to represent a very early stage of stellar evolution lasting no more than a few million years. The Earth was probably not yet fully formed when the Sun went through this stage, and so the Canuto *et al.* calculations are not relevant to early atmospheric evolution. However, more recent calculations using realistic stellar UV fluxes also predicted high O_2 and O_3 concentrations in some simulations, particularly for high atmospheric CO_2 levels (Selsis *et al.*, 2002). This could become an important issue for future exoplanet characterization missions because it represents a possible "false positive" for life. But, neither the Canuto *et al.* models nor the Selsis model included a self-consistent treatment of the atmospheric hydrogen budget. When one includes the hydrogen budget self-consistently, the predicted high O_2 and O_3 concentrations disappear (Kasting *et al.*, 1984b; Segura *et al.*, 2007).

The model of Segura *et al.* (2007) explains why O_2 would be low. These authors modeled an early Earth-type planet orbiting EK Draco, which is a young, solar-analog star with a high UV flux, as shown in Fig. 9.7. For their standard atmospheric model, Segura *et al.* assumed a 1-bar surface pressure with 0.2 bar of CO_2 (roughly 700 PAL). The assumed total hydrogen outgassing rate was 4.2×10^{10} cm^{-2} s^{-1}, or 1.1×10^{13} mol yr^{-1}, which is about three times higher than the total modern outgassing rate of H_2 and H_2 equivalents listed in Table 8.2. This is a reasonable outgassing rate for an early Earth on which plate tectonics and volcanic outgassing were more vigorous than today. Results of the Segura *et al.* calculation are shown in Fig. 9.8. Panel (a) shows calculated O_2 and H_2 profiles for the present solar UV flux and for the higher flux that would be present around EK Draco. The O_2 profiles are similar in both cases to those shown earlier in this chapter (Figs. 9.1

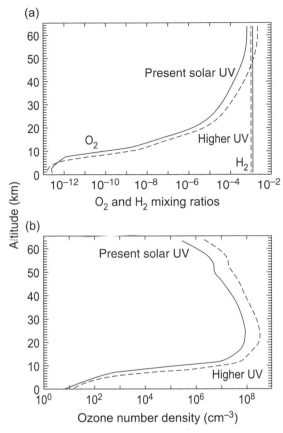

Figure 9.8 Vertical profiles of O_2 and H_2 mixing ratio (a) and O_3 number density (b) for a young planet illuminated by the Sun (solid curves) and by EK Dra (dashed curves). The assumed surface pressure is 1 bar and the atmospheric CO_2 mixing ratio is 0.2 (From Segura *et al.* (2007). Reproduced with permission from ESO. Copyright 2007.)

and 9.4). The calculated O_3 column depths were 1.9×10^{14} cm^{-2} and 6.8×10^{14} cm^{-2} for the low- and high-UV cases, respectively. These may be compared with an average O_3 column depth of 8.6×10^{18} cm^{-2} (~0.32 atm-cm) for the present Earth (Sec. 1.1.2.7). Thus, these calculations suggest that the early Earth – and, by implication, any early Earth-type planet orbiting EK Draco – should have low concentrations of both O_2 and O_3.

The same conclusion holds in the Segura *et al.* model even if volcanic outgassing of reduced gases is completely turned off: neither O_2 nor O_3 is able to accumulate to appreciable levels. The reason has already been discussed: at low outgassing rates H_2 produced from rainout of photochemical oxidants – mainly sulfate aerosols in this high-SO_2 model – builds up to high enough concentrations to limit the accumulation of O_2. The underlying assumption here is that planetary surfaces are reduced, so that oxidants are consumed when they contact them. If

this assumption is correct, then high UV fluxes alone cannot produce a *false positive* for life on a planet orbiting a young star. False positives can occur but they owe their existence to other factors (see Ch. 15).

That said, some recent papers suggest that O_2 concentrations of ~10^{-3} could be produced abiotically on rocky planets with high CO_2 concentrations and low H_2 outgassing rates (Domagal-Goldman *et al.*, 2014; Hu *et al.*, 2012; Tian *et al.*, 2014). The Hu *et al.* result is for an Earth-like planet orbiting the Sun; the Tian *et al.* results are for a planet orbiting an active M star, and Domagal-Goldman *et al.* looked at both cases. We will postpone the discussion of M-star planets until Ch. 15. Their atmospheric photochemistry is quite different from G-star planets because of the much lower fluxes of near-UV stellar radiation, and this makes them somewhat more susceptible to abiotic O_2 buildup. These models all include an atmospheric hydrogen budget, which is appropriate, and the latter two models balance the global redox budget, as well. However, Hu *et al.* assumed an O_2 surface deposition velocity (see Sec. 3.2) of 0, so they implicitly assume that the ocean is not a sink for O_2. That is what allows their model to build up O_2. (Domagal-Goldman *et al.* get the same result if they make this same assumption.) But, if the ocean is filled with ferrous iron, as we think was true for the Archean Earth, then the dissolved O_2 concentration should be negligible, and one can use the methodology of Kharecha *et al.* (2005) to show that the O_2 deposition velocity should be much larger, ~1.5×10^{-4} cm s^{-1} (see Sec. 9.7 below.) This would create a large downward flux of O_2 (or, equivalently, a correspondingly large source of H_2) and that, in turn, would depress the atmospheric O_2 concentration. So, for planets with reduced surfaces like the early Earth, abiotic O_2 concentrations should be small.

9.3   Prebiotic Synthesis of Organic Compounds in Weakly Reduced Atmospheres

The type of prebiotic atmosphere described in the previous section is often termed *weakly reduced* because, while O_2 is virtually absent near the surface, highly reduced gases such as CH_4 and NH_3 are also present only in low concentrations. To researchers pursuing the classical lightning discharge pathway for prebiotic chemical evolution, this type of atmosphere has appeared unsatisfactory because the expected yield of biologically relevant organic compounds from this process is relatively small (Chameides and Walker, 1981; Miller and Schlesinger, 1984; Stribling and Miller, 1987). However, contrary to early reports, modest yields of amino acids, ammonia, and hydrogen cyanide (HCN) from atmospheres dominated by CO_2 and N_2 are possible if ferrous iron is present in a somewhat acidic ocean to prevent the oxidation of organics in solution (Cleaves et al., 2008; Plankensteiner et al., 2006). Also, reducing conditions and lightning are possible locally from particularly reducing volcanic plumes bearing in mind that the average oxidation state of volcanic gases conceals a large range (e.g., Fig. 8.4) (Johnson et al., 2008a; Parker et al., 2011).

Moreover, the atmosphere was not the only place where prebiotic compounds might have come from on the early Earth. Complex organic compounds could have been synthesized abiotically within mid-ocean ridge hydrothermal vents (Corliss et al., 1981; McCollom and Seewald, 2007; Shock, 1990) or by chemistry occurring in the icy mantles of interstellar dust grains (Dworkin et al., 2001) that were later delivered to Earth via cometary impacts or as interstellar dust particles, IDPs (Chyba and Sagan, 1992; Chyba et al., 1990; Delsemme, 2001; Oro, 1961). It is nonetheless interesting to consider whether prebiotic organic synthesis could have been initiated by photochemical reactions in a weakly reduced atmosphere.

9.3.1 Synthesis of RNA Building Blocks: H_2CO and HCN

Whether atmospheric synthesis was possible depends on which types of organic compounds would have been most important to the origin of life. Miller and Urey were excited about the abiotic synthesis of amino acids, which are the building blocks of proteins (Sec. 7.1, Fig. 7.1). But while proteins are important catalysts in modern biochemistry, they are not the fundamental information-bearing molecules. In the past 50 years, researchers have focused on ribonucleic acid (RNA) as the first self-replicating molecule (Higgs and Lehman, 2015; Joyce, 1989; Orgel, 1986; Robertson and Joyce, 2012; Ruiz-Mirazo et al., 2014). RNA has catalytic properties like those of proteins, and it also has information-encoding capability like that of double-stranded DNA, although it is less stable and is therefore limited to much shorter molecules that encode less information. RNA consists of single-stranded chains of monomers called nucleotides. Each nucleotide contains a phosphate molecule, a ribose (sugar) molecule, and a nitrogen-containing base. In nature, the phosphate molecule would likely be derived from phosphate released by weathering of rocks or meteorites, whereas the other components could have had an atmospheric origin.

The sugar part of RNA is in one sense easy to make, yet in another sense it is very hard. Ribose, $C_5H_{10}O_5$, can be formed from five molecules of formaldehyde, H_2CO, if you put them together in the proper way. As we have already seen, photochemical reactions in weakly reduced, CO_2-rich atmospheres are predicted to produce large quantities of formaldehyde by the reactions (9.9)–(9.11). Formaldehyde is soluble in water and, hence, would have rained out of the early atmosphere. Once in the ocean, it could have suffered several different fates. In the surface ocean, H_2CO could have been photolyzed back to CO + H_2. It could also have disproportioned to form methyl alcohol and formic acid (Abelson, 1966)

$$2H_2CO + H_2O \rightarrow CH_3OH + HCOOH \qquad (9.15)$$

Finally, it could have reacted with HCN to form amino acids, or it could have polymerized to form various sugars, including ribose, via the formose reaction. This, of course, would be a useful step in originating life, because ribose is a key ingredient in RNA. Unfortunately, ribose is only one of many sugars that might have formed, and this has been thought to limit the utility of the formose reaction in life's origin.

One possible way of channeling formaldehyde polymerization into ribose was suggested by Ricardo et al. (2004). These authors showed that the borate ion, BO_3^{3-}, forms complexes with pentose sugars and can therefore stabilize ribose with respect to other sugars. Whether or not this observation has significance for the origin of life problem remains to be determined. Still, this is one step of prebiotic synthesis that might have worked in a weakly reduced environment.

Synthesizing the bases needed for RNA is even more difficult. The simplest nitrogen-containing base is adenine, $C_5H_5N_5$, which is a pentamer of hydrogen cyanide, HCN (Sutherland, 2016). Substantial amounts of HCN can be formed in a weakly reduced atmosphere, provided that small amounts of methane are present, as well (Kasting and Brown, 1998; Tian et al., 2011; Zahnle, 1986). The mechanism works as follows. N atoms can be produced by ionization of N_2 in the ionosphere

$$N_2 + h\nu \rightarrow N_2^+ + e \qquad (9.16)$$

followed by dissociative recombination of N_2^+

$$N_2^+ + e \rightarrow N + N \qquad (9.17)$$

The N atoms then flow down into the stratosphere where they combine with methylene (CH_2) and methyl (CH_3) radicals produced by CH_4 photolysis:

$$CH_2 + N \rightarrow HCN + H \qquad (9.18)$$

$$CH_3 + N \rightarrow HCN + H_2 \qquad (9.19)$$

For the above mechanism to work effectively, at least tens of parts per million of methane must have been present in the prebiotic atmosphere (Tian *et al.*, 2011; Zahnle, 1986)). In Ch. 7 (Table 7.3) we saw that a substantial amount of abiotic methane, as much as 4×10^{11} mol yr^{-1}, or 1.5×10^{9} cm^{-2} s^{-1}, is being released today from continental hydrothermal systems and off-axis hydrothermal vents. If released into a weakly reduced atmosphere, this same flux would produce an atmospheric CH$_4$ mixing ratio of ~20 ppmv (see Fig. 9.9). In the model of Tian *et al.* (2011), this would produce an HCN rainout rate of just 10^6–10^7 cm^{-2} s^{-1}, depending on the CO$_2$ concentration (Fig. 9.10). That, by itself, is not much. However, plausible increases in abiotic methane production caused

by a more ultramafic early crust (Ch. 7) might increase this rate by factor of 1000 or more, as shown by the figure. Increases in atomic N production caused by a higher solar EUV flux early in Earth's history (Ribas *et al.*, 2005) are accounted for in this model. So, an HCN production rate of 10^{10} cm^{-2} s^{-1} (2.7×10^{12} mol yr^{-1}) is not implausible for the primitive Earth.

This rate of HCN production, although large, would not have created a prebiotic soup. A flux of 10^{10} HCN molecules cm^{-2} s^{-1} is equivalent to ~5×10^{-7} mol cm^{-2} yr^{-1}. HCN hydrolyzes to ammonium hydroxide, NH$_3$OH, and formic acid, HCOOH, in 70 years at 15 °C (Abelson, 1966). In the absence of continents, the average ocean depth would be 3 km, so the volume of water in a 1 cm^2 column is 300 l. Putting these numbers together yields a dissolved HCN concentration of ~1.2×10^{-7} mol l^{-1}. These numbers apply to the oceans as a whole. The uppermost ocean of down to 100 m depth could have had higher concentrations because the mixing time for the deep oceans is slow, roughly 1000 years. At most, this would have increased dissolved HCN concentrations in the surface ocean by a factor of $1000/70 \cong 15$, so the upper limit on HCN is still only about 1–2 μM. These concentrations are lower by a factor of about 10^4 than those required by modern chemists to perform laboratory equivalents of prebiotic synthesis (Shapiro, 1995). The numbers for production of formaldehyde given earlier imply a similar conclusion. This does not necessarily indicate a failure of the atmospheric models. Rather, it shows that the concept of the entire early oceans as prebiotic soup is probably incorrect. Various concentration mechanisms (e.g., evaporating tide pools (Brasier *et al.*, 2011)) would have been required to create conditions for prebiotic synthesis even if monomers like H$_2$CO and HCN were raining out of the atmosphere more or less continuously.

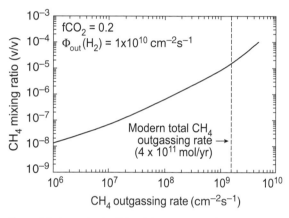

Figure 9.9 Atmospheric CH$_4$ mixing ratio as a function of CH$_4$ outgassing rate, as calculated using the same model as in Figs. 9.5 and 9.6. The dashed line shows the estimated modern *abiotic* CH$_4$ flux. A CO$_2$ mixing ratio of 0.2 and an H$_2$ outgassing rate of 1×10^{10} cm^{-2}s^{-1} (2.7×10^{12} mol yr^{-1}) were assumed.

9.3.2 CO as a Prebiotic Compound

Other origin-of-life researchers have emphasized the primacy of metabolism, as opposed to information storage. It is easy to show that the weakly reduced atmospheres described here are quite favorable for some of these theories. For example, Wächtershäuser (1988a, b, 1990, 1992) has proposed that life originated by way of autotrophic carbon fixation (i.e., making organic matter from CO$_2$) on the surfaces of Fe–S minerals, such as pyrite, FeS$_2$. Following up on this idea, Huber and Wächtershäuser (1997, 1998) showed that amino acids can be activated, enabling them to form *peptides* (components of proteins), by reaction with CO in the presence of (Ni,Fe)S surfaces. Carbon monoxide, which should be abundant in

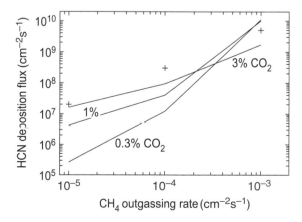

Figure 9.10 Calculated HCN deposition fluxes for different atmospheric CO$_2$ and CH$_4$ mixing ratios. A 1-bar surface pressure is assumed. Crosses are from Zahnle (1986) at 3% CO$_2$. (From Tian *et al.* (2011). Reproduced with permission. Copyright 2011, Elsevier.)

a weakly reduced atmosphere (Fig. 9.3), plays a major role in this theory because it has a high thermodynamic free energy. One can see this by writing the reaction (called the *water–gas shift reaction*)

$$CO + H_2O \rightarrow CO_2 + H_2 \qquad (9.20)$$

According to the NIST webbook database, the Gibbs free energy change, ΔG_R, for this reaction at 25 °C is -20.48 kJ mol^{-1}. (Here, we have used the liquid phase value for H_2O and gas phase values for CO, CO_2, and H_2.) The activity of seawater is close to unity, so the equilibrium constant for this reaction is

$$K_{eq} = \frac{pCO_2 \cdot pH_2}{pCO} = \exp\left(\frac{-\Delta G_R}{RT}\right) \cong 3900 \qquad (9.21)$$

Rearranging terms, and assuming a CO_2 partial pressure of 0.3 bar, gives

$$\frac{pCO}{pH_2} = \frac{pCO_2}{K_{eq}} = \frac{0.3}{3900} \approx 10^{-4} \qquad (9.22)$$

Thus, at equilibrium, the ratio of gaseous CO to H_2 should be about 10^{-4}. But, in the model atmospheres shown in Fig. 9.3, the ratio of CO to H_2 is ~1. That means that these two gases are far away from equilibrium, and so reaction (9.20) should have proceeded strongly to the right, provided that there was some mechanism by which it could be catalyzed. We discuss in Sec. 9.7 how certain modern organisms can do this. If Huber and Wächtershäuser are right, the very earliest organisms might also have taken advantage of the energy available in CO.

This abundant source of thermodynamic free energy should have existed on the early Earth because of a quirk of atmospheric chemistry. The ultimate source of the free energy residing in CO is UV radiation from the Sun, which dissociates CO_2 by reaction (9.1). That energy would be lost if CO recombined rapidly to form CO_2. It doesn't do so, however, because the direct recombination reaction with ground-state atomic O

$$CO + O + M \rightarrow CO_2 + M \qquad (9.23)$$

is spin-forbidden and therefore slow (Clark, 1971). Instead, recombination of CO occurs by the reaction

$$CO + OH \rightarrow CO_2 + H \qquad (9.24)$$

The OH is produced by photolysis of H_2O at a rate that is limited by the availability of UV radiation shortward of ~240 nm. The net result is that CO has a lifetime of several years, or longer. H_2 has a long photochemical lifetime, as well, but unlike CO it also escapes to space; hence, its predicted concentration is relatively low. So, the early atmosphere could not help but have a strong free energy gradient between CO and H_2.

Before leaving this section, we should note that CO can become extremely abundant in some hypothetical model atmospheres. Indeed, in some model calculations, CO can increase until it is more abundant than CO_2 – a process termed *CO runaway* (Kasting *et al.*, 1983; Zahnle, 1986). The runaway phenomenon occurs when the assumed production rate of CO is greater than the production rate of the OH and O radicals that consume it. This can happen, for example, as a result of large impact events (Kasting, 1990). The tendency towards CO runaway is partially offset by uptake of CO by the ocean, followed by hydrolysis to the formate ion (Van Trump and Miller, 1973)

$$CO + OH^- \rightarrow HCOO^- \qquad (9.25)$$

This process can be parameterized by assuming an effective deposition velocity for CO of 10^{-9} to 10^{-8} cm s^{-1} (Kharecha *et al.*, 2005, their Appendix 3). These values do not preclude CO runaway in situations where CO is produced at rapid rates, so a CO-rich atmosphere could, in principle, have been present very early in Earth history. Once life had evolved, however, dissolved CO was probably taken up by organisms, and high CO concentrations would no longer have been possible (Kharecha *et al.*, 2005). Thus, a high abundance of CO in an exoplanet atmosphere might be considered an "anti-biosignature," indicating a lifeless planet (Sec. 15.4).

9.4 When Did Life Originate?

We next wish to consider how the presence of life might have affected early atmospheric composition. Before doing so, however, we first ask: when did life originate? As we will see, the answer to this question is poorly known, although some constraints can be derived from the geologic record (e.g. Hallmann and Summons, 2014; Nisbet and Fowler, 2014). Below, we briefly summarize four lines of evidence for early life: fossils, stromatolites, isotopic indicators, and biomarkers.

9.4.1 Evidence from Microfossils and Stromatolites

Until the 1950s, the earliest evidence for life came from the beginning of the Cambrian Period, currently dated to start at 541.0 ± 1.0 Ma (Gradstein, 2012). That is when many multicellular organisms developed the ability to make shells of calcium carbonates, calcium phosphates, and opal (silica), which are easily preserved as fossils (Bengtson, 1994). In the 1950s, paleontologists such as Elso Barghoorn of Harvard University discovered microfossils – the remnants of single-celled organisms that lacked hard parts – in rocks that were much older than this. One particularly well preserved set of microfossils came from the Gunflint Formation of northwestern Ontario, dated at ~2.1 Ga (Fig. 9.11). These microfossils include both coccoid (spheroidal) and filamentous forms. These

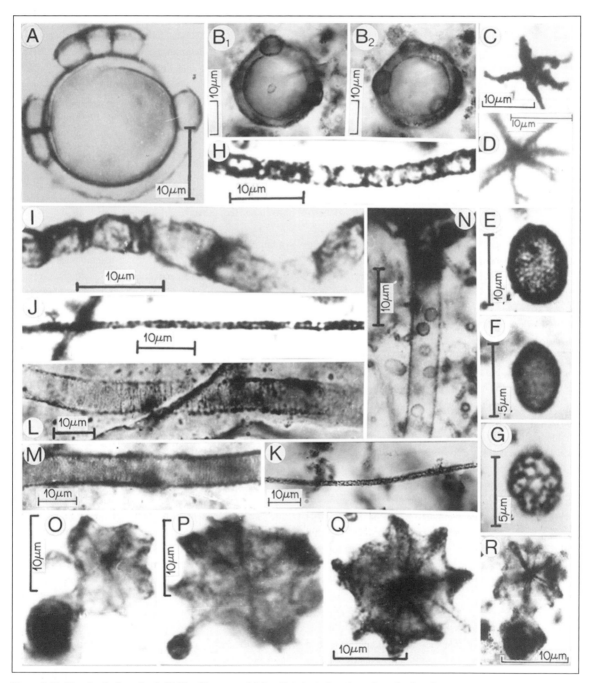

Figure 9.11 Microfossils from the (≈2100-million-year-old) Gunflint chert of southern Canada. A and B: *Eosphaera*, in B shown in two views of the same specimen. C and D: star-shaped filamentous structures named *Eoastrion*. E–G: simple spherical-elliptical forms, *Huroniospora*. H–K: narrow filaments of cylindrical cells, *Gunflintia*. L and M: broader tubes, *Animikiea*. N: *Entosphaeroides*. O–R: unusual forms with slender stalk, basal bulb and umbrella-like cap, *Kakabekia*. The affinities of these fossils remain obscure. (From Schopf (2000). Reproduced with permission. Copyright 2000, National Academy of Sciences.)

are all now considered to be the remnants of *prokaryotic* organisms (Awramik and Barghoorn, 1977), which are microbes that lack a nucleus and other functional subunits called organelles, as opposed to *eukaryotic* organisms that possess these entities. The original interpretation was that some of the microfossils were eukaryotic (Schopf *et al.*, 1983; Tyler and Barghoorn, 1954). Micropaleontologists, though, are in universal agreement that these microfossils have a biological origin.

In more recent years, paleontologists have reported putative microfossils as old as 3.5 Ga (Schopf, 1993; Schopf and Packer, 1987; Westall, 2005; Westall *et al.*, 2001). The most infamous of these microfossils are those from the Apex Chert in the Warrawoona Group of northwestern Australia (Schopf, 1993). Sketches drawn by Schopf included filamentous forms that were said to resemble modern cyanobacteria, but which have been strongly disputed (Brasier *et al.*, 2004; Brasier *et al.*, 2002; Brasier *et al.*, 2005). As we will discuss in more detail later, *cyanobacteria* were almost certainly the first organisms that were capable of making O_2 by oxygenic photosynthesis. Brasier *et al.* argue that these putative microfossils are *not* cyanobacteria. First, the cherts in which they are found are of hydrothermal origin; second, the morphology of the putative microfossils branches in three dimensions and is too indefinite to be ascribed to microbial types. Indeed, Brasier *et al.* argue that they may not have been biogenic at all. In rebuttal, laser–Raman spectroscopy demonstrates the presence of kerogen (Schopf *et al.*, 2002). Other workers have suggested that fractures filled with dark-colored hematite were mistakenly identified as microfossils (Marshall *et al.*, 2011) or that the microfossils are phyllosilicates grains (Brasier *et al.*, 2015). While there is now acceptance of the presence organic carbon, data suggest multiple populations of such material (Marshall *et al.*, 2014; Sforna *et al.*, 2014) perhaps from episodes of fluid infiltration. Overall, the Apex Chert microfossils are a continuing controversy (Marshall *et al.*, 2014; Marshall and Marshall, 2013; Pinti *et al.*, 2013; Schopf and Kudryavtsev, 2012; Kudryavtsev, 2013).

Critiques have also been directed at other reported early Archean microfossils (Buick, 2007b; Wacey, 2009). The oldest ones that can be confidently ascribed to cyanobacteria are at 1.85 Ga in Belcher Islands, Canada (Hoffmann, 1976). The oldest convincing microfossils are from 2.55 Ga in the Transvaal Supergroup, southern Africa; these show behavioral patterns such as cell division and matting, and are also composed of kerogen with carbon isotopic values characteristic of life (Altermann and Schopf, 1995; Klein *et al.*, 1987). Possible

kerogenous microfossils exist in the ~3.23–3.47 Ga Swaziland Supergroup of southern Africa (Walsh, 1992) but they lack behavioral morphologies. Also, ~3.35 Ga sandstone in the Strelley Pool Formation in northwest Australia has plausible silica-replaced microfossil morphologies (Wacey *et al.*, 2012) and isotopically light carbonaceous material, albeit metamorphosed (Wacey *et al.*, 2011). However, proving that these microfossils are as old as the deposit itself (i.e., *syn-depositional*) is difficult because metamorphism extended over a long interval for these rocks. In summary, a question mark hangs over microfossil evidence from all but the latest Archean.

Stromatolites are laminated sedimentary structures accreted as a result of microbial growth, movement or metabolism. They are found today in areas such as in the tidal waters of the Bahamas and Australia, where salinity precludes the presence of fish or mollusks that would otherwise eat them. But in the Precambrian the absence of animals meant no such restriction on abundance. Fossilized structures bearing a strong resemblance to stromatolites are found as early as 3.48 Ga in the Dresser Formation of the Warrawoona Group of northwestern Australia (Buick *et al.*, 1981) (Fig. 9.12) and higher in the stratigraphy in the Strelley Pool Formation in an age bracket of 3.38–3.33 Ga (Allwood *et al.*, 2009; Allwood *et al.*, 2006; Hofmann *et al.*, 1999). The structures described by Buick *et al.* show many biogenic-like features, such as wrinkly lamination that thickens over flexure crests consistent with a phototrophic habit, several orders of flexures that are not self-similar, lateral variation in laminar thickness, and kerogenous microlaminae (Fig. 9.12). The Strelley Pool structures are better preserved and similar to many structures in the Proterozoic that are accepted as biogenic; in addition, the chert contains isotopically light carbon in kerogen. Such stromatolitic structures, which become abundant in the later Archean and Proterozoic, are thought to represent the remnants of microbial mats formed by cyanobacteria or other, anoxygenic photosynthetic bacteria. (As we will see below, photosynthesis does *not* have to produce O_2.)

Some researchers have argued that stromatolite-like structures can form by abiotic processes (Grotzinger and Knoll, 1999; Grotzinger and Rothman, 1996; McLoughlin *et al.*, 2008), while others have disputed the dismissal of all Archean stromatolites as unjustified (Buick *et al.*, 1995a). Hence, as with microfossils, the earliest fossilized structures should be taken as being consistent with the presence of early microbial life, but not absolute proof (Wacey, 2009). Nonetheless, the opinion of most geobiologists is that the Dresser Formation conical stromatolites display morphological characteristics that are almost certainly biogenic

Figure 9.12 Upper: Cross-section of 3.49–3.48 Ga conical stromatolite, Dresser Formation, North Pole, Pilbara, Western Australia. Lens cap is 5 cm diameter. Lower: Plan view. (Photos by D. Catling.) (A black and white version of this figure will appear in some formats. For the color version, please refer to the plate section.)

(e.g., Van Kranendonk, 2006, 2011). In contrast, reports of 3.7 Ga stromatolites in Greenland based solely on limited morphology (Nutman *et al.* 2016) are controversial.

9.4.2 Carbon Isotopic Evidence for Early Life

A second line of evidence that has been used to argue for an early origin of life is the presence of isotopically fractionated organic carbon in rocks. As discussed in the previous chapter, when organisms fix CO_2 to make organic carbon, they typically fractionate the carbon isotopes because ^{12}C, the most abundant isotope, is kinetically favored for fixation compared to ^{13}C.

Graphite inclusions with a biogenic $\delta^{13}C$ value of -24 ± 5‰ are found in a single grain of zircon (zirconium silicate) from Jack Hills, Western Australia, dated at an Hadean age of 4.10 ± 0.1 Ga (Bell *et al.*, 2015). This graphite may originally have been biogenic carbon.

Near Isua, West Greenland, and from nearby Akilia Island, graphite inside apatite with $\delta^{13}C$ values as low as -60‰ have been reported in rocks dated as early as 3.8 Ga and presumed organic (Manning *et al.*, 2006a; Mojzsis *et al.*, 1996). However, the results are controversial and currently disputed by most others in the field. Van Zuilen *et al.* (2002) have argued that the graphitic carbon at Isua was most likely produced by abiotic disproportionation of siderite, $FeCO_3$. Meanwhile, Fedo and Whitehouse (2002) have argued that the carbon-containing rocks at Isua, instead of representing a sedimentary banded iron formation, are igneous in origin. Others have failed to find graphite within apatite from Akilia (Lepland et al., 2005) or have found graphite only in inclusions in gneiss formed from fluids, so the carbon could be both much younger than 3.8 Ga and abiotic (Lepland et al., 2011; Manning et al., 2006a).

Isotopically light organic carbon ($\delta^{13}C = -19$‰) has also been found in 3.7 Ga Isua rocks that are of undisputed sedimentary origin (Rosing, 1999). Some authors have suggested that the organic carbon is meteoritic in origin (Schoenburg et al., 2002) and that the isotopically light carbon is therefore an ambiguous biosignature (Brasier et al., 2004). Detailed spectroscopic and geochemical analysis has, at least, shown that the ^{13}C-depleted carbon came from organics in marine sediments (Ohtomo et al., 2014). Nonetheless, the carbon isotope data from the Hadean or very early Archean lack a complementary $\delta^{13}C$ from marine carbonates to strengthen the argument for biogenic fractionation between carbonate and organic phases.

At 3.52 Ga, there is $\delta^{13}C$ in kerogen of -24‰ and carbonate of -2‰ in chert and sedimentary carbonate of the Coonterunah Group in northwestern Australia, which is persuasive evidence of biological fractionation (Buick, 2007b; Harnmeijer, 2009). We also note that fractionation of sulfur isotopes in pyrite from ~3.47 Ga sedimentary rocks in the Warrawoona Group in northwestern Australia is permissive of biological fractionation (see Sec 9.6.3). Overall, isotopes suggest that life was present by 3.5 Ga. But it remains an open question as to whether life originated prior to 3.5 Ga, given the uncertainties over data from Greenland.

9.4.3 Molecular Biomarkers

A fourth line of evidence for early life comes from *biomarkers*, which are recognizable derivatives of biological molecules, usually in the form of certain hydrocarbons found in kerogen or oil (Gaines et al., 2009; Summons and Brocks, 2004). The process of burial and diagenesis strips biological molecules of functional groups but can often leave a distinctive molecular backbone, in a way that is analogous to the more familiar macroscopic process that strips away flesh to leave fossilized skeletons that we can identify with particular creatures, such as dinosaurs or mammals. Characterizing biomarkers in very ancient rocks is difficult because of contamination either

from fluid migration over geologic time or from modern chemicals.

Early work seemed to indicate the presence of biomarkers in 2.72–2.56 Ga rocks of the Hamersley in northwest Australia (Brocks et al., 1999; Eigenbrode et al., 2008) and 2.67–2.46 Ga rocks from the Transvaal of southern Africa (Waldbauer et al., 2009); and these putative biomarkers were interpreted as evidence for microbes that fixed carbon, produced oxygen and methane, and possibly fixed nitrogen, too. However, a multi-lab study using strict protocols to avoid hydrocarbon contamination has failed to find any biomarkers in cores of the same age from the Hamersley (French et al., 2015). Consequently, contamination likely compromised previous biomarker studies from the Archean.

9.5 The Molecular Phylogenetic Record of Life

To determine the effect of early organisms on the atmosphere, one must also understand what types of organisms may have been present. This is difficult to do by studying the fossil record, as most microbes have very simple morphologies: coccoid, rod-shaped, or filamentous. Stromatolites suggest that the organisms that made them were photosynthetic because their modern analogs are photosynthetic microbial mats. That said, they do not generally tell us whether those organisms were producing O_2 or whether they carried out anoxygenic photosynthesis.

An entirely different approach to studying biological evolution is to construct evolutionary trees based on sequencing of nucleotides in RNA, DNA, or amino acids in proteins. The molecule that provides the most information about early evolution is ribosomal RNA (rRNA). Ribosomes are organelles contained within all cells that are responsible for protein synthesis. The rate of mutation in rRNA is slow because most changes to an organism's protein-manufacturing apparatus are fatal. Conversely, other segments of the DNA molecule may change rapidly because they code for non-essential characteristics, for example, hair color in humans.

The way that rRNA is studied is by locating the rRNA-encoding genes in an organism's DNA and sequencing them. The rRNA molecule contains two subunits of different sizes that used to be separated by centrifugation in the late 1970s and early 1980s when the sequencing was performed on RNA itself. This technique gave rise to an "S" or Svedberg sedimentation coefficient for subunits related to the rate of their movement in the centrifuge. The smaller of these two subunits, labeled 16S in prokaryotes and 18S in eukaryotes, is the one for which the most data are available and that has been used for most evolutionary studies. The 16S-rRNA tree is also sometimes called the *universal tree of life* because it is thought to most faithfully portray actual evolutionary relationships (Woese, 2005). Even so, the interpretation of this tree is not straightforward because genes can be transferred laterally particularly between single-celled organisms and even between different branches, thereby confusing the issue of which organism was descended from whom. Some scientists argue that there is a *web of life* rather than a tree (Doolittle, 2009).

An example of a 16S-rRNA tree is shown in Fig. 9.13. As can be readily seen, extant organisms can be divided into three major *domains*: the Archaea, the Bacteria, and the Eukarya. The Archaea and the Bacteria contain mostly single-celled organisms that lack cell nuclei. Both of these domains were classified simply as "bacteria" prior to the development of molecular phylogenetic schemes. Some bacteria consist of filaments of cells that act together, so it is not strictly true that all of these organisms are single-celled. The Eukarya consist of single-celled and multicellular *eukaryotic* organisms whose cells do contain a nucleus. Single-celled eukarya

include, for example, the intestinal parasite *Giardia* in the Diplomonads, and were originally classified as a separate kingdom called *protists*. (The classical, or Linnaean, taxonomic scheme included 5 *kingdoms*: plants, animals, fungi, bacteria, and protists.)

The tree of life based on rRNA contains valuable information about early evolution, and we shall refer to it repeatedly throughout this chapter and the next. For now, however, we content ourselves with pointing out two particular features. First, at the far tip of the Eukarya domain one sees animals, fungi and plants. Based on rRNA, these three kingdoms diverged fairly recently. This provides some perspective concerning the antiquity of the relationships observed elsewhere in the tree.

Second a phylum near the intersection of the three domains, Euryarcheota, includes *hyperthermophiles*, that is, organisms with optimal growth temperatures in excess of 80 °C. We should point out that the tree drawn here is an unrooted tree, meaning that it is not obvious which organisms are the most ancient. Arguments based on gene duplication – genes that have apparently been duplicated and then evolved differently in different species – suggest that the root lies near the base of the Bacterial domain but

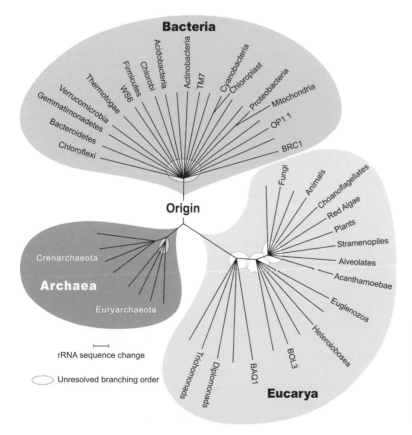

Figure 9.13 The "universal tree of life," based on 16s ribosomal RNA. Distances between nodes are proportional to the number of nucleotide changes. The tree is shown in unrooted form. Various arguments place the root near the foot of the Bacterial lineage. (Courtesy of Norm Pace, University of Colorado.)

not all researchers agree. Distance from the root is proportional to the number of changes in the nucleotide sequence. Thus, if the mutational rate has remained constant with time – a big "if" – then the most ancient organisms are those that lie closest to the root. On this basis, it appears as if some very ancient organisms were hyperthermophiles. Some more recent analyses of various protein sequences support this conclusion (Gaucher *et al.*, 2008; Gaucher *et al.*, 2003), although another suggests that the last universal common ancestor – LUCA for short – was a mesophile (preferred growth temperature of 0–40 °C), while the last common ancestor of both Bacteria and Archaea/Eukarya was a thermophile (40–80 °C) (Boussau *et al.*, 2008).

Increasingly, discoveries in molecular biology may support a view that the eukaryotic lineage originated within the archaeal domain (Spang *et al.*, 2015). If this garners further support from discovery of more close archaeal relatives of eukaryotes, the three-domain tree may be replaced by a two-domain picture where Archaea and Eukarya are grouped together as a single domain (Embley and Williams, 2015; Forterre, 2015; Raymann *et al.*, 2015). This is similar to an idea originally suggested by James Lake and colleagues (Lake *et al.*, 1984). In that case, the eukaryotes will be viewed either as a specialized branch within the new grouped domain or a type of Archaea.

In any case, we can ask the question: What, if anything, does the tree of life tell us about the origin of life and/or about the early environment? There are several possible explanations for the observation that the base of the rRNA tree is populated by hyperthermophiles. The simplest is that this is simply an artifact that arises because G–C bonds are slightly more stable at high temperatures than are A–U bonds (Galtier *et al.*, 1999). (G, C, A, and U are the four nucleic acids in RNA: guanine, cytosine, adenine, and uracil.) But this fails to explain the corresponding evidence from protein sequences, mentioned above. The second is that life itself arose in a hot environment. This explanation is consistent with the hydrothermal vent model for life's origin, discussed in Sec. 9.3. But, it is also consistent with a model in which the entire surface was hot because of the presence of a dense, CO_2-rich early atmosphere. On the other hand, some prebiotic chemists (e.g., Miller and Bada, 1988) view a hot origin of life as being unlikely, because high temperatures reduce the stability and lifetime of complex organic molecules. These authors argue that the most likely environmental temperature for originating life is just above the freezing point of water. Such environments could not have existed if the mean surface temperature was above 80 °C. Proponents of a cold origin of life point

out that there other plausible explanations for the position of hyperthermophiles in the evolutionary tree, even if life did not originate at high temperatures.

This last class of explanations begins with an origin of life in some cold environment such as the surface ocean or a lake. In one variation, this event is assumed to have happened during the time when Earth was still being bombarded by large impactors. Recall from Sec 6.7 that ~500 km diameter impactors capable of ocean sterilization should not have been present after ~3.8 Ga. Impactors larger than ~100 km in diameter would have released enough energy to vaporize the entire photic zone of the ocean (the uppermost 100 m), thereby wiping out most organisms. If life had already spread to the mid-ocean ridge vents, however, and if the oceans were already deep enough to provide an insulating blanket, then the hyperthermophilic organisms living within the vent systems, or in the hot, deep continental subsurface, might have been the only survivors (Gogarten-Boekels *et al.*, 1995; Sleep and Zahnle, 1998; Sleep *et al.*, 1989). Organisms living in the (hot) deep continental subsurface might have been even better protected. After the effects of the impact had disappeared, these organisms could have recolonized the entire planet, thus explaining why all extant organisms appear to be the descendants of hyperthermophiles.

In summary, the jury is still out as to how the phylogenetic evidence for ancient hyperthermophiles should be interpreted. However, there are good reasons to suggest that the early environment was cool, rather than hot, at the time when life originated, as cool temperatures are more conducive to prebiotic synthesis.

9.6 Early Anaerobic Metabolisms and Their Effect on the Atmosphere

Let's return to the question of how life would have affected the atmosphere once it had originated. The answer to this question clearly depends on which organisms originated first and on their metabolism. (For more detail on microbial metabolisms, see books by Konhauser (2007) and by Canfield *et al.* (2005) for comprehensive reviews.) As pointed out earlier, oxygenic photosynthesis was probably not an early invention. This inference is supported by the location of cyanobacteria well up on the Bacterial branch of the rRNA tree (Fig. 9.13). Thus, the earliest organisms must have all been *anaerobes* – organisms that can live in the absence of oxygen. Some of these organisms are discussed here and in the next section. Our discussion is not meant to be exhaustive; rather, we will focus on those metabolic processes that are likely to have had the greatest effect on geochemical cycling of elements and on atmospheric composition.

9.6.1 Heterotrophy and Fermentation

Until the 1980s, the prevailing scientific wisdom was that the earliest organisms were *heterotrophs*, that is, they consumed pre-existing organic matter and converted it to species with lower thermodynamic free energies. An example of such a heterotrophic reaction is the conversion of the carbohydrate glucose ($C_6H_{12}O_6$) to acetate (CH_3COO^-) by *fermentation*. Glucose is equivalent chemically to six molecules of CH_2O, which we have used to represent generic forms of organic carbon. If we stick with this notation, fermentation can be represented by the reaction

$$2\,CH_2O \rightarrow CH_3COO^- + H^+ \tag{9.26}$$

or simply as

$$2\,CH_2O \rightarrow CH_3COOH \tag{9.27}$$

where we have combined the acetate and hydrogen ion to form acetic acid.

If heterotrophic organisms did indeed originate first, their effect on the atmosphere should have been small. The main environmental function of these organisms would have been simply to change the composition of organic matter that was buried in sediments. Biological productivity in this case could not have been very high because it was limited by the rate of abiotic production of organics. An estimate for how fast this process operated can be made by considering the model atmospheres shown in Figs. 9.4 and 9.5. These atmospheres had an assumed CO_2 partial pressure of ~0.2 bar and H_2 mixing ratios ranging from 10^{-5} to 10^{-2}. The maximum rate of formaldehyde production in these simulations was about 2×10^9 cm^{-2} s^{-1}, or 5×10^{11} mol C yr^{-1} on a global basis. This can be compared with a modern (photosynthetic) primary productivity of 4×10^{15} mol yr^{-1} (Prentice *et al.*, 2001) and a modern organic carbon burial rate of about 5×10^{12} mol C yr^{-1} (Ch. 8). So, if these estimates are correct, a completely heterotrophic ecosystem should have been about 10^4 times less productive than the modern ecosystem and should have buried organic carbon at only one-tenth the modern rate, even if none of the material produced was recycled.

9.6.2 Methanogenesis

Fermentation is an example of *chemotrophy*, because the energy needed to drive metabolism comes from rearranging chemical bonds. Another form of chemotrophic metabolism that is thought to have originated early in life's history is *methanogenesis* – the biological production of methane. Modern methanogens can utilize a variety of short-chain carbon compounds, including CO_2, formate, acetate, propionate, and various amines, alcohols, and organic sulfur compounds (Zinder, 1993). About two-thirds of the CH_4 produced on the modern Earth comes from methanogens that utilize acetate (acetic acid, in our notation), according to the reaction

$$CH_3COOH \rightarrow CH_4 + CO_2 \tag{9.28}$$

Methanogens are thought to have evolved early, as discussed below. Once they did, reaction (9.28) would have been a likely fate for much of the acetate produced by fermentation (reaction (9.27)). The net result of an ecosystem consisting of fermenters and acetotrophic (acetate-consuming) methanogens can be seen by adding reactions (9.27) and (9.28) together to get

$$2\,CH_2O \rightarrow CH_4 + CO_2 \tag{9.29}$$

Such a reaction, in which one part of the reactant compound is oxidized and another part is reduced, is termed *disproportionation*. Production of methane by this process would have reduced the amount of organic carbon buried in sediments and increased the concentration of atmospheric CH_4. The maximum production rate of methane would have been half the (abiotic) production rate of formaldehyde, or 2.5×10^{11} mol yr^{-1}, which is roughly half the modern abiotic CH_4 production rate. So, again, this would have had minimal effect on atmospheric composition.

Methanogens, though, are also capable of *chemoautotrophic* metabolism (reducing CO_2 to organic carbon directly). In particular, virtually all methanogens can use hydrogen to reduce CO_2 via

$$CO_2 + 4\,H_2 \rightarrow CH_4 + 2\,H_2O \tag{9.30}$$

This reaction does not produce organic matter. Rather, it yields energy that can be stored for metabolism, typically by converting adenosine diphosphate (ADP) into adenosine triphosphate (ATP). For every ten CO_2 molecules that such an autotrophic methanogen converts to CH_4, it converts approximately one CO_2 molecule into organic carbon (Fardeau and Belaich, 1986; Morii *et al.*, 1987; Schonheit *et al.*, 1980). This reaction would have increased both primary productivity and methane production dramatically. Because methane plays an integral part in climate regulation in Archean models (Sec. 11.6), the last section of this chapter will explore this type of ecosystem in some detail.

The idea that methanogens are ancient is based on two different lines of biological reasoning, supported by some geochemical data. First, methanogens are found within the Archaeal domain, not far in terms of evolutionary distance from the presumed origin of life (Fig. 9.13). Indeed, methanogens are exclusively confined to one of the two main branches of the Archaea, the Euryarchaeota. If the root of the rRNA tree lies within the Bacterial domain as suggested earlier (Sec. 9.5), this would imply that the Last Universal Common Ancestor (LUCA) was not a methanogen. However, analysis using a two domain tree, in which eukaryotes arise within the archaea, allows a root within the Euryarchaeota where LUCA is a methanogen (Weiss et al., 2016). A second argument that methanogens originated early is that they utilize substrates, e.g., CO_2 and H_2, which are thought to have been present in abundance on the early Earth. If methanogens did not evolve early, then the evolutionary process somehow overlooked an obvious opportunity.

The geochemical evidence for early methanogenesis is the discovery of isotopically light ($\delta^{13}C \leq -56‰$) methane in fluid inclusions from 3.5 b.y.-old cherts from the Pilbara craton in Australia (Ueno et al., 2006). As discussed in Sec. 8.5, methanogenesis fractionates carbon isotopes strongly, so that the methane that is produced is highly depleted in ^{13}C relative to ^{12}C. Some abiotic processes, e.g., Fischer–Tropsch-type reactions, can produce large carbon isotope fractionations, as well (Horita, 2005; Horita and Berndt, 1999); however, Ueno et al. argue that the appropriate conditions for such reactions would not have existed in the environments where these cherts formed. Their observation is consistent with the arguments just given that methanogenesis was an early metabolic invention. In the future, nickel (Ni) isotopes might provide further evidence for early methanogenesis because methanogens preferentially assimilate the light isotopes of Ni (Cameron et al., 2009).

9.6.3 Sulfur Metabolism and Sulfate Reduction

Another probable participant in early metabolic processes was elemental sulfur produced from SO_2 photolysis. Phylogenetic studies indicate that sulfur metabolism originated very early (House et al., 2003a). The metabolism that appears to be most ancient is elemental sulfur reduction

$$S^0 + H_2 \rightarrow H_2S \tag{9.31}$$

Molecular hydrogen, as we have seen, should have been widely available. Photochemical calculations (Kasting

et al., 1989b; Pavlov and Kasting, 2002; Pavlov et al., 2001; Zahnle et al., 2006) indicate that elemental sulfur should also have been available from photochemical reduction of outgassed SO_2. The reaction sequence by which sulfur is formed is quite complicated and not well understood. It has been modeled by the following set of reactions

$$SO_2 + h\nu \rightarrow SO + O \tag{9.32}$$
$$SO + SO \rightarrow S + SO_2 \tag{9.33}$$
$$S + S + M \rightarrow S_2 + M \tag{9.34}$$
$$S_2 + S_2 + M \rightarrow S_4 + M \tag{9.35}$$
$$S_4 + S_4 + M \rightarrow S_8 + M \tag{9.36}$$

In reality, sulfur chain formation may involve other allotropes (S_3–S_7) as well as *sulfanes* (sulfur chains with hydrogen atoms attached at the ends). All of these allotropes, including the thermodynamically stable form, S_8, are implicitly included in the generic "S^0" shown in reaction (9.31). Figure 9.14 shows a schematic diagram of sulfur photochemistry, while Fig. 9.15 shows calculated sulfur gas speciation in a simulated low-O_2, Archean atmosphere. Gaseous S_8 is not seen, as its vapor pressure is low at the assumed low surface temperature (15 °C), but elemental sulfur should have been present in particulate form.

Figure 9.14 shows that not all of the outgassed SO_2 is reduced; rather, some of it should have been oxidized to sulfuric acid, H_2SO_4. The reaction sequence is

$$SO_2 + O + M \rightarrow SO_3 + M \tag{9.37}$$
$$SO_2 + OH + M \rightarrow HSO_3 + M \tag{9.38}$$
$$HSO_3 + O \rightarrow SO_3 + OH \tag{9.39}$$
$$SO_3 + H_2O \rightarrow H_2SO_4 \tag{9.40}$$

The O and OH radicals needed to oxidize SO_2 come from photolysis of CO_2 and H_2O, respectively. The sulfuric acid that is formed in reaction (9.40) dissociates to sulfate, SO_4^{2-}, in solution. Surprisingly large amounts of sulfate are formed in model calculations, because O and OH radicals are relatively abundant in the stratosphere and because H_2SO_4, once formed, is very resistant to photolysis.

Early organisms could have used the sulfate formed in this manner to oxidize organic carbon. Today, *biological sulfate reduction (BSR)* is a major metabolic pathway for organisms living in marine sediments. The reason is that the modern ocean is rich in sulfate (28 mM). Consequently, free energy can be derived from the reaction

$$2CH_2O + SO_4^{2-} \rightarrow 2CO_2 + S^{2-} + 2H_2O \tag{9.41}$$

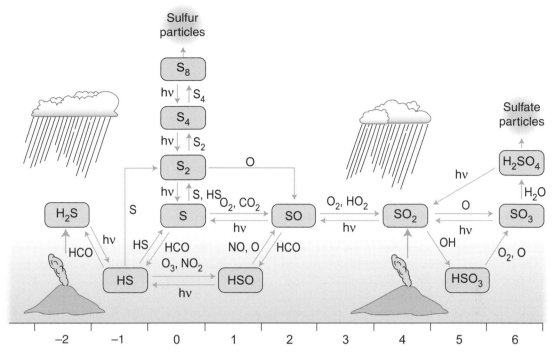

Figure 9.14 Sulfur photochemistry in an anoxic atmosphere. Horizontal scale shows the oxidation state of sulfur. Most sulfur is emitted to the atmosphere as sulfur dioxide, SO_2. Subsequent photochemical reaction can lead to either oxidation or reduction. (From Kasting (2001). Reproduced with permission of Springer. Copyright 2001, Kluwer Academic Publishers.)

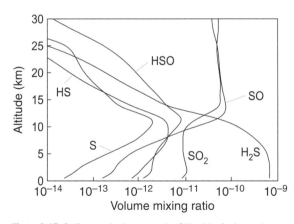

Figure 9.15 Sulfur species in an anoxic, CH_4-rich, Archean atmosphere. The assumed CO_2 mixing ratio is 0.025 and the CH_4 mixing ratio is 0.001. (From Pavlov et al. (2001).)

Sulfide, S^{2-}, then typically combines with iron to make pyrite, FeS_2. The BSR process fractionates sulfur isotopes quite strongly: pyrite that is formed in this manner is typically depleted in ^{34}S relative to ^{32}S by 20‰–40‰, whereas residual sulfate (often preserved as barite, $BaSO_4$) is enriched in ^{34}S by somewhat smaller amounts (Shen and Buick, 2004). Isotopic evidence for BSR is well documented by 2.7 Ga, and microscopic inclusions

in rocks from the Pilbara block in northwestern Australia dated at ~3.5 Ga suggest that BSR was already active by this time (Shen and Buick, 2004; Shen et al., 2001). In criticism, some have suggested that the isotopic fractionation could be caused by thermochemical reactions (Runnegar et al., 2002), but subsequent work has shown that the minor isotopes of sulfur (^{36}S and ^{33}S) are present in ratios that are more consistent with the original interpretation of BSR (Ueno et al., 2008). That said, $\delta^{34}S$ fractionation in rocks remains relatively small until ~2.4 Ga, suggesting that dissolved sulfate remained scarce up until that time, as described in the next chapter (Sec. 10.4.2).

9.6.4 Nitrogen Fixation and Nitrate Respiration

Other anaerobic metabolic pathways are also thought to have originated during the Archean Era. One of the most important was *nitrogen fixation*. Nitrogen, along with phosphorus, is one of two key limiting nutrients in modern marine and terrestrial ecosystems. Nitrogen gas, N_2, is of course the main constituent of Earth's present atmosphere. N_2 is not directly useful to most organisms, however, because it is difficult for them to break the N≡N triple bond. Some abiotic processes are capable of doing so. (1) Impacts with reducing chemistry can generate

HCN (Fegley *et al.*, 1986). (2) In the modern atmosphere, lightning causes N_2 and O_2 to recombine as NO (reaction (9.14)). (3) In the early atmosphere, dissociative recombination of N_2^+ in the ionosphere, followed by downward flow of N atoms, may have resulted in HCN formation in the stratosphere (reactions (9.18) and (9.19)).

The last process, first studied by Zahnle (1986), was probably the dominant abiotic mechanism for nitrogen fixation. Its rate was estimated earlier (Sec. 9.3.1) at 10^{10} cm^{-2} s^{-1}, or ~3×10^{12} mol yr^{-1}. Converting to units used by marine biologists, this is equivalent to about 40 Tg(N) yr^{-1}. By comparison, the modern biological nitrogen fixation rate is ~110 Tg(N) yr^{-1} (Gruber and Sarmiento, 1997). So, the abiotic N fixation rate was substantial if this analysis is correct. These numbers for the abiotic rate, however, are for 3.8 Ga (Tian *et al.*, 2011). By the late Archean, the solar EUV flux would have declined by another factor of 3 or more (Claire *et al.*, 2012), and so fixed N would have become more limiting, perhaps precipitating a fixed nitrogen "crisis." Navarro-Gonzalez *et al.* (2001) suggested a similar scenario based on fixed nitrogen production from lightning, but that process appears less important than the HCN channel, as discussed below.

If the atmosphere contained more CO_2 than CH_4 (which should generally have been the case, for reasons discussed in Ch. 11), lightning would have created NO rather than HCN (Chameides and Walker, 1981). The dominant source of O atoms should have been CO_2, so the relevant reaction can be written as

$$CO_2 + \frac{1}{2} N_2 \leftrightarrow CO + NO \qquad (9.42)$$

Once generated, NO would have been converted to the nitroxyl molecule, HNO, by the reactions

$$H + NO + M \rightarrow HNO + M \qquad (9.43)$$
$$HCO + NO \rightarrow HNO + CO \qquad (9.44)$$

HNO is soluble and would have been rained out of the early atmosphere. Once in the ocean, it would have disproportionated to form a mix of nitrous oxide, nitrite, and nitrate (Mancinelli and McKay, 1988).

The rate of nitrogen fixation from this process can be estimated based on calculations presented earlier in this chapter. As discussed in Sec. 9.2.1, the estimated abiotic NO production rate for a CO_2 partial pressure of 0.2 bar is ~5×10^8 cm^{-2}s^{-1}, or 1.3×10^{11} mol yr^{-1}. Based on redox considerations, only about one-third of this, 4×10^{10} mol yr^{-1} or 0.6 Tg(N) yr^{-1}, would have ended up as biologically useful nitrate or nitrite; thus, this would not have been a significant source of fixed nitrogen

compared to Zahnle's HCN mechanism. It would, however, have enabled another anaerobic metabolism, namely, *nitrate respiration*

$$CH_2O + 2 NO_3^- \rightarrow CO_2 + 2 NO_2^- + H_2O \qquad (9.45)$$

Some authors have suggested that nitrate respiration was a precursor to aerobic respiration (e.g., Egami, 1974, 1976; Takahashi *et al.*, 1963). Broda (1975, 1977) disputed this claim, but his criticism was based on the perceived lack of nitrate in the early oceans, which as we have seen is not a valid objection. So, from our atmospheric standpoint, nitrate respiration could have been an early invention.

The relatively high rates of abiotic nitrogen fixation predicted here are in apparent conflict with previous interpretations of the early biological record. The widespread capability amongst prokaryotes to fix nitrogen biologically has been taken to indicate that fixed nitrogen was in short supply on the early Earth, as it is today (Falkowski, 1997; Walker, 1977). The abiotic N fixation rates predicted here, though, would have been more than enough to keep up with the anaerobic primary productivity rates discussed so far, as well as with those discussed in the next section of this chapter. So, either we have overestimated the abiotic N fixation rate, for reasons that are not clear, or abiotically fixed nitrogen was destroyed at a faster rate than it is today.

9.6.5 Anoxygenic Photosynthesis

The innovation that would have done the most to increase biological productivity, short of oxygenic photosynthesis itself, was the invention of *anoxygenic photosynthesis*, in which organisms used the energy from sunlight to reduce CO_2 to organic carbon. By doing so, they overcame the free energy constraints. This innovation occurred very early in biological evolution, as evidenced by the presence of stromatolites in rocks dating back to almost 3.5 Ga (Sec. 9.4).

Bacteria have invented many different types of anoxygenic photosynthesis. The simplest, and probably one of the most quantitatively important on the early Earth, is the reduction of CO_2 by H_2

$$CO_2 + 2 H_2 \rightarrow CH_2O + H_2O \qquad (9.46)$$

This is the same reaction that we used in Sec. 8.4 to show how organic carbon burial should affect the global redox budget. The hydrogen needed to drive this reaction could have been provided by volcanic gases or, more generally, from the atmosphere itself. Section 9.7 describes a simple model used to estimate biological productivity in an anaerobic biosphere based on this process.

Hydrogen sulfide, H_2S, can also be used to reduce CO_2 photosynthetically. The reaction in this case is

$$CO_2 + 2H_2S(+hv) \rightarrow CH_2O + 2S + H_2O \qquad (9.47)$$

Many modern cyanobacteria are capable of performing this reaction when placed in an environment that is low in O_2 and rich in sulfide. Because of the limited availability of H_2S, the global rate of this process was probably slower than that of other forms of anoxygenic photosynthesis (Canfield et al., 2006; Kharecha et al., 2005). But this reaction was probably part of an early sulfur cycle that may have reprocessed sulfur isotopes on the early Earth, and hence complicated the interpretation of the mass-independent fractionation (MIF) signal (Halevy et al., 2010) (see also Sec. 10.5).

Just as hydrogen and sulfide can be used to reduce CO_2, so too can ferrous iron, Fe^{+2}. Biologists have demonstrated the existence of phototrophic, iron-oxidizing bacteria that fix CO_2 (Heising et al., 1999; Heising and Schink, 1998; Widdel et al., 1993). The iron is precipitated as various oxyhydroxides, often ferrihydrite, $Fe(OH)_3$:

$$4Fe^{+2} + CO_2 + 11H_2O(+hv) \rightarrow 4Fe(OH)_3 + 8H^+$$
$$(9.48)$$

Once in sediments, ferrihydrite dehydrates via goethite (FeOOH) to hematite (Cornell and Schwertmann, 1996)

$$2Fe(OH)_3 \rightarrow Fe_2O_3 + 3H_2O \qquad (9.49)$$

Reactions of this nature have been suggested as the source of early Archean banded iron formations, or BIFs (Ehrenreich and Widdel, 1994; Kappler and Newman, 2004; Konhauser et al., 2002). Magnetite, Fe_3O_4, is more common than hematite in Archean BIFs, probably because of reaction with organic carbon and other reductants (e.g., H_2) in Archean seawater (Klein and Beukes, 1992; Walker, 1984).

Iron-based photosynthesis on the early Earth may have been more or less important than H_2-based photosynthesis, depending on how the analysis is done. To estimate how fast this reaction might have gone, we follow Kharecha et al. (2005) who assume that the deep ocean was filled with 3 ppm ferrous iron (by weight), and that the average upwelling rate was $4\,m\,yr^{-1}$, or $\sim 0.01\,m\,d^{-1}$, as it is today. (This number is just the average ocean depth, 4 km, divided by the deep ocean turnover time, 1000 yr.) Taking into account the surface area of the ocean, $3.6 \times 10^{14}\,m^2$, along with the stoichiometry of reaction (9.48), yields a carbon fixation rate of $\sim 2 \times 10^{13}$ mol yr^{-1}. This number is roughly 40 times faster than the rate of abiotic carbon fixation estimated

in Sec. 9.6.1, but still ~ 200 times slower than the modern rate of net primary productivity (NPP).

An alternative estimate by Canfield (2005) and Canfield et al. (2006) for iron-based photosynthesis is roughly 10 times higher than the above, or 8–24 times slower than modern NPP. This estimate treats Fe as being analogous to P, which is efficiently recycled within marine surface ecosystems. Iron was probably recycled as well by (ferric) iron-reducing bacteria, but we assume that this was a slower process that would have taken place within sediments, as photosynthesizers ballasted with ferric iron coatings (Kappler and Newman, 2004) would have quickly fallen out of the surface ocean.

It should be noted that, prior to the origin of iron-based photosynthesis, upwelled ferrous iron may have been oxidized abiotically by UV radiation (Braterman et al., 1983). The reaction in this case may be written as

$$Fe^{+2} + 3H_2O(+hv) \rightarrow Fe(OH)_3 + 2H^+ + \frac{1}{2}H_2 \quad (9.50)$$

Because of this possibility, there has been a long argument about whether BIFs could have been formed abiotically. More recently, the Braterman et al. experiments have been criticized because they were not conducted with a realistic multicomponent solution that would be expected for Archean seawater (Konhauser et al., 2007a). In such a solution, the formation of hydrous iron silicates has been shown to preclude iron photo-oxidation. So, it seems likely that most BIFs were formed biologically (Pecoits et al., 2015). The debate, then, is whether they were precipitated by anoxygenic photosynthesizers in the manner described above, or whether they were precipitated by cyanobacterially generated O_2.

9.7 Detailed Modeling of H_2-Based Ecosystems

Let us return now to organisms that utilized molecular hydrogen, H_2, either as a reactant for chemosynthesis (reaction (9.30)) or as a reductant for photosynthesis (reaction (9.46)). Because they end up producing methane, either directly or indirectly, these organisms may have had a powerful influence on Earth's early climate. Furthermore, if similar organisms are present on early Earth-like planets around other stars, their effects on atmospheric composition may well be detectable. Thus, it is worth considering such ecosystems in some detail.

Considerable effort has gone into modeling the production of methane and in examining its possible effects on the early atmosphere. In Sec. 7.5.2, we saw that the present abiotic flux of methane is $\sim 1\%$ of the biological flux. By contrast, the predicted flux of methane from an anoxic Archean ecosystem is comparable to the present biological flux (Kharecha *et al.*, 2005). Below, we briefly describe how such an H_2-based ecosystem may have worked.

9.7.1 Atmosphere–Ocean Gas Exchange: the Stagnant Film Model

In the model of Kharecha *et al.*, the factor that limits biological productivity is the transfer of H_2 and CH_4 through the atmosphere–ocean interface. This transfer can be parameterized using the *stagnant film* model (Broecker and Peng, 1982; Liss and Slater, 1974), which is illustrated in Fig. 9.16. Away from the interface, the transport occurs by eddies in the atmospheric boundary layer or in the surface ocean. Near the interface, however, the eddy scale goes to zero, so transport occurs by molecular diffusion. Other processes such as wave breaking and bubble-formation are also important; however, their effects can be subsumed into the stagnant film model with an empirical film thickness.

In the film model with a thickness of the layer z_{film} and the thermal diffusivity D_t, the effective velocity at which gas transfer can take place is given by the piston velocity, $V_p = D_t/z_{film}$. This term acquires its name from the concept of an imaginary piston that moves through the water column at constant velocity, forcing gas either upwards or downwards depending on its concentration gradient (Broecker and Peng, 1982). Aqueous solubilities, thermal diffusivities, and piston velocities for a few key gases are listed in Table 9.1. At the top of the layer, the dissolved gas concentration, $[H_2]$ in this case, is assumed

to be in equilibrium with the atmosphere: $[H_2] = \alpha_H \cdot pH_2$, where α_H is the Henry's Law coefficient and pH_2 is the H_2 partial pressure just above the surface. The dissolved gas concentration at the bottom of the layer, $[H_2]_s$, is equal to the dissolved gas concentration in the surface ocean. The downward flux of H_2 across the interface is given by

$$\Phi_{dep}(H_2) = \frac{D_t \left(\alpha_H \cdot pH_2 - [H_2]_s \right)}{z_{film}} = V_p \cdot \left(\alpha_H \cdot pH_2 - [H_2]_s \right)$$

(9.51)

We have written this as a *deposition flux* (positive downward) because, in the system we are considering, H_2 flows downward from the atmosphere into the surface ocean, where it is consumed by methanogens or other H_2-using anaerobes.

9.7.2 Models of H_2-Based Archean Ecosystems

The stagnant film approach can be used to quantify the rate of primary production in anaerobic ecosystems in which H_2 from the atmosphere is the principal reductant. Both methanogenesis (reaction (9.30)) and H_2-based photosynthesis (reaction (9.46)) qualify as such systems. Both systems should generate methane because, in the absence of dissolved O_2 or sulfate, the organic matter produced by H_2-based photosynthesis would likely have decayed by fermentation and methanogenesis, (combined as reaction (9.29)).

The numerical model of Kharecha *et al.* (2005) of an anaerobic marine ecosystem accounts for primary production by both methanogens and anoxygenic phototrophs (Fig. 9.17). In both cases, H_2 flows from the atmosphere into the surface ocean at a rate determined by its piston velocity. Anoxygenic phototrophs are presumed to pull dissolved H_2 down to negligible concentrations in the surface ocean, so the downward flux from eq. (9.51) is just $V_{dep} \cdot \alpha \cdot pH_2$. Methanogens, like other H_2-using

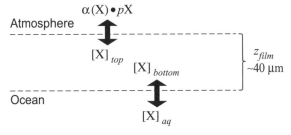

Figure 9.16 The stagnant boundary layer model. The flux of gas X across the atmosphere–ocean interface with film thickness z_{film} is equal to the product of its piston velocity ($v_p(X) = D_t(X)/z_{film}$, where $D_t(X)$ is thermal diffusivity of gas X) and the concentration difference between the top and bottom of the boundary layer ($[X]_{top} - [X]_{bottom}$). In the diagram, $\alpha(X)$ is the Henry's Law solubility constant for gas X, such that $[X]_{top} = \alpha(X)p(X)$, where $p(X)$ is the partial pressure of gas X in the atmosphere. (From Kharecha *et al.* (2005). Reproduced with permission of John Wiley and Sons. Copyright 2005, Blackwell Publishing Ltd.)

Table 9.1 Solubilities, thermal diffusivities, and piston velocities for various gases.

Gas	Solubility (mol l^{-1} bar^{-1})	Diffusivity, D_t (cm^2 s^{-1})	Piston velocity, V_p^* (cm s^{-1})
H_2	7.8×10^{-4}	5.1×10^{-5}	1.3×10^{-2}
CH_4	1.4×10^{-3}	1.8×10^{-5}	4.5×10^{-3}
CO	1.0×10^{-3}	1.9×10^{-5}	4.8×10^{-3}
O_2	1.3×10^{-3}	2.4×10^{-5}	6.0×10^{-3}

[*] Assuming a film thickness, z_{film}, of 40 μm.
Data from the NIST Chemistry WebBook 2009 and Lide (2011). All values at 25 °C.

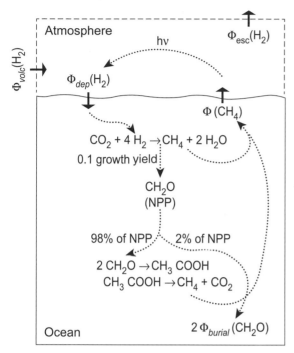

Figure 9.17 Diagram of a methanogen-based ecosystem showing the relevant biochemical reactions and chemical fluxes (see text for term definitions). As in the text, reaction (9.30) represents H₂-using methanogens, reaction (9.27) represents acetogenic bacteria, and reaction (9.28) represents acetotrophic methanogens. (From Kharecha et al. (2005). Reproduced with permission of John Wiley and Sons. Copyright 2005, Blackwell Publishing Ltd.)

chemotrophs, are able to consume dissolved H_2 only until it becomes thermodynamically infeasible to do so. This limit corresponds to a free energy change of about -30 kJ mol^{-1} in reaction (9.30) – approximately the amount of energy needed to convert one mole of ADP into ATP (Conrad, 1996; Kral et al., 1998). The thermodynamic energy yield, ΔG, can be calculated from the free energy form of the Nernst equation

$$\Delta G = \Delta G^0 + RT \ln Q \tag{9.52}$$

where

ΔG = free energy change of the reaction,
ΔG^0 = free energy change of the reaction at STP,
R = universal gas constant = 0.008 314 kJ mol^{-1} K^{-1},
T = water column temperature.
The reaction quotient is

$$Q = \frac{p'CH_4 \cdot a(H_2O)^2}{p'CO_2 \cdot p'H_2^4} \tag{9.53}$$

The partial pressures in eq. (9.53) are not the atmospheric pressures; rather, they are the partial pressures that would be in Henry's Law equilibrium with the dissolved gas

concentrations in the surface ocean, e.g., $p'H_2 = \alpha_H \cdot [H_2]_s$. The activity of water, $a(H_2O)$, is assumed to be 1.

Surprisingly, the effect of this ecosystem on the atmosphere is nearly the same, regardless of whether the assumed primary producers are methanogens or anoxygenic photosynthesizers. (See Table 9.2.) In both cases, much of the available H_2 is converted into CH_4, as predicted previously on heuristic grounds (Walker, 1977). The rate at which conversion occurs is limited by downward diffusion of H_2 through the atmosphere-ocean interface. This happens slightly faster in the photosynthetic ecosystem because dissolved H_2 concentrations are assumed to be negligible. However, the methanogens pull dissolved H_2 down well below its equilibrium concentration with the atmosphere, so this system works in much the same manner.

An example of a coupled photochemical-ecosystem model calculation for an H_2-based photosynthetic biosphere is shown in Fig. 9.18. The three panels show vertical profiles of H_2, CH_4, and CO for three different values of the total hydrogen mixing ratio, $f_{tot}(H_2) \cong f(H_2) + 2 f(CH_4)$, ranging from 200 ppmv to 5000 ppmv. The atmospheric CO_2 mixing ratio was fixed at 0.025 (~80 PAL) in all simulations. The total hydrogen mixing ratio is determined by balancing the global redox budget, as discussed in the previous chapter. The Kharecha et al. model did not explicitly consider seafloor oxidation or pyrite burial. Thus, eq. (8.40) is replaced with a simpler equation

$$\Phi_{out}(Red) = \Phi_{esc}(H_2) + 2\Phi_{burial}(CH_2O) \tag{9.54}$$

Recall that $\Phi_{out}(Red)$ is the total volcanic outgassing rate of reduced species, and $\Phi_{esc}(H_2)$ is the escape rate of hydrogen to space. The coefficient of "2" in the organic carbon burial term comes from the stoichiometry of reaction (9.46). The burial rate of organic carbon was set equal to 2% of NPP (Net Primary Productivity) in these calculations. This is the value for the modern Black Sea, which is anaerobic and sulfidic, i.e. euxinic, at depth (Arthur et al., 1994). By comparison, the estimated burial efficiency for the (mostly oxygenated) modern oceans is ~0.2% (Berner, 1982). The organic carbon that is not buried in this model (98% of the total) was assumed to be recycled by fermentation and methanogenesis (reaction (9.29)).

According to the numbers given in Table 8.2 in the previous chapter, the modern total hydrogen outgassing rate (not including submarine H₂S, which would not have made it out to the atmosphere during the Archean because of the formation of iron sulfides) is just under 3×10^{12} mol yr^{-1}, or ~1×10^{10} cm^{-2} s^{-1}. Because this flux

Table 9.2 Parameters for H_2-based ecosystems.

				Methanogens only				
$f_T(H_2)^*$	$f(H_2)^*$	$f(CH_4)^*$	$\Phi_{esc}(H_2)^{**}$	$\Phi_{burial}(CH_2O)^{**}$	$\Phi(CH_4)^{**}$	NPP**	NPP***	
200	60	70	5.0×10^9	1.4×10^8	7.0×10^{10}	7.0×10^9	1.9	
800	100	350	2.0×10^{10}	2.8×10^8	1.4×10^{11}	1.4×10^{10}	3.6	

H_2-based photosynthesizers, methanogens, and acetogens

$f_T(H_2)^*$	$f(H_2)^*$	$f(CH_4)^*$	$\Phi_{esc}(H_2)^{**}$	$\Phi_{burial}(CH_2O)^{**}$	$\Phi(CH_4)^{**}$	NPP**	NPP***
200	22	83	5.0×10^9	1.4×10^9	3.8×10^{10}	7.0×10^{10}	19
800	42	372	2.0×10^{10}	2.6×10^9	8.9×10^{10}	1.3×10^{11}	35
5000	88	2442	2.3×10^{11}	5.8×10^9	2.3×10^{11}	2.9×10^{11}	76

[*] Mixing ratios in ppmv.
[**] Flux in molecules $cm^{-2}s^{-1}$ (Modern $\Phi(CH_4) = 1.3\times10^{11}$).
[***] Flux in 10^{12} mol yr^{-1} (Modern Net Primary Productivity (NPP) = 4000).

$\Phi_{burial}(CH_2O) = 0.02 \times$ NPP.
Total H_2 outgassing rate = $\Phi_{esc}(H_2) + 2\,\Phi_{burial}(CH_2O)$.
From Kharecha *et al.* (2005).

includes CH_4 from off-axis vents, it implicitly accounts for seafloor oxidation. The net rate of H_2 outgassing is a little over the amount needed to support the 200 ppmv total hydrogen model shown in panel (a). Note that just over half of the outgassed hydrogen escapes to space in this case; the other half goes into organic matter that is buried in sediments (Table 9.2). Even in this case, the calculated CH_4 concentration is 83 ppmv, or more than 100 times higher than the preindustrial CH_4 concentration of ~0.8 ppmv. If hydrogen outgassing rates were higher than today, or if hydrogen escaped to space at less than the diffusion limit, the atmosphere may have resembled panels (b) or (c) in Fig. 9.18, and CH_4 concentrations could have reached ~1000 ppmv.

The calculated net primary productivity and methane flux for two different H_2-based ecosystems are shown in Fig. 9.19 and Table 9.2 as a function of total hydrogen mixing ratio. Calculated values of NPP for the methanogen-only ecosystem are low, roughly $(2-4)\times10^{12}$ mol yr^{-1}, or 1000–2000 times lower than today. NPP is low because these organisms use most of the down-flowing H_2 for metabolism, as opposed to CO_2 reduction. NPP for the H_2-based photosynthetic ecosystem is much higher, $(20-80)\times10^{12}$ mol yr^{-1}, or only 50–200 times lower than today. The lower end of this range is the same value as our estimate for Fe-based photosynthetic productivity given in the previous section. So, if our assumptions are correct, H_2-based photosynthesis was comparable to or faster than Fe-based photosynthesis on the Archean Earth.

Surprisingly, the calculated CH_4 fluxes for the two different ecosystems are comparable, despite the much lower productivity of the methanogen-based ecosystem. For reasonable values of $f_T(H_2)$, calculated methane fluxes are in the range 0.3–2 times the present value. (The present methane flux is 3.6×10^{13} mol yr^{-1}, or 1.3×10^{11} cm^{-2} s^{-1} (Prather *et al.* (2001).) This result is quite remarkable, as there is no *a priori* reason why the methane flux from the Archean ecosystem should be anywhere near the modern methane flux. The Archean Earth was entirely anaerobic, so methanogens could have lived virtually everywhere. The modern Earth, by contrast, is mostly aerobic, so methanogens are confined to restricted environments such as the intestines of cows or the flooded soils beneath rice paddies and other anoxic muds. However, the much greater biological productivity of the modern Earth, caused by oxygenic photosynthesis, makes up for this difference by providing more organic material that methanogens can metabolize in consortia with other microbes.

Finally, we should note that the simulations of the H_2-based photosynthetic ecosystem also included *acetogenic bacteria* (*acetogens*) that metabolize according to the reaction (Genthner and Bryant, 1982; Kerby *et al.*, 1983; Kerby and Zeikus, 1983; Lynd *et al.*, 1982):

$$4\,CO + 2\,H_2O \rightarrow 2\,CO_2 + CH_3COOH \qquad (9.55)$$

If biological CO uptake is not included in this model, the calculated concentration of CO "runs away," as discussed in Sec. 9.3.2, because photochemical production of CO from CH_4 oxidation overwhelms the destruction of CO by reaction with OH and O. This would happen in the methanogen-only ecosystem as well if the calculations were extended to higher values of $f_T(H_2)$. It is not likely

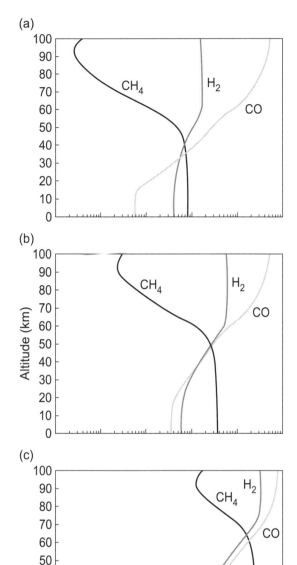

(a)

(b)

(c)

Volume mixing ratio

Figure 9.18 Vertical mixing ratio profiles of H_2, CH_4, and CO in an H_2-based photosynthetic ecosystem. The three panels are for three different assumed values of the total hydrogen mixing ratio, $f_T(H_2)$: (a) 200 ppmv, (b) 800 ppmv, and (c) 5000 ppmv. The CO_2 concentration was kept constant at 25 000 ppmv (2.5%) for each case. (From Kharecha *et al.* (2005). Reproduced with permission of John Wiley and Sons. Copyright 2005, Blackwell Publishing Ltd.)

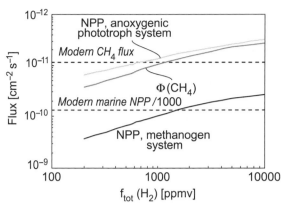

Figure 9.19 Upper two curves: global net primary productivity (NPP) and CH_4 flux as a function of $f_T(H_2)$ for an H_2-based photosynthetic ecosystem. NPP for a methanogen-based ecosystem is also shown (lower curve). Shown for reference are the modern value of the global CH_4 flux and marine NPP (scaled down by 1000). (From Kharecha *et al.* (2005). Reproduced with permission of John Wiley and Sons. Copyright 2005, Blackwell Publishing Ltd.)

to have happened on the post-biotic Earth, however, because the CO would probably have been consumed by acetogens. When these organisms are included in the model, CO mixing ratios rarely rise above 100 ppmv in either of the two H_2-based ecosystems.

9.8   Comparing With the Carbon Isotope Record

In this chapter we have made some fairly detailed predictions about anaerobic biological productivity and about rates of organic carbon burial. Can we test these predictions against the carbon isotope record shown in the last chapter (Fig. 8.2)?

The heterotrophic, sulfur-based, and methanogenic ecosystems all have very low primary productivities compared to today's value. Hence, unless the efficiency of organic carbon burial was much higher than assumed, very little organic carbon would have been buried in such ecosystems, and resulting carbonate $\delta^{13}C$ values should be roughly equal to the mantle value of $-6‰$. There is no evidence for such a signal in the carbonate data shown in Fig. 8.2 back to at least 3.6 Ga; hence, we can infer that life was not only present, but that it has had relatively high productivity since that time.

Now consider our most productive anaerobic ecosystem: the one organized around H_2-based photosynthesis. The estimated organic carbon burial flux for this ecosystem, from Table 9.2, is $(1.4–6)\times10^9$ cm^{-2} s^{-1}, or $(0.4–1.6)\times10^{12}$ mol yr^{-1}. Dividing by the present volcanic CO_2 emission rate, $(8.5\pm2)\times10^{12}$ mol yr^{-1}, gives an organic carbon burial fraction: $f_{org} = (0.06\text{–}0.2)$. This is within the 2-sigma range of f_{org} values (from ~0.05 to ~0.25) estimated for the Archean by Krissansen-Totton et al. (2015), which is shown in Fig. 8.3. But the good agreement does not mean that the Kharecha et al. model is necessarily correct or that our knowledge of Archean ecosystems is secure. The fractionation between carbonate and organic carbon, Δ_B, is lower for anoxygenic photosynthesis than for oxygenic photosynthesis (e.g., House et al., 2003b), whereas the average Δ_B in Archean rocks ranges from $26 \pm 1‰$ in the early Archean (3.8–2.8 Ga) to $31 \pm 3‰$ in the late Archean (2.8–2.5 Ga) (Krissansen-Totton et al., 2015). Thus, to achieve agreement with the carbon isotope record we would need to elaborate on the model discussed above by, for example, allowing some isotopically light carbon derived from methane to become incorporated in sediments. This seems plausible, given the potential importance of methanogens in recycling organic carbon at that time. So, we conclude that we have constructed a self-consistent, but far from definitive, model for how the Archean ecosystem may have worked.

10 The Rise of Oxygen and Ozone in Earth's Atmosphere

In Chapters 6–9, we considered Earth's prebiotic atmosphere followed by how the uptake and release of gases by early life would have changed atmospheric composition. In this chapter, we deal with one of the most significant events in the history of the Earth's atmosphere and biosphere: the advent of *oxygenic photosynthesis*. This single metabolism caused O_2 to become the second most abundant gas after N_2.

An inferred long-term average trend of growth of the oxygen partial pressure, pO_2, over geologic time raises the question of whether pO_2 set the tempo for biological evolution. Animals, in particular, rely on abundant O_2 for respiration and for the synthesis of critical molecules. Consequently, understanding when and why O_2 became plentiful is essential for interpreting the fossil record and our deepest ancestral roots. We start with a brief overview of the co-evolution of O_2 and life, before turning to the geologic constraints on past pO_2 and theories for how and why O_2 levels changed.

10.1 Co-evolution of Life and Oxygen: an Overview

The atmosphere started out with a mixing ratio of O_2 below 10^{-11} (0.1 parts per trillion) before life existed (Fig. 9.1) and now contains 21% O_2 by volume or pO_2 ~0.21 bar (see Fig. 10.1). The first person to discover that levels of O_2 were once negligible was the geologist Alexander MacGregor. In the 1920s, MacGregor deduced that Archean sedimentary rocks in Zimbabwe must have been deposited under an atmosphere devoid of O_2 because the rocks show no sign of reddish-colored, oxidized iron minerals that are common today (MacGregor, 1927). Specifically, MacGregor measured much more ferrous (Fe^{2+}) relative to ferric (Fe^{3+}) iron in Archean sedimentary rocks compared to their modern counterparts. MacGregor's conclusion about a lack of ancient

O_2 strongly influenced the American geologist, Preston Cloud, who championed the view that the geologic record recorded long-term increases in the levels of atmospheric O_2 (Cloud, 1976; Cloud, 1968). Cloud also adopted the suggestion of biologist John Nursall that the late appearance of fossilized macroscopic animals reflected the eventual emergence of an O_2-rich atmosphere capable of supporting animal metabolism (Nursall, 1959). However, the Canadian geologist, Stuart Roscoe, noted that the atmosphere first became oxygenated around 2 Ga, based on levels of oxidation in the Huronian sequence of rocks in Canada (Roscoe, 1969). These ideas suggested that the evolution of atmospheric O_2 concentrations was spread over an interval exceeding two billion years.

A wide variety of geochemical data now show that there were two ancient times when O_2 levels underwent very significant changes (Farquhar *et al.*, 2014) (Fig. 10.1). Two increases occurred near the beginning and end of the Proterozoic eon (2.5–0.541 Ga), respectively. A transition to an oxygenated atmosphere and shallow ocean – named the *Great Oxidation Event* (GOE) by Heinrich Holland (Bekker *et al.*, 2004) – occurred from 2.4–2.2 Ga, within the Paleoproterozoic era (2.5–1.6 Ga). In this chapter, we will discuss evidence that as a result of the GOE, O_2 mixing ratios increased from <1 ppmv to within the absolute range of probably 0.06–2% by volume (also expressed as ~0.3% to 10% PAL, where PAL is present atmospheric level) during much of the Proterozoic, although the lower limit is contentious. At the end of the GOE, at 2.2–2.1 Ga, O_2 may have overshot to some higher level, which is poorly known, before settling to typical middle Proterozoic levels (e.g., Lyons *et al.*, 2014). Around 750–520 Ma, within the Neoproterozoic era (1.0–0.541 Ga) and Cambrian Period (541–485 Ma), O_2 concentrations rose a second time to levels that were probably somewhere between 3% and the modern level of 21% (Frei *et al.*, 2009). Evidence suggests that the second rise of O_2

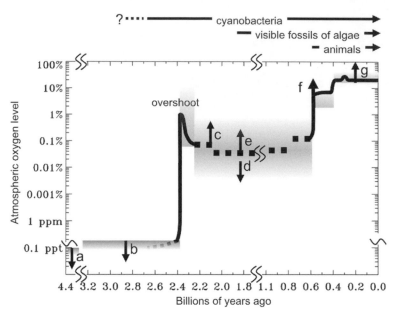

Figure 10.1 Schematic history of the mixing ratio of atmospheric O_2 and the appearance of certain types of organism. The O_2 level in the middle Proterozoic is uncertain with conflicting estimates from geochemical evidence that range from <0.02 and to <2%. Some upper and lower bounds on O_2 are shown as downward and upward-pointing arrows, respectively, labeled a–g. The arrows represent the following constraints. (a) A photochemical model limit for the prebiotic atmosphere (see Fig. 9.1). (b) A constraint of <0.2 ppmv O_2 from the photochemistry required to generate elemental poly-sulfur (S_8) aerosols and preserve mass-independent isotope fractionation in sulfur compounds (Zahnle et al., 2006). (c) Constraint of >0.09% from iodine incorporation into carbonates (Hardisty et al., 2014). (d) Constraint of <0.02% O_2 from the proposed absence of oxidative weathering of chromium (Cr) and its affect on Cr isotopes (Planavsky et al., 2014). (e) Constraint of >0.06% required for an atmosphere with an ozone layer to be photochemically stable with reasonable fluxes (Zahnle et al., 2006); this bound conflicts with arrow c. (f) Constraint of >2% O_2 when macroscopic Ediacaran biota appear. (g) Constraint of >15% from the presence of charcoal since 0.4 Ga (Glasspool and Scott, 2010).

occurred in stages or steps and that oxygenation of the deep sea[1] was a key transition (Canfield, 1998; Canfield et al., 2007; Canfield and Teske, 1996; Fike et al., 2006; Hurtgen et al., 2005; Johnston et al., 2010). Both Precambrian O_2 increases appear in the same eras as substantial changes in the Earth's biota, biogeochemistry (Bekker et al., 2004; Kaufman, 1997; Lindsay and Brasier, 2002; Melezhik et al., 1999) and climate (Hoffman, 2013; Hoffman et al., 1998).

The isotopic composition of Mo in seawater changes in a way that indicates that the area of oxygenated deep seawater was relatively low (<80%) at the beginning of the Cambrian but had increased by c. 535 Ma and that >97% of the seafloor was oxygenated by ~520 Ma (Chen et al., 2015; Wang et al., 2012). From 535–520 Ma,

fossils appear for nearly all the major clades of animals (Erwin et al., 2011; Kouchinsky et al., 2012).

A third rise of O_2 has been hypothesized at around ~400 Ma in the Devonian Period (419–359 Ma). This is correlated with the diversification and spread of vascular plants on land and, possibly, with an increase in the size of predatory fish (Dahl et al., 2010). An increased concentration of molybdenum (Mo) in marine sediments provides key evidence for this third rise of O_2. Mo is released from greater oxidation of molybdenum-containing minerals on land. As in the Cambrian, the isotopic composition of Mo in seawater suggests more oxygenated oceans (see Sec. 10.4.4).

As mentioned above, the importance of O_2 for biological evolution is a compelling reason for an interest in past O_2 levels. Because animals and multicellular plants require O_2 to live (Catling et al., 2005; Margulis et al., 1976), there was no possibility of large animals or plants before an O_2-rich atmosphere. Modern paleontological evidence about organismal size is broadly consistent with

---

[1] In geology, *deep water* means below storm wave base where sediments on the seabed are unaffected by surface waves. The storm wave base is ~30–200 m depth, depending on the strength of winds in storms.

the idea that stepwise oxygenation enabled subsequent biological evolution (Payne *et al.*, 2009; Payne *et al.*, 2011). Field data from oxygen minimum zones in the modern ocean shows that animal size and diversity declines greatly once dissolved O_2 levels fall below ~0.5 ml l^{-1}, which is ~7% of present levels of ~7 ml l^{-1} in the surface ocean (Levin, 2002, 2003). (For comparison, O_2 in air is 210 ml l^{-1}.)

After the GOE, larger organisms appeared than before. The inference that a temporary pulse of relatively high pO_2 over the interval 2.2–2.1 Ga accompanied the GOE (Bekker and Holland, 2012; Canfield *et al.*, 2013; Kump *et al.*, 2011; Planavsky *et al.*, 2012) is consistent with a 2.32–2.1 Ga record of shallow marine sulfate evaporites, given that such of sulfate comes from oxidative weathering of continental sulfides (Schroder *et al.*, 2008). But ocean redox chemistry only settled into a relative stasis after ~1.9–1.8 Ga (Frei *et al.*, 2009; Poulton and Canfield, 2011; Poulton *et al.*, 2004; Slack *et al.*, 2007) (Sec. 10.3.4), which is roughly coincident with the ~1.87 Ga appearance of some of the oldest macroscopic fossils, i.e., those that are visible to the naked eye (Han and Runnegar, 1992).[2] These fossils are found in shales in Michigan and consist of spirally coiled organisms (*Grypania spiralis*), which were probably filamentous algae, i.e., seaweed (Fig. 10.2(a)). However, at 2.1 Ga, centimeter-scale colonies of coordinated aerobic cells (perhaps eukaryotic) in black shales of Gabon may represent the oldest macroscopic life (El Albani *et al.*, 2010). The Gabon assemblage includes possible *acritarchs* (El Albani *et al.*, 2014), which are defined as small, organic-walled fossils that cannot easily be classified into groups of modern organisms. Acritarchs are often interpreted as probable eukaryotes, nonetheless.

Signs of apparent eukaryotic life appear in younger Proterozoic rocks. Moderately diverse acritarchs are found in ~1.8 Ga shales of the Changzhougou Formation in N. China (Lamb *et al.*, 2009). Although there are reports of seaweed-like fossils from 1.6–1.7 Ga Tuanshanzi Formation in China (Zhu and Chen, 1995), these might instead be fragments of microbial mats (Knoll *et al.*, 2006a). However, acritarchs become abundant in younger Proterozoic rocks (Fig. 10.2(b)), generally with a size range of 40–200 µm, which is typical of eukaryotes (Javaux *et al.*, 2004; Knoll, 1992). Then, by 1.4 Ga, marine sedimentary rocks contain carbonaceous fossils of multicellular algae on a worldwide basis (Knoll, 1992). Later, fossils of filamentous red algae at 1.2 Ga are notable as the

(a)

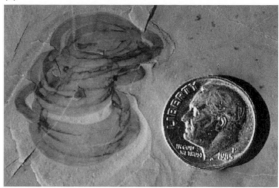

(b)

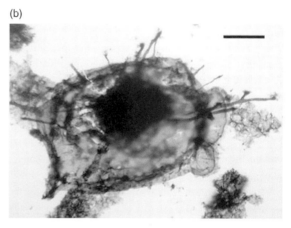

Figure 10.2 (a) *Grypania Spiralis*, which is interpreted as an extinct seaweed, dating to ~1.87 Ga (Courtesy of Bruce Runnegar). (b) *Tappania plana*, which is an acritarch from 1.5 Ga rocks in the Roper Group of northern Australia. Cylindrical branches protruding from this fossil organism suggests the presence of a cytoskeleton, which is an internal protein scaffold characteristic of single-celled eukaryotes. Scale bar = 35 µm. (From Javaux *et al.* (2001). Reproduced with permission from Macmillan Publishers Ltd. *Nature* (Javaux, E.J., *et al.* 412, 66–69) Copyright 2001.)

oldest known sexual organisms (Butterfield, 2000). Meiotic cell division to produce gametes for sexual reproduction, may have originated in eukaryotes as a repair mechanism for oxidative damage (Horandl and Hadacek, 2013; Nedelcu *et al.*, 2004) and so may possibly be related to environmental redox.

The second rise of O_2 near the end of the Proterozoic probably acted as a precursor to the appearance of animals, although the suggestion that it was a direct O_2 trigger for animal evolution is controversial (Budd, 2008; Butterfield, 2009; Mills and Canfield, 2014; Zhang *et al.*, 2014). The presence of animal fossils (possibly sponges) from 635–659 Ma rocks (Maloof *et al.*, 2010) is consistent with independent evidence from biomarkers, which

---

[2] Revised dating from 2.1 Ga originally reported by Han and Runnegar is given by Schneider *et al.* (2002).

indicate that ancient sponges existed before ~635 Ma (Love *et al.*, 2009) and possible proto-sponge structures in ~650 Ma carbonates (Wallace *et al.*, 2014). Later, embryo-like forms inside ornamented organic vesicles are found in rocks just older than 632.5 ± 0.5 Ma (Yin *et al.*, 2007).

In the late Neoproterozoic, the deep oceans were generally anoxic. Sometimes they were sulfidic, but most of the time they were rich in dissolved iron, as Fe^{2+} (Farquhar *et al.*, 2008; Johnston *et al.*, 2010). Periods of ocean anoxia may have lasted until after a short-lived (<1 m.y.) glaciation at ~582 Ma, called the *Gaskiers* (Bowring *et al.*, 2007; Halverson, 2006).

Deep ocean oxygenation was probably a necessary prelude to colonization of the seafloor by macroscopic organisms (Canfield *et al.*, 2007; Fike *et al.*, 2006; Xiao, 2014). Soft-bodied fossils that appear after the Gaskiers were first clearly identified in the Ediacaran hills in the Flinders Range of S. Australia. Similar fossils that were subsequently discovered on six continents are collectively known as the *Ediacaran biota* (Xiao and Laflamme, 2009). The Ediacaran fossils fall within the second half of the Ediacaran Period, 635–541 Ma (Knoll *et al.*, 2004, 2006b), with the oldest at ~575–565 Ma (Narbonne, 2005; Narbonne and Gehling, 2003). Subsequently, fossil evidence for animals with twofold (*bilaterian*) symmetry is found at 555 Ma (Martin *et al.*, 2000). Bilaterians continue to the present day, of course, and include us. Note that some literature uses an old term *Vendian* for the Ediacaran or the latter Ediacaran from 565–541 Ma.

The *Cambrian Explosion* is the relatively fast appearance of animals with new body plans in a 10–30 Myr interval after the beginning of the Cambrian at 541 Ma. Sufficient body size, diverse carnivorous taxa, and extensive food webs apparently allowed an escalatory predator–prey "arms race" that enabled the Cambrian Explosion (Butterfield, 2011; Sperling *et al.*, 2013). As mentioned above, oxygenation of the Cambrian seafloor may have been an important precursor (Chen *et al.*, 2015; Wang *et al.*, 2012). However, molecular genetics suggests that animal ancestors originated during the early Ediacaran (Benton and Ayala, 2003; Peterson *et al.*, 2008). This would be consistent with the first animal fossils (Maloof *et al.*, 2010), geochemical evidence for an increase in marine Mo concentrations at 635–632 Ma (Sahoo *et al.*, 2012), and other geochemical inferences that the second rise in O_2 played out over 750–535 Ma (Canfield *et al.*, 2007; Fike *et al.*, 2006; Och and Shields-Zhou, 2012).

Finally, during the Phanerozoic eon (0.541 Ga to present), pO_2 has probably always been 0.20 ± 0.15 bar (see Sec. 10.10). As mentioned previously, pO_2 increased in concert with a diversification of vascular plants on land at ~400 Ma (Dahl *et al.*, 2010). From ~400 Ma, a record of charcoal in continental rocks indicates that there was sufficient O_2 to burn wood (Chaloner, 1989; Cope and Chaloner, 1980; Scott, 2000). The amount of O_2 required to make charcoal depends on moisture, but generally an O_2 mixing ratio >15% is necessary (Belcher and McElwain, 2008; Belcher *et al.*, 2010; Lenton and Watson, 2000; Wildman *et al.*, 2004). The percentage of charcoal in coals or lignites derived from peat (which controls for the influence of moisture) has been used to infer that pO_2 ranged between 15% and 30% since ~400 Ma (Glasspool and Scott, 2010). The higher end of the range is uncertain due to poor calibration of the proxy at high pO_2. Another proxy for Phanerozoic pO_2 is the stable carbon isotope composition of plant resins such as amber. The $\delta^{13}C$ of plant resins becomes more negative with increasing pO_2. This proxy has been applied back to the Triassic but inferred O_2 concentrations are systematically lower than those derived from charcoal studies, ranging from 12% to modern since ~220 Ma (Tappert *et al.*, 2013).

10.2 Controls on O_2 Levels

10.2.1 Redox Budgeting for the Modern O_2-Rich System

In order to understand the evidence for the Precambrian history of O_2, let us start by considering controls on O_2 levels in recent geological time. As with every atmospheric gas, the abundance of O_2 is set by kinetics – a competition between sources and sinks (Sec. 3.1.2). O_2 exchanges freely with the ocean as well, but we will consider the sources and sinks of O_2 for the combined atmosphere–ocean system for long-term changes. In Ch. 8, we referred to this balance sheet as the *global redox budget*. There, we kept track of inputs and outputs in terms of moles of H_2. For the present O_2-rich atmosphere, it makes sense to consider fluxes in Teramoles (10^{12} mol) of O_2. As pointed out already (eq. 8.2), units of H_2 or O_2 moles can be inter-converted with the stoichiometry of the reaction:

$$2H_2 + O_2 \leftrightarrow 2\,H_2O \qquad (10.1)$$

So, two moles of H_2 correspond to one mole of O_2, albeit with a minus sign, because H_2 is an electron donor, whereas O_2 is an electron acceptor.

We make two additional changes from the redox budgeting described in Ch. 8. There, SO_2 and ferrous iron were treated as redox "neutral" species, meaning that we did not keep track of their fluxes into and out of the atmosphere-ocean system. This makes sense when the system is anoxic, because much of the ferrous iron entering the Archean system was never oxidized and because SO$_2$ is an oxidant relative to an atmosphere rich in H$_2$ and CH$_4$. But, for the modern atmosphere-ocean system, these definitions lead to confusion, in addition to being inconsistent with much published literature. In today's atmosphere, oxidative weathering of sulfides creates sulfate, as does oxidation of volcanic SO$_2$. So, it makes sense to define *sulfate* as the redox neutral sulfur compound, giving SO$_2$ a stoichiometric redox coefficient of –0.5 relative to O$_2$ (SO$_2$ + ½O$_2$ + H$_2$O → H$_2$SO$_4$) or 1 relative to H$_2$. The stoichiometric coefficient for H$_2$S outgassing is then –2 relative to O$_2$ (H$_2$S + 2O$_2$ → H$_2$SO$_4$) or 4 relative to H$_2$ and sulfate (H$_2$S + 3H$_2$O → H$_2$SO$_4$ + 4H$_2$), whereas in Ch. 8 it was 3 relative to H$_2$ and SO$_2$ (H$_2$S + 2H$_2$O → SO$_2$ + 3H$_2$).

Similarly, picking ferrous iron as the reference redox state for the present oxidized system leads to problems. Virtually all ferrous iron is oxidized to ferric iron during weathering. Some of ferric iron stays in the rock and some of it is transported to the oceans in particulate form where it is ultimately used to make pyrite. If we took ferrous iron as the reference state, we would have to determine how much ferric iron left the system in each way. So, instead, we choose ferric iron as the redox reference state. Hydrothermal ferrous iron is then an O$_2$ sink, as is oxidative weathering of ferrous iron on the continents. One picks redox standards for convenience. But the choice of redox standard does not influence the outcome of the budget analysis because our interest lies how sources and sinks change relative to each other, which is fine if we stick to the same unit convention.

10.2.2 The "Net" Source Flux of O$_2$

The subtlety of how photosynthesis influences long-term atmospheric O$_2$ levels was not widely appreciated until the 1970s, even though a French geologist, Jacques Ebelmen, correctly described the key ideas in 1845 (Berner, 2004; Berner and Maasch, 1996; Holland, 2011). Ebelmen realized that oxygenic photosynthesis is nearly a zero-sum activity, which can be summarized schematically, as follows (see also Sec. 8.1):

$$CO_2 + H_2O \underset{\text{respiration, decay}}{\overset{\text{photosynthesis}}{\rightleftharpoons}} CH_2O + O_2 \tag{10.2}$$

Here, the photosynthesis of one mole of organic carbon with average stoichiometry CH$_2$O (i.e., hydrated carbon or carbohydrate) is accompanied by production of one mole of O$_2$. Respiration and decay almost completely reverse the above reaction over ~10^2 years.

The burial of organic carbon turns out be critical because, from eq. (10.2), the burial of one mole of organic carbon means that one mole of O$_2$ escapes the fate of oxidizing this carbon. Consequently, on geologic timescales, the burial of organic carbon is often called the "net source" of O$_2$. An estimated ~0.1–0.2% of organic carbon escapes oxidation through burial in sediments, mainly on continental shelves (Berner, 1982). The assumption here is that buried organic matter is not fossil organic carbon that has been eroded off the continents and reburied, which is confirmed today by a measured fossil carbon flux of only ~0.06 Tmol C yr^{-1} (Dickens *et al.*, 2004), compared with a total burial flux ~10 Tmol C yr^{-1} (Holland, 2002).

The word "net" in the aforementioned O$_2$ source is somewhat misleading because the net O$_2$ is eventually lost to numerous other sinks for O$_2$ besides a reversal of eq. (10.2), as illustrated in Fig. 10.3. In fact, a common interpretation of the geologic record, which we will examine in Secs. 10.7 and 10.8, is that O$_2$ did not accumulate in the Archean atmosphere immediately after the origin of oxygenic photosynthesis because a flux of kinetically rapid O$_2$ sinks exceeded the flux of organic burial providing the source of O$_2$.

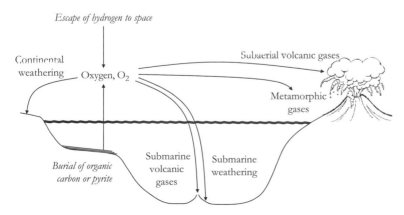

Figure 10.3 Schematic diagram of the source fluxes of oxygen (*in italics*) and sink fluxes of oxygen that operate over geologic timescales.

Table 10.1 The modern source flux of O_2. Data from Holland (1978). The pyrite burial flux in the second row differs by a factor of 2 from a stoichiometric coefficient for O_2 that was published in error in Holland (2002).

Burial Process	Wt% in new sediments	Stoichiometry	O_2 gain (Tmol yr^{-1})
Organic carbon burial	0.6 ± 0.1 as C^0	$CO_2 \rightarrow C + O_2$	10 ± 1.7
Pyrite burial	0.4 ± 0.1 as $S^=$	$Fe(OH)_3 + 2H_2SO_4 \rightarrow FeS_2 + 15/4O_2 + 7/2H_2O$	4.7 ± 1.2
Fe^{2+} burial	1.6 ± 0.6 as FeO	$1/2Fe_2O_3 \rightarrow FeO + 1/4O_2$	1.1 ± 0.4
		Total rate of oxygen production =	**15.8 ± 3.3**

In the twentieth century, with Ebelmen forgotten, Berkner and Marshall (1965) assumed that ancient O_2 concentrations were directly tied to the origin of oxygenic photosynthesis and subsequent changes in the *gross* rate of global photosynthesis rather than the rate of organic burial. However, Van Valen (1971) re-invented Ebelmen's concept. He realized that on geologic timescales almost all the O_2 produced in photosynthesis is used up in oxidizing the accompanying organic carbon back to carbon dioxide, so that rate of burial of organic carbon – not instantaneous photosynthetic production – controls the long-term size of the atmospheric O_2 reservoir.

The net source of O_2 is affected by the burial of other redox-sensitive species besides organic carbon. The most quantitatively significant are sulfur and iron minerals. In seafloor sediments, bacteria use organic matter to chemically reduce sulfate and ferric iron (Fe^{3+}) to pyrite (FeS_2). Effectively, the reducing power of photosynthesized organic carbon is transferred to pyrite, so that pyrite buried in sediments is ultimately balanced by "net" O_2 production. The chemistry can be written as follows (Berner, 2004):

the sinks of O_2 where O_2 reacts with H_2S and SO_2 gases. Otherwise, we would be doing double accounting of the same O_2 molecules. So, in Table 10.1, we use the same geological data as Holland (2002) but our redox accounting is strictly according to the convention we discussed in Sec. 10.2.1, which is critical for consistency.

Consider the sulfur burial flux as an example of how O_2 fluxes are inferred. Average sedimentary rock has 0.4 ± 0.1 wt% S in sulfide (from Table 6–8 in Holland (1978), p. 285). Given an average erosion flux of $\sim 2 \times 10^{16}$ g yr^{-1} (Holland, 1978, pp. 277–278), the burial flux of sulfide sulfur is:

$$\frac{(0.4 \pm 0.1) \times 10^{-2}}{(32 \text{ g S/mol S})} \left(\frac{\text{g S}}{\text{g rock}} \right) \times 2 \times 10^{16} \left(\frac{\text{g rock}}{\text{yr}} \right)$$

$$= (2.50 \pm 0.63) \text{ Tmol S yr}^{-1} \tag{10.4}$$

where, as before, 1 Tmol $= 10^{12}$ mol. Since FeS_2 has 2 moles of S and each mole of FeS_2 buried generates 15/4 mole of O_2 from the pyrite burial reaction in Table 10.1, the O_2 flux is 15/8 times the S flux (as in eq. (10.3)), which is 4.7 ± 1.2 Tmol O_2 yr^{-1}.

$$\begin{aligned} \text{oxygenic photosynthesis :} \quad & 15H_2O + 15CO_2 \rightarrow 15CH_2O + 15O_2 \\ \text{sulfate reduction :} \quad & 15CH_2O + 2Fe_2O_3 + 16H^+ + 8SO_4^{2-} \rightarrow 4FeS_2 + 23H_2O + 15CO_2 \\ \hline \text{net :} \quad & 2Fe_2O_3 + 16H^+ + 8SO_4^{2-} \rightarrow 4FeS_2 + 8H_2O + 15O_2 \end{aligned}$$

$$\tag{10.3}$$

For every mole of S buried in pyrite this way, 15/8 mole of O_2 is generated. The budgeting here takes sulfate and ferric iron as reference oxidation states for sulfur and iron, for reasons pointed out in Sec. 10.2.1.

We follow the approach of Holland (1978, 2002) by using the average composition of fresh sedimentary rocks to estimate burial fluxes of reduced species and the corresponding O_2 fluxes that are released. However, unlike Holland, we use our convention that sulfate is the neutral redox reference state, as discussed above. So burying sulfate, by our convention, does not lead to loss of O_2. Instead, consumption of O_2 to produce sulfate in the first place is considered below in Sec. 10.2.3 in our budget of

With this method, organic carbon and pyrite burial provide 10.0 ± 1.7 Tmol O_2 yr^{-1} and 4.7 ± 1.2 Tmol O_2 yr^{-1}, respectively. These fluxes are about a factor of two larger than those of Berner who estimated 5.3 Tmol O_2 yr^{-1} from burial of organic carbon (Berner, 2004, p. 42) and 2.3 Tmol O_2 yr^{-1} from sulfide burial (converting to O_2 equivalents from 39 Mt S yr^{-1} given in Berner (1982)). However, we tabulate estimates from Holland's data for consistency because we also use his data to estimate oxidative weathering fluxes. In Table 10.1, the reduction of oxidized iron and the burial of ferrous iron adds a minor flux of O_2. In total, the modern O_2 "net source" is 15.8 ± 3.3 Tmol O_2 yr^{-1} (Table 10.1).

Table 10.2 The modern weathering sink flux of O_2. Data from Holland (1978). The pyrite burial flux in the second row differs by a factor of 2 from a stoichiometric coefficient for O_2 that was published in error in Holland (2002).

Loss process	Wt% in average rock undergoing weathering	Stoichiometry	O_2 loss (Tmol yr^{-1})
Carbon weathering	0.45 ± 0.1 as C^0	$C + O_2 \rightarrow CO_2$	7.5 ± 1.7
Sulfide weathering	0.3 ± 0.1 as $S^=$	$FeS_2 + 15/4O_2 + 7/2H_2O \rightarrow Fe(OH)_3 + 2H_2SO_4$	3.5 ± 1.2
Fe^{2+} weathering	1.9 ± 0.6 as FeO	$FeO + 1/4O_2 \rightarrow 1/2Fe_2O_3$	1.3 ± 0.4
		Total rate of continental oxidative weathering =	**12.0 ± 3.3**

10.2.3 The O₂ Sink Fluxes

The modern net O_2 source of Table 10.1 is balanced by losses involving reactions of O_2 in the air or water (Fig. 10.3). The O_2 sinks include reactions with many species: reduced minerals on the continents, gases released from hot rocks that melt (volcanism), gases from hot rocks that do not melt (metamorphism), seafloor minerals encountered by percolating seawater, and gases and dissolved minerals released from hot seafloor vents into the oceans. By examining the minerals that are undergoing weathering in average continental rock, an estimated 12.0 ± 3.3 Tmol O_2 yr^{-1} is lost to continental weathering (Table 10.2). Note that the concentrations of sulfide sulfur and elemental carbon are larger in sediments (Table 10.1) than average rock undergoing weathering (Table 10.2) because the latter includes metamorphic and igneous rocks, while sediments accumulate sulfur and carbon from outgassing sources.

The non-weathering O_2 losses require detailed consideration. In Table 7.3, we considered the fluxes of reducing gases emitted from the Earth. In the atmosphere, reducing gases are oxidized through photochemical reactions that sum to a net oxidation by O_2, which effectively is equivalent to combustion. The main gases that consume O_2 in this way are H_2, CO, CH_4, H_2S, and SO_2. Table 10.3 recasts the fluxes of Table 7.3 with the others, in terms of O_2 sinks (see also Table 8.2.)

However, once again, the sulfur-bearing fluxes require care. Ferrous iron in the seafloor is oxidized to magnetite by seawater sulfate in hydrothermal circulation. If we assume that the sulfide vents to the sea and gets oxidized by O_2, there's an O_2 sink as follows, tied directly to the flux of magnetite made.

$$12FeO + H_2SO_4 \rightarrow 4Fe_3O_4 + H_2S$$
$$\underline{2O_2 + H_2S \rightarrow H_2SO_4}$$
$$net: 6FeO + O_2 \rightarrow 2Fe_3O_4 \qquad (10.5)$$

In eq. (10.5), one batch of seawater sulfate is replaced with another, while O_2 enters oceanic crust in the Fe^{3+} component of magnetite. However, if the S^{2-} derived

from sulfate does not vent but stays in the rock as FeS_2, then seawater sulfate is being reduced and iron oxidized, such that there is no net O_2 sink if we take sulfate as our redox reference state of sulfur (i.e., the redox states of the FeS_2 and Fe_3O_4 balance out). Only the portion of the flux of 0.35 Tmol H_2S yr^{-1} listed under "submarine volcanism" in Table 7.3 that comes from the reduction of sulfate acts as an O_2 sink into magnetite. The rest of the H_2S flux is derived from sulfide that is leached from basaltic sulfides by hydrothermal vent fluid and is a qualitatively different O_2 sink. Shanks and Seyfried (1987) suggest that ~30% is an upper limit to the molar amount of H_2S from seafloor vents derived from sulfate based on $\delta^{34}S$ data, while Ono et al. (2007) suggest ~10%–30% based on more comprehensive sulfur isotope systematics. We assume that $20 \pm 10\%$ of H_2S from vents is derived from sulfate, i.e., 0.070 ± 0.035 Tmol H_2S yr^{-1} equivalent to 0.14 ± 0.07 Tmol O_2 yr^{-1} consumption listed under seafloor oxidation in Table 10.3.

Actually, a considerable amount of magnetite is made from the oxidation of new seafloor by sulfate (Lecuyer and Ricard, 1999). Sleep (2005) suggests that the total oxidation may be equivalent to full oxidation of FeO to 0.5 km depth. With 10wt% FeO, the new moles of FeO yr^{-1} = [0.1 × 3 km^2 yr^{-1} spreading rate × (0.5 km depth) × 3000 kg m^{-3} × 10^9 m^3 km^{-3}]/(0.072 kg/(mol of FeO)) = 6.25×10^{12} mol FeO yr^{-1}. With the stoichiometry of the oxidation of FeO to magnetite (eq. (10.5)), the O_2 sink is potentially 1×10^{12} mol O_2 yr^{-1}. However, if we use sulfate as our redox reference state for sulfur, the O_2 sink is equivalent only to the sulfate-derived H_2S that vents and consumes O_2, which is ~0.14 Tmol O_2 yr^{-1}, as described above.

The result of adding up all the sinks in Table 10.3 is 5.7 ± 1.2 Tmol O_2 yr^{-1} loss. We can add this to the weathering sink of 12.0 ± 3.3 Tmol O_2 yr^{-1} to give a total sink flux of 17.7 ± 4.5 Tmol O_2 yr^{-1}, which agrees with the total O_2 source flux of 15.8 ± 3.3 Tmol O_2 yr^{-1} within a large uncertainty. To put it another way, today roughly 70% of the O_2 is lost to continental weathering and ~30%

Table 10.3 The O_2 sink from volcanic, seafloor and metamorphic reductant fluxes.

Species	Flux (Tmol yr^{-1})	Stoichiometry for O_2 loss	O_2 Loss flux (Tmol yr^{-1})
Surface volcanism[a]			
H_2	1 ± 0.5	$H_2 + 1/2O_2 \rightarrow H_2O$	0.5 ± 0.3
CO	0.1 ± 0.05	$CO + 1/2O_2 \rightarrow CO_2$	0.05 ± 0.03
H_2S	0.03 ± 0.015	$H_2S + 2O_2 \rightarrow H_2SO_4$	0.06 ± 0.03
SO_2	1.8 ± 0.6	$H_2O + SO_2 + 1/2O_2 \rightarrow H_2SO_4$	0.9 ± 0.3
		Subtotal for surface volcanism	**1.5±0.7**
Submarine volcanism[a]			
H_2 (vents)	0.1 ± 0.05	$H_2 + 1/2O_2 \rightarrow H_2O$	0.05 ± 0.03
H_2S	0.28 ± 0.10	$H_2S + 2O_2 \rightarrow H_2SO_4$	0.6 ± 0.2
CH_4 (axial)	0.01 ± 0.005	$CH_4 + 2O_2 \rightarrow CO_2 + H_2O$	0.02 ± 0.01
CH_4 (off-axis)	0.03 ± 0.02	$CH_4 + 2O_2 \rightarrow CO_2 + H_2O$	0.06 ± 0.04
		Subtotal for submarine volcanic gases	**0.7±0.3**
Surface metamorphism and serpentinization			
CH_4 (abiotic)[*]	0.3	$CH_4 + 2O_2 \rightarrow CO_2 + H_2O$	0.6
CH_4 (thermogenic)[**]	1.25	$CH_4 + 2O_2 \rightarrow CO_2 + H_2O$	2.5
		Subtotal for metamorphic gases	**3.1**
Seafloor oxidation, serpentinization, and hydrothermal iron			
Fe^{2+} oxidation to magnetite sulfate		See eq. (10.5) and discussion in the text	0.14 ± 0.07
Fe^{2+} conversion to magnetite from serpentinization[***]		$3FeO + H_2O \rightarrow Fe_3O_4 + H_2$	0.2 ± 0.1
Fe^{2+}(aq) hydrothermal flux (axial + off-axis)[#]	0.25 ± 0.09	$FeO + 1/4O_2 \rightarrow 1/2Fe_2O_3$	0.06 ± 0.02
		Subtotal for seafloor oxidation	**0.4±0.2**
		Total sink from reductant fluxes	**5.7±1.2**

Sources: [a]Table 7.3 for H_2, CO, H_2S, where submarine H_2S only includes that sourced from basaltic sulfide, not seawater sulfate (see text). Table 7.1 for SO_2.
[*]Estimate from Fiebig *et al.* (2009), although others note that abiotic CH_4 fluxes are poorly constrained (Etiope and Lollar, 2013).
[**]Lower estimate of 1.25–2.5 Tmol CH_4 yr^{-1} from Etiope *et al.* (2009).
[***] Sleep (2005) notes that significant mantle rock is exposed only at ultraslow spreading centers at a rate of 0.3 km^2 yr^{-1}, which for 50% serpentinization to ~2 km depth generates enough H_2 to consume 0.2 Tmol O_2 yr^{-1}. Bottoms-up estimates of serpentinization fluxes of H_2 are somewhat lower at ~0.1 Tmol O_2 yr^{-1} (Cannat *et al.*, 2010; Keir, 2010). These estimates may be missing H_2 from systems like "Lost City" that are difficult to detect, so we take Sleep's estimate. CH_4/H_2 molar ratios in ultramafic-hosted hydrothermal fluids are ~1:15 (Cannat *et al.*, 2010), hence the CH_4 (off-axis) flux of 0.4/15 = 0.03 Tmol yr^{-1}.
[#]From Poulton and Raiswell (2002), which includes a review of seafloor hydrothermal iron sources.

is lost to a flux of reductants from volcanism, metamorphism, and seafloor weathering.

In Table 10.3, we see that the four largest sinks of modern O_2 come from metamorphic thermogenic CH_4, subaerial volcanic SO_2, seafloor H_2S, and abiotic CH_4, respectively. We also note that we lack an estimate of the diffuse flux of reducing volcanic gases that diffuses up from areas of igneous intrusions but it may be important (Sec. 10.7.3.2).

The thermogenic CH_4 flux is based on a bottoms-up inventory of sources of natural microseeps and macroseeps of CH_4 from onshore hydrocarbon reservoirs and breakdown of disseminated sedimentary organic carbon (Etiope *et al.*, 2009; Etiope *et al.*, 2008). We can check its credibility using an independent method. A comparison of 0.6wt% of C in average new sediments to 0.45 wt% of C in uplifted rock exposed to weathering (Holland, 2002) implies that a 0.15 wt% difference has been lost in metamorphic reactions such as $C + 2H_2O = CO_2 + 2H_2$ (Mason, 1990) or thermogenic cracking, such as $C_nH_{n+2} = CH_4 + C_{n-1}H_{n-2}$. Given the organic burial flux of ~10 Tmol yr^{-1} (Table 10.1), an estimate of today's metamorphic reducing flux due to carbon is $(0.15/0.6) \times$ 10 Tmol yr^{-1} = 2.5 Tmol O_2 consumption yr^{-1}. A consistency with Etiope *et al.*'s (2009) lower estimate of 1.25 Tmol CH_4 yr^{-1}, equivalent to 2.5 Tmol O_2 consumption yr^{-1}, is why we list this number in Table 10.3 rather than the upper limit of 2.5 Tmol CH_4 yr^{-1} from Etiope *et al.* (2009).

10.2.4 Generalized History of Atmosphere–Ocean Redox

We have just considered the O_2 balance sheet in modern geologic times, but the history of O_2 can be described generally. We need to consider O_2-consuming or O_2-producing fluxes in and out of the atmosphere-ocean reservoir (Fig. 10.3). As in Tables 10.1–10.3, a convenient unit for the relevant fluxes is Tmol yr^{-1} ($=10^{12}$ mol O_2 yr^{-1}). In consistent units, the modern atmosphere–ocean reservoir of oxygen is $R_{O2} = 3.8 \times 10^7$ Tmol O_2. Dividing the atmosphere–ocean reservoir (3.8×10^7 Tmol O_2) by the source flux from Table 10.1 of ~15.8 Tmol O_2 yr^{-1} gives 2.4 Myr for the average amount of time an O_2 molecule spends in the atmosphere–ocean system today (the *residence time*).

The rate of change of the reservoir of O_2, R_{O2}, in the atmosphere–ocean system is

$$\frac{d(R_{O2})}{dt} = F_{\text{source}} - F_{\text{sink}}$$
$$= F_{\text{source}} - (F_{\text{volcanic}} + F_{\text{metamorphic}} + F_{\text{weathering}})$$
$$(10.6)$$

Here, F_{sink} is the removal flux of O_2 from the atmosphere–ocean (in Tmol yr^{-1}) due to numerous oxidation reactions. F_{source} is the source flux of oxygen (in Tmol yr^{-1}) due to burial of organic carbon and pyrite. The flux of hydrogen to space from the top of the atmosphere is also an oxidizing source flux. The sink fluxes are the reaction of O_2 with reduced volcanic gases (both subaerial and submarine) (F_{volcanic}), reduced metamorphic gases ($F_{\text{metamorphic}}$), and reduced material on the continents and seafloor ($F_{\text{weathering}}$). If R_{O2} is in "steady-state," F_{sink} will match F_{source} and $d(R_{O2})/dt \approx 0$. Oxygen levels increase when the terms on the right-hand side of eq. (10.6) are in temporary imbalance with $F_{\text{source}} > F_{\text{sink}}$. Before attempting to answer how this might have happened, we will consider the evidence for past changes in O_2, some of which is summarized in Fig. 10.4.

Before we leave the issue of controls on O_2, we will mention an alternative hypothesis that the evolution of different forms of rubisco, and their efficiencies in producing various gases and fixing carbon, may have regulated the proportions of atmospheric O_2, CO_2, and CH_4 over time (Nisbet and Nisbet, 2008). *Rubisco* is the enzyme ribulose-1,5-bisphosphate carboxylase/oxygenase. This protein comes in different forms. Rubisco I, found in cyanobacteria and plants, is the most abundant enzyme on Earth (Ellis, 1979). In contrast, rubisco II occurs in some anoxygenic photosynthetic bacteria. Rubisco III is found in methanogens and some other archaea, while rubisco IV (also known as rubisco-like protein or RLP) occurs in some bacteria and archaea. Forms I, II, and III catalyze the attachment of CO_2 or O_2 to ribulose biphosphate, a five-carbon molecule, whereas rubisco IV does not catalyze either reaction.

Rubisco I works both ways: depending upon the O_2:CO_2 ratio, rubisco I either fixes carbon from CO_2 (making O_2) or produces CO_2 in photorespiratory uptake of O_2. An *O_2 compensation point* is the O_2 level, for a given temperature and CO_2 level, at which net O_2 exchange is zero. Similarly, a *CO_2 compensation point* is the CO_2 level, with a given temperature and O_2 level, at which net CO_2 fixation is zero. In a closed system, net photosynthetic uptake of CO_2 reaches equilibrium where it balances the gross O_2 uptake, which can be expressed by the relationship (Andre, 2011):

$$f_{O2} \text{ [in \%]} = 0.5(Sp)f_{CO2}\text{[in ppmv]} \qquad (10.7)$$

Here, f_{O2} and f_{CO2} are mixing ratios of O_2 and CO_2 in % and ppmv, respectively, while Sp (in units of % per ppm CO_2) is an empirical *specificity* representing the relative preference of rubisco for CO_2 versus O_2 that incorporates the gradient of CO_2 from the air to rubisco. At the concentrations of O_2 and CO_2 in eq. (10.7), O_2 and CO_2 uptakes are equal. At an average, specificity for C3 plants is 0.132 % ppmv^{-1} (Andre, 2011).

$$20\% \text{ } O_2 \approx 0.5 \times \left(0.132 \text{ \% ppmv}^{-1}\right) \times \left(300 \text{ ppmv } CO_2\right)$$

These levels of O_2 and CO_2 correspond roughly to preindustrial Holocene values.

Because of the above relationship, some have hypothesized that evolutionary selection of rubisco variants might control the O_2:CO_2 ratio of the modern atmosphere (Tolbert *et al.*, 1995) and moderate Earth's climate (Nisbet *et al.*, 2012; Nisbet and Nisbet, 2008). A possible negative feedback is that if ambient O_2 were low, CO_2 would be reduced to organic matter preferentially, and once buried, O_2 would increase.

The hypothesis faces some challenges. If photosynthetic specificity had been the same in the past, O_2 and CO_2 levels should correlate, but evidence suggests that a time of high O_2 at about 300 Ma was associated with low CO_2 levels (compare Fig. 10.10 and Fig. 11.17 later in the

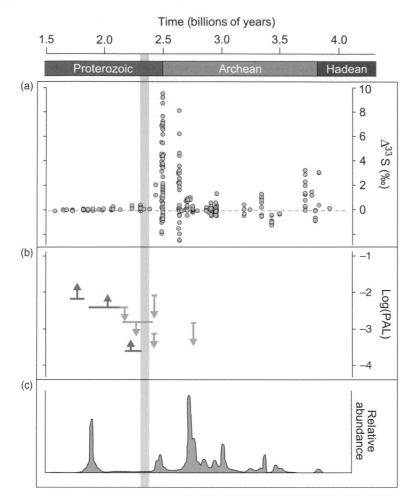

Figure 10.4 Various proxies that indicate changes in atmospheric O_2 at ~2.4 Ga (indicated by the grey vertical bar) coupled to changes in oceanic redox. (From Sessions *et al.* (2009).) (a) Mass-independent fractionation of sulfur isotopes, which largely disappears after the Great Oxidation Event (see Sec. 10.5). (b) Constraints on O_2 levels from ancient, lithified soils (paleosols), expressed in terms of present atmospheric level (PAL). (c) Relative abundance of Banded Iron Formations.

book). Also, specificity presumably adapted to ambient gas levels that have been likely been affected by external factors. Ultimately, even if specificity is involved, we must frame a model of the time-history of O_2 in terms of global redox fluxes, i.e., eq. (10.6).

10.3 Evidence for a Paleoproterozoic Rise of O_2

10.3.1 Continental Indicators: Paleosols, Detrital Grains, and Redbeds

Paleosols, detrital grains, and redbeds all suggest very low levels of O_2 before ~2.4 Ga.

Paleosols are lithified soils, which can provide estimates of pO_2 based on iron and rare-earth element geochemistry (see Rye and Holland (1998), Murakami (2011), and Fig. 10.4(b)). Modern soils generally contain more iron per kg than underlying rock because oxygenated rainwater washes away soluble elements and leaves

behind insoluble ferric iron, Fe^{3+}, in the form of iron oxides (e.g., Murakami *et al.*, 2011). But paleosols older than ~2.4 Ga are deficient in iron (e.g., Utsunomiya *et al.*, 2003). Evidently, iron was flushed from soils because ferrous iron, Fe^{2+}, was soluble in anoxic rainwater. Sometimes, the Fe^{2+} ended up in clay minerals near the base of the paleosol, which was converted into high-Fe^{2+} chlorite and quartz during metamorphism. Changes in a paleosol's cerium concentration profile have also been used to similarly trace oxygen because Ce^{3+} is soluble whereas Ce^{4+} is insoluble (Murakami *et al.*, 2001). However, some argue that the Ce profile is more sensitive to pH than redox (Alfimova *et al.*, 2011).

Other evidence for redox change comes from *detrital grains*, which are sedimentary minerals that never completely dissolve in weathering. Detrital grains in pre-2.4 Ga riverbeds commonly contain reduced minerals that would only survive at low pO_2 (England *et al.*, 2002; Hofmann *et al.*, 2009; Rasmussen and Buick, 1999).

The roundness of such grains shows that they were transported long distances in aerated rivers.

Detrital grains of pyrite (FeS_2), uraninite (UO_2), and siderite ($FeCO_3$) place upper bounds on Archean pO_2. The bounds are debated because they depend on assumptions about pCO_2, the time to transport grains along rivers, and initial grain size. The dissolution kinetics of uraninite and pyrite suggest pO_2 upper bounds of ~0.004 atm (Grandstaff, 1980; Holland, 1984, p. 331) and ~2×10^{-4} atm (Ono, 2001), respectively. However, very fine pyrite sand (~100 μm sidelength cubes) could dissolve at lower pO_2 ~2.1×10^{-6} atm (Anbar et al., 2007; Reinhard et al., 2009), and so a continental source of sulfate to late Archean oceans is possible if sulfides were in proximity to cyanobacterial mats in lakes and relatively high local O_2 concentrations (Stüeken et al., 2012). A chemical–physical model applied to detrital grains in the 2.415 Ga Koegas Subgroup of southern Africa, which was a ~1000 km-long system where transport took ~10^5 years, suggests $pO_2 < 1 \times 10^{-4}$ atm and $< 3.2 \times 10^{-6}$ atm from the survival of pyrite and uraninite, respectively (Johnson et al., 2014). These trace O_2 levels are consistent with estimates of $pO_2 < 0.2$ ppmv inferred from mass independent sulfur isotope fractionation discussed in Sec. 10.5.

As originally noted by Cloud (1968), the appearance of continental redbeds in the early Proterozoic suggests that free O_2 became present in the atmosphere at that time. Redbeds are sedimentary sandstones derived from wind-blown or river-transported particles coated with red-colored iron oxides, usually hematite (Fe_2O_3), which formed after there was sufficient atmospheric O_2 to make the iron oxides (Eriksson and Cheney, 1992; Melezhik et al., 1999; Prasad and Roscoe, 1996).

Post-depositional alteration is a potential pitfall with ancient mineralogical indicators and explains occasional oxidation in Archean rocks. One example is the ~3 Ga Steep Rock succession of Canada, where "buckshot" iron ore, which looks like an Archean redbed, is found 300–350 m underground. However, Machado (1987) found microscopic features in the ore characteristic of termites, and subsequently woody fragments were identified (Nisbet, 2002). The iron oxide was apparently made by Cretaceous or Tertiary termites that tunneled 500 m downwards from a paleosurface to the water table, bringing down wood and O_2. Another claim is that hematite (Fe_2O_3) in the 3.46 Ga Marble Bar Chert Member and Apex Basalt in the Pilbara, Australia, formed in the presence of Archean-age oxygenated seawater (Hoashi et al., 2009) or groundwater (Kato et al., 2009). However, U–Th–Pb isotope systematics indicate the addition of U from oxygenated Phanerozoic surface waters to these rocks (Li et al., 2013; Li et al., 2012) while examination of the petrology shows that hematite was formed from post-depositional oxidation (Rasmussen et al., 2014).

10.3.2 Banded Iron Formations

As well as the continental indicators described above, ancient redox change is indicated by banded iron formations (BIFs), which appear from the start of the sedimentary geologic record at 3.7–3.8 Ga but decline in abundance through the Paleoproterozoic and disappear after ~1.8 Ga (Klein, 2005) (Fig. 10.4(c)). BIFs reappear briefly around 720-600 Ma, but in association with glacial deposits, where they are taken as evidence for possible "Snowball Earth" episodes (see Sec. 11.10). BIFs are defined as laminated marine sedimentary rocks with greater than 15 wt% iron, which contain iron-rich and iron-poor (usually silica-rich) layers (Bekker et al., 2010). BIFs are categorized as either Algoma or Superior types. *Algoma-type BIFs* occur in restricted basins closely associated with volcanic rocks, whereas *Superior-type BIFs* are from a range of water depths on continental shelves.

BIFs provide evidence that the deep Archean and Paleoproterozoic oceans were anoxic and probably had 40–120 μM (2–7 ppm by wt) of dissolved ferrous iron (Fe^{2+}), unlike the ~2 ppb today (Holland, 1973b; 1984, p. 387). Trace and rare earth element patterns in BIFs indicate hydrothermal input of Fe^{2+}, which would have remained in solution if the Archean deep ocean had been anoxic (Bau et al., 1997; Holland, 1984). For example, trace europium (Eu) is leached from the calcium–feldspar component ($CaAl_2Si_2O_8$) of basalts when seawater flows through hydrothermal vents. Vent fluid consequently has a so-called *positive europium anomaly*. Such anomalies are large in Algoma-type BIFs of the early Archean and present but smaller in Superior-type BIFs.

In Superior-type BIFs, Fe^{2+} emanating from deep-sea hydrothermal vents was transported to shallow continental shelves where oxidation precipitated ferric iron (Fig. 10.5(a)). Microscopic studies show that magnetite from a large number of BIFs contains hematite cores (Han, 1988), so that the original precipitate was likely ferrihydrite ($Fe(OH)_3$), which became hematite (Fe_2O_3) and then was chemically reduced to magnetite in the sediments. Iron isotopes suggest only partial oxidation of Fe^{2+} in surface waters of 3.7–3.8 Ga BIFs of Isua, but complete oxidation of Fe^{2+} in Paleoproterozoic BIFs (Czaja et al., 2013). These inferences are consistent with oxidation by anoxygenic photosynthesis (Garrels et al., 1973; Widdel et al., 1993) and later by oxygenic photosynthesis once it evolved (e.g., Posth et al., 2014).

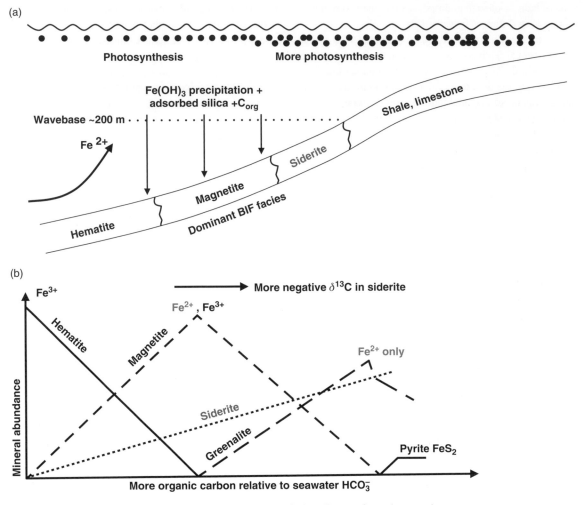

Figure 10.5 (a) The schematic relationship between dominant facies, distance from shore, and productivity for Superior-type banded iron formations (BIFs), e.g., the Kuruman BIF in southern Africa, following Klein (2005). (b) The iron chemistry corresponding to the diagram in (a), following the theory that the primary precipitate is ferric iron (Fe^{3+}), which is biologically reduced to ferrous iron (Fe^{2+}) minerals to an extent depending upon the availability of organic carbon (Walker, 1984). More organic carbon is delivered to sediments closer to shore where there is greater productivity.

As mentioned in Sec. 9.6.5, experiments have cast doubt on old proposals (Braterman *et al.*, 1983; Cairns-Smith, 1978) that UV light could have caused abiotic precipitation of ferric iron (Konhauser *et al.*, 2007a). UV oxidation is far slower than abiotic precipitation of ferrous silicate or carbonate minerals, or, alternatively, than precipitation of iron oxides from anoxygenic or oxygenic photosynthesis.

Many iron minerals in BIFs were likely made during post-depositional chemical reactions (e.g., Melnik, 1982; Posth *et al.*, 2013). Walker (1984) proposed that if organic carbon was originally co-deposited with ferric iron, bacteria would use the organic matter to reduce the iron in the

hematite to make magnetite (Fe_3O_4), iron silicates or siderite ($FeCO_3$) with increasingly available quantities of organic carbon (Fig. 10.5(b)). Lab experiments (Kohler *et al.*, 2013) and isotopes of Fe, O, and C in BIFs confirm Walker's hypothesis (Heimann *et al.*, 2010; Johnson *et al.*, 2013a). Additional support for this idea comes from the microscopic studies (Han, 1988; Sun *et al.*, 2015). However, others argue that iron silicates were the dominant primary precipitates in BIFs (Rasmussen *et al.*, 2016). We return to this issue in Sec. 11.4.3 because it bears on the question of whether BIFs provide constraints on atmospheric CO_2.

The cessation of BIF deposition at ~1.8 Ga can be explained by a change in seawater chemistry that prevented iron from upwelling to coastal areas. Three competing ideas about the ocean chemistry have been put forward. The first is that the deep ocean became oxic after the GOE so that iron precipitated on the ocean floor near its hydrothermal source, as it does today (Holland, 1984, 2006). Exhalative sedimentary aprons called *exhalites* around volcanic massive sulfide (VMS) deposits associated with hydrothermal vents provide some support (Slack and Cannon, 2009; Slack *et al.*, 2007). The depth of VMS deposits with >1 wt% Cu exceeds 850 m because Cu is only soluble at >300 °C, which requires sufficient hydrostatic pressure to raise the boiling point of water. Before ~1.85 Ga, VMS-related exhalites generally consist of minerals characteristic of anoxic conditions, such as iron sulfides or pyritic chert, or iron silicates or carbonates. After 1.85 Ga, exhalites contain jasper (hematitic chert), hematite or magnetite, which are minerals that may be characteristic of deep seawater with trace oxygen levels, ~1 µM. However, the data have exceptions, e.g., a magnetite and jasper-containing exhalite at 2.96 Ga in Scuddles, W. Australia (Slack and Cannon, 2009). Indeed, magnetite and goethite can coexist at very low oxygen fugacity below 10^{-50} bar (Farquhar *et al.*, 2010), while hematite is increasing favored over goethite at higher temperatures (Diakonov *et al.*, 1994).

A second proposal is that the ocean became sulfidic at depth, which scavenged ferrous iron into iron sulfides. The proposed source of the sulfide is sulfate-reducing bacteria in the surface ocean, which provided a flux of sulfide to the deep ocean (Canfield, 1998). Data from molybdenum isotopes (see Sec 10.4.4), as well as Cr and Mo abundances, suggest that several percent of the Proterozoic ocean seafloor area was probably *euxinic*, meaning oxygen-free with dissolved H$_2$S (Reinhard *et al.*, 2013b). This is far larger than today's 0.3% area of euxinic ocean, but is not the whole deep ocean.

A third proposal is that the Proterozoic deep ocean generally remained rich in ferrous iron (*ferruginous*), even after the cessation of BIFs, while only the middle depth continental margins were sulfidic (Poulton and Canfield, 2011). Analysis of iron-bearing minerals (see Sec 10.3.4) at 1.8 Ga shows sulfidic conditions in continental slope settings out to ~100 km from the paleoshore, whereas the deep Paleoproterozoic ocean remained ferruginous (Poulton *et al.*, 2010). Iron minerals in mid-Proterozoic shales also suggest a ferruginous deep sea (Planavsky *et al.*, 2011), while relatively low uranium enrichments in shales indicate a small marine uranium reservoir and long-lived ocean anoxia (Partin *et al.*, 2013). Under such conditions, upwelling ferrous iron would be removed as pyrite on offshore ocean

margins, preventing BIF deposition. Using the world's modern coastline length of 3.5×10^5 km and assuming 100 km of offshore euxinia, one derives a euxinic area of 3.5×10^7 km^2, which is ~10% of the global ocean area of 3.61×10^8 km^2. Assuming smaller continents, the implied area of euxinia is consistent with the aforementioned few percent based on molybdenum isotopes.

In at least one Mesoproterozoic basin, iron-bearing minerals indicate oxic deep waters (Sperling *et al.*, 2014). Such oxygenated conditions are expected in places of low productivity (Holland, 2006), so we must keep in mind that the Proterozoic deep ocean likely had spatial variability in redox. Nonetheless, the deep ocean could have been predominantly ferruginous from the late Paleoproterozoic (~1.8 Ga) to the late Neoproterozoic (~0.58 Ga) (Poulton and Canfield, 2011).

10.3.3 Concentration of Redox-Sensitive Elements and the Rise of Oxygen

Some metals and non-metals are sensitive to redox, and so their concentrations in marine sediments through time can be related to environmental levels of oxygen. In particular, the abundance of many metals in black shales is roughly proportional to their concentration in seawater, which reflects a source often dominated by rivers (Holland, 1979). In particular, transition metals (those in the third to twelfth columns of the Periodic Table)[3] trace the history of Earth's O$_2$ through their concentrations and isotopes (reviewed by Anbar and Rouxel, 2007). Of those, molybdenum and rhenium are highly *soluble* under oxygenated conditions as the oxyanions MoO$_4^{2-}$ and ReO$_4^{2-}$, while uranium (an actinide) behaves similarly as the soluble uranyl oxyanion UO$_2^{2+}$. Archean black shales have low abundances of these elements, which is consistent with a low concentration in seawater under anoxic conditions (Yang and Holland, 2002). Later in time, their concentrations increased. Amongst non-metals, selenium and iodine are two elements that are also useful as paleoredox indicators.

Molybdenum (Mo) has a redox-sensitive cycle as follows. Oxygen dissolved in rainwater or rivers reacts with continental molydenum sulfides to make soluble molybdate (MoO$_4^{2-}$), which is washed to the ocean (Miller *et al.*, 2011). In the oceans, some Mo is incorporated into Mn-oxyhydroxides, while a major Mo loss is to anoxic sediments. The Mo accumulation increases when MoO$_4^{2-}$ reacts with sulfidic bottom or pore waters to form

---

[3] As defined by the IUPAC. Some definitions exclude column 12, which contains Zn, Cd, and Hg.

thiomolybdates ($MoO_{4-x}S_x^{2-}$), which get taken up by organic matter. Modern euxinic basins, such as the Cariaco Basin off the coast of Venezuela, have sediments enriched in Mo. So more extensive euxinia after the late Paleoproterozoic would efficiently consume the Mo flowing in from rivers and produce low marine Mo levels. With an Archean anoxic atmosphere, little Mo would be sourced in the first place.

Mo concentrations through time show that before 2.15 Ga, the accumulation of Mo in black shales was small, consistent with little oxidative continental weathering (Cheng *et al.*, 2015; Scott *et al.*, 2008), although a modest Mo increase occurred after 2.7–2.8 Ga probably because of oxidative weathering associated with photosynthetic life in lakes (see Sec. 10.6 and Stüeken *et al.* (2012)). Afterwards, Mo enrichments remain moderate during the Proterozoic, suggesting oxidative weathering at rates lower than modern. A consequence of the new availability of molydenum may have been increased dominance of Mo-based enzymes for nitrogen fixation, in contrast with nitrogen-fixing enzymes based on iron and vanadium (Boyd *et al.*, 2011). Later, after 551 Ma, large Mo enrichments imply a highly oxygenated atmosphere and ocean (Dahl *et al.*, 2010; Scott *et al.*, 2008).

Unlike molybdenum, manganese forms an *insoluble* oxide, MnO_4, under oxic conditions. Manganese enrichment is found after the last (Gowganda) Huronian glaciation at 2.3–2.2 Ga and although now found in metamorphic minerals was likely sourced from Mn-oxides (Sekine *et al.*, 2011c). The largest occurrence of manganese ore is the ~2.22 Ga Kalahari manganese field, which accounts for half of the world's manganese reserves (Maynard, 2010). Manganese was deposited in O_2-rich conditions at ~2.4-2.3 Ga (Kirschvink *et al.*, 2000). Oxygenation is confirmed by cerium. Cerium is soluble as Ce^{3+} in reducing fluids but scavenged into insoluble seafloor CeO_2 under oxidizing conditions, leaving residual seawater with a so-called *negative Ce anomaly* relative to other rare earth elements. A negative cerium anomaly in the Kalahari manganese deposit is comparable to that in modern shallow seawater (Tsikos and Moore, 1997). On the other hand, scavenging of the cerium into BIFs after the GOE created positive Ce anomalies in those deposits (Wang *et al.*, 2014).

Another sign of the GOE exists in sediment-hosted stratiform copper deposits. These appear in the geologic record after 2.25 Ga (Hitzman *et al.*, 2010). Such deposits form when highly saline fluids leach copper and transport it in colorful chloride complexes of the form $CuCl_{2+x}^{x-}$(aq), e.g., $CuCl_3^-$ (aq). The copper deposits are generally stratigraphically above redbed clastics. The copper accumulates as an insoluble sulfide from redox reactions at the interface between an oxidizing redbed aquifer that contains the dissolved Cu and is overlain by organic-rich shale.

Amongst non-metals, changes in marine selenium (Se) (with a geologic cycle discussed below in Sec. 10.4.4) and iodine concentrations are consistent with oxygenation. The weathering of sulfides on the continents provides a flux of Se because Se can substitute for S in sulfides. Selenium concentrations in marine shales increase around 2.7 Ga (Stüeken *et al.*, 2015b), consistent with a growth in oxidative weathering inferred from Mo and S concentrations (Stüeken *et al.*, 2012). No statistical change in Se concentrations is seen across the GOE. However, Se concentrations increase at the end of Proterozoic into the Phanerozoic, consistent with Neoproterozoic and Devonian increases in oxygen (Pogge von Strandmann *et al.*, 2015; Stüeken *et al.*, 2015b).

Iodine exists as iodate (IO_3^-) in oxic waters and influences iodine-to-calcium ratios (I/Ca) in carbonates. An increase in I/Ca in shallow water carbonates suggests a transition after ~2.4 Ga to oxic surface waters exceeding 1 μM dissolved O_2 around 2.1–2.2 Ga (Hardisty *et al.*, 2014). This compares with ~250 μM equilibrium concentration in modern surface seawater at 15 °C, i.e., a constraint of >0.4% present atmospheric level (PAL) O_2 after the GOE, or >0.08% O_2 in terms of absolute mixing ratio, at least for ages 2.2–2.1 Ga (Fig. 10.1).

10.3.4 Iron Speciation: Ocean Anoxia or Euxinia, and the Rise of Oxygen

Iron speciation is an important oceanic paleoredox tracer (Lyons *et al.*, 2009; Lyons and Severmann, 2006). Three redox states can be distinguished: (1) *ferruginous*, which is anoxic and rich in Fe^{2+}(aq), (2) *euxinic*, which is anoxic and rich in H_2S, and (3) oxygenated. The anoxia of the first two states is indicated by a high mass ratio of highly reactive iron (Fe_{HR}) to total iron (Fe_{TOT}) in sediments. Fe_{HR} consists of iron phases formed in the water column or during early diagenesis, including pyrite (Fe_{pyrite}), iron carbonate (Fe_{carb}), magnetite (Fe_{mag}) and ferric oxides (Fe_{oxide}). Total iron (Fe_{TOT}) includes Fe_{HR} phases plus Fe in silicate minerals (Poulton *et al.*, 2004), i.e.,

$$\text{highly reactive iron} = Fe_{HR} = \left(Fe_{pyrite} + Fe_{carb} + Fe_{mag} + Fe_{oxide}\right)$$
$$\text{total iron} = Fe_{TOT} = Fe_{HR} + Fe_{silicate} \tag{10.8}$$

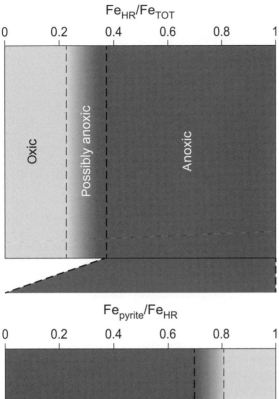

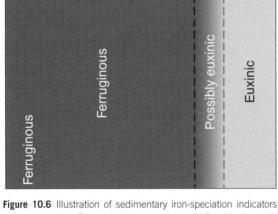

Figure 10.6 Illustration of sedimentary iron-speciation indicators of oceanic redox. Fe$_{HR}$ = highly reactive iron (defined in the text); Fe$_{TOT}$ = total iron sediments; Fe$_{pyrite}$ = pyrite iron. (From Poulton and Canfield (2011). Reproduced with permission. Copyright 2011, The Minerological Society of America.)

Sediments deposited under oxic conditions have Fe$_{HR}$/Fe$_{TOT}$ < 0.22, whereas those deposited in anoxic bottom waters have Fe$_{HR}$/Fe$_{TOT}$ > 0.38 (Fig 10.6). High Fe$_{HR}$/Fe$_{TOT}$ ratios arise from the water column formation of Fe$_{HR}$ either as pyrite in euxinic basins, or magnetite, siderite or ferric oxides in ferruginous basins. The Fe$_{pyrite}$/Fe$_{HR}$ ratio can then be used to distinguish between ferruginous (<0.8) and euxinic (>0.8) types of anoxia (Fig 10.6). In the latter case, most of the reactive iron is

pyrite, as one would expect from reaction of dissolved sulfide with Fe^{2+}(aq).

On the basis of iron speciation relationships in Fig. 10.6, ~2.6–2.5 Ga ocean margins generally had shallow waters that were mildly oxygenated with sediments characterized by a low Fe$_{HR}$/Fe$_{TOT}$ ratio (Kendall *et al.*, 2010; Poulton *et al.*, 2010; Reinhard *et al.*, 2009), even though the global atmosphere was still anoxic based on sulfur isotopes (see Sec. 10.5). An explanation is that microbes produced O$_2$ in the surface ocean before O$_2$ substantially accumulated in the atmosphere. Simple models of such localized *oxygen oases* predict that dissolved O$_2$ concentrations could reach as high as 0.02 mM, or ~8% of present dissolved O$_2$ levels (Kasting, 1992), while more sophisticated models suggest 0.001–0.01 mM (Olson *et al.*, 2013).

In the simple model of an oxygen oasis, the dissolved O$_2$ concentration is estimated by balancing production of O$_2$ at rates comparable to those in modern, highly productive upwelling regions, 1 g C m^{-2} day^{-1} = 0.083 mol O$_2$ m^{-2} day^{-1}, with loss of O$_2$ to the atmosphere at its piston velocity of 6×10^{-3} cm s^{-1}, or ~5 m day^{-1} (Table 9.1). A slightly lower piston velocity of 4 m day^{-1} (4.8×10^{-3} cm s^{-1}) was used in Kasting (1992). With the piston velocity from Ch. 9, the dissolved O$_2$ concentration would be (0.083 mol O$_2$ m^{-2} day^{-1})/(5 m day^{-1}) = 0.017 mol O$_2$ m^{-3} = 0.017 mM. Given a Henry's Law coefficient of 1.3 ×10^{-3} mol l^{-1} atm^{-1}, the equivalent pO$_2$ is 0.013 atm or 6% PAL.

Sediments deposited in the deep ocean with low Fe$_{pyrite}$/Fe$_{HR}$, high Fe$_{TOT}$ (>10 wt%) and elevated Fe$_{HR}$/Fe$_{TOT}$ indicate that the ~2.6–2.5 Ga deep ocean was ferruginous, consistent with BIF models that we described earlier (Fig. 10.5). However, the middle depths of late Archean seawater column were transiently sulfidic, according to high Fe$_{pyrite}$/Fe$_{HR}$ and Fe$_{HR}$/Fe$_{TOT}$ data, as well as a significant accumulation of molybdenum (Kendall *et al.*, 2010; Poulton *et al.*, 2010; Reinhard *et al.*, 2009). Such euxinic episodes may have occurred during times or places with relatively high fluxes of organic matter.

Poulton *et al.* (2010) have also used iron speciation to infer marine redox chemistry in the 1.8 Ga Animikie Basin near Lake Superior. In this basin, an oxic surface ocean overlay a deep ocean that was ferruginous, while in near shore environments, euxinic conditions formed a wedge between oxic surface and deep zones. As a result of this and other work, Poulton and Canfield (2011) propose an evolution of oceanic redox state depicted in Fig. 10.7.

A very different analysis of the redox state of iron allows the deduction of pO$_2$ < 10^{-4} bar in the Archean from the composition of impact spherules, which are

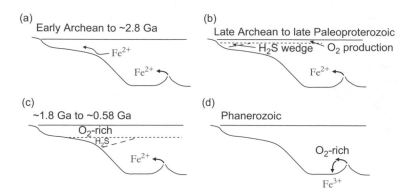

Figure 10.7 Schematic diagram of the evolution of ocean redox conditions from the Precambrian to the Phanerozoic, following Poulton and Canfield (2011).

formed by the condensation of rock vapor in impact plumes. Spinels ($A^{2+}B^{3+}_2O^{2-}_4$, where A and B represent cations) in 3.24 Ga impact spherules from the Barberton greenstone belt, South Africa, have Fe^{3+}/Fe_{TOT} ~0.29-0.69 by atoms compared to impact spherules from the Phanerozoic with Fe^{3+}/Fe_{TOT} ~0.8-0.99. Experimental measurements of Fe^{3+}/Fe_{TOT} for melts at particular pO_2 constrains the pO_2 at 3.24 Ga to $< 10^{-4}$ bar (Krull-Davatzes et al., 2010).

10.4 Mass-Dependent Stable Isotope Records and the Rise of Oxygen

Above, we have discussed concentration data; but, in addition, isotopes of carbon, sulfur, nitrogen, iron, molybdenum, chromium, and selenium in marine sediments are all consistent with major redox change of the atmosphere in the Paleoproterozoic and/or Neoproterozoic. However, the strongest evidence for redox change at ~2.4 Ga comes from sulfur isotopes that are mass-independently fractionated, which we discuss in Sec. 10.5.

10.4.1 Carbon Isotopes

As described in Sec. 8.5, photosynthesis concentrates ^{12}C into organic matter, leaving seawater – recorded in ancient carbonates – relatively enriched in ^{13}C. Thus, the relative fluxes of carbon buried in the past as carbonate or organic carbon can be deduced by examining carbon isotopes in sedimentary rocks. Figure 8.2 shows that the average carbon isotope composition of marine carbonates has been roughly close to 0‰ throughout much of geologic time. In Sec. 8.5, we discussed how some have interpreted this fact as indicating that, since ~3.5 Ga, ~20% of the carbon in CO_2 entering the ocean–atmosphere system has exited as buried organic carbon (a fraction designated as f_{org}) and ~80% as carbonate (Schidlowski, 1988). On the other hand, more detailed

statistical analysis suggests that there was an increase by a factor of 1.2–2 in fractional organic carbon f_{org} burial over geologic time if a simple carbon cycle model like that of Sec. 8.5 is considered (Krissansen-Totton et al., 2015). If the data are interpreted in terms of a carbon cycle model that includes uncertain ranges in the fraction of carbon that gets sequestered into carbonatized seafloor (Bjerrum and Canfield, 2004) and the amount that forms as authigenic carbonate from the oxidation of organic matter in sediments (Schrag et al., 2013), the trend in f_{org} has unknown sign, however (Krissansen-Totton et al., 2015).

Irrespective of long-term trends, there are very positive $\delta^{13}C$ excursions in carbonates during the Paleoproterozoic and Neoproterozoic, the same eras as significant increases of O_2. In particular, after oxygen rose, $\delta^{13}C$ excursions occur during 2.4–2.06 Ga, including the global *Lomagundi Event*, which began 2.31–2.22 Ga and ended 2.06 Ga (Melezhik et al., 2013a). The event is named after a province in Zimbabwe where ^{13}C-rich dolomites were first studied (Bekker et al., 2008; Bekker et al., 2006; Maheshwari et al., 2010; Melezhik and Fallick, 2010; Melezhik et al., 2007; Schidlowski et al., 1976). Doubts about the number of excursions arise because currently available radiometric dates have error bars that are too big for exact chronological comparisons between different locations.

The Lomagundi Event has been interpreted in different ways. One explanation concerns global scavenging of ^{12}C into organic matter associated with enhanced organic burial (Karhu and Holland, 1996; Kerr et al., 2015). Alternatively, the isotope composition may have been the result of diagenesis involving methanogens (Hayes and Waldbauer, 2006). Methanogens convert CO_2 to methane, which can escape from sediments, leaving behind ^{13}C-enriched CO_2 that ends up in carbonates (Mazzullo, 2000). The idea is that methanogens, which are poisoned by O_2, were pushed into sediments during the GOE while, prior to the GOE, organic carbon was metabolized into methane or CO_2 in the water column (Walker, 1987). However, $\delta^{13}C$ data for Lomagundi carbonates and $\delta^{34}S$ of carbonate-associated sulfate that matches contemporaneous evaporites appear inconsistent with a diagenetic origin of the carbonates (Planavsky et al., 2012). The first abundant diagenetic carbonates, such as ^{12}C-rich carbonate concretions,[4] appear at ~2.0 Ga, after the Lomagundi Event (Fallick et al., 2008).

Certainly, the Lomagundi $\delta^{13}C$ excursions cannot be the cause of the GOE, as originally suggested (Karhu and Holland, 1996), because radiometric dating now shows that the excursions occur after the GOE (Bekker et al., 2004). Recognizing this, Holland (2002) suggested that continental oxidative weathering of pyrites after the GOE produced sulfuric acid, which, in turn, enhanced weathering and released of high nutrient fluxes to the ocean that stimulated productivity and organic burial. From 2.48–2.32 Ga, Cr enrichment in iron-rich sedimentary rocks without a large $\delta^{53}Cr$ that would be associated with manganese oxide-catalysed oxidation (Sec. 10.4.4) suggests that a pulse of acidity associated with pyrite oxidation was sufficient to mobilize chromium in its reduced form, Cr^{3+}, from continents (Konhauser et al., 2011). Such a transitional sedimentary cycle of oxidative weathering might be related to the initiation of the ~2.3–2.1 Ga Lomagundi Event. In turn, later oxidative weathering of uplifted organic-rich sediments deposited during the Lomagundi Event would be a negative feedback on an O_2 overshoot (Bekker and Holland, 2012). A subsequent negative excursion in marine $\delta^{13}C$ of −14‰ at c. 2.06 Ga has been attributed to a high O_2 pulse that weathered organic matter buried during the Lomagundi Event (Kump et al., 2011).

Other possible evidence for an overshoot of Paleoproterozoic pO_2 comes from phosphorites, which are sedimentary rocks with enough phosphorus to be economically useful. The absence of phosphorites in the Archean, their appearance in the Paleoproterozoic at ~2.1–1.9 Ga (Lepland et al., 2013), and their subsequent disappearance until the Neoproterozoic (Papineau, 2010) are consistent with a postulated supply of phosphorus associated with acidic oxidative weathering and significant deposition and remineralization of organic material in oxygenated deep waters. Evidence for temporary oxygenation of deep waters (Canfield et al., 2013), the appearance of 2.2–2.1 Ga sulfate-rich evaporites (Reuschel et al., 2012;

Schroder et al., 2008), and an analysis of sulfur isotope and Mo, U, and Fe proxies, all support a pulse of oxidation and subsequent decline (Asael et al., 2013; Scott et al., 2014).

Another curious aspect of the ancient carbon isotope record concerns very ^{13}C-depleted organic carbon in 2.8–2.7 Ga sediments in Canada, South Africa and Western Australia, which is interpreted to arise from methanotrophs, i.e., bacteria that consume methane (Hayes, 1983, 1994; Hinrichs, 2002; Schoell and Wellmer, 1981). Hayes (1983, 1994) favored aerobic methanotrophs that oxidize methane with O_2. But Hinrichs suggested so-called reverse methanogens that operate in consortia with sulfate reducers, effectively reacting methane with sulfate in anaerobic oxidation of methane.

Methanogens fractionate the stable isotopes of carbon strongly in favor of ^{12}C. When methanogen CH_4 is consumed by methanotrophs, the carbon is further fractionated, leading to organic carbon with $\delta^{13}C$ around −60‰. CH_4 could have been present at higher trace levels throughout the Archean but would have been invisible in the carbon isotope record in the absence of significant methanotrophy, which is postulated to have become conspicuous once marine SO_4^{2-} became more prevalent. For example, organic carbon with $\delta^{13}C$ ranging from −33‰ to −51‰ in the 2.76 Ga Mt. Roe paleosol in Western Australia has also been attributed to methanotrophs living in an ephemeral pond under an anoxic atmosphere relatively rich in methane (Rye and Holland, 2000). Methanotrophic signatures declined after 2.6 Ga perhaps because of increasing competition with other aerobes for oxidants.

10.4.2 Sulfur Isotopes

Sulfur has four stable isotopes 32, 33, 34, and 36, of which ^{32}S and ^{34}S represent ~85% and 4.2% of the total sulfur on Earth, respectively. Hence, mass dependent isotope fractionation is commonly expressed in terms of the most abundant isotopes as $^{34}S/^{32}S$ ratios. In the delta notation, $\delta^{34}S = (R/R_{standard} - 1) \times 1000$, where $R = {^{34}S}/{^{32}S}$ and $R_{standard}$ is the ratio for the standard.

The record of mass-dependently fractionated sulfur isotopes in ancient pyrites and sulfates (Fig. 10.8) suggests that the Archean oceans contained little sulfate compared to ~28 mM today, and that marine sulfate concentrations increased in the Paleoproterozoic (Habicht et al., 2002). Sulfur isotopes and marine sulfate concentrations are related. Some microbes reduce sulfate for metabolism, using either organic matter or hydrogen

$$SO_4^{2-} + 2H^+ + 2CH_2O \rightarrow H_2S + 2CO_2 + 2H_2O \quad (10.9)$$

$$SO_4^{2-} + 2H^+ + 4H_2 \rightarrow H_2S + 4H_2O \quad (10.10)$$

---

[4] Concretions are hard, compact mineral accumulations of different composition from the surrounding rock. They form from groundwater that carries dissolved minerals through soft sediments or porous rock. A concretion grows when the dissolved minerals precipitate and partially incorporate or replace surrounding sediment.

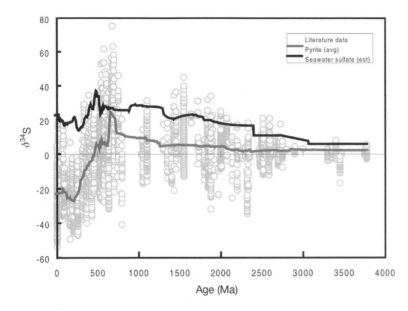

Figure 10.8 The variation of $\delta^{34}S$ in marine pyrites over time, shown as gray symbols. The *gray line* is the running mean of the pyrite values. The *black line* is the estimated running mean for seawater sulfate $\delta^{34}S$. (From Farquhar *et al.* (2010).)

The H_2S that is produced reacts with ferrous iron in sediments to make pyrite (FeS_2). If there is sufficient sulfate, sulfate-reducing microbes preferentially use ^{32}S rather than ^{34}S and produce sulfides that are enriched in ^{32}S relative to seawater sulfate. However, at low sulfate concentrations <0.05–0.2 mM, there is no significant isotope fractionation (Habicht *et al.*, 2002). Sulfides with significant ^{32}S enrichment become widespread only after 2.3 Ga, which is interpreted as an increase in sulfate levels above a 0.2 mM threshold (Farquhar *et al.*, 2010; Huston and Logan, 2004). In turn, increased sulfate implies more atmospheric O_2 because the sulfate ultimately comes from sulfate in rivers, which is produced when continental sulfides react with oxygenated rainwater. So a low concentration of sulfate in the Archean ocean indicates a lack of atmospheric O_2. A stronger constraint on pO_2 comes from examining the relative fractionation of three or all four stable isotopes of sulfur, as described in Sec. 10.5.

Debate persists about the exact sulfate levels in the Archean ocean and how they tie to $\delta^{34}S$ data. Photochemical calculations show that even an anoxic atmosphere can provide a sulfate flux to the ocean of ~10^9 SO_4^{2-} cm^{-2} s^{-1} (0.3 Tmol S yr^{-1}) (Ono *et al.*, 2003; Zahnle *et al.*, 2006), almost 10% of the modern oxidative weathering sulfur flux (Table 10.2). Radicals produced by water and CO_2 photolysis oxidize sulfurous volcanic gases. Huston and Logan (2004) suggest that the presence of bedded marine sulfates before 3.2 Ga but not afterwards indicates early Archean seawater with higher sulfate concentrations than during late Archean. Possibly, greater biological productivity in the late Archean allowed for more sulfate reduction in marine sediments. Perhaps increased productivity was associated with evolutionary innovation in photosynthesis.

10.4.3 Nitrogen Isotopes

On the modern Earth, microbes extract nitrogen from the air, and when it is returned to the atmosphere by other microbes, ^{14}N is preferentially converted into N_2, leaving behind dissolved nitrogen compounds enriched in ^{15}N (Sigman *et al.*, 2008). The fractionation is expressed as $\delta^{15}N$ where $\delta^{15}N = (R/R_{air} -1) \times 1000$, with $R = ^{15}N /^{14}N$ and $R_{air} = 0.003\,67$. Limited fractionation of −3‰ to +4‰ occurs when microbial *nitrogen fixation* converts atmospheric N_2 to ammonia and then ammonium (NH_4^+) in organic matter. Some fractionation occurs in *nitrification*, when ammonium is oxidized to nitrate. However, strong fractionation occurs during *anammox*, which is anaerobic ammonium oxidation ($NH_4^+ + NO_2^- \rightarrow N_2 + 2H_2O$) (Brunner *et al.*, 2013), and *denitrification*, when dissolved nitrate is reduced by bacteria in oxygen minimum zones to N_2 via intermediates of NO_2^-, NO, and N_2O (Casciotti, 2009). Today, seawater nitrate and organic matter derived from nitrate has average $\delta^{15}N$ ~+5‰ (Peters *et al.*, 1978), which is similar to the mean $\delta^{15}N$ in sedimentary organic matter of +(7 ± 1)‰ (Altabet and Francois, 1994). Such nitrogen isotope fractionation reflects the role of redox-sensitive pathways and so can be used to trace the evolution of ancient environmental redox (Canfield *et al.*, 2010).

Nitrogen fixation is strictly anaerobic and presumably existed before the advent of oxygenic photosynthesis

(Boyd and Peters, 2013; Postgate, 1987; Zehr *et al.*, 1995). Genetic coding for *nitrogenase*, the enzyme for nitrogen fixation, is absent in eukaryotes but found in several phyla of bacteria (including cyanobacteria) and one methanogen phylum of archaea (Dos Santos *et al.*, 2012). Nitrogen fixation may have evolved only once but a suggestion that it pre-dates the universal common ancestor (LUCA) of the three domains of life (Fani *et al.*, 2000) is disputed (Raymond *et al.*, 2004). Nitrogenase genes in the deeply rooted thermophilic bacteria *Aquificales* appear to have been acquired by lateral gene transfer rather than evolving prior to the divergence of bacteria and archaea (Boyd *et al.*, 2011; Boyd and Peters, 2013). In any case, once microbes developed nitrogen fixation, nitrogen should have been lost to sediments and buried in the form of clay minerals because NH_4^+ can substitute for K^+ as a result of similar ionic radii. Of course, even in an anoxic world, we would expect some recycling: first, geologic recycling back to N_2 via mid-ocean ridges and subduction zones, and second, biological cycling through the release of NH_3, which would photolyze to N_2.

The Precambrian record of nitrogen isotopes shows some relationship to paleoredox. We can interpret this record because atmospheric $\delta^{15}N$ appears to have changed little: it was ~0‰ in the Archean, as it is today (Marty *et al.*, 2013; Nishizawa *et al.*, 2007; Sano and Pillinger, 1990). Proterozoic kerogens have average $\delta^{15}N$ ~+5‰, similar to modern values, which suggests that oxidative pathways operated (Thomazo and Papineau, 2013). Archean $\delta^{15}N$ values are variable and can reflect alteration of rocks by metamorphism or hydrothermal fluids. The $\delta^{15}N$ in bulk rocks and kerogen ranges from −7‰ to +27‰ between 4.0 and 2.7 Ga (Pinti *et al.*, 2007; Thomazo and Papineau, 2013), while $\delta^{15}N$ after 2.7 Ga is generally positive. Very positive $\delta^{15}N$ values in kerogen of highly metamorphosed Archean shales (Jia and Kerrich, 2004; Kerrich *et al.*, 2006) are unlikely to be representative of global seawater.

In early work, Beaumont and Robert (1999) reported an upward trend in marine $\delta^{15}N$ from negative in the mid-Archean to positive in the Proterozoic, which they attributed to increasing oxygenation. More recent results from 3.2–2.8 Ga rocks of relatively low metamorphic grade show a few per mil range of $\delta^{15}N$ centered near ~0‰ (Stueken *et al.*, 2015c), which suggests anaerobic cycling of nitrogen prior to ~2.7 Ga.

After 2.7 Ga, mildly metamorphosed rocks provide a record of generally positive $\delta^{15}N$ values, which suggest some aerobic cycling (Busigny *et al.*, 2013; Thomazo and Papineau, 2013). However, extreme $\delta^{15}N$ values up to +50‰ in the 2.7 Ga Tumbiana Formation of northwest

Australia (Thomazo *et al.*, 2011) may not simply reflect aerobic cycling. The Tumbiana rocks were formed in lakes (Buick, 1992), and extremely positive $\delta^{15}N$ can be produced by loss of ammonia if the lakes were highly alkaline (Stueken *et al.*, 2015d). Around 2.5 Ga, there are also reports of temporary departures from anaerobic to aerobic nitrogen cycling when $\delta^{15}N$ increases in association with other indications of oxygen availability (Garvin *et al.*, 2009; Godfrey and Falkowski, 2009).

In general, some negative $\delta^{15}N$ values down to −7‰ for marine organic matter in the early Archean (similar to samples from Phanerozoic ocean anoxic events) have inspired three different explanations. One is a greater contribution from chemosynthetic hydrothermal microbes (Pinti *et al.*, 2001; Pinti *et al.*, 2009). The organic matter made by chemolithoautotrophs (with C extracted from CO_2 using energy released from the oxidation of inorganic substances such as NH_4^+) has $\delta^{15}N$ of −9.6‰ to +0.9‰. However, attributing all negative $\delta^{15}N$ to chemolithotrophy is difficult to reconcile with the shallow-water setting of some samples. A second explanation involves the benthic uptake of isotopically light nitrogen produced during *ammonification* when organic matter is broken down to release ammonium in anoxic settings (Papineau *et al.*, 2009). Ammonification can produce fractionation of down to −5‰ (Sigman *et al.*, 2008). A third possibility is that a greater contribution to nitrogen fixation before ~3.2 Ga from anoxygenic photosynthesis, which tends to produce ^{15}N-depleted organic matter. Increasingly positive $\delta^{15}N$ in the late Archean would then represent a bigger contribution from cyanobacteria and aerobic pathways.

10.4.4 Transition Metal (Iron, Chromium, and Molybdenum) and Non-Metal Isotopes (Selenium)

Changes in iron isotopes (reported as $^{56}Fe/^{54}Fe$) provide circumstantial evidence for oxygenation. Going back before ~2.3 Ga, the $\delta^{56}Fe$ in pyrites changes from mostly positive to negative (Rouxel *et al.*, 2005). A plausible explanation is based on bacterial dissimilatory iron reduction (DIR), which reduces iron (III) oxides with organic carbon and enriches Fe^{2+}(aq) with ^{54}Fe during sediment diagenesis (Archer and Vance, 2006; Johnson *et al.*, 2008b). Indeed, we saw earlier that DIR is the likely explanation for the variety of iron minerals in Superior-Type banded iron formations (BIFs) (Sec. 10.3.2). A negative excursion in $\delta^{56}Fe$ in 2.7–2.5 Ga BIFs could be a signature of enhanced DIR because of greater iron oxide and organic carbon deposition generated by oxygenic photosynthesis (Czaja *et al.*, 2010; Johnson *et al.*, 2008b). However, oxygenic photosynthesis may be indicated earlier. Moderatively positive δ^{56}Fe combined with elevated uranium concentrations in shallow water, Fe-poor facies of the Manzimnyama banded iron formation, Fig Tree Group, South Africa, have been interpreted as suggesting a redox-stratified ocean at 3.2 Ga (Satkoski *et al.*, 2015).

Chromium (Cr) isotopes in BIFs also appear to be correlated with changes in atmospheric O_2 (Frei *et al.*, 2009). When manganese oxide catalyzes the oxidation of insoluble Cr^{3+} in soils to a soluble Cr^{6+} chromate ion, the heavy isotope ^{53}Cr is preferentially oxidized compared to ^{52}Cr. After transport from rivers to the ocean, the chromate ion is reduced back to Cr^{3+} oxyhydroxide by reaction with ferrous iron. Ferric oxyhydroxides coprecipitate with the Cr^{3+} preserving marine Cr isotopes in BIF. An increase of ^{53}Cr / ^{52}Cr in marine sediments after 2.4 Ga is therefore consistent with new oxidative weathering. A large increase in the abundance of marine ^{53}Cr / ^{52}Cr after 750 Ma suggests a second rise of O_2 in the Neoproterozoic.

Molybdenum (Mo) isotopes also trace both oxygenation and ocean euxinia (Arnold *et al.*, 2004; Lyons *et al.*, 2009). Oxic uptake of Mo on marine solids preferentially removes ^{95}Mo relative to ^{98}Mo, leaving the ocean isotopically heavy. In contrast, under conditions of low oxygenation, seawater is ^{95}Mo-enriched. Highly euxinic deep waters scavenge Mo efficiently into sulfides that tend to capture the Mo isotope composition of seawater. Rocks that are ^{95}Mo-enriched from the 1.7–1.4 Ga MacArthur Basin in Australia suggest that the deep ocean in the mid-Proterozoic was commonly oxygen-poor, with an area of euxinia that was probably several percent compared to today's 0.3% (Arnold *et al.*, 2004; Kendall *et al.*, 2009). This extent of euxinia may have been enough to significantly deplete trace metals in the Mesoproterozoic ocean by scavenging them into insoluble sulfides, which may have affected marine ecology during the Proterozoic (Anbar and Knoll, 2002).

Selenium (Se) isotopes are sensitive to redox as well as to biological processing. In the Periodic Table, selenium is below sulfur and so has some chemical parallels. Selenium has four oxidation states, +6, +4, 0 and –2, and, unlike sulfur, six stable isotopes, ^{74}Se, ^{76}Se, ^{77}Se, ^{78}Se, ^{80}Se, and ^{82}Se. In the modern surface ocean, Se exists as selenite (Se^{4+}) or selenate (Se^{6+}) anions (collectively SeO_x^{2-}), and dissolved organic Se (denoted Se_{org}), where the latter usually dominates (Cutter and Bruland, 1984). Marine biota incorporate selenium. Then, particulate biogenic detritus carries Se to the deep sea, where today it is

oxidized back to selenite and selenate. In sediments, reduction of selenium ($Se^{6+} \rightarrow Se^{4+}$, $Se^{6+} \rightarrow Se^0$, or $Se^{4+} \rightarrow Se^0$) concentrates the lighter isotopes in Se^0 (Johnson and Bullen, 2004). In turn, Se^0 can be further reduced to Se^{2-} that associates with pyrite in minerals such as $FeSe_2$.

On the early Earth, there was surely a different Se cycle than today because of the lack of deep-sea oxygenation and different delivery to the deep sea via dead planktonic matter instead of zooplankton fecal pellets. While Se isotopes in marine shales appear to have changed little across the GOE (Stueken *et al.*, 2015b), there is a significant spread towards lighter isotopes in the Neoproterozoic. This change suggests deep-sea oxygenation (Pogge von Strandmann *et al.*, 2015) and the increasing ecological influence of eukaryotes in the water column (Lenton *et al.*, 2014; Logan *et al.*, 1995).

10.5 Mass-Independent Fractionation of Sulfur Isotopes and the Rise of Oxygen

By far the most unambiguous evidence for the Great Oxidation Event comes from the *mass-independent fractionation* (MIF) of sulfur isotopes, which is large before 2.4 Ga but small afterwards. Because sulfur MIF (S-MIF) can be linked to the photochemistry of an anoxic atmosphere, it is widely considered to be the "smoking gun" evidence for a rise of O_2 around 2.4 Ga (Bekker *et al.*, 2004; Farquhar *et al.*, 2000a).

To understand S-MIF, recall that sulfur has four stable isotopes: ^{32}S, ^{33}S, ^{34}S, and ^{36}S. Two classes of fractionation can be defined for these isotopes. Most processes produce *mass dependent fractionation* (MDF), where the fractionation is proportional to the mass difference between isotopes. In such cases (e.g., diffusion of a sulfur-bearing gas or bacterial sulfate reduction), ^{33}S is fractionated relative to ^{32}S by only half as much as ^{34}S because the mass difference of ^{33}S and ^{32}S is half of that between ^{34}S and ^{32}S. In the delta notation ($\delta^x S = [(^x S/^{32}S)_{sample}/[(^x S/^{32}S)_{standard}]-1]$), variations in δ^{33}S are usually about half as large as those in δ^{34}S for MDF. Mathematically, a slope of roughly half defines a MDF straight line for MDF isotope data, whereas MIF is simply a deviation from that line.

$$\delta^{33}S \approx 0.5(\delta^{34}S) \quad \text{mass dependent fractionation} \quad (10.11)$$

$$\delta^{33}S \neq 0.5(\delta^{34}S) \quad \text{mass independent fractionation}$$

$$(10.12)$$

For eq. (10.11), we've written the proportionality constant as 0.5 but a more precise value is 0.515, i.e., $\delta^{33}S = 0.515\delta^{34}S$. Similarly, an MDF straight line defining the relationship of $\delta^{36}S$ and $\delta^{34}S$ is roughly $\delta^{36}S \approx 2\delta^{34}S$ or more precisely $\delta^{36}S = 1.90\delta^{34}S$. We express MIF as how much the $\delta^{33}S$ or $\delta^{36}S$ of measured samples lies off the expected MDF lines using the following notation:

$$\Delta^{33}S = \delta^{33}S - 0.515\delta^{34}S \tag{10.13}$$

$$\Delta^{36}S = \delta^{36}S - 1.9\delta^{34}S \tag{10.14}$$

Here $\Delta^{33}S$ is "cap delta 33 S", i.e., excess or depleted ^{33}S relative to the MDF line. Such $\Delta^{33}S$ is generally correlated with $\Delta^{36}S$. Figure 10.4(a) shows the remarkable time-history of $\Delta^{33}S$ and the obvious step change at 2.4 Ga. Pre-2.4 Ga pyrites (FeS$_2$) mostly have positive $\Delta^{33}S$, although some have negative $\Delta^{33}S$. Similarly, Archean sulfates preserved as barite (BaSO$_4$) predominately have negative $\Delta^{33}S$ but some have positive $\Delta^{33}S$.

Photochemistry is the only known process that produces S-MIF greater than +10‰, as shown in Fig. 10.4(a). Laboratory experiments show that S-MIF is present in elemental sulfur produced from the photolysis of SO$_2$ and associated photochemistry (Farquhar et al., 2001). Other possible reactions that induce S-MIF are photolysis of the free radical SO (Ono et al., 2003), sulfur polymerization S + S$_2$ → S$_3$ (Du et al., 2011), and SO$_3$ photolysis, where the latter two are perhaps of unlikely significance. Photolysis of carbon disulfide (CS$_2$) creates S-MIF (Zmolek et al., 1999) but CS$_2$ was probably very minor in the early atmosphere (Domagal-Goldman et al., 2011). Negligible S-MIF is seen in the photolysis products of H$_2$S (Farquhar et al., 2000b) and OCS (carbonyl sulfide) (Hattori et al., 2011; Lin et al., 2011). Measureable S-MIF ($\Delta^{33}S \approx 2‰$) has been reported from thermochemical reduction of sulfate by powdered amino acids (Watanabe et al., 2009); but this can be attributed to a magnetic isotope effect. Since even-numbered isotopes of all elements have zero net nuclear spin, magnetic isotope effects cannot produce MIF in ^{36}S (Oduro et al., 2011) and hence cannot account for the substantial $\Delta^{36}S$ values seen in Archean rocks.

The creation of S-MIF in the Archean depends on the penetration of UV photons into the troposphere. UV photons with wavelength less than about 230 nm cause photolysis reactions in sulfur oxide gases, as follows.

$$SO_2 + h\nu \ (\lambda < 217 \text{ nm}) \rightarrow SO + O \tag{10.15}$$

$$SO_2 + h\nu \ (\lambda < 208 \text{ nm}) \rightarrow S + O_2 \tag{10.16}$$

$$SO + h\nu \ (\lambda < 231 \text{ nm}) \rightarrow S + O \tag{10.17}$$

On the modern Earth, O$_2$ and ozone absorb wavelengths shorter than ~300 nm, so that reactions (10.15) –(10.17)

do not occur in the troposphere. S-MIF can be generated by photolysis of SO$_2$ in the upper stratosphere, and this signal is observed in sulfate aerosols preserved in polar ice cores (Alexander et al., 2003; Whitehill et al., 2015). But when those sulfate particles enter the ocean they become part of the vast oceanic dissolved sulfate reservoir, and any MIF signal is erased before sulfur is deposited in sediments.

In an anoxic atmosphere, SO$_2$ and SO can be photolyzed within the troposphere. Moreover, the elemental sulfur (S) polymerizes to S$_8$ aerosols (Gunning and Strausz, 1966; Khare and Sagan, 1975) which is another process that may induce MIF. The net result is that both sulfate and S$_8$ aerosols fall out of the sky in an anoxic atmosphere with S-MIF signatures of complementary sign (Pavlov and Kasting, 2002 , also Fig. 9.14). An anoxic environment thus allows isotopic heterogeneity in sulfur to be transferred to the surface and then the geologic record.

Conventionally, it has been assumed that sulfate aerosols form with negative $\Delta^{33}S$ values, whereas the S$_8$ aerosols tend to have positive ones. In this picture, the $\Delta^{33}S$ of sulfate aerosols gets directly transferred to sedimentary sulfates, while pyrites are derived from S$_8$ aerosols. This idea receives support from the geologic record where these are the predominant signs of $\Delta^{33}S$ in sulfates and sulfides. Early measurements of $\Delta^{33}S$ at a single photolysis wavelength of 193.0 nm also made sulfates with negative $\Delta^{33}S$ (Farquhar et al., 2001; Farquhar et al., 2010). However, such photolysis experiments cannot be used to directly interpret the sign of S-MIF in the rock record because S-MIF depends upon photochemistry at all relevant wavelengths acting at plausible gas concentrations (Claire et al., 2014). Pathways of sulfur processing after sulfur exits the atmosphere are also important (Halevy, 2013; Halevy et al., 2010).

Presently, the generation of S-MIF at the molecular level is not fully understood. Two hypotheses are: (1) shielding (e.g., Lyons, 2009; Ono et al., 2013), and (2) kinetic isotope effects associated with excited molecules and/or atoms.

In shielding, sulfur-bearing gases in the lower atmosphere are protected from part of the UV spectrum. In *self-shielding* by SO$_2$, ^{32}SO$_2$ absorbs incoming solar photons of wavelength λ down to a certain altitude, precluding further ^{32}SO$_2$ photolysis at wavelength λ at lower altitudes. However, photons that affect $^{33/34/36}$SO$_2$ can reach lower heights and induce S-MIF by photolysis of $^{33/34/36}$SO$_2$ but not ^{32}SO$_2$. The problem with the self-shielding hypothesis is that photochemical models do not produce optically thick SO$_2$ in the early atmosphere for plausible volcanic inputs (Claire *et al.*, 2014), except possibly for short time intervals following explosive eruptions. Other gases might provide shielding, which is called *mutual shielding* (Ueno *et al.*, 2009).

In the kinetic isotope effect hypothesis, photochemistry generates long-lived excited molecules and/or atoms. The absorption cross-section of sulfur-bearing gases depends on the isotopologue, i.e., in the case of SO_2 whether the gas is $^{32}SO_2$ or $^{33}SO_2$, etc. (Danielache *et al.*, 2008; Endo *et al.*, 2015; Lyons, 2009)). The SO_2 absorption spectrum is highly structured because of rotational and vibrational excitations. Excited species can create S-MIF by reacting through more than one pathway.

Regardless of the molecular mechanism of S-MIF, basic atmospheric chemistry is sufficient to deduce that the presence or absence of S-MIF distinguishes anoxic and oxic atmospheres. In the modern atmosphere, volcanic or biogenic sulfur gases are all oxidized to sulfate, which dissolves in rainwater, rains out, and homogenizes the sulfur into a single reservoir of isotopically-uniform oceanic sulfate with no MIF. For S-MIF to be generated and preserved, photochemical models show that an atmosphere must have three different properties, as follows (Zahnle *et al.*, 2006).

(1) *Ground-level O_2 < 1 ppmv.* Because ozone derives from O_2 (Sec. 3.3.1), there is no stratospheric ozone layer in an anoxic atmosphere, which allows UV photons of less than 220 nm wavelength to penetrate the troposphere and cause SO_2 and SO photolysis (eqs. (10.15)–(10.17)). Also, in an anoxic atmosphere, both soluble SO_4^{2-} and insoluble sulfur aerosols (S_8) are produced rather than just sulfate. Sulfur can also exit the atmosphere as soluble SO_2, which would have yet another isotopic signature. Such species can preserve S-MIF in different minerals (Farquhar and Wing, 2003; Ono *et al.*, 2003; Pavlov and Kasting, 2002). Pavlov and Kasting suggested that the presence of S-MIF constrained O_2 to $<10^{-5}$ PAL or <2 ppmv. Updated models that generate transitional chemistry between reducing and oxidizing atmospheres suggest that the transition to an atmosphere with an S_8 exit channel occurs at $<10^{-6}$ PAL O_2 or <0.2 ppm (Zahnle *et al.*, 2006).

(2) *Relatively abundant methane and/or hydrogen.* An atmosphere must also have a sufficiently abundant reducing gas, such as methane or molecular hydrogen, at concentrations exceeding tens of ppmv. This enables reducing chemistry in the upper troposphere, which is the altitude where S_8 is produced. At such heights, photochemical models show that the O_2 concentration is exceedingly small (<1 pptv).

(3) *Sufficient input fluxes of sulfur-bearing gases.* Finally, there must be a sufficient flux of sulfur-bearing gases to the atmosphere, which is satisfied for expected levels of Precambrian volcanism. However, variations in the volcanic flux and biogenic sources of sulfur gases may have modulated the MIF magnitude (Claire *et al.*, 2014; Ono *et al.*, 2003; Zahnle *et al.*, 2006). It has been proposed that the proportion of SO_2 vs. H_2S in sulfur-bearing emissions might also be important (Halevy *et al.*, 2010) but photochemical models show that SO_2 and H_2S exchange so rapidly that the ratio in volcanic gases makes little difference to S-MIF (Claire *et al.*, 2014). Atmospheric redox state is instead dominated by larger redox fluxes such as H_2, CO, and CH_4 relative to O_2.

Exactly how the MIF was transferred from the atmosphere to sediments is uncertain. Current photochemical models that use measured UV absorption cross-sections of SO_2 isotopologues (and assume that they are the source of S-MIF) produce sulfate with positive $\Delta^{33}S$ and S_8 aerosols with negative $\Delta^{33}S$, which is opposite to the conventional interpretation of S-MIF signs in Archean sulfates and pyrites (Claire *et al.*, 2014; Ueno *et al.*, 2009). Possibly the input assumptions are incorrect. The SO_2 cross-sections used to calculate the $\Delta^{33}S$ could be inaccurate, while the assumption of SO_2 photolysis as the only means to generate S-MIF make be incorrect if other reactions are important, such as SO photolysis. Alternatively, the SO_2 cross-sections could be correct, but the quantum yield of SO + O may differ for the various isotopologues (M. Thiemans, private communication). Further laboratory work and theoretical models could resolve this issue.

The record of $\Delta^{33}S$ vs. time is structured (Fig. 10.4(a)), which may tell us about concentrations of gases that could offer broadband shielding of sulfur-bearing gases from photolysis. For example, the presence of S-MIF could set an upper limit on amount of CO_2. If CO_2 concentrations had been too high, they would have shielded the lower atmosphere from shortwave ultraviolet because CO_2 is efficient at Rayleigh scattering. (Absorption by CO_2 is small at wavelengths >200 nm.) Farquhar *et al.* (2001) suggested that CO_2 levels exceeding ~0.8 bar prohibits S-MIF. Ueno *et al.* (2009) suggested that carbonyl sulfide (OCS) might be potentially important for broadband shielding of SO_2 because, in their model, integrated SO_2 photolysis between 200 and 220 nm produced sulfate with positive $\Delta^{33}S$, while that between 180 and 200 nm produced sulfate with negative $\Delta^{33}S$. Consequently, the presence of ppm levels of OCS, which has high opacity over 200–220 nm, was posited to be responsible for sulfate with negative $\Delta^{33}S$ and S_8 aerosols with positive $\Delta^{33}S$. However, because OCS photolyzes quickly, concentrations above 10 ppb are improbable, according to more complete photochemical models (Claire *et al.*, 2014; Domagal-Goldman *et al.*, 2011). So, the formation of positive $\Delta^{33}S$ sulfides and negative $\Delta^{33}S$ sulfates probably requires another explanation. Mutual shielding by an organic haze may have also left a trace in the structure of S-MIF through time (Zerkle *et al.*, 2012). Similarly, fluxes and concentrations of CH_4 and O_2 affect the flux of S_8 leaving the atmosphere and so could affect the structure of S-MIF over time (Kurzweil *et al.*, 2013; Zahnle *et al.*, 2006).

10.6   When Did Oxygenic Photosynthesis Appear?

At this point, we have described how changes in atmospheric O_2 levels are indicated by geologic (Sec. 10.3) and isotopic (Secs. 10.4–10.5) evidence, including a dramatic increase in O_2 at ~2.4–2.3 Ga. But when did oxygenic photosynthesis appear?

Various lines of evidence suggest that oxygenic photosynthesis evolved before 2.4 Ga, perhaps by 2.7 Ga and possibly even earlier (Buick, 2008; Farquhar *et al.*, 2011). Physical evidence comes in two main forms. First, certain oxidized geochemical species suggest that some O_2 was locally available in the late Archean. Second, fossil evidence perhaps implies O_2 production. We consider the combination of these types of evidence (described below) to be reasonably convincing. However, a minority opinion disputes the evidence and argues that O_2-producing organisms arose virtually at the same time as the GOE at 2.4 Ga (Kirschvink and Kopp, 2008; Kopp *et al.*, 2005).

Another potential line of evidence concerns molecular remnants of organisms – *biomarkers* – of microbes that made and used O_2 (as mentioned in Sec. 9.4.3). However, earlier literature reports (e.g., Brocks *et al.*, 1999) evidently suffered contamination by more recent organic material (French *et al.*, 2015), so the validity of most late Archean biomarkers is in doubt (Rasmussen *et al.*, 2008).

Finally, a potentially useful technique that can give insight into the history of cyanobacteria and environment is phylogeny (Schirrmeister *et al.*, 2013). For example, phylogenetic inferences suggest that aerobic metabolisms arose around ~3.0 Ga (Kim *et al.*, 2012). However, the interpretation of phylogeny in terms of absolute rather than relative time must ultimately rely on physical evidence to provide calibration points and a reality check. Consequently, divergence times of microbial lineages in the Archean based on phylogeny have very large uncertainties (David and Alm, 2011).

10.6.1   Geochemical Evidence for O_2 Before the Great Oxidation Event

Evidence of small amounts of O_2 before the GOE comes in several varieties: enrichments of redox-sensitive elements (e.g., uranium, molybdenum, rhenium, and sulfur) and changes in their isotopes in marine sediments; iron speciation in marine sediments; and extremely ^{12}C-enriched organic carbon attributed to microbes that oxidize methane with oxidants derived from O_2 or with O_2 itself. We discuss these proxies in chronological order of their indicating the possible advent of oxygenic photosynthesis.

Most lines of evidence proposed for early oxygenic photosynthesis concern the late Archean, but more ancient oxygenic photosynthesis has been proposed. Analysis of lead isotopes and U–Th–Pb systematics in metasediments from Isua, Greenland, has been interpreted to infer that oxygenic photosynthesis existed at 3.7 Ga (Rosing and Frei, 2004). By examining Pb derived from the decay of radioactive U, Rosing and Frei concluded that U was in initially high abundance relative to Th. Uranium is mobile in oxidizing fluids, whereas thorium is not, so they suggested that the sediments derived from oxidizing upper waters. However, doubts arise because such U–Th–Pb patterns are not seen in late Archean sediments until 3.2 Ga. At that time, elevated U concentrations and fractionation of iron isotopes in the Manzimnyama Banded Iron Formation of the Fig Tree Group, South Africa are possible evidence of shallow water O_2 (Satkoski *et al.*, 2015).

Geochemical proxies from the ~3.0 Ga Pongola Supergroup, South Africa, provide later evidence for a change in redox potentially associated with O_2. As mentioned in Sec. 10.4.4, the oxidative weathering of chromium enriches Cr^{4+} in the heavy ^{53}Cr isotope relative to residual Cr^{3+}, and heavy Cr^{4+} can be carried to the oceans where its isotopic signature is captured in sediments. The ~3 Ga Ijzermyn iron formation records heavy $^{53}Cr/^{52}Cr$, while the nearly contemporaneous Nsuze paleosol has light $^{53}Cr/^{52}Cr$, consistent with oxidative weathering (Crowe *et al.*, 2013). Molybdenum isotopes from Pongola rocks also suggest the presence of O_2 in nearshore waters. Molydenum adsorbed as molybdate (MoO_4^{2-}) anions onto nodules of manganese-rich oxides and oxyhydroxides induces light $\delta^{98}Mo$ relative to seawater, whereas sorption onto iron oxyhydroxides causes less isotope fractionation. Consequently, if manganese (Mn) is oxidized during sediment deposition then light $\delta^{98}Mo$ should correlate with an increasing Mn/Fe concentration ratio. This pattern is found in nearshore sediments of the 2.95-billion-year-old Sinqeni Formation and consistent with the presence of O_2 in overlying water (Planavsky et al., 2014a). Modern oxidation of Mn^{2+} requires O_2 and is microbially mediated, given slow abiotic kinetics. It has been postulated that Mn-oxidizing anoxygenic photosynthesis existed in the Archean (Johnson et al., 2013b), but no extant organisms are known, making the hypothesis controversial (Johnson et al., 2013c; Jones and Crowe, 2013).

From 2.8 Ga, a variety of proxies indicate changes associated with the possible presence of O_2. Nisbet et al. (2007b) argue that the difference of ~25‰ in marine $\delta^{13}C$ in carbonates and organic matter from ~2.8 Ga Steep Rock, Canada, imply global partitioning of carbon by photosynthesis, and that abundant stromatolites (up to meter-scale) corroborate the presence of a highly productive biosphere likely to be based on oxygenic photosynthesis. The carbonates are stacked in thick beds (whereas older beds are thin) and there is a negative cerium anomaly (as defined in Sec. 10.3.3) in the limestones, which suggests the presence of O_2 in shallow seawater (Fralick and Riding, 2015; Riding et al., 2014). The inferred dissolved O_2 at ~0.01 mM is consistent with an oxygen oasis and inferences of late Archean shallow oxic waters from iron speciation measurements (Sec. 10.3.4).

Analyses of carbonates and shales from late Archean South Africa also suggests the presence of some O_2. As mentioned above, isotopes of $\delta^{98}Mo$ are affected by adsorption onto iron oxyhydroxides, such as ferrihydrite, $Fe(OH)_3$. Correlation of Mo and Fe isotopes in carbonates and shales from the ~2.7–2.5 Ga Campbellrand-Malmani platform of the Kaapvaal Craton, South Africa, combined with modeling, suggest photic zone O_2 concentrations of 0.02–0.1 mM (Czaja et al., 2012). On the other hand, the lack of a negative cerium anomaly in the rare earth element patterns of Archean BIFs indicates only 0.5–5 µM O_2 if the water redox allows Fe to be oxidized but not Ce or Mn (Planavsky et al., 2010). These contrasting inferences might be explained by spatial heterogeneity and differing sedimentary environments.

A statistically significant upsurge in the concentration of Mo and S in marine shales occurs after 2.8–2.7 Ga, indicating an approximately ten-fold increase in the delivery of S from rivers to the oceans (Stüeken et al., 2012). The global atmosphere was anoxic at the time based on sulfur mass independent isotope fractionation (Sec. 10.5), but oxidation of continental sulfides could have occurred next to O_2-producing cyanobacteria on land, such as in lakes, consistent with the model of Lalonde and Konhauser (2015). The colonization of early land surfaces by cyanobacteria is consistent with molecular genomics, which indicates that cyanobacteria originated in freshwater or endolithic environments before spreading to the ocean (Blank and Sanchez-Baracaldo, 2010). An alternative hypothesis is that the entire atmosphere may have been oxidizing for brief periods during the Late Archean, without noticeably affecting the sulfur MIF record. Recycling of MIF-rich sediments allows S-MIF to persist 10–100 Myr (Reinhard et al., 2013a).

In Sec. 10.4.1, we noted how a worldwide appearance of remarkably ^{12}C-enriched organic carbon ($\delta^{13}C \approx -60‰$) at ~2.8–2.7 Ga can be explained by the presence of methanotrophs. Such microbes consume ^{12}C-enriched methane using sulfate, nitrate or oxygen as oxidants. While both sulfate and nitrate are produced photochemically in anoxic atmospheres (Sec. 9.6.3 and 9.6.4), the fluxes and resulting oceanic concentrations would be small. As is the case today, significant fluxes of sulfate or nitrate have to be derived from the reaction of O_2 with sulfides or reduced nitrogen. Evidence for an increased sulfate flux to the ocean after 2.8–2.7 Ga (Stüeken et al., 2012), along with a greater production of organic matter from marine photosynthesis, would be consistent with increased methanotrophy in the late Archean.

At the end of the Archean, several geochemical signs of increased redox state occur. Over the interval 2.6–2.5 Ga, increased fractionation of Mo isotopes is seen in black shales of the Transvaal Supergroup, southern Africa, and Hamersley, Australia, which is possible if some Mo was oxidized and present as oxyanions (Kurzweil et al., 2015b; Wille et al., 2007). Enrichments of Re concentrations, iron speciation (Sec. 10.3.4), and nitrogen isotopes from Transvaal rocks from 2.6–2.5 Ga are consistent with photic zone ocean margins that were mildly oxygenated (Godfrey and Falkowski, 2009; Kendall et al., 2010; Poulton et al., 2010; Reinhard et al., 2009).

Also, at 2.5 Ga, a pulse of trace levels of Mo, Se, and Re along with mass-dependent sulfur isotope fractionation in pyrite and isotopic excursions in Mo and Se, are found in the Mt. McRae Shale of the Hamersley Basin, Australia, and attributed to a so-called "whiff" of oxygen (Anbar et al., 2007; Kaufman et al., 2007; Stueken et al., 2015a). Mobilization of Mo, Se, and Re and sulfur through oxidation on the continents are suggested. Lack of a U enrichment is attributed to U being hosted in silicates and requiring more O_2 to liberate. However, critics have suggested that the spikes in element concentrations and isotope fractionations could be caused by post-depositional oxidizing ground water (Fischer et al., 2014).

10.6.2 Fossil and Biomarker Evidence for O$_2$ Before the Great Oxidation Event

Fossil evidence for O$_2$ production comes from stromatolites dating back to 2.7 Ga with particular features that suggest they were built by oxygenic photosynthesis. *Stromatolites* are laminated sedimentary structures produced by microbial communities. Abundant stromatolites in the 2.7 Ga Tumbiana Formation, Australia, grew on a lake shore, according to field geology (Buick, 1992). There is no sign of abundant hydrothermal reductants needed for anoxygenic photosynthesis, so Buick suggested that the stromatolites grew by oxygenic photosynthesis. A counter-argument is that the early atmosphere should have contained enough H$_2$ to support H$_2$-based anoxygenic photosynthesis (Sec. 9.6.5). However, additional support for Buick's hypothesis comes from isotopic and structural evidence. The biomass in the Tumbiana stromatolites is extremely enriched in ^{12}C with δ^{13}C $\approx$ –50‰ to –60‰, suggesting assimilation of carbon from methane oxidation, which requires O$_2$ or oxidants derived from O$_2$ (Eigenbrode and Freeman, 2006). Disrupted, curled and contorted laminae along with enmeshed millimeter-scale bubbles in conical stromatolites are consistent with cyanobacterial production of O$_2$ bubbles (Bosak *et al.*, 2009). Such features are seen in the 2.7 Ga Tumbiana stromatolites as well as 2.7 Ga stromatolites in Africa and North America. Also, centimeter-scale tufts, pinnacles and ridges in Tumbian stromatolites are made only in modern mats by filamentous cyanobacteria with gliding motility and phototaxis (Flannery and Walter, 2012). Thus, stromatolite evidence is consistent with O$_2$ production by 2.7 Ga.

In principle, *biomarkers*, which are recognizable remnants of specific biological molecules, could provide evidence for the presence of cyanobacteria and O$_2$. In modern rocks, 2α-methylhopanes are commonly derived from cyanobacterial cell membranes. One strain of a non-marine photosynthetic bacterium (*Rhodopseudomonas palustris*) is also known to produce abundant 2α-methylhopanes (Rashby *et al.*, 2007), but it should also produce other diagnostic biomarkers (gammaceranes). Steranes are derived from eukaryotic sterols – the most well known example being cholesterol – which are generally considered to require O$_2$ in their biosynthesis (Summons *et al.*, 2006) (but see Kirschvink and Kopp (2008) for a contrasting view). In the 1990s, sterane biomarkers found together with 2-α-methylhopanes in 2.7 Ga black shales from the Pilbara, northwest Australia, suggested the presence of O$_2$ and cyanobacteria (Brocks *et al.*, 2003; Brocks *et al.*, 1999). However, subsequently Brocks argued that his own data were compromised by contamination

(Rasmussen *et al.*, 2008). His skepticism has been vindicated by rigorous sterile drilling procedures that found no biomarkers in the same stratigraphy (French *et al.*, 2015). Because of this case of contamination, other late Archean biomarker work is in doubt (Eigenbrode *et al.*, 2008; Waldbauer *et al.*, 2009).

Cyanobacteria biomarkers and diverse steranes have been detected in fluids inclusions in the 2.45 Ga Matinenda Formation at Elliot Lake, Canada (Dutkiewicz *et al.*, 2006; George *et al.*, 2008), which is a sandstone deposited prior to the rise of O$_2$. Rasmussen *et al.* (2008) state that the Elliot Lake biomarkers contain contaminant plant waxes, but this statement is contested by George *et al.* In any case, these biomarkers, if valid, do not much pre-date the rise of oxygen.

Despite the uncertainty about Archean biomarkers, we conclude that overall, geochemical and fossil data suggest that oxygenic photosynthesis originated by ~2.7–2.8 Ga and possibly by 3.0 Ga. The apparent paradox of oxygenic photosynthesis hundreds of millions of years before the 2.4 Ga rise of atmospheric O$_2$ is resolved if the O$_2$ was scavenged efficiently in the air and in seawater. Consequently, many researchers have argued that the Earth had to undergo a global redox titration before O$_2$ could rise. Essentially, oxygenic photosynthesis was necessary but insufficient for O$_2$ to rise. One version of this hypothesis was described in Sec. 8.6. We now return to this problem.

10.7 Explaining the Rise of O$_2$

10.7.1 General Conditions for an Anoxic Versus Oxic Atmosphere

Before describing theories for why O$_2$ rose, let's define an *anoxic* versus *oxic* atmosphere. An oxidizing atmosphere is poor in hydrogen-bearing reducing gases (e.g., H$_2$, CH$_4$, and NH$_3$) whereas a reducing atmosphere is not. We see examples in the Solar System (Sec. 1.1.2.2). Earth's atmosphere (21% O$_2$, 1.8 ppmv CH$_4$) and that of Mars (0.13% O$_2$, 15 ppmv H$_2$) are oxidizing, whereas Titan's atmosphere (95% N$_2$, 5% CH$_4$) is reducing. A small excess of hydrogen relative to oxygen (or vice versa) can flip the redox state. For example, an atmosphere with a steady-state abundance of ~0.1% H$_2$ or methane would be anoxic, while one with 0.1% O$_2$ would be oxic.

The Archean atmosphere could have remained anoxic even with a relatively large net flux of O$_2$ from burial of photosynthesized organic carbon (Sec 10.2.2) if the flux of rapidly reactive reductants (such as reducing gases and hydrothermal cations) was greater. We express this quantitatively with an oxygenation parameter, K_{OXY}, defined as follows (Catling and Claire, 2005; Claire et al., 2006):

reducing fluxes of organic matter, which is an O_2 sink directly or indirectly if the organic matter is converted into methane.

$$K_{OXY} = \frac{O_2 \text{ source flux}}{\text{non-weathering } O_2 \text{ sink flux}} = \frac{F_{reductant\_burial}}{F_{metamorphic} + F_{volcanic}}; \quad \begin{array}{l} K_{OXY} > 1 \text{ gives an oxic atmosphere} \\ K_{OXY} < 1 \text{ gives an anoxic atmosphere} \end{array} \quad (10.18)$$

This formulation for K_{OXY} looks different from that in Ch. 8 (eq. (8.44)) but is really the same. The anaerobic iron oxidation term in eq. (8.46), Φ_{burial} (Fe_3O_4), is included implicitly in the denominator in eq. (10.18). When $K_{OXY} < 1$, efficient O_2 sinks are larger than O_2 sources and so an atmosphere has trace quantities of O_2 that are probably less than part per million levels. As a consequence, H_2 and hydrogen-bearing reducing gases such as CH_4 builds up and hydrogen escapes to space. When $K_{OXY} > 1$, oxygen sources outweigh efficient sinks and O_2 levels rise so that oxidative weathering becomes a big sink.

In eq. (10.18), $F_{reductant\_burial}$ (in Tmol O_2 yr^{-1}) is the net source flux of O_2 that accompanies the burial of reductants: organic carbon from oxygenic photosynthesis and reductants whose reducing power has been derived from organic carbon, such as pyrite (eq. (10.3)). As in eq. (10.6), $F_{metamorphic}$ is the flux of metamorphic gases and $F_{volcanic}$ is the flux subaerial volcanic gases, both expressed as the rate that they consume oxygen in Tmol O_2 yr^{-1}. Conceptually, a serpentinization flux of H_2 or CH_4 is included in "$F_{metamorphic} + F_{volcanic}$" (see Kasting, 2013). Serpentinization occurs on the seafloor and on continents when water oxidizes ferrous iron in ultramafic rocks with a net reaction, $3FeO + H_2O \rightarrow Fe_3O_4 + H_2$. The hydrogen may be converted into methane abiotically or by microbes, but it remains an O_2 sink. In the Archean, Fe^{2+} in seawater that upwelled from hydrothermal vents consumed O_2 if it was used to form oxide BIFs; thus, for an Archean calculation, this flux can be considered part of the denominator of eq. (10.18).

Anaerobic photosynthesis and burial of ferric iron converts a metamorphic, volcanic, or hydrothermal source of reducing power into an equivalent amount of organic carbon, which in the Archean should have been microbially degraded into methane that vented to the atmosphere (e.g. Walker, 1987). For example, the net reactions of iron and hydrogen photosynthesis, respectively, can be written as follows.

$$CO_2 + 4FeO + H_2O + hv \rightarrow CH_2O + 2Fe_2O_3 \quad (10.19)$$

$$2H_2 + CO_2 + hv \rightarrow CH_2O + 2H_2O \quad (10.20)$$

In eqs. (10.19) and (10.20), the reducing fluxes of ferrous iron ("FeO") and H_2, respectively, are converted into

If we took the total O_2 sources and efficient sinks from Tables 10.1 and 10.3, as $F_{reductant\_burial}$ ~15.8 Tmol O_2 yr^{-1} and "$F_{metamorphic} + F_{volcanic}$" of 5.7 Tmol O_2 yr^{-1} (consumed), we would obtain a K_{OXY} of 15.8/5.7 $\approx$ 2.8. However, a ~0.2 Tmol O_2 yr^{-1} seafloor sink in Table 10.3 depends on sulfate from oxidative weathering, while the pyrite and ferrous O_2 sources of Table 10.1 also depend on an oxidative weathering supply of sulfate and ferric iron (pyrite was mostly detrital in the Archean). To try to examine how O_2 would be consumed in an anoxic world, let us neglect the seafloor oxidation from sulfate in the sink and neglect the aforementioned sources of O_2. Then K_{OXY} ~10/5.5 $\approx$ 1.8. The denominator could be even bigger if BIF deposition is included as an O_2 sink, although the estimates here are highly uncertain, e.g., Kasting (2013) lists values in the range 0.1–12.5 Tmol O_2 yr^{-1}. If we use a lower estimate of 5.3 Tmol O_2 yr^{-1} for burial of organic carbon (Berner, 2004, p. 42), then K_{OXY} ~5.3/5.5 $\approx$ 0.96, which corresponds to an anoxic atmosphere.

Our values of modern K_{OXY} are somewhat smaller than previously published (Catling and Claire, 2005; Claire et al., 2006; Kasting, 2013) because we have revised the O_2 source flux in Table 10.1, as described in the table caption, and because Table 10.3 contains a more exhaustive list of O_2 sinks than before. In any case, today $K_{OXY} > 1$ and so the O_2 source is mostly balanced by oxidative weathering, which permits high concentrations of O_2 and marine sulfate. In the past, if reductant fluxes in the denominator of eq. (10.18), were larger by a factor of ~2 relative to the numerator, then K_{OXY} would have been < 1, and the atmosphere would have been redox-dominated by hydrogen-rich reducing gases, such as CH_4 and H_2. Thus, the redox state of the atmosphere depends as much on the magnitude of the rapid sink flux as it does on the O_2 source. This is illustrated schematically in Fig. 10.9. Kasting (2013) reviews how various contributions to K_{OXY} may have varied in the past, using the form of K_{OXY} in eq. (8.46).

A further nuance is that the source of carbon for organic burial is not just volcanic and metamorphic CO_2 but also weathering of carbonates and organic carbon. Thus, if we consider the source of O_2 from organic carbon

(a)

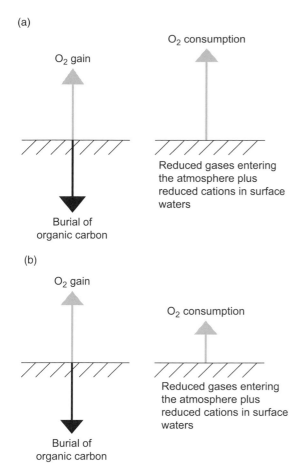

O2 gain

O2 consumption

Reduced gases entering the atmosphere plus reduced cations in surface waters

Burial of organic carbon

(b)

O2 gain

O2 consumption

Reduced gases entering the atmosphere plus reduced cations in surface waters

Burial of organic carbon

Figure 10.9 (a) Relative size of fluxes in an *anoxic* atmosphere, where the length of arrows indicates the magnitude of the flux in units of moles O_2 per year either produced or consumed. Here, it is assumed that the flux of organic burial is the only significant source of O_2. In the anoxic atmosphere, the efficient consumption fluxes of O_2 exceed the source flux of O_2. (b) Relative sizes of fluxes in an *oxic* atmosphere. Here source flux of O_2 exceeds the efficient consumption fluxes of O_2; oxygen levels will come into balance as a result of weathering fluxes, which are not shown.

burial, the numerator in eq. (10.18) could be written (Krissansen-Totton *et al.*, 2015)

$$F_{reductant\_burial,Corg} = f_{org}(F_{volc+meta.CO2} + F_{weather.CO2})$$

$$(10.21)$$

Here, f_{org} is the fraction of carbon buried as organic carbon, as given in eq. (8.42). Today, $F_{volc+meta.CO2}$ ~8.5 Tmol CO_2 yr^{-1} (Table 7.1). If the modern value of is f_{org} ~0.25 (Fig. 8.3(a)) and $F_{reductant\_burial,Corg}$ ~10 Tmol C_{org} yr^{-1} (Holland, 2002) then $F_{weather.CO2}$ ~31.5 Tmol CO_2 yr^{-1}. This estimate of modern $F_{weather.CO2}$ might be too large (see Sec. 7.2.2) and it would be ~11.5 Tmol CO_2 yr^{-1} if we used the estimate

of $F_{reductant\_burial,Corg}$ ~5 Tmol C_{org} yr^{-1} from Berner (2004). In the Archean, the flux of carbon from weathering may have been lower still for two reasons: the lack of oxidative weathering of organic carbon and the possibility that continental area was smaller (Derry, 2014) and that continents had less coverage of carbonate. If $F_{weather.CO2}$ had been lower in the Archean, then the numerator in eq. (10.18) would be smaller and an anoxic atmosphere ($K_{OXY} < 1$) more readily achievable.

Another general consideration of anoxic atmospheres concerns the escape of hydrogen to space. Anoxic atmospheres on small, Earth-like planets, by definition, have sufficient fluxes of reducing gases from their surface to be maintained as relatively hydrogen-rich and so they are prone to lose hydrogen to space relatively rapidly. There are always some upward directed H atoms traveling fast enough to escape into interplanetary space, in contrast to very slow or negligible rates for heavier atoms such as O (Ch. 5). If an entity loses net hydrogen, it becomes oxidized, *by definition*. So, from a redox standpoint, hydrogen escape oxidizes a planet as a whole over time.

In fact, hydrogen escape is the widely accepted explanation for atmospheric oxygen and surface oxidation on other bodies in the Solar System. Hydrogen escape is why the surfaces of Mars and Venus are red with ferric iron. On Mars, hydrogen has been lost preferentially from an amount of water that was the equivalent of at least tens of meters deep spread globally (Ch. 12), while on Venus the quantity of hydrogen may have been an ocean's worth (Ch. 13). Hydrogen escape that leaves oxygen behind also creates very thin, collisionless atmospheres (i.e., exospheres) of O_2 on some icy moons: Europa and Ganymede (Hall *et al.*, 1998; Smyth and Marconi, 2006), which are satellites of Jupiter, and Saturn's moons Rhea (Teolis *et al.*, 2010) and Dione (Tokar *et al.*, 2012).

In general, hydrogen on rocky planets originates within water, hydrated silicates, or hydrocarbons (Ch. 6). Hydrogen that escapes has ultimately split from one of these compounds and left behind an oxidized product to balance the redox reaction. It doesn't matter whether the hydrogen that escapes is vertically transported through the atmosphere within CH_4, H_2, H_2O or some other H-bearing compound. For example, if volcanic hydrogen is released on Earth and subsequently escapes to space, the Earth's upper mantle is oxidized, schematically as $3FeO + H_2O = Fe_3O_4 + H_2 = Fe_3O_4 + 2H(\uparrow space)$ (Kasting *et al.*, 1993a). Similarly, when escaped hydrogen derives from methane generated by serpentinization in continental rocks, Earth's crust is oxidized (Catling *et al.*, 2001; Claire *et al.*, 2006).

In terms of eq. (10.18), we see that to transition from an anoxic to oxic atmosphere the O_2 source flux must exceed the efficient O_2 sink flux. In relative terms, either the O_2 source ($F_{reductant\_burial}$) increases or the O_2 sink ($F_{metamorphic} + F_{volcanic}$) decreases. Although both of these changes can happen together, theories that attempt to explain the Great Oxidation Event of 2.4 Ga usually focus on changes in the source or sink flux individually, which we now consider.

10.7.2 Hypotheses for an Increasing Flux of O_2

Some have argued that the GOE occurred because the rate of organic carbon burial (net O_2 production) increased. The most extreme variant of this idea is that the advent of oxygenic photosynthesis caused the GOE almost immediately in geologic time (Kirschvink and Kopp, 2008). However, the majority of geoscientists are persuaded by the various lines of evidence described in Sec. 10.6 that oxygenic photosynthesis arose hundreds of millions of years before the rise of O_2.

On idea for causing oxygenation is a pulse of organic burial, although we argue that this mechanism is incompatible with the timing of the GOE and would not cause a permanent shift in O_2 levels given the short geologic residence time of O_2 (Sec. 10.2.4). Organic matter extracts the light carbon isotope, ^{12}C, from seawater, so seawater carbon, recorded in ancient carbonates, should decline in ^{12}C content if the rate of organic burial increases. Consequently, some have attributed the rise of O_2 to a pulse of organic burial indicated by ^{12}C-depleted carbonates. The hypothesis that the 2.2–2.06 Ga Lomagundi pulse in organic carbon burial caused the rise of O_2 was mentioned in Sec. 10.4.1. But it cannot be the cause of the GOE, only its consequence, because it happened *after* the ~2.4 Ga GOE as indicated by mass independent fraction of sulfur isotopes (Sec. 10.5).

Rather than a pulse in the organic burial rate, a more attractive possibility is a long-term increase. Early models with box-averaging of $\delta^{13}C$ data over time suggested a secular increase in organic carbon burial rates (DesMarais et al., 1992). However, the result was influenced by starting at 2.6 Ga, which is a time with an anomalously negative $\delta^{13}C$ for organic carbon that has been interpreted as a signature of methanotrophs in sediments (Hayes, 1994; Hayes and Waldbauer, 2006; Hinrichs, 2002; Hinrichs and Boetius, 2002; Schoell and Wellmer, 1981; Thomazo et al., 2009). Another model suggested an increase in the fraction of organic burial from the mid-Archean onwards based on the idea that there was a significant loss of ^{12}C-depleted carbon into hydrothermal

carbonates (Bjerrum and Canfield, 2004). If hydrothermal carbonates sequestered ^{12}C relative to sedimentary marine carbonates at a greater rate in the Archean, this would allow for a smaller rate of organic burial. The model assumed a biologically pumped gradient in the $^{13}C/^{12}C$ ratio between the surface and deep seawater. Measurements of 3.46 Ga seafloor do not support such a gradient (Nakamura and Kato, 2004), however.

Suffice it to say that differing views persist about whether the fraction of carbon buried as organic carbon has changed or not, given the noise in the carbon isotope record (see Sec. 8.5). As mentioned in Sec. 10.4.1, a comprehensive statistical analysis shows that determining a meaningful trend in organic burial from the carbon isotope data depends on the assumed carbon cycle model (Krissansen-Totton et al., 2015; Derry, 2014).

Another recurring variant on the theme of increased organic burial concerns continental growth. Organic carbon is buried on continental shelves (Knoll, 1979), while continental weathering is a source of phosphorus, which is a critical nutrient. One numerical model ties organic burial to phosphorus delivery, which is set proportional to continental area, so that increasing O_2 levels parallels continental growth (Godderis and Veizer, 2000). Clearly, such models are prescriptive. In another approach, bursts of continental growth have been inferred from zircon U–Pb ages, and linked to increased O_2 (Campbell and Allen, 2008). However, the inferred growth spurts are probably not real but reflect preferential preservation of crust during supercontinent formation (Hawkesworth et al., 2009). Indeed, most of the present continental crust volume was probably generated by ~3 Ga (Hawkesworth et al., 2010) after which crustal growth slowed down (Dhuime et al., 2012; Kemp and Hawkesworth, 2014).

Some models suggest that most of the continental area was submerged below seawater at the end of the Archean (Flament et al., 2008), which might have limited organic burial. However, marine Hf–Nd isotope data from ~2.7 Ga require weathering of substantial landmass above sea level for selective mobilization of radiogenic Hf (Viehmann et al., 2014). Overall, it is hard to make a convincing case that changes in organic carbon burial caused the GOE, although this hypothesis cannot be ruled out (e.g., Krissansen-Totton et al., 2015).

10.7.3 Hypotheses for a Decreasing Sink of O_2

A decreasing O_2 sink is an alternative hypothesis for the rise of O_2. Many authors have taken a conservative view that the average $\delta^{13}C$ value of marine carbonates has remained roughly ~0‰ since 3.5 Ga (Schidlowski, 1988), while the average $\delta^{13}C$ for organic carbon has been about $-25‰$. This suggests that roughly the same proportion of ^{12}C has been preferentially extracted from seawater into organic sediments, so that relative burial rates of organic and carbonate carbon have remained fairly constant over geologic time (Holland, 1984, 2002, 2009; Kump et al., 2001; Kump and Barley, 2007). As shown in Sec. 8.5, the inferred typical ratio of carbon burial is $\approx 1:4$ organic to carbonate. If we also accept the evidence for an origin of oxygenic photosynthesis well before the GOE (Sec. 10.6) and assume that this process dominated productivity, the *only* way that O_2 could have remained a trace atmospheric constituent in the Archean is if the flux of reduced species to the atmosphere from volcanoes and metamorphism exceeded the flux of O_2 from organic carbon and sulfide burial (Catling et al., 2001; Holland, 2002; Kasting et al., 1993a; Kump et al., 2001). Subsequently, for O_2 to rise, the reductant flux must have decreased to a tipping point when K_{OXY} reached unity in eq. (10.18). Once K_{OXY} exceeded unity, O_2 would have flooded the atmosphere and reached a new equilibrium concentration balanced by a new, big loss to oxidative weathering. Of course, we see this new sink directly in the form of redbeds and oxidized paleosols, as described in Sec. 10.3.1.

10.7.3.1 Decreasing Sink from Reducing Volcanic Gases

One variant of the above idea is that prior to the GOE there was a greater amount of reducing gases from *volcanism*, which subsequently declined to the point where biogenic O_2 fluxes overwhelmed them. As Earth's interior cooled, the flux of volcanic reducing gases surely dwindled (Sec. 7.6). However, this does not necessarily lessen the sink on O_2 because increased past volcanic outgassing would have provided more CO_2 for photosynthetic fixation. If carbon isotopes imply that ~20% of the CO_2 input was buried as organic carbon, increased past outgassing, on its own, cannot explain the oxic transition because going back in time, O_2 production arising from organic burial would have paralleled O_2 losses. The problem can be overcome if the *proportion* of reducing gases in the total volcanic plus metamorphic flux has decreased. Volcanic gases that had an increasingly oxidizing composition with time could explain the GOE (Holland, 2002, 2009; Kasting et al., 1993a; Kump et al., 2001).

The upper mantle is the source region for volcanic gases and so mantle geochemistry generally determines the redox state of the gases. In Sec. 7.4, we saw that the more reducing the upper mantle, the more reducing are volcanic emissions. So whether Archean volcanic gases were less oxidized than today hinges on the question of whether the upper mantle used to be more reduced. As discussed in Sec. 7.3, analyses of redox-sensitive chromium and vanadium abundances in basalts (Canil, 2002; Delano, 2001; Li and Lee, 2004) have addressed this question and found that the mantle's average oxidation state has remained roughly constant since ~3.5 Ga, presumably because the mantle is big and strongly redox buffered (Lee et al., 2005). The studies indicate that the average oxygen fugacity f_{O2} (defined in Sec. 7.3) in the Archean upper mantle differed from the modern mantle by no more than 0.3 $\log_{10}$ units. To appreciate how little this means for the redox state of volcanic gas emissions, we can examine the relationship between f_{O2} and gas composition.

To first order, let's assume that the ratios of the emitted volcanic gases scale with their fugacities. For simplicity, let's consider the H_2 flux (a sink for O_2) relative to the CO_2 flux (which is related to the O_2 source via carbon fixation and organic burial):

$$\frac{\text{volcanic flux of } H_2}{\text{volcanic flux of } CO_2} \approx \text{fugacity ratio} = \left(\frac{f_{H_2}}{f_{H_2O}}\right)\left[\frac{f_{H_2O}}{f_{CO_2}}\right] \tag{10.22}$$

Let's assume that the ratio in the square brackets is proportional to the flux ratio of H_2O/CO_2 and remains constant (we'll consider the validity of this assumption later). Then the f_{H2}/f_{H2O} ratio, r_{H2}, is given in terms of f_{O2} by eq. (7.13):

$$r_{H2} = \frac{f_{H_2}}{f_{H_2O}} = \frac{K_{11}}{f_{O_2}^{0.5}} \tag{10.23}$$

From Sec. 7.4, the modern value of r_{H2} is ~0.02, using K_{11} $= 1.31 \times 10^{-6}$ $atm^{0.5}$ for the equilibrium constant and $f_{O2} =$ $10^{-8.5}$ atm as an average f_{O2} for the modern upper mantle (for a typical magmatic temperature of 1200 °C and pressure of 5 atm). If f_{O2} were 0.3 $\log_{10}$ units lower, r_{H2} would only increase to 0.03, i.e, ~3% of the hydrogen would come out of volcanoes as H_2 compared to H_2O instead of ~2% today. For illustrative purposes, if the Archean mantle f_{O2} had been drastically lower by ~3 $\log_{10}$ units at $10^{-11.5}$ atm, then r_{H2} would have been ~0.74, i.e., almost three-quarters of volcanic hydrogen would have been emitted as H_2 rather than water vapor, providing a huge sink for O_2. But the geochemical data do not allow such f_{O2} changes. However, there are ways that the redox speciation of volcanic gases could have changed even if the mantle redox state remained roughly constant.

First, as discussed in Sec. 8.6.1, Holland (2009, 2011) suggested that the CO_2/H_2O flux ratio increased with time, whereas in eq. (10.22) we assumed it was constant. If past geologic processes had recycled significantly less CO_2 then the proportion of H_2 relative to the carbon outgassing would have been greater. Holland argued that mantle-derived carbon accumulated in the surface reservoir so that more carbon was geologically recycled as time went on. He also proposed that the SO_2/H_2O flux ratio increased because of more geologic recycling of sulfur. If less SO_2 had been outgassed in the past, less H_2 would have been required to reduce SO_2 to sulfide via

$$5H_2 + FeO + 2SO_2 = FeS_2 + 5H_2O \qquad (10.24)$$

Then H_2 would be in excess, creating an anoxic atmosphere.

Unfortunately, data are lacking to confirm Holland's proposed secular increase in CO_2/H_2O and SO_2/H_2O. It is reasonable that the volcanic S fluxes increased after the GOE once marine sulfate became abundant and entered subducted oceanic crust, but this would be an effect rather than a cause. It is also possible that carbon accumulated on the continents at the expense of the mantle, but carbon cycle models establish this balance early, by about 3 Ga (Zahnle and Sleep, 2002). Finally, Holland assumes that the fraction of carbon buried as organics is fixed at 4:1, which gives him a smaller O_2 source with a smaller recycling rate of carbon. Whether his mechanism could trigger an O_2 rise at 2.4 Ga is unclear.

Kump and Barley (2007) and Gaillard et al. (2011) have argued that there was a shift from predominantly submarine to subaerial volcanism with time, which caused the GOE. Gaillard (2011) modeled the composition of gases from thermodynamic equilibrium of a melt and showed a slightly more reducing mixture at seafloor pressures ~100 bar compared to subaerial pressures ~1 bar. This difference is insufficient to flip the atmosphere to anoxic unless the O_2 source flux is also changed (Kasting et al., 2012).

Certainly, the thermodynamics of relatively low temperature submarine emissions tends to favor more reducing gases (Sec. 7.5). Also in the absence of significant seawater sulfate, and possibly shallower mid-ocean ridges, the Fe^{2+} flux of hydrothermal vent fluids could have been much greater than today (Kump and Seyfried, 2005). Today the submarine sink on O_2 from volcanic gases, ferrous iron exhalation, and serpentinization sums to ~1 Tmol O_2 yr^{-1}, which is a factor of ~1.5 less than the subaerial volcanic sink (Table 10.3), and together (2.5 Tmol O_2 yr^{-1}) these are a factor of ~4 smaller than the O_2 flux released from organic burial. So, very substantial changes in the balance of volcanic emissions between subaerial and submarine are required for this proposal to

work. One idea is that seafloor serpentinization may have been much larger in the past, which is discussed at the end of the next section.

10.7.3.2 Decreasing Sink from Reducing Metamorphic Gases and Diffuse Volcanic Gases

Another possibility is that the flux of reducing, metamorphic gases decreased from the Archean to the Proterozoic (Catling et al., 2001; Claire et al., 2006; Zahnle et al., 2013). Of the total geologic reducing gas flux entering the atmosphere today, metamorphic methane accounts for about half of the total sink on net O_2 (Table 10.3).

Much of the methane is thermogenic, i.e., created from disseminated organic carbon in rocks. If we assume that the carbon isotope record shows a proportion of organic carbon buried that has not varied drastically, then this flux may have been relatively important in the past also. Metamorphic chemistry in the crust includes reactions that generate hydrogen or methane, such as,

$$C + H_2O \rightarrow CO_2 + 2H_2 \qquad (10.25)$$
$$2C + 2H_2O \rightarrow CO_2 + CH_4 \qquad (10.26)$$

When such hydrogen is lost to space through the atmosphere, the crust will oxidize.

Another source of hydrogen is serpentinization in hydrothermal systems. This generates hydrogen from water, which we can write schematically as:

$$3FeO + H_2O \rightarrow Fe_3O_4 + H_2 \qquad (10.27)$$

Microbes can then make methane by using such hydrogen released from serpentinization.

There is evidence that serpentinization was important in Archean. *Greenstone belts* are folded, structurally distinct, generally elongate regions containing abundant dark-green, altered mafic to ultramafic igneous rocks. These igneous rocks were more mafic than the modern crust and were prone to serpentinize. Hydrogen escape to space from H_2 and CH_4 intermediaries would oxidize the crust. Any subsequent metamorphic processing of the resultant crust would release less reducing gases, as a matter of redox conservation, and each geologic cycle would give further decline.

Volcanic volatiles emerge on the continents from areas of intrusive igneous rocks and associated hydrothermal systems that are better described as a *diffuse volcanic volatiles* rather than metamorphic per se. In arcs, data suggests that ~80% of the igneous volatile flux to the crust is emplaced at plutonic depths, implying that diffusive transport in hydrothermal and groundwater systems

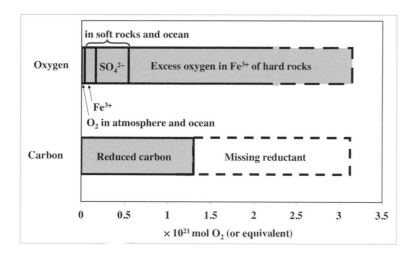

Figure 10.10 The inventory of oxygen in the Earth's crust. "Hard rocks" are high grade metamorphic rocks and igneous rocks. "Soft rocks" are sedimentary rocks. In units of 10^{21} mol O_2 equivalent, there is 0.037 in the atmosphere and ocean, 0.13 in ferric iron of soft rocks, and 0.384 in oceanic and sedimentary sulfate. Reduced carbon in the crust is <1.3. Hard rocks contain 1.6–2.5, with the upper limit illustrated. The imbalance between oxygen and reduced carbon inventories shows that there is missing reductant, which can plausibly be attributed to the escape of hydrogen to space. Quantities are based on tabulated values in Catling *et al.* (2001) and references therein; see also Sleep (2005) and Hayes and Waldbauer (2006) for similar estimates.

potentially supplies a large amount of volatiles to the surface (Carmichael, 2002). The fluxes and redox state of diffusive volcanic volatiles are very poorly constrained. However, theoretical calculations suggest that low temperature systems in reducing crust generally favor reducing volatiles (Claire *et al.*, 2006), unlike high temperature direct volcanic emissions.

One piece of evidence in support of ancient continental oxidation comes from inventories showing that "excess" oxygen has accumulated in the Earth's continental crust (Fig. 10.10). The continental crust, ocean and atmosphere contain ~2.5×10^{21} mol of O_2 equivalents (Catling *et al.*, 2001), which is mostly in the form of ferric iron that started out as ferrous iron, i.e., oxygen has been added in the net redox reaction, $FeO + \frac{1}{4}O_2 \rightarrow \frac{1}{2}Fe_2O_3$. Consequently, O_2 cannot have all come from organic carbon burial because otherwise there would be equivalent number of moles of reduced carbon in the crust but instead there is <1.3×10^{21} mol (Wedepohl, 1995). Alternatively, to explain Fig. 10.10, organic carbon could have been preferentially subducted into the mantle (Hayes and Waldbauer, 2006) relative to iron oxides, but this seems unlikely given the refractory and dense nature of ferric oxides. Furthermore, old Archean cratons contain some of the excess ferric iron, which was not formed after the GOE by oxidative weathering. So instead of organic carbon burial, perhaps the escape of hydrogen to space was responsible for *net* oxidation of the crust (Catling *et al.*, 2001). If the modern metamorphic outgassing of 1.55 CH_4 Tmol yr^{-1} (Table 10.3) were scaled up by a factor of 3.2, and all of the hydrogen was lost to space, then the O_2 equivalent production rate would have been 5×10^{12} mol yr^{-1}. Over 0.5 b.y. this would produce a net

irreversible quantity of 2.5×10^{21} mol O_2 – enough to account for the observed oxidation of the crust.

Some other data suggest that crustal redox evolved. Yang *et al.* (2014c) deduce a more reduced Hadean and early Archean continental crust from cerium anomalies in zircons. The data are very scattered but Yang *et al.* suggest that they imply a trend of increasing oxygen fugacity, which they attribute to a declining bombardment of reducing impactors. However, the crust will also oxidize with hydrogen loss. Further similar work may shed more light on the duration and extent of crustal redox evolution.

Furthermore, isotopic evidence suggests that considerable hydrogen did escape from the Archean Earth. Pope *et al.* (2012) found that the D/H ratio of seawater was lower than present by analyzing hydrated minerals (serpentine) in 3.8 Ga rocks from Greenland. Chlorite and amphibolite in 3.5 Ga seafloor basalts from Barberton, South Africa, confirm lower Archean D/H (Ueno *et al.*, unpublished). The preferential escape of H versus D raises the D/H ratio and the data imply the early ocean had ~1/4 more volume. The modern ocean is 1.4×10^{21} kg so the water lost was (1/4 × 1.4 × 10^{24} g)/(18 g mol H_2O^{-1}) = 1.9×10^{22} mol H_2O, which would produce 9.5×10^{21} mol O_2 using eq. (10.1). This O_2 could account for all of the excess oxidation of the crust as well as the export of considerable oxidized material (e.g., iron oxides) to the mantle.

The escape of hydrogen may be also be indicated indirectly by xenon isotopes. Fluid inclusions in Archean barites and quartz show that the pattern of xenon's nine stable isotopes was lighter going back in time (Hebrard and Marty, 2014; Pujol *et al.*, 2011). This relationship is explained if xenon (Xe), the heaviest gas in the atmosphere, escaped to space. Xenon ionizes easily and so could have escaped along open magnetic field lines at the poles if it were dragged by rapid hydrogen escape in a *polar wind* (see Ch. 5). However, the argument is uncertain: the ancient xenon data might be mixtures of modern air with an unfractionated mantle component (Pepin, 2013).

All of the above data (crustal redox, D/H evolution, and a possible interpretation of Xe isotopes) are inconsistent with claims that appear in the literature from time to time that hydrogen escape was far slower in the Archean than the diffusion-limited escape rate described in Sec. 5.2.4 and that a highly reducing atmosphere built up. Tian *et al.* (2005) modeled slow H escape but Kuramoto *et al.* (2013) have argued that the slow escape was an artifact of their numerical diffusion. More importantly, geological evidence from the Archean Earth conflicts with the idea of a strongly reducing atmosphere. If the Archean atmosphere had been strongly reducing, minerals that are abundant in Archean sediments, such as magnetite, should have been replaced with more reduced minerals such as siderite (Abelson, 1966; Rosing and Bird, 2007). Even pyrite would decompose to siderite and H_2S (Krupp *et al.*, 1994). Similarly, iron in Archean impact spherule spinels (formed from condensed rock vapor in impact plumes) would be more reduced than is observed (Krull-Davatzes *et al.*, 2010) if the Archean atmosphere had been highly reducing.

In a biogeochemical box model, Claire *et al.* (2006) estimated the evolution of the sink on O_2 with time by tracking redox fluxes in and out of the crust (and diffusion-limited hydrogen escape to space), and calculating the metamorphic gas speciation according to a crustal oxygen fugacity specified in terms of a ferric/ferrous iron content. For numerical simplicity, the model converted organic carbon to Fe^{2+} equivalents via $6Fe_2O_3 + C \rightarrow 4Fe_3O_4 + CO_2$, which is a process that happens in greenschist metamorphism. Figure 10.11 shows model results assuming that reducing metamorphic gases and volcanic gases dominated O_2 sinks in the Archean.

The Claire *et al.* model is obviously a big simplification of the Earth system, but some general characteristics are illuminating. We would expect negligible oxidative weathering prior to the GOE and a redox tipping point when K_{OXY} (eqn. (10.18)) reaches unity. The model illustrates these features nicely. However, because the model lumps the atmosphere and ocean together, the duration of the oxic transition in Fig. 10.11 is certainly too long. Photochemical models show that a transition from a weakly reducing atmosphere to an oxic state as a result of a redox flux imbalance would actually take only ~10^4 years (Claire, 2008).

(a)
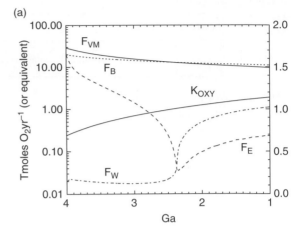

(b)
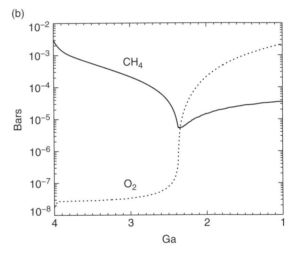

Figure 10.11 Results following the box model of Claire *et al.* (2006). (a) Redox fluxes are as follows: F_W = continental oxidative weathering, F_E = hydrogen escape to space, F_{VM} = sink on O_2 from volcanic and metamorphic reducing gases. K_{OXY} is the oxygenation parameter defined by eq. (10.17). Note how K_{OXY} reaches unity at the oxic transition when F_{VM} falls below F_B. (b) The oxygen and methane concentrations as a function of time corresponding to the effect of the fluxes in (a).

Finally, we should point out that the Archean seafloor, like the continents, may also have been more mafic, and hence more prone to serpentinization (Kasting, 2013 and references therein). According to Table 10.3, the total amount of O_2 consumed by seafloor oxidation today is 0.4 ± 0.2 Tmol yr^{-1}. Half of this sink is sourced from the reduction of sulfate in modern seawater, but sulfate would have been much less important on the Archean Earth, while oxidation by dissolved O_2 would also have been negligible. However, about half of modern seafloor oxidation in Table 10.3, ~0.2 Tmol yr^{-1} of O_2 loss, is caused by serpentinization when warm (~300–400 K) water interacts with ultramafic seafloor. As discussed in Sec. 7.6, some models suggest that the seafloor could have been both thicker and more Mg-rich as a consequence of higher mantle temperatures and a correspondingly greater degree of partial melting at midocean ridges (Herzberg *et al.*, 2010). Exactly how big an O_2 sink this would have produced is unknown, but factors of 10 or more above the modern serpentinization rate cannot be ruled out. Table 10.1 shows that a factor of ~75 increase in this O_2 sink would be required to allow it to match the current O_2 source and make $K_{OXY} < 1$ in eq. (10.18). This is a potential factor whose contribution to early O_2 sinks deserves further examination.

10.7.3.3 Biological Trace Element Hypothesis for Declining Reducing Gases

One other type of hypothesis about declining reducing gases is that trace elements may have controlled biological emissions of reducing gases. However, in ideas of this kind it is important to remember that metabolisms (on their own) do not change net redox state because they create equal amounts of reductants and oxidants. What matters is whether physical or chemical processes can remove the reductant component preferentially in order to oxidize the atmosphere.

It has been postulated that a decline in the amount of ultramafic seafloor in the Archean led to lower concentrations of nickel (Ni) in the ocean, which, in turn, caused a decrease in biogenic methane fluxes because methanogens depend upon Ni for their enzyme cofactors (Konhauser *et al.*, 2009; Konhauser *et al.*, 2015). Most methane is produced from fermentation of organic matter that derives from oxygenic photosynthesis. So a decline in the importance of methanogens should allow more organic matter to be remineralized by reacting with oxygen in the water column. Less methane would decrease the rate of hydrogen escape, diminishing the rate of oxidation (Kasting, 2013, Sec. 3.6). So, this mechanism does not provide an obvious trigger for the GOE.

10.8 Atmospheric Chemistry of the Great Oxidation Event

10.8.1 A Great Collapse of Methane

The Great Oxidation Event has also been described as a great collapse of methane because of the shift in the global redox budget (Zahnle *et al.*, 2006). Atmospheric O_2 and CH_4 destroy each other. Consequently, with <1 ppmv O_2 in the Archean, it is possible to have ~10^2–10^3 ppmv CH_4 compared to today's 1.8 ppmv, if there had been a global

methane source in the range 10%–100% of today's flux. As we discuss in Ch. 11, the early Sun was 25%–30% fainter than today and so the atmosphere likely had more greenhouse gases to prevent Earth's surface from being permanently frozen. In Earth's water-vapor rich atmosphere, abundant methane, along with ethane, which is a photochemical product of methane, could have provided at least part of the required greenhouse warming, with the rest from CO_2 (Haqq-Misra *et al.*, 2008).

Once O_2 began to rise, even to ppm levels at the surface, atmospheric CH_4 levels should have fallen dramatically. In photochemical models, a lack of CH_4 and H_2, rather than the rise of O_2 to high levels, initiates a failure to generate S_8 aerosols, which causes the disappearance of mass-independent fractionation of sulfur isotopes (Zahnle *et al.*, 2006). The loss of CH_4 could have cooled the Earth by ~10 °C, enough to explain the ~2.4 Ga Paleoproterozoic "Snowball Earth" (Sec. 11.6).

To understand the consequences of the GOE on atmospheric chemistry, consider the biogeochemistry of O_2 and CH_4. In the models of Kasting *et al.* (2001), Pavlov *et al.* (2001) and Claire *et al.* (2006), the most important redox-sensitive biogenic gases emitted to the Archean atmosphere are assumed to be CH_4 and O_2 from methanogenesis and photosynthesis, respectively. Organic burial produces a net flux of O_2 (eq. (10.2)), but an important fate for a much greater organic carbon flux is methanogenesis. A net reaction describes the result of the processing of organic carbon by fermenters and methanogens:

$$2CH_2O \rightarrow CH_4 + CO_2 \qquad (10.28)$$

The overall effect of recycling of photosynthetically-produced organic carbon is found by summing twice eq. (10.2) and eq. (10.28), to give:

$$CO_2 + 2H_2O \rightarrow CH_4 + 2O_2 \qquad (10.29)$$

Thus, the combination of oxygenic photosynthesis and methanogenesis produces fluxes of O_2 (ϕ_{O2}) and CH_4 (ϕ_{CH4}) in the ratio $\phi_{O2}/\phi_{CH4} = 2$.

Now consider the chemical consequences of the oxic transition, which we assume will be driven by a shift in the balance of the global redox budget such that K_{OXY} reaches unity; proposed mechanisms were discussed in Sec. 10.7. Once there was enough O_2 around that a small amount of sulfate or nitrate became available, ϕ_{CH4} should have diminished relative to ϕ_{O2} because methanotrophs oxidize methane using either SO_4^{2-} (Valentine, 2002) or NO_3^- (Raghoebarsing *et al.*, 2006).

Photochemical models show that with an increased ϕ_{O2}/ϕ_{CH4} ratio due to the GOE, stable steady state atmospheric compositions are oxic. Figure 10.12 shows results

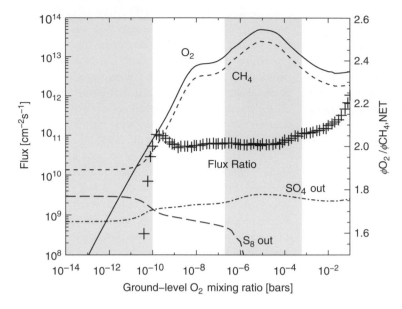

Figure 10.12 Important atmospheric fluxes of chemical species as a function of the ground-level O_2 mixing ratio. Mixing ratio is equivalent to a partial pressure in bars for a 1 bar atmosphere. Results are shown from a photochemical thought experiment using a fixed CH_4 mixing ratio of 100 ppmv and fixed volcanic outgassing fluxes of sulphur gases and reducing gases (see text). Plus symbols ("+") map to the right-hand vertical axis and show the net ratio of the biogenic O_2 flux (ϕ_{O2}) to CH_4 flux (ϕ_{CH4}) as a function of the ground-level O_2 mixing ratio. Deposition fluxes of atmospheric sulfate (SO_4) and elemental sulfur (S_8) are also shown. The shaded regions represent cases when the biogenic O_2 fluxes are either implausibly high or low. (From Catling *et al.* (2007).)

from a photochemical thought experiment using a fixed CH_4 mixing ratio of 100 ppmv and fixed outgassing of ~ 1 Tmol S yr^{-1} (with SO_2:H_2S in 10:1 ratio), 2.7 Tmol H_2 yr^{-1}, and 0.3 Tmol CO yr^{-1}. This outgassing sink on O_2 adds up to only ~ 2.15 Tmol O_2 yr^{-1} in this particular model, but one can think of it as ignoring the zero sum of an organic burial flux that would might have been of order ~ 10 Tmol O_2 yr^{-1} and a complementary ~ 10 Tmol O_2 yr^{-1} geologic gas sink. The graph shows the O_2, CH_4, SO_4^{2-} and S_8 fluxes needed to produce ground-level O_2 mixing ratios shown on the horizontal axis (remembering that O_2 will not be well mixed at higher levels in the anoxic atmospheres).

The decline in the S_8 production for O_2 mixing ratios >0.2 ppm, is related to the loss of mass-independent fractionation (MIF) of sulfur isotopes, as discussed in Sec. 10.5. The plus symbols ("+") are the ratio of the O_2 and CH_4 flux curves, which map to the right-hand vertical axis. Shaded regions in Fig. 10.12 require biogenic fluxes that are unreasonably high or low in the presence of oxygenic photosynthesis (Zahnle *et al.*, 2006). On the anoxic part of the graph, ϕ_{O2}/ϕ_{CH4} is not exactly 2 because there are also relatively large fluxes of redox-sensitive rainout species, principally hydrogen peroxide and formaldehyde, as described in Sec. 8.3, all of which have to be tallied up to describe the full redox budget of the atmosphere.

One observation is that oxic solutions on the right of Fig. 10.12 have similar but slightly higher ϕ_{O2}/ϕ_{CH4} ratios than anoxic solutions on the left. The change in the flux ratio ϕ_{O2}/ϕ_{CH4} is less than 3% between the most oxic anoxic atmosphere (O_2 mixing ratio $\sim 2 \times 10^{-7}$) and

the least oxic oxygenated atmosphere (O_2 mixing ratio $\sim 10^{-3}$) either side of the shaded zone. Effectively, this increase in ϕ_{O2}/ϕ_{CH4} is equivalent to creating an imbalanced O_2 flux that exceeds the outgassing flux of reductants. This would ultimately be caused by a shift in the global redox budget, such as a decline in H_2 outgassing from the solid Earth making less H_2 available to methanogens. Given the tiny size of the geologic outgassing fluxes compared to biogenic fluxes (e.g., 1×10^{13} O_2 molecules cm^{-2} s^{-1} = 2670 Tmol O_2 yr^{-1}), a take-home message is that *the redox state of the atmosphere is determined by small differences between very large fluxes.* A second take-home message is that atmospheres with ground-level mixing ratios of O_2 between ~ 0.2 ppmv and $\sim 0.1\%$, which is a shaded region in Fig. 10.12 separating anoxic and oxic atmospheres, require physically implausible fluxes of O_2 greater than that of today's biosphere. Thus, the oxic transition should have been a rapid shift from a weakly reducing, anoxic atmosphere to an oxidizing atmosphere where levels of O_2 were $\sim 0.1\%$ or more. Essentially, the atmosphere should have transitioned quickly ($\sim 10^4$ yr) because the intervening redox states are unstable. This transition was effectively an *oxygen runaway* until stabilized by new negative feedbacks, such as oxidative weathering.

One counter-intuitive effect of the rise of O_2 was that CH_4 levels may have reached a minimum at the inception of the GOE and then risen afterwards to concentrations that were less than during the Archean but nonetheless climatologically significant. The mid-Proterozoic ocean probably had a large biogenic source of CH_4, given widespread areas of deep ocean anoxia (Fig. 10.6). The behavior of CH_4 is captured in the parameterized photochemistry of the model results shown in Fig. 10.11. The creation of a stratospheric ozone layer (see below) shields the troposphere from shortwave ultraviolet (UV) photons, which slows down the net reaction between CH_4 and O_2 (Catling *et al.*, 2004; Claire *et al.*, 2006; Goldblatt *et al.*, 2006). This allows tropospheric CH_4 to increase and regain a climatologically significant abundance after the stratospheric ozone layer forms.

Possibly, Mesoproterozoic CH_4 persisted at levels of tens of ppmv or ~10^2 ppmv until the Neoproterozoic if the global biogenic CH_4 flux were larger than today (Catling *et al.*, 2002; Pavlov *et al.*, 2003). Today, a vast flux of seafloor CH_4 is consumed by microbial SO_4^{2-} reduction at the CH_4–SO_4^{2-} transition zone in sediments (D'Hondt *et al.*, 2002), so that the ocean is a very minor source of CH_4. But in a Proterozoic ocean with extensive euxinia, considerable methane could have fluxed to the atmosphere. We consider the climatic consequences in Sec. 11.9.

10.8.2 The Formation of a Stratospheric Ozone Shield

The creation of the stratospheric ozone layer was an important consequence of the 2.4–2.3 Ga rise of O_2 because the new layer shielded Earth's surface from biologically harmful solar UV radiation. As discussed in Sec. 3.3.1, stratospheric ozone (O_3) derives from O atoms produced when O_2 is photolyzed (eqs. (3.40) and (3.43)), so the abundance of O_3 is tied to the concentration of O_2. But it doesn't require much O_2 to establish an effective ozone shield. Photochemical models show that an O_2 concentration $\geq 1\%$ of the present atmospheric level (i.e, roughly $\geq 10^{-3}$ bar) would be sufficient to protect the surface from radiation in the harmful range of 200–300 nm (Kasting, 1987; Kasting and Donahue, 1980; Levine *et al.*, 1979; Segura *et al.*, 2003), as depicted in Fig. 10.13.

Because solutions to photochemical models are difficult to construct with plausible biogenic fluxes for ground level O_2 mixing ratios that range ~10^{-7} to ~10^{-3} (Zahnle *et al.*, 2006) (Fig. 10.13), from a modeling perspective, around ~10^{-3} O_2 or 1%PAL is theoretically predicted as minimum concentration of O_2 during the *Boring Billion* interval of 1.8–0.8 Ga (discussed below in Sec. 10.9.1).

10.8.3 Did the Rise of O_2 Affect Atmospheric N_2 Levels?

Some evidence suggests that the Archean atmosphere had relatively low pN_2, and so N_2 levels may have increased

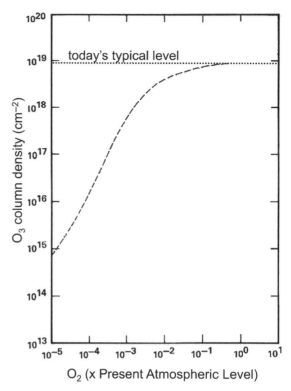

Figure 10.13 The ozone (O_3) layer column abundance (dashed line) as a function of the concentration of atmospheric oxygen. The level of O_2 is expressed as a ratio to the Present Atmospheric Level (PAL). A typical ozone layer column abundance is shown by the horizontal dotted line. The ozone layer's protective absorption of biologically harmful ultraviolet radiation becomes significant at ~0.01 PAL.

after the GOE. How the GOE affected the nitrogen cycle requires further constraints from data, so the following ideas are somewhat incomplete and speculative.

Fossil raindrop imprints (Som *et al.*, 2012) combined with studies of the size distribution of vesicles in ancient lava flows (Som *et al.*, 2016) suggest that Archean air had surface pressure < 0.5 bar at 2.7 Ga. Other work on the $N_2/^{36}Ar$ ratio from fluid inclusions trapped in 3 to 3.5 Ga hydrothermal quartz shows that the partial pressure of nitrogen (pN_2) was lower than 1.1 bar and possibly lower than 0.5 bar (Marty *et al.*, 2013). The latter value is consistent with Som *et al.* (2012, 2016). Low pN_2 might be explained if, in the anoxic Archean environment, aqueous seawater ammonium allowed nitrogen to be sequestered into buried minerals. Indeed, elevated amounts of ammonium are found in Archean sediments.

After the GOE, the dominant form of marine nitrogen would be nitrate, so sequestration of ammonium into seafloor clays would diminish. In particular, once O_2 reached concentrations ~15 μM in the water column, nitrifying chemoautotrophs would have been able to oxidize NH_4^+ to nitrite and then NO_3^- (Falkowski, 1997). In turn, in anaerobic settings, denitrifying bacteria should have cycled the oxidized N-species back to atmospheric N_2. N_2 levels likely rose after the GOE because a new source of N_2 would have been oxidative weathering of organics on the continents, followed by rapid denitrification. Today, this flux is comparable to volcanic degassing of N_2 (Sec. 7.2.3).

10.9 The Neoproterozoic Oxidation Event (NOE) or Second Rise of Oxygen

10.9.1 Evidence for Neoproterozoic Oxygenation

The Great Oxidation Event was apparently not great enough for animal life because the first large (centimeter- to meter-scale) fossils and signs of animals appeared over 1.5 billion years later, in the Ediacaran Period (635–541 Ma), after *another* rise in O_2. As discussed below, there could, of course, have been other factors holding back animal life. However, the middle Proterozoic was an era of comparative biogeochemical and climatic stasis (Brasier and Lindsay, 1998; Buick *et al.*, 1995b). In fact, the interval 1.8–0.8 Ga is called the *Boring Billion*. Buick *et al.* state, "... it may be fair to say about the Mesoproterozoic that never in the course of Earth's history did so little happen to so much for so long." During the Mesoproterozoic, atmospheric O_2 levels probably stayed somewhere between ~0.1% and 2%, and levels of marine sulfate remained only ~1–2 mM compared with 28 mM today (Canfield, 2014; Kah *et al.*, 2004; Shen *et al.*, 2003).

The Neoproterozoic era (1.0–0.541 Ga) that followed the Mesoproterozoic is divided up into three Periods: the Tonian (1.0–0.85 Ga), Cryogenian (850–635 Ma), and Ediacaran (635–541 Ma). The last two Periods have great changes in climate, atmospheric and oceanic chemistry, and biological evolution (Fig. 10.14). The Cryogenian is so-called because it contains two major glaciations, the Sturtian and Marinoan. Because these extended to low latitudes, they are so-called *Snowball Earth* phenomena (see Sec. 11.10). The Sturtian has glacial deposits with interbedded volcanic ash from 716.5 Ma (Macdonald *et al.*, 2010) while the Marinoan has glacial deposits below carbonates containing volcanic ash at ~635 Ma (Condon *et al.*, 2005). The Sturtian glaciation ends by 662 Ma, so that its duration may ~55 million years, while the Marinoan begins by ~641 Ma and ends by ~632 Ma and so its duration appears to be <9 million years (Rooney *et al.*, 2014; Rooney *et al.*, 2015).

After the Marinoan, the first widespread evidence for animals appears (Sec. 10.1 and Fig. 10.14(A)). There is a general consensus that sufficient O_2 must at least have been a precursor for animals, particularly for large biota that appeared in the Ediacaran and Cambrian. Exactly what the Ediacaran biota were is still debated (Liu *et al.*, 2015), but the deep-sea environment is generally considered to require aerobic heterotrophy. The O_2 demand of animals of centimeter-scale thickness is pO_2 >0.02 bar absolute, or ~10% PAL (Catling *et al.*, 2005; Knoll and Carroll, 1999; Margulis *et al.*, 1976; Raff and Raff, 1970; Runnegar, 1982, 1991), which exceeds the O_2 level that apparently existed in the Mesoproterozoic. During the Mesoproterozoic, low O_2 may also have frustrated the evolution of metazoans through nutrient limitation in an ocean with wide areas of euxinia because sulfide precipitation would have removed trace elements needed for critical enzymes (Anbar and Knoll, 2002).

The first reports of geochemical evidence for a second rise of O_2 came from Neoproterozoic sedimentary sulfides with [32]S-enrichment exceeding the isotope discrimination of sulfate-reducing bacteria (Canfield and Teske, 1996; Halverson and Hurtgen, 2007; Hurtgen *et al.*, 2005; McFadden *et al.*, 2008; Strauss *et al.*, 2001). The spread in isotopic fractionation (Fig. 10.8) can be explained by sulfide produced by bacterial sulfate reduction being re-oxidized at the sediment-water interface and cyclically re-reduced by sulfur disproportionating bacteria. Canfield and Teske originally suggested that O_2 rose above a threshold of 5% PAL necessary for *Beggiatoa*, a sulfide-oxidizing bacteria. However, evidence from multiple isotopes of sulfur indicates that sulfide oxidizers appeared 1.3–1.45 Ga (Johnston *et al.*, 2005). In fact, by 1.2 Ga, enough O_2 was around to see the effects of sulfide-oxidizing bacteria in lake sediments (Parnell *et al.*, 2010). Later on, the churning of seafloor sediments by animals such as worms, *bioturbation*, should have facilitated an increase in the rate of sulfide oxidation (Canfield and Farquhar, 2009). The first evidence for organism locomotion is ~565 Ma (Liu *et al.*, 2010), while that for burrowing is at 545.1–548.8 Ma (Jensen *et al.*, 2000). Consequently, bioturbation is a process that affected the late Ediacaran sulfur cycle.

Other lines of geochemical evidence have bolstered the reality of a Neoproterozoic Oxygenation (or Oxidation) Event (NOE) (Fig. 10.14), although the exact timing and magnitude of O_2 increases are still a matter of research (Shields-Zhou and Och, 2011). Besides sulfur isotopes, the NOE is indicated by changes in the isotopic composition of marine molybdenum (Mo), chromium (Cr), and selenium (Se), as well as increases in the

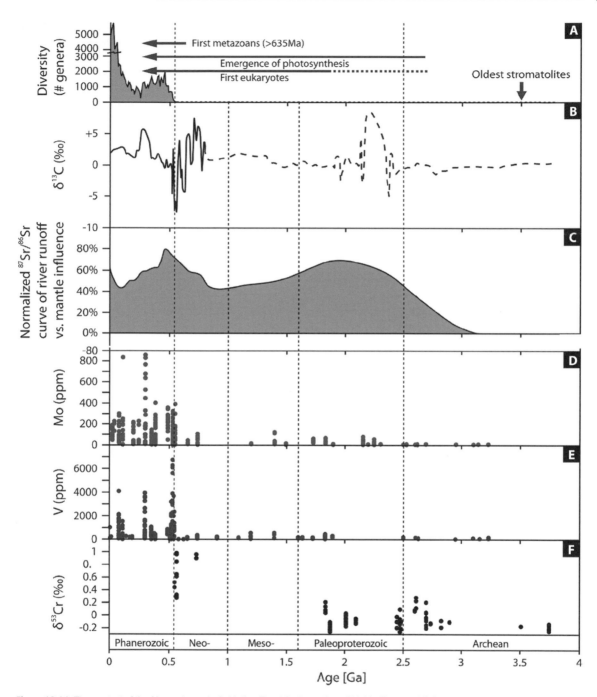

Figure 10.14 The context of the Neoproterozoic Oxidation Event (redrawn from Shields-Zhou and Och (2011). Reproduced with permission. Copyright 2011, Geological Society of America.) (A) Biological innovation. (B) Carbon isotopes in marine carbonates. (C) Normalized strontium isotopes in seawater. (D) Molybdenum concentrations in black shales. (E) Vanadium concentrations in black shales. (F) Chromium isotope fractionation recorded in banded iron formations.

abundance of Mo and vanadium (V) in black shales (Fig. 10.14(D,E)). One similarity with the Great Oxidation Event is that increased oxygenation of the atmosphere (or, at least, oxidation on land) may have preceded

that of the ocean. Cr isotope evidence suggests that atmospheric O_2 levels were increasing as early as ~750 Ma (Frei *et al.*, 2009) (Sec. 10.4.4 and Fig. 10.14 (F)) and indeed thick sulfate evaporites are found as early as just prior to 811 Ma (Turner and Bekker, 2015). However, marine sulfate levels increased significantly only after the Marinoan glaciation, ~635 Ma (Halverson and Hurtgen, 2007), along with geochemical tracers of greater oxygen (Sahoo *et al.*, 2012) and a recovered productivity (Kunzmann *et al.*, 2013). Yet even at this stage and before the Marinoan, the deep ocean generally remained anoxic (Hood and Wallace, 2014; Kurzweil *et al.*, 2015a; Shen *et al.*, 2008) and it stayed rich in Fe^{2+} until the end of the Gaskiers glaciation, ~582 Ma (Farquhar *et al.*, 2008; Johnston *et al.*, 2010). Then the deep ocean became more oxidized (Canfield *et al.*, 2007; Fike *et al.*, 2006), although oxygenation may still have been episodic, with more persistent and extensive oxygenation occurring within the period ~551 Ma (McFadden *et al.*, 2008) to 520 Ma (Chen *et al.*, 2015; Wen *et al.*, 2015). Seafloor colonization by soft-bodied Ediacaran organisms occurred soon after the Gaskiers, however.

Canfield *et al.* (2007) estimated that atmospheric O_2 increased to at least 15% PAL by 551 Ma (i.e., >3% in terms of absolute O_2 concentration). They assumed an O_2 demand for 500–1500 m water depths of sediments at the Avalon Peninsula, Newfoundland. By 551 Ma, a significant jump in Mo and V enrichments in black shales (Scott *et al.*, 2008) along with $\delta^{98}Mo$ (Dahl *et al.*, 2010) provide evidence for increased O_2 levels. Isotopes of Mo also indicate oxygenation of large areas of the deep sea by 535–520 Ma, consistent with the Cambrian Explosion (Chen *et al.*, 2015; Wen *et al.*, 2015).

Another similarity between the GOE and the NOE concerns large excursions in the isotopic composition of marine carbon (Fig. 10.14(B)). Long before the Sturtian and Marinoan glaciations, marine carbonates generally have positive $\delta^{13}C$, averaging several per mil, but there are negative excursions down to –6‰ before and after the glaciations. Negative $\delta^{13}C$ excursions also occur at other times in the Neoproterozoic (Swanson-Hysell *et al.*, 2010).

The most extraordinary $\delta^{13}C$ excursion went down to –12‰, and has an age bracketed between ~600 Ma and 551 Ma (Bowring *et al.*, 2007; Condon *et al.*, 2005; Macdonald *et al.*, 2013), although some stratigraphic evidence places the excursion just before the Gaskiers glaciation at 582 Ma (Bowring *et al.*, 2007; Halverson *et al.*, 2010). The $\delta^{13}C$ excursion is known as the *Shuram anomaly* after the Shuram Formation, Oman, where it was first identified (Burns and Matter, 1993). The Shuram anomaly is a biogeochemical enigma. One hypothesis is that it was caused by a global diagenetic event (Grotzinger *et al.*, 2011). Increased O_2 at this time could have dumped mineral oxidants (such as sulfate or ferric iron) into sediments, which created a pulse of ^{12}C-depleted

inorganic carbon through microbial oxidation of organic carbon (Rothman *et al.*, 2003; Swanson-Hysell *et al.*, 2010).

After the Shuram anomaly, significant changes in biology occurred. Large ornamented Ediacaran microfossils are interpreted as the egg-resting stages of early animals (Yin *et al.*, 2007) and they decline after ~560 Ma, when oxygenation became more extensive. A leading hypothesis is that such cysts evolved in response to episodes of deep sea anoxia in shelf and platform environments before the late Ediacaran (Cohen *et al.*, 2009; Li *et al.*, 2010).

In the fully oxygenated seawater of the late Ediacaran, peculiar self-similar branching organisms – the canonical Ediacarans, such as the frond-like *Charnia* – were replaced by segmented and mobile animals. For example, in the late Ediacaran, *Cloudina* appears, which is a millimeter-scale fossil consisting of nested calcite cones. *Cloudina* provides one of first examples of animals with biomineralization and is named after Preston Cloud, the early champion of connections between biology and oxygenation.

10.9.2 What Caused the Second Rise of Oxygen?

We lack a mechanistic understanding of the NOE, but hypotheses about its cause divide up into biological and geological categories, which the reader will notice as being similar to ideas about the causes of the GOE.

One group of biological proposals relies upon an increase in organic burial, either as a pulse or long-term trend. Large positive carbonate $\delta^{13}C$ excursions in the Neoproterozoic could have resulted from a pulse of organic burial and associated O_2 release (Derry *et al.*, 1992). However, given the short residence time of O_2 (~2 million years, even today (Sec. 10.2.4)), first order kinetics means that a pulse of organic burial should merely cause a parallel pulse in O_2. O_2 should return to its previous low levels once burial and oxidation of previously buried carbon have re-equilibrated over uplift timescales of ~10^8 years. One variant of enhanced organic burial includes the evolution of zooplankton with fecal pellets that dragged a greater amount of organic carbon to the seafloor (Logan *et al.*, 1995). However, fecal pellets are found near the Cambrian boundary (Walter, 1995), long after the ~580 Ma deep-sea oxygenation inferred by S isotopes and Fe mobilization (Canfield *et al.*, 2007; Canfield and Teske, 1996; Fike *et al.*, 2006; Hurtgen *et al.*, 2005). So, data argue against fecal pellets being the cause of the second rise of O_2. A logical alternative is that aerobic organisms large enough to produce significant ballast for organic carbon burial evolved in response to higher O_2.

With different biological proposals involving nutrients or other ways to stimulate organic burial, the problem remains that greater organic burial should extract more ^{12}C and cause marine carbonates to become permanently ^{13}C-enriched. Composite $\delta^{13}C$ data in the Neoproterozoic show a series of complicated fluctuations, including negative $\delta^{13}C$ excursions before and after glacial periods (Halverson et al., 2005; Hoffman and Schrag, 2002). Fluctuations in $\delta^{13}C$ have been modeled as non-steady-state interactions involving a deep-sea carbon pool (Rothman et al., 2003) but negative $\delta^{13}C$ excursions may require oxidation of a dissolved organic carbon pool that is too large to be consistent with steady or increasing marine sulfate and O_2 levels (Bristow and Kennedy, 2008). Statistical analysis shows that the average fraction of carbon buried as marine organic carbon, f_{org}, increased somewhat from ~18% to ~23% from the mid-Proterozoic to the Phanerozoic (Krissansen-Totton et al., 2015), perhaps reflecting the emergent dominance of photosynthetic eukaryotes (Butterfield, 2015; Lenton et al., 2014).

An idea for an infusion of nutrients in the ocean released from continental weathering by terrestrial plants (Knauth and Kennedy, 2009) – a "worldwide greening of the land" – is subject to the problem of an overall lack of a clear secular trend in $\delta^{13}C$ of marine carbonates from the Mesoproterozoic to Neoproterozoic. The hypothesis for a continental greening starting at ~850 Ma is based on the idea that meteoric waters depleted in ^{18}O carried isotopically light photosynthetic C to carbonates. Currently, there is a lack of fossil evidence for a cover of land plants prior to ~470 Ma (noting some evidence for lichens at 600 Ma (Yuan et al., 2005) and microbial crusts (Prave, 2002)). A further challenge is that negative $\delta^{13}C$ excursions in marine carbonates occur either side of Cryogenian glaciations, while during clement Neoproterozoic times, marine C is moderately isotopically heavy. Presumably, the opposite would be expected if light carbon were contributed from continental biotic weathering. Models that rely on a greatly reduced phosphorus flux to limit O_2 production before the NOE (Laakso and Schrag, 2014) are also challenged by the carbon isotope record.

Geological proposals for explaining the NOE include a jump in the production of clays that absorbed organics (Kennedy et al., 2006), continental re-configurations more conducive to platforms for organic burial (Campbell and Allen, 2008), or greater sedimentation from erosion of continental mountains (Campbell and Squire, 2010). One problem is that these proposals are still in apparent disagreement with the lack of associated trend in $\delta^{13}C$ of marine carbonates (Schidlowski, 1988). Nonetheless, "Wilson cycles" of continent amalgamation and breakup occurred. The supercontinent of Rodinia was established around 900 Ma and then began to break up around 800 Ma. By 700 Ma, a scattering of tropical microcontinents dominated the global geography (Hoffman and Li, 2009). Landmass collisions eventually resulted in a new supercontinent of Gondwana around 530 Ma, which extended from the South Pole to the tropics. Strontium isotopes provide support for an average increase in continental weathering from the Cryogenian to the early Cambrian. The radiogenic isotope ^{87}Sr is produced by the decay of ^{87}Rb that is more prevalent in continental crust than seafloor. Consequently, the $^{87}Sr/^{86}Sr$ ratio of seawater should increase if continental weathering increases, which is a trend observed from the Sturtian to the early Cambrian (Shields, 2007; Shields-Zhou and Och, 2011). It has also been proposed that gases released from subduction zones became more oxidizing in the Neoproterozoic (Macouin et al., 2015),

Going back to basics, the answer to the Neoproterozoic rise of O_2 must ultimately be rooted in an imbalance of O_2 source and sink fluxes (Fig. 10.3), comparable to what we discussed for the GOE in Sec. 10.2.4. If it is true that the middle Proterozoic deep ocean remained anoxic and ferruginous (Poulton and Canfield, 2011), the flux of reductants from seafloor emissions, including oxidative weathering of the seafloor, must have buffered the redox state of the deep ocean and atmosphere to their relatively low level prior to the Cryogenian and Ediacaran. To titrate such a buffer and reach a tipping point, large irreversible losses of reductant are required. Two possibilities are subduction of seafloor sulfides deposited in euxinic areas or the escape of hydrogen to space via moderate levels of Proterozoic atmospheric methane (Catling et al., 2002). If the Proterozoic ocean was at least partly deeply sulfidic, seafloor sulfide would be subducted, given its refractory nature (Canfield, 2004), and would provide a long-term, net source of O_2 (eq. (10.3)). Loss of isotopically light sulfide would solve the puzzle of having an average $\delta^{34}S$ of both sulfate and sulfide that is positive in the Mesoproterozoic (Fig. 10.8). Also, if atmospheric methane persisted from ~2.2 Ga to ~0.8 Ga at ~10^2 ppmv (Catling et al., 2002; Pavlov et al., 2003), it would be decomposed in the upper atmosphere to release hydrogen that escapes to space. Such hydrogen escape operating over 1.4 b.y. would oxidize the Earth and produce the equivalent of ~25 times of all the O_2 in the modern atmosphere and ocean, following the theory of Catling et al. (2001). Of course, such irreversible oxidation would induce feedbacks, including changes in the biosphere that then affect the atmosphere and climate. So the viewpoint of a global redox titration does not necessarily exclude some of the ideas mentioned above.

Finally, it could be hypothesized that the NOE at 0.8–0.6 Ga and the GOE at 2.4 Ga were both caused by the same phenomenon: a gradual cooling of the mantle, accompanied by a progressive decrease in the amount of ultramafic rock on the seafloor and continents that was subject to serpentinization. In this view, a high CH_4 content of the Mesoproterozoic atmosphere was a consequence of redox balance, rather than a driver of O_2 increase (Kasting, 2013). In this scenario, the Earth oxidized as a consequence of continued hydrogen escape but most of the oxidation was in the mantle where it buffered. By the beginning of the Neoproterozoic, the composition of seafloor and continental basalts had become more modern (less mafic), the flux of reduced gases from serpentinization decreased, and so less hydrogen needed to escape to space in order to balance the global redox budget. Deep ocean O_2 increased, shutting down the production of methane from marine sediments, and establishing the modern redox regime of high atmospheric O_2 and negligible escape of hydrogen to space. However, data to support such change of seafloor composition so late in Earth history are currently lacking.

10.10 Phanerozoic Evolution of Atmospheric O_2

The Phanerozoic evolution of atmospheric O_2 has been described in detail by Berner (2004), so we will just give a summary. Figure 10.15 shows two models of pO_2 over the past 500 million years. These models use carbon and sulfur isotopic mass balance to deduce the comings and goings of O_2 primarily according to eqs. (10.2) and (10.3). These equations operate from left-to-right to produce O_2 from buried organic carbon or sulfide and they operate from right-to-left to consume O_2 in weathering or through oxidation or reducing gases produced by the thermal decomposition of sulfides or organics. The model of Bergman et al. (2004) also includes a number of biogeochemical feedbacks. The salient features of Fig. 10.15 are that in the Berner (2006a, 2009) model, O_2 rises in the Devonian and peaks in the Carboniferous, whereas in the Bergman et al. model, O_2 rises in the Devonian and peaks in both the Carboniferous and Cretaceous.

Various lines of evidence support Phanerozoic changes of O_2 that are broadly consistent with the geochemical models. In marine sediments, an increase in the concentration of Mo at ~400 Ma indicates that O_2 levels increased within the Devonian (416–359 Ma) (Dahl et al., 2010). Greater pO_2 correlates with a higher diversity of vascular plants and it may have enabled the evolution of large, O_2-hungry predatory fish. Plants appear to have originated earlier in the Silurian Period (485–444 Ma) on Gondwana (Steemans et al., 2009) where plant spores are found at ~470 Ma (Rubinstein et al., 2010; Wellman et al., 2003). But fossils of substantial parts of plants do not occur until ~425 Ma, in the late Silurian (Gensel, 2008).

It has been suggested that once forests became prevalent in the late Devonian, large amounts of organic carbon were buried on the continents because lignin, an insoluble

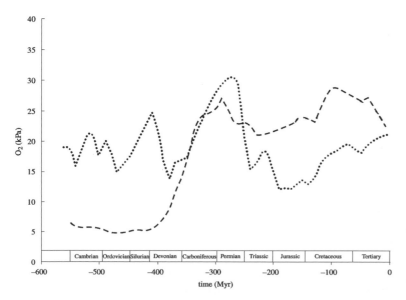

Figure 10.15 Two models for Phanerozoic changes in atmospheric O_2. Dotted line: Berner (2006a). Dashed line: Bergman et al. (2004). (Adapted from Harrison et al. (2010). Reproduced with permission from the Royal Society. Copyright 1990.)

structural compound found in wood, persisted until organisms such as fungi evolved the means to decompose lignin (Robinson, 1990). The stoichiometric release of O_2 from buried wood via eq. (10.2) may have produced a peak pO_2 of ~0.3 bar at ~300 Ma (Berner et al., 2003). At this time, fossils of giant Carboniferous insects occur, e.g., dragonflies with 70-cm wingspans. Because insects rely on diffusion through holes in their cuticles for their respiration, high levels of O_2 may have been required for such insects to exist that were ~10 times bigger than those today (Dudley, 1998; Harrison et al., 2010). But this argument may not be definitive: Birds had not evolved at the time, so there was also an ecological niche for large flying insects. This does not explain giant arachnids, however.

Further biological change may be related to oxygen. Long after fish evolved into amphibians by 365 Ma (Clack, 2000), ancestors of our own lineage of placental mammals originated by ~125 Ma (Ji et al., 2002). Placental mammals require high O_2 concentrations because veins distribute O_2 to a fetus. So, it has been speculated that a further boost in O_2 levels during the Cenozoic (since 65 Ma) could have enabled the radiation of placental mammals. This O_2 surge may have been associated with greater organic burial along the passive continental margins of the Atlantic Ocean, enabled also by greater sedimentation rates resulting from new diatom species (Falkowski et al., 2005). However, a pO_2 record using the proxy of $\delta^{13}C$ of plant resins does not support any linkage of pO_2 to the radiation of placental mammals (Tappert et al., 2013).

Although there have been fluctuations, Phanerozoic O_2 has probably always remained within roughly 0.20 ± 0.15 bar, which must require negative feedbacks to stabilize O_2 to these high levels. Trees have burnt naturally since ~400 Ma. The presence of charcoal suggests that the atmosphere contained at least ~15% O_2 needed to ignite forest fires (Chaloner, 1989; Wildman et al., 2004), while charcoal concentrations constrain O_2 levels to 15%–30% since 400 Ma (Glasspool and Scott, 2010). These inferred O_2 levels are consistently above those deduced from plant resin proxies, which range from ~12% to 21% since ~220 Ma (Tappert et al., 2013). Nonetheless, O_2 level did not swing wildly. Yet, the average amount of time an O_2 molecule spends in the atmosphere–ocean system is only ~2 m.y. (Sec. 10.2.4).

Respiration and decay cannot regulate O_2: they can modulate no more than 1% of the total amount of atmospheric O_2 because the amount of organic carbon contained in the terrestrial and marine biota at the Earth's surface (~4.4×10^5 Tmol) is ~10^2 times smaller than the surface reservoir of O_2 (3.8×10^7 Tmol) (Walker, 1980). Instead, respiration and decay only control the small size of the surface organic reservoir, as we commonly see in seasonal cycles of vegetation.

The general principles by which O_2 is regulated on geologic timescales today involve counteracting feedbacks. When O_2 goes up, the consumption of O_2 increases and/or its production goes down. Likewise, when O_2 goes down, feedbacks increase pO_2. Oxidation during continental weathering is thought to be relatively complete at today's O_2 level or even lower (Holland, 1984), so that a negative feedback on the O_2 source, i.e., organic burial, is generally favored. However, what increases the organic burial rate most appears to be a greater overall sedimentation rate rather than a lack of O_2 in the water column (Betts and Holland, 1991). So, indirect controls on the supply of nutrients have been suggested.

Most geochemists favor phosphorus as the limiting nutrient for marine productivity because its only source is from continental weathering, which allows for a potential negative feedback on O_2. Phosphorus is lost by burial in sediments, but under anoxic conditions, phosphorus is not so easily bound to iron hydroxides in sediments (Colman and Holland, 2000; Van Cappellen and Ingall, 1996). So if O_2 decreases, more phosphorus is then retained in the ocean to support productivity, which is a stabilizing negative feedback against significant decreases in O_2.

Large increases in O_2 are prohibited by the negative feedback of increased frequency of forest fires, which removes O_2 (Lenton and Watson, 2000). The feedback here is rather subtle, as forest regrowth would normally produce as much O_2 as was consumed by fire. However, some organic material that is produced by forests is carried by streams and rivers down to the ocean where it is buried, and that terrestrial organic material has a lower P:C ratio than does typical marine organic matter (Kump et al., 2010). If phosphorus is limiting, then, the burning of forests reduces organic carbon burial in the sea, lowering O_2 production, and thus providing a negative feedback. We note, however, that definitive data and a high level of confidence in our understanding of the controls on modern O_2 levels are still lacking.

10.11 O₂ and Advanced Life in the Cosmos

Finally, turning to astrobiology, high O_2 levels may be profoundly important in for the prevalence of complex life in the Universe. Earth's high pO_2 atmosphere is unique amongst the planets of the Solar System. Indeed, in Ch. 15, we argue that the remote detection of high pO_2

in exoplanet atmospheres could be used as a sign of life, at least in some circumstances. We might go further and speculate that high O_2 is generally required for life comparable to advanced animals (Catling *et al.*, 2005). For people unfamiliar with the details of this argument, the idea may seem biased and geocentric. However, life is made from the limited toolkit of the Periodic Table. Normalized "per electron transfer," fluorine is the most energetic oxidant in the Periodic Table, while chlorine is comparable to oxygen. But fluorine is useless because it spontaneously explodes on contact with organics while aqueous chlorine forms bleach that also destroys organics. O_2 has stronger internal bonding than the halogens. Weak, singly bonded F_2 or Cl_2 could never realistically accumulate in a planetary atmosphere to high abundance because they are so reactive. As well as being more chemically stable, oxygen is also cosmologically abundant: it comes third behind H and He, and is a component of water.

For the above reasons, molecular oxygen likely provides the most feasible, most energetic metabolism available for advanced life anywhere. Thus, the prevalence of high pO_2 in Earth-like exoplanet atmospheres could be the key to whether animal-like life exists elsewhere and, indeed, whether there is extraterrestrial intelligent life. For this reason, understanding how and why an oxic atmosphere appeared on the Earth, which is the focus of this chapter, is a general consideration relevant to the question of "Are we alone?" in the most interesting case of other sentient organisms.

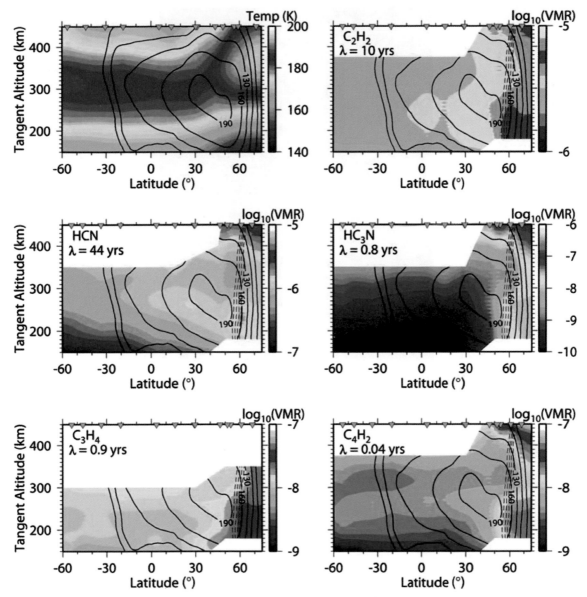

Figure 4.33 Titan zonal mean temperatures (K), cyclostrophic zonal mean winds (m s^{-1}), and trace constituent concentrations (as $\log_{10}$ volume mixing ratio (VMR)) above 150 km altitude. The photochemical lifetimes at 300 km are denoted by λ. The zonal mean winds exhibit strong equatorial superrotation as well as seasonal effects, and the constituents exhibit strong downward transport in the region of the temperature maximum near the winter pole. Blue dashed lines denote a dynamical mixing barrier (a region of steepest horizontal potential vorticity gradient) that confines a polar vortex. (Source: Teanby *et al* (2008). Reproduced with permission. Copyright 2008, John Wiley and Sons.)

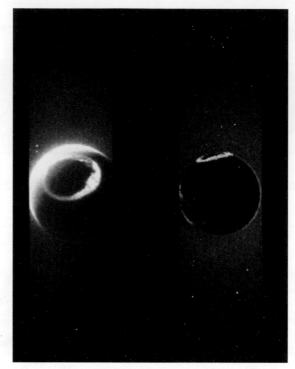

Figure 5.1 Earth imaged in the vacuum ultraviolet (VUV) by NASA's *Dynamics Explorer 1* (Rairden *et al.*, 1986). *Left panel*: View with the spacecraft at 16 500 km altitude above 67° N latitude at 2017 UT on October 14, 1981. Glow beyond the limb of the planet (red false color) is due to Lyman-α (121 nm) solar radiation resonantly scattered by Earth's extended hydrogen atmosphere or *geocorona*. Energetic hydrogen atoms in the geocorona are escaping to space. Features on the Earth's disk (dayglow from the sunlit atmosphere, a northern auroral oval, and equatorial airglow) are due to the emission of atomic oxygen at 130.4 and 135.6 nm and emission in the Lyman–Birge–Hopfield band of N_2 (140–170 nm). Isolated points of light are background stars that are bright in the VUV. *Right panel:* A view of Earth's dark hemisphere at 0222 UT on February 16, 1982, with the Sun behind Earth. Spacecraft altitude and latitude are 19,700 km and 13° N, respectively. Equatorial airglow straddles the magnetic equator in the pre-midnight sector. (Image credit: NASA.)

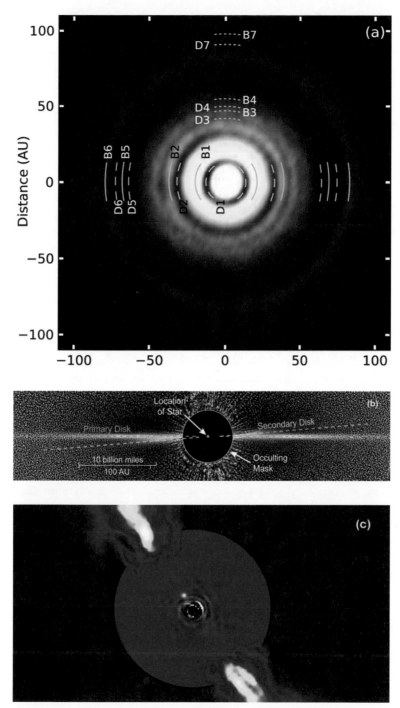

Figure 6.1 (a) A pattern of bright concentric rings (labeled B1, B2, etc.) separated by dark rings (labeled D1, D2, etc.) around the star HL Tauri, imaged by the *Atacama Large Millimeter/submillimeter Array* (ALMA) at 1 mm wavelength. (Source: ALMA-Partnership (2015).) (b) The disk of Beta Pictoris seen in visible light by the *Hubble Space Telescope*. The central star is blocked out in the photo and a faint secondary disk, inclined at 4°, is seen in scattered light. (Courtesy of NASA, ESA.) (c) Near infrared photograph of Beta Pictoris taken by the *Very Large Telescope* (VLT) in Chile. The star is again blocked out. The white dot to the upper left of the star is an 8-Jupiter-mass planet aligned with the disk at 8 AU from the star. A separate disk image from ESO's 3.6 m telescope has been grafted onto the central VLT image in this photo. (Courtesy of ESO/ A.-M. Lagrange *et al.*)

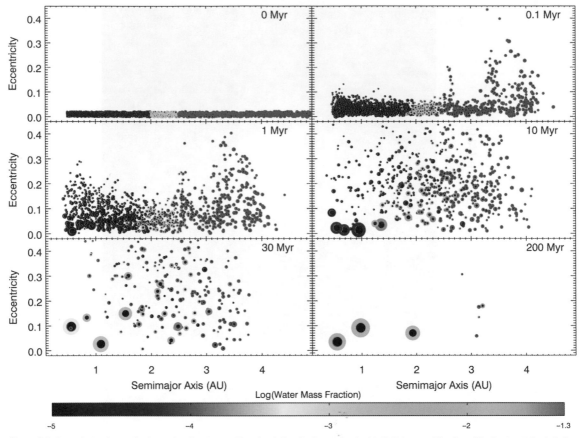

Figure 6.6 Snapshots of a particular rocky planet accretion simulation for the region inside 5 AU around the Sun. The horizontal axis is the planet's semi-major axis, i.e., its mean distance from the Sun. The dots represent large planetesimals, some of which will grow into planetary embryos and planets. The position of the dot on the vertical axis indicates the planetesimal's eccentricity. The size of the dot indicates the mass of the planetesimal or planet, and its color shows the fraction of its total mass that is made up of water. The simulation was terminated after 200 million years. (From Raymond *et al.* (2006).)

Figure 7.4 The top of a black smoker chimney on the south-eastern edge of the iguanas vent field imaged on the Galapagos 2005 expedition. (NOAA image ID: expl1233, courtesy of UCSB, University of S. Carolina, NOAA, WHOI.)

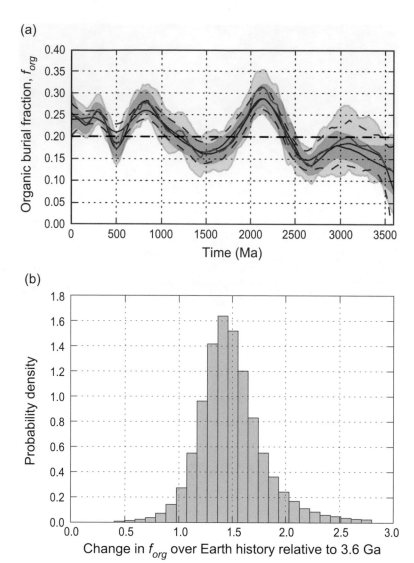

Figure 8.3 (a) Time series of the organic carbon burial fraction, f_{org}, based on the isotopic data shown in Fig. 8.2. The curves show different statistical analyses. *Blue*: locally weighted scatterplot smoothing (LOWESS); *red*: kernel regression; *green*: Kalman smoothing. Shading or dashed lines of corresponding color show 95% confidence intervals. All statistical techniques indicate the same general pattern. (b) The probability distribution of the change in organic carbon burial fraction, f_{org}, from 3.6 Ga to present based on a generalized least squares regression bootstrap analysis. The most probable change is a factor of ~1.5 increase and the 95% confidence range is an increase of ~1.2 to ~2.0. However, the increase in f_{org} relative to 3.6 Ga is mainly in the past ~400 Ma with no statistical difference between average f_{org} in the Archean (3.8–2.5 Ga) versus "boring billion" (1.8–0.8 Ga). From Krissansen-Totton *et al.* (2015).

Figure 9.12 Upper: Cross-section of 3.49–3.48 Ga conical stromatolite, Dresser Formation, North Pole, Pilbara, Western Australia. Lens cap is 5 cm diameter. Lower: Plan view. (Photos by D. Catling.)

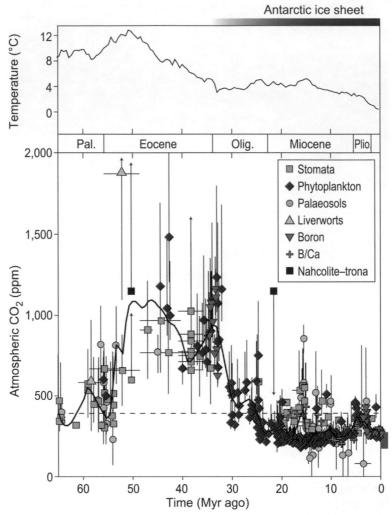

Figure 11.18 The Cenozoic history of atmospheric CO_2 and climate. Deep-sea temperatures (upper panel) generally track the estimates of atmospheric CO_2 (lower panel) reconstructed from terrestrial and marine proxies. Errors represent reported uncertainties. Symbols with arrows indicate either upper or lower limits. The vertical grey bar on the lower right axis indicates the glacial–interglacial CO_2 range from ice cores. The top blue bar indicates ice-sheet development on Antarctica. Horizontal dashed line indicates the atmospheric CO_2 concentration (390 ppm) in 2011. (From Beerling and Royer (2011). Reproduced with permission from Macmillan Publishers Ltd. Copyright 2011, Nature Publishing Group.)

Figure 12.1 A map of the topography of Mars, using Mars Orbiter Laser Altimeter (MOLA) data. (Source: NASA/MOLA Team.)

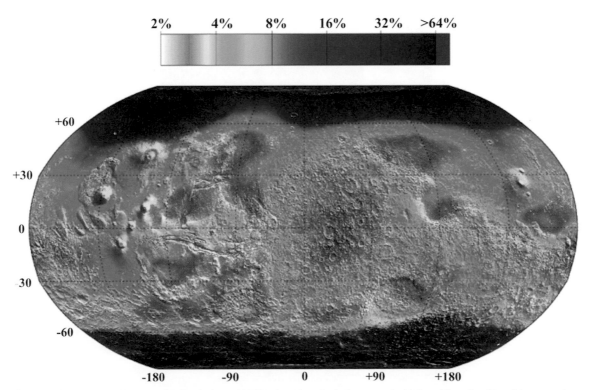

Figure 12.6 Water-equivalent mass fraction in the subsurface based on the hydrogen content inferred from the *Mars Odyssey* neutron spectrometer (From Feldman *et al.* (2004). Reproduced with permission from John Wiley and Sons. Copyright 2004, American Geophysical Union.)

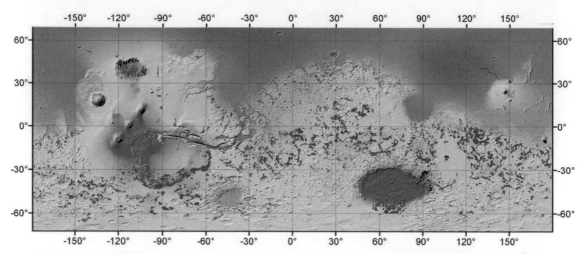

Figure 12.9 The global distribution of valley networks on a background contour map where highlands are shaded red and lowlands are shaded blue. Valleys on Noachian terrain are red, while those on younger terrain are blue for Amazonian and purple for Hesperian (From Hynek *et al.* (2010). Reproduced with permission from John Wiley and Sons. Copyright 2010, American Geophysical Union.)

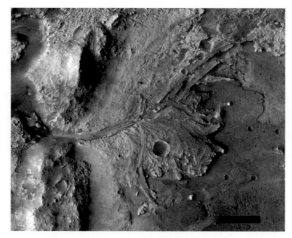

Figure 12.10 Deltaic deposit in western Jezero crater. The deposit is a positive relief feature, indicating that the material was more resistant to erosion than surrounding material. Phyllosilicate-bearing areas are green, while yellow and blue colors indicate basaltic minerals. Purple-brown surfaces have no distinctive spectral features. Scale bar = 2 km. (Courtesy of NASA/JPL/JHUAPL/MSSS/Brown University.)

Figure 13.4 An image of Venus taken on February 14, 1990, by the *Galileo* spacecraft Solid State Imager (SSI) camera through its violet filter (centered at 404 nm with 45 nm band-pass (Belton *et al.*, 1992)). The image is shown in false blue color to emphasize contrast. North is at the top and the evening terminator is to the left, i.e., the sunlit part of the planet is rotating towards the observer. The image shows markings in the sulfuric acid clouds such as west-to-east cloud bands. The bright region is the subsolar point. (Image P-37218; courtesy of NASA/ JPL.)

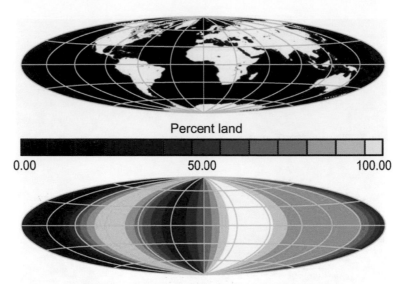

Percent land

0.00   50.00   100.00

Figure 15.13 Inferred land–sea distribution for Earth, obtained from analysis of the *EPOXI* color time series. (From Cowan *et al.* (2009). Reproduced with permission. Copyright 2009, American Astronomical Society.)

11 Long-Term Climate Evolution

We have avoided examining climate evolution in the past three chapters in order to place our discussion here in one coherent narrative. Of course, climate depends on the variations in atmospheric composition that we described in Chapters 9 and 10, which included changes in levels of methane, a greenhouse gas. Variations in carbon dioxide concentrations over time were probably even more important, though, and planetary albedo changes associated with glacial episodes and greenhouse gas decreases are known to have dominated climate during cold times. Other greenhouse gases such as nitrous oxide, ammonia, and molecular hydrogen may also have played a climatic role in the Precambrian. Our goal is to understand how climate and its drivers have varied as the Sun brightened and the atmospheric redox state changed.

11.1 Solar Evolution

A fundamental problem in Earth's long-term climate evolution is to explain how it managed to compensate for the gradual brightening of the Sun (Feulner, 2012). Stellar physicists realized many years ago that main sequence stars, including the Sun, get brighter as they age. Three pioneers were Arthur Eddington, Hans Bethe, and Martin Schwarzschild. Eddington suggested in 1920 that the Sun produces energy by nuclear fusion in its core, while Bethe (1939) recognized that fusion of protons was the key energy source. Schwarzschild (1958) elaborated these concepts by developing detailed models of stellar evolution. The actual nuclear fusion reactions can be quite complex, and different processes apply to stars of various masses. Nevertheless, the results for a star like our Sun are relatively simple: 4 hydrogen nuclei (protons) combine to form one nucleus of helium-4 (2 protons and 2 neutrons)

$$4\,^{1}\text{H} \rightarrow \,^{4}\text{He} + 2\text{c}^{+} + 2\nu \qquad (11.1)$$

$$4\,^{1}\text{H} + \text{e}^{-} \rightarrow \,^{4}\text{He} + \text{e}^{+} + 2\nu \qquad (11.2)$$

Here, e^{+} is a positron, e^{-} is an electron, and ν is a nearly massless neutrino. Equation (11.1) is the net effect of chain reactions called ppI, ppIII, and the CNO cycle, while eq. (11.2) is the net reaction of the ppII chain ("pp" means proton–proton fusion) (LeBlanc, 2010). There is a difference in mass, Δm, between the reactants and products, e.g., in (11.1), the mass of the helium nucleus and two positrons (anti-electrons) is less than the four individual hydrogen nuclei. Thus, energy ΔE is released, following Einstein's formula, $\Delta E = \Delta mc^{2}$, where c is the speed of light. The positrons release energy, as well, when they annihilate ambient electrons.

Besides releasing energy, reaction (11.1) also increases the density of the solar core, as one helium-4 nucleus (^{4}He) occupies less volume than four individual protons. The Sun is a self-gravitating sphere, so the increase in density causes the core to contract, and the resultant release of gravitational potential energy produces heat. Alternatively, one can demonstrate that an increase in core temperature is required in order for the core pressure to be able to balance the pressure bearing down from the increased gravitational attraction of the overlying material. Regardless of how one considers it, as the Sun converts hydrogen to helium, the core temperature gradually increases. This rise in temperature, in turn, causes an increase in the rate of nuclear fusion. This latter result makes sense because fusion occurs only when two particles come close enough for the strong nuclear force to act between them. Because positively charged protons (or other atomic nuclei) repel each other electrostatically, they can only fuse when by chance they are moving towards each other at sufficiently high velocity to overwhelm this repulsion. High temperatures increase the likelihood of such event so the rate of fusion depends

300

Long-Term Climate Evolution

on core temperature as $\propto T_{core}^4$ for the net pp chains near $T_{core}= 15\times10^6$ K (Carroll and Ostlie, 2007, p. 311). The extra energy produced within the Sun's core must eventually be released at its surface, and so the Sun and other H-burning stars get brighter as they age.

Gough (1981) expressed the results of his own solar evolution modeling in a parametric form that remains reasonably accurate when compared to later models (e.g., Bahcall et al. (2001))

$$\frac{L(t)}{L_0} = \frac{1}{1 + 0.4(t/4.6)} \quad (11.3)$$

Here, $L(t)$ is the solar luminosity at time t, L_0 is the present solar luminosity, and t is time before the present in units of Ga. Equation (11.3) predicts that the Sun was $1/1.4 = 0.71$ times as bright as today at 4.6 Ga and that it has brightened roughly linearly with time since then. In reality, the rate of brightening is accelerating with time. The current rate of increase, according to eq. (11.3), is about 1% every hundred million years. Gough's formula is represented by the solid curve in Fig. 11.1.

An alternative to the faint young Sun is that the young Sun shed considerable mass in a rapid solar wind so that the Sun's core wasn't compressed as much as assumed above, but this idea has not stood up to observational scrutiny. While the theory of early mass loss has been suggested numerous times (Boothroyd et al., 1991; Graedel et al., 1991; Guzik et al., 1987; Sackmann and Boothroyd, 2003; Willson et al., 1987), empirical estimates of the mass loss rate from young stars appear too small for the theory to hold (Wood et al., 2002; Wood et al., 2005).

Wood et al. used the Hubble Space Telescope to measure excess absorption on the blue-shifted side of the Lyman-α emission (at 121 nm) from eight nearby young stars, including ε-Eridani – a 400–500 million-year-old K dwarf. They attributed this absorption to the buildup of neutral hydrogen at the boundary of a stellar *astrosphere*. The astrosphere, like the solar heliosphere, is the region of space that is swept free of interstellar material by the outflowing stellar wind. By using a 2-D hydrodynamic model, Wood et al. were able to relate the intensity of astrospheric absorption to the strength of the stellar wind. They then correlated these data with data on x-ray emissions to estimate stellar ages.

The bottom line is that young, solar-type stars do indeed have winds that are ~ 1000 times stronger than the solar wind; however, these winds decrease in intensity rapidly as the star ages. Figure 11.2 shows cumulative solar mass loss with time. As one can see, the best estimate for the total mass loss for the Sun is about 0.6% of a solar mass (M_o), with an upper limit of about 0.03 M_o. The Sun's luminosity is roughly proportional to $M^{4.5}$ (eq. (2.4)). Furthermore, if the mass loss is not too rapid, a planet's semi-major axis, a, should expand so as to conserve angular momentum, giving $a \sim 1/M$. The solar flux, F, incident at a planet's orbit obeys the inverse square law: $F \sim a^{-2}$. Putting all of this together yields $F \sim M^{6.5}$. Thus, a young Sun that was 3% more massive than today should have been brighter than the standard solar model by $1.03^{6.5} \approx 1.19$, i.e. a 19% increase. This is roughly half the amount needed to compensate for the

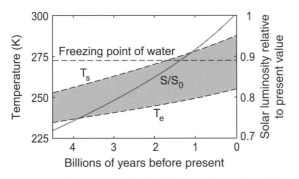

Figure 11.1 Diagram illustrating the faint young Sun problem. The solid curve labeled "S/S_0" represents solar luminosity S relative to today's value S_0 (right-hand scale). The two dashed curves represent Earth's effective radiating temperature, T_{eff}, and its global mean surface temperature, T_s, as calculated using a 1-D climate model. The temperature difference between T_s (upper dashed curve) and T_e (lower dashed curve) at any time is a measure of the greenhouse effect. A constant CO_2 concentration of 300 ppmv was assumed. (From Kasting et al. (1988).)

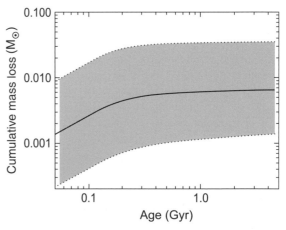

Figure 11.2 Estimated cumulative mass loss from the Sun, based on observations of Lyman-α absorption in the astrospheres of nearby solar-type stars. The shaded area represents the range of uncertainty in the calculations. (From Wood et al. (2002). Reproduced with permission. Copyright 2002, American Astronomical Society.)

decrease in luminosity predicted by the standard model. Note, however, that nearly all of the predicted mass loss, and hence all of the luminosity difference from the standard model, occurs within the first 200 million years of the Sun's history. As we shall see, the geologic record provides few constraints on climate history during that very early time. Then, after 4.4 Ga, the faint young Sun problem remains as previously described.

11.2 Implications for Planetary Surface Temperatures: Sagan and Mullen's Model

The theory of solar evolution was well developed by the mid-1960s, and its implications for planetary climates were soon pointed out by Sagan in Shklovskii and Sagan (1966, pp. 221–223) and then by Sagan and Mullen (1972). The latter authors showed how the effects of such a solar luminosity change on Earth's surface temperature can be readily estimated if one assumes that Earth's greenhouse effect and albedo have remained unchanged. We showed in Sec. 2.2.4 (eq. (2.15)) that the equilibrium temperature, T_{eq}, of the planet is given by

$$\sigma T_{eq}^4 = \frac{S}{4}(1 - A) \tag{11.4}$$

Here, S is the solar constant, A is the planet albedo, and σ is the Stefan-Boltzmann constant. The surface temperature, T_s, is higher than the equilibrium temperature by an amount, ΔT_g, that we defined as the greenhouse effect (eq. (2.16)).

$$T_s = T_{eq} + \Delta T_g \tag{11.5}$$

The climate problem for Earth is illustrated graphically in Fig. 11.1. For a planet like the Earth, where the internal heat flux is negligible compared to the solar flux, the equilibrium temperature is equal to the effective blackbody radiating temperature, T_{eff}, as discussed in Sec. 2.2.6. As mentioned earlier, the solid curve in Fig. 11.1 shows Gough's solar luminosity curve, eq. (11.3), while the two dashed curves show T_{eff} and T_s, as calculated by a 1-D radiative convective climate model (Kasting et al., 1989a). In this calculation, a constant CO_2 concentration of 300 ppm was assumed, along with a fixed relative humidity profile. Fixing the relative humidity in this manner allows the total water vapor abundance to increase with increasing surface temperature – a positive feedback that is widely recognized to be an important part of Earth's climate system. This is why the greenhouse effect increases with time in Fig. 11.1. The results of this calculation are similar to those of Sagan and Mullen: the mean surface

temperature drops below the freezing point of water prior to about 2 Ga. We know, however, that both liquid water and life have been present for much longer than that (see Ch. 9). Thus, the results of this type of calculation have become known as the *faint young Sun problem*, or sometimes the *faint young Sun paradox*. It is only a paradox, however, if both the greenhouse effect and Earth's albedo remained constant with time. Neither of these assumptions is likely to have been true; hence, we'll avoid the term "paradox" from here on.

Sagan and Mullen suggested that high concentrations of ammonia, NH_3, in Earth's early atmosphere would resolve this problem. Preston Cloud had already published his theory for the rise of oxygen in Earth's atmosphere, discussed in the previous chapter, so Sagan and Mullen knew that Earth's primitive atmosphere should have contained little free O_2. Sagan leaned towards the highly reduced, CH_4–NH_3 atmospheres favored by Oparin and Urey which, as we have seen in Sec. 7.1, are now out of favor, having been replaced by models in which the dominant gases were CO_2 and N_2. Ammonia is an extremely effective greenhouse gas, as it absorbs strongly across the thermal-IR spectrum. Sagan and Mullen showed that NH_3 concentrations of 10–100 ppmv would have been enough to compensate for a 30% reduction in solar luminosity.

The trouble with their theory is that ammonia can be photochemically destroyed by the following reaction sequence (Kasting, 1982; Kuhn and Atreya, 1979).

$$NH_3 + hv \, (\lambda < 230 \, nm) \rightarrow NH_2 + H \tag{11.6}$$

$$NH_2 + NH_2 + M \rightarrow N_2H_4 + M \tag{11.7}$$

$$N_2H_4 + H \rightarrow N_2H_3 + H_2 \tag{11.8}$$

$$N_2H_4 + hv \rightarrow N_2H_3 + H \tag{11.9}$$

$$\begin{aligned} N_2H_3 + N_2H_3 &\rightarrow N_2H_4 + N_2H_2 \\ &\rightarrow N_2H_4 + N_2 + H_2 \end{aligned} \tag{11.10}$$

Here, N_2H_4 is *hydrazine*, best known for its use as a liquid propellant on spacecraft. This reaction sequence is fast in a low-O_2 atmosphere because other gases do not block solar UV radiation between 200 nm and 230 nm. (CO_2 absorbs strongly below ~200 nm.) The net result of the reactions above is that NH_3 is irreversibly converted to N_2 and H_2. The H_2 escapes to space by mechanisms discussed in Ch. 5, while the N_2 remains in the atmosphere. Because it is triply bonded, however, N_2 can be converted back into NH_3 only with great difficulty. We return to this issue in Sec. 11.6.

Despite these objections, Sagan never gave up on his original idea. In a paper published posthumously, he pointed out that ammonia might have been protected from photolysis by high-altitude organic haze formed from polymerization of methane (Sagan and Chyba, 1997). A "thin slab" model parameterized the haze, which effectively assumed infinitely small particles. This hypothesis was tested in a more elaborate photochemical model by Pavlov et al. (2001) and found to be lacking. When the effects of finite particle size were included, the ratio of the UV optical depth of the haze to its visible optical depth was lower than Sagan and Chyba had estimated (see Sec. 2.4.2.2 for the definition of *optical depth*). According to Pavlov et al., an effective UV screen could have been established only if the visible optical depth of the haze was high. This would have lowered the surface temperature by creating an anti-greenhouse effect (as discussed below), and so once again the ammonia greenhouse model was deemed impossible. The Pavlov et al. model was itself oversimplified, however, as it assumed spherical haze particles with a single size at each altitude. More elaborate particle models yield different results, so we return to this issue in Sec. 11.6.

11.3 Geological Constraints on Archean and Hadean Surface Temperatures

Before discussing other solutions to the faint young Sun problem, we should ask the question: What do we actually know about surface temperatures during the Archean and Hadean eons? Unfortunately, the answer is "not much." The environment in the Hadean eon (4.5–4.0 Ga) is particularly poorly constrained because the geologic record from this time period is almost nonexistent. That said, some rocks, such as those found in fossilized gravel in the Jack Hills of western Australia, contain zirconium silicate ($ZrSiO_4$) crystals that are much older than the rocks themselves (Compston and Pidgeon, 1986; Harrison, 2009; Mojzsis et al., 2001; Peck et al., 2001; Wilde et al., 2001). These *zircons*, as they are called, have such high melting points that they occasionally survive when rocks are subducted into the mantle, and then reappear and grow new layers when fresh igneous rocks are created. Dating of the interiors of zircons show that some of them started to form as early as 4.4 Ga (*ibid.*). Furthermore, many of these zircons are significantly enriched in ^{18}O relative to ^{16}O compared with mantle rocks, indicating that they formed from magmas which were themselves derived from rocks that had undergone low-temperature aqueous alteration at Earth's surface (Valley et al., 2002). Based on this result, Valley et al. titled their paper "A cool early Earth." But one needs to examine carefully what this means. The conclusion of the paper is that the temperature of the water with which

zircon protoliths interacted was less than about 200 °C. Although novel, this conclusion is fully consistent with numerical models of Earth's accretion (Matsui and Abe, 1986a, b; Zahnle et al., 1988) that were discussed in Chapters 6 and 7. These models indicate that Earth's surface temperature should have fallen well below 200 °C once the main accretion period ended and solar energy became the main source of heating for the atmosphere. Had that not been the case, the oceans would likely have been lost by the "moist greenhouse" process that will be explored in Chapter 13.

After the Hadean, our knowledge of surface temperatures improves because a sedimentary rock record exists. As a result, Sagan and Mullen knew that the oceans were not frozen during the Archean and that life itself existed, at least by 3.5 Ga, if not before. The upper temperature limit on modern life (at the time of writing in 2016) is 122 °C (Kashefi and Lovley, 2003; Takai et al., 2008). If early organisms were no more heat tolerant than modern ones, this gives an upper limit on Archean surface temperature.

11.3.1 Glacial Constraints on Surface Temperature

During the mid- to late Archean the geologic record improves, and climatic constraints become better defined. In particular, glacial episodes appear. Some of these glaciations were severe and have been known about for many years. Convincing evidence for low-latitude glaciation is found at 2.4–2.2 Ga (Evans et al., 1997), as well as in the Neoproterozoic around 0.6–0.8 Ga (Hoffman et al., 1998) (see Fig. 9.1). These intervals are considered to be possible "Snowball Earth" episodes, as discussed in Sec. 11.10. Other Precambrian glacial episodes have been discovered more recently. Most of these lack paleolatitude constraints and thus could represent more conventional polar glaciation. The oldest of these are 3.5 Ga rocks of the Barberton Super group, South Africa (De Wit and Furnes, 2016). The next oldest are 2.9 Ga rocks from the Pongola Supergroup in South Africa (Young et al., 1998). Later, 2.7 Ga glacial rocks are found in the Dharwar Supergroup, India (Ojakangas et al., 2014). In the Paleoproterozoic, a 1.8 Ga glaciation is recorded in the King Leopold Formation in northwestern Australia (Schmidt and Williams, 2008; Williams, 2005). At roughly this same time, in central Sweden, a single 5-m-scale 1.9 Ga outcrop suggests a local, sub-zero climate because of evidence for the formation of polygonal cracks in the ground that form from freezing and thawing (Kuipers et al., 2013). Finally, the glacial

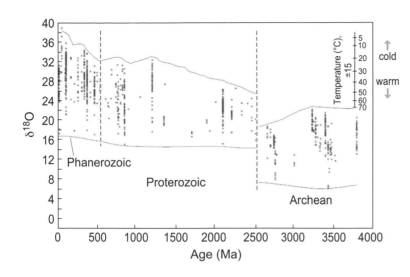

Figure 11.3 Oxygen isotopic composition of cherts (SiO_2) of different ages. The isotope standard is SMOW (Standard Mean Ocean Water). (From Knauth (2005). Reproduced with permission from Elsevier. Copyright 2005.)

Vazante Group in Brazil, which had previously been considered Neoproterozoic in age, has been re-dated at 1.1-1.3 Ga (Geboy *et al.*, 2013), placing it in the mid-Proterozoic, which had previously been thought to be uniformly warm. (See further discussion in Sec. 11.9.) Taken as a whole, these data suggest that the Precambrian was not persistently warmer than the ensuing Phanerozoic.

Phanerozoic glaciations (not shown in Fig. 9.1) are discussed in Sec. 11.10. These glacial episodes, when one or both poles are covered with ice (as they are today), appear to require a mean global surface temperature below ~ 20 °C (Kasting, 1987). This surface temperature limit is inferred from deep-sea oxygen isotopes around 35 Ma, when Antarctica first became glaciated (Zachos *et al.*, 2008). At about this time, the Drake Passage opened up between Antarctica and South America, and so the onset of glaciation may have been affected by changes in ocean heat transport, as well as by changes in mean global surface temperature. The modern circumpolar current, which goes through Drake Passage, restricts the poleward flow of warm water from lower latitudes, thereby making Antarctica colder. Mean surface temperature during a glacial period must also be above a certain threshold, except during the Snowball Earth events discussed later in this chapter. The mean surface temperature during the Last Glacial Maximum of the Pleistocene Ice Age, at about 20 ka, is estimated to have been between 5 °C and 10 °C, or 5–10 °C cooler than today's mean global temperature of 15 °C (e.g. Crowley, 2000, and references therein). At that time, polar ice in the northern hemisphere extended down to $\sim 40°$ latitude in North America and $\sim 50°$ in Europe. Sea ice probably did not reach this far equatorward, but it would likely have done

so if the climate were just a few degrees cooler. Once sea ice reaches about 25° latitude, ice–albedo feedback is predicted to take over, creating a Snowball Earth (see Sec. 11.9). Hence, a generous lower limit on mean global surface temperature during times of polar glaciation is ~ 0 °C, and 5 °C is probably a better estimate.

11.3.2 Isotopic Constraints on Surface Temperature

The evidence for glaciation during the Archean and Paleoproterozoic stands in stark contrast to paleotemperatures reported from oxygen isotopes in ancient rocks. It has been known for decades that both cherts (SiO_2) and carbonates become progressively depleted in ^{18}O relative to ^{16}O in rocks of increasing age (Knauth, 2005; Knauth and Epstein, 1976; Knauth and Lowe, 2003; Shields and Veizer, 2002). The chert record, which is thought to be less subject to later, diagenetic alteration is shown in Fig. 11.3. A straightforward, but likely incorrect, interpretation of this pattern would suggest that Earth's mean surface temperature was much warmer in the distant past than it is today. Silica that precipitates from seawater concentrates less ^{18}O relative to the water as temperature increases. The ^{18}O isotope forms stronger bonds with Si than does ^{16}O and so, at equilibrium, warmer temperatures are needed to remove ^{18}O from a bond and replace the ^{18}O with ^{16}O. On the basis of chert data from the Barberton greenstone belt, South Africa, Knauth and Lowe (2003) argued that the mean surface temperature at 3.3 Ga was 70 ± 15 °C. This is certainly warm, but not so warm as to preclude prokaryotic life and possibly even cyanobacteria, some of which can survive at temperatures almost this high.

A high temperature interpretation of the O isotope record seems implausible, though, given the evidence for glaciation at 3.5 Ga, 2.9, 2.7, and 2.4 Ga (Fig. 9.1), and so alternative explanations for the isotope data have been offered. The simplest is that all of the oxygen isotope ratios in ancient rocks have been reset during diagenesis and metamorphism, and that the temperatures being measured are those at depth within the sediments rather than those of the seawater from which they precipitated (Degens and Epstein, 1962; Land, 1995). Other possible explanations include gradual changes in the oxygen isotope composition of seawater (Kasting *et al.*, 2006; Perry *et al.*, 1978; Walker and Lohmann, 1989; Wallmann, 2004) or widespread hydrothermal alteration of the seafloor (van den Boorn *et al.*, 2007). The latter explanation could also account for the observed trend in silicon isotopes in cherts, which had been previously cited in support of the high paleo-surface temperatures (Robert and Chaussidon, 2006). Even in the Phanerozoic, the data are difficult to explain unless the O isotope composition of seawater changed gradually in time (Veizer and Prokoph, 2015), so we continue to think that this process must have played a role.

New isotope techniques have been applied to the Archean recently. Hren *et al.* (2009) concluded that Archean surface temperatures were <40 °C, based on combined analysis of O and H isotope ratios. Blake *et al.* (2010) analyzed O isotopes in Archean phosphates and derived upper limits of 26 °C–35 °C.

On the whole, the isotopic data are more subject to interpretation, and hence less convincing, than the glacial evidence, and so the mean surface temperature during the Archean was probably nowhere near as warm as the original oxygen isotopic data would suggest. Instead the average Archean climate was likely as warm or warmer than today, but that is probably all that can be said based on the available data.

In the future, the *clumped isotope* method of determining paleotemperatures might help. This new technique examines the extent to which rare isotopes bond to each other rather than to the abundant light isotopes. In principle, it can be used as a paleothermometer independent of assumptions about the $\delta^{18}O$ of ancient seawater (Came *et al.*, 2007; Eiler, 2007). However, currently the method has limitations and can yield a wide range of paleotemperatures (Eiler, 2011; Henkes *et al.*, 2014). The carbonate clumped isotope thermometer preserves primary temperatures through metamorphism up to $\sim 200\ °C$, provided that the carbonate does not dissolve and reprecipitate. There are no Archean carbonates of sufficiently low metamorphic grade to satisfy this temperature criterion, but there may be trends with metamorphic grade that point to an original

low temperature of deposition. At the very least, the technique is promising for later times, such as the Mesoproterozoic, where low-grade carbonates are available.

11.4 Solving the Faint Young Sun Problem with CO_2

Let's return now to theories for how the faint young Sun problem might be resolved. When the ammonia greenhouse warming mechanism ran into problems, researchers turned their attention to other greenhouse gases. The first to be looked at carefully was CO_2. Along with H_2O, CO_2 is one of the two most important greenhouse gases in the present atmosphere, and there are good reasons to think that it could have been more abundant in the distant past. Hart (1978) was among the first to realize this. He constructed an elaborate computer model of Earth's evolution which predicted that that atmospheric CO_2 concentrations were initially as much as 1000 times higher than today. (We will return to Hart's model in Ch. 15 when we discuss the concept of the circumstellar habitable zone.) Following this, Owen *et al.* (1979) performed radiative–convective climate modeling using Hart's proposed evolutionary sequence for atmospheric CO_2 and solar luminosity. According to their calculation, a CO_2 partial pressure of 0.31 bar, or about 1000 PAL (times the Present Atmospheric Level), could have produced a mean surface temperature of 310 K at 4.25 Ga. This, of course, is well above the present mean temperature of 288 K. We will henceforth define 1 PAL of CO_2 as 3×10^{-4} bar, or 300 ppmv in a 1-bar atmosphere. This is slightly higher than the preindustrial CO_2 level of 280 ppmv.

11.4.1 The Carbonate–Silicate Cycle

Could atmospheric CO_2 have actually been this high, though? To answer this question, one must examine the processes that control atmospheric CO_2 on long timescales. It is, of course, the carbon cycle that does this. However, it is not the part of the carbon cycle with which most readers are familiar. The more familiar part of the carbon cycle is the *organic carbon cycle*, which we discussed in some detail in the previous chapter. That cycle is important because it generates O_2. On timescales greater than 10^5–10^6 years, however, atmospheric CO_2 is controlled mostly by the *inorganic carbon cycle*, also known as the *carbonate–silicate cycle* (see Fig. 11.5).

To understand how this cycle works, let's begin on the left-hand side of this diagram. CO_2 dissolves in rainwater to form a weak acid, carbonic acid (H_2CO_3). Chemically, this can be written as

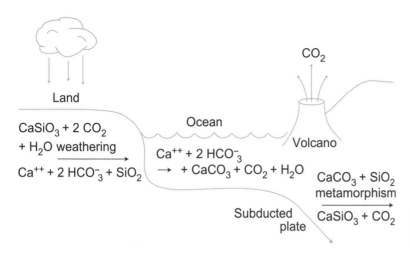

Figure 11.4 Diagram illustrating the carbonate–silicate cycle. (From Catling and Kasting (2007).)

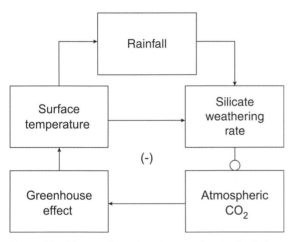

Figure 11.5 Diagram illustrating the negative feedback loop between atmospheric CO_2 and climate that is part of the carbonate–silicate cycle. (From Kasting (2010). Reproduced with permission. Copyright 2012, Princeton University Press.)

$$CO_2 + H_2O \leftrightarrow H_2CO_3 \quad (11.11)$$

Carbonic acid is not a strong acid, i.e., it does not dissociate completely to yield two H^+ ions plus carbonate ion, CO_3^{2-}. Rather, when CO_2 is dissolved in pure water, the H_2CO_3 partly dissociates into one H^+ ion plus a bicarbonate ion, HCO_3^-. (An exception occurs in alkaline-buffered solutions with $pH > 10.3$, where dissociation proceeds so that the carbonate ion, CO_3^{2-} dominates). When CO_2 dissolves in rainwater, the fluid is acidic and strong enough to dissolve silicate rocks on the continents. If we represent all silicates by the simple silicate mineral, wollastonite ($CaSiO_3$), we can write *silicate weathering* as follows.

$$CaSiO_3 + 2\,CO_2 + H_2O \rightarrow Ca^{2+} + 2\,HCO_3^- + SiO_2 \quad (11.12)$$

Here, SiO_2 represents dissolved silica. Once formed, these byproducts of weathering are carried by rivers to the ocean. There, organisms that live in the surface ocean, e.g., planktonic foraminifera, use the calcium and bicarbonate to make shells out of calcium carbonate ($CaCO_3$). This process is *carbonate precipitation*.

$$Ca^{2+} + 2\,HCO_3^- \rightarrow CaCO_3 + CO_2 + H_2O \quad (11.13)$$

When the organisms die, they sink into the deep ocean. Most of the calcium carbonate redissolves, because the deep ocean is more acidic than the surface ocean as a result of the downward flux and subsequent oxidation of organic matter. Some of the calcium carbonate survives, however, and forms sediments on the seafloor. The net result of silicate weathering and carbonate precipitation can be determined by adding equations (11.12) and (11.13) together to get

$$CaSiO_3 + CO_2 \rightarrow CaCO_3 + SiO_2 \quad (11.14)$$

What we care about here is the fate of the carbon. The carbonate sediments are mobile because the seafloor moves. It is continually created at the midocean ridges, spreads slowly away from there at a few centimeters per year, and then at certain plate boundaries it is *subducted* back into the mantle. Some of the carbonate is scraped off, but some of it is carried to great depths with the subducted slab where temperatures and pressures are much higher. Under these conditions, calcium (and magnesium) carbonates recombine with SiO_2, which by this time is the mineral *quartz*, to reform calcium and magnesium silicates. The name given to this process is *carbonate metamorphism*. Schematically, it can be represented as follows.

$$CaCO_2 + SiO_2 \rightarrow CaSiO_3 + CO_2 \quad (11.15)$$

Gaseous CO_2 is given off in this process. This CO_2 makes its way back to the surface through cracks and vents and is emitted to the atmosphere in volcanic plumes or diffusely (e.g., Mather, 2015), thereby completing the cycle. Indeed, reaction (11.15) is just the opposite of reaction (11.14), indicating that the carbonate–silicate cycle is closed.

Carbonate can also be incorporated into seafloor by way of low-temperature alteration of oceanic crust. Weathering of the continents probably dominates the CO_2 sink currently because the total CO_2 concentration in seawater is such that the flux of CO_2 into sediments is bigger than the flux into the oceanic crust (Caldeira, 1995). But high concentrations of carbonate in Mesozoic oceanic crust suggest that the Mesozoic sink for CO_2 into oceanic crust needs to be added to that from silicate weathering on the continents followed by deposition of carbonates on the seafloor (Coogan and Dosso, 2015; Coogan and Gillis, 2013). Other authors have argued that this process was important for CO_2 removal on the Archean Earth when the amount of exposed continental area was smaller and rates of seafloor production were likely higher (Sleep and Zahnle, 2001; Walker, 1990). While our discussion here will concentrate on the traditional carbonate-silicate cycle, i.e., that involving continental weathering, we should bear in mind that the less studied process of seafloor weathering may also be important in controlling atmospheric CO_2 in the Precambrian.

11.4.2 Feedbacks in the Carbonate–Silicate Cycle and a Possible Solution to the Faint Young Sun Problem

The carbonate–silicate cycle acts as a thermostat for the Earth's climate on timescales of about 0.5 Myr in response to changes in solar forcing (Berner et al., 1983; Walker et al., 1981). Consider first the extreme case where the oceans freeze over entirely. (Later in the chapter, we'll discuss whether this may have actually happened several times in Earth's history in so-called "Snowball Earth" episodes.) If the oceans froze, evaporation and rainfall would cease. Some sublimation would occur from the ice, and some snowfall would occur at high latitudes. But the hydrologic cycle would be extremely slow, and liquid water would be absent at the surface. The oceans would, however, remain liquid at depth. Under these conditions, silicate weathering should slow down or stop, and volcanic CO_2 should accumulate in the atmosphere. An atmosphere with ~ 300 ppmv CO_2 contains about 6×10^{16} moles of CO_2, and the CO_2 outgassing rate is a little over 6×10^{12} mol yr^{-1} (Table 7.1).

So, in about 10 000 years, assuming no transfer of gas to the iced-over ocean, the atmospheric CO_2 concentration would double, and in ten million years it would reach values a thousand times higher – enough to melt the ice on even a hard, white Snowball Earth (Caldeira and Kasting, 1992b).

Now consider how the system would work if forced in a less drastic manner. The interplay between atmospheric CO_2 and climate constitutes a classic negative feedback loop, as shown in Fig. 11.5. In this figure a box represents a system *component*, and the arrows between them represent *couplings*. The notation here is similar to that used by Kump et al. (2010). An arrow with a normal arrowhead represents a positive coupling, meaning that an increase in component A causes a corresponding *increase* in component B. An arrow with a circular arrowhead represents a negative coupling, meaning that an increase in component A causes a corresponding *decrease* in component B. With this notation in place, we can explain the CO_2-climate feedback loop as follows. An increase in surface temperature, T_s, causes an increase in evaporation, and hence an increase in rainfall. The increased rainfall, along with the higher temperature itself, causes the silicate weathering rate to increase. This decreases atmospheric CO_2, which decreases the greenhouse effect, and this in turn decreases surface temperature. After one cycle around the loop, the sign of the initial perturbation to surface temperature is reversed, indicating that the system acts as a *negative feedback loop*.

If one accepts this reasoning, then it is possible to imagine a solution to the faint young Sun problem that involves only CO_2 and H_2O. Because of its participation in the carbonate–silicate cycle, atmospheric CO_2 varies in such a way that it keeps surface temperatures high enough to permit the presence of liquid water over significant areas of the Earth's surface. Inverting this logic, one can then ask the question: How much CO_2 would have been required to compensate for lower solar luminosity early in Earth's history?

An answer to this question is given in Fig. 11.6. These calculations were performed by one of the authors (JK) using a *radiative–convective climate model*. (The principles behind such models are described in Sec. 2.4.5 and details for a 1-D model are given in Appendix A.) This model assumed a moist adiabatic lapse rate (see Sec.1.1.3.6) and a fixed vertical distribution of relative humidity. The shaded area on the figure represents the range of atmospheric CO_2 concentrations that are consistent with the solar forcing and with the observed climate record if CO_2 and H_2O were the only important

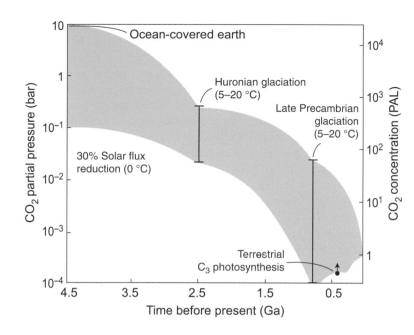

Figure 11.6 Diagram illustrating the amount of CO$_2$ required to compensate for reduced solar luminosity in the past, assuming that CO$_2$ and H$_2$O were the only greenhouse gases. The shaded area shows CO$_2$ concentrations that would have kept the mean surface temperature between 5 °C and 20 °C. The scale on the right shows the CO$_2$ concentration in terms of PAL (Present Atmospheric Level). 1 PAL is defined as 300 ppmv of CO$_2$. (From Kasting (1993). Reproduced with permission. Copyright 1993, American Association for the Advancement of Science.)

greenhouse gases. At 4.5 Ga, when the Sun was ∼30% dimmer, a minimum of 0.1 bar of CO$_2$, or about 330 PAL, was needed to keep the mean surface temperature above freezing. On the other hand, the actual atmospheric CO$_2$ concentration could conceivably have been much higher than this. The upper limit of 10 bar of CO$_2$ shown in the figure comes from Walker (1985), who argued that atmospheric CO$_2$ levels should have been extremely high prior to the formation of the continents because of the lack of land area on which silicate weathering might take place. The predicted mean surface temperature for a 10-bar CO$_2$ atmosphere at this time is about 85 °C (Kasting and Ackerman, 1986). (See also the discussion in Sec. 11.5.) This is warm, but not sufficiently so to boil the oceans. According to Kasting and Ackerman, the oceans never boil when CO$_2$ is added to Earth's atmosphere, because the increased surface pressure always compensates for the increased temperature. This surface temperature is also well below the 200 °C limit derived from zircons, discussed earlier in the chapter, and so it is theoretically possible that Earth was this warm in the Hadean.

In contrast to the hypothesis of high CO$_2$ levels on the earliest Earth, Sleep and Zahnle (2001) argue that the CO$_2$ concentration on the early Earth was *low* and the Hadean relatively cold because of enhanced weathering of seafloor on a tectonically more active young planet. A significant flux of carbon into the early oceanic crust is supported by the considerable carbonatization of thick, mid-Archean greenstone belts in the Pilbara, Australia,

that have whole rock composition similar to modern sea-floor basalt (Shibuya *et al.*, 2012). The Late Mesozoic oceanic crust also contains several times higher CO$_2$ concentrations (2.5 wt%) than Cenozoic upper oceanic crust (0.5 wt%) (Coogan and Gillis, 2013). Based on this observation, along with the knowledge that atmospheric CO$_2$ concentrations were several times higher at that time, these latter authors suggest that carbonatization of the oceanic crust is another strong negative feedback that helps to stabilize CO$_2$ over long timescales.

Later on in the Archean, the glacial data mentioned earlier become relevant. Atmospheric CO$_2$ could not have been too high when continental glaciation occurred. Figure 11.6 assumes that the mean surface temperature at the time of the 2.4-Ga and 0.6-Ga glaciations must have been below 20 °C – the mean surface temperature when Antarctica first became glaciated (Sec. 11.3). The lower limit on surface temperature at these times is taken to be 5 °C – the approximate mean surface temperature during the Last Glacial Maximum. If the surface temperature was much lower than this, a Snowball Earth scenario would likely have ensued, from which the Earth would have recovered by the feedback mechanism described above.

11.4.3 Geochemical Constraints on Past CO$_2$ Concentrations

The hypothesis that the faint young Sun problem was solved mostly by CO$_2$ is attractive because it can be justified on theoretical grounds, as we have just explained. It has faced major hurdles, however, as geochemists have attempted to place observational constraints on pCO_2. Some results suggest that paleo-CO_2 concentrations were significantly lower than those shown in Fig. 11.6.

One type of evidence comes from *paleosols*. As we discussed in Sec. 10.3.1, paleosols are ancient soils that formed in contact with the atmosphere and, thus, can be used to estimate atmospheric O_2 concentrations. Similarly, they can also be used to quantify atmospheric CO_2 concentrations (e.g., Sheldon and Tabor, 2009).

In an early study, Rye *et al.* (1995) looked for the presence of the mineral siderite, $FeCO_3$, in paleosols ranging in age up to 2.8 Ga, which they related to the pCO_2. Such paleosols formed prior to the initial rise in O_2. This is important because siderite is a reduced mineral that is unstable in the presence of free O_2 and, hence, is not usually found in modern soils (although see Sheldon (2006) for some exceptions). Surprisingly, siderite was not found in the paleosols that Rye *et al.* studied, either. Instead iron precipitated as a silicate mineral, rather than as a carbonate. Rye *et al.* used the following equilibrium to place an upper limit of ~0.03 bar, or 100 PAL, on pCO_2 at 2.8 Ga:

$$\underset{\text{greenalite}}{Fe_3Si_2O_5(OH)_4} + 3CO_2 + 2H_2O = \underset{\text{siderite}}{3FeCO_3} + \underset{\text{silicic acid}}{2H_4SiO_4}$$

(11.16)

This pCO_2 is roughly equal to the lower limit estimated in Fig. 11.6.

The Rye *et al.* paper has been criticized by Sheldon (2006) who points out that the iron-silicate mineral used in their analysis, greenalite, is a metamorphic mineral, not one formed directly in soils. When the analysis is repeated with a more appropriate equilibrium and updated thermodynamic data, it yields a much lower limit on pCO_2, equal

to or below today's value. That, of course, is far below the value needed to compensate for the faint young Sun. Indeed, Sheldon argues that any thermodynamic constraint on pCO_2 derived from a single chemical reaction is suspect. So, he approached the problem in another way. He performed a mass-balance study of weathering in three different paleosols dated at 2.2 Ga and derived an estimate of $23^{\times 3}_{\div 3}$PAL of CO_2. The same approach yields an estimate of 10–50 PAL of CO_2 at 2.7 Ga (Driese *et al.*, 2011). The CO_2 levels in these papers are referenced to a value of about 370 ppmv, so the Driese *et al.* estimates correspond to a CO_2 partial pressure of 0.0037–0.019 bar. Sheldon also looked at some younger, Proterozoic paleosols and concluded that pCO_2 dropped significantly between 1.8 Ga and 1.0 Ga (Fig. 11.7). This conclusion is supported by an independent analysis based on calcification of cyanobacterial sheaths that yields an upper limit of ~10 PAL of CO_2 (0.36%) at 1.2 Ga (Kah and Riding, 2007). Thus, the trend in Sheldon's estimated CO_2 concentrations parallels that predicted by the climate model. However, Sheldon's pCO_2 values are still lower by a factor of 5–10 than the climate model values, suggesting that some additional warming mechanism besides CO_2 was probably needed. Sheldon's analysis is not definitive, however. Several parameters in the mass balance model are subject to choice (Holland and Zbinden, 1988), especially the time required for soil formation and the diffusion coefficient for CO_2 within the soil. The model also assumes that all of the CO_2 that goes into the soil results in mineral dissolution. If this is not true, then pCO_2 could have been higher.

In fact, Kanzaki and Murakami (2015) apply a different method to analyze eight paleosols from the Archean and Paleoproterozoic and obtain higher pCO_2. Their calculations estimate the composition of aqueous solutions at the time of weathering; the pCO_2 also depends upon

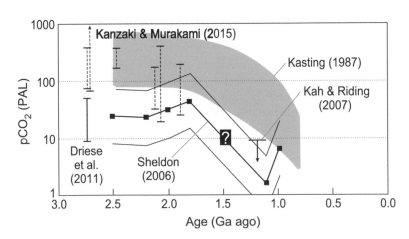

Figure 11.7 CO_2 estimates from paleosols compared to those from the climate model shown in Fig. 11.7. Modified from Sheldon (2006). Sheldon's estimates (with error bars) are shown by the dark squares connected by the solid black line. The downward arrow at 1.2 Ga is the upper limit on pCO_2 from cyanobacterial sheath calcification (Kah and Riding, 2007). The bar at the middle left is the paleosol estimate at 2.7 Ga from Driese *et al.* (2011). Bars in the upper left are paleosol estimates from Kanzaki and Murakami (2015).

inferences of the local temperature. Their best estimates for pCO_2 are 85–510 PAL at 2.77 Ga, 78–2500 PAL at 2.75 Ga, 160–490 PAL at 2.46 Ga, 30–190 PAL at 2.15 Ga, 20–620 PAL at 2.08 Ga, and 23–210 PAL at 1.85 Ga. In general, the lower bounds are too low for CO_2 to have been the only important greenhouse gas, whereas the upper bounds are within or above the CO_2–H_2O climate model of Kasting (1987).

More recently, Rosing *et al.* (2010) have attempted to place constraints on Archean pCO_2 using banded iron formations, or BIFs. As discussed in Sec. 10.3.2, BIFs were deposited throughout the Archean and Paleoproterozoic, ending at ~ 1.7 Ga. BIFs contain many iron-bearing minerals, including hematite (Fe_2O_3), magnetite (Fe_3O_4), siderite ($FeCO_3$), and various iron silicates, represented here by fayalite (Fe_2SiO_3). Rosing *et al.* base their pCO_2 constraint on phase stability relationships between these different minerals (Fig. 11.8). The hydrogen partial pressure, pH_2 (shown on the vertical axis), is assumed to

have been controlled by consumption by methanogens. Rosing *et al.* argue that in order to prevent magnetite from converting fully to siderite, CO_2 must have been to the left of this phase boundary at this particular pH_2. At 25 °C, the corresponding upper limit on pCO_2 is 10^{-3} bar, or ~ 3 PAL.

This upper limit on pCO_2 has been questioned (Dauphas and Kasting, 2011; Reinhard and Planavsky, 2011). The primary precipitate in BIFs was probably ferrihydrite ($Fe(OH)_3$) (see Sec. 10.3.2). To convert the ferric iron to magnetite ($Fe_2^{3+} Fe^{2+}O_4$) and then Fe^{2+}-containing siderite, iron must be reduced. Given that carbon in siderite is isotopically enriched in ^{12}C, the reductant used in this process was likely organic carbon, and iron isotopes suggest microbially mediated reduction (Becker and Clayton, 1972; Heimann *et al.*, 2010). If organic carbon was in short supply in some localities, then magnetite might have remained stable, which casts doubt on the low estimate for Archean pCO_2.

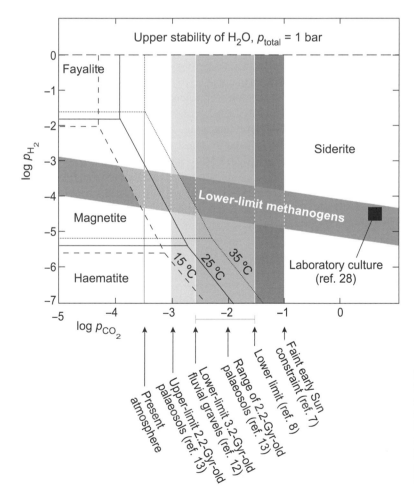

Figure 11.8 Phase diagram for various iron minerals in banded iron formations, or BIFs. The axes represent logs of the H_2 and CO_2 partial pressures that would be in equilibrium with the water in which the BIFs formed. (From Rosing *et al.* (2010). Reproduced with permission from Macmillan Publishers Ltd. Copyright 2010, Nature Publishing Group.)

11.5 Clouds and the Faint Young Sun Problem

Another factor that could help to resolve the faint young Sun problem is the possible effect of changes in fractional cloud cover or cloud albedo. Equation (11.4) shows that any decrease in planetary albedo, A, from its current value of ~ 0.3 would warm the early Earth. Rosing et al. (2010) speculated that Earth's albedo was lower during the Archean because there were fewer cloud condensation nuclei (CCNs) in the atmosphere. Sparse CCNs would cause cloud particles to be less abundant and larger, creating a decrease in cloud albedo. In their view, the scarcity of CCNs was a consequence of reduced continental area, leading to less wind-blown dust, along with smaller fluxes of biogenic sulfur gases such as dimethyl sulfide (DMS) (see p. 36), which today oxidize to form sulfate aerosol particles. But the effect of variations in CCNs appears to have been overestimated in the Rosing et al. model (Goldblatt and Zahnle, 2011) and in any case their mechanism is just barely able to bring the Archean mean surface temperature up to 0 °C. As discussed earlier, a more realistic goal would be 15 °C or higher. A similar mechanism involving reduced CCN availability in the past was suggested by Svensmark (2007), who speculated that cosmic rays trigger cloud formation and that cosmic rays were shielded by a magnetically active young Sun. However, reflective clouds over the ocean are limited by sulfate CCN, not ions, so there is no in situ evidence that cosmic rays have any significant influence. Second, Svensmark's additional claim about a causal link between cosmic rays, decreased clouds, and contemporary climate change is false because decades of direct measurements of the average cosmic ray since 1953 show no trend (Shea and Smart, 2000), while a global warming trend is unambiguous (Stocker et al., 2013).

Other cloud-based solutions to the faint young Sun problem have been suggested. Several of these involve negative feedback between cloud cover (or cloud radiative properties) and surface temperature. Rossow et al. (1982) and Rondanelli and Lindzen (2010) have both suggested that clouds respond to changes in surface temperature in such a way that they tend to damp external forcings, including that of lowered solar luminosity. In the Rondanelli and Lindzen model, which is based on the "Iris" hypothesis of Lindzen et al. (2001), high-altitude tropical cirrus clouds become more widespread as the ocean surface cools, and this leads to increased greenhouse warming. The Iris hypothesis, though, has been strongly criticized by other meteorologists on the basis of both theory and observations (Chambers et al., 2002;

Fu et al., 2002; Lin et al., 2002; Su et al., 2008). If such a strong cloud feedback existed, it would be difficult to explain more recent changes in Earth's climate, such as the very cold Last Glacial Maximum at 20 ka and the Mid-Cretaceous climatic optimum at 100 Ma (Hoffert and Covey, 1992).

Because the spatial coverage of clouds varies greatly across the globe, 3-D GCMs are required to quantify the effect of clouds on the faint young Sun problem. Goldblatt and Zahnle find that 1-D climate models in which the clouds are parameterized as a surface albedo change (e.g., Haqq-Misra et al., 2008; Kasting, 1987; Pavlov et al., 2000) may underestimate the amount of greenhouse warming needed to compensate for reduced solar luminosity. Clouds are still a major problem for 3-D models, as their sizes are typically sub-gridscale; however, with GCMs one can at least quantitatively explore the effects of different cloud parameterizations. Early GCM results suggested that cloudiness decreased with increased rotation rate, which would help with the faint young Sun problem, as the early Earth rotated faster (Sec. 3.8) (Jenkins, 1993; Jenkins et al., 1993). Subsequent calculations, however, failed to reproduce this behavior (Jenkins, 1996; Kienert et al., 2012). However, even more recent GCMs suggest that the various impacts of clouds may mitigate the faint young Sun problem and allow for open water at relatively low CO_2 levels of ~ 7–70 PAL at 2.5–3.8 Ga if CH_4 is also at levels of 1000–2000 ppmv (Charnay et al., 2013; Wolf and Toon, 2013). For example, with lower solar radiation absorbed by the ocean, the GCM of Charnay et al. (2013) has weaker evaporation and fewer clouds in the Archean. The GCMs also make spatial predictions about the Archean climate. Faster rotation tends to reduce meridional heat transport, as expected from Hadley cell theory (Sec. 4.2), so models suggest a bigger equator-to-pole temperature differential in the Archean than today (Charnay et al., 2013; Hunt, 1979).

11.6 Effect of Reducing Gases on Archean Climate

Although Sagan and Mullen's original suggestion of a greenhouse of reducing gases for the early Earth was initially discarded, it may have been partly correct. One reason for falling back on this hypothesis is the apparent difficulty with getting enough warming from CO_2 and from cloud feedbacks on their own, as discussed above. Another consideration is the glacial record.

The Paleoproterozoic glaciations begin at ~ 2.4 Ga when atmospheric O_2 concentrations first rose, strongly

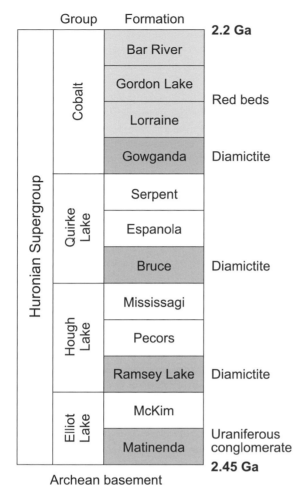

Figure 11.9 Stratigraphic sequence from the Huronian Super-group in southern Canada showing the three glacial diamictites. This sequence also shows the initial rise in atmospheric O_2, as documented by the presence of detrital uraninite and pyrite at the bottom and redbeds at the top. (Modified from Young (1991). Reproduced with permission from the Geological Association of Canada. Copyright 1991.)

suggesting a causal connection between these two events. Some evidence on which this idea is based is shown in Fig. 11.9, which depicts a stratigraphic section through the Huronian Supergroup in southern Canada. The section is bounded by volcanic rocks above and below, which have ages of 2.2 Ga and 2.45 Ga, respectively (Young, 1991). It also contains three diamictites,[1] which are interpreted as glacial deposits. From oldest to youngest,

---

[1] A *diamictite* is a sedimentary rock derived from continental material composed of unsorted sandsized or larger particles dispersed throughout a fine-textured matrix. It is a non-genetic term, unlike *tillite*, which presupposes that the unsorted particles were *till*, i.e., sediment transported by glacier ice and subsequently deposited.

these formations are the Ramsey Lake, the Bruce, and the Gowganda. The lowermost diamictite is underlain by the Matinenda Formation, which contains abundant detrital uraninite and pyrite. As discussed previously (Sec. 10.3.1), these minerals are considered evidence for very low atmospheric O_2 levels. Above the uppermost diamictite lies the Lorraine Formation, a redbed. Redbeds, as we have seen, are evidence for high levels of atmospheric O_2. So, this suggests that the glaciations occurred very close to when O_2 levels went up, even if the details of the transition – why are there three glacial layers? – remain to be understood.

If reduced gases such as methane and ammonia were part of the atmospheric greenhouse during the Archean, then their disappearance when O_2 levels rose provides a natural, but perhaps not unique, explanation for the glaciations. To the Canadian geologist, Stuart Roscoe, who mapped the Huronian sequence originally, the correlation between the rise of O_2 and glaciation appeared coincidental (Roscoe, 1969). More recent evidence suggests that after the second glaciation, atmospheric O_2 levels became high enough to weather osmium (Os) from continents and deliver Os to shallow marine sediments, which is indicated by a high, radiogenic $^{187}Os/^{188}Os$ ratio (Sekine *et al.*, 2011b). Alternatively, the glaciations triggered the rise of O_2 by releasing trapped oxidants such as H_2O_2 that then stimulated the origin of oxygenic photosynthesis (Liang *et al.*, 2006). But this latter hypothesis contradicts evidence, discussed previously (Sec.10.6), that O_2 was being generated photosynthetically well before 2.4 Ga. (See Melezhik *et al.* (2013b) for a review of the glaciations and hypotheses for triggering them.)

11.6.1 Methane and Climate: Greenhouse and Anti-Greenhouse Effects

Sagan and Mullen (1972) focused on ammonia because it is such a strong greenhouse gas, but their model also included methane. Methane has a strong absorption band centered at 7.6 μm and weaker bands spread across the thermal-IR spectrum. Methane is sometimes described as being 34 times stronger than CO_2 on a molecule-per-molecule basis on a century timescale (Stocker *et al.*, 2013); however, this statement applies only to the present atmosphere, in which CH_4 is much less abundant than CO_2. At comparably high concentrations, CO_2 is a better greenhouse gas (Kiehl and Dickinson, 1987, their Fig. 11).

Methane has two other properties that make it a viable candidate for warming the Archean Earth. First, it is much more stable photochemically than is ammonia because

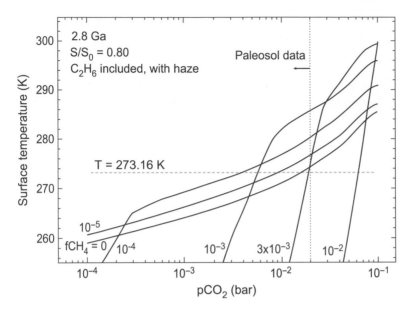

Figure 11.10 Diagram illustrating the late Archean methane greenhouse. Solid curves represent global mean surface temperature for different atmospheric CH_4 mixing ratios, as calculated using a 1-D climate model. The greenhouse effect of CO_2, H_2O, CH_4, and C_2H_6 are included, along with anti-greenhouse cooling by organic haze that becomes dominant at high CH_4/CO_2 ratios. The dotted line labeled "Paleosol data" is the estimated upper limit of 50 PAL of CO_2 at 2.7 Ga from Driese *et al.* (2011). (Modified from Haqq-Misra *et al.* (2008).)

CH_4 photolyzes only below 145 nm, whereas NH_3 photolyzes out to about 230 nm. The solar photon flux, of course, is much weaker at short UV wavelengths, and shielding by atmospheric gases is stronger. The other factor favoring methane is that it should have had a substantial biological source during the Archean. We discussed this source in some detail in Sec. 9.7. The estimated CH_4 flux from our simulated Archean ecosystem was $(0.3–3) \times 10^{11}$ cm^{-2}s^{-1}, which brackets the modern biological methane flux of 1.3×10^{11} cm^{-2}s^{-1}. These fluxes would have been capable of sustaining atmospheric CH_4 mixing ratios ranging from 100 ppmv to well over 1000 ppmv (see Fig. 9.19).

The greenhouse effect of methane in a reduced atmosphere is complicated by the byproducts of its photochemistry. In atmospheres in which the ratio of CH_4:CO_2 is greater than ~0.1, the photochemical products of CH_4 tend to polymerize, both in theoretical models (Haqq-Misra *et al.*, 2008) and in laboratory experiments (Trainer *et al.*, 2006). The reactions begin with

$$CH_4 + hv \rightarrow CH_3 + H \qquad (11.17)$$

$$CH_4 + OH \rightarrow CH_3 + H_2O \qquad (11.18)$$

$$CH_3 + CH_3 + M \rightarrow C_2H_6 + M \qquad (11.19)$$

This sequence leads to ethane, C_2H_6, and eventually to higher alkanes such as propane, C_3H_8, and butane, C_4H_{10} (Kasting *et al.*, 1983; Pavlov *et al.*, 2001; Yung *et al.*, 1984; Zahnle, 1986). Photolysis of ethane then leads to formation of ethylene, C_2H_4, and acetylene, C_2H_2 (Yung *et al.*, 1984). Acetylene can then polymerize to form

polyacetylenes, which become long enough to condense out as particles.

The net result of methane photochemistry on Archean climate depends on the effects of hydrocarbon gases and particles. The ethane formed from reaction (11.19) is itself an excellent greenhouse gas, as it has an absorption band centered at about 12 μm that extends into Earth's window region (Haqq-Misra *et al.*, 2008) (see Fig. 2.13). Other alkanes should exhibit thermal-IR absorption, as well, although their concentrations are expected to be smaller than that of ethane. But the haze particles generated from the polyacetylenes should have just the opposite effect. They would have created an *anti-greenhouse effect*, similar to that found on Saturn's moon, Titan (Sec. 12.4.2). On Titan the anti-greenhouse effect reduces the surface temperature by about 9 K (McKay *et al.*, 1991). As its name implies, the anti-greenhouse effect is just the opposite of the greenhouse effect. In the greenhouse effect, the atmosphere allows most of the incoming solar energy to penetrate, while absorbing much of the outgoing thermal-IR radiation to create warm air, which radiates IR upwards and downwards by virtue of its finite temperature. The additional downward IR flux warms the planet's surface. In the anti-greenhouse effect, the organic haze particles absorb much of the incoming solar radiation high up in the stratosphere and radiate energy back to space from that level. This reduces the solar flux at the surface, thereby cooling it.

When warming by ethane and cooling by organic haze are included, the effect of methane on surface temperature is that shown in Fig. 11.10. These calculations were performed for a solar flux of 0.8 times the present value, which is appropriate for 2.8 Ga, and 1 bar surface pressure. The solid curves represent global mean surface temperature as a function of CO_2 partial pressure and CH_4 mixing ratio. The dashed line at 0.019 bar shows the upper end of the range of pCO_2 values (10–50 PAL) given by Driese *et al.* (2011). If pCO_2 was this low, then the highest surface temperature that can be obtained from this climate model is ~ 286 K at a CH_4 mixing ratio of 10^{-3}, or 1000 ppmv. Wolf and Toon (2013) get an almost identical solution from their 3-D climate model for this same set of conditions (see their Fig. 8a). Kanzaki and Murakami (2015) allow for much more CO_2, in which case surface temperatures could have been more than the present value.

The fact that climate can be both warmed and cooled by methane suggests that methane might have regulated Archean climate. A feedback loop exists because methane production itself depends on climate. This cannot help but be true in a broad sense, as global surface temperatures below freezing would kill off most of the methanogens and other near-surface biota, thereby reducing methane production. It has also been observed, though, that thermophilic methanogens have shorter doubling times (i.e., faster growth rates) than methanogens that live at lower temperatures (Cooney, 1975). Thus, at low methane concentrations, the methane/climate system might have contained a positive feedback loop: high surface temperatures would have led to more methane production, thereby increasing the greenhouse effect, and causing still higher surface temperatures. However, at some point, the atmospheric CH_4:CO_2 ratio would have approached the limit at which organic haze began to form. This would have cooled the climate, creating a negative feedback loop. CO_2 concentrations should also have been linked to climate through a negative feedback loop, as discussed in Sec. 11.4.

The combined CH_4–CO_2 climate system can be described by the feedback diagram shown in Fig. 11.11. An increase in surface temperature, T_s, should cause an increase in atmospheric CH_4 and a decrease in CO_2 by enhancing the silicate weathering rate. The combination of higher CH_4 and lower CO_2 increases the CH_4:CO_2 ratio, eventually leading to formation of organic haze. The organic haze then cools the surface by creating an anti-greenhouse effect. The overall feedback loop is negative; thus, if this feedback loop actually operated during the Archean, it should have stabilized the surface temperature at a value where a thin organic haze existed. We will return to this thought towards the end of this book, as planets like the Archean Earth may one day be detected around nearby stars, and the presence of such a haze layer might be one way to identify them.

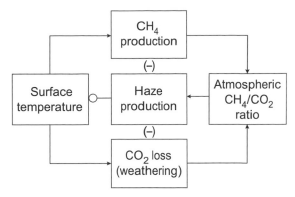

Figure 11.11 Diagram illustrating the combined CO_2 and CH_4 negative feedback loops that may have stabilized Archean climate. (From Kump *et al.* (2010). Reproduced with permission. Copyright 2010, Pearson Education Inc., New York.)

11.6.2 Fractal Organic Haze and UV Shielding of Ammonia

According to Fig. 11.10, we may still require some greenhouse warming in addition to that from CH_4 and CO_2 in order to produce a non-glacial Archean climate. This, of course, depends on what one believes about pCO_2 and cloud albedo, or, indeed, whether the late Archean climate was non-glacial. Two additional mechanisms could have warmed the early Earth further. The first of these comes from recent studies of Titan. Researchers trying to reproduce the albedo spectrum (including polarization) of Titan have found that they can generate better fits if the haze particles are assumed to be *fractals* rather than spheres (Cabane *et al.*, 1993; Rannou *et al.*, 1997; Rannou *et al.*, 1995; Rannou *et al.*, 2003; Tomasko *et al.*, 2008). Organic haze, like soot in Earth's modern atmosphere, is known to be composed of small "flakes" of organic material, just as cirrus clouds are composed of tiny snowflakes, or ice crystals. The fine structures that form the soot particle, or ice crystal, are able to interact strongly with wavelengths of light that are much smaller than the particle's overall diameter. A robust prediction of Mie theory, which describes scattering by spherical particles, is that particles interact most strongly with radiation at wavelengths comparable to their diameter (Sec. 2.4.2.4). Their extinction efficiency, Q_e (see eq. (2.54)) is highest at these wavelengths.

It is difficult, and computationally expensive, to calculate scattering from nonspherical particles. That is where fractals are useful. Fractals are structures whose patterns repeat on different spatial scales. A haze particle with intricate fine structure can be approximated as a large sphere composed of hundreds or thousands of much tinier spheres. The same technique has been used to model cirrus cloud particles (ice crystals) on Earth

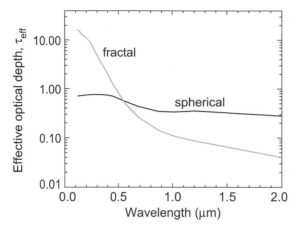

Figure 11.12 Effective optical depth versus wavelength for a hypothetical Archean organic haze layer, as calculated for spherical and fractal haze particles. See text for definition of the term "effective." (From Wolf and Toon (2010). Reproduced with permission. Copyright 2010, The American Association for the Advancement of Science.)

(Grenfell and Warren, 1999). This allows Mie theory to be used to describe scattering by the entire structure. The small spheres within the larger sphere interact much more strongly with short-wavelength radiation than would the large sphere itself.

These insights from the Titan and cirrus cloud literature have been applied to organic haze on the early Earth (Wolf and Toon, 2010). Recall from Sec. 11.2 that Sagan and Chyba (1997) had suggested that organic haze produced from methane photolysis could have provided a UV shield for ammonia, and that Pavlov *et al.* (2001) refuted this argument by showing that the haze should have become optically thick in the visible before the UV shield was fully developed. With fractal particles, however, the shielding mechanism becomes considerably more efficient. Figure 11.12 shows the effective optical depth of a representative haze layer for both spherical and fractal particles. At 0.2 μm, where NH_3 photolyzes, the ratio of UV optical depth to visible optical depth is about 15. By contrast, for spherical particles, this ratio is of order unity, because the particles are much larger than either of these wavelengths. The fractal haze in this particular model would still probably be too thick to be consistent with a warm Archean climate. However, a somewhat thinner fractal haze would have had only a small effect on climate and might still have provided an effective UV shield for ammonia.

That said, whether or not ammonia could have been present in appreciable concentrations on the early Earth depends on its sources and on whether there were other important sinks. The main source of ammonia may not

have been biological nitrogen fixation, as it is today. Today, organisms fix an estimated total of 110 Tg N yr^{-1}, or 3×10^{10} N atoms $cm^{-2}s^{-1}$ (Gruber and Sarmiento, 1997). Biological productivity on the Archean Earth before oxygenic photosynthesis is estimated to have been lower than today by a factor of roughly 100 (see Sec. 9.7), and rates of N fixation may have been scaled down correspondingly. But the abiotic nitrogen fixation mechanism described in Sec. 9.3.1 could have been relatively efficient. In this mechanism, N atoms flowing down from the ionosphere recombine with the byproducts of methane photolysis in the stratosphere to form HCN. The HCN diffuses downwards through the atmosphere into the oceans, where it hydrolyzes to form NH_3. The rate at which this happens depends on the amount of methane present and on the flux of solar EUV radiation that is capable of ionizing N_2. The most recent calculations suggest that the NH_3 production rate from this process should have been in the range of 6-30 Tg N yr^{-1}, which compares with a modern non-anthropogenic N fixation rate of ~ 200 Tg N yr^{-1} (Tian *et al.*, 2011)(see Sec. 7.2.3). If these estimates hold up, ammonia might have contributed several Kelvin of surface warming during the Archean, but only *after* methanogens began producing substantial amounts of methane and *if* this produced a haze with suitable UV shielding.

11.6.3 Effect of H_2 on Archean Climate

Sagan and Mullen (1972) presented an alternative evolutionary path for early Earth in which the planet began its life with a captured 1-bar H_2 atmosphere. They were aware that H_2 can absorb in the infrared (see Sec. 2.5.6), and they calculated that such a planet would have started out with a surface temperature well above 100 °C. Indeed, they discounted this model because it would likely have been too hot for life to have originated. However, Sagan (1977) later revived the idea of an H_2 greenhouse as a means to keep early Mars and early Earth warm under the faint young Sun; he noted that collision-induced dipole and forbidden quadrupole transitions give H_2 a high opacity in the thermal infrared.

Wordsworth and Pierrehumbert (2013) also suggested that high concentrations of H_2, about 10% by volume, could have made an important contribution to the greenhouse effect on the early Earth. Whether this was actually important depends on how fast hydrogen and other reduced gases were outgassed (as discussed in Ch. 7), on how fast hydrogen escaped to space, and the conversion of H_2 to CH_4 by the biosphere. The models discussed in Ch. 9 suggest that the H_2 mixing ratio on the prebiotic

Earth should have been of the order of 0.1% – much lower than the value assumed by Wordsworth and Pierrehumbert. We also showed in Ch. 9 that methanogens and/or anoxygenic photosynthetic bacteria would have converted much of Earth's H_2 into CH_4 once they evolved. So, it seems unlikely that H_2 was a major climate factor on Earth, except perhaps during its early prebiotic history. Greenhouse warming of early Mars with H_2 has also been suggested, as we discuss in Sec. 12.5.2.2.

11.7 The Gaia Hypothesis

An alternative concept for planetary climate stability is called the *Gaia hypothesis*, which "proposes that organisms contribute to self-regulating feedback mechanisms that have kept the Earth's surface environment stable and habitable for life" (Lenton, 1998). The idea was first advocated by Lovelock (1972) and has been followed by numerous papers (Lovelock, 1989; Margulis and Lovelock, 1974) and books (Lovelock, 1988; Lovelock, 1979, 1991). In one sense, the idea that Earth's climate has remained suitable for life for 3.5 billion years is a truism because otherwise we would not be here. Thus, it is not controversial to accept that negative feedbacks involving coupling of the biosphere to the inorganic components of the Earth system (such as the carbonate–silicate cycle) have operated over Earth history. For example, the climate regulation suggested in Sec. 11.6.1 by methane and an organic haze could be described as Gaian.

A key problem is that the Gaia hypothesis has been stated in different forms, including strong ones where the biosphere maintains "homeostasis at an optimum by and for the biosphere" (Lovelock and Watson, 1982). It is difficult to see how this version of Gaia can be true given evidence in the geologic record for huge changes in the climate and atmospheric composition. For example, although beneficial for some forms of life, the rise of atmospheric oxygen at 2.4 Ga greatly limited the future habitats of anaerobic microbes. Possible Snowball Earth events discussed in Sec. 11.6 and Sec. 11.10 are also hard to reconcile with a strong form of the Gaia hypothesis. Thus, although the Gaia hypothesis has stimulated debate, produced some interesting conjectures, and is popular in the media, it has remained at the fringes of mainstream professional science. Criticism of Gaia stems from its variable definition between strong and weak forms, questions about the testability of Gaia, and whether the hypothesis is actually needed at all (Free and Barton, 2007; Kirchner, 1989, 2002, 2003).

11.8 N₂, Barometric Pressure, and Climate

Thus far, we have ignored the major constituent of Earth's present atmosphere, which is N_2, molecular nitrogen. N_2 is not a greenhouse gas, meaning that it does not absorb or emit thermal-IR radiation effectively. But the presence of N_2 can increase absorption of thermal-IR radiation by greenhouse gases such as CO_2 and H_2O by helping to pressure-broaden their absorption lines (see Sec. 2.5.6). So, a high partial pressure of N_2 in the early atmosphere could have helped to compensate for the faint young Sun.

Goldblatt *et al.* (2009) have suggested that pN_2 was higher on the Archean Earth by a factor of 2 or more (although data that we discuss below cast doubt on this idea). Their argument is based partly on observations of Venus' atmosphere, which contains around three times as much N_2 as does Earth's atmosphere by mass (Table 11.1). Earth and Venus have similar amounts of

Table 11.1 The nitrogen inventory of Earth compared with Venus.*

Planet	Reservoir	Nitrogen (10^{18} kg N)	(bar N₂)
Venus	Atmosphere	10.9 ± 2.5	3.3 ± 0.7 (on Venus)
	Atmosphere scaled to Earth mass	13.4 ± 3.0	4.0 ± 0.9 (on Venus)
Earth	Atmosphere	4.0	0.78 (on Earth)
	Mantle	24 ± 16	
	Continental crust	1.7 ± 0.1	
	Oceanic crust	0.21 ± 0.07	
	Total	31 ± 20	6 ± 4 (on Earth)

* For calculating Venus' inventory, we used a total surface pressure of 93.3 bar at the mean radius of 6051.84 km and a mole fraction of N_2 of $3.5 \pm 0.8\%$. Other values are from Johnson and Goldblatt (2015).

carbon at their surfaces if we account for the carbon locked up in terrestrial carbonates, so Earth was probably originally endowed with substantially more nitrogen than is present in its atmosphere today. Goldblatt *et al.* argue that, over time, significant amounts of nitrogen were incorporated into sediments and subducted into the mantle. But when did this process occur, and how much N_2 was present in the early atmosphere? For the latter question, an evaluation of nitrogen reservoirs suggests that about 5 ± 3 bar of terrestrial nitrogen is non-atmospheric, making the total N inventory of Earth and Venus similar (Table 11.1). According to Goldblatt *et al.*, a doubling of pN_2 could have contributed 4–5 K of greenhouse warming in the late Archean. That is somewhat smaller than the expected contribution from CO_2 and reduced greenhouse gases, but it would help explain the climate record under a fainter Sun.

Recent constraints on Archean barometric pressure suggest that the pN_2 at 2.7–3.5 Ga was no greater and probably lower than present. One line of evidence is analysis of fossil raindrop imprints from 2.7 Ga (Som *et al.*, 2012). The size of raindrop craters depends on the terminal velocity of raindrops, which scales inversely with the square root of air density. Inferences from raindrop imprints are not especially precise, but they restrict the 2.7 Ga surface air density to <2.3 kg m^{-3} and probably less than 0.6–1.3 kg m^{-3} compared with today's 1.2 kg m^{-3}. This suggests a total pressure less than ~ 0.5–1.1 bar (Som *et al.*, 2012).

A second line of evidence concerns the size of vesicles (i.e., frozen gas bubbles) in lavas from ~ 2.7 Ga. The size of vesicles at the top and bottom of a basaltic lava flow is constrained by air pressure at the top of the flow and lithostatic plus air pressure at the base (Sahagian and Maus, 1994), which allows air pressure to be inferred. Analysis of vesicles in 2.7 Ga basaltic lavas erupted at sea level implies a remarkably low sea-level barometric pressure of <0.5 bar (Som *et al.*, 2016).

Finally, a third line of evidence is the $N_2/^{36}Ar$ ratio from fluid inclusions trapped in 3–3.5 Ga hydrothermal quartz (Marty *et al.*, 2013). The inferred isotopic composition of atmospheric N_2 from this data is similar to today's, while the partial pressure is lower than 1.1 bar, possibly lower than 0.5 bar.

If the empirical results are correct, it would imply that nitrogen accumulated in the crust and mantle in the early Archean or even Hadean, given that biological nitrogen fixation likely evolved early (Raymond *et al.*, 2004), at least by 3.2 Ga (Stüeken *et al.*, 2015c) or even by 3.8 Ga (Stüeken, 2016). Thus, enhanced pressure-broadening might have helped counter the faint young sun in the

early Archean or Hadean, but not during the late Archean. The data also imply that pN_2 must have risen later. Perhaps N_2 levels increased after the Great Oxidation Event (GOE) ~ 2.4 Ga because oxidative weathering of organic material should have released nitrate and, after rapid denitrification, the input flux of N_2 to the atmosphere would be larger than before the GOE (Sec. 10.8.3). The timescale over which N_2 may have increased would be coincident with evidence for a pulse of oxygen to the atmosphere around 2.2–1.9 Ga (Sec. 10.4.1).

11.9 The Warm and Stable Mid-Proterozoic Climate

The problem of the Archean climate has been debated vigorously, because solar luminosity was lowest then and because atmospheric composition was so different from today. But the Proterozoic climate is also a worthy topic of discussion. As Fig. 9.1 shows, the Proterozoic began with a series of glaciations from 2.4–2.2 Ga, and it ended with a series of deep, possibly Snowball Earth glaciations around 0.7–0.6 Ga. Until just recently, the time period in between was considered to have been uniformly warm and ice-free. But, there are now reports of a glaciation at 1.8–1.9 Ga and another at 1.1–1.3 Ga (see Sec. 11.3.1), so this story is no longer secure. This may actually make the mid-Proterozoic somewhat easier to understand, as this putative climate stability had always been considered puzzling when compared to the variable Phanerozoic climate (see Section 11.10).

The carbon isotope record is also stable during the mid-Proterozoic (Fig. 8.2), so that the period 1.8–0.8 Ga has been called *the Boring Billion* (Holland, 2006) (see 10.9.1). But, if glaciations were occurring sporadically during this time interval, maybe it wasn't really that boring after all. In fact, during the Boring Billion, eukaryotes appear to have diversified and, in red algae, the first evidence for sex is found (Butterfield, 2000).

11.9.1 Greenhouse Warming by CH_4

Mid-Proterozoic warmth and (maybe) stability may well be related to the redox evolution described in the previous chapter (see Section 10.9). Compared with today, much larger portions of the deep oceans are thought to have been anoxic and either sulfidic or ferruginous (see Sec. 10.3). Under such conditions, the processes by which organic matter decayed in sediments should have been significantly different from today. These contrasts are illustrated in Fig. 11.13. Panel (a) shows the modern ocean, which contains roughly 200 μM dissolved

(a) Modern ocean

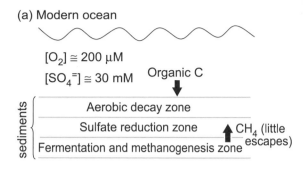

$[O_2] \cong 200 \ \mu M$

$[SO_4^=] \cong 30 \ mM$ Organic C

sediments
- Aerobic decay zone
- Sulfate reduction zone   → CH₄ (little
- Fermentation and methanogenesis zone   escapes)

(b) Proterozoic "Canfield ocean"

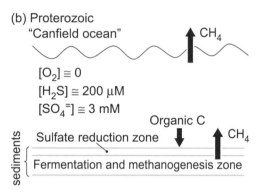

CH₄

$[O_2] \cong 0$

$[H_2S] \cong 200 \ \mu M$

$[SO_4^=] \cong 3 \ mM$

Organic C   CH₄

sediments
- Sulfate reduction zone
- Fermentation and methanogenesis zone   CH₄

Figure 11.13 Cartoon comparing the modern ocean/sediment system with the hypothesized Proterozoic ocean/sediment system. In Proterozoic sediments organic matter may have decayed largely by fermentation and methanogenesis.

O_2 and 30 mM sulfate. The sediments underlying the modern ocean typically include three different regimes where organic matter is decaying: a topmost *aerobic decay* layer ($CH_2O + O_2 \rightarrow CO_2 + H_2O$), a layer below that in which *sulfate reduction* occurs ($2 \ CH_2O + SO_4^{2-} \rightarrow 2 \ CO_2 + 2 \ H_2O + S^{2-}$), and a deeper layer in which decay occurs by *fermentation and methanogenesis* ($2 \ CH_2O \rightarrow CO_2 + CH_4$). The current global methane flux generated in this bottommost layer is estimated to be a few tens of Tmol CH_4 yr^{-1} (Catling *et al.*, 2007), which can be compared to the total global flux of methane from all sources that actually reaches the modern atmosphere of 31–38 Tmol CH_4 yr^{-1} (Prather *et al.*, 2001). However, virtually none of the methane flux in modern ocean sediments makes its way into the air, and so the modern ocean is a negligible source of atmospheric methane. The reason is that methane diffusing up from below must pass through the sulfate reduction zone and the aerobic decay zone before it can escape. But microbes within these zones consume methane by reacting it with either sulfate or O_2 (Hinrichs *et al.*, 1999).

By contrast, the Proterozoic deep oceans, or at least some Proterozoic ocean basins, contained no dissolved O_2. Sulfate concentrations were also only about one-tenth

that of today, based on trace sulfate measurements in carbonates (Fike *et al.*, 2015; Hurtgen *et al.*, 2002; Kah *et al.*, 2004). Also, rapid fluctuations in marine $\delta^{34}S$ values would not have been possible if the marine sulfate reservoir was large (Anbar and Knoll, 2002; Shen *et al.*, 2002). Thus, much of the organic matter that was deposited in marine sediments at this time must have decayed by fermentation and methanogenesis. For these reasons, Catling *et al.* (2002) first suggested that methane was probably an important greenhouse gas not just in the Archean but also in the mid-Proterozoic. The amount of organic matter that decomposed in Proterozoic marine sediments could have been quite large, as marine primary productivity was probably within a factor of 10 of today's value (Pavlov *et al.*, 2003), while the amount of organic carbon that was buried was probably no higher than today. (The evidence for this comes from the constant, near-zero $\delta^{13}C$ values of marine carbonates during this time (Fig. 8.2).) A simple calculation shows that the methane flux from marine sediments during the Proterozoic could have been 10–20 times higher than the global methane flux today (Pavlov *et al.*, 2003).[2] Proterozoic methane was coming from the oceans, though, whereas modern methane comes mostly from the land surface.

The climatic effect of this increased methane flux is even higher than one would think, because the photochemistry of CH_4 is nonlinear, even in an O_2-rich atmosphere. The main sink for CH_4 in today's troposphere is its reaction with OH (see Sec. 3.3.2):

$$CH_4 + OH \rightarrow CH_3 + H_2O \tag{11.20}$$

But reaction (11.20) is also a major sink for OH; hence, the concentration of OH decreases as the methane flux increases (Chameides and Walker, 1975; Pavlov *et al.*, 2003; Prather, 1996; Thompson and Cicerone, 1986). According to Pavlov *et al.* (2003), a factor of 10 increase in the methane flux would produce an increase in the methane mixing ratio by a factor of 60. Hence, instead of the paltry 1.8 ppmv of CH_4 that exists in today's atmosphere, the Proterozoic atmosphere might have had a CH_4 mixing ratio of 100 ppmv – enough to produce 6–7 K of greenhouse warming (Fig. 11.14(a)). The warming shown here is a factor of 2 lower than predicted

---

[2] Bjerrum and Canfield (2011, Supp. Info.) predict a much smaller Proterozoic methane flux (20–30 Tmol CH_4 yr^{-1}), or about double the preanthropogenic methane flux. Their much lower estimate for CH_4 production from marine sediments is based on modeling by Habicht *et al.* (2002). Two other models also produce low Proterozoic methane fluxes (Daines and Lenton, 2016; Olson *et al.*, 2016). How biology is parameterized is a key uncertainty.

by Pavlov *et al.* because we have corrected the mistake, mentioned earlier, in their methane absorption coefficients.

11.9.2 Greenhouse Warming by N_2O

Nitrous oxide, N_2O (or "laughing gas"), may also have been relatively abundant in the Proterozoic atmosphere. This idea was suggested by Buick (2007a), who called it the "laughing gas greenhouse." N_2O is a minor greenhouse gas today (0.3 ppmv), which contributes a few tenths of a degree to surface warming. Its source is almost entirely from *denitrification* by anaerobic bacteria. In denitrification, nitrite or nitrate is reduced to either N_2O or N_2, with associated release of energy that fuels the metabolism of these bacteria. Today, most denitrification proceeds all the way to the terminal end product, N_2. But about 5%–10% of the total denitrification flux is released as N_2O. However, Buick pointed out that the final step of denitrification, from N_2O to N_2, requires the metalloenzyme NOS (nitric oxide synthase), which contains 12 Cu atoms at its active sites. Cu is relatively abundant in today's oxidized oceans, but it reacts readily with sulfide, and hence would have been largely stripped out of the euxinic parts of a Proterozoic ocean of the type suggested by Poulton and Canfield (2011) (see Sec. 10.9). Hence, most denitrification at that time may have been incomplete, and the resulting biogenic N_2O flux might have been higher than today by a factor of 10–20.

The main loss process for N_2O is photolysis, so its photochemistry is not nearly as nonlinear as that of CH_4. Nevertheless, a factor of 10–20 increase in N_2O could have raised its concentration by the same amount, providing an additional 3–5 K of greenhouse warming (Fig. 11.14(b)). This, by itself, probably cannot explain the warm mid-Proterozoic climate. But when added to the effect of CH_4, the combined warming of greenhouse gases during the Proterozoic is substantial.

11.10 The Neoproterozoic "Snowball Earth" Episodes

As mentioned earlier, the Proterozoic was warm during its middle part but extremely cold on both ends. Indeed, both the Paleoproterozoic and Neoproterozoic glaciations are thought by some geologists to have been "Snowball Earth" episodes, in which the oceans froze over entirely (Evans *et al.*, 1997; Hoffman *et al.*, 1998; Hoffman and Schrag, 2002; Kirschvink, 1992). A putative 1.8-Ga glaciation from the King Leopold Formation of the Kimberley Basin, in western Australia (Schmidt and Williams, 2008; Williams, 2005) and a ~1.9 Ga glaciation from central Sweden

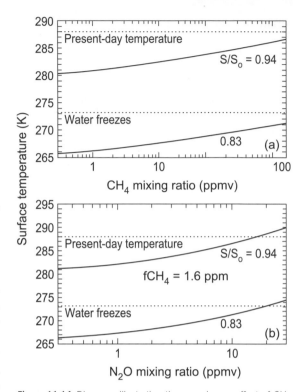

Figure 11.14 Diagrams illustrating the greenhouse effect of CH_4 (a) and N_2O (b) on Proterozoic mean surface temperatures. Calculations were performed for solar luminosities of 94% and 83% of today's value, corresponding to ages of 700 Ma and 2.3 Ga, respectively (Gough, 1981). A fixed CO_2 mixing ratio of 320 ppmv was assumed. (From Roberson *et al.* (2011). Reproduced with permission. Copyright 2011, John Wiley and Sons.)

(Kuipers *et al.*, 2013) are also both reported to be low latitude, but a Snowball event around this time remains doubtful due to the geographical sparseness of these rocks. For the other periods, the Snowball Earth Hypothesis (SEH) continues to be extensively debated in the geological and climatological literature (Pierrehumbert *et al.*, 2011).

Here, we focus on five issues. (1) What is the basic evidence for Snowball events? (2) Are there alternative models for the glaciations? (3) What triggered them? (4) How did the climate system recover from them? And, (5) how did photosynthetic life survive these events?

11.10.1 Geologic Evidence for Snowball Earth

The evidence for Snowball glaciations has been compiled by numerous geologists, notably Joseph Kirschvink, who coined the name *Snowball Earth* (Kirschvink, 1992), and Paul Hoffman, who led the team of geologists and geochemists that gave credibility to the hypothesis (Hoffman *et al.*, 1998; Hoffman and Schrag, 2002). The initial concept originated much earlier, however. In the 1960s, Brian Harland championed the idea of low-latitude Neoproterozoic glaciation, demonstrating that glacial deposits could be found on all seven continents (e.g., Harland and Rudwick, 1964). He tried using paleomagnetic studies to support a low-latitude location, but at the time such measurements were unreliable. Three decades later, Kirschvink and his then-student, Dawn Sumner, using better equipment, succeeded at this effort. They took samples of laminated sediments from the Elatina Formation of Southeast Australia (~650 Ma). These laminations were initially interpreted by the Australian geologist George Williams as sedimentation modulated by the solar cycle (Williams, 1981). Williams (1989) later changed his interpretation of these deposits to tidal rhythmites, but their close association with other glaciogenic deposits remained secure. (*Rhythmites* are successive sedimentary layers with clear periodic frequencies.) Australia at that time was situated directly on the equator, so the presence of glacial sediments was surprising. Williams had noted this puzzle previously (see, e.g. Williams, 1975) and had explained it by postulating that the Earth at that time had a very high obliquity (70°), which made the equatorial regions colder than the poles. Earth's present obliquity is 23.5°, which is safely within the regime where the poles are colder than the equator. The crossover occurs at an obliquity of ~54° (Ward, 1974).

The Elatina rhythmites contained detrital hematite that had a field of natural remnant magnetization (NRM) oriented parallel to the bedding plane of the sediments (Embleton and Williams, 1986). The presence of this signal in detrital hematite suggested strongly that the NRM was primary, i.e., it was not induced by some later remagnetization event. Kirschvink and Sumner, and later others (Schmidt and Williams, 1995; Schmidt *et al.*, 1991; Sohl *et al.*, 1999), demonstrated that these deposits also passed the "fold test", i.e., the magnetic field vectors followed folds in the sediments induced by later tectonic activity. Indeed, the folds formed while the sediments were still unlithified (unsolidified), indicating a tight constraint on the age of the NRM. This reinforced the conclusion that the paleomagnetism was primary. Because Earth's (surface) magnetic field is predominantly a dipole field aligned more or less with the geographic poles, horizontally oriented magnetic field vectors indicate that the deposits formed near the equator. So, their analysis confirmed claims that a glacial climate existed near Earth's equator at this time. Other glacial indicators are found in Australia, as well, including dropstones in marine sediments. (*Dropstones* are isolated fragments of rock in finely bedded sedimentary rocks, which fell into the sediments from melting icebergs, or ice shelves,

originating from the glaciers.) Along with the Elatina rhythmites, a marine environment justified the idea that sea ice was present at low latitude.

Hoffman and his colleagues added several important pieces to this story. They studied glacial diamictites in Namibia, which also formed near the equator. Here, and elsewhere during this time, the glacial deposits are covered with thick layers of carbonate rocks. Normally, carbonates form in warm waters because the solubility of calcium carbonate is lower at high temperatures. Namibian carbonates did not seem to fit this pattern. Furthermore, the bottommost few meters of the carbonate layers, termed *cap carbonates*, had a fine texture, indicating rapid deposition. Hoffman *et al.* argued that the cap carbonates were formed from CO_2 that accumulated in the atmosphere during the Snowball Earth events. This, of course, is just what would be expected based on the carbonate–silicate cycle feedback discussed earlier in Sec. 11.4. With the oceans completely frozen over, silicate weathering should have ceased almost entirely during these episodes, and volcanic CO_2 should have accumulated in the atmosphere, eventually creating enough greenhouse effect to melt the ice (see Sec. 11.3.4).

One other piece of geological evidence for Snowball Earth deserves mention here because it ties back to indicators for the rise of O_2 story mentioned in the previous chapter. Banded iron formations, or BIFs, reappear briefly during the Neoproterozoic, always in association with glacial deposits (Hoffman and Schrag, 2002; Kirschvink, 1992). BIFs, as we saw earlier, require an anoxic deep ocean to transport dissolved ferrous iron long distances from the midocean ridges (Sec. 10.3.2). If the ocean had been largely cut off from exchange with the atmosphere, dissolved O_2 should have been quickly depleted underneath the ice layer. The supply of sulfide must also have been lowered, or else sulfide would have titrated out the iron needed to form the BIFs. Various mechanisms have been proposed to do this, most of them associated in some way with glaciation (Canfield and Raiswell, 1999; Kump and Seyfried, 2005; Mikucki *et al.*, 2009). So, the reappearance of BIFs during the Neoproterozoic is consistent with the Snowball Earth hypothesis, but does not by itself require it to be correct.

11.10.2 Alternative Models to Explain Low-Latitude Glaciation

Not all researchers are convinced that the Snowball Earth Hypothesis (SEH) is correct. Indeed, probably the majority of geologists question whether the evidence supports a completely frozen Earth because the perceived consequences for life are very severe and because the sedimentology allows or requires open water (Chumakov, 2008). However, in general, it is difficult to tell what part of a Neoproterozoic glacial period a glacial sequence represents. As in Quaternary glacial sequences, the bulk of surviving evidence dates from the end of the glaciation. So evidence for open water does not necessarily rule out the SEH.

Nevertheless, skepticism has led to the models in which land areas near the equator were glaciated but the tropical oceans remained ice-free (Baum and Crowley, 2001; Crowley et al., 2001; Hyde et al., 2000). Schrag and Hoffman (2001) refer to this as the *slushball* or *soft Snowball* hypothesis, while Pierrehumbert et al. (2011) prefer the term *waterbelt state*. The sea–ice line in these models was at 25–30° latitude. The same is true for the model of Voigt et al., but this model does not predict ice in the tropics (Voigt et al., 2011). More recently, Abbot and colleagues have described a so-called *Jormungand* climate state, in which the sea–ice line is closer to 10–20° latitude – effectively a "narrow waterbelt" state (Abbot et al., 2011; Rodehacke et al., 2013). (The term Jormungand refers to a mythical Norwegian sea serpent that encircled the Earth and grasped his own tail.) The key to the narrow waterbelt is that the net evaporation in the descent region of the Hadley cell keeps subtropical ice bare with low albedo, unlike high-albedo snow-covered ice at high latitudes. Other recent climate models have produced ice lines near 50° latitude (Ferreira et al., 2011; Rose and Marshall, 2009), but these models are inconsistent with geologic evidence of tropical ice.

Two specific aspects of the SEH that have been questioned are whether the hydrological cycle remained active and the timing of the events. Neoproterozoic open marine deposits sandwiched between glacial ones have suggested that there were multiple glacial advance-retreat intervals rather than hard Snowballs (Allen and Etienne, 2008; Condon et al., 2002; Leather et al., 2002). In the Flinders Range of South Australia, hummocky cross-bed deposits found between diamictite and ice-rafted debris require open water because they are the signature of turbulent oceans caused by storms (Le Heron et al., 2010). But perhaps these glacial deposits post-date the collapse of tropical sea glaciers. Also, Neoproterozoic glaciogenic rocks in South Australia are overlain by thick mudstones (~1 km) that contain abundant dropstones (Young and Gostin, 1989), which may suggest a slowly warming climate with persistent floating ice rather than the rapid ~10^4 year meltback of the SEH. The main problem is that the duration of the mudstone deposition is unconstrained. However, geomagnetic reversals within cap carbonates

(Kilner et al., 2005; Raub et al., 2007; Trindade et al., 2003) suggest a timescale that approached ~10^6 years rather than the 10^3–10^4 years meltback of the SEH.

Some geologists also question the SEH origin of Neoproterozoic BIFs. They argue that the distribution of Neoproterozoic BIFs is far sparser than that of glaciogenic rocks, whereas the SEH might predict global BIFs. In cases, such as the Rapitan Group, North America, BIFs occur below rather than above diamictites. Yet the SEH suggests deposition of iron oxides after the melting of ice (Kirschvink, 1992) unless sub-ice meltwater discharge at ice-sheet grounding lines delivered O_2 from air bubbles in the ice. Because Neoproterozoic BIFs occur in rift basins (Chumakov, 2008; Young, 2002), some favor an older hypothesis that such BIF resulted from hydrothermal activity in restricted basins, similar to the Red Sea, where glaciers descended to sea level (Yeo, 1981). However, the rare earth element chemistry of BIFs in the Rapitan Group, North America, suggests that any hydrothermal input was highly diluted (Klein and Beukes, 1993). Also, dropstones and faceted pebbles occur in the Rapitan iron formation itself, consistent with glaciomarine conditions.

Post-glacial conditions have also been questioned. Chemical analysis of post-glacial Paleoproterozoic mudstones of the upper Gowganda Formation (Fig. 11.9) show an upward increase in chemical weathering (Young and Nesbitt, 1999), similar to the end of a Pleistocene-style glaciation when the climate warmed and weathering accelerated. Young and Nesbitt argue that Snowball Earth post-glacial siliciclastic sediments should show the opposite weathering trend, with initially extreme weathering from a high CO_2 atmosphere followed by less intensity. However, an upward increase in chemical weathering may just track progressive soil development rather than atmospheric CO_2. Also, the timescale of deposition of the mudstones is poorly constrained and there is no paleomagnetic evidence to indicate that the Gowganda glaciation occurred at low latitude.

Various alternative models have been suggested for the origin of cap carbonates. One possibility concerns differences in carbonate chemistry between the Phanerozoic oceans and the Precambrian oceans (Ridgwell et al., 2003). Carbonate-shell-forming metazoans originated in the Cambrian, and calcareous algae in the Mesozoic; thus, carbonate formation in the Proterozoic oceans was restricted to shallow shelf areas, creating a greater feedback between atmospheric CO_2 levels and carbonate ion concentrations in surface water. Whether or not this mechanism can account for the observed mean thickness of cap carbonates, ~18 m (Hoffman and Li, 2009), is unclear. The predicted cap thicknesses in Ridgwell et al. (2003) are only 0.8–2.1 m. Other suggestions include the

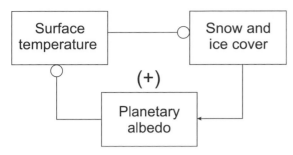

Figure 11.15 Diagram illustrating the snow/ice–albedo feedback loop.

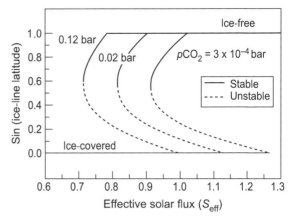

Figure 11.16 Diagram illustrating climate relationships for a hard Snowball Earth. Solid curves show the extent of the ice line (the equatorward extent of the polar caps) as a function of effective solar flux, $S_{eff} = S/S_o$, and CO_2 partial pressure. Dashed curves show regions where the climate system is unstable, indicating that global glaciation should occur. (From Caldeira and Kasting (1992b).)

idea that atmospheric oxygen that dissolved in the cold glacial oceans would have remineralized considerable organic carbon, creating a CO_2-charged ocean that was a precursor to cap carbonates (Peltier *et al.*, 2007). Alternatively, cap carbonates are posited to originate from CO_2 produced by the oxidation of methane released from clathrates that were destabilized during post-glacial sea-level rise (Kennedy *et al.*, 2008). However, carbon isotopic evidence of methane is lacking in cap carbonates, except for two locations in South China where the isotopically depleted carbon in late-stage cement is apparently hydrothermal (Bristow *et al.*, 2011).

11.10.3 Triggering a Snowball Earth

One of the great mysteries of Snowball Earth episodes is the question of what triggers them. In one sense, the answer is easy: ice–albedo feedback (Budyko, 1969; Sellers, 1969). As the climate cools, the polar ice caps extend further and further towards the equator. Ice and snow are highly reflective, so the albedo increases, causing the planet to absorb less and less solar radiation. This creates a positive feedback loop (Fig. 11.15), which is known to have been a key factor in amplifying the glacial-to-interglacial cycles of the Pleistocene Epoch (Lorius *et al.*, 1990). As long as the ice caps remain sufficiently small, this feedback does not destabilize the climate system. But if the edge of the ice caps reaches roughly 25° latitude, in simple models at least, ice albedo feedback takes over and the oceans should freeze entirely (Fig. 11.16). The calculations shown here are from Caldeira and Kasting (1992b), but they are similar to those carried out originally by Budyko and Sellers. All of these calculations were performed using energy-balance climate models (EBMs), the theory of which is well described by North (1975). In more complicated global climate models, in which ice albedo is allowed to vary with latitude, a stable waterbelt state may exist (Abbot *et al.*, 2011; Voigt and Abbot, 2012). This solution is

similar in some ways to the "thin-ice" solution, which we discuss in Sec. 11.10.5. In general, GCMs show that sea-ice dynamics can promote Snowball Earth episodes by equatorward movement of ice, while meridional heat transport by the atmosphere and ocean, along with the latitude-dependent albedo of ice, act to prevent a Snowball (Voigt and Abbot, 2012; Yang *et al.*, 2012a, b).

To trigger the ice–albedo catastrophe, something needs to make the climate cold enough to allow the polar caps to extend down to 25° latitude. Hoffman *et al.* (1998) suggested that this climatic cooling was initiated by drawdown of atmospheric CO_2 caused by increased burial of organic carbon. The evidence for this hypothesis was positive $\delta^{13}C$ values in carbonates found below the glacial deposits. As we saw in Sec. 8.5, positive $\delta^{13}C$ values in carbonates correspond to increased burial of isotopically light organic carbon. In Hoffman's model, the increase in organic carbon burial was caused, at least in part, by the creation in new continental shelf area during the breakup of the Rodinia supercontinent, starting at ~800 Ma. The carbon isotope story is complicated, though, because in many locations the $\delta^{13}C$ values decline sharply *below* the actual glacial horizons, indicating either that the marine ecosystem had already collapsed prior to the glaciation or that isotopically light carbon, possibly methane, was being added to the system (Hoffman and Schrag, 2002). Tziperman *et al.* (2011) hypothesize a mechanism involving enhanced export of organic matter (driven by diversification of eukaryotes) and its anoxic remineralization in the deep ocean by sulfate or iron-reducing bacteria. The process lowers

CO_2 by promoting carbonate deposition, and may be consistent with the isotopic record. While this hypothesis is speculative, it has testable consequences.

If CO_2 drawdown was relatively slow, the above hypotheses have a problem: why did the carbonate–silicate cycle not compensate for this effect? Normally, one would expect that silicate weathering should have slowed down as the global climate cooled. That, after all, is the basis of the solution to the faint young Sun problem described in Sec. 11.4. Proponents of geological drivers of CO_2 drawdown note that the continents were clustered at low latitudes (Donnadieu et al., 2004; Marshall et al., 1988). In that case, continental weathering might have continued unabated, even as the ice caps extended down to lower and lower latitudes. Other tectonic/geologic factors could have enhanced CO_2 drawdown. The breakup of Rodinia was evidently accompanied by the eruption of large, basaltic provinces, particularly the Laurentian traps, dated at 780 Ma. Laurentia was in the dry subtropics at this time, but it drifted into the much wetter tropics over the next few tens of millions of years (Godderis et al., 2007; Godderis et al., 2003). These latter authors have suggested that rapid weathering of these traps might have lowered atmospheric CO_2 enough to trigger the Sturtian glaciation, even though the traps themselves must have emitted CO_2. But, if CO_2 drawdown was indeed the trigger, the carbonate–silicate negative feedback must not have happened on the Neoproteozoic Earth, or else it happened so slowly that it was overwhelmed by faster processes, perhaps those associated with the organic carbon cycle.

An alternative way of triggering a Snowball Earth is to have a sudden decrease in greenhouse gases other than CO_2. For example, Schrag et al. (2002) suggested that large quantities of submarine methane hydrates released their CH_4 prior to the glaciations, causing transient climatic warming. This increased silicate weathering, thereby pulling down atmospheric CO_2. When the clathrates were exhausted, there was too little CO_2 remaining in the atmosphere to keep the climate warm, and so the climate fell into a Snowball state. This mechanism probably does not work, because the postulated methane fluxes from clathrate decomposition are too small to have created a significant climate perturbation (Pavlov et al., 2003). But the idea of using alternative greenhouse gases to trigger Snowball Earth is a good one. We have discussed how the concentrations of both CH_4 and N_2O may have been elevated during the mid-Proterozoic because of large gas fluxes originating from euxinic oceans. All that is needed, then, is for these fluxes to have been shut off relatively quickly during the Neoproterozoic, and the climate might indeed have been thrown into a Snowball state. An

obvious way to do this would be if atmospheric O_2 were to have increased suddenly near the end of the Proterozoic, as suggested in Sec. 10.9. This could have caused the deep ocean basins to become oxidized, thereby greatly reducing the sources for both CH_4 and N_2O. So, a Neoproterozoic rise in O_2 might account for the extreme climate of that time, in addition to paving the way for the rise of animal life. This hypothesis could conceivably be tested by exploring its implications for isotope fractionation in carbon, sulfur, and nitrogen.

11.10.4 Recovery from Snowball Earth

As mentioned earlier, the problem of recovery from a Snowball Earth has been studied with numerical models. Caldeira and Kasting (1992b) used an EBM to show that high concentrations of atmospheric CO_2, built up in response to continued volcanism on a frozen Earth, could have deglaciated a frozen planet even if the snow-covered surface was very clean (albedo = 0.663). Their model included parameterized thermal-IR fluxes from the 1-D model of Kasting and Ackerman (1986), as modified by Kasting (1991). They reported that a CO_2 partial pressure of 0.12 bar could have deglaciated the planet, given the modern solar flux. At the present subaerial volcanic outgassing rate of CO_2 of 5.5×10^{12} mol yr^{-1} (Table 7.1), this would have required ~ 4 million years. Indeed, the Caldeira and Kasting estimate assumed present solar luminosity, so the required CO_2 amount and corresponding recovery time for a Neoproterozoic Earth could have been even greater.

Other researchers (Le Hir et al., 2010; Pierrehumbert, 2004) have reported difficulties in deglaciating a frozen planet by way of CO_2 buildup. Both of these studies used 3-D climate models, which should in principle be more accurate than the Caldeira and Kasting EBM. However, both models assumed a fixed 1-bar surface pressure, and so the lack of increased pressure broadening at high CO_2 partial pressures may have influenced the results. Indeed, it can be readily shown by using 1-D climate models that deglaciation of a hard Snowball is possible when the buildup in surface pressure is accounted for, although it could require up to 0.4 bar of CO_2 to do so if the ice remained clean.

Constraints on pCO_2 during deglaciation may be possible from geochemical proxies. The stable isotopes of oxygen (^{16}O, ^{17}O, ^{18}O) potentially provide the information. Atmospheric ozone is enriched in ^{17}O. Thus, $O(^1D)$ that originates from the photolysis of ozone can transfer oxygen to CO_2 via the exchange reaction $O(^1D) + CO_2 \rightarrow CO_3^* \rightarrow O(^3P) + CO_2$, where CO_3^* is an excited intermediate. The oxygen isotope composition is measured as $\Delta^{17}O = \delta^{17}O - 0.528\delta^{18}O$, in which $\delta = (R_{sample}/R_{standard}) -1$, where R is the ratio of $^{18}O/^{16}O$, or $^{17}O/^{16}O$. The net result of photochemical reactions is that CO_2 ends up with positive $\Delta^{17}O$ (i.e., enriched in ^{17}O from ozone) while mass balance requires that O_2 has negative $\Delta^{17}O$. Bao et al. (2008) find that, whereas marine sulfate from the Phanerozoic generally has negative $\Delta^{17}O$, there is a negative $\Delta^{17}O$ spike in barite ($BaSO_4$) contained within 635 Ma cap carbonates from South China. Sulfate (SO_4^{2-}) derives from weathering of continental pyrites (S^-) using O_2 dissolved in rainwater, so the ^{17}O negative anomaly is assumed to derive from O_2 in a high pCO_2 atmosphere. The higher pCO_2 implies a greater reservoir of ^{17}O-enriched CO_2, and a correspondingly more negative $\Delta^{17}O$ for O_2. Bao et al. estimate that $pCO_2 \cong 0.012$ bar or 40 PAL, which falls well short of the 0.1–0.4 bar levels discussed above. A similar analysis on sulfate extracted from carbonates within a 635 Ma glacial diamictite suite from Svalbard resulted in even more negative $\Delta^{17}O$, which, if interpreted in the same way, would give $pCO_2 \sim 270$ PAL (Bao et al., 2009). However, the pCO_2 depends on many uncertain parameters, such as the size of the O_2 reservoir and stratosphere-troposphere fluxes of O_2, so these results should be viewed as semi-quantitative.

Another insight into pCO_2 comes from proxies for ocean pH. Boron exists as trigonal $B(OH)_3$ and tetrahedral $B(OH)_4^-$ species in a ratio of $\sim$4:1 in modern seawater. But boron speciation is pH-dependent, and boron isotopes are thus pH-modulated by an exchange reaction (Hemming and Honisch, 2007):

$$^{11}B(OH)_3 + {}^{10}B(OH)_4^- \rightarrow {}^{11}B(OH)_3 + {}^{11}B(OH)_4^-$$

(11.21)

Inferences of pH from boron isotopes in Namibia suggest a largely constant ocean pH and no significantly elevated pCO_2 in the post-glacial period after the older (Sturtian) Neoproterozoic Snowball glaciation, whereas a decrease in pH marks the deglaciation of the younger (Marinoan) Snowball and is compatible with elevated post-glacial pCO_2 (Kasemann et al., 2010).

Calcium isotopes also have excursions after the glaciations and are related to weathering and pH (Kasemann et al., 2005; Silva-Tamayo et al., 2010). In the modern ocean, isotopic fractionation during carbonate precipitation keeps seawater $\delta^{44}Ca$ heavier, at 1.9‰, than an input of $\sim$0.8‰ mainly from continental weathering. A negative excursion to $\sim$0‰ followed by a rise to $\sim$1‰ is found in post-Sturtian carbonates, whereas after the Marinoan, $\delta^{44}Ca$ oscillates from 0 to 2‰ before returning to $\sim$1‰. Such variability in marine $\delta^{44}Ca$ exceeds that in the Phanerozoic. It records an imbalance between Ca input and output, as would accompany large changes in weathering and/or seawater pH.

11.10.5 Survival of the Photosynthetic Biota: the Thin-Ice Model and Narrow Waterbelt State

The salient feature of the Neoproterozoic Snowball Earth episodes from an astrobiological standpoint is that life, including photosynthetic life, made it through these catastrophes. Indeed, Kunzmann et al. (2013) report that primary productivity returned to normal values shortly after the end of the glaciations, based on evidence from Zn isotopes, which are fractionated during biological Zn assimilation. That said, biodiversity may well have been severely impacted in the short term by a "bottleneck" during the Neoproterozoic that may have paved the way biologically for the explosion in biodiversity that began just before the Cambrian (Hoffman and Schrag, 2002; Knoll, 2003). Indeed, Boyle et al. (2007) argue that extreme ecological isolation created by Snowball conditions may have fostered altruism, and hence the evolution of multicellularity. However, evidence for a direct cause-and-effect relationship between the glaciations and the first evidence for multicellularity remains unclear. On the one hand, biomarkers for sponges occur by $\sim$635 Ma, i.e., before the Marinoan glacial sequence (Love et al., 2009). On the other hand, metazoan embryos within diapause cysts (which are dormant forms of embryos that develop in response to adverse environmental conditions) appear at 632.5 ± 0.5 Ma, about 3 m.y. after the termination of the Marinoan glaciation at 635 Ma (Yin et al., 2007). We note also that if the Snowball episodes were triggered by an O_2 rise, as suggested above, it is difficult to know whether the evolutionary breakthroughs were caused by the glaciation or by the rise in O_2.

Irrespective of the answer to this question, just the fact that photosynthetic life survived at all is remarkable, as the sea ice thickness on the type of hard Snowball Earth envisioned by Hoffman's group is at least a half a kilometer everywhere. Fortunately, the Snowball Earth model has variants that could help to resolve the biological issues, and perhaps some of the geological ones, as well. Two such models that look different superficially, but are actually closely related, are the thin-ice model of McKay (2000) and Pollard and Kasting (2005), and the Jormungand model of Abbot et al. (2011). These models both take advantage of the fact that the albedo of snow-free ice is much lower than that of snow-covered ice, especially if the ice is thin so that the ocean can be seen through it. Snow-free ice is expected in the tropics, even in the hard

Snowball model, because the weak hydrological cycle should result in a net excess of evaporation over precipitation at low latitudes. If the surface albedo is low, and if meridional heat transport is sufficiently slow, then open water or thin ice can be maintained in the tropics even if the rest of the planet's surface is frozen solid.

The Jormungand state is the easiest to understand, because the tropical oceans remain ice free up to 5–15° latitude. Open water has a very low albedo, and the flux of sunlight hitting the tropics is large, so the possibility that open water could exist can be demonstrated analytically (Abbot *et al.*, 2011).

In the thin-ice model, by contrast, tropical temperatures are slightly cooler, so that ice forms at the ocean surface. The ice then thickens until the flux of sunlight transmitted through the ice equals the heat flux, F, that can be carried back up through it by conduction. According to Fourier's law of heat transfer

$$F = k \cdot \frac{\Delta T}{\Delta z} \tag{11.22}$$

Here, Δz is the ice thickness, ΔT is the temperature difference between the top and bottom of the ice layer, and k is the thermal conductivity (~ 2 W m^{-1} K^{-1}). As the bottom of the ice is in contact with seawater, the temperature at the base of the ice is approximately 0 °C, or -2°C if one corrects for salinity. In the tropics, the surface temperature of the thin-ice model is about -25 °C (Pollard and Kasting, 2005); thus, the temperature difference between the top and bottom of the ice layer, ΔT, is about 23 K. Now, assume for the moment that 10% of the sunlight hitting the surface makes it through the ice. The globally averaged solar flux, after accounting for 30% reflection by clouds, is 240 W m^{-2}. Average insolation at the equator is 20% higher than this, or roughly 300 W m^{-2}, so 10% of this gives a transmitted solar flux of 30 W m^{-2}. Plugging this value into eq. (11.22) and solving for Δz yields an ice thickness of ~ 1.5 m. That is thin enough to allow 10% of the sunlight to penetrate, if that the ice is clear and bubble-free (McKay, 2000); thus, the solution is self-consistent. Thin, clear ice also has a low albedo because one can see the water beneath it, so the analytical analysis of Abbot *et al.* (2011) applies to this class of models, as well.

If, on the other hand, the ice is thick, as in the hard Snowball model, then eq. (11.22) still applies, but the relevant value of F is the geothermal heat flux, ~ 0.09 W m^{-2}. The surface temperature is also somewhat lower, about -33 °C in the model of Hyde *et al.* (2000). Solving for Δz then yields an ice thickness of just under 700 m. This is thick enough to completely block any incident sunlight, so the question of how photosynthetic

organisms survived is immediately raised. Hoffman argues that there would have been cracks in the ice where sunlight might penetrate, particularly in the subtropics where flow rates of floating ice are estimated to be maximal (Li and Pierrehumbert, 2011). Examples, include channels in sea ice, polynyas (pockets of open water found in modern pack ice), tidal cracks along ice grounding lines, transient meltwater ponds, and perhaps most plausibly, shallow hot springs around volcanic islands, like Iceland. Continental hot springs like those in Yellowstone in the USA would not have made good refugia, as they are supplied by meteoric water, and hence would quickly have dried out under Snowball Earth conditions. Sea glacier flow, discussed further below, might also have helped open up cracks through which sunlight might penetrate.

Thus, while the hard Snowball model may be consistent with the survival of photosynthetic life, a Jormungand or thin-ice Snowball Earth would ensure this automatically. These models should also deglaciate more easily than a hard Snowball, because the surface is always somewhat darker. The thin-ice model of Pollard and Kasting (2005) deglaciates at about 30–60 PAL of CO_2, which is consistent with the oxygen isotope evidence mentioned in the previous section. Deglaciation can also be facilitated if the ice gets dusty (Abbot and Pierrehumbert, 2010), although this process may be self-limiting because the ice surface should cleanse itself as the climate becomes warmer and precipitation increases.

None of these models is generally accepted. Both the Jormungand model and the thin-ice model encounter potential problems from flowing *sea glaciers* – thick sea ice formed near the poles that flows equatorward at speeds of hundreds of meters per year (Goodman and Pierrehumbert, 2003; Voigt and Abbot, 2012). In the Goodman and Pierrehumbert model, which assumes an entirely ocean-covered Earth, this sea glacier flow prevents thin ice from remaining stable at the equator. Other calculations indicate that the results depend on the parameters chosen for the model, particularly the albedo of ice and snow (Goodman, 2006; Pollard and Kasting, 2006; Warren and Brandt, 2006). The albedo of snow goes down rapidly when it gets dusty, while the albedo of ice increases rapidly if it gets bubbly. So, it is not clear at this point whether or not thin ice could have been maintained near the equator in the open oceans. That said, even if the ice cover was thick in most places, open water or thin ice might still have existed regionally in some low-latitude, long narrow sea, analogous to the modern Red Sea, where the flow of sea glaciers from higher latitudes would have been impeded by friction and sublimation (Campbell *et al.*, 2011). In this case, the algae might have had a refugium, but the amount of CO_2 needed to deglaciate would have been similar to the hard Snowball case.

11.11 Phanerozoic Climate Variations

This brings us finally to the Phanerozoic Eon, 541 Ma to present. Although it represents just over a tenth of the Earth's history, far more has been written about the biological and climate evolution of the Phanerozoic than the Precambrian. The reason, of course, is that we have a much better geologic record in the Phanerozoic.

Here, we will not attempt to do justice to extensive research on Phanerozoic climate. Instead, we refer the reader to books and reviews by Frakes (1979), Imbrie and Imbrie (1979), Crowley and North (1991), Crowell (1999), Berner (2004, 2006a), and Beerling and Royer (2011). During most of this time, Berner and others have argued, convincingly, that climate has been driven largely by variations in atmospheric CO_2, along with gradually increasing solar luminosity (Royer *et al.*, 2007). The CO_2 variations were caused by changes in plate tectonics, including formation and breakup of supercontinents, and by biological innovations such as the emergence of life on land (which increased silicate weathering rates and also led to coal formation). Another important factor influencing climate was changes in paleogeography. The opening of the Drake Passage at 35 Ma, mentioned earlier in this chapter, was one such paleogeographic change that may well have helped initiate glaciation on Antarctica.

Phanerozoic climate is marked by three great Ice Ages: in increasing age, the Late Cenozoic (0–35 Ma), the Permo-Carboniferous glaciation (260–340 Ma), and the Late Ordovician glaciation (443–445 Ma). These glacial episodes are indicated by the shaded areas in Fig. 11.17. The curves shown in the figure are three different model calculations of the atmospheric CO_2 in units of Present Atmospheric Level, PAL, actually meaning the preindustrial level of 280 ppmv. According to Berner's calculations, atmospheric CO_2 levels could have been as much as 50 times higher than this during the early Phanerozoic. The high CO_2 levels predicted at this time are a response to changes in solar luminosity and the silicate weathering rate feedback on CO_2 described earlier in this chapter. Berner's models include both the organic and inorganic carbon cycles and are driven by many of the climate forcing mechanisms mentioned in the preceding paragraph. Encouragingly, atmospheric CO_2 levels predicted by Berner's calculations tend to be low during glacial periods, as one would expect if changes in CO_2 were driving the glaciations. Berner's models include a large number of parameters, though, and so it is hard to

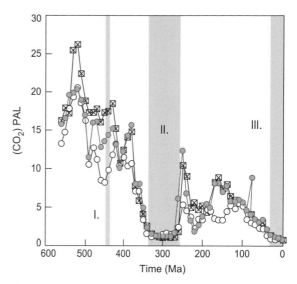

Figure 11.17 Diagram showing calculated values of the atmospheric CO_2 level relative to Present Atmospheric Level, PAL, equivalent to the pre-industrial level of 280 ppmv. (From Berner (2006a). Reproduced with permission from Elsevier. Copyright 2006.)

determine to what extent this agreement was achieved by simply adjusting these parameters.

If Berner's model is correct, then the Permo-Carboniferous glaciation was largely a result of the growth of forests during the Carboniferous Period (359–299 Ma) and the consequent sequestration of CO_2 in coal beds. The warm Mesozoic Era, 250–67 Ma, was caused by increased seafloor spreading rates (leading to increased volcanic release of CO_2), along with lower emergent continental area. These factors are linked because higher rates of seafloor spreading lead to higher sea level. The young, warm oceanic crust is less dense, midocean ridges rise, and the ocean volume spreads upward relative to continents (isostasy). Some of the highest sea levels and most extensive shelf seas are in the Cretaceous Period (145–66 Ma) at the end of the Mesozoic. The cooling during the latter part of the Cenozoic Era, 66 Ma up to the present, has been caused by decreases in seafloor spreading rates, possibly aided by the collision of India with Asia, starting at about 40 Ma (Broecker, 2015). This increased silicate weathering rates by exposing fresh rock in the Himalayan Mountains and creating a monsoonal circulation that dumped rain on them and helped them to erode (Raymo and Ruddiman, 1992). One needs to state this last hypothesis carefully, as the rate of weathering cannot exceed the rate at which CO_2 is supplied from volcanism (Caldeira *et al.*, 1993). But this does not contradict the hypothesis, as atmospheric CO_2 levels and global temperatures should have

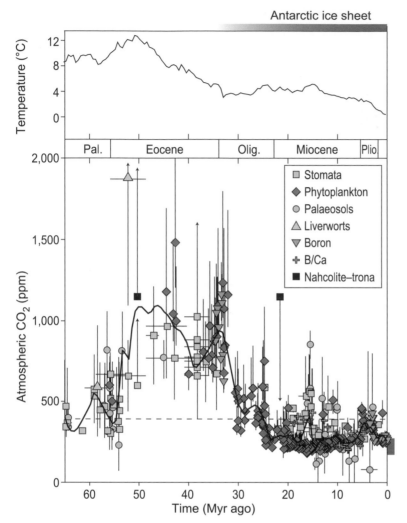

Figure 11.18 The Cenozoic history of atmospheric CO_2 and climate. Deep-sea temperatures (upper panel) generally track the estimates of atmospheric CO_2 (lower panel) reconstructed from terrestrial and marine proxies. Errors represent reported uncertainties. Symbols with arrows indicate either upper or lower limits. The vertical grey bar on the lower right axis indicates the glacial–interglacial CO_2 range from ice cores. The top blue bar indicates ice-sheet development on Antarctica. Horizontal dashed line indicates the atmospheric CO_2 concentration (390 ppm) in 2011. (From Beerling and Royer (2011). Reproduced with permission from Macmillan Publishers Ltd. Copyright 2011, Nature Publishing Group.) (A black and white version of this figure will appear in some formats. For the color version, please refer to the plate section.)

fallen in just such a way as to keep the carbonate–silicate cycle balanced. Thus, the negative feedback loop described in Section 11.4 appears to have been operating throughout the Phanerozoic.

Finally, a goal in understanding Phanerozoic climate history is to develop models that are consistent with isotopic records of temperature and with multi-proxy records of atmospheric CO_2. A large amount of work continues to be done in this area, and researchers are getting closer to self-consistent paleoclimate theories. Both the data and the models improve as one gets closer to the present. In particular, because most seafloor older than ~ 70 Ma has been subducted, ocean temperature estimates from O isotopes in benthic and planktonic foraminifera and other shell forming organisms only become available after that time.

Estimates for Cenozoic deep-sea temperatures (top panel) and CO_2 concentrations (bottom panel) are shown

in Fig. 11.18. Temperatures peaked during the Eocene Optimum, around ~ 50 Ma (there was an earlier Paleocene–Eocene Thermal Maximum, which was a 0.2 m.y. spike around 56 Ma, but not shown in the figure). Temperatures then declined over the next 15–20 million years, culminating in the onset of Antarctic glaciation around 35 Ma. During this time, atmospheric CO_2 fell from ~ 1000 ppmv to just over 500 ppmv and then to ~ 300 ppmv by the early Miocene, around 20 Ma. CO_2 levels in the current ~ 100 k.y. cycles of interglacial–glacial periods have been about 280 ppmv during interglacials. Because of the burning of fossil fuels, a mean annual CO_2 level of ~ 400 ppmv was reached in 2015, which may be higher than at any time during the past 20 million years. This, of course, has implications for future global warming. But we will leave those consequences for others to discuss (Stocker *et al.*, 2013) and for all of us to experience.

PART III

Atmospheres and Climates on Other Worlds

12 Mars

12.1 Introduction to Mars

The most interesting question about Mars is: Did it ever have life, and if not, why not? Because all Earth-like biology requires liquid water, the question of life on Mars is linked to whether Mars' climate once supported widespread liquid water on the planet's surface. Today, Mars is a cold, planet-wide desert, with a mean global emission temperature of –58 °C (215 K) (see Sec. 2.2.5), which is similar to the temperature of the Antarctic interior. But the ancient surface of Mars has dried-up river valley networks, including deltas at the mouths of some valleys (e.g., Fassett and Head, 2005). Also, minerals are detected such as clays that were produced when liquid water chemically altered volcanic rocks (e.g., Ehlmann *et al.*, 2013; Ehlmann and Edwards, 2014) and *in situ* sedimentology on the surface of Mars at the *Curiosity Rover* site suggests the presence of an ancient lake in Gale Crater (Grotzinger *et al.*, 2015b).

In the previous chapter, we discussed climate change over Earth's history and how to reconcile abundant liquid water on the early Earth with a faint young Sun. The same problem applies to Mars when we try to understand early (>3.7 Ga) geomorphic and geochemical signs of liquid water. But the severity of the problem is greater for Mars than for Earth because Mars is an average of 1.5 times farther from the Sun, and inherently colder. In fact, exactly how the early atmosphere on Mars produced a different environment and the nature of that environment – whether persistently or intermittently wetter – remain open questions.

An early age produced *fluvial* (stream-related) erosion, which made *valley networks*, i.e., valleys that spread out in dendritic branches with tributaries like terrestrial river valleys. But since then, Mars has likely been predominantly cold and dry. Geological features on the surface, such as large outflow channels, which are more recent than the valley networks, were probably formed by mechanisms that did not depend upon a warm climate. Small-scale channels have also formed on steep slopes in the geologically recent past (Dundas *et al.*, 2012; Malin *et al.*, 2006), but these do not require a different climate from the modern one. Seasonal streaks with spectral evidence of hydrated salts even occur in the present cold climate (Ojha *et al.*, 2015),

In this chapter, we review the geology and atmosphere of Mars. We are particularly concerned with long-term changes in the quantities of volatiles, i.e., substances that tend to form gases or vapors at the temperature of a planet's surface. We also examine the evidence for a different climate on early Mars, especially the hypothesis known as *a warm, wet early Mars*. While there is wide agreement that early Mars was different, one idea is that the early climate was more conducive to the presence of liquid water on the surface than today's climate.

Another relevant issue is that throughout its history, Mars has undergone large variations in its orbital parameters analogous to the Milankovitch cycles that paced the terrestrial ice ages. On Mars, such cycles had much larger amplitudes. The influence of orbital variability will also be discussed. Finally, we note that the evidence suggests that there has been considerable modification of the Martian surface by wind erosion and deposition, which complicates the interpretation of Mars' climate history.

12.1.1 Overview of Mars

Mars is often said to be the most Earth-like planet in the Solar System. This statement is true in some respects but in other ways Mars' geophysics and atmosphere are very different from Earth's. Early Venus might arguably be a better candidate for the most Earth-like status.

The smallness of Mars accounts for many key geophysical differences with Earth. Mars is roughly half the diameter and one-ninth the mass of Earth. Because Mars is small, it cooled relatively quickly after it formed, so that its surface heat flow declined from an estimated 60–70 mW m^{-2} at 4.4 Ga to $\leq$ ~10–20 m W m^{-2} today (Hauck and Phillips, 2002; Ruiz *et al.*, 2011). Mars has no active volcanoes, but it does have many geologically young lava flows, so the planet is probably volcanically dormant today. Upper limits on atmospheric SO$_2$ of 0.3 ppb (Krasnopolsky, 2012) and HCl of <0.2 ppb (Hartogh *et al.*, 2010), which are expected volcanic gases, corroborate the lack of current volcanism. Furthermore, no thermal hotspots have been detected from orbital infrared mapping. However, Mars has extinct volcanoes, including the largest in the Solar System, Olympus Mons. Age relationships indicate that there was prolonged volcanism that became less intense and more sporadic after about 3.5 Ga (Werner, 2009). The overall morphology of Martian volcanoes is similar to that of hotspot shield volcanoes on Earth, such as Mauna Loa in Hawaii. The Martian crust's mineralogy is also dominated by basalt (Bandfield, 2002).

Like the Earth, Mars is differentiated into core, mantle, and crust; but, unlike the Earth, Mars has no intrinsic global magnetic field. In the absence of seismic data, the size of the core is poorly defined, but its radius is estimated to be 1300–1500 km (Stevenson, 2001). Although Mars lacks a magnetic field today, surface magnetic anomalies in ancient terrain suggest that Mars had one in the past. When old parts of the crust formed, their minerals captured the direction of a contemporaneous global field (Acuna *et al.*, 1999). Generally, it is thought that the magnetic dynamo on Mars only operated prior to ~4.1–3.9 Ga, which is the age of the Hellas basin (Werner, 2008). Ancient impact basins of this age or younger commonly have no magnetic anomalies, which can be explained if impacts occurred in the absence of a magnetic field and destroyed pre-existing magnetization. However, a minority view is that the magnetic field turned on late, after ~4 Ga, and that the magnetic anomalies were caused by magmatic intrusions after the large impact basins formed (Schubert *et al.*, 2000). In particular, some younger magnetic anomalies exist, such as Apollinaris Mons (8°S, 174°E), dated at ~3.7 Ga (Schubert and Soderlund, 2011). One observational constraint is that around ~3.9 Ga, the magnetic field could have been comparable in strength to Earth's modern field if remnant magnetism in minerals embedded in carbonate globules of the ALH84001 Martian meteorite captured a global field rather than a local one (Weiss *et al.*, 2008).

One of the most striking features of Mars is the difference between the northern and southern hemispheres (Fig. 12.1), which is evident in elevation, crustal thickness,

Figure 12.1 A map of the topography of Mars, using Mars Orbiter Laser Altimeter (MOLA) data. (Source: NASA/MOLA Team.) (A black and white version of this figure will appear in some formats. For the color version, please refer to the plate section.)

and crater density. Northern lowlands cover about one-third of the planet and are ~4 km lower on average than the heavily cratered highlands elsewhere. (In Fig. 13.6, this dichotomy is displayed in roughly bimodal hypsometry.) Although the northern lowlands are superficially less cratered than the southern highlands, a comparably dense number of craters in the northern plains appear to be buried at shallow depth (Frey, 2006a). The hemispheric dichotomy is also reflected in estimates of the thickness of the crust from topographic and gravity data. Gravity data suggest that the global average crustal thickness is ~45 km. In the northern lowlands the mean thickness is ~30 km, while in the south it is ~60 km (Neumann et al., 2004).

Excavation associated with a giant impact is one popular mechanism thought to have caused the dichotomy, and initially an impact was proposed for the northern hemisphere (Marinova et al., 2008; Nimmo et al., 2008; Wilhelms and Squyres, 1984). However, the impacted hemisphere could be the southern hemisphere, because the impact melt thickens the crust and produces a basin centered at the antipode (Leone et al., 2014; Reese et al., 2010). Subsequent extraction of heat in the impacted hemisphere by mantle convection may have generated a long-lived mantle plume that built the Tharsis volcanic province (Golabek et al., 2011). Tharsis is a ~10 km-high equatorial bulge in the overall shape of Mars centered around 265° E.

An issue for atmospheric evolution is volatile recycling by tectonics. Mars currently lacks plate tectonics and has shield volcanoes, which we associate with plume volcanism. Sleep (1994) proposed that the smooth northern plains may have been created by early plate tectonics. This cannot have happened in the Late Noachian or later, as originally proposed, because an older sub-surface exists below the present surface (Frey et al., 2002). However, an earlier episode of plate tectonics would have avoided massive internal melting on early Mars and such cooling would have helped drive a dynamo and magnetic field (Nimmo and Stevenson, 2000). Some models switch from early plate tectonics (mobile lid) to later stagnant lid mantle convection (Breuer and Spohn, 2003; Lenardic et al., 2004). But evidence for Earth-like plate tectonics on Mars (linear mountain chains, oceanic trenches, rifts at spreading centers, and very folded rocks) is scant (e.g., Nimmo and Tanaka, 2005). Elongated magnetic anomalies in the southern hemisphere of Mars are much wider than more symmetrical terrestrial lineations caused by seafloor spreading, and may be due to intrusion of dikes (Nimmo, 2000) rather than plate tectonics (Fairen et al., 2002).

The atmosphere of Mars (discussed further in Sec. 12.3), is thin and is composed mostly of carbon dioxide (Table 12.1). Equation (2.15) with Mars' Bond albedo of 0.25 (Pleskot and Miner, 1981) gives an effective temperature for Mars of 210 K. This would be the blackbody surface emission temperature if Mars lacked an atmosphere. The actual global mean surface temperature of emission is ~215 K (see Sec. 2.2.5, Haberle (2013)), so that Mars's thin, dry atmosphere provides only ~5 K of greenhouse warming. Mars' atmosphere is largely transparent to solar radiation, except to UV photons with wavelength <200 μm, which comprise only ~0.07% of total solar radiation. The greenhouse effect is small because the air is very dry. Without much water vapor, the air absorbs little thermal-IR radiation from the surface except within the 15 μm CO_2 band. Here, the atmosphere absorbs and radiates ~15% of the total surface radiation. An additional factor affecting the atmosphere's energy budget is dust, which absorbs thermal radiation emanating from Mars and absorbs and scatters incoming visible radiation from the Sun (Madeleine et al., 2011).

Phase changes from gas to solid and vice versa for CO_2 and H_2O occur as a result of temperature changes. The winter poles are so cold that the main atmospheric gas, CO_2, condenses into deposits 1–2 m thick. During the northern summer, this CO_2 ice sublimates (turns from solid to gas) from the north polar cap to expose underlying water ice, which reaches temperatures ~205 K. In contrast, the southern polar cap is covered in some CO_2 ice all year round. The reason for the southern cap's behavior is a regional climate with a permanent low-pressure zone that produces CO_2 snow, a high albedo, and a self-sustaining low temperature (Colaprete et al., 2005).

Because of the cold temperatures and low pressures, water generally consists of ice or vapor. Although surface temperatures range from 140 K to 310 K (Kieffer et al., 1992), temperatures above the freezing point of water occur only temporarily during summer late afternoon conditions and only in a desiccated upper layer of a few millimeters depth in low-latitude regions of low *thermal inertia*.[1] Such surfaces have loose material that warms and cools quickly. The pressure over much of highland surface is below the triple point of water of 611 Pa, or 6.11 mbar (Haberle et al., 2001), because the annual and spatial average surface pressure on Mars is only ~600 Pa (Haberle et al., 2008). Persistent liquid water on the surface of Mars today can only exist in the form of highly saline solutions or thin adsorbed layers on soil particles. No liquid water has been conclusively identified.

---

[1] Thermal inertia is $(\kappa \rho c)^{1/2}$, the square root of the product of thermal conductivity (κ), bulk density (ρ), and specific heat capacity (c). It represents the ability of a material to store and conduct heat.

Table 12.1 Basic planetary parameters for Mars and its present atmosphere.

Parameter	Value on Mars
Mass relative to Earth's 5.97×10^{24} kg	$0.107 \approx 1/9$
Mean radius relative to Earth's 6371 km	$0.532 \approx 1/2$
Orbital eccentricity	0.093
Semimajor axis of orbit (AU)	1.523 66 (ranges 1.3815 to 1.666)
Obliquity (°)	25.19
Mean orbital period (a Mars year)	669.60 Mars solar days (sols)
	686.98 Earth days $\approx$ 1.88 Earth years
Gravitational acceleration, equator (m s^{-2})	3.711
Mean solar day (noon-to-noon period, sol)	88 775.2 s $\approx$ 24.66 hr
Sidereal day (axial rotation period)	88 642.7 s $\approx$ 24.6229 hr
Mean atmospheric surface pressure[a] (Pa)	$\approx$ 600 Pa
Mean global surface emission temperature[b] (K)	215
Near-surface scale height[c] (km)	10.3
Atmospheric composition[d] (by volume) below 120 km	CO_2 95.7±1.6%, N_2 2.03±0.3%,
(ppm = parts per million	Ar 2.07±0.02%, O_2 0.173±0.006%, CO 749±26 ppm,
ppb = parts per billion)	H_2O 0.03% (varies), H_2 15±5 ppm, He 10 ppm, Ne 2.5
	ppm, Kr 0.3 ppm, Xe 0.08 ppm, O_3 0.04-0.2 ppm (varies),
	H_2O_2 0–40 ppb, NO <1.7 ppb, SO_2 <0.3 ppb, HCl <0.2
	ppb
Column dust content of the atmosphere[e]	0.1–5 visible optical depth

[a] Haberle et al. (2008). [b] Haberle (2013). [c] Leovy (2001). [d] Main constituents from Mars Science Lab results in Franz et al. (2015); see also Owen et al. (1977) and Krasnopolsky (2011), and references therein. [e] Kahn et al. (1992).

However, spectral images suggest that brines rich in perchlorate (ClO_4^-) salts were concentrated into patches at NASA's *Phoenix Lander* site (Cull et al., 2010). Salts such as $Mg(ClO_4)_2$ can depress the freezing point of brines down to eutectic temperatures of –67 °C to –64 °C (Pestova et al., 2005; Toner et al., 2015). Above the eutectic temperature, liquid formation is spontaneous if the salt is in contact with ice. In fact, slowly cooled $Mg(ClO_4)_2$ and $Ca(ClO_4)_2$ solutions remain liquid and solidify near ~153 K as amorphous solids (i.e., as glasses) (Toner et al., 2014a).

12.1.2 The Geologic Timescale for Mars

Later in this chapter, we discuss the evolution of the Martian atmosphere over geologic time. Martian geologic time is divided into the Pre-Noachian, Noachian, Hesperian and Amazonian. Each of these stages extends over hundreds of millions of years, so they are analogous to the eons on Earth: the Hadean, Archean, etc. (Table 12.2). We use the term "period" to describe these stages following Tanaka et al. (2014), who have mapped the ages of units on Mars, although in the literature the stages are variously called periods, eras, or epochs. Surfaces on

Mars are placed within each period according to the density of impact craters. Older surfaces have accumulated more impact craters. A Noachian surface, named after the Noachis region "type locality," is one that has accumulated at least 200 impact craters with diameters exceeding 5 km within an area of 10^6 km^2. Over the same area, a Hesperian surface, named after Hesperia Planum, is defined according to lower crater densities listed in Table 12.2. The Amazonian, which follows the Hesperian, refers to the most recent period with surfaces with the lowest cratering densities.

A Pre-Noachian period has also been defined, although the term was not used prior to 2006. The largest visible impact craters are all estimated to have formed after ~4.1 Ga, when the Hellas basin (Fig. 12.1) was excavated by a large impact. Before Hellas formed, huge visually obscure craters called Quasi-Circular Depressions (QCDs), such as Chryse and Acidalia, were created from very large impacts. These QCDs formed in the Pre-Noachian, while Hellas dates the base of the Noachian (Carr and Head, 2010; Frey, 2006b).

The absolute dates for Martian geologic time are estimated using knowledge from the Moon. Lunar craters have been dated from radioisotopes in rocks brought back

Table 12.2 The geologic timescale for Mars.

Time-stratigraphic system	Estimated age[a] (Ga)	Boundaries[b] (craters per 10^6 km^2)
Pre-Noachian	Before ~4.1	Before the formation of Hellas[c]
Noachian	~4.1 to 3.74–3.57	$\geq$200 craters $\geq$5 km diameter
Hesperian	~3.74–3.57 to 3.46–3.0	$\geq$2100 craters $\geq$1 km diameter[d]
Amazonian	3.46–3.0 to present	$\leq$2100 craters $\geq$1 km diameter

[a] Noachian–Hesperian and Hesperian-Amazonian ages were summarized by Hartmann and Neukum (2001) and updated by Werner and Tanaka (2011) [b] Crater density definitions as proposed by Tanaka (1986) and updated by Werner and Tanaka (2011).
[c] Frey (2006b) proposed that the formation of the Hellas impact basin should be the base of the Noachian, which has an age 4.1–3.9 Ga (Werner, 2008). [d] Werner and Tanaka (2011).

by Apollo astronauts. From these, a correlation between lunar cratering density and absolute age has been made. To extrapolate lunar ages to Mars, a correction is made for a Mars:Moon cratering rate ratio >1, which depends on various factors such as the gravitational focusing for Mars versus the Moon. From such methods, the Noachian is estimated to have ended at ~3.7–3.6 Ga, while the Hesperian-Amazonian boundary is placed at 3.46–3.0 Ga (Werner and Tanaka, 2011). In general, the southern highlands are Noachian, while the geologic units that have resurfaced the northern plains are Hesperian or Amazonian.

We also note that the three chronostratigraphic periods after the Pre-Noachian are divided into eight epochs: Early, Middle, and Late Noachian; Early and Late Hesperian, and Early, Middle, and Late Amazonian, with crater density designations given by Tanaka (1986).

12.1.3 The Basis of our Knowledge: Spacecraft Data and Martian Meteorites

Spacecraft have given us a scientific understanding of Mars, so no discussion of Mars can be complete without appreciating their key findings (Table 12.3). In brief, revelations began with the flyby of *Mariner 4* in 1965, which photographed a cratered, Moon-like surface. Analysis of radio-occultation data indicated a thin atmosphere (Kliore *et al.*, 1965), while later Earth-based spectroscopy suggested that a predominantly CO_2 atmosphere (Young, 1971). *Mariner 6* and *7* flybys in 1969 extended visual coverage from 1% to 10% of the planet, but again showed mainly old, cratered terrain.

Mariner 9 transformed our knowledge of Martian landforms. When *Mariner 9* went into orbit in 1971, there was a global dust storm, which persisted until the following year. But after the dust cleared, *Mariner 9* cameras revealed gigantic volcanoes, enormous canyons, layered sedimentary terrain, and intriguing channels and valley networks. The latter suggested fluvial erosion, and therefore a past when water flowed freely on the surface.

Following their launch in 1975, NASA's *Viking* spacecraft surpassed all previous Mars missions in the extent of their data (Snyder, 1979). *Viking* consisted of two identical spacecraft, each composed of an orbiter and a lander. The orbiters' purpose was to aid landing site selection and provide a global perspective. *Viking*'s main objective was to look for life using instruments on the landers. However, in retrospect, orbital data proved just as essential for understanding Mars. Cameras mapped the entire surface at 200 m resolution, infrared radiometry mapped the albedo and seasonal temperatures of the surface, and infrared spectroscopy measured atmospheric column water abundances for 1.5 Mars years.

To great disappointment, the Viking biology experiments failed to find life (**Box 12.1**). Partly because of this result, a hiatus in the exploration of Mars ensued until the mid-1990s. Since then, there have been numerous missions to Mars that have made many discoveries (Table 12.3). Key new findings include the presence of extensive subsurface ice at mid to high latitudes and the widespread occurrence of types of minerals that were produced by the action of past liquid water on volcanic rocks. The detection of past or present life, however, remains elusive and may be difficult at the surface given the degradation of organics by radiation (Pavlov *et al.*, 2012).

Another crucial discovery came in the 1980s, when it was discovered that some meteorites are from Mars. These pieces of rock were ejected from the shallow subsurface of Mars during impacts and, after going into orbit around the Sun, were captured by Earth's gravity. The analysis of Martian meteorites continues to provide a wide range of constraints on the evolution of Mars (Bogard *et al.*, 2001; McSween and McLennan, 2014). Roughly 60 000 meteorites exist in various collections

Table 12.3 Important spacecraft missions to Mars since the 1970s.

Mission	Operational at Mars	Most notable results
Mariner 9	1971–72	imaged volcanoes and apparent fluvial features, such as channels and valley networks
Viking (two Viking Orbiters (VOs), two Viking Landers (VLs)	1976–78/80 (VO-2, VO-1) 1976–79/82 (VL-2, VL-1)	VOs mapped the surface (visually to 200 m resolution) and atmosphere; VLs failed to find organics; VLs measured surface physical properties, soil elemental geochemistry, and atmospheric properties
Mars Pathfinder	1997	engineering demo; additional soil and rock elemental chemistry
Mars Global Surveyor (MGS)	1997–2006	topography, magnetic data, images down to 1.5 m pixel^{-1}; thermal infrared mineral mapping.
Mars Odyssey	2001–present*	mapped subsurface water ice in high latitudes; found halide salts
ESA Mars Express	2003–present*	revealed layered sulfates and clay minerals; measured hydrogen and oxygen ion escape rates
2 Mars Exploration Rovers (MERs): *Spirit* and *Opportunity*	2004–2010 (*Spirit*) 2004–present*	identified sulfate-rich sedimentary rocks, concretions indicative of groundwater, fossilized ripples from small streams, mud cracks, hydrated silica, and a carbonate-rich outcrop
Mars Reconnaissance Orbiter (MRO)	2006–present*	found outcrops of sulfates and carbonates and widespread clay minerals; ice exposed in new, mid-latitude craters. Images to 20 cm pixel^{-1}
Phoenix Lander	May–Nov. 2008	subsurface ice confirmed; first identification of soluble soil salts, including low eutectic perchlorate (ClO_4^-) salts
Curiosity Rover (Mars Science Laboratory)	2012–present*	found mudstone lakebed deposits, including clay minerals deposited in near-neutral pH fluid; found stream deposits; first *in situ* radiometric dating; first *in situ* detection of organics and nitrogen oxides (likely nitrates) in surface samples
Mars Atmosphere and Volatile EvolutioN (MAVEN) orbiter	2014–present*	enhanced ion escape and production of diffuse aurora by solar storms; *in situ* measurement of thermosphere; detection of infalling interplanetary dust particles
Mars Orbiter Mission (India)	2014–present*	yet to report major results*

* At present time of writing, 2016.

worldwide and, by 2016, over 100 samples were identified as coming from Mars,[2] although the number grows continually.

Martian meteorites were originally called SNC (pronounced "snik") meteorites after the locations where meteorites that are typical of three subdivisions were found. Shergotty in India is the type specimen of Martian meteorites called *shergottites*, Nakhla in Egypt gives its

name to the *nakhlites*, while Chassigny in France is one of a couple of *chassignites*. However, it now makes sense simply to call the whole group *Martian meteorites* because some of the meteorites do not fit within the SNC classification. Misfits include "Black Beauty" (NWA 7034), which is a *breccia* (i.e., composed of cemented fragments) from the southern highlands, and ALH84001, which crystallized at a much older age from molten rock than the SNCs.

Together, three lines of evidence prove that Martian meteorites are from Mars. First, such meteorites are

[2] In 2016, following the list of Professor A. J. Irving: http://www.imca.cc/mars/martian-meteorites-list.htm

Box 12.1 The Viking Lander biology experiments.

Three biology experiments looked for metabolic activity (Klein, 1979, 1998).

(1) The **Gas Exchange Experiment** (GEx) examined changes in the makeup of gases in a test chamber that would indicate biological activity after Martian soil was moistened with a solution containing organic material. Several gases were given off, but these are believed to have been atmospheric gases that had been adsorbed on the soil. O_2 was evolved, but this probably arose from decomposition of inorganic oxidants in the soil. Repeated addition of water resulted in decreased production of gases, unlike biological systems, where an increase would be expected.

(2) The **Labeled Release Experiment** (LR) aimed to detect the uptake of ^{14}C radioactively tagged liquid nutrient by life. $^{14}CO_2$ emitted by Martian life would show the tagging. Initial results showed an increase in $^{14}CO_2$, but this has generally been interpreted as a product of the oxidation of the organic compounds by soil oxidants (Oyama and Berdahl, 1979).

(3) The **Pyrolytic Release Experiment** (PR) (or Carbon Assimilation Experiment) incubated soil samples that had been exposed to $^{14}CO_2$ or ^{14}CO to see if carbon was fixed by organisms into organic compounds. A little fixation was found, but was not considered biological because carbon-fixation still occurred after heating to 175 °C. Iron-containing minerals are believed to have taken up the carbon.

A further experiment relevant to life was the **Gas Chromatograph Mass Spectrometer** (GCMS). This found no organic material in the soil to ppb levels. A plausible explanation is that organic compounds were absent because their destruction rate in Mars' present oxidizing environment exceeds their production rate. Alternatively, organics were destroyed by oxidation in the pyrolysis protocol required for the GCMS.

Overall, the experiments demonstrated the importance of understanding the inorganic chemistry of the environment before trying to look for life.

volcanic rocks with relatively young crystallization ages compared to the ~4.6 Ga age of the Solar System.[3] For example, the shergotittes are basalt that crystallized at ~150–570 Ma, the nakhlites all formed at ~1.3 Ga, NWA 7034 (paired with NWA 7533 and other samples) has zircons with radiometric ages of 4.4 Ga (Humayun et al., 2013), and ALH84001 formed at 4.1 Ga (Borg and Drake, 2005; Lapen et al., 2010). So, such extraterrestrial rocks must come from a celestial body that has produced lavas as recently as ~150 Ma. Mars is the prime candidate. We can compare the very youngest lavas on the Moon, which date from ~1 Ga according to crater counts.

Second, the silicate minerals in Martian meteorites possess unique triple oxygen isotope (^{16}O, ^{17}O, ^{18}O) compositions. Different bodies in the Solar System have dissimilar oxygen isotope ratios in their bulk rock depending on where they formed; thus, the fact that every Martian meteorite has the same ratio indicates a single parent body (see Ch. 6).

Finally, the definitive proof that the meteorites came from Mars is that atmospheric gases were trapped in the

meteorites. Specifically, gases that had been in the pores of the rocks on Mars were sealed into basaltic glass made from the shock of impact (Wiens and Pepin, 1988). (*Basaltic glass* is the amorphous product created when a basaltic melt cools rapidly). Figure 12.2 shows how trapped gases match the composition of the Martian atmosphere that was measured by the *Viking* landers (Becker and Pepin, 1984; Bogard and Johnson, 1983).

While there is no doubt that Martian meteorites are from Mars, trying to identify source craters for the meteorites from matches of orbital spectra to meteorite composition is inexact and remains a topic of debate (e.g., Ody et al., 2015; Werner et al., 2014).

12.2 The Present-Day Atmosphere and Climate of Mars

12.2.1 Composition and Thickness of the Present Atmosphere

The thinness of the Martian atmosphere was first securely established by the small refraction of *Mariner 4* radio signals that passed through the atmosphere during its 1965 flyby (Kliore et al., 1965). Measurements from pressure sensors on the *Viking Landers*, which landed on Mars in 1976, allow the global annual mean surface pressure to be estimated as ~600 Pa (Haberle et al., 2008).

---

[3] There has been controversy about the ages of Martian meteorites, with one research group claiming that the shergotittes have very old ages of 4.3 Ga and 4.1 Ga (Bouvier et al., 2009). However, this is disputed by most others (e.g., Bogard and Park, 2008) and we stick with the consensus.

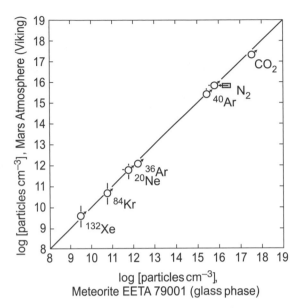

Figure 12.2 Gases trapped in the basaltic glass of the shergottite meteorite EETA 79001 match to those measured at ground level by the Viking landers over a range of nine orders of magnitude. (Adapted from Pepin (2006).)

This compares to $\sim 10^5$ Pa on Earth's surface. Because CO_2 condenses at the poles each winter, the surface pressure varies seasonally by $\sim 25\%$. The annual pressure cycle is very repeatable except for stochastic perturbations from planetary-scale dust storms that occur in some Martian years but not others.

Carbon dioxide dominates the atmosphere near $\sim 96\%$ by volume. The next major atmospheric constituents are N_2 and Ar, each $\sim 2\%$ (Table 12.1). The minor gases are a variable amount up to 0.1% of water vapor (H_2O), small concentrations of gases derived from the photodissociation of carbon dioxide (e.g., CO, O_2, and O_3), and trace amounts of noble gases, including neon (Ne), krypton (Kr), and xenon (Xe). Mass spectroscopy measurements made during the entry of the Viking probes show that bulk gases are well mixed up to ~ 120 km height.

Trace amounts of methane (CH_4) averaging around 10 ppbv have been reported (Formisano et al., 2004; Geminale et al., 2011; Krasnopolsky, 2012; Mumma et al., 2009; Webster et al., 2015). The potential significance of methane is that it could have geothermal or even biological sources (Krasnopolsky, 2006). However, reports that the methane is variable both spatially and on timescales of days to months are difficult to reconcile with the photochemical stability of CH_4, which should have a residence time of ~ 300 years and be well mixed by atmospheric turbulence (Lefevre and Forget, 2009; Zahnle et al., 2011). Consequently, skeptics question the validity of

models used to interpret the spectral data and suggest that an intermittent local source reported for MSL (Webster et al., 2015) is actually the rover itself (Zahnle, 2015).

12.2.2 Climate and Meteorology

Mars today has a climate that differs from Earth's primarily because of its thin, dry atmosphere combined with its greater distance from the Sun (Leovy, 2001). Because the atmosphere has little heat capacity, daily changes in the surface temperature of up to 80 K are observed – more than double the maximum range of ~ 30 K found in terrestrial deserts. The response of atmospheric temperatures to variations in planetary radiation is faster for Mars than Earth because the Martian atmosphere has a radiative time constant of ~ 2–3 sols (Goody and Belton, 1967), where a sol is one Martian solar day (Table 12.1). The radiative time constant is the $1/e$ decay time of a small thermal perturbation (Sec. 2.3.2). Mars also has seasons that are analogous to terrestrial ones because the tilt of the rotation axis of Mars to its orbital plane – the *obliquity* – is 25.2°, which is similar to the Earth's 23.45°. However, because the Martian year is about 1.9 Earth years, the seasons are longer. Perihelion is 31 sols before the northern winter solstice and the eccentricity of Mars' orbit means that the northern winter is warmer and shorter than the southern winter.

Suspended atmospheric dust strongly affects atmospheric and surface temperatures. The Martian atmosphere always has some airborne dust, as demonstrated by the fact that the visible optical depth (defined in Sec 2.4.2.2) never drops below ~ 0.1–0.2 (Kahn et al., 1992). This dust load is constantly replenished by dust devils (Whelley and Greeley, 2008) that are typically ~ 2 km tall but can occasionally extend throughout the daytime convective layer (Balme and Greeley, 2006; Fisher et al., 2005), even up to ~ 20 km height. During a Martian year, the atmosphere becomes dustier as perihelion and southern summer solstice approaches. The dust absorbs sunlight, which warms the air and can diminish the diurnal temperature changes on the surface to only ~ 5 K (Pollack et al., 1979).

Martian dust storms are initiated by strong winds that scour particles from the ground. Speeds above a typical threshold of ~ 30 m s^{-1} propel ~ 100 μm diameter particles into a bouncing motion along the surface called *saltation* (Greeley and Iversen, 1985). Saltating particles dislodge smaller particles and possibly acquire a negative charge relative to the ground, causing electrostatic loosening (Kok and Renno, 2009). Local dust storms with areas smaller than $\sim 10^6$ km^2 occur during all seasons, especially along the edges of retreating polar caps or in association

Box 12.2 Areocentric longitude: Seasons on Mars

Seasonal periods are described in terms of *areocentric longitude*, L_s, which is a measure of the orbital position of Mars. $L_s = 0°$ corresponds to spring equinox in the northern hemisphere (or southern hemisphere autumn equinox), while $L_s = 90°$, $180°$, and $270°$, are northern summer solstice, autumn equinox, and winter solstice, respectively. Perihelion occurs at $L_s = 250.87°$ in the current epoch, which is shortly before southern summer solstice.

with topography, such as Valles Marineris (Cantor *et al.*, 2001). Regional storms exceeding 10^6 km^2 in area appear during the dusty season, $L_s = 130\text{–}340°$ (See **Box 12.2**). Around perihelion ($L_s = 180\text{–}360°$), planet-encircling dust storms occur in some years but not others, averaging about once every 3 Mars years. They grow from regional dust storms and appear to be triggered by the increased solar insolation close to perihelion. It takes a couple of months for a return to nominal optical depths after a planet-encircling storm subsides.

The Coriolis force, polar condensation and sublimation of CO_2, and seasonal temperature variations control the general circulation of the atmosphere. (By *general circulation* we mean the large-scale wind systems and the associated temperature and pressure distribution when averaged around latitude circles and/or for periods long enough to include many daily cycles and individual weather systems, but still short compared to seasonal changes.) Orbital infrared remote sensing, supplemented by microwave remote sensing from the Earth and radio occultation, has provided three-dimensional distributions of atmospheric temperature. The thermal or gradient wind equations described in Sec. 4.1.2.2 can be used to calculate global winds from temperature distributions (Fig. 4.13), and numerical circulation models provide their physical interpretation (Haberle *et al.*, 1993b; Leovy and Mintz, 1969; Pollack *et al.*, 1990; Read and Lewis, 2003).

Both data and models indicate that a Hadley circulation occurs on Mars (Sec. 4.2.4). On the Earth, one Hadley cell dominates for most of the year, where warm air ascends in the summer hemisphere near the equator and moves towards the opposite hemisphere at upper levels, falling in the subtropics (Lindzen and Hou, 1988). Two Hadley cells occur briefly at the Earth's equinoxes, one in each hemisphere, both with ascending branches at the equator (Hartmann, 1994). Mars follows a similar pattern, except that the single cross-equatorial Hadley cell at the solstices is stronger, overturns faster, and is wider in latitude (Fig. 12.3). Also the higher topography in the south causes the southern summer Hadley cell to dominate the annual average (Richardson and Wilson, 2002). Near the surface, the Coriolis deflection of air causes trade winds similar to terrestrial counterparts.

At mid and high latitudes, a different circulation regime prevails, again similar to the Earth's with predominantly eastward winds. Wind speed increases with height and jet streams blow at 25–40 km altitude. On the Earth, the tropospheric jet streams are stronger in winter but blow perennially because oceans prevent a significant seasonal variability in the north–south temperature gradient that drives the jets. On Mars, the wintertime jet stream is nearly four times faster than its terrestrial counterparts, driven by the big temperature gradient between the CO_2 ice and adjacent bare ground. In the summer hemisphere, the meridional temperature gradient away from the residual ice cap is small, so the jet stream and associated storms disappear. In wintertime midlatitudes, however, the Martian atmosphere is strongly baroclinic (i.e., isotherms and isobars are strongly inclined to one another), which is an unstable state that leads to the growth of wavelike disturbances on top of the eastward flow (Sec. 4.3.5). In these baroclinic waves, crests (excursions towards the poles) are associated with regions of high pressure and troughs (excursions towards the equator) with low-pressure systems. *Viking Lander* measurements show that in the northern hemisphere, storms migrate eastward, following the path of the overlying jet. On Earth, midlatitude weather is difficult to predict more than four days in advance but Martian weather systems are remarkably repeatable. For example, a ~1500 km diameter, annular-shaped water ice cloud appears every Martian year, around $L_s \sim 120°$, in Vastitas Borealis to the south of the northern cap (Cantor *et al.*, 2002; Malin *et al.*, 2008).

One noticeable difference with the Earth is the strength of thermal tides. Thermal tides, which form a family of waves, are global oscillations of the atmosphere caused by day-night differences in solar insolation (Sec. 4.4.3). As Mars rotates, "migrating" thermal tides follow the apparent westward motion of the overhead Sun, but interactions with Mars' large-scale topography also generate "non-migrating" waves that do not. Unlike Earth, these tides are present as large oscillations in atmospheric temperature profiles, as measured by *Viking* (Seiff and Kirk, 1977) and *Phoenix* (Withers and Catling, 2010). Indeed, energy from thermal tides

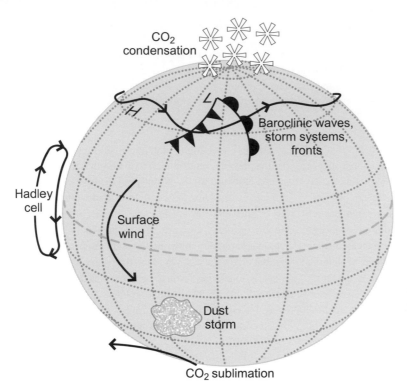

Figure 12.3 A schematic diagram showing the global circulation in northern winter on Mars, where two Hadley cells that exist during autumn give way to a single cross-equatorial cell. In winter mid-latitudes, a jet stream guides storm systems at the surface but in the summer hemisphere jets and storms vanish, while dust storms are prevalent.

significantly affects the circulation of Mars' middle atmosphere (Forbes *et al.*, 2004).

12.2.3 Atmospheric Chemistry

We have just considered the current Martian climate, but what controls the atmospheric composition? Fortunately, the photochemistry of Mars' atmosphere is simpler than that of either Earth or Venus. Essentially, the photolysis products of carbon dioxide and water dominate: CO, O, O_2, and O_3 from CO_2, and H, OH, HO_2, H_2O_2, and H_2 from H_2O. Of these, the most quantitatively significant are CO, O_2 and H_2, while H_2O_2 and O_3 are key minor species (Table 12.1). Products derived from N_2 (N, NO, and NO_2) have only trace abundance and relatively little influence.

A schematic overview of the photochemistry of the Martian atmosphere is shown in Fig. 12.4. CO_2 is unstable to photodissociation by solar UV photons that have wavelengths less than 225 nm. Photolysis of CO_2 gives rise to O, O_2, and O_3, shown in the boxes, as follows.

$$CO_2 + h\nu \rightarrow CO + \boxed{O} \tag{12.1}$$

$$O + O + M \rightarrow M + \boxed{O_2} \tag{12.2}$$

$$O + O_2 + M \rightarrow \boxed{O_3} + M \tag{12.3}$$

Here, M is any molecule that takes up the excess energy released when two atoms combine to form a single molecule. Because two O atoms from reaction (12.1) are needed to produce an O_2, we might naively expect O_2 and CO to exist in a 1:2 ratio. However, the observed O_2: CO abundance ratio is ~1.5:1. Also, if there were no mechanisms to reform CO_2, all Martian CO_2 would convert to CO and O_2 in ~3500 years. (A similar stability issue exists for CO_2 on Venus (Sec. 11.5)). Despite being present only at small levels, atmospheric water vapor and its photochemistry "save" the CO_2 from destruction, as we shall see below.

Photodissociation of water vapor provides the source of oxygen that accounts for why O_2 exceeds CO in abundance. The reaction sequence starts with:

$$H_2O + h\nu \ (\leq 200 \ nm) \rightarrow OH + H \tag{12.4}$$

$$H_2O + O(^1D) \rightarrow OH + OH \tag{12.5}$$

Here, O-singlet-D ($O(^1D)$) is a metastable excited state of the oxygen atom that originates from ozone photolysis. The products of water vapor photolysis give rise to the hydroperoxyl radical, HO_2, via the following reaction.

$$OH + O + M \rightarrow HO_2 + M \tag{12.6}$$

The species H, OH, and HO_2 (denoted HO_x) contain a single hydrogen atom and are called *odd hydrogen*

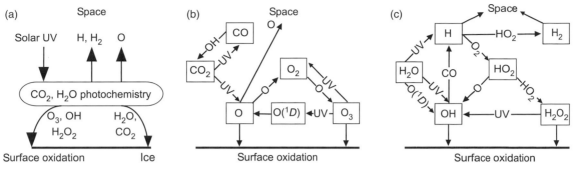

Figure 12.4 (a) Atmospheric chemical reactions that dominate on Mars. Oxygen either escapes to space or is lost to the surface primarily through ozone and hydrogen peroxide intermediaries. Reaction with surface minerals oxidizes them and this process over time has created the reddish color of Mars in iron oxides. (b) CO_2 photochemistry. Ultraviolet photons split CO_2, which produces oxygen atoms. Oxygen molecules and ozone are produced by subsequent reactions. (c) H_2O photochemistry. Water vapor is also spilt by solar ultraviolet to generate oxygen and hydrogen. The hydrogen ultimately escapes to space.

species. When these species react with each other, they can produce O_2 or hydrogen peroxide (H_2O_2).

$$H + HO_2 \rightarrow H_2O + O$$
$$\rightarrow H_2 + O_2 \quad (12.7)$$

$$HO_2 + HO_2 \rightarrow H_2O_2 + O_2 \quad (12.8)$$

$$HO_2 + OH \rightarrow H_2O + O_2 \quad (12.9)$$

Hydrogen peroxide has been detected on Mars and is an important oxidant. Microwave (Clancy *et al.*, 2004; Hartogh *et al.*, 2010) and infrared data (Encrenaz *et al.*, 2004; Encrenaz *et al.*, 2008) prove that H_2O_2 is present. Loss pathways of H_2O_2 include photodissociation ($H_2O_2 + hv \rightarrow OH + OH$), polar condensation, or dry deposition and oxidation of the surface. H_2O_2 was also proposed as a possible oxidant responsible for the results of the *Viking Lander* biology experiments (Hunten, 1979b). With a metal catalyst, H_2O_2 decomposes to O_2, which could account for O_2 evolved in the Labeled Release experiment (**Box 12.1**). H_2O_2 can also oxidize organics to CO_2 keeping the Martian surface devoid of organics (Zent and McKay, 1994) (**Box 12.1**).

The importance of odd hydrogen species (H, OH, and HO_2) on Mars is that they catalyze the recombination of CO and O in cycles such as the following (McElroy and Donahue, 1972; Parkinson and Hunten, 1972).

$$OH + CO \rightarrow CO_2 + H \quad (12.10)$$

$$H + O_2 + M \rightarrow HO_2 + M \quad (12.11)$$

$$\underline{HO_2 + O \rightarrow OH + O_2} \quad (12.12)$$

$$= CO + O \rightarrow CO_2 \quad \text{net} \quad (12.13)$$

Thus, CO_2 on Mars is regenerated, analogous to a proposed maintenance of CO_2 on Venus by H, Cl and N

catalysts (reviewed by Mills and Allen (2007)). Another aspect of this process is that O, CO, and O_2 are mixed downwards from the upper atmosphere so that they can associate and reform CO_2. The turbulent mixing is strong on Mars, probably because of the dissipation of atmospheric waves in a way that is relatively more important than on Earth (see Sec. 4.6.2).

We have assumed that water vapor is the source of all hydrogen species in the Martian atmosphere. This may be incorrect if there is geothermal H_2 or CH_4, or if a subsurface biosphere feeds on geothermal H_2 and exhales non-negligible quantities of CH_4. However, water vapor photolysis can explain the observed H_2 abundance of ~15 ppmv (Table 12.1) if balanced by hydrogen escape to space at the *diffusion-limited* rate described in Secs. 5.2.4 and 5.8–5.9 (Zahnle *et al.*, 2008).

Another effect of the water-derived odd hydrogen species (H, OH and HO_2) is that catalytic cycles control ozone concentrations, e.g.,

$$H + O_3 \rightarrow OH + O_2$$
$$\underline{O + OH \rightarrow H + O_2}$$
$$= O + O_3 \rightarrow O_2 + O_2 \quad \text{net} \quad (12.14)$$

This cycle with an H catalyst is analogous to the destruction of ozone by OH catalysis in the Earth's stratosphere (Sec. 3.3.1). On Mars, the result is that ozone often anti-correlates with the water vapor from which odd hydrogen (HO_x) derives (Barth *et al.*, 1973; Perrier *et al.*, 2006). Today, ozone is most abundant at the cold, winter pole where water vapor is low because of its tiny saturation vapor pressure. However, ozone levels remain rather constant in the tropics during aphelion ($L_s = 40$–$130°$), despite increasing water vapor abundance as northern spring progresses. A possible explanation is heterogeneous loss of odd hydrogen onto water ice in clouds (Krasnopolsky, 2006; Lefevre *et al.*, 2008).

In the past, atmospheric chemistry should have been important for producing salts on Mars. Volcanic gases containing S or Cl oxidize to form salts. For example, SO_2 and H_2S oxidize to form sulfate aerosols (Settle, 1979; Smith *et al.*, 2014). Mass-independent isotopic signatures, which are characteristic of photochemical reactions, show that sulfates produced in Mars' atmosphere have passed isotopic fractionation to both sulfate and sulfide phases found in Martian meteorites (Farquhar *et al.*, 2007; Farquhar *et al.*, 2000b; Franz *et al.*, 2014). The discovery of ~0.6 wt% of perchlorate in soil (ClO_4^-, probably as $KClO_4$, $NaClO_4$, or $Mg(ClO_4)_2$) by NASA's *Phoenix Lander* (Hecht *et al.*, 2009) led to the suggestion that perchlorate might have been produced globally in the past from atmospheric oxidation of chlorine-bearing gases as happens on Earth (Catling *et al.*, 2010). However, purely gas phase chemistry is insufficient because Mars' atmosphere is colder and contains less ozone than Earth's (Smith *et al.*, 2014). Gas–solid (heterogeneous) reactions are required and although suggestions exist (Carrier and Kounaves, 2015), the exact mechanism is yet to be proven. In any case, perchlorate is widespread, given its detection at the MSL site (Glavin *et al.*, 2013) and continuous generation is required given that perchlorate is decomposed over time by ionizing radiation (Quinn *et al.*, 2013).

Because perchlorate was likely present in *Viking Lander* soils, it might have been important for destroying organics. Experiments suggest that perchlorate would destroy organics during the pyrolysis protocol that was used by the *Viking Lander* for detecting organics (Navarro-Gonzalez *et al.*, 2010, 2011) (but see Biemann and Bada (2011) for a different view). When heated, perchlorate decomposes and oxidizes organic matter. The same principle is used in fireworks, in which O_2 from heated perchlorate combusts sulfur and charcoal. On Mars, reaction with perchlorate and chlorate is also a plausible source of chlorinated hydrocarbons detected by MSL and *Viking* (Freissinet *et al.*, 2015; Glavin *et al.*, 2013; Ming *et al.*, 2014).

12.2.4 The escape of H, O, C, and N

The atmosphere of Mars leaks to interplanetary space. Key atoms, including hydrogen, oxygen, carbon, and nitrogen are slowly lost from Mars today, in processes that we now consider.

Hydrogen escape. In the upper Martian atmosphere, H_2 molecules react with O_2^+ and CO_2^+ ions to make H atoms, which escape to space and form a halo of hydrogen surrounding Mars (see Fig. 5.1 for a similar halo around Earth). Hydrogen absorbs Lyman-α (121.6 nm) photons from the Sun. These photons have the right energy to promote an electron to its first excited state, from which it decays back to the ground state, releasing 121.6 nm photons in random directions. Such resonant scattering has been observed around Mars by UV spectrometers on *Mariner 6, 7*, and *9* (Anderson, 1974). It has also been seen by the *Rosetta* spacecraft (Feldman *et al.*, 2011), the SPICAM instrument (Spectroscopy for Investigation of Characteristics of the Atmosphere of Mars) on *Mars Express* (Chaufray *et al.*, 2008), the *Hubble Space Telescope* (Clarke *et al.*, 2009), and a UV-sensitive detector on *Mars Express* (Galli *et al.*, 2006b). Measurements provide the altitude profile of exospheric hydrogen from which hydrogen escape rates can be inferred. Zahnle *et al.* (2008) conclude from the data that hydrogen probably escapes at the diffusion-limited rate of ~3.5×10^8 atoms $cm^{-2} s^{-1}$.

Barth *et al.* (1972) realized that the inferred hydrogen escape rate, if extended over geologic time, would imply a build-up of oxygen to far higher abundances than observed unless oxygen is lost to space or to the Martian surface. At the same time, McElroy (1972) proposed a mechanism for oxygen to escape. Heavy atoms, O, N, and C cannot escape thermally because the temperature of Mars' exobase, ~200 K (Bougher *et al.*, 2009), is too low. But photochemical and ionic reactions in Mars' ionosphere can boost such atoms to the escape velocity in so-called *nonthermal escape* (see Sec. 5.2.2).

Oxygen escape. The escape of oxygen is tied to the photochemistry of CO_2 in the ionosphere. *Photoionization* of CO_2, which remains the main gas at the exobase level, is the principal source of ions.

$$CO_2 + hv \rightarrow CO_2^+ + e^- \tag{12.15}$$

These CO_2^+ ions are removed by reaction with O (which derives from CO_2 photodissociation (eq. (12.1)).

$$CO_2^+ + O \rightarrow O_2^+ + CO \tag{12.16}$$

Below 250 km altitude, O_2^+ exceeds CO_2^+ in abundance because reaction (12.16) is more efficient than (12.15) (Fig. 12.5). Oxygen atoms that are energetic enough to escape are generated as follows.

$$O_2^+ + e^- \rightarrow O + O \tag{12.17}$$

$$CO_2^+ + e^- \rightarrow CO + O \tag{12.18}$$

Reactions of this type are called *dissociative recombination* because an ion and an electron recombine into a

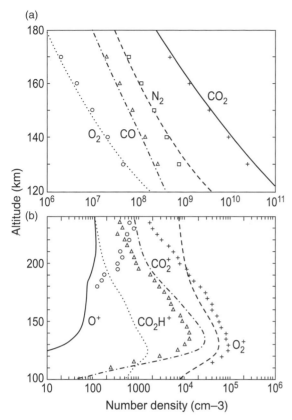

Figure 12.5 Species in the upper atmosphere of Mars: (a) neutral species and (b) major ions. Points are based on measurements from mass spectrometry on the *Viking 1* probe. Smooth lines are from the model of Nair *et al.* (1994).

The sum of eqs. (12.1), (12.15), (12.16), and (12.17) is $2CO_2 + 2hv \rightarrow 2CO + O + O$. So every O atom that escapes leaves behind an unbalanced CO. But oxygen derived from water vapor photolysis can oxidize CO (eq. (12.20)), and the removal of $\frac{1}{2}O_2$ generates net, unbalanced hydrogen, H_2 (eq. (12.19)), which can subsequently escape. If you add up the atoms, H and O escape in a ratio of 2:1 in this scheme. Overall, water is lost and the atmosphere and surface are neither oxidized by a net gain of oxygen nor chemically reduced by the accumulation of hydrogen. This idea of *stoichiometric escape* of water has been influential. In many photochemical models, the O escape flux is fixed at the upper boundary to balance two times the H escape flux (Krasnopolsky, 1993; Nair *et al.*, 1994).

However, oxygen emission measurements failed to identify the numbers of fast oxygen atoms hypothesized to form from dissociative recombination (Feldman *et al.*, 2011), while detailed calculations have produced a large spread of O escape rates that are often far less than that needed for a 2:1 H-to-O escape ratio (Fox and Hac, 2009; Groller *et al.*, 2014; Lammer *et al.*, 2003a) (although not always (Valeille *et al.*, 2010)). The dayside escape rate of Lammer *et al.* (2003a) is $\sim 1 \times 10^7$ O atoms cm^{-2} s^{-1}, which is much less than the $\sim 1.8 \times 10^8$ O atoms cm^{-2} s^{-1} that corresponds to half of the diffusion-limited H escape rate.

McElroy's (1972) insight about redox balance must remain true somehow, however, because oxygen associated with hydrogen escape must go somewhere. If stoichiometric *escape* is not happening, oxygen must react with the surface, which we illustrate in Fig. 12.4 with surface reactions of hydrogen peroxide and ozone. Oxidation of the surface would generate products such as ferric iron from ferrous iron or sulfate from sulfide. Alternatively, O_2 could be accumulating in the Martian atmosphere. However, this accumulation would somehow need to reverse itself periodically, as the doubling time for atmospheric O_2, assuming no loss, is only $\sim 10^5$ yr.

Oxygen can also escape by other nonthermal mechanisms besides dissociative recombination. *Ion pickup* involves ionization above the ionopause, i.e., the upper boundary of the ionosphere, which separates planetary plasma from that which is induced by the solar wind. The ions are picked up by electric fields associated with the solar wind. This escape flux is small on its own, but some of the ions re-impact the atmosphere and cause sputtering. *Atmospheric sputtering* is the ejection of previously gravitationally bound species by ions incident on the upper atmosphere. Estimates of the rates in the current epoch suggest that escape from dissociative recombination is the most important (Valeille *et al.*, 2010).

product that dissociates or breaks up. An excess of kinetic energy is produced because the positive ion and negative electron attract each other. Half of the O atoms generated by eqs. (12.17) and (12.18) move upward and half move downward. The upward moving O atoms can exceed Mars' 5 km s^{-1} escape velocity. However, the more massive CO molecules generated in eq. (12.18) cannot. The velocities of the reaction products can be calculated from the ionization energy, using conservation of momentum and energy.

McElroy (1972) proposed that a coupled chemistry operates on Mars in which one oxygen atom escapes for every two hydrogen atoms, i.e., the same ratio as in H_2O. If true, Mars would be in redox stasis. McElroy's idea was that nonthermal escape of O regulates H escape. One can understand the coupled chemistry in terms of net reactions for water vapor dissociation and CO oxidation.

$$H_2O \rightarrow H_2 + \frac{1}{2}O_2 \tag{12.19}$$

$$CO + \frac{1}{2}O_2 \rightarrow CO_2 \tag{12.20}$$

Carbon escape. Carbon atoms also escape through dissociative recombination.

$$CO^+ + e^- \rightarrow C + O \qquad (12.21)$$
$$CO_2^+ + e^- \rightarrow C + O_2 \qquad (12.22)$$

Another source of hot (i.e., meaning fast) carbon atoms is UV photolysis of carbon monoxide.

$$CO + hv \; (<111.6 \, nm) \rightarrow C + O \qquad (12.23)$$

According to Nagy *et al.* (2001), photolysis of CO is more important than the dissociative recombination of its ion, eq. (12.21). Estimated photochemical losses of carbon are $(2–6) \times 10^5$ C atoms cm^{-2} s^{-1} (Fox and Bakalian, 2001; Nagy *et al.*, 2001). Other sources of escaping carbon include collisions of carbon with fast oxygen atoms, as well as solar wind sweeping of carbon dioxide ions. Sputtering of CO_2 is estimated as $(6–17) \times 10^5$ molecules cm^{-2} s^{-1} (reviewed by Lammer *et al.* (2008)). Total loss rates range $(0.6–2.2) \times 10^6$ C atoms cm^{-2} s^{-1} (Groller *et al.*, 2014). The 6 mbar Martian atmosphere today contains $\sim 2.2 \times 10^{23}$ CO_2 molecules cm^{-2}, so that the present rate of carbon escape would take 3-12 Gyr $((1–3.7) \times 10^{17}$ s) to remove the entire atmosphere. Rates of ion escape (summed O^+, O_2^+, and CO_2^+) inferred from MAVEN are ~ 0.1 kg s^{-1}, increasing to ~ 1 kg s^{-1} during a solar coronal mass ejection (Jakosky *et al.*, 2015). The latter would take ~ 1 Gyr to remove the current Martian atmosphere or ~ 100 Gyr to remove 1 bar, so the effect is small. However, rates of CO_2 escape were probably much higher in the past because of hydrodynamic and impact erosion processes (Sec 12.3.3 below).

Nitrogen escape. Nitrogen molecules photolyze or photo-ionize under extreme UV, allowing N to escape by dissociative recombination. The branches of photolysis include (Fox, 2007):

$$
\begin{aligned}
N_2 + hv &\rightarrow N + N^* \\
&\rightarrow N^* + N^* \\
&\rightarrow N_2^+ + e^-
\end{aligned}
\qquad (12.24)
$$

Here, N^* is excited nitrogen (in a quantum state of either $N(^2D)$ or $N(^2P)$ compared to the ground state $N(^4S)$). The N_2^+ ions can undergo dissociative recombination through two branches that each produce excited atoms but leave enough energy for escape:

$$N_2^+ + e^- \rightarrow N + N(^2D) \qquad (12.25)$$
$$N_2^+ + e^- \rightarrow N + N(^2P) \qquad (12.26)$$

If such dissociative recombination takes place above the exobase, where collisions are negligible, N atoms can be lost. However, the escape associated with dissociative recombination is only about half the total nitrogen escape because of other photochemical reactions (Fox, 2007), e.g.,

$$
\begin{aligned}
N_2 + hv &\rightarrow N^+ + N + e^- \; \text{(photodissociative ionization)} \\
N_2 + e^- &\rightarrow N^+ + N + 2e^- \; \text{(electron impact dissociative ionization)}
\end{aligned}
\qquad (12.27)
$$

Fox (1993) estimates the escape flux of ^{14}N to be $\sim 5 \times 10^5$ cm^{-2} s^{-1} averaged over high and low solar activity. This slow rate would take 6 b.y. to remove the amount of N_2 in the present Martian atmosphere.

The creation of electronically excited neutral atoms in eqs. (12.25) and (12.26) leaves N atoms with barely enough energy to escape, so that escape is sensitive to isotopic mass. In addition, ^{14}N is more abundant at the exobase than ^{15}N because of diffusive separation in the upper atmosphere. So, the escape of ^{15}N is about a factor of ~ 0.6 times that of ^{14}N (Fox and Hac, 1997). Thus, photochemical escape over time is plausibly responsible for the high $^{15}N/^{14}N$ nitrogen isotope ratio on Mars ($\delta^{15}N \approx 620 \pm 160‰$), which is a factor of 1.6 times greater than Earth's (Nier and McElroy, 1977). One can explain the enrichment in ^{15}N by assuming a model in which primordial nitrogen started with a terrestrial $^{15}N/^{14}N$ ratio. Because ^{15}N was escaping 0.6 times as fast as ^{14}N, the predicted enrichment in ^{15}N is $1/0.6 \cong 1.6$. Actually, the $^{15}N/^{14}N$ ratio in Martian meteorite silicates suggests that the primordial $\delta^{15}N$ may have been $-30‰$ rather than the 0‰ of terrestrial air (Bogard *et al.*, 2001), so this might be the starting value from which the nitrogen isotope ratio has evolved. However, this correction is small compared to the total fractionation observed

12.3 Volatile Inventory: Present and Past

Before turning to Mars' atmospheric evolution, we should consider the volatile reservoirs on Mars today (Table 12.4). We start with the amount of CO_2, which by convention is expressed as an equivalent surface pressure. Then we turn to the H_2O inventory, which is described in terms of a global equivalent layer (GEL) of an ocean.

12.3.1 The Present-Day Volatile Inventories

Carbon dioxide. Today, the total abundance of CO_2 is probably small. The inventory includes 6 mbar in the atmosphere, 4–5 mbar in the south polar residual cap (Phillips *et al.*, 2011; Thomas *et al.*, 2009), ~ 1 mbar in the seasonal caps (Smith *et al.*, 2009a), and less than 40 mbar adsorbed in the *regolith* because of competition

Table 12.4 Current volatile inventories on Mars.

Carbon Dioxide (CO₂)	Surface Pressure Equivalent (mbar)	Sources
Atmosphere	6	Haberle *et al.* (2008)
Polar caps	4–5	Thomas *et al.* (2009); Phillips *et al.* (2011)
Seasonal caps	1	Smith (2009a)
Carbonate in weathered dust or soil	~200 per 100 m global average layer of dust or soil	Bandfield *et al.* (2003); Boynton *et al.* (2009)
Adsorbed in regolith	<40	Zent and Quinn (1995)
Carbonate rocks	0.25–12	Edwards and Ehlmann (2015).
Subsurface bedrock	<250 per km depth	From carbonate abundance in meteorites.
Water (H₂O)	**Global Equivalent Layer, GEL (m)**	
Atmosphere	10^{-5} m	Jakosky and Farmer (1982)
Polar caps and layered terrains	<20 m	Plaut *et al.* (2007); Zuber *et al.* (2007)
Ice, adsorbed water, and/or hydrated salts stored in the regolith	<20 m	See text.
Covered glaciers	<3 m	Levy *et al.* (2014); Karlsson *et al.* (2015)
Alternation minerals in the crust	150–900 m	1wt% hydration to 5 km depth to 3wt% hydration to 10 km depth (Mustard *et al.*, 2012).
Deep aquifers	None known	Radar finds no aquifers to ~5 km depth.
Sulfur	**Global Equivalent**	
Weathered soil	$<0.9 \times 10^{16}$ kg SO₃	Assumes <8wt% SO₃
Sedimentary rock reservoirs	per meter of global average soil ~10^{17} kg SO₃	Assumes ~20 vol% SO₃ in observed layered deposits. See also Michalski and Niles (2012).

with water vapor for adsorption (Zent and Quinn, 1995). *Regolith* is the geologic unit consisting of dust, sand, loose rocks, and rocky fragments, but excluding bedrock.

The quantity of CO₂ locked up in carbonate rocks is unknown but appears small. Despite an extensive search from orbital spectroscopy, only a few carbonate outcrops have been found, restricted to Noachian terrain (Ehlmann *et al.*, 2008b; Ehlmann *et al.*, 2009; Michalski and Niles, 2010; Wray *et al.*, 2016) and dominated by those in Nili Fossae. The carbonate content in Nili Fossae outcrops is ≤ 20%, and the area ranges from 6800 km² of the carbonate outcrops themselves to the regional extent of the olivine unit that contains the carbonates, 300 000 km² (Edwards and Ehlmann, 2015). The thickness of carbonation is (generously) 500 m maximum, so the rock volume ranges from 3400–150 000 km³. If we assume

magnesite (MgCO₃), the CO₂ inventory[4] would be 0.25 to ~12 mbar (Edwards and Ehlmann, 2015).

Carbonate is also found in dust, which contains 3–5 wt% magnesite (MgCO₃) (Bandfield *et al.*, 2003). This is similar to estimates of the amount of carbonate in the soil (Boynton *et al.*, 2009; Christensen *et al.*, 2004; Palomba *et al.*, 2009). If this mass fraction were representative of the crust down to 1–3 km depth then 1–3 bars could be sequestered. However, such an extrapolation is likely unjustified. First, processes operating in the current cold Martian climate can enrich dust and soil in carbonates (Boynton *et al.*, 2009) relative to the subsurface. Second, actual samples of the subsurface contain less carbonate than dust. We can make an estimate using the Martian meteorite, ALH84001, which has 1 vol% carbonate. If we assume ~0.5 wt% carbonate, extrapolation through 1 km of subsurface would suggest <0.25 bar of sequestered CO₂. Younger meteorites have less carbonate (e.g., Nakhla has ~0.02wt% (Carr *et al.*, 1985)) but they crystallized at a time when perhaps there was less carbonate to sequester.

---

[4] Using eq. (1.19), $P = [((0.34 \text{ to } 15) \times 10^{13} \text{ m}^3 \times 0.2 \times 2960 \text{ kg MgCO}_3 \text{ m}^{-3} \times g \times 0.044 \text{ kg CO}_2 \text{ mol}^{-1})/(0.084 \text{ kg MgCO}_3 \text{ mol}^{-1} \times A_{\vartheta})]$, where $A_{\vartheta} = 1.44 \times 10^{14} \text{ m}^2$ is the surface area of Mars, and g is Martian gravity of 3.72 m s^{-2}.

Water. Water is present on Mars as atmospheric water vapor or aerosols, as ice in the permanent north polar caps and its surrounding layered terrains, and as ice in the layered terrains around the permanent southern CO_2 cap. Water is also buried in midlatitudes as massive ice, and buried between thin debris covers in midlatitude glaciers (confirmed by radar) and perhaps also equatorial glaciers (suspected from morphology, but unconfirmed). Water also resides within the regolith as adsorbed water and hydrous salts.

Consider each reservoir in turn. The extremely arid Martian atmosphere only contains water equivalent to a global layer of $\sim 10^{-5}$ m depth (Jakosky and Farmer, 1982). The volume of the water-rich layered terrains around the CO_2 south polar ice cap has been estimated from radar as $\sim 1.6 \times 10^6$ km^3 or 11 ± 1.4 m GEL, assuming pure water ice (Plaut *et al.*, 2007). Since the dust content is $\sim 15\%$ (Zuber *et al.*, 2007), a better estimate of the water GEL is 9.4 ± 1.2 m. The permanent north polar cap is $\geq 95\%$ water ice (Grima *et al.*, 2009) with a total volume of $\sim 1.4 \times 10^6$ km^3, implying 9–10 m of global water. Neutron spectroscopy from orbit has been used to infer the presence of abundant hydrogen in the top 1–2 m of the regolith. Detection is possible because cosmic rays enter the surface and neutrons are ejected with an energy spectrum that depends upon the distribution of elements and indicates the presence of abundant hydrogen. Neutrons lose energy efficiently by collisions with H atoms because a neutron and a proton have nearly the same mass. By contrast, collisions with heavier atoms, e.g., O or Si, allow the neutrons to rebound without losing much energy. (Think of analogies with two tennis balls, or a tennis ball and a basketball.) At high latitudes, the hydrogen is assumed to be in the form of subsurface ice. i.e., *ground ice*, whereas

at low latitudes, hydrogen is assumed to be within mineral phases such as hydrated salts (Fig. 12.6), although some have argued that there might be ice deposited in past climatic excursions, which is still sublimating (Mouginot *et al.*, 2010). The robotic arm and thrusters on the *Phoenix* lander exposed ice at 68°N, which was 5–10 cm below the surface, consistent with inferences from orbiter data (Smith *et al.*, 2009b) (Fig. 12.7(a)). Ice has also been exposed in meter-scale craters created by impacts that have occurred on Mars during observational operations from orbit (Byrne *et al.*, 2009; Dundas *et al.*, 2014) (Fig. 12.7 (b)) and sublimation can alter crater morphology (Viola *et al.*, 2015). The total depth to which the ice extends in the high latitudes is unknown but is probably no more than ~ 20–30 m depth (Campbell *et al.*, 2008; Catling *et al.*, 2012; Putzig *et al.*, 2014). The GEL depends strongly on the ice:rock ratio (Mouginot *et al.*, 2010), but is probably <20 m. The glacier contribution to the ice inventory is modest, 1–3 m GEL (Karlsson *et al.*, 2015; Levy *et al.*, 2014). In summary, the water inventory is probably not more than ~ 150–900 m GEL and dominated by hydrated minerals rather than ice, where the latter totals no more than ~ 35 m GEL (Carr and Head, 2015; Christensen, 2006; Lasue *et al.*, 2013).

Sulfur. Sulfur-bearing gases are undetectable in today's atmosphere, while the surface inventory of sulfur can only be quantified crudely. Measurements by all landers show that 4–8 wt% SO_3 or ~ 1.6-3.2 wt% S is present in soils (**Box 12.3**), with minor exceptions. Near-infrared spectroscopy and the *Opportunity* rover have also revealed sulfate-rich sedimentary deposits (Gendrin *et al.*, 2005; Murchie *et al.*, 2009b; Squyres *et al.*, 2006). Additional sulfate is present in northern circumpolar dunes (Langevin *et al.*, 2005). To estimate the total sulfate

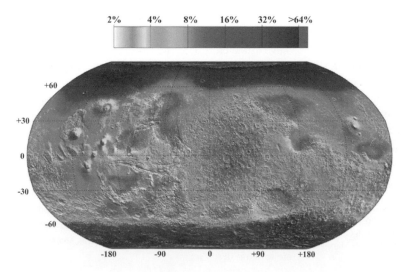

Figure 12.6 Water-equivalent mass fraction in the subsurface based on the hydrogen content inferred from the *Mars Odyssey* neutron spectrometer (From Feldman *et al.* (2004). Reproduced with permission from John Wiley and Sons. Copyright 2004, American Geophysical Union.) (A black and white version of this figure will appear in some formats. For the color version, please refer to the plate section.)

(a)

(b)

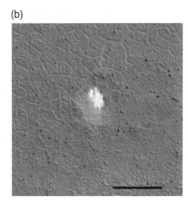

Figure 12.7 (a) The ice table underneath NASA's *Phoenix Lander* exposed by descent thrusters, imaged by the robotic arm camera. (Courtesy of NASA/ Univ. of Arizona/ MPI/ Marco Di Lorenzo/ Ken Kremer). (b) Subsurface ice exposed by a recent small crater in the northern plains of Mars. Scale bar = 50 m. (HiRISE image PSP_001380_2520 at 71.7° N, 189.9° E; Courtesy of NASA/ University of Arizona.)

Box 12.3 Oxide representation of major elements in soils and rocks

A convention in soil chemistry and petrology, which can confuse the uninitiated, is to report major elements as if they were in their simple oxide form, e.g., K_2O for K, SiO_2 for Si, SO_3 for S, etc. Mars rocks and soil compositions are commonly tabulated in such a way. The convention has the advantage of adding about the right amount of oxygen overall for many systems. Conversion is based on atomic mass. Thus, wt% S = wt% $SO_3 \times 0.4$, where $0.4 = 32/ (32 + (3 \times 16))$.

---

inventory would require knowing the volume and sulfate fraction of every deposit. For a rough estimate, let us assume ~20% sulfate by volume in all sulfate-rich sediments. The sum of various known deposits[5] gives $>7.1 \times 10^{16}$ kg SO_3. Because we have not quantified the volume of all sulfate deposits, the total sulfur is probably of order ~10^{17} kg or more. This is approaching within an order of magnitude of Earth's oceanic sulfate, which contains $(1.4 \times 10^{21} \text{ l}) \times (28.2 \text{ mM SO}_4^{-2}) \times (0.08 \text{ kg/mol SO}_3) = 3.2 \times 10^{18}$ kg SO_3. But Earth's total sulfur inventory (in the form of pyrite and gypsum) is much larger than this, ~3×10^{20} mol S, or 2.4×10^{19} kg SO_3 equivalent (Holser *et al.*, 1988). Even accounting for the factor of 4 difference

in surface area between the two planets, Mars' surface appears to be deficient in sulfur compared to Earth presumably because of less efficient outgassing.

12.3.2 Past Volatile Inventory

We have just considered the present volatile inventory but what was it initially? The original inventory is tied to the way that Mars accreted, as described in Ch. 6. Data imply that Mars formed quickly, within a few million years (Dauphas and Pourmand, 2011). The radioactive decay of ^{182}Hf (with a 9 m.y. half-life) produces ^{182}W, which has an affinity for iron. Martian meteorites have excess ^{182}W, which implies that the core formed before much ^{182}W was produced. The small size of Mars also suggests that it formed from planetesimals and did not experience late-stage accretion of planetary embryos (see Sec. 6.1). Consequently, Mars may be a planetary embryo that is poor in volatiles because, unlike the Earth, which is thought to be a combination of many embryos, Mars did not incorporate volatile-rich embryos formed at ice-rich distances of 2.5 AU or beyond (Lunine *et al.*, 2003). However, others argue that contribution from volatile-rich planetesimals could have made Mars rich in volatiles (Drake and Righter, 2002).

---

[5] The polar dune sulfate is 7500 km² in area (Fishbaugh *et al.*, 2007), and assuming ~10 m average thickness, this is 1.5×10^{10} m³ or 3.5×10^{13} kg sulfate, using parameters for gypsum, which is ~1.6×10^{13} kg SO_3. Other known sulfate volumes are generally thought to be kieserite-dominated ($MgSO_4.H_2O$, 0.134 kg mol⁻¹, 56.6 cm³ mol⁻¹). They include Aram Chaos (Masse *et al.*, 2008) with an oblate ellipsoid dome deposit with total volume ~1.5×10^{13} m³, equivalent to ~0.42×10^{16} kg SO_3 at 20 vol% sulfate. Meridiani Planum probably has at least 1.1×10^{14} m³ deposits (McCollom *et al.*, 2007), equivalent to 3.1×10^{16} kg SO_3 at 20 vol% sulfate. The interior layered deposits of Valles Marineris have at least 1.3×10^{14} m³ volume (Nedell *et al.*, 1987), which is 3.6×10^{16} kg SO_3 at 20 vol% sulfate.

Carbon. One way to estimate Mars' initial CO_2 inventory to an order of magnitude assumes that Mars accreted carbon with a fractional abundance similar to that of Earth or Venus. Venus has ~90 bars of CO_2 in its atmosphere, while the Earth has a comparable amount (estimates range up to 90 bars) locked up in carbonate rocks. By mass, Mars should have 1/9 of the mass inventory. The equivalent CO_2 pressure is (90 bars) $\times$ (1/9) $\times$ 4 $\times$ (1/2.6) $\approx$ 15 bars, accounting for surface area and gravity. The amount in bars doesn't scale as 1/9 because surface pressure is (column mass) $\times$ g, where g is gravitational acceleration (eq. (1.19)). So when you put a mass of CO_2 onto Mars, the column mass is increased by a factor of 4 relative to Earth because Mars has only ¼ of the area. But Mars' gravity is 2.6 times less than Earth's, which reduces the CO_2 pressure.

Another way to estimate the early inventory of carbon, albeit after heavy bombardment, is to examine the loss of carbon relative to krypton. Carbon can escape today (Sec. 12.2.4) but krypton is too heavy to have escaped since the end of heavy bombardment. The ratio of $C/^{84}Kr$ is $\approx 4 \times 10^7$ for both Earth (Pepin, 1991) and Venus (Donahue and Pollack, 1983). But on Mars, the ratio $C/^{84}Kr \approx (4.4$–$6) \times 10^6$ (Owen *et al.*, 1977; Pepin, 1991). Owen and Bar-Nun (1995) suggest that the much smaller $C/^{84}Kr$ ratio on Mars represents a ~85–90% depletion of carbon on Mars since the end of heavy bombardment from nonthermal escape processes that have affected carbon but not krypton. The current carbon reservoirs on Mars that we can accurately quantify are the atmosphere and polar caps, which together contain ~12 mbar (Table 12.4). Thus, accounting for 85%–90% loss, the amount of CO_2 present at the end of heavy bombardment could have been 80–120 mbar.

We can also examine the quantity of Kr per kg of planet, assuming that all the krypton resides in the atmosphere to estimate how much krypton (and similarly heavy gas) escaped early. In the Martian atmosphere, Kr is present at 0.3 ppm, whereas on Earth it is 1 ppm (Lodders and Fegley, 1998). These amounts are equivalent to 1.5×10^{13} kg of Kr on Earth and 1.4×10^{10} kg of Kr on Mars. Normalizing to the mass of each planet, the Kr concentration is 2.5×10^{-12} kg Kr kg^{-1} on Earth and 2.2×10^{-14} kg Kr kg^{-1} on Mars. Thus, Mars has only ~1% of the expected krypton concentration if we assume that Mars should have had a fractional endowment of volatiles similar to the Earth's. Possibly, ~99% of the krypton was lost early in hydrodynamic escape and impact erosion of the atmosphere (Sec. 12.3.3, below). If Kr escaped, CO_2 would have escaped also. If there had been 80–120 mbar of CO_2 at the end of heavy bombardment, then a prior, hundredfold-depletion would mean an initial CO_2 inventory of 8–12 bar. This agrees with our earlier estimate of ~15 bar to within a factor of two, which is reasonable considering that we're just using rough scaling arguments.

Water. Estimates of the initial amount of water on Mars come from four sources: mass scaling, accretion models, inferences from geomorphology, and calculations of water loss from escape. The unit commonly used is global equivalent layer (GEL).

Mass scaling is the simplest. As we did above for CO_2, assume that the amount of water per unit mass of planet was the same for Mars as for Earth (with the caveat that this is doubtful because of different impactors during accretion (Horner *et al.*, 2009)). Earth's oceans are 4 km deep, on average. Continents cover 30% of the Earth's surface, so if these were removed, the ocean would be

Table 12.5 Estimates of the water inventory on Mars (global equivalent layer, GEL).

Estimated water (m)	Basis	Reference
>30	Loss of water to account for D/H ratio.	Krasnopolsky (2002)
130	Bulk geochemistry and accretion	Dreibus and Wänke (1987)
284±86	D/H, hydrodynamic escape fractionation.	This text
120	Outgassed during Tharsis formation	Phillips *et al.* (2001)
>(300–500)	Outflow channel incision	Carr (1986), Carr (1996, p. 165), Baker (2006)
~40	Outflow channel incision	Carr and Head (2015)
50–500	To incise 146 000 km of Noachian valleys	Carr and Malin (2000)
30	To incise 81 000 km of Hesperian valleys	
130–1000	Putative ocean shorelines	Parker *et al.* (1993)
130	Deuteronilus shoreline enclosure	Clifford and Parker (2001)
1000	Noachian ocean	Clifford and Parker (2001)

about 3 km deep. Mars' surface to volume ratio is twice that of Earth, so ocean depth on Mars (ignoring mantle water) would be reduced by a factor of 2, yielding 1.5 km. This estimate is in crude factor of 1.5–10 agreement with some other estimates made on the basis of geomorphology (Table 12.5). For example, if there was a Noachian ocean that covered about one-third of Mars, as some have suggested (see Sec. 12.4.1), the total volume would be equivalent to ~1 km depth (Clifford and Parker, 2001).

In contrast to simple mass scaling, most accretion models, which are based on the geochemistry of Martian meteorites or dynamical simulations, suggest that Mars formed dry compared to Earth. Dreibus and Wänke (1987) proposed that Mars accreted from two components, which were volatile-rich and volatile-poor, respectively. They assumed homogeneous accretion because chalcophile (sulfur-loving) elements, such as Cu, Co, and Ni, are depleted, which implies efficient mixing and removal by FeS segregation to the core. In their model, water was chemically reduced by iron during core formation, which oxidized mantle iron and produced a large amount of hydrogen that escaped to space. At the end, only ~130 m GEL was left. A key assumption is that mixing of the water with the undifferentiated planet was complete. In contrast, dynamical models that allow for water delivery after core formation can result in 600–2700 m GEL (Lunine et al., 2003). The proportion of this water that outgasses would then determine the surface reservoir.

As described below (Sec. 12.4), liquid water is thought to have carved various geomorphic features on the surface of ancient Mars, and the amount of water can be estimated from the scale of the erosion. The features include valley networks, large outflow channels, and hypothesized ocean shorelines. The existence of past oceans is controversial (Carr and Head, 2003), and some also dispute the dominant view that outflow channels were eroded by water, instead favoring incision by inviscid lavas (Hopper and Leverington, 2014; Leone, 2014; Leverington, 2011).

The amount of water needed to cut Noachian valleys has been estimated as a 50–500 m global depth layer, while the Hesperian valleys require 30 m (Carr and Malin, 2000). These are not really estimates for the total water inventory but estimates for the amount of runoff needed to form the valleys, which could be done with a smaller inventory if water is recycled. We should compare these numbers with estimated rainfall requirements for valley formation on Earth. An example is Colorado's Grand Canyon. When the Grand Canyon was carved has been debated. Overall, geomorphic and thermochronological data favor a younger age of 5–6 Ma (Darling and Whipple, 2015; Karlstrom et al., 2014; Young and Crow, 2014) rather than older ages up to 70 Ma (Flowers and Farley,

2012; Polyak et al., 2008). Since 6 Ma, ~1.8 million meters of runoff from the plateau probably drained down the canyon, assuming ~0.3 m yr^{-1} rainfall. This value is higher than Carr and Malin's estimate for Noachian Martian valley formation by a factor of 10^3–10^4. Other analyses of Martian valleys based on modeling of sediment transport also suggest higher runoff figures. Hoke et al. (2011) estimate that episodic runoff rates of 0.5 cm day^{-1}, along with sustained runoff rates of ~10 cm yr^{-1} for $(3–4)\times10^7$ yr, were needed to form the larger valleys such as the Naktong east valley (2° N, 34° E). The annual runoff of ~10 cm comes from multiplying a daily runoff rate of 0.5 cm d^{-1} by 365 d yr^{-1} and assuming that rainfall occurs only 1 out of every 20 days, as in many arid regions on Earth. In any case, the total local runoff is ~10^6 m in this model. Martian valleys are generally narrower and shallower by a factor of ~10 than the Grand Canyon's 29 km width and 1.8 km depth, and the Martian surface could be more easily eroded than is Earth's; nevertheless, Carr and Malin's estimate for the amount of water needed to form the valleys may be too small. High runoff estimates, of course, do *not* imply high initial water inventories on Mars. Instead, they demonstrate the need for active hydrological recycling through evaporation or sublimation and precipitation.

For outflow channels, early estimates suggested a minimum amount of water for fluvial erosion of 300–500 m (Table 12.5), assuming that there was 60% water to 40% sediment load. Such sediment loads are high compared to terrestrial systems, so that more realistic loads would require recycling of the water back to areas of hypothesized outflow bursts for further flooding. More recent estimates of outflow channel discharge are much smaller than earlier assessments (Kleinhans, 2005; Williams et al., 2000). By taking into account volume that is tectonic or due to collapse rather than erosion per se, Carr and Head (2015) estimate ~40 m.

Estimates of the total water loss from escape processes set a lower bound on the initial water reservoir. The D/H ratio (deuterium/hydrogen) in atmospheric water vapor on Mars has been deduced from remote spectroscopy as 5–6 times higher than in Earth's oceans (Krasnopolsky, 2000; Owen, 1992; Owen et al., 1988), which is consistent with MSL in situ measurements of 6±1 (Webster et al., 2013). However, high spatial resolution remote spectroscopy suggests a factor of ~7 D/H ratio relative to Earth (Villanueva et al., 2015). Models have used the enhanced escape of H relative to D from the upper atmosphere of Mars to account for the large D/H ratio. Calculated losses include those from sputtering, ion pickup, nonthermal, and thermal escape of hydrogen and oxygen. These have generated estimates of water loss ranging between 12 m (Lammer et al., 2003a) and 30 m (Krasnopolsky and Feldman, 2001).

Linear extrapolation of the current escape rates of D and H, assuming that the initial D/H ratio was the same as Standard Mean Ocean Water (SMOW) on Earth, would imply that Mars has lost 3 m of water over the past 4 Gyr, and that the remainder will be gone in the next 320 m.y. (Yung et al., 1988). Such extrapolations are unrealistic, however. The assumption of a uniformitarian escape rate conflicts with data. Specifically, the Los Angeles, Shergotty, Zagami, and Larkman Nunatak 06319 shergotitte Martian meteorites have D/H ratios (relative to SMOW) of 4.5–5.1, 5.6, 5.4, and 6 (Usui et al., 2012). These values are similar to that in today's Martian atmospheric water vapor, which implies little change in D/H since these meteorites formed at ~170 Ma (Greenwood et al., 2008). The ALH84001 meteorite, which formed at 4.1 Ga, contains both low and high D/H of ~4 SMOW and 5.6 SMOW, respectively. The higher ratio is similar to modern, whereas the lower value may be a mix of mantle and atmospheric components, where magmatic D/H is probably 1.28 ± 0.01 SMOW (Usui et al., 2012). A D/H ratio of 3.0 ± 0.2 SMOW has also been found in ~3 Ga mudstones from the Yellowknife Bay lakebed on the floor of Gale Crater, which suggests considerable early loss of water (Mahaffy et al., 2015b).

One possible explanation of the elevated D/H ratio by ~3 Ga is that hydrogen was Rayleigh fractionated, mostly by *hydrodynamic escape* in the Pre-Noachian (Batalha et al., 2015; Greenwood et al., 2008; Zahnle et al., 1990). Earlier, in Sec. 5.11.1, we considered the theory of Rayleigh fractionation of the inventory of a heavy gas, denoted N_2, from a lighter gas inventory, denoted N_1, which we developed through eqs. (5.81)–(5.84). Let us multiply eq. (5.84) by N_1^0/N_1 and substitute $f = 1/(1+y)$. Then we have:

relative efficieny of D escape to H escape. To use this equation and calculate $M_{initial}$, we must know the fractionation factor and the initial D/H ratio, (D/H)$_{initial}$.

The inventory that undergoes Rayleigh fractionation must be exchangeable surface water, i.e., water that can give rise to atmospheric hydrogen, in order for us to calculate a past inventory relevant to fluvial features. Usually, photolysis of water vapor is assumed to be the source of hydrogen (Donahue, 1995; Yung et al., 1988). The factor f would be 1 if deuterium escaped as easily as H, so $f < 1$. A value of $f = 0.4$ is a reasonable estimate for diffusion-limited escape. In diffusion-limited escape, escape rates are proportional to mixing ratios in the upper atmosphere and today hydrogen appears to escape at the diffusion limit (see Sec. 5.9). The observed D/H ratio in H_2 is 0.4 that in water vapor, i.e., $(HD/H_2)/(HDO/H_2O)$ $\approx 2.4/6 = 0.4$ (Krasnopolsky, 2002).

An alternative scenario is that early H_2 comes from surface water that is tectonically cycled, broken down into H_2 in the upper mantle, and released as volcanic H_2 that escapes, leaving oxygen behind (Batalha et al., 2015). By coincidence, a fractionation factor of $f \sim 0.4$ is also calculated for hydrodynamic escape of a H_2-rich atmosphere. Let us assume that H_2 concentrations of ~5% by volume were created on early Mars if volcanic outgassing of H_2 had been sufficient (see Sec. 12.5.2.2.). If hydrogen escaped at the diffusion-limited rate, the volcanic outgassing rate (and escape rate) would be 8×10^{11} H_2 molecules cm^{-2} s^{-1}. This rate yields a crossover mass of 3.6 a.m.u. from eq. (5.78) and a fractionation factor, f, of 0.4 from eq. (5.84), assuming that HD and H_2 were the two escaping atmospheric constituents.

What was the initial D/H value? Usui et al. (2012) argue that δD of (275 ± 10)‰ in olivine-hosted melt inclusions in the shergottite Yamoto 980459 represents

$$\left(\frac{N_2}{N_2^0}\right) = \left(\frac{N_1}{N_1^0}\right)^f \Rightarrow \left(\frac{N_2}{N_2^0}\right)\left(\frac{N_1^0}{N_1}\right) = \left(\frac{N_1}{N_1^0}\right)^f \left(\frac{N_1^0}{N_1}\right) \Rightarrow \frac{(N_2/N_1)}{(N_2^0/N_1^0)} = \left(\frac{N_1}{N_1^0}\right)^{f-1} \qquad (12.28)$$

Here, the "0" superscript indicates an initial inventory. If N_2/N_1 is interpreted as the D/H ratio of an inventory, then we can write an equation for the Rayleigh fractionation of hydrogen between an initial D/H and later D/H, as follows, by rearranging eq. (12.28) (see also Donahue, 1995):

$$\left(\frac{M_{initial}}{M_{later}}\right)^{1-f} = \left[\frac{(D/H)_{later}}{(D/H)_{initial}}\right] \qquad (12.29)$$

Here, $M_{initial}$ is the initial hydrogen reservoir mass, which is assumed to be water, M_{later} is the water reservoir at later time, and f is a fractionation factor that indicates the

an upper limit for undegassed primordial mantle water. This is equivalent[6] to (D/H)$_{initial} = 1.275 \pm 0.01$ SMOW, which is comparable to other estimates of the primordial D/H that are close to SMOW (Hallis et al., 2012).

Taking (D/H)$_{initial} = 1.28 \pm 0.01$ SMOW, (D/H)$_{later}$ as 5.5 ± 1 SMOW, and $M_{later} = 25$ m of H_2O, eq. (12.29)

---

[6] Recall that δD = $(R - 1) \times 1000$ for a sample where ratio $R = $ (D/H)$_{sample}$/ (D/H)$_{SMOW}$. Here, (D/H)$_{SMOW}$ is the standard mean ocean water value of 155.8×10^{-6}. Consequently, $R = $ (δD/1000) +1, so that a D/H of 6 SMOW on Mars is a δD = 5000‰, for example.

gives $M_{initial} = 284\pm86$ m of H_2O. This initial inventory is roughly consistent with estimates from bulk planet geochemistry (Taylor, 2013; Wänke and Dreibus, 1994) and calculated values of 100–450 m water by others from D/H (Batalha *et al.*, 2015; Kurokawa *et al.*, 2014; Villanueva *et al.*, 2015).

Our estimate of $M_{initial}$ is only valid if diffusion-limited escape with $f = 0.4$ was the dominant fractionating process. However, the idea is testable. Dominant fractionation from early escape predicts that polar ice should have a D/H similar to present atmospheric water vapor. A model in which D/H fractionation happens on time-scales $\sim10^5$ yr involving a small exchangeable reservoir (Carr, 1990) would mean that polar ice could have a D/H ratio similar to SMOW. Currently, spectroscopy suggests that the D/H in the polar caps is ~8 SMOW (Villanueva *et al.*, 2015), consistent with some variability between caps and atmosphere due to modern climatic factors (Montmessin *et al.*, 2005) and consistent with early escape causing the fractionation. However, the D/H ratio in polar ice remains to be measured *in situ*.

12.3.3   Noachian and Pre-Noachian Atmospheric Escape: Theory and Evidence

If CO_2 was lost from the early atmosphere then nitrogen and other comparably heavy gases would also have been lost. Above, we discussed the possibility that Mars started with 8–15 bar of CO_2, while the current inventory of Mars is perhaps only a few times 0.1 bar (Table 12.4). This is a $\sim99\%$ loss. Similarly, Mars' atmospheric N_2 inventory is merely 2×10^{-4} bar, compared to 0.78 bar and 3.3 bar N_2 in the atmospheres of Earth and Venus, respectively. We can also compare 6 ± 4 bar of N_2 in Earth's mantle, crust and atmosphere together (Table 11.1). The measured abundance of nitrogen on Mars (which is probably in nitrates) by MSL ranges from ~200 ppm in eolian deposits up to ~1000 ppm in mudstones (Stern *et al.*, 2015a). These low levels suggest big loss of nitrogen to space if Mars was endowed with similar proportions of nitrogen as the Earth. In Sec. 12.2.4, we discussed ways in which O, H, C, and N atoms can slowly escape from Mars today. But in the ancient past, evidence and theory suggests that *impact erosion* and *hydrodynamic escape* were much more efficient at removing atmospheric gases than any process operating after the Noachian.

12.3.3.1   *Impact Erosion*

Many impacts hitting early Mars would have been big and fast enough that they vaporized themselves and a similar

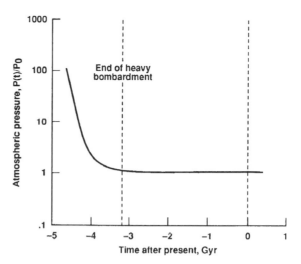

Figure 12.8 Calculated impact erosion of the Martian atmosphere, expressed as the surface pressure as a function of time, $P(t)$, which is normalized to the present surface pressure, P_0. (From Melosh and Vickery (1989). Reproduced with permission from Macmillan Publishers Ltd. Copyright 1989, Nature Publishing Group.)

mass on the surface. The hot vapor plume from such events can expand faster than the escape velocity and drive off the overlying atmosphere into space (Sec. 5.12). Mars, with its small escape velocity, is much more prone to atmospheric impact erosion than Earth or Venus.

Models of impact erosion that use the estimated cratering rate on early Mars suggest that the early atmosphere lost 99% of its mass to impact erosion (Fig. 12.8) (Melosh and Vickery 1989). Thus, if Mars started with a 10 bar atmosphere after accretion, it would have had a ~100 mbar atmosphere by the end of the Noachian (~3.7 Ga), when bombardment had subsided. This model assumed that the heavy bombardment period represented an exponential tail to the main accretion process – a view that continues to undergo revision (see Sec. 6.7).

More recent models generally support atmospheric erosion on Mars, with the extent depending upon assumptions about the impactor flux and fraction of volatiles within impactors (Manning *et al.*, 2006b; Pham *et al.*, 2011; Svetsov, 2007). Another caveat is that if CO_2 was locked up as carbonates it would have been protected and not subject to impact erosion (Zahnle, 1993b).

Although the efficiency of impact erosion on Mars is debated (e.g., Lammer *et al.*, 2013), the abundance of some noble gas isotopes provides possible evidence for impact erosion. Argon has a radiogenic form, ^{40}Ar, which derives primarily from the radioactive decay of ^{40}K with a half-life of 1.3 Gyr. But there is also a nonradiogenic form, ^{36}Ar, which is primordial. Although the absolute amount of ^{40}Ar (on a kg Ar per kg planet basis) is much

Table 12.6 The isotopic composition of the atmosphere of Mars, as measured by the *Viking Landers*, MSL, and in trapped gas in Martian meteorites. Protosolar values (for primordial gases) and ratios for the Earth are listed for comparison. VLs = *Viking Landers*. MSL = *Mars Science Laboratory*.

Ratio	Mars (landers)	Mars (meteorites)	Earth	Proto-Solar	Mars isotope ratio relative to terrestrial	Inferences
$D/H \times 10^{-4}$ in H_2O	9.3 ± 1.7 (MSL)[a]	~10.8[b]	1.56[c]	0.25[d] ~1.6[e]	~6±1	Water loss
$^{12}C/^{13}C$ in CO_2	$85.1\pm0.3a$ (MSL)[a]	not reported	89.9[f]	89[g]	0.95	Exchangeable reservoir
$^{14}N/^{15}N$	173 ± 9 (MSL)[a]	>181	272[c]	440±6[c] 272[e]	0.64	Nonthermal escape
$^{20}Ne/^{22}Ne$	Not yet measured	~10	9.8	14.0[h]	~1	Possible hydrodynamic escape
$^{16}O/^{18}O$ in CO_2	476 ± 4 (MSL)[a]	~490	499[f]	499[g]	0.95	Exchangeable reservoir
$^{40}Ar/^{36}Ar$	1900 ± 300 (MSL)[a]	1800 ± 100	296	$\sim10^{-3b}$	~6	Impact erosion
$^{36}Ar/^{38}Ar$	4.2 ± 0.1 (MSL)[a]	3.9–4.2	5.3	5.50[h]	0.79	Hydrodynamic escape
$^{129}Xe/$ ^{132}Xe	2.5^{+2}_{-1} (VLs)[i]	2.4–2.6	0.97	–	~2.5	Hydrodynamic escape or impact erosion of ^{132}Xe

[a] Mahaffy (2015a) [b] Bogard (2001). The $^{40}Ar/^{36}Ar$ ratio is for 4.56 Ga. [c] Marty (2012). [d] Not the present Sun but pre-deuterium burning (Marty, 2012). [e] Primitive meteorites (Marty, 2012). [f] Henderson and Henderson (2009). [g] Lodders (2010b). [h] Pepin (2012). [i] Owen (1992). The Xe isotope composition measured by MSL is unavailable at the time of writing, 2016.

less on Mars than on Earth (see Fig. 6.13), the $^{40}Ar/^{36}Ar$ ratio is 1900±300 as measured by MSL or ~1800 in the trapped gas of Martian meteorites compared to 296 on Earth (Table 12.6). A possible explanation for the relative enrichment of ^{40}Ar on Mars is removal of ^{36}Ar by impact erosion early in Martian history and later accumulation of ^{40}Ar as it was released from radioactive decay and volcanism. We also noted earlier (Sec. 12.3.2) how krypton appears to have suffered a ~99% loss. Because krypton isotopes are not strongly fractionated, impact erosion, which is non-fractionating, is a plausible cause.

12.3.3.2 Hydrodynamic Escape

The second form of efficient escape operating early in Martian history was hydrodynamic escape (see Sec. 5.10). This thermal escape mechanism applies to a hydrogen-rich atmosphere, which may have existed from 4.5 Ga to ~4.2 Ga, in the Pre-Noachian. The atmosphere during accretion would have been steam from impact volatilization and outgassing from a magma ocean (Elkins-Tanton, 2008, 2012). Hydrogen would also have been very abundant from volcanism during internal differentiation of Mars, albeit for the first few million years (Dauphas and

Pourmand, 2011), when water in the mantle reacted with metallic iron (Dreibus and Wänke, 1987).

How long Mars' atmosphere remained hydrogen-rich depends on the redox state of Mars' mantle. If Mars' mantle is highly reducing (Wadhwa, 2001), the emission of hydrogen-rich volcanic gases could plausibly have continued long after core formation. For example, a mantle oxygen fugacity 2 log units below that of Earth, which may be possible (Sec. 7.3), would have increased the H_2:H_2O ratio in volcanic gases by a factor of 10, according to eq. (7.13). During the Pre-Noachian, high fluxes of extreme ultraviolet radiation (EUV) from the young Sun should have driven a hydrodynamic outflow of hydrogen that dragged away heavy species such as carbon and nitrogen. In hydrodynamic escape, heavy atoms are carried upward by collisions with hydrogen faster than they fall down under gravity.

The escape of heavy gases from hydrodynamic escape causes isotope fractionation (Sec. 5.11). Martian noble gases have isotopic patterns consistent with fractionation by early hydrogen-driven hydrodynamic escape. The xenon isotope ^{129}Xe is derived from radioactive decay of extinct iodine ^{129}I with a half-life 16 Myr and the fission of extinct ^{244}Pu with a half-life of 82 Myr. Xenon-129 is somewhat less abundant on Mars than Earth, but the relative abundance of ^{129}Xe compared to nonradiogenic xenon, ^{132}Xe, is large. *Viking* measured ^{129}Xe/^{132}Xe ~2.5 compared with ~1 on Earth and in most Solar System materials. A possible explanation is very early loss of the ^{132}Xe isotope is hydrodynamic escape (Jakosky and Jones, 1997; Zahnle, 1993b). The stable isotopes of argon on Mars, ^{36}Ar, and ^{38}Ar may also be fractionated by hydrodynamic escape, as discussed in Sec. 5.11.

Even after the atmosphere was no longer hydrogen-rich, high EUV fluxes may have removed considerable CO_2 on a short timescale (Tian *et al.*, 2009) or considerable hydrodynamic escape of the atmosphere could have happened shortly during a steam atmosphere phase (Maindl *et al.*, 2015). Under conditions of $20\times$ present EUV, the upper thermosphere of Mars reaches temperatures >2500 K, and carbon from CO_2 photolysis may escape hydrodynamically (Tian *et al.*, 2009). According to Tian *et al.* (2009), 1 bar of CO_2 could have escaped in 10 Myr. However, this model did not include significant amounts of hydrogen. If much of the EUV energy went into escape of H_2, then perhaps the amount of CO_2 lost would be less. More modeling is needed to test this possibility.

Certainly carbon isotopes on Mars with are fractionated compared to Earth with δ^{13}C = 46±4‰ in CO_2 (Mahaffy *et al.*, 2015a). Carbonates in ALH84001 formed at ~3.9 Ga and have δ^{13}C = 40‰ (apart from a Ca-rich phase with +20‰) (Niles *et al.*, 2013; Shaheen *et al.*, 2015). This could mean that much of the carbon was fractionated early but the interpretation depends upon an initial, primordial δ^{13}C value, which is unknown for Mars.

It is important to realize that escape losses are cumulative. For example, if impacts and hydrodynamic escape removed 99% of the earliest atmosphere, while later sputtering and photochemical escape removed 50%–90% of the remainder (Jakosky and Phillips, 2001), the total loss comes to 99.5%–99.9%.

Finally, these arguments for early escape of Mars' atmosphere must be weighed against evidence that fluvial features were produced at or near 3.8 Ga, as discussed below. That may be difficult if Mars lost too much of its atmosphere too fast. This is part of what makes understanding Mars interesting. Different lines of evidence sometimes appear to point in different directions.

12.4 Evidence for Past Climate Change and Different Atmospheres

12.4.1 Geomorphic Evidence of Possible Water Flow

Geomorphic features suggest that liquid water has affected the surface of Mars during the Noachian,

Hesperian, and Amazonian. Four different classes of features are valley networks (including their associated deltas), outflow channels, gullies, and recurring slope lineae, which we discuss in turn. Suggestions that Mars had oceans in the Noachian or Hesperian are also relevant, of course, and heavily eroded Noachian craters are germane if they were eroded by liquid water.

Valley networks. Valley networks consist of dendritic, interconnected valleys that have some similarity to terrestrial river networks. They occur mainly on Noachian or early Hesperian surfaces (Fassett and Head, 2008a) and are 1–4 km wide and 50–200 m deep (Fig. 12.9). Their cross-section tends to be V-shaped near their heads and more rectangular downstream. Many valley networks are less dense than terrestrial counterparts (Aharonson *et al.*, 2002; Stepinski *et al.*, 2004), which might be explained if they formed from subsurface release of water, called *groundwater sapping* (Baker, 1990; Carr and Clow, 1981; Gulick, 1998; Lamb *et al.*, 2006; Pieri, 1980), although many question whether sapping is sufficient and invoke precipitation (Craddock and Howard, 2002; Lamb *et al.*, 2008; Matsubara *et al.*, 2013). In any case, groundwater would need to be charged by precipitation or hydrothermal fluids (Goldspiel and Squyres, 2000; Gulick, 2001). Some Noachian networks have drainage densities that lie within the range of terrestrial river valleys, which suggests that they were fed by rain or snowmelt (Hynek *et al.*, 2010). Also, the characteristics of some valleys have been used to estimate peak discharge rates of 10^2–10^4 m^3 s^{-1}, which are comparable to terrestrial rivers (Burr *et al.*, 2010; Howard *et al.*, 2005; Irwin *et al.*, 2005; Jerolmack *et al.*, 2004; Moore *et al.*, 2003). Because the overall landscape erosion is relatively immature, most valleys terminate in local topographic depressions, where lakes may have formed temporarily. Sedimentary layers are common in such lows (Grant *et al.*, 2008) and some have exit breaches implying partial drainage (Fassett and Head, 2008b). Some simulations of erosional degradation suggest that valley networks were formed by repeated and modest fluvial events over timescales of 10^5–10^7 years (Barnhart *et al.*, 2009; Hoke *et al.*, 2011; Howard, 2007).

Most valleys do not terminate in deltaic deposits, but some do. Along the boundary between the northern plains and southern highlands, 52 deltas have been identified, with a mean elevation of –(1.8±1.2) km (Di Achille and Hynek, 2010). An example of a deltaic deposit is found in Eberswalde crater (24° S, 327° E), which contains clay minerals that were probably detrital, i.e., transported to the delta rather than forming in place (Milliken and Bish, 2010; Pondrelli *et al.*, 2011). The dimensions of the delta

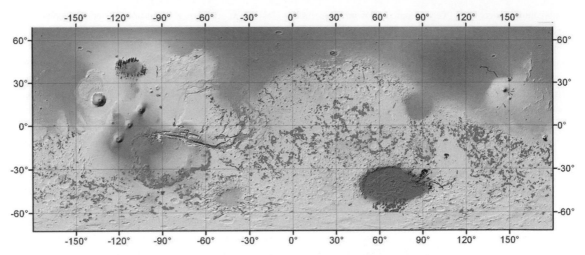

Figure 12.9 The global distribution of valley networks on a background contour map where highlands are shaded red and lowlands are shaded blue. Valleys on Noachian terrain are red, while those on younger terrain are blue for Amazonian and purple for Hesperian (From Hynek *et al.* (2010). Reproduced with permission from John Wiley and Sons. Copyright 2010, American Geophysical Union.) (A black and white version of this figure will appear in some formats. For the color version, please refer to the plate section.)

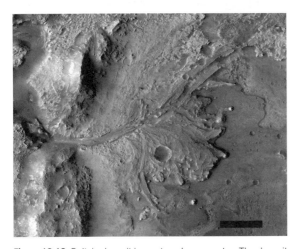

Figure 12.10 Deltaic deposit in western Jezero crater. The deposit is a positive relief feature, indicating that the material was more resistant to erosion than surrounding material. Phyllosilicate-bearing areas are green, while yellow and blue colors indicate basaltic minerals. Purple-brown surfaces have no distinctive spectral features. Scale bar = 2 km. (Courtesy of NASA/JPL/JHUAPL/MSSS/Brown University.) (A black and white version of this figure will appear in some formats. For the color version, please refer to the plate section.)

suggest a discharge rate comparable to terrestrial rivers operating over 10^3–10^6 years (Malin and Edgett, 2003; Moore *et al.*, 2003); intermittent rainfall ~1 cm day^{-1} and seasonal snowmelt could have produced an Eberswalde paleolake over 10^3–10^6 years (Irwin *et al.*, 2015). However, high impact-triggered fluvial discharges may have

formed the delta in as little as 10–10^3 years (Jerolmack *et al.*, 2004; Kleinhans, 2010; Kleinhans *et al.*, 2010; Mangold *et al.*, 2012). Other deltaic deposits such as the one in Jezero crater (18.4° N, 282.4° W) also contain clay minerals (Ehlmann *et al.*, 2008a) (Fig. 12.10). Sometimes there is no link between past warm global climates and deltas because the presence of late Amazonian-aged deltas indicates that water must have been mobilized locally from subsurface ice in these cases (Hauber *et al.*, 2013). The valleys associated with these young deltas are generally very short.

Shallow-floored and eroded craters. Craters of Noachian age are frequently heavily eroded with degraded rims and shallow floors (Fig. 12.11). The valley networks are incised into this degraded landscape and so cannot be responsible for the inferred erosion. Nonetheless, the estimated rates of Noachian erosion are modest and comparable to the cold, dry polar regions of the modern Earth (Golombek *et al.*, 2006). Erosion rates drop by a factor of ~10^3 after the Noachian, down to ~1 m Gyr^{-1}, suggesting a possible link between high erosion rates and impact bombardment. For comparison, current erosion rates in the Antarctic mountains are ~100 m Gyr^{-1} (Nishiizumi *et al.*, 1991). Some models give fluvial activity a significant role in producing shallow crater floors (Howard, 2007). Images suggest that craters were also degraded or obscured by impacts, impact ejecta, wind, mass wasting, or volcanic ash, which complicates specific attribution to fluvial processes, however. It is even possible for impacts into layered, icy terrain to produce shallow crater floors without post-impact modification (Senft and Stewart, 2008). Whatever process produced shallow crater floors, it shut down around the end of the Noachian because younger craters are systematically deeper. The global distribution of crater depth to diameter ratios is bimodal, reflecting this transition (Robbins and Hynek, 2012).

Some of the younger, deeper-floored craters have alluvial fans (Moore and Howard, 2005), but these fans

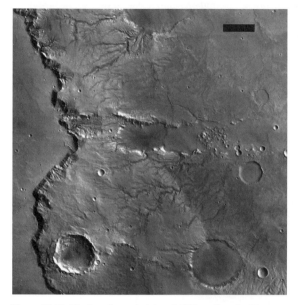

Figure 12.11 The heavily eroded rim of 456 km diameter Huygens Crater and smaller craters above the rim. Valley networks are imposed on the high ground and drain into local topographic depressions. The image is centered at 14° S, 61° E, north upwards, with scale bar = 20 km. (High Resolution Stereo Camera image, from *Mars Express*. (Courtesy of ESA/DLR/FU Berlin (G. Neukum).)

are post-Noachian (Grant and Wilson, 2011; Kraal *et al.*, 2008; Mangold *et al.*, 2012).

Outflow channels. A system of enormous channels, which formed during the Hesperian to early Amazonian, is another class of features indicating fluid flow. The dimensions of outflow channels dwarf the valley networks. Channels are 20–100 km in width, up to ~1000 km in length, and as much as several km deep (Fig. 12.12). The source of the outflow channels are areas of chaotic terrain, which are places where the ground collapsed and broke (Carr, 2006; Sharp, 1973), mainly within ±20° latitude and on the periphery of volcanic provinces such as Tharsis and Elysium. Most outflow channels drain into the northern plains. The dominant view is that the channels formed when groundwater was expelled from areas of chaos (Baker, 2006; Baker *et al.*, 2015).

Some estimates of the total amount of water needed to produce all the channels on Mars are equivalent to a global ocean several hundreds of meters deep (Table 12.5). The idea that floodwaters pooled provides the motivation for suggesting that an ocean existed in the northern lowlands during the late Hesperian, which is the cratering age of the channels (Kreslavsky and Head, 2002). However, some discharge rate estimates involve smaller amounts of water in events spread over time (Kleinhans, 2005; Williams *et al.*, 2000), with no single event big enough to fill the northern plains. The peak discharge rate required to form the largest channel system, Kasei Valles, which extends 3000 km from a source in Echus Chasma, is comparable to that of the most studied terrestrial analog, the channeled scablands of eastern Washington State, USA (Williams *et al.*, 2000). The terrestrial channels formed when floodwaters were released from Lake Missoula at the end of the last ice

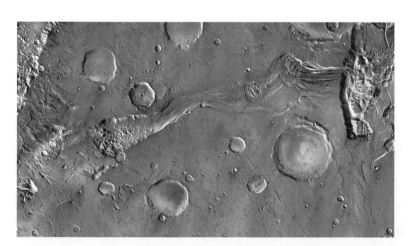

Figure 12.12 The outflow channel Ravi Vallis (0.5° S, 318° E), north upwards. The image is a mosaic of infrared frames from the Thermal Emission Imaging System on NASA's *Mars Odyssey* orbiter. The scene is 237 × 415 km. At the western end of the channel is the source region for the channel, Aromatum Chaos in a 90-km long depression. On the eastern end, the channel drops into Hydraoates Chaos, which connects to further channels that terminate in the northern plains. (Courtesy: NASA/JPL/ASU.)

age during periodic ruptures of an ice dam that reformed after each flood event.

An alternative school of thought is that the outflow channels were not carved by water at all but by very fluid lavas (Leone, 2014; Leverington, 2011). Certainly, the floodwater interpretation has several problems including the lack of correlation of outflow regions with evaporite minerals, the possible inconsistency of the large amounts of water required against the small water inventory on Mars inferred from geochemistry, and doubtful hydrological assumptions needed to reconcile large outbursts from relatively small areas. The latter includes very large permeabilities and headwaters at high elevations where the hydraulic pressure should be least (Leverington, 2011), unless volcanism melted ice sheets. Some outflow channels (e.g., Marte Vallis and Athabasca Valles) have lava flows on their floors (Jaeger *et al.*, 2007) or have geomorphology that resembles lava-carved lunar and Venusian rilles (Leverington, 2004). The sources of some outflow channels are also sources of lava (e.g., Cerberus Fossae for Athabasca Valles and Memnonia Fossae for Mangala Valles (Leverington, 2007)). Furthermore, ridged and platy flows at the heads of channels are evident at Athabasca, Mangala, and the largest channel, Kasei (Chapman *et al.*, 2010a, b). Geochemically inferred viscosities of Martian lavas are very low (Greeley *et al.*, 2005). This favors turbulent flow, which is required for

significant thermal or mechanical erosion of a channel floor (Jaeger *et al.*, 2010; Williams *et al.*, 2005). Analysis of Kasei Valles landforms suggests that lava there was erosional, but that lava alone cannot account for the full size of the channel (Dundas and Keszthelyi, 2014). Moreover, the presence of delta fans at the termination of Okavango Valles outflow channels is compelling evidence that at least some channels had a fluvial origin (Mangold and Howard, 2013).

Our view, given the above evidence, is that both erosional lava flows and water outbursts occurred on Mars. This conclusion should be unsurprising: magmatic intrusions should have melted permafrost, triggering floods in association with lava outpourings (e.g., McKenzie and Nimmo, 1999). For example, liquid water might be sourced from icy southern highlands (Cassanelli *et al.*, 2015).

Gullies. Gullies are incisions of tens to hundreds of meters length found on the walls of craters, plateaus, canyons or dunes, at latitudes 30–70°, predominantly on poleward-facing slopes with higher abundance in the southern hemisphere (Fig. 12.13) (Heldmann and Mellon, 2004). Tens of thousands of these features indicate flow on Mars during the Late Amazonian (Kneissl *et al.*, 2010; Malin and Edgett, 2000a). Slope inclinations vary 5–40° (Dickson *et al.*, 2007; Heldmann *et al.*, 2007) and gullies have the form of alcoves above small channels of lengths

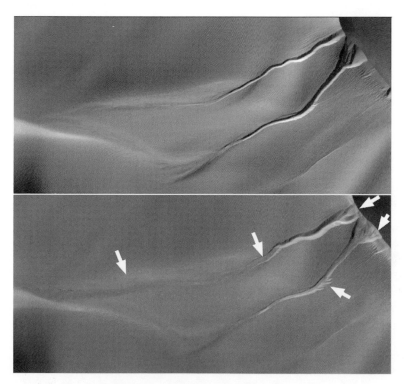

Figure 12.13 Gullies in Matara Crater (49.4° S, 34.7° E) run downhill from the crest of a sand dune. The image is 1.2 km across. The upper image was taken during mid-autumn in the southern hemisphere in 2008, while the lower image was taken during the beginning of the following summer, in 2009. Arrows at the upper right point to alcoves that widened over winter and other arrows point to changes in the channel structure, including a lengthening of the apron at the bottom of the channels. (HiRISE images PSP_007650_1300 and ESP_013834_1300, NASA/JPL-Caltech/University of Arizona.)

~0.01–3 km that terminate in debris aprons. Because gullies formed so recently, they cannot require a much thicker atmosphere to form. Their setting on slopes initially led to suggestions that they were all produced from sediment-rich flows similar to those produced in terrestrial alpine regions when ice or snow melts. One idea was that on Mars, such ice would have been deposited in Martian mid-latitudes during past "ice ages", corresponding to periods of high obliquity in Milankovitch-type cycles (Sec. 12.6) (Christensen, 2003; Costard *et al.*, 2002; Head *et al.*, 2003; Williams *et al.*, 2009).

Subsequent high-resolution images have shown that gullies are forming on today on Mars when CO_2 frost vaporizes, which presumably releases a dry flow of rocks and soil (Dundas *et al.*, 2012; Dundas *et al.*, 2015; Raack *et al.*, 2015; Vincendon, 2015). An association with seasonal CO_2 explains the predominantly poleward facing slopes and the higher abundance of gullies in the southern hemisphere, which experiences a longer winter. On timescales of millions of years over Milankovitch-type cycles, all gullies might form in this way, making a role for liquid water unnecessary (Diniega *et al.*, 2013; McEwen, 2013). Nonetheless, models for gully formation that invoke the melting of water ice continue to be discussed (e.g., Jouannic *et al.*, 2015).

The influence of Milankovitch-type cycles is also evident in a thin (~10 m), patchy mantle of material in a 30–55° latitude band in each hemisphere that is made of cemented dust (Mustard *et al.*, 2001). This *latitude-dependent mantle* (LDM) is assumed to have originated as an ice–dust mixture deposited during times of higher obliquity (>~35°) when ice spread to lower latitudes. Presently, much of the ice has sublimated. The LDM can bury gullies and its removal can exhume them, which complicates our understanding of gully formation (Dickson *et al.*, 2015).

Recurring Slope Lineae (RSL). On Mars today, narrow (0.5–5 m), dark linear features appear on steep slopes (25–40°) and lengthen during the warm season; these are RSL found in equatorial and mid-latitude regions (McEwen *et al.*, 2014; McEwen *et al.*, 2011). The RSL disappear in the cold season, but reappear annually. Currently, how RSL form is unknown. However, an association with equatorward slopes and temperatures of 250 K to >273 K suggests seepage or flow of water (Grimm *et al.*, 2014; Stillman *et al.*, 2014). RSL on equatorial dunes perhaps makes groundwater an unlikely source, so they may result from deliquescence of very hydroscopic salts (e.g., perchlorates). The presence of brines is supported by near-infrared detection of hydrous salts, possibly perchlorates, in the RSL (see Mushkin

et al., 2010; Ojha *et al.*, 2015). If brines cause RSL, they might be excellent places to look for extant life.

Oceans. Whether there were once oceans on Mars is controversial. Different oceans have been proposed for various times in Martian history. As mentioned earlier, a late Hesperian ocean in the northern plains has been proposed because floodwaters from outflow channels are presumed to have pooled there. Loss of floodwaters is hypothesized to have left behind a ~10^2 m deep layer of sediments in the northern lowlands, called the Vastitas Borealis Formation (VBF) (Head *et al.*, 2002). However, small impact craters that have penetrated the VBF produce ejecta consisting of apparently dark boulders and such boulders elsewhere on Mars are interpreted as igneous (Catling *et al.*, 2012). Also, Hesperian material underlying the VBF that is exposed by impact craters has basaltic mineral signatures in the near-infrared (Salvatore *et al.*, 2010).

Based on hypothesized shoreline features and common drainage elevations of some deltas, some have argued that an older, different ocean existed during the Noachian (Clifford and Parker, 2001; Di Achille and Hynek, 2010) or that oceans were episodic from the Noachian to post-Hesperian times (Baker, 2001; Fairen *et al.*, 2003). (Some of these deltas, however, are now dated as later than Noachian). Inner (*Deuteronilus*), center (*Arabia*) and outer (*Meridiani*) "shorelines" of different putative ocean stands have been identified in orbiter images, with much of the inner "shoreline" corresponding to the edge of the VBF (Clifford and Parker, 2001; Parker *et al.*, 1993; Parker *et al.*, 2010).

Shorelines were originally mapped from *Viking* orbiter images, but extensive examination of newer high-resolution imagery has failed to find convincing shoreline geomorphology (Ghatan and Zimbelman, 2006; Malin and Edgett, 1999, 2001). Indeed, many "shorelines" have been reinterpreted as mass-wasting and volcanic flow features (Carr and Head, 2003). It is also thought that finding evidence of a possible Noachian ocean without very deep drilling is unlikely because of burial or erosion (Head *et al.*, 2002). Proof of past oceans thus remains elusive.

12.4.2 Mineralogy and Sedimentology

Apart from geomorphlogy, another indication of liquid water in the past is the presence of minerals produced by water. *Aqueous alteration* is the process by which minerals change because of chemical reactions with water. The surface of Mars is predominantly basalt (Bandfield, 2002; Bibring *et al.*, 2005) and basaltic minerals can react with water. Ions are released into solution, which give rise to *alteration minerals*, such as clay minerals, carbonates, sulfates, silica, or iron oxides, depending upon the subsequent chemical and physical environment. The reddish dust on Mars is the most obvious example of material altered from basalt. It has 20–25 wt% ferric iron in a poorly crystalline phase described as nanophase ferric oxide (Morris *et al.*, 2006). The origin of the dust composition remains unknown but it appears to have geochemical similarity to meteorite NWA 7533/7034, a Noachian basaltic breccia that probably came from the highlands (Agee *et al.*, 2013; Beck *et al.*, 2015; Humayun *et al.*, 2013).

As a global average, Mars appears to have a physically processed regolith with minor aqueous alteration. For example, only ~3% of the Noachian surface has hydrous minerals (Carter *et al.*, 2013). Nonetheless, many alteration minerals have been detected (e.g., Ehlmann and Edwards, 2014) (Table 12.7). The minerals most convincingly associated with liquid water are phyllosilicates,[7] which are found in Noachian terrains. Sulfates are found in late Noachian or Hesperian areas. Iron oxides are common on Amazonian surfaces but are not necessarily associated with formation in liquid water.

The observations of predominant types of alteration mineral through time led Bibring *et al.* (2006) to suggest an alternative Martian timeline consisting of three mineralogical epochs. The oldest one has a prevalence of phyllosilicates and some carbonates; this is followed by a sulfate era; and finally, there is a long period up to the present day when anhydrous iron oxides dominate. Bibring *et al.* called these epochs the *Phyllocian*, when alkaline or near-neutral waters weathered basalt into clay minerals, the *Theiikian*, which was an acidic sulfur-dominated environment, and the *Siderikian*, marked by a cold, dry environment with iron oxides. Roughly, the mineralogical epochs are equivalent to the Noachian, the late Noachian to Hesperian, and Amazonian, respectively. However, dividing up time on Mars into mineralogical epochs has problems, and so the names have not caught

on. Sulfates and phyllosilicates are sometimes found interbedded, e.g., Columbus crater (30° S, 158° W) in Terra Sirenum (Wray *et al.*, 2009) or mixed together, e.g., Melas Chasma (Weitz *et al.*, 2015).

Carbonates. Carbonates have generated much discussion concerning their implications for the past atmosphere because of their connection to CO_2 (Niles *et al.*, 2013). Pollack *et al.* (1979) proposed that Mars must have once had a thicker CO_2 atmosphere with a sufficiently large greenhouse effect to allow liquid water to exist. Calculations suggested that 5 bar of CO_2 would be needed (Pollack *et al.*, 1987). But it was soon realized that CO_2 above a ~1 bar threshold condenses into clouds at the distance of early Mars under a fainter young sun (Kasting, 1991). Calculations that include the climatic effect of CO_2 ice clouds show that any warming from such clouds is limited by the precipitation of CO_2 ice and self-dissipation upon warming, so that mean global temperatures above freezing appear unachievable (Colaprete and Toon, 2003) (see Sec. 12.5.2.1 below).

Because CO_2 dissolves in water and produces solutions during weathering reactions that precipitate carbonates (e.g., see Sec. 11.4.1), a prediction of an early Mars model with a thick CO_2 atmosphere is the production of abundant carbonates. Using up 1 bar of CO_2 in weathering reactions would produce a planet-wide cover on Mars of ~20 m of pure calcite. However, carbonate-bearing rocks are very geographically restricted. They are found in Mg/Fe form in layered rock in Nili Fossae (some large fissures northwest of Isidis basin) (Ehlmann *et al.*, 2008b), in layered rocks excavated by an impact from ~6 km depth east of Syrtis Major (Michalski and Niles, 2010), and in magnesite outcrops in Gusev Crater (Morris *et al.*, 2010; Ruff *et al.*, 2014). Additionally, carbonates are found in small volumes (<1% vol) in Martian meteorites (Bridges *et al.*, 2001), in a 2–5 wt% magnesite component of the global Martian dust (Bandfield *et al.*, 2003), and as a 3–5 wt% component in soil at the *Phoenix Lander* site (Boynton *et al.*, 2009; Sutter *et al.*, 2012).

The amount of CO_2 locked up in the above occurrences depends upon assumptions but is probably small (Table 12.4). For example, the carbonate percentage in dust probably can't be extrapolated into the surface, as explained in Sec. 12.3. Lab experiments show that slow weathering of dust (involving thin films of water) over hundreds of millions of years could account for the amount of carbonate in the dust (Stephens, 1995a, b). Also, the fact that outcrops are magnesite-dominated may provide clues to their formation. Hydrous magnesites usually form from hydrothermal alteration of mafic or ultramafic rocks or alteration of serpentinites that were, in turn, originally

---

[7] The term *clay* alone refers to materials with grain size less than 1/256 mm or 4 μm. In contrast, *clay mineral* is a compositional term taken to mean phyllosilicates and aluminosilicate minerals with a crystal structure that derives from phyllosilicates. Phyllosilicates are minerals consisting of parallel sheets of $[SiO_4]^{4-}$ tetrahedra linked to sheets with Al–O or (Mg,Fe)–O chemistry. Phyllosilicates are placed in groups according to the layering. The 1:1 group, which includes *kaolinite* ($Al_2Si_2O_5(OH)_4$) have a layer of silicate tetrahedra linked by hydrogen bonds to OH groups on an octahedral aluminum layer. The 2:1 group, exemplified by *smectite*, has an octahedral sheet sandwiched between two tetrahedral sheets and interlinked to them.

Table 12.7 Alteration minerals on Mars. (Adapted from Ehlmann (2010) and Murchie *et al.* (2009b).) With the exception of gypsum circumpolar dunes and the *Phoenix* site, occurrences tend to be concentrated in low latitudes.

Class	Minerals	Chemistry	Example locations on Mars
Non-silicates			
Carbonates	Ca/Mg carbonate	$(Ca.Mg)CO_3$	*Phoenix* site soil Large impact crater east of Syrtis Major
	Magnesite	$MgCO_3$	Nili Fossae; Gusev crater; global dust
	Siderite	$FeCO_3$	Large impact crater east of Syrtis Major
Sulfates	Gypsum	$CaSO_4 \cdot 2H_2O$	Iani Chaos; north circumpolar dunes; Noctis Labyrinthus; perhaps Mawrth Vallis and Meridiani Gusev; veins at Endeavour and Gale craters.
	Bassanite	$CaSO_4 \cdot 0.5H_2O$	Mawrth Vallis; Gale crater
	Mono and poly-hydrated sulfates	$(Fe,Mg)SO_4 \cdot nH_2O$ (e.g., kieserite, $MgSO_4 \cdot H_2O$)	Possibly Terby; Valles Marineris region: Ius, Hebes, Capri, Candor, Melas, Juventae Chasma, Iani Chaos, Aram Chaos, Meridiani, Gale crater
	Jarosite	$KFe^{3+}_3(SO_4)_2(OH)_6$ $(H_3O)Fe^{3+}_3(SO_4)_2(OH)_6$	Mawrth Vallis; Meridiani; Juventae plateau Melas Chasma
	Alunite	$KAl_3(SO_4)_2(OH)_6$	Terra Sirenum craters
Halides (chlorides?)	Metal-halides	Probably NaCl-dominated	Hundreds of Noachian and Hesperian locations; *Phoenix* site
Perchlorates	No mineral names exist	$Mg(ClO_4)_2$, $(Na,K)ClO_4$	*Phoenix* site; MSL; North polar cap
Iron oxides	Hematite	Fe_2O_3	Meridiani; Aram Chaos; Candor, Melas, Juventae, Tithonium and Eos Chasmata
	Goethite	$FeOOH$	Gusev Crater rocks
	Akaganeite	$FeO(OH)Cl$	Gale, Antoniadi basin, various craters
Sheet silicates			
Phyllosilicates	Fe/Mg- smectites	$(Ca, Na)_{0.3--0.5}(Fe,Mg, Al)_{2-3}(Si, Al)_4O_{10}(OH)_2$ (e.g., Fe-rich nontronite $Na_{0.3}Fe^{3+}_2(Si,Al)_4 O_{10}(OH)_2$, Mg-rich saponite $Ca_{1/6} Mg_3(Al,Si)_4O_{10}(OH)_2$)	Mawrth Vallis; Nili Fossae; Holden, Eberswalde, Jezero, Terby, Gale craters; Walls of Valles Marineris; Terra Sirenum; Meridiani; S. highlands
	Kaolinite group	$Al_2Si_2O_5(OH)_4$ (kaolinite) (e.g., halloysite $Al_2Si_2O_5(OH)_4 \cdot (0-2) H_2O$)	Mawrth Vallis; Nili Fossae
	Montmorillonite	$(Na,Ca)_{0.33}(Al,Mg)_2(Si_4O_{10}) (OH)_2$	Mawrth Vallis
	Al,K-phyllosilicates	$KAl_2AlSi_3O_{10}(OH)_2$	Nili Fossae
	Chlorite	$(Mg,Fe^{2+})_5Al(Si_3Al) O_{10}(OH)_8$	Large N. Plains or S. highland craters
	Serpentine	$(Mg,Fe)_3Si_2O_5(OH)_4$	Claritas Rise; Nili Fossae; highlands impact craters, regional olivine-rich unit near Isidis
	Prehnite	$Ca_2Al(AlSi_3O_{10})(OH)_2$	Large N. Plains craters; S. highlands

Table 12.7 (*cont.*)

Class	Minerals	Chemistry	Example locations on Mars
Framework silicates			
Silica	Opaline silica	$SiO_2 \cdot 2H_2O$	Gusev; Valles Marineris Hesperian plains; Ius/ Melas Chasmata; Noctis Labyrinthus; Mawrth Vallis
Analcime (a.k.a. analcite)	Analcime/ analcite	$NaAlSi_2O_6 \cdot H_2O$	Highlands bordering Isidis or Hellas; Terra Cimmeria

produced by hydrothermal activity (Pohl, 1989; Russell, 1996). However, some thermodynamic models suggest that magnesite could form on Mars at modest temperatures of ~278 K (van Berk and Fu, 2011).

The lack of extensive carbonate outcrops has been used as evidence that Mars never had a thick CO_2 atmosphere, but some dispute this interpretation. One proposal to reconcile a thicker CO_2 atmosphere with lack of carbonates is that tiny amounts of volcanic SO_2 ~10^{-5} bar in the atmosphere could have inhibited carbonate formation by making surface waters acidic (Bullock and Moore, 2007; Fairen et al., 2004; Halevy et al., 2007). However, eventually, the CO_2 has to go somewhere after volcanism subsides and sulfur-related acidity is neutralized by a basaltic regolith. There are many post-Noachian craters with many kilometers of stratigraphic uplift in central peaks, but no carbonates. So, in the scenario involving the SO_2-induced suppression of carbonates, the CO_2 has to escape to space.

Another idea is that carbonates were not stable on Mars' surface because rainfall was acidic from high atmospheric CO_2 levels (Kasting, 2010, Ch. 8). Rainfall on Holocene Earth with ~300 ppmv, or 3×10^{-4} bar of CO_2, has a pH of ~5.7. The pH drops by 1 unit for each factor of 100 increase in pCO_2. Thus, rainwater on early Mars with an assumed pCO_2 ~3 bar would have a pH ~3.7, even in the absence of SO_2, and would have dissolved near-surface carbonates. Both bicarbonate and the associated Ca^{2+} and Mg^{2+} ions would be transported to the subsurface and precipitated once acidity was neutralized by reaction with basalt. Thus, this model predicts that carbonates should be abundant in the deep Martian subsurface. However, the amounts of carbonate required to remove 1 bar of CO_2 is equivalent to a layer of pure carbonate ~20 m thick. Martian meteorites, some of which represent samples from depth, have very small concentrations of carbonate (see above), which does not support this weathering hypothesis. Also, post-Noachian craters do not reveal large quantities of carbonate.

Phyllosilicates. Near-infrared orbital spectra show diverse phyllosilicates, with thousands of occurrences in the southern highlands and incidences associated with northern lowlands' craters that penetrate through kilometer-thick lava flows (Ehlmann et al., 2011; Ehlmann et al., 2013). In contrast, young volcanic provinces do not exhibit clay mineral spectra. Three geomorphic settings of phyllosilicates exist: crustal, sedimentary and stratigraphic (Ehlmann et al., 2011). *Crustal* (or deep) ones occur in igneous rocks, including areas degraded by impact craters. *Sedimentary* deposits are fans or layered materials within basins. *Stratigraphic* clays exist within multiple clay-bearing units but without a crustal or in-basin sedimentary designation. In all settings, Fe/Mg smectites are the most common clay minerals, found in 75% of occurrences. The smectites range from the iron endmember, nontronite, to the magnesium endmember, saponite (see Table 12.7 for their chemical formulae). Chlorite is the second most common phyllosilicate. Less common minerals, which nonetheless are often present in crustal settings, include prehnite, serpentine, illite or muscovite, analcime, and epidote, which generally indicate alteration by at elevated temperatures. Hydrated silica is also found. Sedimentary clay mineral assemblages include Fe/Mg smectites, kaolinite, and also non-clays such as salts (carbonate, sulfate and chloride (Glotch et al., 2010; Osterloo et al., 2010)). Stratigraphic clay-bearing units often have a lower unit with Fe/Mg-clays and an upper unit with Al-clays.

The location of phyllosililcates in exposed Noachian crust suggests that clay-forming aqueous chemistry mostly occurred during the Noachian. However, diverse settings for the phyllosilicates indicate multiple origins for the phyllosilicates such as subsurface fluids, near-surface weathering, and in ice-dominated near-surface systems. For example, phyllosilicates excavated from craters include minerals such as serpentine that form in environments up to 400 °C (Ehlmann et al., 2010). In contrast, higher temperature alteration minerals appear absent in sedimentary phyllosilicates. In Jezero crater (18.4° N, 282.4° W) there are Fe,Mg-smectite clays and Mg-carbonate (Ehlmann et al., 2008a), while Terby crater (27.7° S, 74.1° E) has layered deltaic deposits containing Fe,Mg smectites (Ansan et al., 2011). These sedimentary phyllosilicates are thought to be detrital because the source areas surrounding the basins contain the same phyllosilicates (Milliken and Bish, 2010). In the third type of setting, stratigraphic, phyllosilicate units sometimes drape pre-existing topography, which suggests in situ alteration of surface materials, airfall deposition, or both. Differences in alteration chemistry and temperature might explain why Al clays are found above Fe/Mg clays.

Sulfates. Other abundant minerals are sulfates. Sulfates occur in many light-toned layered deposits (LLDs) based on spectroscopic evidence (Bibring et al., 2006; Gendrin et al., 2005; Murchie et al., 2009b) but not all LLDs (e.g., mounds in W Arabia Terra). Soils also contain 4–8 wt% sulfur budgeted as SO_3, based on in situ x-ray spectroscopy (Clark, 1993; Clark et al., 2005; Wänke et al., 2001; Yen et al., 2005), with 1.3 ± 0.5 wt% soluble sulfate identified in *Phoenix* soils (Kounaves et al., 2010; Toner et al., 2014b). More specifically, basaltic materials mixed with Fe^{3+}/Mg/Ca-sulfates, silica, and Ca-phosphates dominate light-toned soil exposed by the wheels of *Spirit Rover* (Yen et al., 2008). Veins of Ca-sulfate also occur at *Opportunity* and MSL rover sites (Arvidson et al., 2014; Grotzinger et al., 2015a; Grotzinger et al., 2014).

In low latitudes, LLDs occur inside chasms, canyons, craters, or chaotic terrain (Bishop et al., 2009; Catling et al., 2006; Glotch and Rogers, 2007; Lichtenberg et al., 2010; Noel et al., 2015; Roach et al., 2010a, b). Generally, LLDs inside chasms contain mono- and polyhydrated sulfates, whereas LLDs in depressions on the plateau around Valles Marineris contain Fe-sulfates and opaline silica (Le Deit et al., 2010; Milliken et al., 2008; Weitz et al., 2010).

Despite their significance, how LLDs formed remains an enigma. Proposals include: (1) sedimentary deposits from lakes (Komatsu et al., 1993; Malin and Edgett, 2000b; McCauley, 1978; Nedell et al., 1987); (2) airfall deposits (Catling et al., 2006; Malin and Edgett, 2000b; Nedell et al., 1987; Niles and Michalski, 2009; Peterson, 1981); (3) mass wasting materials (Lucchitta et al., 1994; Nedell et al., 1987); or (4) volcanic deposits (Chapman et al., 2003; Hynek et al., 2003; Komatsu et al., 2004; Lucchitta et al., 1994; Peterson, 1981). Proposal (2) also includes an idea that LLDs could be sublimation residues left behind from ice and sulfate aerosols co-deposited during times of active volcanism and high obliquity when snow fell preferentially in the tropics (Catling et al., 2006). Dust may have been aqueously altered to produce sulfates when ice-trapped dust was heated by sunlight (Michalski and Niles, 2012; Niles and Michalski, 2009).

Observations in Meridiani Planum by *Opportunity Rover* provide insights into the origin of LLDs there, which are several hundred meters thick and cover >500 000 km^2 (Squyres et al., 2006). There, the LLD is sandstone composed of ~40 wt% salts (mainly sulfates), ~30 wt% hydrated silicate alteration products, ~10 wt% oxides, and ~20 wt% basaltic debris (Clark et al., 2005). LLDs in West Candor Chasma may be comparable because orbital spectra show that they are dominated by dust and basaltic sand, accompanied by hydrated sulfates and ferric minerals (Bibring et al., 2007; Murchie et al., 2009a). At Meridiani, millimeter-scale spherules of hematite (Fe_2O_3) embedded within the sulfate-rich sandstone (Weitz et al., 2006) are interpreted as concretions, which are compact mineral accumulations of different composition from the surrounding rock formed by precipitation from groundwater (Fig. 12.14(a)). Meridiani LLDs are of Noachian–Hesperian boundary age, ~3.7 Ga (Hynek and Phillips, 2008), although most LLDs are Hesperian (Carr and Head, 2010). Meter-scale cross-bedding (where the deposition of layers is at an angle to the original deposition surface) in some LLD exposures (Squyres et al., 2009) show that the sediments were formerly migrating dunes (Grotzinger et al., 2005).

The following sequence of events probably gave rise to the Meridiani LLDs. Sulfates are presumed to be deposited originally in playas. Then wind erosion and re-deposition produced sulfate-rich dunes (McLennan and Grotzinger, 2008; Squyres et al., 2006). Subsequently, groundwater upwelled through the dunes, carrying dissolved iron from which hematite precipitated into concretions on a timescale of ~10^2–10^3 years (Sefton-Nash and Catling, 2008), possibly under hydrothermal conditions (Golden et al., 2008). The presence of jarosite (generically $(K,Na,H_3O)(Fe,Al)(OH)_6(SO_4)_2$) in the LLD (Klingelhofer et al., 2006), implies acidic and oxidizing conditions (Catling, 2004; King and McSween, 2005; Squyres and Knoll, 2005). The upper ~1 m of the Burns Formation (part of the regional LLD) contains evidence that upwelling groundwater ponded between dunes. Centimeter-scale cross-laminations that intersect in concave-upwards sets called *festoons* indicate ripples formed by gentle subaqueous flow (Grotzinger et al., 2005) (Fig. 12.14(b)). Initial aqueous alteration was limited and subsequent liquid water has been minimal (Elwood-Madden et al., 2009; Tosca and Knoll, 2009). Wind has eroded some of the sandstone at

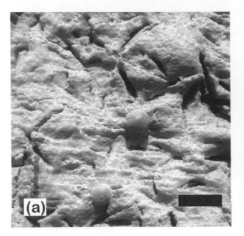

Figure 12.14 (a) A microscopic image mosaic from the Mars *Opportunity Rover* taken of a rock called "McKittrick" in Eagle Crater (the landing site) on sol 29 of the mission (scale bar = 6 mm). The two millimeter-scale spherules in positive relief are hematite (Fe_2O_3) concretions. There are also needle-shaped crystal molds in negative relief. (b) Microscopic imager mosaic from the Mars *Opportunity Rover* taken on sols 39 and 40 of a rock called "Last chance" in Eagle Crater. These swooping cross-beds (indicated by the lines drawn on) are known as festoon cross-laminations. On Earth, they are formed in shallow water that flows toward or away from the viewer. (Courtesy: NASA/JPL/USGS.)

Meridiani, leaving a lag of hematite concretions that is detectable from orbit in thermal infrared spectra (Christensen *et al.*, 2001).

Inferences about the past climate from Meridiani are that ponding between dunes and the formation of concretions required groundwater but not direct rainfall per se, otherwise the surface would show dissolution textures (McLennan and Grotzinger, 2008). However, liquid water would be required to charge regional-scale groundwater somehow, perhaps through snowmelt (e.g., Kite *et al.* (2012) and Sec. 12.5.2.3). Alternatively, some models assume that aquifers were supplied by rainfall cosine-distributed about the equator, which upwelled later (Andrews-Hanna *et al.*, 2007; Andrews-Hanna *et al.*, 2010). Sulfates at high latitudes may have their origin in eolian transport, periglacial or impact processes (Ackiss and Wray, 2014)

Abundant sulfates hint at the chemistry of volcanic gases. Wänke and Dreibus (1994) used the high abundance of sulfur relative to water in Martian meteorites to argue that volcanic gases were more sulfur-rich on Mars than on Earth. Essentially, with a higher ratio of sulfur to water, Martian volcanoes would have effectively spewed sulfuric acid. In contrast, terrestrial volcanic gases are dominated by steam (Sec. 7.2.3). Outgassing models suggest that Martian volcanoes released 10–100 times more S than typical terrestrial volcanoes (Gaillard and Scaillet, 2009). Tharsis, for example, could have cumulatively produced the equivalent of a 20–60 m global layer of sulfate. Although SO_2 is a greenhouse gas, the net climatic effect of SO_2 volcanic gas is probably to cool early Mars because SO_2 oxidizes into aerosols that reflect sunlight (Tian *et al.*, 2010) (see Sec. 12.5.2.1).

12.5 Explaining the Early Climate of Mars

12.5.1 The Faint Young Sun Problem

The evidence in favor of liquid water on early Mars just discussed in Sec. 12.4 has to be reconciled with the problem that the ancient Sun was fainter. In Sec. 11.1, we discussed the theory behind the gradual brightening of the Sun over its lifetime and supporting astronomical data. The Sun should have been ~25% less luminous around 3.7 Ga at the end of Noachian when fluvial erosion and the formation of large drainage networks were prevalent (Fassett and Head, 2008a; Howard *et al.*, 2005). Using eq. (2.15), the equilibrium mean global emission temperature of early Mars – equivalent to the temperature in the absence of an atmosphere – would have been 195.5 K, using a Bond albedo of 0.25 (the same as the modern value) and a solar flux that was 75% of present. Consequently, 77.5 K of greenhouse warming would have been needed on Mars to reach a mean temperature above 273 K, the melting point of pure water. This amount of warming is very large compared to the ~33 K greenhouse effect of the Holocene Earth (Sec. 2.2.4).

A Bond albedo on early Mars that was lower than today's (perhaps because the surface was not as oxidized and as bright) does not solve the problem. An albedo of ~0.1 is typical of basalt. It would only raise the equilibrium temperature to ~205 K, which would still require a 68 K greenhouse effect to reach 273 K.

12.5.2 Mechanisms for Producing Early Climates Conducive to Fluvial Erosion

For Mars, three ideas have been proposed to reconcile evidence of widespread fluvial activity with the fainter Sun. One is that the greenhouse effect was bigger in the past because a thicker atmosphere contained higher concentrations of greenhouse gases (reviewed by Haberle et al. (2017)). A second concept rejects the assumption of a persistently warmer climate and advocates that fluvial features were caused by many temporary warm episodes associated with impacts. The idea is that the energy released from impacts would have heated the surface of early Mars, vaporized ice into steam, and produced rainfall that eroded river valleys (Toon et al., 2010). A third proposal is that the fluvial and sedimentary features were produced in an environment that was never substantially warmer, only marginally so. Fluvial erosion might be produced in response to fortuitous combinations of orbital parameters, allowing localized snowmelt (Kite et al., 2013; McKay, 2004). The freezing point depression of concentrated brines might also have produced fluvial erosion in a cold environment. Let us consider each of these three concepts in turn after considering empirical constraints on the atmospheric paleopressure of Mars.

12.5.2.1 Empirical Constraints on the Paleopressure of Early Mars

Because the Martian atmosphere is postulated to have been thicker in the past, geomorphic and geochemical data have been used to quantify paleopressure p_{surf} on early Mars.

The results for paleopressure are as follows: <0.4 bar at ~4.1 Ga, <0.9±0.1 bar at 3.7-3.5 Ga, and >0.12 bar in the late Noachian/early Hesperian. The lower bound at 4.1 Ga comes from a $^{40}Ar/^{36}Ar$ ratio of 626±100 in trapped argon in ALH84001 (plausibly the atmospheric ratio at the time), compared with ~10^{-3} when Mars formed and the present ratio of ~1800 (Table 12.6). The large early value of $^{40}Ar/^{36}Ar$ before ^{40}Ar could accumulate from radioactive decay implies that ^{36}Ar must have been lost early. A range of models for the evolution of $^{40}Ar/^{36}Ar$ implies the upper limit on p_{surf} (Cassata et al., 2012). The constraint at 3.5–3.7 Ga comes from examining the size distribution of craters (which depends on

atmospheric pressure) in an area of Mars, Aeolis Dorsa, with highly dense and well-preserved craters (Kite et al., 2014). The area also contains river valleys, so was contemporaneous with a "wet" Mars. The lower limit of 0.12 bar was derived from an estimate of the terminal velocity of a volcanic bomb sag in Gusev crater (Manga et al., 2012). A volcanic bomb is a rock ejected in a volcanic eruption that falls and creates a bomb sag, or crater, where it lands. Earlier we mentioned an inference of a ~0.1 bar atmosphere after the end of late heavy bombardment, from the C/^{84}Kr ratio (Sec. 12.3.2), which is consistent with Kite et al. (2014). It is also possible to infer a CO_2 partial pressure, pCO_2, for the formation of carbonates, e.g., the Comanche outcrop found by Spirit Rover (van Berk et al., 2012). However, pCO_2 may not represent the atmosphere if the carbonates are hydrothermal (Morris et al., 2010).

12.5.2.2 An Enhanced Greenhouse Effect on Early Mars?

Greenhouse gases that have been considered for warming early Mars include carbon dioxide (CO_2), water vapor (H_2O), methane (CH_4), ammonia (NH_3), sulfur dioxide (SO_2), and hydrogen (H_2). As mentioned in Sec. 12.4.2, early work proposed that a thick CO_2 atmosphere together with water vapor warmed early Mars (Pollack et al., 1987). However, there are two problems with high CO_2 providing a lot of greenhouse warming on early Mars (Kasting, 1991). The first is that under the low radiation flux from the early Sun, CO_2 condenses into dry ice in the Martian upper troposphere. As CO_2 condenses, enthalpy (latent heat) is released, which fixes the lapse rate in the upper troposphere to the CO_2 frost point. More outgoing infrared from a warmer upper atmosphere means that the surface remains cool to maintain the balance between the outgoing infrared flux to space and the incoming solar flux. Adding more CO_2 simply causes a greater vertical extent of the atmospheric temperature profile to follow the CO_2 frost point. The second problem is Rayleigh scattering of sunlight by CO_2. As the amount of CO_2 increases, more sunlight returns to space by Rayleigh scattering, which counteracts the greenhouse warming. An updated parameterization for collision-induced absorption (see Sec. 2.5.6) in dense CO_2 atmospheres (Wordsworth et al., 2010) has been applied, which provides less absorption than previous calculations. The result is that at ~2 bars surface pressure, a 1-D radiative-convective model for the early Martian atmosphere attains a surface temperature of only ~212 K at 3.8 Ga, which is barely different than today (Fig. 12.15).

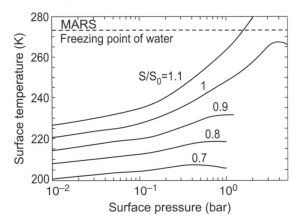

Figure 12.15 Global mean surface temperature as a function of surface pressure for a CO_2–H_2O atmosphere. S/S_0 is the solar luminosity relative to today's value. (Source: R. Ramirez.)

In principle, the formation of CO_2 ice clouds might scatter infrared radiation back to the surface and warm it. However, calculations show insufficient warming to solve the faint young Sun problem. Forget and Pierrehumbert (1997) used a model with 100% cloud cover and comparatively large, 10-μm radius ice particles, in order to obtain surface temperatures of 273 K in atmospheres at >0.5 bar CO_2. But a more sophisticated 1-D model with microphysics (i.e., the physics of growth and precipitation of ice particles) produced only 5-10 K of greenhouse warming, which would have left early Mars below freezing (Colaprete and Toon, 2003). Three-dimensional (3-D) climate model simulations confirm that CO_2 ice clouds do not provide major surface warming (Forget et al., 2013; Wordsworth et al., 2013). Lack of 100% cloud cover allows infrared radiation to escape through "holes," thereby cooling the surface.

Because CO_2–H_2O atmospheric models fail to warm early Mars above freezing, other greenhouse gases have been considered. However, CH_4, NH_3, and SO_2 gases also have various physical and chemical limitations.

There are two problems with methane. The first is that climatically significant concentrations of atmospheric methane (~10^2 ppmv or more) require a considerable source flux because methane is destroyed continuously by oxidation and photolysis. The source flux of methane would have to have been comparable to the flux from the modern terrestrial biosphere, but there is no evidence for such an extensive biosphere on early Mars. Alternatively, very reduced magmas on early Mars (which likely contained reduced carbon (Steele et al., 2012)) may have produced geothermal methane. But the second problem is more severe. Net greenhouse warming from even comparatively large amounts of methane (~500 ppmv) is too

small to keep early Mars above freezing because of competing anti-greenhouse effects (Ramirez et al., 2014a). As methane increases, an early Mars stratosphere would warm from the absorption of infrared radiation by methane. Such a warm stratosphere emits more radiation to space, which would cancel out some of the expected greenhouse warming (McKay et al., 1991). The net result is <5°C of surface warming from methane (Ramirez et al., 2014a).

Ammonia is also an unlikely explanation because of its susceptibility to destruction from photolysis on short timescales (Kuhn and Atreya, 1979). Protection by ultraviolet shielding by an organic haze (Wolf and Toon, 2010) suffers the problem that it requires a very large source of methane to maintain an organic haze.

Ideas that SO_2 could serve as an effective greenhouse gas on early Mars (Halevy et al., 2007; Postawko and Kuhn, 1986) have generally omitted the consideration that sulfur-bearing gases are rapidly converted into aerosols through atmospheric chemistry. Specifically, H_2SO_4 aerosols are generated in oxidizing atmospheres and a mixture of H_2SO_4 and elemental sulfur in reducing atmospheres. Overall, such aerosols increase planetary albedo and would cause net cooling of early Mars (Tian et al., 2010). Another problem with SO_2 is its high solubility. A warm climate with rainfall would quickly remove SO_2 (McGouldrick et al., 2011) unless an ocean was saturated with dissolved S(IV) species (Kasting et al., 1989). It is possible to construct a model that warms early Mars above freezing intermittently with volcanic pulses of SO_2 but it requires H_2SO_4 aerosols made of absorptive dust with variable H_2SO_4 coatings (Halevy and Head, 2014). Whether such particles represent reality is open to question. Also, rainout and surface deposition of SO_2 was ignored; when included in a photochemical model, these decrease atmospheric SO_2 by orders of magnitude, removing the warming effect (Batalha et al., 2015).

H_2 has also been proposed as a greenhouse gas for early Mars. Sagan (1977) suggested that a 1-bar H_2 atmosphere could have warmed the planet. Ramirez et al. (2014a, b) propose a variant of this idea. These authors rely on data from Martian meteorites (Sec. 12.1.3) that suggest that the Martian mantle may be more reduced than that of Earth. Recall from Sec. 7.3 that the oxygen fugacity (f_{O2}) of Earth's mantle is near the QFM redox buffer (and that QFM ≈ IW + 3.5 log units at magmatic conditions). The inferred f_{O2} from Martian meteorites has a large range from –3.5 log units below QFM to QFM (Table 12.8). If low f_{O2} values are representative, taking –3 log units, the ratio of H_2:H_2O in volcanic gases would have been ~30 times higher than it

Table 12.8 Oxygen fugacities inferred from Martian meteorites (n.r. = not recorded).

Mars Meteorite	Type	Age (Ga)	Oxygen fugacity, f_{O2} ($\log_{10} \Delta$QFM)	Reference
ALH84001	orthopyroxenite	4.1	−2.7	Herd (2003)
Nakhla	Nakhlite	1.3	−0.0	Szymanski et al. (2010)
NWA 998	Nakhlite	1.29	−0.8	Treiman and Irving (2008)
SAU 130[a]	Shergottite	0.81	−3.5	Herd (2003)
Dhofar 019[b]	Shergottite	0.56	−3.7	Herd (2003)
DaG 476[c]	Shergottite	0.47	−2.5	Herd (2003)
NWA 1110[d]	Shergottite	0.19	−2.5	Herd (2006)
Zagami	Shergottite	0.18	−1.2	Steele et al. (2012)
NWA 1950[e]	Shergottite	n.r.	−2.7	Steele et al. (2012)
NWA 6234[f]	Shergottite	n.r.	−2.5	Gross et al. (2013)

[a] Paired with SAU 005, 008, 051, 060, 090, 094, 120, 125, 150, 687. [b] Paired with Dhofar 1668, 1674. [c] Paired with DaG 489, 670, 735, 876, 975, 1037, 1051. [d] Paired with NWA 1068, 1183, 1175, 2373, 2969. [e] Paired with NWA 7721. [f] Paired with NWA 2990, 5960, 6710, 10170.

is on Earth today, so that more than half of outgassed hydrogen should have been emitted as H_2. If Mars had vigorous volcanism at 3.8 Ga, it might have been sufficient to maintain an atmospheric H_2 mixing ratio of several percent, or higher. As shown in Fig. 12.16, an atmosphere containing ~4 bar of CO_2 and 5% H_2 would have brought Mars' average surface temperature up to the freezing point of water. At 20% H_2, the required CO_2 partial pressure falls to 1.3 bar. However, the mechanism requires outgassing fluxes larger than modern Earth to maintain the H_2 against escape to space (Batalha et al., 2015). Also, the assumed pCO_2 levels are higher than empirical constraints on Noachian/Hesperian paleopressure described in Sec. 5.2.1.1. Nonetheless, revised collision-induced absorption coefficients allow lower H_2 and CO_2 levels (Wordsworth et al., 2016).

In summary, it remains difficult to make Mars persistently warm and wet, like Earth, so that not all researchers accept this hypothesis. Consequently, other explanations for Martian fluvial erosion have been suggested, as we now describe.

12.5.2.3 Impact-Induced Climates

An alternative scenario to a persistently warm climate is that the fluvial features on early Mars were produced by many temporary warm episodes resulting from impact events. The idea dates back to the 1970s (e.g., Leovy (1977) favored it) but only recently have the climatic consequences of impacts been modeled (Segura et al., 2008; Segura et al., 2002; Segura et al., 2013; Toon et al., 2010). Between large impacts, Mars is envisaged to be largely cold and dry. There is no doubt that the

surface of early Mars is heavily cratered and that impacts occurred. The energy released from a sufficiently large impact and its molten ejecta would flash-heat the surface of Mars, which would create steam atmospheres from surface and subsurface ices. The steam would cool, making hot rain followed by cold rain. Such precipitation integrated over many such impact events could be sufficient to carve river valleys. Previously formed impact craters might also be eroded, diminishing the crater rims and removing central peaks. A second effect is that water added to the atmosphere by an impact would be recycled back into the atmosphere creating a climatic episode lasting 10^2–10^3 years, with mean temperatures below freezing but in the range where many brines remain liquid, ~250 K.

The amount of water mobilized depends on the impactor size. Rainfall amounts are plotted in Fig. 12.17 for single impactors of various diameters. Rainfall rates are similar to terrestrial values, around 2–4 m yr^{-1}. The estimated cumulative amount of rainfall generated by all observed impact craters is hundreds of meters, which is comparable to Carr and Malin's (2000) estimate of the amount of water needed to erode valley networks (Table 12.5). We showed earlier, however (Sec. 12.3.2), that Carr and Malin's number is lower than some other estimates.

There have been several criticisms of the impact hypothesis. Initial suggestions were that the timing of the impacts and the formation of valley networks might be too far apart (Jakosky and Mellon, 2004). However, ages of the last flows in river valleys cluster around 3.7-3.9 Ga (Fassett and Head, 2008a), which is similar to the ages of 15 impact craters larger than 250 km, within

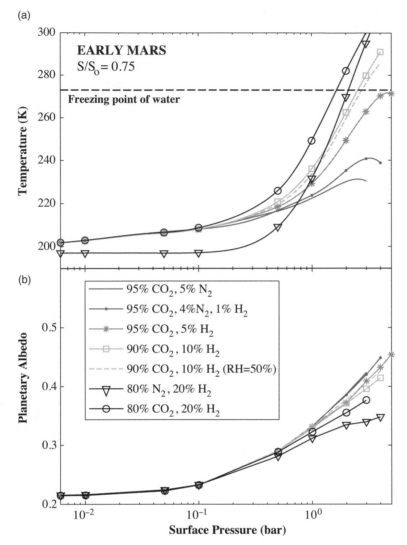

(a)

EARLY MARS
$S/S_0 = 0.75$

Freezing point of water

Temperature (K)

(b)

Planetary Albedo

Surface Pressure (bar)

95% CO_2, 5% N_2
95% CO_2, 4% N_2, 1% H_2
95% CO_2, 5% H_2
90% CO_2, 10% H_2
90% CO_2, 10% H_2 (RH=50%)
80% N_2, 20% H_2
80% CO_2, 20% H_2

Figure 12.16 One-dimensional (1-D) climate model simulations for early Mars with an assumed solar flux 75% of present. (A) Surface temperature variation with surface pressure. (B) Planetary albedo variation with surface pressure. The symbols in (B) show the atmospheric compositions that were assumed with any remaining gas taken as N_2. Fully saturated atmospheres (solid lines) and 50% tropospheric relative humidity (RH) atmospheres (dashed lines) were considered. In the model, greenhouse warming is mainly caused by H_2 and CO_2 collision-induced absorption. (From Ramirez *et al.* (2014a).)

uncertainty (Werner, 2008). Valley networks simulated in landscape evolution models are consistent with dry environments with sporadic rainfall (Barnhart *et al.*, 2009; Hoke *et al.*, 2011). The number of flood events is large, however: about half a million in the model of Barnhart *et al.* One caveat is that Earth-based assumptions about erodability, evaporative loss, and infiltration may be different for Mars. For example, if the landscape is more easily infiltrated and erodible (perhaps because there is a considerable component of impact ejecta, dunes, salts, and tuffs), less water and fewer events are required. Impacts can also raise the surface temperature to ~100°C, which would increase chemical erosion. The erodability of Mars remains controversial, though. It is difficult for impacts to provide rainfall equivalent to

500 000 Colorado plateau mean annual floods (Barnhart *et al.*, 2009).

12.5.2.4 *Fluvial Erosion in Cold Climates*

Because of the problems of greenhouse warming on early Mars and uncertainty about whether impacts can account for observed fluvial geomorphology, a third idea is that fluvial erosion on early Mars was possible in a predominantly cold climate. Three general concepts lie behind a hydrological cycle on a cold early Mars. One concerns seasonal melting of snow or ice under moderately more massive atmospheres with surface pressures ~100 mbar or greater (Kite *et al.*, 2013; McKay, 2004). A second proposal is that aquifers could have erupted on a cold early

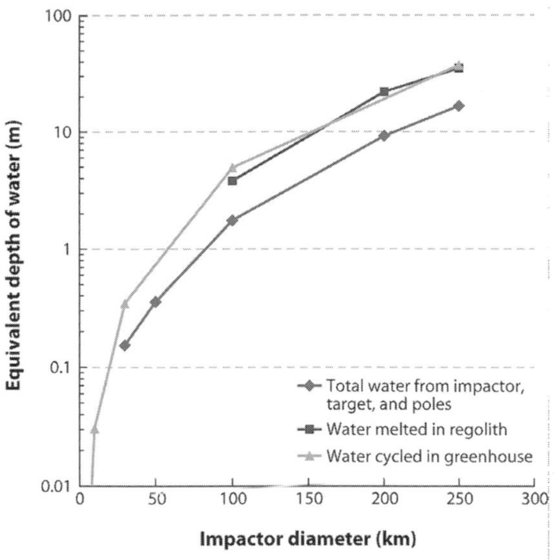

Figure 12.17 The total amount of water (given as global equivalent depth) injected from an impactor, its target and polar caps on early Mars as a function of impactor diameter. The curve with squares shows water melted in the regolith from hot impact ejecta. The curve with triangles shows water recycled through the atmosphere according to the model of Segura *et al.* (2008). (Adapted from Toon *et al.* (2010). Reproduced with permission of Annual Review. Copyright 2010, Annual Review of Earth and Planetary Sciences.)

Table 12.9 Possible salts on Mars and their eutectic temperatures.

Salt or substance	Eutectic temperature (K)
$H_2SO_4 + H_2O$	199
$Fe_2(SO_4)_3$	246
$Ca(ClO_4)_2$	198
$Mg(ClO_4)_2$	206–209
$CaCl_2$	223
$MgBr_2$	230
$NaClO_4$	239
$NaCl$	252
$MgSO_4$	269

Mars, possibly in the form of brines that exist as liquids at temperatures below 273 K (Gaidos and Marion, 2003). A third, related suggestion emphasizes the importance of low temperature brines (Fairen, 2010; Fairen *et al.*, 2009).

Cold environments with mean annual temperatures as low as −20°C can support rivers and lakes through seasonal melting, as we know from the Antarctic (McKay, 2004), and the formation of Martian sedimentary deposits might be possible. One hypothesis mentioned earlier for the origin of low-latitude sedimentary deposits is that they accumulated as a result of snowfall that incorporated aerosols and dust during times of high obliquity (Catling *et al.*, 2006). Interaction of dust with thin films of water in such deposits could account for alteration minerals (Michalski and Niles, 2012; Niles and Michalski, 2009). Energy-balance simulations suggest that the preferential low latitude location of layered sedimentary rocks (64% at $<10°$ latitude (Malin *et al.*, 2010)) can be explained by localized snowmelt in a sub-zero climate (Kite *et al.*, 2013). Simulations take into account Milankovitch cycles, which are discussed below (Sec. 12.6). Sedimentary deposits are ~2 km elevation below average ancient terrain, including those in Valles Marineris and Gale Crater (4.5° S, 137.4° E). Atmospheric pressures of order ~0.1 bar are needed to prevent evaporative cooling of

snowpack. At pressures below 6.1 mbar – the triple point of water – liquid water is unstable. At 6.1 mbar, water boils at 0°C, whereas only slightly higher, at 10 mbar, water boils at 7°C and so still evaporates and cools quickly (Hecht, 2002; Ingersoll, 1970). For localized snowmelt, the atmospheric pressure must not be too high, though. As pressure increases, melting is favored but so is sublimation of snow through an increased greenhouse effect, which reduces the likelihood that snow will be available for melting.

If brines rather than pure water were prevalent on early Mars then temperatures in excess of ~245 K can cause melting (Brass, 1980; Fairen, 2010; Fairen *et al.*, 2009; Zent and Fanale, 1986). Any contact of a salt with ice spontaneously generates a liquid if the temperature exceeds the eutectic temperature, which is the temperature below which a salty solution freezes into ice and solid salt(s). Several relevant salts, including calcium chloride and several perchlorates, have eutectic temperatures that are lower than ~245 K (Table 12.9). However, ~245 K on early Mars would still require a substantial degree of greenhouse warming. For example, in the model of Colaprete and Toon (2003), which included the effects of CO_2 ice clouds, ~2 bar of CO_2 was needed to achieve a mean global temperature of 245 K. Fluids at lower temperatures are possible with perchlorate brines because of eutectic temperatures down to 198 K and supercooling (Toner *et al.*, 2014a).

One problem with dry, moderately thick atmospheres of ~0.1 bar CO_2 or greater in fairly cold climates is photochemical instability (Zahnle *et al.*, 2008). Around this pressure level, carbon monoxide (CO) becomes as abundant as CO_2. Because CO is not a good greenhouse gas, the global mean temperature tends to be limited to ~200 K, although under high obliquity, water ice at the equator may still be important for seasonal hydrology. The reason that CO becomes abundant in thick cold atmospheres is that the amount of water vapor, which depends on temperature, is relatively small and so there is a limited concentration of OH. The OH radical is the key species controlling CO through CO destruction, CO + OH → H + CO_2. Surface sinks on CO might be a way to mitigate CO build-up. For example, CO reacts with metals to form metal carbonyls, such as $Ni(CO)_4$. If CO-rich atmospheres once existed, some geochemical trace may exist because of carbonyls.

12.5.2.5 Intermittent Greenhouse Warming Paced by Limit Cycles

Finally, another idea is that early Mars could have experienced intermittent warm periods caused by volcanically released greenhouse gases, but still spent most of its time in a cold, frozen state. Volcanic CO_2 could build up on frozen planets and cause them to deglaciate, for reasons described in Sec. 11.4 and 11.10.4. Energy balance climate models (EBMs) show that Mars-like planets could experience "limit cycles" in which the climate oscillates between brief warm episodes and longer, frozen Snowball Earth-like periods (Kadoya and Tajika, 2014, and references therein; Menou, 2015; Mills et al., 2011). In carbon cycle models of abiotic planets, the rate of silicate weathering depends on CO_2, so if pCO_2 is too high, then CO_2 cannot be maintained for long under a warm climate, so that the climate can oscillate.

Based on earlier discussion, CO_2 and H_2O are not capable by themselves of producing even transiently warm climates on early Mars, so one would have to argue that these gases are supplemented by volcanic H_2 at sufficient levels. Cycle lengths would depend on the assumed volcanic outgassing rates and the amount of land area available for silicate weathering. Preliminary models suggest that warm periods last much longer than those

following impact events, making it easier to explain the amounts of rainfall thought by some authors to be needed to erode the valleys.

12.6 Effect of Orbital Change on Past Martian Climate

Compared with the Earth, Mars has large variations in its orbital elements, the obliquity, eccentricity and longitude of perihelion. (See Sec. 2.2.1 for a discussion of orbits.) The oscillations in orbital elements over geologic time are Milankovitch cycles, which have periods shown in Table 12.10. On Earth, these cycles are responsible for the ice ages during the Quaternary Period (i.e, since ~2.588 Ma). On Mars, the average obliquity over geologic time was ~40°, but sometimes the obliquity may have been ~80° (Laskar et al., 2004). In recent geologic history, Martian eccentricity and obliquity have oscillated with periods of ~10^5 years, modulated with envelopes of 2.4 Myr and 1.2 Myr respectively (Fig. 12.18). The orientation of the spin axis, which determines the season of perihelion, also precesses every 51 000 years.

Milankovitch cycles change the climate on Mars in a two-step process. First, they alter the annual average solar insolation and its seasonal distribution with latitude. Second, the changes in insolation cause carbon dioxide ice and water ice to migrate. Obliquity variations exert the largest influence by altering the latitudinal distribution of sunlight. At low obliquity, <10–20° depending on the precise values of polar cap albedo and thermal emissivity, the CO_2 atmosphere collapses onto permanent CO_2 ice polar caps and for the current inventory, pressure may drop to ~1 mbar (Manning et al., 2006b). Calculations indicate that this collapse occurs ~1%–2% of the time. At obliquity above 54°, the poles receive more annual average insolation than the equator (Ward, 1974, 1992). But

Table 12.10 The orbital elements of Mars and the Earth and their variability.

Parameter	Present Mars	Martian variability		Present Earth	Terrestrial variability	
		Range	Cycle (years)		Range	Cycle (years)
Obliquity (°)	25.19	0–85*	120 000**	23.45	22–24	41 000
Eccentricity	0.093	0–0.12	120 000***	0.017	0.01–0.04	100 000
Longitude of perihelion (°)	250	0–360	51 000	285	0–360	21 000

* Before ~10 Ma, obliquity variations are chaotic. While unpredictable at an exact time, statistically they would have varied between 0 and 85° (Laskar et al., 2004; Touma and Wisdom, 1993).
** The amplitude of obliquity oscillation is modulated with a ~1.2 Myr period envelope.
*** The amplitude of eccentricity oscillation is modulated with a ~2.4 Myr period envelope.

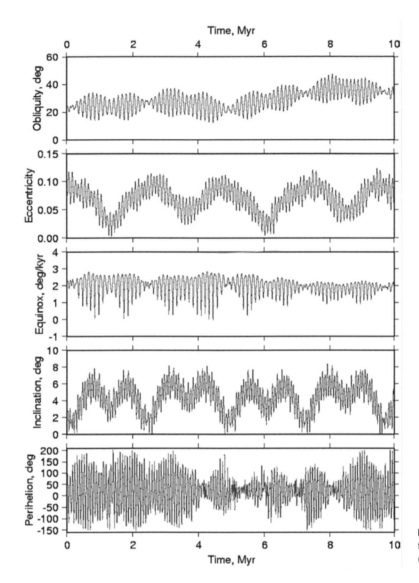

Figure 12.18 Simulated variations of Martian orbital elements over the last 10 Myr. (From Armstrong *et al.* (2004).)

ice can accumulate in the tropics at obliquity above ~35° because of the nonlinear dependence of sublimation rate on temperature (Kite *et al.*, 2013). The global annual mean surface temperature is calculated to drop at high obliquity because of the increased extent of high albedo, associated with large seasonal CO_2 ice caps (Haberle *et al.*, 2003). At high obliquity, atmospheric pressure may increase because of warming and the release of adsorbed carbon dioxide from high-latitude regolith. Calculations suggest, however, that the maximum pressure increase is only a few millibars. If the obliquity were 45° (with other orbital parameters the same as today's), the southern cap would reach the equator of Mars during southern winter (Haberle *et al.*, 2003). Consequently, surface ice may have reached the equator even within the past 10 Myr.

12.7 Wind Modification of the Surface

Finally, we note that the interpretation of Mars' climate history is complicated by wind modification of the surface. In particular, there are good reasons that we discuss below why wind erosion and deposition would have been more important than today if the atmosphere had been thicker in the past.

Today, wind-modified geology is widespread. Bedforms, such as dunes and ripples, litter the Martian surface. *Yardangs* are common, which are positive relief features in coherent materials sculpted by wind on scales from tens of meters to kilometers. *Ventifacts*, which are wind-modified objects, are evident in faceted and grooved surfaces (Bridges *et al.*, 2004). We have already discussed the ubiquity of dust devils, the frequent presence of dust storms in some areas, and great dust storms that occur in some years (Sec. 12.2.2).

Observations suggest that wind erosion, dust transport, and dust deposition have repeatedly buried and exhumed surfaces. Based on the heights of erosionally resistant mesas, parts of Meridiani Planum appear to have been exhumed from beneath at least several hundred meters of material (Edgett, 2005; Edgett and Malin, 2002). In fact, many of the tropical, sulfate-rich sedimentary layered deposits appear to be undergoing exhumation. From stratigraphy, we suspect that they are old, but their surfaces are young with very few small craters. Burial, exposure, and re-burial over time complicates how we determine the real ages of geomorphic units and sequence of events.

Models of atmospheric circulation suggest that net erosion must have taken place in lowlands, particularly in the Hellas basin and the northern lowlands, with net deposition in uplands and in some places of moderate elevation, such as parts of Arabia Terra and southern Amazonis Planitia (Anderson et al., 1999). Regions of high thermal inertia, where there are consolidated or coarse-grained soils, surface rocks, or bedrock, are found where the models predict net erosion over many Milankovitch cycles. Conversely, dust-laden regions of low thermal inertia are found where net deposition is predicted (Haberle et al., 2003).

The potential of Noachian atmospheres to cause wind erosion is perhaps under-appreciated. Saltation (the hopping motion of fine sand) results from the high velocity tail of the wind speed distribution and is very sensitive to changes in the surface pressure. Models indicate that an increase in atmospheric pressure up to only 40 mbar would intensify potential surface erosion rates by two orders of magnitude (Armstrong and Leovy, 2005). If Mars had a surface pressure ~100 mbar or higher during the Noachian, rates of surface modification by wind could have been far greater than today, depending on the surface erosional resistance. Noachian surfaces experienced relatively rapid resurfacing that has sometimes been attributed solely to fluvial processes earlier. But modification by winds under a denser atmosphere is also worthy of consideration. In particular, wind abrasion can also modify the morphology of fluvial channels (albeit parallel to the wind direction) if the substrate is suitably erodible (Perkins et al., 2015).

12.8 Unanswered Questions of Mars' Astrobiology and Atmospheric Evolution

It is appropriate to return to the question at the start of this chapter that motivated our desire to decipher the atmospheric evolution on Mars: Did life arise on Mars? As our understanding of Mars has improved, the possibilities for life on Mars have narrowed from pre-Space Age speculations to a more sober, nuanced view. The hypothesis that life might still exist in today's subsurface of Mars faces the challenge that organisms would have needed to survive in an exceptionally long-lived aquifer or somehow moved within the subsurface over billions of years to hotspots that supported temporary aquifers. Because of these problems, many consider the best hope for finding

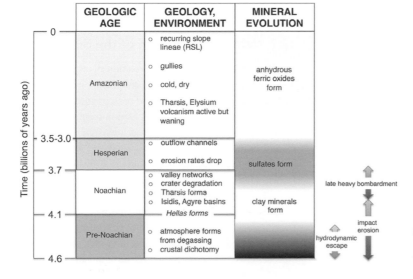

Figure 12.19 Overview of the Martian geologic time and major events in the history of surface and atmosphere of Mars.

life on Mars to be fossil or biogeochemical remnants dating from the Noachian or Hesperian.

When we consider the evidence as a whole (summarized in Fig. 12.19), it seems unlikely that early Mars had an uninterrupted "warm, wet" climate for vast periods extending over hundreds of millions of years. The explanation for fluvial features on early Mars, despite the fainter Sun, could be a combination of repeated, temporary impact-induced climates (Sec. 12.5.2.2), hydrology under prolonged cold-based climates (Sec. 12.5.2.3), and periods of greenhouse warming separated by longer cold periods (Sec. 12.5.2.5). In the last case, however, it remains to be demonstrated that rates of volcanic outgassing of CO_2 and H_2 were sufficient.

What all of this discussion implies about the possibility of life on early Mars remains unclear. Life can certainly exist within cold-based hydrology, as demonstrated in the Antarctic Dry Valleys. If Mars was warmed intermittently by a CO_2–H_2 greenhouse, then life (counterintuitively) may have been *less* likely to have been present, because microbes tend to consume H_2 in the presence of oxidants. It can be argued that a CO_2 atmosphere rich in H_2 is inconsistent with the presence of a biosphere (see Sec. 9.7). That said, such arguments should not stand in the way of continuing to search for life on Mars. Exploration of Mars with novel instruments and higher resolution always throws up new puzzles and surprises, and such advances we hope will continue.

13 Evolution of Venus' Atmosphere

13.1 Current State of Venus' Atmosphere

In the previous chapter, we looked at Mars, which is the most Earth-like of the other planets in our Solar System with respect to its potential for supporting life. But, in other ways, Venus is the most Earth-like planet. Venus has nearly the same mass as Earth (Table 13.1) and, as we shall see, similar inventories of volatile elements. But Venus has a very un-Earth-like surface temperature, ~740 K, and a vastly different, 93-bar, CO_2 atmosphere.[1] Venus also has an extremely slow rotation period, 243 Earth days, and it rotates in a *retrograde* direction with respect to the planet's orbit – just the opposite of what is observed for Earth and most of the other planets. The combination of this slow retrograde rotation with its orbital period of 224 days gives Venus a solar daylength (noon to noon) of just under 117 Earth days. So, the interesting story about Venus is not how like the Earth it is, but rather how it came to be so different. Before addressing this question, though, we briefly look at what Venus' atmosphere is like today.

13.1 Current State of Venus' Atmosphere

Venus is the most-visited planet in our Solar System, having been explored by some 24 different spacecraft, several of which have dropped probes into its atmosphere. These include the Soviet *Venera* spacecraft, ten of which descended to Venus' surface from 1970 to 1985, and NASA's *Pioneer Venus*, which made a successful visit in 1978. *Pioneer Venus* consisted of two spacecraft, an orbiter, and a multiprobe mission that included a descent bus, a large descent probe, and three smaller probes. Much of what we know about Venus' lower atmosphere comes from data obtained by this mission. *Magellan* sent

back orbital data from 1990 to 1994, mapping the surface with radar. *Venus Express*, which went into orbit around Venus in 2006, has contributed new atmospheric measurements and also discovered evidence that Venus may be volcanically active. Near-infrared emissivity anomalies around volcanic areas suggest that Venus has been volcanically active in the past 0.25 million years (Smrekar *et al.*, 2010) while transient bright spots seen through an atmospheric window at ~1 µm are consistent with active lava flows (Shalygin *et al.*, 2015).

13.1.1 Atmospheric Temperature and Composition: the Concept of "Excess Volatiles"

Figure 13.1 shows a composite vertical temperature profile obtained from the four *Pioneer Venus* descent vehicles. The atmospheric temperature increases from about 200 K at the 100-km level to roughly 730 K at the surface. The surface temperature is hot enough to melt lead or tin, and no probe has survived there for more than 2 h. The surface temperature exceeds the critical point of water, 647 K, above which water can only exist in the vapor phase.

Both the *Pioneer Venus Large Probe* and the *Orbiter* contained neutral mass spectrometers, and the *Large Probe* contained a gas chromatograph, as well. Between them, they were able to measure the concentrations of a number of species in Venus' atmosphere, several of which are shown in Table 13.2. As noted above, CO_2 is the most abundant gas, comprising some 96.5% of the atmosphere. N_2, with a mixing ratio of 3.5%, makes up most of the rest.

It is instructive to compare the N_2 column mass in Venus' atmosphere with that on Earth. As pointed out in Ch. 1 (eq. (1.19)), the mass of an atmospheric column is equal to the surface pressure divided by the gravitational acceleration. To get the mass of an individual species, one

---

[1] Values are given at the mean radius of 6051.84 km (Fegley, 2014), noting that temperature varies ~100 K and pressure varies several tens of bar across the surface because of topography.

Table 13.1 Planetary characteristics of Venus. (Source: Fegley (2014).)

Parameter	Value	Value relative to Earth
Mass	4.8685×10^{24} kg	0.815
Mean radius	6052 km	0.95
Mean density	5243 kg m^{-3}	0.95
Mean surface gravity	8.87 m s^{-2}	0.904
Semi-major axis	108 208 930 km	0.723
Orbital eccentricity	0.0067	
Sidereal orbital period	224.701 d[*]	0.615
Rotation period	243 d[*] (retrograde)	243
Daylength (noon to noon)	116.75 d[*]	116.75
Surface temperature at the mean radius	737 K	2.6
Surface pressure at the mean radius	93.3 bar	93.3
Bond albedo	0.76	2.53

[*] Units are in Earth days.

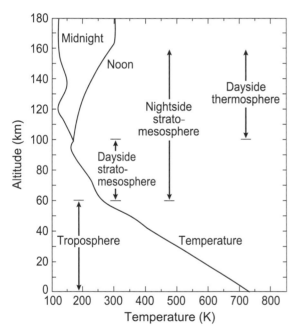

Figure 13.1 Vertical temperature at 30° latitude on Venus as determined from the *Pioneer Venus* mission. (From Prinn and Fegley (1987). Reproduced with permission Annual Reviews, Inc. Copyright 1987, Annual Review of Earth and Planetary Sciences.)

Table 13.2 Composition of Venus' atmosphere. (Source: Fegley (2014) and references therein.)

Gas	Abundance by volume
CO_2	$96.5 \pm 0.8\%$
N_2	$3.5 \pm 0.8\%$
SO_2	150 ± 30 ppm (22–42 km)
	30–150 ppm (12–22 km)
Ar	66 (+40, −20) ppm
H_2O	30 ± 15 ppm (0–45 km)
CO	17 ± 1 ppm (12 km)

must then multiply this result by the species' mass mixing ratio, μ_i given by eq. (1.5), $\mu_i = m_i n_i / \rho = (m_i / m) \cdot f_i$, where m_i, n_i, and f_i are the molecular mass, number density, and volume mixing ratio of species i, and m and ρ are the molecular mass and mass density of the atmosphere itself. Performing this calculation for N_2 on Earth and Venus yields 7.8×10^3 kg m^{-2} and 23.7×10^3 kg m^{-2},

respectively. Thus, Venus has around three times more atmospheric mass of N_2 than Earth. Recall from Ch. 11 that some have used this disparity to argue that Earth once had greater amounts of N_2 in its atmosphere (Goldblatt *et al.*, 2009; Lecuyer *et al.*, 2000). Table 11.1 shows that most of Earth's nitrogen is locked in minerals.

The CO_2 content of Venus' atmosphere is, of course, vastly greater than the amount of CO_2 in Earth's atmosphere or even in Earth's combined atmosphere–ocean system. Neglecting recent anthropogenic additions, Earth's atmosphere contains about 300 ppmv of CO_2, and the ocean contains about 60 times this amount, mostly as bicarbonate ions. Together, these yield a CO_2 column mass of ~280 kg m^{-2}. By comparison, Venus has a column CO_2 mass of ~1.0×10^6 kg m^{-2}, or more than 3500 times higher. As many authors have pointed out, this discrepancy is easily explained, as most of Earth's CO_2 is tied up in carbonate minerals and organic carbon in rocks. According to Walker (1985), the total amount of carbon in these reservoirs is $\sim 10^{20}$ kg C, which corresponds to 7×10^5 kg m^{-2} of CO_2. If one were to convert all of this stored carbon back into CO_2, the resulting CO_2 partial pressure would be 70 ± 10 bar. On this basis, Venus and Earth have comparable CO_2 inventories. Venus has no water, however, and thus no effective way of forming carbonate minerals, so virtually all of its lithospheric CO_2 is present in its atmosphere.

By comparison with the Earth, Venus' deep interior might retain carbon in elemental form. Earth's upper mantle contains ~ 100 ppm C, the lower mantle ~ 2000 ppm C, and core ~ 2000–3000 ppm C (as Fe_3C)), by mass (Wood et al., 1996). If this carbon were released and oxidized, it would constitute an inventory of hundreds of bar.

Sulfur dioxide, SO_2, is similar to CO_2 in the sense of being much more abundant in the atmosphere of Venus than in Earth's atmosphere. The SO_2 concentration on Venus is ~ 150 ppmv below the clouds (i.e., below ~ 45 km altitude), compared to only ~ 1 ppb or less in Earth's troposphere. The ~ 150 ppmv value on Venus was measured by the Pioneer Venus and Venera 11–12 gas chromatographs and is consistent with Venus Express remote measurements (Marcq et al., 2008). According to UV spectroscopic measurements made by the Vega 1 and Vega 2 descent probes, SO_2 decreases towards the surface, down to ~ 25 ppmv (Bertaux et al., 1996). This gradient would require the presence of another major sulfur species, perhaps OCS, to take up the sulfur from SO_2, or an unusually large sink for SO_2 at the surface. Such a vertical gradient requires confirmation. Above the clouds, the SO_2 level drops to ~ 10 ppbv (Belyaev et al., 2012; Sandor et al., 2010) because of photooxidation, and the SO_2 concentration is somewhat variable with altitude, perhaps because of evaporating aerosols (Zhang et al., 2012).

On Earth, SO_2 levels are low because SO_2 is weakly soluble in water. Furthermore, in the presence of O_2 and H_2O, sulfur dioxide is quickly oxidized to sulfuric acid, H_2SO_4, which is highly soluble. Thus, any volcanic SO_2 that is outgassed on Earth soon ends up in the oceans as sulfate. From there, it makes its way into the rocks where it forms large reservoirs of gypsum, $CaSO_4 \cdot 2H_2O$, and pyrite, FeS_2. The reservoir sizes of these minerals on Earth are each about 2.5×10^{20} moles S (Berner and Raiswell, 1983; Lasaga et al., 1985) for a combined surface sulfur reservoir of 5×10^{20} moles S, or $\sim 1.6 \times 10^{19}$ kg S. Canfield (2004) lists a slightly smaller total sulfur inventory of 3.5×10^{20} moles S. This terrestrial sulfur inventory vastly exceeds the amount of sulfur stored as SO_2 in Venus' atmosphere which, by the

calculation method outlined above, is 114 kg m^{-2} S, or 5.3×10^{16} kg S total. So, Earth's surficial sulfur inventory is 200–300 times larger. As the carbon and nitrogen surface inventories of Earth and Venus are roughly comparable, this suggests that large amounts of sulfur may reside in minerals on Venus' surface. We return to this question later in the chapter.

This brings us to water. Pioneer Venus was not able to measure the water vapor content of Venus' lower atmosphere, partly because the inlet leak of the Large Probe mass spectrometer became clogged with a sulfuric acid particle as it descended through the clouds (see below). But H_2O has been measured spectroscopically from the Venera 11, 13, and 14 landers (Ignatiev et al., 1997), from ground-based telescopes (Bailey, 2009; Meadows and Crisp, 1996; Pollack et al., 1993), and from Venus Express (Bezard et al., 2009; Marcq et al., 2006) (Fig. 13.2). Seeing through the clouds and dense atmosphere involves taking advantage of various near-IR windows. Based on an analysis of many measurements, the H_2O concentration is estimated to be 30 ± 10 ppm from 15–45 km and 30 ± 15 ppm below 15 km (Taylor et al., 1997). Bezard et al. (2009) derive a slightly higher value of 44 ± 9 ppm averaged over the lower atmosphere. A comparison with Earth is again helpful. The 30 ppm of H_2O in Venus' atmosphere is equivalent to a column mass of ~ 13 kg m^{-2}. By comparison, the column mass of Earth's oceans, which would be roughly 3 km deep if spread out over the entire planet, is about 3×10^6 kg m^{-2}. So, Venus is deficient in water compared to Earth by a factor of 4×10^{-6}.

We can generalize the discussion above as follows. The American geochemist, William Rubey, defined excess volatiles as compounds at Earth's surface that cannot be derived simply by weathering of igneous rocks and conversion to sedimentary rocks (Rubey, 1951, 1955). These compounds include H_2O, CO_2, N_2, and SO_2, along with Cl. Because these compounds are not present in appreciable quantities in igneous rocks, Rubey assumed that they must have been released from Earth's interior as constituents of volcanic gases. Alternatively, based on what we believe now about the violent process of planetary accretion, these compounds could have been emplaced directly into the atmosphere by impact degassing (Sec. 6.5). In any case, using Rubey's terminology, what we have been comparing above are the excess volatile inventories of Venus and Earth. The conclusion is that, based on what are able to directly observe, Earth has much more water and surficial sulfur, and the two planets have comparable inventories of carbon and nitrogen when you account for amounts locked up in the solid Earth.

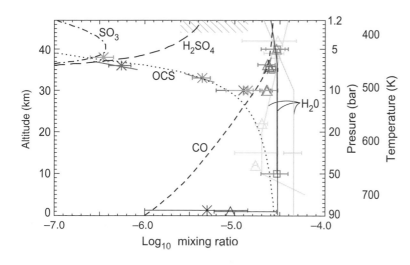

Figure 13.2 Predicted and observed composition of the Venus lower atmosphere. Calculated abundances from a chemical kinetics model for CO, OCS, H_2SO_4, and SO_3. Calculated abundances from a thermochemical equilibrium model for the conditions believed to exist at the surface are shown for CO (triangle) and OCS (asterisk). Retrieved H_2O abundances are the medium gray solid curve (Meadows and Crisp, 1996), the light gray solid curve (Ignatiev *et al.*, 1997), and the dark solid line with squares (Pollack *et al.*, 1993). The range of H_2SO_4 vapor abundances retrieved from observations is indicated by the hatched region near the top of the figure. The other symbols represent additional measurements. (From Mills *et al.* (2007). See their paper for further details and references. Reproduced with permission from John Wiley and Sons. Copyright 2007, American Geophysical Union.)

13.1.2 Cloud Composition and Photochemistry

Venus is entirely shrouded in clouds that make it impossible to see the surface at visible wavelengths, although it is possible to see through them at radio wavelengths and in some near-IR windows. The clouds are found at altitudes of 30–70 km (Fig. 13.3). *Pioneer Venus* measured the cloud composition to be a mixture of 75% sulfuric acid, H_2SO_4, and 25% water. The sulfuric acid is produced photochemically from photolysis of SO_2 and CO_2.

$$SO_2 + h\nu \rightarrow SO + O \qquad (13.1)$$

$$CO_2 + h\nu \rightarrow CO + O \qquad (13.2)$$

$$SO_2 + O + M \rightarrow SO_3 + M \qquad (13.3)$$

$$SO_3 + H_2O \rightarrow H_2SO_4 \qquad (13.4)$$

H_2SO_4 is formed near the top of the cloud deck, around 70 km, where UV photons are available. Because of the low temperatures at these altitudes, H_2SO_4 condenses out to create particles which gradually fall through the cloud layer. H_2SO_4 becomes thermodynamically unstable, however, below about 45 km, causing the cloud deck to thin out by evaporation.

The presence of 100% upper cloud cover on Venus illustrates an important distinction between photochemically produced clouds such as these and condensation clouds on Earth. Photochemistry (reactions (13.1)–(13.4)) occurs everywhere where SO_2 is exposed to solar UV; hence, clouds form more or less uniformly over the sunlit hemisphere of Venus. By contrast, condensation clouds, such as the H_2O clouds on Earth, typically form in updrafts

as a result of adiabatic cooling. Updrafts can occur over only a fraction, such as ~50% of Earth's surface – what goes up must come down – and so condensation clouds typically do not reach 100% cloud cover. Some clouds, e.g., marine stratus clouds, do span large areas, but such clouds do not occur uniformly over Earth's surface. Extremely thick atmospheres like those of the giant planets are 100% filled with condensation clouds, as viewed from above, but this happens because convection occurs over multiple scale heights and because different constituents condense at different altitudes, thereby producing multiple cloud layers. Saturn's moon, Titan, is similar to Venus: it is completely enveloped in photochemically produced haze. Titan's haze, of course, is composed of organic compounds, whereas the Venus haze is composed of sulfuric acid. By contrast, just like water clouds on Earth, the methane condensation clouds that exist in Titan's troposphere are spotty (e.g., Griffith, 2009).

Unlike the upper cloud layer on Venus, the middle and lower cloud layers appear patchy, based on contrasts in brightness temperatures measured at different near-IR wavelengths (McGouldrick and Toon, 2007). These authors argue that the lower cloud layers on Venus are maintained by radiative-dynamical feedbacks similar to those that maintain stratocumulus clouds on Earth.

Venus' planet-encircling cloud cover gives it a high planetary albedo of 0.76 ± 0.01 (Moroz *et al.*, 1985). At visible wavelengths, Venus appears through a telescope as a bright, featureless disk. At violet and UV wavelengths, however, cloud markings become visible (Fig. 13.4). The markings are caused by the presence of some compound

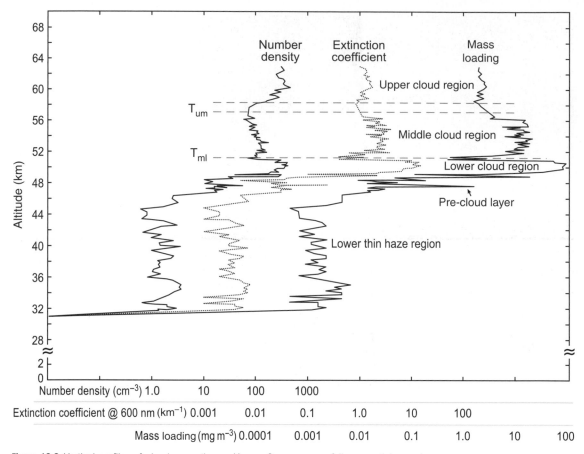

Figure 13.3 Vertical profiles of cloud properties on Venus. Curves are as follows: particle number density (cm^{-3}), the extinction coefficient at 600 nm wavelength, and mass loading of the cloud particles (mg m^{-3}). Labels T_{um} and T_{ml} are the upper-middle and lower-middle cloud transitions, respectively. (From Mills *et al.* (2007). Reproduced with permission from John Wiley and Sons. Copyright 2007, American Geophysical Union. Data from Knollenberg and Hunten, (1979).)

that absorbs from 0.32–0.5 nm. In fact, about half of the solar flux received by Venus is absorbed in the upper clouds (58–65 km) by the UV absorber and the CO_2 column (Titov *et al.*, 2007; Titov *et al.*, 2013). The absorber's identity has remained unresolved for many years.

Early suspicion about the nature of UV absorber centered on various compounds related to elemental sulfur (Toon *et al.*, 1982). SO_2 does not just oxidize to sulfate when it is photolyzed. (This would not make sense, anyway, from a redox standpoint, as little free O_2 is thought to exist in Venus' atmosphere.) Rather, the SO_2 *disproportionates*, that is, some of it is oxidized to sulfate by reactions (13.1)–(13.4), while some of it is reduced to elemental sulfur. The reduction is thought to occur by the sequence of reactions described in Ch. 9, Sec. 9.6.3 (reactions (9.32)–(9.36)). The end product of this reaction sequence is S_8, which is the stable form of elemental sulfur at room temperature on Earth. As

mentioned in Ch. 9, other sulfur *allotropes*, such as S, S_2, and S_4, are also formed during this process. In reality, odd-numbered sulfur allotropes should also form, e.g. by the following reaction.

$$S + S_2 + M \rightarrow S_3 + M \qquad (13.5)$$

The sulfur molecules that are created by these reactions can be either linear chains or ring molecules (for S_3 and higher allotropes). Some of this sulfur is present as vapor, while some of it should condense to form particles of *amorphous* (noncrystalline) *sulfur* (Toon *et al.*, 1982). Sulfur can also react with hydrogen to form *sulfanes* (sulfur chains with hydrogen atoms attached to one end). These sulfanes can then play a role in chain growth by way of the following reactions.

$$H_2 + S_n \rightarrow HS_n + H \qquad (13.6)$$

$$HS_n + S \rightarrow S_{n+1} + H \qquad (13.7)$$

Figure 13.4 An image of Venus taken on February 14, 1990, by the *Galileo* space-craft Solid State Imager (SSI) camera through its violet filter (centered at 404 nm with 45 nm band-pass (Belton *et al.*, 1992)). The image is shown in false blue color to emphasize contrast. North is at the top and the evening terminator is to the left, i.e., the sunlit part of the planet is rotating towards the observer. The image shows markings in the sulfuric acid clouds such as west-to-east cloud bands. The bright region is the subsolar point. (Image P-37218; courtesy of NASA/ JPL.) (A black and white version of this figure will appear in some formats. For the color version, please refer to the plate section.)

Some, or all, of these sulfur chain molecules and sulfanes absorb near-UV radiation and are candidates for the markings in the Venus clouds (Hapke and Nelson, 1975).

Other UV absorbers have also been considered. Photochemical models generate elemental sulfur in the lower atmosphere, which condenses in the middle cloud layer but may have insufficient abundance to account for near-UV absorption at ~60 km (Krasnopolsky and Lefevre, 2013; Mills *et al.*, 2007; Zhang *et al.*, 2012). So, another leading candidate for the UV absorber is ~1% $FeCl_3$ dissolved in sulfuric acid (Markiewicz *et al.*, 2014; Zasova *et al.*, 1981).

13.1.3 Atmospheric Circulation

Our brief tour of the modern Venus atmosphere would be incomplete without commenting on its atmospheric circulation (e.g., see reviews by Limaye and Rengel, 2013; Read, 2013). Long ago it was noticed that the markings in the clouds – for example, a characteristic sideways "Y" pattern that can be seen in Fig. 13.4 – persist for long enough to allow a measurement of the wind velocities near the cloud tops. The wind velocities at this level are directed in a retrograde sense, like the planet's rotation, but much faster. The apparent rotation period at cloud-top level is about 4 days, which corresponds to wind velocities of ~120 m s^{-1}. Atmospheric rotation at equatorial latitudes that exceeds the rotation of the surface is *superrotation*, which is a basic feature of Venus' atmospheric circulation (see Sec. 4.5).

The *Pioneer Venus* probes made it possible to measure wind velocities as a function of altitude (Fig. 13.5). Doppler tracking of the probes by Earth-based radar gave their velocities as they descended through Venus' atmosphere. According to these measurements, wind velocities decrease monotonically with decreasing altitude, reaching values of 1–2 m s^{-1} near the surface, which is consistent with similar Doppler tracking of *Venera* and *Vega* entry probes. At all locations where it has been observed, however, the atmosphere appears to be superrotating. This rotation extends up to high latitudes and so is readily observed in images taken by *Venus Express* as it went over Venus' poles (e.g. Taylor, 2010, his Sec. 8.3). There, the superrotation produces a strong *polar vortex* that carries air downward at its center, much as the wintertime circumpolar vortex does in Earth's polar stratosphere.

As discussed in Sec. 4.5, why the atmosphere circulates in this way is not entirely understood, even though numerous papers and book chapters have been written about this topic (see, e.g., Gierasch *et al.*, 1997; Lellouch *et al.*, 1997; Schubert, 1983). Below the clouds the circulation should be dominated by large Hadley cells in each

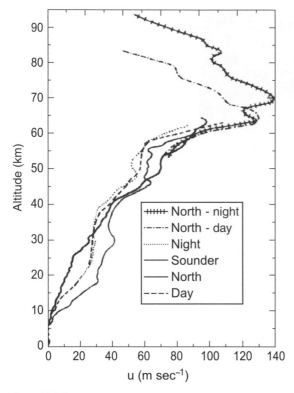

Figure 13.5 Retrograde zonal wind velocity, *u*, versus altitude, as determined from *Pioneer Venus*. (From Prinn and Fegley, (1987). Reproduced with permission from Annual Reviews, Inc. Copyright 1987, Annual Review of Earth and Planetary Sciences.)

hemisphere caused by preferential heating of air near the equator and cooling near the poles (Sec. 4.2.5). Support for the existence of a planet-wide Venus Hadley circulation comes from observations of CO abundance made by *Venus Express* and summarized by Taylor and Grinspoon (2009). Above the cloud tops, Venus' atmospheric circulation is different: evidence suggests that both subsolar-to-antisolar and equator-to-pole circulations are present. In between, within the clouds, the circulation appears to be driven primarily by strong thermal tides (Gierasch *et al.*, 1997). In general, as discussed in Sec. 4.5, some mechanism, either thermal tides or eddies associated with atmospheric waves, must transfer angular momentum to the 65–70 km level of maximum wind speed in the equatorial atmosphere.

13.2 The Solid Planet: Is Plate Tectonics Active on Venus?

As discussed in Ch. 11, a critical aspect of Earth's long-term climate control system is its ability to recycle carbonate rocks back into gaseous CO_2. On Earth, this recycling

occurs largely because of plate tectonics. Carbonate sediments on the seafloor, and carbonate veins that form within mid-ocean ridge hydrothermal vent circulation systems, are carried down into the upper mantle when seafloor is subducted. Volatilization of those carbonate minerals as they are heated up leads to outgassing of CO_2 through volcanoes, thereby resupplying CO_2 to the atmosphere and helping to stabilize climate over long timescales.

Do other rocky planets within our Solar System show evidence for plate tectonics? One way to tell is through *hypsometry* – measuring the relative elevation of surfaces. Hypsometric curves for the Moon and three terrestrial planets are shown in Fig. 13.6. The distribution of surface elevation on Earth is distinctly bimodal. Roughly 30% of the surface lies at elevations of 2 ± 1 km, measured with respect to sea level. This, of course, represents the continents. The other 70% of Earth's surface – the seafloor – lies at elevations of -4 ± 2 sea-level difference in elevation between continents and seafloor is a direct consequence of the action of plate tectonics. Indeed, granitic continents might not even exist if plate tectonics were not active (Campbell and Taylor, 1993).

According to Fig. 13.6, Mercury, the Moon, Venus, and Titan have unimodal elevation distributions. The data for Venus were collected by radar altimetry from NASA's *Magellan Mission*. This, by itself, suggests that plate tectonics is not active on these bodies. (The topography of Mars was discussed in Sec. 12.1.1.)

Other data from *Magellan* support the idea that plate tectonics does not operate on Venus today. *Magellan* used *synthetic aperture radar*, or SAR, to map surface features on Venus down to a spatial resolution of ~120–280 m. SAR makes images by bouncing radar beams off the surface and combining signals from different points along the spacecraft's orbit. The effective aperture of the radar becomes the distance between these points, rather than the diameter of the antenna itself, thereby allowing the SAR system to achieve high spatial resolution. With SAR, *Magellan* mapped about 900 impact craters on Venus, all of them over 2 km in diameter (Herrick *et al.*, 1997; Schaber *et al.*, 1992). The reason for the lack of smaller impact craters is easy to determine: incoming bolides smaller than ~1 km are unable to penetrate Venus' dense atmosphere, so they fragment and produce airbursts rather than craters. Intriguingly, impact craters on Venus appear to be randomly spaced over Venus' surface (but see further discussion below). Since impacts occur sporadically over long periods of time, this implies that the mean age of Venus' surface is approximately the same everywhere, initially estimated as 0.5 ± 0.3 billion years (Schaber *et al.*, 1992). Subsequent models that take account of

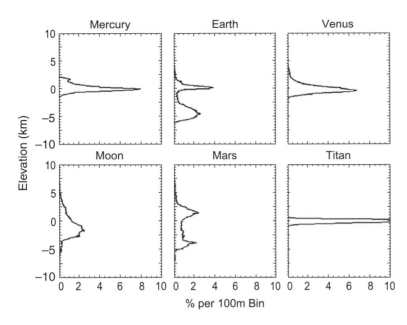

Figure 13.6 Diagram showing hypsometric curves, i.e. the fraction of a planetary surface in a given elevation range, for the terrestrial planets, the Moon, and Titan. The unimodal distribution of surface elevation on Venus shows lack of evidence for plate tectonics. (From Lorenz *et al.* (2011).)

the deceleration and flattening effects of Venus' dense atmosphere on impactors put the age at 0.73 ± 0.1 billion years (Korycansky and Zahnle, 2005).

Unlike Venus, the age distribution of Earth's surface, like its elevation, is bimodal. The seafloor has a mean age of 60 million years, whereas the continents exhibit a variety of ages but are typically about 1 billion years old. The difference is caused by the fact that the seafloor is continually being subducted, whereas the continents are not.

The fact that plate tectonics does not operate on Venus does not necessarily imply that its surface is volcanically inactive. Time-dependent changes in the measured SO_2 concentration above the clouds have been interpreted as evidence of periodic emission of volcanic gases (Esposito, 1984). Alternative explanations include time-dependent changes in dynamical transport of SO_2 from the lower atmosphere (Mills *et al.*, 2007). As mentioned earlier, anomalously high near-IR emissivities measured in volcanic regions, including transient brightening, support the idea that Venus is volcanically active (Shalygin *et al.*, 2015; Smrekar *et al.*, 2010).

What happens inside Venus' interior if plate tectonics is not active? Upward transport of heat by conduction through the crust is inefficient, so heat released by radioactive decay of elements within Venus' mantle should cause it to become extremely hot. One suggestion is that Venus experiences episodic, planet-wide volcanism (Turcotte, 1993). Instead of releasing its interior heat more or less continuously, as Earth does by creating new seafloor, Venus stores it up until the upper mantle

becomes so hot that it melts over wide regions. Global volcanism results, resurfacing much or all of the planet's surface, and thereby accounting for the uniform surface age inferred from crater counts. Once the heat is released, the upper mantle freezes solid again and volcanism slows or stops until the cycle is repeated.

Not all authors agree that Venus experiences episodic global resurfacing. Phillips *et al.* (1992) argue that the distribution of impact craters is not completely random; rather, some areas have slightly fewer craters, and these craters are partially embayed (filled) with material that could represent fresh lava flows. But they acknowledge that the hypothesis of random crater spacing for Venus cannot be ruled out. Recent crater density studies remain consistent with global resurfacing scenarios (Kreslavsky *et al.*, 2015).

If the catastrophic resurfacing hypothesis is correct then one must ask why Venus behaves this way, while Earth does not, given that Venus is almost the same size as Earth. The difference between the two planets is probably related to Venus' lack of water (Moresi and Solomatov, 1998). On Earth, the presence of water in the upper mantle lowers the viscosity of rocks, thereby helping to create the ductile *asthenosphere* on which the *lithospheric* plates ride. Without water, the asthenosphere would be much more viscous, making it more difficult for plates to slide around. On Earth, water weakens rock and lowers the strength of the lithosphere (Stern, 2007), and water substantially lowers the melting point below arcs from ~1400 to 800 °C (Till *et al.*, 2012). Finally, on Earth, the presence of liquid water helps remove CO_2 from the atmosphere, limiting the magnitude of the greenhouse effect, and keeping the surface relatively cool. On Venus, the dense CO_2 atmosphere that built up in the absence of water (see discussion below) creates such high surface temperatures that the temperature distribution within the mantle is affected. The conversion of seafloor basalt into *eclogite*, which occurs once subducting slabs reach ~35 km depth on Earth, may not occur on Venus. Eclogite is denser than basalt, and so this transition is thought to increase "slab pull," which is one of the forces driving plate motions on Earth. For all of these reasons, the lack of water on Venus could be the reason why plate tectonics does not operate there. Thus, how Venus lost its water, which we discuss below, may be the key to understanding the operation of the entire planet.

13.3 Formation of Venus' Atmosphere: Wet or Dry?

As already discussed, a key difference between Venus and Earth is that Earth has large oceans on its surface, while Venus is extremely dry. Most of the remaining parts of this chapter are concerned with why that is the case. The first question, obviously, is: Did Venus form dry, or did it form wet and then lose its water over time? This question was debated vigorously during the 1960s and 1970s (see, e.g., Donahue *et al.*, 1982; Lewis and Prinn, 1984).

Over the past 30 years, a consensus has emerged that Venus probably started wet, for two reasons. The first was discussed in Sec. 6.2.1: John Lewis's equilibrium condensation model – which had predicted that Venus formed dry – is no longer accepted as a viable model for planetary formation. Although large portions of Earth's mass may indeed have been derived from volatile-poor material that condensed at or near 1 AU, significant quantities of volatile-rich material are thought to have come from the asteroid belt region or beyond. If these volatile-rich planetesimals were hitting the early Earth, then Venus must have gotten hit, too. Venus need not have received the same amount of water as Earth, but forming Venus with as little water as we observe there today is considered to be highly unlikely, given that Venus has roughly similar inventories of other volatiles, such as carbon and nitrogen.

It was a fortuitous measurement of the deuterium/hydrogen (D/H) ratio from *Pioneer Venus*, though, that finally seemed to clinch the argument for a wet early Venus (Donahue *et al.*, 1982). During its descent through the clouds, the inlet leak on the *Large Probe* mass spectrometer became clogged with a sulfuric acid particle. While this meant that the mass spectrometer results below the

cloud deck were largely useless, it provided a great enhancement in the hydrogen abundance (because H is included in H_2SO_4), enough to enable the detection of deuterium. The D/H ratio returned from the mass spectrometer was 0.019, or about 120 times higher than the ratio in Earth's oceans (1.56×10^{-4}). Since then, the D/H ratio in Venus' atmosphere has been measured even more accurately using near-IR spectroscopic observations from Earth (DeBergh *et al.*, 1991), which yield an estimate of 150 times the terrestrial value. A similar D/H ratio was obtained from the ion mass spectrometer aboard the *Pioneer Venus Orbiter* (McElroy *et al.*, 1982). The investigators from the latter instrument reported the presence of a mass-2 positive ion in Venus' ionosphere, which was initially considered to be H_2^+. After some thought, and following comparison with the probe data, it was realized that the ion must be D^+.

How can the high D/H ratio on Venus be explained? As discussed in Sec. 6.2.3, both Venus and Earth are thought to have received most of their water from the outer asteroid belt. Water-rich carbonaceous chondrites, which are believed to originate from that region, have D/H ratios that scatter widely, but that are on average about the same as that of Earth's ocean (Robert, 2001). As described in detail in Sec. 6.2.3, the D/H values in Oort Cloud and Kuiper Belt comets range from around one to three times the terrestrial ocean value. So, neither comets nor asteroids could have provided water with a D/H ratio as high as that observed on Venus. The only plausible mechanism for producing such an enhancement is by having an initially much larger H_2O endowment, and then by enriching D/H by way of preferential escape of the lighter hydrogen isotope. Not only is this idea plausible, but the mechanism to enable it – the runaway greenhouse – had already been proposed some 13 years before Venus' D/H ratio was measured (Ingersoll, 1969; Rasool and DeBergh, 1970). Thus, the dry early Venus model quickly lost support after the D/H data were published.

Before leaving this discussion, it should be acknowledged that the D/H data do not by themselves require a large initial water endowment on Venus. In theory, if only H atoms escaped to space, and all D atoms were left behind, the required initial inventory of water is just 150 times the present inventory, which is still less than 0.1% of Earth's oceans. Furthermore, it is possible that Venus' water has been resupplied by comet impacts (Grinspoon, 1987; Grinspoon, 1993), in which case the present D/H ratio tells us little about the initial water inventory. But it is also possible that Venus was endowed with much larger quantities of water. If the initial escape of hydrogen was hydrodynamic, as seems likely (Kasting and Pollack, 1983; Kumar *et al.*, 1983), deuterium should have been carried away along with hydrogen. In this case, the fractionation between the two isotopes would have been small (Sec. 5.11), leaving little or no signature, and so the amount of water lost could have been very large.

13.4 The Runaway Greenhouse

This brings us to the runaway greenhouse hypothesis. Before getting into the details of this theory, we should say what we mean by *runaway greenhouse*, as various definitions have been used. A useful review is given in Goldblatt and Watson (2012) and briefly summarize some of their discussion here.

Gold (1964) defined a runaway greenhouse as a positive feedback loop that causes surface temperatures to increase almost without bound. That feedback loop was described earlier in the book (Sec. 2.3) and is illustrated in Fig. 13.7. An increase in surface temperature causes an increase in atmospheric water vapor, which then increases the greenhouse effect, causing a further increase in surface temperature. This feedback loop amplifies perturbations, e.g., those caused by CO_2 increases, by about a factor of two in Earth's present climate system, but it does not destabilize the system. The feedback becomes more powerful, though, as the climate warms, and can eventually reach a point at which it makes the climate system unstable.

We can define the runaway greenhouse more precisely by identifying what happens when the temperature "runs away." A consequence of wet atmospheres is that spectral atmospheric windows close up, and thermal infrared radiation cannot escape to cool the planet. In this case, a limit to the outgoing radiation is set by the physics of water vapor, as pointed out initially by Simpson (1927) and later by Komabayashi (1967, 1968) and Ingersoll (1969). If the absorbed solar flux exceeds the outgoing infrared limit, then the planet's surface heats up until rocks melt. The glow of the hot atmosphere at near-infrared wavelengths, where water vapor is more transparent, shines through to space and restores balance in incoming and outgoing fluxes. The temperature of molten rocks, ~1500 K, is well above the critical temperature for water, 647 K, and so in this state liquid water cannot exist on the planet's surface. This distinction is important because one can also identify climate states in which surface liquid water is present, but in which the stratosphere is moist, so that water vapor is rapidly photodissociated and the hydrogen is lost to space. Kasting *et al.* (1984a) termed this a *moist greenhouse* state, following an earlier suggestion by Towe (1981). As we will see, this latter mechanism is how Venus may actually have lost its

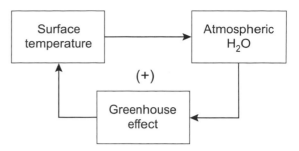

Figure 13.7 Diagram illustrating the positive feedback loop caused by water vapor.

water, although the topic is debated (Hamano *et al.*, 2013; Kurosawa, 2015).

To summarize, we define the *runaway greenhouse effect* as an upper limit on the outgoing thermal infrared radiation flux from a dense atmosphere dominated by water vapor, which is set by the physics of water vapor, so that under an incoming solar flux that exceeds the outgoing limit, all water enters the vapor state and a planet's surface melts. We note, parenthetically, that such runaway states could apply to other condensable species (e.g., methane on a Titan-like planet), as well. However, we restrict our discussion here to water on Earth-like planets because our interest is in habitability.

13.4.1 The Classical Runaway Greenhouse

One of the earliest papers on the runaway greenhouse, and conceptually one of the easiest to understand, was by Rasool and DeBergh (1970). These authors performed a simple climate calculation for three young planets, Venus, Earth, and Mars, as they formed atmospheres from volcanic outgassing. The planets were assumed to begin their existence as airless bodies with their present masses, and variations in solar luminosity were neglected. (This should not be surprising, as Sagan and Mullen had not yet published their 1972 paper on the faint young Sun.) In Rasool and DeBergh's model, H_2O and CO_2 were assumed to have been outgassed in a 4:1 ratio. A slightly simpler, pure-H_2O model was published later by Goody and Walker (1972), also without consideration of solar luminosity changes. Their model illustrates the point well and is easier to understand, so we use it as our example.

Both Rasool and DeBergh and Goody and Walker assumed gray, radiative-equilibrium atmospheres, and they used the Eddington approximation to obtain an analytic solution to the radiative transfer equations. This approach is a variation of the two-stream method described in Sec. 2.4.5.3. The key step is to assume that the specific intensity, I, in Schwarzchild's equation of radiative transfer in a plane parallel atmosphere, eq, (2.81), is independent of zenith angle (Goody, 1964, p.52). In Sec. 2.4.5.3, we used a diffusivity factor D, of 1.66, whereas in this treatment, it is 3/2. When this approximation is used, the temperature within the atmosphere is given by

$$\sigma T^4(\tau) = \frac{F_S}{2}\left(1 + \frac{3}{2}\tau\right) \tag{13.8}$$

This is just eq. (2.106) with $D = 3/2$. Similarly, the surface temperature is given by

$$\sigma T^4(\tau_{ground}) = \frac{F_S}{2}\left(2 + \frac{3}{2}\tau_{ground}\right) \tag{13.9}$$

In these equations, τ represents vertical thermal infrared optical depth at an arbitrary altitude, τ_{ground} is the total atmospheric optical depth, and F_S is the absorbed solar flux, $S(1 - A)/4$, which (in equilibrium) is equal to the outgoing infrared flux at the top of the atmosphere.

For these pure water vapor atmospheres, the optical depth, measured vertically downwards, is given by

$$d\tau = -\kappa_v \rho_v \, dz = \kappa_v \frac{dP}{g} \tag{13.10}$$

Here, κ_v is the absorption coefficient of water vapor (assumed to be 0.01 m^2 kg^{-1}), ρ_v is mass density of water vapor, g is gravity, z is altitude, and P is pressure. We have used the barometric law, $dP/dz = -g\rho$, to change from altitude to pressure coordinates. The absorption coefficient is one that is relevant to the 8–12 μm window region where water vapor is most transparent (Goody, 1964, p. 195). This is a reasonable assumption, as most of the outgoing radiation should escape from within that interval. The optical depth at the surface is found by integrating equation (13.10) downward starting from $\tau = 0$ at the top.

$$\int_{\tau_{ground}}^{0} d\tau = \int_{P_{ground}}^{0} \frac{\kappa_v P}{g} \Rightarrow \tau_{ground} = \frac{\kappa_v P_{ground}}{g} \tag{13.11}$$

The above equations can be used in the following way. A given temperature defines a saturation vapor pressure, according to the Clausius–Clapeyron equation (Sec. 1.1.3.5). If we vary the vapor pressure, assumed to be P_{ground}, we can calculate an optical depth from eq. (13.11) and a surface temperature from eq. (13.9), if we specify the solar flux, F_S. The results of such calculations are shown in Fig. 13.8. The dark, solid curve represents the saturation vapor pressure curve for water. The three dashed curves show the mean surface temperatures of the three terrestrial planets, as calculated from eq. (13.9). All three planets were given surface albedos of 0.17, similar

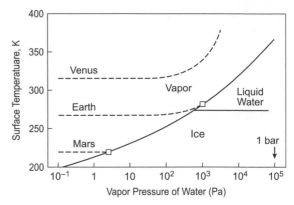

Figure 13.8 Diagram illustrating the "classical" runaway greenhouse effect. The solid curve represents the saturation vapor pressure of water. The three dashed curves show surface temperature of three initially airless planets as they outgas an atmosphere of pure water vapor. Atmospheric radiative transfer is gray, i.e., with a single broadband infrared absorption coefficient. (From Goody and Walker, (1972). Reproduced with permission of Pearson Education, Inc., New York. Copyright 1973. Adapted originally from Rasool and DeBergh, 1970.)

to that of Mars today. Mars itself, being furthest from the Sun, starts out the coldest, ~220 K. As vapor pressure is increased, which we can think of as water being outgassed, one moves from left to right in the diagram. Once the surface pressure of Mars' atmosphere reached about 2 Pa (0.02 mbar), where the solid and dashed curves intersect, water vapor began to exceed its saturation vapor pressure, and so it condensed out to form ice. To a first approximation, this explains what we see today: Mars is basically a frozen desert with a very thin atmosphere.

Earth, being somewhat closer to the Sun, started out a little warmer than Mars, about 270 K, which is fairly warm because the model used an assumed surface albedo that's low and a modern solar luminosity. Hence, water vapor was able to accumulate to higher concentrations before it began to condense. In Earth's case, enough water vapor accumulated to produce an optical depth in eq. (13.11) greater than unity and a greenhouse effect. That is why the curve for Earth bends slightly upwards as one follows it towards the right. As a result, when water vapor began to condense on Earth, the surface temperature was >0 °C, and so it formed liquid water instead of ice. Once again, to a first approximation, this is what we observe. Earth's large inventory of surface water is almost all contained in its oceans.

For Venus, the calculated surface temperature evolution is quite different. Because Venus is even closer to the Sun than is Earth, its surface started out hotter still, ~315 K. This allowed large amounts of water vapor to accumulate in Venus' atmosphere – enough so that the greenhouse effect became large well before saturation was reached. Thus, the surface temperature curve for Venus bends sharply upwards at pressures above about 0.01 bar, and so it never intersects the saturation vapor pressure curve. According to this calculation, all of the water released from Venus' interior would have remained in the vapor phase, forming a dense steam atmosphere. If one assumes that Venus started out with the same amount of water as did Earth, the surface pressure of this atmosphere would have been about 270 bar. Once in the atmosphere, the water vapor would have been lost by photodissociation followed by escape of hydrogen to space. We have not yet described that process in detail but will do so below. The net result would be a dry planet similar to the one we observe today.

Although the calculation described above illuminates the basic idea of a runaway greenhouse very well, in other ways it is unsatisfying. As described in Ch. 6, we no longer believe that planets begin their lives as airless bodies; instead, they form atmospheres and oceans as they accrete. The variation in solar luminosity with time was ignored, whereas in reality the Sun was approximately 30% less bright when the planets formed. And the climate model used was a gray, radiative-equilibrium model, which ignores convection and offers only a crude approximation of the detailed line absorption in a real CO_2–H_2O atmosphere. We will describe more complicated atmospheric models in Section 13.4.4. For now, though, let's retain the gray atmosphere approximation and look at the problem from a different perspective.

13.4.2 A Simple Approximation to the Outgoing Infrared Flux from a Runaway Greenhouse Atmosphere

Another way to approach the runaway greenhouse problem is to treat it as a static phenomenon and to calculate the solar flux needed to trigger it. This is the approach that has been taken historically by most authors. Two different types of static *radiation limits* have been derived: one based on emission of radiation from the convective troposphere (Simpson, 1927), the other based on emission from a radiative-equilibrium stratosphere (Ingersoll, 1969; Komabayashi, 1967, 1968). The tropospheric limit was the first one to be studied and is actually the more physically relevant limit, so we discuss it first. The treatment is approximate. A more rigorous calculation is described in the next section.

We consider a pure H_2O atmosphere, as before. As explained in Sec. 2.4.4, most of the outgoing infrared radiation should be emitted from the region around optical depth unity. Let's start by assuming that the mass absorption coefficient depends on pressure linearly, as follows.

$$\kappa_v = \kappa_0 \left(\frac{P}{P_0}\right)$$ (13.12)

Here, P_0 is a reference pressure where the opacity is κ_0. Such pressure-dependence of the mass absorption coefficient is appropriate for *pressure-broadened* or *collision-induced absorption*, which is the situation in tropospheres (see Sec. 2.5.7). The vertical optical depth at any height in a pure H_2O atmosphere is given by eq. (13.10) at some pressure, P, within the atmosphere, as follows.

$$d\tau = -\kappa_v \rho_v \, dz = \kappa_0 \left(\frac{P}{P_0}\right)\frac{dP}{g} \Rightarrow \tau = \frac{\kappa_0 P^2}{2gP_0}$$ (13.13)

In the last step, we have done a simple integration. Remember that the subscript "v" stands for "vapor", not frequency. There is no frequency dependence because the atmosphere is gray.

Now, assume that thermal-infrared radiation is being emitted from within the saturated troposphere. Then, $P = P_v$, the saturation vapor pressure. Most of the radiation emerges from the region where the optical depth along a slant path is equal to unity, i.e., $\tau^* \simeq D\tau = 1.66\,\tau = 1$. The *diffusivity factor* $D = 1.66$ accounts for the increase in pathlength along an average slant path, from Rodgers and Walshaw (1966). The vapor pressure at the $\tau^* = 1$ radiation level will therefore be:

$$P_v(\text{at } \tau^* = 1) \approx \sqrt{\frac{2gP_0}{\kappa_0 D}}$$ (13.14)

But, the vapor pressure P_v is uniquely defined by temperature through the Clausius–Clapeyron equation (1.49). Combining this equation with eq. (13.14) yields

$$P_v = \left(\frac{2gP_0}{\kappa_0 D}\right)^{1/2} = P_{ref} \exp\left[\frac{l_c}{R_c}\left(\frac{1}{T_{ref}} - \frac{1}{T}\right)\right]$$ (13.15)

Here, P_{ref} and T_{ref} are the triple point values for water of 611 Pa and 273.15 K, and l_c and R_c are the latent heat of vaporization and the specific gas constant for H_2O. Inverting the previous equation and solving for T gives

$$T(\text{at } \tau^* = 1) = \frac{T_{ref}}{\left[1 - T_{ref}\frac{R_c}{l_c}\ln\left(\left(\frac{2gP_0}{\kappa_0 D}\right)^{1/2}\frac{1}{P_{ref}}\right)\right]}$$ (13.16)

We can think of T as the temperature of an infrared "photosphere" for a planet in a runaway greenhouse.

We can write the outgoing infrared radiation from the planet as $F_{IR} = \sigma(CT)^4$, where σ is the Stefan–Boltzmann constant and C is a constant of order unity, which accounts for the fact that the blackbody temperature of emission will be slightly less than that found at optical depth of unity and that the ratio l_c / R_c (which is dimensionally a temperature) in eq. (13.16) is not really constant but depends on temperature and pressure. We take $R_c = 462$ J K^{-1} kg^{-1}, $l_c = 2.425 \times 10^6$ J kg^{-1}, and $g = 9.8$ m s^{-2}. The gray body water vapor absorption coefficient is less certain. A reasonable fit to non-gray numerical 1-D modeling results is given by $C = 0.928$ and $\kappa_0 = 0.05$ m^2 kg^{-1} at $P_0 = 10^4$ Pa. This gives $CT = 266.3$ K from eq. (13.16), and a limit on the outgoing radiative flux of $F_{IR} = \sigma(CT)^4 = 285$ W m^{-2}, where σ is the Stefan–Boltzmann constant. The value of $\kappa_0 = 0.05$ m^2 kg^{-1} approximately matches the absorption around 10 μm at a temperature of ~320 K in the mid-runaway troposphere at a saturation pressure ~10^4 Pa.

We can gain a better appreciation of this limit by comparing it to the radiative fluxes in Earth's present atmosphere. In equilibrium, the outgoing infrared flux, F_{IR}, must equal the absorbed solar flux, $S(1- A)/4$. For Earth's present albedo of ~0.3, the absorbed solar flux is about 239 W m^{-2}. The runaway greenhouse limit derived above is higher than this by a factor of 1.2. So, this analysis predicts that a pure water vapor atmosphere on Earth (assuming a gray opacity and no change in planetary albedo) would turn into a runaway greenhouse if solar luminosity were to increase by ~20%.

The analytical form of eq. (13.16) allows us to generalize this result to planets other than Earth. The dependence on g indicates that a planet that is more massive than Earth will have a higher runaway limit, while low-mass planets will have smaller limits and be more susceptible to runaway greenhouses. The reason is that the atmospheres of high-mass planets are more compressed. The level at which the emitting optical depth is unity occurs at higher pressure, according to eq. (13.14). It must lie along the saturation vapor curve, as well, so the emitting temperature must be higher, along with the emitted flux. To give an example, a 10-Earth-mass, water vapor-saturated planet with a surface gravity of ~18 m s^{-2} should have a runaway greenhouse limit of 305 W/m^2, about 7% higher than a 1-Earth-mass planet. This prediction is borne out by non-gray numerical models, which show that the runaway greenhouse threshold changes by about $\pm 8\%$ for planets that are, respectively, 10 times more massive or 10 times less massive than Earth (Kasting *et al.*, 1993b, Table 1; see also Pierrehumbert, 2010, his Fig. 4.37).

13.4.3 More Rigorous Limits on Outgoing Infrared Radiation from Gray Atmospheres

Kombabayashi (1967, 1968) and Ingersoll (1969) independently developed another way of illustrating the runaway greenhouse mathematically. Their analyses focused on the radiative-equilibrium stratosphere. Their approach is elegantly explained by Nakajima *et al.* (1992), who also showed how the radiation limit derived in this manner is related to the one derived by considering emission from the convective troposphere. Hence, we follow Nakajima *et al.*

The simplest version of the story once again concerns a pure water vapor atmosphere. Assume further that the atmosphere consists of a stratosphere, which is in radiative equilibrium, along with a convective troposphere. By the same logic as used for eq. (13.11), the opacity at the tropopause is

$$\tau_{TP} = \frac{\kappa_v P_{TP}}{g} = \frac{\kappa_v P_v(T_{TP})}{g} \tag{13.17}$$

Here, we have set the pressure at the tropopause equal to the saturation vapor pressure, $P_v(T_{TP})$, at the tropopause temperature. But P_v is given by the Clausius–Clapeyron equation, as before; hence, eq. (13.17) gives $\tau_{TP} = \tau_{TP}(T_{TP})$. Meanwhile, evaluating eq. (13.8) at the tropopause gives $T_{TP} = T_{TP}(\tau_{TP})$. These two equations are displayed graphically in Fig. 13.9. The dashed curve represents eq. (13.17), while the solid curves represent eq. (13.8) for three different values of F_{IR}, the outgoing flux at the top of the atmosphere. It can be seen immediately that the curves do not intersect for $F_{IR} > 385$ W m^{-2}, meaning that there is no solution in this region. Physically, that's because if the outgoing flux is to be any higher, the tropopause temperature must also be higher. But this, in turn, would increase the amount of water vapor in the stratosphere, and thus its opacity, which would make it even harder for the radiation to get out. This is the *Komabayashi–Ingersoll (KI) limit* on the outgoing infrared radiation flux (name coined by Nakajima *et al.*). Note that it is about 35% higher than the approximate radiation limit derived in the previous section. The KI limit would be achieved for a net absorbed solar flux increase by a factor of 1.61, relative to today's value of 239 W m^{-2}.

One can do more with the gray atmosphere approximation. A real atmosphere would contain not just water vapor, but also one or more non-condensable components. It would also have a convective troposphere linking the stratosphere with the surface. Both Ingersoll and Nakajima *et al.* have treated this more general case. In a

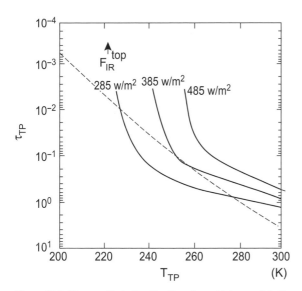

Figure 13.9 Diagram illustrating the Komabayashi–Ingersoll limit on the runaway greenhouse outgoing thermal infrared flux, F_{IR}. The dashed line represents the saturation vapor pressure of water. The three solid curves show the variation of temperature with altitude calculated from eq. (13.8). A gray atmosphere is assumed. (From Nakajima et al. (1992). Reproduced with permission. Copyright 1992, American Meteorological Society.)

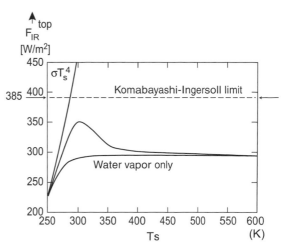

Figure 13.10 Outgoing infrared flux F_{IR}, as a function of surface temperature T_s for a gray atmosphere consisting of pure water vapor (bottom curve) or water vapor plus 1 bar of non-condensable, non-absorbing gas (middle curve). The flux calculated from the Stefan–Boltzmann law is shown for comparison. (From Nakajima et al. (1992). Reproduced with permission. Copyright 1992, American Meteorological Society.)

two-component model, the opacity along a vertical path is given by

$$dτ = -(κ_v x_v m_v + κ_n x_n m_n)\frac{dP}{mg}$$ (13.18)

Here, m_v and m_n are the molecular masses of water vapor and the non-condensable species, m is the mean molecular mass, and x_v and x_n are the respective volume mixing ratios. The variation of temperature with pressure in the troposphere is assumed to be given by the moist adiabatic lapse rate (eq. (1.56)). In the model considered here, $m_n = m_v = m$, and $κ_n = 0$, i.e., the non-condensable component has the same molecular mass as water vapor and does not absorb radiation. Results of two separate calculations are shown in Fig. 13.10. The lower curve shows the outgoing infrared flux as a function of surface temperature for a pure water vapor atmosphere, while the middle curve shows the flux for an atmosphere with a non-condensable gas partial pressure of 1 bar, similar to modern Earth. The flux from a blackbody, $σT^4$, is shown for comparison. In neither case does the outgoing infrared flux reach the Komabayashi–Ingersoll limit (dashed line). Other model atmospheres with larger non-condensable components can reach this limit, but in thin atmospheres this does not happen. Instead, the radiation being emitted to space comes mostly from the troposphere. This tropospheric

radiation reaches an asymptotic limit of ~293 W m^{-2} at surface temperatures >400 K. The physical reason for this behavior is straightforward: As the surface temperature becomes higher and higher, the lower atmosphere becomes increasingly opaque to infrared radiation. Eventually, none of the outgoing radiation comes from the surface; rather, it all comes from the upper troposphere. The amount of radiation being emitted depends on the absorption coefficient of water vapor, along with the slope of the moist adiabat. This same asymptotic behavior is exhibited in non-gray models, as well (Abe and Matsui, 1988; Kasting, 1988; Pollack, 1971; Watson et al., 1984). Although all of these models preceded the paper of Nakajima et al. (1992), it was the latter authors who most clearly elucidated the physics behind this tropospheric emission limit and, as mentioned earlier, it was Simpson (1927) who discovered it. Hence, following Goldblatt and Watson (2012), we will term this the *Simpson–Nakajima (SN) radiation limit*

The SN limit on the outgoing infrared flux is substantially lower than the KI limit and similar to the approximate runaway greenhouse threshold derived in the previous section. Under the same assumptions as above (no albedo change) the calculated value of 293 W m^{-2} corresponds to an increase in solar flux by a factor of 1.23 compared to present Earth. This result depends on the assumed atmospheric composition because one needs to get over the "hump" in the middle curve in Fig. 13.10, and that hump can be as high as the KI limit. As we shall

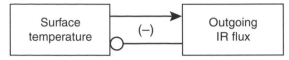

Figure 13.11 Diagram illustrating the negative feedback between surface temperature and the outgoing infrared flux. The line terminating in a circle is negative coupling. This feedback loop breaks down when the atmosphere becomes optically thick at all infrared wavelengths.

see, however, the SN limit shown here is probably closer to the real limit for triggering a runaway greenhouse.

Finally, Fig. 13.10 provides a visual way of thinking about feedback loops, or lack thereof, in the runaway greenhouse calculation. As discussed in Ch. 1, the feedback loop that stabilizes Earth's climate on short time scales (months to years) today is that between surface temperature and the outgoing infrared flux (Fig. 13.11). An increase in the outgoing infrared flux always cools the planet, so that coupling is always negative. And, under normal circumstances, an increase in surface temperature causes an increase in the outgoing infrared flux, so that coupling is positive. Once one gets over the hump in Fig. 13.10, though, this coupling changes sign, becoming strongly or weakly negative, depending on how much background gas is present and how well that gas absorbs in the infrared. As a result, the stabilizing negative feedback loop disappears, and the system is free to "run away" to very high surface temperatures.

13.4.4 Radiation Limits from Non-Gray Models

The runaway greenhouse has been studied with non-gray atmospheric models, as well. Two early studies were those by Abe and Matsui (1988) and Kasting (1988). Abe and Matsui looked at possible steam atmospheres that might have existed during Earth's accretion. They found that a runaway greenhouse was caused when the accretional heat flux exceeded 150 W m^{-2}. Their model included solar energy, as well, and with a little effort, one can calculate that the SN limit in their model[2] was ~304 W m^{-2}. This is slightly higher than the SN limit calculated by the gray model in the previous section (293 W m^{-2}), but it shows that the same atmospheric physics is coming into play;

namely, most of the emitted infrared flux is coming from the convective troposphere. One can also calculate that their planetary albedo was ~0.37. As we will see, both the outgoing IR flux and the planetary albedo are close to values derived independently by Kasting (1988) (discussed below), but both have now been superseded by more up-to-date calculations.

The non-gray model of the runaway greenhouse by Kasting (1988) was driven by solar radiation only. The calculation that he performed was equivalent to taking the present Earth, with its existing atmosphere and ocean, and sliding it slowly in toward the Sun. Like the Rasool and DeBergh simulation discussed earlier, this calculation does not correspond directly to how a real runaway greenhouse atmosphere might evolve. But it demonstrates under what conditions an Earth-like atmosphere can become unstable.

Because of the strong positive feedback from water vapor, climate model calculations can be difficult to control once the critical threshold for a runaway greenhouse is passed. Kasting avoided this problem by performing inverse calculations. The stratospheric temperature was fixed at 200 K, a value appropriate for Earth without an ozone layer, and the surface temperature, T_s, was raised sequentially to higher and higher values. By-products of H_2O photolysis would presumably catalytically destroy an ozone layer if H_2O became a major constituent in the stratosphere. The surface pressure was set equal to 1 bar of non-condensable gas ($N_2 + O_2$) plus the saturation vapor pressure of water. The tropospheric temperature profile was assumed to follow a moist pseudo-adiabat as long as the surface temperature was below the critical point for water, 647 K (Fig. 13.12(a)). As discussed in Sec. 1.1.3.6, the pseudo-adiabat is one in which the condensed water is assumed to immediately rain out. The formulation of Ingersoll (1969), which remains valid in the region near the critical point, was used to calculate the pseudo-adiabatic lapse rate. This assumption makes physical sense, because the vapor pressure of (pure) water at the critical point, 220.6 bar, is lower than the pressure of a fully vaporized ocean, ~270 bar. In Kasting's model, once the surface temperature exceeded 647 K, the surface pressure immediately jumped to 271 bar (= 270 bar H_2O + 1 bar N_2/O_2), and the lower atmosphere became unsaturated. The tropospheric lapse rate was assumed to follow a dry adiabat up to the point where water vapor began to condense, after which it followed a moist adiabat up to the tropopause (Fig. 13.12(b)).

Temperature and water vapor profiles for the low-temperature simulations ($T_s < 420$ K) are shown in Fig. 13.13. As the surface temperature is increased, the convective troposphere extends to higher and higher

---

[2] Abe and Matsui (1988) assumed a solar constant of 960 W m^{-2}, appropriate for early Earth. Parameters for their E150 model, which is just slightly below the runaway greenhouse threshold, are given in their Table 4. They used spherical geometry, and they listed global fluxes, which we normalize here to Earth's surface area of 5.1×10^{14} m^2. The outgoing infrared flux in their E150 model, L_{pl}, was 155.1×10^{15} W, which is equivalent to a flux per unit area of 304 W m^{-2}.

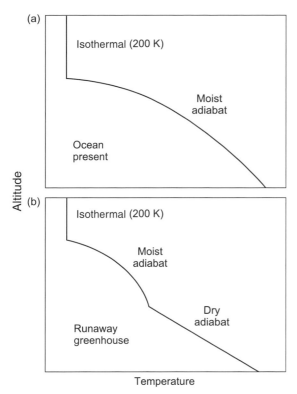

Figure 13.12 Schematic vertical temperature profiles for moist (a) and runaway (b) atmospheres. (From Kasting (1988). Reproduced with permission from Elsevier. Copyright 1988.)

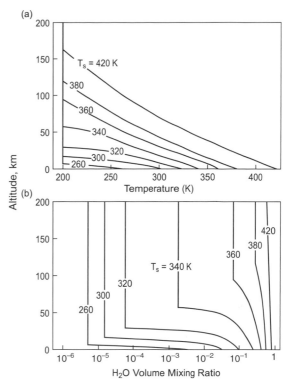

Figure 13.13 Vertical profiles of temperature (a) and water vapor mixing ratio (b) for atmospheres with different surface temperatures, T_s. A 1-bar N_2/O_2 background atmosphere is assumed. (From Kasting, (1988). Reproduced with permission from Elsevier. Copyright 1988.)

altitudes, reaching values >150 km for surface temperatures above 400 K. The reason, as explained originally by Ingersoll (1969), is because of the large amount of latent heat released when water vapor condenses. At high surface temperatures, the atmosphere becomes water-dominated, and so the tropospheric P–T profile approaches the saturation vapor pressure profile given by the Clausius–Clapeyron equation (1.49). The lapse rate in this water-dominated regime can be derived from the differential form of this eq. (1.48). If, for clarity, we replace the meteorologists' symbol for saturation vapor pressure, e_s, with the symbol we have been using in this chapter, P_v, eq. (1.48) becomes

$$\frac{dP_v}{dT} = \frac{l_c P_v}{R_c T^2} \tag{13.19}$$

Inverting this equation and using the barometric law and the perfect gas law gives

$$\frac{dT}{dz} = \rho g \frac{dT}{dP_v} = \frac{R_c T^2 \rho g}{l_c P_v} = \frac{gT}{l_c} \tag{13.20}$$

The latent heat of condensation, $l_c \cong 2.5 \times 10^6$ J kg^{-1}, so at 288 K the predicted lapse rate from eq. (13.20) is

~1.1 K km^{-1}. By comparison, the dry adiabatic lapse rate for Earth's present atmosphere is: $\Gamma_{dry} = g/c_p \cong$ 10 K km^{-1}. The moist adiabatic lapse rate for a water-dominated atmosphere is clearly very much smaller, so the troposphere must extend to very high altitudes before the temperature drops to the assumed stratospheric temperature (T_{strat}) of 200 K. One could argue that it is artificial to keep T_{strat} fixed during such a calculation. However, by analogy to the gray atmosphere solution, as the tropopause moves higher and higher, T_{strat} should approach the skin temperature of the planet, which is given by eq. (2.97) as $T_{skin} = T_{eff}/2^{1/4}$, where T_{eff} is the effective radiating temperature. Thus, T_{strat} varies only as the fourth root of the outgoing infrared flux. As the stratosphere becomes higher and thinner it emits less radiation, and so any deviations in T_{strat} make little difference to the overall planetary radiation balance (Kasting, 1988).

The assumption that an upper stratosphere approaches the skin temperature has been challenged. Leconte et al. (2013) point out (following Pierrehumbert, 2010, p. 289 ff.) that the temperature at low optical depths in a non-gray atmosphere can be much lower than the gray atmosphere skin temperature. In their 3-D numerical model, the temperature at the top of the troposphere reaches 115 K. This produces a much more effective cold trap and may limit the amount of water that can be lost from such planets. However, Venus *does* seem to have lost lots of water, so this prediction may be in conflict with observations. The tropopause moves up to high altitudes (low pressures) in all of these models, so non-LTE effects may also become important, which need to be considered with further models.

The water vapor profiles corresponding to the vertical temperature profiles of Fig. 3.13(a) are shown in Fig. 13.13(b). At low surface temperatures, T_s, the water vapor mixing ratio, f_{H2O}, drops off rapidly with altitude, as it does in Earth's troposphere today. As a consequence, the stratosphere is dry (<10 ppmv for $T_s = 280$ K). The 1-D model overestimates stratospheric water vapor slightly because it lacks a cold, high, equatorial tropopause. The actual stratospheric H_2O mixing ratio is 3–5 ppmv.

As T_s rises above about 320 K, a dramatic change takes place: the water vapor mixing ratio in the upper troposphere and stratosphere increases dramatically, reaching values near unity for $T_s > 400$ K. Once again, Ingersoll (1969) predicted such behavior. He pointed out that the critical transition occurs when the volume mixing ratio of water vapor at the surface exceeds ~0.2. The saturation vapor pressure of water at 340 K is ~0.3 bar, and so for the Kasting (1988) model, with its 1 bar of non-condensable gas, this is about where the critical transition to a wet stratosphere occurs.

The calculated outgoing infrared flux, F_{IR}, and absorbed solar flux, F_S, for these simulations are shown in Fig. 13.14. Here, F_S is the absorbed solar flux *for the present solar constant*. In reality, of course, the outgoing infrared flux must equal the absorbed solar flux. This is true at only one particular surface temperature in the figure, $T_s = 288$ K, the present value for Earth. So, at other surface temperatures, the required, or "effective," solar flux, S_{eff}, would be either higher or lower than the present flux, according to

$$F_{IR} = S_{eff} \cdot F_S \qquad (13.21)$$

Turning this relationship around, $S_{eff} = F_{IR}/F_S$. The significance of S_{eff} will be discussed below.

Let's focus on the behavior of F_{IR} and F_S. The outgoing infrared flux, F_{IR}, is similar to that shown in the gray atmosphere model of Nakajima *et al.* (1992) (Fig. 13.10), except that the "hump" is much smaller. According to Nakajima *et al.*, this is probably because the non-condensable gas in this model contains CO_2, which is an infrared absorber, whereas the calculations

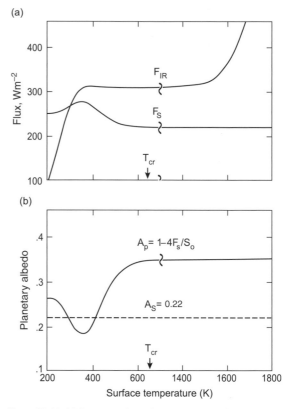

Figure 13.14 (a) Outgoing infrared flux, F_{IR}, and absorbed solar flux, F_S, as a function of surface temperature. The incident solar flux is assumed to be the same as that at Earth today. T_{cr} is the critical temperature, 647 K, above which water only exists as a vapor. (b) Planetary albedo, A_p. The surface albedo is A_s. (From Kasting, (1988). Reproduced with permission from Elsevier. Copyright 1988.)

shown in Fig. 13.10 assume that the background gas is transparent in the infrared. The asymptotic limit on the outgoing infrared flux, termed the SN limit above, is about 310 W m^{-2} (versus 293 W m^{-2} for the gray model). More recent models put the SN limit at 282 W m^{-2} (Goldblatt *et al.*, 2013) to 290 W m^{-2} (Kopparapu *et al.*, 2013; Pierrehumbert, 2010, his Fig. 4.37).

Whereas the gray atmosphere was assumed to be completely transparent in the visible, the non-gray Kasting model atmosphere absorbs and scatters incident solar radiation. As seen from Fig. 13.14(a), the absorbed solar flux F_S first increases slightly then decreases as the surface temperature is raised. The initial increase in F_S is caused by the presence of more water vapor in the model atmosphere. As mentioned in Sec 2.4.2, water vapor has several strong absorption bands in the visible/near-IR region, which absorb some of the incident solar flux. At higher surface temperatures, the atmospheric density increases because of the additional H_2O, and this causes increased Rayleigh scattering. More sunlight is reflected back to space, causing F_S to decrease. The behavior of the solar radiation is easier to visualize if one realizes that $F_S = S_0(1-A)/4$, where S_0 is the present solar constant, 1360 W m^{-2} in this particular calculation. Turning this relationship around allows one to calculate the albedo from the following.

$$A = 1 - \frac{4F_S}{S_0} \qquad (13.22)$$

The albedo can be seen to dip slightly at first and then increase to an asymptotic value of 0.35 as the surface temperature increases and the atmosphere becomes water-dominated (Fig. 13.14(b)). The dashed line in the figure shows the assumed surface albedo of 0.22. The surface albedo was adjusted in this cloud-free model to yield the correct surface temperature for modern Earth, given the present solar constant.

The planetary albedo of an H_2O-dominated atmosphere has also changed in newer models. Goldblatt *et al.* (2013) calculate an asymptotic value of 0.17 at high surface temperatures, and Kopparapu *et al.* (2013) calculate 0.19. As explained by Goldblatt *et al.*, the reason why these albedos are much lower than previously found is because the absorption coefficients for H_2O are being calculated from the new HITEMP database, which includes weak absorption lines shortward of 500 nm that are not found in the older HITRAN database.

13.4.5 Evolution of Venus' Atmosphere: the "Moist Greenhouse"

Figure 13.15 summarizes the Kasting (1988) results. The runaway greenhouse threshold is determined by the relationship between surface temperature T_s, and a solar flux S_{eff}, normalized to the solar constant. Although S_{eff} was calculated as a function of T_s in the model, the axes have been inverted in Fig. 13.15 to show T_s as a function of S_{eff} (solid curve). Again, imagine that the Earth is slowly sliding toward the Sun. S_{eff} will of course increase according to the inverse square law, $S_{eff} \equiv \frac{S}{S_0} = \left(\frac{r_0}{r}\right)^2$.

As this happens, T_s increases, slowly at first, then more rapidly, as the atmosphere becomes more and more water-dominated. At $S_{eff} \cong 1.4$, T_s shoots up sharply from ~600 K to ~1600 K. The critical temperature is 647 K, so any surface temperature in excess of this is too hot to allow liquid water. The reason that T_s levels out around 1600 K is that the moist convective layer is now sufficiently optically thin that radiation from the dry convective region below it can now radiate to space. Note that 1600 K is above the melting temperature for typical silicate rocks, so a planet in this state should have a magma ocean at its surface, similar to that predicted for the accretional models of Matsui and Abe (1986a, b), Abe and Matsui (1988), and Zahnle *et al.* (1988). This runaway greenhouse threshold occurs at $S_{eff} \cong 1.06$ in the newer model of Kopparapu *et al.* (2013).

Figure 13.15 shows something else, though, that is even more important in terms of planetary habitability. The dashed curve in the figure, which goes with the scale on the right, shows the stratospheric water vapor mixing ratio as a function of S_{eff}. Stratospheric water vapor increases rapidly with surface temperature, as we have already seen in Fig. 13.13(b). When expressed in terms of S_{eff}, the behavior is quite dramatic: the stratosphere goes from being dry to water-dominated as S_{eff} increases beyond 1.1 (or 1.015 in the Kopparapu *et al.* (2013) model). Once this happens, the planet is likely to become

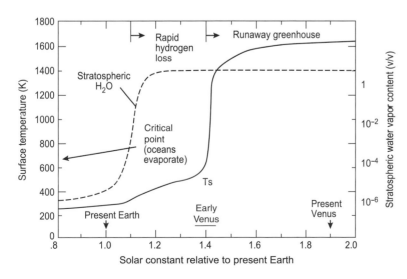

Figure 13.15 Surface temperature T_s (solid curve, mapping to the left-hand vertical axis), and stratospheric water vapor volume mixing ratio (dashed curve, mapping to the right-hand axis) as a function of effective solar flux, $S_{eff} = F_{IR}/F_S$. (From Kasting, (1988).)

uninhabitable, for two reasons. (1) If the planet had oxygen and ozone in its atmosphere, the ozone layer would be destroyed by catalytic cycles involving the byproducts of H_2O photolysis. (2) More importantly, the presence of large amounts of H_2O in the stratosphere would overcome the "cold-trap bottleneck" for hydrogen escape (described in Sec. 5.8.4). Photodissociation of this water vapor, followed by escape of hydrogen to space, could lead to loss of the planet's oceans on time scales of tens to hundreds of millions of years (Kasting, 1988; Kasting and Pollack, 1983; Kasting et al., 1984a). This "water loss" limit on planetary habitability is much more stringent than the runaway greenhouse limit. According to the inverse square law, the (non-gray) runaway greenhouse limit corresponds to an orbital distance of $(1/1.4)^{1/2} \cong 0.85$ AU in the Kasting (1988) model. By comparison, the water loss limit corresponds to a distance of $(1/1.1)^{1/2} \cong 0.95$ AU in that model. In the Kopparapu et al. (2013) model, the corresponding limits are 0.97 AU and 0.99 AU. In reality, these limits derived from 1-D models are probably too pessimistic, because clouds could provide negative feedback and because real tropospheres are not fully saturated.

A 3-D climate model has tested how clouds may affect the runaway greenhouse limit (Leconte et al., 2013). As mentioned earlier, these authors do not predict a wet stratosphere because their cold trap temperatures are very low. However, they do compute a runaway greenhouse limit of 0.95 AU, or $S_{eff} \cong 1.1$. Their model includes H_2O absorption coefficients from HITEMP, so it should in principle give similar results to Kopparapu et al. (2013) except for 3-D effects. Decreased relative humidity is indeed a stabilizing influence in this model, causing the runaway greenhouse limit to move in slightly, Surprisingly, clouds give *positive* feedback in this model, contrary to previous predictions (Kasting, 1988). It remains to be seen whether other 3-D climate models produce the same result.

What do these calculations imply for the evolution of Venus' atmosphere? Venus' orbital radius is 0.723 AU, so the solar flux at Venus' orbit today is 1.91 times that at Earth. Taken at face value, this should put present Venus well into the runaway greenhouse regime, according to any of the published 1-D or 3-D calculations. But the solar flux in Venus' early history was about 30% lower, yielding $S_{eff} \cong 1.34$. That is below the (older) runaway greenhouse limit, but well above the (older) water loss limit ($S_{eff} = 1.1$). This led Kasting (1988) to conclude that early Venus should have been in a *moist greenhouse state* in which liquid water remained present at the surface while at the same time water was being rapidly depleted

by photolysis and hydrogen escape. The newer calculations, however, suggest that Venus was always well inside the runaway greenhouse limit, so that it never had liquid water on its surface. This prediction is consistent with other recent studies, which predict that Venus formed a steam atmosphere during accretion, and that this steam atmosphere persisted until almost all of Venus' water was lost (Hamano et al., 2013; Kurosawa, 2015).

Finally, let's consider how Venus' atmosphere and surface might have evolved from an early water-rich state to its present condition. As water was lost, any ocean at its surface – if it existed at all – would have become smaller and smaller until it eventually disappeared. H_2O photolysis and hydrogen escape would have continued, however, until some other process limited it. On Venus, that process was probably the buildup of SO_2 and the formation of the hygroscopic sulfuric acid clouds. These would have locked up H_2O, severely limiting the H_2O vapor pressure above the clouds and thereby shutting down hydrogen escape. SO_2 would have built up in the atmosphere because volcanoes were still active at this early stage in Venus' history – indeed, they may still be active today, as discussed earlier in Sec. 13.2. However, loss of SO_2 to the surface would have become slow once the liquid water disappeared. The same statement holds true for CO_2. Formation of carbonate minerals on Earth requires liquid water. So, once Venus lost its water, volcanic CO_2 should have accumulated, leading to the very dense CO_2 atmosphere that we observe today. And, of course, if water was important to plate tectonics, as suggested in Sec. 13.2, then plate tectonics would have shut down, as well.

One last, but important, question is: what happened to the oxygen that was left behind, as H_2O was lost? This question has several possible answers. If much of the hydrogen loss occurred early in Solar System history when the solar EUV flux was very high, then some of the oxygen may have been dragged off to space along with the escaping hydrogen (see Sec. 5.10.4). Quantifying this loss would require better models of hydrodynamic escape than currently exist. Some oxygen is being lost to space today by interactions with the solar wind (Barabash et al., 2007), although not enough to explain the loss of an entire ocean. Some oxygen could have combined with outgassed CO to form CO_2. How much oxygen could have been lost depends on how much of Venus' CO_2 was initially outgassed as CO. If the Earth is a good analogy (see Ch. 7, eq. (7.12)), the CO should only have been a small amount, ~5% of the outgassed carbon. That is enough to soak up only a small amount of oxygen, less than 1% of the amount contained in water in Earth's oceans.

A more likely sink for the oxygen is crustal oxidation. This has previously been considered to be a major issue for water loss (Lewis and Prinn, 1984, p. 190). Suppose that Venus started off with a 3 km-deep ocean. That is equivalent to 3×10^5 g H_2O cm^{-2}, or 1.7×10^4 moles H_2O cm^{-2}. Assume that crustal density is 2.5 g cm^{-3}, the crust is 10% FeO, and that the FeO is oxidized to magnetite.

$$3\, FeO + \frac{1}{2} O_2 \rightarrow Fe_3O_4 \qquad (13.23)$$

Three moles of FeO are required for each mole of water, so the total FeO demand is 5×10^4 moles cm^{-2}, or 3.6×10^6 g cm^{-2}. Using the crustal density and FeO content from above, this requires oxidizing the topmost 140 km of the crust. That is difficult to do if the crust was solid, as it is on Earth today. The global rate of crustal generation at Earth's mid-ocean ridges is ~21 km^3 yr^{-1}. Earth's surface area is ~5×10^8 km^2, so the average rate of crust generation is about 4.2×10^{-8} km yr^{-1}. If Venus had the same crustal generation rate, and if all of this crust was oxidized, the time required to consume Venus' leftover oxygen would be 3.3 b.y. Thus, this mechanism could conceivably have consumed a large amount of oxygen, but the time scale for doing so would probably have been quite long.

The oxygen loss problem largely goes away if the more recent climate models are correct. If Venus' atmosphere remained in a runaway greenhouse state throughout the water loss process, its surface would have remained molten. Oxygen could thus have been directly taken up by the molten mantle, which would have overturned much more quickly than Earth's mantle does today (Hamano et al., 2013; Kurosawa, 2015). If so, then Venus-like planets around other stars might not show evidence of O_2 buildup. So, observations of young exoplanets may eventually provide a test for this hypothesis.

13.5 Stability of Venus' Present Atmosphere

Other researchers have proposed alternative explanations for how Venus' atmosphere evolved. Bullock and Grinspoon (1996) suggested that the atmospheric pressure on present Venus is buffered by chemical interactions between the atmosphere and the planet's surface. In their model, atmospheric CO_2 was assumed to be in equilibrium with carbonate minerals in Venus' crust. Specifically, they considered the following reaction.

$$CaSiO_3 + CO_2 \leftrightarrow CaCO_3 + SiO_2 \qquad (13.24)$$

Here, $CaSiO_3$ is the mineral wollastonite, $CaCO_3$ is calcium carbonate (limestone), and SiO_2 is silica, or quartz.

This reaction is called the wollastonite equilibrium, or sometimes the Urey equilibrium, as Harold Urey proposed this long ago as a buffer for atmospheric CO_2 on Earth. According to Fegley (2012, 2014), the equilibrium CO_2 partial pressure (in bar) for this reaction is given by

$$\log_{10} pCO_2 = 7.97 - \frac{4456}{T} \qquad (13.25)$$

Equation (13.25) predicts that $pCO_2 \cong 88$ bar at 740 K, which is close to the observed surface pressure. So, this does indeed suggest that the Urey equilibrium might apply. However, the consequences of adopting this assumption are probably unphysical, because the equilibrium defined by eqs. (13.24) and (13.25) is unstable: carbonates are favored at lower temperatures, whereas gaseous CO_2 is favored at higher temperatures. Thus, the hotter the surface gets, the more CO_2 goes into the atmosphere, and this makes the surface still hotter. Conversely, if the surface cools, atmospheric CO_2 goes into the rocks, making the surface still colder.

This is a classic positive feedback loop, similar to the one involved in the runaway greenhouse. Hence, if this model were correct, Venus' surface pressure would be poised precariously at an unstable equilibrium point. This seems unlikely, however, because the surface pressure ought to "run away" in one direction or the other. Furthermore, a simpler explanation is available: Chemical reaction rates between gases and dry rocks are exceedingly slow. (That is why the hydrated silicates predicted by John Lewis' equilibrium condensation model should not have actually formed.) Such reactions can be further inhibited by the buildup of weathering products on the surfaces of mineral grains. Once a silicate grain becomes coated with calcium carbonate, no further reaction with atmospheric CO_2 should occur. On Earth, the products of weathering are removed by liquid water, but on Venus this would not happen. So, it seems simpler to assume that surface interactions are indeed slow and that Venus' atmosphere has served as a simple collector for CO_2 that was released from its interior. If so, then the near coincidence between the surface pressure and the CO_2 partial pressure predicted from the Urey equilibrium must be just that – a coincidence.

Control of atmospheric SO_2 by surface–atmosphere interactions is much more plausible, as SO_2 is present in much lower concentrations (~150 ppm), and so less surface rock would be needed to control its concentration. Subsequently, Bullock and Grinspoon (2001) suggested that SO_2 in Venus' atmosphere is buffered by the following equilibrium reaction.

$$SO_2 + CaCO_3 \leftrightarrow CaSO_4 + CO \qquad (13.26)$$

Here, $CaSO_4$ is the mineral *anhydrite*. This reaction could proceed both ways (as implied by thermodynamic equilibrium) if both calcite and anhydrite were present on Venus' surface. The equilibrium SO_2 concentration predicted from this reaction, though, is only about 1% of the observed SO_2 concentration in the atmosphere (Hashimoto and Abe, 2005), and the predicted lifetime of SO_2 with respect to this reaction is relatively short, only 20 million years. Hence, if this reaction occurs, it probably goes only in one direction, from left to right, and continued volcanic outgassing of SO_2 right up to the present day would be needed to maintain its atmospheric concentration.

Alternatively, Hashimoto and Abe (2005) proposed that Venus' SO_2 is buffered by reaction with pyrite, FeS_2. The reaction in this case is as follows.

$$3FeS_2 + 16CO_2 \leftrightarrow Fe_3O_4 + 6SO_2 + 16CO \qquad (13.27)$$

Fe_3O_4 is the mineral *magnetite*. As discussed in Sec. 10.3, pyrite is one of the reduced minerals whose presence (in detrital form) is considered to be an indicator of low O_2 levels on the early Earth. Although we do not know the exact O_2 concentration in Venus' lower atmosphere, it is thought to be well below the upper limit of 0.3 ppm that is present above the cloud tops (Mills and Allen, 2007; Trauger and Lunine, 1983). So, this reaction is plausible, and it predicts an equilibrium atmospheric SO_2 concentration close to the various observations discussed in Sec. 13.1.1. Pyrite might also account for the anomalously high radar reflectivity of high-altitude regions of Venus' surface (Klose *et al.*, 1992). So, it may be that the SO_2 in Venus' atmosphere is actively controlled by surface interactions, even if CO_2 is not.

13.6 Implications for Earth and Earth-Like Planets

Although Venus is an interesting planet in its own right, more interesting to us is what we can learn from Venus about the limits of climatic stability on Earth and on other Earth-like planets that may exist elsewhere in the galaxy. The latter topic will be discussed at some length in Ch. 15. Below, we look at how the runaway/moist greenhouse theory might apply to our own planet Earth.

13.6.1 Can CO₂ Cause a Runaway Greenhouse on Earth?

A question that has arisen more than once in the context of the modern global warming problem is whether or not

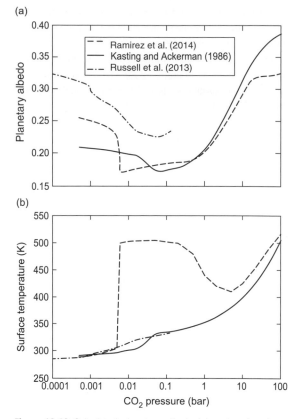

Figure 13.16 Calculated planetary albedo (a) and surface temperature (b) as a function of CO_2 pressure for present Earth. Results from three different calculations are shown. Ramirez *et al.* (2014b) use the new HITEMP absorption coefficients for H_2O, whereas the other two groups used older coefficients. (Reproduced with permission. Copyright 2014, Mary Ann Liebert, Inc.)

CO_2 increases could create a runaway greenhouse on present Earth. Kasting and Ackerman (1986) concluded that the answer was "no." These older calculations are shown by the solid curves in Fig. 13.16. In the Kasting and Ackerman model, Earth's albedo increases rapidly as CO_2 is added to the atmosphere because of the high Rayleigh scattering cross section of CO_2 (about 2.5 times that of air) (Fig. 13.16(a)). Consequently, the calculated surface temperature never exceeds the critical temperature for water, 647 K, and so the ocean never boils (Fig. 13.16 (b)). People often mistakenly assume that the oceans would boil at 373 K, or 100 °C, as this is the normal boiling point for water. But the boiling temperature is a function of the overlying atmospheric pressure: water boils at 100 °C today because the corresponding saturation vapor pressure is just over 1 bar. But if the entire planetary surface was at 100 °C, the atmospheric pressure would be 2 bar (1 bar N_2/O_2 plus 1 bar H_2O), and the ocean would not boil.

This problem has been revisited because of the newer HITEMP absorption database for H_2O mentioned earlier in the chapter. Stronger absorption of incoming solar radiation in the visible causes a warm Earth to have a lower albedo, thereby making the climate more susceptible to runaway (Goldblatt *et al.*, 2013). These authors argue that a CO_2 increase up to 100 PAL might trigger a runaway. But Ramirez *et al.* (2014b) have challenged their conclusion with results shown in Fig. 13.16 (dashed curves). In these 1-D calculations, surface temperature increases sharply to over 500 K at a CO_2 pressure of 6×10^{-3} bar, or ~12 PAL.[3] The sharp uptick in surface temperature is caused by an assumed relative humidity feedback that was present in the original Kasting and Ackerman (1986) model, as well. Its effects are magnified in the new model, however, because of the stronger absorption of sunlight by H_2O. A second reason why Ramirez *et al.* find higher surface temperatures is that their model also predicts an increase in surface relative humidity as the climate warms (Fig. 13.17). By contrast, Kasting and Ackerman assumed a fixed surface relative humidity of 0.8. Actual surface relative humidity *must* increase at high surface temperatures in order to balance the surface energy budget. (If the surface relative humidity is too low, evaporation will remove energy from the surface ocean faster than solar heating can provide it (Ramirez *et al.*, 2014b).)

This problem is admittedly *not* well handled by 1-D climate models. Such models *can* be used to estimate surface relative humidity, as discussed above, but they cannot calculate vertical relative humidity distributions self-consistently. A recent 3-D simulation of CO_2 increases up to 256 PAL by Russell *et al.* (2013) is also shown in Fig. 13.16(b). Their results agree more closely with the old Kasting and Ackerman (1986) calculations. These authors make no mention of HITEMP, however, and so they presumably used older absorption coefficients for H_2O. Finally, Popp *et al.* (2016) find that CO_2 can drive the planet into a moist greenhouse even with the old HITRAN coefficients, likely because of positive cloud feedback in their GCM.

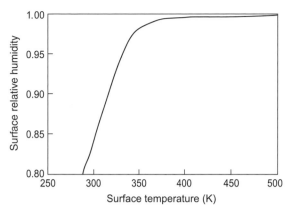

Figure 13.17 Calculated surface relative humidity as a function of surface temperature for the high-CO_2 runs from Ramirez *et al.* (2014b) plotted in Fig. 13.16. Surface relative humidity increases in order to balance the surface energy budget. (Reproduced with permission. Copyright 2014, Mary Ann Liebert, Inc.)

13.6.2 Future Evolution of Earth's Climate

While Earth thus appears to be protected against a runaway greenhouse today, this will not remain true in the distant future. The Sun has brightened by about 30% since it formed, some 4.6 billion years ago, and it continues to brighten today at a rate of roughly 1% per hundred million years. If one extrapolates forward in time using eq. (11.3), the Sun should be about 10% brighter one billion years from now (see Fig. 13.18(a)). As we have seen, that is the same value of the solar flux that causes rapid water loss in models of early Venus. Hence, one might expect that Earth will follow Venus' path and lose its water as well, starting at about that time.

The future evolution of Earth's atmosphere is complicated by the fact that atmospheric CO_2 concentrations should change as the Sun brightens. According to the discussion in Chapter 11, silicate weathering should speed up as the climate warms, and so atmospheric CO_2 should decrease with time, thereby partially offsetting the increase in solar luminosity. This prediction may seem counter-intuitive, as CO_2 in the atmosphere is currently *increasing* at a rapid rate as a consequence of fossil fuel burning. The timescale being discussed here, though, is much longer. Fossil fuels will likely be exhausted within a few hundred years unless we start consuming them more slowly or we find large new reserves. The CO_2 produced from burning them will remain in the Earth system for much longer; indeed, the last dregs will linger for up to a million years (Archer, 2005; Walker and Kasting, 1992). But this is still only a blip compared to the time scale for solar evolution. In the long run, the increase in silicate

---

[3] One has to be careful when measuring atmospheric gas inventories in terms of pressure because, contrary to what one would expect based on Dalton's Law, the partial pressure of a gas in a planetary atmosphere *does* depend on the other gases that are present. If a light gas is added to a heavier one, for example, it mixes with the heavier gas, increasing the pressure scale height (eq. (1.12), $H = kT/\overline{m}g$, where $\overline{m}$ is lower), causing the heavier gas to spread away from the planet's surface, thereby lowering its partial pressure. The pressure depicted on the horizontal axis of Fig. 13.16 is the pressure that CO_2 would exert if it was present by itself.

(a)

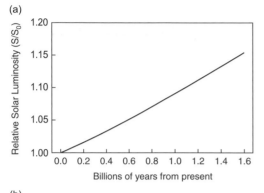

(b)

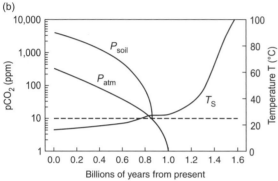

Figure 13.18 Predicted evolution of the Earth system in the distant future. (a) Solar flux relative to today. (b) Surface temperature, T_s, and CO_2 concentration in the atmosphere, P_{atm}, and in the soil, P_{soil}. (Originally from Caldeira and Kasting, (1992a).)

weathering will win out, and atmospheric CO_2 should begin to decrease.

The future evolution of the Earth system has been simulated by several groups using coupled climate/geochemical cycling models (Caldeira and Kasting, 1992a; Franck et al., 1999; Lovelock and Whitfield, 1982; Ramirez et al., 2014b; Von Bloh et al., 2003). In the earliest model, atmospheric CO_2 was predicted to fall below 150 ppm about 100 million years from now. A CO_2 concentration of 150 ppm represents a critical level, called the *CO2 compensation point*, below which C_3 plants cannot live, because they respire faster than they photosynthesize. (The name C_3 refers to the length of the carbon chain that is created during the initial step of photosynthesis.) C_3 plants account for about 95% of all plants on Earth, including trees and most agricultural crops. Hence, Lovelock and Whitfield concluded that Earth might be able to support an active biosphere for only another 100 million years.

Caldeira and Kasting (1992a) revisited this calculation by using an improved climate model and a different

set of assumptions about how weathering rates vary with temperature. The results are shown in Fig. 13.18(b). Their model was based on older absorption coefficients for H_2O and CO_2, but the results are not greatly affected compared to using newer coefficients, because surface temperatures remain relatively low during the next billion years. The model included the so-called *terrestrial biological pump*, which enhances pCO_2 in soils through root respiration and decay of organic matter. In the calculation shown, C_4 plants like corn and sugar cane are assumed to dominate productivity once atmospheric CO_2 concentrations decline below 10 ppm. These plants go extinct about 900 million years from now in the simulation. And even C_3 plants last for 500 million years – five times longer than predicted by Lovelock and Whitfield. Of course, photosynthetic life could even last longer if it evolves more efficient ways of fixing carbon, which has a precedent in the evolution of C_4 plants.

In the Caldeira and Kasting (1992a) model, once atmospheric CO_2 declines below 10 ppm, its greenhouse effect is greatly diminished, its stabilizing effect on climate vanishes, and the surface temperature climbs rapidly over the next few hundred million years (Fig. 13.18(b)). Unlike the CO_2-induced greenhouse described earlier, water should be rapidly lost during this process because the stratosphere will become water-dominated and the warming will be permanent. It might take a few hundred million years to lose the oceans, but the Earth will eventually be left completely dry, as Venus is today, unless descendants of humans (if they exist) somehow intervene. Indeed, this should be the fate of any Earth-like planet that absorbs a stellar flux more than about 10% higher than Earth does today.

To sum up, Venus today is hot and inhospitable to life. The planet may well have had surface liquid water early in its history, but it lost that water through a runaway or moist greenhouse. Whether a terrestrial planet evolves more like Venus or more like Earth depends primarily on the amount of energy it absorbs from its parent star, and to a lesser extent on how much CO_2 its atmosphere contains. These inferences will be used in Ch. 15 to help delineate the region around a star in which Earth-like planets can exist. Indeed, it seems reasonable to predict that that one day astronomers will observe an early Venus-like exoplanet shedding its water under the intense glow of extreme ultraviolet light from its young parent star. Studying such a phenomenon would help confirm the theory of atmospheric evolution presented in this chapter.

14 Giant Planets and their Satellites

An overall theme in this book is the importance of atmospheric evolution for astrobiology. Kinetics determines the composition of atmospheres, but the chemistry of the giant planet atmospheres in the Solar System has probably always been much closer to thermodynamic equilibrium than the atmospheres of rocky planets such as Earth or Mars. When an atmosphere is near thermodynamic equilibrium, little free energy is available for biology and no sign is seen of significant perturbation by metabolic waste gases. Consequently, thermodynamic equilibrium suggests that the atmospheres of giant planets are not favorable habitats for life. Giant planets also have no solid or liquid surface at moderate pressures and temperatures and hence no stable environment in which life might originate and evolve. In contrast, giant planet satellites can be icy and they may have stages in their evolution where liquid water resides at the surface or in the subsurface. So, in this chapter, we devote more discussion to the giant planet moons than the giant planets themselves.

14.1 Giant Planets

The giant planets in our Solar System are just a few of a vast number throughout the galaxy. An estimated $17^{+6}_{-9}\%$ of stars possess planets with mass 0.3–10 M_J, where M_J is the mass of Jupiter. Neptune-mass exoplanets (10–30$M_\oplus$, where $M_\oplus$ is an Earth mass) are even more common, orbiting about half of all stars (Cassan *et al.*, 2012) (and smaller exoplanets are probably even more common still (Howard, 2013)). Meanwhile, unbound Jupiter-mass planets float between stars. Estimates for their abundance in the Milky Way range between 20–50 times less numerous than main sequence stars (Scholz *et al.*, 2012) to twice as common (Sumi *et al.*, 2011).

14.1.1 Current Atmospheres

Basic physical data about the giant plants of the Solar System is given in Table 14.1, while Table 14.2 provides the composition of their atmospheres. More than 99% of the observable atmospheres of giant planets consist of H_2, He, and CH_4 (Table 14.2). But in the composition of the planets as a whole, ice-forming elements are much more abundant in Uranus and Neptune compared to Jupiter and Saturn. Ice-forming elements are O, C, and N, which are found in H_2O, CH_4, NH_3, CO_2, and CO ices. Because of their inferred internal structure and composition (Fig. 14.1), Uranus and Neptune are called *ice giants*, while Jupiter and Saturn are *gas giants*. In bulk, the ice giants have less than 50% by mass H and He.

The atmospheric composition of Jupiter and Saturn is similar to that of the Sun but slightly enriched in elements heavier than helium (Table 14.2). The $^{13}C/^{12}C$ ratio on both Jupiter and Saturn is close to the protosolar value of 0.011, while the D/H ratios are also similar to protosolar (Atreya *et al.*, 2003). One element that bucks the trend is neon (Ne), which is only ~1/10 solar abundance on Jupiter. A possible reason is that neon is soluble in He drops that have "rained out" of the deep atmosphere towards the center of the planet (Stevenson and Salpeter, 1977b; Wilson and Militzer, 2010).

Apart from helium and neon, other noble gases (Ar, Kr, and Xe) are enriched by a factor of ~3, which is puzzling. These elements don't form solids at 5 AU and Ar and H cannot be separated except at <30 K. Either the noble gases were brought in later in ices from the cold outer solar system (Owen and Encrenaz, 2006) or external photoevaporation from the far-UV of a nearby massive star fractionated elements in the solar nebula (Monga and Desch, 2015).

Clouds characterize the visual appearance of the giant planets. Figure 1.11 shows the saturation vapor curves for

Table 14.1 Basic data on the giant planets. (Source: Lodders and Fegley (1998).)

Property	Jupiter	Saturn	Uranus	Neptune
Mass (10^{26} kg)	18.986	5.6846	0.868 25	1.0243
(Earth masses)	~318	~95	~14.5	~17
Radius at 1 bar (km)	71 492	60 268	24 973	24 764
(Earth radii)	~11	~9.5	~4	~4
Mean distance to Sun (AU)	5.2026	9.5719	19.194	30.066
Mean density (g cm^{-3})	1.326	0.6873	1.318	1.638
Sidereal rotation period (hr)	9.925	10.65	17.24 (retro)	16.11
Bond albedo	0.343	0.342	0.300	0.290
Effective temperature (K)	124.4 ± 0.3	95.0 ± 0.4	59.1 ± 0.3	59.3 ± 0.8
Temperature at 1 bar (K)	165 ± 5	134 ± 4	76 ± 2	71.5 ± 2
Internal energy flux (W m^{-2})	5.44 ± 0.43	2.01 ± 0.14	0.042 ± 0.047	3.22 ± 0.34
(relative to absorbed solar)	~1.7	~1.8	~1	~2.6

Table 14.2 Composition of giant planet atmospheres. Gases are given as volume mixing ratios. (Sources: Lodders (2010a); Atreya *et al.* (2003); Irwin (2009).)

Species	Jupiter	Saturn	Uranus	Neptune
H_2	86.4 ± 2.6 %	88 ± 2 %	82.5 ± 3.3 %	80 ± 3.2 %
He	13.6 ± 0.3 %	12 ± 2 %	15.2 ± 3.3 %	19.0 ± 3.2 %
*CH_4	0.204 ± 0.049 %	0.47 ± 0.2 %	~2.3 %	~1–2%
			(below CH_4 clouds)	(below CH_4 clouds)
$^{13}CH_4$	$1.9 \pm 0.1 \times 10^{-5}$	$5.1 \pm 0.2 \times 10^{-5}$		
*NH_3	$5.74 \pm 0.22 \times 10^{-4}$	~1.1×10^{-4}	$<1 \times 10^{-7}$	$<6 \times 10^{-7}$
	(8.9-11.7 bar, below	($p > 1$ bar, below		(at 6 bar)
	NH_3 clouds)	NH_3 clouds)		
H_2O	$4.23 \pm 1.38 \times 10^{-4}$	2.3×10^{-7}	$6\text{-}14 \times 10^{-9}$	$1.7\text{-}4.1 \times 10^{-9}$
	(at 17.6-20.9 bar)	(upper troposphere)	(at $p < 0.03$ mbar)	(at $p < 0.6$ mbar)
*H_2S	$6.7 \pm 0.4 \times 10^{-5}$	$\leq 0.4 \times 10^{-6}$	$<8 \times 10^{-7}$	$<3 \times 10^{-6}$
Ne	1.99×10^{-5}			
Ar	$1.57 \pm 0.35 \times 10^{-5}$			
C_2H_6	$5.8 \pm 1.5 \times 10^{-6}$	$7.0 \pm 1.5 \times 10^{-6}$	1.5×10^{-5}	1.5×10^{-6}
	(stratosphere)	(stratosphere)	(at 0.5–1 mbar)	(stratosphere)
PH_3	$1.04 \pm 0.1 \times 10^{-6}$	$5.9 \pm 0.2 \times 10^{-6}$	$<8.3 \times 10^{-7}$	
		($p > 0.5$ bar)		
C_2H_2	~4×10^{-8}	$3.0 \pm 0.2 \times 10^{-7}$	~1.0×10^{-8}	$6.0^{+14.0}_{-4.0} \times 10^{-9}$
	(at 5 mbar)	(at 1 mbar)	(stratosphere)	
				(stratosphere)
CO_2	3.5×10^{-10}	3.4×10^{-10}	$4.0 \pm 0.5 \times 10^{-11}$	6×10^{-8}
	(at $p < 10$ mbar)	(at $p < 10$ mbar)	(at 0.1 mbar)	(at $p < 5$ mbar)
CO	$1.6 \pm 0.3 \times 10^{-9}$	$< 10^{-7}$	$<4 \times 10^{-8}$	$6.5 \pm 3.5 \times 10^{-6}$
	(at 5–8 bar)	(stratosphere)	(stratosphere)	(stratosphere)
D/H	$2.3\text{-}2.6 \times 10^{-5}$	$1.7^{+0.75}_{-0.45} \times 10^{-5}$		
$^{13}C/^{12}C$	0.0108 ± 0.0005	0.011		

* Condensable species, so levels are indicated. CH_4 condenses on Uranus and Neptune. H_2O and NH_3 condense on the giant planets. H_2S reacts with NH_3 to form ammonium hydrosulfide (NH_4SH) clouds.

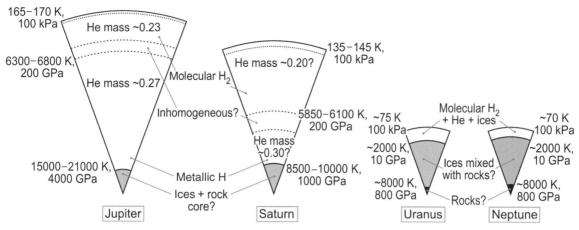

Figure **14.1** Schematic diagram showing giant planet interiors. (Adapted from Perryman (2014, p. 296).)

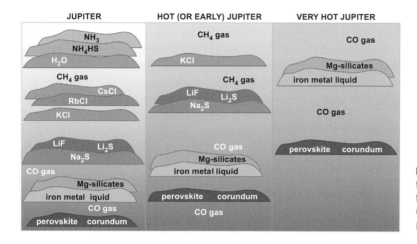

Figure **14.2** Schematic diagram showing the vertical distribution of clouds in Jupiter (left), early Jupiter before 4.4 Ga (middle), and a very hot Jupiter-like planet. (Redrawn from Marley (2010).)

various condensable species and the observed temperature profiles of the giant planets. A cloud base for each species should form where the saturation curve intersects the temperature profile (Sanchez-Lavega *et al.*, 2004). Jupiter and Saturn are predicted to have clouds of NH_3, NH_4SH (ammonium hydrosulfide, which forms from the reaction of H_2S and NH_3), and H_2O. Spectra supports these compositions except for the deep H_2O clouds (West *et al.*, 2004). In the colder atmospheres of Uranus and Neptune, clouds of the aforementioned compounds form at greater depths, while clouds of CH_4 and H_2S condense in the cold upper tropospheres, unlike on Jupiter or Saturn. Going in the opposite direction of temperature, very hot giant exoplanets are expected to have clouds of exotic substances such as droplets of molten silicates or liquid iron (Fig. 14.2) (Marley, 2010). Clouds of these

substances probably exist on Jupiter today but cannot be detected because they would exist much deeper than the levels that emit radiation to space.

Further details of the formation, atmospheric structure, chemistry, and dynamics of giant planet atmospheres are summarized elsewhere, e.g., books by Irwin (2009) or Sanchez-Lavega (2010) and reviews (e.g., Guillot and Gautier, 2015). Here, we focus on how such atmospheres evolve.

14.1.2 Thermal Evolution of Giant Planets and their Atmospheres

The general evolutionary behavior of giant planets is to start out warm and then cool and shrink. Jupiter had an effective temperature T_{eff}, around 600–1000 K when it was first formed, compared to ~124 K today (Burrows *et al.*, 2001). So, Jupiter's atmosphere was initially too hot for ammonia or water clouds. Instead, alkali metals may have formed halide or sulfide particles (Fig. 14.2). By 4.4 Ga, Jupiter's T_{eff} had dropped to ~400 K and water clouds appeared in the upper troposphere. By 3.5 Ga, T_{eff} was ~160 K, and ammonia clouds were possible. At that point, Jupiter's cloud structure would have been similar to that of today (Burrows, 2005). We speculate that another possible consequence of a warm, early Jupiter might be to drive off early atmospheres of the Galilean moons by hydrodynamic escape. But this has not yet been studied.

Simple convective cooling models predict the current T_{eff} of Jupiter in a state where it radiates twice the energy it absorbs because contraction still releases gravitational energy, but they do not model Saturn correctly. In the absence of other physics, Saturn is about 1.5 times more luminous than it should be (Pollack *et al.*, 1977; Stevenson and Salpeter, 1977a). The generally accepted explanation is that He can form immiscible drops in metallic hydrogen and sink, which converts gravitational potential energy into heat (Stevenson, 1975). However, an outstanding puzzle is that He in Saturn's atmosphere does not appear to be significantly depleted compared to protosolar abundance (Table 14.2). The H/He ratio might be resolved in the future by sending a probe into Saturn.

Data for Uranus and Neptune rely upon *Voyager 2* encounters and are sparse. Even so, convective cooling models for these planets have the opposite problem as for Saturn and tend to overestimate T_{eff}. The models also cannot easily explain the heat flow from these planets (Table 14.1), which is especially low for Uranus. Possibly, stable stratification exists at depth (Hubbard *et al.*, 1995). Stability against convection would result in less efficient removal of internal heat and a lower heat flux. Alternatively, the observed infrared emission on Uranus may be affected by the planet's high obliquity and thus may not be equal to its orbit-averaged value.

In the long term, very old Neptune-like planets may cool down enough that their interiors enter a pressure–temperature regime where oceans of liquid water become possible (Wiktorowicz and Ingersoll, 2007). At the moment, the depth at which the interior gases of Neptune become compressed to water-like densities is too hot for the liquid state. But on a colder Neptune, liquid water can occur when the pressure exceeds the critical pressure of water if the temperature is lower than water's critical temperature of 647 K. Calculations suggest that oceans are favored if the bulk mole fraction of water on a Neptune-like planet is greater than ~1/3, compared with ~0.27 on Neptune. Even so, if Neptune cooled down to $T_{eff} < 30$ K, the water in its atmosphere might condense into a deep ocean. Such a situation should occur when the Sun ages and becomes a white dwarf. Unbound ice giants floating between stars could also potentially have deep internal oceans of liquid water. The relevance of such conditions for astrobiology depends on the pressure limits for life. Some terrestrial deep-sea piezophiles have pressure optima >100 MPa (Bartlett, 2002), while other studies show active cells at ~1 GPa (Sharma *et al.*, 2002). Oceans on old Neptunes would have ~GPa pressures.

14.1.3 Thermal (Hydrodynamic) Escape on Hot Giant Exoplanets

The discovery of giant planets near their host stars – *hot Jupiters* or *hot Neptunes* – suggests that some giant planets might have ended up so close in the process of orbital migration (Sec. 3.1.3) that their atmospheres strongly evaporate. As discussed in Sec. 5.10, hydrodynamic escape is a limiting case of thermal escape that occurs when a planet's upper atmosphere is sufficiently warm to expand, accelerate through the speed of sound, and attain escape velocity *en masse*. Hot, hydrogen-rich atmospheres are prone to hydrodynamic escape. It may be difficult to evaporate the entire atmosphere of a Jupiter-mass planet unless it's extremely close to its parent star (Kurokawa and Nakamoto, 2014; Murray-Clay *et al.*, 2009), but the atmospheres of hot Neptunes or objects with masses ~$15M_\oplus$ or less have been hypothesized to disappear and leave behind remnant rocky cores (Garcia-Munoz, 2007; Lecavalier des Etangs *et al.*, 2004; Lopez and Fortney, 2013; Owen and Wu, 2013).

Hydrodynamic escape is not just a theory given that observations suggest such escape from hot Jupiters HD 209458b and WASP-12b (Fossati *et al.*, 2013; Fossati *et al.*, 2010; Haswell *et al.*, 2012; Vidal-Madjar *et al.*, 2003). HD 209458b ($0.71M_J$) orbits a G0V star at ~0.05 AU and UV observations show a hydrogen nebula around the planet, covering an area of ~2.5 planetary radii (Ben-Jaffel, 2007, 2008; Vidal-Madjar *et al.*, 2003). Oxygen and carbon are detected in this corona (Vidal-Madjar *et al.*, 2004). Because these atoms are too heavy to escape on their own, drag from hydrodynamically escaping hydrogen has been inferred with a hydrogen escape rate of ~10^7 kg s^{-1} (Linsky *et al.*, 2010; Vidal-Madjar *et al.*, 2008). This flux is far too small to have a significant effect on the planet's mass even over billions of years. However, smaller, highly irradiated exoplanets are likely vulnerable to big atmospheric loss.

In general, an atmosphere of initial mass M_{atm} will be lost over time $\tau_{lifetime}$ if

$$\frac{M_{atm}}{\left(\overline{dM_{atm}/dt}\right)} < \tau_{lifetime} \qquad (14.1)$$

Here, $\overline{(dM_{atm}/dt)}$ is the time-averaged atmospheric mass escape rate.

A useful upper limit on escape is the energy-limited rate (Ch. 4). In this case, stellar energy heats the upper atmosphere and lifts a particle out of the gravity well, which requires $\frac{1}{2}v_{esc}^2$ energy per unit mass. The stellar flux that induces escape is usually the flux in the extreme UV, F (in units of W m^{-2}). We can write the energy input as:

$$\text{energy per unit time} = \left(\pi R_p^2\right)\varepsilon F \qquad (14.2)$$

Here R_p is the radius of the planet and ε is a numerical factor (range 0 to 1) that takes account of heating efficiency, albedo, and geometric factors. Thus, ε represents the fraction of incoming energy carried away by the escaping gas. A typical value for hot Jupiters is ε is <0.6 and probably 0.1–0.25 (Lammer et al., 2009). Because the energy in eq. (14.2) is required to supply $\frac{1}{2}v_{esc}^2$ per unit mass for escape and because $\frac{1}{2}v_{esc}^2 = GM_p/R_p$, where M_p is the planetary mass and G is the universal gravitational constant, it follows that the escape flux is

$$\frac{dM_{atm}}{dt} = \frac{\left(\pi R_p^3\right)\varepsilon F}{GM_p} \qquad (14.3)$$

A further correction factor, K_{tide}, is needed to account for tidal effects at small orbital distances, as follows (Erkaev et al., 2007):

$$\frac{dM_{atm}}{dt} = \frac{\left(\pi R_p^3\right)\varepsilon F}{GM_p K_{tide}}, \qquad K_{tide} = 1 - \frac{3}{2\xi} + \frac{1}{2\xi^3}, \qquad \xi = \frac{(M_p/3M_*)^{1/3}a}{R_p} \qquad (14.4)$$

Effectively, K_{tide} is a weight between 0 and 1 that accounts for the reduction in escape energy caused by the need for an atmospheric particle to reach only the Roche lobe radius $=(M_p/3M_*)^{1/3}a$, where M_* is stellar mass and a is the semi-major axis of the planetary orbit. Beyond the Roche radius, the gravity of the planet is no longer important compared to that of the host star. Note that the planetary radius, planetary mass, stellar flux, and possibly the heating efficiency in eq. (14.4) are all functions of time.

Candidates for the remnant cores of evaporated giant planets have been discussed but their identification remains uncertain. CoRoT-7b is a rocky exoplanet ($4.8 \pm 0.8M_{\oplus}$) that orbits a G9V yellow dwarf star at a distance of only 0.0172 AU and Kepler-10b is a $4.6 \pm 1.2M_{\oplus}$ exoplanet orbiting a G star at 0.017 AU. Application of eq. (14.4) suggests that CoRoT-7b and Kepler-10b might both be giant planets that lost their atmospheres (Jackson et al., 2010; Kurokawa and Kaltenegger, 2013). However, Leitzinger et al. (2011) deduced that the rate of escape is too small for CoRoT-7b or Kepler-10b to have once been ice giants; we note that they used a different function for the evolution of XUV luminosity. Currently, definitive identification of the scorched remnants of evaporated ice giants remains elusive.

It is important to realize that exoplanets in tight orbits such as Kepler-10b or CoRoT-7b will also be subject to impact erosion of their atmospheres. Median impact velocities scale with orbital velocities, which are faster for planets close to their host star. Consequently, impact erosion is a mechanism that might also create remnant cores of worlds that formerly had atmospheres (Catling and Zahnle, 2013).

14.2 Tenuous Atmospheres on Icy Worlds

14.2.1 Overview of Outer Satellite Atmospheres

Moons of giant planets likely form within *circumplanetary disks*, which is ultimately where the atmospheres of such bodies comes from. As discussed in Sec. 6.1, formation models for giant planets include *disk instability* and *core accretion* beyond the snow line, with evidence mostly favoring the latter. So-called *regular satellites* of the giant planets orbit in the same plane as the planets' equator and in the same direction as the giant planet spin, which suggests their formation from a circumplanetary disk. As discussed later, geochemical evidence suggests that icy moons such as Titan formed from icy planetesimals in the disk (Sec. 14.4.4). Large moons (e.g., Europa) were evidently able to retain water ice, even though they differentiated. Smaller moons did not get hot enough to differentiate. The main ice in icy objects in the outer solar system is water ice, which essentially forms the "bedrock" on objects ranging from Europa to Enceladus to Pluto.

Irregular satellites have eccentric orbits at high inclination to the giant planets' equators and these were almost certainly captured. Triton, the largest moon of Neptune, is a prime example.

Apart from Titan, the atmospheres on the outer planet satellites are all extremely tenuous (Table 14.3). In Ch. 1, we mentioned three compositional groups of

Table 14.3 Properties of Pluto and outer planet satellites that have tenuous atmospheres or exospheres.

	Io	Europa	Ganymede	Callisto	Enceladus	Dione	Rhea	Triton	Pluto
Mass[a] (10^{23} kg)	0.893 54	0.480 17	1.4824	1.0763	0.001 08	0.010 95	0.0231	0.215	0.1305
Mean radius[a] (km)	1821	1561	2631	2410	252.1	533.0	764.3	1353	1180
Mean density[a] (g/cm^{-3})	3.528	3.014	1.942	1.834	1.608	1.476	1.233	2.059	2.050
Orbital semi-major axis[a] (10^5 km)	4.22	6.71	10.7	18.8	2.38	3.77	5.27	3.55	5.9064 $\times 10^5$
Orbital period	1.77d	3.55d	7.16d	16.689d	1.370d	2.737d	4.518d	5.877d	247.92 yr
Rotation period[b]	Syn.	Syn.	Syn.	Syn.	Syn.	Syn.	Syn.	Syn.	6.3872d
Surface temperature (K)	90 (night), 130 (day)	50 (pole), 110 (equator)	90 (night), 150 (day)	100 (night), 150 (day)	65–85, (~120–155 in plumes)	50 (night), 90 (day)	53 (night), 99 (day)	37–38	35 (bright areas) 55 (dark areas)
Surface pressure[c] (bar)	<10^{-9} (away from plumes)	10^{-12}	10^{-12}	10^{-9}	variable	10^{-12}	10^{-12}	14–20$\times 10^{-6}$	~(8-15) $\times 10^{-6}$
Bond albedo[d]	≈0.52	0.68 ± 0.05	0.32 ± 0.04	0.2 ± 0.4	0.89 ± 0.02	0.63 ± 0.15	0.6 ± 0.2	0.85 ± 0.05	0.4–0.6
Atmospheric composition	SO_2	O_2	O_2	O_2 with some CO_2	H_2O	O_2	O_2	N_2, trace CH_4, CO	N_2, trace CH_4, CO

[a] Weiss (2004); D. Yeomans, online "Planetary Satellite Physical Parameters," JPL Solar System Dynamics. [b] Syn. means "synchronous rotation," in which the rotational period is equal to the orbital period. [c] Cunningham et al. (2015) for Callisto's atmosphere; Zalucha et al. (2011) for Pluto. [d] Grundy et al. (2007) for Europa's albedo; Howett et al. (2010) for albedo of Saturn's satellites. Brown et al. (1991) for Triton's albedo.

atmospheres, as follows (with surface pressures in bar in parentheses).

(1) The N_2-rich atmospheres of Titan (1.5), Triton ($\sim(14-20)\times10^{-6}$), and Pluto ($\sim15\times10^{-6}$).

(2) Volcanogenic atmospheres: an SO_2 atmosphere of Io ($\sim10^{-7}-10^{-9}$) and a water vapor–CO_2 atmosphere that emanates from the south pole of Enceladus.

(3) The oxidized, tenuous atmospheres of icy moons, including Europa ($\sim10^{-12}-10^{-13}$ O_2), Ganymede ($\sim10^{-12}$ O_2) and Callisto ($\sim10^{-9}$ bar of O_2 with some CO_2) around Jupiter, and Rhea ($\sim10^{-12}$ O_2 and CO_2) and Dione ($\sim10^{-12}$ O_2) around Saturn.

Molecules are collisional on Io, in the atmospheres in the first group, within Callisto's atmosphere (Cunningham *et al.*, 2015), and within plumes very near the surface of Enceladus. In contrast, all the other atmospheres are nearly collisionless and are exospheres at the surfaces of each body.

Below, we briefly describe the atmospheres of Io and Enceladus (Sec. 14.2.2) and the tenuous oxidized atmospheres of icy moons (14.2.3). We then consider the N_2 atmospheres of Pluto and Triton (Sec. 14.2.4) before that of Titan (Sec. 14.4).

14.2.2 Tenuous Volcanic or Cryovolcanic Atmospheres

Two moons have very thin atmospheres that are generated by volcanism. Io's volcanoes produce silicate magmas and secondary melting of sulfur deposits, while Enceladus's *cryovolcanism* melts ice that drives jets of icy particles and gas.

Tidal heating ultimately produces the volcanism and atmosphere on Io. Gravitational forces from Jupiter vary because Io's orbit has *forced eccentricity* caused by periodic gravitational prodding from Europa and Ganymede. This three-body relationship is the famous *Laplace resonance*,[1] reflecting a 4:2:1 period ratio for the orbits of Ganymede, Europa and Io. The resulting global heat flux of 3 ± 1 W m^{-2} on Io (Veeder *et al.*, 1994) is higher than that of any other solid body in the Solar System.

Io's large heat flux has various consequences. Over 400 calderas and volcanoes rise up to 17 km above the surrounding plains. Temperatures up to 1600 K indicate the melting of silicates whose composition is *ultramafic* (defined in Sec. 7.2.1). Given the heat flux, Io should be resurfaced at a rate of 0.5–1 cm yr^{-1}, which is consistent with the absence of impact craters (McEwen *et al.*, 2004).

Typically, about ten or more volcanoes are active. They supply SO_2 at a rate of $\sim10^5$ kg s^{-1} (Lellouch *et al.*, 2003), which freezes in the atmosphere and settles on Io's surface. Except near hot vents, SO_2 frost is at least several millimeters thick (Schmitt *et al.*, 1994). Additional atmospheric constituents are SO, present at 3%–10% of SO_2, and NaCl, which is present at 0.4% of SO_2. Volcanoes are thought to emit the NaCl (Fegley and Zolotov, 2000; Lellouch *et al.*, 2003). Apart from active volcanoes, SO_2 should also be supplied by sublimation of SO_2 frost (McGrath *et al.*, 2004). For example, at a dayside temperature of 130 K, SO_2 frost would be in equilibrium with a 0.1 µbar atmosphere. Photochemistry predicts the formation of other atmospheric constituents, including sulfur allotropes, S_2, S_3,...S_8 (Moses *et al.*, 2002). These plausibly explain the brilliant colors on Io, such as a red ring around the volcano Pele. Elemental sulfur also appears to emanate directly from volcanoes because a ratio of SO_2/S_2 of 3–12 has been detected in Pele's volcanic plume (Spencer *et al.*, 2000).

Io's atmosphere escapes above an exobase at $\sim100-300$ km altitude, so that a banana-shaped cloud of neutral gas extends both ahead of and behind Io's orbit. O and S atoms that derive from SO_2 are the primary constituents, along with minor Na and K. There is also a donut-shaped plasma torus along Io's orbital path, which mainly consists of O and S ions.

The probable eventual fate of Io's SO_2 atmosphere is to vanish. The current heat loss from Io can be explained by tidally driven frictional dissipation, but Io is spiraling inwards towards Jupiter, and thus the Laplace resonance will eventually end (Lainey *et al.*, 2009). When the resonance breaks, Io's volcanism will cease and its atmosphere will no longer be replenished. Exactly when this breakage will occur depends on poorly constrained mechanical details of tidal dissipation. In the past, the orbits of Io, Europa, and Ganymede were originally closer to Jupiter but expanded because of tides raised by Jupiter on the moons (similar to the expansion of the lunar orbit in the Earth–Moon system). However, it is also not known exactly when the Laplace resonance came into existence (Schubert *et al.*, 2010).

Similar to Io, gravitational tides are also ultimately responsible for producing the atmosphere of Enceladus. Enceladus is in a 2:1 mean motion resonance with Dione (Fig. 14.3), i.e., Enceladus orbits Saturn twice for every orbit of Dione. Periodic gravitational nudges from Dione cause forced eccentricity of Enceladus's orbit, resulting in tidal forces from Saturn that flex and heat Enceladus. Thus, tidal heat probably explains why the surface temperature of ~90 K at the south pole exceeds the equilibrium temperature of 60 K (Spencer *et al.*, 2006).

---

[1] The star Gliese 876, which is a red dwarf ~15 light years away, has a 4:2:1 Laplace resonance amongst its e, b, and c planets.

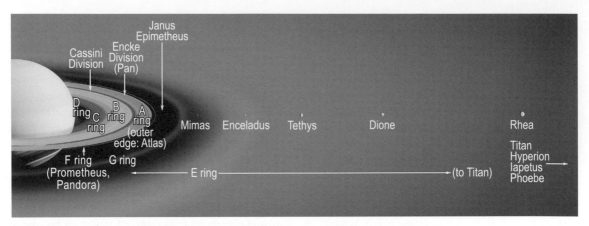

Figure 14.3 Saturn with its inner satellites and rings. The known Pre-Space-Age satellites of Saturn – Mimas to Phoebe – can be remembered with the mnemonic "MET DR. THIP." Two of these moons, Hyperion and Phoebe, are not large enough to be spherical. (Adapted from a NASA diagram.)

There, in Enceladus's south polar region, jets emanate from parallel lineaments dubbed *tiger stripes* (~500 m deep, ~130 km long, 2 km wide) (Porco *et al.*, 2006). The jets release water vapor and other gases at a rate of 200 ± 30 kg s^{-1} (Hansen *et al.*, 2011) and tiny icy particles at ~50 kg s^{-1} (Ingersoll and Ewald, 2011). The gas composition is a ~90% mixing ratio of water vapor, ~5% CO_2, some fraction of CO, 0.9% CH_4, 0.8% NH_3, <0.5% N_2, and trace organic compounds (Hansen *et al.*, 2011; Waite *et al.*, 2009a). The icy particles contain 1% NaCl (Postberg *et al.*, 2011). Because the jets are restricted to the south polar region, the modeled exosphere of Enceladus is extremely asymmetric, with higher density and vertical extent in the south (Tenishev *et al.*, 2010; Tenishev *et al.*, 2014).

Enceladus's jets explain the outermost ring of Saturn, the E-ring. Enceladus orbits within the peak density of this ring (Fig. 14.3), which extends from 3 to 8 Saturn radii and is mostly made up of icy particles with an average size of ~1 μm (Hillier *et al.*, 2007). Destruction of ~1 μm grains by sputtering limits their lifetime to <50 years, so that the E-ring would disappear without resupply. A flux of water vapor that originates from Enceladus can also explain H_2O observed in Saturn's atmosphere, as well as the O in atmospheric CO on Titan (Hartogh *et al.*, 2011a). Remote measurements of H_2O, combined with photochemical models, suggest that the supply from Enceladus and interplanetary dust particles can explain the presence of atmospheric H_2O on Titan. However, explaining both H_2O and CO_2 abundances requires a variable H_2O source, because of very different lifetimes of H_2O and CO_2 (Dobrijevic *et al.*, 2014; Horst *et al.*, 2008; Lara *et al.*, 2014; Moreno *et al.*, 2012).

Presently, the details of how Enceladus is heated and its evolutionary history are not understood. The overall composition of Enceladus' plumes is similar to that of cometary comae. This suggests that Enceladus was accreted from comet-like material. But why tidal heating is concentrated only at the south pole is unsolved (Schubert *et al.*, 2010). The heating is of great interest because the most interesting aspect of Enceladus is a global liquid water, salty sea between its silicate core and outer ice shell (Spencer and Nimmo, 2013; Thomas *et al.*, 2016). Thus, current debates concern possible hydrothermal activity on the seafloor and the nature of the seawater chemistry (Glein *et al.*, 2015; Hsu *et al.*, 2015).

14.2.3 Tenuous O_2-Rich and CO_2-Rich Atmospheres

Icy moons that lack volcanism can have tenuous oxidized atmospheres. The explanation behind such compositions is the breakdown of water ice and the preferential escape of hydrogen to space. Earlier, we discussed the influence of hydrogen escape in oxidizing the Earth (Sec. 10.7), Venus (Sec. 11.3.6), and Mars (Sec. 12.2.4). The same effect occurs on giant planet satellites. On the surfaces of Ganymede, Europa, Rhea, and Dione, molecules in water ice are split by energetic particles or UV photons. Hydrogen escapes rapidly, but the oxygen, which is heavier, lags behind.

Charged particles generate oxidized atmospheres on the icy satellites through *radiolysis* – the chemical alteration of ices by charged particles, and surface *sputtering*, which is erosion by charged particle impacts. The harsh irradiation experienced by gas giant moons is caused by the capture and acceleration of particles in the gas giant magnetospheres. Europa, for example, orbits through Jupiter's magnetic field and plasma and receives $\sim 10^2$–10^3 times the dose rate of charged particles that the Moon gets from the solar wind. Charged particle interactions dissociate water in ice into H + OH. The H can diffuse out of the ice but the OH is trapped. Further reactions inside the ice can produce O_2 molecules, as follows.

$$OH + OH \rightarrow H_2O + O \qquad (14.5)$$

$$O + OH \rightarrow O_2 + H \qquad (14.6)$$

Molecular hydrogen can also be formed.

$$H + OH \rightarrow H_2 + O \qquad (14.7)$$

$$H + H + M \rightarrow H_2 + M \qquad (14.8)$$

Sputtering releases H_2O molecules and trapped gases heavier than hydrogen. It was originally thought that dayside H_2O sublimation from icy satellites would generate O_2 atmospheres through photochemistry (Yung and McElroy, 1977). However, temperatures are so cold that H_2O molecules are easily removed from an atmosphere by sticking to icy surfaces. In contrast, O_2 molecules bounce off the ice and can form atmospheres after sputtering releases them. Thus, radiolysis and sputtering prove critical for forming oxidized atmospheres on icy satellites (Johnson et al., 2009).

An O_2-dominated exosphere of Europa has been detected by *Cassini* from UV emission during a Jupiter flyby (Porco et al., 2003) and from the Hubble Space Telescope (Hall et al., 1998). The surface pressure is $\sim 10^{-13}$–10^{-12} bar and the O_2 scale height is ~ 20 km, assuming a temperature of ~ 100 K. Oxygen is ionized by solar photons or by interaction with magnetospheric plasma. Oxygen ions are "picked up" by Jupiter's field lines and swept away (Johnson et al., 2009). The net result is a balance between the release of O_2 from the surface (mainly by sputtering, which removes ~ 10–100 m of surface per Gyr (Cooper et al., 2001)) and the escape of the oxygen to space.

Trace constituents in tenuous atmospheres may provide a way to detect subsurface oceans. The cratering age of Europa is only 20–180 Myr (Schenk et al., 2004) because of resurfacing associated with tidal heating, so this means that material from a hypothesized subsurface ocean might have been transported to the surface in the recent geologic past. In passing through Jupiter's magnetic field, Europa acquires an induced magnetic field with a strength that requires an electrically conducting fluid within 200 km of Europa's icy surface (Chyba and Phillips, 2007; Khurana et al., 2009). The most likely explanation is a salty ocean. Measurements of induced magnetic fields on Ganymede (Kivelson et al., 2002) and Callisto (Zimmer et al., 2000) also imply somewhat deeper subsurface oceans below roughly 200 km and 300 km depth, respectively. Sodium and potassium have been detected in Europa's extended atmosphere through UV emission suggesting that oceanic salts might be sputtered off the surface of Europa (Brown, 2001; Brown and Hill, 1996). An inferred Na/K ratio of 20 ± 3 at Europa's surface and 25 ± 2 in Europa's atmosphere is consistent with models of the Na/K ratio in Europa's ocean that is fractionated on freezing and upwelling; the ratio is difficult to explain as implantation of Io sodium because the Na/K ratio in Io's atmosphere is 10 ± 3 (Brown, 2001; Johnson et al., 2002; Zolotov and Shock, 2001).

The *Hubble Space Telescope* detected two 200-km-tall plumes of water vapor from Europa in 2012 and 2014 from UV emission (Roth et al., 2014; Sparks et al., 2016). The presence of such plumes is yet to be verified and, if they're common, whether the plumes are a surface phenomenon or eruptions of deeper water.

Radiolysis, sputtering and sublimation also produce an O_2 atmosphere on Ganymede, which is seen in auroral emission at the poles (Hall et al., 1998), analogous to aurorae on Earth. But unlike on Europa, sodium has not been detected in Ganymede's atmosphere (Brown, 1997). Ganymede's surface age is ~ 0.5 Ga (Zahnle et al., 1998), so its subsurface ocean has probably not interacted with its surface recently.

Callisto, the Galilean satellite with the oldest surface of ~ 4 Ga, also has an O_2 atmosphere with some CO_2 (Carlson, 1999; Cunningham et al., 2015; Kliore et al., 2002; Liang et al., 2005). The CO_2 may originate from radiolysis and oxidation of carbonaceous matter in Callisto's dark, icy surface (Johnson et al., 2004). The dark material itself could be a remnant of a primordial atmosphere (Griffith and Zahnle, 1995).

In the Saturnian system, O_2 atmospheres have been detected on Rhea (Teolis et al., 2010) and indirectly on Dione from measurements of molecular oxygen ions, O_2^+ (Tokar et al., 2012) (see Fig. 14.3 for the relative location of these moons). Overall, it would not be surprising if further tenuous O_2 atmospheres on other icy satellites were detected from future data.

14.2.4 The Nitrogen Atmospheres of Triton and Pluto

At and beyond Neptune, we find a different composition of atmosphere on icy bodies, which is dominated by N_2.

The "bedrock" is still water ice but N_2 ice and air sits above.

Triton and Pluto have the characteristics of Kuiper Belt Objects (KBOs). As discussed in Sec. 6.2, the *Kuiper Belt* is a region of icy bodies within the plane of the Solar System at 30–50 AU left over from Solar System formation (see Fig. 6.7). A scattered disk also extends to ~1000 AU, including bodies such as Eris with ~1.3 Pluto masses (Brown and Schaller, 2007). Triton, the largest of 13 moons of Neptune, is a captured KBO because of its retrograde orbit (opposite in sense to Neptune's rotation) with an inclination of 157° with respect to Neptune's equator (Agnor and Hamilton, 2006; McKinnon *et al.*, 1995; Nogueira *et al.*, 2011).

Pluto is now recognized as a KBO that just happens to be relatively large and near the inner edge of the Kuiper Belt. However, Pluto is big only in comparison to most other KBOs: it is smaller than the "Big Seven" satellites of the Solar System: the Moon, the four Galilean satellites, Titan, and Triton. Pluto is in a stable 3:2 mean motion resonance with Neptune, i.e., Pluto orbits the Sun twice for every three orbits of Neptune, and was likely captured into this state when Neptune migrated outwards early in Solar System history (Brown, 2012). Both Triton and Pluto have similar mean density (Table 14.3) and N_2 atmospheres with minor CH_4 and CO components.

14.2.4.1 *Triton*

Detailed knowledge about Triton's environment comes from the *Voyager* 2 flyby in 1989. Because of its high Bond albedo (~0.85), Triton's surface is extremely cold at 38 K. Nonetheless, Triton is geologically active, relatively organic rich, and may have a subsurface ammonia–water ocean at a depth of over 100 km (Gaeman *et al.*, 2012; Hussmann *et al.*, 2006).

As discussed previously, the global atmospheric pressure on certain bodies is sometimes modulated by the vapor pressure of an ice – for example, CO_2 ice on Mars or SO_2 ice on Io – and N_2 behaves like this on Triton and Pluto. Thus, seasonal N_2 frost migrates on Triton on top of a water ice crust (Cruikshank *et al.*, 2000; Spencer, 1990).

The atmosphere is mainly N_2 with a thin haze. The vapor pressure of N_2 ice at 38 K is 14 μbar (Brown and Ziegler, 1979), which matches the surface pressure inferred from *Voyager* 2 radio occultation data of 14 ± 1 μbar. In the 1990s, Triton's surface pressure apparently increased to ~20 μbar because of a ~1–2 K increase in mean surface temperature associated with greater insolation of the

southern cap (Elliot *et al.*, 1998). Atmospheric CH_4 and CO are present at ~0.03% and 0.06% of N_2, respectively (Lellouch *et al.*, 2011b). The methane is destroyed by photolysis and so must be replenished from surface sources. There is a thin haze of particles with a mean size ~0.1–0.2 μm throughout the lower tens of kilometers in Triton's atmosphere, probably as a result of methane photochemistry and production of hydrocarbon aerosols. If photochemistry has been operating at the same rate since 4.5 Ga, about 1 m of solid organics should have fallen out onto Triton's surface. Frosts can cover organics but a reddish tint of the surface perhaps indicates the presence of organics.

Triton's atmosphere is dynamic (Yelle *et al.*, 1995). *Voyager* 2 imaged clouds around a bright southern polar cap and at least two plumes. The plumes had maximum heights of ~8 km and tails that extended laterally over 100 km (Soderblom *et al.*, 1990). The plumes carried vapor and dark particles, perhaps organics. The leading hypothesis for their origin is that solar heating of ice (a *solid state greenhouse*) vaporizes N_2, which is supported by the location of the plumes in the sub-solar southern hemisphere (Brown *et al.*, 1990; Kirk *et al.*, 1995). Another hypothesis is that geothermal energy in Triton's crust causes the plumes (Duxbury and Brown, 1997).

Triton must be geologically active for several reasons. Its surface age is only ~10 Ma (Schenk and Zahnle, 2007), and there are cryovolcanic structures made of ice that are equivalent to vents, fissures, and lavas (Croft *et al.*, 1995). The internal heat that must be responsible presumably derives from both primordial heat left over from Triton's capture by Neptune and radiogenic sources.

When Triton evolved during its capture by Neptune, it may have had a much thicker atmosphere than today. Triton was captured either by gas drag in a Neptunian subnebula, or by collision with another satellite (McKinnon *et al.*, 1995). In any case, the heat derived from gravitational potential energy would have melted Triton completely. Triton would then have differentiated into an icy outer surface and inner rocky core. As Triton's orbit circularized during the first 100 million years or more, tidal heat flow plausibly reached ~2–4 W m^{-2}. Volatilization from such a large heat flow could have produced an atmosphere with a surface pressure ranging from one to tens of bar (Lunine and Nolan, 1992). The greenhouse effect from collision-induced absorption (see Sec. 2.5.6) in such an atmosphere could have generated a surface temperature ~100–200 K.

The sad fate of a satellite in a retrograde orbit is to spiral inward and either crash into its host or to disintegrate into a ring of debris at the distance from its host

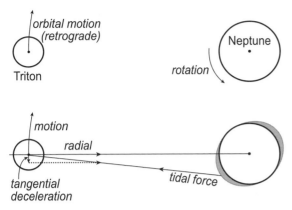

Figure 14.4 Triton has a retrograde orbit around Neptune and so suffers drag from tidal bulges raised on Neptune. As a result, Triton's orbit must gradually decay.

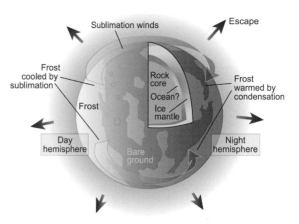

Figure 14.5 A pre-*New Horizons* schematic diagram of Pluto showing volatile migration on the surface and possible internal structure. On the dayside, an atmosphere is created from the sublimation of ice according to vapor pressure equilibrium. On the nightside, frost condenses. The atmosphere also slowly escapes. An interior ocean is drawn but uncertain. (Adapted from McKinnon (2002). Reproduced with permission from Macmillan Publishers Ltd. Copyright 2002, Nature Publishing Group.)

where the Roche limit is reached and tidal forces exceed the satellite's cohesion. Tidal bulges on Neptune are dragged ahead of the position of Triton by Neptune's rotation and Neptune's nearside bulge exerts a gravitational force on Triton that slows Triton's pace along its orbit (Fig. 14.4). In the 1960s, it was projected that Triton's orbit would decay in ~0.1 Gyr. However, this calculation incorporated unrealistic assumptions about tidal dissipation. The decay timescale is now estimated to exceed the remaining main sequence lifetime of the Sun (McKinnon *et al.*, 1995).

14.2.4.2 *Pluto*

Pluto has some similarities to Triton. Like Triton, the main surface ice is N_2, while other surface ices are far less abundant (CH_4/N_2 ice ~0.4% and CO/N_2 ice ~0.01%) (Lellouch *et al.*, 2011b). The *New Horizons* spacecraft imaged mountains of several kilometers height, and structural integrity requires that the "bedrock" be water ice (Stern *et al.*, 2015b). In contrast, N_2 appears to form glaciers. Vapor pressure equilibrium causes Pluto to possess an N_2 atmosphere with ~8–15 μbar surface pressure determined prior to *New Horizons* (Zalucha *et al.*, 2011), while *New Horizons* measured ~10 μbar (Stern *et al.*, 2015b). Atmospheric CH_4 and CO are also present at abundances of ~0.5% and 0.05% of N_2, respectively (Lellouch *et al.*, 2011a), which is consistent with the relative volatility of $N_2 > CO > CH_4$ at ~40 K (Owen *et al.*, 1993; Young *et al.*, 1997). Prior to *New Horizons*, bright regions were attributed to N_2 ice at temperatures down to 35 K caused by sublimation cooling (Stern, 2008) (Fig. 14.5), but *New Horizons* found spectral features of CO in one bright, flat area, Sputnik Planum. Pluto has geologically young surfaces and so must have been recently geologically active (Stern *et al.*, 2015b). Three ices of N_2, CO, and CH_4 have also been found on big KBOs: Eris, Makemake, Sedna, and Quaoar (Brown, 2012).

Photochemistry destroys atmospheric CH_4 on Pluto, as on Triton, and produces a global hydrocarbon haze up to 150 km altitude (Stern *et al.*, 2015b). Again, this implies re-supply of CH_4. Dark areas on Pluto's variegated surface may be rich in hydrocarbons arising either from photochemistry or the radiolysis of CH_4 ice by cosmic rays. *New Horizons* found distinct haze layers in the atmosphere, including one as high as ~80 km altitude.

The global atmospheric pressure should vary greatly as Pluto orbits the Sun every 248 years. Pluto is tipped on its side with an obliquity of 122°, causing N_2 frost to migrate seasonally (Grundy and Buie, 2001; Hansen and Paige, 1996). Also, high orbital eccentricity (0.25) means that the solar flux changes substantially during Pluto's orbit. The semi-major axis is 39.5 AU, but perihelion (which occurred last in 1989) and aphelion (which occurs next in 2114) are at distances of 29.6 AU and 49.3 AU, respectively. Observations suggest that Pluto's atmosphere roughly doubled in surface pressure from 1988 to 2002 despite receding from the Sun; the explanation is that the southern cap entered sunlight (Sicardy *et al.*, 2003). However, models suggest that the atmosphere should condense by ~2020 as heliocentric distance increases. Eris, which is a Pluto-like KBO at 96 AU with

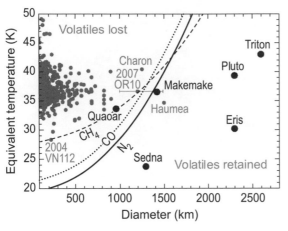

Figure 14.6 Boundaries for volatile retention on Kuiper Belt Objects, including Pluto, Charon, and Triton. The bodies shown as black dots have had CH$_4$ detected on their surfaces. The bodies to the left of the CH$_4$, CO, and N$_2$ lines have either too little mass or are too warm to retain these volatiles as result of volatilization and Jeans escape. Objects to the left of all the boundaries have had no volatiles detected (redrawn from Brown (2012). Reproduced with permission Annual Reviews. Copyright 2012, Annual Review of Earth and Planetary Sciences.)

perihelion at 38 AU, appears to be in a state with such a collapsed atmosphere (Sicardy *et al.*, 2011).

Vapor pressure and susceptibility to atmospheric escape may explain the detection of ices and atmospheres on some KBOs, such as Pluto, but not on others. Schaller and Brown (2007) suggest that KBOs retain volatiles if they are either cold enough to prevent large vapor pressure of the frosts or massive enough to prevent significant long-term Jeans escape. On this basis, Pluto is able to retain its ices and atmosphere (Fig. 14.6). Some authors argued that Pluto's atmosphere was in a "slow hydrodynamic escape" regime rather than a Jeans escape regime (Strobel, 2008a; Tian and Toon, 2005; Trafton *et al.*, 1997). But the assumption of a purely fluid hydrodynamic model for Pluto gives an incorrect structure for the atmosphere because the problem requires a combined fluid-kinetic model (Tucker *et al.*, 2012) (see discussion in Sec. 5.10.1 and Fig. 5.13). Kinetics (in the form of the Boltzmann equation or its approximations) needs to be applied to the rarefied atmosphere at high altitudes because hydrodynamic assumptions break down due to the lack of collisions between molecules or atoms. However, the escape rates are in the Jeans regime and only ~(a few)$\times 10^{23}$ N$_2$ molecules s^{-1} (Gladstone *et al.*, 2016). If constant, such escape would remove only ~6 cm of ice over 4.5 Gyr.

Pluto is gravitationally coupled to Charon, the largest of Pluto's five moons, which is half Pluto's size and

around one-ninth its mass. Pluto and Charon have rotational and orbital periods of 6.4 days and orbit around a center of mass that lies outside of Pluto. Charon's surface is different from Pluto's, however. The ice is mainly H$_2$O (Buie and Grundy, 2000). Ammonia–water ice is also present (Brown and Calvin, 2000) and has been taken as evidence that ammonia-rich water once flowed on Charon's surface (Cook *et al.*, 2007). There are other KBOs with surfaces containing ammonia–water ice, so perhaps past flows of ammonia-rich water are common (Brown, 2012). In any case, the presence of NH$_3$ rather than N$_2$ on Charon is also different from Pluto. Charon might be unlike Pluto because Charon probably formed by an impact on Pluto analogous to the Moon-forming impact on Earth (Canup, 2005). Such a process could explain why Charon is less dense than Pluto, and Chiron's different evolutionary fate may have allowed the preservation of ammonia-rich cryovolcanic units (McKinnon *et al.*, 2008). Unlike Pluto, there is no evidence that Charon has an atmosphere (Gulbis *et al.*, 2006; Sicardy *et al.*, 2006); lack of impact craters, however, implies recent geologic activity.

14.3 The Dense Atmosphere on Titan versus the Barren Galilean Satellites

Before discussing Titan's atmosphere in detail, it is worth considering a major puzzle, which is how bodies of nearly identical mass such as the Titan or Ganymede end up with completely different atmospheres. Ganymede's exosphere exerts a surface pressure of 10^{-12} bar (Table 14.3), while Titan has a thick atmosphere with 1.5 bar surface pressure. Although differences in the initial volatile inventories may have played some role, even if Ganymede had started with a thick atmosphere it would have been subject to atmospheric losses that never existed on Titan. As mentioned in Ch. 5, two processes that cause big atmospheric losses are impact erosion and hydrodynamic escape. In comparing Titan to the Galilean moons, impact erosion is a probable factor in explaining the differences.

Impacts supply or diminish volatiles depending on three factors: (1) the volatile content of impactors, (2) the mass distribution of impactors, and (3) the probability that impactors hit slowly enough that volatiles are retained rather than blasted away. Factors (1) and (2) should have been similar for the Galilean moons and Titan. However, there is a big difference in the third factor because the Galilean moons sit in the intrinsically deeper gravitational well of Jupiter. The effectiveness of impact erosion of atmospheres depends on the impactor energy, which goes as the square of the velocity of the impactors.

Roughly speaking, the orbital velocity is a good measure of impactor speed. Thus, the mean orbital velocity of Titan, 5.6 km s^{-1}, compared to that of Ganymede, 10.9 km s^{-1}, accounts for a factor of roughly 4 decrease in impactor energy per unit mass. In more detail, a minimum impact velocity is the moon's escape velocity v_{esc}, by energy conservation. Then there is an additional energy contribution from the orbital velocity of the satellite being struck, $v_{sat\text{-}orb}$, and the velocity of the impactor with respect to the satellite when far away, v_{∞}. It can be shown that the mean impact velocity, v_{impact}, is given by (Zahnle *et al.*, 1992):

$$v_{impact}^2 \approx 3v_{sat\text{-}orb}^2 + v_{\infty}^2 + v_{esc}^2 \qquad (14.9)$$

The escape velocity term is generally the smallest. Also, for asteroids and Kuiper Belt comets, $v_{\infty}^2 \approx (\sqrt{2} - 1)V_{planet\text{-}orb}$, where $V_{planet\text{-}orb}$ is the orbital velocity of Saturn or Jupiter, giving, e.g., $v_{\infty} \approx 5$ km s^{-1} for Jupiter. On this basis, asteroids and Kuiper Belt comets have $v_{impact} \approx 11$ km s^{-1} at Titan but 21 km s^{-1} at Ganymede, which is a factor of 3.6 in J kg^{-1}.

Monte Carlo simulations of impact histories show that Ganymede ends up barren while Titan usually retains a thick atmosphere (Griffith and Zahnle, 1995). Interestingly, in some trials, Callisto (which has slower orbital velocity than Ganymede) acquires an atmosphere. If Callisto really did once have an atmosphere, sputtering and thermal escape must have removed it since. Other simulations show that Titan's N$_2$ atmosphere can be formed by impact conversion of ammonia (Sekine *et al.*, 2011a). However, one study suggests that Late Heavy Bombardment would erode Titan's atmosphere (Marounina *et al.*, 2015).

As mentioned previously, similar considerations of impact erosion are likely important for determining whether exoplanets or exomoons have atmospheres or not, particularly on smaller bodies (Catling and Zahnle, 2013).

14.4 Titan

14.4.1 Overview

Titan is a special moon. It is the biggest of Saturn's 62 moons,[2] the second largest satellite in the Solar System (after Ganymede), the only moon with a dense atmosphere, and the sole object in the Solar System other than Earth that has standing bodies of liquid. The Spanish

astronomer José Comas-Solá first inferred the presence of an atmosphere from limb darkening around Titan's disk (Comas-Solá, 1908). Later, near-infrared absorption spectra indicated the presence of methane (Kuiper, 1944). Initial spacecraft knowledge about Titan came from *Voyager 1* and *Voyager 2* flybys in 1980 and 1981, respectively, but this has been surpassed in quantity and detail by data from the *Cassini–Huygens* mission. *Cassini–Huygens* was launched in 1997 and arrived at the Saturnian system in 2004 (Brown *et al.*, 2009). *Huygens* is a probe that landed on Titan in 2005, while *Cassini* is a Saturn orbiter.

Basic characteristics of Titan are listed in Table 14.4, several of which bear upon Titan's bulk structure. Titan's mean density of ~1.9 g cm^{-3} suggests that it is a mixture of ice and 55%–65% rock, like Ganymede and Callisto. Unlike Ganymede, however, no detection has been made of an intrinsic dipolar magnetic field or an induced field on Titan. There are two confounding factors in measuring such fields. First, Saturn's magnetic field is not inclined, so there is not a simple and strong modulation of the field at Titan as there is at Europa or Ganymede. Second, currents in Titan's ionosphere partly shield Titan's interior from spacecraft observation (and the thick atmosphere prevents a flyby or orbiter from getting close). In any case, the presence of a subsurface ocean on Titan cannot be deduced from a magnetic field.

However, other properties suggest a subsurface ocean. Titan experiences gravitational tides. The eccentricity of Titan's orbit causes flexure under Saturn's gravity. Tidal distortions changes Titan's gravity field enough that Titan cannot be entirely solid. Thus, a subsurface ocean is inferred to exist below an icy crust of <100 km thickness (Iess *et al.*, 2012). This deduction is consistent with electric-field measurements made from the *Huygens* probe that hint at a conductive layer 45 ± 15 km below Titan's surface (Béghin *et al.*, 2010). However, unlike Europa, where the subsurface ocean has a rocky seafloor, the seafloor of Titan's ocean is probably dense ice because Titan's size allows high-pressure ice phases (Fig. 14.7). This situation might be less conducive to supplying any hypothetical life in Titan's ocean with nutrients or energy (Sec. 14.4.5 discusses life on Titan further). Another interesting finding is that Titan's gravity field suggests that the ice and rock inside Titan did not completely differentiate into a rocky core and icy mantle (Iess *et al.*, 2010). However, others think that full differentiation is consistent with observations (Baland *et al.*, 2014).

The current atmosphere of Titan, discussed further below, has some analogy to that on Earth. Titan's

---

[2] The number of moons at the time of writing. Many are small, with distant eccentric orbits at 187–383 Saturn radii.

Table 14.4 Some characteristics of Titan.

Property	Value	Comment
Radius	2575 km	6% bigger than Mercury; ~1.5 times the Moon
Mass	1.346×10^{23} kg	~2% of Earth's mass and twice the lunar mass
Mean density	1882 ± 1 kg m^{-3}	Similar to Ganymede and Callisto
Distance from Saturn	1.2×10^9 m	~20 Saturn radii
Surface gravity	1.35 m s^{-1}	Falls appreciably in the extended atmosphere
Mean distance to Sun[*]	9.537 AU	The solar flux at Titan's orbit (S_{Titan} ~15 Wm^{-2}) is ~1.1% of Earth's solar constant (1361 Wm^{-2})
Orbital period (Saturn)	15.945 days	Synchronous rotation is observed from Hubble
Orbital period (Sun)	29.458 years	Same as Saturn's orbital period around the Sun. Seasons are ~30 times longer than Earth's
Titan orbital eccentricity	0.028	So, Titan has periodic tidal stress
Saturn orbital eccentricity	0.0557	~3 times Earth's. Modulates insolation by ~20%; phasing with respect to equinoxes creates Milankovitch-like cycles, affecting lake distribution
Saturn's obliquity	26.73°	Titan's distribution of sunlight is dominated by Saturn's obliquity, which causes seasonality on Titan
Inclination of Titan's orbital plane with respect to Saturn	0.33°	
Titan's obliquity with respect to its orbital plane	0.3°	
Surface temperature, T_s	93.7 ± 0.6 K at 10° S ~2–3 K cooler at the poles	CO_2, O_2, H_2O, and NH_3 are solid at Titan's surface temperature
Near-surface scale height, H	~21 km	$H = k\overline{T}/\overline{m}g$ (eq. (1.12)), where surface gravity $g = 1.35$ m s^{-2}, $T = T_s = 94$ K, $\overline{m} = 4.533 \times 10^{-26}$ kg
Dry lapse rate, Γ_a	1.3 K km^{-1}	$\Gamma_a = g/c_p$ (eq. 1.27) with $c_p = 1040$ J kg^{-1} K^{-1}
Actual lapse rate, Γ	1 K km^{-1}	From Strobel et al. (2009)
Bond albedo, A	0.265 ± 0.03	As adopted by Li et al. (2011)
Surface pressure, p_s (bar)	1.467 ± 0.03 @Huygens site	~50% higher than Earth's surface pressure
Column mass of air, M_{col}	~10^5 kg m^{-2}	$M_{col} = p_s/g$ (eq. (1.19)), ~10 times terrestrial
Surface escape velocity	2.65 km s^{-1}	The thermosphere is sufficiently warm at 110–188 K that there is small Jeans-like escape of heavier constituents, such as CH_4
Lower atmosphere composition[§] (mole fractions)	94.2% N_2, 5.65% CH_4 (1.48% CH_4 in stratosphere), 0.099%H_2, CO ~ 4.7×10^{-5}, ^{40}Ar~3.4×10^{-5}, ^{36}Ar~2.1×10^{-7}, $C_2H_6 \leq 1 \times10^{-5}$, $C_2H_2 \leq 2 \times10^{-6}$, ^{20}Ne $<2 \times 10^{-5}$, ^{22}Ne ~3×10^{-7}, CO_2 ~ 10^{-8} (above the stratospheric condensation level), $H_2O < 10^{-8}$ Mean molar mass from Huygens abundances:$\overline{M} = 27.3$ g mol^{-1}	

[*] Titan was 9.44 AU during the Voyager 1 flyby and 9.04 AU at the time of Huygens probe entry.
[§] Niemann et al. (2010).

atmosphere is predominantly N_2 and methane. The latter plays a role similar to water on Earth by condensing into clouds and rain, and pooling into rivers that have carved channels in the surface. Methane and its chemical products also form lakes of liquid hydrocarbons on Titan. Nearly all the lakes are in Titan's north polar region (Stofan et al., 2007); only one large lake, Ontario Lacus, exists near the southern pole (Turtle et al., 2009).

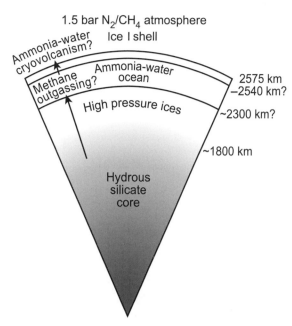

1.5 bar N$_2$/CH$_4$ atmosphere

Ammonia-water cryovolcanism?

Ice I shell

Methane outgassing?

Ammonia-water ocean

High pressure ices

2575 km
-2540 km?

~2300 km?

~1800 km

Hydrous silicate core

Figure 14.7 Schematic diagram of Titan's internal structure. The surface is icy, but an ocean probably exists at a depth of tens of kilometers.

Like the Earth, Titan experiences seasons. *Huygens* measured a surface temperature of 94 K at a pressure of 1.5 bar. Because of the dependence of density on temperature ($\rho = P/\overline{R}T$(eq. 1.1)), the surface air is around four times denser than on Earth. The axial tilt of Titan with respect its orbital plane is 0.3° (Stiles *et al.*, 2008), while the orbital plane is inclined 0.33° to Saturn's equator. However, Saturn's obliquity with respect to the orbital plane around the Sun is ~27°. This last angle dominates the seasonal modulation of Titan's sunlight at high latitudes. The atmospheric radiative time constant (see Sec. 2.3.2) can be estimated with an equation similar to eq. (2.29) through the ratio of the column heat capacity to the average solar energy input:

One fundamental chemical difference between the environments on Titan and the Earth is redox. Because Titan's troposphere contains roughly 94% N$_2$ and nearly 6% CH$_4$ (Niemann *et al.*, 2010) (Table 14.4), it is chemically reducing. In Earth's water-rich oxidizing atmosphere, the OH radical removes gases such as methane and other hydrocarbons (Sec. 3.3.2). This process makes the terrestrial atmosphere mostly transparent to sunlight even in the Earth's stratosphere, which, unlike Earth's troposphere, does not experience the additional cleansing process of rainfall. In contrast, an orange-brown haze surrounds Titan and obscures its surface at visual wavelengths.

The early, anoxic Earth would have had similar redox chemistry to Titan except for another key difference, which is the extreme cold of Titan. H$_2$O and CO$_2$ are basically frozen out of Titan's atmosphere (although see Sec. 14.4.2 below). Thus, the oxidizing radicals produced from their photolysis (and from the escape of hydrogen) do not exist. Also, many volatiles, including ethane (C$_2$H$_6$) and higher-order hydrocarbons, would have existed only as gases in the early Earth's atmosphere but on Titan they either condense or are involatile. For example, ethane becomes solid in the lower stratosphere or liquid at Titans' surface. *Gaseous* ethane – a by-product of methane photochemistry – is considered important for the greenhouse effect on the Archean Earth because it absorbs within Earth's infrared window at wavelengths just below 12 μm (Sec. 11.6.1).

14.4.2 Titan's Atmosphere: Structure, Climate, Chemistry, and Methane Cycle

14.4.2.1 *Structure and Climate*

Titan's atmospheric structure and climate have several features that make them particularly informative for understanding planetary atmospheres in general. The

$$\tau_{rad} = \frac{(T_s p_s c_p/g)}{(S_{\text{Titan}}/4)(1-A)} = \frac{(94 \text{ K})(1.5 \times 10^5 \text{ Pa})(1040 \text{ J K}^{-1}\text{kg}^{-1})/1.35 \text{ m s}^{-2}}{\left((150 \text{ W m}^{-2})/4\right)(1-0.265)}$$

$$= 125 \text{ Earth years}$$

(14.10)

Variables are those given in Table 14.4 plus c_p, the specific heat capacity at constant pressure. This radiative time constant is much longer than seasons that last ~7.4 Earth years on average (Flasar *et al.*, 2009). But enough solar radiation (~10%) makes it down to the surface to cause seasonal variations in ground temperature that affect the atmosphere and volatiles (Lorenz *et al.*, 2009).

physical consequences of Titan's photochemical haze are especially interesting. Three key effects include the strongest stratospheric temperature inversion amongst stratospheres in the Solar System, stabilization of a significant portion of the lower atmosphere against convection, and an *anti-greenhouse effect* (discussed earlier in Sec. 11.6.1).

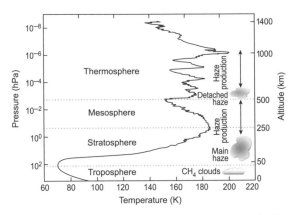

Figure 14.8 Titan's atmospheric structure as measured by the *Huygens* probe during atmospheric entry. Also, a schematic representation of haze and cloud layers.

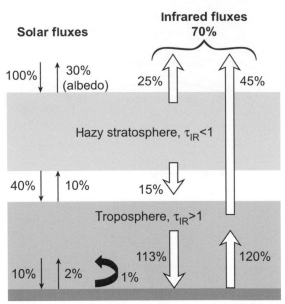

Figure 14.9 An estimate of the global average energy balance of Titan based on numerical radiative-convective calculations by McKay *et al.* (1991). Here, τ_{IR} is the thermal infrared optical depth. The diagram assumes a two-layer model, where the "troposphere" is optically thick in the infrared and the stratosphere is optically thin. Fluxes are percentages of the globally averaged solar flux at the top of the atmosphere, which is ~3.7 W m^{-2}. Solid arrows are solar fluxes, while open arrows are infrared fluxes. The curved arrow is a convective flux ~0.04 W m^{-2}. The net infrared emission from the surface is 120% − 113% = 7%, which is balanced by incoming sunlight, 10%, minus losses from the surface to scattering and convection (2%+1%).

Titan's atmosphere has layers analogous to those on Earth: a troposphere, stratosphere, mesosphere, and thermosphere (Fig. 14.8). The haze causes a very steep temperature inversion in Titan's stratosphere because it has a large ratio of shortwave to infrared opacity. At the top of Titan's atmosphere, through the inverse-square law, the solar insolation is only ~1.1%, so $S_{Titan}/4$ ~3.7 W m^{-2} compared to the terrestrial time-averaged solar flux of 340 W m^{-2}, given the Sun–Titan distance of almost 10 AU. Titan's haze absorbs and scatters ~90% of the incoming radiation, so that a solar flux of only ~0.4 W m^{-2} reaches the ground as a global average (Fig. 14.9).

Voyager radio occultation profiles (Flasar, 1983, 1998; McKay *et al.*, 1997), *Huygens* probe data (Fulchignoni *et al.*, 2005), and analytic radiative–convective models (Robinson and Catling, 2012) show that Titan's atmosphere is typically unstable to convection from the surface at ~1.5 bar up to a radiative–convective boundary at ~1.3 bar. This boundary is remarkably shallow, considering that the 70 K tropopause is located at ~0.11 bar (Fig. 14.9) (Fulchignoni *et al.*, 2005; Schinder *et al.*, 2011). Shallow convection arises because there is enough haze even in Titan's middle troposphere to absorb sunlight and create a stable lapse rate where radiation dominates. The troposphere absorbs ~30% of the incoming top-of-atmosphere sunlight (Fig. 14.9).

Titan's atmosphere shows the difficulty of using terrestrial nomenclature for other worlds. The ancient Greek root *tropos* for "troposphere" comes from the "turning" of convection, but radiative heat transfer dominates most of Titan's troposphere, making "troposphere" something of a misnomer. Shallow convection in a troposphere generally occurs when the ratio of the tropospheric

shortwave to gray infrared optical depth exceeds ~0.1 (Robinson and Catling, 2012). On Titan, it is ~0.2.

Titan's atmosphere has competing *greenhouse* and *anti-greenhouse* effects. Infrared opacity is dominated by collision-induced absorption (CIA) of N_2–N_2, N_2–CH_4, N_2–H_2, and minor CH_4–CH_4 (see Sec. 2.5.6 for a description of CIA). Analogous to the Earth 8–12 μm infrared window (Sec. 2.2.5), Titan's cooling is dominated by a window ~16.5–25 μm where H_2 is a key absorber. This window is closed by abundant H_2 on the giant planets. But on Titan, the H_2 abundance is ~0.1% and H_2 plays a climatic role similar to low levels of CO_2 on Earth (Table 14.5). The calculated greenhouse effect is ~20 K. But this is counterbalanced by an anti-greenhouse effect of ~10 K caused by absorption of sunlight by the high altitude haze, which lies in an optically thin region of the atmosphere (in the infrared) where energy can be radiated to space (McKay *et al.*, 1991). The result is a net warming at the surface of ~10 K. This warming is consistent with an effective temperature of

Table 14.5 Analogies between Earth and Titan: Volatiles and the atmosphere.

Earth	Titan analog	Phenomena
$H_2O(l)$ or $H_2O(g)$	$CH_4(l)$ or $CH_4(g)$	The main condensable for clouds, rain and rivers. The vapor phase is also a greenhouse gas
Silicate rock and silica	Water ice, solid organics	Solids on the surface that are weathered
CO_2 greenhouse absorption (through pressure-broadened absorption)	H_2 greenhouse absorption (through N_2–H_2 collision-induced absorption)	Control of the greenhouse effect by a minor, non-condensable gas
8–12 μm infrared window	16.5–25 μm infrared window	Where thermal infrared photons escape directly from the surface to space
Stratospheric ozone layer	Organic haze	Shortwave absorption creates a temperature inversion

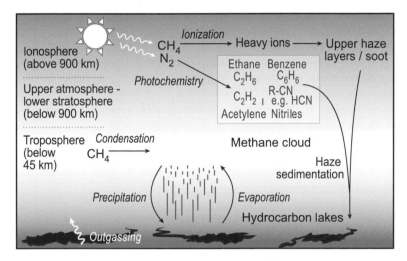

Figure 14.10 A schematic diagram showing chemical processes operating on Titan. (Based on a diagram by Atreya (2010).)

84.08 ± 0.09 K measured by *Cassini* (Li *et al.*, 2011) and a surface temperature of 94 K.

Titan's atmospheric properties may also provide a sort of thermostat (McKay *et al.*, 1989). Were Titan's surface to cool, its blackbody emission peak, which is currently ~31 μm by Wien's Law (eq. (2.47)) would shift into a region of strong opacity from N_2–N_2 CIA. Were Titan to warm, the peak would move into the infrared window.

14.4.2.2 Chemistry

The atmospheric chemistry on Titan is extremely diverse (see, e.g., Bezard *et al.*, 2014). Besides the bulk gases, there are trace constituents at part per million levels or less, which we can group as follows.

(1) Hydrocarbons, such as acetylene (C_2H_2), ethane (C_2H_6), propane (C_3H_8), and benzene (C_6H_6).
(2) Nitriles, which are carbon–nitrogen–hydrogen compounds with general formula R–C≡N, where "R" is a side group. The simplest and most abundant is HCN.
(3) Noble gases, including argon as ^{36}Ar and ^{40}Ar, and neon detectable as ^{22}Ne.
(4) Macromolecular organic haze particles.

The photochemistry of neutral species and ion–molecule reactions in the ionosphere are the sources of the hydrocarbons, nitriles, and haze particles (Fig. 14.10). In particular, the destruction of methane and the reaction of its products lead to the production of heavier hydrocarbons.

Methane is lost from Titan's atmosphere in two photochemical pathways and by direct escape to space (Sec. 14.4.3). The two photochemical pathways are

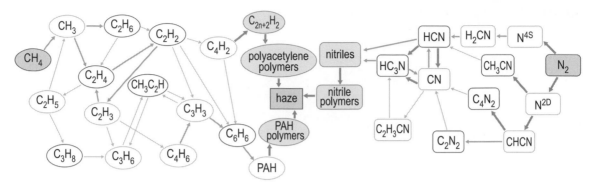

Figure 14.11 A schematic diagram showing the coupled nitrogen–carbon–hydrogen chemistry in the atmosphere of Titan. The shaded species and those bordered in bold have been detected on Titan. (Redrawn from Atreya *et al.* (2006).)

photolysis and catalytic destruction by C_2H radicals that are derived from photolysis of acetylene (C_2H_2). Methane photolysis is driven by the relatively large solar flux above ~700 km altitude at the Lyman-α wavelength of 121.6 nm. About half of the photolysis produces methyl radicals (CH_3) while the other half produces methylidyne (CH) and methylene (CH_2) radicals. Radicals combine to form more complex molecules. For example, methyl radicals give rise to ethane, as follows, through both methane photolysis and acetylene-catalyzed destruction of methane.

$$
\begin{array}{ll}
2(CH_4 + h\nu \rightarrow CH_3 + H) & 2(C_2H_2 + h\nu \rightarrow C_2H + H) \\
CH_3 + CH_3 + M \rightarrow C_2H_6 + M & 2(C_2H + CH_4 \rightarrow C_2H_2 + CH_3) \\
& CH_3 + CH_3 + M \rightarrow C_2H_6 + M \\
\hline
Net: 2CH_4 \rightarrow C_2H_6 + 2H & Net: 2CH_4 \rightarrow C_2H_6 + 2H
\end{array}
$$

(14.11)

Excluding atmospheric escape, direct photolysis of methane accounts for about one-third of the methane loss, while the remaining approximate two-thirds are lost by catalytic destruction (Wilson and Atreya, 2004, 2009).

Acetylene (a.k.a. ethyne) is the simplest of the $C\equiv C$ bond-containing alkynes. It is also ultimately produced from photolysis of methane, with a net reaction, as follows.

$$2CH_4 + photons \rightarrow C_2H_2 + 3H_2 \qquad (14.12)$$

When acetylene photolyzes, its products enter a series of reactions that produce other hydrocarbons. The net result is that Titan's atmosphere contains less acetylene (≤ 2 ppmv) than ethane (≤ 10 ppmv) (Niemann *et al.*, 2010).

Some of Titan's organic molecules incorporate nitrogen. Atmospheric N_2 is dissociated by UV photons of <100 nm wavelength and by impacts of electrons in

Saturn's magnetosphere. Figure 14.11 summarizes the network of methane–nitrogen neutral chemistry on Titan. The most abundant nitrile is hydrogen cyanide, HCN, but several others have been detected in Titan's stratosphere including C_2N_2 (cyanogen or ethanedinitrile), HC_3N (cyanoacetylene or propynenitrile), and CH_3CN (acetonitrile or ethanenitrile) (reviewed by Strobel *et al.* (2009)).

Titan's heavy hydrocarbons are built up through photochemistry. They include a series of alkanes, polyacetylenes and polyynes ($C_{2n+2}H_2$, where n is an integer).

Two C_3H_3 (propargyl) radicals can form benzene in the stratosphere (Lavvas *et al.*, 2008a; Vuitton *et al.*, 2008). Fused benzene rings can then form polycyclic aromatic hydrocarbons (PAHs), in a schematic overall reaction, as follows.

$$6nCH_4 + photons \rightarrow \underset{\text{polyaromatic hydrocarbons}}{(C_6H_6)_n} + 9nH_2$$

(14.13)

However, neutral chemistry cannot account for why there is greater benzene abundance in the ionosphere than in the stratosphere.

The photochemistry of neutral species is not the whole story because Titan's ionosphere is a source of hydrocarbons and nitriles as ions (Waite *et al.*, 2009b). Benzene is ~10^3 times more abundant in the ionosphere than in the neutral atmosphere below, which implicates high altitude ion–neutral reactions that are generally faster than reactions between neutrals (Waite *et al.*, 2007). The $C_6H_7^+$ ion is produced efficiently by ion reactions from smaller hydrocarbons so that benzene can be formed by electron recombination (Vuitton *et al.*, 2008).

$$C_6H_7^+ + e^- \rightarrow C_6H_6 + H \qquad (14.14)$$

Other surprising species in the ionosphere include massive negative ions up to 13 800 Daltons at ~950–1400 km altitude (Coates *et al.*, 2007; Coates *et al.*, 2009). A tentative suggestion for how heavy negative ions are produced is that ions and atoms assemble into large, so-called *cluster ions*.

The products of organic polymerization are refractory macromolecular organic compounds that make up Titan's haze. Such substances were originally investigated in laboratory experiments by irradiating mixtures of gases. Carl Sagan and Bishun Khare named these reddish-brown organics *tholins* from the Greek *tholos* for "muddy" (Khare *et al.*, 1984; Sagan and Khare, 1979). They also predicted the presence of PAHs within tholins (Sagan *et al.*, 1993a). Simulated tholins are made of C, H, and N atoms, which is consistent with Titan aerosol particles that released NH_3 and HCN when they were heated and examined by *Huygens* gas chromatograph-mass spectrometry (Israel *et al.*, 2005). Aerosols in the lower atmosphere are also interpreted to be aggregates of a few thousand monomers of 0.05–1 μm radius (Tomasko *et al.*, 2009).

Surprisingly for a cold, reducing atmosphere, three oxygen-containing molecules are also found in Titan's stratosphere: CO at ~50 ppmv, CO_2 at 16 ± 2 ppbv, and H_2O at 0.1–0.4 ppbv. Because these substances are involatile ices on the surface of Titan, the source of the oxygen must be space (Horst *et al.*, 2008). This idea is supported by the observed precipitation of O^+ ions into Titan's atmosphere (Hartle *et al.*, 2006; Sillanpaa *et al.*, 2011).

All the complex products of methane photolysis and subsequent photochemistry are less volatile than methane and tend to sediment out of the atmosphere (Fig. 14.10). Ethane will solidify in the lower stratosphere and melt at 91 K (at 1.5 bar pressure), which is colder than the 94 K surface temperature. Consequently, ethane forms a volatile liquid on Titan's surface. Propane (C_3H_8) is also liquid at the surface, but all the other products of photochemistry, e.g., acetylene, will form solids. Liquids have evidently accumulated in polar lakes, while ~20% of Titan's surface appears to be covered in organic-rich, tropical dunes (Lorenz *et al.*, 2008). Dunes are probably made of organic sand or organic-coated sand because they are dark in visible, near infrared, and *Cassini* radar (Radebaugh *et al.*, 2008).

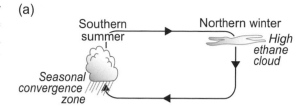

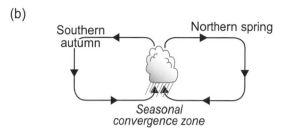

Figure 14.12 The meridional circulation on Titan affects the distribution of clouds and the cycling of hydrocarbons. Whereas the Earth has an intertropical convergence zone at the rising branch of the Hadley circulation that is near the equator, Titan has a seasonal convergence zone with a much larger range of latitude. In southern summer (a), the seasonal convergence zone is near the south pole and the descending branch produces a high cloud of ethane over the opposite pole. At the equinox (b), the seasonal convergence zone moves to the tropics.

14.4.2.3 *Atmospheric Circulation*

Many key features of Titan's atmospheric dynamics are understood in broad outline but the details of the troposphere are elusive (as reviewed by Tokano (2009) and Flasar *et al.* (2009)). With a rotational period of 16 days, Titan is slowly rotating and so its atmospheric circulation should be characterized by a large Hadley circulation from pole-to-pole at the solstices with winds that are in approximate cyclostrophic balance in the upper troposphere, similar to Venus (see Sec. 4.2.5). The most obvious manifestations of such dynamics include more stratospheric haze in the winter hemisphere (Fig. 14.12) and a weak equator-to-pole temperature gradient of 2–3 K at the surface associated with light winds of ~1 m s^{-1} (Bird *et al.*, 2005).

The rising branch of the Hadley circulation should acquire methane moisture and redistribute it to other latitudes, and when the Sun moves to the opposite hemisphere the circulation should reverse at the equinox[3] (Fig. 14.12). A further consequence of the poleward transport of angular momentum is a polar vortex, like that on Venus and Earth, which prevents mixing of chemical

---

[3] Southern summer solstice occurred October 2002 while northern spring equinox was in August 2009.

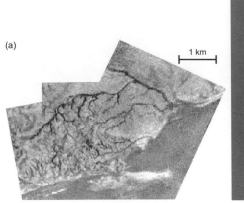

Figure 14.13 (a) Images taken by the *Huygens* probe during descent projected to a common altitude of 6.5 km show a network of channels that appear to flow into a plain. (b) Image of the surface at the *Huygens* landing site. Stones in the foreground are 10–15 cm in size and sit on a darker, fine-grained substrate. (Courtesy of ESA/NASA/University of Arizona.)

species, somewhat like the Antarctic vortex in the terrestrial stratosphere (Teanby *et al.*, 2008).

Seasonal changes in circulation have been observed. After the mid-2009 northern spring equinox, a reversal of circulation was seen through chemical tracers in the middle atmosphere (Teanby *et al.*, 2012). At high altitude an HCN cloud and a vortex at the south pole developed from 2011 (de Kok *et al.*, 2014; West *et al.*, 2015).

Meridional temperature gradients of 3 K at the surface and ~20 K in the stratosphere combined with the thermal wind equation (Sec. 4.1.2.3) imply *superrotation*, meaning that stratospheric winds travel faster than the equatorial surface. (The surface speed is at the equator is $\Omega a = (4.56 \times 10^{-6}$ rad s$^{-1}) \times (2575$ km$) \sim 12$ m s^{-1}, where Ω is the angular rotation rate and a is Titan's radius.) *Cassini* data indicate a single eastward jet with speeds up to 190 m s^{-1} between (winter) latitudes 30–50° N at the 1 mbar level or ~200 km altitude (Achterberg *et al.*, 2008), which is consistent with ground-based observations (Flasar *et al.*, 2009). A weakening of the meridional temperature gradient during 2005–2009 (up to equinox) caused a slower decay of the jet with increasing altitude (Achterberg *et al.*, 2011).

14.4.2.4 *Methane Cycle and Hydrocarbon Lakes*
Dendritic channels akin to dried-up rivers, lakes of hydrocarbons, and methane clouds suggest that Titan has an active hydrological cycle based around the condensation and evaporation of methane (Lunine and Lorenz, 2009; Hayes, 2016).

The *Huygens* landing site provides evidence for liquids. The region nearby has a network of channels reminiscent of those on Earth that are caused by fluvial erosion (Fig. 14.13(a)). At the *Huygens* landing site, there

are a number of cobbles, presumably made of water or CO$_2$ ice, which have been rounded as if by a liquid (Tomasko *et al.*, 2005) (Fig. 14.13(b)). After it landed, *Huygens* measured a flux of methane and ethane, which can be explained by damp ground that was heated by the warmth of the spacecraft (Lorenz *et al.*, 2006; Niemann *et al.*, 2010). Surface darkening at low latitude after a cloud system had been seen there also suggests wetting and actual rainfall (Turtle *et al.*, 2011a).

Cassini has also imaged more than 500 lakes at high latitudes using radar (Aharonson *et al.*, 2009; Stofan *et al.*, 2007; Turtle *et al.*, 2009). Glint indicates that the lakes are filled with liquid (Soderblom *et al.*, 2012; Stephan *et al.*, 2010). Observed polar lakes cover an area of ~400 000 km^2 and probably have a depth on the order of tens of meters, suggesting a liquid volume exceeding 10^4 km^3 (Lorenz *et al.*, 2008). For example, the deepest part of the second largest northern sea, Ligeia Mare, is 160 m based on radar (Mastrogiuseppe *et al.*, 2014); the same lake is also smooth and lacks waves given specular radar reflection (Zebker *et al.*, 2014). Thermodynamic equilibrium predicts mixtures of liquid ethane, methane, and propane (Cordier *et al.*, 2009; Cordier *et al.*, 2012; Glein and Shock, 2013), but uncertainty remains until a future probe measures the composition. The lakes also have "bathtub ring" features, suggesting variable lake levels (Turtle *et al.*, 2011b; Wall *et al.*, 2010), which is consistent with seasonal migration of liquids caused by cold-trapping at the winter pole (Schneider *et al.*, 2012). In 2013, transient bright features (waves, bubbles, or solids) in Ligeia Mare implied change as summer approaches (Hofgartner *et al.*, 2014).

The concentration of most lakes in the north polar region might be explained by Milankovitch-type cycles of insolation (Aharonson *et al.*, 2009). Seasonal redistribution between poles is improbable given the large mass of liquid that would need to be moved. On longer timescales, the eccentricity of Saturn's orbit (Table 14.4) means a change of the phasing of perihelion and aphelion with respect to equinoxes over ~50 kyr. Currently, northern summer occurs at aphelion and so is long. Summer is the rainy season (see Fig. 14.12), so currently the north might accumulate liquid over a longer period. Also evaporation is less because aphelion summer is cooler.

Titan has different cloud types composed of methane, ethane, and HCN. Typically, Titan has ~1%–10% cloud cover (Griffith, 2009). Cumulus convection clouds near the summer pole and sporadic clouds in mid-latitudes are composed of methane droplets with dissolved nitrogen (Rodriguez et al., 2011). In contrast, a more stratiform cloud persisting over the northern winter pole is attributed to the condensation of ethane from subsiding strato-spheric air (Griffith et al., 2006). An HCN cloud has also been seen post-equinox over the south pole (de Kok et al., 2014; West et al., 2015). Tropical clouds were more frequent near the equinox. Such behavior is consistent with a seasonal reversal of the Hadley circulation (Rannou et al., 2006).

During Huygens' descent to the surface, measure-ments suggested that methane was saturated at ~8 km altitude. At the surface, the methane relative humidity was ~50%. But the methane abundance began to follow a saturation vapor curve (100% relative humidity) for a CH_4–N_2 solution at 8 km, consistent with a cloud base or lifting condensation level (see Sec. 1.1.3.5, Fig. 1.10).

On Earth, O_2 and N_2 are poorly soluble in liquid water, but on Titan, N_2 is highly soluble in liquid methane, and so the saturation vapor pressure of CH_4 depends on the relative composition of a solution as well as temperature. Figure 14.14(a) shows the calculated sat-uration vapor pressure for CH_4 and a CH_4–N_2 mixture, assuming the temperature profile derived from Voyager 1 radio occultation data. The right-hand vertical axis refers to the total pressure at altitude, not to the saturation vapor pressure. The horizontal axis shows how the saturation vapor pressure (the maximum allowed partial pressure of methane relative to total pressure) varies from ~10% of the total pressure at the surface to ~1.5% at several tens of km altitude. Liquid methane can also mix with ethane, propane and heavier hydrocarbons, so there are potential further complications. Figure 14.14(b) shows the Huy-gens measurements. Departure from the CH_4–N_2 satur-ation curve at ~14 km height or 0.7 bar is consistent with temperatures falling below freezing and the formation of methane ice particles.

The behavior of methane on Titan can be understood through its phase diagram. The triple point is 90.7 K at a pressure of 0.117 bar (Setzmann and Wagner, 1991) (Fig. 14.15). Titan's surface pressure of 1.5 bar far exceeds the triple point pressure, but the surface tempera-ture of 94 K is close to the methane freezing point. However, Titan's poles are close to the triple point tem-perature with measurements of 90.5 ± 0.8 K at 87° N and 91.7 ± 0.7 K at 88° S (Jennings et al., 2009). This is analogous to Earth, where the total atmospheric pressure far exceeds the 6.1 mbar triple point of water but tem-peratures at the poles allow water to freeze. In general, it has been argued that methane–nitrogen–ethane seas cannot freeze because of mutual solution effects, i.e., freezing point depression, and that dense ice would sink. However, ice can float if porous (Hofgartner and Lunine, 2013) or if enough N_2 is dissolved in the liquid (Roe and Grundy, 2012).

Methane rainfall is important for fluvial features on Titan's surface, and thermodynamics gives some insight into its magnitude (Griffith et al., 2008; Lorenz, 2000). The specific enthalpy of vaporization is $l_v = 2.3 \times 10^6$ J kg^{-1} for H_2O and 5.19×10^5 J kg^{-1} for CH_4. On Earth, globally averaged evaporation is powered by ~80 W m^{-2} (around 30% of the 240 W m^{-2} net absorbed sunlight) (Trenberth et al., 2009). The mass of water evaporated per year on Earth is therefore

$$\text{mass} = (80 \, \text{W}) \times (3.15 \times 10^7 \text{s} \ \ \text{yr}^{-1}) \ / l_v = 1096 \, \text{kg m}^{-2}$$

Rainfall must balance evaporation. Given the density of water, 1000 kg m^{-3}, we deduce ~1.1 m of rainfall, con-sistent with the observed annual average global precipita-tion of ~1.1 m (Hulme, 1995). On Titan, a global average of ~0.4 W m^{-2} solar energy reaches the ground. If this were all used to evaporate methane, it would produce ~5 cm of methane rain per Earth year, assuming a liquid methane density of 447 kg m^{-3} at 94 K. This seems difficult to reconcile with the amount of methane rainfall needed to erode channels such as those in Fig. 14.13(a). Instead, rare heavy rainfall is required (Lorenz, 2000) that perhaps has decades and centuries between downpours.

However, a global average calculation is too simplis-tic because of latitudinal heat imbalance and horizontal heat transport. In Earth's tropics, the local radiation imbalance makes 115 W m^{-2} available for evaporation, while horizontal heat transport away from the equator accounts for an additional 40 W m^{-2} of evaporation (Trenberth and Stepaniak, 2003). On Titan, the outgoing longwave flux remains constant seasonally because of the atmosphere's long radiative time constant, but the incom-ing solar flux varies. This allows ~2 W m^{-2} of imbalance

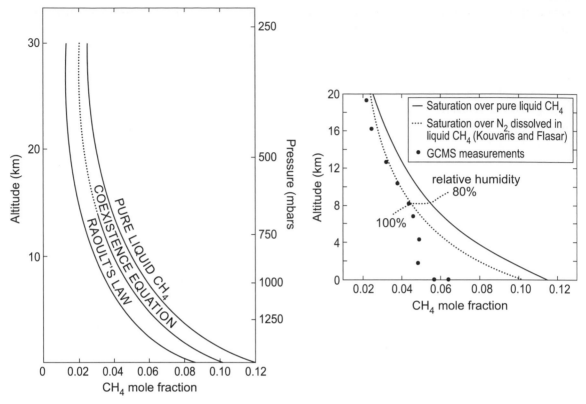

Figure 14.14 (a) The saturation vapor pressure for CH_4 and a CH_4–N_2 mixture, using Titan's temperature profile derived from *Voyager 1* measurements according to Kovaris and Flasar (1991). Raoult's
Law assumes an ideal solution of CH_4 and N_2 with the partial pressure proportional to the mole fraction
at a given temperature and pressure. The coexistence solution is a more sophisticated treatment where
vapor and liquid phases coexist in a two-component non-ideal solution. (b) Measurements from the
Huygens probe Gas Chromatography Mass Spectrometer (GCMS). 100% CH_4 saturation with a
CH_4–N_2 mixture suggests a lifting condensation level (CH_4 cloud base) at 8 km. Departure from the
liquid equilibrium at ~16 km altitude suggests frozen particles of CH_4. (From Atreya *et al.* (2009).
Reproduced with permission of Springer. Copyright 2009, Springer Science + Business Media B.V.)

at the summer poles that should significantly enhance the
methane evaporation–condensation cycle.

The net effect of methane photolysis and hydrogen
escape to space is to generate organic solids or liquids,
which accumulate on the surface over time. One can think
of this overall effect as follows.

$$CH_4 + \text{photons} \rightarrow C_{\text{organic}} + 4H(\uparrow \text{space}) \qquad (14.15)$$

Thus, the rate of hydrogen escape to space provides an
upper limit on the net rate of organic production. The
hydrogen escape rate inferred from *Cassini* data is
~2.0×10^{10} atoms cm^{-2} s^{-1} or ~1.7×10^{28} atoms s^{-1} (Bell
et al., 2010a; Cui *et al.*, 2008), which is similar to the
diffusion-limited escape flux (Sec. 5.8). Thus, the rate of
organic carbon production would be 5×10^{13} C atoms m^{-2}
sec^{-1}, accounting for a factor of four in the stoichiometry
of eq. (14.15). If the same rate has been maintained since

4.5 Ga, then ~10^7 mol C m^{-2} would have been generated.
If most of the carbon had ended up in the form of benzene
(serving to approximate the chemistry of PAHs) packed at
an assumed density of ~0.5 g cm^{-3}, it would be equivalent
to a thickness of 260 m over the whole of Titan. Of course,
because methane is being used up via the scheme of
eq. (14.15), it should be exhausted over time, unless there
is a source, as discussed below in Sec. 14.4.4.

In the 1980s, there were suggestions of a ~1 km deep
ocean of ethane on Titan's surface because it was
assumed that ethane would be the main end-product of
Titan's photochemistry (Lunine *et al.*, 1983). However,
there is no ocean. The volume of dunes is probably
equivalent to a thickness of several meters over the whole
of Titan, which is an inventory several times that of the
lakes. Possible answers to the *missing ethane* problem are
that atmospheric ethane is sequestered into solid haze

particles (Hunten, 2006), that ethane gets trapped inside ices in the form of clathrates, or that ethane soaks through the subsurface (Mousis *et al.*, 2014). In any case, the "ocean" on Titan could be an "ocean" of solid hydrocarbons.

14.4.3 Atmospheric Escape

Current rates of atmospheric escape from Titan (summarized in Table 14.6) are important for understanding Titan's past and future. Titan's escape velocity *at the exobase* is ~2.1 km s^{-1}, which allows hydrogen, carbon and nitrogen to escape. As discussed above, the escape of hydrogen ultimately drives the net generation of large hydrocarbons. Although there has been debate about putative high rates of methane escape, in general the evidence and theory imply that methane escapes at a rate

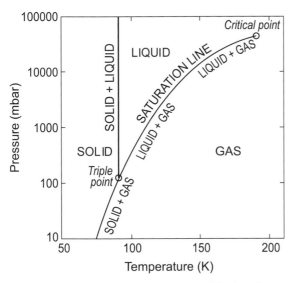

Figure 14.15 Phase diagram for pure methane. Titan's surface at ~1500 mbar and 94 K is in the liquid part of the diagram. The boiling temperature for CH$_4$ at this pressure is around 111 K. This is analogous to Earth's surface being in the liquid part of the pure H$_2$O phase diagram with a boiling temperature of 373 K.

some ~20 times less than its destruction from photolysis. Nitrogen escapes very slowly by nonthermal processes.

Hydrogen. The escape of hydrogen is indicated by three observations: a steep H$_2$ density profile above the exobase detected by *Cassini*'s ion-neutral mass spectrometer (Waite *et al.*, 2005) (Fig. 14.16), the presence of a hydrogen corona detected through energetic neutral atoms produced by charge exchange reactions (Garnier *et al.*, 2008), and exospheric hydrogen atoms detected by resonant scattering of solar UV (Hedelt *et al.*, 2010). In Fig. 14.17, the exobase occurs at a change in slope at ~1400–1500 km altitude. Bell *et al.* (2010a, 2010b) find that hydrogen escapes at the Jeans rate, i.e., individual hydrogen molecules are warm enough to evaporate into space from the exobase. Others have suggested that the rate is sometimes enhanced with energy input by Saturn's magnetospheric particles (Cui *et al.*, 2011). The inferred escape rate is ~2.0×10^{10} atoms cm^{-2} s^{-1} or ~1.7×10^{28} atoms s^{-1} (Bell *et al.*, 2010a; Cui *et al.*, 2008) (Table 14.6). Most hydrogen escapes as H$_2$, with probably ~30% as H (Yung *et al.*, 1984). Calculations also suggest that hydrogen may be lost downwards to Titan's surface at a rate similar to the escape flux because there is a gradient where thermospheric hydrogen is elevated in mixing ratio compared to the tropopause (Strobel, 2010). The production of hydrogen at altitude beyond that from neutral photochemistry might be explained by interactions with magnetospheric particles (Cui *et al.*, 2011).

Methane. Titan's rate of CH$_4$ escape has been debated. Prior to *Cassini*, the escape of CH$_4$ was assumed to be dominantly nonthermal (Michael *et al.*, 2005). Initially, *Cassini* data were interpreted to indicate a big escape flux of 2–3×10^9 CH$_4$ molecules cm^{-2} s^{-1} (44–66 kg s^{-1}) driven by hydrodynamic escape (Strobel, 2008b, 2009, 2010; Yelle *et al.*, 2008). However, such a high rate of escape is inconsistent with very low densities of carbon ions at a few Saturn radii from Titan, which limits escape to <10^7 CH$_4$ molecules cm^{-2} s^{-1} (Crary *et al.*, 2010). Hydrodynamic escape of methane would be strange

Table 14.6 Estimates of current rates of atmospheric escape on Titan referenced to the surface area (rather than exobase area). These can be converted into global rates by multiplying by the surface area of Titan, 8.33×10^{17} cm^2. See text for sources.

Species	Escape rate	Type of escape process
Hydrogen as H$_2$ and H	~2.0×10^{10} atoms cm^{-2} s^{-1}	Jeans modified by energy input from magnetospheric particles
CH$_4$	~1.0×10^6 molecules cm^{-2} s^{-1}	Jeans modified for non-Maxwellian behavior
N$_2$	(4–6)×10^7 N atoms cm^{-2} s^{-1}	Mainly sputtering and minor photochemical

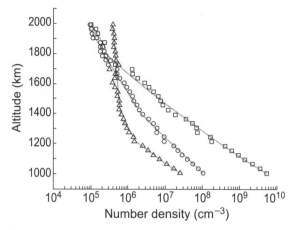

Figure 14.16 Profiles of H_2 (triangles), CH_4 (circles), and N_2 (squares) that are averages of measurements made by *Cassini*'s Ion and Neutral Mass Spectrometer (INMS) from 18 passes through Titan's upper atmosphere. The exobase lies at 1400–1500 km. (From Magee (2009).)

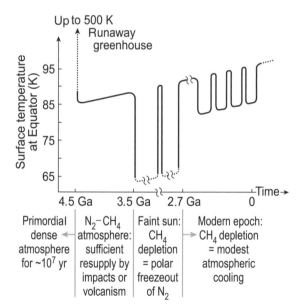

Figure 14.17 Schematic diagram of the climate evolution of Titan. (Following Lunine *et al.* (1998).)

because a heavy molecule, CH_4, would have to be selectively energized given that H_2 escape is Jeans-like and N_2 escape is slow and nonthermal.

Monte Carlo kinetic models and others suggest that the escape rate is ~1.0×10^6 CH_4 molecules cm^{-2} s^{-1}, which is Jeans-like (Bell *et al.*, 2011; Tucker and Johnson, 2009). The effects of H_2 escape might explain why

the density profile of CH_4 is different from a simple hydrostatic case, which could be misconstrued as CH_4 hydrodynamic escape. H_2 escape from Titan is borderline with hydrodynamic (the Jeans escape parameter, λ (see eq. (5.25)) is ~ 3 at the exobase) so H_2 escape extracts heat from heavier molecules, which alters the CH_4 density profile (Tucker *et al.*, 2013).

Nitrogen. Nitrogen also escapes one atom at a time from Titan by atmospheric sputtering and photochemical escape.

Photochemical escape is mainly through impact dissociation by fast electrons, which can be photoelectrons released from ionization or Saturn's magnetospheric electrons (Cravens *et al.*, 1997; Lammer and Bauer, 1991).

$$N_2 + e^-(\text{fast}) \rightarrow N(^2D) + N(^4S) + e^- \quad 0.693\,\text{eV}$$
$$(14.16)$$

Here, N(^{4}S) is ground state nitrogen, N(^{2}D) is metastable (excited state) nitrogen, and 0.693 eV is the excess energy released in the reaction. In the literature, the escape energy, E_{esc}, is often expressed in eV (where 1 eV = 1.6 $\times$ 10^{-19} J), so that $E_{esc} \approx 0.024 m_{Da}$, where m_{Da} is the mass of a species in Daltons. On Titan, E_{esc} is 0.34 eV for N and 0.68 eV for N_2. In eq. (14.16), the energy in eV per N atom of 0.693/2 only slightly exceeds the escape energy but models show that the reaction contributes ~70% of the escaping nitrogen atoms (Cravens *et al.*, 1997).

Dissociative recombination produces more energetic atoms, as follows, but proves less effective than impact dissociation as a result of dependencies on temperature and chemistry.

$$N^+_2 + e^- \rightarrow N(^2D) + N(^4S)\,3.45\,\text{eV}$$
$$N^+_2 + e^- \rightarrow N(^2D) + N(^2D)\,1.04\,\text{eV}$$
$$(14.17)$$

The global nitrogen escape from photochemistry is estimated as ~1.2×10^7 N atoms cm^{-2} s^{-1} (Cravens *et al.*, 1997).

Sputtering on Titan occurs when an energetic ion collides with an atmospheric molecule or atom and initiates collisions that lead to atmospheric loss. Energetic ions can be supplied either by the solar wind or Saturn's magnetosphere, where the latter dominates the nitrogen loss. Both *Voyager 1* and *Cassini* data indicate that Saturn's outer magnetospheric plasma contains a few percent N$^+$ ions as a result of atmospheric escape from Titan (Sittler *et al.*, 2006). Models suggest that sputtering at rates of $(3–5) \times 10^7$ N atoms cm^{-2} s^{-1} dominates the current escape of nitrogen (Michael *et al.*, 2005; Shematovich *et al.*, 2003). Whether the loss rates have been greater in the past is important for understanding the isotopic composition of nitrogen and also atmospheric evolution in general – topics to which we now turn.

14.4.4 Origin and Evolution of Titan's Atmosphere

Titan is considered to have accreted from a disk around Saturn called Saturn's subnebula (Lunine *et al.*, 2009; Prinn and Fegley, 1989). During accretion, a proto-atmosphere may have formed from the melting and vaporization of an outer icy layer of Titan, possibly producing a water-vapor runaway greenhouse (Kuramoto and Matsui, 1994). (See Sec. 11.4 for a general discussion of the runaway greenhouse.) In this model, the composition of Titan's early atmosphere is determined by three factors: the subnebula composition, the chemistry associated with accretionary processes such as impacts, and photochemistry.

14.4.4.1 The Origin and Initial Inventory of Nitrogen

Perhaps the main question for Titan's atmosphere is the origin of N_2, for which there are two options. Nitrogen was either "primordial," accreting as N_2, or it arrived in some other compound, generally considered to be ammonia, that was converted into N_2. Earlier, we discussed Pluto and the Kuiper Belt Objects, which have surfaces of N_2 ice. These objects presumably formed in a region of the Solar System where it was sufficiently cold, below ~40 K, for N_2 ice to form[4] (Owen and Bar-Nun, 1998). Saturn's subnebula was at a somewhat higher pressure than the solar nebula, so nitrogen should have condensed in the subnebula as ammonia ice (Prinn and Fegley, 1989). Thus, it is generally thought that nitrogen was captured as ammonia ice ($NH_3 \cdot H_2O$) or in N-rich organic compounds.

Data from the Gas Chromatograph Mass Spectrometer on *Huygens* support the notion that nitrogen was not originally accreted as N_2 ice. The molar ratio of $^{36}Ar/^{14}N$ in Titan's atmosphere is ~10^{-7}, compared with a solar value of 0.06 (Grevesse *et al.*, 2007) (Table 14.4). Argon and nitrogen would condense into ice from the solar nebula at similar temperatures, so if nitrogen had

been accreted as N_2 ice, we would expect $^{36}Ar/^{14}N$ to be comparable to the solar value, or even larger given probable nitrogen loss from Titan. These arguments favor accretion of NH_3 (Atreya *et al.*, 2009) but they do not exclude the possibility that N was gathered in solid organic compounds or in HCN. The abundance of neon similarly suggests that Titan did not accrete gas directly from Saturn's subnebula. In Titan's atmosphere, ^{20}Ne and ^{22}Ne have mole fractions of $<2\times10^{-5}$ and ~3×10^{-7}, respectively (Niemann *et al.*, 2010). Comparing the solar molar ratio of $^{20}Ne/^{14}N$ of ~1.6 (Grevesse *et al.*, 2007), we see that Titan's current atmosphere was not captured directly as gas from a subnebula around Saturn, because this ought to have a similar composition to the solar nebula (Owen and Niemann, 2009).

Photolysis or the shock chemistry of impacts can convert nitrogen accreted as NH_3 into N_2. If photolysis was responsible, Titan needs to have been warmer in the past because the vapor pressure of NH_3 at 94 K is negligible, ~2.6×10^{-6} Pa. If temperatures exceed ~150 K, enough NH_3 photolysis can proceed through a pathway such as that involving hydrazine (N_2H_4) and radicals of amidogen (NH_2) and hydrazinyl (N_2H_3) (Atreya *et al.*, 1978).

$$4NH_3 + 4hv \rightarrow 4NH_2 + 4H \tag{14.18}$$

$$2NH_2 + 2NH_2 + M \rightarrow 2N_2H_4 + M \tag{14.19}$$

$$2N_2H_4 + 2H \rightarrow 2N_2H_3 + 2H_2 \tag{14.20}$$

$$\underline{N_2H_3 + N_2H_3 \rightarrow N_2H_4 + N_2 + H_2} \tag{14.21}$$

$$Net: 4NH_3 + 4hv \rightarrow N_2 + N_2H_4 + 3H_2 + 3H \tag{14.22}$$

On the other hand, according to Atreya *et al.* (2009), the temperature must not be higher than ~250 K because then water ice produces a significant vapor pressure and the reaction of hydroxyl (OH) and NH_2 radicals chokes off the conversion of ammonia to N_2. (This assumes that there are not other photochemical pathways that might effect conversion, e.g., those involving N atoms.)

Accretionary heat could maintain the surface temperature of early Titan between 100–300 K (Lunine *et al.*, 2009), but once Titan reached a threshold radius of 1000–1500 km, a water-vapor rich, proto-atmosphere could have formed, increasing the surface temperatures to 500 K through a runaway greenhouse effect (Kuramoto and Matsui, 1994). Within the right temperature regime, up to 20 bars of ammonia could convert to N_2 in ~2 Myr according to calculations assuming an isothermal atmosphere (Atreya, 1986; Atreya *et al.*, 1978). But calculations using an atmosphere with cold-trapping of NH_3 model suggest a longer timescale of ~20–100 Myr for conversion to N_2 (Adams, 2006), which is big compared to

---

[4] Triton also has N_2 ice on its surface but the N_2 is not necessarily primordial because Triton is sufficiently large that the shock of impacts can convert NH_3 ice to N_2. Pluto or the KBOs have low impact velocities so that their N_2 may not have been produced from NH_3 (Sekine *et al.*, 2011a).

accretionary timescales of ~10^5–10^6 years (Lunine et al., 2009). The photolysis model may be untenable if there was only a short-lived, warm proto-atmosphere (Kuramoto and Matsui, 1994).

An alternative hypothesis for the origin of Titan's nitrogen is that it was delivered in comets during the early history of the Solar System (Griffith and Zahnle, 1995). In this model, impact shocks produce high temperatures that convert ammonia to N_2 Calculations and lab experiments support such a conversion, which could have happened during the initial accretion from impactors or later (Jones and Lewis, 1987; Marounina et al., 2015; McKay et al., 1988; Sekine et al., 2011a).

One testable consequence of shock chemistry from impacts is that chemical species will be produced that would not otherwise be abundant. For example, CO and H_2 should convert to CH_4 and CO_2 (Kress and McKay, 2004). Also impact conversion of NH_3 to N_2 is efficient only in a CO_2-rich atmosphere (Ishimaru et al., 2011). Thus, CO_2 ice should exist on Titan if there has been

difficulty is that hydrodynamic escape is not strongly mass fractionating (Sec. 5.11), while highly fractionating escape mechanisms such as nonthermal or Jeans escape operate at such low rates that they cannot shift Titan's large N_2 inventory. Thus, the data suggest that Titan's atmospheric nitrogen was derived from a cometary source and has remained with a nearly primordial ratio because plausible escape scenarios provide little fractionation. We also note that if this is the case, the large error bars on the primordial value of $^{14}N/^{15}N$ in Jupiter-family comets mean that the data do not (currently) constrain the amount of nitrogen that has escaped.

We can make a crude estimate that the original nitrogen inventory of Titan was 40 times larger. We assume that the fraction of nitrogen that Titan accreted was similar to that of ammonia and HCN in Jupiter-family comets. The molecular abundance of NH_3 in the Jupiter family comets 10/Tempel 2 and 103P/Hartley 2 is 0.5% (Biver et al., 2012; Meech et al., 2011). Thus, the accreted nitrogen would be:

$$(\text{Titan ice mass, } 0.5 \times 1.346 \times 10^{23}\,\text{kg}) \times (0.005) \times (17/18\,NH_3\text{: } H_2O \text{ mole ratio})$$
$$\approx 3.2 \times 10^{20}\,\text{kg}\,NH_3 = 2.6 \times 10^{20}\,\text{kg}\,N$$

shock chemistry conversion of NH_3 to N_2. This is consistent with spectral detection of CO_2 ice on Titan's surface (McCord et al., 2008).

Isotope ratios favor a cometary origin for Titan's nitrogen. Stable isotope ratios of hydrogen and carbon on Titan are within uncertainty of those measured in Jupiter-family comets and meteorites (Table 14.7). But isotopes of nitrogen particularly constrain the origin of Titan's nitrogen. The salient point is that Titan's $^{14}N/^{15}N$ of ~167 is similar to that in Jupiter-family comets, with values in Table 14.7 averaging 160 ± 47. (The nitrogen isotope ratio is also similar to that on Mars of 173 ± 9 (Table 12.6).) In contrast, Titan's $^{14}N/^{15}N$ is very dissimilar to the terrestrial and meteorite value of ~272, and even more different from the proto-solar or Jupiter ratio of ~440 (Marty et al., 2011).

There are three alternatives for explaining the nitrogen isotopes. (1) Nitrogen started from a solar-like value of ~440 and has been fractionated by enormous escape. (2) Nitrogen started at a meteorite-like value ~272 and had been fractionated by large escape. (3) Nitrogen started from a value similar to that in Jupiter-family comets and has been little fractioned despite escape. The most thorough models of escape rule out the first idea and also cannot achieve the large ^{15}N enrichment in the second option (Mandt et al., 2009). The reason for the

where 1.346×10^{23} kg is Titan's mass. A similar calculation gives 1×10^{20} kg N from HCN for the contribution of HCN, which is ~0.2% abundance in Jupiter family comets (Meech et al., 2011). This gives a total of 3.6×10^{20} kg N, which is about 40 times the current atmospheric inventory of nitrogen of 8.7×10^{18} kg N.

14.4.4.2 The Origin of Methane and its Evolution

As with nitrogen, there are two options for the origin of Titan's methane. Either methane is primordial or carbon was accreted in some other chemical form, such as CO or CO_2, and has since been converted into methane. There is also the need for methane to be replenished on 10–100 Myr timescales.

The main form of carbon in Saturn's circumplanetary disk was probably either CH_4 or CO_2. In one model, the subnebula is relatively hot and dense compared to the solar nebula and gas-phase and gas–solid reactions convert N_2 and CO into NH_3 and CH_4, respectively (Prinn and Fegley, 1989; Sekine et al., 2005). In another model, called the gas-starved disk, Saturn gathers icy planetesimals into a disk and Titan gains NH_3 but more CO_2 than CH_4 (Alibert and Mousis, 2007; Mousis et al., 2009b).

Table 14.7 A summary of Titan's isotopic ratios compared with other reservoirs.

Ratio	Titan	Protosolar	Earth	Jupiter-family comets
D/H ($\times 10^{-4}$)	1.35 ± 0.3 in H_2 [a]	0.25 ± 0.05 (nebula) [d]	1.56 [d]	1.6 ± 0.2 [g]
	$1.32^{+0.15}_{-0.11}$ in CH_4	~1.6 (meteorites) [d]		(103P/Hartley 2)
$^{12}C/^{13}C$	91 ± 1.4 in CH_4 [a]	89 [e]	89.9 [f]	95 ± 15 [h]
				(103P/Hartley 2)
$^{14}N/^{15}N$	167 ± 0.6 [a]	440 ± 6 [d] (nebula)	272 [d]	155 ± 25 [h]
		272 [e] (meteorites)		(103P/Hartley 2)
				160 ± 40 [i] (CN in 17P/Holmes)
$^{16}O/^{18}O$	~250 in CO [b]	499 [e]	499 [f]	490–520
	346 ± 110 in CO_2 [c]			(81P/Wild2) [j]

[a] Niemann *et al.* (2010). [b] Owen *et al.* (1999). [c] Nixon *et al.* (2008). [d] Marty (2012) and references therein. [e] Lodders (2010b). [f] Henderson and Henderson (2009) [g] Hartogh *et al.* (2011b). [h] Meech *et al.* (2011). [i] Bockelee *et al.* (2008). [j] McKeegan *et al.* (2006).

Isotopic data help distinguish which model is more likely (Table 14.7). The D/H ratio and $^{12}C/^{13}C$ ratio measured by *Huygens* are indistinguishable within uncertainty from those found in Jupiter family comets and meteorites. This argues in favor of the accretion of Titan's carbon from a circum-Saturn disk of planetesimals – the gas-starved disk. The nitrogen isotope ratio also favors a Jupiter-family comet source, as discussed previously.

In the current atmosphere, methane must be replenished because its destruction is geologically rapid. Making the crude approximation that methane is well mixed, eq. (1.21) gives its column abundance N_{col}, as:

atmospheric methane for ~10 Myr before exhaustion (Lorenz *et al.*, 2008). However, a small (60×40 km^2) lake in low latitudes may have been detected, which is unexpected because lakes should be unstable in the tropics (Griffith *et al.*, 2012). This finding suggests that methane might be seeping from deep within the subsurface, but needs confirmation with high resolution imaging.

Others suggestions for a source of methane are subsurface serpentinization reactions between water, rock and carbon dioxide like those reactions discussed in Sec. 7.5.2 (Atreya *et al.*, 2006). But a serpentinization source would conflict with the D/H of CH_4, if an Enceladus-like D/H ratio

$$N_{col} \approx n_{surf} H = \left(\frac{P_{CH4}}{kT}\right)\left(\frac{kT}{mg}\right) = \left(\frac{P_{CH4}}{mg}\right)$$

$$= \left(\frac{\left(1.467 \times 10^5 \text{ Pa}\right) \times (0.0565 \text{ mol fraction})}{(27.3 \text{ dalton}) \times (1.35 \text{ m s}^{-2})}\right) = 1.35 \times 10^{25} \text{ molecules cm}^{-2} \tag{14.23}$$

Here, n_{surf} is the surface number density, H is the scale height, P_{CH4} is the partial pressure of CH_4, g is gravitational acceleration, and $\overline{m}$ is the mean molar mass. Using a net methane destruction rate of 5×10^9 CH_4 cm^{-2} s^{-1} derived from the hydrogen escape rate, the lifetime of CH_4 is $\approx 1.35 \times 10^{25}/5 \times 10^9 = 2.7 \times 10^{15}$ s = 85 million years. This is within estimates in the range 30–100 million years in the literature (Yung *et al.*, 1984).

Suggestions to explain why methane persists are that the subsurface releases methane continuously or that we are seeing Titan at a special time after a discrete event that puffed methane into the atmosphere through outgassing or cometary impact. The amount of methane estimated in polar hydrocarbon lakes could only buffer loss of

in H_2O is assumed (Glein *et al.*, 2009; Mousis *et al.*, 2009a). An alternative hypothesis proposes that methane is episodically released from subsurface methane clathrates (Tobie *et al.*, 2006). Yet another possibility is that occasional cometary impacts release enough methane to saturate Titan's atmosphere (Kress and McKay, 2004).

If methane was destroyed faster than it was resupplied in the past, Titan's nitrogen atmosphere could have collapsed into N_2 ice or N_2 lakes so that the partial pressure of N_2 was controlled by vapor pressure equilibrium, similar to the situation on Pluto or Triton (Charnay *et al.*, 2014; Lorenz *et al.*, 1997; McKay *et al.*, 1993). Today, the removal of methane and its photochemical product H_2 from Titan's atmosphere would cause the temperature to fall by 5 K because CH_4–CH_4, CH_4–N_2 and N_2–H_2 collision-induced absorption would cease. However, when the Sun's luminosity was ~80% less than present, a freeze-out of N_2 could drive Titan's surface temperatures below ~65 K.

A highly schematic diagram of the evolutionary history is shown in Fig. 14.17. An early proto-atmosphere is hot but lasts probably $<10^7$ years. Then impacts keep Titan resupplied with methane until ~3.5 Ga when complete freeze-out may have occurred. After the sun was ~80% of present luminosity (2.7 Ga), N_2 could no longer freeze out but temperatures may have oscillated if methane levels fluctuated.

14.4.4.3 Noble Gases

The noble gases are always especially helpful in determining the origin and evolution of atmospheres because these elements ought to end up in the gas phase and the heavier ones should have difficulty escaping.

The amount of radiogenic ^{40}Ar in Titan's atmosphere indicates that that only 6% of the argon has outgassed from Titan's interior (McKinnon, 2010). This estimate derives from a potassium ^{40}K inventory on Titan, assuming that Titan is made of ~55% rock of CI carbonaceous chondrite composition. Possibly, Titan has experienced little outgassing and cryovolcanism during its history (Moore and Pappalardo, 2011).

Huygens did not detect krypton or xenon. The explanation could simply be that *Huygens* instrumentation did not have sufficient sensitivity (Owen and Niemann, 2009). However, there are two other possibilities. One is that krypton and xenon are trapped in clathrates on Titan (Mousis *et al.*, 2011; Thomas *et al.*, 2007). The other is that noble gases actually take part in chemistry and can be captured by aerosols under the cold conditions of Titan (Jacovi and Bar-Nun, 2008). We also note that noble gases should be very soluble in Titan's hydrocarbon lakes.

14.4.5 Life on Titan: "Weird Life" or Liquid Water Life

Of great interest is whether there is life on Titan. The two possibilities are that Earth-like life that uses liquid water lives in Titan's interior ocean (Fortes, 2000) or that exotic, so-called "weird life," that uses hydrocarbon solvents exists on Titan's surface.

From an evolutionary point of view, a very early water ocean on the surface of Titan would have been in contact with the atmosphere. Possibly water-based life evolved and survived as the ocean retreated into the subsurface.

A suggestion has also been made for metabolism of weird life, which is the reduction of acetylene to methane (McKay and Smith, 2005):

$$C_2H_2 + 3H_2 \rightarrow 2CH_4 \qquad (14.24)$$

This yields free energy, given the concentrations of gases at the surface of Titan. But presently it must be seen as highly speculative. Also, a problem with weird life on Titan is that that biochemical complexity is limited by the tiny availability of oxygen-containing molecules; on Earth, O atoms are in nucleotides, amino acids, and sugars.

Hypothetical weird life that uses very cold hydrocarbon solvents has been discussed by Benner *et al.* (2004). One idea is that molecules that use hydrogen bonding could be immersed in non-hydrogen-bonding solvents such as liquid ethane. Overall, there are problems in dissolving large molecules such as those needed for genomes. Genomes are required for life: to provide a recipe for reproduction and for an organism's characteristics. The very low temperature on Titan implies insolubility of large molecules. For example, alkanes larger than propane have triple-point temperatures that are higher than surface temperatures on Titan. The liquefied gas literature shows that larger molecules are generally less soluble than smaller ones (Kohn *et al.*, 1976; Kurata, 1975; Szczepanieccieciak *et al.*, 1978). We can understand this semi-quantitatively,[5] through an ideal solubility equation for a solid (Preston and Prausnit, 1970):

$$\ln (x) = \frac{\Delta H_{fus}}{RT_t} \left(1 - \frac{T_t}{T} \right) \qquad (14.25)$$

Here, x is the mole-fraction solubility of the solid, ΔH_{fus} is the enthalpy of fusion of the solid, R is the universal gas constant, T_t is the triple-point temperature of the solute of interest, and T is the ambient temperature. According to this equation, we should not expect large molecules to be appreciably soluble in liquid hydrocarbons on Titan because (1) the ambient temperature is low, and (2) large molecules have large enthalpies of fusion because of a large number of intermolecular interactions that stabilize the solid, which makes $\ln(x)$ small, given a negative term in the parentheses. The low temperature on Titan implies that liquid methane on Titan is a much worse solvent than is n-hexane under typical laboratory conditions on Earth. Dissolution of a solid is thermodynamically equivalent to melting the solid at a temperature below the triple point (Prausnitz *et al.*, 1999).

---

[5] Although author DCC has argued before that solubility is a problem for weird life on Titan (e.g., Catling (2013, p. 106)), he is indebted to Christopher Glein for providing this semi-quantitative argument and associated references.

In hydrocarbon systems, this means that solids that are more difficult to melt are also more difficult to dissolve into solvents. Thus, a "weird life" genetic molecule on Titan falls victim to solution thermodynamics. This treatment neglects non-ideal effects, but nonpolar mixtures generally exhibit positive deviations from Raoult's law (activity coefficients greater than unity), meaning that the actual solubility is lower than ideal (Prausnitz et al., 1999).

Perhaps it is important to remember that, even if there is no life, Titan would still remain as a remarkable crucible of organic chemistry that should inform us about prebiotic chemistry (Raulin et al., 2012).

14.5 The Exoplanet Context for Outer Planets and their Satellites

Although the diversity of atmospheres on planets and moons can seem bewildering, in this chapter we have seen several common principles at work. The atmospheres of ice giants may be subject to hydrodynamic escape and impact erosion that could affect whether they maintain atmospheres or not. There is a trend going outwards from the Sun in ice composition from the water ice of moons of Jupiter or Saturn to nitrogen surface ice on Triton and Pluto. Icy satellites of gas giants can acquire tenuous O_2-rich atmospheres from sputtering of water ice and hydrogen escape. Possibly, thicker atmospheres could be acquired from sublimation or melting of icy moons around Jupiter-like exoplanets that migrate inwards in their orbits if the moons are big enough to prevent atmospheric escape. Given the thickness of icy shells on such satellites, if water melts, deep oceans could provide habitats on the surfaces of such exomoons.

Finally, understanding Titan may help us recognize exo-Titans or related variants. There may be many Titan-like worlds around other stars. The most common type of star is an M-dwarf and at a distance of about 1 AU around a typical M-dwarf the stellar flux is similar to that received by Titan (Lunine, 2010). At that distance, compared with the ~0.1 AU range of the habitable zone, a planet avoids tidal locking, the flare activity of M-dwarf stars, and potential impact erosion of volatiles (Lissauer, 2007). But surface temperatures would be conducive to liquid hydrocarbons rather than liquid water. Thus, if we were to discover weird life on Titan, the inference might be a universe teeming with life quite different from our own, where we're the weird ones.

15 Exoplanets: Habitability and Characterization

Until recently, the study of planetary atmospheres was largely confined to the planets within our Solar System plus the one moon (Titan) that has a dense atmosphere. But, since 1991, thousands of planets have been identified orbiting stars other than our own. Lists of exoplanets are currently maintained on the *Extrasolar Planets Encyclopedia* (http://exoplanet.eu/), NASA's *Exoplanet Archive* (http://exoplanetarchive.ipac.caltech.edu/) and the *Exoplanets Data Explorer* (http://www.exoplanets .org/). At the time of writing (2016), over 3500 exoplanets have been detected using a variety of methods that we discuss in Sec. 15.2. Furthermore, NASA's *Kepler* telescope mission has reported a few thousand additional planetary "candidates" (unconfirmed exoplanets), most of which are probably real. Of these detected exoplanets, a handful that are smaller than 1.5 Earth radii, and probably rocky, are at the right distance from their stars to have conditions suitable for life (Batalha, 2014; Batalha *et al.*, 2013; Borucki *et al.*, 2011). Describing the different planetary detection techniques and summarizing the data collected so far would require a book in itself (which would immediately be out-of-date), so we shall not attempt to do that. Rather, we will focus on general issues related to planetary habitability and the detection of life, both of which build on concepts discussed earlier within this book.

For general discussions of exoplanets, several good reviews are available. Kasting (2010) covers some of the material discussed here, but at a lower level. Tutorial books at approximately the upper undergraduate level include Scharf (2009) for exoplanet astrobiology and Haswell (2010) on transiting exoplanets; both have great clarity. Seager (2010), Perryman (2014), and Winn and Fabrycky (2015) review exoplanet detection and characterization at the researcher level.

15.1 The Circumstellar Habitable Zone

We focus our attention on Earth-like planets and on the possibility of remotely detecting life. We begin by defining what life is and how we might look for it. As we shall see, life may need to be defined differently for an astronomer using a telescope than for a biologist looking through a microscope or using other *in situ* techniques.

15.1.1 Requirements for Life: the Importance of Liquid Water

Biologists have offered various definitions of life, none of them entirely satisfying (Benner, 2010; Tirard *et al.*, 2010). One is that given by Gerald Joyce, following a suggestion by Carl Sagan: "Life is a self-sustained chemical system capable of undergoing Darwinian evolution" (Joyce, 1994). This definition is useful for laboratory scientists: If they make a self-replicating, mutating chemical system, they will hopefully recognize it, regardless of its construction. The same applies for astrobiologists who look for life on Mars or Titan, although efforts to date have focused on looking for signs of metabolism rather than a genome, e.g., the *Viking* mission to Mars (Sec. 12.1.3). We have not yet seen evidence for life in the Solar System, but as exploration becomes more sophisticated, we could possibly identify life – even exotic life – if it exists.

Astronomers are not so fortunate because they have no hope of actually visiting exoplanets in the foreseeable future. Instead, they must use remote detection. One way to find life involves taking spectra of exoplanet atmospheres and looking for gases such as O_2 or CH_4 that might be produced by life. Other, more general, life-detection criteria have been proposed, e.g., finding extreme thermodynamic disequilibrium (Lederberg, 1965; Lovelock,

1965), which we discuss later (Sec. 15.4). But such criteria may be difficult to interpret if they involve chemical signatures that are alien from those we find on Earth. For this reason, it makes sense to concentrate initially on planets that could support life similar to that found on Earth. A key requirement of Earth-like life is the availability of liquid water. It will be much easier to believe that some possible biosignature is indeed evidence for life if the planet on which it is observed supports liquid water. Of course, the most credible remote biosignature of all would be a radio or visible light transmission from an extraterrestrial civilization. But if such signals exist, the well-known Drake Equation implies that they should be rarer than the atmospheric signatures of biospheres (e.g., Vakoch and Dowd, 2015), which is our focus here.

Exoplanets with hidden subsurface life are not practical candidates for remote life detection. For example, Mars is dry and lifeless at its surface, but might harbor liquid water at depth and, at best, a meager subsurface biosphere. But even though Mars is our planetary neighbor, a contemporary debate has erupted about whether trace levels of methane are actually present in its atmosphere, irrespective of whether such methane is biogenic or abiotic (see Sec. 12.2.1). Detecting similar miniscule signals from an exoplanet atmosphere is beyond the realm of feasibility for the foreseeable future. For life to influence a planet's atmosphere in a way that is readily detectable from interstellar distances, it needs to exist on the planet's surface. For life as we know it, this means that liquid water must also be present at the surface. Hence, we focus our interest on planets that could support surface liquid water. These planets orbit in the region that we now call the *circumstellar habitable zone*.

15.1.2 Historical Treatment of the Habitable Zone

The idea of a habitable zone has a long history. William Whewell was the first to note, in a book about extraterrestrial life, how Earth's orbit is in a *temperate zone* between a "central torrid zone" and external "frigid zone" (Whewell, 1853). A century later, the astronomer Harlow Shapley (1953) defined a *liquid water belt* as the region in a planetary system where liquid water could exist at a planet's surface. Also at that time, Strughold (1953, 1955) defined an analogous *ecosphere* around the Sun. The astronomer Su-Shu Huang (1959, 1960) then identified a variety of issues bearing on habitability. He pointed out that binary or multiple star systems are less likely to harbor habitable planets than single stars because the planetary orbits would be unstable in most cases. He also

concluded, perhaps correctly, that stars that are similar in mass to our Sun are the most likely to have habitable planets. But his most lasting contribution, probably, was to coin the term *habitable zone* as a synonym for Shapley's liquid water belt.

Shortly after this, in *Habitable Planets for Man*, Dole (1964) considered a more focused question: How many nearby stars might harbor planets suitable for human colonization? His anthropocentric conditions for habitability included mean annual temperatures of 0–30 °C over 10% of a planet's surface, an O_2-rich atmosphere, and a surface gravity less than 1.5 times that of Earth, so that humans could walk upright. Dole's climate models were crude: black planets with no greenhouse effect and 45% cloud cover. Nevertheless, Dole made a number of valid points, reiterating Huang's concerns about orbital stability in binary star systems, and pointing out the problem of tidal locking of planets orbiting red dwarfs, which we discuss later.

None of the early researchers attempted to define the habitable zone using a realistic climate model until Hart (1978, 1979). In retrospect, Hart's climate models were flawed, but his papers were influential and provoked further scientific interest. In particular, Hart defined a *continuously habitable zone*, or CHZ, for short, as the region around a star where a planet could support liquid water for some specified period, usually the star's main sequence lifetime. The conventional *habitable zone*, then, can be abbreviated as "HZ," and is defined at a single instant in time. The HZ must move outwards with time because main sequence stars brighten as they age (Sagan and Mullen, 1972), as indicated in Fig. 15.1. Suppose the Sun's initial HZ at time t_0 covered the range of distances shown in the figure, and that by some later time t_1 it had moved further out, as indicated. Then the CHZ is represented by the overlap between the two regions.

Hart's approach to calculating the boundaries of the HZ was ambitious. He performed time-dependent calculations for planetary atmospheres and climates. His models included a simplified greenhouse effect, along with numerous physical processes, including outgassing of volcanic CO_2, reactions of CO_2 with surface minerals, the presence of reduced greenhouse gases (CH_4 and NH_3) early in Earth's history, organic carbon burial, the rise of O_2, changes in solar luminosity, and ice albedo feedback.

A key flaw in Hart's climate model was that he underestimated the CO_2 greenhouse effect, making the outer edge of his HZ too small. Volcanic CO_2 could build up in a planet's atmosphere if the planet became entirely ice-covered, but a planet could not deglaciate (Levenson, 2015). Consequently, Hart determined that the outer edge

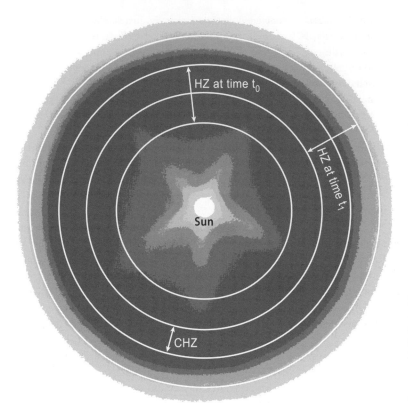

Figure 15.1 Diagram illustrating the habitable zone and continuously habitable zone (HZ), which shifts between times t_0 and t_1, and the continuously habitable zone (CHZ) around a star. (From Kasting, (2010). Reproduced with permission of Princeton University Press. Copyright 2012.)

of the HZ for our own Solar System was at 1.01 AU. If the Earth had formed farther from the Sun than this, it would have remained perpetually frozen. We have already argued, though, that a Snowball Earth planet *should* be able to deglaciate (Sec. 11.10). Hart also found that a runaway greenhouse would have occurred if the Earth had formed inside of 0.95 AU. This prediction turns out to have been reasonable, as the best current estimate for this threshold based on 3-D climate modeling is at that same distance (see next section). In any case, the overall conclusion of Hart's (1978) paper was that the 4.6 b.y. continuously habitable zone around the Sun was quite narrow, 0.95-1.01 AU. Then, in his 1979 paper, Hart argued that the CHZ around other main sequence stars was even narrower, or nonexistent. The bottom line was that habitable planets were extremely rare. Indeed, pessimists who believed Hart's papers might have concluded that Earth was the only one in the galaxy.[1]

---

[1] Hart also repeatedly argued that humans are special and that advanced life is absent from the rest of the Universe (Hart, 1975, 1982). These views fit rather snugly with the results of his habitable zone models.

15.1.3 Modern Limits on the Habitable Zone Around the Sun

Estimating the inner and outer boundaries of the habitable zone for our own Sun is closely related to the problem of understanding long-term climate evolution on Venus and Mars, respectively. In Sec. 13.4, we discussed Venus' susceptibility to a runaway greenhouse, which sets a hard inner edge to the HZ, while in Sec 12.5.2, we considered how a CO_2–H_2O greenhouse can fail to warm early Mars, which is an issue pertinent to the outer edge of the HZ. Given these earlier discussions, we summarize the ideas here.

The inner edge of the HZ is determined by when a planet develops a wet stratosphere and loses its water through photodissociation, followed by escape of hydrogen to space (Kasting *et al.*, 1993b; Kopparapu *et al.*, 2013). In the model of Kopparapu *et al.* (2013), this happens when the solar flux became more than ~2% higher than its present value at Earth or, equivalently, at a distance of ~0.99 AU. This is the *water-loss limit* of Kasting *et al.* (1993b). The *runaway greenhouse limit* in the Kopparapu *et al.* model occurs at ~0.97 AU, corresponding to an effective solar flux, S_{eff}, of ~1.05.

(The effective solar flux is defined as $S_{eff} = S/S_0$, where S is the solar flux at some distance r from the Sun and S_0 is the solar flux at Earth's orbital distance of 1 AU.) As pointed out in Sec. 13.4.5, a 3-D calculation by Leconte *et al.* (2013) does *not* predict a wet stratosphere because their tropopause is extremely cold. The inner edge of the HZ in this model is determined by the last stable climate simulation, which occurs at $S_{eff} = 1.1$, or 0.95 AU.[2] Henceforth, we use 0.95 AU as the conservative estimate for the inner edge of the habitable zone, as this 3-D calculation is arguably better than the 1-D model estimates. By accident, this limit agrees with that calculated over 20 years or so earlier by Kasting *et al.* (Kasting, 1988; Kasting *et al.*, 1993b).

Following this same line of reasoning, the outer edge of the HZ can be estimated by modeling the effect of slowly sliding the Earth out toward Mars' orbit. As this happens, the climate should get colder, silicate weathering should slow down, and volcanic CO_2 should build up in the planet's atmosphere. This negative feedback is sufficient to prevent global glaciation, provided that the solar flux is above a critical value. However, at some distance, CO_2 clouds begin to form in the planet's atmosphere (the *first condensation limit*), and at some greater distance CO_2 condensation becomes so extensive that further greenhouse warming is prohibited (the *maximum greenhouse limit*). The relevance of the first condensation limit depends on whether CO_2 clouds can warm a climate or not, which recent models don't favor (see Sec. 12.5.2.2). The *maximum greenhouse* limit is a firmer but more generous limit for the outer edge of a CO_2–H_2O greenhouse. In the Kasting *et al.* (1993b) model, this limit was reached at 1.67 AU, or $S_{eff} = 0.36$, whereas Kopparapu *et al.* (2013) place it at 1.69 AU, or $S_{eff} = 0.35$. Both estimates are well outside of Mars' orbital distance of 1.52 AU. So, as pointed out in Ch. 12, Mars might well be habitable today if it was able to recycle its atmospheric CO_2.

These estimates for the boundaries of the HZ were calculated for cloud-free atmospheres and fully saturated tropospheres and may thus be too conservative. In Leconte *et al.* (2013), the unsaturated troposphere is a stabilizing influence, but cloud feedback is *positive*, so these effects tend to cancel. Cloud feedback is notoriously difficult to calculate, and other models predict that it is

negative near the inner edge of the HZ (Wolf and Toon, 2014). Thus, Wolf and Toon predict that the inner edge is at 0.93 AU, or even closer, although their model does not account for water vapor as a major constituent near the inner edge of the HZ, unlike Leconte *et al.* (2013). The outer edge of the HZ could be much farther out if there are other greenhouse gases, e.g., H_2, that provide additional warming, which we discuss further below.

The uncertainty on the outer limit of the HZ may not matter too much because even the conservative estimate for the HZ width is fairly broad. If the outer edge is near 1.7 AU and the inner edge is near 0.95 AU, then the width of the HZ is ~0.75 AU. Our own Solar System contains four terrestrial planets between 0.4 AU and 1.5 AU, and the mean spacing between them is ~0.35 AU. This suggests that two of them ought to be in the HZ, which is exactly what we observe, because both Earth and Mars are in it, according to Kopparapu *et al.* (2013). If rocky planets in other planetary systems are spaced as they are in our Solar System, then the chance that at least one of them will be in the habitable zone is reasonable. As discussed further below, our own Solar System is dynamically "packed," meaning that it contains as many dynamically stable planets as it can. Originally, there were many more, smaller planets, and they continued to collide with each other until they reached a configuration in which collisions became infrequent on Gyr timescales, producing a stable, packed system. But it remains to be seen if such packing holds true for other planetary systems.

We can also use climate modeling results to derive an estimate for the width of the 4.6-billion-year (b.y.) continuously habitable zone (CHZ) around the Sun. The Sun is brighter today than it has been in the past. Hence, the inner edge of the CHZ is the same as the inner edge of the modern HZ: 0.95 AU. The outer edge, though, must be closer in because at 4.6 Ga the Sun was only about 70% as bright as it is today. Hence, a planet would need to have been closer to the Sun by a factor of $0.7^{1/2} = 0.84$ in order to receive the same flux that it does at present. If the outer edge of the modern HZ is at 1.7 AU, then at 4.6 Ga it should have been at $1.7 \times 0.84 \cong 1.4$ AU. Our CHZ is thus relatively wide, ~0.4 AU, which may be compared with the value of 0.06 AU in Hart's model. Once again, this illustrates the importance of the carbonate–silicate cycle and the negative feedback it provides on atmospheric CO_2 and climate. Without this feedback, habitable planets might indeed be rare.

A caveat should be added here: The CO_2–climate feedback on Earth depends on the presence of exposed continents. A "water world" with little or no continental area would not be subject to this same feedback

---

[2] Whether or not a moist greenhouse atmosphere can occur is currently a matter of contention. A recent 3-D climate simulation by Wolf and Toon (2015) suggests that it will. Their stratosphere is not quite as cold as that of Leconte *et al.* (2013), and they are able to extend their calculations to higher surface temperatures. A 1-D study by Kasting *et al.* (2015) supports the Wolf and Toon result.

(Abbot *et al.*, 2012). As discussed in Sec. 11.4, CO_2 would likely be removed by weathering of the seafloor on such a planet. The dependence of this process on surface temperature is uncertain. Coogan and Gillis (2013) argue that carbonatized seafloor shows that it provided a stabilizing negative feedback on high atmospheric CO_2 concentration on the Late Mesozoic Earth.

Other authors have proposed modifications to the inner edge of the HZ. Abe *et al.* (2011) used a 3-D climate model to show that hot, rocky planets with small water endowments and low obliquities might remain habitable in their polar regions, because the lack of a large ocean would reduce the positive feedback of water vapor on climate. They called such planets "*Dune* planets" after Frank Herbert's eponymous novel, which describes a desert planet with small habitable regions near its poles. Their estimated inner edge is at 0.77 AU, or $S_{eff} \cong 1.7$. But the water on such *Dune* planets might combine chemically with the surface, so it is unclear if they would remain habitable (Kasting *et al.*, 2014). Similarly, Zsom *et al.* (2013) estimated 0.5 AU for the inner edge using 1-D calculations of low-relative-humidity planets. But surface energy balance suggests that liquid water would quickly evaporate under such conditions (Kasting *et al.*, 2014).

As mentioned earlier, the outer edge of the HZ can expand if greenhouse gases other than H_2O and CO_2 are considered. Of the various candidates, H_2 is very effective because of its broad infrared absorption spectrum (Sec. 2.5.6) and low condensation temperature. As discussed in Sec. 12.5.2.2, Ramirez *et al.* (2014a) invoked several percent of H_2 to produce a warm climate on early Mars, which would otherwise have resided outside of the conventional HZ. Other authors have considered much larger amounts of H_2. Following Pierrehumbert and Gaidos (2011), Seager (2013) suggested that the outer edge of the HZ around our Sun might be as far out as 10 AU for a 3-Earth-mass planet with a 40-bar H_2 atmosphere that was captured during accretion. Stevenson (1999) had demonstrated earlier that H_2-rich rocky planets could maintain liquid water on their surfaces even if they were wandering freely in interstellar space. In that case, geothermal heat is sufficient to keep a planet warm if its H_2-rich atmosphere is thick enough.

The debate about HZ boundaries has practical consequences. Kasting *et al.* (2014) argue that a space telescope to look for Earth-like planets should be designed using a conservative definition of the HZ. That's because the frequency of Earth-size planets around stars, $\eta_\oplus$ (said as "eta sub Earth" or "eta Earth"), depends on the assumed HZ width. If $\eta_\oplus$ is high, then fewer stars need to be searched and the telescope can be made smaller. If $\eta_\oplus$ is low, then a larger telescope is needed. (Note that there are varying definitions of $\eta_\oplus$ in the literature that have different choices for the meaning of "Earth-like" or size of the HZ, which we discuss in Sec. 15.2.3.) Because both *Dune*-like planets and H_2-rich super-Earths are speculative, and they may or may not exist in reality, wider HZ limits for these objects should probably *not* be used when designing an Earth-finding telescope, lest the instrument be undersized. Once such a telescope has been launched, however, then broadening one's definition of the HZ makes sense, as one would not want to overlook any potentially habitable planets.

15.1.4 Empirical Estimates of Habitable Zone Boundaries

Theoretical calculations, regardless of how sophisticated they might be, must always be viewed skeptically, as nature is often subtler than our imagination. This statement applies to estimates of habitable zone boundaries, as well. Fortunately, our Solar System provides some *empirical* estimates for these boundaries. As discussed in Ch. 13 and above, Venus appears to have lost its water through either a runaway or moist greenhouse effect. The young cratering age of Venus' surface ranges 0.35–1.5 Ga, based on uncertainties in modeling, with a typical value near ~0.7 Ga (Korycansky and Zahnle, 2005). Consequently, Venus had probably already lost its water prior to 0.7 Ga, at which time the Sun was about 6% dimmer than it is today, using eq. (11.3) (Gough, 1981). Venus' semi-major axis is 0.723 AU, so the effective solar flux at that distance is $S_{eff} = (1/0.723)^2 = 1.91$. Taking into account the lower solar luminosity at 0.7 Ga, the "recent Venus" empirical estimate for the HZ inner edge is $S_{eff} = 1.91 \cdot (0.94) = 1.8$, or 0.75 AU. Note that this is slightly higher than the *Dune*-planet limit of $S_{eff} = 1.7$ estimate by Abe *et al.* (2011).

At the other end of the habitability spectrum, Mars appears (to some, at least) to have been habitable back around 3.8 Ga during the heavy bombardment period (Sec. 12.4). Mars' semi-major axis is 1.52 AU, and the solar flux at that time was about 25% lower than today by eq. (11.3), yielding $S_{eff} = (1/1.52)^2 \cdot 0.75 = 0.32$, which corresponds to a distance of 1.76 AU. This is just slightly lower than the maximum greenhouse limit of $S_{eff} = 0.35$ from the Kopparapu *et al.* (2013) model, which is why warming early Mars is a challenge (see Sec. 12.5.2). Thus, the empirical and theoretical limits on the outer edge of the HZ are in fairly close agreement. Only if large amounts of H_2 are present in a planet's atmosphere should it remain habitable beyond about 1.7 AU.

15.1.5 Habitable Zones Around Other Main Sequence Stars

The same types of climate calculations that are described above can be done for planets orbiting other types of stars. (The stellar classification scheme was described in Sec. 2.1.2.) The calculations change in two ways, however. Most obviously, the stellar flux is much higher around bright blue stars and much lower around dim red ones; hence, the habitable zone must move either out or in depending on the stellar type. But it is not just the star's luminosity that changes. The spectral distribution of its radiation changes as well, because of the change in the star's surface temperature. This is illustrated in Fig. 15.2 (a), which shows the distribution of radiation for a planet orbiting an F2 star, a G2 star (the Sun), and a K2 star. The radiation from an F star is relatively bluer, while that from a K star is redder. These color shifts affect the climate calculations because blue light is more easily reflected from a planet due to increased Rayleigh scattering, whereas red and near-infrared radiation is scattered less and is also partly absorbed by the planet's atmosphere. Ice is also more reflective at shorter wavelengths (Warren et al., 2002). Hence, according to 1-D climate simulations, the HZ around a blue star is slightly closer in than one would expect based solely on its luminosity, while the HZ for a red star is slightly farther out. (See below for how these predictions can change in 3-D.)

A graphical summary of HZ calculations is shown in Fig. 15.3(a). Here, the horizontal axis represents the distance from the star, and the vertical axis is the star's mass, relative to that of our Sun. The habitable zone is the strip running from the lower left-hand part of the diagram to the upper right. Also shown are the eight planets of our Solar System. Because the HZ moves outwards with time at different rates for stars of different masses, one has to choose a particular time in the star's lifetime in order to make a plot like this. In Fig. 15.3(a), the HZ has been plotted at the time when each star first enters the main sequence.

Figure 15.3(a) shows various relationships. First, the habitable zone lies farther out for more massive stars and closer in for less massive ones. That is to be expected, as a planet must receive roughly the same amount of starlight as does Earth in order to be habitable. Second, Fig. 15.3 (a) demonstrates a point made earlier: The HZ in our own Solar System is relatively wide, compared to the spacing between the planets. The orbital distance is shown on a

(a)

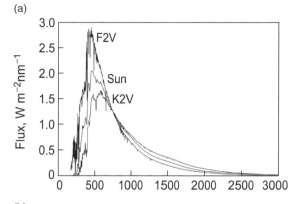

(b)

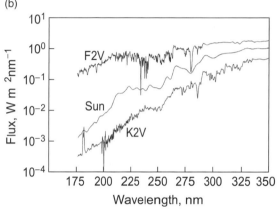

Figure 15.2 Incident stellar flux distribution for a planet orbiting an F2 star, a G2 star (the Sun), and a K2 star. The planet is assumed to receive the same total amount of sunlight as the present Earth. Panel (a) shows the entire wavelength spectrum; panel (b) shows the far ultraviolet portion of the spectrum. (From Kasting, (2010). Reproduced with permission of Princeton University Press; originally from Segura et al. (2003). Reproduced with permission from Mary Ann Liebert, Inc. Copyright 2003.)

log scale because the planets in our own Solar System are spaced logarithmically. This is "geometric" spacing, because each planet's orbital distance is larger than that of its inner neighbor by an average factor of ~1.7. Of course, the asteroid belt occupies the gap between Mars and Jupiter, where it appears as if a planet should exist. This observation is nothing new. The relationship between planetary orbital distances was noticed a long time ago and is referred to as *Bode's Law* or the *Titius–Bode Law*.

Bode's Law has frequently been dismissed as nothing more than a simple empirical fit, but a Bode's Law spacing of planetary orbits can arise naturally from the right initial mass distribution within the solar nebula in order for orbits to be stable (Laskar, 2000). More specifically, separate power-law fits for the four inner and four

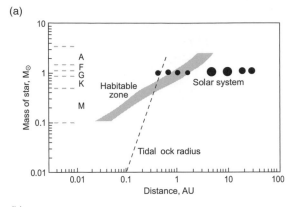

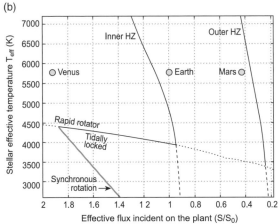

Figure 15.3 (a) Diagram illustrating the extent of the habitable zone around different types of stars. The vertical scale represents the mass of the star in units of the mass of our Sun, $M_\odot$, and the figure is drawn for the time when the stars have just reached the main sequence. The eight planets of our own Solar System are shown. The dashed line shows the orbital distance at, and below which, a planet experiences tidal locking of its rotation rate. (b) Diagram showing the habitable zone (HZ) in units of stellar flux S relative to today's solar flux, S_0. The runaway greenhouse "inner HZ" (IHZ) limit is shown for an Earth-mass planet, corresponding to 3-D climate model results (Leconte *et al.*, 2013). For cool stars ($T_{eff} < 4500$ K), the INZ jumps inward along the "synchronous rotation" line within the tidal locking limit (dotted curve) because 3-D models suggest a high planetary albedo develops on the dayside from thick water clouds at the subsolar point (Yang *et al.*, 2013). The outer HZ is the *maximum greenhouse* limit. (Adapted from Kopparapu *et al.* (2014). Reproduced with permission. Copyright 2014, American Astronomical Society.)

outer planets describe the orbits fairly accurately (Laskar, 2000). In any case, Bode's Law may have a physical basis. At least for the giant planets, it has been argued that you cannot pack planets more closely together without making their orbits unstable (Chambers *et al.*, 1996; Gladman, 1993). Whether or not *terrestrial* planets should be spaced geometrically is debated, however.

Resonances with giant planets may influence terrestrial planet spacing more than mutual interactions between the terrestrial planets. This question should be answered when we are able to observe other inner planetary systems.

The issue of planetary spacing becomes important in considering stars that are significantly different from the Sun. As can be seen from Fig. 15.3(a), the HZ is roughly constant in width when plotted against the logarithm of orbital distance. In actual distance units, the HZ around an M star is quite narrow compared to the Sun's HZ, whereas the HZ around an F star is quite large. This caused Huang (1959) to conclude that the chance of finding habitable planets around M stars was small. But such a conclusion seems premature (or it may be correct for other reasons – see Sec. 15.1.5.2 below). When expressed in terms of log distance, there is just as much habitable space around an M star as there is around a G star. Whether one can populate this space with Earth-like planets remains to be seen.

Although Fig. 15.3(a) depicts the conventional habitable zone, it represents that zone only at the time a star enters the main sequence and it is only valid for rapidly rotating, 1-Earth-mass planets. A better way of defining habitable zones is in terms of stellar flux derived from 1-D or 3-D climate models. In particular, it makes much more sense to delineate the HZ boundaries in terms of stellar fluxes than in terms of planetary effective temperatures, which some astronomers have done. Calculating an effective temperature for a planet requires that one assume a value for its Bond albedo (eqs. (2.15) and (2.20)). But the albedo of a planet depends both on the composition of its atmosphere, which should be quite different near the two habitable zone boundaries, and on the stellar spectrum. It is not safe to assume that a planet's albedo is near Earth's value of ~0.3 (Table 2.2). Thus, defining the habitable zone in terms of stellar flux is better.

Kopparapu *et al.* (2013, 2014) have derived a useful parameterization from 1-D climate modeling results. This parameterization is now being continually updated, including using results from 3-D models.[3] A fit was performed as follows: Theoretical stellar spectra (non-blackbody) were used. Then, HZ boundaries for our Solar System were defined in terms of effective solar flux, $S_{eff} = S/S_0$. Correction factors for stars of other types are expressed in terms of their effective radiating temperature, T_*.

---

[3] Currently, a stellar flux HZ calculator is available here: depts. washington.edu/naivpl/content/hz-calculator

$$S_{eff} = S_{eff}^0 + aT_* + bT_*^2 + cT_*^3 + dT_*^4 \qquad (15.1)$$

Here, S_{eff}^0 is the HZ boundary for our Sun. The "recent Venus" and "early Mars" limits are estimated by making the fluxes proportional to those for water loss and the maximum greenhouse, respectively. The corresponding orbital distances can then be obtained by applying the inverse square law, where L is the stellar luminosity.

$$r = 1 \, \text{AU} \cdot \left(\frac{L/L_S}{S_{eff}} \right)^{0.5} \qquad (15.2)$$

Some results are obvious: The habitable zone moves inward for M stars and outward for F stars because of their vastly different luminosities (Fig. 15.3(a)). But there are also subtle differences. When expressed in terms of S_{eff}, the boundaries shift inward by an additional 10%–30% for F stars because their relatively bluer radiation is more effectively scattered (and more poorly absorbed) by a planet's atmosphere, thereby raising the planet's albedo. Just the opposite happens for M stars: the HZ boundaries shift outward because the planet's albedo is lowered (Fig. 15.3(b)).

Three-dimensional (3-D) climate models generate alternative boundaries. Leconte *et al.* (2013) considered rapidly rotating planets orbiting a Sun-like star, as discussed previously. Yang *et al.* (2013) have used a different 3-D climate model (the NCAR CCSM3) to look at synchronously rotating planets around late K and M stars. As discussed further below, these are planets on which tidal forces are sufficiently strong to slow the planet's rotation so that one side permanently faces the star as the Moon does to the Earth. Yang *et al.* find that the sunlit side of such planets should be permanently cloud-covered, greatly increasing the planet's albedo, and allowing $S_{eff} \cong 1.85$ for the inner edge of the HZ. Cloud cover is enhanced on slowly rotating planets where long daytime illumination and a weak Coriolis force promote strong convergence in the substellar area (Yang *et al.*, 2014a). So, even planets inside the *Dune* planet limit could remain habitable if they are tidally locked. These limits derived from 3-D models are incorporated in the version of the habitable zone shown in Fig. 15.3(b). The inner edge depends on planet mass, for reasons discussed in Sec. 13.4.2 (it will move inwards with bigger mass), and it depends as well on the planet's assumed rotation rate.

Tidal locking is also affected by atmospheric tides, which were described in Sec. 4.4.3. Some planets within the HZs of M and late K stars should *not* rotate synchronously because their rotation is affected by atmospheric thermal tides (Cunha *et al.*, 2015; Laskar and Correia, 2004; Leconte *et al.*, 2015). Stellar heating creates a bulge in atmospheres, which is not aligned with the sub-stellar point because of thermal inertia (Fig. 4.31). The star's gravitation acts on the pressure bulge and accelerates the atmosphere; then frictional coupling to the surface produces a torque on the planet that can prevent synchronous rotation. The torque is stronger on planets with thick atmospheres. Planets near the outer edge of the HZ that build up dense, CO_2-rich atmospheres could have strong thermal tides and rotate non-synchronously, whereas planets near the inner edge with thin atmospheres and weak thermal tides should rotate synchronously. The inner edge of the HZ could thus still be close to the star because of cloud feedback, as shown by Yang *et al.* (2013), but some planets towards the middle or outer parts of K- and M-star HZs may, in theory, be rotating more rapidly.

15.1.5.1 *Problems for Planets Orbiting Early-Type Stars*

Planets orbiting stars that are significantly more or less massive than the Sun face a variety of problems that may affect their habitability. Early-type stars have short main sequence lifetimes, and emit large UV fluxes. The first problem makes the O and B stars, and many of the A stars as well, not very interesting for astrobiology. The F stars, though, are a different matter. The main sequence lifetime of an F0 star is about 2 billion years. This is enough time for life to originate and evolve, as we know from our own planet's history. It may not be enough time to develop complex, multicellular life, based on our experience here on Earth (e.g., Catling (2005)), but that is a separate issue.

High stellar UV may appear detrimental to life, but it is not necessarily so. The problem is illustrated in Fig. 15.2(b). A planet at 1-AU equivalent distance around an F star would receive around four times as much UV radiation as does Earth at wavelengths less than 315 nm (see Table 15.1). This wavelength region includes biologically damaging UVB (290–320 nm) and UVC radiation (100–290 nm). Relative fluxes at 250 nm, where absorption by DNA peaks, are even higher – a factor of 10 or more. For a planet like the early Earth, that lacks an ozone layer, this could lead to higher rates of mutation for near-surface life. To astronomers, this appeared to be a serious problem (Sagan, 1973) but not to biologists, because organisms can avoid UV damage by forming mats, such as the stromatolites mentioned in Chapters 9 and 10 (Margulis *et al.*, 1976; Rambler and Margulis, 1980). If a mat-forming strategy worked on Earth, perhaps it would work on F-star planets as well.

Table 15.1 Relative values of UV flux (< 315 nm), ozone, and DNA dose rate on Earth-like planets around different stars

Stellar type	Incident UV flux	Ozone column depth	Surface UV flux	Relative dose rate[*]
G2 (Sun)	1	1	1	1
K2	0.26	0.79	0.43	0.5
F2	3.7	1.87	0.68	0.38

[*] Dose rate for DNA damage.
All values from Segura *et al.* (2003).

Furthermore, for F-star planets with O_2-rich atmospheres, a thick ozone layer should develop and provide UV shielding (Segura *et al.*, 2003). So, if life on such planets could make it through the anoxic–oxic transition, the stellar UV should cease to be an issue.

15.1.5.2 *Problems for Planets Orbiting Late-Type Stars*

Planets orbiting late-type stars face a completely different set of problems related to habitability. One issue that was recognized very early (Dole, 1964) and was discussed by Kasting *et al.* (1993b),[4] is the tidal locking problem (see the dashed curves in Figs. 15.3(a) and (b)). The HZ for an M star, and for a late K star as well, lies within the *tidal locking radius* of the star. This problem was discussed already in the previous section because it affects the HZ boundaries. But it can also affect a planet's climate. The potential danger with a tidally locked planet is that the planet's atmosphere and oceans could freeze out to form a giant ice cap on the dark side, thereby rendering the entire planet uninhabitable. Fortunately, this particular problem can be circumvented in several ways. One of these is illustrated by the planet Mercury in our own Solar System. As can be seen from Fig. 15.3(a), Mercury is within the Sun's tidal locking radius, yet it does not rotate synchronously. Instead, it spins three times on its axis for every two times it orbits the Sun because it is in a *spin–orbit resonance*. This probably happened because Mercury's mass distribution is slightly non-spherical as a consequence of violent, large impacts that occurred during its formation process, and because Mercury's orbit is highly eccentric ($e \cong 0.21$). Consequently, the energy associated with Mercury's rotation is lowest when its long

axis (the one associated with the lowest moment of inertia) is aligned with its radius vector at perihelion. Other close-in planets in highly eccentric orbits may also avoid synchronous rotation in a similar manner.

Another way out of the tidal locking problem is atmospheric or oceanic advection of heat. If a synchronously rotating planet has an atmosphere with at least 30 mbar of CO_2 – about 100 times the amount in Earth's present atmosphere – 3-D climate simulations show that the atmosphere transports sufficient heat from the dayside to the nightside to prevent the atmosphere from freezing out (Joshi, 2003; Joshi *et al.*, 1997). High CO_2 concentrations facilitate such heat transfer by lengthening the time required to radiate heat off to space. Similarly, if a planet has a deep ocean, like Earth, then ocean currents can also carry heat from the dayside to the nightside (Edson *et al.*, 2011; Merlis and Schneider, 2011). Pierrehumbert (2011) also identifies an alternative *eyeball Earth* state in which the front side of the planet is warm while the back side is cold and frozen. So, from a climatic standpoint, M-stars should not be excluded as candidates for harboring habitable planets.

Several other potential problems for M-star planets exist, though, and these may be more serious than the tidal locking issue. One problem concerns the ability of planets to retain an atmosphere. M stars have much more flare activity than does our Sun, and they have correspondingly enhanced stellar winds. Furthermore, tidally locked planets around all but the least massive M stars should be rotating fairly slowly, 10–100 days, and hence might be unable to generate strong magnetic fields. In such a situation, a planet's atmosphere could conceivably be stripped away by the intense stellar wind (Lammer *et al.*, 2007). Other problems with M-star planets could result from differences in the nebular environment in which they form (Lissauer, 2007). The habitable zone around an M star is very close in; hence, the amount of material available to be swept up by a growing planet is relatively small, and so the planets that form there may be significantly less massive than Earth. Furthermore,

---

[4] Dobrovolskis (2009, p. 9) has pointed out that the tidal despinning rate from equation (8) of Peale (1977) is a factor of two too fast, while the corresponding timescale from his equation (9) is a factor of two too short. Thus, the tidal locking distance should be closer by a factor of $0.5^{1/6} \sim 0.8909$, and the coefficient in equation (10) of Kasting *et al.* (1993b) should be 0.024 instead of 0.027.

because the orbital times at these distances are short, accretion of planets should occur rapidly, giving the nebula less time to cool. This means that terrestrial planets, as they form, would be further removed from the *snow line* – the distance where icy planetesimals can form (Sec. 6.2.1). Delivery of water could also be inhibited by the apparent dearth of Jovian-size planets orbiting M stars (Bonfils *et al.*, 2013), which should result in reduced radial mixing of water-rich planetesimals from outer planetary systems. The tight orbits within M-star habitable zones also mean that the planetesimal relative velocities would have been high, and so impacts would have been even more violent than those that formed the Earth. Such energetic impacts may have swept away, or eroded, more atmosphere than they delivered (Catling and Zahnle, 2013; Melosh and Vickery, 1989).

The biggest problem for M-star habitability, though, is probably that the pre-main sequence luminosity of M stars is high and can cause a runaway greenhouse on M-star planets (Luger and Barnes, 2015; Ramirez and Kaltenegger, 2014). The luminosity, of course, comes from gravitational energy released by the collapsing protostellar cloud. The Sun was about twice as bright during its pre-main sequence phase as it was later on (Baraffe *et al.*, 1998; Baraffe *et al.*, 2002) (illustrated in Fig. 6.5). The corresponding ratio for an M star is 20–180, with the higher values applying to the later (smaller) M stars. The formation time for an M star is also slow – a few hundred million years, as compared with less than 50 million years for the Sun (Luger and Barnes, 2015). At the same time, the expected accretion time for a planet within the HZ of an M star is shorter than for our Sun because the orbital periods are shorter. Earth is thought to have accreted in 10–100 million years, whereas the corresponding time for a planet accreting in situ in the HZ of an M star is ~a few million years (Lissauer, 2007). Thus, whereas the latter part of Earth's accretion occurred *after* the Sun had reached the main sequence, the same is not true for M-star planets. A planet forming *in situ* within the HZ of an M star would likely suffer a runaway greenhouse and lose its water at that time. If this water was not later replenished, then the planet might never be habitable. The best bet for creating a habitable M-star planet might thus be to form the planet well outside of the HZ, beyond the ice-line of the stellar nebula where it would accrete lots of volatiles, then migrate it into the HZ (Luger *et al.*, 2015). If the planet starts with an overabundance of water, perhaps it could lose a large percentage of it and still retain enough water to be habitable. Alternatively, an M-star *Dune* planet might avoid a runaway.

Although none of the above points are necessarily show-stoppers, M stars appear to have more issues working against their habitability than planets around F, G, and early K stars. Eventually, exoplanet data will help determine the validity of this idea. Certainly, over the past 50 years, the suggestion that one should focus primarily on F, G, and K stars has been made consistently (Dole, 1964; Huang, 1960; Kasting *et al.*, 1993b). Fortunately, ~20% of stars in the solar neighborhood have spectral classifications between F0 and K5, so this limitation is not too restrictive. On the other hand, most of the closest stars are M stars, and the small sizes and even smaller luminosities of these stars also make it easier to search for and characterize Earth-like planets using transit observations, as discussed further below. So, for the next few years, M-star exoplanets are likely to remain high-priority targets.

15.1.5.3 *Limit Cycling in the Outer Habitable Zone*
While planets in the outer parts of a HZ may not be able to maintain stable, warm climates, they may instead oscillate between short-lived warm periods and longer periods of global glaciation (Haqq-Misra *et al.*, 2016; Kadoya and Tajika, 2014; Menou, 2015). This behavior is referred to as *limit cycling*. On such planets, CO_2 is consumed by weathering during the warm periods faster than it can be resupplied by volcanism. The resulting, time-dependent behavior depends on the planet's volcanic outgassing rate of CO_2, the availability of water to form snow and ice, and on the type of host star. Planets around F and early-G stars are more prone to limit cycling than are planets around K and M stars because the stellar radiation is shifted towards the blue, where the albedo of water ice is higher (Joshi and Haberle, 2012; Shields *et al.*, 2013; Warren *et al.*, 2002). This type of climate behavior may or may not pose problems for simple (unicellular) life, as we know that simple life was able to make it through repeated Snowball Earth episodes on Earth (Sec. 11.10.5). But complex, multicellular life – in particular, animal life on the continents – would be challenged (Haqq-Misra *et al.*, 2016). Humans are a subset of such life, of course, and so this may imply that other forms of land based, intelligent life are limited to the inner region of the HZ around hotter, bluer stars. Earth is in the inner part of the Sun's HZ (Fig. 15.3), and perhaps we owe our own existence partly to this fortuitous circumstance.

15.1.6 Other Concepts of the Habitable Zone
There is obviously more to habitability than the habitable zone because planetary properties matter. An extension of the habitable zone concept is to take the properties of a

planet into account (Franck *et al.*, 2000a, b; Franck *et al.*, 1999). For example, different continental growth rates lead to differences in CO_2 uptake by silicate weathering, and so the greenhouse effect varies in a time-dependent manner in the calculations of Franck *et al.* Thus, the width of the HZ varies with time for different assumed planetary properties. Similarly, the length of time that a planet can remain habitable depends on its distance from its parent star. For an Earth-like planet orbiting a star like the Sun, Franck *et al.* calculate an optimal orbital distance of 1.08 AU. To use these types of calculations, it is necessary to have lots of information about the planet being considered. But if we already know that much about the planet, then we shouldn't need to predict whether or not it is in the HZ. So, it is not clear that these extensions of the HZ concept are practically useful.

Another extension of the habitable zone concept is for planets with significant tidal heating (Barnes *et al.*, 2009; Barnes *et al.*, 2013; Barnes *et al.*, 2010; Jackson *et al.*, 2008a; Jackson *et al.*, 2008b; Jackson *et al.*, 2010). The conventional (insolation-determined) HZ for late-K and M stars lies within their tidal locking radius. Planets that are beyond this orbital distance but for which tidal drag is strong have an internal heat source that can add to the stellar insolation and widen the HZ. Such planets must be on eccentric orbits to generate this tidal heating, and their eccentricities must be forced by other planets within the system; otherwise, their orbits would rapidly circularize. This factor should be included when studying planets orbiting such late-type stars.

15.1.7 The Galactic Habitable Zone

There may be a region around the center of our galaxy that is optimal for finding habitable planetary systems, which is analogous to a stellar habitable zone. The name given to this concept is the *galactic habitable zone*, or GHZ (Gonzalez *et al.*, 2001; Ward, 2000). The idea can be elaborated to include time as well as space (Lineweaver *et al.*, 2004), but, in essence, the concept is that not all stars in the Milky Way are equally likely to harbor habitable planets. Planets orbiting stars that are too close to the center of the galaxy could have their orbits perturbed by close stellar encounters, and they are also more likely to experience catastrophic events such as nearby supernovae and gamma ray bursts. Stars too far out towards the rim of the Milky Way spiral galaxy are less metal-rich than the Sun and may be less likely to be accompanied by rocky planets. (Recall that a "metal" to an astronomer is any element heavier than hydrogen and helium.) Similarly, stars that form too early in the history of the galaxy are likely to be metal-poor, because not enough hydrogen and helium will have been reprocessed through stars to form the heavy elements. In contrast, our Sun formed at 4.6 Ga and is located $27\,200\pm1100$ light years from the galactic center (Gillessen *et al.*, 2009) – about half of the radius of the Milky Way, whose stellar disc is considered to be $100\,000$–$130\,000$ light years in diameter (e.g., Schneider, 2014).

The spatial GHZ concept is challenged by recent simulations showing that stars migrate within the galaxy as a result of scattering off of spiral arms (Roskar *et al.*, 2008). Thus, habitable planets could theoretically exist anywhere. In any case, this concept is more relevant to the distant future of galactic exploration than it is to the near-term search for habitable planets. For practical reasons, the exoplanet systems that we hope to study within the foreseeable future are all located relatively nearby, within ~50 light years. The host stars appear to have roughly the same metal content as the Sun, provided that one restricts the comparison to F–G–K stars (Boone *et al.*, 2006). (Some early work suggested otherwise, but it included M stars, which tend to be older on average, and hence less metal-rich (Gonzalez, 1999).) Nearby stars are also subject to the same background level of supernovae and gamma ray bursts that the Solar System experiences. So, even if the potentially habitable area of the galaxy is indeed limited, this should have little effect on the probability of finding habitable planets in our stellar neighborhood.

15.2 Finding Planets Around Other Stars

Before examining exoplanets to see if they harbor life, we must find suitable exoplanets. Here we give an overview of exoplanet detection methods. More technical detail is given in the review by Wright and Gaudi (2013).

Astronomers have been searching for exoplanets for a long time. In the 1960s to 1980s, Peter van de Kamp thought he had detected a wobble in the position of Barnard's star as it moved along its track, i.e., deviations of the star's *proper motion*, which is movement relative to the Sun. Barnard's star is an M star six light years away, which is the closest individual star after the triple α-Centauri system. At the end of a series of papers, van de Kamp concluded that the perturbations were the effect of two gas giants orbiting the star with periods of 12 and 20 years (van de Kamp, 1969, 1975; van de Kamp, 1982). The data were photographic plates recorded between 1938 and 1981. However, subsequently systematic errors were found, including shifts when the telescope lens was adjusted. Further observations, using a variety of methods, show no evidence for large planets orbiting Barnard's star (Benedict *et al.*, 1999; Choi *et al.*, 2013; Dieterich *et al.*, 2012; Kurster *et al.*, 2003).

15.2.1 The Astrometric Method

Van de Kamp's technique was an example of the *astrometric method*. This finds exoplanets by looking for perturbations in the motion of a star, which occurs because the star moves around a common center of mass with the exoplanets. Although the modern astrometric method uses CCDs (charge-coupled devices) with high spatial resolution, ground-based astrometry still has limited accuracy – generally not enough to find planets.[5] The problem is that the Earth's atmosphere *scintillates*, or twinkles, because of the disruption of the wavefronts coming from stars by atmospheric turbulence. Some of this interference can be removed by *adaptive optics*, which are techniques to sense and correct the distortions caused by turbulence. However, it remains challenging to measure stellar positions accurately enough to find exoplanets.

An obvious solution to atmospheric effects is to do astrometry from space. ESA's Gaia Mission, which has operated since 2014, measures stellar positions to an accuracy enough to detect Jovian plantes orbiting Sun-like stars. NASA's proposed Space Interferometry Mission, SIM (later "SIM Lite"), would have done even better. The estimated error in a single angular measurement from SIM would have been ~1 μas (micro-arcsecond) (Unwin *et al.*, 2008). Earth's motion around the Sun causes the Sun's position to wobble by ~0.3 μas, as viewed from a distance of 10 pc. This follows from the definition of a parsec, and from the ratio of Earth's mass to the Sun's mass, $M_\oplus/M_\odot \simeq 3\times10^{-6}$. One parsec is the distance at which the Earth's orbital radius, 1 AU, subtends an angle of 1 arcsec. Viewed from 10 pc, Earth's orbital radius would subtend one-tenth of this angle, or 0.1 arcsec. If we ignore the effects of the other planets, Earth and the Sun would orbit at distances $r_\oplus$ and $r_\odot$ from a common barycenter defined by

$$M_\oplus r_\oplus - M_\odot r_\odot \qquad (15.3)$$

As Earth orbits the Sun, the Sun moves by a distance $r_\odot = r_\oplus \cdot (M_\oplus/M_\odot) = 3\times10^{-6}$ AU, which would subtend an angle of 0.3 μas, as viewed from 10 pc. Random errors decrease as $1/\sqrt{N}$, where N is the number of measurements. Hence, within ~100 measurements, SIM

Lite should have been capable of detecting an Earth-like planet. Unfortunately, SIM Lite was cancelled in 2010. We will have to find the Earth-like planets using some other method.

15.2.2 The Radial Velocity Method

The planet-finding technique that has proved most successful from ground-based telescopes is the *radial velocity* (*RV*) or *Doppler method* (e.g., Mayor *et al.*, 2014). Instead of measuring the position of the star on the sky, one measures its motion back and forth in the line of sight by looking at the Doppler shifts of multiple spectral absorption lines. The Doppler shift, $\Delta\lambda$, of a line centered at wavelength λ is given by $\Delta\lambda/\lambda \cong v_S/c$, where v_S is the star's radial velocity and c is the speed of light. (See Beaugé *et al.* (2008) for a technical review of RV theory.)

If we consider the Earth moving around the Sun, ignore its eccentricity (which is small), and ignore the other planets, as we did above, conservation of momentum ensures that

$$M_\oplus v_\oplus = M_\odot v_\odot \qquad (15.4)$$

The average velocity of the Earth around the Sun, $v_\oplus$, is about 30 km s^{-1}, or 3×10^4 m s^{-1}; hence, the Sun's maximum radial velocity, assuming an edge-on view of the system, is: $v_\odot = v_\oplus \cdot (M_\oplus/M_\odot) \cong 0.1$ m s^{-1}, or 10 cm s^{-1}. Current RV precision around quiet, bright, Sun-like stars is ~1 m s^{-1}. So, with sufficient measurements it is possible, in principle, to find an Earth-mass planet around a Sun-like star using this technique. But, in practice, it currently takes a very bright star (e.g., α-Centauri), along with lots of time on a very big telescope.

The RV technique works best for massive planets in tight orbits around their parent stars. (By contrast, astrometry works best for massive planets at large distances, because their lever arm is longer.) As such, the first exoplanets to be discovered by the RV method were so-called *hot Jupiters* – which are loosely defined as Jovian gas giant planets with orbital radii less than roughly 0.1 AU (e.g. Perryman, 2014, p. 103).[6] The first such planet to be discovered was 51 Peg b (Mayor and Queloz, 1995). The notation follows that of binary star systems, where the companion (in this case a planet) is designated by the star's name, followed by "b". 51 Peg b has a minimum

---

[5] By 2013, the *Extrasolar Planets Encyclopedia* website reported just one planet candidate identified by PHASES – The Palomar High-precision Astrometric Search for Exoplanet Systems.

[6] Exoplanet descriptions such as "hot Jupiter," "very hot Jupiter," "mini-Neptune," "sub-Neptune," "Super-Earth," and so on, currently have different definitions in the literature, which is still evolving. In this chapter, we use definitions that are close to the current consensus.

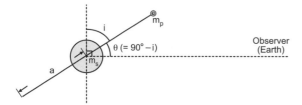

Figure 15.4 Diagram illustrating the geometry of an exoplanet observation. Here, i is the inclination of the planet's orbit with respect to the plane of the sky, and θ (= 90° – i) is the angle of the planet's orbit with respect to the observer on Earth. The star's mass is m_s and the planets mass is m_p. The planet orbits the star at a semi-major axis of a. (From Kasting (2010). Reproduced with permission of Princeton University Press. Copyright 2012.)

mass of 0.44 Jupiter masses and orbits its (Sun-like) star in just over 4 days.

With the RV technique, one cannot measure the planet's mass, m_p, directly because one does not generally know the inclination, i, of the planet's orbit with respect to the plane of the sky (see Fig. 15.4). So, the reported mass is actually $m_p \sin i$. The mass of an ensemble of such RV planets can be estimated statistically from

$$\langle \sin i \rangle = \frac{\int_0^{\pi/2} \sin i \cdot \sin i \, di}{\int_0^{\pi/2} \sin i \, di} = \frac{\frac{1}{2}\int_0^{\pi/2}(1 - \cos 2i)\, di}{1} = \frac{\pi}{4} \cong 0.785 \tag{15.5}$$

Thus, the average exoplanet detected by the RV method is heavier than its measured mass by a factor of $1/0.785 \cong 1.27$. However, applying this factor to any individual planet is risky.

The RV method has been astounding successful, with almost 700 detected exoplanets by late 2016. As the time series lengthens, planets with longer orbital periods emerge from the data. New instruments are also aiming for precision that would allow detection of Earth-mass planets, in principle. For example, the Eschelle SPectrograph for Rocky Exoplanet and Stable Spectroscopic Observations (ESPRESSO) on the European Southern Observatory's Very Large Telescope array aims for a precision of RV variations of a few centimeters per second, compared to Earth's 9 cm s^{-1} effect on the Sun (Pepe *et al.*, 2014). The practical limit of RV measurements may depend on inherent noise levels of the stars themselves. Stars like the Sun have vertical motions on their surfaces of the order of several meters per second, and they also have sunspots that rotate with the star, along with irregular flare activity. So, even with essentially perfect spectral calibration techniques (laser combs) that

are currently being developed, there may well be a mass limit below which planets cannot be detected.

The most famous RV detection is a 1.3 (minimum) Earth mass planet in the HZ of the nearest star, Proxima Centauri (Anglada-Escude *et al.*, 2016). This planet could conceivably be imaged using 30–40 m ground-based telescopes.

15.2.3 The Transit Method and Results from NASA's *Kepler* Mission

Another exoplanet detection technique that has been extremely successful is the *transit method* (Cameron, 2016). When a planet *transits* (passes in front) of its parent star, as seen from Earth, it blocks out some starlight, and can be detected. To give an example from our own Solar System, the diameters of the Sun, Jupiter, and Earth are approximately in the ratio 100:10:1. The projected area of a planet is just πr_p^2, where r_p is the planet's radius. Hence, if Jupiter or Earth were to pass in front of the Sun, as viewed from a great distance, the Sun's brightness would diminish by 1% and 0.01%, respectively. A 1% change in stellar brightness can be readily detected using a ground-based telescope. Indeed, the first such planet detected, the hot Jupiter planet HD209458b, was found from a 1.6% dip in starlight using a small "backyard" telescope equipped with an accurate CCD-based photometer (Charbonneau *et al.*, 2000). The light curve from his measurements is reproduced in Fig. 15.5(a), along with later, precise measurements from the *Hubble Space Telescope* (Brown *et al.*, 2001) (Fig. 15.5(b)).

Charbonneau realized that he could measure such a transit because the planet itself had already been detected by radial velocity searches. It was one of about ten such hot Jupiter planets that were known at the time. The probability that such a planet will transit is approximately equal to the radius of the star divided by the semi-major axis of the planet's orbit. This probability can be computed by integrating the inclination angle, i, weighted by $\sin i$, over the range of angles for which a transit occurs. Equivalently, if we let $\theta = 90° - i$ (see Fig. 15.6) and integrate over angles from face-on, $\theta = 0$, to some actual angle $\theta = \theta_0$, then the probability, P, that a transit will occur is given by this integral divided by an integral over all angles to $\theta = 90°$, i.e.

$$P = \frac{\int_0^{\theta_0} \cos \theta \, d\theta}{\int_0^{\pi/2} \cos \theta \, d\theta} = \sin \theta_0 = \frac{R_S}{a} \tag{15.6}$$

Here, R_S is the radius of the star and a is the semi-major axis of the planet. The calculation is more complicated if

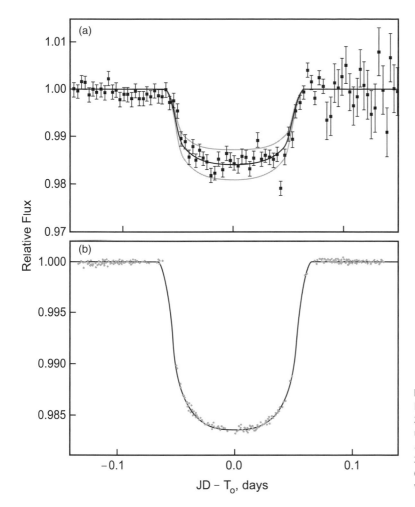

Figure 15.5 (a) Light curve for the star HD 209458. (From Charbonneau *et al.*, 2000.) (b) The same light curve measured a few months later using the *Hubble Space Telescope*. (From Brown *et al.*, 2001). (From Kasting (2010). Reproduced with permission of Princeton University Press. Copyright 2012.)

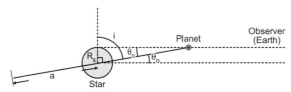

Figure 15.6 Diagram illustrating the geometry of a transiting planet relative to an observer on Earth. Here *a* is the semi-major axis of the planet's orbit, *i* is the orbit's inclination, and R_s is the stellar radius. Here θ (= $90° - i$) is the angle of the planet's orbit with respect to the observer on Earth. (From Kasting (2010). Reproduced with permission of Princeton University Press. Copyright 2012.)

one allows for eccentric orbits, but the results do not change much unless the eccentricity is very high. Now, the radius of the Sun is about 7×10^5 km, while the semi-major axis of a typical hot Jupiter is about 0.05

AU $\cong 0.05 \cdot (1.5 \times 10^8$ km$) = 7.5 \times 10^6$ km. Hence, the probability that a hot Jupiter will transit is ~10%. So, Charbonneau knew that if he observed 10 hot Jupiters each for long enough to see a transit, the odds suggested that he would see one.

Seeing Earth-like planets by the transit method from ground-based telescopes is more difficult. An Earth-like planet passing in front of a Sun-like star would only block one part in 10^4 of the star's light. Because of scintillation, this signal is too small to detect. The chances of seeing an Earth-like planet transit are also much smaller. For Earth around the Sun, the probability is $R_S/(1$ AU$) \cong 7 \times 10^5$ km/1.5×10^8 km = 5×10^{-3}, or 0.5%. Hence, one would need to observe ~200 such systems in order to find one transiting Earth. Consequently, ground-based surveys, such as *MEarth*, have focused on the easier task of searching for short-orbit planets around nearby M-dwarfs, which are smaller, dimmer stars (Berta *et al.*, 2012). This has discovered excellent candidates for follow-up detailed study, such as a $1.2R_\oplus$ Super-Venus 12 parsecs away (Berta-Thompson *et al.*, 2015).

Fortunately, there is a way to observe transits of Earth-size planets. NASA's *Kepler* space telescope monitored the brightness of more than 150 000 stars in the region of the Cygnus and Lyra constellations for almost four years, starting in June, 2009. *Kepler* is capable of measuring stellar brightness to one part in 10^5; hence, it is able to find Earth-sized planets, particularly around stars that are smaller than the Sun. (G stars, it turns out, are more difficult to study than anticipated, because most of them are noisier (have more spots) than the Sun.) By mid-2015, the analysis of *Kepler* data had revealed over 3900 unconfirmed planet candidates and over 1000 confirmed planets orbiting more than 2000 stars. If a planetary size range of 0.75–$2.5R_\oplus$ is considered, the occurrence rate of planets per star for *Kepler* GK stars is 0.77 with an range of 0.3–1.9 (Christopher *et al.*, 2015). Basically, GK stars on average have planets.

Some *Kepler* planets orbit within the habitable zones of their parent stars, allowing statistical inferences to be extrapolated to the rest of the galaxy (albeit with the caution of residual sample bias even after attempts have been made to correct for bias). A parameter of interest is $\eta_\oplus$, which has slightly varying definitions but is generally described as *the frequency of Earth-size planets in a circumstellar habitable zone*. One needs to pay attention to what authors mean by "Earth-size" and which boundaries they choose for the HZ in order to compare different estimates of $\eta_\oplus$. For example, assumptions of "Earth-size" include 0.5–1.4 $R_\oplus$ (Kopparapu *et al.*, 2013), 0.5–2 $R_\oplus$ (Silburt *et al.*, 2015) or 1–2 $R_\oplus$ (Petigura *et al.*, 2013). Observations where radius and mass are both available, so that mean density can be deduced, suggest that the upper size limit for rocky planets is 1.5–$1.75R_\oplus$ (Lopez and Fortney, 2014; Rogers, 2015; Weiss and Marcy, 2014). Meanwhile, 0.5 $R_\oplus$ (the size of Mars) is a reasonable lower limit for an Earth-size planet, as planets much smaller than that would have trouble holding onto their atmospheres. So, to be conservative, we suggest 0.5–1.5 $R_\oplus$ as the appropriate size range to define $\eta_\oplus$. The range of spectral types of star is also another parameter because one defines $\eta_\oplus$ for a particular stellar class.

Interestingly, $\eta_\oplus$ is perhaps not greatly different for G, K, or M stars. Early estimates indicated that $\eta_\oplus$ for M stars is between 0.4 and 0.6 (Gaidos, 2013; Kopparapu *et al.*, 2013). Bonfils *et al.* (2013) also estimated $\eta_\oplus =$ $0.41^{+0.54}_{-0.13}$ for M stars using data from the High Accuracy Radial velocity Planet Searcher (HARPS) instrument on the 3.6 m telescope at La Silla, Chile. However, both of these estimates were made before the upper size limit on rocky planets had been estimated. When 1.5 $R_\oplus$ is taken as that upper limit and we use the conservative estimate for the HZ discussed in Sec. 15.1.3, $\eta_\oplus$ for M stars is $0.16^{+0.17}_{-0.07}$ (Dressing and Charbonneau, 2015). More massive stars appear to have a similar $\eta_\oplus$ value. Although Petigura *et al.* (2013) estimated 0.22 ± 0.08 for K and G stars, they used a HZ inner edge of 0.5 AU (or its stellar flux equivalent), following Zsom *et al.* (2013), along with an outer edge of 2.0 AU. As discussed earlier in this chapter, the first of these two limits is almost certainly too close to the star. A better estimate for the HZ inner edge is 0.95 AU (Leconte *et al.*, 2013). Thus, in terms of log distance, the HZ is only about half as wide as Petigura *et al.* assumed, and their estimate for $\eta_\oplus$ should be reduced by this same factor, putting it at 0.11 ± 0.04. Results from the *Kepler* team suggest GK $\eta_\oplus$ values that are in this same range (N. Batalha, private communication).

A mean density that indicates whether an exoplanet is rocky or gaseous can sometimes be obtained using just the transit method alone. In multi-planet systems, gravitational perturbations of one exoplanet on another cause changes in the transit times of the transiting exoplanet, even if the perturbing planet is non-transiting. This is the method of *transit timing variations* (or TTVs) (Agol *et al.*, 2005; Holman and Murray, 2005; Sam and Yoram, 2014) (covered as Ch. 7 of Haswell (2010)). Such mean density information is not only critical for planet mass–radius relationships but also our understanding of how planetary systems form.

15.2.4 Gravitational Microlensing

A fourth technique for finding planets is *gravitational microlensing*. Light is bent as it passes around a star, a planet, or a galaxy. Equivalently, spacetime is curved by massive objects, and light follows a straight line in curved space. This general relativistic effect is easiest to discern over large distances; hence, *gravitational lensing* has been used for some time to study distant galaxies and clusters of galaxies. Only more recently has this technique been applied to stars and planets.

Microlensing is a regime of gravitational lensing where the additional light bent forward to an observer brightens the source but without multiple images or distortion of the source image that happens in the two other cases of lensing, called *strong* and *weak* lensing, respectively. In microlensing, when a planet orbits a lens star, the background star brightens more than once because of the lens star and its planet. The technique provides no direct information about planetary atmospheres or nearby planets, so we will not dwell on it here. Space-based microlensing can, however, provide statistics, and potentially an independent estimate of $\eta_\oplus$. Surveys suggest every star has $1.6^{+0.72}_{-0.89}$ planets in the $5M_\oplus$ to $10M_J$ range orbiting at 0.5-10 AU (Cassan *et al.*, 2012); also ~1.8 Jupiter-mass planets per star exist towards the galactic bulge (Sumi *et al.*, 2011).

Microlensing is a planned component of NASA's Wide Field Infrared Telescope (WFIRST), which is currently a top priority mission for launch in the 2020s (Spergel *et al.*, 2015). So, we should obtain statistical information about Earth-like planets and outer planetary systems within the next 10–15 years. In addition, WFIRST's baseline design has a coronagraph (see below) to allow direct imaging of exoplanets, which we now discuss.

15.2.5 Direct Detection Methods: *Terrestrial Planet Finder* (TPF) and *Darwin*

All of the exoplanet detection methods discussed so far are termed *indirect methods* because they look for light emitted by the star itself (or by another star behind it). Astrobiologists are most interested in *direct detection methods* for finding exoplanets because such methods can potentially provide detailed spectra that may be used to characterize their atmospheres and surfaces. (The transit method can yield planetary spectra, as well, as discussed in Sec. 15.3.1.) As its name suggests, direct detection means looking directly for light from the planet itself.

Direct detection is difficult for several reasons. For one, planets are assumed to be embedded within a cloud of exozodiacal dust, analogous to the zodiacal dust in our own Solar System, which comes from the collisions of asteroids and comets. Resolving the planet in the midst of this background requires a large telescope (> 4 m for a dust cloud similar to our own). Furthermore, the planet being observed is very close to a much brighter star, and so one needs to be able to separate the planet's image from that of the star and to block out any diffracted light from the star that might otherwise obscure the planet. This can also place lower limits on telescope size, as discussed further below.

Direct detection has already been done using both large telescopes on the ground and the *Hubble Space Telescope*. So far the imaged planets are all massive planets, Jupiter-sized or larger, with semi-major axes

ranging from tens to hundreds of AU (e.g. Delorme *et al.*, 2013; Kuzuhara *et al.*, 2013; Lagrange *et al.*, 2010; Macintosh *et al.*, 2015; Rameau *et al.*, 2013). To see planets, the telescope must have a *coronagraph*, an instrument that can block out the light from the star and retain the light from the planets around it. The term "coronagraph" comes from attachments that were originally used to block out the light from the Sun in order to expose the tenuous corona surrounding it. Alternatively, the telescope could be flown in combination with an external occulter, or *starshade*, that blocks out the star's light before it enters the telescope (see below).

Finding Earth-like planets by direct imaging will require either large (≥ 4 m) telescopes in space or extremely large telescopes on the ground. Several 30 m-plus ground-based telescopes are being considered, however (Kasper *et al.*, 2010; Matsuo and Tamura, 2010). These may or may not be able to image planets like Earth, depending on how well they are able to remove the effects of the atmosphere. To search for spectral biosignatures (e.g., O_2, O_3, CH_4, N_2O) from the ground one must look through an atmosphere that contains all of these gases and high dispersion spectroscopy has been proposed for detecting biogenic O_2 on Earth twins around M stars (Snellen *et al.*, 2013; Rodler and Lopez-Morales, 2014). But the best hope for characterizing Earth-like planets and for looking for signs of life comes from space-based telescopes.

NASA and ESA (the European Space Agency) have been developing space mission concepts to achieve this goal. One idea is to observe in the thermal-IR, where the *contrast ratio* between the planet and the star is more

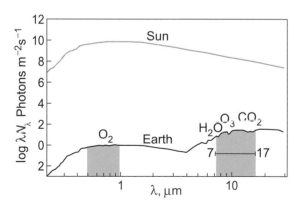

Figure 15.7 Diagram illustrating how the flux of photons from the Earth and Sun would appear if viewed from a distance of 10 pc. The shaded area is the wavelength region that would be observed by a telescope operating in the visible/near-IR (at left) or in the thermal-IR (at right). (From Kasting (2010). Reproduced with permission of Princeton University Press. Copyright 2012.)

favorable (Fig. 15.7). The Earth, for example, is 10^{10} times dimmer than the Sun in the mid-visible, but only 10^7 times dimmer in the thermal-IR. This arises from evaluating the Planck function (eq. (2.43)) in the visible and thermal-IR at emission temperatures of 255 K and 5780 K for the Earth and Sun, respectively, and also accounting for the two bodies' different size.

Apart from contrast ratio, another effect of wavelength must be considered, which is diffraction. The angular resolution, θ (in radians) of a telescope obeys

$$\theta \geq 1.22 \frac{\lambda}{D} \tag{15.7}$$

Here, λ is the wavelength at which the observations are being made, and D is the diameter of the telescope mirror (or lens). Equality applies in eq. (15.7) at the *diffraction limit*, which is the smallest size (a so-called *Airy disk*) to which an optical system can focus. As mentioned before, a planet orbiting at 1 AU from star located at 10 pc distance forms an angle of ~0.1 arcsec or ~5×10^{-7} radians. For observations in the visible to near-IR, i.e., out to 1 μm = 1×10^{-6} m, we have $D = 1.22\lambda/\theta \approx 2$ m. Real telescopes involving coronagraphs operate at some multiple of this limit, typically 4 times, requiring a mirror diameter ≥ 8 m. Such large telescope missions have been advocated, e.g., a proposed *High Definition Space Telescope* (HDST) (Dalcanton *et al.*, 2015). Turned around, for a given mirror size, the smallest angle from a star at which an exoplanet can be detected is known as the *inner working angle*. In the thermal-IR, a wavelength of 10 μm is ~10 times longer, and so, according to eq. (15.7), you need an 80-m telescope. An 80 m mirror is too big for a single dish space telescope because it wouldn't fit in a launch vehicle, so thermal-IR designs for an Earth-finding telescope combine multiple detectors using *interferometry*. Both NASA and ESA have studied this concept. TPF-I (*Terrestrial Planet Finder – Interferometer*) is a NASA mission concept, and *Darwin* is an ESA mission concept. Both missions would be expensive, and both are currently on indefinite hold.

Doing a direct imaging mission in the visible is considered slightly easier, although this is still costly and no full-scale missions are currently being pursued. However, preliminary studies are anticipated for a (10–12 m) *Large UV-optical-IR telescope* (LUVOIR) and a somewhat smaller (4–8 m) *Habitable Planets Explorer* (HabEx). Some smaller probe-class (<$1B) missions have also been studied (see http://exep.jpl.nasa.gov/).

Visible direct imaging missions come in two varieties, depending on how one blocks the light from the star. One idea is to suppress the starlight internally within the telescope using a coronograph. This idea was studied under the name TPF-C (where the "C" stands for coronograph) (see http://exep.jpl.nasa.gov/TPF-C/tpf-C_index .cfm). The other idea is to use an external occulter, or *starshade*, to block out the light from the star. This technique is similar to what happens during a solar eclipse, when the Moon passes in front of the Sun. To provide adequate starlight suppression, an occulter would typically need to be 50–70 m in diameter, and it would need to fly at a distance of ~50 000 km from the telescope. So, this mission, like the interferometer, requires multiple spacecraft. The demands on the telescope mirror, however, are much less stringent than for the internal coronagraph, so this mission, sometimes called TPF-O (where the "O" stands for occulter), is under serious consideration. One such concept that has been studied in the USA is called *New Worlds Observer* (http://newworlds.color ado.edu/). NASA's smaller, probe-class analogs to these missions are concepts called Exo-C (Stapelfeldt *et al.*, 2015) and Exo-S (Seager *et al.*, 2015), where the "S" stands for "starshade."

15.3 Characterizing Exoplanet Atmospheres and Surfaces

We hope not only to find Earth-like planets, if they exist, but also to characterize their atmospheres and surfaces (see reviews by Crossfield, 2015; Seager, 2010, 2013). After all, if an Earth-mass planet is discovered within the habitable zone of some nearby star, the first question that will be asked is: "Is it habitable?" And the second will be: "Is there any evidence that it is indeed inhabited?"

A number of different characterization techniques can be imagined, ranging from relatively crude to highly detailed. The methods used will depend on what type of stellar system is being investigated and on when the measurements are being made. Today, for most systems, it is not possible to separate out the light from a planet from that of its parent star (except for a few super-Jupiters at large distances from their parent stars, which have already been directly imaged using ground-based telescopes); thus, our options are currently rather limited. Below, we look to the future instead.

15.3.1 The Near Term: Transit Spectra of Planets Around Low-Mass Stars

In the next 10–15 years, our best chance of characterizing habitable planets will be to look at nearby low-mass (late K and M) stars. The habitable zones around such stars are relatively close in, and so the probability that a potentially habitable planet will transit is higher, according to eq. (15.6). The decreased orbital distances within the HZ are offset to some extent by the fact that a star's diameter tends to decrease with its mass, although not as rapidly as its habitable zone shrinks. A NASA mission called TESS (the *Transiting Exoplanet Survey Satellite*), scheduled for launch in 2017, will search for such transiting planets around nearby G and K stars. If such planets can be identified, then techniques that have already been used to study hot Jupiters can be used to characterize their atmospheres and surfaces. Later, an ESA mission, *PLATO* (Planetary Transits and Oscillations of stars) is planned for launch in 2024 to find further rocky exoplanets.

A transit spectrum can be obtained in two ways. When the planet is in front of the star, called *primary transit*, light passes through an annulus of the planet's atmosphere on its way to Earth (Fig. 15.8). By measuring the spectrum of the star when the planet is in front of it, and then subtracting the spectrum of the star when the planet has passed by, one can obtain a spectrum of the planet's atmosphere.

This primary transit allows *transmission spectroscopy*. During the transit, the whole spectrum gets dimmer, but at wavelengths where atmospheric gases absorb, the dimming is bigger, which can give insight into atmospheric composition and structure. The technique was first applied to Solar System planets (Smith and Hunten, 1990). As discussed in Sec.15.2.3, the flux decreases according to the planet : star area ratio as $(r_p/R_s)^2$; so if a planet's atmospheric opacity varies with wavelength, λ, the planet's apparent size changes. Thus, measuring $(r_p(\lambda)/R_s)^2$ gives a transmission spectrum. As the transit decrement is small, high precision is needed. Nonetheless, the technique has been used to study the atmospheres of hot Jupiters (Charbonneau *et al.*, 2002; Lecavalier Etangs *et al.*, 2008), Neptune-mass planets (Knutson *et al.*, 2014), and super-Earths (Bean *et al.*, 2010; Kreidberg *et al.*, 2014).

Clouds and hazes can confound transmission spectroscopy. In the Solar System, *all* thick planetary atmospheres have clouds, while the giant planets, Venus, and Titan have considerable haze. Haze, in particular, obscures deeper layers and, through absorption and multiple scattering, hazes can render transit spectra flat and featureless (Marley *et al.*, 2013; Robinson *et al.*, 2014b; Seager and Sasselov, 2000).

Another technique, which works well for some hot Jupiter-type planets, is called *secondary transit spectroscopy*. In this method, which works best in the thermal infrared where the planet/star contrast ratio is lowest, one first measures the spectrum of the planet + star when the planet is beside the star. Then, one takes another spectrum when the planet the planet is behind the star (secondary transit) and subtracts that one from the first (Fig. 15.9). Such spectroscopy has been performed successfully for hot Jupiter planets using the *Spitzer Space Telescope*, which operates at thermal-infrared wavelengths and has a roughly 1-m diameter mirror (e.g., Agol *et al.*, 2010).

The *James Webb Space Telescope* (JWST), scheduled to launch in 2018, has a 6.5-m mirror, so it just might be

Figure 15.8 Diagram illustrating the technique of primary transit spectroscopy. During transit, some of the light reaching the observer passes through the atmosphere of the planet. The planet blocks more light in a transit at wavelengths at which the atmosphere absorbs. (From Kasting (2010). Reproduced with permission of Princeton University Press. Copyright 2012.)

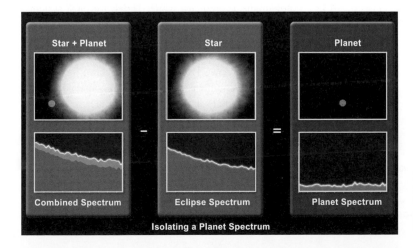

Figure 15.9 Diagram illustrating the technique of secondary transit spectroscopy. (From Kasting (2010). Reproduced with permission of Princeton University Press. Copyright 2012.)

able to obtain a spectrum of an Earth-like planet around a nearby M star (although not for any of the currently known transiting systems). That, of course, would be extremely exciting. We will not dwell on this topic here, though, because it remains to be seen if this can actually be done.

15.3.2 The Future: Direct Detection of Habitable Planets

Much more can be learned once we are able to image planets directly, that is, to separate out their light from that of the parent star. The TPF missions described earlier in this chapter would be able to do this. What can be discovered depends on how much of a planet's orbit can be observed and on how many photons are available. Let's consider the simplest observations and progress to the more complex.

15.3.2.1 *Searching for Liquid Water on a Planet's Surface*

If one can measure the planet's brightness over a significant fraction of its orbit, one can obtain information about its atmosphere and surface. Figure 15.10 illustrates the geometry of the observation for a planet whose orbital plane is aligned with the observer. (Such a planet would

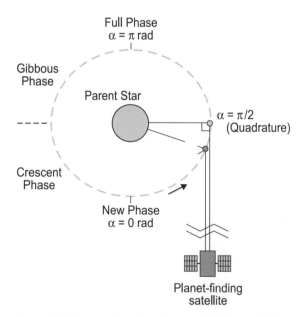

Figure 15.10 Diagram illustrating the geometry of an observation of an exoplanet out of transit. The orbital longitude, α, is defined as 0° when the planet is in front of the star. The phases of the planet are indicated, which are analogous to those we see of the Moon.

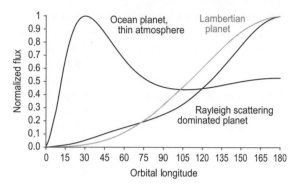

Figure 15.11 Light curves for different types of planets for an edge-on viewing geometry. (From Zugger *et al.* (2010). Reproduced with permission. Copyright 2010, American Astronomical Society.)

transit its star, but in this case we are interested in observing it out of transit.) In practice, it may be difficult to obtain a complete light curve because the planet gets harder to see as it moves closer to the star. Figure 15.11, though, illustrates the kind of information that might be obtained from even a partial orbit. If a planet behaves like a Lambertian sphere, i.e., if it scatters light equally in all directions (see Fig 2.7), then its brightness relative to its parent star is given by the following (Russell, 1916; Sobolev, 1975).

$$C(\alpha) = \frac{2}{3} A \left(\frac{r_p}{a}\right)^2 \left(\frac{\sin \alpha + (\pi - \alpha) \cos \alpha}{\pi}\right) \quad (15.8)$$

Here, A is the planet's Bond albedo, r_p is its radius, a is its orbital distance, and α is the orbital longitude. $C(\alpha)$ is wavelength-dependent, so technically A needs to be defined for the wavelength *range* of interest. At quadrature ($\alpha = \pi/2$), $C = (2/3)A(r/a)^2/\pi$. For an Earth-like planet orbiting a Sun-like star, $A \cong 0.3$ (Palle *et al.*, 2003), $r = 6371$ km, and $a = 1.5 \times 10^8$ km, the contrast ratio C is $\sim 1.15 \times 10^{-10}$. This is why a visible TPF mission is so difficult: the planet is dimmer by a factor of $\sim 10^{10}$ than its parent star.

As one can see from Fig. 15.11, the light curve for a planet depends on its type of atmosphere and surface. A Lambertian planet would appear brightest right before it passed behind the star. (It would, of course, disappear from view somewhat before that.) Raleigh scattering mimics Lambertian scattering, so a planet with a thick atmosphere (and little absorption) would do the same thing. An ocean planet with a thin atmosphere would much different, however: its brightness would peak at an orbital longitude of ~30°. That's because the albedo of a liquid water surface depends strongly on the angle of the incident starlight. Light that is vertically incident is strongly absorbed, but light that strikes the surface at a grazing angle is efficiently reflected. A *glint* spot is present where the angle of incidence equals the angle of reflection; it is spatially extended because of the finite size of the host star and the waviness of the liquid surface (Robinson *et al.*, 2010; Sagan *et al.*, 1993b). In principle, polarization could also be used to look for liquid water, but this appears not to be useful unless the planet's atmosphere is exceedingly thin (Zugger *et al.*, 2010).

15.3.2.2 *Using Color to Characterize Exoplanets*

To get a better estimate of an exoplanet's nature, the next step will be to measure its reflectance spectrum in several broad wavelength bands, at a resolution $R \equiv \lambda/\Delta\lambda \cong 5$. Photometric observations can be decomposed into three bands, and a planet can be placed on a color–color diagram, as shown in Fig. 15.12. In Fig. 15.12, the reflectance ratios of Solar System bodies were obtained from photometric observations of the EPOXI emission and other spacecraft with filter bands of 100 nm at the center wavelengths indicated (Crow *et al.*, 2011). (The EPOXI mission was what became of the *Deep Impact* spacecraft once it launched its probe into the comet 9P/Tempel.) In Fig. 15.12, the Earth occupies a special position compared to other Solar System bodies. Rocky planets with little or no atmosphere (Mercury, Moon, Mars) cluster in the red or near-IR corner of the diagram, while methane-rich gas giants (Uranus, Neptune) cluster on the opposite "blue" corner. Earth has a blue up-turn at short

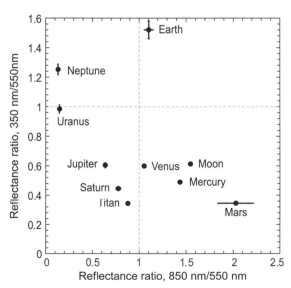

Figure 15.12 Color–color diagram for planets within our own Solar System using filters from the EPOXI mission. (From Crow *et al.* (2011). Reproduced with permission. Copyright 2011, American Astronomical Society.)

wavelengths and so is separated from the other planets. Earth is a *Pale Blue Dot* with a U-shaped spectrum in the visible to near-IR because of high reflectance in the blue (<450 nm) from Rayleigh scattering, weak absorption across the visible from ozone Chappuis bands (450–850 nm), white clouds reflecting across the visible, and an upward slope at 600–900 nm caused by continents and vegetation (Arnold *et al.*, 2002; Robinson *et al.*, 2011; Tinetti *et al.*, 2006).

A study examining optimal photometric bins over 350–1000nm for identifying an exo-Earth shows that it can be difficult to distinguish an Earth from potential false positives. Icy worlds with thick atmospheres are similar in color because of Rayleigh scattering in the blue (Krissansen-Totton *et al.*, 2016b). Consequently, somewhat higher spectral resolution, $R{\sim}10$ is needed. The same study also shows that an early Earth-type planet shrouded in organic haze (an Archean analog) cannot be distinguished by color alone from a Titan-like planet. So, while color could provide some information, ultimately spectra are required to reveal the true nature of exoplanets and remove ambiguities.

15.3.2.3 *Measuring Planetary Rotation Rates, Land–Sea Distributions, and Planetary Obliquities*

Color can be used to derive other planetary characteristics if one can measure how a planet's color changes with time. Ocean surfaces are dark, whereas land surfaces are somewhat reddish, and clouds are gray. So, if one can measure the color variation as the planet rotates, it is possible to learn something about the planet's rotation rate (Oakley and Cash, 2009; Palle *et al.*, 2008) and about land–sea distributions (Cowan *et al.*, 2009).

NASA demonstrated the capability to examine color change with its EPOXI spacecraft. Figure 15.13 shows a map of planet Earth, along with its inferred geography from EPOXI data. All latitudinal information is lost because the planet was viewed as a single pixel, as would be the case for an exoplanet. But the longitudinal distribution of continents is readily apparent from the time series analysis of different color bands. Knowing whether a planet has continents as well as oceans could be valuable in evaluating its ability to harbor complex, or intelligent, life. This technique is limited to relatively bright planets (or very big direct imaging telescopes), however, because the integration time must be substantially shorter than the planet's rotation period.

Finally, if one is able to perform such a rotation rate analysis at a variety of orbital phase angles, one can in

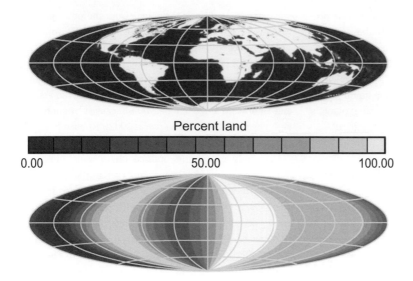

Figure 15.13 Inferred land–sea distribution for Earth, obtained from analysis of the *EPOXI* color time series. (From Cowan *et al.* (2009). Reproduced with permission. Copyright 2009, American Astronomical Society.) (A black and white version of this figure will appear in some formats. For the color version, please refer to the plate section.)

principle determine an exoplanet's obliquity and obtain a 2-D map of continental geometry (Kawahara and Fujii, 2010). We say "in principle" because this probably requires being able to view the planet when it is relatively close to the star, and that, in turn, needs sufficient angular resolution. But even if resolution proved difficult with a first-generation direct imaging mission, it might be possible to do this at a later time by repeating the observation with a bigger space telescope.

15.3.2.4 *Visible/Near-IR Spectra*

Much more information is available if one can obtain a reasonably detailed ($R \cong 70$) spectrum of the planet's atmosphere. A spectrum of the Earth's atmosphere at UV/visible/near-IR wavelengths is shown in Fig. 15.14. This spectrum was taken using "Earthshine" data – light reflected from the dark side of the Moon (from Woolf *et al.*, 2002). These photons originated from the Sun, bounced off the Earth, and then back off the Moon. If one subtracts a spectrum of the bright side of the Moon, one obtains a reflection spectrum of the Earth. In the figure, the curve at the top represents the Earthshine data. The smooth curve running through the data, which matches the "clear sky" curve from lower down, is a model fit to the data. From this fit, one can detect three different gases: O_2, O_3, and H_2O. They are all seen as absorption bands. The brightness within these bands is less than it is elsewhere in the spectrum because the gases are absorbing some of the incident sunlight. O_2 itself has three different absorption bands that can be seen at this

spectral resolution. The brightest of these is the O_2 "A" band at 760 nm. The "A" band is easy to observe – it would span two pixels in a spectrograph at $R = 70$ – and was singled out almost 30 years ago as a possible indicator of life on exoplanets (Owen, 1980). The O_2 "B" band at 690 nm is also easy to pick out, as are the three H_2O absorption bands at 720 nm, 820 nm, and 940 nm.

Ozone has broad absorption bands in the visible (the Chappuis bands) that extend from 450 nm to 850 nm and peak in the yellow-orange at ~600 nm (as mentioned in Sec. 2.4.2). O_3 absorbs even more strongly at shorter UV wavelengths, between 200 and 310 nm (the Huggins and Hartley bands), although that wavelength region is not shown in Fig. 15.14. The O_3 band in the visible might be harder to observe on an exoplanet than are the O_2 bands because it is easily masked by clouds (Segura *et al.*, 2003) and because at wavelengths below 600 nm its effects might be confused with those of Rayleigh scattering (labeled "ray" in the model curves near the bottom of the figure). The Earthshine spectrum in Fig. 15.14 is also very "blue," as shown by the increase in intensity at shorter wavelengths.

Also possibly visible in Fig. 15.14 is the so-called *red edge* of chlorophyll at 700 nm. It is difficult to pick out, however, because it is partially masked by an H_2O absorption band at 720 nm. Plants and algae absorb sunlight effectively shortward of this wavelength and reflect most of the sunlight longward of this wavelength (Kiang *et al.*, 2007a; Kiang *et al.*, 2007b; Sagan *et al.*, 1993b; Seager *et al.*, 2005). The red edge is a questionable biosignature for extraterrestrial life, partly because it

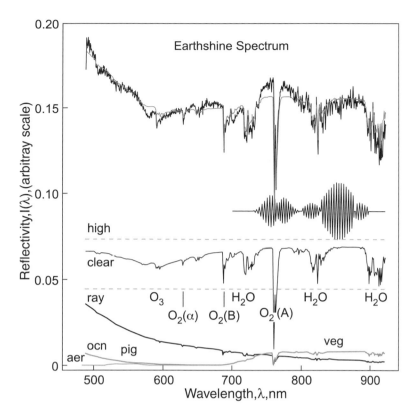

Figure 15.14 Visible/near-IR reflectivity spectrum of the Earth, taken from "Earthshine" data. The wiggly curve at the top shows the reflectivity data, which is the ratio of earthshine to moonshine corrected for phase, but with arbitrary scale. The smooth curve running through it shows a model fit (from seven components below). Five CCD interference fringes (inset and offset to the right) were subtracted from the data. The seven components were: "high" for reflectivity from a high cloud; "clear" for clear atmosphere transmission; "ray" for Rayleigh-scattered light; "veg" for reflected light from vegetated land; "pig" for reflected light from green-pigmented marine phytoplankton; "aer" for aerosol scattered light (negligible); and "ocn" for reflected light from the ocean (From Kasting, (2010). Reproduced with permission of Princeton University Press.) (Originally from Woolf *et al.* (2002). Reproduced with permission from The American Astronomical Society, Copyright 2002.)

is only marginally visible in disk-averaged spectra like Fig. 15.14 and partly because it is not certain that an alien biota would exhibit a spectral signature at this same wavelength. But we would certainly look for if we did obtain the spectrum of an extrasolar Earth.

15.3.2.5 *Thermal-IR Spectra*

Useful biosignatures are also available in the thermal-infrared. Figure 15.15 shows thermal-IR spectra of Venus, Earth, and Mars. Looking first at Venus and Mars, one can see that at this relatively low spectral resolution only a single feature is clearly visible: the 15-μm band of CO_2. This band is created by the ν_2 "bending" mode of the CO_2 molecule, and it is the primary reason why CO_2 is a strong greenhouse gas (see Sec. 2.5.4).

If we look at Earth's spectrum, the 15-μm CO_2 band is also clearly visible, even though Earth's CO_2 concentration is relatively low. Thus, the thermal-IR is an excellent place to look for CO_2. It is also a good way of distinguishing terrestrial (rocky) planets like Venus, Earth, and Mars from gas giant planets like those in the outer Solar System, because the gas giants lack CO_2. The

atmospheres of some hot Jupiters contain measurable amounts of CO_2, as a consequence of rapid photochemistry, but these objects can be readily distinguished from terrestrial planets on the basis of their orbital distances. Earth's thermal-IR spectrum contains additional information. The absorption at short wavelengths (<8 μm) is caused by H_2O, as is the absorption at long wavelengths (>17 μm). The short-wavelength feature is the 6.3-μm rotation-vibration band, and the long-wavelength feature is the H_2O pure rotation band, which extends all the way out to the microwave region where it is used for heating food. So, as in the visible/near-IR, it should be possible to determine whether a planet has abundant water vapor in its atmosphere, but not necessarily on its surface.

Even more interesting is the strong band of O_3 centered at 9.6 μm. This band is clearly visible even though ozone is only a trace constituent of Earth's atmosphere, because the ozone resides mostly in the stratosphere. Ozone is formed photochemically from O_2 (see Sec. 3.3.1) so its presence indicates that the planet must have O_2 in its atmosphere. But, one also finds that O_3 can be detected even if only small amounts of O_2 are present (Leger *et al.*, 1993; Segura *et al.*, 2003). The reasons have

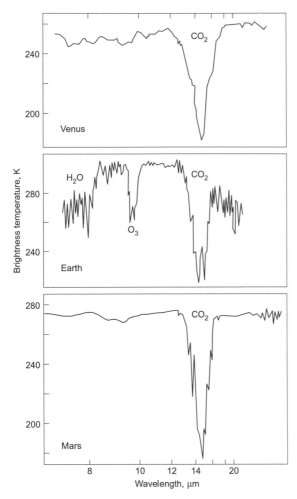

Figure 15.15 Thermal-IR spectra of Venus, Earth, and Mars. (From Kasting (2010). Original figure courtesy of Robert Hanel, NASA GSFC.)

to do partly with the nonlinear nature of the ozone photochemistry and partly with the fact that the stratospheric temperature drops as the amount of ozone decreases, making the absorption feature appear stronger. Thus, O_3 is, in some ways, an even more sensitive indicator of O_2 than is O_2 itself. By comparison, the O_2 "A" band becomes difficult to see below about 10% of Earth's O_2 abundance (Segura et al., 2003).

15.4 Interpretation of Possible Biosignatures

Once a planet is found to be habitable, the next step will be to try to determine if it is actually inhabited. To do this, we will need to look for biosignature gases, some of which (e.g., O_2) have already been mentioned. But the

question of what exactly to look for in a planet's atmosphere and how to interpret the data, once obtained, remains an active area of investigation.

15.4.1 The Criterion of Extreme Thermodynamic Disequilibrium

The idea of looking for remote biosignatures was suggested many decades ago in the context of Solar System exploration (Lederberg, 1965; Lovelock, 1965). Lederberg suggested that "kinetic instability in the context of local physical and chemical conditions" could indicate life. In retrospect, his concern for possible abiotic influences was very wise. Lovelock argued something more specific: we should "search for the presence of compounds in [a] planet's atmosphere which are incompatible on a long-term basis." In fact, Earth's atmosphere–ocean system is in a disequilibrium state because of the presence of biogenic gases.

However, a complication with the idea of chemical disequilibrium as a biosignature is that all planetary atmospheres are in disequilibrium to some degree as a consequence of atmospheric chemistry driven by the free energy of sunlight, particle radiation, and perhaps internal or tidal heat for some exoplanets. Some hypothetical, abiotic, CO-rich atmospheres could even be extremely out of equilibrium. For example, Kasting (1990) pointed out that such an atmosphere could have been produced by impacts on the early Earth. Zahnle et al. (2008) generated CO-rich atmospheres by way of photochemistry in a cold, dense early atmosphere on early Mars. As discussed in Sec. 12.2.3, the underlying reason why high-CO atmospheres are possible has to with a quirk of photochemistry: photolysis of CO_2 is relatively rapid, but recombination of (ground-state) O and CO to reform CO_2 is slow because the reaction is spin-forbidden; consequently, O atoms are more likely to recombine with each other to form O_2. That said, if CO ever was abundant in Earth's early atmosphere, it should have quickly fallen to low concentrations once life evolved. Thus, if anything, the presence of highly disequilibrium CO in a planet's atmosphere should be considered an "anti-biosignature" (Zahnle et al., 2011).

15.4.1.1 The Simultaneous Presence of O_2 and Reduced Gases

Certain types of chemical disequilibrium, however, may be reliable biosignatures. Lovelock later reiterated and developed his original argument by suggesting that the simultaneous presence of O_2 and CH_4 would be a good indication of life (Hitchcock and Lovelock, 1967;

Lovelock, 1975). It is indeed difficult to see how these gases could be present simultaneously in appreciable concentrations without having a biological source for both, as the lifetime of CH_4 in today's O_2-rich atmosphere is only ~12 yr. Although the CH_4–O_2 couple is minor in terms of thermodynamic free energy (see below), kinetics requires a large flux of CH_4 into the atmosphere, and that flux is ~90% biogenic. Conversely, an atmosphere with a large abiogenic CH_4 flux would be unlikely to build up free O_2 because the global redox budget (eq. (8.40)) would be heavily weighted towards the reduced side. Kinetically speaking, not a single molecule of CH_4 should exist in the presence of so much O_2 were it not for the continuous input of CH_4 (as shown in Sec. 3.1.1). So, the CH_4–O_2 couple is important if we think in terms of kinetics and lifetimes rather than available free energy (Krissansen-Totton et al., 2016a; Simoncini et al., 2013).

O_2 and N_2O should also constitute a good biosignature. N_2O has a longer photochemical lifetime than CH_4, about 150 yr; however, it has even fewer natural sources and may therefore provide even stronger evidence for life. The simultaneous detection of O_2 and either CH_4 or N_2O is thus sometimes cited as the "Holy Grail" of remote life detection (Kasting, 2010).

Unfortunately, it would not be easy to observe these combinations of gases simultaneously on a planet like modern Earth because the concentrations of CH_4 and N_2O are only ~1.7 ppmv and 0.3 ppmv, respectively. Consequently, both CH_4 and N_2O are missing from the relatively low resolution spectra shown in Figs. 15.14 and 15.15.

These gases were seen, however, in a somewhat higher resolution, near-IR spectrum of the Earth obtained by the NIMS (Near Infrared Mapping Spectrometer) instrument on the Galileo spacecraft as it swung by Earth on its way out to Jupiter (Sagan et al., 1993b). The bands observed by NIMS were between 2 μm and 5 μm, which are difficult to access for exoplanets because of the low photon flux at these wavelengths (Fig. 15.7). But if we saw an interesting Earth-analog planet on a first-generation direct imaging mission, one can imagine sending a bigger, follow-up mission that would be capable of taking spectra in this wavelength range. So, the Holy Grail may be accessible if one invests enough effort in looking for it.

There may also be Earth-like planets on which the O_2–CH_4 pair is naturally more observable. As discussed in Chapters 10 and 11, both gases may have been present in appreciable concentrations during the Proterozoic as a consequence of the very different nature of the marine biosphere. So, a planet resembling the Proterozoic Earth might be a good candidate for remote life detection.

O_2 and CH_4 might also be observed simultaneously on a modern Earth-like planet orbiting an M star. As mentioned in the previous section, atmospheric photochemistry on an M-star planet would be very different from that of Earth because of the low flux of visible and near-UV radiation. On Earth today, CH_4 is primarily destroyed in the troposphere by the following reaction sequence (see Sec. 3.1.2).

$$O_3 + h\nu \rightarrow O_2 + O(^1D) \qquad \lambda \leq 310\,nm \qquad (15.9)$$

$$O(^1D) + H_2O \rightarrow 2\,OH \qquad (15.10)$$

$$CH_4 + OH \rightarrow CH_3 + H_2O \qquad (15.11)$$

But the rate of reaction (15.9) would be slow on an M-star planet; thus, if methanogenic microbes were producing methane at the same rate as on modern Earth, the concentration of methane in the planet's atmosphere could be as high as 500 ppm (Segura et al., 2005). So, perhaps M-star planets, if they can harbor life at all, will end up being the place where extraterrestrial life is first definitively detected.

15.4.1.2 The O_2–N_2 Pair

Somewhat surprisingly, the degree of thermodynamic disequilibrium in Earth's atmosphere (ignoring any interaction with the ocean) is not that large. Indeed, when disequilibrium is quantified as the Gibbs free energy available from reacting all the gases in a mole of air to equilibrium, it turns out that Earth's air has nearly a 100-fold smaller thermodynamic disequilibrium than does Mars (Table 15.2). (Mars' disequilibrium is the presence of CO and O_2 created by photochemistry.) The O_2–CH_4 pair is the dominant source of the disequilibrium in

Table 15.2 Comparison of the available Gibbs free energy in Solar System atmospheres defined here as the difference in Gibbs free energy between the observed composition and a theoretical equilibrium chemical composition calculated at the surface pressure–temperature conditions or 1 bar conditions for giant planets (Krissansen-Totton et al., 2016a).

Planetary atmosphere	Available Gibbs free energy (Joules per mole of atmosphere)
Venus	0.06
Earth (atmosphere only)	1.5
Earth (atmosphere and ocean)	2326
Mars	136
Jupiter	0.001
Titan	1.2
Uranus	<0.1 (upper limit)

Earth's atmosphere, but the free energy is small because the concentration of CH_4 is small.

It is only when you allow Earth's atmosphere to react with the ocean water in such a calculation that Earth stands out as having a far larger disequilibrium compared to other bodies in the Solar System. In the air–ocean calculation, nearly all of Earth's disequilibrium is provided by O_2, N_2, and water. The equilibrium state of these substances is dilute nitric acid, i.e.,

$$2N_2(g) + 5O_2(g) + 2H_2O(l) \rightleftharpoons 4H^+(aq) + 4NO_3^-(aq) \tag{15.12}$$

In fact, the idea that this disequilibrium is the largest one at the Earth's surface was first noted by the chemist Gilbert Lewis (who discovered the covalent bond), although he didn't quantify it (Lewis and Randall, 1923, pp. 567–568).

Earth's disequilibrium is certainly biogenic, including the contribution from O_2–N_2–water. Oxygenic photosynthesis obviously produces the O_2, but the N_2 is also maintained by photosynthesis because the large quantities of photosynthetic organic carbon generate the anoxic conditions in sediments that allow denitrifying microbes to turn nitrate back into N_2.

We conclude that the concept of extreme thermodynamic chemical disequilibrium as a biosignature applies well to gases in a photosynthetic world, such as the O_2–CH_4 kinetic instability and the N_2–O_2–water disequilibrium, but the extent to which it can be generalized to other biospheres is yet to be fully investigated. The pairs identified so far are, in principle, detectable. Even N_2 has spectral features: N_2–N_2 dimers, best seen at 4.15 μm on Earth-like worlds (Schwieterman *et al.*, 2015).

15.4.2 Classification of Biosignature Gases

Seager *et al.* (2013a, b; 2012) and Seager and Bains (2015) have proposed a classification of potential biosignature gases into three categories. *Type I biosignatures* are by-products of existing thermodynamic redox gradients, such as CH_4 on the Archean Earth. Such biosignatures are equivocal because abiotic processes might synthesize them, if suitable kinetic pathways exist. Methane, for example, can be formed by serpentinization of ultramafic rocks by water containing dissolved CO_2 (Berndt *et al.*, 1996). Seager *et al.* suggest that NH_3 might be a Type I biosignature on a cold "Haber world" with a dense N_2–H_2 atmosphere, because it is difficult to produce NH_3 by gas-phase photochemistry. But NH_3 might also be produced by photolytically catalyzed reaction of

N_2 with H_2O on surfaces containing TiO_2 (Schrauzer and Guth, 1977).

Type II biosignatures are biomass-building by-products, such as O_2 produced from photosynthesis. Anoxygenic photosynthesis also produces biomass-building by-products, as discussed in Sec. 9.6.5. H_2-, H_2S-, and Fe^{2+}-based photosynthesis yield H_2O, S, and $Fe(OH)_3$, respectively. Other forms of anoxygenic photosynthesis yield SO_4^{2-} and NO_3^{2-} (Seager *et al.*, 2012). Looking at this list, however, shows that most of these other by-products are either solids, liquids, or dissolved ions; thus, none of them would be likely to create an atmospheric biosignature. O_2 may be the only useful one.

Type III biosignatures are secondary metabolic products, such as dimethyl sulfide (DMS), OCS, CS_2, CH_3Cl, and higher hydrocarbons such as isoprene. Many of these species are volatile. But on Earth today, they are produced in such small amounts and/or are photolyzed sufficiently rapidly that they do not build up to high concentrations. If this is true on other Earth-like planets then they will likely not be observable by a first-generation direct imaging mission. But they might be useful targets for larger follow-up missions in the more distant future. Overall, although less obvious biosignature gases are interesting, the issue of whether they are ever likely to be detectable is surely paramount.

15.4.3 Is O_2 by Itself a Reliable Biosignature?

Let's return for a moment to Fig. 15.14, the visible/near-IR spectrum of Earth obtained from Earthshine data. As the figure shows, the only molecules that can be identified from these moderate-resolution data are O_2, O_3, and H_2O. O_3, of course, is formed photochemically from O_2, and so its information content is closely related. Should O_2 (and/or O_3) by itself be considered a reliable biosignature?

This question receives continued attention because it is not that easy to come up with a definitive answer. Nearly all of the O_2 in Earth's present atmosphere has been produced by photosynthesis followed by organic carbon burial. Furthermore, models of early prephotosynthetic atmospheres show that it is difficult or impossible to build up significant abiotic O_2 levels on a planet with a large ocean like Earth orbiting a star like the Sun (Sec. 9.2). However, it is easy to conceive of abiotic methods for producing O_2 on planets that do not satisfy these conditions. The most obvious one is by way of a Venus-like runaway, or moist, greenhouse. Suppose that Venus started out with as much water as is present in Earth's oceans today, 1.4×10^{21} kg, and that Venus lost most of its water within the first few hundred million years of its history by photodissociation, followed by escape of hydrogen to space (see Sec. 13.4.5). That process could conceivably have left an enormous amount of oxygen behind – enough to produce about 240 bars of surface pressure if it was all converted to O_2. But most of this oxygen was probably lost during accretion through contact between a steam atmosphere and an underlying magma ocean (Hamano *et al.*, 2013; Kurosawa, 2015). Alternatively, if the *moist greenhouse* model of Kasting (1988) is correct, then Venus' atmosphere might have remained O_2-rich for a billion years or more following the loss of its water. This "false positive" for life would probably not fool us, however, because the planet would be located suspiciously close to the inner edge of the habitable zone and the H_2O bands in the planet's atmosphere would be weak or nonexistent.

A second "false positive" that is easy to identify ahead of time would be a frozen planet like Mars near or outside the outer edge of the habitable zone (Kasting, 1997; Kasting, 2010). As noted in Ch. 12, Mars itself has about 0.17% O_2 in its atmosphere. This is too little to see spectroscopically from a distance because Mars' atmosphere is very thin; however, the process that produces it could operate on other planets as well. Mars' O_2 comes from photodissociation of H_2O and CO_2, followed by escape of hydrogen to space. The net source of O_2 is small. However, the sinks for O_2 are also very small. Volcanic outgassing is negligible, given undetectable SO_2 in the atmosphere (Krasnopolsky, 2012). Furthermore, Mars' surface is cold and dry, and so the rate of surface oxidation is slow. Mars' surface *is* highly oxidized, to be sure, but most of this oxidation may have happened a long time ago. In the absence of liquid water, surface erosion is relatively slow, and so fresh, non-oxidized rocks that serve as an O_2 sink are exposed only gradually.

In addition to surface oxidation, escape of oxygen atoms could limit the O_2 concentration in Mars' atmosphere. Mars is so small (~0.1 Earth mass) that oxygen escapes from Mars' atmosphere at a slow rate. As discussed in Sec. 12.2.4, O atom losses include dissociative recombination (eqs. (12.17) and (12.18)) and sputtering by the solar wind. However, if Mars were a little bigger, neither of these oxygen escape mechanisms would be likely to operate. The higher gravity would prevent O atom escape, and the warmer interior might be sufficient to generate a magnetic dynamo. If the planet remained too small to outgas hydrogen, then O_2 might conceivably accumulate to higher levels.

A more interesting question – because it might lead to confusion – is whether high abiotic concentrations of O_2 and/or O_3 could develop on a planet within the habitable zone. Once again, some false positives can be readily

identified. Wordsworth and Pierrehumbert (2014) pointed out that planets that are deficient in N_2 might build up O_2 abiotically because there would be little background gas to prevent a moist greenhouse from developing. This type of pathological situation may or may not occur in reality, as it would be difficult to accrete water and carbon without bringing in nitrogen at the same time. However, skeptics could still point to such an explanation as an alternative to the existence of life. It may eventually be possible to rule out such this type of false positive by looking for the spectral signature of O_2–O_2 and N_2–N_2 dimers (Misra *et al.*, 2014; Schwieterman *et al.*, 2015).

Another false positive may exist for planets orbiting M stars. This one has already been introduced in Sec. 15.1.5.2, because it poses a fundamental problem for M-star planet habitability. Because of the long, bright, pre-main-sequence phase of the host star, planets around M stars may lose their water by a runaway or moist greenhouse. But this implies that they might also build up O_2-rich atmospheres, which could constitute false positives (Luger and Barnes, 2015). Like other post-runaway-greenhouse planets, however, these planets might be identified by their lack of H_2O absorption.

Finally, some recent calculations suggest that O_2 could accumulate on less extreme planets within the HZ, under certain circumstances. Hu *et al.* (2012) calculated that abiotic O_2 concentrations in a CO_2-rich atmosphere might be as high as 10^{-3} by volume, if surface outgassing of H_2 was low. This would likely be detectable spectroscopically, either using the O_2 0.76-μm band or from the associated O_3 signal (Segura *et al.*, 2003). An O_2 mixing ratio of 10^{-3} is above the Proterozoic O_2 level of 2×10^{-4} atm (10^{-3} PAL) estimated by Planavsky *et al.* (2014b), so such a detection would constitute a false positive. But Harman *et al.* (2015) do not get this result, for reasons that are not entirely clear.

Other calculations for abiotic O_2 on M stars have been debated. Tian *et al.* (2014) calculated that abiotic O_2 mixing ratios as high as 4×10^{-3} could be produced on planets with CO_2-rich atmospheres orbiting M stars, even when H_2 outgassing is included. In their model, the dearth of stellar visible/near-UV radiation on these planets lowers the efficiency of catalytic O_2 destruction mechanisms that operate on Solar System planets. Their prediction was not borne out in parallel calculations by Domagal-Goldman *et al.* (2014), but it was supported by Harman *et al.* (2015). All three of these models balanced both the atmospheric redox budget (eq. (8.30)) and the global redox budget (eq. (8.36)). As Harman *et al.* show, whether or not abiotic O_2 can accumulate in a planet's atmosphere depends critically on chemical reactions that occur within the planet's ocean. The presence of dissolved ferrous iron, similar to Earth's oceans during the Archean (Ch. 12), will suppress atmospheric O_2. Direct reaction between dissolved O_2 and CO, perhaps catalyzed by metal ions, can also suppress O_2. Laboratory experiments are needed to study this reaction. So, while this question is not totally resolved, it remains possible that O_2 by itself is a good biosignature for planets within the habitable zones of FGK stars.

15.5 Parting Thoughts

This brings us to the end of our book. But it is certainly *not* the end of our study of atmospheric evolution. As a society, we are just partway through our exploration of the Solar System, and we are only beginning to study exoplanets. Thus, much of what we have written here, particularly in this last chapter, will be modified in the future. We hope, though, that we have provided a framework that will allow current and future generations of atmospheric, planetary, and Earth scientists, along with astronomers and astrobiologists, to place their discoveries in the context of the work that has preceded them. And we look forward to the construction of the future ground- and space-based telescopes that will eventually allow us to look for other planets like Earth and to test our theories of atmospheric, planetary, and biological evolution.

Appendix A: One-Dimensional Climate Model

A tool used frequently in this book and in other studies of planetary atmospheres is a one-dimensional *radiative–convective climate model*, or RCM. Such models were pioneered by Manabe and Strickler (1964) and Manabe and Wetherald (1967). The basic premise is that a planet's globally averaged surface temperature can be estimated using a model in which incident solar insolation and surface temperature are averaged over the planet's surface, and only the vertical variation in atmospheric temperature structure is calculated. Such models were described conceptually in Sec. 2.4.5 and in Fig. 2.20.

A.1 Numerical Method

Numerically, an RCM is very simple. The basic equation to be solved is that of energy conservation

$$\rho c_{\mathrm{p}} \frac{\partial T(z,t)}{\partial t} = \frac{\partial F(z)}{\partial z} \tag{A.1}$$

Here, ρ is mass density, c_{p} is the specific heat at constant pressure, $T(z,t)$ is the vertical temperature profile, z is altitude, and $F(z)$ is the globally averaged total radiative flux at each altitude, defined in our model as being positive downwards. Typically, $F(z)$ is divided into two components

$$F(z) = F_S(z) + F_{\mathrm{IR}}(z) \tag{A.2}$$

$F_S(z)$ is the solar flux and $F_{\mathrm{IR}}(z)$ is the thermal-infrared flux. $F_S(z)$ includes both the direct solar beam plus a diffuse (scattered) flux, while $F_{\mathrm{IR}}(z)$ is entirely diffuse and includes thermal emission as part of the source function. The apparent simplicity of eq. (A.2) may be misleading, as both $F_S(z)$ and $F_{\mathrm{IR}}(z)$ involve integrations over zenith angle and over wavelength, both of which can be approximated in different ways. Below, we will describe the approach used here, which is only one of many different methods that have been applied to this problem.

It is convenient to work in pressure coordinates, p, rather than in altitude so that, as the temperature changes and the atmosphere expands or contracts, the vertical grid moves with it. Thus, by dividing eq. (A.1) by ρc_{p} and using the barometric law, $dp = -\rho g dz$, we obtain

$$\frac{\partial T(p,t)}{\partial t} = -\frac{g}{c_{\mathrm{p}}} \frac{dF(p)}{dp} \tag{A.3}$$

Equation (A.3) is a single partial differential equation, first-order in both t and p, that can be readily solved using finite differences. An initial guess at a temperature profile is required. As was the case with the 1-D photochemical model, we are typically only interested in the steady-state solution; hence, first-order time differencing is adequate. An unevenly spaced grid in log p is assumed, with 100 layers extending from the surface pressure, p_s, to the top-of-atmosphere pressure, $p_0 \cong 3 \times 10^{-5}$ bar. The grid is numbered downward from the top. Temperatures are calculated at the midpoints of the layers, and fluxes are computed at the layer boundaries. An extra temperature grid point is defined at the surface. In finite difference form, eq. (A.3) becomes

$$T_j^{n+1} = T_j^n - \frac{g}{c_{\mathrm{p}j}} \left(\frac{F_{j+1/2}^n - F_{j-1/2}^n}{\Delta p_j} \right) \Delta t \tag{A.4}$$

Here, Δt is the time step, Δp_j is the thickness of layer j, and the superscript n designates the time step. A variable time-stepping method is used, starting from small time steps and progressing to larger ones, based on how fast the temperature changes at each altitude. The time-stepping here is explicit, because the fluxes are evaluated at the current time step n. Hence, unlike the photochemical model, in which Δt increases to very large values as steady state is approached, Δt in this model is always limited to rather low values, typically 10^4 s, or ~3 h. This presents issues with reaching convergence in dense atmospheres, for which the heat capacity $\rho c_p \Delta z$ of the lowermost layers may become high. As such dense atmospheres are of great interest to us, we have developed methods to achieve rapid convergence in these cases.

When the model atmosphere is thin, and its heat capacity is low, we use a standard, energy-conserving approach. Equation (A.4) is advanced from time step to time step and the temperature, T_g, is found from

$$T_g^{n+1} = T_g^n - \left(\frac{F_g^n}{C_g}\right)\Delta t \qquad (A.5)$$

Here, F_g^n is the net upward flux at the surface, and C_g is the surface heat capacity, which is usually taken to be that of a 0.5-m thick ocean. The ocean thickness is intentionally made small to allow the model to converge quickly. In steady state, $F_g^n = 0$, and the solution is independent of the magnitude of C_g.

As described in Sec. 2.4.5.3, within the troposphere the radiative equilibrium solution calculated from eq. (A.4) often leads to layers that are unstable against convection. To deal with this, we employ a convective adjustment scheme broadly similar to the one described by Manabe and Strickler (1964). We define a critical lapse rate, which we take to be a moist pseudo-adiabat, i.e., an adiabat in which the condensate (usually liquid water or water ice) is assumed to immediately fall out. An expression for the moist pseudo-adiabatic lapse rate was given in eq. (1.56). Our climate model, however, was designed to operate in regions near the critical points of H_2O and CO_2, where the gas is nonideal and the Clausius–Clapeyron equation is no longer valid. Thus, we use the more complicated lapse rate expressions derived by Ingersoll (1969), Kasting (1988), and Kasting (1991).

When the model atmosphere is thin, as for present Mars, we perform the convective adjustment as follows. We begin by cycling through the vertical grid from bottom to top looking for pairs of layers that are unstable. (Pairs of layers are involved because the temperatures are defined at the layer midpoints.) When such a pair is found, the temperatures of both layers are adjusted in such a way that energy is conserved and the temperature gradient is equal to the moist adiabat. Multiple cycles (typically 20) through the grid are required because as each pair of layers is adjusted, the pair below it may once again become unstable. Fortunately, a simple iteration procedure converges nicely.

For thick atmospheres, the heat capacity of the lower atmosphere becomes large, and it becomes impractical to time march to equilibrium in the manner described above, so we take a shortcut to the answer. What we are trying to achieve is radiative equilibrium at all levels within the

stratosphere, along with flux balance at the top of the atmosphere. At steady state, the net absorbed solar flux must equal the net outgoing infrared flux. The outgoing IR flux, though, is a monotonically increasing function of surface temperature, T_g. Hence, if we simply ignore energy conservation during the time-stepping procedure, we can adjust the temperature of the bottommost atmospheric layer, J, according to

$$T_J^{n+1} = T_J^n + \frac{g}{c_{pJ}}\left(\frac{F_1^n}{\Delta p_J}\right)\Delta t \qquad (A.6)$$

Here, F_1^n is the net downward flux (solar plus IR) at the *top* of the model atmosphere. The surface temperature is then calculated by extending a moist adiabat from the midpoint of the first layer down to the surface. Temperatures at other layers within the model atmosphere are adjusted by extending a moist adiabat upward from the bottommost layer. In this case, energy conservation is not required, and so iterations of the convective adjustment scheme are not needed: a single pass through the troposphere achieves the desired vertical profile.

The scheme works in the following way: If the net downward flux is positive then, according to eq. (A.6), the troposphere and surface will warm, causing an increase in the upward infrared flux. Conversely, if the net downward flux is negative, then the troposphere and surface will cool, causing a decrease in the upward infrared flux. In either case, the atmosphere quickly converges (usually in <50 time steps) to one in which flux balance has been achieved at the top of the atmosphere. Further iterations are often necessary to allow the rest of the stratosphere to achieve radiative balance, but this problem is tractable because the heat capacity of the stratosphere is much less than that of the troposphere. Additional energy-conserving iterations can be performed, as well, to check that the solution is self-consistent. These are generally not necessary, however, because even though energy conservation was ignored during the time integration, the steady state solution still satisfies both radiative–convective equilibrium within the atmosphere and top-of-atmosphere flux balance, which are the criteria for a converged solution. Thus, energy balance is achieved in the final, steady-state solution.

A.2 Calculation of Radiative Fluxes

As mentioned above, many different methods are available to calculate radiative fluxes. We have adopted a *delta two-stream method* based on the numerical approach of Toon *et al.* (1989). When particles are involved, the "delta" scaling of Joseph *et al.* (1976) is used, as well.

Two-stream methods are ones in which the integrations of diffuse radiation over the upward and downward hemispheres are approximated by a single stream of radiation in each direction at some average zenith angle. A two-stream approach was described in Sec. 2.4.5.3. The Toon *et al.* model can be used with three different two-stream models (Eddington, quadrature, and hemispheric mean) by specifying different coefficients from their Table 1. Following their recommendation we use the quadrature method at solar wavelengths and the hemispheric mean method at thermal-infrared wavelengths. The numerics of their rather complex method will not be described here except to point out that there is a mistake in the last line of their equation (42), which should read

$$E_l = \left[C_{n+1}^+(0) - C_n^+(\tau_n)\right]e_{2n+1} + \left[C_n^-(\tau_n) - C_{n+1}^-(0)\right]e_{4n+1}$$
$$(A.7)$$

The correction is that the order of the two terms in the second square bracket is reversed compared with the expression given in their paper.

The delta scaling is a technique that is often applied in low angular resolution methods, such as the two-stream approximation, to better account for forward scattering by particles. The forward scattered radiation is effectively removed from the diffuse beam by expressing the phase function as the sum of three terms, one of which is a Dirac delta function in the forward direction. The scaling is done within each layer using the asymmetry factor, g, of the combined gases plus particles in that layer. In the description that follows, the layer subscript j is eliminated for clarity. The asymmetry factor represents the fraction of radiation scattered in the forward hemisphere minus the fraction scattered backwards. The gases scatter by Rayleigh scattering, so their asymmetry factor, g_g, is zero. Thus, $g = \left(\tau_g g_g + \tau_p g_p\right)/\left(\tau_g + \tau_p\right) = g_p \cdot \tau_p/\left(\tau_g + \tau_p\right)$, where τ indicates the optical depth within the layer and the subscripts g and p refers to gases and particles, respectively. The delta scaling for g, τ, and single scattering albedo ω can then be defined by the expressions

$$f = g^2 \qquad \tau' = (1 - \omega f)\tau$$
$$\omega' = \frac{(1-f)\omega}{1 - \omega f} \qquad g' = \frac{g}{1 + g} \qquad (A.8)$$

Geometry must be considered when using a 1-D climate model. The globally averaged solar flux hitting a planet such as Earth is $S/4$, where S is the flux at the planet's orbit. In our 1-D model, we typically simulate this by putting the sun at a solar zenith angle, θ, of $60°$ (so that $\cos\theta = \frac{1}{2}$) and by multiplying all fluxes by another factor of $\frac{1}{2}$ to account for the day–night variation. Alternatively,

one can perform multiple solar calculations at different zenith angles over the sunlit hemisphere and use Gaussian integration to calculate the total direct plus diffuse solar flux.

The real difficulty in performing radiative transfer calculations in planetary atmospheres is to account for absorption by various gaseous species, including CO_2 and H_2O. The most accurate radiative transfer models, such as LBLRTM (Clough and Iacono, 1995; Clough *et al.*, 1992) and SMART (Crisp, 1997; Meadows and Crisp, 1996) resolve the individual vibration–rotation lines of each molecule using a Voigt profile to capture the variation in line broadening with altitude. Such models typically require of the order of tens of minutes to an hour or more to perform a complete wavelength integration, however, and this is too slow for our climate model, for which many flux calculations must be performed in succession. Hence, we employ a broadband approximation in which the thermal-infrared is divided into 55 spectral intervals between 0.7 and 500 μm and the solar spectrum is broken into 38 intervals between 0.25 and 4.5 μm. The infrared code extends to short wavelengths in order to accurately simulate runaway greenhouse atmospheres of the type described in Ch. 13. Within each interval, absorption by each individual gas is parameterized in terms of *correlated-k coefficients* (Kato *et al.*, 1999). In this technique, the (broadband) transmission over some pathlength, P, is parameterized as $\overline{T}(P) = \sum_{i=1}^{N} \alpha_i e^{-k_i P}$. The weights, α_i, are taken to be the usual Gaussian weights, and the absorption coefficients, k_i, are derived by binning the line-by-line coefficients derived from a line-by-line model. Prior to 2013, we used coefficients derived by R. Freedman at NASA Ames, using a model similar to LBLRTM. Since that time, we have used new coefficients for CO_2 and H_2O derived by Kopparapu *et al.* (2013). The new CO_2 coefficients are based on the updated HITRAN 2008 database, while the H_2O coefficients are from HITEMP, 2010 version. The HITEMP database contains many more H_2O lines than does HITRAN, particularly in the visible/near-UV part of the spectrum. The binning method is described by Liou (2002, pp. 127–129). A more complicated double-Gauss procedure is used by Kopparapu *et al.* (2013). Different sets of ks are derived for different temperatures and pressures, and then interpolation is used to find values appropriate for the model pressure-temperature grid. We use eight-term sums for CO_2 and H_2O and six-term sums for CH_4, C_2H_6, and N_2O. By treating each gas separately, we effectively assume that the individual absorption lines of each gas are uncorrelated with those of the other gases within each spectral interval. This assumption may or may not be true and should be tested by comparing with detailed line-by-line calculations.

If the absorption of each gas is uncorrelated, then the transmission through a combination of gases can be found from

$$\overline{T}_{tot} = \overline{T}_{H_2O} \cdot \overline{T}_{CO_2} \cdots = \sum_{i=1}^{N} a_i e^{-k_i P_{H_2O}} \cdot \sum_{j=1}^{N} a_j e^{-k_j P_{CO_2}} \cdots$$

(A.9)

Mathematically, the terms in each individual summation must be convolved with each other, that is, each separate term of gas 1 must be multiplied by each term of gas 2. Thus, for a model including just CO_2 and H_2O, we must perform $8 \times 8 = 64$ separate calculations within each spectral interval. For the 55 spectral intervals in the thermal-IR routine, this requires $64 \times 55 = 3520$ separate calls to the two-stream routine at each time step. A slightly smaller number of two-stream calls are required in the solar code. Despite all this work, a typical time step for a two-gas model requires only a fraction of a second on a 3-GHz Linux processor. Thus, full calculations involving hundreds of time steps can be performed in a few tens of seconds. The addition of each new gas, however, slows the code down by a factor proportional to the number of k-coefficients, as these too must be convolved with those from CO_2 and H_2O. Thus, the method remains tractable for up to three or four gases but becomes impractical for more complex mixtures.

An alternative k-coefficient method, which we have used for generating spectra (Segura et al., 2005; Segura et al., 2003), is that of Mlawer et al. (1997). These authors have derived k-coefficients for combined CO_2–H_2O mixtures over 16 spectral intervals in the thermal-IR. They use 16-term sums, with eight terms clustered around the Doppler cores of the lines and eight terms covering the pressure-broadened wings. As such, their calculation of absorption is accurate in the stratosphere as well as in the troposphere. Newer correlated k-coefficients (Kopparapu et al., 2013) employ a similar double-Gauss method and should exhibit a similar degree of accuracy. The Mlawer et al. model is accurate and fast for simulating atmospheres like that of modern Earth, but it fails when pCO_2 is more than about 50 times the current terrestrial value. Their approach can be generalized, however, and this may be an appropriate way to handle gaseous absorption in more complex, 3-D climate models.

A.3 Treatment of Water Vapor

Water vapor is an important greenhouse gas in most terrestrial planet atmospheres, so it is necessary to treat it as self-consistently as possible in climate models. Three-dimensional general circulation models (GCMs) do a good job here because they typically include a prognostic equation for water vapor that takes into account evaporation, precipitation, and transport. One-dimensional RCMs lack the spatial resolution to simulate the hydrologic cycle properly; hence, one is forced to make various approximations. Following Manabe and Wetherald (1967), the approach we take in the troposphere is to assume a fixed vertical distribution of relative humidity, RH. These authors suggested an empirical profile of the following form.

$$RH(z) = \frac{Q(z) - 0.02}{1 - 0.02}$$

(A.10)

Here, $Q(z) = p(z)/p_g$ represents the ratio of the ambient pressure at altitude z to the surface pressure, p_g. Although crude, this expression is appropriate for a planet like modern Earth, for which it was derived. It reflects the observation that relative humidity declines in the upper troposphere as a consequence of condensation in localized updrafts, followed by subsidence of cold, dry air over broader regions of the troposphere. Fixing the relative humidity in a climate model means that water vapor acts as a positive feedback on surface temperature, because its absolute abundance increases with temperature according to the Clausius–Clapeyron equation (1.49). Equation (A.10) is often referred to as a "Manabe–Wetherald" relative humidity distribution.

Whether or not eq. (A.10) works for any atmosphere other than modern Earth is unknown. It is almost certainly not a valid approximation for atmospheres that are much thinner or denser or for which the mean surface temperature is greatly different from the surface temperature of modern Earth. Consequently, we make different assumptions about relative humidity depending on what we are trying to show. For example, for early Mars we often assume that $RH = 1$ in the troposphere, as this allows us to calculate an upper limit on surface temperature. But no one would argue that this assumption is realistic.

Finally, a key aspect of how water vapor behaves in a planetary atmosphere is the concept of the tropopause cold trap. Earth's stratosphere is dry today because the mixing ratio of water vapor in the upper troposphere is low. This would be true even if the relative humidity there was high, simply because the air is cold and the saturation vapor pressure is low. We simulate the tropopause cold trap in two different ways in our model. When we attempt to reproduce the Earth's current atmosphere, we use a Manabe–Wetherald relative humidity distribution in the troposphere and simply define a minimum H_2O mixing

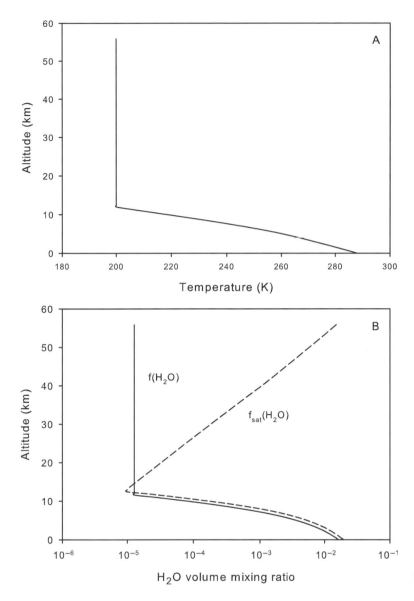

Figure A.1

ratio of 4 ppmv. The stratospheric H_2O mixing ratio is then assumed to be a constant 4 ppmv above this height. Again, this is an empirical approach that is not likely to be valid for any atmosphere other than that of modern Earth. For other types of atmospheres, we follow the approach illustrated in Figure A.1. Consider an atmosphere with a surface temperature of 288 K, a moist adiabatic lapse rate, and an isothermal stratosphere with a temperature of 200 K (Fig. A.1(a)). Assume, for simplicity, that the troposphere is fully saturated ($RH = 1$). Define the saturation mixing ratio of water vapor as $f_{sat}(H_2O) = p_{sat}(T)/p(z)$,

where $p_{sat}(T)$ is the saturation vapor pressure at temperature $T(z)$ and $p(z)$ is the ambient pressure at height z. Then, $f_{sat}(H_2O)$ goes through a minimum near the top of the troposphere, as shown in Fig. A.1(b). Below that height f_{sat} decreases with altitude because of the steep decrease in temperature; above it, $p_{sat}(T)$ becomes constant (because $T(z)$ is constant), so the ratio $p_{sat}(T)/p(z)$ must increase with altitude. The cold trap occurs where $f_{sat}(H_2O)$ reaches its minimum value. In the model, the H_2O mixing ratio $f(H_2O)$ is held constant above this level, as shown in the figure.

A.4 Treatment of Clouds

Clouds are difficult to treat self-consistently in one-dimensional models, not because they are hard to include, but rather because there is no good, non-empirical way to parameterize them. Various authors, e.g., Goldblatt and Zahnle (2011), have used observations of real clouds in Earth's atmosphere to derive a statistical ensemble of cloud height, cloud optical properties, and fractional cloud cover. But this ensemble is valid only for Earth's modern atmosphere, and it is unclear how one would change it to simulate some other atmosphere. To avoid this problem, we use the planet's surface albedo as a tuning parameter to mimic the effect of clouds. This is equivalent to "painting the clouds on the surface" (Goldblatt and Zahnle, 2011). If we are simulating the Earth, then the surface albedo is adjusted such that the calculated surface temperature, given present solar insolation, is equal to the observed mean temperature of 288 K. This required a surface albedo of ~0.22 in our old model, but this value rises to 0.3 in the new model because of increased atmospheric absorption. Both values are much higher than the actual surface albedo, which is closer to 0.05. When combined with Rayleigh scattering by the atmosphere, we get a planetary albedo of ~0.25. This is lower than the actual planetary albedo of 0.3, implying that the model Earth is absorbing too much solar radiation. But by placing the cloud layer at the surface, we have ignored the greenhouse effect of clouds, which in the real world would have helped to keep the surface warm. Thus, in our model the errors cancel: we underestimate the planetary albedo and we underestimate the greenhouse effect, but by doing so we achieve the correct surface temperature for modern Earth. If we want to then simulate the early Earth, we hold the surface albedo constant and hope that the errors will cancel there, as well.

Goldblatt and Zahnle (2011) have performed a comparison of how this procedure works compared to an approach that uses prescribed, statistical clouds. They conclude that putting the clouds at the surface may result in an overestimate of greenhouse warming by CO_2 in the distant past by as much as 25%. The reason is that the 8–12 μm window region is too clear in our cloud-free model; hence, putting in species that absorb there (which CO_2 does at high concentrations) overestimates their greenhouse effect. The Goldblatt and Zahnle analysis is useful in pointing out the possible errors, but it does not offer an alternative way to proceed in modeling past Earth atmospheres or atmospheres of other planets. Three-dimensional GCMs are needed to improve on the cloud feedback problem, and even these models are problematical because the clouds are almost always sub-grid-scale. We conclude that climate modelers will be kept busy for many years trying to develop self-consistent models of generalized planetary atmospheres. Simulating climate realistically is not an easy problem.

Appendix B: Photochemical Models

One of the useful tools for studying the evolution of planetary atmospheres is a one-dimensional (1-D) photochemical model. Such a model takes advantage of the fact that variations in atmospheric composition with altitude are typically more pronounced than those with latitude or longitude, for two reasons. (1) The penetration depth of solar UV photons depends on the amount of overlying gas, and hence on altitude. (2) Vertical transport is much slower than horizontal transport in most planetary atmospheres; hence, all atmospheres show some degree of lateral homogeneity. The 1-D approximation is clearly less appropriate for tidally locked planets with permanent day and night sides, although even there one can do separate day- and night-side calculations (see, e.g., Line *et al.*, 2010).

B.1 Photochemical Model Equations

The equations solved in a photochemical model are generally the following.

Continuity

$$\frac{\partial n_i}{\partial t} = P_i - l_i n_i - \frac{\partial \Phi_i}{dz} \tag{B.1}$$

Flux

$$\Phi_i = -Kn \frac{\partial f_i}{\partial z} - D_i n_i \left(\frac{1}{n_i} \frac{\partial n_i}{\partial z} + \frac{1}{H_i} + \frac{1 + \alpha_{Ti}}{T} \frac{\partial T}{\partial z} \right) \tag{B.2}$$

Here, t = time, z = altitude, n_i = number density of species i (molecules cm^{-3}), P_i = chemical production rate (molecules cm^{-3} s^{-1}), l_i = chemical loss frequency (s^{-1}), Φ_i = flux of species i, $f_i = n_i/n$ = mixing ratio of species i, K = eddy diffusion coefficient (cm^2 s^{-1}), n = total number density, D_i = diffusion coefficient between species i and the background atmosphere, H_i ($= kT/m_i g$) is the scale heights of species i, and α_{Ti} is the thermal diffusion coefficient of species i with respect to the background atmosphere. Within these definitions, k = Boltzmann's constant, m_i = molecular mass of species i, and m = mean molecular mass of the atmosphere. All quantities are given in CGS units, based largely on the fact that the kinetic theory of gases (see, e.g., Chapman and Cowling, 1939; Chapman *et al.*, 1990) was originally developed in these units. A derivation of eq. (B.2) can be found in Banks and Kockarts (1973, Ch. 15). See also the derivation of eq. (5.43) in Sec. 5.8.1.

Some simplifications have already been made in deriving eq. (B.2). Equation (B.2) is strictly valid only for a minor constituent diffusing through a more abundant background gas. This may lead to errors when these equations are used to model gases that become major constituents at some altitudes. Second, the molecular diffusion coefficient between two gases, 1 and 2, is of the form

$$D_{12} = \frac{b_{12}}{n_1 + n_2} \tag{B.3}$$

Here, n_1 and n_2 are the number densities of the two gases. The binary diffusion parameter, b_{12}, is different for each pair of gases. In a multi-component mixture, it is convenient to approximate these relationships by the formula (Banks and Kockarts, 1973, eq. 15.29):

$$D_i = 1.52 \times 10^{18} \left(\frac{1}{M_i} + \frac{1}{M} \right) \frac{T^{1/2}}{n} \tag{B.4}$$

Here, M_i = molecular weight of species i, M = mean molecular weight of the atmosphere, and n = total number density. Furthermore, because the thermal diffusion coefficient depends in a complex way on the other species that are present, one typically sets $\alpha_{Ti} = 0$ in lower atmosphere photochemical models. In Earth's atmosphere thermal diffusion is most important for light species like H and

H_2 in Earth's thermosphere, where molecular diffusion dominates and the vertical temperature gradient is strong. This is not an important issue in lower atmospheres where transport is dominated by eddy diffusion.

It is convenient to work in mixing ratio units, so we recast the second term on the right-hand side of eq. (B.2) by using

$$\frac{\partial f_i}{\partial z} = \frac{1}{n}\frac{\partial n_i}{\partial z} - \frac{n_i}{n^2}\frac{\partial n}{\partial z} \tag{B.5}$$

The ideal gas law says that pressure, p, is given by $p = nkT$. So,

$$\frac{1}{p}\frac{\partial p}{\partial z} = \frac{1}{n}\frac{\partial n}{\partial z} + \frac{1}{T}\frac{\partial T}{\partial z} \tag{B.6}$$

But $\dfrac{1}{p}\dfrac{\partial p}{\partial z} = -\dfrac{1}{H_a}$, where H_a is the pressure scale height, $kT/(mg)$. Substituting this relationship into eq. (B.6) and then plugging the result back into eq. (B.5) yields

$$\frac{\partial f_i}{\partial z} = \frac{1}{n}\frac{\partial n_i}{\partial z} + \frac{n_i}{n}\left(\frac{1}{T}\frac{\partial T}{\partial z} + \frac{1}{H_a}\right) \tag{B.7}$$

By solving eq. (B.7) for $\partial n_i/\partial z$ and by defining

$$\zeta_i \equiv D_i\left(\frac{1}{H_i} - \frac{1}{H_a} + \frac{\alpha_{Ti}}{T}\frac{\partial T}{\partial z}\right) \tag{B.8}$$

we can recast eq. (B.2) as

$$\Phi_i = -(K + D_i)n\frac{\partial f_i}{\partial z} - \zeta_i n f_i \tag{B.9}$$

Finally, by differentiating eq. (B.9) with respect to altitude, and noting that in steady state $\partial n/\partial t = 0$, we can rewrite eq. (B.1) as

$$\frac{\partial f_i}{\partial t} = \frac{1}{n}\frac{\partial}{\partial z}\left[(K + D_i)n\frac{\partial f_i}{\partial z} + \zeta_i n f_i\right] + \frac{P_i}{n} - l_i f_i \tag{B.10}$$

In taking the derivative of Φ_i, we have intentionally *not* differentiated term by term in eq. (B.9); instead, we have left the entire right-hand side within the square brackets in eq. (B.10). This is referred to as the *conservation form* of the equation because it allows explicit mass (or number density) conservation when the equation is cast in finite difference form. To see how this works when the solution is in steady state, set the left-hand side of eq. (B.10) equal to zero, multiply the right-hand side by n, and rewrite it as

$$-\frac{\partial}{\partial z}[\Phi_i] + P_i - L_i = 0 \tag{B.11}$$

Here, $L_i = l_i n_i =$ chemical loss rate of species i in units of molecules cm^{-3} s^{-1}. Integration with respect to z from the lower to the upper boundary of the model then yields

$$\int_{z_{low}}^{z_{up}} (P_i - L_i)dz + \Phi_{low} - \Phi_{up} = 0 \tag{B.12}$$

In practice, the integral is replaced by a summation over all levels, and the column-integrated production and loss rates for each species are stored in vectors TP(I) and TL(I). The sum of the terms on the left-hand side of eq. (B.12) is then stored in a vector CON(I), which is never actually zero, but which should be small compared to either TP(I) or TL(I). Mass conservation is a useful property for these models, particularly when one is interested in accurately tracking redox balance, as we often are in doing early Earth calculations.

B.2 Finite Differencing the Model Equations

Equation (B.10) represents a set of partial differential equations (PDEs) for the mixing ratios of each species i that is being tracked in the model. To solve this system of PDEs, we use a finite difference method. In most of the model calculations described in this book, the vertical grid is evenly spaced with a 1-km step size (Δz) up to 100 km (see Fig. B.1). The pressure scale height of Earth's atmosphere is of the order of 6–8 km, so this gives roughly eight points per scale height, allowing accurate calculation of the solar UV flux. Species mixing ratios are defined at the midpoints of the grid levels, whereas fluxes are defined at the grid boundaries. An evenly spaced grid allows one to approximate the spatial derivatives by centered, second-order finite difference equations of the form

$$\left.\frac{\partial(\zeta_i n f_i)}{\partial z}\right|_{j+1/2} = \frac{(\zeta_i n f_i)^{j+1} - (\zeta_i n f_i)^j}{\Delta z} \tag{B.13}$$

$$\left.\frac{\partial\Phi_i}{\partial z}\right|_j = \frac{\Phi_i^{j+1/2} - \Phi_i^{j-1/2}}{\Delta z} \tag{B.14}$$

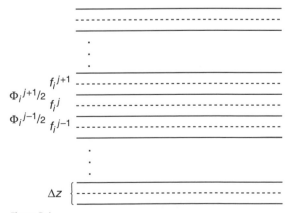

Figure B.1

Here, the superscript j indicates the vertical level. The fluxes at the half grid points are evaluated from

$$\Phi_i^{j+1/2} = (K + D_i)^{j+1/2} n^{j+1/2} \frac{f_i^{j+1} - f_i^j}{\Delta z}$$
$$+ \zeta^{j+1/2} n^{j+1/2} f_i^{j+1/2} \tag{B.15}$$

and similarly for $\Phi_i^{j-1/2}$. Thus, by replacing the spatial derivatives with finite difference approximations, the system of PDEs (eq. (B.10)) has been reduced to a set of ODEs (ordinary differential equations) of the form

$$\frac{df_i^j}{dt} = A f_i^{j+1} + B f_i^j + C f_i^{j-1} + D_i^j \tag{B.16}$$

with $A = \dfrac{(K + D_l)^{j+1/2}}{n^j (\Delta z)^2} + \dfrac{(\zeta_i n)^{j+1}}{2 n^j \Delta z}$

$$B = \frac{(K + D_i)^{j+1/2}}{n^j (\Delta z)^2} + \frac{(K + D_i)^{j-1/2}}{n^j (\Delta z)^2} - l_i^j$$

$$C = \frac{(K + D_i)^{j-1/2}}{n^j (\Delta z)^2} - \frac{(\zeta_i n)^{j-1}}{2 n^j \Delta z}, \text{ and } D_i^j = \frac{P_i^j}{n^j}.$$

To solve the system of ODEs represented by eq. (B.16), it is convenient to renumber the equations. Suppose that one has I chemical species and J vertical layers. Let

$$x_k \equiv f_i^j \quad k = i + (j - 1) \cdot I \qquad i = 1, ..., I; \; j = 1, ..., J \tag{B.17}$$

One then has $K (= I \cdot J)$ equations for K unknowns, x_k.

These equations may be written as

$$\frac{dx_k}{dt} = A x_{k+I} + B x_k + C x_{k-I} + D_k \tag{B.18}$$

A typical photochemical model may have 30 species and 100 vertical layers, so the number of ODEs to be solved is $30 \times 100 = 3000$. Note that we have varied the species number first and layer number second (meaning that we have numbered all the species in the first row, then those in the second row, etc.). That is important because it leads to a Jacobian matrix that is easily inverted (see below). Note also that D_k (or D_i^j) is not a constant, as it contains chemical production terms involving other species on the jth level. So, we should properly write $D_k = D_k(x_l), \; l = k - I, ..., k + I$.

B.3   Solving the System of Ordinary Differential Equations (ODEs)

At this point one could solve this system of equations with a standard ODE solver. One needs to use a special type of package, though – one that is designed to handle *stiff equations*. Stiff equations are ones in which some of the dependent variables change their values much more quickly than do others. In atmospheric chemistry, short-lived species such as OH have photochemical lifetimes of the order of 1 s, whereas long-lived species such as CH_4 have lifetimes of 10 yr or more. Vertical transport times are also slow, of the order of tens of years. Hence, to reach steady state in a model of Earth's present atmosphere one might need to integrate for 100 yr or more, or $\sim 3 \times 10^9$ s. With a standard *explicit* ODE solver one would be forced to take time steps shorter than the lifetime of the shortest-lived chemical species; thus, it would require at least 3×10^9 tiny time steps to reach a converged solution. An explicit method is one in which the right-hand side of the equations being integrated (eq. (B.18)) is evaluated at the current time step.

Numerical analysts bypass this limitation by using *implicit* integration methods in which the right-hand side is evaluated at the forward time step. Doing this requires that one develop an approximation for what those function values will be. Many different methods have been developed to do this. A book by C. W. Gear (Gear, 1971) describes some of the best variable-order predictor-corrector method; as such, they are sometimes called *Gear codes*. Many different computer libraries, for example, the one accompanying the book *Numerical Recipes* (Press, 2007), contain implementations of these codes.

We used Gear codes at one time to solve the equations for our model. Doing so has its disadvantages, though. If the code fails to converge, it can be difficult to determine what is going wrong if one does not understand the details of the algorithm. (We will see below that there are several potential numerical problems that can be encountered.) Furthermore, we are generally only interested in steady-state solutions, so maintaining accuracy during the time-stepping sequence is not essential. If the model converges, it must converge to the right answer. We assume, without proof, that solutions to these systems of equations are unique. In such a situation, any fully implicit method will suffice, and we choose the simplest one – the *reverse Euler* method. A brief description follows.

Rewrite the system of ODEs (B.18) in vector form

$$\frac{d\mathbf{x}}{dt} = \mathbf{F}(\mathbf{x}) \qquad \mathbf{x} = (x_1, x_2, ..., x_K) \tag{B.19}$$

Approximate the time derivative by a simple forward difference

$$\frac{d\mathbf{x}}{dt} \simeq \frac{\mathbf{x}^{n+1} - \mathbf{x}^n}{\Delta t} \tag{B.20}$$

Here, the superscript n represents the time step. For a fully implicit method the function values $\mathbf{F}(\mathbf{x})$ must be evaluated at the forward time step, i.e.

$$\mathbf{x}^{n+1} = \mathbf{x}^n + \mathbf{F}(\mathbf{x}^{n+1})\,\Delta t \qquad (B.21)$$

But we do not yet know the value of $\mathbf{x}$ at this time step, so how do we evaluate the function values? To do so, we expand $\mathbf{F}(\mathbf{x})$ around the point $\mathbf{F}(\mathbf{x}^n)$ using one step of a Taylor series

$$\mathbf{F}(\mathbf{x}^{n+1}) \cong \mathbf{F}(\mathbf{x}^n) + \frac{d\mathbf{F}}{d\mathbf{x}}\bigg|_n (\mathbf{x}^{n+1} - \mathbf{x}^n) + \cdots \qquad (B.22)$$

Define $\Delta\mathbf{x} \equiv \mathbf{x}^{n+1} - \mathbf{x}^n$. The derivative in eq. (B.22) is called the Jacobian matrix, $\mathbf{J}$

$$\frac{d\mathbf{F}}{d\mathbf{x}}\bigg|_n \equiv \mathbf{J} = \begin{pmatrix} \dfrac{\partial F_1}{\partial x_1} & \dfrac{\partial F_1}{\partial x_2} & \cdots & \dfrac{\partial F_1}{\partial x_n} \\[2mm] \dfrac{\partial F_2}{\partial x_1} & \dfrac{\partial F_2}{\partial x_2} & \cdots & \dfrac{\partial F_2}{\partial x_n} \\[2mm] \vdots & & & \vdots \\[2mm] \dfrac{\partial F_n}{\partial x_1} & \dfrac{\partial F_n}{\partial x_2} & \cdots & \dfrac{\partial F_n}{\partial x_n} \end{pmatrix} \qquad (B.23)$$

Making use of these definitions and substituting back into eq. (B.21) yields

$$\Delta\mathbf{x} = [\mathbf{F}(\mathbf{x}^n) + \mathbf{J}\,\Delta\mathbf{x}]\cdot\Delta t \qquad (B.24)$$

Rearranging terms and dividing through by Δt then gives

$$\left(\frac{\mathbf{I}}{\Delta t} - \mathbf{J}\right)\Delta\mathbf{x} = \mathbf{F}(\mathbf{x}^n) \qquad (B.25)$$

Here, $\mathbf{I}$ is the identity matrix, which has 1s along the diagonal and 0s everywhere else. Equation (B.25) is a linear matrix equation that can be solved using standard techniques. So, this provides a two-step procedure for advancing the solution to the next time step. First, one solves eq. (B.25) for $\Delta\mathbf{x}$. Then, $\mathbf{x}$ is advanced according to

$$\mathbf{x}^{n+1} = \mathbf{x}^n + \Delta\mathbf{x} \qquad (B.26)$$

Importantly, the matrix $\mathbf{A} = (\mathbf{I}/\Delta t - \mathbf{J})$ on the left hand side of eq. (B.25) has a special form. Each species within layer j of the vertical grid is related chemically to other species in layer j and by diffusion to its own mixing ratio in layers $j-1$ and $j+1$ (see eqs. (B.16) and (B.18)). It is *not* related directly to chemical mixing ratios in any other layer or to its own mixing ratio more than one layer distant. Hence, most of the partial derivatives in the Jacobian matrix are zero. The $\mathbf{A}$ matrix therefore has a banded form, as shown in Fig. B.2. The dimensions of the matrix are still $K \times K$, typically 3000×3000, but the

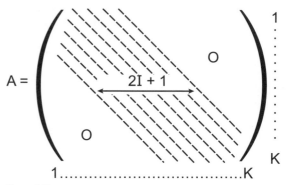

Figure B.2

bandwidth is only $2I + 1 = 61$ for the example given earlier. Such a matrix can be inverted much more quickly than could full 3000×3000 matrix. So, this system of equations can be advanced from time step to time step in only a small fraction of a second on a single-processor computer. In our model, the transport terms in the Jacobian matrix are computed analytically by inspection of eq. (B.16), while the chemistry terms are computed numerically by performing repeated calls to the chemistry routine. (See further discussion below.)

To find a solution, one makes an initial guess for $\mathbf{x}$, generally starting from a previous solution, and eqs. (B.25) and (B.26) are then iterated using a variable time step procedure. The time step is initially set to an extremely small value (10^{-8} s), then is allowed to increase as first the short-lived species and then the longer-lived species approach equilibrium. Our time-stepping algorithm tries to keep the maximum percentage change in any species at any height at ~20% of its value at the previous time step. Typically, several hundred to several thousand time steps are required to achieve a steady state, generally taken as an integration time of 10^{17} s, or ~3×10^9 yr. Wall clock time is a few tens of seconds to a few minutes on a 3-GHz Linux processor. Integrating for this long is overkill in some sense, but it enables the program to conserve mass to about one part in 10^{10}. All calculations are run in single precision on a 64-bit processor, allowing about 14 significant digits of precision. Note that as the time step approaches infinity, the first term in the brackets in eq. (B.25) vanishes, and so the method reverts to Newton's method for solving systems of algebraic equations.

B.4 Boundary Conditions

Boundary conditions must be applied to each species at both the top and bottom of the model grid. Obviously, eq. (B.16) cannot be used because it would involve quantities that lie outside the model grid. To develop the boundary conditions, we go back to eq. (B.10) and recast it in finite difference form as

$$\frac{df_i^j}{dt} = \frac{-1}{n^j \Delta z}\left(\Phi_i^{j+1/2} - \Phi_i^{j-1/2}\right) + \frac{p_i^j}{n^j} - l_i^j f_i^j \qquad (B.27)$$

Consider first the lower boundary, $j = 1$. Molecular diffusion is negligible near the surface, and the flux at the $j - 1/2$ gridpoint is just the flux at the ground, Φ_g. So, eq. (B.27) reduces to

$$\frac{df_i^1}{dt} = \frac{1}{n^1 \Delta z}\left[(Kn)^{3/2}\left(\frac{f_i^2 - f_i^1}{\Delta z}\right) + \Phi_g\right] + \frac{p_i^1}{n^1} - l_i^1 f_i^1 \qquad (B.28)$$

When recast in terms of the variables x_k, this equation can be used to impose flux boundary conditions. Two different types of flux boundary conditions are useful. (1) For a species like CH_4 in the modern atmosphere, which has a specified biogenic flux, one can simply fix Φ_g at this value. Alternatively, as the surface mixing ratio of CH_4 is known (~1.6 ppmv), one can specify fixed mixing ratio boundary conditions by setting $\partial f_i^1/\partial t = 0$, in which case the model will calculate the surface flux needed to sustain CH_4 at this observed concentration. (2) For species such as OH or O, which are removed chemically at the surface, or for soluble species such as SO_2 or H_2S, which dissolve in the oceans, we set $\Phi_g = -v_d n_i^1 = -v_d n^1 f_i^1$, where v_d is a *deposition velocity* (see Sec. 3.2). Deposition velocities are limited to values between 0 and 1 cm s^{-1} by boundary layer theory within the atmosphere (Lee and Schwartz, 1981; Slinn et al., 1978), with more reactive species having higher velocities. A deposition velocity of 1 cm s^{-1} would be appropriate over land for a species that is destroyed every time it hits the ground. Deposition velocities over the ocean are typically much smaller than this because they are limited by diffusion through the uppermost layer of the ocean and they depend on the concentration of the species in solution. See further discussion in Sec. 9.7.1.

Upper boundary conditions are treated in a similar manner, except that one cannot neglect molecular diffusion; hence, the full form of $\Phi_i^{j-1/2}$ must be used (see eq. (B.15)). The flux at the top of the grid, Φ_t, is set to zero for most species. In doing so, however, we first recast it in terms of an effusion velocity, v_{eff}, defined such that $\Phi_t = v_{eff} n_i^J = v_{eff} n^J f_i^J$; then, v_{eff} is set equal to zero. For H and H_2, v_{eff} is set equal to the diffusion-limited value, $v_{eff} = D_i/H_a$ (see eq. (5.53) in Ch. 5). Species such as O_2 that can be photodissociated above the model grid are

given upward velocities that balance the column-integrated loss above this level, i.e., $v_{eff} = J_{O_2} \cdot H_a$, where J_{O_2} is the photorate. Atomic O is then given a fixed downward flux equal to twice the upward flux of O_2. Likewise, photolysis of CO_2 above the model grid leads to downward fluxes of CO and O.

B.5   Including Particles

Particles can be treated in a manner directly analogous to gases if one assumes a *monodisperse* particle size distribution, i.e., one size at each altitude. To remain within the same numerical framework we solve for the effective "mixing ratio," f_{ip}, of the particle species, that is, the mixing ratio that would occur if the particle were gaseous. The relationship between f_{ip} and particle number density is determined by mass conservation, i.e., how many molecules of the species can fit within a sphere of radius r_i and density ρ_i. Molecular diffusion does not apply to particles, but they can be transported by eddy diffusion and they have a fall velocity w_i that depends on their radius and density through Stokes' Law. Thus, the flux of a given particle species must satisfy the relation

$$\Phi_{ip} = -Kn\frac{\partial f_{ip}}{\partial z} - w_i n f_{ip} \qquad (B.29)$$

which is analogous to eq. (B.9) for gaseous species. One can see by inspection that the fall velocity w_i simply replaces the molecular diffusion term ζ_i (eq. (B.8)) that occurs for gaseous species.

To estimate the particle size at each altitude we use a procedure described by Toon and Farlow (1981). At each time step, and beginning at the top of the model grid, lifetimes of the particle against sedimentation, diffusion, and rainout are computed, to determine which is the most important. The minimum of these three loss lifetimes is then compared against the particle coagulation lifetime. If coagulation is fast compared to the other time scales, the particles are allowed to grow. Particles are not allowed to decrease in size at low altitudes in this approximation, nor are they allowed to evaporate and convert back to gases. Because of this, and because some particles (e.g., sulfate and S_8) are not present in the upper part of the model, we generally do *not* include them in the Jacobian matrix. Instead, each particle equation is solved separately using a tridiagonal matrix analogous to that implied by eq. (B.16), but with the time derivative set to zero. Recall again that we are interested in finding steady-state solutions, so ignoring the time dependence of the particles is an acceptable approximation.

B.6 Setting up the Chemical Production and Loss Matrices

A significant challenge in solving large chemical systems is to compute the chemical production and loss rates for each species. These are shown by the terms $P_i/n - l_i f_i$ in eq. (B.10). Writing these out manually would be extremely tedious and would likely lead to errors, as each species can be involved in many different chemical reactions. To handle this, we developed a method for computing these rates automatically. To begin, the list of chemical reactions is written in a file in the form shown in Table B.1.

Here, the first column contains the reaction number, the next two columns contain the reactants, and the last three columns contain the products. Most photochemical reactions, including 3-body reactions and photolysis reactions can be put in this form. The rate of each reaction is $Rate_j = k_j n_1 n_2$, where k_j is the rate constant for reaction j, and n_1 and n_2 are the number densities of the two reactants. In the examples shown, reaction 1 is a standard reaction with two reactants, OH and O, and two products, O_2 and H. Reaction 2 is a three-body reaction; the third body density (not shown) is folded into the rate constant. Reaction 3 is a photolysis reaction, which can be handled in the same way if one equates the photolysis rate with the rate constant and assigns a number density of unity to the dummy variable "hv."

To make use of this reaction list, we use the following procedure. Within the Fortran 77 code we define a Hollerith sequence in which each chemical species is assigned a number. Once the data table has been read in, the symbols in each of the last five columns of the reaction table are replaced with species numbers. Then, the table is scanned to create production and loss matrices. The production matrix has the form shown in Table B.2.

Thus, for each of the N_{sp} different species, a list of reaction numbers, R_j, that form it is given in the corresponding row of the matrix. The production rate of species i can then be computed by using the information stored in Table B.1. The chemical loss matrix is similar, but it contains an additional layer that holds the species number of the species with which species i reacts.

B.7 Long- and Short-Lived Species and Ill-Conditioned Matrices

In principle, one could solve for all chemical species by the procedure outlined in Sections B.2 and B.3. In practice, doing so may not be optimal, for two reasons. (1) For very short-lived species such as the singlet D oxygen atom, $O(^1D)$, transport is not important. Such species are well described by assuming photochemical equilibrium: $P_i^j = l_i^j n_i^j$. So, one can reduce the size of the Jacobian matrix by removing such species from it and solving them by photochemical equilibrium. (2) More importantly, one can encounter situations in which the Jacobian matrix becomes ill conditioned, i.e., nearly singular, in which case errors may occur when solving the matrix eq. (B.25). In practice, the program becomes "lost" and is unable to increase the time step in the normal manner. This can happen, for example, if a particular species becomes decoupled from the other species because its abundance goes nearly to zero at some height within the model grid. The Jacobian matrix can then contain multiple rows in which most of the partial derivatives, $\partial F_i/\partial x_j$, go to zero.

When this latter problem occurs, it can be solved in several ways. One solution is to modify the chemical scheme in such a way that it remains more tightly coupled. That, however, may be unphysical. A second option is to simply ignore the errors that occur in the calculated mixing ratio of the decoupled species over

Table B.1 List of chemical reactions.

Reaction #	Reactants		Products		
R_1	OH	O	O_2	H	
R_2	H	O_2	HO_2		
R_3	O_2	hv	O	O	
R_4	HO_2	O_3	OH	O_2	O_2
$\ldots$					

Table B.2 Chemical production matrix.

1	R_3	R_7	R_8	R_9	$\ldots$														
2	R_1	R_4	R_{10}	$\ldots$															
$\ldots$																			
N_{sp}	R_5	R_7	R_{11}	$\ldots$															

some height range, that is, to not let such species influence the time step. We use this option frequently, particularly in large chemical systems such as those that occur in trying to solve for multiple hydrocarbon species in a weakly reduced atmosphere (see, e.g., Pavlov et al., 2001). A third option is to remove some species from the Jacobian matrix and solve for their densities by assuming photochemical equilibrium, as described above. Very long-lived species such as CO_2 can be removed from the matrix and solved by assuming diffusive equilibrium. This does not preclude including them in chemical reactions that affect other species. Only "neutral" species such as CO_2 and N_2 should be treated in this way in weakly reduced atmospheres in which redox balance is an issue. For Earth's modern atmosphere, O_2 is usually treated in this manner.

If one does choose to treat some species as short lived, one must be careful to avoid "dependencies." Short-lived species are solved sequentially in our model; hence, one must have already computed the densities of other short-lived species involved in their production. Consider the system of reactions

$$A + B \rightarrow C + D$$

$$C + E \rightarrow A + G$$

Suppose that "A" and "C" are both treated as short-lived species. To find the density of "C," one must first know the density of "A." But, likewise, to find "A" one must first know "C." This would therefore constitute a dependency. So, either species "A" or "C" must be treated as a long-lived species. One could get around this problem by solving for the short-lived species simultaneously using a Newton step. But then one would essentially be doing a Newton step each time one evaluated a partial derivative within the Jacobian matrix, and this would be inefficient. Another obvious dependency would be a species that reacted with itself, e.g.

$$A + A \rightarrow B + C$$

To evaluate the loss frequency of "A," one needs to already know its density; hence, "A" cannot be treated as a short-lived species. Similarly, if species "A" reacts to form "B," but "B" is not involved in producing "A," then they can both be treated as short-lived, but only if "A" is calculated ahead of "B."

We emphasize the problem of dependencies because, in practice, this is the easiest way to go wrong in creating a photochemical model. Because the chemical terms in the Jacobian matrix are evaluated numerically, any dependency within the short-lived species will result in an incorrect Jacobian matrix. Typically, this problem manifests itself, as before, by the program appearing to become "lost," and thus failing to converge. The safest way to check for dependencies, if one's compiler allows it, is to initialize all variables as negative indefinites. (The old Cray supercomputers used to do this automatically, which was very convenient.) A negative indefinite will cause the program to crash if it is used in an arithmetic operation; hence, all dependencies are automatically detected. Some modern Fortran compilers may only allow variables to be initialized to zero. In that case, the safest option is to check for zeros the first time one runs the chemistry routine to see if this is being caused by a dependency.

B.8 Rainout, Lightning, and Photolysis

Three additional physical processes that are included in the model are rainout, lightning, and photolysis. Rainout is treated by the method described in Giorgi and Chameides (1985). Each long-lived species is assigned a solubility. For most species, e.g., O_2, the solubility is just that corresponding to Henry's Law. Species that dissociate in solution, though, like SO_2, are given pH-dependent "effective" solubilities that account for the total dissolved abundance of the species and its dissociation products. This requires estimating the pH of raindrops. So, at each layer within the model troposphere, raindrop chemistry is computed by taking the relevant gaseous and dissolved species, typically CO_2, SO_2, and H_2CO, and using a Newton–Raphson step to solve a system of ten coupled algebraic equations. Once the raindrop pH is known, effective solubilities are computed, and then the formalism of Giorgi and Chameides is used to calculate rainout rates. Highly soluble species such as H_2SO_4 rain out on a time scale of about 5 days, whereas less soluble species (SO_2, H_2S) can have lifetimes of several weeks.

Lightning is also included in the model, as discussed in Sec. 9.2. Lightning is the main natural source of odd nitrogen compounds in Earth's present troposphere and of tropospheric O_2 in the prebiotic atmosphere. Lightning production of NO in the modern atmosphere is assumed to occur at a column-integrated rate of 10^9 cm^{-2}s^{-1} (Borucki and Chameides, 1984; Martin et al., 2002). Production rates of other gases are scaled by assuming that thermodynamic equilibrium occurs at an effective freeze-out temperature of 3500 K. This represents a simplification of the more elaborate model of Chameides et al. (1977) in which different species are removed sequentially at different freeze-out temperatures. The errors involved in making this approximation have not been quantified, although they are presumably no larger than other uncertainties in early atmosphere photochemical models. Species produced in this manner are assumed to be concentrated in the lower troposphere, following the rainout profile of Fishman and Crutzen (1977). Redox balance is ensured in low-O_2 atmospheres by requiring that all species be produced in redox-balanced proportions.

Finally, photolysis is obviously a key process in photochemical models, as this is ultimately what makes all atmospheric photochemistry work. Photorates in our model are calculated using the two-stream approximation of Toon *et al.* (1989), which accounts for both absorption and multiple scattering. This is the same algorithm that is used in the 1-D radiative-convective model described in Appendix A. Photolysis can occur as a consequence of photons arriving from any direction, and so the photochemical model uses the calculated mean intensity rather than the net flux. We assume 118 wavelength bins, spanning the wavelength range from 121.6 nm (Ly-α) to 850 nm. Rayleigh scattering is included at all wavelengths. Radiative transfer within the O_2 Schumann–Runge bands, 175–205 nm, is handled using an exponential sum fit to the Allen and Frederick (1982) band model. Again, the accuracy of this scheme has not been tested at low O_2 levels – yet another area where the model could use further improvement.

Appendix C: Atomic States and Term Symbols

Electronic states are expressed in atmospheric chemistry as ground state and electronically excited states. For example, the *ground electronic state* of O atoms is given the notation $O(^3P)$, and the *first electronically excited state* is designated $O(^1D)$. Inside the brackets is a *Russell–Saunders*[1] *term symbol*, which is a compact way of describing the angular momentum quantum numbers in a multi-electron atom. Because term symbols are commonly used in writing equations of atmospheric chemistry, here we describe where term symbols come from, which is the quantum physics of atoms with many electrons. Moreover, the total angular momentum determines the transition probabilities in radiative absorption or emission and the magnetic properties of an atom. The reader is referred to various quantum textbooks for more detail (e.g., Atkins and Friedman, 2005; Ellis *et al.*, 2005; McQuarrie and Simon, 1997).

Let us take an oxygen atom as our example. Oxygen has an atomic number of 8. Consequently, O atoms have two inner shell electrons and six outer shell electrons. Individual electronic orbitals in O atoms are given labels with electron occupancy as the superscript:

$$1s^2 2s^2 2p^4$$

Here, the outer p orbital is not full (it fills to six electrons in neon) and so the distribution of electrons and their spins in an O atom can have a variety of configurations, which give rise to the different atomic electronic states.

To understand electronic states, we need to review the quantum description of electrons given in terms of five quantum numbers. Electrons orbit in shells, defined by a particular energy level designated by a principal (or Bohr) quantum number, i.e., the numbers 1 and 2 in the above

electronic configuration: 1s, 2s, 2p. Electrons also have their own spin. The orbit gives rise to orbital angular momentum, which has magnitude and direction, and the spin gives rise to spin angular momentum, which has magnitude and direction. Magnitude is quantized and requires a quantum number, and direction is also quantized and needs another quantum number. Thus, quantum numbers for orbital and spin angular momentum (two each) are added to the principal quantum number to make five quantum numbers in total. Letters designating these quantum numbers are n, l, m_l, s, and m_s, as follows.

(1) **Principal quantum number**, n, where $n = 1, 2, 3, ...$

This quantum number governs the total energy of the electron and corresponds to the Bohr quantum number or electron shell number.

(2) **Orbital (or azimuthal) quantum number**, l, where $l = 0, 1, 2, ..., (n-1)$.

This quantum number governs the magnitude of the electron's angular momentum around the nucleus and corresponds to an *atomic orbital* or *subshell*. Early spectroscopists had names for alkali metal spectroscopic lines, and these historical names cause the electron angular-momentum states to be designated by lower-case letters. The letter s ("sharp") corresponds to $l = 0$, p ("principal") to $l = 1$, d ("diffuse") to $l = 2$, and f ("fundamental") to $l = 3$. After f, the letters become alphabetical, as follows:

$l =$	0	1	2	3	4	5	6...
	s	p	d	f	g	h	i...

If an electron has orbital quantum number l, then it has angular momentum of magnitude $L_e = \sqrt{l(l+1)}(h/2\pi)$.

(3) **Magnetic quantum number**, m_l, where $m_l = 0, \pm 1, \pm 2, \pm 3, ..., \pm l$. There are $2l + 1$ values for any given l.

---

[1] Henry Norris Russell, who also gives his name to the *Hertzsprung–Russell diagram* of stars (Fig. 2.2).

The magnetic quantum number, m_l, governs the direction of the angular momentum. Thus, while quantum number l determines the magnitude of L_e of electron momentum, $\mathbf{L_e}$, we also need direction specification from m_l.

(4) **Spin quantum number**, s, where $s = \frac{1}{2}$ only.

The spin quantum number, s, describes the magnitude S_e of the spin angular momentum of the electron, $\mathbf{S_e}$. The magnitude of the angular momentum due to the spin is given by $S_e = \sqrt{s(s+1)}(h/2\pi) = (\sqrt{3}/2)(h/2\pi)$.

(5) **Spin magnetic quantum number**, m_s, where $m_s = +1/2$ and $-1/2$.

The spin magnetic quantum number describes the direction of vector $\mathbf{S_e}$.

The total orbital and spin angular momenta of the electrons in a closed subshell is zero. Thus, for the O atom, we ignore electrons in the inner closed subshells ($1s$ and $2s$) and just consider the $2p$ electrons to obtain the atomic electronic state. For p electrons, $l = 1$, as noted above, so the magnetic quantum number, m_l, can equal 1, 0, –1. To understand all the possible atomic states, we have to work out all of the ways in which four p electrons can be arranged with the six parameters ($m_l = 1,0,-1$) and $m_s = +1/2$ and $-1/2$.

We also need to be aware of *Hund's Rules*, which state that for the lowest and most stable energy state of an atom, the electrons in a subshell remain unpaired, i.e., have parallel spins whenever possible. This is because of the mutual repulsion of electrons: the farther apart the electrons in an atom are, the lower the energy of the atom. Thus, the electronic state with the largest total spin angular momentum has the lowest energy.

We can calculate the combinations of p electrons in the O atom. There are six different combinations corresponding to two m_s (spin) values and 3 m_l values. One can also think of the p subshell with six possible individual orbitals ($2p_x\alpha$, $2p_x\beta$; $2p_y\alpha$, $2p_y\beta$; $2p_z\alpha$, $2p_z\beta$). For our four electrons, there are six choices for the first spin orbital quantum state, five for the second, four for the third, and three for the fourth, giving $6\times5\times4\times3$ ways. But the electrons are indistinguishable and the order we choose the orbitals is irrelevant, so we divide $6\times5\times4\times3$ by $4\times3\times2\times1$, giving 15 ways. Expressing this as a formula, we arrange the four electrons in $6!/(4!2!) = 15$ ways. More generally, the number of distinct ways to assign N electrons to G spin orbital quantum states is:

$$\text{number of ways} = \frac{G!}{N!(G-N)!} \tag{C.1}$$

Table C.1 Fifteen different arrangements of two electrons in $2p$ orbital. An up or down arrow indicates the spin state of the electron, $m_s = +1/2$ or $-1/2$.

Magnetic quantum number (m_l) value		
$m_l = 1$	$m_l = 0$	$m_l = -1$
↑↓		
	↑↓	
		↑↓
↑	↑	
↑		↑
↑	↓	
↑		↓
↓	↑	
↓		↑
↓	↓	
↓		↓
	↑	↑
	↑	↓
	↓	↑
	↓	↓

To simplify matters for the O atom, we note that choosing four items from six is the same as choosing the two to be left behind. Thus, we can start by considering the possible arrangements of two "left-behind" electrons. Table C.1 shows combinations of two electrons in a $2p$ orbital. Then we fill in the blanks of Table C.1 to generate Table C.2, which shows the actual four electrons for the O atom.

In the arrangements of electrons shown in each row of Table C.2, we sum all the m_l values to give M_L, which is the quantum number associated with the z-component of the total orbital angular momentum,

$$L_z = M_L \hbar \quad M_L = \pm L, \pm(L-1), ..., 0 \tag{C.2}$$

For example, in the first row of the table, $M_L = (2\times0) + (2\times-1) = -2$. Essentially, the orbital angular momenta of the electrons couple together and $\mathbf{L}$ is the vector sum of all the $\mathbf{L_e}$s due to each electron. The value of M_L describes the direction of the vector $\mathbf{L}$. We also sum the m_s values of each electron to give M_S, which is the quantum number associated with the z-component of the total spin angular momentum of the state,

$$S_z = M_S \hbar \quad M_S = \pm S, \pm(S-1), ..., 0 \tag{C.3}$$

For example, in the last row of the table, $M_S = \frac{1}{2}+(-\frac{1}{2}) +\frac{1}{2}+\frac{1}{2} = 1$.

The vector $\mathbf{J} = \mathbf{L} + \mathbf{S}$ gives the total angular momentum for a given configuration, which is called *spin–orbit*

Table C.2 Electronic states of an O atom given the $2p$ orbital electronic configuration. Shaded columns are values derived from the left-hand columns.

Magnetic quantum number (m_l) value			Total orbital quantum number, $M_L = \sum m_l$	Total spin quantum number, $M_S = \sum m_s$	O atom's electronic state	L	M_L
$m_l = 1$	$m_l = 0$	$m_l = -1$					
	↑↓	↑↓	−2	0	^1D	2	−L
↑↓	↑↓		0	0	^1S	0	L
↑↓		↑↓	2	0	^1D	2	L
↓	↓	↑↓	−1	−1	^3P	1	−L
↓	↑↓	↓	0	−1	^3P	1	L−1
↓	↑	↑↓	−1	0	^1D	2	−L+1
↓	↑↓	↑	0	0	^3P	1	L−1
↑	↓	↑↓	−1	0	^3P	1	−L
↑	↑↓	↓	0	0	^1D	2	L−2
↑	↑	↑↓	−1	1	^3P	1	−L
↑	↑↓	↑	0	1	^3P	1	L−1
↑↓	↓	↓	1	−1	^3P	1	L
↑↓	↓	↑	1	0	^1D	2	L−1
↑↓	↑	↓	1	0	^3P	1	L
↑↓	↑	↑	1	1	^3P	1	L

coupling. The total angular momentum has a magnetic quantum number $M_J = M_L + M_S$, with a range:

$$M_J = \pm J, \pm (J-1), ..., 0$$
$$J = L + S, (L+S) - 1, (L+S) - 2, ..., |L - S| \quad \text{(C.4)}$$

Thus, the state of the atom is specified by a set L, S, and J quantum numbers, with J depending on L and S. An atomic electronic configuration with a given L and S is called a *term* and can be represented by a *term symbol*:

$$\text{term symbol} = {}^{(2S+1)}_{\substack{\text{the number}}} L_{\substack{\text{corresponding letter}}} \quad \text{(C.5)}$$

States of a term having the same L- and S-values but different J-values are close in energy and are a *multiplet*. Sometimes the term symbol has a J subscript, i.e., $^{(2S+1)}L_J$ to specify a particular state within a multiplet. The total number of such states is $2S+1$, which is called the *multiplicity* and is the superscript in the term symbol. More generally, for multiplicity, we have the following.

When $S = 0$, the multiplicity is 1, called a *singlet state*, and $J = L$.

When $S = ½$, the multiplicity is 2, called a *doublet state*, and $J = L \pm ½$.

When $S = 1$, the multiplicity is 3, called a *triplet state*, and $J = L \pm 1$, L, or $L-1$. And so on.

The total orbital angular-momentum quantum number L is given by capitalized letters analogous to the letters used for electron orbitals, as follows:

$L =$	0	1	2	3	4	5	6...
	S	P	D	F	G	H	$I...$

Before, we saw that the value of the orbital quantum number, l, was equal to the largest m_l quantum number value. For the O atom, the largest possible M_L value is equal to the L, which is the *total orbital angular momentum* of the atom. But we must bear in mind that we will see combinations of $M_L = L, L-1, ..., -L$, and $M_S = S, S-1, ..., -S$.

To work out electronic states of an atom (e.g., in Table C.2), we follow a procedure, which works for all but the heaviest atoms:

Step 1. We find the largest M_S and, for it, the largest M_L, which will be equal to maximum S for this state. This largest spin state is the most stable state by Hund's rules.

In our case, in Table C.2, the largest M_S is 1. Then for $M_S = 1$, the largest M_L is 1. With $L = M_L = 1$ and $S = M_S = 1$, the lowest energy state of the O atom is designated ^3P, using eq. (C.5). There are three (multiplicity) J values of 0, 1, and 2. In equations for atmospheric chemistry, O(^3P) represents the ground state O atom, e.g., eq. (3.44) describes absorption of shortwave ultraviolet by stratospheric ozone that generates O(^3P) after an excited O atom state is quenched.

With $L = 1$ and $S = 1$, the possible values of both M_L and M_S from eqs. (3.2) and (3.3) are −1, 0, 1. As a result, ^3P term accounts for nine of the rows in Table C.2 with so-called basis states

$$(M_L, M_S) = (-1, 1), (-1, 0), (-1, 1), (0, -1), (0, -1), (0, 0), (0, 1), (1-, -1), (1, 0), and (1, 1).$$

We do not label all $(0, 0)$ states as ^3P in Table C.2 because a particular (M_L, M_S) combination can only appear once. The Pauli Exclusion Principle states that identical fermions (e.g., electrons) cannot occupy the same quantum state.

Step 2. We find the next largest M_S and, for it, the largest M_L. We repeat the identification of the next term.

In Table C.2, the next largest M_S is $M_S = 0$, which is found in all six rows of Table C.2 that are not in the ^3P ground state. Then for $M_S = 0$, the largest M_L is 2. With $L = M_L = 2$ and $S = M_S = 0$, the next highest energy state of the O atom is designated ^1D, using eq. (C.5) with one J value of 2. This *first electronically excited state*, denoted O(^1D), is said as "O singlet D." Looking at the up and down arrows in Table C.2, we see that O(^1D) corresponds to the excitation of one electron in the $2p$ orbital. With $L = 2$ and $S = 0$, the possible values of both M_L and M_S from eqs. (C.2) and (C.3), respectively, are $M_L = -2, -1, 0, 1, 2$ and $M_S = 0$. This makes five basis states: $(M_L, M_S) = (-2, 0)$,

$(-1, 0)$, $(0, 0)$, $(1, 0)$, and $(2, 0)$, and these rows are labeled ^1D in Table C.2.

Step 3. Repeat Step 2.

In Table C.2, one row with $(M_L, M_S) = (0, 0)$ remains unassigned. This has $L = S = 0$, so has the term symbol ^1S with $J = 0$. This "O singlet S" state is the *second excited state* of atomic oxygen.

In this way, we see that the states of atomic oxygen are a ground state ^3P, and nearby excited states ^1D and ^1S. Of course, if electrons were excited into higher orbitals such as the $3s$ orbital, higher excitations states would be produced with different terms.

Changes between different states of O atoms are important for emission and absorption of radiation in the atmosphere. In the Earth's thermosphere, O atoms are excited by charged solar particles. A transition to the ground state, $^1D \rightarrow ^3P_2$, produces red auroral emission at 630 nm, while $^1S \rightarrow ^1D$ produces green auroral emission at 557.7 nm. Quantum selection rules determine the full range of such electric dipole transitions ($\Delta L = \pm 1$, $\Delta S = 0$, $\Delta J = 0$ or ± 1 (but $J=0$ to $J'= 0$ is forbidden), and $\Delta M_J = 0$ or ± 1).

Bibliography

Abbot, D. S., *et al.* (2012). Indication of insensitivity of planetary weathering behavior and habitable zone to surface land fraction. *Astrophys. J.* **756**, 178.

Abbot, D. S. and Pierrehumbert, R. T. (2010). Mudball: Surface dust and Snowball Earth deglaciation. *J. Geophys. Res.* **115**, D03104, doi: 10.1029/2009JD012007.

Abbot, D. S., *et al.* (2011). The Jormungand global climate state and implications for Neoproterozoic glaciations. *J. Geophys. Res.* **116**, D18103, doi:10.1029/2011JD015927.

Abe, Y. (2011). Protoatmospheres and surface environment of protoplanets. *Earth Moon Planets* **108**, 9–14.

Abe, Y., *et al.* (2011). Habitable zone limits for dry planets. *Astrobiology* **11**, 443–460.

Abe, Y. and Matsui, T. (1988). Evolution of an impact-generated H2O-CO2 atmosphere and formation of a hot proto-ocean on Earth. *J. Atmos. Sci.* **45**, 3081–3101.

Abelson, P. H. (1966). Chemical events on the primitive Earth. *Proc. Nat. Acad. Sci.* **55**, 1365.

Achterberg, R. K., *et al.* (2008). Titan's middle-atmospheric temperatures and dynamics observed by the Cassini Composite Infrared Spectrometer. *Icarus* **194**, 263–277.

Achterberg, R. K., *et al.* (2011). Temporal variations of Titan's middle-atmospheric temperatures from 2004 to 2009 observed by Cassini/CIRS. *Icarus* **211**, 686–698.

Ackiss, S. E. and Wray, J. (2014). Occurrences of possible hydrated sulfates in the southern high latitudes of Mars. *Icarus* **243**, 311–324.

Acuna, M. H., *et al.* (1999). Global distribution of crustal magnetization discovered by the Mars Global Surveyor MAG/ER experiment. *Science* **284**, 790–793.

Adams, E. Y. (2006). *Titan's thermal structure and the formation of a nitrogen atmosphere*. University of Michigan, Ph.D. thesis, Ann Arbor, MI.

Agee, C. B., *et al.* (2013). Unique meteorite from early Amazonian Mars: Water-rich basaltic breccia Northwest Africa 7034. *Science* **339**, 780–785.

Agnor, C. and Asphaug, E. (2004). Accretion efficiency during planetary collisions. *Astrophys. J.* **613**, L157–L160.

Agnor, C. B. and Hamilton, D. P. (2006). Neptune's capture of its moon Triton in a binary-planet gravitational encounter. *Nature* **441**, 192–194.

Agol, E., *et al.* (2010). The climate of HD 189733b from fourteen transits and eclipses measured by Spitzer. *Ap. J.* **721**, 1861–1877.

Agol, E., *et al.* (2005). On detecting terrestrial planets with timing of giant planet transits. *Mon. Not. R. Astron. Soc.* **359**, 567–579.

Aharonson, O., *et al.* (2009). An asymmetric distribution of lakes on Titan as a possible consequence of orbital forcing. *Nat. Geosci.* **2**, 851–854,

Aharonson, O., *et al.* (2002). Drainage basins and channel incision on Mars. *P. Natl. Acad. Sci. USA* **99**, 1780–1783.

Ahrens, T. J. (1993). Impact erosion of terrestrial planetary atmospheres. *Annu. Rev. Earth Planet. Sci.* **21**, 525–555.

Alexander, B., *et al.* (2003). East Antarctic ice core sulfur isotope measurements over a complete glacial–interglacial cycle. *J. Geophys. Res.* **108**, 4786, doi:10.1029/2003JD003513.

Alexander, R. D., et al. (2006). Photoevaporation of protoplanetary discs - II. *Evolutionary models and observable properties. Mon. Not. R. Astron. Soc.* **369**, 229–239.

Alfimova, N. A., et al. (2011). Mobility of cerium in the 2.8–2.1 Ga exogenous environments of the Baltic Shield: data on weathering profiles and sedimentary carbonates. *Lithol. Miner. Resour.* **46**, 397–408.

Alibert, Y. and Mousis, O. (2007). Formation of Titan in Saturn's subnebula: constraints from Huygens probe measurements. *Astron. Astrophys.* **465**, 1051–1060.

Allard, P. (1997). Endogenous magma degassing and storage at Mount Etna. *Geophys. Res. Lett.* **24**, 2219–2222.

Allegre, C. J., et al. (1987). Rare gas systematics: Formation of the atmosphere, evolution and structure of the Earths mantle. *Earth Planet. Sci. Lett.* **81**, 127–150.

Allen, M. and Frederick, J. E. (1982). Effective photo-dissociation cross sections for molecular oxygen and nitric oxide in the Schumann–Runge bands. *J. Atmos. Sci.* **39**, 2066–2075.

Allen, P. A. and Etienne, J. L. (2008). Sedimentary challenge to Snowball Earth. *Nature Geosc.* **1**, 817–825.

Allwood, A. C., et al. (2009). Controls on development and diversity of Early Archean stromatolites. *P. Natl. Acad. Sci. USA* **106**, 9548–9555.

Allwood, A. C., et al. (2006). Stromatolite reef from the Early Archaean era of Australia. *Nature* **441**, 714–718.

ALMA-Partnership, et al. (2015). The 2014 ALMA Long Baseline Campaign: First Results from High Angular Resolution Observations toward the HL Tau Region. *Astrophys. J. Lett.* **808**, L3, doi:10.1088/2041-8205/808/1/L3.

Alt, J. C. (1995). Sulfur isotopic profile through the oceanic crust: Sulfur mobility and seawater-crustal sulfur exchange during hydrothermal alteration. *Geology* **23**, 585–588.

Altabet, M. A. and Francois, R. (1994). Sedimentary nitrogen isotopic ratio as a recorder for surface ccean nitrate utilization. *Global Biogeochemical Cycles* **8**, 103–116.

Altermann, W. and Schopf, J. W. (1995). Microfossils from the Neoarchean Campbell Group, Griqualand West Sequence of the Transvaal Supergroup, and their paleoenvironmental and evolutionary Implications. *Precambrian Res.* **75**, 65–90.

Altwegg, K. et al. (2015). 67P/Churyumov–Gerasimenko, a Jupiter family comet with a high D/H ratio. *Science* **347**, doi: 10.1126/science.1261952.

Amelin, Y., et al. (2010). U–Pb chronology of the Solar System's oldest solids with variable 238U/235U. *Earth Planet. Sci. Lett.* **300**, 343–350.

Anbar, A. D., et al. (2007). A whiff of oxygen before the Great Oxidation Event? *Science* **317**, 1903–1906.

Anbar, A. D. and Knoll, A. H. (2002). Proterozoic ocean chemistry and evolution: a bioinorganic bridge? *Science* **297**, 1137–1142.

Anbar, A. D. and Rouxel, O. (2007). Metal stable isotopes in paleoceanography. *Annu. Rev. Earth Planet. Sci.* **35**, 717–746.

Anders, E. and Grevesse, N. (1989). Abundances of the elements – meteoritic and solar. *Geochim. Cosmochim. Acta* **53**, 197–214.

Anderson, D. E. (1974). Mariner 6, 7, and 9 ultraviolet spectrometer experiment: Analysis of hydrogen Lyman alpha data. *J. Geophys. Res.* **79**, 1513–1518.

Anderson, D. E. and Hord, C. W. (1971). Mariner 6 and Mariner 7 ultraviolet spectrometer experiment: analysis of hydrogen Lyman-alpha data. *J. Geophys. Res.* **76**, 6666–6673.

Anderson, F. S., et al. (1999). Assessing the Martian surface distribution of aeolian sand using a Mars general circulation model. *J. Geophys. Res.* **104**, 18 991–19 002.

Anderson, G. M. (2005). *Thermodynamics of Natural Systems*. New York: Cambridge University Press.

Anderson, G. M. and Crerar, D. A. (1993). *Thermodynamics in Geochemistry: The Equilibrium Model*. New York: Oxford University Press.

Andre, M. J. (2011). Modelling $^{18}O_2$ and $^{16}O_2$ unidirectional fluxes in plants: I. Regulation of pre-industrial atmosphere. *Biosystems* **103**, 239–251.

Andrews, D. G. (2010). *An Introduction to Atmospheric Physics*. New York: Cambridge University Press.

Andrews, D. G., et al. (1987). *Middle Atmosphere Dynamics*. Orlando: Academic Press.

Andrews-Hanna, J. C., et al. (2007). Meridiani Planum and the global hydrology of Mars. *Nature* **446**, 163–166.

Andrews-Hanna, J. C., *et al.* (2010). Early Mars hydrology: Meridiani playa deposits and the sedimentary record of Arabia Terra. *Journal of Geophysical Research-Planets* **115**.

Anglada-Escude, G., *et al.* (2016). A terrestrial planet candidate in a temperate orbit around Proxima Centauri. *Nature* **536**, 437–440.

Ansan, V., *et al.* (2011). Stratigraphy, mineralogy, and origin of layered deposits inside Terby crater, Mars. *Icarus* **211**, 273–304.

Archer, C. and Vance, D. (2006). Coupled Fe and S isotope evidence for Archean microbial Fe(III) and sulfate reduction. *Geology* **34**, 153–156.

Archer, D. (2005). Fate of fossil fuel CO_2 in geologic time. *J. Geophys. Res.* **110**.

Armstrong, B. H. (1968). Theory of diffusivity factor for atmospheric radiation. *J. Quant. Spectros. Radiat. Transfer* **8**, 1577–1599.

Armstrong, J. C. and Leovy, C. B. (2005). Long term wind erosion on Mars. *Icarus* **176**, 57–74.

Armstrong, J. C., *et al.* (2004). A 1 Gyr climate model for Mars: new orbital statistics and the importance of seasonally resolved polar processes. *Icarus* **171**, 255–271.

Arnold, G. L., *et al.* (2004). Molybdenum isotope evidence for widespread anoxia in mid-Proterozoic oceans. *Science* **304**, 87–90.

Arnold, L., *et al.* (2002). A test for the search for life on extrasolar planets-Looking for the terrestrial vegetation signature in the Earthshine spectrum. *Astron. Astrophys.* **392**, 231–237.

Artemieva, N. and Lunine, J. I. (2005). Impact cratering on Titan - II. Global melt, escaping ejecta, and aqueous alteration of surface organics. *Icarus* **175**, 522–533.

Arthur, M. A. (2000). Volcanic contributions to the carbon and sulfur geochemical cycles and global change. In: *Encyclopedia of Volcanoes*, ed. H. Sigurdsson, Academic Press, pp. 1045–1056.

Arthur, M. A., *et al.* (1994). Varve-calibrated records of carbonate and organic carbon accumulation over the last 2000 years in the Black Sea. *Glob. Biogeochem. Cycles* **8**, 195–217.

Arvidson, R. E., *et al.* (2014). Ancient aqueous environments at Endeavour Crater. *Mars. Science* **343**, doi:10.1126/science.1248097.

Asael, D., *et al.* (2013). Coupled molybdenum, iron and uranium stable isotopes as oceanic paleoredox proxies during the Paleoproterozoic Shunga Event. *Chem. Geol.* **362**, 193–210.

Aston, F. W. (1924). The rarity of the inert gases on Earth. *Nature* **114**, 786.

Atkins, P. W. and Friedman, R. S. (2005). *Molecular Quantum Mechanics*. New York: Oxford University Press.

Atkinson, D. H., *et al.* (1997). Deep winds on Jupiter as measured by the Galileo probe. *Nature* **388**, 649–650.

Atreya, S. K. (1986). *Atmospheres and Ionospheres of the Outer Planets and their Satellites*. Heidelberg: Springer-Verlag.

Atreya, S. K. (2010). Atmospheric moons Galileo would have loved. *Proc. IAU* **6**, 130–140. doi:10.1017/S1743921310007349.

Atreya, S. K., *et al.* (2006). Titan's methane cycle. *Planet. Space Sci.* **54**, 1177–1187.

Atreya, S. K., *et al.* (1978). Evolution of a nitrogen atmosphere on Titan. *Science* **201**, 611–613.

Atreya, S. K., *et al.* (2009). Volatile origin and cycles: Nitrogen and methane. In: *Titan from Cassini-Huygens*, ed. R. H. Brown, *et al.*, New York: Springer, pp. 77–99.

Atreya, S. K., *et al.* (2003). Composition and origin of the atmosphere of Jupiter - an update, and implications for the extrasolar giant planets. *Planet. Space Sci.* **51**, 105–112.

Atreya, S. K., *et al.* (1991). Photochemistry and vertical mixing. In: *Uranus*, ed. J. T. Bergstralh, *et al.*, Tucson, AZ: Univ. of Arizona Press, pp. 110–146.

Atreya, S. K., *et al.* (2013). Primordial argon isotope fractionation in the atmosphere of Mars measured by the SAM instrument on Curiosity and implications for atmospheric loss. *Geophys. Res. Lett.* **40**, 5605–5609.

Atreya, S. K., *et al.* (1999). A comparison of the atmospheres of Jupiter and Saturn: deep atmospheric composition, cloud structure, vertical mixing, and origin. *Planet. Space Sci.* **47**, 1243–1262.

Aulbach, S. and Stagno, V. (2016). Evidence for a reducing Archean ambient mantle and its effects on the carbon cycle. *Geology* **44**, 751–754.

Awramik, S. M. and Barghoorn, E. S. (1977). Gunflint Microbiota. *Precambrian Res.* **5**, 121–142.

Axford, W. I. (1968). The polar wind and the terrestrial helium budget. *J. Geophys. Res.* **73**, 6855–6859.

Bahcall, J. N., *et al.* (2001). Solar models: Current epoch and time dependences, neutrinos, and helioseismological properties. *Ap. J.* **555**, 990–1012.

Bailey, J. (2009). A comparison of water vapor line parameters for modeling the Venus deep atmosphere. *Icarus* **201**, 444–453.

Baker, N. L. and Leovy, C. B. (1987). Zonal winds near Venus cloud top level: a model study of the interaction between the zonal mean circulation and the semidiurnal tide. *Icarus* **69**, 202–220.

Baker, V. R. (1990). Spring sapping and valley network development, with case studies by R. C. Kochel, J. E. Laity, and A. D. Howard. In: C. G. Higgins, D. R. Coates, ed., *Groundwater Geomorphology: The Role of Subsurface Water in Earth-Surface Processes and Landforms*, Geol. Soc. Am. Spec. Pap., Vol. 252, pp. 235–265.

Baker, V. R. (2001). Water and the martian landscape. *Nature* **412**, 228–236.

Baker, V. R. (2006). Geomorphological evidence for water on Mars. *Elements* **2**, 139–143.

Baker, V. R., *et al.* (2015). Fluvial geomorphology on Earth-like planetary surfaces: A review. *Geomorphology* **245**, 149–182.

Baland, R.-M., *et al.* (2014). Titan's internal structure inferred from its gravity field, shape, and rotation state. *Icarus* **237**, 29–41.

Balme, M. and Greeley, R. (2006). Dust devils on Earth and Mars. *Rev. Geophys.* **44**, RG3003 doi:10.1029/2005RG000188.

Bandfield, J. L. (2002). Global mineral distributions on Mars. *J. Geophys. Res.* **107**.

Bandfield, J. L., *et al.* (2003). Spectroscopic identification of carbonate minerals in the martian dust. *Science* **301**, 1084–1087.

Banks, P. M. and Kockarts, G. (1973). *Aeronomy: Part B.* New York: Academic Press.

Bannon, P. R., *et al.* (1997). Does the surface pressure equal the weight per unit area of a hydrostatic atmosphere? *Bull. Am. Met. Soc.* **78**, 2637–2642.

Bao, H. M., *et al.* (2009). Stretching the envelope of past surface environments: Neoproterozoic glacial lakes from Svalbard. *Science* **323**, 119–122.

Bao, H. M., *et al.* (2008). Triple oxygen isotope evidence for elevated CO_2 levels after a Neoproterozoic glaciation. *Nature* **453**, 504–506.

Barabash, S., *et al.* (2007). The loss of ions from Venus through the plasma wake. *Nature* **450**, 650–653.

Baraffe, I., *et al.* (1998). Evolutionary models for solar metallicity low-mass stars: Mass-magnitude relationships and color-magnitude diagrams. *Astron. Astrophys.* **337**, 403–412.

Baraffe, O., *et al.* (2002). Evolutionary models for low-mass stars and brown dwarfs: Uncertainties and limits at very young ages. *Astron. Astrophys.* **382**, 563–572.

Barakat, A. R. and Lemaire, J. (1990). Monte Carlo study of the escape of a minor species. *Phys. Rev. A* **42**, 3291–3302.

Barnes, R., *et al.* (2009). Tidal limits to planetary habitability. *Astrophys. J. Lett.* **700**, L30–L33.

Barnes, R., *et al.* (2013). Tidal Venuses: Triggering a climate catastrophe via tidal heating. *Astrobiology* **13**, 225–250.

Barnes, R., *et al.* (2010). CoRoT-7b: Super-Earth or Super-Io? *Astrophys. J. Lett.* **709**, L95–L98.

Barnhart, C. J., *et al.* (2009). Long-term precipitation and late-stage valley network formation: Landform simulations of Parana Basin, *Mars. J. Geophys. Res.* **114**.

Barr, A. C. and Canup, R. M. (2010). Origin of the Ganymede-Callisto dichotomy by impacts during the late heavy bombardment. *Nat. Geosci.* **3**, 164–167.

Barr, A. C. and Citron, R. I. (2011). Scaling of melt production in hypervelocity impacts from high-resolution numerical simulations. *Icarus* **211**, 913–916.

Barth, C. A. (1974). Atmosphere of Mars. *Annu. Rev. Earth Planet. Sci.* **2**, 333–367.

Barth, C. A., *et al.* (1973). Mariner 9 ultraviolet spectrometer experiment: Seasonal variation of ozone on Mars. *Science* **179**, 795–796.

Barth, C. A., *et al.* (1972). Mariner 9 ultraviolet spectrometer experiment: Mars airglow spectroscopy and variations in Lyman alpha. *Icarus* **17**, 457–468.

Bartlett, D. H. (2002). Pressure effects on in vivo microbial processes. *Biochim. Biophys. Acta* **1595**, 367–381.

Batalha, N., *et al.* (2015). Testing the early Mars H$_2$–CO$_2$ greenhouse hypothesis with a 1-D photochemical model. *Icarus* **258**, 337–349.

Batalha, N. M. (2014). Exploring exoplanet populations with NASA's Kepler Mission. *P. Natl. Acad. Sci. USA* **111**, 12647–12654.

Batalha, N. M., *et al.* (2013). Planetary candidates observed by Kepler. Iii. Analysis of the first 16 months of data. *Astrophys. J. Supp. S.* **204**.

Bau, M., *et al.* (1997). Sources of rare-earth elements and iron in Paleoproterozoic iron-formations from the Transvaal Supergroup, South Africa: Evidence from neodymium isotopes. *J. Geol.* **105**, 121–129.

Baum, S. K. and Crowley, T. J. (2001). GCM response to late precambrian (similar to 590 Ma) ice-covered continents. *Geophysical Research Letters* **28**, 583–586.

Bean, J. L., *et al.* (2010). A ground-based transmission spectrum of the super-Earth exoplanet GJ 1214b. *Nature* **468**, 669–672.

Beaugé, C., *et al.* (2008). Planetary masses and orbital parameters from radial velocity measurements. In: *Extrasolar Planets: Formation, Detection and Dynamics*, Weinheim, Germany: Wiley-VCH Verlag GmbH & Co. KGaA, pp. 1–25.

Beaumont, V. and Robert, F. (1999). Nitrogen isotope ratios of kerogens in Precambrian cherts: A record of the evolution of atmosphere chemistry? *Precambrian Res.* **96**, 63–82.

Beck, P., *et al.* (2015). A Noachian source region for the "Black Beauty" meteorite, and a source lithology for Mars surface hydrated dust? *Earth Planet. Sc. Lett.* **427**, 104–111.

Becker, R. H. and Clayton, R. N. (1972). Carbon isotopic evidence for origin of a Banded Iron Formation in Western Australia. *Geochim. Cosmochim. Acta* **36**, 577.

Becker, R. H. and Pepin, R. O. (1984). The case for a martian origin of the Shergottites: Nitrogen and noble gases in EETA79001. *Earth Planet. Sc. Lett.* **69**, 225–242.

Beer, J. (2000). Long-term indirect indices of solar variability. *Space Science Reviews* **94**, 53–66.

Beerling, D. J. and Royer, D. L. (2011). Convergent Cenozoic CO$_2$ history. *Nat. Geosci.* **4**, 418–420.

Béghin, C., *et al.* (2010). Titan's native ocean revealed beneath some 45 km of ice by a Schumann-like resonance. *C. R. Geosci.* **342**, 425–433.

Bekker, A. and Holland, H. D. (2012). Oxygen overshoot and recovery during the early Paleoproterozoic. *Earth Planet. Sc. Lett.* **317**, 295–304.

Bekker, A., *et al.* (2004). Dating the rise of atmospheric oxygen. *Nature* **427**, 117–120.

Bekker, A., *et al.* (2008). Fractionation between inorganic and organic carbon during the Lomagundi (2.22–2.1 Ga) carbon isotope excursion. *Earth Planet. Sci. Lett.* **271**, 278–291.

Bekker, A., *et al.* (2006). Carbon isotope record for the onset of the Lomagundi carbon isotope excursion in the Great Lakes area, North America. *Precam. Res.* **148**, 145–180.

Bekker, A., *et al.* (2010). Iron Formation: The sedimentary product of a complex interplay among mantle, tectonic, oceanic, and biospheric processes. *Econ. Geol.* **105**, 467–508.

Belcher, C. M. and McElwain, J. C. (2008). Limits for combustion in low O$_2$ redefine paleoatmospheric predictions for the Mesozoic. *Science* **321**, 1197–1200.

Belcher, C. M., *et al.* (2010). Baseline intrinsic flammability of Earth's ecosystems estimated from paleoatmospheric oxygen over the past 350 million years. *P. Natl. Acad. Sci. USA* **107**, 22 448–22 453.

Bell, D. R. and Anbar, A. D. (2013). *Oxygen titration of continental lithosphere and the rise of atmospheric O$_2$.* American Geophysic Union Fall Mtg., p. V41E-05.

Bell, E. A., *et al.* (2015). Potentially biogenic carbon preserved in a 4.1 billion year old zircon. *P. Natl. Acad. Sci. USA*, doi: 10.1073/pnas.1517557112.

Bell, J. M., *et al.* (2010a). Simulating the one-dimensional structure of Titan's upper atmosphere: 1. Formulation of the Titan Global Ionosphere-Thermosphere Model and benchmark simulations. *J. Geophys. Res.* **115**, E12002, doi:10.1029/2010JE003636.

Bell, J. M., *et al.* (2011). Simulating the one-dimensional structure of Titan's upper atmosphere: 3. Mechanisms determining methane escape. *J. Geophys. Res.* **116**.

Bell, J. M., *et al.* (2010b). Simulating the one-dimensional structure of Titan's upper atmosphere: 2. Alternative scenarios for methane escape. *J. Geophys. Res.* **115**.

Bell, J. M., *et al.* (2014). Developing a self-consistent description of Titan's upper atmosphere without hydrodynamic escape. *J. Geophys. Res.* **119**, 4957–4972.

Belton, M. J. S., *et al.* (1991). Images from Galileo of the Venus cloud deck. *Science* **253**, 1531–1536.

Belton, M. J. S., *et al.* (1992). The Galileo Solid-State Imaging experiment. *Space Sci. Rev.* **60**, 413–455.

Belyaev, D. A., *et al.* (2012). Vertical profiling of SO_2 and SO above Venus' clouds by SPICAV/SOIR solar occultations. *Icarus* **217**, 740–751.

Ben-Jaffel, L. (2007). Exoplanet HD 209458b: Inflated hydrogen atmosphere but no sign of evaporation. *Astrophysical Journal Letters* **671**, L61–L64.

Ben-Jaffel, L. (2008). Spectral, spatial, and time properties of the hydrogen nebula around exoplanet HD 209458b. *Astrophys. J.* **688**, 1352–1360.

Benedict, G. F., *et al.* (1999). Interferometric astrometry of Proxima Centauri and Barnard's star using Hubble Space Telescope Fine Guidance Sensor 3: Detection limits for substellar companions. *Astron. J.* **118**, 1086–1100.

Bengtson, S. (1994). The advent of animal skeletons. In: *Early Life on Earth*, ed. S. Bengtson, New York: Columbia University Press, pp. 412–425.

Benner, S. A. (2009). *Life, the Universe...and the Scientific Method*. Gainesville, FL: FfAME Press.

Benner, S. A. (2010). Defining life. *Astrobiology* **10**, 1021–1030.

Benner, S. A., *et al.* (2004). Is there a common chemical model for life in the universe? *Curr. Opin. Chem. Biol.* **8**, 672–689.

Bent, H. A. (1965). *The Second Law: An Introduction to Classical and Statistical Thermodynamics*. New York: Oxford University Press.

Benton, M. J. and Ayala, F. J. (2003). Dating the tree of life. *Science* **300**, 1698–1700.

Bergin, E. A., *et al.* (2015). Tracing the ingredients for a habitable Earth from interstellar space through planet formation. *P. Natl. Acad. Sci. USA* **112**, 8965–8970.

Bergman, N. M., *et al.* (2004). COPSE: A new model of biogeochemical cycling over Phanerozoic time. *Am. J. Sci.* **304**, 397–437.

Berkner, L. V. and Marshall, L. C. (1965). On the origin and rise of oxygen concentration in the Earth's atmosphere. *J. Atmos. Sci.* **22**, 225–261.

Berkner, L. V. and Marshall, L. L. (1964). The history of oxygenic concentration in the Earth's atmosphere. *Disc. Faraday Soc.* **34**, 122–141.

Berkner, L. V. and Marshall, L. L. (1966). Limitation on oxygen concentration in a primitive planetary atmosphere. *J. Atmos. Sci.* **23**, 133–143.

Berkner, L. V. and Marshall, L. L. (1967). The rise of oxygen in the Earth's atmosphere with notes on the martian atmosphere. *Adv. Geophys.* **12**, 309–331.

Bernal, J. D. (1951). *The Physical Basis of Life*. London: Routledge and Paul.

Berndt, M. E., *et al.* (1996). Reduction of CO_2 during serpentinization of olivine at 300 °C and 500 bar. *Geology* **24**, 351–354.

Berner, R. A. (1982). Burial of organic-carbon and pyrite sulfur in the modern ocean - Its geochemical and environmental significance. *Am. J. Sci.* **282**, 451–473.

Berner, R. A. (2004). *The Phanerozoic Carbon Cycle: CO_2 and O_2*. Oxford: Oxford University Press.

Berner, R. A. (2006a). GEOCARBSULF: A combined model for Phanerozoic atmospheric O_2 and CO_2. *Geochim. Cosmochim. Acta* **70**, 5653–5664.

Berner, R. A. (2006b). Geological nitrogen cycle and atmospheric N_2 over Phanerozoic time. *Geology* **34**, 413–415.

Berner, R. A. (2009). Phanerozoic atmospheric oxygen: new results using the Geocarbsulf Model. *Am. J. Sci.* **309**, 603–606.

Berner, R. A., *et al.* (2003). Phanerozoic atmospheric oxygen. *Ann. Rev. Earth Planet. Sci.* **31**, 105–134.

Berner, R. A. and Canfield, D. E. (1989). A new model for atmospheric oxygen over Phanerozoic time. *Amer. J. Sci.* **289**, 333–361.

Berner, R. A., *et al.* (1983). The carbonate-silicate geochemical cycle and its effect on atmospheric carbon dioxide over the past 100 million years. *Amer. J. Sci.* **283**, 641–683.

Berner, R. A. and Maasch, K. A. (1996). Chemical weathering and controls on atmospheric O_2 and CO_2: Fundamental principles were enunciated by J. J. Ebelmen in 1845. *Geochim. Cosmochim. Acta* **60**, 1633–1637.

Berner, R. A. and Raiswell, R. (1983). Burial of organic-carbon and pyrite sulfur in sediments over Phanerozoic time – A new theory. *Geochim. Cosmochim. Acta* **47**, 855–862.

Berresheim, H. and Jaeschke, W. (1983). The contribution of volcanoes to the global atmospheric sulfur budget. *J. Geophys. Res.* **88**, 3732–3740.

Berta, Z. K., *et al.* (2012). Transit detection in the MEarth survey of nearby M dwarfs: Bridging the clean-first, search-later divide. *Astronom. J.* **144**, 145.

Berta-Thompson, Z. K., *et al.* (2015). A rocky planet transiting a nearby low-mass star. *Nature* **527**, 204–207.

Bertaux, J. L. (1975). Observed variations of the exospheric hydrogen density with the exospheric temperature. *J. Geophys. Res.* **80**, 639–642.

Bertaux, J. L., *et al.* (1978). Lyman-alpha observations of Venera-9 and Venera-10 .1. Nonthermal hydrogen population in exosphere of Venus. *Planet. Space Sci.* **26**, 817–831.

Bertaux, J. L., *et al.* (1982). Altitude profile of H in the atmosphere of Venus from Lyman alpha observations of Venera 11 and Venera 12 and origin of the hot exospheric component. *Icarus* **52**, 221–244.

Bertaux, J. L., *et al.* (1996). VEGA 1 and VEGA 2 entry probes: An investigation of local UV absorption (220–400 nm) in the atmosphere of Venus (SO_2, aerosols, cloud structure). *J. Geophys. Res.* **101**, 12 709–12 745.

Bessell, M. S. (2005). Standard photometric systems. *Annu. Rev. Astron. Astrophys.* **43**, 293–336.

Bethe, H. A. (1939). Energy production in stars. *Phys. Rev.* **55**, 0103–0103.

Betts, J. N. and Holland, H. D. (1991). The oxygen-content of ocean bottom waters, the burial efficiency of organic-carbon, and the regulation of atmospheric oxygen. *Global Planet. Change* **97**, 5–18.

Bézard, B., *et al.* (1990). The deep atmosphere of Venus revealed by high-resolution nightside spectra. *Nature* **345**, 508–511.

Bezard, B., *et al.* (2009). Water vapor abundance near the surface of Venus from Venus Express/VIRTIS observations. *J. Geophys. Res.* **114**.

Bezard, B., *et al.* (2014). The composition of Titan's atmosphere. In: *Titan: Surface, Atmosphere and Magnetosphere*, ed. I. Muller-Wodarg, *et al.*, New York: Cambridge University Press, pp. 158–189.

Bibring, J. P., *et al.* (2007). Coupled ferric oxides and sulfates on the Martian surface. *Science* **317**, 1206–1210.

Bibring, J. P., *et al.* (2005). Mars surface diversity as revealed by the OMEGA/Mars Express observations. *Science* **307**, 1576–1581.

Bibring, J. P., *et al.* (2006). Global mineralogical and aqueous mars history derived from OMEGA/Mars express data. *Science* **312**, 400–404.

Bickle, M. J. (1986). Implications of melting for stabilization of the lithosphere and heat loss in the Archean. *Earth Planet. Sci. Lett.* **80**, 314–324.

Biemann, K. and Bada, J. (2011). Comment on "Reanalysis of the Viking results suggests perchlorate and organics at midlatitudes on Mars" by Rafael Navarro-Gonzalez *et al. J. Geophys. Res.* **116**, E12001.

Birch, F. (1964). Density and composition of mantle and core. *J. Geophys. Res.* **69**, 4377–4388.

Bird, G. A. (1994). *Molecular gas dynamics and the direct simulation of gas flows*. New York: Oxford University Press.

Bird, M. K., *et al.* (2005). The vertical profile of winds on Titan. *Nature* **438**, 800–802.

Bishop, J. L., *et al.* (2009). Mineralogy of Juventae Chasma: Sulfates in the light-toned mounds, mafic minerals in the bedrock, and hydrated silica and hydroxylated ferric sulfate on the plateau. *J. Geophys. Res.* **114**, E00D09, doi:10.1029/2009JE003352.

Biver, N., *et al.* (2012). Ammonia and other parent molecules in comet 10P/Tempel 2 from Herschel/HIFI and ground-based radio observations. *Astron. Astrophys*, **539**.

Bjerrum, C. J. and Canfield, D. E. (2002). Ocean productivity before about 1.9 Gyr ago limited by phosphorus adsorption onto iron oxides. *Nature* **417**, 159–162.

Bjerrum, C. J. and Canfield, D. E. (2004). New insights into the burial history of organic carbon on the early Earth. *Geochem. Geophys. Geosys.* **5**, Q08001, doi:10.1029/2004GC000713.

Bjerrum, C. J. and Canfield, D. E. (2011). Towards a quantitative understanding of the late Neoproterozoic carbon cycle. *P. Natl. Acad. Sci. USA* **108**, 5542–5547.

Blake, G. A. and Bergin, E. A. (2015). Prebiotic chemistry on the rocks. *Nature* **520**, 161–162.

Blake, R. E., *et al.* (2010). Phosphate oxygen isotopic evidence for a temperate and biologically active Archaean ocean. *Nature* **464**, 1029–U89.

Blank, C. E. and Sanchez-Baracaldo, P. (2010). Timing of morphological and ecological innovations in the cyanobacteria – a key to understanding the rise in atmospheric oxygen. *Geobiology* **8**, 1–23.

Bockelee-Morvan, D., *et al.* (2008). Large excess of heavy nitrogen in both hydrogen cyanide and cyanogen from comet 17P/Holmes. *Astrophys. J. Lett.* **679**, L49–L52.

Bogard, D. D. (1995). Impact ages of meteorites: A synthesis. *Meteoritics* **30**, 244–268.

Bogard, D. D. (1997). A reappraisal of the Martian Ar-36/Ar-38 ratio. *J. Geophys. Res.* **102**, 1653–1661.

Bogard, D. D., *et al.* (2001). Martian volatiles: Isotopic composition, origin, and evolution. *Space Sci. Rev.* **96**, 425–458.

Bogard, D. D. and Garrison, D. H. (1998). Relative abundances of argon, krypton, and xenon in the Martian atmosphere as measured in Martian meteorites. *Geochim. Cosmochim. Acta* **62**, 1829–1835.

Bogard, D. D. and Johnson, P. (1983). Martian gases in an Antarctic meteorite. *Science* **221**, 651–654.

Bogard, D. D. and Park, J. (2008). Ar-39–Ar-40 dating of the Zagami Martian shergottite and implications for magma origin of excess Ar-40. *Meteorit. Planet. Sci.* **43**, 1113–1126.

Bohren, C. F. and Albrecht, B. A. (1998). *Atmospheric Thermodynamics*. New York: Oxford University Press.

Bohren, C. F. and Fraser, A. B. (1985). Colors of the sky. *Phys. Teach.* **23**, 267–272.

Bondi, H. (1952). On spherically symmetrical accretion. *Mon. Not. R. Astron. Soc.* **112**, 195–204.

Bonfils, X., *et al.* (2013). The HARPS search for southern extra-solar planets XXXI. *The M-dwarf sample. Astron. Astrophys.* **549**, A109, doi: 10.1051/0004-6361/201014704.

Bonnefoy, M., *et al.* (2011). High angular resolution detection of β Pictoris b at 2.18 μm. *Astron. Astrophys*, **528**.

Boone, R. H., *et al.* (2006). Metallicity in the solar neighborhood out to 60 pc. *New Astronomy Reviews* **50**, 526–529.

Boothroyd, A. I., *et al.* (1991). Our Sun II. Early mass loss of 0.1 Mo and the case of the missing lithium. *Ap. J.* **377**, 318–329.

Bordoni, S. and Schneider, T. (2008). Monsoons as eddy-mediated regime transitions of the tropical overturning circulation. *Nat. Geosci.* **1**, 515–519.

Bordoni, S. and Schneider, T. (2010). Regime transitions of steady and time-dependent Hadley circulations: Comparison of axisymmetric and eddy-permitting simulations. *J. Atmos. Sci.* **67**, 1643–1654.

Borg, L. and Drake, M. J. (2005). A review of meteorite evidence for the timing of magmatism and of surface or near-surface liquid water on Mars. *J. Geophys. Res.* **110**, E12S03, doi: 10.1029/2005JE002402.

Borucki, W. J. and Chameides, W. L. (1984). Lightning: Estimates of the rates of energy dissipation and nitrogen fixation. *Rev. Geophys.* **22**, 363–372.

Borucki, W. J., *et al.* (2011). Characteristics of planetary candidates observed by Kepler. I. Analysis of the first four months of data. *Astrophys. J.* **736**, 19, doi: 10.1088/0004-637X/736/1/19.

Bosak, T., *et al.* (2009). Morphological record of oxygenic photosynthesis in conical stromatolites. *P. Natl. Acad. Sci. USA* **106**, 10 939–10 943.

Boss, A. P. (2005). Evolution of the solar nebula. *VII. Formation and survival of protoplanets formed by disk instability. Astrophys. J.* **629**, 535–548.

Boss, A. P. (2006). Gas giant protoplanets formed by disk instability in binary star systems. *Astrophys. J.* **641**, 1148–1161.

Boss, A. P. (2008). Flux-limited diffusion approximation models of giant planet formation by disk instability. *Astrophys. J.* **677**, 607–615.

Boss, A. P. (2012). Giant planet formation by disc instability: flux-limited radiative diffusion and protostellar wobbles. *Mon. Not. R. Astron. Soc.* **419**, 1930–1936.

Boss, A. P. and Ciesla, F. J. (2014). The solar nebula. In: *Treatise on Geochemistry*, ed. H. D. Holland, K. K. Turekian, New York: Elsevier, pp. 1–23.

Bottke, W. F., *et al.* (1995). Collisional lifetimes and impact statistics of near-Earth asteroids. In: *Hazards Due to Comets and Asteroids*, ed. T. Gehrels, Tucson: University of Arizona Press, pp. 337–357.

Bottke, W. F., *et al.* (2015). Dating the Moon-forming impact event with asteroidal meteorites. *Science* **348**, 321–323.

Bottke, W. F., *et al.* (2012). An Archaean heavy bombardment from a destabilized extension of the asteroid belt. *Nature* **485**, 78–81.

Bougher, S. W., *et al.* (2000). Comparative terrestrial planet thermospheres 3. Solar cycle variation of global structure and winds at solstices. *J. Geophys. Res.* **105**, 17 669–17 692.

Bougher, S. W., *et al.* (2009). Solar cycle variability of Mars dayside exospheric temperatures: Model evaluation of underlying thermal balances. *Geophysical Research Letters* **36**.

Boussau, B., *et al.* (2008). Parallel adaptations to high temperatures in the Archaean eon. *Nature* **456**, 942–U74.

Bouvier, A., *et al.* (2009). Martian meteorite chronology and the evolution of the interior of Mars. *Earth Planet. Sc. Lett.* **280**, 285–295.

Bowring, S. A., *et al.* (2007). Geochronologic constraints on the chronostratigraphic framework of the neoproterozoic Huqf Supergroup, Sultanate of Oman. *American Journal of Science* **307**, 1097–1145.

Boyd, E. S., *et al.* (2011). A late methanogen origin for molybdenum-dependent nitrogenase. *Geobiology* **9**, 221–232.

Boyd, E. S. and Peters, J. W. (2013). New insights into the evolutionary history of biological nitrogen fixation. *Front. Microbiol.* **4**.

Boyle, R. A., *et al.* (2007). Neoproterozoic 'snowball Earth' glaciations and the evolution of altruism. *Geobiology* **5**, 337–349.

Boynton, W. V., *et al.* (2009). Evidence for calcium carbonate at the Mars Phoenix landing site. *Science* **325**, 61–64.

Brain, D. A. and Jakosky, B. M. (1998). Atmospheric loss since the onset of the Martian geologic record: Combined role of impact erosion and sputtering. *J. Geophys. Res.* **103**, 22 689–22 694.

Brantley, S. L. and Koepenick, K. W. (1995). Measured carbon dioxide emissions from Oldoinyo-Lengai and the skewed distribution of passive volcanic fluxes. *Geology* **23**, 933–936.

Brasier, M., *et al.* (2004). Earth's oldest (similar to 3.5 Ga) fossils and the 'Early Eden hypothesis': Questioning the evidence. *Orig. Life Evol. Biosph.* **34**, 257–269.

Brasier, M. D., *et al.* (2015). Changing the picture of Earth's earliest fossils (3.5–1.9 Ga) with new approaches and new discoveries. *P. Natl. Acad. Sci. USA* **112**, 4859–4864.

Brasier, M. D., *et al.* (2002). Questioning the evidence for the Earth's oldest fossils. *Nature* **416**, 76–81.

Brasier, M. D., *et al.* (2005). Critical testing of Earth's oldest putative fossil assemblage from the similar to 3.5 Ga Apex Chert, Chinaman Creek, western Australia. *Precam. Res.* **140**, 55–102.

Brasier, M. D. and Lindsay, J. F. (1998). A billion years of environmental stability and the emergence of eukaryotes: New data from northern Australia. *Geology* **26**, 555–558.

Brasier, M. D., *et al.* (2011). Pumice as a remarkable substrate for the origin of life. *Astrobiology* **11**, 725–735.

Brass, G. W. (1980). Stability of brines on Mars. *Icarus* **42**, 20–28.

Brasseur, G. and Solomon, S. (2005). *Aeronomy of the Middle Atmosphere: Chemistry and Physics of the Stratosphere and Mesosphere*. Dordrecht: Springer.

Braterman, P. S., *et al.* (1983). Photooxidation of hydrated Fe^{+2} – significance for banded iron formations. *Nature* **303**, 163–164.

Breuer, D. and Spohn, T. (2003). Early plate tectonics versus single-plate tectonics on Mars: Evidence from magnetic field history and crust evolution. *J. Geophys. Res.* **108**.

Brewer, A. W. (1949). Evidence for a world circulation provided by the measurements of helium and water vapour distribution in the stratosphere. *Q. J. Roy. Meteor. Soc.* **75**, 351–363.

Bridges, J. C., *et al.* (2001). Alteration assemblages in martian meteorites: Implications for near-surface processes. *Space Science Reviews* **96**, 365–392.

Bridges, J. C. and Warren, P. H. (2006). The SNC meteorites: basaltic igneous processes on Mars. *J. Geol. Soc. London* **163**, 229–251.

Bridges, N. T., *et al.* (2004). Insights on rock abrasion and ventifact formation from laboratory and field analog studies with applications to Mars. *Planet. Space Sci.* **52**, 199–213.

Brinkman, R. T. (1969). Dissociation of water vapor and evolution of oxygen in the terrestrial atmosphere. *J. Geophys. Res.* **74**, 5355–5368.

Brinkmann, R. T. (1970). Departure from Jeans escape rate for H and He in the Earth's atmosphere. *Planet. Space Sci.* **18**, 449–478.

Brinkmann, R. T. (1971). More comments on the validity of Jeans escape rate. *Planet. Space Sci.* **19**, 791–794.

Bristow, T. F., *et al.* (2011). A hydrothermal origin for isotopically anomalous cap dolostone cements from south China. *Nature* (advance online publication).

Bristow, T. F. and Kennedy, M. J. (2008). Carbon isotope excursions and the oxidant budget of the Ediacaran atmosphere and ocean. *Geology* **36**, 863–866.

Brocks, J. J., *et al.* (2003). A reconstruction of Archean biological diversity based on molecular fossils from the 2.78 to 2.45 billion-year-old Mount Bruce Supergroup, Hamersley Basin, Western Australia. *Geochim. Cosmochim. Acta* **67**, 4321–4335.

Brocks, J. J., *et al.* (1999). Archean molecular fossils and the early rise of eukaryotes. *Science* **285**, 1033–1036.

Broda, E. (1975). Beginning of photosynthesis. *Origins Life Evol. Biosph.* **6**, 247–251.

Broda, E. (1977). Evolution of photosynthesis. *Precambrian Res.* **4**, 117–132.

Broecker, W. S. (2015). The collision that changed the world. *Elem. Sci. Anth.* **3**, 000061, doi: 10.12952/journal.elementa.000061.

Broecker, W. S. and Peng, T. H. (1982). *Tracers in the Sea*. Palisades, New York: Lamont Doherty Geol. Obs.

Brown, G. N. and Ziegler, W. T. (1979). Vapor pressure and heats of sublimation of liquids and solids of interest in cryogenics below 1-atm pressure. In: *Advances in Cryogenic Engineering*, ed. K. Timmerhaus, H. A. Snyder, New York: Plenum Press, pp. 662–670.

Brown, M. E. (1997). A search for a sodium atmosphere around Ganymede. *Icarus* **126**, 236–238.

Brown, M. E. (2001). Potassium in Europa's atmosphere. *Icarus* **151**, 190–195.

Brown, M. E. (2012). The compositions of Kuiper Belt Objects. *Ann. Rev. Earth Planet. Sci.* **40**, 467–495.

Brown, M. E. and Calvin, W. M. (2000). Evidence for crystalline water and ammonia ices on Pluto's satellite Charon. *Science* **287**, 107–109.

Brown, M. E. and Hill, R. E. (1996). Discovery of an extended sodium atmosphere around Europa. *Nature* **380**, 229–231.

Brown, M. E. and Schaller, E. L. (2007). The mass of dwarf planet Eris. *Science* **316**, 1585–1585.

Brown, R. H., *et al.* (1991). Triton's global heat budget. *Science* **251**, 1465–1467.

Brown, R. H., *et al.* (1990). Energy-Sources for Tritons Geyser-Like Plumes. *Science* **250**, 431–435.

Brown, R. H., *et al.* (2009). *Titan from Cassini-Huygens*. New York: Springer.

Brown, T. M., *et al.* (2001). Hubble Space Telescope time-series photometry of the transiting planet of HD 209458. *Astrophysical Journal* **552**, 699–709.

Brunner, B., *et al.* (2013). Nitrogen isotope effects induced by anammox bacteria. *P. Natl. Acad. Sci. USA* **110**, 18 994–18 999.

Buchhave, L. A., *et al.* (2014). Three regimes of extrasolar planet radius inferred from host star metallicities. *Nature* **509**, 593–595.

Buchhave, L. A., *et al.* (2012). An abundance of small exoplanets around stars with a wide range of metallicities. *Nature* **486**, 375–377.

Budd, G. E. (2008). The earliest fossil record of the animals and its significance. *Phil. Trans. R. Soc. Lond. B* **363**, 1425–1434.

Budyko, M. I. (1969). The effect of solar radiation variations on the climate of the Earth. *Tellus* **21**, 611–619.

Buick, R. (1992). The antiquity of oxygenic photosynthesis: Evidence from stromatolites in sulphate-deficient Archaean lakes. *Science* **255**, 74–77.

Buick, R. (2007a). Did the Proterozoic 'Canfield Ocean' cause a laughing gas greenhouse? *Geobiology* **5**, 97–100.

Buick, R. (2007b). The earliest records of life on Earth. In: *Planets and Life: The Emerging Science of Astrobiology*, ed. W. T. Sullivan, J. Baross, Cambridge: Cambridge University Press, pp. 237–264.

Buick, R. (2008). When did oxygenic photosynthesis evolve? *Phil. Trans. R. Soc. Lond. B* **363**, 2731–2743.

Buick, R., *et al.* (1981). Stromatolite recognition in ancient rocks: An appraisal of irregularly laminated structures in an early Archean chert-barite unit from North Pole, Western Australia. *Alcheringa* **5**, 161–181.

Buick, R., *et al.* (1995a). Abiological origin of described stromatolites older than 3.2 Ga – Comment. *Geology* **23**, 191–191.

Buick, R., *et al.* (1995b). Stable isotopic compositions of carbonates from the Mesoproterozoic Bangemall Group, Northwestern Australia. *Chem. Geol.* **123**, 153–171.

Buie, M. W. and Grundy, W. M. (2000). The distribution and physical state of H_2O on Charon. *Icarus* **148**, 324–339.

Bullock, M. A. and Grinspoon, D. H. (1996). The stability of climate on Venus. *J. Geophys. Res.* **101**, 7521–7529.

Bullock, M. A. and Grinspoon, D. H. (2001). The recent evolution of climate on Venus. *Icarus* **150**, 19–37.

Bullock, M. A. and Moore, J. M. (2007). Atmospheric conditions on early Mars and the missing layered carbonates. *Geophys. Res. Lett.* **34**.

Bunch, T. E. and Chang, S. (1980). Carbonaceous chondrites .2. Carbonaceous chondrite phyllosilicates and light-element geochemistry as indicators of parent body processes and surface conditions. *Geochim. Cosmochim. Acta* **44**, 1543–1577.

Burgasser, A. J., *et al.* (2002). The spectra of T dwarfs. I. Near-infrared data and spectral classification. *Astrophys. J.* **564**, 421–451.

Burgisser, A. and Scaillet, B. (2007). Redox evolution of a degassing magma rising to the surface. *Nature* **445**, 194–197.

Burke, K., *et al.* (1976). In: *Dominance Of Horizontal Movements, Arc, And Microcontinental Collisions During The Late Permobile Regime*, New York: Wiley, pp. 113–130.

Burns, S. J. and Matter, A. (1993). Carbon isotopic record of the latest Proterozoic from Oman. *Eclogae Geol. Helv.* **86**, 595–607.

Burr, D. M., *et al.* (2010). Inverted fluvial features in the Aeolis/Zephyria Plana region, Mars: Formation mechanism and initial paleodischarge estimates. *J. Geophys. Res.* **115**.

Burrows, A. (2005). A theoretical look at the direct detection of giant planets outside the Solar System. *Nature* **433**, 261–268.

Burrows, A., *et al.* (2001). The theory of brown dwarfs and extrasolar giant planets. *Rev. Mod. Phys.* **73**, 719–765.

Burton, M. R., *et al.* (2013). Deep carbon emissions from volcanoes. *Rev. Mineral. Geochem.* **75**, 323–354.

Busigny, V., *et al.* (2013). Nitrogen cycle in the Late Archean ferruginous ocean. *Chem. Geol.* **362**, 115–130.

Busse, F. H. (1976). Simple model of convection in Jovian atmosphere. *Icarus* **29**, 255–260.

Busse, F. H. (1983). A model of mean zonal flows in the major planets. *Geophys. Astro. Fluid.* **23**, 153–174.

Butler, S. and Peltier, W. (2002). Thermal evolution of Earth: Models with time-dependent layering of mantle convection which satisfy the Urey ratio constraint. *J. Geophys. Res.* **107**, 3-1–3-15.

Butterfield, N. J. (2000). *Bangiomorpha pubescens* n. gen., n. sp.: Implications for the evolution of sex, multicellularity, and the Mesoproterozoic/Neoproterozoic radiation of eukaryotes. *Paleobiology* **26**, 386–404.

Butterfield, N. J. (2009). Oxygen, animals and oceanic ventilation: an alternative view. *Geobiology* **7**, 1–7.

Butterfield, N. J. (2011). Animals and the invention of the Phanerozoic Earth system. *Trends Ecol. Evol.* **26**, 81–87.

Butterfield, N. J. (2015). Proterozoic photosynthesis – a critical review. *Palaeontology* **58**, 953–972.

Byrne, S., *et al.* (2009). Distribution of mid-latitude ground ice on Mars from new impact craters. *Science* **325**, 1674–1676.

Caballero, R. and Huber, M. (2013). State-dependent climate sensitivity in past warm climates and its implications for future climate projections. *P. Natl. Acad. Sci. USA* **110**, 14 162–14 167.

Cabane, M., *et al.* (1993). Fractal aggregates in Titan atmosphere. *Planet. Space Sci.* **41**, 257–267.

Cairns-Smith, A. G. (1978). Precambrian solution photochemistry, inverse segregation, and banded iron formations. *Nature* **276**, 807–808.

Caldeira, K. (1995). Long-term control of atmospheric carbon dioxide: Low temperature seafloor alteration or terrestrial silicate rock weathering. *Am. J. Sci.* **295**, 1077–1114.

Caldeira, K., *et al.* (1993). Cooling in the Late Cenozoic. *Nature* **361**, 123–124.

Caldeira, K. and Kasting, J. F. (1992a). The life span of the biosphere revisited. *Nature* **360**, 721–723.

Caldeira, K. and Kasting, J. F. (1992b). Susceptibility of the early Earth to irreversible glaciation caused by carbon dioxide clouds. *Nature* **359**, 226–228.

Came, R. E., *et al.* (2007). Coupling of surface temperatures and atmospheric CO_2 concentrations during the Palaeozoic era. *Nature* **449**, 198–202.

Cameron, A. C. (2016). Extrasolar planetary transists. In: *Methods of Detecting Exoplanets: 1st Advanced School on Exoplanetary Science*, ed. V. Bozza *et al.* Springer, Cham, pp. 89–131.

Cameron, V., *et al.* (2009). A biomarker based on the stable isotopes of nickel. *P. Natl. Acad. Sci. USA* **106**, 10 944–10 948.

Campbell, A. J., *et al.* (2011). Refugium for surface life on Snowball Earth in a nearly-enclosed sea? A first simple model for sea-glacier invasion. *Geophys. Res. Lett.* **38**.

Campbell, B., *et al.* (2008). SHARAD radar sounding of the Vastitas Borealis Formation in Amazonis Planitia. *J. Geophys. Res.* **113**, 1–10.

Campbell, I. H. and Allen, C. M. (2008). Formation of supercontinents linked to increases in atmospheric oxygen. *Nature Geosci.* **1**, 554–558.

Campbell, I. H. and Squire, R. J. (2010). The mountains that triggered the Late Neoproterozoic increase in oxygen: The Second Great Oxidation Event. *Geochim. Cosmochim. Acta* **74**, 4187–4206.

Campbell, I. H. and Taylor, S. R. (1993). No water, no granites no continents, no oceans. *Geophys. Res. Lett.* **10**, 1061–1064.

Canfield, D. E. (1998). A new model for Proterozoic ocean chemistry. *Nature* **396**, 450–453.

Canfield, D. E. (2004). The evolution of the Earth surface sulfur reservoir. *Am. J. Sci.* **304**, 839–861.

Canfield, D. E. (2005). The early history of atmospheric oxygen: homage to Robert M. *Garrels. Ann. Rev. Earth Planet. Sci.* **33**, 1–36.

Canfield, D. E. (2014). Proterozoic atmospheric oxygen. In: *Treatise on Geochemistry*, ed. H. D. Holland and K. K. Turekian, New York: Elsevier, pp. 197–216.

Canfield, D. E. and Farquhar, J. (2009). Animal evolution, bioturbation, and the sulfate concentration of the oceans. *P. Natl. Acad. Sci. USA* **106**, 8123–8127.

Canfield, D. E., *et al.* (2010). The evolution and future of Earth's nitrogen cycle. *Science* **330**, 192–196.

Canfield, D. E., *et al.* (2005). *Aquatic Geomicrobiology*. San Diego, CA: Elsevier Academic Press.

Canfield, D. E., *et al.* (2013). Oxygen dynamics in the aftermath of the Great Oxidation of Earth's atmosphere. *P. Natl. Acad. Sci. USA* **110**, 16 736–16 741.

Canfield, D. E., *et al.* (2007). Late-Neoproterozoic deep-ocean oxygenation and the rise of animal life. *Science* **315**, 92–95.

Canfield, D. E. and Raiswell, R. (1999). The evolution of the sulfur cycle. *Amer. J. Sci.* **299**, 697–723.

Canfield, D. E., *et al.* (2006). Early anaerobic metabolisms. *Phil. Trans. R. Soc. Lond. B* **361**, 1819–1834.

Canfield, D. E. and Teske, A. (1996). Late-Proterozoic rise in atmospheric oxygen concentration inferred from phylogenetic and sulfur isotope studies. *Nature* **382**, 127–132.

Canil, D. (1997). Vanadium partitioning and the oxidation state of Archaean komatiite magmas. *Nature* **389**, 842–845.

Canil, D. (2002). Vanadium in peridotites, mantle redox and tectonic environments: Archean to present. *Earth Planet. Sci. Lett.* **195**, 75–90.

Cannat, M., *et al.* (2010). Serpentinization and associated hydrogen and methane fluxes at slow spreading ridges. *Geophysical Monograph Series*, Vol. 188, pp. 241–264.

Cantor, B., *et al.* (2002). Multiyear Mars Orbiter Camera (MOC) observations of repeated Martian weather phenomena during the northern summer season. *J. Geophys. Res.* **107**, doi:10.1029/2001JE001588.

Cantor, B. A., *et al.* (2001). Martian dust storms: 1999 Mars Orbiter Camera observations. *J. Geophys. Res.* **106**, 23653–23687.

Canup, R. M. (2005). A giant impact origin of Pluto–Charon. *Science* **307**, 546–550.

Canuto, V. M., *et al.* (1982). UV radiation from the young Sun and oxygen levels in the pre-biological paleoatmosphere. *Nature* **296**, 816–820.

Canuto, V. M., *et al.* (1983). The young Sun and the atmosphere and photochemistry of the early Earth. *Nature* **305**, 281–286.

Carlson, R. W. (1999). A tenuous carbon dioxide atmosphere on Jupiter's moon Callisto. *Science* **283**, 820–821.

Carmichael, I. S. E. (2002). The andesite aqueduct: perspectives on the evolution of intermediate magmatism in west-central (105-99 degrees W) Mexico. *Contrib. Mineral. Petr.* **143**. 641–663.

Carr, M. H. (1986). Mars: a water-rich planet. *Icarus* **68**, 187–216.

Carr, M. H. (1990). D/H on Mars: Effects of floods, volcanism, impacts and polar processes. *Icarus* **87**, 210–27.

Carr, M. H. (1996). *Water on Mars*. New York: Oxford University Press.

Carr, M. H. (2006). *The Surface of Mars*. Cambridge: Cambridge University Press.

Carr, M. H. and Clow, G. D. (1981). Martian channels and valleys – Their characteristics, distribution, and age. *Icarus* **48**, 91–117.

Carr, M. H. and Head, J. W. (2003). Oceans on Mars: An assessment of the observational evidence and possible fate. *J. Geophys. Res.* **108**, 5042.

Carr, M. H. and Head, J. W. (2010). Geologic history of Mars. *Earth Planet. Sci. Lett.* **294**, 185–203.

Carr, M. H. and Head, J. W. (2015). Martian surface/near-surface water inventory: Sources, sinks, and changes with time. *Geophys. Res. Lett.* **42**, 726–732.

Carr, M. H. and Malin, M. C. (2000). Meter-scale characteristics of martian channels and valleys. *Icarus* **146**, 366–386.

Carr, R. H., *et al.* (1985). Martian atmospheric carbon-dioxide and weathering products in SNC meteorites. *Nature* **314**, 248–250.

Carrier, B. L. and Kounaves, S. P. (2015). The origins of perchlorate in the Martian soil. *Geophys. Res. Lett.* **42**, 3739–3745.

Carroll, B. W. and Ostlie, D. A. (2007). *An Introduction to Modern Astrophysics*. San Francisco: Pearson Addison-Wesley.

Carter, J., *et al.* (2013). Hydrous minerals on Mars as seen by the CRISM and OMEGA imaging spectrometers: Updated global view. *Journal of Geophysical Research-Planets* **118**, 831–858.

Casciotti, K. L. (2009). Inverse kinetic isotope fractionation during bacterial nitrite oxidation. *Geochim. Cosmochim. Acta* **73**, 2061–2076.

Cassan, A., *et al.* (2012). One or more bound planets per Milky Way star from microlensing observations. *Nature* **481**, 167–169.

Cassanelli, J. P., *et al.* (2015). Sources of water for the outflow channels on Mars: Implications of the Late Noachian "icy highlands" model for melting and groundwater recharge on the Tharsis rise. *Planet. Space Sci.* **108**, 54–65.

Cassata, W. S., *et al.* (2012). Trapped Ar isotopes in meteorite ALH 84001 indicate Mars did not have a thick ancient atmosphere. *Icarus* **221**, 461–465.

Catling, D. and Kasting, J. F. (2007). Planetary atmospheres and life. In: *Planets and Life: The Emerging Science of Astrobiology*, ed. W. T. Sullivan and J. A. Baross, Cambridge: Cambridge University Press, pp. 91–116.

Catling, D. C. (2004). Planetary science: On Earth, as it is on Mars? *Nature* **429**, 707–708.

Catling, D. C. (2013). *Astrobiology: A Very Short Introduction*. Oxford: Oxford University Press.

Catling, D. C. (2014). The Great Oxidation Event Transition. In: *Treatise on Geochemistry*, ed. H. D. Holland and K. K. Turekian, 2nd edn. New York: Elsevier, pp. 191–233.

Catling, D. C. (2015). Planetary Atmospheres. In: *Treatise on Geophysics*, ed. G. Schubert, 2nd edn. Oxford: Elsevier, pp. 429–472.

Catling, D. C. and Claire, M. W. (2005). How Earth's atmosphere evolved to an oxic state: A status report. *Earth Planet. Sci. Lett.* **237**, 1–20.

Catling, D. C., *et al.* (2004). Understanding the evolution of atmospheric redox state from the Archaean to the

Proterozoic. In: *Field Forum on Processes on the Early Earth*, ed. W. U. Reimold and A. Hofmann, Kaapvaal Craton, S. Africa: University of Witwatersrand, pp. 17–19.

Catling, D. C., *et al.* (2007). Anaerobic methanotrophy and the rise of atmospheric oxygen. *Phil. Trans. R. Soc. Lond. A* **365**, 1867–1888.

Catling, D. C., *et al.* (2010). Atmospheric origins of perchlorate on Mars and in the Atacama. *J. Geophys. Res.* **115**, E00E11.

Catling, D. C., *et al.* (2005). Why O_2 is required by complex life on habitable planets and the concept of planetary "oxygenation time". *Astrobiology* **5**, 415–438.

Catling, D. C., *et al.* (2012). Does the Vastitas Borealis Formation contain oceanic or volcanic deposits? *Third Int. Conf. on Early Mars*, 7031.

Catling, D. C., *et al.* (2006). Light-toned layered deposits in Juventae Chasma, Mars. *Icarus* **181**, 26–51.

Catling, D. C. and Zahnle, K. J. (2013). An impact erosion stability limit controlling the existence of atmospheres on exoplanets and solar system bodies. *44th Lunar Planet. Sci. Conf.* 2665.

Catling, D. C., *et al.* (2001). Biogenic methane, hydrogen escape, and the irreversible oxidation of early Earth. *Science* **293**, 839–843.

Catling, D. C., *et al.* (2002). What caused the second rise of O_2 in the late Proterozoic? Methane, sulfate, and irreversible oxidation. *Astrobiology* **2**, 569 (Abstract).

Catling, D. C. and Zahnle, K. L. (2009). The Planetary Air Leak. *Sci. Am.* **300**, 36–43.

Cavosie, A. J., *et al.* (2005). Magmatic delta O-18 in 4400–3900 Ma detrital zircons: A record of the alteration and recycling of crust in the Early Archean. *Earth Planet. Sci. Lett.* **235**, 663–681.

Chaffin, M. S., *et al.* (2014). Unexpected variability of Martian hydrogen escape. *Geophys. Res. Lett.* **41**, 314–320.

Chaloner, W. G. (1989). Fossil charcoal as an indicator of paleoatmospheric oxygen level. *J. Geol. Soc. London* **146**, 171–174.

Chamberlain, J. W. (1963). Planetary coronae and atmospheric evaporation. *Planet. Space Sci.* **11**, 901–960.

Chamberlain, J. W. and Campbell, F. J. (1967). Rate of evaporation of a non-Maxwellian atmosphere. *Ap. J.* **149**, 687–705.

Chamberlain, J. W. and Hunten, D. M. (1987). *Theory of Planetary Atmospheres*. Orlando: Academic Press.

Chamberlain, J. W. and Smith, G. R. (1971). Rate of evaporation of a non-Maxwellian atmosphere. *Planet. Space Sci.* **19**, 675–684.

Chambers, J. E. (2007). On the stability of a planet between Mars and the asteroid belt: Implications for the Planet V hypothesis. *Icarus* **189**, 386–400.

Chambers, J. E. (2009). Planetary migration: What does it mean for planet formation? *Annu. Rev. Earth. Pl. Sc.* **37**, 321–344.

Chambers, J. E. (2014). Planet Formation. In: *Treatise on Geochemistry*, ed. H. D. Holland and K. K. Turekian, New York: Elsevier, pp. 55–72.

Chambers, J. E., *et al.* (1996). The stability of multi-planet systems. *Icarus* **119**, 261–268.

Chambers, L. H., *et al.* (2002). Examination of new CERES data for evidence of tropical Iris feedback. *Journal of Climate* **15**, 3719–3726.

Chameides, W. and Walker, J. C. G. (1975). Possible variation of ozone in troposphere during course of geologic time. *Am. J. Sci.* **275**, 737–752.

Chameides, W. L., *et al.* (1977). NOx production in lightning. *J. Atmos. Sci.* **34**, 143–149.

Chameides, W. L. and Walker, J. C. G. (1981). Rates of fixation by lightning of carbon and nitrogen in possible primitive terrestrial atmospheres. *Origins of Life* **11**, 291–302.

Chandrasekhar, S. (1960). *Radiative Transfer*. New York: Dover Publications.

Chang, S., *et al.* (1983). Prebiotic organic syntheses and the origin of life. In: *Earth's Earliest Biosphere: Its Origin and Evolution*, ed. J. W. Schopf, Princeton, New Jersey: Princeton University Press, pp. 53–92.

Chapman, G. A., *et al.* (2012). Comparison of TSI from SORCE TIM with SFO Ground-Based Photometry. *Sol. Phys.* **276**, 35–41.

Chapman, M. G., *et al.* (2003). Possible Juventae Chasma subice volcanic eruptions and Maja Valles ice outburst floods on Mars: Implications of Mars Global Surveyor crater densities, geomorphology, and topography. *J. Geophys. Res.* **108**, 5113.

Chapman, M. G., *et al.* (2010a). Noachian–Hesperian geologic history of the Echus Chasma and Kasei Valles system on Mars: New data and interpretations. *Earth Planet. Sc. Lett.* **294**, 256–271.

Chapman, M. G., *et al.* (2010b). Amazonian geologic history of the Echus Chasma and Kasei Valles system on Mars: New data and interpretations. *Earth Planet. Sc. Lett.* **294**, 238–255.

Chapman, S. and Cowling, T. G. (1939). *The Mathematical Theory Of Non-Uniform Gases: An Account of the Kinetic Theory of Viscosity, Thermal Conduction, and Diffusion In Gases*. Cambridge: Cambridge University Press.

Chapman, S., *et al.* (1990). *The Mathematical Theory of Non-Uniform Gases: An Account of the Kinetic Theory of Viscosity, Thermal Conduction and Diffusion In Gases* (3rd edition). Cambridge: Cambridge University Press.

Chapman, S. and Lindzen, R. S. (1970). *Atmospheric Tides: Thermal and Gravitational*. New York: Gordon and Breach.

Charbonneau, D., *et al.* (2000). Detection of planetary transits across a sun-like star. *Ap. J.* **529**, L45–L49.

Charbonneau, D., *et al.* (2002). Detection of an extrasolar planet atmosphere. *Astrophys. J.* **568**, 377–384.

Charlson, R. J., *et al.* (1987). Oceanic phytoplankton, atmospheric sulfur, cloud albedo and climate. *Nature* **326**, 655–661.

Charnay, B., *et al.* (2015). Methane storms as a driver of Titan's dune orientation. *Nat. Geosci.* **8**, 362–366.

Charnay, B., *et al.* (2014). Titan's past and future: 3D modeling of a pure nitrogen atmosphere and geological implications. *Icarus* **241**, 269–279.

Charnay, B., *et al.* (2013). Exploring the faint young Sun problem and the possible climates of the Archean Earth with a 3-D GCM. *J. Geophys. Res.* **118**, 10 414–10 431.

Chatterjee, S. and Tan, J. C. (2014). Inside-out Planet Formation. *Astrophys. J.* **780**.

Chaufray, J., *et al.* (2008). Observation of the hydrogen corona with SPICAM on Mars Express. *Icarus* **195**, 598–613.

Chaufray, J. Y., *et al.* (2012). Hydrogen density in the dayside venusian exosphere derived from Lyman-alpha observations by SPICAV on Venus Express. *Icarus* **217**, 767–778.

Chen, G. Q. and Ahrens, T. J. (1997). Erosion of terrestrial planet atmosphere by surface motion after a large impact. *Physics of the Earth and Planetary Interiors* **100**, 21–26.

Chen, X., *et al.* (2015). Rise to modern levels of ocean oxygenation coincided with the Cambrian radiation of animals. *Nat. Commun.* **6**.

Cheng, M., *et al.* (2015). Mo marine geochemistry and reconstruction of ancient ocean redox states. *Sci. China Earth Sci.*, 1–11.

Cho, J. Y. K. and Polvani, L. M. (1996). The morphogenesis of bands and zonal winds in the atmospheres on the giant outer planets. *Science* **273**, 335–337.

Choi, J., *et al.* (2013). Precise Doppler monitoring of Barnard's Star. *Astrophys. J.* **764**, 131, doi: 10.1088/0004-637X/764/2/131.

Christensen, P. R. (2003). Formation of recent martian gullies through melting of extensive water-rich snow deposits. *Nature* **422**, 45–48.

Christensen, P. R. (2006). Water at the poles and in permafrost regions of Mars. *Elements* **2**, 151–155.

Christensen, P. R., *et al.* (2001). Global mapping of Martian hematite mineral deposits: Remnants of water-driven processes on early Mars. *J. Geophys. Res.* **106**, 23 873–23 885.

Christensen, P. R., *et al.* (2004). Initial results from the Mini-TES experiment in Gusev crater from the Spirit rover. *Science* **305**, 837–842.

Christopher, J. B., *et al.* (2015). Terrestrial planet occurrence rates for the Kepler GK dwarf sample. *Astrophys. J.* **809**, 8, doi: 10.1088/0004-637X/809/1/8.

Chumakov, N. M. (2008). A problem of total glaciations on the Earth in the Late Precambrian. *Stratigr. Geo. Correl.* **16**, 107–119.

Chyba, C. and Sagan, C. (1992). Endogenous production, exogenous delivery and impact-shock synthesis of organic molecules: an inventory for the origins of life. *Nature* **355**, 125–132.

Chyba, C. F. (1987). The cometary contribution to the oceans of primitive Earth. *Nature* **330**, 632–635.

Chyba, C. F. (1990). Extraterrestrial amino acids and terrestrial life. *Nature* **348**, 113–114.

Chyba, C. F. and Phillips, C. B. (2007). Europa. In: *Planets and Life: The Emerging Science of Astrobiology,* ed. W. T. Sullivan and J. A. Baross, Cambridge: Cambridge University Press.

Chyba, C. F., *et al.* (1990). Cometary delivery of organic molecules to the early Earth. *Science* **249**, 366–373.

Cicerone, R. (1989). Analysis of sources and sinks of atmospheric nitrous oxide (N_2O). *J. Geophys. Res.* **94**, 18 265–18 271.

Clack, J. A. (2000). The origin of the tetrapods. In: *Amphibian Biology*, ed. H. Heatwole and R. L. Carroll, Chipping Norton, Australia: Surrey Beatty, pp. 979–1029.

Claire, M., *et al.* (2006). Biogeochemical modelling of the rise in atmospheric oxygen. *Geobiology* **4**, 239–269.

Claire, M. W. (2008). *Quantitative modeling of the rise in atmospheric oxygen*. University of Washington, Ph.D., Seattle.

Claire, M. W., *et al.* (2014). Modeling the signature of sulfur mass-independent fractionation produced in the Archean atmosphere. *Geochim. Cosmochim. Acta* **141**, 365–380.

Claire, M. W., *et al.* (2012). The evolution of solar flux from 2 nm to 160 microns: Quantitative estimates for planetary studies. *Astrophys. J.* **757**, 95 doi:10.1088/0004-637X/757/1/95.

Clancy, R., *et al.* (2004). A measurement of the 362 GHz absorption line of Mars atmospheric H_2O_2. *Icarus* **168**, 116–121.

Clark, B. C. (1993). Geochemical components in Martian soil. *Geochim. Cosmochim. Acta* **57**, 4575–4581.

Clark, B. C., *et al.* (2005). Chemistry and mineralogy of outcrops at Meridiani Planum. *Earth Planet. Sc. Lett.* **240**, 73–94.

Clark, I. D. (1971). Chemical kinetics of CO_2 atmospheres. *J. Atmos. Sci.* **28**, 847–858.

Clarke, J. T., *et al.* (2009). HST observations of the extended hydrogen corona of Mars. *Bull. Am. Astron. Soc.* **41**, 49.11 (abstract).

Clarke, J. T., *et al.* (2014). A rapid decrease of the hydrogen corona of Mars. *Geophys. Res. Lett.* **41**, 8013–8020.

Clayton, R. N. (2002). Solar System : Self-shielding in the solar nebula. *Nature* **415**, 860–861.

Cleaves, H. J., *et al.* (2008). A reassessment of prebiotic organic synthesis in neutral planetary atmospheres. *Origins Life Evol. Biosph.* **38**, 105–115.

Clifford, S. M. and Parker, T. J. (2001). The evolution of the martian hydrosphere: Implications for the fate of a primordial ocean and the current state of the northern plains. *Icarus* **154**, 40–79.

Cloud, P. (1976). Beginnings of biospheric evolution and their biogeochemical consequences. *Paleobiol.* **2**, 351–387.

Cloud, P. E. (1968). Atmospheric and hydrospheric evolution on the primitive earth. Both secular accretion and biological and geochemical processes have affected earth's volatile envelope. *Science* **160**, 729–736.

Clough, S. A. and Iacono, M. J. (1995). Line-by-line calculation of atmospheric fluxes and cooling rates. 1. Application to carbon dioxide, ozone, methane, nitrous oxide, and the halocarbons. *J. Geophys. Res.* **100**, 16 519–16 535.

Clough, S. A., *et al.* (1992). Line-by-line calculations of atmospheric fluxes and cooling rates: Application to water vapor. *J. Geophys. Res.* **97**, 15 761–15 785.

Coates, A. J., *et al.* (2007). Discovery of heavy negative ions in Titan's ionosphere. *Geophys. Res. Lett.* **34**.

Coates, A. J., *et al.* (2009). Heavy negative ions in Titan's ionosphere: Altitude and latitude dependence. *Planet. Space Sci.* **57**, 1866–1871.

Cohen, B. A., *et al.* (2000). Support for the lunar cataclysm hypothesis from lunar meteorite impact melt ages. *Science* **290**, 1754–1756.

Cohen, P. A., *et al.* (2009). Large spinose microfossils in Ediacaran rocks as resting stages of early animals. *P. Natl. Acad. Sci. U.S.A.* **106**, 6519–6524.

Colaprete, A., *et al.* (2005). Albedo of the south pole on Mars determined by topographic forcing of atmosphere dynamics. *Nature* **435**, 184–188.

Colaprete, A. and Toon, O. B. (2003). Carbon dioxide clouds in an early dense Martian atmosphere. *J. Geophys. Res.* **108**, 5025, doi:10.1029/2002JE001967.

Colburn, D., *et al.* (1989). Diurnal variations in optical depth at Mars. *Icarus* **79**, 159–89.

Colman, A. S. and Holland, H. D. (2000). The global diagenetic flux of phosphorus from marine sediments to the oceans: Redox sensitivity and the control of atmospheric oxygen levels. In: *Marine Authigenesis: From Global to Microbial*, ed. C. R. Glenn, *et al.*: SEPM Special Pub. No. 66.

Comas-Solá, J. (1908). Observationes des satellites principapaux de Jupiter et de Titan. *Astron. Nachr.* **179**.

Compston, W. and Pidgeon, R. T. (1986). Jack Hills, Evidence of more very old detrital zircons in Western Australia. *Nature* **321**, 766–769.

Condon, D., *et al.* (2005). U–Pb ages from the neoproterozoic Doushantuo Formation, China. *Science* **308**, 95–98.

Condon, D. J., *et al.* (2002). Neoproterozoic glacial-rainout intervals: Observations and implications. *Geology* **30**, 35–38.

Connelly, J. N., *et al.*, Pb–Pb chronometry and the early Solar System. *Geochim. Cosmochim. Acta*, in press, doi:10.1016/j.gca.2016.10.044.

Conrad, R. (1996). Soil microorganisms as controllers of atmospheric trace gases (H_2, CO, CH_4, OCS, N_2O, and NO). *Microbiol. Rev.* **60**, 609–640.

Coogan, L. A. and Dosso, S. E. (2015). Alteration of ocean crust provides a strong temperature dependent feedback on the geological carbon cycle and is a primary driver of the Sr-isotopic composition of seawater. *Earth Planet. Sc. Lett.* **415**, 38–46.

Coogan, L. A. and Gillis, K. M. (2013). Evidence that low-temperature oceanic hydrothermal systems play an important role in the silicate–carbonate weathering cycle and long-term climate regulation. *Geochem. Geophys. Geosys.* **14**, 1771–1786.

Cook, J. C., *et al.* (2007). Near-infrared spectroscopy of Charon: Possible evidence for cryovolcanism on Kuiper Belt objects. *Astrophys. J.* **663**, 1406–1419.

Cook, K. H. (2004). Hadley circulation dynamics. In: *The Hadley Circulation: Past, Present and Future*, ed. H. F. Diaz and R. S. Bradley, Dordrecht: Kluwer, pp. 61–83.

Cooney, C. L. (1975). Thermophilic anaerobic digestion of solid waste for fuel gas production. *J. Biotech. Bioengineer.* **17**, 1119–1135.

Cooper, J. F., *et al.* (2001). Energetic ion and electron irradiation of the icy Galilean satellites. *Icarus* **149**, 133–159.

Cope, M. J. and Chaloner, W. G. (1980). Fossil charcoal as evidence of past atmospheric composition. *Nature* **283**, 647–649.

Cordier, D., *et al.* (2009). An estimate of the chemical composition of Titan's lakes. *Ap. J.* **707**, L128–L131.

Cordier, D., *et al.* (2012). Titan's lakes chemical composition: Sources of uncertainties and variability. *Planet. Space Sci.* **61**, 99–107.

Corliss, J. B., *et al.* (1981). An hypothesis concerning the relationship between submarine hot springs and the origin of life on Earth. *Ocean Acta*, 59–69.

Cornell, R. M. and Schwertmann, U. (1996). *The iron oxides : structure, properties, reactions, occurrences and uses*. Weinheim ; Cambridge: VCH.

Correia, A. C. M. and Laskar, J. (2001). The four final rotation states of Venus. *Nature* **411**, 767–770.

Correia, A. C. M. and Laskar, J. (2003). Long-term evolution of the spin of Venus II. *Numerical simulations*. *Icarus* **163**, 24–45.

Correia, A. C. M., *et al.* (2003). Long-term evolution of the spin of Venus I. Theory. *Icarus* **163**, 1–23.

Costard, F., *et al.* (2002). Formation of recent Martian debris flows by melting of near-surface ground ice at high obliquity. *Science* **295**, 110–113.

Coughenour, C. L., *et al.* (2009). Tides, tidalites, and secular changes in the Earth-Moon system. *Earth Sci. Rev.* **97**, 59–79.

Cowan, N. B., *et al.* (2009). Alien maps of an ocean-bearing world. *Astrophys. J.* **700**, 915–923.

Craddock, R. A. and Howard, A. D. (2002). The case for rainfall on a warm, wet early Mars. *J. Geophys. Res.* **107**, 5111, doi:10.1029/2001JE001505.

Crary, F., *et al.* (2010). Upper limits on carbon group ions near the orbit of Titan: Implications for methane escape from Titan. *38th COSPAR Scientific Assembly*, 5.

Cravens, T. E., *et al.* (1997). Photochemical sources of non-thermal neutrals for the exosphere of Titan. *Planetary and Space Science* **45**, 889–896.

Crespin, A., *et al.* (2008). Diagnostics of Titan's stratospheric dynamics using Cassini/CIRS data and the 2-dimensional IPSL circulation model. *Icarus* **197**, 556–571.

Crisp, D. (1997). Absorption of sunlight by water vapor in cloudy conditions: A partial explanation for the cloud absorption anomaly. *Geophys. Res. Lett.* **24**, 571–574.

Croft, S. K., *et al.* (1995). The geology of Triton. In: *Neptune and Triton*, ed. D. P. Cruikshank, Tucson: University of Arizona Press, pp. 879–947.

Crossfield, I. J. M. (2015). Observations of exoplanet atmospheres. *Publ. Astron. Soc. Pac.* **127**, 941–960.

Crow, C. A., *et al.* (2011). Views from EPOXI: colors in our solar system as an analog for extrasolar planets. *Astrophys. J.* **729**, 130.

Crowe, S. A., *et al.* (2013). Atmospheric oxygenation three billion years ago. *Nature* **501**, 535.

Crowell, J. C. (1999). *Pre-Mesozoic Ice Ages: Their Bearing on Understanding the Climate System.* Vol. GSA Memoir 192. Geological Society of America.

Crowley, T. J. (2000). CLIMAP SSTs re-revisited. *Climate Dyn.* **16**, 241–255.

Crowley, T. J., *et al.* (2001). CO_2 levels required for deglaciation of a "Near-Snowball" Earth. *Geophys. Res. Lett.* **28**, 283–286.

Crowley, T. J. and North, G. R. (1991). *Paleoclimatology.* New York: Oxford University Press.

Cruikshank, D. P., *et al.* (2000). Water ice on Triton. *Icarus* **147**, 309–316.

Crutzen, P. J. (1976). Possible importance of CSO for sulfate layer of stratosphere. *Geophys. Res. Lett.* **3**, 73–76.

Crutzen, P. J. (1979). Role of NO and NO_2 in the chemistry of the troposphere and stratosphere. *Annu. Rev. Earth Planet. Sci.* **7**, 443–472.

Crutzen, P. J. and Zimmermann, P. H. (1991). The changing photochemistry of the troposphere. *Tellus A* **43**, 136–151.

Cui, J., *et al.* (2011). The implications of the H_2 variability in Titan's exosphere. *J. Geophys. Res.* **116**.

Cui, J., *et al.* (2008). Distribution and escape of molecular hydrogen in Titan's thermosphere and exosphere. *J. Geophys. Res.* **113**, E10004, doi:10.1029/2007JE003032.

Cull, S. C., *et al.* (2010). Concentrated perchlorate at the Mars Phoenix landing site: Evidence for thin film liquid water on Mars. *Geophys. Res. Lett.* **37**, L22203.

Cunha, D., *et al.* (2015). Spin evolution of Earth-sized exoplanets, including atmospheric tides and core–mantle friction. *Int. J. Astrobiol.* **14**, 233–254.

Cunningham, N. J., *et al.* (2015). Detection of Callisto's oxygen atmosphere with the Hubble Space Telescope. *Icarus* **254**, 178–189.

Curtis, A. R. and Goody, R. M. (1956). Thermal radiation in the upper atmosphere. *Proc. R. Soc. Lond. A* **236**, 193–206.

Cushing, M. C., *et al.* (2011). The discovery of Y Dwarfs using data from the Wide-Field Infrared Survey Explorer (WISE). *Astrophys. J.* **743**.

Cutter, G. A. and Bruland, K. W. (1984). The marine biogeochemistry of selenium: a re-evaluation. *Limnol. Oceanogr.* **29**, 1179–1192.

Cuzzi, J. N., *et al.* (2008). Toward planetesimals: dense chondrule clumps in the protoplanetary nebula. *Astrophys. J.* **687**, 1432–1447.

Czaja, A. D., *et al.* (2010). Iron and carbon isotope evidence for ecosystem and environmental diversity in the similar to 2.7 to 2.5 Ga Hamersley Province, Western Australia. *Earth Planet. Sci. Lett.* **292**, 170–180.

Czaja, A. D., *et al.* (2013). Biological Fe oxidation controlled deposition of banded iron formation in the ca. 3770 Ma Isua Supracrustal Belt (West Greenland). *Earth Planet. Sc. Lett.* **363**, 192–203.

Czaja, A. D., *et al.* (2012). Evidence for free oxygen in the Neoarchean ocean based on coupled iron-molybdenum isotope fractionation. *Geochim. Cosmochim. Acta* **86**, 118–137.

D'Hondt, S., *et al.* (2002). Metabolic activity of subsurface life in deep-sea sediments. *Science* **295**, 2067–2070.

Dahl, T. W., *et al.* (2010). Devonian rise in atmospheric oxygen correlated to the radiations of terrestrial plants and large predatory fish. *P. Natl. Acad. Sci. U.S.A.* **107**, 17 911–17 915.

Dai, A. and Wang, J. (1999). Diurnal and semidiurnal tides in global surface pressure fields. *J. Atmos. Sci.* **56**, 3874–3891.

Daines, S. J. and Lenton, T. M. (2016). The effect of widespread early aerobic marine ecosystems on methane cycling and the Great Oxidation. *Earth Planet. Sci. Lett.* **434**, 42–51.

Dalcanton, J., *et al.* (2015). From Cosmic Birth to Living Earths: The Future of UVOIR Space Astronomy. *arXiv preprint arXiv:1507.04779.*

Danielache, S. O., *et al.* (2008). High-precision spectroscopy of ^{32}S, ^{33}S, and ^{34}S sulfur dioxide: Ultraviolet absorption cross sections and isotope effects. *J. Geophys. Res.* **113**, D17314.

Darling, A. and Whipple, K. (2015). Geomorphic constraints on the age of the western Grand Canyon. *Geosphere* **11**, 958–976.

Dauphas, N. (2003). The dual origin of the terrestrial atmosphere. *Icarus* **165**, 326–339.

Dauphas, N., *et al.* (2014). Geochemical arguments for an Earth-like Moon-forming impactor. *Phil. Trans R. Soc. Lond. A* **372**.

Dauphas, N. and Kasting, J. F. (2011). Low pCO$_2$ in the pore water, not in the Archean atmosphere. *Nature* **474**, E1, doi:10.1038/nature09960.

Dauphas, N. and Pourmand, A. (2011). Hf-W-Th evidence for rapid growth of Mars and its status as a planetary embryo. *Nature* **473**, 489–U227.

Dauphas, N., *et al.* (2000). The late asteroidal and cometary bombardment of Earth as recorded in water deuterium to protium ratio. *Icarus* **148**, 508–512.

David, L. A. and Alm, E. J. (2011). Rapid evolutionary innovation during an Archaean genetic expansion. *Nature* **469**, 93–96.

Davies, G. F. (1980). Thermal histories of convective Earth models and constraints on radiogenic heat production in the Earth. *J. Geophys. Res.* **85**, 2517–2530.

Davies, G. F. (2002). Stirring geochemistry in mantle convection models with stiff plates and slabs. *Geochim. Cosmochim. Acta* **66**, 3125–3142.

Davies, G. F. (2006). Gravitational instability of the early Earth's upper mantle and the viability of early plate tectonics. *Earth Planet. Sci. Lett.* **243**, 376–382.

de Kok, R. J., *et al.* (2014). HCN ice in Titan's high-altitude southern polar cloud. *Nature* **514**, 65.

de Wit, M. J. and Furnes, H. (2016). 3,5-Ga hydrothermal fields and diamictites in the Barberton Greenstone Belt – Paleoarchean crust in cold environments. *Sci. Adv.* **2**, e1500368.

DeBergh, C., *et al.* (1991). Deuterium on Venus – Observations from Earth. *Science* **251**, 547–549.

Degens, E. T. and Epstein, S. (1962). Relationship between O18/O16 ratios in coexisting carbonates, cherts and diatomites. *Amer. Assoc. Petrol. Geol. Bull.* **46**, 534–542.

Del Genio, A. D., *et al.* (2009). Saturn atmospheric structure and dynamics. In: *Saturn from Cassini-Huygens*, ed. M. Dougherty, *et al.*, New York: Springer, pp. 113–160.

Del Genio, A. D. and Suozzo, R. J. (1987). A comparative study of rapidly and slowly rotating dynamic regimes in a terrestrial General Circulation Model. *J. Atmos. Sci.* **44**, 973–986.

Del Genio, A. D. and Zhou, W. (1996). Simulations of superrotation on slowly rotating planets: Sensitivity to rotation and initial condition. *Icarus* **120**, 332–343.

Delano, J. W. (2001). Redox history of the Earth's interior: implications for the origin of life. *Orig. Life Evol. Biosph.* **31**, 311–341.

Delorme, P., *et al.* (2013). Direct-imaging discovery of a 12–14 Jupiter-mass object orbiting a young binary system of very low-mass stars. *Astronomy & Astrophysics* **553**, L5.

Delsemme, A. H. (2001). An argument for the cometary origin of the biosphere. *Am. Sci.* **89**, 432–442.

Derry, L. A., *et al.* (1992). Sedimentary cycling and environmental change in the Late Proterozoic: Evidence from stable and radiogenic isotopes. *Geochim. Cosmochim. Acta* **56**, 1317–1329.

Derry, L. A. (2014). Organic carbon cycling and the lithosphere. In: *Treatise on Geochemistry (Second Edition)*, ed. H. D. Holland and K. K. Turekian. Oxford: Elsevier, pp. 239–249.

Des Marais, D. J., *et al.* (1992). Carbon isotope evidence for the stepwise oxidation of the Proterozoic environment. *Nature* **359**, 605–609.

DesMarais, D. J., *et al.* (1992). Carbon isotope evidence for the stepwise oxidation of the Proterozoic environment. *Nature* **359**, 605–609.

Dessler, A. E., *et al.* (2013). Stratospheric water vapor feedback. *P. Natl. Acad. Sci. USA* **110**, 18087–18091.

Dessler, A. E., *et al.* (2008). Water-vapor climate feedback inferred from climate fluctuations, 2003–2008. *Geophys. Res. Lett.* **35**, L20704, doi:10.1029/2008GL035333.

Dhuime, B., *et al.* (2012). A Change in the Geodynamics of Continental Growth 3 Billion Years Ago. *Science* **335**, 1334–1336.

Di Achille, G. and Hynek, B. M. (2010). Ancient ocean on Mars supported by global distribution of deltas and valleys. *Nature Geoscience* **3**, 459–463.

Diakonov, I., *et al.* (1994). Thermodynamic properties of iron-oxides and hydroxides .1. Surface and bulk thermodynamic properties of goethite (alpha-FeOOH) up to 500 K. *Eur. J. Mineral.* **6**, 967–983.

Dickens, A. F., *et al.* (2004). Reburial of fossil organic carbon in marine sediments. *Nature* **427**, 336–339.

Dickey, J. O., *et al.* (1994). Lunar laser ranging – a continuing legacy of the Apollo Program. *Science* **265**, 482–490.

Dickson, J. L., *et al.* (2015). Recent climate cycles on Mars: Stratigraphic relationships between multiple generations of gullies and the latitude dependent mantle. *Icarus* **252**, 83–94.

Dickson, J. L., *et al.* (2007). Martian gullies in the southern mid-latitudes of Mars: Evidence for climate-controlled formation of young fluvial features based upon local and global topography. *Icarus* **188**, 315–323.

Dieterich, S., B., *et al.* (2012). The Solar Neighborhood. XXVIII. The multiplicity fraction of nearby stars from 5 to 70 AU and the brown dwarf desert around M Dwarfs. *Astron. J.* **144**, 64.

Dima, I. M. and Wallace, J. M. (2003). On the seasonality of the Hadley cell. *J. Atmos. Sci.* **60**, 1522–1527.

Diniega, S., *et al.* (2013). A new dry hypothesis for the formation of martian linear gullies. *Icarus* **225**, 526–537.

Dobrijevic, M., *et al.* (2014). Coupling of oxygen, nitrogen, and hydrocarbon species in the photochemistry of Titan's atmosphere. *Icarus* **228**, 324–346.

Dobrovolskis, A. R. (1980). Atmospheric tides and the rotation of Venus .2. Spin evolution. *Icarus* **41**, 18–35.

Dobrovolskis, A. R. (2009). Insolation patterns on synchronous exoplanets with obliquity. *Icarus* **204**, 1–10.

Dobrovolskis, A. R. and Ingersoll, A. P. (1980). Atmospheric tides and the totation of Venus .1. Tidal theory and the balance of torques. *Icarus* **41**, 1–17.

Dobson, G. M. B. (1956). Origin and distribution of the polyatomic molecules in the atmosphere. *Proc. R. Soc. Lond. A* **236**, 187–193.

Dohnanyi, J. S. (1972). Interplanetary objects in review: Statistics of their masses and dynamics. *Icarus* **17**, 1.

Dole, S. H. (1964). *Habitable Planets for Man.* New York: Blaisdell Publishing.

Domagal-Goldman, S. D., *et al.* (2011). Using biogenic sulfur gases as remotely detectable biosignatures on anoxic planets. *Astrobiology* **11**, 419–441.

Domagal-Goldman, S. D., *et al.* (2014). Abiotic ozone and oxygen in atmospheres similar to prebiotic Earth. *Astrophys. J.* **792**, doi: 10.1088/0004-637X/792/2/90.

Donahue, T. M. (1995). Evolution of water reservoirs on Mars from D/H ratios in the atmosphere and crust. *Nature* **374**, 432–434.

Donahue, T. M., *et al.* (1982). Venus was wet: A measurement of the ratio of deuterium to hydrogen. *Science* **216**, 630–633.

Donahue, T. M. and Pollack, J. B. (1983). Origin and evolution of the atmosphere of Venus. In: *Venus*, ed. D. M. Hunten, *et al.*, Tucson: University of Arizona Press, pp. 1003–1036.

Donnadieu, Y., *et al.* (2004). A 'snowball Earth' climate triggered by continental break-up through changes in runoff. *Nature* **428**, 303–306.

Doolittle, W. F. (2009). The practice of classification and the theory of evolution, and what thedemise of Charles Darwin's tree of life hypothesis means for both of them. *Phil. Trans. R. Soc. Lond. B* **364**, 2221–2228.

Dos Santos, P. C., *et al.* (2012). Distribution of nitrogen fixation and nitrogenase-like sequences amongst microbial genomes. *BMC Genomics* **13**, 162.

Dowling, T. E. (1995). Dynamics of Jovian atmospheres. *Annu. Rev. Fluid Mech.* **27**, 293–334.

Drake, M. J. and Righter, K. (2002). Determining the composition of the Earth. *Nature* **416**, 39–44.

Dreibus, G. and Wänke, H. (1987). Volatiles on Earth and Mars, A comparison. *Icarus* **71**, 225–40.

Dressing, C. D. and Charbonneau, D. (2015). The occurrence of potentially habitable planets orbiting M dwarfs estimated from the full Kepler dataset and an empirical measurement of the detection sensitivity. *Astrophys. J.* **807**, 45.

Driese, S. G., *et al.* (2011). Neoarchean paleoweathering of tonalite and metabasalt: Implications for reconstructions of 2.69 Ga early terrestrial ecosystems and paleoatmospheric chemistry. *Precambrian Res.* **189**, 1–17.

Du, S. Y., *et al.* (2011). The kinetics study of the S + S-2 -> S-3 reaction by the chaperone mechanism. *Journal of Chemical Physics* **134**.

Dudley, R. (1998). Atmospheric oxygen, giant Paleozoic insects and the evolution of aerial locomotor performance. *J. Exp. Biol.* **201**, 1043–1050.

Dundas, C. M., *et al.* (2014). HiRISE observations of new impact craters exposing Martian ground ice. *J. Geophys. Res.* **119**, 109–127.

Dundas, C. M., *et al.* (2012). Seasonal activity and morphological changes in martian gulllies. *Icarus*, doi:10.1016/j.icarus.2012.04.005.

Dundas, C. M., *et al.* (2015). Long-term monitoring of martian gully formation and evolution with MRO/ HiRISE. *Icarus* **251**, 244–263.

Dundas, C. M. and Keszthelyi, L. P. (2014). Emplacement and erosive effects of lava in south Kasei Valles, Mars. *J. Volcanol. Geoth. Res.* **282**, 92–102.

Dutkiewicz, A., *et al.* (2006). Biomarkers from Huronian oil-bearing fluid inclusions: An uncontaminated record of life before the Great Oxidation Event. *Geology* **34**, 437–440.

Duxbury, N. S. and Brown, R. H. (1997). The role of an internal heat source for the eruptive plumes on Triton. *Icarus* **125**, 83–93.

Dworkin, L. P., *et al.* (2001). Self-assembling amphiphilic molecules: Synthesis in simulated interstellar/precometary ices. *P. Natl. Acad. Sci. USA* **98**, 815–819.

Dyudina, U. A., *et al.* (2010). Detection of visible lightning on Saturn. *Geophys. Res. Lett.* **37**, doi:10.1029/2010GL043188.

Edgett, K. S. (2005). The sedimentary rocks of Sinus Meridiani: Five key observations from data acquired by the Mars Global Surveyor and Mars Odyssey orbiters. *Mars* **1**, 5–58.

Edgett, K. S. and Malin, M. C. (2002). Martian sedimentary rock stratigraphy: Outcrops and interbedded craters of northwest Sinus Meridiani and southwest Arabia Terra. *Geophys. Res. Lett.* **29**.

Edson, A., *et al.* (2011). Atmospheric circulations of terrestrial planets orbiting low mass stars,. *Icarus* **212**, 1–13.

Edwards, C. S. and Ehlmann, B. L. (2015). Carbon sequestration on Mars. *Geology*, doi:10.1130/G36983.1.

Egami, F. (1974). Inorganic types of fermentation and anaerobic respirations in evolution of energy-yielding metabolism. *Origins Life Evol. Biosph.* **5**, 405–413.

Egami, F. (1976). Comment on position of nitrate respiration in metabolic evolution. *Origins Life Evol. Biosph.* **7**, 71–72.

Ehlmann, B. (2010). Diverse aqueous environments during Mars' first billion years: The emerging view from orbital visible-near infrared spectroscopy. *Geochem. News* **142**.

Ehlmann, B., *et al.* (2011). Subsurface water and clay mineral formation during the early history of Mars. *Nature* **479**, 53–60.

Ehlmann, B. L., *et al.* (2013). Geochemical consequences of widespread clay mineral formation in Mars' ancient crust. *Space Sci. Rev.* **174**, 329–364.

Ehlmann, B. L. and Edwards, C. S. (2014). Mineralogy of the martian surface. *Annu. Rev. Earth Pl. Sc.* **42**, 291–315.

Ehlmann, B. L., *et al.* (2008a). Clay minerals in delta deposits and organic preservation potential on Mars. *Nature Geoscience* **1**, 355–358.

Ehlmann, B. L., *et al.* (2010). Geologic setting of serpentine deposits on Mars. *Geophysical Research Letters* **37**, L06201, doi:10.1029/2010GL042596.

Ehlmann, B. L., *et al.* (2008b). Orbital identification of carbonate-bearing rocks on Mars. *Science* **322**, 1828–1832.

Ehlmann, B. L., *et al.* (2009). Identification of hydrated silicate minerals on Mars using MRO-CRISM: Geologic context near Nili Fossae and implications for aqueous alteration. *J. Geophys. Res.* **114**.

Ehrenreich, A. and Widdel, F. (1994). Anaerobic oxidation of ferrous iron by purple bacteria, a new type of phototrophic metabolism. *Appl. Environ. Microbiol.* **60**, 4517–4526.

Eigenbrode, J. L. and Freeman, K. H. (2006). Late Archean rise of aerobic microbial ecosystems. *P. Natl. Acad. Sci. USA* **103**, 15 759–15 764.

Eigenbrode, J. L., *et al.* (2008). Methylhopane biomarker hydrocarbons in Hamersley Province sediments provide evidence for Neoarchean aerobiosis. *Earth Planet. Sci. Lett.* **273**, 323–331.

Eiler, J. M. (2007). "Clumped-isotope" geochemistry–The study of naturally-occurring, multiply-substituted isotopologues. *Earth Planet. Sci. Lett.* **262**, 309–327.

Eiler, J. M. (2011). Paleoclimate reconstruction using carbonate clumped isotope thermometry. *Quaternary Sci. Rev.* **30**, 3575–3588.

El Albani, A., et al. (2010). Large colonial organisms with coordinated growth in oxygenated environments 2.1 Gyr ago. *Nature* **466**, 100–104.

El Albani, A., et al. (2014). The 2.1 Ga old Francevillian biota: Biogenicity, taphonomy and biodiversity. *PLoS ONE* **9**, e99438 doi:10.1371/journal.pone.0099438.

Elachi, C. and Van Zyl, J. (2006). *Introduction to the Physics and Techniques of Remote Sensing.* Hoboken, N.J.: Wiley.

Elkins-Tanton, L. T. (2008). Linked magma ocean solidification and atmospheric growth for Earth and Mars. *Earth Planet. Sc. Lett.* **271**, 181–191.

Elkins-Tanton, L. T. (2012). Magma oceans in the inner Solar System. *Ann. Rev. Earth Planet. Sci.* **40**, 113–139.

Elliot, J. L., et al. (1998). Global warming on Triton. *Nature* **393**, 765–767.

Ellis, A. M., et al. (2005). *Electronic and Photoelectron Spectroscopy: Fundamentals and Case Studies.* New York: Cambridge University Press.

Ellis, R. J. (1979). Most abundant protein in the world. *Trends Biochem. Sci.* **4**, 241–244.

Elwood-Madden, M. E., et al. (2009). How long was Meridiani Planum wet? Applying a jarosite stopwatch to determine the duration of aqueous diagenesis. *Geology* **37**, 635–638.

Embleton, B. J. and Williams, G. E. (1986). Low paleolatitude of deposition for the late Precambrian periglacial varvitesin South Australia: implications for paleoclimatology. *Earth Planet. Sci. Lett.* **79**, 419–430.

Embley, T. M. and Williams, T. A. (2015). Steps on the road to eukaryotes. *Nature* **521**, 169–170.

Encrenaz, T., et al. (2004). Hydrogen peroxide on Mars: Evidence for spatial and seasonal variations. *Icarus* **170**, 424–429.

Encrenaz, T., et al. (2008). Simultaneous mapping of H_2O and H_2O_2 on Mars from infrared high-resolution imaging spectroscopy. *Icarus* **195**, 547–556.

Endo, Y., et al. (2015). Photoabsorption cross-section measurements of S-32, S-33, S-34, and S-36 sulfur dioxide from 190 to 220 nm. *J. Geophys. Res.* **120**, 2546–2557.

England, G. L., et al. (2002). Paleoenvironmental significance of rounded pyrite in siliclastic sequences of the Late Archean Witwatersrand Basin: Oxygen-deficient atmosphere or hydrothermal alteration? *Sedimentol.* **49**, 1133–1156.

Eriksson, P. G. and Cheney, E. S. (1992). Evidence for the transition to an oxygen-rich atmosphere during the evolution of red beds in the lower Proterozoic sequences of southern Africa. *Precamb. Res.* **54**, 257–269.

Erkaev, N. V., et al. (2007). Roche lobe effects on the atmospheric loss from "Hot Jupiters". *Astron. Astrophys.* **472**, 329–334.

Erwin, D. H., et al. (2011). The Cambrian Conundrum: Early divergence and later ecological success in the early history of animals. *Science* **334**, 1091–1097.

Erwin, J., et al. (2013). Hybrid fluid/kinetic modeling of Pluto's escaping atmosphere. *Icarus* **226**, 375–384.

Esposito, L. W. (1984). Sulfur dioxide: Episodic injection shows evidence for active Venus volcanism. *Science* **223**, 1072–1074.

Etiope, G., et al. (2009). Terrestrial methane seeps and mud volcanoes: A global perspective of gas origin. *Mar. Petrol. Geol.* **26**, 333–344.

Etiope, G. and Klusman, R. W. (2002). Geologic emissions of methane to the atmosphere. *Chemosphere* **49**, 777–789.

Etiope, G., et al. (2008). Reappraisal of the fossil methane budget and related emission from geologic sources. *Geophys. Res. Lett.* **35**.

Etiope, G. and Lollar, B. S. (2013). Abiotic methane on Earth. *Rev. Geophys.* **51**, 276–299.

Eugster, H. P. and Wones, D. R. (1962). Stability relations of the ferruginous biotite, annite. *J. Petrol.* **3**, 82–125.

Evans, D. A., et al. (1997). Low-latitude glaciation in the Proterozoic era. *Nature* **386**, 262–266.

Evans, K. A., et al. (2012). Oxidation state of subarc mantle. *Geology* **40**, 783–786.

Fahr, H. J. (1976). Reduced hydrogen temperatures in the transition region between thermosphere and exosphere. *Ann. Geophys.* **32**, 277–282.

Fahr, H. J. and Weidner, B. (1977). Gas evaporation from collision determined from planetary exospheres. *Mon. Not. R. Astron. Soc.* **180**, 593–612.

Fairen, A. G. (2010). A cold and wet Mars. *Icarus* **208**, 165–175.

Fairen, A. G., et al. (2009). Stability against freezing of aqueous solutions on early Mars. *Nature* **459**, 401–404.

Fairen, A. G., et al. (2003). Episodic flood inundations of the northern plains of Mars. *Icarus* **165**, 53–67.

Fairen, A. G., et al. (2004). Inhibition of carbonate synthesis in acidic oceans on early Mars. *Nature* **431**, 423–426.

Fairen, A. G., et al. (2002). An origin for the linear magnetic anomalies on Mars through accretion of terranes: Implications for dynamo timing. *Icarus* **160**, 220–223.

Falkowski, P. G. (1997). Evolution of the nitrogen cycle and its influence on the biological CO_2 pump in the oceans. *Nature* **387**, 272–275.

Falkowski, P. G., et al. (2005). The rise of oxygen over the past 205 million years and the evolution of large placental mammals. *Science* **309**, 2202–2204.

Falkowski, P. G., et al. (1992). Natural versus anthropogenic factors affecting low-level cloud albedo over the North Atlantic. *Science* **256**, 1311–1313.

Fallick, A. E., et al. (2008). The ancient anoxic biosphere was not as we know it. In: *Biosphere Origin and Evolution*, ed. N. Dobretsov, et al., New York: Springer, pp. 169–188.

Fani, R., et al. (2000). Molecular evolution of nitrogen fixation: the evolutionary history of the nifD, nifK, nifE, and nifN genes. *J Mol. Evol.* **51**, 1–11.

Fardeau, M. L. and Belaich, J. P. (1986). Energetics of the growth of Methanococcus-Thermolithotrophicus. *Arch. Microbiol.* **144**, 381–385.

Farley, K. A. and Neroda, E. (1998). Noble gases in the Earth's mantle. *Ann. Rev. Earth Planet. Sci.* **26**, 189–218.

Farley, K. A. and Poreda, R. J. (1993). Mantle neon and atmospheric contamination. *Earth Planet. Sc. Lett.* **114**, 325–339.

Farman, J. C., et al. (1985). Large losses of total ozone in Antarctica reveal seasonal ClO_x/NO_x interaction. *Nature* **315**, 207–210.

Farmer, C. B. and Houghton, J. T. (1966). Collision-induced absorption in Earth's atmosphere. *Nature* **209**, 1341.

Farquhar, J., et al. (2000a). Atmospheric influence of Earth's earliest sulfur cycle. *Science* **289**, 756–758.

Farquhar, J., et al. (2008). Sulfur and oxygen isotope study of sulfate reduction in experiments with natural populations from Faellestrand, Denmark. *Geochim. Cosmochim. Acta* **72**, 2805–2821.

Farquhar, J., et al. (2007). Implications from sulfur isotopes of the Nakhla meteorite for the origin of sulfate on Mars. *Earth Planet. Sci. Lett.* **264**, 1–8.

Farquhar, J., et al. (2001). Observation of wavelength-sensitive mass-independent sulfur isotope effects during SO_2 photolysis: application to the early atmosphere. *J. Geophys. Res.* **106**, 1–11.

Farquhar, J., et al. (2000b). Evidence of atmospheric sulfur in the martian regolith from sulphur isotopes in meteorites. *Nature* **404**, 50–52.

Farquhar, J. and Wing, B. A. (2003). Multiple sulfur isotopes and the evolution of the atmosphere. *Earth Planet. Sci. Lett.* **213**, 1–13.

Farquhar, J., et al. (2010). Connections between sulfur cycle evolution, sulfur isotopes, sediments, and base metal sulfide deposits. *Econ. Geol.* **105**, 509–533.

Farquhar, J., et al. (2011). Geological constraints on the origin of oxygenic photosynthesis. *Photosyn. Res.* **107**, 11–36.

Farquhar, J., et al. (2014). Geologic and geochemical constraints on Earth's early atmosphere. In: *Treatise on Geochemistry*, ed. H. D. Holland and K. K. Turekian, New York: Elsevier, pp. 91–138.

Fassett, C. I. and Head, J. W. (2005). Fluvial sedimentary deposits on Mars: Ancient deltas in a crater lake in the Nili Fossae region. *Geophys. Res. Lett.* **32**.

Fassett, C. I. and Head, J. W. (2008a). The timing of martian valley network activity: Constraints from buffered crater counting. *Icarus* **195**, 61–89.

Fassett, C. I. and Head, J. W. (2008b). Valley network-fed, open-basin lakes on Mars: Distribution and implications for Noachian surface and subsurface hydrology. *Icarus* **198**, 37–56.

Fassett, C. I. and Minton, D. A. (2013). Impact bombardment of the terrestrial planets and the early history of the Solar System. *Nat. Geosci.* **6**, 520–524.

Fedo, C. M. and Whitehouse, M. J. (2002). Metasomatic origin of quartz-pyroxene rock, Akilia, Greenland, and implications for Earth's earliest life. *Science* **296**, 1448–1452.

Fegley, B. (2012). *Practical Chemical Thermodynamics for Geoscientists*. Academic Press.

Fegley, B. (2014). Venus. In: *Treatise on Geochemistry*, ed. H. D. Holland and K. K. Turekian, New York: Elsevier, pp. 127–148.

Fegley, B. and Osborne, R. (2013). *Practical Chemical Thermodynamics for Geoscientists*. London: Academic Press.

Fegley, B. and Schaefer, L. (2010). Cosmochemistry of the Biogenic Elements C, H, N, O, and S. In: *Astrobiology: Emergence, Search and Detection of Life*, ed. V. A. Basuik, Stevenson Ranch, CA: Am. Sci. Publishers, pp. 23–49.

Fegley, B. and Zolotov, M. Y. (2000). Chemistry of sodium, potassium, and chlorine in volcanic gases on Io. *Icarus* **148**, 193–210.

Feldman, P., *et al.* (2011). Rosetta-Alice observations of exospheric hydrogen and oxygen on Mars. *Icarus* **214**, 394–399.

Feldman, W. C., *et al.* (2004). Global distribution of near-surface hydrogen on Mars. *J. Geophys. Res.* **109**, E09006.

Ferraz-Mello, S., *et al.* (2008). Tidal friction in close-in satellites and exoplanets: The Darwin theory revisited. *Celestial Mech. Dyn. Ast.* **101**, 171–201.

Ferreira, D., *et al.* (2011). Climate determinism revisited: Multiple equilibria in a complex climate model. *J. Climate* **24**, 992–1012.

Feulner, G. (2012). The faint young Sun problem. *Rev. Geophys.* **50**.

Feynman, R. P., *et al.* (1963). *The Feynman Lectures On Physics*. Reading, Mass: Addison-Wesley.

Fiebig, J., *et al.* (2009). Excess methane in continental hydrothermal emissions is abiogenic. *Geology* **37**, 495–498.

Fike, D. A., *et al.* (2015). Rethinking the ancient sulfur cycle. *Annu. Rev. Earth Pl. Sc.* **43**, 593–622.

Fike, D. A., *et al.* (2006). Oxidation of the Ediacaran Ocean. *Nature* **444**, 744–747.

Fink, U., *et al.* (1972). Water vapor in the atmosphere of Venus. *Icarus* **17**, 617–631.

Finlayson-Pitts, J., B. J. and Pitts, J. N. (2000). *Chemistry of the Upper and Lower Atmosphere : Theory, Experiments, and Applications*. San Diego: Academic Press.

Fischer, D. A. and Valenti, J. (2005). The planet-metallicity correlation. *Astrophys. J.* **622**, 1102–1117.

Fischer, T. P. (2008). Fluxes of volatiles (H_2O, CO_2, N_2, Cl, F) from arc volcanoes. *Geochem. J.* **42**, 21–38.

Fischer, W. W., *et al.* (2014). Archean "whiffs of oxygen" tied to post-depositional processes. *Mineral. Mag.* **78**.

Fishbaugh, K. E., *et al.* (2007). On the origin of gypsum in the Mars north polar region. *J. Geophys. Res.* **112**.

Fisher, J. A., *et al.* (2005). A survey of martian dust devil activity using Mars Global Surveyor Mars Orbiter Camera images. *J. Geophys. Res.* **110**.

Fishman, J. and Crutzen, P. J. (1977). Numerical study of tropospheric photochemistry using a one-dimensional model. *J. Geophys. Res.* **82**, 5897–5906.

Flament, N., *et al.* (2008). A case for late-Archaean continental emergence from thermal evolution models and hypsometry. *Earth Planet. Sc. Lett.* **275**, 326–336.

Flannery, D. T. and Walter, M. R. (2012). Archean tufted microbial mats and the Great Oxidation Event: new insights into an ancient problem. *Australian Journal of Earth Sciences* **59**, 1–11.

Flasar, F. M. (1983). Oceans on Titan. *Science* **221**, 55–57.

Flasar, F. M. (1998). The dynamic meteorology of Titan. *Planet. Space Sci.* **46**, 1125–1147.

Flasar, F. M., *et al.* (2009). Atmosperic dynamics and meteorology. In: *Titan from Cassini–Huygens*, ed. R. H. Brown, *et al.*, New York: Springer.

Flasar, F. M., *et al.* (1981). Titan's atmosphere: Temperature and dynamics. *Nature* **292**, 693–698.

Fleagle, R. G. and Businger, J. A. (1980). *An Introduction to Atmospheric Physics*. New York: Academic Press.

Flowers, R. M. and Farley, K. A. (2012). Apatite He-4/He-3 and (U-Th)/He Evidence for an Ancient Grand Canyon. *Science* **338**, 1616–1619.

Forbes, J., *et al.* (2004). Tides in the middle and upper atmospheres of Mars and Venus. *Adv. Space Res.* **33**, 125–131.

Forbes, J. M. (2002). Wave coupling in terrestrial planetary atmospheres. In: *Atmospheres in the Solar System: Comparative Aeronomy* ed. M. Mendillo, *et al.*, Washington, D. C.: AGU, pp. 171–190.

Forget, F. and Pierrehumbert, R. T. (1997). Warming early Mars with carbon dioxide clouds that scatter infrared radiation. *Science* **278**, 1273–1276.

Forget, F., *et al.* (2013). 3D modelling of the early martian climate under a denser CO_2 atmosphere: Temperatures and CO_2 ice clouds. *Icarus* **222**, 81–99.

Formisano, V., *et al.* (2004). Detection of methane in the atmosphere of Mars. *Science* **306**, 1758–1761.

Forterre, P. (2015). The universal tree of life: An update. *Front. Microbiol.* **6**, Article 717, doi: 10.3389/fmicb.2015.00717.

Fortes, A. D. (2000). Exobiological implications of a possible ammonia-water ocean inside Titan. *Icarus* **146**, 444–452.

Fossati, L., *et al.* (2013). Absorbing gas around the WASP-12 planetary system. *Astrophys. J. Lett.* **766**, L20 doi:10.1088/2041-8205/766/2/L20.

Fossati, L., *et al.* (2010). Metals in the exosphere of the highly irradiated planet WASP-12b. *Astrophys. J. Lett.* **714**, L222–L227.

Fowler, D., *et al.* (2013). The global nitrogen cycle in the twenty-first century. *Phil. Trans. R. Soc. Lond. B* **368**.

Fowler, D., *et al.* (2009). Atmospheric composition change: Ecosystems-atmosphere interactions. *Atmos. Env.* **43**, 5193–5267.

Fox, J. and Hac, A. (2009). Photochemical escape of oxygen from Mars: A comparison of the exobase approximation to a Monte Carlo method. *Icarus* **204**, 527–544.

Fox, J. L. (1993). The production and escape of nitrogen atoms on Mars. *J. Geophys. Res.* **98**, 3297–3310.

Fox, J. L. (2007). Comment on the papers "Production of hot nitrogen atoms in the martian thermosphere" by F. Bakalian and "Monte Carlo computations of the escape of atomic nitrogen from Mars" by F. Bakalian and R.E. Hartle. *Icarus* **192**, 296–301.

Fox, J. L. and Bakalian, F. M. (2001). Photochemical escape of atomic carbon from Mars. *J. Geophys. Res.* **106**, 28 785–28 795.

Fox, J. L. and Hac, A. (1997). The $^{15}N/^{14}N$ isotope fractionation in dissociative recombination of N_2^+. *J. Geophys. Res.* **102**, 9191–9204.

Frakes, L. A. (1979). *Climates Throughout Geologic Time*. New York: Elsevier.

Fralick, P. and Riding, R. (2015). Steep Rock Lake: Sedimentology and geochemistry of an Archean carbonate platform. *Earth Sci. Rev.* **151**, 132–175.

Franck, S., *et al.* (2000a). Habitable zone for Earth-like planets in the solar system. *Planet. Space Sci.* **48**, 1099–1105.

Franck, S., *et al.* (2000b). Reduction of biosphere life span as a consequence of geodynamics. *Tellus B* **52**, 94–107.

Franck, S., *et al.* (1999). Modelling the global carbon cycle for the past and future evolution of the earth system. *Chem. Geol.* **159**, 305–317.

Francois, L. M. and Walker, J. C. G. (1992). Modelling the Phanerozoic carbon cycle and climate: Constraints from the $^{87}Sr/^{86}Sr$ isotopic ratio of seawater. *Amer. J. Sci.* **292**, 81–135.

Franz, H., *et al.* (2015). Reevaluated martian atmospheric mixing ratios from the mass spectrometer on the Curiosity rover. *Planet. Space Sci.* **109–110**, 154–158.

Franz, H. B., *et al.* (2014). Isotopic links between atmospheric chemistry and the deep sulphur cycle on Mars. *Nature* **508**, 364–368.

Free, A. and Barton, N. H. (2007). Do evolution and ecology need the Gaia hypothesis? *Trends Ecol. Evol.* **22**, 611–619.

Frei, R., *et al.* (2009). Fluctuations in Precambrian atmospheric oxygenation recorded by chromium isotopes. *Nature* **461**, 250–U125.

Freissinet, C., *et al.* (2015). Organic molecules in the Sheepbed Mudstone, Gale Crater. *Mars. J. Geophys. Res.* **120**, 495–514.

French, K. L., *et al.* (2015). Reappraisal of hydrocarbon biomarkers in Archean rocks. *P. Natl. Acad. Sci. USA* **112**, 5915–5920.

Frey, H. V. (2006a). Impact constraints on the age and origin of the lowlands of Mars. *Geophys. Res. Lett.* **33**, 1–4.

Frey, H. V. (2006b). Impact constraints on, and a chronology for, major events in early Mars history. *J. Geophys. Res.* **111**, 1–11.

Frey, H. V., *et al.* (2002). Ancient lowlands on Mars. *Geophys. Res. Lett.* **29**.

Friedson, A. J., *et al.* (2009). A global climate model of Titan's atmosphere and surface. *Planet. Space Sci.* **57**, 1931–1949.

Frierson, D. M. W., *et al.* (2006). A gray-radiation aquaplanet moist GCM. Part I: Static stability and eddy scale. *J. Atmos. Sci.* **63**, 2548–2566.

Frierson, D. M. W., *et al.* (2007). Width of the Hadley cell in simple and comprehensive general circulation models. *Geophys. Res. Lett.* **34**, doi:10.1029/2007GL031115.

Frost, B. R. (1991). Introduction to oxygen fugacity and its petrologic importance. In: *Reviews In Mineralogy*, ed. D. H. Lindsley, Washington, D.C.: Mineralogical Society of America, pp. 1–9.

Frost, D. J., *et al.* (2004). Experimental evidence for the existence of iron-rich metal in the Earth's lower mantle. *Nature* **428**, 409–412.

Frost, D. J. and McCammon, C. A. (2008). The redox state of Earth's mantle. *Ann. Rev. Earth Planet. Sci.* **36**, 389–420.

Fu, Q., *et al.* (2002). Tropical cirrus and water vapor: an effective Earth infrared iris feedback? *Atmos. Chem. Phys.* **2**, 31–37.

Fujii, Y., *et al.* (2010). Colors of a second Earth: Estimating the fractional areas of ocean, land, and vegetation of Earth-Like exoplanets. *Astrophys. J.* **715**, 866–880.

Fulchignoni, M., *et al.* (2005). In situ measurements of the physical characteristics of Titan's environment. *Nature* **438**, 785–791.

Gaeman, J., *et al.* (2012). Sustainability of a subsurface ocean within Triton's interior. *Icarus* **220**, 339–347.

Gaidos, E. (2013). Candidate planets in the habitable zones of Kepler stars. *Astrophys. J.* **770**, 90, doi: 10.1088/0004-637X/770/2/90.

Gaidos, E. and Marion, G. (2003). Geological and geochemical legacy of a cold early Mars. *J. Geophys. Res.* **108**.

Gaillard, F. and Scaillet, B. (2009). The sulfur content of volcanic gases on Mars. *Earth Planet. Sci. Lett.* **279**, 34–43.

Gaillard, F., *et al.* (2011). Atmospheric oxygenation caused by a change in volcanic degassing pressure. *Nature* **478**, 229–233.

Gaines, S. M., *et al.* (2009). *Echoes of Life: What Fossil Molecules Reveal about Earth History.* New York: Oxford University Press.

Galli, A., *et al.* (2006a). Energetic hydrogen and oxygen atoms observed on the nightside of Mars. *Space Sci. Rev.* **126**, 267–296.

Galli, A., *et al.* (2006b). The hydrogen exospheric density profile measured with ASPERA-3/NPD. *Space Sci. Rev.* **126**, 447–467.

Galtier, N., *et al.* (1999). A non-hyperthermophilic common ancestor to extant life forms. *Science*, **283**, 220–221.

Ganguli, S. B. (1996). The polar wind. *Rev. Geophys.* **34**, 311–348.

Garcia, R. R. and Solomon, S. (1985). The effect of breaking gravity waves on the dynamics and chemical composition of the mesosphere and lower thermosphere. *J. Geophys. Res.* **92**, 3850–3868.

Garcia-Munoz, A. (2007). Physical and chemical aeronomy of HD 209458b. *Planet. Space Sci.* **55**, 1426–1455.

Garnier, P., *et al.* (2008). The lower exosphere of Titan: Energetic neutral atoms absorption and imaging. *J. Geophys. Res.* **113**, A10216, doi:10.1029/2008JA013029.

Garrels, R. M., *et al.* (1973). Genesis of Precambrian iron formations and development of atmospheric oxygen. *Econ. Geol.* **68**, 1173–1179.

Garvin, J., *et al.* (2009). Isotopic evidence for an aerobic nitrogen cycle in the latest Archean. *Science* **323**, 1045–1048.

Gaucher, E. A., *et al.* (2008). Palaeotemperature trend for Precambrian life inferred from resurrected proteins. *Nature* **451**, 704–707.

Gaucher, E. A., *et al.* (2003). Inferring the palaeoenvironment of ancient bacteria on the basis of resurrected proteins. *Nature* **425**, 285–288.

Gear, C. W. (1971). *Numerical Initial Value Problems In Ordinary Differential Equations.* Englewood Cliffs, N.J.: Prentice-Hall.

Geboy, N. J., *et al.* (2013). Re-Os age constraints and new observations of Proterozoic glacial deposits in the Vazante Group, Brazil. *Precambrian Research* **238**, 199–213.

Geminale, A., *et al.* (2011). Mapping methane in Martian atmosphere with PFS-MEX data. *Planet. Space Sci.* **59**, 137–148.

Genda, H. and Abe, Y. (2003). Survival of a proto-atmosphere through the stage of giant impacts: The mechanical aspects. *Icarus* **164**, 149–162.

Genda, H. and Abe, Y. (2005). Enhanced atmospheric loss on protoplanets at the giant impact phase in the presence of oceans. *Nature* **433**, 842–844.

Gendrin, A., *et al.* (2005). Sulfates in martian layered terrains: the OMEGA/Mars Express view. *Science* **307**, 1587–1591.

Gensel, P. G. (2008). The Earliest Land Plants. *Ann. Rev. Ecol. Evol. Sys.* **39**, 459–477.

Genthner, B. R. S. and Bryant, M. P. (1982). Growth of eubacterium: Limosum with carbon monoxide as the energy source. *Appl. Environ. Microbiol.* **43**, 70–74.

George, S. C., *et al.* (2008). Preservation of hydrocarbons and biomarkers in oil trapped inside fluid inclusions for > 2 billion years. *Geochim. Cosmochim. Acta* **72**, 844–870.

Gerasimov, M. V. and Mukhin, L. M. (1984). Studies of the chemical composition of gaseous phase released from laser pulse evaporated rocks and meteorite materials. *Proc. Lunar Planet. Sci. Conf. XV*, 298–299.

Gerlach, T. M. (2011). Volcanic versus anthropogenic carbon dioxide. *EOS Trans. AGU* **92**, 201–202.

German, C. R. and Seyfried, W. E. (2014). Hydrothermal processes. In: *Treatise on Geochemistry*, 2nd Edn, ed. H. D. Holland and K. K. Turekian, New York: Elsevier, pp. 181–222.

Ghatan, G. J. and Zimbelman, J. R. (2006). Paucity of candidate coastal constructional landforms along proposed shorelines on Mars: Implications for a northern lowlands-filling ocean. *Icarus* **185**, 171–196.

Gierasch, P. J. (1975). Meridional circulation and maintenance of Venus atmospheric rotation. *J. Atmos. Sci.* **32**, 1038–1044.

Gierasch, P. J., *et al.* (1997). The general circulation of the Venus atmosphere: An assessment. In: *Venus II*, ed. S. W. Bougher, *et al.*, Tucson: University of Arizona Press, pp. 459–500.

Giggenbach, W. (1997). Relative importance of thermodynamic and kinetic processes in governing the chemical and isotopic composition of carbon gases in high-heatflow sedimentary basins. *Geochim. Cosmochim. Acta* **61**, 3763–3785.

Giggenbach, W. F. and Matsuo, S. (1991). Evaluation of results from Second and Third IAVCEI Field Workshops on Volcanic Gases. *Appl. Geochem.* **6**, 125–141.

Giggenbach, W. H. (1996). Chemical composition of volcanic gases. In: *Monitoring and Mitigation of Volcano Hazards*, ed. R. Scarpa and R. I. Tilling, Berlin: Springer-Verlag, pp. 221–256.

Gillessen, S., *et al.* (2009). Monitoring stellar orbits around the massive black hole in the galactic center. *Astrophys. J.* **692**, 1075–1109.

Giorgi, F. and Chameides, W. L. (1985). The rainout parameterization in a photochemical model. *J. Geophys. Res.* **90**, 7872–7880.

Gladman, B. (1993). Dynamics of systems of 2 close planets. *Icarus* **106**, 247–263.

Gladstone, G. R., *et al.* (2016). The atmosphere of Pluto as observed by New Horizons. *Science* **351**, 1280, doi:10.1126/science.aad8866.

Glasspool, I. J. and Scott, A. C. (2010). Phanerozoic concentrations of atmospheric oxygen reconstructed from sedimentary charcoal. *Nat. Geosci.* **3**, 627–630.

Glavin, D. P., *et al.* (2013). Evidence for perchlorates and the origin of chlorinated hydrocarbons detected by SAM at the Rocknest aeolian deposit in Gale Crater. *J. Geophys. Res.* **118**, 1955–1973.

Glein, C. R., *et al.* (2015). The pH of Enceladus' ocean. *Geochim. Cosmochim. Acta* **162**, 202–219.

Glein, C. R., *et al.* (2009). The absence of endogenic methane on Titan and its implications for the origin of atmospheric nitrogen. *Icarus* **204**, 637–644.

Glein, C. R. and Shock, E. L. (2013). A geochemical model of non-ideal solutions in the methane-ethane-propane-nitrogen-acetylene system on Titan. *Geochim. Cosmochim. Acta* **115**, 217–240.

Glotch, T. D., *et al.* (2010). Distribution and formation of chlorides and phyllosilicates in Terra Sirenum, Mars. *Geophysical Research Letters* **37**,.

Glotch, T. D. and Rogers, A. D. (2007). Evidence for aqueous deposition of hematite- and sulfate-rich light-toned layered deposits in Aureum and Iani

Chaos, *Mars. J. Geophys. Res.* **112**, E06001, doi:10.1029/2006JE002863.

Godderis, Y., *et al.* (2007). Coupled modeling of global carbon cycle and climate in the Neoproterozoic: links between Rodinia breakup and major glaciations. *Comptes Rendus Geoscience* **339**, 212–222.

Godderis, Y., *et al.* (2003). The Sturtian 'snowball' glaciation: fire and ice. *Earth and Planetary Science Letters* **211**, 1–12.

Godderis, Y. and Veizer, J. (2000). Tectonic control of chemical and isotopic composition of ancient oceans: The impact of continental growth. *Am. J. Sci.* **300**, 434–461.

Godfrey, L. V. and Falkowski, P. G. (2009). The cycling and redox state of nitrogen in the Archaean ocean. *Nature Geosci.*, doi: 10.1038/NGEO633.

Gogarten-Boekels, M., *et al.* (1995). The effects of heavy meteorite bombardment on the early evolution: The emergence of the three domains of life. *Orig. Life and Evol. Biosph.* **25**, 251–264.

Golabek, G. J., *et al.* (2011). Origin of the martian dichotomy and Tharsis from a giant impact causing massive magmatism. *Icarus* **215**, 346–357.

Gold, T. (1964). Outgassing processes on the Moon and Venus. In: *The Origin and Evolution of Atmospheres and Oceans*, ed. P. J. Brancazio and A. G. W. Cameron, New York: Wiley, pp. 249–256.

Gold, T. and Soter, S. (1969). Atmospheric tides and resonant rotation of Venus. *Icarus* **11**, 356–366.

Gold, T. and Soter, S. (1971). Atmospheric tides and 4-day circulation on Venus. *Icarus* **14**, 16–20.

Goldblatt, C., *et al.* (2009). Nitrogen-enhanced greenhouse warming on early Earth. *Nat. Geosci.* **2**, 891–896.

Goldblatt, C., *et al.* (2006). Bistability of atmospheric oxygen and the Great Oxidation. *Nature* **443**, 683–686.

Goldblatt, C., *et al.* (2013). Low simulated radiation limit for runaway greenhouse climates. *Nat. Geosci.* **6**, 661–667.

Goldblatt, C. and Watson, A. J. (2012). The runaway greenhouse: implications for future climate change, geoengineering and planetary atmospheres. *Phil. Trans. R. Soc. Lond. A* **370**, 4197–4216.

Goldblatt, C. and Zahnle, K. J. (2011). Clouds and the faint young Sun paradox. *Clim. Past* **7**, 203–220.

Golden, D. C., *et al.* (2008). Hydrothermal synthesis of hematite spherules and jarosite: Implications for diagenesis and hematite spherule formation in sulfate outcrops at Meridiani Planum, *Mars. Am. Mineral.* **93**, 1201–1214.

Goldreich, P. (1966). History of the lunar orbit. *Rev. Geophys.* **4**, 411–439.

Goldreich, P. and Soter, S. (1966). Q in Solar System. *Icarus* **5**, 375–389.

Goldspiel, J. M. and Squyres, S. W. (2000). Groundwater sapping and valley formation on Mars. *Icarus* **148**, 176–192.

Golombek, M. P., *et al.* (2006). Erosion rates at the Mars Exploration Rover landing sites and long-term climate change on Mars. *J. Geophys. Res.* **111**.

Gomes, R., *et al.* (2005). Origin of the cataclysmic Late Heavy Bombardment period of the terrestrial planets. *Nature* **435**, 466–469.

Gonzalez, G. (1999). Is the Sun anomalous? *Astron. Geophys.* **40**, 25–29.

Gonzalez, G., *et al.* (2001). The Galactic Habitable Zone: Galactic chemical evolution. *Icarus* **152**, 185–200.

Goodman, J. C. (2006). Through thick and thin: Marine and meteoric ice in a "Snowball Earth" climate. *Geophys. Res. Lett.* **33**.

Goodman, J. C. and Pierrehumbert, R. T. (2003). Glacial flow of floating marine ice in 'Snowball Earth'. *J. Geophys. Res.* **108**, 3308.

Goody, R. M. (1964). *Atmospheric Radiation*. Oxford: Clarendon Press.

Goody, R. M. and Belton, M. J. S. (1967). Radiative relaxation times for Mars: a discussion of Martian atmospheric dynamics. *Planet. Space Sci.* **15**, 247–56.

Goody, R. M. and Walker, J. C. G. (1972). *Atmospheres*. Englewood Cliffs, NJ: Prentice Hall.

Goody, R. M. and Yung, Y. L. (1989). *Atmospheric Radiation: Theoretical Basis*. New York: Oxford University Press.

Gough, D. O. (1981). Solar interior structure and luminosity variations. *Solar Phys.* **74**, 21–34.

Grad, H. (1949). On the kinetic theory of rarefied gases. *Comm. Pure Appl. Math.* **2**, 331–407.

Gradie, J. and Tedesco, E. (1982). Compositional structure of the asteroid belt. *Science* **216**, 1405–1407.

Gradstein, F. M. (2012). *The Geologic Time Scale 2012*. Boston: Elsevier.

Graedel, T. E. and Crutzen, P. J. (1993). *Atmospheric Change: An Earth System Perspective*. New York: W.H. Freeman.

Graedel, T. E., *et al.* (1991). Early solar mass loss: A potential solution to the weak sun paradox. *Geophys. Res. Lett.* **18**, 1881–1884.

Graham, D. W. (2002). Noble gas isotope geochemistry of mid-ocean ridge and ocean island basalts: Characterization of mantle source reservoirs. *Rev. Mineral. Geochem.* **47**, 247–317.

Grandstaff, D. E. (1980). Origin of uraniferous conglomerates at Elliot Lake, Canada and Witwatersrand, South-Africa – Implications for oxygen in the Precambrian atmosphere. *Precam. Res.* **13**, 1–26.

Grant, J. A., *et al.* (2008). HiRISE imaging of impact megabreccia and sub-meter aqueous strata in Holden Crater, Mars. *Geology* **36**, 195–198.

Grant, J. A. and Wilson, S. A. (2011). Late alluvial fan formation in southern Margaritifer Terra, *Mars. Geophy. Res. Lett.* **38**.

Greeley, R., *et al.* (1997). Aeolian processes and features on Venus. In: *Venus II*, ed. S. W. Bougher, *et al.*, Tucson: University of Arizona Press, pp. 547–589.

Greeley, R., *et al.* (2005). Fluid lava flows in Gusev crater. *Mars. J. Geophys. Res.* **110**, E05008.

Greeley, R. and Iversen, J. D. (1985). *Wind as a Geological Process on Earth, Mars, Venus, and Titan*. New York: Cambridge University Press.

Greenwood, J. P., *et al.* (2008). Hydrogen isotope evidence for loss of water from Mars through time. *Geophys. Res. Lett* **35**.

Grenfell, T. C. and Warren, S. G. (1999). Representation of a nonspherical ice particle by a collection of independent spheres for scattering and absorption of radiation. *J. Geophys. Res.* **104**, 31 697–31 709.

Grevesse, N., *et al.* (2007). The solar chemical composition. *Space Sci. Rev.* **130**, 105–114.

Griessmeier, J. M., *et al.* (2005). Cosmic ray impact on extrasolar earth-like planets in close-in habitable zones. *Astrobiology* **5**, 587–603.

Griffith, C. A. (2009). Storms, polar deposits and the methane cycle in Titan's atmosphere. *Phil. Trans. R. Soc. Lond. A* **367**, 713–728.

Griffith, C. A., *et al.* (2012). Possible tropical lakes on Titan from observations of dark terrain. *Nature* **486**, 237–239.

Griffith, C. A., *et al.* (2008). Titan's tropical storms in an evolving atmosphere. *Astrophys. J. Lett.* **687**, L41–L44.

Griffith, C. A., *et al.* (2006). Evidence for a polar ethane cloud on Titan. *Science* **313**, 1620–1622.

Griffith, C. A. and Zahnle, K. (1995). Influx of cometary volatiles to planetary Moons: The atmospheres of 1000 possible Titans. *J. Geophys. Res.* **100**, 16907–16922.

Grima, C., *et al.* (2009). North polar deposits of Mars: Extreme purity of the water ice. *Geophysical Research Letters* **36**.

Grimm, R. E., *et al.* (2014). Water budgets of martian recurring slope lineae. *Icarus* **233**, 316–327.

Grinspoon, D. (1987). Was Venus wet? Deuterium reconsidered. *Science* **238**, 1702–1704.

Grinspoon, D. H. (1993). Implications of the high D/H ratio for the sources of water in Venus' atmosphere. *Nature* **363**, 428–431.

Groller, H., *et al.* (2014). Hot oxygen and carbon escape from the martian atmosphere. *Planet. Space Sci.* **98**, 93–105.

Gross, J., *et al.* (2013). Petrography, mineral chemistry, and crystallization history of olivine-phyric shergottite NWA 6234: A new melt composition. *Meteoritics & Planetary Science* **48**, 854–871.

Gross, S. H. (1972). Exospheric temperature of hydrogen-dominated planetary atmospheres. *J. Atmos. Sci.* **29**, 214–218.

Gross, S. H. (1974). Atmospheres of Titan and Galilean satellites. *J. Atmos. Sci.* **31**, 1413–1420.

Grotzinger, J. P., *et al.* (2005). Stratigraphy and sedimentology of a dry to wet eolian depositional system, Burns formation, Meridiani Planum, Mars. *Earth and Planetary Science Letters* **240**, 11–72.

Grotzinger, J. P., *et al.* (2015a). Curiosity's mission of exploration at Gale Crater, Mars. *Elements* **11**, 19–26.

Grotzinger, J. P., *et al.* (2011). Enigmatic origin of the largest-known carbon isotope excursion in Earth's history. *Nature Geoscience* **4**, 285–292.

Grotzinger, J. P., et al. (2015b). Deposition, exhumation, and paleoclimate of an ancient lake deposit, Gale crater. Mars. Science **350**, doi: 10.1126/science.aac7575.

Grotzinger, J. P. and Knoll, A. H. (1999). Stromatolites in Precambrian carbonates: Evolutionary mileposts or environmental dipsticks? Annu. Rev. Earth Planet. Sci. **27**, 313–358.

Grotzinger, J. P. and Rothman, D. H. (1996). An abiotic model for stromatolite morphogenesis. Nature **383**, 423–425.

Grotzinger, J. P., et al. (2014). A habitable fluvio-lacustrine environment at Yellowknife Bay, Gale Crater, Mars. Science **343**, 1242777, doi: 10.1126/science.1242777.

Gruber, N. and Sarmiento, J. L. (1997). Global patterns of marine nitrogen fixation and denitrification. Global Biogeochem. Cyc. **11**, 235–266.

Grundy, W. M. and Buie, M. W. (2001). Distribution and evolution of CH_4, N_2, and CO ices on Pluto's surface: 1995 to 1998. Icarus **153**, 248–263.

Grundy, W. M., et al. (2007). New horizons mapping of Europa and Ganymede. Science **318**, 234–237.

Guillot, T. (1999). Interiors of giant planets inside end outside the solar system. Science **286**, 72–77.

Guillot, T. and Gautier, D. (2015). Giant planets. In: Treatise on Geophysics (Second Edition), ed. G. Schubert, Oxford: Elsevier, pp. 529–557.

Guillot, T., et al. (2004). The interior of Jupiter. In: Jupiter: The Planet, Satellites, and Magnetosphere, ed. F. Bagenal, et al., New York: Cambridge University Press, pp. 35–57.

Gulbis, A. A. S., et al. (2006). Charon's radius and atmospheric constraints from observations of a stellar occultation. Nature **439**, 48–51.

Gulick, V. C. (1998). Magmatic intrusions and a hydrothermal origin for fluvial valleys on Mars. J. Geophys. Res. **103**, 19365–19387.

Gulick, V. C. (2001). Origin of the valley networks on Mars: a hydrological perspective. Geomorphology **37**, 241–268.

Gunning, H. E. and Strausz, O. P. (1966). The reactions of sulfur atoms. In: Advan. Photochem., ed. W. A. Noyes, et al., Hoboken, NJ: Wiley, pp. 143–194.

Gutowsky, H. S. (1976). Halocarbons: Effects on Stratospheric Ozone. National Academy of Sciences.

Guzik, J. A., et al. (1987). A comparison between mass-losing and standard solar models. Ap. J. **319**, 957–965.

Haberle, R. M. (2013). Estimating the power of Mars' greenhouse effect. Icarus **223**, 619–620.

Haberle, R. M., et al. (2017). Early Mars. In: The Atmosphere and Climate of Mars, ed. R. M. Haberle et al., New York: Cambridge University Press.

Haberle, R. M., et al. (2008). The effect of ground ice on the Martian seasonal CO_2 cycle. Planet. Space Sci. **56**, 251–255.

Haberle, R. M., et al. (1993a). Atmospheric effects on the utility of solar power on Mars In: Resources of Near-Earth Space, ed. J. Lewis, M. S. Matthews, Tucson: University of Arizona Press, pp. 845–885.

Haberle, R. M., et al. (2001). On the possibility of liquid water on present-day Mars. J. Geophys. Res. **106**, 23 317–23 326.

Haberle, R. M., et al. (2003). Orbital change experiments with a Mars general circulation model. Icarus **161**, 66–89.

Haberle, R. M., et al. (1993b). Mars atmospheric dynamics as simulated by the NASA Ames General Circulation Model. 1. The zonal-mean circulation. J. Geophys. Res. **98**, 3093–3123.

Habicht, K. S., et al. (2002). Calibration of sulfate levels in the Archean ocean. Science **298**, 2372–2374.

Hadley, G. (1735). Concerning the cause of the general trade-winds. Phil. Trans. R. Soc. **39**, 58–62.

Haisch, K. E., et al. (2001). Disk frequencies and lifetimes in young clusters. Astrophys. J. **553**, L153–L156.

Haldane, J. B. S. (1928). Possible Worlds and Other Papers. London: Harper & Brothers Publishers.

Haldane, J. B. S. (1929). The origin of life. Rationalist Annual **148**, 3–10.

Halevy, I. (2013). Production, preservation, and biological processing of mass-independent sulfur isotope fractionation in the Archean surface environment. P. Natl. Acad. Sci. USA **110**, 17 644–17 649.

Halevy, I. and Head, J. W. (2014). Episodic warming of early Mars by punctuated volcanism. Nat. Geosci. **7**, 865–868.

Halevy, I., *et al.* (2010). Explaining the structure of the Archean mass-independent sulfur isotope record. *Science* **329**, 204–207.

Halevy, I., *et al.* (2007). A sulfur dioxide climate feedback on early Mars. *Science* **318**, 1903–1907.

Hall, D. T., *et al.* (1998). The far-ultraviolet oxygen airglow of Europa and Ganymede. *Astrophys. J.* **499**, 475–481.

Halliday, A. N. (2013). The origins of volatiles in the terrestrial planets. *Geochim. Cosmochim. Acta* **105**, 146–171.

Hallis, L. J., *et al.* (2012). Magmatic water in the martian meteorite Nakhla. *Earth Planet. Sc. Lett.* **359**, 84–92.

Hallmann, C. and Summons, R. E. (2014). Paleobiological clues to early atmospheric evolution. In: *Treatise on Geochemistry*, ed. H. D. Holland and K. K. Turekian, 2nd edn, New York: Elsevier, pp. 139–155.

Halmer, M. M., *et al.* (2002). The annual volcanic gas input into the atmosphere, in particular into the stratosphere: a global data set for the past 100 years. *J. Volcanol. Geotherm. Res.* **115**, 511–528.

Halverson, G. P. (2006). A Neoproterozoic chronology. In: *Neoproterozoic Geobiology and Paleobiology*, ed. S. Xiao and A. J. Kaufman, New York: Kluwer, pp. 231–271.

Halverson, G. P., *et al.* (2005). Toward a Neoproterozoic composite carbon-isotope record. *GSA Bull.* **117**, 1181–1207.

Halverson, G. P. and Hurtgen, M. T. (2007). Ediacaran growth of the marine sulfate reservoir. *Earth Planet. Sci. Lett.* **263**, 32–44.

Halverson, G. P., *et al.* (2010). Neoproterozoic chemostratigraphy. *Precambrian Research* **182**, 337–350.

Hamano, K., *et al.* (2013). Emergence of two types of terrestrial planet on solidification of magma ocean. *Nature* **497**, 607–610.

Han, T. M. (1988). Origin of magnetite in Precambrian iron-formations of low metamorphic grade. In: *Proc. Seventh IAGOD Symposium*, ed. E. Zachrisson, Stuttgart: E. Schweizerbart, pp. 641–656.

Han, T. M. and Runnegar, B. (1992). Megascopic eukaryotic algae from the 2.1 billion-year-old Negaunee iron-formation, Michigan. *Science* **257**, 232–235.

Han, T. T., *et al.* (2011). Global features and trends of the tropopause derived from GPS/CHAMP RO data. *Sci. China: Phys. Mech. Astron.* **54**, 365–374.

Hanel, R. A. (1981). Fourier spectroscopy on planetary missions including Voyager. *Proc. Soc. Photo-Opt. Instrum. Eng.* **289**, 331–344.

Hanel, R. A., *et al.* (1981). Albedo, internal heat, and energy balance of Jupiter: Preliminary results of the Voyager infrared investigation. *J. Geophys. Res.* **86**, 8705–8712.

Hanel, R. A., *et al.* (1972). Nimbus 4 infrared spectroscopy experiment. 1. Calibrated thermal emission spcctra. *J. Geophys. Res.* **77**, 2629–2641.

Hansen, B. M. S. (2009). Formation of the terrestrial planets from a narrow annulus. *Astrophys. J.* **703**, 1131–1140.

Hansen, C. J. and Paige, D. A. (1996). Seasonal nitrogen cycles on Pluto. *Icarus* **120**, 247–265.

Hansen, C. J., *et al.* (2011). The composition and structure of the Enceladus plume. *Geophys. Res. Lett.* **38**.

Hansen, J., *et al.* (2008). Target atmospheric CO_2: Where should humanity aim? *Open Access Atmos. Sci. J.* **2**, 217–231.

Hapke, B. (2012). *Theory of Reflectance and Emittance Spectroscopy*. New York: Cambridge University Press.

Hapke, B. and Nelson, R. (1975). Evidence for an elemental sulfur component of clouds from Venus spectrophotometry. *J. Atmos. Sci.* **32**, 1212–1218.

Haqq-Misra, J. D., *et al.* (2008). A revised, hazy methane greenhouse for the Archean Earth. *Astrobiol.* **8**, 1127–1137.

Haqq-Misra, J., *et al.* (2016). Limit cycles can reduce the width of the habitable zone. *Astrophys. J.* **827**, doi:10.3847/0004-637x/827/2/120.

Hardisty, D. S., *et al.* (2014). An iodine record of Paleoproterozoic surface ocean oxygenation. *Geology* **42**, 619–622.

Harland, W. B. and Rudwick, M. J. S. (1964). The great infra-cambrian ice age. *Sci. Am.* **211**, 28–36.

Harman, C. E., *et al.* (2015). Abiotic O2 levels on planets aroudn F, G, K, and M stars: Possible false positives for life? *Astrophys. J.*, in press.

Harnmeijer, J. P. (2009). *Squeezing blood from a stone: Inferences into the life and depositional environments of the Early Archaean*. University of Washington, PhD thesis, Seattle.

Harper, C. L. and Jacobsen, S. B. (1996). Noble gases and Earth's accretion. *Science* **273**, 1814–1818.

Harries, J., *et al.* (1996). On the distribution of mesospheric molecular hydrogen inferred from HALOE measurements of H_2O and CH_4. *Geophy. Res. Lett.* **23**, 297–300.

Harrison, J. F., *et al.* (2010). Atmospheric oxygen level and the evolution of insect body size. *Proc. R. Soc. Lond. B* **277**, 1937–1946.

Harrison, T. M. (2009). The Hadean crust: Evidence from > 4 Ga zircons. *Ann. Rev. Earth Planet. Sci.* **37**, 479–505.

Hart, M. H. (1975). Explanation for absence of extraterrestrials on Earth. *Q. J. Roy. Astron. Soc.* **16**, 128–135.

Hart, M. H. (1978). The evolution of the atmosphere of the Earth. *Icarus* **33**, 23–39.

Hart, M. H. (1979). Habitable zones around main sequence stars. *Icarus* **37**, 351–357.

Hart, M. H. (1982). Atmospheric evolution, the Drake Equation, and DNA: Sparse life in an infinite Universe. In: *In Extraterrestrials: Where Are They?*, ed. M. H. H. a. B. Zuckerman, New York: Pergamon Press, pp. 154–164.

Harteck, P. and Jensen, J. H. D. (1948). Uber Den Sauerstoffgehalt Der Atmosphare (On the oxygen content of the atmosphere). *Z. Naturforsch. A* **3**, 591–595.

Hartle, R. E. and Grebowsky, J. M. (1990). Upward ion flow in ionospheric holes on Venus. *J. Geophys. Res.* **95**, 31–37.

Hartle, R. E., *et al.* (2006). Initial interpretation of Titan plasma interaction as observed by the Cassini plasma spectrometer: Comparisons with Voyager 1. *Planet. Space Sci.* **54**, 1211–1224.

Hartman, H. and McKay, C. P. (1995). Oxygenic photosynthesis and the oxidation state of Mars. *Planet. Space Sci.* **43**, 123–128.

Hartmann, D. L. (1994). *Global Physical Climatology*. San Diego: Academic Press.

Hartmann, L., *et al.* (2005). IRAC observations of Taurus pre-main-sequence stars. *Astrophys. J.* **629**, 881–896.

Hartmann, W. K. and Davis, D. R. (1975). Satellite-sized planetesimals and lunar origin. *Icarus* **24**, 504–515.

Hartmann, W. K., *et al.* (2000). The time-dependent intense bombardment of the primordial Earth/Moon system. In: *Origin of the Moon and Earth*, ed. R. M. Canup and K. Righter, Tucson, AZ: University of Arizona Press, pp. 493–512.

Hartmann, W. K. and Neukum, G. (2001). Cratering chronology and the evolution of Mars. *Space Sci. Rev.* **96**, 165–194.

Hartogh, P., *et al.* (2010). Herschel/HIFI observations of Mars: First detection of O_2 at submillimetre wavelengths and upper limits on HCl and H_2O_2. *Astron. Astrophys.* **521**.

Hartogh, P., *et al.* (2011a). Direct detection of the Enceladus water torus with Herschel. *Astron. Astrophys.* **532**, article ID L2.

Hartogh, P., *et al.* (2011b). Ocean-like water in the Jupiter-family comet 103P/Hartley 2. *Nature* **478**, 218–220.

Hashimoto, G. L. and Abe, Y. (2005). Climate control on Venus: Comparison of the carbonate and pyrite models. *Planet. Space Sci.* **53**, 839–848.

Hashimoto, G. L., *et al.* (2007). The chemical composition of the early terrestrial atmosphere: Formation of a reducing atmosphere from CI-like material. *J. Geophys. Res.* **112**, E05010.

Haskin, L. A. (1998). The Imbrium impact event and the thorium distribution at the lunar highlands surface. *J. Geophys. Res.* **103**, 1679–1689.

Haswell, C. A. (2010). *Transiting Exoplanets*. Cambridge: Cambridge University Press.

Haswell, C. A., *et al.* (2012). Near-ultraviolet absorption, chromospheric activity, and star-planet interactions in the WASP-12 system. *Astrophys. J.* **760**, 79: doi:10.1088/0004-637X/760/1/79.

Hattori, S., *et al.* (2011). Ultraviolet absorption cross sections of carbonyl sulfide isotopologues (OCS)-S-32, (OCS)-S-33, (OCS)-S-34 and (OCS)-C-13: isotopic fractionation in photolysis and atmospheric implications. *Atmos. Chem. Phys.* **11**, 10 293–10 303.

Hauber, E., *et al.* (2013). Asynchronous formation of Hesperian and Amazonian-aged deltas on Mars and implications for climate. *J. Geophys. Res.* **118**, 1529–1544.

Hauck, S. A. and Phillips, R. J. (2002). Thermal and crustal evolution of Mars. *J. Geophys. Res.* **107**.

Hawkesworth, C., *et al.* (2009). A matter of preservation. *Science* **323**, 49–50.

Hawkesworth, C. J., *et al.* (2010). The generation and evolution of the continental crust. *J. Geol. Soc.* **167**, 229–248.

Hayashi, C., *et al.* (1979). Earths melting due to the blanketing effect of the primordial dense atmosphere. *Earth. Planet. Sc. Lett.* **43**, 22–28.

Hayashi, C., *et al.* (1985). Formation of the Solar System. In: *Protostars and Protoplanets II*, ed. D. C. Black and M. S. Matthews, Tucson: University of Arizona Press, pp. 1100–1153.

Hayes, A. G. (2016). The lakes and seas of Titan. *Annu. Rev. Earth Planet. Sci.* **44**, 57–83.

Hayes, J. M. (1983). Geochemical evidence bearing on the origin of aerobiosis, a speculative hypothesis. In: *Earth's Earliest Biosphere: Its Origin and Evolution*, ed. J. W. Schopf, Princeton, New Jersey: Princeton University Press, pp. 291–301.

Hayes, J. M. (1994). Global methanotrophy at the Archean-Preoterozoic transition. In: *Early Life on Earth*, ed. S. Bengtson, New York: Columbia University Press, pp. 220–236.

Hayes, J. M. and Waldbauer, J. R. (2006). The carbon cycle and associated redox processes through time. *Phil. Trans. R. Soc. Lond. B* **361**, 931–950.

Hays, P. B. and Liu, V. C. (1965). On loss of gases from a planetary atmosphere. *Planet. Space Sci.* **13**, 1185–1211.

Head, J. W., *et al.* (2002). Northern lowlands of Mars: Evidence for widespread volcanic flooding and tectonic deformation in the Hesperian Period. *J. Geophys. Res.* **107**, 5003.

Head, J. W., *et al.* (2003). Recent ice ages on Mars. *Nature* **426**, 797–802.

Hebrard, E. and Marty, B. (2014). Coupled noble gas-hydrocarbon evolution of the early Earth atmosphere upon solar UV irradiation. *Earth Planet. Sc. Lett.* **385**, 40–48.

Hecht, M. H. (2002). Metastability of liquid water on Mars. *Icarus* **156**, 373–386.

Hecht, M. H., *et al.* (2009). Detection of perchlorate and the soluble chemistry of martian soil at the Phoenix Lander site. *Science* **325**, 64–67.

Hedelt, P., *et al.* (2010). Titan's atomic hydrogen corona. *Icarus* **210**, 424–435.

Heimann, A., *et al.* (2010). Fe, C, and O isotope compositions of banded iron formation carbonates demonstrate a major role for dissimilatory iron reduction in similar to 2.5 Ga marine environments. *Earth Planet. Sci. Lett.* **294**, 8–18.

Heimpel, M. and Aurnou, J. (2007). Turbulent convection in rapidly rotating spherical shells: A model for equatorial and high latitude jets on Jupiter and Saturn. *Icarus* **187**, 540–557.

Heimpel, M., *et al.* (2005). Simulation of equatorial and high-latitude jets on Jupiter in a deep convection model. *Nature* **438**, 193–196.

Heising, S., *et al.* (1999). *Chlorobium ferrooxidans* sp. nov., a phototrophic green sulfur bacterium that oxidizes ferrous iron in coculture with a "Geospirillum" sp. strain. *Arch. Microbiol.* **172**, 116–124.

Heising, S. and Schink, B. (1998). Phototrophic oxidation of ferrous iron by a *Rhodomicrobium vannielii* strain. *Microbiol.* **144**, 2263–2269.

Held, I. M. (1975). Momentum transport by quasi-geostrophic eddies. *J. Atmos. Sci.* **32**, 1494–1497.

Held, I. M. (2000). General circulation of the atmosphere. In: *2000 Program in Geophysical Fluid Dynamics*, ed. J.-L. Thiffeault, Woods Hole, MA: Woods Hole Oceanographic Institution, pp. 1–54 (http://www.whoi.edu/page.do?pid=13076).

Held, I. M. and Hou, A. Y. (1980). Non-linear axially-symmetric circulations in a nearly inviscid atmosphere. *J. Atmos. Sci.* **37**, 515–533.

Held, I. M. and Phillips, P. J. (1990). A barotropic model of the Interaction between the Hadley-Cell and a Rossby wave. *J. Atmos. Sci.* **47**, 856–869.

Heldmann, J. L., *et al.* (2007). Observations of martian gullies and constraints on potential formation mechanisms II. The northern hemisphere. *Icarus* **188**, 324–344.

Heldmann, J. L. and Mellon, M. T. (2004). Observations of martian gullies and constraints on potential formation mechanisms. *Icarus* **168**, 285–304.

Helled, R., *et al.* (2011). Jupiter's moment of inertia: A possible determination by Juno. *Icarus* **216**, 440–448.

Helled, R., *et al.* (2014). Giant planet formation, evolution, and internal structure. In: *Protostars and*

Planets VI, ed. H. Beuther, *et al.*, Tucson, AZ: Univ. of Arizona Press, pp. 643–665.

Helled, R., *et al.* (2008). Grain sedimentation in a giant gaseous protoplanet. *Icarus* **195**, 863–870.

Helling, C. and Casewell, S. (2014). Atmospheres of brown dwarfs. *Astron. Astrophys. Rev.* **22**, 80.

Hemming, N. G. and Honisch, B. (2007). Boron isotopes in marine carbonate sediments adn the pH of the ocean. In: *Proxies in Late Cenozoic Paleoceanography*, ed. A. de Vernal, C. Hillaire-Marcel, Amsterdam: Elsevier, pp. 717–734.

Henderson, P. and Henderson, G. M. (2009). *The Cambridge Handbook of Earth Science Data*. New York: Cambridge University Press.

Heng, K., *et al.* (2014). Analytical models of exoplanetary atmospheres. II. Radiative transfer via the two-stream approximation. *Astrophys. J. Suppl. S.* **215**, 4, doi:10.1088/0067-0049/215/1/4.

Henkes, G. A., *et al.* (2014). Temperature limits for preservation of primary calcite clumped isotope paleotemperatures. *Geochim. Cosmochim. Acta* **139**, 362–382.

Herd, C. D. K. (2003). The oxygen fugacity of olivine-phyric martian basalts and the components within the mantle and crust of Mars. *Meteorit. Planet. Sci.* **38**, 1793–1805.

Herd, C. D. K. (2006). Insights into the redox history of the NWA 1068/1110 martian basalt from mineral equilibria and vanadium oxybarometry. *Am. Min.* **91**, 1616–1627.

Herrick, R. R., *et al.* (1997). Morphology and morphometry of impact craters. In: *Venus II: Geology, Geophysics, Atmospghere, and Solar Wind Environment*, Tucson: University of Arizona Press, pp. 1015–1046.

Herwartz, D., *et al.* (2014). Identification of the giant impactor Theia in lunar rocks. *Science* **344**, 1146–1150.

Herzberg, C., *et al.* (2010). Thermal history of the Earth and its petrological expression. *Earth Planet. Sc. Lett.* **292**, 79–88.

Hide, R. (1969). Dynamics of atmospheres of major planets with an appendix on viscous boundary layer at rigid bounding surface of an electrically-conducting rotating fluid in presence of a magnetic field. *J. Atmos. Sci.* **26**, 841–853.

Hide, R. (1970). Equatorial jets in planetary atmospheres. *Nature* **225**, 254–255.

Higgs, P. G. and Lehman, N. (2015). The RNA world: molecular cooperation at the origins of life. *Nat. Rev. Genet.* **16**, 7–17.

Hillier, J. K., *et al.* (2007). The composition of Saturn's E ring. *Mon. Not. R. Astron. Soc.* **377**, 1588–1596.

Hinrichs, K. U. (2002). Microbial fixation of methane carbon at 2.7 Ga: Was an anaerobic mechanism possible? *Geochem. Geophys. Geosys.* **3**, doi:10.1029/2001GC000286.

Hinrichs, K. U. and Boetius, A. (2002). The anaerobic oxidation of methane: New insights in microbial ecology and biogeochemistry. In: *Ocean Margin Systems*, ed. G. Wefer, *et al.*, Berlin: Springer-Verlag, pp. 457–477.

Hinrichs, K. U., *et al.* (1999). Methane-consuming archaebacteria in marine sediments. *Nature* **398**, 802–805.

Hitchcock, D. R. and Lovelock, J. E. (1967). Life detection by atmospheric analysis. *Icarus* **7**, 149.

Hitzman, M. W., *et al.* (2010). Formation of sedimentary rock-hosted stratiform copper deposits through Earth history. *Econ. Geol.* **105**, 627–639.

Hoashi, M., *et al.* (2009). Primary haematite formation in an oxygenated sea 3.46 billion years ago. *Nat. Geosci.* **2**, 301–306.

Hobbs, P. V. (2000). *Introduction to Atmospheric Chemistry*. New York: Cambridge University Press.

Hodges, R. R. (1999). An exospheric perspective of isotopic fractionation of hydrogen on Venus. *J. Geophys. Res.* **104**, 8463–8471.

Hodges, R. R. and Tinsley, B. A. (1981). Charge exchange in the Venus ionosphere as the source of exospheric hydrogen. *J. Geophys. Res.* **86**, 7649–7656.

Hoeppe, G. (2007). *Why the Sky is Blue? Discovering the Color of Life*. Princeton, N.J.: Princeton University Press.

Hoffert, M. I. and Covey, C. (1992). Deriving global climate sensitivity from paleoclimate reconstructions. *Nature* **360**, 573–576.

Hoffman, P. F. (2013). The Great Oxidation and a Siderian snowball Earth: MIF-S based correlation of Paleoproterozoic glacial epochs. *Chemical Geology* **362**, 143–156.

Hoffman, P. F., *et al.* (1998). A Neoproterozoic Snowball Earth. *Science* **281**, 1342–1346.

Hoffman, P. F. and Li, Z. X. (2009). A palaeogeographic context for Neoproterozoic glaciation. *Palaeogeogr. Palaeocl.* **277**, 158–172.

Hoffman, P. F. and Schrag, D. P. (2002). The Snowball Earth hypothesis: Testing the limits of global change. *Terra Nova* **14**, 129–155.

Hoffmann, H. J. (1976). Precambrian microflora, Belcher Islands, Canada: Significance and systematics. *J. Paleontol.* **50**, 1050–1073.

Hofgartner, J. D., *et al.* (2014). Transient features in a Titan sea. *Nature Geoscience* **7**, 493–496.

Hofgartner, J. D. and Lunine, J. I. (2013). Does ice float in Titan's lakes and seas? *Icarus* **223**, 628–631.

Hofmann, A., *et al.* (2009). Multiple sulphur and iron isotope composition of detrital pyrite in Archaean sedimentary rocks: A new tool for provenance analysis. *Earth Planet. Sci. Lett.* **286**, 436–445.

Hofmann, H. J., *et al.* (1999). Origin of 3.45 Ga coniform stromatolites in Warrawoona Group, Western Australia. *Geol. Soc. Am. Bull.* **111**, 1256–1262.

Hoinka, K. P. (1998). Statistics of the global tropopause pressure. *Mon. Weather Rev.* **126**, 3303–3325.

Hoke, M. R. T., *et al.* (2011). Formation timescales of large Martian valley networks. *Earth Planet. Sc. Lett.* **312**, 1–12.

Holland, G., *et al.* (2009). Meteorite Kr in Earth's mantle suggests a late accretionary source for the atmosphere. *Science* **326**, 1522–1525.

Holland, H. D. (1962). Model for the evolution of the Earth's atmosphere. In: *Petrologic Studies: A Volume to Honor A.F. Buddington*, ed. A. E. J. Engel, *et al.*, New York: Geol. Soc. Am., pp. 447–477.

Holland, H. D. (1973a). Ocean water, nutrients, and atmospheric oxygen through geologic time. In: *Proceedings of Symposium of Hydrogeochemistry and Biogeochemistry*, ed. E. Ingerson, Washington, DC: The Clark Co., pp. 68–81.

Holland, H. D. (1973b). Oceans: Possible source of iron in iron-formations. *Econ. Geol.* **68**, 1169–1172.

Holland, H. D. (1978). *The Chemistry of the Atmosphere and Oceans*. New York: Wiley.

Holland, H. D. (1979). Metals in black shales: Reassessment. *Econ. Geol.* **74**, 1676–1680.

Holland, H. D. (1984). *The Chemical Evolution of the Atmosphere and Oceans*. Princeton: Princeton University Press.

Holland, H. D. (2002). Volcanic gases, black smokers, and the Great Oxidation Event. *Geochim. Cosmochim. Acta* **66**, 3811–3826.

Holland, H. D. (2006). The oxygenation of the atmosphere and oceans. *Phil. Trans. R. Soc. Lond. B* **361**, 903–915.

Holland, H. D. (2009). Why the atmosphere became oxygenated: A proposal. *Geochim. Cosmochim. Acta* **73**, 5241–5255.

Holland, H. D. (2011). Discovering the history of atmospheric oxygen. In: *Frontiers in Geochemistry: Contribution of Geochemistry to the Study of the Earth*, ed. R. S. Harmon and A. Parker, Oxford: Blackwell Publishing Ltd., pp. 43–60.

Holland, H. D. and Zbinden, E. A. (1988). Paleosols and the evolution of the atmosphere: part I. In: *Physical and Chemical Weathering in Geochemical Cycles*, ed. A. Lerman and M. Meybeck, Dordrecht: Reidel, pp. 61–82.

Hollingsworth, J. L. and Barnes, J. R. (1996). Forced stationary planetary waves in Mars's winter atmosphere. *J. Atmos. Sci.* **53**, 428–448.

Hollingsworth, J. L., *et al.* (2007). A simple-physics global circulation model for Venus: Sensitivity assessments of atmospheric superrotation. *Geophys. Res. Lett.* **34**.

Holloway, J. R. (2004). Redox reactions in seafloor basalts: possible insights into silicic hydrothermal systems. *Chem. Geol.* **210**, 225–230.

Holman, M. J. and Murray, N. W. (2005). The use of transit timing to detect terrestrial-mass extrasolar planets. *Science* **307**, 1288–1291.

Holser, W. T., *et al.* (1988). Geochemical cycles of carbon and sulfur. In: *Chemical Cycles in the Evolution of the Earth*, ed. C. B. Gregor, *et al.*, New York: Wiley, pp. 105–173.

Holton, J. R. (2004). *An Introduction to Dynamic Meteorology*. New York: Elsevier Academic Press.

Holton, J. R. and Hakim, G. J. (2013). *An Introduction to Dynamic Meteorology* (5th Edition). Amsterdam: Elsevier, Academic Press.

Holzer, T. E., *et al.* (1971). Comparison of kinetic and hydrodynamic models of an expanding ion exosphere. *J. Geophys. Res.* **76**, 2453.

Hood, A. V. S. and Wallace, M. W. (2014). Marine cements reveal the structure of an anoxic, ferruginous Neoproterozoic ocean. *J. Geol. Soc. London* **171**, 741–744.

Hopkins, M., *et al.* (2008). Low heat flow inferred from > 4 Gyr zircons suggests Hadean plate boundary interactions. *Nature* **456**, 493–496.

Hopper, J. P. and Leverington, D. W. (2014). Formation of Hrad Vallis (Mars) by low viscosity lava flows. *Geomorphology* **207**, 96–113.

Horandl, E. and Hadacek, F. (2013). The oxidative damage initiation hypothesis for meiosis. *Plant Reprod.* **26**, 351–367.

Horita, J. (2005). Some perspectives on isotope biosignatures for early life. *Chem. Geol.* **218**, 171–186.

Horita, J. and Berndt, M. E. (1999). Abiogenic methane formation and isotopic fractionation under hydrothermal conditions. *Science* **285**, 1055–1057.

Horner, J., *et al.* (2009). Differences between the impact regimes of the terrestrial planets: Implications for primordial D:H ratios. *Planet. Space Sci.* **57**, 1338–1345.

Horst, S. M., *et al.* (2008). Origin of oxygen species in Titan's atmosphere. *J. Geophys. Res.* **113**.

Houghton, J. T. (2002). *The Physics of Atmospheres.* New York: Cambridge University Press.

Houghton, J. T., *et al.* (1995). *Climate Change, 1994: Radiative Forcing Of Climate Change And An Evaluation Of The IPCC IS92 Emission Scenarios.* New York: Cambridge University. Press.

House, C. H., *et al.* (2003a). Geobiological analysis using whole genome-based tree building applied to the Bacteria, Archea, and Eukarya. *Geobiology* **1**, 15–26.

House, C. H., *et al.* (2003b). Carbon isotopic fractionation by Archaeans and other thermophilic prokaryotes. *Org. Geochem.* **34**, 345–356.

Howard, A. D. (2007). Simulating the development of Martian highland landscapes through the interaction of impact cratering, fluvial erosion, and variable hydrologic forcing. *Geomorphology* **91**, 332–363.

Howard, A. D., *et al.* (2005). An intense terminal epoch of widespread fluvial activity on early Mars: 1. Valley network incision and associated deposits. *J. Geophys. Res.* **110**.

Howard, A. W. (2013). Observed properties of extrasolar planets. *Science* **340**, 572–576.

Howett, C. J. A., *et al.* (2010). Thermal inertia and bolometric Bond albedo values for Mimas, Enceladus, Tethys, Dione, Rhea and Iapetus as derived from Cassini/CIRS measurements. *Icarus* **206**, 573–593.

Hren, M. T., *et al.* (2009). Oxygen and hydrogen isotope evidence for a temperate climate 3.42 billion years ago. *Nature* **462**, 205–208.

Hsu, H. W., *et al.* (2015). Ongoing hydrothermal activities within Enceladus. *Nature* **519**, 207–210.

Hu, R. Y., *et al.* (2012). Photochemistry in terrestrial exoplanet atmospheres. *I. Photochemistry model and benchmark cases. Astrophys. J.* **761**.

Huang, S. S. (1959). Occurrence of life in the universe. *Amer. Scientist* **47**, 397–402.

Huang, S. S. (1960). Life outside the solar system. *Scientific American* **202**, 55–63.

Hubbard, W. B., *et al.* (1995). The interior of Neptune. In: *Neptune and Triton*, ed. D. P. Cruikshank, Tucson: Univ. of Arizona Press, pp. 109–140.

Hubbard, W. B., *et al.* (1993). The occultation of 28 Sgr by Titan. *Astronomy & Astrophysics* **269**, 541–563.

Huber, C. and Wachtershauser, G. (1997). Activated acetic acid by carbon fixation on (Fe,Ni)S under primordial conditions. *Science* **276**, 245–247.

Huber, C. and Wachtershauser, G. (1998). Peptides by activation of amino acids with CO on (Ni,Fe) surfaces: implications for the origin of life. *Science* **281**, 670–672.

Hulburt, E. O. (1953). Explanation of the brightness and color of the sky, particularly the twilight sky. *J. Opt. Soc. Am.* **43**, 113–118.

Hulme, M. (1995). Estimating global change in precipitation. *Weather* **50**, 34–42.

Humayun, M., *et al.* (2013). Origin and age of the earliest Martian crust from meteorite NWA 7533. *Nature* **503**, 513–516.

Hunt, B. G. (1979). Effects of past variations of the Earth's rotation rate on climate. *Nature* **281**, 188–191.

Hunten, D. M. (1973). The escape of light gases from planetary atmospheres. *J. Atmos. Sci.* **30**, 1481–1494.

Hunten, D. M. (1975). Estimates of stratospheric pollution by an analytic model. *Proc. Nat. Acad. Sci.* **72**, 4711.

Hunten, D. M. (1979a). Capture of Phobos and Deimos by protatmospheric drag. *Icarus* **37**, 113–123.

Hunten, D. M. (1979b). Possible oxidant sources in the atmosphere and surface of Mars. *J Mol. Evol.* **14**, 71–78.

Hunten, D. M. (1979c). Possible oxidant sources in the atmosphere and surface of Mars. *J. Mol. Evol.* **14**, 71–78.

Hunten, D. M. (1990). Kuiper Prize Lecture: Escape of atmospheres, ancient and modern. *Icarus* **85**, 1–20.

Hunten, D. M. (2002). Exospheres and planetary escape. In: *Aeronomic systems on planets, moons, and comets*, ed. M. Mendillo, *et al.*, Washington, D. C.: AGU, pp. 191–202.

Hunten, D. M. (2006). The sequestration of ethane on Titan in smog particles. *Nature* **443**, 669–670.

Hunten, D. M. and Donahue, T. M. (1976). Hydrogen loss from the terrestrial planets. *Ann. Rev. Earth Planet Sci.* 4, 265–292.

Hunten, D. M., *et al.* (1989). Escape of atmospheres and loss of water. In: *Origin And Evolution Of Planetary And Satellite Atmospheres*, ed. S. K. Atreya, *et al.*, Tucson, AZ: University of Arizona.

Hunten, D. M., *et al.* (1987). Mass fractionation in hydrodynamic escape. *Icarus* **69**, 532–549.

Hunten, D. M. and Strobel, D. F. (1974). Production and escape of terrestrial hydrogen. *J. Atmos. Sci.* **31**, 305–317.

Hunten, D. M. and Watson, A. J. (1982). Stability of Pluto's atmosphere. *Icarus* **51**, 665–667.

Hurtgen, M. T., *et al.* (2005). Neoproterozoic sulfur isotopes, the evolution of microbial sulfur species, and the burial efficiency of sulfide as sedimentary pyrite. *Geology* **33**, 41–44.

Hurtgen, M. T., *et al.* (2002). The sulfur isotopic composition of Neoproterozoic seawater sulfate: implications for a snowball Earth? *Earth Planet. Sci. Lett.* **203**, 413–429.

Hussmann, H., *et al.* (2006). Subsurface oceans and deep interiors of medium-sized outer planet satellites and large trans-neptunian objects. *Icarus* **185**, 258–273.

Huston, D. L. and Logan, G. A. (2004). Barite, BIFs and bugs: evidence for the evolution of the Earth's early hydrosphere. *Earth Planet. Sci. Lett.* **220**, 41–55.

Hutchins, K. S. and Jakosky, B. M. (1996). Evolution of Martian atmospheric argon: Implications for sources of volatiles. *J. Geophys. Res.* **101**, 14933–14949.

Hutchins, K. S., *et al.* (1997). Impact of a paleomagnetic field on sputtering loss of Martian atmospheric argon and neon. *J. Geophys. Res.* **102**, 9183–9189.

Hyde, W. T., *et al.* (2000). Neoproterozoic 'snowball Earth' simulations with a coupled climate/ice-sheet model. *Nature* **405**, 425–429.

Hynek, B. M., *et al.* (2010). Updated global map of martian valley networks and implications for climate and hydrologic processes. *J. Geophys. Res.* **115**, E09008, doi:10.1029/2009JE003548.

Hynek, B. M. and Phillips, R. J. (2008). The stratigraphy of Meridiani Planum, Mars, and implications for the layered deposits' origin. *Earth Planet. Sci. Lett.* **274**, 214–220.

Hynek, B. M., *et al.* (2003). Explosive volcanism in the Tharsis region: Global evidence in the Martian geologic record. *J. Geophys. Res.* **108**, 5111, doi:10.1029/2003JE002062.

Iess, L., *et al.* (2012). The tides of Titan. *Science* **337**, 457–459.

Iess, L., *et al.* (2010). Gravity field, shape, and moment of inertia of Titan. *Science* **327**, 1367–1369.

Ignatiev, N. I., *et al.* (1997). Water vapour in the lower atmosphere of Venus: A new analysis of optical spectra measured by entry probes. *Adv. Space Res.* **19**, 1159–1168.

Iizuka, T., *et al.* (2015). Meteorite zircon constraints on the bulk Lu-Hf isotope composition and early differentiation of the Earth. *P. Natl. Acad. Sci. USA* **112**, 5331–5336.

Imbrie, J. and Imbrie, K. P. (1979). *Ice Ages: Solving the Mystery*. Cambridge, MA: Harvard University Press.

Inaba, S. and Ikoma, M. (2003). Enhanced collisional growth of a protoplanet that has an atmosphere. *Astron. Astrophys.* **410**, 711–723.

Ingersoll, A. B. and Cuzzi, J. N. (1969). Dynamics of Jupiter's cloud bands. *J. Atmos. Sci.* **26**, 981–985.

Ingersoll, A. P. (1969). The runaway greenhouse: A history of water on Venus. *J. Atmos. Sci.* **26**, 1191–1198.

Ingersoll, A. P. (1970). Mars: Occurrence of liquid water. *Science* **168**, 972–973.

Ingersoll, A. P. (2013). *Planetary Climates*. Princeton, N.J.: Princeton Uniersity Press.

Ingersoll, A. P. and Dobrovolskis, A. R. (1978). Venus rotation and atmospheric tides. *Nature* **275**, 37–38.

Ingersoll, A. P., *et al.* (2004). Dynamics of Jupiter's atmosphere. In: *Jupiter: The Planet, Satellites, and Magnetosphere*, ed. F. Bagenal, *et al.*, New York: Cambridge University Press.

Ingersoll, A. P. and Ewald, S. P. (2011). Total particulate mass in Enceladus plumes and mass of Saturn's E ring inferred from Cassini ISS images. *Icarus* **216**, 492–506.

Irwin, P. (2009). *Giant Planets of Our Solar System: Atmospheres, Composition, and Structure*. Chichester, UK: Springer.

Irwin, R. P., *et al.* (2005). An intense terminal epoch of widespread fluvial activity on early Mars: 2. Increased runoff and paleolake development. *J. Geophys. Res.* **110**.

Irwin, R. P., *et al.* (2015). Paleohydrology of Eberswalde crater, Mars. *Geomorphology* **240**, 83–101.

Ishimaru, R., *et al.* (2011). Oxidizing proto-atmosphere on Titan: Constraint from N-2 formation by impact shock. *Astrophys. J. Lett.* **741**.

Isley, A. E. (1995). Hydrothermal plumes and the delivery of iron to banded iron-formation. *J. Geol.* **103**, 169–185.

Israel, G., *et al.* (2005). Complex organic matter in Titan's atmospheric aerosols from in situ pyrolysis and analysis. *Nature* **438**, 796–799.

Jackson, B., *et al.* (2008a). Tidal heating of terrestrial extrasolar planets and implications for their habitability. *Mon. Not. R. Astron. Soc.* **391**, 237–245.

Jackson, B., *et al.* (2008b). Tidal heating of extrasolar planets. *Astrophys. J.* **681**, 1631–1638.

Jackson, B., *et al.* (2010). The roles of tidal evolution and evaporative mass loss in the origin of CoRoT-7 b. *Mon. Not. R. Astron. Soc.* **407**, 910–922.

Jacob, D. J. (1999). *Introduction to Atmospheric Chemistry*. Princeton, N.J.: Princeton University Press.

Jacobsen, S. B., *et al.* (2008). Isotopes as clues to the origin and earliest differentiation history of the Earth. *Phil. Trans. R. Soc. Lond. A* **366**, 4129–4162.

Jacobson, M. Z. (2005). *Fundamentals of Atmospheric Modeling* (2nd Edn). New York: Cambridge University Press.

Jacovi, R. and Bar-Nun, A. (2008). Removal of Titan's noble gases by their trapping in its haze. *Icarus* **196**, 302–304.

Jaeger, W. L., *et al.* (2007). Athabasca Valles, Mars: A lava-draped channel system. *Science* **317**, 1709–1711.

Jaeger, W. L., *et al.* (2010). Emplacement of the youngest flood lava on Mars: A short, turbulent story. *Icarus* **205**, 230–243.

Jakosky, B. M. and Farmer, C. B. (1982). The seasonal and global behaviour of water vapour in the Mars atmosphere: Complete global results of the Viking atmospheric water detector experiment. *J. Geophys. Res.* **87**, 2999–3019.

Jakosky, B. M., *et al.* (2015). MAVEN observations of the response of Mars to an interplanetary coronal mass ejection. *Science* **350**, doi: 10.1126/science.aad0210.

Jakosky, B. M. and Jones, J. H. (1997). The history of Martian volatiles. *Rev. Geophys.* **35**, 1–16.

Jakosky, B. M. and Mellon, M. T. (2004). Water on Mars. *Phys. Today* **57**, 71–76.

Jakosky, B. M., *et al.* (1994). Mars atmospheric loss and isotopic fractionation by solar-wind-induced sputtering and photochemical escape. *Icarus* **111**, 271–288.

Jakosky, B. M. and Phillips, R. J. (2001). Mars' volatile and climate history. *Nature* **412**, 237–244.

Janssen, M. A. (1993). *Atmospheric Remote Sensing by Microwave Radiometry*. New York: Wiley.

Jansson, K. W. and Johansen, A. (2014). Formation of pebble-pile planetesimals. *Astron. Astrophys.* **570**.

Jarrard, R. D. (2003). Subduction fluxes of water, carbon dioxide, chlorine, and potassium. *Geochem. Geophys. Geosys.* **4**, doi: 10.1029/2002gc000392.

Jaumann, R., *et al.* (2009). Geology and surface processes on Titan. In: *Titan from Cassini-Huygens*, ed. R. H. Brown, *et al.*, New York: Springer, pp. 75–140.

Javaux, E. J., *et al.* (2001). Morphological and ecological complexity in early eukaryotic ecosystems. *Nature* **412**, 66–69.

Javaux, E. J., *et al.* (2004). TEM evidence for eukaryotic diversity in mid-Proterozoic oceans. *Geobiology* **2**, 121–132.

Javoy, M., *et al.* (2010). The chemical composition of the Earth: Enstatite chondrite models. *Earth Planet. Sci. Lett.* **293**, 259–268.

Jeans, J. H. (1954). *The Dynamical Theory of Gases.* New York: Dover.

Jenkins, G. S. (1993). A general circulation model study of the effects of faster rotation, enhanced CO2 concentrations, and solar forcing: implications for the Faint Young Sun Paradox. *J. Geophys. Res.* **98**, 20 803–20 811.

Jenkins, G. S. (1996). A sensitivity study of changes in Earth's rotation rate with an atmospheric general circulation model. *Global and Planet. Change* **11**, 141–154.

Jenkins, G. S., *et al.* (1993). Precambrian climate: the effects of land area and Earth's rotation rate. *J. Geophys. Res.* **98**, 8785–8791.

Jennings, D. E., *et al.* (2009). Titan's surface brightness temperatures. *Astrophys. J. Lett.* **691**, L103–L105.

Jensen, S., *et al.* (2000). Complex trace fossils from the terminal Proterozoic of Namibia. *Geology* **28**, 143–146.

Jerolmack, D. J., *et al.* (2004). A minimum time for the formation of Holden Northeast fan, *Mars. Geophys. Res. Lett.* **31**.

Ji, Q., *et al.* (2002). The earliest known eutherian mammal. *Nature* **416**, 816–822.

Jia, Y. F. and Kerrich, R. (2004). Nitrogen 15-enriched Precambrian kerogen and hydrothermal systems. *Geochem. Geophys. Geosys.* **5**, doi:10.1029/2004GC000716.

Johansen, A., *et al.* (2014). The multifaceted planetesimal formation process. In: *Protostars and Planets VI*, ed. H. Beuther, *et al.*, Tucson, AZ: University of Arizona Press, pp. 571–594.

Johansen, A., *et al.* (2015). Growth of asteroids, planetary embryos, and Kuiper belt objects by chondrule accretion. *Sci. Adv.* **1**, e1500109, doi:10.1126/sciadv.1500109.

Johnson, A. P., *et al.* (2008a). The Miller volcanic spark discharge experiment. *Science* **322**, 404–404.

Johnson, B. and Goldblatt, C. (2015). The nitrogen budget of Earth. *Earth Sci. Rev.* **148**, 150–173.

Johnson, B. C. and Melosh, H. J. (2012). Impact spherules as a record of an ancient heavy bombardment of Earth. *Nature* **485**, 75–77.

Johnson, C. M., *et al.* (2008b). The iron isotope fingerprints of redox and biogeochemical cycling in the modern and ancient Earth. *Ann. Rev. Earth Planet. Sci.* **36**, 457–493.

Johnson, C. M., *et al.* (2013a). Iron formation carbonates: Paleoceanographic proxy or recorder of microbial diagenesis? *Geology* **41**, 1147–1150.

Johnson, H. E. and Axford, W. I. (1969). Production and loss of He^3 in Earth's atmosphere. *J. Geophys. Res.* **74**, 2433.

Johnson, J. A., *et al.* (2010). Giant planet occurrence in the stellar mass-metallicity plane. *Publ. Astron. Soc. Pac.* **122**, 905–915.

Johnson, J. E., *et al.* (2014). O$_2$ constraints from Paleoproterozoic detrital pyrite and uraninite. *Bull. Geol. Soc. Am.* **126**, 813–830.

Johnson, J. E., *et al.* (2013b). Manganese-oxidizing photosynthesis before the rise of cyanobacteria. *P. Natl. Acad. Sci. USA* **110**, 11 238–11 243.

Johnson, J. E., *et al.* (2013c). Reply to Jones and Crowe: Correcting mistaken views of sedimentary geology, Mn-oxidation rates, and molecular clocks. *P. Natl. Acad. Sci. USA* **110**, E4119–E4120.

Johnson, R. E., *et al.* (2009). Composition and detection of Europa's sputter-induced atmosphere. In: *Europa*, ed. R. T. Pappalardo, *et al.*, Tucson: University of Arizona Press, pp. 507–527.

Johnson, R. E., *et al.* (2004). Radiation effects on the surfaces of the Galilean satellites. In: *Jupiter: The Planet, Satellites, and Magnetosphere*, ed. F. Bagenal, *et al.*, New York: Cambridge University Press.

Johnson, R. E., *et al.* (2008c). Exospheres and atmospheric escape. *Space Sci. Rev.* **139**, 355–397.

Johnson, R. E., *et al.* (2002). Energy distributions for desorption of sodium and potassium from ice: The Na/K ratio at Europa. *Icarus* **156**, 136–142.

Johnson, R. E., *et al.* (2013d). Molecular-kinetic simulations of escape from the ex-planet and exoplanets: Criterion for transonic flow. *Astrophys. J. Lett.* **768**.

Johnson, T. M. and Bullen, T. (2004). Mass-dependent fractionation of selenium and chromium isotopes in low-temperature environments. *Rev. Mineral. Geochem.* **55**, 289–317.

Johnston, D. T., *et al.* (2010). An emerging picture of Neoproterozoic ocean chemistry: Insights from the Chuar Group, Grand Canyon, USA. *Earth Planet. Sci. Lett.* **290**, 64–73.

Johnston, D. T., *et al.* (2005). Active microbial sulfur disproportionation in the Mesoproterozoic. *Science* **310**, 1477–1479.

Jones, C. and Crowe, S. A. (2013). No evidence for manganese-oxidizing photosynthesis. *P. Natl. Acad. Sci. USA* **110**, E4118.

Jones, T. D. and Lewis, J. S. (1987). Estimated impact shock production of N_2 and organic compounds on early Titan. *Icarus* **72**, 381–393.

Jöns, H.-P. (1985). Late sedimentation and late sediments in the northern lowlands on Mars. *Lunar Planet. Sci. Conf. XVI*, 414–415.

Jorgensen, U. G., *et al.* (2009). The Earth-Moon system during the late heavy bombardment period - Geochemical support for impacts dominated by comets. *Icarus* **204**, 368–380.

Joseph, J. H., *et al.* (1976). Delta-Eddington approximation for radiative flux transfer. *J. Atmos. Sci.* **33**, 2452–2459.

Joshi, M. (2003). Climate model studies of synchronously rotating planets. *Astrobiology* **3**, 415–427.

Joshi, M. M. and Haberle, R. M. (2012). Suppression of the water ice and snow albedo feedback on planets orbiting red dwarf stars and the subsequent widening of the habitable zone. *Astrobiology* **12**, 3–8.

Joshi, M. M., *et al.* (1997). Simulations of the atmospheres of synchronously rotating terrestrial planets orbiting M dwarfs: Conditions for atmospheric collapse and the implications for habitability. *Icarus* **129**, 450–465.

Jouannic, G., *et al.* (2015). Laboratory simulation of debris flows over sand dunes: Insights into gully-formation (Mars). *Geomorphology* **231**, 101–115.

Joyce, G. F. (1989). RNA evolution and the origins of life. *Nature* **338**, 217–224.

Joyce, G. F. (1994). Foreward. In: *Origins of life : the central concepts*, ed. D. W. Deamer, G. R. Fleischaker, Boston: Jones and Bartlett Publishers, pp. xi–xii.

Junge, C. E., *et al.* (1975). Model calculations for the terrestrial carbon cycle: Carbon isotope geochemistry and evolution of photosynthetic oxygen. *J. Geophys. Res.* **80**, 4542–4552.

Kadoya, S. and Tajika, E. (2014). Conditions for oceans on Earth-Like planets orbiting within the habitable zone: Importance of volcanic CO_2 degassing. *Astrophys. J.* **790**, 107.

Kah, L. C., *et al.* (2004). Low marine sulphate and protracted oxygenation of the Proterozoic biosphere. *Nature* **431**, 834–838.

Kah, L. C. and Riding, R. (2007). Mesoproterozoic carbon dioxide levels inferred from calcified cyanobacteria. *Geology* **35**, 799–802.

Kahn, R. A., *et al.* (1992). The martian dust cycle. In: *Mars*, ed. H. H. Kieffer, *et al.*, Tucson: University of Arizona Press, pp. 1017–1053.

Kaib, N. A. and Chambers, J. E. (2015). The fragility of the terrestrial planets during a giant planet instability. *Mon. Not. R. Astron. Soc.*, in press.

Kanzaki, Y. and Murakami, T. (2015). Estimates of atmospheric CO2 in the Neoarchean-Paleoproterozoic from paleosols. *Geochim. Cosmochim. Acta* **159**, 190–219.

Kappler, A. and Newman, D. K. (2004). Formation of Fe(III)-minerals by Fe(II)-oxidizing photoautotrophic bacteria. *Geochim. Cosmochim. Acta* **68**, 1217–1226.

Kargel, J. S. and Lewis, J. S. (1993). The composition and early evolution of Earth. *Icarus* **105**, 1–25.

Karhu, J. A. and Holland, H. D. (1996). Carbon isotopes and the rise of atmospheric oxygen. *Geology* **24**, 867–870.

Karkoschka, E. (1994). Spectrophotometry of the Jovian planets and Titan at 300 nm to 1000-nm wavelength: The Methane Spectrum. *Icarus* **111**, 174–192.

Karkoschka, E. (1998). Methane, ammonia, and temperature measurements of the Jovian planets and Titan from CCD-spectrophotometry. *Icarus* **133**, 134–146.

Karlsson, N. B., *et al.* (2015). Volume of Martian midlatitude glaciers from radar observations and ice flow modeling. *Geophys. Res. Lett.* **42**, 2627–2633.

Karlstrom, K. E., *et al.* (2014). Formation of the Grand Canyon 5 to 6 million years ago through integration of older palaeocanyons. *Nat. Geosci.* **7**, 239–244.

Karttunen, H. (2007). *Fundamental astronomy*. Berlin: Springer-Verlag.

Kasemann, S. A., *et al.* (2005). Boron and calcium isotope composition in Neoproterozoic carbonate rocks from Namibia: evidence for extreme environmental change. *Earth Planet. Sc. Lett.* **231**, 73–86.

Kasemann, S. A., *et al.* (2010). Neoproterozoic ice ages, boron isotopes, and ocean acidification: Implications for a snowball Earth. *Geology* **38**, 775–778.

Kashefi, K. and Lovley, D. R. (2003). Extending the upper temperature limit for life. *Science* **301**, 934–934.

Kasper, M., *et al.* (2010). *EPICS: direct imaging of exoplanets with the E-ELT*. SPIE Astronomical Telescopes+ Instrumentation. International Society for Optics and Photonics, pp. 77352E–77352E-9.

Kaspi, Y. and Showman, A. P. (2015). Atmospheric dynamics of terrestrial exoplanets over a wide range of orbital and atmospheric parameters. *Astrophys. J.* **804**, 60, doi:10.1088/0004-637x/804/1/60.

Kasting, J. F. (1982). Stability of ammonia in the primitive terrestrial atmosphere. *J. Geophys. Res.* **87**, 3091–3098.

Kasting, J. F. (1985). Photochemical consequences of enhanced CO_2 levels in Earth's early atmosphere. In: *The Carbon Cycle and Atmospheric CO_2: Natural Variations Archean to Present*, ed. E. T. Sundquist and W. S. Broecker, Washington D.C.: American Geophysical Union, pp. 612–622.

Kasting, J. F. (1987). Theoretical constraints on oxygen and carbon dioxide concentrations in the Precambrian atmosphere. *Precambrian Res.* **34**, 205–229,

Kasting, J. F. (1988). Runaway and moist greenhouse atmospheres and the evolution of Earth and Venus. *Icarus* **74**, 472–494.

Kasting, J. F. (1990). Bolide impacts and the oxidation state of carbon in the Earth's early atmosphere. *Origins of Life* **20**, 199–231.

Kasting, J. F. (1991). CO_2 condensation and the climate of early Mars. *Icarus* **94**, 1–13.

Kasting, J. F. (1992). Models relating to Proterozoic atmospheric and oceanic chemistry. In: *The Proterozoic Biosphere: A Multidisciplinary Study*, ed. J. W. Schopf and C. Klein, Cambridge: Cambridge University Press, pp. 1185–1187.

Kasting, J. F. (1993). Earth's early atmosphere. *Science* **259**, 920–926.

Kasting, J. F. (1997). Habitable zones around low mass stars and the search for extraterrestrial life. *Origins of Life* **27**, 291–307.

Kasting, J. F. (2001). The rise of atmospheric oxygen (Perspective). *Science* **293**, 819–820.

Kasting, J. F. (2010). *How to Find a Habitable Planet*. Princeton, N.J.: Princeton University Press.

Kasting, J. F. (2013). What caused the rise of atmospheric O_2? *Chem. Geol.* **362**, 13–25.

Kasting, J. F. and Ackerman, T. P. (1986). Climatic consequences of very high CO_2 levels in the Earth's early atmosphere. *Science* **234**, 1383–1385.

Kasting, J. F. and Brown, L. L. (1998). Setting the stage: the early atmosphere as a source of biogenic compounds. In: *The Molecular Origins of Life: Assembling the Pieces of the Puzzle*, ed. A. Brack, New York: Cambridge University. Press, pp. 35–56.

Kasting, J. F. and Canfield, D. E. (2012). The global oxygen cycle. In: *Fundamentals of Geobiology*, ed. K. O. Konhauser, *et al.*, Oxford: Wiley-Blackwell, pp. 93–104.

Kasting, J. F. and Catling, D. (2003). Evolution of a habitable planet. *Ann. Rev. Astron. Astrophys.* **41**, 429–463.

Kasting, J. F., *et al.* (2012). Atmospheric oxygenation and volcanism. *Nature* **487**, E1.

Kasting, J. F., *et al.* (2015). Stratospheric temperatures and water loss from moist greenhouse atmospheres of Earth-like planets. *Astrophys. J. Lett.* **813**, L3, doi: 10.1088/2041-8205/813/1/13.

Kasting, J. F. and Donahue, T. M. (1980). The evolution of atmospheric ozone. *J. Geophys. Res.* **85**, 3255–3263.

Kasting, J. F., *et al.* (1993a). Mantle redox evolution and the oxidation state of the Archean atmosphere. *J. Geol.* **101**, 245–257.

Kasting, J. F., *et al.* (1985). Oxidant abundances in rainwater and the evolution of atmospheric oxygen. *J. Geophys. Res.* **90**, 10 497–10 510.

Kasting, J. F., *et al.* (2006). Paleoclimates, ocean depth, and the oxygen isotopic composition of seawater. *Earth Planet. Sci. Lett.* **252**, 82–93.

Kasting, J. F., *et al.* (2014). Remote life-detection criteria, habitable zone boundaries, and the frequency of Earth-like planets around M and late K stars. *P. Natl. Acad. Sci. USA* **111**, 12 641–12 646.

Kasting, J. F., *et al.* (2001). A coupled ecosystem-climate model for predicting the methane concentration in the Archean atmosphere. *Origins Life Evol. Biosph.* **31**, 271–285.

Kasting, J. F. and Pollack, J. B. (1983). Loss of water from Venus. I. Hydrodynamic escape of hydrogen. *Icarus* **53**, 479–508.

Kasting, J. F., *et al.* (1984a). Response of Earth's atmosphere to increases in solar flux and implications for loss of water from Venus. *Icarus* **57**, 335–355.

Kasting, J. F., *et al.* (1984b). Effects of high CO_2 levels on surface temperature and atmospheric oxidation state of the early earth. *J. Atmos. Chem.* **1**, 403–428.

Kasting, J. F., *et al.* (1989a). Climate evolution on the terrestrial planets. In: *Origin and Evolution of Planetary and Satellite Atmospheres*, ed. M. S. Matthews, Tucson, Arizona: University. of Arizona Press, pp. 423–449.

Kasting, J. F., *et al.* (1988). How climate evolved on the terrestrial planets. *Scientific Am.* **256**, 90–97.

Kasting, J. F. and Walker, J. C. G. (1981). Limits on oxygen concentration in the prebiological atmosphere and the rate of abiotic fixation of nitrogen. *J. Geophys. Res.* **86**, 1147–1158.

Kasting, J. F., *et al.* (1993b). Habitable zones around main sequence stars. *Icarus* **101**, 108–128.

Kasting, J. F., *et al.* (1989b). Sulfur, ultraviolet radiation, and the early evolution of life. *Origins Life Evol. Biosph.* **19**, 95–108.

Kasting, J. F., *et al.* (1983). Photochemistry of methane in the Earth's early atmosphere. *Precambrian Res.* **20**, 121–148.

Kato, S., *et al.* (1999). The k-distribution method and correlated-k approximation for a shortwave radiative transfer model. *J. Quant. Spectrosc. Radiat. Transf.* **62**, 109–121.

Kato, Y., *et al.* (2009). Hematite formation by oxygenated groundwater more than 2.76 billion years ago. *Earth Planet. Sc. Lett.* **278**, 40–49.

Kaufman, A. J. (1997). Palaeoclimatology – An ice age in the tropics. *Nature* **386**, 227–228.

Kaufman, A. J., *et al.* (2007). Late Archean biospheric oxygenation and atmospheric evolution. *Science* **317**, 1900–1903.

Kawahara, H. and Fujii, Y. (2010). Global mapping of Earth-Like exoplanets from scattered light curves. *Astrophys. J.* **720**, 1333–1350.

Keir, R. S. (2010). A note on the fluxes of abiogenic methane and hydrogen from mid-ocean ridges. *Geophys. Res. Lett.* **37**.

Keller, C. B. and Schoene, B. (2012). Statistical geochemistry reveals disruption in secular lithospheric evolution about 2.5 Gyr ago. *Nature* **485**, 490–495.

Kelley, D. S., *et al.* (2001). An off-axis hydrothermal vent field near the Mid-Atlantic Ridge at 30° N. *Nature* **412**, 145–149.

Kelley, D. S., *et al.* (2005). A serpentinite-hosted ecosystem: the Lost City hydrothermal vent field. *Science* **307**, 1428–1434.

Kelley, K. A. and Cottrell, E. (2009). Water and the oxidation state of subduction zone magmas. *Science* **325**, 605–607.

Kemp, A. I. S. and Hawkesworth, C. J. (2014). Growth and differentiation of the continental crust from isotope studies of accessory minerals. In: *Treatise on Geochemistry*, ed. H. D. Holland and K. K. Turekian, New York: Elsevier, pp. 379–421.

Kendall, B., *et al.* (2009). Re-Os and Mo isotope systematics of black shales from the Middle Proterozoic Velkerri and Wollogorang Formations, McArthur Basin, northern Australia. *Geochim. Cosmochim. Ac.* **73**, 2534–2558.

Kendall, B., *et al.* (2010). Pervasive oxygenation along late Archaean ocean margins. *Nature Geosci.* **3**, 647–652.

Kennedy, M., *et al.* (2006). Late Precambrian oxygenation; inception of the clay mineral factory. *Science* **311**, 1446–1449.

Kennedy, M., *et al.* (2008). Snowball Earth termination by destabilization of equatorial permafrost methane clathrate. *Nature* **453**, 642–645.

Kenyon, S. J. and Bromley, B. C. (2006). Terrestrial planet formation. I. The transition from oligarchic growth to chaotic growth. *Astron. J.* **131**, 1837–1850.

Kerby, R., *et al.* (1983). Single-carbon catabolism in acetogens: Analysis of carbon flow in Acetobacterium-Woodii and Butyribacterium-Methylotrophicum by fermentation and C-13 nuclear magnetic-resonance measurement. *J. Bacteriol.* **155**, 1208–1218.

Kerby, R. and Zeikus, J. G. (1983). Growth of Clostridium-Thermoaceticum on H_2/CO_2 or CO as energy source. *Curr. Microbiol.* **8**, 27–30.

Kerr, G. B., *et al.* (2015). The Palaeoproterozoic global carbon cycle: insights from the Loch Maree Group, NW Scotland. *J. Geol. Soc.*, doi:10.1144/jgs2014-042.

Kerrich, R., *et al.* (2006). Secular variations in N-isotopes in terrestrial reservoirs and ore deposits. *Geol. Soc. Am. Mem.* **198**, 81–104.

Kerrick, D. (2001). Present and past nonanthropogenic CO_2 degassing from the solid Earth. *Rev. Geophys.* **39**, 565–585.

Kerrick, D. M., *et al.* (1995). Convective hydrothermal CO_2 emission from high heat-flow regions. *Chem. Geol.* **121**, 285–293.

Kerridge, J. F. (1985). Carbon, hydrogen and nitrogen in carbonaceous chondrites – Abundances and isotopic compositions in bulk samples. *Geochim. Cosmochim. Acta* **49**, 1707–1714.

Khare, B. N. and Sagan, C. (1975). Cyclic octatomic sulfur: Possible infrared and visible chromophore in clouds of Jupiter. *Science* **189**, 722–723.

Khare, B. N., *et al.* (1984). Optical constants of organic tholins produced in a simulated Titanian atmosphere: From soft X-ray to microwave frequencies. *Icarus* **60**, 127–137.

Kharecha, P., *et al.* (2005). A coupled atmosphere-ecosystem model of the early Archean Earth. *Geobiology* **3**, 53–76.

Khurana, K. K., *et al.* (2009). Electromagnetic induction from Europa's ocean and deep interior. In: *Europa*, ed. R. T. Pappalardo, *et al.*, Tucson: University of Arizona Press.

Kiang, N. Y., *et al.* (2007a). Spectral signatures of photosynthesis. II. Coevolution with other stars and the atmosphere on extrasolar worlds. *Astrobiology* **7**, 252–274.

Kiang, N. Y., *et al.* (2007b). Spectral signatures of photosynthesis. I. Review of Earth organisms. *Astrobiology* **7**, 222–251.

Kieffer, H. H., *et al.* (1992). The planet Mars from antiquity to present. In: *Mars*, ed. H. H. Kieffer, *et al.*, Tucson: University of Arizona Press, pp. 1–33.

Kiehl, J. T. and Dickinson, R. E. (1987). A study of the radiative effects of enhanced atmospheric CO_2 and CH_4 on early earth surface temperatures. *J. Geophys. Res.* **92**, 2991–2998.

Kiehl, J. T. and Trenberth, K. E. (1997). Earth's annual global mean energy budget. *Bull. Am. Meteorol. Soc.* **78**, 197–208.

Kienert, H., *et al.* (2012). Faint young Sun problem more severe due to ice-albedo feedback and higher rotation rate of the early Earth. *Geophys. Res. Lett.* **39**, L23710.

Kilner, B., *et al.* (2005). Low-latitude glaciation in the Neoproterozoic of Oman. *Geology* **33**, 413–416.

Kim, K. M., *et al.* (2012). Protein domain structure uncovers the origin of aerobic metabolism and the rise of planetary oxygen. *Structure* **20**, 67–76.

King, P. L. and McSween, H. Y. (2005). Effects of H_2O, pH, and oxidation state on the stability of Fe minerals on Mars. *J. Geophys. Res.* **110**, 1–15.

Kirchner, J. W. (1989). The Gaia Hypothesis – can it be tested? *Reviews of Geophysics* **27**, 223–235.

Kirchner, J. W. (2002). The Gaia hypothesis: Fact, theory, and wishful thinking. *Climatic Change* **52**, 391–408.

Kirchner, J. W. (2003). The Gaia hypothesis: Conjectures and refutations. *Climatic Change* **58**, 21–45.

Kirk, R. L., *et al.* (1995). Triton's plume: Discovery, characteristics and models. In: *Neptune and Triton*, ed. D. P. Cruikshank, Tucson: University of Arizona Press, pp. 807–877.

Kirkpatrick, J. D., *et al.* (1999). Dwarfs cooler than "M": The definition of spectral type "L" using discoveries from the 2-Micron All-Sky Survey (2MASS). *Astrophysical Journal* **519**, 802–833.

Kirschke, S., *et al.* (2013). Three decades of global methane sources and sinks. *Nat. Geosci.* **6**, 813–823.

Kirschvink, J. L. (1992). In: *Late Proterozoic low-latitude global glaciation: the snowball Earth*, ed. J. W. Schopf and C. Klein, Cambridge: Cambridge University Press, pp. 51–52.

Kirschvink, J. L., *et al.* (2000). Paleoproterozoic snowball Earth: Extreme climatic and geochemical global change and its biological consequences. *P. Natl. Acad. Sci. USA* **97**, 1400–1405.

Kirschvink, J. L. and Kopp, R. E. (2008). Palaeoproterozoic ice houses and the evolution of oxygen-mediating enzymes: the case for a late origin of photosystem II. *Phil. Trans. R. Soc Lond..* **363**, 2755–2765.

Kite, E. S., *et al.* (2013). Seasonal melting and the formation of sedimentary rocks on Mars, with predictions for the Gale Crater mound. *Icarus* **223**, 181–210.

Kite, E. S., *et al.* (2013). Seasonal melting and the formation of sedimentary rocks on Mars, with predictions for the Gale Crater mound. *Icarus* **223**, 181–210.

Kite, E. S., *et al.* (2014). Low palaeopressure of the martian atmosphere estimated from the size distribution of ancient craters. *Nature Geoscience* **7**, 335–339.

Kivelson, M. G., *et al.* (2002). The permanent and inductive magnetic moments of Ganymede. *Icarus* **157**, 507–522.

Kivelson, M. G. and Russell, C. T. (1995). *Introduction to Space Physics*. New York: Cambridge University Press.

Klein, C. (2005). Some Precambrian banded iron-formations (BIFs) from around the world: Their age, geologic setting, mineralogy, metamorphism, geochemistry, and origin. *Am. Mineral.* **90**, 1473–1499.

Klein, C. and Beukes, N. J. (1992). Time distribution, stratigraphy, and sedimentologic setting, and geochemistry of Precambrian iron formations. In: *The Proterozoic Biosphere: A Multidisciplinary Study*, ed. J. W. Schopf and C. Klein, Cambridge: Cambridge University Press, pp. 139–146.

Klein, C. and Beukes, N. J. (1993). Sedimentology and geochemistry of the glaciogenic Late Proterozoic Rapitan iron-formation in Canada. *Econ. Geol. Bull. Soc.* **88**, 542–565.

Klein, C., *et al.* (1987). Filamentous microfossils in the Early Proterozoic Transvaal Supergroup: Their morphology, significance, and paleoenvironmental setting. *Precambrian Res.* **36**, 81–94.

Klein, H. P. (1979). Viking Mission and the Search for Life on Mars. *Rev. Geophys.* **17**, 1655–1662.

Klein, H. P. (1998). The search for life on Mars: What we learned from Viking. *J. Geophys. Res.* **103**, 28 463–28 466.

Kleine, T. and Rudge, J. F. (2011). Chronometry of meteorites and the formation of the Earth and Moon. *Elements* **7**, 41–46.

Kleinhans, M. G. (2005). Flow discharge and sediment transport models for estimating a minimum timescale of hydrological activity and channel and delta formation on Mars. *J. Geophys. Res.* **110**.

Kleinhans, M. G. (2010). A tale of two planets: geomorphology applied to Mars' surface, fluvio-deltaic processes and landforms. *Earth Surf. Proc. Land.* **35**, 102–117.

Kleinhans, M. G., *et al.* (2010). Palaeoflow reconstruction from fan delta morphology on Mars. *Earth Planet. Sc. Lett.* **294**, 378–392.

Klingelhofer, G., *et al.* (2006). Two earth years of Mossbauer studies of the surface of Mars with MIMOS II. *Hyperfine Interactions* **170**, 169–177.

Kliore, A., *et al.* (1965). Occultation experiment: Results of first direct measurement of Mars atmosphere and ionosphere. *Science* **149**, 1243.

Kliore, A. J., *et al.* (2002). Ionosphere of Callisto from Galileo radio occultation observations. *J. Geophys. Res.* **107**.

Klose, K. B., *et al.* (1992). Mineral equilibria and the high radar reflectivity of Venus mountaintops. *J. Geophys. Res.* **97**, 16 353–16 369.

Knauth, L. P. (2005). Temperature and salinity history of the Precambrian ocean: implications for the course of microbial evolution. *Palaeogeog., Palaeoclimat., Palaeoecol.* **219**, 53–69.

Knauth, L. P. and Epstein, S. (1976). Hydrogen and oxygen isotope ratios in nodular and bedded cherts. *Geochim. Cosmochim. Acta* **40**, 1095–1108.

Knauth, L. P. and Kennedy, M. J. (2009). The late Precambrian greening of the Earth. *Nature* **460**, 728–732.

Knauth, P. and Lowe, D. R. (2003). High Archean climatic temperature inferred from oxygen isotope geochemistry of cherts in the 3.5 Ga Swaziland Supergroup, South Africa. *GSA Bull.* **115**, 566–580.

Kneissl, T., *et al.* (2010). Distribution and orientation of northern-hemisphere gullies on Mars from the evaluation of HRSC and MOC-NA data. *Earth Planet. Sc. Lett.* **294**, 357–367.

Knoll, A. H. (1979). Archean photoautotrophy – some alternatives and limits. *Orig. Life Evol. Biosph.* **9**, 313–327.

Knoll, A. H. (1992). The early evolution of the eukaryotes: a geological perspective. *Science* **256**, 622–627.

Knoll, A. H. (2003). *Life on a Young Planet: The First Three Billion Years of Evolution on Earth*. Princeton, N.J.; Oxford: Princeton University Press.

Knoll, A. H. and Carroll, S. B. (1999). Early animal evolution: Emerging views from comparative biology and geology. *Science* **284**, 2129–2137.

Knoll, A. H., *et al.* (2006a). Eukaryotic organisms in Proterozoic oceans. *Phil. Tran. R. Soc. Lond. B* **361**, 1023–1038.

Knoll, A. H., *et al.* (2004). A new period for the geologic time scale. *Science* **305**, 621–622.

Knoll, A. H., *et al.* (2006b). The Ediacaran Period: a new addition to the geologic time scale. *Lethaia* **39**, 13–30.

Knollenberg, R. G. and Hunten, D. M. (1979). Clouds of Venus: Particle-size distribution measurements. *Science* **203**, 792–795.

Knutson, H. A., *et al.* (2014). A featureless transmission spectrum for the Neptune-mass exoplanet GJ 436b. *Nature* **505**, 66–68.

Knutson, H. A., *et al.* (2007). A map of the day–night contrast of the extrasolar planet HD 189733b. *Nature* **447**, 183–186.

Koeberl, C. (2006). The record of impact processes on the early Earth: A review of the first 2.5 billion years. *GSA Special Papers* **405**, 1–22.

Kohler, I., *et al.* (2013). Biological carbon precursor to diagenetic siderite with spherical structures in iron formations. *Nat. Commun.* **4**, 1741, doi :10.1038/ncomms2770.

Kohn, J. P., *et al.* (1976). 3-Phase solid-liquid-vapor equilibria of binary-n-alkane systems (ethane-n-octane, ethane-n-decane, ethane-n-dodecane). *J. Chem. Eng. Data* **21**, 360–362.

Kok, J. F. and Renno, N. O. (2009). Electrification of wind-blown sand on Mars and its implications for atmospheric chemistry. *Geophys. Res. Lett.* **36**.

Komabayashi, M. (1967). Discrete equilibrium temperatures of a hypothetical planet with the atmosphere and the hydrosphere of one component-two phase system under constant solar radiation. *J. Meteor. Soc. Japan* **45**, 137–139.

Komabayashi, M. (1968). Conditions for the existence of the atmosphere and oceans. *Shizen* **23**, 24–31 (in Japanese).

Komatsu, G., *et al.* (1993). Stratigraphy and erosional landforms of layered deposits in Valles Marineris, *Mars. J. Geophys. Res.* **98**, 11 105–11 121.

Komatsu, G., *et al.* (2004). Interior layered deposits of Valles Marineris, Mars. Analogous subice volcanism related to Baikal rifting, southern Siberia. *Planet. Space Sci.* **52**, 167–187.

Komiya, T., *et al.* (1999). Plate tectonics at 3.8-3.7 Ga: Field evidence from the Isua Accretionary Complex, southern West Greenland. *J. Geol.* **107**, 515–554.

Konhauser, K. (2007). *Introduction to Geomicrobiology*. Malden, MA: Blackwell Publishing.

Konhauser, K. O., *et al.* (2007a). Decoupling photochemical Fe(II) oxidation from shallow-water BIF deposition. *Earth Planet. Sci. Lett.* **258**, 87–100.

Konhauser, K. O., *et al.* (2002). Could bacteria have formed the Precambrian banded iron formations? *Geology* **30**, 1079–1082.

Konhauser, K. O., *et al.* (2007b). Was there really an Archean phosphate crisis? *Science* **315**, 1234.

Konhauser, K. O., *et al.* (2011). Aerobic bacterial pyrite oxidation and acid rock drainage during the Great Oxidation Event. *Nature* **478**, 369.

Konhauser, K. O., *et al.* (2009). Oceanic nickel depletion and a methanogen famine before the Great Oxidation Event. *Nature* **458**, 750–754.

Konhauser, K. O., *et al.* (2015). The Archean nickel famine revisited. *Astrobiology* **15**, 804–815.

Konn, C., *et al.* (2015). The production of methane, hydrogen, and organic compounds in ultramafic-hosted hydrothermal vents of the mid-Atlantic Ridge. *Astrobiology* **15**, 381–399.

Kopp, G. and Lean, J. L. (2011). A new, lower value of total solar irradiance: Evidence and climate significance. *Geophys. Res. Lett.* **38**, L01706, doi:10.1029/2010GL045777.

Kopp, R. E., *et al.* (2005). The Paleoproterozoic Snowball Earth: A climate disaster triggered by the evolution of oxygenic photosynthesis. *P. Natl. Acad. Sci. USA* **102**, 11 131–11 136.

Kopparapu, R. K., *et al.* (2013). Habitable zones around main-sequence stars: New estimates. *Astrophys. J.* **765**, doi: 10.1088/0004-637X/765/2/131.

Kopparapu, R. K., *et al.* (2014). Habitable zones around main-sequence stars: Dependence on planetary mass. *Astrophys. J. Lett.* **787**.

Korenaga, J. (2006). Archean geodynamics and the thermal evolution of Earth. In: *Archean Geodynamics and Environments*, ed. K. Benn, Washington DC: American Geophys. Union, pp. 7–32.

Korenaga, J. (2007). Eustasy, supercontinental insulation, and the temporal variability of terrestrial heat flux. *Earth Planet. Sci. Lett.* **257**, 350–358.

Korenaga, J. (2008a). Plate tectonics, flood basalts and the evolution of Earth's oceans. *Terra Nova* **20**, 419–439.

Korenaga, J. (2008b). Urey ratio and the structure and evolution of Earth's mantle. *Reviews of Geophysics* **46**.

Korycansky, D. G. and Zahnle, K. J. (2005). Modeling crater populations on Venus and Titan. *Planet. Space Sci.* **53**, 695–710.

Korycansky, D. G. and Zahnle, K. J. (2011). Titan impacts and escape. *Icarus* **211**, 707–721.

Koskinen, T. T., *et al.* (2014). Thermal escape from extrasolar giant planets. *Phil. Trans. R. Soc. A* **372**, 20130089.

Kostiuk, T., *et al.* (2006). Stratospheric global winds on Titan at the time of Huygens descent. *J. Geophys. Res.* **111**, E07S03.

Kouchinsky, A., *et al.* (2012). Chronology of early Cambrian biomineralization. *Geol. Mag.* **149**, 221–251.

Kounaves, S. P., *et al.* (2010). Soluble sulfate in the martian soil at the Phoenix landing site. *Geophy. Res. Lett.* **37**, L09201, doi:10.1029/2010GL042613.

Kouvaris, L. C. and Flasar, F. M. (1991). Phase-equilibrium of methane and nitrogen at low-temperatures – application to Titan. *Icarus* **91**, 112–124.

Kraal, E. R., *et al.* (2008). Catalogue of large alluvial fans in martian impact craters. *Icarus* **194**, 101–110.

Kral, T. A., *et al.* (1998). Hydrogen consumption by methanogens on the early Earth. *Origins of Life and Evol. of the Biosph.* **28**, 311–319.

Krasnopolsky, V. (2000). On the deuterium abundance on Mars and some related problems. *Icarus* **148**, 597–602.

Krasnopolsky, V. A. (1993). Photochemistry of the martian atmosphere (mean conditions). *Icarus* **101**, 313–332.

Krasnopolsky, V. A. (1999). Hydrodynamic flow of N_2 from Pluto. *J. Geophys. Res.* **104**, 5955–5962.

Krasnopolsky, V. A. (2002). Mars' upper atmosphere and ionosphere at low, medium, and high solar activities: Implications for evolution of water. *J. Geophys. Res.* **107**.

Krasnopolsky, V. A. (2006). Photochemistry of the martian atmosphere: Seasonal, latitudinal, and diurnal variations. *Icarus* **185**, 153–170.

Krasnopolsky, V. A. (2010). The photochemical model of Titan's atmosphere and ionosphere: A version without hydrodynamic escape. *Planet. Space Sci.* **58**, 1507–1515.

Krasnopolsky, V. A. (2011). Atmospheric chemistry on Venus, Earth, and Mars: Main features and comparison. *Planet. Space Sci.* **59**, 952–964.

Krasnopolsky, V. A. (2012). Search for methane and upper limits to ethane and SO_2 on Mars. *Icarus* **217**, 144–152.

Krasnopolsky, V. A. and Feldman, P. D. (2001). Detection of molecular hydrogen in the atmosphere of Mars. *Science* **294**, 1914–1917.

Krasnopolsky, V. A. and Lefevre, F. (2013). Chemistry of the atmospheres of Mars, Venus, and Titan. In: *Comparative Climatology of Terrestrial Planets*, ed. S. J. Mackwell, *et al.*, Tucson: University of Arizona Press, pp. 231–275.

Krasnopolsky, V. A. and Pollack, J. B. (1994). H_2O–H_2SO_4 system in Venus' clouds and OCS, CO, and H_2SO_4 profiles in Venus' troposphere. *Icarus* **109**, 58–78.

Kraus, R. G., *et al.* (2011). Impacts onto H_2O ice: Scaling laws for melting, vaporization, excavation, and final crater size. *Icarus* **214**, 724–738.

Kreidberg, L., *et al.* (2014). Clouds in the atmosphere of the super-Earth exoplanet GJ 1214b. *Nature* **505**, 69–72.

Kreslavsky, M. A. and Head, J. W. (2002). Fate of outflow channel effluents in the northern lowlands of Mars: The Vastitas Borealis Formation as a sublimation residue from frozen ponded bodies of water. *J. Geophys. Res.* **107**, 5121.

Kreslavsky, M. A., *et al.* (2015). The resurfacing history of Venus: Constraints from buffered crater densities. *Icarus* **250**, 438–450.

Kress, M. E. and McKay, C. P. (2004). Formation of methane in comet impacts: implications for Earth, Mars, and Titan. *Icarus* **168**, 475–483.

Krimigis, S. M., *et al.* (2013). Search for the exit: Voyager 1 at heliosphere's border with the galaxy. *Science* **341**, 144–147.

Krissansen-Totton, J., *et al.* (2015). A statistical analysis of the carbon isotope record from the Archean to Phanerozoic and implications for the rise of oxygen. *Amer. J. Sci.* **315**, 275–316.

Krissansen-Totton, J., *et al.* (2016a). On detecting biospheres from chemical disequilibrium in planetary atmospheres. *Astrobiology*, **16**, 39–67.

Krissansen-Totton, J., *et al.* (2016b). Is the Pale Blue Dot unique? Optimized photometric bands for identifying Earth-like exoplanets. *Astrophys. J.* **871**, 31, doi: 10.3847/0004-637X/817/1/31.

Kruijer, T. S., *et al.* (2014). Protracted core formation and rapid accretion of protoplanets. *Science* **344**, 1150–1154.

Krull-Davatzes, A. E., *et al.* (2010). Evidence for a low-O_2 Archean atmosphere from nickel-rich chrome spinels in 3.24 Ga impact spherules, Barberton greenstone belt, South Africa. *Earth Planet. Sci. Lett.* **296**, 319–328.

Krupp, R., *et al.* (1994). The Early Precambrian atmosphere and hydrosphere: Thermodynamic constraints from mineral deposits. *Econ. Geol.* **89**, 1581–1598.

Kuhn, W. R. and Atreya, S. K. (1979). Ammonia photolysis and the greenhouse effect in the primordial atmosphere of the Earth. *Icarus* **37**, 207 213.

Kuiper, G. P. (1944). Titan: A satellite with an atmosphere. *Astrophys. J.* **100**, 378–383.

Kuipers, G., *et al.* (2013). Periglacial evidence for a 1.91–1.89 Ga old glacial period at low latitude, Central Sweden. *Geol. Today* **29**, 218–221.

Kumar, S., *et al.* (1983). Nonthermal escape of hydrogen and deuterium from Venus and implications for loss of water. *Icarus* **55**, 369–389.

Kump, L. R. and Barley, M. E. (2007). Increased subaerial volcanism and the rise of atmospheric oxygen 2.5 billion years ago. *Nature* **448**, 1033–1036.

Kump, L. R., *et al.* (2011). Isotopic evidence for massive oxidation of organic matter following the Great Oxidation Event. *Science* **334**, 1694–1696.

Kump, L. R., *et al.* (2001). The rise of atmospheric oxygen and the "upside-down" Archean mantle. *Geol. Geochem. Geophys. (online)* **2**.

Kump, L. R., *et al.* (2010). *The Earth System*. Upper Saddle River, NJ: Pearson.

Kump, L. R. and Seyfried, W. E. (2005). Hydrothermal Fe fluxes during the Precambrian: Effect of low oceanic sulfate concentrations and low hydrostatic pressure on the composition of black smokers. *Earth Planet. Sc. Lett.* **235**, 654–662.

Kunzmann, M., *et al.* (2013). Zn isotope evidence for immediate resumption of primary productivity after snowball Earth. *Geology* **41**, 27–30.

Kuramoto, K. and Matsui, T. (1994). Formation of a hot proto-atmosphere on the accreting giant icy satellite: Implications for the origin and evolution of Titan, Ganymede, and Callisto. *J. Geophys. Res.* **99**, 21 183–21 200.

Kuramoto, K. and Matsui, T. (1996). Partitioning of H and C between the mantle and the core during the core formation in the Earth: its implications for the atmospheric evolution and redox state of the early mantle. *J. Geophys. Res.* **101**, 14 909–14 932.

Kuramoto, K., *et al.* (2013). Effective hydrodynamic hydrogen escape from an early Earth atmosphere inferred from high-accuracy numerical simulation. *Earth Planet. Sci. Lett.* **375**, 312–318.

Kurata, F. (1975). *Solubility of heavier hydrocarbons in liquid methane*. Research Report, RR-14. Gas Processors Association.

Kurokawa, H. and Kaltenegger, L. (2013). Atmospheric mass-loss and evolution of short-period exoplanets: the examples of CoRoT-7b and Kepler-10b. *Mon. Not. R. Astron. Soc.* **433**, 3239–3245.

Kurokawa, H. and Nakamoto, T. (2014). Mass-loss evolution of close-in exoplanets: evaporation of hot Jupiters and the effect on population. *Astrophysical Journal* **783**.

Kurokawa, H., *et al.* (2014). Evolution of water reservoirs on Mars: Constraints from hydrogen isotopes in martian meteorites. *Earth Planet. Sc. Lett.* **394**, 179–185.

Kurosawa, K. (2015). Impact-driven planetary desiccation: The origin of the dry Venus. *Earth Planet. Sci. Lett.* **429**, 181–190.

Kurster, M., *et al.* (2003). The low-level radial velocity variability in Barnard's star (= GJ 699): Secular acceleration, indications for convective redshift, and planet mass limits. *Astron. Astrophys.* **403**, 1077–1087.

Kurzweil, F., *et al.* (2013). Atmospheric sulfur rearrangement 2.7 billion years ago: Evidence for oxygenic photosynthesis. *Earth Planet. Sci. Lett.* **366**, 17–26.

Kurzweil, F., *et al.* (2015a). Coupled sulfur, iron and molybdenum isotope data from black shales of the Teplá-Barrandian unit argue against deep ocean oxygenation during the Ediacaran. *Geochim. Cosmochim. Acta* **171**, 121–142.

Kurzweil, F., *et al.* (2015b). Continuously increasing δ^{98}Mo values in Neoarchean black shales and iron formations from the Hamersley Basin. *Geochim. Cosmochim. Acta* **164**, 523–542.

Kuzuhara, M., *et al.* (2013). Direct imaging of a cold Jovian exoplanet in orbit around the Sun-like star GJ 504. *The Astrophysical Journal* **774**, 11.

Laakso, T. A. and Schrag, D. P. (2014). Regulation of atmospheric oxygen during the Proterozoic. *Earth Planet. Sci. Lett.* **388**, 81–91.

Lacis, A. A., *et al.* (2013). The role of long-lived greenhouse gases as principal LW control knob that governs the global surface temperature for past and future climate change. *Tellus B* **65**, 19734, http://dx.doi.org/10.3402/tellusb.v65i0.19734.

Lacis, A. A., *et al.* (2010). Atmospheric CO_2: Principal control knob governing Earth's temperature. *Science* **330**, 356–359.

Lagrange, A. M., *et al.* (2010). A giant planet imaged in the disk of the young star β pictoris. *Science* **329**, 57–59.

Lainey, V., *et al.* (2009). Strong tidal dissipation in Io and Jupiter from astrometric observations. *Nature* **459**, 957–959.

Lake, J. A., *et al.* (1984). Eocytes: A new ribosome structure indicates a kingdom with a close relationship to eukaryotes. *P. Natl. Acad. Sci. USA* **81**, 3786–3790.

Lalonde, S. V. and Konhauser, K. O. (2015). Benthic perspective on Earth's oldest evidence for oxygenic photosynthesis. *P. Natl. Acad. Sci. USA* **112**, 995–1000.

Lamb, D. M., *et al.* (2009). Evidence for eukaryotic diversification in the similar to 1800 million-year-old Changzhougou Formation, North China. *Precam. Res.* **173**, 93–104.

Lamb, M. P., *et al.* (2008). Formation of Box Canyon, Idaho, by megaflood: Implications for seepage erosion on Earth and Mars. *Science* **320**, 1067–1070.

Lamb, M. P., *et al.* (2006). Can springs cut canyons into rock? *J. Geophys. Res.* **111**.

Lambrechts, M. and Johansen, A. (2012). Rapid growth of gas-giant cores by pebble accretion. *Astron. Astrophys.* **544**.

Lammer, H. (2013). *Origin and Evolution of Planetary Atmospheres: Implications for Habitability*. New York: Springer.

Lammer, H. and Bauer, S. J. (1991). Nonthermal atmospheric escape from Mars and Titan. *J. Geophys. Res.* **96**, 1819–1825.

Lammer, H., *et al.* (2013). Outgassing history and escape of the Martian atmosphere and water inventory. *Space Sci. Rev.* **174**, 113–154.

Lammer, H., *et al.* (2008). Atmospheric escape and evolution of terrestrial planets and satellites. *Space Sci. Rev.* **139**, 399–436.

Lammer, H., *et al.* (2006). Loss of hydrogen and oxygen from the upper atmosphere of Venus. *Planet. Space Sci.* **54**, 1445–1456.

Lammer, H., *et al.* (2003a). Loss of water from Mars: Implications for the oxidation of the soil. *Icarus* **165**, 9–25.

Lammer, H., *et al.* (2007). Coronal Mass Ejection (CME) activity of low mass M stars as an important factor for the habitability of terrestrial exoplanets. II. CME-induced ion pick up of Earth-like exoplanets in close-in habitable zones. *Astrobiology* **7**, 185–207.

Lammer, H., *et al.* (2009). Determining the mass loss limit for close-in exoplanets: what can we learn from transit observations? *Astron. Astrophys.* **506**, 399–410.

Lammer, H., *et al.* (2003b). Atmospheric loss of exoplanets resulting from stellar X-ray and extreme-ultraviolet heating. *Astrophys. J.* **598**, L121–L124.

Land, L. S. (1995). Oxygen and carbon isotopic composition of Ordovician brachiopods: Implications for Coeval seawater – Comment. *Geochim. Cosmochim. Acta* **59**, 2843–2844.

Lange, M. A. and Ahrens, T. J. (1982). The evolution of an impact generated atmosphere. *Icarus* **51**, 96–120.

Lange, M. A. and Ahrens, T. J. (1986). Shock-induced CO_2 loss from $CaCO_3$ – Implications for early planetary atmospheres. *Earth Planet. Sci. Lett.* **77**, 409–418.

Langevin, Y., *et al.* (2005). Sulfates in the north polar region of Mars detected by OMEGA/Mars express. *Science* **307**, 1584–1586.

Lapen, T. J., *et al.* (2010). A younger age for ALH84001 and its geochemical link to shergottite sources in Mars. *Science* **328**, 347–351.

Lara, L. M., *et al.* (2014). A time-dependent photochemical model for Titan's atmosphere and the origin of H_2O. *Astron. Astrophys.* **566**.

Lasaga, A. C., *et al.* (1971). Primordial oil slick. *Science* **174**, 53–55.

Lasaga, A. C., *et al.* (1985). An improved geochemical model of atmospheric CO2 fluctuations over the past 100 million years. In: *The Carbon Cycle and Atmospheric CO2: Natural Variations Archean to Present*, ed. E. T. Sundquist and W. S. Broecker, Washington, DC: American Geophysical Union, pp. 397–411.

Laskar, J. (2000). On the spacing of planetary systems. *Phys. Rev. Lett.* **84**, 3240–3243.

Laskar, J. and Correia, A. C. M. (2004). The rotation of extra-solar planets. In: *Extrasolar Planets: Today and Tomorrow*, ed. J.-P. Beaulieu, *et al.*: Astronom. Soc. of the Pacific, pp. 401–409.

Laskar, J., *et al.* (2004). Long term evolution and chaotic diffusion of the insolation quantities of Mars. *Icarus* **170**, 343–364.

Lasue, J., *et al.* (2013). Quantitative assessments of the martian hydrosphere. *Space Sci. Rev.* **174**, 155–212.

Lavvas, P. P., *et al.* (2008a). Coupling photochemistry with haze formation in Titan's atmosphere, part I: Model description. *Planet. Space Sci.* **56**, 27–66.

Lavvas, P. P., *et al.* (2008b). Coupling photochemistry with haze formation in Titan's atmosphere. Part II: Results and validation with Cassini/Huygens data. *Planet. Space Sci.* **56**, 67–99.

Le Deit, L., *et al.* (2010). Morphology, stratigraphy, and mineralogical composition of a layered formation covering the plateaus around Valles Marineris, Mars: Implications for its geological history. *Icarus* **208**, 684–703.

Le Heron, P., *et al.* (2010). Sea ice−free conditions during the Sturtian glaciation (early Cryogenian), South Australia. *Geology* **39**, 1–34.

Le Hir, G., *et al.* (2010). Toward the snowball Earth deglaciation. *Climate Dynamics* **35**, 285–297.

Lean, J. and Rind, D. (1998). Climate forcing by changing solar radiation. *Journal of Climate* **11**, 3069–3094.

Leather, J., *et al.* (2002). Neoproterozoic snowball Earth under scrutiny: Evidence from the Fiq glaciation of Oman. *Geology* **30**, 891–894.

LeBlanc, F. (2010). *An Introduction to Stellar Astrophysics*. Chichester, UK: Wiley.

Lebonnois, S., *et al.* (2003). Atomic and molecular hydrogen budget in Titan's atmosphere. *Icarus* **161**, 474–485.

Lebonnois, S., *et al.* (2012). Angular momentum budget in General Circulation Models of superrotating atmospheres: A critical diagnostic. *J. Geophys. Res.* **117**.

Lebonnois, S., *et al.* (2010). Superrotation of Venus' atmosphere analyzed with a full general circulation model. *J. Geophys. Res.* **115**.

Lecavalier des Etangs, A., *et al.* (2004). Atmospheric escape from hot Jupiters. *Astron. Astrophys.* **418**, L1–L4.

Lecavalier Etangs, L. D., *et al.* (2008). Rayleigh scattering in the transit spectrum of HD189733b. *Astron. Astrophys.* **481**, L83–L86.

Leconte, J., *et al.* (2013). Increased insolation threshold for runaway greenhouse processes on Earth-like planets. *Nature* **504**, 268–271.

Leconte, J., *et al.* (2015). Asynchronous rotation of Earth-mass planets in the habitable zone of lower-mass stars. *Science* **347**, 632–635.

Lecuyer, C., *et al.* (1998). The hydrogen isotope composition of seawater and the global water cycle. *Chem. Geol.* **145**, 249–261.

Lecuyer, C. and Ricard, Y. (1999). Long-term fluxes and budget of ferric iron: implication for the redox states of the Earth's mantle and atmosphere. *Earth Planet. Sci. Lett.* **165**, 197–211.

Lecuyer, C., *et al.* (2000). Comparison of carbon, nitrogen and water budgets on Venus and the Earth. *Earth Planet. Sc. Lett.* **181**, 33–40.

Lederberg, J. (1965). Signs of life: criterion-system of exobiology. *Nature* **207**, 9–13.

Lee, C. T. A., *et al.* (2005). Similar V/Sc systematics in MORB and arc basalts: Implications for the oxygen fugacities of their mantle source regions. *J. Petrol.* **46**, 2313–2336.

Lee, C. T. A., *et al.* (2010). The redox state of arc mantle using Zn/Fe systematics. *Nature* **468**, 681–685.

Lee, H., *et al.* (2016). Massive and prolonged deep carbon emissions associated with continental rifting. *Nat. Geosci.* **9**, 145–149.

Lee, J. Y., *et al.* (2006). A redetermination of the isotopic abundances of atmospheric Ar. *Geochim. Cosmochim. Acta* **70**, 4507–4512.

Lee, S. and Kim, H. K. (2003). The dynamical relationship between subtropical and eddy-driven jets. *J. Atmos. Sci.* **60**, 1490–1503.

Lee, Y. N. and Schwartz, S. E. (1981). Evaluation of the rate of uptake of nitrogen dioxide by atmospheric and surface liquid water. *J. Geophys. Res.* **86**, 11 971–11 983.

Lefevre, F., *et al.* (2008). Heterogeneous chemistry in the atmosphere of Mars. *Nature* **454**, 971–975.

Lefevre, F. and Forget, F. (2009). Observed variations of methane on Mars unexplained by known atmospheric chemistry and physics. *Nature* **460**, 720–723.

Leger, A., *et al.* (1993). Search for primitive life on a distant planet: Relevance of O_2 and O_3 dectections. *Astron. Astrophys.* **277**, 309–313.

Leighton, R. B. and Murray, B. C. (1966). Behaviour of carbon dioxide and other volatiles on Mars. *Science* **153**, 136–144.

Leitzinger, M., *et al.* (2011). Could CoRoT-7b and Kepler-10b be remnants of evaporated gas or ice giants? *Planet. Space Sci.* **59**, 1472–1481.

Lellouch, E., *et al.* (1997). Monitoring of mesospheric structure and dynamics. In: *Venus II: Geology, Geophysics, Atmosphere, and Solar Wind Environment,* ed. S. W. Bougher, *et al.*, Tucson, AZ: University of Arizona Press, pp. 295–324.

Lellouch, E., *et al.* (2011a). High resolution spectroscopy of Pluto's atmosphere: detection of the 2.3 μm CH_4 bands and evidence for carbon monoxide. *Astron. Astrophys.* **530**.

Lellouch, E., *et al.* (2011b). The tenuous atmospheres of Pluto and Triton explored by CRIRES on the VLT. *ESO Messenger* **145**, 20–23.

Lellouch, E., *et al.* (2003). Volcanically emitted sodium chloride as a source for Io's neutral clouds and plasma torus. *Nature* **421**, 45–47.

Lemaire, J. F., *et al.* (2007). History of kinetic polar wind models and early observations. *J. Atmos. Sol-Terr. Phys.* **69**, 1901–1935.

Lenardic, A., *et al.* (2004). Growth of the hemispheric dichotomy and the cessation of plate tectonics on Mars. *J. Geophys. Res.* **109**, E02003, doi:10.1029/2003JE002172.

Lenton, T. M. (1998). Gaia and natural selection. *Nature* **394**, 439–447.

Lenton, T. M., *et al.* (2014). Co-evolution of eukaryotes and ocean oxygenation in the Neoproterozoic era. *Nature Geoscience* **7**, 257–265.

Lenton, T. M. and Watson, A. J. (2000). Redfield revisited: II. What regulates the oxygen content of the atmosphere? *Global Biogeochem. Cyc.* **14**, 249–268.

Leone, G. (2014). A network of lava tubes as the origin of Labyrinthus Noctis and Valles Marineris on Mars. *J. Volcanol. Geoth. Res.* **277**, 1–8.

Leone, G., *et al.* (2014). Three-dimensional simulations of the southern polar giant impact hypothesis for the origin of the Martian dichotomy. *Geophys. Res. Lett.* **41**, 2014GL062261.

Leovy, C. (1982a). Martian meteorological variability. *Adv. Space Res.* **2**, 19–44.

Leovy, C. (2001). Weather and climate on Mars. *Nature* **412**, 245–249.

Leovy, C. B. (1964). Simple models of thermally driven mesospheric circulation. *J. Atmos. Sci.* **21**, 327–341.

Leovy, C. B. (1973). Rotation of upper-atmosphere of Venus. *J. Atmos. Sci.* **30**, 1218–1220.

Leovy, C. B. (1977). The atmosphere of Mars. *Sci. Am.* **237**, 34–43.

Leovy, C. B. (1982b). Control of the homopause level. *Icarus* **50**, 311–321.

Leovy, C. B. (1987). Zonal winds near Venus cloud top level: An analytic model of the equatorial wind speed. *Icarus* **69**, 193–201.

Leovy, C. B. and Mintz, Y. (1969). Numerical simulation of the weather and climate of Mars. *Journal of Atmospheric Sciences* **26**, 1167–90.

Leovy, C. B., *et al.* (1973). Mechanisms for Mars dust storms. *J. Atmos. Sci.* **30**, 749–762.

Lepland, A., *et al.* (2013). The earliest phosphorites: Radical change in the phosphorus cycle ruring the Palaeoproterozoic. In: *Reading the Archive of Earth's Oxygenation*, ed. V. A. e. a. Melezhik, Berlin: Springer, pp. 1275–1296.

Lepland, A., *et al.* (2005). Questioning the evidence for Earth's earliest life - Akilia revisited. *Geology* **33**, 77–79.

Lepland, A., *et al.* (2011). Fluid-deposited graphite and its geobiological implications in early Archean gneiss from Akilia, Greenland. *Geobiology* **9**, 2–9.

Lesniak, M. V. and Desch, S. J. (2011). Temperature structure of protoplanetary disks undergoing layered accretion. *Astrophys. J.* **740**, 118.

Levenson, B. P. (2015). Why Hart found narrow ccospheres: A minor science mystery solved. *Astrobiology* **15**, 327–330.

LeVeque, R. J. (2002). *Finite Volume Methods for Hyperbolic Problems*. Cambridge: Cambridge University Press.

Leverington, D. W. (2004). Volcanic rilles, streamlined islands, and the origin of outflow channels on Mars. *J. Geophys. Res.* **109**, 1–14.

Leverington, D. W. (2007). Was the Mangala Valles system incised by volcanic flows? *J. Geophys. Res.* **112**, 1–22.

Leverington, D. W. (2011). A volcanic origin for the outflow channels of Mars: Key evidence and major implications. *Geomorphology* **132**, 51–75.

Levin, L. A. (2002). Deep-ocean life where oxygen is scarce. *Am. Sci.* **90**, 436–444.

Levin, L. A. (2003). Oxygen minimum zone benthos: Adaptation and community response to hypoxia. *Oceanogr. Mar. Biol.* **41**, 1–45.

Levine, J. S., *et al.* (1979). The evolution and variability of atmospheric ozone over geologic time. *Icarus* **39**, 295–309.

Levine, X. J. and Schneider, T. (2011). Response of the Hadley Circulation to climate change in an aquaplanet GCM coupled to a simple representation of ocean heat transport. *J. Atmos. Sci.* **68**, 769–783.

Levison, H. F. and Dones, L. (2014). Comet populations and cometary dynamics. In: *Encyclopedia of the Solar System*, ed. T. Spohn, *et al.*, 3rd edn. Boston: Elsevier, pp. 705–719.

Levison, H. F., *et al.* (2001). Could the lunar "Late heavy bombardment" have been triggered by the formation of Uranus and Neptune? *Icarus* **151**, 286–306.

Levison, H. F., *et al.* (2015). Growing the gas-giant planets by the gradual accumulation of pebbles. *Nature* **524**, 322–324.

Levy, H. (1971). Normal atmosphere – large radical and formaldehyde concentrations predicted. *Science* **173**, 141–143.

Levy, J. S., *et al.* (2014). Sequestered glacial ice contribution to the global Martian water budget: Geometric constraints on the volume of remnant, midlatitude debris-covered glaciers. *J. Geophys. Res.* **119**, 2188–2196.

Lew, S. K. (1967). *The problem of hydrogen escape in the Earth's upper atmosphere – a reappraisal*. University of California, Los Angeles, PhD, Los Angeles, CA.

Lewis, G. N. and Randall, M. (1923). *Thermodynamics and the Free Energy of Chemical Substances*. New York: McGraw-Hill.

Lewis, J. S. (2004). *Physics and Chemistry of the Solar System* (2nd Edn). Boston: Elsevier.

Lewis, J. S. and Prinn, R. G. (1984). *Planets and Their Atmospheres: Origin and Evolution*. Orlando, Florida: Academic Press.

Lewis, S. R., *et al.* (2007). Assimilation of thermal emission spectrometer atmospheric data during the Mars Global Surveyor aerobraking period. *Icarus* **192**, 327–347.

Li, C., *et al.* (2010). A stratified redox model for the Ediacaran ocean. *Science* **328**, 80–83.

Li, D. W. and Pierrehumbert, R. T. (2011). Sea glacier flow and dust transport on Snowball Earth. *Geophys. Res. Lett.* **38**.

Li, L. M., *et al.* (2011). The global energy balance of Titan. *Geophy. Res. Lett.* **38**.

Li, W. Q., *et al.* (2013). An anoxic, Fe(II)-rich, U-poor ocean 3.46 billion years ago. *Geochim. Cosmochim. Acta* **120**, 65–79.

Li, W. Q., *et al.* (2012). U-Th-Pb isotope data indicate phanerozoic age for oxidation of the 3.4 Ga Apex Basalt. *Earth Planet. Sci. Lett.* **319**, 197–206.

Li, Z. X. A. and Lee, C. T. A. (2004). The constancy of upper mantle fO(2) through time inferred from V/Sc ratios in basalts. *Earth Planet. Sci. Lett.* **228**, 483–493.

Lian, Y. and Showman, A. P. (2010). Generation of equatorial jets by large-scale latent heating on the giant planets. *Icarus* **207**, 373–393.

Liang, M. C., *et al.* (2006). Production of hydrogen peroxide in the atmosphere of a Snowball Earth and the origin of oxygenic photosynthesis. *P. Natl. Acad. Sci. USA* **103**, 18 896–18 899.

Liang, M. C., *et al.* (2005). Atmosphere of Callisto. *Journal of Geophysical Research-Planets* **110**.

Lichtenberg, K. A., *et al.* (2010). Stratigraphy of hydrated sulfates in the sedimentary deposits of Aram Chaos, Mars. *J. Geophys. Res.* **115**, E00D17, doi:10.1029/2009JE003353.

Lide, D. R. (2011). *Handbook of Chemistry and Physics*. Boca Raton, FL: CRC Press.

Limaye, S. S. and Rengel, M. (2013). Atmospheric circulation and dynamics. In: *Towards Understanding the Climate of Venus*, ed. L. Bengtsson, *et al.*, New York: Springer, pp. 55–70.

Lin, B., *et al.* (2002). The iris hypothesis: A negative or positive cloud feedback? *Journal of Climate* **15**, 3–7.

Lin, Y., *et al.* (2011). Multiple-sulfur isotope effects during photolysis of carbonyl sulfide. *Atmos. Chem. Phys.* **11**, 10 283–10 292.

Lindemann, T. A. and Dobson, G. M. B. (1923). A theory of meteors, and the density and temperature of the outer atmosphere to which it leads. *Proc. R. Soc. Lond. A* **102**, 411–437.

Lindsay, J. F. and Brasier, M. D. (2002). Did global tectonics drive early biosphere evolution? Carbon isotope record from 2.6 To 1.9 Ga carbonates of Western Australian basins. *Precam. Res.* **114**, 1–34.

Lindzen, R. S. (1971). Atmospheric tides. In: *Mathematical Problems in the Geophysical Sciences, No. 2 : Inverse Problems, Dynamo Theory and Tides*, ed. W. H. Reid, Providence, RI: Amer. Math. Soc., pp. 293–362.

Lindzen, R. S., *et al.* (2001). Does the earth have an adaptive infrared iris? *Bull. Am. Met. Soc.* **82**, 417–432.

Lindzen, R. S. and Hou, A. Y. (1988). Hadley circulations for zonally averaged heating centred off the equator. *J. Atmos. Sci.* **45**, 2416–2447.

Line, M. R., *et al.* (2010). High temperature photochemistry in the atmosphere of HD 189733b. *Astrophys. J.* **717**, 496–502.

Lineweaver, C. H., *et al.* (2004). The Galactic habitable zone and the age distribution of complex life in the Milky Way. *Science* **303**, 59–62.

Linsky, J. L., *et al.* (2010). Observations of mass loss from the transiting exoplanet HD 209458b*. *Astrophys. J.* **717**, 1291–1299.

Liou, K.-N. (2002). *An Introduction to Atmospheric Radiation*. Amsterdam; London: Academic Press.

Liss, P. S., *et al.* (1997). Marine sulphur emissions. *Phil. Trans. R. Soc. Lond. B* **352**, 159–168.

Liss, P. S. and Slater, P. G. (1974). Flux of gases across air-sea interface. *Nature* **247**, 181–184.

Lissauer, J. J. (2007). Planets formed in habitable zones of M dwarf stars probably are deficient in volatiles. *Astrophys. J.* **660**, L149–L152.

Little, B., *et al.* (1999). Galileo images of lightning on Jupiter. *Icarus* **142**, 306–323.

Liu, A. G., *et al.* (2015). Remarkable insights into the paleoecology of the Avalonian Ediacaran macrobiota. *Gondwana Res.* **27**, 1355–1380.

Liu, A. G., *et al.* (2010). First evidence for locomotion in the Ediacara biota from the 565 Ma Mistaken Point Formation, Newfoundland. *Geology* **38**, 123–126.

Liu, S. C. and Donahue, T. M. (1974). The aeronomy of hydrogen in the atmosphere of the earth. *J. Atmos. Sci.* **31**, 1118–1136.

Lodders, K. (2003). Solar system abundances and condensation temperatures of the elements. *Astrophys. J.* **591**, 1220–1247.

Lodders, K. (2010a). Atmospheric chemistry of the gas giant planets. *Geochem. News* **142**, 1–11.

Lodders, K. (2010b). Solar system abundances of the elements. In: *Principles and Perspectives in Cosmochemistry*, ed. A. Goswami and B. E. Reddy, Berlin: Springer, pp. 379–417.

Lodders, K. and Fegley, B. (1998). *The Planetary Scientist's Companion*. New York: Oxford University Press.

Loeb, N. G., *et al.* (2009). Toward optimal closure of the Earth's top-of-atmosphere radiation budget. *J. Climate* **22**, 748–766.

Logan, G. A., *et al.* (1995). Terminal Proterozoic reorganization of biogeochemical cycles. *Nature* **376**, 53–56.

Lopez, E. D. and Fortney, J. J. (2013). The role of core mass in controlling evaporation: The Kepler radius distribution and the Kepler-36 density dichotomy. *Astrophysical Journal* **776**.

Lopez, E. D. and Fortney, J. J. (2014). Understanding the mass-radius relation for sub-Neptunes: Radius as a proxy for composition. *Astrophys. J.* **792**, 1, doi: 10.1088/0004-637X/792/1/1.

López-Puertas, M. and Taylor, F. W. (2001). *Non-LTE Radiative Transfer in the Atmosphere*. London: World Scientific.

Lora, J. M., *et al.* (2015). GCM simulations of Titan's middle and lower atmosphere and comparison to observations. *Icarus* **250**, 516–528.

Lorenz, R. D. (2000). The weather on Titan. *Science* **290**, 467–468.

Lorenz, R. D., *et al.* (2009). Seasonal change on Titan. In: *Titan from Cassini-Huygens*, ed. R. H. Brown, *et al.*, New York: Springer.

Lorenz, R. D., *et al.* (1997). Photochemically driven collapse of Titan's atmosphere. *Science* **275**, 642–644.

Lorenz, R. D., *et al.* (2008). Titan's inventory of organic surface materials. *Geophys. Res. Lett.* **35**.

Lorenz, R. D., *et al.* (2006). Titan's damp ground: Constraints on Titan surface thermal properties from the temperature evolution of the Huygens GCMS inlet. *Meteorit. Planet. Sci.* **41**, 1705–1714.

Lorenz, R. D., *et al.* (2011). Hypsometry of Titan. *Icarus* **211**, 699–706.

Lorius, C., *et al.* (1990). The ice-core record: Climate sensitivity and future greenhouse warming. *Nature* **347**, 139–145.

Love, G. D., *et al.* (2009). Fossil steroids record the appearance of Demospongiae during the Cryogenian period. *Nature* **457**, 718–721.

Lovelock, J. (1988). *The Ages of Gaia*. New York: W.W. Norton.

Lovelock, J. E. (1965). A physical basis for life detection experiments. *Nature* **207**, 568–570.

Lovelock, J. E. (1972). Gaia as seen through atmosphere. *Atmos. Environ.* **6**, 579–&.

Lovelock, J. E. (1975). Thermodynamics and the recognition of alien biospheres. *Proc. R. Soc. Lond.B* **189**, 167–181.

Lovelock, J. E. (1979). *Gaia: A New Look at Life on Earth*. Oxford: Oxford University Press.

Lovelock, J. E. (1989). Geophysiology, the science of Gaia. *Rev. Geophys.* **27**, 215–222.

Lovelock, J. E. (1991). *Gaia: The Practical Science of Planetary Medicine*. London: Gaia Books.

Lovelock, J. E. and Watson, A. J. (1982). The regulation of carbon dioxide and climate: Gaia or Geochemistry. *Planet. Space Sci.* **30**, 795–802.

Lovelock, J. E. and Whitfield, M. (1982). Life span of the biosphere. *Nature* **296**, 561–563.

Lowe, D. R. and Byerly, G. R. (2015). Geologic record of partial ocean evaporation triggered by giant asteroid impacts, 3.29–3.23 billion years ago. *Geology* **43**, 535–538.

Lowe, D. R., *et al.* (2014). Recently discovered 3.42-3.23 Ga impact layers, Barberton Belt, South Africa: 3.8 Ga detrital zircons, Archean impact history, and tectonic implications. *Geology* **42**, 747–750.

Lu, J., *et al.* (2009). Cause of the widening of the tropical belt since 1958. *Geophys. Res. Lett.* **36**, L03803, doi:10.1029/2008GL036076.

Lucchitta, B. K., *et al.* (1994). Topography of Valles-Marineris - Implications for erosional and structural history. *J. Geophys. Res.* **99**, 3783–3798.

Luger, R. and Barnes, R. (2015). Extreme water loss and abiotic O_2 buildup on planets throughout the habitable zones of M dwarfs. *Astrobiology* **15**, 119–143.

Luger, R., *et al.* (2015). Habitable evaporated cores: Transforming mini-Neptunes into Super-Earths in the habitable zones of M dwarfs. *Astrobiology* **15**, 57–88.

Lunine, J., *et al.* (2009). The origin and evolution of Titan. In: *Titan from Cassini-Huygens*, ed. R. H. Brown, *et al.*, New York: Springer, pp. 35–59.

Lunine, J. I. (2010). Titan and habitable planets around M-dwarfs. *Faraday Discuss.* **147**, 405–418.

Lunine, J. I., *et al.* (2003). The origin of water on Mars. *Icarus* **165**, 1–8.

Lunine, J. I. and Lorenz, R. D. (2009). Rivers, lakes, dunes, and rain: crustal processes in Titan's methane cycle. *Ann. Rev. Earth Planet. Sci.* **37**, 299–320.

Lunine, J. I., *et al.* (1998). Some speculations on Titan's past, present and future. *Planetary and Space Science* **46**, 1099–1107.

Lunine, J. I. and Nolan, M. C. (1992). A massive early atmosphere on Triton. *Icarus* **100**, 221–234.

Lunine, J. I., *et al.* (2011). Dynamical models of terrestrial planet formation. *Adv. Sci. Lett.* **4**, 325–338.

Lunine, J. I., *et al.* (1983). Ethane ocean on Titan. *Science* **222**, 1229–1230.

Lupu, R. E., *et al.* (2014). The atmospheres of Earth-like planets after giant impact events. *Astrophys. J.* **784**.

Luth, R. W. and Canil, D. (1993). Ferric iron in mantle-derived pyroxenes and a new oxybarometer for the mantle. *Contrib. Mineral. Petrol.* **113**, 236–248.

Lynd, L., *et al.* (1982). Carbon monoxide metabolism of the methylotrophic acidogen *Butyribacterium Methylotrophicum. J. Bacteriol.* **149**, 255–263.

Lyons, J. R. (2009). Atmospherically-derived mass-independent sulfur isotope signatures, and incorporation into sediments. *Chem. Geol.* **267**, 164–174.

Lyons, J. R. and Young, E. D. (2005). CO self-shielding as the origin of oxygen isotope anomalies in the early solar nebula. *Nature* **435**, 317–320.

Lyons, T. W., *et al.* (2009). Tracking euxinia in the ancient ocean: A multiproxy perspective and Proterozoic case study. *Ann. Rev. Earth Planet. Sci.* **37**, 507–534.

Lyons, T. W., *et al.* (2014). The rise of oxygen in Earth's early ocean and atmosphere. *Nature* **506**, 307–315.

Lyons, T. W. and Severmann, S. (2006). A critical look at iron paleoredox proxies: New insights from modern euxinic marine basins. *Geochim. Cosmochim. Acta* **70**, 5698–5722.

Macdonald, F. A., *et al.* (2010). Calibrating the Cryogenian. *Science* **327**, 1241–1243.

Macdonald, F. A., *et al.* (2013). The stratigraphic relationship between the Shuram carbon isotope excursion, the oxygenation of Neoproterozoic oceans, and the first appearance of the Ediacara biota and bilaterian trace fossils in northwestern Canada. *Chem. Geol.* **362**, 250–272.

MacGregor, A. M. (1927). The problem of the Precambrian atmosphere. *S. African J. Sci.* **24**, 155–172.

Machado, A. D. (1987). On the origin and age of the Steep Rock Buckshot, Ontario, Canada. *Chem. Geol.* **60**, 337–349.

Macintosh, B., *et al.* (2015). Discovery and spectroscopy of the young jovian planet 51 Eri b with the Gemini Planet Imager. *Science* **350**, 64–67.

Macouin, M., *et al.* (2015). Is the Neoproterozoic oxygen burst a supercontinent legacy? *Front. Earth Sci.* **3**, 44, doi:10.3389/feart.2015.00044.

Madeleine, J. B., *et al.* (2011). Revisiting the radiative impact of dust on Mars using the LMD Global Climate Model. *J. Geophys. Res.* **116**.

Magalhaes, J. A., *et al.* (2002). The stratification of Jupiter's troposphere at the Galileo probe entry site. *Icarus* **158**, 410–433.

Magee, B. A., *et al.* (2009). INMS-derived composition of Titan's upper atmosphere: Analysis methods and model comparison. *Planet. Space Sci.* **57**, 1895–1916.

Mahaffy, P. R., *et al.* (2015a). Volatile and Isotopic Imprints of Ancient Mars. *Elements* **11**, 51–56.

Mahaffy, P. R., *et al.* (2015b). The imprint of atmospheric evolution in the D/H of Hesperian clay minerals on Mars. *Science* **347**, 412–414.

Maher, L. J. and Tinsley, B. A. (1977). Atomic hydrogen escape rate due to charge-exchange with hot plasmaspheric ions. *J. Geophys. Res.* **82**, 689–695.

Maheshwari, A., *et al.* (2010). Global nature of the Paleoproterozoic Lomagundi carbon isotope excursion A review of occurrences in Brazil, India, and Uruguay. *Precam. Res.* **182**, 274–299.

Maindl, T. I., *et al.* (2015). Impact induced surface heating by planetesimals on early Mars. *Astron. Astrophys.* **574**, A22, doi: 10.1051/0004-6361/201424256.

Mak, M. (2011). *Atmospheric Dynamics*. Cambridge ; New York: Cambridge University Press.

Malin, M. C., *et al.* (2008). Climate, weather, and north polar observations from the Mars Reconnaissance Orbiter Mars Color Imager. *Icarus* **194**, 501–512.

Malin, M. C. and Edgett, K. S. (1999). Oceans or seas in the Martian northern lowlands: High resolution imaging tests of proposed coastlines. *Geophys. Res. Lett.* **26**, 3049–3052.

Malin, M. C. and Edgett, K. S. (2000a). Evidence for recent groundwater seepage and surface runoff on Mars. *Science* **288**, 2330–2335.

Malin, M. C. and Edgett, K. S. (2000b). Sedimentary rocks of early Mars. *Science* **290**, 1927–1937.

Malin, M. C. and Edgett, K. S. (2001). Mars Global Surveyor Mars Orbiter Camera: Interplanetary cruise through primary mission. *J. Geophys. Res.* **106**, 23429–23570.

Malin, M. C. and Edgett, K. S. (2003). Evidence for persistent flow and aqueous sedimentation on early Mars. *Science* **302**, 1931–1934.

Malin, M. C., *et al.* (2010). An overview of the 1985–2006 Mars Orbiter Camera science investigation. *Mars* **5**, 1–60.

Malin, M. C., *et al.* (2006). Present-day impact cratering rate and contemporary gully activity on Mars. *Science* **314**, 1573–1577.

Mallmann, G. and O'Neill, H. S. C. (2009). The crystal/melt partitioning of V during mantle melting as a function of oxygen fugacity compared with some other elements (Al, P, Ca, Sc, Ti, Cr, Fe, Ga, Y, Zr and Nb). *J. Petrol.* **50**, 1765–1794.

Maloof, A. C., *et al.* (2010). Possible animal-body fossils in pre-Marinoan limestones from South Australia. *Nature Geosci.* **3** (9), 653–659.

Manabe, S. and Strickler, R. F. (1964). Thermal equilibrium of the atmosphere with a convective adjustment. *J. Atmos. Sci.* **21**, 361–385.

Manabe, S. and Wetherald, R. T. (1967). Thermal equilibrium of the atmosphere with a given distribution of relative humidity. *J. Atmos. Sci.* **24**, 241–259.

Mancinelli, R. L. and McKay, C. P. (1988). The evolution of nitrogen cycling. *Origins of Life* **18**, 311–325.

Mandt, K. E., *et al.* (2009). Isotopic evolution of the major constituents of Titan's atmosphere based on Cassini data. *Planetary and Space Science* **57**, 1917–1930.

Manga, M., *et al.* (2012). Wet surface and dense atmosphere on early Mars suggested by the bomb sag at Home Plate, *Mars. Geophys. Res. Lett.* **39**.

Mangold, N. and Howard, A. D. (2013). Outflow channels with deltaic deposits in Ismenius Lacus, Mars. *Icarus* **226**, 385–401.

Mangold, N., *et al.* (2012). The origin and timing of fluvial activity at Eberswalde crater, Mars. *Icarus* **220**, 530–551.

Manning, C. E., *et al.* (2006a). Geology, age and origin of supracrustal rocks at Akilia, West Greenland. *American Journal of Science* **306**, 303–366.

Manning, C. V., *et al.* (2006b). Thick and thin models of the evolution of carbon dioxide on Mars. *Icarus* **180**, 38–59.

Marchi, S., *et al.* (2013). High-velocity collisions from the lunar cataclysm recorded in asteroidal meteorites. *Nat. Geosci.* **6**, 303–307.

Marchi, S., *et al.* (2014). Widespread mixing and burial of Earth's Hadean crust by asteroid impacts. *Nature* **511**, 578–582.

Marcq, E., *et al.* (2008). A latitudinal survey of CO, OCS, H_2O, and SO_2 in the lower atmosphere of Venus: Spectroscopic studies using VIRTIS-H. *J. Geophys. Res.* **113**, doi: 10.1029/2008JE003074.

Marcq, E., *et al.* (2006). Remote sensing of Venus' lower atmosphere from ground-based IR spectroscopy: Latitudinal and vertical distribution of minor species. *Planet. Space Sci.* **54**, 1360–1370.

Margulis, L. and Lovelock, J. E. (1974). Biological modulation of the Earth's atmosphere. *Icarus* **21**, 471–489.

Margulis, L., *et al.* (1976). Reassessment of roles of oxygen and ultraviolet light in Precambrian evolution. *Nature* **264**, 620–624.

Marinova, M. M., *et al.* (2008). Mega-impact formation of the Mars hemispheric dichotomy. *Nature* **453**, 1216–1219.

Markiewicz, W. J., *et al.* (2014). Glory on Venus cloud tops and the unknown UV absorber. *Icarus* **234**, 200–203.

Marley, M. S. (2010). The atmospheres of extrasolar planets. *EAS Publications Series* **41**, 411–428.

Marley, M. S., *et al.* (2013). Clouds and hazes in exoplanets. In: *Comparative Climatology of Terrestrial Planets*, ed. S. J. Mackwell, *et al.*, Tucson: University of Arizona Press, pp. 367–391.

Marounina, N., *et al.* (2015). Evolution of Titan's atmosphere during the Late Heavy Bombardment. *Icarus* **257**, 324–335.

Marrero, T. R. and Mason, E. A. (1972). Gaseous diffusion coefficients. *J. Phys. Chem. Ref. Data* **1**, 3–118.

Marshall, A. O., *et al.* (2014). Multiple generations of carbonaceous material deposited in Apex chert by basin-scale pervasive hydrothermal fluid flow. *Gondwana Res.* **25**, 284–289.

Marshall, A. O. and Marshall, C. P. (2013). Comment on "Biogenicity of Earth's earliest fossils: A resolution of the controversy" by J. W. Schopf and A. B. Kudryavtsev, Gondwana Research, Volume 22, Issue 3–4, Pages 761–771. *Gondwana Res.* 23, 1654–1655.

Marshall, C. P., *et al.* (2011). Haematite pseudomicrofossils present in the 3.5-billion-year-old Apex Chert. *Nat. Geosci.* **4**, 240–243.

Marshall, H. G., *et al.* (1988). Long-term climate change and the geochemical cycle of carbon. *J. Geophys. Res.* **93**, 791–802.

Marshall, J. and Plumb, R. A. (2008). *Atmosphere, Ocean, and Climate Dynamics: An Introductory Text*. Boston: Academic Press.

Martin, R. V., *et al.* (2002). An improved retrieval of tropospheric nitrogen dioxide from GOME. *J. Geophys. Res.* **107**, 4437, doi: 10.1029/2001JD001027.

Martin, W., *et al.* (2008). Hydrothermal vents and the origin of life. *Nat. Rev. Microbiol.* **6**, 805–814.

Marty, B. (2012). The origins and concentrations of water, carbon, nitrogen and noble gases on Earth. *Earth Planet. Sc. Lett.* **313–314**, 56–66.

Marty, B., *et al.* (2011). A ^{15}N-poor isotopic composition for the Solar System as shown by Genesis solar wind samples. *Science* **332**, 1533–1536.

Marty, B. and Dauphas, N. (2003). The nitrogen record of crust-mantle interaction and mantle convection from Archean to present. *Earth Planet. Sci. Lett.* **206**, 397–410.

Marty, B. and Yokochi, R. (2006). Water in the early Earth. *Rev. Mineral. Geochem.* **62**, 421–450.

Marty, B. and Zimmermann, L. (1999). Volatiles (He, C, N, Ar) in mid-ocean ridge basalts: Assesment of shallow-level fractionation and characterization of source composition. *Geochim. Cosmochim. Acta* **63**, 3619–3633.

Marty, B., *et al.* (2013). Nitrogen isotopic composition and density of the Archean atmosphere. *Science* **342**, 101–104.

Martyn, D. F. and Pulley, O. O. (1936). The temperature and constituents of the upper atmosphere. *Proc. R. Soc. Lond. A* **154**, 455–486.

Mason, R. (1990). *Petrology of the Metamorphic Rocks*. Boston: Unwin Hyman.

Masse, M., *et al.* (2008). Mineralogical composition, structure, morphology, and geological history of Aram Chaos crater fill on Mars derived from OMEGA Mars Express data. *J. Geophys. Res.* **113**, E12006.

Massie, S. T. and Hunten, D. M. (1981). Stratospheric eddy diffusion coefficients from tracer data. *J. Geophys. Res.* **86**, 9859–9868.

Mastrogiuseppe, M., *et al.* (2014). The bathymetry of a Titan sea. *Geophys. Res. Lett.* **41**, 1432–1437.

Mather, T. A. (2015). Volcanoes and the environment: Lessons for understanding Earth's past and future from studies of present-day volcanic emissions. *J. Volcanol. Geoth. Res.*, doi:10.1016/j.jvolgeores.2015.08.016.

Matsubara, Y., *et al.* (2013). Hydrology of early Mars: Valley network incision. *J. Geophys. Res.* **118**, 1365–1387.

Matsui, T. and Abe, Y. (1986a). Evolution of an impact-induced atmosphere and magma ocean on the accreting Earth. *Nature* **319**, 303–305.

Matsui, T. and Abe, Y. (1986b). Impact-induced atmospheres and oceans on Earth and Venus. *Nature* **322**, 526–528.

Matsuo, T. and Tamura, M. (2010). *Second-earth imager for TMT (SEIT): a proposal and concept Description*. SPIE Astronomical Telescopes + Instrumentation. International Society for Optics and Photonics, pp. 773584–773584-9.

Mattey, D. P. (1987). Carbon isotopes in the mantle. *Terra Cognita* **7**, 31–37.

Mattioli, G. S. and Wood, B. J. (1986). Upper mantle oxygen fugacity recorded by spinel lherzolites. *Nature* **322**, 626–628.

Maynard, J. B. (2010). The chemistry of manganese ores through time: A signal of increasing diversity of Earth-surface environments. *Econ. Geol.* **105**, 535–552.

Mayor, M., *et al.* (2014). Doppler spectroscopy as a path to the detection of Earth-like planets. *Nature* **513**, 328–335.

Mayor, M. and Queloz, D. (1995). A Jupiter-mass companion to a solar-type star. *Nature* **378**, 355–359.

Mazzullo, S. J. (2000). Organogenic dolomitization in peritidal to deep-sea sediments. *J. Sediment. Res.* **70**, 10–23.

McBride, N. and Gilmour, I. (2004). *An Introduction to the Solar System*. New York: Cambridge University Press.

McCammon, C. (2005). The paradox of mantle redox. *Science* **308**, 807–808.

McCauley, J. F. (1978). Geologic Map of the Coprates Quadrangle of Mars, scale 1:5,000,000. *U.S. Geol. Surv. Misc. Inv. Series Map I-897*.

McCleese, D. J., *et al.* (2010). Structure and dynamics of the Martian lower and middle atmosphere as observed by the Mars Climate Sounder: Seasonal variations in zonal mean temperature, dust, and water ice aerosols. *J. Geophys. Res.* **115**.

McCollom, T. M., *et al.* (2007). Could erosion of Meridiani Planum represent a significant contributer to global sulfate-rich martian soils. *38th Lunar Planet. Sci. Conf.*, no. 2151 (abstract).

McCollom, T. M. and Seewald, J. S. (2007). Abiotic synthesis of organic compounds in deep-sea hydrothermal environments. *Chem. Rev.* **107**, 382–401.

McCord, T. B., *et al.* (2008). Titan's surface: Search for spectral diversity and composition using the Cassini VIMS investigation. *Icarus* **194**, 212–242.

McDermott, J. M., *et al.* (2015). Pathways for abiotic organic synthesis at submarine hydrothermal fields. *P. Natl. Acad. Sci. USA* **112**, 7668–7672.

McDonough, W. F. (2003). Compositional model for the Earth's core. In: *Treatise on Geochemistry*, ed. H. D. Holland and K. K. Turekian, Oxford: Pergamon, pp. 547–568.

McElroy, M. B. (1972). Mars: An evolving atmosphere. *Science* **175**, 443–445.

McElroy, M. B. and Donahue, T. M. (1972). Stability of the Martian atmosphere. *Science* **177**, 986–988.

McElroy, M. B., *et al.* (1982). Escape of hydrogen from Venus. *Science* **215**, 1614–1615.

McEwen, A. S. (2013). Mars in Motion: The surface of Mars changes all the time. Is flowing water one of the causes? *Sci. Am.* **308**, 58–65.

McEwen, A. S., *et al.* (2014). Recurring slope lineae in equatorial regions of Mars. *Nature Geoscience* **7**, 53–58.

McEwen, A. S., *et al.* (2004). The lithosphere and surface of Io. In: *Jupiter: The planet, satellites, and magnetosphere*, ed. F. Bagenal, *et al.*, New York: Cambridge University Press, pp. 307–328.

McEwen, A. S., *et al.* (2011). Seasonal flows on warm martian slopes. *Science* **333**, 740–743.

McFadden, K. A., *et al.* (2008). Pulsed oxidation and bioloical evolution in the Ediacaran Doushantuo Formation. *P. Natl. Acad. Sci. USA* **105**, 3197–3202.

McGouldrick, K. and Toon, O. B. (2007). An investigation of possible causes of the holes in the condensational Venus cloud using a microphysical cloud model with a radiative-dynamical feedback. *Icarus* **191**, 1–24.

McGouldrick, K., *et al.* (2011). Sulfuric acid aerosols in the atmospheres of the terrestrial planets. *Planet. Space Sci.* **59**, 934–941.

McGovern, W. E. (1973). Potential atmospheric composition of smaller bodies in Solar System and some aspects of planetary evolution. *J. Geophys. Res.* **78**, 274–280.

McGrath, M. A., *et al.* (2004). Satellite atmospheres. In: *Jupiter: The planet, satellites, and magnetosphere*, ed. F. Bagenal, *et al.*, New York: Cambridge University Press.

McIlveen, J. F. R. (1992). *Fundamentals of Weather and Climate*. London: Chapman & Hall.

McKay, C. P. (2000). Thickness of tropical ice and photosynthesis on a snowball Earth. *Geophys. Res. Lett.* **27**, 2153–2156.

McKay, C. P. (2004). Wet and cold thick atmosphere on early Mars. *J. Phys. IV France* **121**, 283–288.

McKay, C. P., *et al.* (1997). Temperature lapse rate and methane in Titan's troposphere. *Icarus* **129**, 498–505.

McKay, C. P., *et al.* (1989). The thermal structure of Titan's atmosphere. *Icarus* **80**, 23–53.

McKay, C. P., *et al.* (1991). The greenhouse and antigreenhouse effects on Titan. *Science* **253**, 1118–1121.

McKay, C. P., *et al.* (1993). Coupled atmosphere ocean models of Titan's past. *Icarus* **102**, 88–98.

McKay, C. P., *et al.* (1988). High temperature shock formation of N_2 and organics on primordial Titan. *Nature* **332**, 520–522.

McKay, C. P. and Smith, H. D. (2005). Possibilities for methanogenic life in liquid methane on the surface of Titan. *Icarus* **178**, 274–276.

McKeegan, K. D., *et al.* (2006). Isotopic compositions of cometary matter returned by Stardust. *Science* **314**, 1724–1728.

McKenzie, D. and Nimmo, F. (1999). The generation of Martian floods by the melting of ground ice above dykes. *Nature* **397**, 231–233.

McKinnon, W. B. (2002). Planetary science – Out on the edge. *Nature* **418**, 135–+.

McKinnon, W. B. (2010). Argon-40 outgassing from Titan and Enceladus: A tale of two satellites. *41st Lunar Planet. Sci. Conf.*, 2718 (abstract).

McKinnon, W. B., *et al.* (1995). Origin and evolution of Triton. In: *Neptune and Triton*, ed. D. P. Cruikshank, Tucson: University of Arizona Press, pp. 807–877.

McKinnon, W. B., *et al.* (2008). Structure and evolution of Kuiper Belt Objects and dwarf planets. In: *The Solar System Beyond Neptune*, ed. M. A. Barucci, *et al.*, Tucson: University of Arizona Press, pp. 213–241.

McLennan, S. M. and Grotzinger, J. P. (2008). The sedimentary rock cycle on Mars. In: *The Martian Surface: Composition, Mineralogy and Physical Properties*, ed. J. Bell, New York: Cambridge University Press, pp. 541–577.

McLoughlin, N., *et al.* (2008). Growth of synthetic stromatolites and wrinkle structures in the absence of microbes: Implications for the early fossil record. *Geobiology* **6**, 95–105.

McNutt, R. L. (1989). Models of Pluto upper atmosphere. *Geophys. Res. Lett.* **16**, 1225–1228.

McQuarrie, D. A. and Simon, J. D. (1997). *Physical Chemistry: A Molecular Approach*. Sausalito: University Science Books.

McSween, H. Y. and McLennan, S. M. (2014). Mars. In: *Treatise on Geochemistry* (2nd Edn), ed. H. D.

Holland and K. K. Turekian, New York: Elsevier, pp. 251–300.

Meadows, V. S. and Crisp, D. (1996). Ground-based near-infrared observations of the Venus nightside: The thermal structure and water abundance near the surface. *J. Geophys. Res.* **101**, 4595–4622.

Meech, K. J., *et al.* (2011). Epoxi: Comet 103P/Hartley 2 observations from a worldwide campaign. *Astrophys. J. Lett.* **734**.

Melezhik, V. A. and Fallick, A. E. (2010). On the Lomagundi-Jatuli carbon isotopic event: The evidence from the Kalix Greenstone Belt, Sweden. *Precam. Res.* **179**, 165–190.

Melezhik, V. A., *et al.* (2013a). The Palaeoproterozoic perturbation of the global carbon cycle: The Lomagundi-Jatuli isotopic event. In: *Reading the Archive of Earth's Oxygenation*, ed. V. A. e. a. Melezhik, Berlin: Springer, pp. 1111–1150.

Melezhik, V. A., *et al.* (1999). Extreme $^{13}C_{carb}$ enrichment in ca. 2.0 Ga magnesite-stromatolite-dolomite- 'red beds' association in a global context: A case for the world-wide signal enhanced by a local environment. *Earth Sci. Rev.* **48**, 71–120.

Melezhik, V. A., *et al.* (2007). Temporal constraints on the Paleoproterozoic Lomagundi-Jatuli carbon isotopic event. *Geology* **35**, 655–658.

Melezhik, V. A., *et al.* (2013b). Huronian-age glaciation. In: *Reading the Archive of Earth's Oxygenation*, ed. V. A. e. a. Melezhik, Berlin: Springer, pp. 1059–1109.

Melnik, Y. P. (1982). *Precambrian Banded Iron-Formations: Physicochemical Conditions Of Formation.* Amsterdam: Elsevier.

Melosh, H. J. and Vickery, A. M. (1989). Impact erosion of the primordial atmosphere of Mars. *Nature* **338**, 487–489.

Menou, K. (2015). Climate stability of habitable Earth-like planets. *Earth Planet. Sc. Lett.* **429**, 20–24.

Menou, K., *et al.* (2003). "Weather" variability of close-in extrasolar giant planets. *Astrophys. J.* **587**, L113–L116.

Menou, K. and Rauscher, E. (2009). Atmospheric circulation of hot Jupiters: A shallow three-dimensional model. *Astrophys. J.* **700**, 887–897.

Mercer, C. M., *et al.* (2015). Refining lunar impact chronology through high spatial resolution $^{40}Ar/^{39}Ar$

dating of impact melts. *Sci. Adv.* **1**, e1400050 doi:10.1126/sciadv.1400050.

Merlis, T. M. and Schneider, T. (2011). Changes in zonal surface temperature gradients and Walker Circulations in a wide range of climates. *J. Climate* **24**, 4757–4768.

Merryfield, W. J. and Shizgal, B. D. (1994). Discrete velocity model for an escaping single-component atmosphere. *Planet. Space Sci.* **42**, 409–419.

Michael, M., *et al.* (2005). Ejection of nitrogen from Titan's atmosphere by magnetospheric ions and pick-up ions. *Icarus* **175**, 263–267.

Michalski, G., *et al.* (2004). Long term atmospheric deposition as the source of nitrate and other salts in the Atacama Desert, Chile: New evidence from mass-independent oxygen isotopic compositions. *Geochim. Cosmochim. Acta* **68**, 4023–4038.

Michalski, J. and Niles, P. B. (2010). Deep crustal carbonate rocks exposed by meteor impact on Mars. *Nature Geosci.* **3**, 751–755.

Michalski, J. and Niles, P. B. (2012). Atmospheric origin of martian interior layered deposits: Links to climate change and global sulfur cycle. *Geology* **40**, 419–422.

Migliorini, A., *et al.* (2012). Investigation of air temperature on the nightside of Venus derived from VIRTIS-H on board Venus-Express. *Icarus* **217**, 640–647.

Mikucki, J. A., *et al.* (2009). A contemporary microbially maintained subglacial ferrous "ocean". *Science* **324**, 397–400.

Miller, C. A., *et al.* (2011). Re-assessing the surface cycling of molybdenum and rhenium. *Geochim. Cosmochim. Acta* **75**, 7146–7179.

Miller, S. L. (1953). A production of amino acids under possible primitive Earth conditions. *Science* **117**, 528–529.

Miller, S. L. (1955). Production of some organic compounds under possible primitive Earth conditions. *J. Am. Chem. Soc.* **77**, 2351–2361.

Miller, S. L. and Bada, J. L. (1988). Submarine hot springs and the origin of life. *Nature* **334**, 609–611.

Miller, S. L. and Schlesinger, G. (1984). Carbon energy yields in atmospheres containing CH_4, CO, and CO_2. *Origins of Life* **14**, 83–90.

Milliken, R. E. and Bish, D. L. (2010). Sources and sinks of clay minerals on Mars. *Philos. Mag.* **90**, 2293–2308.

Milliken, R. E., *et al.* (2008). Opaline silica in young deposits on Mars. *Geology* **36**, 847–850.

Mills, B., *et al.* (2011). Timing of Neoproterozoic glaciations linked to transport-limited global weathering. *Nature Geosci.* **4**, 861–864.

Mills, D. B. and Canfield, D. E. (2014). Oxygen and animal evolution: Did a rise of atmospheric oxygen trigger the origin of animals? *Bioessays* **36**, doi:10.1002/bies.201400101.

Mills, F. P. and Allen, M. (2007). A review of selected issues concerning the chemistry in Venus' middle atmosphere. *Planet. Space Sci.* **55**, 1729–1740.

Mills, F. P., *et al.* (2007). Atmospheric composition, chemistry, and clouds. In: *Exploring Venus as a Terrestrial Planet*, American Geophysical Union, pp. 73–100.

Ming, D. W., *et al.* (2014). Volatile and organic compositions of sedimentary rocks in Yellowknife Bay, Gale Crater, Mars. *Science* **343**, 1245267, doi: 10.1126/science.1245267.

Misra, A., *et al.* (2014). Using dimers to measure biosignatures and atmospheric pressure for terrestrial exoplanets. *Astrobiology* **14**, 67–86.

Mitchell, J. L. and Vallis, G. K. (2010). The transition to superrotation in terrestrial atmospheres. *J. Geophys. Res.* **115**, E12008, doi:10.1029/2010JE003587.

Mlawer, E. J., *et al.* (1997). Radiative transfer for inhomogeneous atmospheres: RRTM, a validated correlated-k model of the longwave. *J. Geophys. Res.* **102**, 16 663–16 682.

Mojzsis, S. J., *et al.* (1996). Evidence for life on Earth before 3,800 million years ago. *Nature* **384**, 55–59.

Mojzsis, S. J., *et al.* (2001). Oxygen isotope evidence from ancient zircons for liquid water at the Earth's surface 4,300 Myr ago. *Nature* **409**, 178–181.

Monga, N. and Desch, S. (2015). External photoevaporation of the solar nebula: Jupiter's noble gas enrichments. *Astrophys. J.* **798**.

Monteith, J. L. (2012). *Principles of Environmental Physics*. Oxford: Academic.

Montmessin, F., *et al.* (2011). A layer of ozone detected in the nightside upper atmosphere of Venus. *Icarus* **216**, 82–85.

Montmessin, F., *et al.* (2005). Modeling the annual cycle of HDO in the Martian atmosphere. *J. Geophys. Res.* **110**, E03006, doi: 10.1029/2004JE002357.

Montzka, S. A., *et al.* (2007). On the global distribution, seasonality, and budget of atmospheric carbonyl sulfide (COS) and some similarities to CO_2. *J. Geophys. Res.* **112**, D09302, doi:10.1029/2006JD007665.

Moorbath, S., *et al.* (1973). Early Archaean age for Isua iron formation, West Greenland. *Nature* **245**, 138–139.

Moore, J. M. and Howard, A. D. (2005). Large alluvial fans on Mars. *J. Geophys. Res.* **110**.

Moore, J. M., *et al.* (2003). Martian layered fluvial deposits: implications for Noachian climate scenarios. *Geophys. Res. Lett.* **30**.

Moore, J. M. and Pappalardo, R. T. (2011). Titan: An exogenic world? *Icarus* **212**, 790–806.

Moore, W. B. and Webb, A. A. G. (2013). Heat-pipe Earth. *Nature* **501**, 501–505.

Moores, E. M. (1986). The Proterozoic ophiolite problem, continental emergence, and the Venus connection. *Science* **234**, 65–68.

Moores, E. M. (1993). Neoproterozoic oceanic crustal thinning, emergence of continents, and origin of the Phanerozoic ecosystem; a model. *Geol.* **21**, 5–8.

Moores, E. M. (2002). Pre-1 Ga (pre-Rodinian) ophiolites: Their tectonic and environmental implications. *GSA Bull.* **114**, 80–95.

Morbidelli, A., *et al.* (2000). Source regions and timescales for the delivery of water to the Earth. *Meteoritics and Planet. Sci.* **35**, 1309–1320.

Moreira, M. (2013). Noble gas constraints on the origin and evolution of Earth's volatiles. *Geochem. Perspectives* **2**, 229–403.

Moreno, R., *et al.* (2012). The abundance, vertical distribution and origin of H2O in Titan's atmosphere: Herschel observations and photochemical modelling. *Icarus* **221**, 753–767.

Moresi, L. and Solomatov, V. (1998). Mantle convection with a brittle lithosphere: thoughts on the global tectonic styles of the Earth and Venus. *Geophys. J. Int.* **133**, 669–682.

Morii, H., *et al.* (1987). Energetic analysis of the growth of Methanobrevibacter-Arboriphilus A2 in hydrogen-limited continuous cultures. *Biotechnology and Bioengineering* **29**, 310–315.

Morner, N. A. and Etiope, G. (2002). Carbon degassing from the lithosphere. *Global And Planetary Change* **33**, 185–203.

Moroz, V. I., *et al.* (1985). Solar and thermal radiation in the Venus atmosphere. *Adv. Space Res.* **5**, 197–232.

Morris, R. V., *et al.* (2006). Mössbauer mineralogy of rock, soil, and dust at Gusev crater, Mars: Spirit's journey through weakly altered olivine basalt on the plains and pervasively altered basalt in the Columbia Hills. *Journal of Geophysical Research-Planets* **111**.

Morris, R. V., *et al.* (2010). Identification of carbonate-rich outcrops on Mars by the Spirit Rover. *Science* **329**, 421–424.

Moses, J. I., *et al.* (2004). The stratosphere of Jupiter. In: *Jupiter: The Planet, Satellites, and Magnetosphere*, ed. F. Bagenal, *et al.*, New York: Cambridge University Press.

Moses, J. I., *et al.* (2011). Disequilibrium carbon, oxygen, and nitrogen chemistry in the atmospheres of HD 189733b and HD 209458b. *Astrophys. J.* **737**.

Moses, J. I., *et al.* (2002). Photochemistry of a volcanically driven atmosphere on Io: Sulfur and oxygen species from a Pele-type eruption. *Icarus* **156**, 76–106.

Moskalenko, N. I., *et al.* (1979). Pressure-induced infrared radiation absorption in atmospheres. *Bull. Acad. Sci. USSR, Atmospheric and Oceanic Physics* **15**, 632–637.

Mottl, M. J. and Wheat, C. G. (1994). Hydrothermal circulation through mid-ocean ridge flanks: fluxes of heat and magnesium. *Geochim. Cosmochim. Acta* **58**, 2225–2237.

Mouginot, J., *et al.* (2010). The 3–5 MHz global reflectivity map of Mars by MARSIS/Mars Express: Implications for the current inventory of subsurface H_2O. *Icarus* **210**, 612–625.

Mousis, O., *et al.* (2014). Equilibrium composition between liquid and clathrate reservoirs on Titan. *Icarus* **239**, 39–45.

Mousis, O., *et al.* (2009a). A primordial origin for the atmospheric methane of Saturn's moon Titan. *Icarus* **204**, 749–751.

Mousis, O., *et al.* (2011). Removal of Titan's atmospheric noble gases by their sequestration in surface clathrates. *Astrophys. J. Lett.* **740**.

Mousis, O., *et al.* (2009b). Clathration of volatiles in the solar nebula and implications for the origin of Titan's atmosphere. *Astrophys. J.* **691**, 1780–1786.

Mueller, R. F. and Saxena, S. K. (1977). *Chemical Petrology.* New York: Springer-Verlag.

Mulders, G., D., *et al.* (2015). The snow line in viscous disks around low-mass stars: Implications for water delivery to terrestrial planets in the habitable zone. *Astrophys. J.* **807**, 9.

Mumma, M. J., *et al.* (2009). Strong release of methane on Mars in northern summer 2003. *Science* **323**, 1041–1045.

Murakami, T., *et al.* (2011). Quantification of atmospheric oxygen levels during the Paleoproterozoic using paleosol compositions and iron oxidation kinetics. *Geochim. Cosmochim. Acta* **75**, 3982–4004.

Murakami, T., *et al.* (2001). Direct evidence of Late Archean to Early Proterozoic anoxic atmosphere from a product of 2.5 Ga old weathering. *Earth Planet. Sci. Lett.* **184**, 523.

Murchie, S., *et al.* (2009a). Evidence for the origin of layered deposits in Candor Chasma, Mars, from mineral composition and hydrologic modeling. *J. Geophys. Res.* **114**, E00D05, doi:10.1029/2009JE003343.

Murchie, S. L., *et al.* (2009b). A synthesis of Martian aqueous mineralogy after 1 Mars year of observations from the Mars Reconnaissance Orbiter. *J. Geophys. Res.* **114**, E00D06, doi:10.1029/2009JE003342.

Murray, C. D. and Dermott, S. F. (2001). *Solar System Dynamics.* New York: Cambridge Univ. Press.

Murray-Clay, R. A., *et al.* (2009). Atmospheric escape from hot Jupiters. *Astrophys. J.* **693**, 23–42.

Mushkin, A., *et al.* (2010). Spectral constraints on the composition of low-albedo slope streaks in the Olympus Mons Aureole. *Geophys. Res. Lett.* **37**, L22201, doi: 10.1029/2010GL044535.

Mustard, J. F., *et al.* (2001). Evidence for recent climate change on Mars from the identification of youthful near-surface ground ice. *Nature* **412**, 411–414.

Mustard, J. F., *et al.* (2012). *Sequestration of volatiles in the martian crust through hydrated minerals: Implications for planetary evolution. Third Int. Conf. on Early Mars,* 7075.

Mutch, T. A., *et al.* (1976). *The Geology of Mars.* Princeton: Princeton University Press.

Nagy, A. F., *et al.* (2001). Hot carbon densities in the exosphere of Mars. *J. Geophys. Res.* **106**, 21565–21568.

Nair, H., *et al.* (1994). A photochemical model of the Martian atmosphere. *Icarus* **111**, 124–150.

Nakajima, S., *et al.* (1992). A study on the "runaway greenhouse effect" with a one-dimensional radiative-convective equilibrium model. *J. Atmos. Sci.* **49**, 2256–2266.

Nakamura, K. and Kato, Y. (2004). Carbonatization of oceanic crust by the seafloor hydrothermal activity and its significance as a CO_2 sink in the Early Archean. *Geochim. Cosmochim. Acta* **68**, 4595–4618.

Nappo, C. J. (2012). *An Introduction to Atmospheric Gravity Waves* (2nd Edn). Waltham, MA: Academic Press.

Nappo, C. J. (2013). *An Introduction to Atmospheric Gravity Waves* (2nd Edn). Waltham, MA: Elsevier.

Narbonne, G. M. (2005). The Ediacara biota: Neoproterozoic origin of animals and their ecosystems. *Ann. Rev. Earth Planet. Sci.* **33**, 421–442.

Narbonne, G. M. and Gehling, J. G. (2003). Life after the Snowball: The oldest complex Ediacaran fossils. *Geology* **31**, 27–30.

National Research Council. (1979). *Stratospheric Ozone Depletion by Halocarbons: Chemistry and Transport.* National Academy of Sciences.

Navarra, A. and Boccaletti, G. (2002). Numerical general circulation experiments of sensitivity to Earth rotation rate. *Clim. Dynam.* **19**, 467–483.

Navarro-Gonzalez, R., *et al.* (2001). A possible nitrogen crisis for Archean life due to reduced nitrogen fixation by lightning. *Nature* **412**, 61–64.

Navarro-Gonzalez, R., *et al.* (2010). Reanalysis of the Viking results suggests perchlorate and organics at midlatitudes on Mars. *J. Geophys. Res.* **115**.

Navarro-Gonzalez, R., *et al.* (2011). Correction to "Reanalysis of the Viking results suggests perchlorate and organics at midlatitudes on Mars". *J. Geophys. Res.* **116**, E08011.

Nedelcu, A. M., *et al.* (2004). Sex as a response to oxidative stress: a twofold increase in cellular reactive oxygen species activates sex genes. *Proc. R. Soc. Lond. B–Biol. Sci.* **271**, 1591–1596.

Nedell, S. S., *et al.* (1987). Origin and evolution of the layered deposits in the Valles Marineris, Mars. *Icarus* **70**, 409–441.

Neumann, G. A., *et al.* (2004). Crustal structure of Mars from gravity and topography. *J. Geophys. Res.* **109**.

Newman, C. E., *et al.* (2011). Stratospheric superrotation in the TitanWRF model. *Icarus* **213**, 636–654.

Newsom, H. E. and Sims, K. W. W. (1991). Core formation during early accretion of the Earth. *Science* **252**, 926–933.

Nicklas, R. W., *et al.* (2017). The redox history of the Archean Mantle: Evidence from komatiites, in preparation.

Niemann, H. B., *et al.* (2010). Composition of Titan's lower atmosphere and simple surface volatiles as measured by the Cassini-Huygens probe gas chromatograph mass spectrometer experiment. *J. Geophys. Res.* **115**.

Nier, A. O. and McElroy, M. B. (1977). Composition and structure of Mars' upper atmosphere: Results from the neutral mass spectrometers on Viking 1 and 2. *J. Geophys. Res.* **82**, 4341–49.

Niles, P. B., *et al.* (2013). Geochemistry of carbonates on Mars: Implications for climate history and nature of aqueous environments. *Space Sci. Rev.* **174**, 301–328.

Niles, P. B. and Michalski, J. (2009). Meridiani Planum sediments on Mars formed through weathering in massive ice deposits. *Nature Geoscience* **2**, 215–220.

Nimmo, F. (2000). Dike intrusion as a possible cause of linear Martian magnetic anomalies. *Geology* **28**, 391–394.

Nimmo, F., *et al.* (2008). Implications of an impact origin for the martian hemispheric dichotomy. *Nature* **453**, 1220–U32.

Nimmo, F. and Stevenson, D. J. (2000). Influence of early plate tectonics on the thermal evolution and magnetic field of Mars. *J. Geophys. Res.* **105**, 11969–11979.

Nimmo, F. and Tanaka, K. (2005). Early crustal evolution of mars. *Annu. Rev. Earth Pl. Sc.* **33**, 133–161.

Nisbet, E., *et al.* (2007a). Creating habitable zones, at all scales, from planets to mud micro-habitats, on earth and on mars. *Space Sci. Rev.* **129**, 79–121.

Nisbet, E. G. (2002). Fermor lecture: The influence of life on the face of the Earth: garnets and moving continents.

In: *The Early Earth: Physical, Chemical and Biological Development*, ed. C. M. R. Fowler, *et al.*, London: Geological Soc. of London.

Nisbet, E. G. and Fowler, C. M. R. (2014). The early history of life. In: *Treatise on Geochemistry* (2nd Edn), ed. H. D. Holland and K. K. Turekian, Oxford: Elsevier, pp. 1–42.

Nisbet, E. G., *et al.* (2012). The regulation of the air: a hypothesis. *Solid Earth* **3**, 87–96.

Nisbet, E. G., *et al.* (2007b). The age of Rubisco: The evolution of oxygenic photosynthesis. *Geobiology* **5**, 311–335.

Nisbet, E. G. and Nisbet, R. E. R. (2008). Methane, oxygen, photosynthesis, rubisco and the regulation of the air through time. *Phil. Trans. R. Soc. Lond. B* **363**, 2745–2754.

Nisbet, E. G. and Sleep, N. H. (2001). The habitat and nature of early life. *Nature* **409**, 1083–1091.

Nishiizumi, K., *et al.* (1991). Cosmic ray produced ^{10}Be and ^{26}Al in Antarctic rocks: Exposure and erosion history. *Earth Planet. Sci. Lett.* **104**, 440–454.

Nishizawa, M., *et al.* (2007). Speciation and isotope ratios of nitrogen in fluid inclusions from seafloor hydrothermal deposits at similar to 3.5 Ga. *Earth Planet. Sci. Lett.* **254**, 332–344.

Nixon, C. A., *et al.* (2008). Isotopic ratios in Titan's atmosphere from Cassini CIRS limb sounding: CO_2 at low and midlatitudes. *Astrophys. J. Lett.* **681**, L101–L103.

Noel, A., *et al.* (2015). Mineralogy, morphology and stratigraphy of the light-toned interior layered deposits at Juventae Chasma. *Icarus* **251**, 315–331.

Nogueira, E., *et al.* (2011). Reassessing the origin of Triton. *Icarus* **214**, 113–130.

Nordstrom, D. K. and Munoz, J. L. (1994). *Geochemical Thermodynamics*. Boston: Blackwell Scientific Publications.

Norman, M. D., *et al.* (2003). Chronology, geochemistry, and petrology of a ferroan noritic anorthosite clast from Descartes breccia 67215: Clues to the age, origin, structure, and impact history of the lunar crust. *Meteorit. Planet. Sci.* **38**, 645–661.

Norman, M. D., *et al.* (2010). Imbrium provenance for the Apollo 16 Descartes terrain: Argon ages and geochemistry of lunar breccias 67016 and 67455. *Geochim. Cosmochim. Acta* **74**, 763–783.

Norman, M. D. and Nemchin, A. A. (2014). A 4.2 billion year old impact basin on the Moon: U-Pb dating of zirconolite and apatite in lunar melt rock 67955. *Earth Planet. Sci. Lett.* **388**, 387–398.

North, G. R. (1975). Theory of energy-balance climate models. *J. Atmos. Sci.* **32**, 2033–2043.

Nursall, J. R. (1959). Oxygen as a prerequisite to the origin of the metazoa. *Nature* **183**, 1170–1172.

Nutman, A. P. (2006). Antiquity of the oceans and continents. *Elements* **2**, 223–227.

Nutman, A. P., *et al.* (1997). Recognition of >=3850 Ma water-lain sediments in West Greenland and their significance for the early Archaean Earth. *Geochim. Cosmochim. Acta* **61**, 2475–2484.

Nutman, A. P., *et al.* (2016). Rapid emergence of life shown by discovery of 3,700-million-year-old microbial structures. *Nature* **537**, 535–538.

O'Brien, D. P., *et al.* (2014). Water delivery and giant impacts in the 'Grand Tack' scenario. *Icarus* **239**, 74–84.

Oakley, P. H. H. and Cash, W. (2009). Construction of an Earth model: Analysis of exoplanet light curves and mapping the next Earth with the New Worlds Observer. *Astrophys. J.* **700**, 1428–1439.

Och, L. M. and Shields-Zhou, G. A. (2012). The Neoproterozoic oxygenation event: Environmental perturbations and biogeochemical cycling. *Earth Sci. Rev.* **110**, 26–57.

Oduro, H., *et al.* (2011). Evidence of magnetic isotope effects during thermochemical sulfate reduction. *P. Natl. Acad. Sci. USA* **108**, 17 635–17 638.

Ody, A., *et al.* (2015). Candidates source regions of martian meteorites as identified by OMEGA/MEx. *Icarus* **258**, 366–383.

Ohtomo, Y., *et al.* (2014). Evidence for biogenic graphite in early Archaean Isua metasedimentary rocks. *Nat. Geosci.* **7**, 25–28.

Ojakangas, R. W., *et al.* (2014). The Tayla conglomerate: an Archean (~2.7 Ga) glaciomarine formation, Western Dharwar Craton, Southern India. *Curr. Sci.* **106**, 287–396.

Ojha, L., *et al.* (2015). Spectral evidence for hydrated salts in recurring slope lineae on Mars. *Nature Geosci.*, doi: 10.1038/ngeo2546.

Ojima, M. and Podosek, F. A. (2002). *Noble gas geochemistry*. New York: Cambridge University Press.

Olson, S. L., *et al.* (2013). Quantifying the areal extent and dissolved oxygen concentrations of Archean oxygen oases. *Chem. Geol.* **362**, 35–43.

Olson, S. L., *et al.* (2016). Limited role for methane in the mid-Proterozoic greenhouse. *P. Natl Acad. Sci. USA* **113**, 11 447–11 452.

Ono, S. (2001). *Detrital uraninite and the early Earth's atmosphere: SIMS analyses of uraninite in the Elliot Lake District and the dissolution kinetics of natural uraninite*. Penn State University, PhD, College Station, PA.

Ono, S., *et al.* (2003). New insights into Archean sulfur cycle from mass-independent sulfur isotope records from the Hamersley Basin, *Australia. Earth Planet. Sci. Lett.* **213**, 15–30.

Ono, S., *et al.* (2007). S-33 constraints on the seawater sulfate contribution in modern seafloor hydrothermal vent sulfides. *Geochim. Cosmochim. Acta* **71**, 1170–1182.

Ono, S., *et al.* (2013). Contribution of isotopologue self-shielding to sulfur mass-independent fractionation during sulfur dioxide photolysis. *J. Geophys. Res.* **118**, 2444–2454.

Oparin, A. I. (1924). *Proiskhozhdenie Zhizni (The Origin of Life)*. Moscow: Izd. Moskovskii Rabochii (in Russian).

Oparin, A. I. (1938). *The Origin of Life*. New York: MacMillan.

Oparin, A. I. (1957). *The Origin of Life*. New York: Academic Press.

Oparin, A. I. (1968). *Genesis and Evolutionary Development of Life*. New York: Academic Press.

Opik, E. J. (1963). Selective escape of gases. *Geophys. J. R. Astron. Soc.* **7**, 490–509.

Oppenheimer, C., *et al.* (2014). Volcanic degassing: Process and impact. In: *Treatise on Geochemistry* (2nd Edn), ed. H. D. Holland and K. K. Turekian, New York: Elsevier.

Orgel, L. E. (1986). RNA catalysis and the origins of life. *Journal of Theoretical Biology* **123**, 127–149.

Oro, J. (1961). Comets and the formation of biochemical compounds on the primitive Earth. *Nature* **190**, 389–390.

Osborn, H. F. (1917). *The Origin and Evolution of Life: On the Theory of Action, Reaction and Interaction of Energy*. New York: Charles Scribner's Sons.

Osterloo, M. M., *et al.* (2010). Geologic context of proposed chloride-bearing materials on Mars. *J. Geophys. Res.* **115**.

Owen, J. E. and Jackson, A. P. (2012). Planetary evaporation by UV & X-ray radiation: basic hydrodynamics. *Mon. Not. R. Astron. Soc.* **425**, 2931–2947.

Owen, J. E. and Wu, Y. Q. (2013). Kepler planets: A tale of evaporation. *Astrophys. J.* **775**, 105, doi: 10.1088/0004-637X/775/2/105.

Owen, T. (1980). The search for early forms of life in other planetary systems: future possibilities afforded by spectroscopic techniques. In: *Strategies for Search for Life in the Universe*, ed. M. D. Papagiannis, Dordrecht: Reidel, pp. 177–185.

Owen, T. (1992). The composition and early history of the atmosphere of Mars. In: *Mars*, ed. H. H. Kieffer, *et al.*, Tucson: University of Arizona Press, pp. 818–834.

Owen, T. and Bar-Nun, A. (1995). Comets, impacts, and atmospheres. *Icarus* **116**, 215–226.

Owen, T. and Bar-Nun, A. (1998). From the interstellar medium to planetary atmospheres via comets. *Faraday Discuss.* **109**, 453–462.

Owen, T., *et al.* (1992). Possible cometary origin of heavy noble gases in the atmospheres of Venus, Earth, and Mars. *Nature* **358**, 43–46.

Owen, T., *et al.* (1977). The composition of the atmosphere at the surface of Mars. *J. Geophys. Res.* **82**, 4635–4639.

Owen, T., *et al.* (1999). Saturn VI (Titan). *IAU Circ.* **7306**.

Owen, T., *et al.* (1979). Early Earth: An enhanced carbon dioxide greenhouse to compensate for reduced solar luminosity. *Nature* **277**, 640–642.

Owen, T. and Encrenaz, T. (2006). Compositional constraints on giant planet formation. *Planet. Space Sci.* **54**, 1188–1196.

Owen, T., *et al.* (1988). Deuterium on Mars: The abundance of HDO and the value of D/H. *Science* **240**, 1767–70.

Owen, T. and Niemann, H. B. (2009). The origin of Titan's atmosphere: some recent advances. *Phil. Trans. R. Soc. Lond. A* **367**, 607–615.

Owen, T. C., *et al.* (1993). Surface ices and the atmospheric composition of Pluto. *Science* **261**, 745–748.

Oyama, V. I. and Berdahl, B. J. (1979). Model of Martian surface chemistry. *J. Mol. Evol.* **14**, 199–210.

Pahlevan, K. and Stevenson, D. J. (2007). Equilibration in the aftermath of the lunar-forming giant impact. *Earth Planet. Sc. Lett.* **262**, 438–449.

Palle, E., *et al.* (2008). Identifying the rotation rate and the presence of dynamic weather on extrasolar earth-like planets from photometric observations. *Astrophys. J.* **676**, 1319–1329.

Palle, E., *et al.* (2003). Earthshine and the Earth's albedo: 2. Observations and simulations over 3 years. *J. Geophys. Res.* **108**, 4710, doi:10.1029/2003jd003611.

Palomba, E., *et al.* (2009). Evidence for Mg-rich carbonates on Mars from a 3.9 µm absorption feature. *Icarus* **203**, 58–65.

Papineau, D. (2010). Global biogeochemical changes at both ends of the Proterozoic: Insights from phosphorites. *Astrobiology* **10**, 165–181.

Papineau, D., *et al.* (2009). High primary productivity and nitrogen cycling after the Paleoproterozoic phosphogenic event in the Aravalli Supergroup, India. *Precam. Res.* **171**, 37–56.

Parker, E. N. (1963). *Interplanetary Dynamical Processes*. New York: Interscience Publishers.

Parker, E. T., *et al.* (2011). Primordial synthesis of amines and amino acids in a 1958 Miller H_2S-rich spark discharge experiment. *P. Natl. Acad. Sci. USA* **108**, 5526–5531.

Parker, T. J., *et al.* (1993). Coastal geomorphology of the martian Northern Plains. *J. Geophys. Res.* **98**, 11 061–11 078.

Parker, T. J., *et al.* (2010). *The northern plains: A martian oceanic basin? In: Lakes on Mars*, eds. N. A. Cabrol, E. A. Grin. New York: Elsevier.

Parkinson, T. D. and Hunten, D. M. (1972). Spectroscopy and aeronomy of O_2 on Mars. *J. Atmos. Sci.* **29**, 1380–1390.

Parmentier, V. and Guillot, T. (2014). A non-grey analytical model for irradiated atmospheres. *I. Derivation. Astron. Astrophys.* **562**, A133, doi:10.1051/0004-6361/201322342.

Parmentier, V., *et al.* (2015). A non-grey analytical model for irradiated atmospheres II. Analytical vs. numerical solutions. *Astron. Astrophys.* **574**, A35, doi:10.1051/0004-6361/201323127.

Parnell, J., *et al.* (2010). Early oxygenation of the terrestrial environment during the Mesoproterozoic. *Nature* **468**, 290–293.

Partin, C. A., *et al.* (2013). Large-scale fluctuations in Precambrian atmospheric and oceanic oxygen levels from the record of U in shales. *Earth Planet. Sci. Lett.* **369**, 284–293.

Pasek, M. and Lauretta, D. (2008). Extraterrestrial flux of potentially prebiotic C, N, and P to the early Earth. *Origins Life Evol. B.* **38**, 5–21.

Pavlov, A. A., *et al.* (2003). Methane-rich Proterozoic atmosphere? *Geology* **31**, 87–90.

Pavlov, A. A. and Kasting, J. F. (2002). Mass-independent fractionation of sulfur isotopes in Archean sediments: strong evidence for an anoxic Archean atmosphere. *Astrobiology* **2**, 27–41.

Pavlov, A. A., *et al.* (2001). UV-shielding of NH_3 and O_2 by organic hazes in the Archean atmosphere. *J. Geophys. Res.* **106**, 23 267–23 287.

Pavlov, A. A., *et al.* (2000). Greenhouse warming by CH_4 in the atmosphere of early Earth. *J. Geophys. Res.* **105**, 11,981–11,990.

Pavlov, A. A., *et al.* (2012). Degradation of the organic molecules in the shallow subsurface of Mars due to irradiation by cosmic rays. *Geophys. Res. Lett.* **39**.

Payne, J. L., *et al.* (2009). Two-phase increase in the maximum size of life over 3.5 billion years reflects biological innovation and environmental opportunity. *P. Natl. Acad. Sci. USA* **106**, 24–27.

Payne, J. L., *et al.* (2011). The evolutionary consequences of oxygenic photosynthesis: a body size perspective. *Photosynth. Res.* **107**, 37–57.

Peale, S. J. (1977). Rotational histories of the natural satellites. In: *Planetary Satellites*, ed. J. A. Burns, Tucson, AZ: University of Arizona Press.

Pearl, J. C. and Conrath, B. J. (1991). The albedo, effective temperature, and energy balance of Neptune, as determined from Voyager data. *J. Geophys. Res.* **96**, 18 921–18 930.

Pearl, J. C., *et al.* (1990). The albedo, effective temperature, and energy balance of Uranus, as determined from Voyager IRIS data. *Icarus* **84**, 12–28.

Pearson, D. G., *et al.* (2003). Mantle samples included in volcanic rocks: Xenoliths and diamonds. In: *Treatise on Geochemistry - Volume 2: The Mantle and Core*, ed. R. W. Carlson, Amsterdam: Elsevier, pp. 171–276.

Pechmann, J. B. and Ingersoll, A. P. (1984). Thermal tides in the atmosphere of Venus: Comparison of model results with observations. *J. Atmos. Sci.* **41**, 3290–3313.

Peck, W. H., *et al.* (2001). Oxygen isotope ratios and rare earth elements in 3.3 to 4.4 Ga zircons: Ion microprobe evidence for high delta O-18 continental crust and oceans in the Early Archean. *Geochim. Cosmochim. Acta* **65**, 4215–4229.

Pecoits, E., *et al.* (2015). Atmospheric hydrogen peroxide and Eoarchean iron formations. *Geobiology* **13**, 1–14.

Pedone, M., *et al.* (2014). Tunable diode laser measurements of hydrothermal/volcanic CO_2 and implications for the global CO_2 budget. *Solid Earth* **5**, 1209–1221.

Peixoto, J. P. and Oort, A. H. (1984). Physics of climate. *Rev. Mod. Phys.* **56**, 365–429. © 1984 by The American Physical Society.

Peltier, W. R., *et al.* (2007). Snowball Earth prevention by dissolved organic carbon remineralization. *Nature* **450**, 813–U1.

Pepe, F., *et al.* (2014). ESPRESSO: The next European exoplanet hunter. *Astronomische Nachrichten* **335**, 8–20.

Pepin, R. O. (1989). Atmospheric compositions: Key similarities and differences. In: *Origin and Evolution of Planetary and Satellite Atmospheres*, ed. S. K. Atreya, *et al.*, Tucson, AZ: University of Arizona Press, pp. 291–305.

Pepin, R. O. (1991). On the origin and evolution of terrestrial planet atmospheres and meteoritic volatiles. *Icarus* **92**, 2–79.

Pepin, R. O. (2000). On the isotopic composition of primordial xenon in terrestrial planet atmospheres. *Space Sci. Rev.* **92**, 371–395.

Pepin, R. O. (2006). Atmospheres on the terrestrial planets: Clues to origin and evolution. *Earth Planet. Sci. Lett.* **252**, 1–14.

Pepin, R. O. (2013). Comment on "Chondritic-like xenon trapped in Archean rocks: A possible signature of the ancient atmosphere" by M. Pujol, B. Marty, R. Burgess [Earth Planet. Sci. Lett. 308 (2011) 298–306]. *Earth Planet. Sci. Lett.* **371**, 294–295.

Pepin, R. O. and Porcelli, D. (2006). Xenon isotope systematics, giant impacts, and mantle degassing on the early Earth. *Earth Planet. Sci. Lett.* **250**, 470–485.

Pepin, R. O., *et al.* (2012). Helium, neon, and argon composition of the solar wind as recorded in gold and other Genesis collector materials. *Geochim. Cosmochim. Acta* **89**, 62–80.

Peralta, J., *et al.* (2007). A reanalysis of Venus winds at two cloud levels from Galileo SSI images. *Icarus* **190**, 469–477.

Peralta, J., *et al.* (2012). Solar migrating atmospheric tides in the winds of the polar region of Venus. *Icarus* **220**, 958–970.

Perez-de-Tejada, H. (1987). Plasma flow in the Mars magnetosphere. *J. Geophys. Res.* **92**, 4713–4718.

Perkins, J. P., *et al.* (2015). Amplification of bedrock canyon incision by wind. *Nat. Geosci.* **8**, 305–310.

Pernice, H., *et al.* (2004). Laboratory evidence for a key intermediate in the Venus atmosphere: Peroxychloroformyl radical. *P. Natl. Acad. Sci. USA* **101**, 14 007–14 010.

Perrier, S., *et al.* (2006). Global distribution of total ozone on Mars from SPICAM/MEX UV measurements. *J. Geophys. Res.* **111**, E09S06.

Perry, E. C., *et al.* (1978). The oxygen isotope composition of 3800 m.y. old metamorphosed chert and iron formation from Isukasia, West Greenland. *J. Geol.* **86**, 223–239.

Perryman, M. A. C. (2014). *The Exoplanet Handbook.* New York: Cambridge University Press.

Pestova, O. N., *et al.* (2005). Polythermal study of the systems $M(ClO_4)(2)-H_2O(M^{2+} = Mg^{2+}, Ca^{2+}, Sr^{2+}, Ba^{2+})$. *Russian J. Appl. Chem.* **78**, 409–413.

Peters, K. E., *et al.* (1978). Correlation of carbon and nitrogen stable isotope ratios in sedimentary organic matter. *Limnol. Oceanogr.* **23**, 598–604.

Peterson, C. (1981). A secondary origin for the central plateau of Hebes Chasma. *Proc. Lunar Planet. Sci. Conf. 11th*, 1459–1471.

Peterson, K. J., *et al.* (2008). The Ediacaran emergence of bilaterians: congruence between the genetic and the geological fossil records. *Phil. Trans. R. Soc. Lond. B* **363**, 1435–1443.

Petigura, E. A., *et al.* (2013). Prevalence of Earth-size planets orbiting Sun-like stars. *P. Natl. Acad. Sci. USA* **110**, 19273–19278.

Petty, G. W. (2006). *A First Course in Atmospheric Radiation.* Madison, Wis.: Sundog Pub.

Pfalzner, S., *et al.* (2015). The formation of the solar system. *Phy. Scr.* **90**, 068001 doi:10.1088/0031-8949/90/6/068001.

Pham, L. B. S., *et al.* (2011). Effects of impacts on the atmospheric evolution: Comparison between Mars, Earth, and Venus. *Planet. Space Sci.* **59**, 1087–1092.

Phillips, R. J., *et al.* (2011). Massive CO(2) ice deposits sequestered in the South Polar layered deposits of Mars. *Science* **332**, 838–841.

Phillips, R. J., *et al.* (1992). Impact craters and Venus resurfacing history. *J. Geophys. Res.* **97**, 15 923.

Phillips, R. J., *et al.* (2001). Ancient geodynamics and global-scale hydrology on Mars. *Science* **291**, 2587–2591.

Pierazzo, E., *et al.* (2008). Validation of numerical codes for impact and explosion cratering: Impacts on strengthless and metal targets. *Meteorit. Planet. Sci.* **43**, 1917–1938.

Pierazzo, E. and Chyba, C. F. (1999). Amino acid survival in large cometary impacts. *Meteorit. Planet. Sci.* **34**, 909–918.

Pieri, D. C. (1980). Geomorphology of valleys on Mars: A summary of morphology, distribution, age and origin. *Science* **210**, 895–897.

Pierrard, V. (2003). Evaporation of hydrogen and helium atoms from the atmospheres of Earth and Mars. *Planet. Space Sci.* **51**, 319–327.

Pierrehumbert, R. and Gaidos, E. (2011). Hydrogen greenhouse planets beyond the habitable zone. *Astrophys. J. Lett.* **734**.

Pierrehumbert, R. T. (2004). High levels of atmospheric carbon dioxide necessary for the termination of global glaciation. *Nature* **429**, 646–649.

Pierrehumbert, R. T. (2010). *Principles of Planetary Climate*. Cambridge: Cambridge University Press.

Pierrehumbert, R. T. (2011). A pallette of climates for Gliese 581g. *Astrophys. J. Lett.* **726**.

Pierrehumbert, R. T., *et al.* (2011). Climate of the Neoproterozoic. *Annu. Rev. Earth Planet. Sci.* **39**, 417–460.

Pierrehumbert, R. T. and Swanson, K. L. (1995). *Baroclinic instability. Annu. Rev. Fluid Mech.* **27**, 419–467.

Pinti, D. L., *et al.* (2001). Nitrogen and argon signatures in 3.8 To 2.8 Ga metasediments: Clues on the chemical state of the Archean ocean and the deep biosphere. *Geochim. Cosmochim. Acta* **65**, 2301–2315.

Pinti, D. L., *et al.* (2007). Biogenic nitrogen and carbon in Fe-Mn-oxyhydroxides from an Archean chert, Marble Bar, Western Australia. *Geochem. Geophys. Geosys.* **8**, doi:10.1029/2006GC001394.

Pinti, D. L., *et al.* (2009). Isotopic fractionation of nitrogen and carbon in Paleoarchean cherts from Pilbara craton, Western Australia: Origin of N-15-depleted nitrogen. *Geochim. Cosmochim. Acta* **73**, 3819–3848.

Pinti, D. L., *et al.* (2013). Comment on "Biogenicity of Earth's earliest fossils: a resolution of the controversy" by J. William Schopf and Anatoliy B. Kudryavtsev, Gondwana Research 22 (2012), 761–771. *Gondwana Res.* **23**, 1652–1653.

Pinto, J. P., *et al.* (1980). Photochemical production of formaldehyde in the earth's primitive atmosphere. *Science* **210**, 183–185.

Planavsky, N., *et al.* (2010). Rare Earth Element and yttrium compositions of Archean and Paleoproterozoic Fe formations revisited: New perspectives on the significance and mechanisms of deposition. *Geochim. Cosmochim. Acta* **74**, 6387–6405.

Planavsky, N. J., *et al.* (2014a). Evidence for oxygenic photosynthesis half a billiion years before the Great Oxidation Event. *Nat. Geosci.* **7**, 283–286.

Planavsky, N. J., *et al.* (2012). Sulfur record of rising and falling marine oxygen and sulfate levels during the Lomagundi event. *P. Natl. Acad. Sci. USA* **109**, 18 300–18 305.

Planavsky, N. J., *et al.* (2011). Widespread iron-rich conditions in the mid-Proterozoic ocean. *Nature* **477**, 448–451.

Planavsky, N. J., *et al.* (2014b). Low Mid-Proterozoic atmospheric oxygen levels and the delayed rise of animals. *Science* **346**, 635–638.

Plankensteiner, K., *et al.* (2006). Amino acids on the rampant primordial Earth: Electric discharges and the hot salty ocean. *Mol. Divers.* **10**, 3–7.

Plaut, J. J., *et al.* (2007). Subsurface radar sounding of the south polar layered deposits of Mars. *Science* **316**, 92–95.

Pleskot, L. K. and Miner, E. D. (1981). Time variability of martian bolometric albedo. *Icarus* **45**, 179–201.

Pogge von Strandmann, P. A. E., *et al.* (2015). Selenium isotope evidence for post-glacial oxygenation trends in the Ediacaran ocean. *Nat. Commun.*, submitted.

Pohl, W. (1989). Comparative geology of magnesite and ocurrences. *Monograph Series on Mineral Deposits* **28**, 1–13.

Poirier, J. P. (1994). Light elements in the Earths outer core: a critical review. *Phys. Earth Planet. In.* **85**, 319–337.

Pollack, J. B. (1971). A nongrey calculation of the runaway greenhouse: implications for Venus' past and present. *Icarus* **14**, 295–306.

Pollack, J. B. (1979). Climatic change on the terrestrial planets. *Icarus* **37**, 479–553.

Pollack, J. B. (1991). Kuiper Prize Lecture: Present and past climates of the terrestrial planets. *Icarus* **91**, 173–198.

Pollack, J. B., *et al.* (1979). Properties and effects of dust particles suspended in the Martian atmosphere. *J. Geophys. Res.* **84**, 2929–2945.

Pollack, J. B., *et al.* (1993). Near-infrared light from Venus nightside: A spectroscopic analysis. *Icarus* **103**, 1–42.

Pollack, J. B., *et al.* (1977). Calculation of Saturn's gravitational contraction history. *Icarus* **30**, 111–128.

Pollack, J. B., *et al.* (1990). Simulations of the general circulation of the Martian atmosphere 1. Polar Processes. *J. Geophys. Res.* **95**, 1447–1473.

Pollack, J. B., *et al.* (1996). Formation of the giant planets by concurrent accretion of solids and gas. *Icarus* **124**, 62–85.

Pollack, J. B., *et al.* (1987). The case for a wet, warm climate on early Mars. *Icarus* **71**, 203–224.

Pollard, D. and Kasting, J. F. (2005). Snowball Earth: A thin-ice model with flowing sea glaciers. *J. Geophys. Res.* **110**, C07010, doi: 10.1029/2004JC002525.

Pollard, D. and Kasting, J. F. (2006). Reply to comment by Stephen G. Warren and Richard E. Brandt on "Snowball earth: A thin-ice solution with flowing sea glaciers". *J. Geophys. Res.* **111**, C09017, doi: 10.1029/2006JC003488.

Polyak, V., *et al.* (2008). Age and evolution of the Grand Canyon revealed by U-Pb dating of water table-type speleothems. *Science* **319**, 1377–1380.

Pondrelli, M., *et al.* (2011). Geological, geomorphological, facies and allostratigraphic maps of the Eberswalde fan delta. *Planet. Space Sci.* **59**, 1166–1178.

Pope, E. C., *et al.* (2012). Isotope composition and volume of Earth's early oceans. *P. Natl. Acad. Sci. USA* **109**, 4371–4376.

Popp, M., *et al.* (2016). Transition to a moist greenhouse with CO_2 and solar forcing. *Nat. Commun.* **7**, 10627, doi:10.1038/ncomms10627.

Porcelli, D. and Elliott, T. (2008). The evolution of He isotopes in the convecting mantle and the preservation of high He-3/He-4 ratios. *Earth Planet. Sci. Lett.* **269**, 175–185.

Porcelli, D. and Wasserburg, G. J. (1995). Mass transfer of helium, neon, argon, and xenon through a steady-state upper mantle. *Geochim. Cosmochim. Acta* **59**, 4921–4937.

Porco, C. C., *et al.* (2006). Cassini observes the active South Pole of Enceladus. *Science* **311**, 1393–1401.

Porco, C. C., *et al.* (2003). Cassini imaging of Jupiter's atmosphere, satellites, and rings. *Science* **299**, 1541–1547.

Postawko, S. E. and Kuhn, W. R. (1986). Effect of the greenhouse gases (CO_2, H_2O, SO_2) on Martian paleoclimate. *J. Geophys. Res. (Proc. Lunar Planet. Sci. Conf. 16th)* **91**, D431–D438.

Postberg, F., *et al.* (2011). A salt-water reservoir as the source of a compositionally stratified plume on Enceladus. *Nature* **474**, 620–622.

Postgate, J. R. (1987). *Nitrogen Fixation.* London: Edward Arnold.

Posth, N. R., *et al.* (2014). Biogenic Fe(III) minerals: From formation to diagenesis and preservation in the rock record. *Earth Sci. Rev.* **135**, 103–121.

Posth, N. R., *et al.* (2013). Simulating Precambrian banded iron formation diagenesis. *Chem. Geol.* **362**, 66–73.

Poulton, S. W. and Canfield, D. E. (2011). Ferruginous conditions: A dominant feature of the ocean through Earth's history. *Elements* **7**, 107–112.

Poulton, S. W., *et al.* (2004). The transition to a sulphidic ocean ~1.84 billion years ago. *Nature* **431**, 173–177.

Poulton, S. W., *et al.* (2010). Spatial variability in oceanic redox structure 1.8 billion years ago. *Nature Geosci.* **3**, 486–490.

Poulton, S. W. and Raiswell, R. (2002). The low-temperature geochemical cycle of iron: From continental fluxes to marine sediment deposition. *Am. J. Sci.* **302**, 774–805.

Prasad, N. and Roscoe, S. M. (1996). Evidence of anoxic to oxic atmosphere change during 2.45–2.22 Ga from lower and upper sub-Huronian paleosols, Canada. *Catena* **27**, 105–121.

Prather, M. (1996). Time scales in atmospheric chemistry: Theory, GWPs for CH4 and CO, and runaway growth. *Geophys. Res. Lett.* **23**, 2597–2600.

Prather, M., *et al.* (2001). Atmospheric chemistry and greenhouse gases. In: *Climate Change 2001, The Scientific Basis*, ed. J. T. Houghton, *et al.*, Cambridge; New York: Cambridge University Press, pp. 239–287.

Prausnitz, J. M., *et al.* (1999). *Molecular Thermodynamics Of Fluid-Phase Equilibria.* Upper Saddle River, NJ: Prentice Hall.

Prave, A. R. (2002). Life on land in the Proterozoic: Evidence from the Torridonian rocks of northwest Scotland. *Geology* **30**, 811–814.

Prentice, I. C., *et al.* (2001). The carbon cycle and atmospheric carbon dioxide. In: *Climate Change 2001: the Scientific Basis* ed. J. T. Houghton, *et al.*, Cambridge; New York: Cambridge University Press, pp. 183–238.

Press, W. H. (2007). *Numerical Recipes: The Art Of Scientific Computing.* New York: Cambridge University Press.

Preston, G. T. and Prausnit, J. M. (1970). Thermodynamics of solid solubility in cryogenic solvents. *Ind. Eng. Chem .Proc. Dd.* **9**, 264–271.

Prinn, R. G. (1971). Photochemistry of HCl and other minor constituents in atmosphere of Venus. *J. Atmos. Sci.* **28**, 1058–1068.

Prinn, R. G. (2014). Ozone, hydroxyl radical, and oxidative capacity. In: *Treatise on Geochemistry*, ed. H. D. Holland and K. K. Turekian, New York: Elsevier.

Prinn, R. G. and Fegley, B. (1987). The atmospheres of Venus, Earth, and Mars: A critical comparison. *Annu. Rev. Earth Planet. Sci.* **15**, 171–212.

Prinn, R. G. and Fegley, B. (1989). Solar nebula chemistry: origin of planetary, satellite and cometary volatiles. In: *Origin and Evolution of Planetary and Satellite Atmospheres*, ed. S. K. Atreya, *et al.*, Tucson, AZ: University of Arizona Press, pp. 78–136.

Prinn, R. G., *et al.* (1995). Atmospheric trends and ilfetime of CH_3CCl_3 and global OH concentrations. *Science* **269**, 187–192.

Proskurowski, G., *et al.* (2008). Abiogenic hydrocarbon production at Lost City hydrothermal field. *Science* **319**, 604–607.

Pujol, M., *et al.* (2011). Chondritic-like xenon trapped in Archean rocks: A possible signature of the ancient atmosphere. *Earth Planet. Sc. Lett.* **308**, 298–306.

Pujol, M., *et al.* (2013). Argon isotopic composition of Archaean atmosphere probes early Earth geodynamics. *Nature* **498**, 87–90.

Putzig, N. E., *et al.* (2014). SHARAD soundings and surface roughness at past, present, and proposed landing sites on Mars: Reflections at Phoenix may be attributable to deep ground ice. *J. Geophys. Res.* **119**, 1936–1949.

Quinn, R. C., *et al.* (2013). Perchlorate radiolysis on Mars and the origin of martian soil reactivity. *Astrobiology* **13**, 515–520.

Raack, J., *et al.* (2015). Present-day seasonal gully activity in a south polar pit (Sisyphi Cavi) on Mars. *Icarus* **251**, 226–243.

Radebaugh, J., *et al.* (2008). Dunes on Titan observed by Cassini Radar. *Icarus* **194**, 690–703.

Raff, R. A. and Raff, E. C. (1970). Respiratory mechanisms and the metazoan fossil record. *Nature* **228**, 1003–1005.

Raghoebarsing, A. A., *et al.* (2006). A microbial consortium couples anaerobic methane oxidation to denitrification. *Nature* **440**, 918–921.

Rairden, R. L., *et al.* (1986). Geocoronal imaging with Dynamics Explorer. *J. Geophys. Res.* **91**, 3613–3630.

Ramanathan, V. and Carmichael, G. (2008). Global and regional climate changes due to black carbon. *Nat. Geosci.* **1**, 221–227.

Ramanathan, V. and Inamdar, A. (2006). The radiative forcing due to clouds and water vapor. In: *Frontiers of Climate Modeling*, ed. J. T. Kiehl, V. Ramanathan, New York: Cambridge University Press, pp. 119–151.

Ramaswamy, V., *et al.* (2001). Radiative forcing of climate change. In: *Climate Change 2001: Working Group I: The Scientific Basis*, ed. J. T. Houghton, et al., New York: Cambridge University Press, pp. 349–416.

Rambler, M. and Margulis, L. (1980). Bacterial resistance to ultraviolet irradiation under anaerobiosis: Implications for pre-Phanerozoic evolution. *Science* **210**, 638–640.

Rameau, J., *et al.* (2013). Discovery of a probable 4–5 Jupiter-mass exoplanet to HD 95086 by direct imaging. *Astrophys. J. Lett.* **772**, L15.

Ramirez, R. M. and Kaltenegger, L. (2014). The habitable zones of pre-main-sequence stars. *Astrophys. J. Lett.* **797**.

Ramirez, R. M., *et al.* (2014a). Warming early Mars with CO_2 and H_2. *Nat. Geosci.* **7**, 59–63.

Ramirez, R. M., *et al.* (2014b). Can increased atmospheric CO_2 levels trigger a runaway greenhouse? *Astrobiology* **14**, 714–731.

Randel, W. J. and Held, I. M. (1991). Phase speed spectra of transient eddy fluxes and critical layer absorption. *J. Atmos. Sci.* **48**, 688–697.

Rannou, P., *et al.* (1997). A new interpretation of scattered light measurements at Titan's limb. *J. Geophys. Res.* **102**, 10 997–11 013.

Rannou, P., *et al.* (1995). Titan's geometric albedo: Role of the fractal structure of the aerosols. *Icarus* **118**, 355–372.

Rannou, P., *et al.* (2004). A coupled dynamics-microphysics model of Titan's atmosphere. *Icarus* **170**, 443–462.

Rannou, P., *et al.* (2003). A model of Titan's haze of fractal aerosols constrained by multiple observations. *Planet. Space Sci.* **51**, 963–976.

Rannou, P., *et al.* (2006). The latitudinal distribution of clouds on Titan. *Science* **311**, 201–205.

Rashby, S. E., *et al.* (2007). Biosynthesis of 2-methylbacteriohopanepolyols by an anoxygenic phototroph. *P. Natl. Acad. Sci. USA* **104**, 15 099–15 104.

Rasmussen, B. and Buick, R. (1999). Redox state of the Archean atmosphere: Evidence from detrital heavy minerals in ca. 3250–2750 Ma sandstones from the Pilbara Craton, Australia. *Geology* **27**, 115–118.

Rasmussen, B., *et al.* (2008). Reassessing the first appearance of eukaryotes and cyanobacteria. *Nature* **455**, 1101–1104.

Rasmussen, B., *et al.* (2014). Hematite replacement of iron-bearing precursor sediments in the 3.46-b.y.-old Marble Bar Chert, Pilbara craton, Australia. *Bull. Geol. Soc. Am.*, doi:10.1130/B31049.1.

Rasmussen, B., *et al.* (2016). Dust to dust: Evidence for the formation of "primary" hematite dust in banded iron formations via oxidation of iron silicate nanoparticles. *Precam. Res.* **284**, 49–63.

Rasool, S. I. and DeBergh, C. (1970). The runaway greenhouse and the accumulation of CO_2 in the Venus atmosphere. *Nature* **226**, 1037–1039.

Raub, T. D., *et al.* (2007). Siliciclastic prelude to Elatina–Nuccaleena deglaciation: Lithostratigraphy and rock magnetism of the base of the Ediacaran system. *Spec. Publ. Geol. Soc. London* **286**, 53–75.

Raulin, F., *et al.* (2012). Prebiotic-like chemistry on Titan. *Chem. Soc. Rev.* **41**, 5380–5393.

Raymann, K., *et al.* (2015). The two-domain tree of life is linked to a new root for the Archaea. *P. Natl Acad. Sci. USA* **112**, 6670–6675.

Raymo, M. E. and Ruddiman, W. F. (1992). Tectonic forcing of Late Cenozoic climate. *Nature* **359**, 117–122.

Raymond, J., *et al.* (2004). The natural history of nitrogen fixation. *Molec. Biol. Evol.* **21**, 541–554.

Raymond, S. N. (2014). Terrestrial planet formation at home and abroad. In: *Protostars and Planets VI*, ed. H. Beuther, *et al.*, Tucson, AZ: University of Arizona Press, pp. 595–618.

Raymond, S. N., *et al.* (2006). High-resolution simulations of the final assembly of Earth-like planets I: Terrestrial accretion and dynamics. *Icarus* **183**, 265–282.

Raymond, S. N., *et al.* (2010). Formation of Terrestrial Planets. In: *Formation and Evolution of Exoplanets*, ed. Rony Barnes, Wiley - VCH Verlag GmbH & Co. KGaA, pp. 123–143.

Read, P. L. (2013). The dynamics and circulation of Venus atmosphere. In: *Towards Understanding the Climate of Venus*, ed. L. Bengtsson, *et al.*, New York: Springer, pp. 73–110.

Read, P. L. and Lewis, S. R. (2003). *The Martian Climate Revisited: Atmosphere and Environment of a Desert Planet*. London: Praxis.

Redfield, A. C. (1958). The biological control of chemical factors in the environment. *Am. Sci.* **46**, 205–221.

Reese, C. C., *et al.* (2010). Impact origin for the Martian crustal dichotomy: Half emptied or half filled? *J. Geophys. Res.* **115**.

Reinhard, C. T. and Planavsky, N. J. (2011). Mineralogical constraints on Precambiran pCO_2. *Nature* **474**, E1, doi:10.1038/nature09959.

Reinhard, C. T., *et al.* (2013a). Long-term sedimentary recycling of rare sulphur isotope anomalies. *Nature* **497**, 100–104.

Reinhard, C. T., *et al.* (2013b). Proterozoic ocean redox and biogeochemical stasis. *P. Natl. Acad. Sci. USA* **110**, 5357–5362.

Reinhard, C. T., *et al.* (2009). A late Archean sulfidic sea stimulated by early oxidative weathering of the continents. *Science* **326**, 713–716.

Reuschel, M., *et al.* (2012). Isotopic evidence for a sizeable seawater sulfate reservoir at 2.1 Ga. *Precambrian Res.* **192–95**, 78–88.

Rhines, P. B. (1975). Waves and turbulence on a beta plane. *J. Fluid Mech.* **69**, 417–443.

Ribas, I., *et al.* (2005). Evolution of the solar activity over time and effects on planetary atmospheres. I. High-energy irradiances (1–1700 ångström). *Astrophys. J.* **622**, 680–694.

Ricardo, A., *et al.* (2004). Borate minerals stabilize ribose. *Science* **303**, 196–196.

Richardson, M. I. and Wilson, R. J. (2002). A topographically forced asymmetry in the martian circulation and climate. *Nature* **416**, 298–301.

Ridgwell, A. J., *et al.* (2003). Carbonate deposition, climate stability, and neoproterozoic ice ages. *Science* **302**, 859–862.

Riding, R., *et al.* (2014). Idenfication of an Archean marine oxygen oasis. *Precam. Res.* **251**, 232–237.

Riedel, K. and Lassey, K. (2008). Detergent of the atmosphere. *Water Atmos.* **16**, 22–23.

Roach, L. H., *et al.* (2010a). Diagenetic haematite and sulfate assemblages in Valles Marineris. *Icarus* **207**, 659–674.

Roach, L. H., *et al.* (2010b). Hydrated mineral stratigraphy of Ius Chasma, Valles Marineris. *Icarus* **206**, 253–268.

Robbins, S. J. and Hynek, B. M. (2012). A new global database of Mars impact craters >= 1 km: 1. Database creation, properties, and parameters. *J. Geophys. Res.* **117**.

Roberson, A. L., *et al.* (2011). Greenhouse warming by nitrous oxide and methane in the Proterozoic Eon. *Geobiology* **9**, 313–320.

Robert, F. (2001). The origin of water on Earth. *Science* **293**, 1056–1058.

Robert, F. and Chaussidon, M. (2006). A palaeotemperature curve for the Precambrian oceans based on silicon isotopes in cherts. *Nature* **443**, 969–972.

Robertson, M. P. and Joyce, G. F. (2012). The origins of the RNA World. *CSH Perspect. Biol.* **4**.

Robin, C. M. I. and Bailey, R. C. (2009). Simultaneous generation of Archean crust and subcratonic roots by vertical tectonics. *Geology* **37**, 523–526.

Robinson, J. M. (1990). Lignin, land plants, and fungi – Biological evolution affecting Phanerozoic oxygen balance. *Geology* **18**, 607–610.

Robinson, T. D. and Catling, D. C. (2012). An analytic radiative-convective model for planetary atmospheres. *Astrophys. J.* **757**, 104 doi:10.1088/0004-637X/757/1/104.

Robinson, T. D. and Catling, D. C. (2014). Common 0.1 bar tropopause in thick atmospheres set by pressure-dependent infrared transparency. *Nat. Geosci.* **7**, 12–15.

Robinson, T. D., *et al.* (2014a). Detection of ocean glint and ozone absorption using LCROSS Earth observations. *Astrophys. J.* **787**, 171.

Robinson, T. D., *et al.* (2014b). Titan solar occultation observations reveal transit spectra of a hazy world. *P. Natl. Acad. Sci. USA* **111**, 9042–9047.

Robinson, T. D., *et al.* (2010). Detecting oceans on extrasolar planets using the glint effect. *Astrophys. J. Lett.* **721**, L67–L71.

Robinson, T. D., *et al.* (2011). Earth as an extrasolar planet: Earth model validation using EPOXI Earth observations. *Astrobiology* **11**, 393–408.

Robock, A. (2000). Volcanic eruptions and climate. *Rev. Geophys.* **38**, 191–219.

Rodehacke, C. B., *et al.* (2013). An open ocean region in Neoproterozoic glaciations would have to be narrow to allow equatorial icesheets. *Geophys. Res. Lett.* **40**, 5503–5507.

Rodgers, C. D. and Walshaw, C. D. (1966). Computation of infra-red cooling rate in planetary atmospheres. *Q. J. R. Met. Soc.* **92**, 67–92.

Rodler, F. and Lopez-Morales, M. (2014). Feasibility studies for the detection of O_2 in an Earth-like exoplanet. *Astrophys. J.* **781**, 54.

Rodriguez, S., *et al.* (2011). Titan's cloud seasonal activity from winter to spring with Cassini/VIMS. *Icarus* **216**, 89–110.

Roe, H. G. and Grundy, W. M. (2012). Buoyancy of ice in the CH_4–N_2 system. *Icarus* **219**, 733–736.

Rogers, L. A. (2015). Most 1.6-Earth-radii planets are not rocky. *Ap. J.* **801**, 41.

Rondanelli, R. and Lindzen, R. S. (2010). Can thin cirrus clouds in the tropics provide a solution to the faint young Sun paradox? *J. Geophys. Res.* **115**.

Rooney, A. D., *et al.* (2014). Re–Os geochronology and coupled Os–Sr isotope constraints on the Sturtian snowball Earth. *P. Natl. Acad. Sci. USA* **111**, 51–56.

Rooney, A. D., *et al.* (2015). A Cryogenian chronology: Two long-lasting synchronous Neoproterozoic glaciations. *Geology* **43**, 459–462.

Roscoe, S. M. (1969). Huronian rocks and uraniferous conglomerates in the Canadian Shield. *Geol. Surv. Can. Pap. 68–40*, 205 pp.

Rose, B. E. J. and Marshall, J. (2009). Ocean heat transport, sea ice, and multiple climate states: Insights from energy balance models. *J. Atmos. Sci.* **66**, 2828–2843.

Rosing, M. T. (1999). ^{13}C-depleted carbon microparticles in >3700-Ma sea-floor sedimentary rocks from West Greenland. *Science* **283**, 674–676.

Rosing, M. T. and Bird, D. K. (2007). Constraints on atmospheric H-2 from banded iron formations. *Geochim. Cosmochim. Acta* **71**, A852.

Rosing, M. T., *et al.* (2010). No climate paradox under the faint early Sun. *Nature* **464**, 744-U117.

Rosing, M. T. and Frei, R. (2004). U-Rich Archaean seafloor sediments from Greenland – Indications of > 3700 Ma oxygenic photosynthesis. *Earth Planet. Sci. Lett.* **217**, 237–244.

Roskar, R., *et al.* (2008). Riding the spiral waves: Implications of stellar migration for the properties of galactic disks. *Astrophys. J. Lett.* **684**, L79–L82.

Rossow, W. B., *et al.* (1982). Cloud feedback: A stabilizing effect for the early Earth? *Science* **217**, 1245–1247.

Roth, L., *et al.* (2014). Transient water vapor at Europa's south pole. *Science* **343**, 171–174.

Rothman, D. H., *et al.* (2003). Dynamics of the Neoproterozoic carbon cycle. *P. Natl. Acad. Sci. USA* **100**, 8124–8129.

Rouxel, O. J., *et al.* (2005). Iron isotope constraints on the Archean and Paleoproterozoic ocean redox state. *Science* **307**, 1088–1091.

Royer, D. L., *et al.* (2007). Climate sensitivity constrained by CO_2 concentrations over the past 420 million years. *Nature* **446**, 530–532.

Rubey, W. W. (1951). Geological history of seawater. An attempt to state the problem. *Geol. Soc. Am. Bull.* **62**, 1111–1148.

Rubey, W. W. (1955). Development of the hydrosphere and atmosphere, with special reference to probable composition of the early atmosphere. In: *Crust of the Earth*, ed. A. Poldervaart, New York: Geol. Soc. Am., pp. 631–650.

Rubie, D. C., *et al.* (2011). Heterogeneous accretion, composition and core-mantle differentiation of the Earth. *Earth Planet. Sc. Lett.* **301**, 31–42.

Rubie, D. C. and Jacobson, S. A. (2015). Mechanisms and geochemical models of core formation. In: *Deep Earth: Physics and Chemistry of the Lower Mantle and Core. AGU Monograph*, ed. R. Fischer and H. Terasaki, Washington DC: Amer. Geophys. Union.

Rubie, D. C., *et al.* (2015a). Accretion and differentiation of the terrestrial planets with implications for the compositions of early-formed Solar System bodies and accretion of water. *Icarus* **248**, 89–108.

Rubie, D. C., *et al.* (2015b). Formation of the Earth's core. In: *Treatise on Geophysics*, ed. G. Schubert, New York: Elsevier, pp. 43–79.

Rubinstein, C. V., *et al.* (2010). Early Middle Ordovician evidence for land plants in Argentina (eastern Gondwana). *New Phytol.* **188**, 365–369.

Ruff, S. W., *et al.* (2014). Evidence for a Noachian-aged ephemeral lake in Gusev crater, Mars. *Geology* **42**, 359–362.

Ruiz, J., *et al.* (2011). The thermal evolution of Mars as constrained by paleo-heat flows. *Icarus* **215**, 508–517.

Ruiz-Mirazo, K., *et al.* (2014). Prebiotic systems chemistry: new perspectives for the origins of life. *Chem. Rev.* **114**, 285–366.

Runnegar, B. (1982). Oxygen requirements, biology and phylogenetic significance of the late Precambrian worm Dickinsonia, and the evolution of the burrowing habit. *Alcheringa* **6**, 223–239.

Runnegar, B. (1991). Oxygen and the early evolution of the Metazoa. In: *Metazoan Life without Oxygen*, ed. C. Bryant, New York: Chapman and Hall, pp. 65–87.

Runnegar, B., *et al.* (2002). Mass-independent and mass-dependent sulfur processing throughout the Archean. *Geochim. Cosmochim. Acta* **66**, A655–A655.

Russell, G. L., *et al.* (2013). Fast atmosphere–ocean model runs with large changes in CO_2. *Geophys. Res. Lett.* **40**, 5787–5792.

Russell, H. N. (1916). On the albedo of the Moon and planets. *Ap. J.* **43**.

Russell, J. M., *et al.* (1996). Satellite confirmation of the dominance of chlorofluorocarbons in the global stratospheric chlorine budget. *Nature* **379**, 526–529.

Russell, M. J. (1996). The generation at hot springs of sedimentary ore deposits, microbialites and life. *Ore Geology Reviews* **10**, 199–214.

Russell, S. S., *et al.* (2006). Timescales of the protoplanetary disk. In: *Meteorites and the Early Solar System II*, ed. D. S. Lauretta and H. Y. McSween, Tuscon, AZ: University of Arizona Press, pp. 233–251.

Ryan, S., *et al.* (2006). Mauna Loa volcano is not a methane source: Implications for Mars. *Geophysical Research Letters* **33**.

Ryder, G., *et al.* (2000). Heavy bombardment on the Earth at ~3.85 Ga: The search for petrographic and geochemical evidence. In: *Origin of the Moon and Earth,* ed. R. M. Canup and K. Righter, Tucson, AZ: University of Arizona Press, pp. 475–492.

Rye, R. and Holland, H. D. (1998). Paleosols and the evolution of atmospheric oxygen: A critical review. *Amer. J. Sci.* **298**, 621–672.

Rye, R. and Holland, H. D. (2000). Life associated with a 2.76 Ga ephemeral pond?: Evidence from Mount Roe #2 paleosol. *Geology* **28**, 483–486.

Rye, R., *et al.* (1995). Atmospheric carbon dioxide concentrations before 2.2 billion years ago. *Nature* **378**, 603–605.

Sackmann, I. J. and Boothroyd, A. I. (2003). Our Sun. V. A bright young Sun consistent with helioseismology and warm temperatures on ancient Earth and Mars. *Astrophys. J.* **583**, 1024–1039.

Safronov, V. S. (1972). *Evolution of the Protoplanetary Cloud and Formation of the Earth and Planets.* NASA.

Sagan, C. (1973). Ultraviolet selection pressure on earliest organisms. *J. Theor. Biol.* **39**, 195–200.

Sagan, C. (1977). Reducing greenhouses and temperature history of Earth and Mars. *Nature* **269**, 224–226.

Sagan, C. (1994). *Pale Blue Dot: A Vision of the Human Future in Space.* New York: Random House.

Sagan, C. and Chyba, C. (1997). The early faint Sun paradox: Organic shielding of ultraviolet-labile greenhouse gases. *Science* **276**, 1217–1221.

Sagan, C. and Khare, B. N. (1979). Tholins: Organic chemistry of interstellar grains and gas. *Nature* **277**, 102–107.

Sagan, C., *et al.* (1993a). Polycyclic aromatic hydrocarbons in the atmospheres of Titan and Jupiter. *Astrophys. J.* **414**, 399–405.

Sagan, C. and Mullen, G. (1972). Earth and Mars: Evolution of atmospheres and surface temperatures. *Science* **177**, 52–56.

Sagan, C., *et al.* (1993b). A search for life on Earth from the Galileo spacecraft. *Nature* **365**, 715–721.

Sahagian, D. L. and Maus, J. E. (1994). Basalt vesicularity as a measure of atmospheric pressure and palaeoelevation. *Nature* **372**, 449–451.

Sahoo, S. K., *et al.* (2012). Ocean oxygenation in the wake of the Marinoan glaciation. *Nature* **489**, 546–549.

Salaris, M. and Cassisi, S. (2005). *Evolution of Stars and Stellar Populations.* Hoboken, NJ: Wiley.

Salvatore, M. R., *et al.* (2010). Definitive evidence of Hesperian basalt in Acidalia and Chryse planitiae. *J. Geophys. Res.* **115**, E07005.

Sam, H. and Yoram, L. (2014). Densities and eccentricities of 139 Kepler planets from transit time variations. *Astrophys. J.* **787**, 80.

Sanchez-Lavega, A. (2010). *An Introduction to Planetary Atmospheres.* Boca Raton, FL: CRC Press/Taylor & Francis.

Sanchez-Lavega, A., *et al.* (2004). Clouds in planetary atmospheres: A useful application of the Clausius–Clapeyron equation. *Am. J. Phys.* **72**, 767–774.

Sandor, B. J., *et al.* (2010). Sulfur chemistry in the Venus mesosphere from SO_2 and SO microwave spectra. *Icarus* **208**, 49–60.

Sano, Y. and Pillinger, C. T. (1990). Nitrogen isotopes and N_2/Ar ratios in cherts: An attempt to measure time evolution of atmospheric delta-N-15 value. *Geochem. J.* **24**, 315–325.

Sanudo, J., *et al.* (1997). Bounds for the mass of the atmosphere. *J. Geophys. Res. – Atmospheres* **102**, 7007–7009.

Sasaki, S. and Nakazawa, K. (1988). Origin of isotopic fractionation of terrestrial Xe: Hydrodynamic fractionation during escape of the primordial H_2–He atmosphere. *Earth Planet. Sc. Lett.* **89**, 323–334.

Satkoski, A. M., *et al.* (2015). A redox-stratified ocean 3.2 billion years ago. *Earth Planet. Sci. Lett.* **430**, 43–53.

Satoh, M. (2004). *Atmospheric circulation dynamics and general circulation models.* Chichester, UK: Praxis Pub.

Saunders, A. and Reichow, M. (2009). The Siberian Traps and the End-Permian mass extinction: a critical review. *Chinese Sci. Bull.* **54**, 20–37.

Sausen, R. and Santer, B. D. (2003). Use of changes in tropopause height to detect human influences on climate. *Meteorol. Z.* **12**, 131–136.

Schaber, G. G. *et al.* (1992). Geology and distribution of impact craters on Venus: what are they telling us? *J. Geophys. Res.* **97**, 13 257–13 301.

Schaefer, L. and Fegley, B. (2010). Chemistry of atmospheres formed during accretion of the Earth and other terrestrial planets. *Icarus* **208**, 438–448.

Schaller, E. L. and Brown, M. E. (2007). Volatile loss and retention on Kuiper Belt objects. *Astrophys. J.* **659**, L61–L64.

Scharf, C. A. (2009). *Extrasolar Planets and Astrobiology*. Sausalito, CA: University Science Books.

Schenk, P. M., *et al.* (2004). Ages and interiors: The cratering record of the Galilean satellites. In: *Jupiter: The Planet, Satellites, and Magnetosphere*, ed. F. Bagenal, *et al.*, New York: Cambridge University Press.

Schenk, P. M. and Zahnle, K. (2007). On the negligible surface age of Triton. *Icarus* **192**, 135–149.

Schidlowski, M. (1988). A 3,800-million-year isotopic record of life from carbon in sedimentary rocks. *Nature* **333**, 313–318.

Schidlowski, M., *et al.* (1976). Carbon isotope geochemistry of Precambrian Lomagundi carbonate province, Rhodesia. *Geochim. Cosmochim. Acta* **40**, 449–455.

Schidlowski, M., *et al.* (1983). Isotopic inferences of ancient biochemistries: carbon, sulfur, hydrogen, and nitrogen. In: *Earth's Earliest Biosphere: Its Origin and Evolution*, ed. J. W. Schopf, Princeton, New Jersey: Princeton University Press, pp. 149–186.

Schinder, P. J., *et al.* (2011). The structure of Titan's atmosphere from Cassini radio occultations. *Icarus* **215**, 460–474.

Schirrmeister, B. E., *et al.* (2013). Evolution of multicellularity coincided with increased diversification of cyanobacteria and the Great Oxidation Event. *P. Natl. Acad. Sci. USA* **110**, 1791–1796.

Schlichting, H. E., *et al.* (2015). Atmospheric mass loss during planet formation: The importance of planetesimal impacts. *Icarus* **247**, 81–94.

Schmidt, G. A., *et al.* (2010). Attribution of the present-day total greenhouse effect. *J. Geophys. Res.* **115**, D20106, doi:10.1029/2010JD014287.

Schmidt, P. W. and Williams, G. E. (1995). The Neoproterozoic climatic paradox - equatorial paleolatitude for marinoan glaciation near sea-level in south Australia. *Earth and Planetary Science Letters* **134**, 107–124.

Schmidt, P. W. and Williams, G. E. (2008). Palaeomagnetism of red beds from the Kimberley Group, Western Australia: Implications for the palaeogeography of the 1.8 Ga King Leopold glaciation. *Precambrian Res.* **167**, 267–280.

Schmidt, P. W., *et al.* (1991). Low paleolatitude of Late Proterozoic glaciation - early timing of remanence in hematite of the Elatina Formation, South-Australia. *Earth Planet. Sci. Lett.* **105**, 355–367.

Schmitt, B., *et al.* (1994). Identification of 3 absorption bands in the 2 micron spectrum of Io. *Icarus* **111**, 79–105.

Schneider, D. A., *et al.* (2002). Age of volcanic rocks and syndepositional iron formations, Marquette Range Supergroup: Implications for the tectonic setting of Paleoproterozoic, iron formations of the Lake Superior region. *Can. J. Earth Sci.* **39**, 999–1012.

Schneider, P. (2014). *Extragalactic Astronomy and Cosmology: An Introduction*. Berlin: Springer.

Schneider, T. (2006). The general circulation of the atmosphere. *Annu. Rev. Earth Planet. Sci.* **34**, 655–688.

Schneider, T. and Bordoni, S. (2008). Eddy-mediated regime transitions in the seasonal cycle of a Hadley circulation and implications for monsoon dynamics. *J. Atmos. Sci.* **65**, 915–934.

Schneider, T., *et al.* (2012). Polar methane accumulation and rainstorms on Titan from simulations of the methane cycle. *Nature* **481**, 58–61.

Schneider, T. and Liu, J. J. (2009). Formation of jets and equatorial superrotation on Jupiter. *J. Atmos. Sci.* **66**, 579–601.

Schoell, M. and Wellmer, F. W. (1981). Anomalous C-13 depletion in Early Precambrian graphites from Superior-Province, Canada. *Nature* **290**, 696–699.

Schofield, J. T. and Taylor, F. W. (1983). Measurements of the mean, solar-fixed temperature and cloud structure of the middle atmosphere of Venus. *Q. J. R. Meteor. Soc.* **109**, 57–80.

Scholz, A., *et al.* (2012). Substellar objects in nearby young clusters (SONYC) VI: The planetary-mass domain of NGC1333. *Astrophys. J.* **756**, 24.

Schonheit, P., *et al.* (1980). Growth: Parameters (Ks, Mu-Max, Ys) of methanobacterium-thermoautotrophicum. *Arch. Microbiol.* **127**, 59–65.

Schopf, J. W. (1993). Microfossils of the Early Archean Apex chert: new evidence for the antiquity of life. *Science* **260**, 640–646.

Schopf, J. W. (2000). Solution to Darwin's dilemma: Discovery of the missing Precambrian record of life. *P. Natl. Acad. Sci. USA* **97**, 6947–6953.

Schopf, J. W., *et al.* (1983). Evolution of Earth's earliest ecosystems: recent progress and unsolved problems. In: *Earth's Earliest Biosphere: Its Origin and Evolution*, ed. J. W. Schopf, Princeton, NJ: Princeton University Press, pp. 361–384.

Schopf, J. W. and Kudryavtsev, A. B. (2012). Biogenicity of Earth's earliest fossils: A resolution of the controversy. *Gondwana Res.* **22**, 761–771.

Schopf, J. W., *et al.* (2002). Laser-Raman imagery of Earth's earliest fossils. *Nature* **416**, 73–76.

Schopf, J. W. and Kudryavtsev, A. B. (2013). Reply to the comments of DL Pinti, R. Mineau and V. Clement, and of AO Marshall and CP Marshall on "Biogenicity of Earth's earliest fossils: A resolution of the controversy" by J. William Schopf and Anatoliy B. Kudryavtsev, Gondwana Research 22 (2012), 761–771. *Gondwana Res.* **23**, 1656–1658.

Schopf, J. W. and Packer, B. M. (1987). Early Archean (3.3 billion to 3.5 billion-year-old) microfossils from Warrawoona Group, Australia. *Science* **237**, 70–73.

Schrag, D. P., *et al.* (2002). On the initiationof a snowball Earth. *Geochem. Geophys. Geosystems* **3**.

Schrag, D. P., *et al.* (2013). Authigenic carbonate and the history of the global carbon cycle. *Science* **339**, 540–543.

Schrag, D. P. and Hoffman, P. F. (2001). Geophysics - Life, geology and snowball Earth. *Nature* **409**, 306.

Schrauzer, G. N. and Guth, T. D. (1977). Photolysis of water and photoreduction of nitrogen on titanium dioxide. *J. Am. Chem. Soc.* **99**, 7189–7193.

Schroder, S., *et al.* (2008). Rise in seawater sulphate concentration associated with the Paleoproterozoic positive carbon isotope excursion: evidence from sulphate evaporites in the similar to 2.2-2.1 Gyr shallow-marine Lucknow Formation, South Africa. *Terra Nova* **20**, 252.

Schubert, G. (1983). General circulation and the dynamical state of the Venus circulation. In: *Venus*, ed. D. M. Hunten, *et al.*, Tucson: University of Arizona Press, pp. 681–765.

Schubert, G., *et al.* (2010). Evolution of icy satellites. *Space Sci. Rev.* **153**, 447–484.

Schubert, G., *et al.* (2000). Geophysics: Timing of the Martian dynamo. *Nature* **408**, 666–667.

Schubert, G. and Soderlund, K. M. (2011). Planetary magnetic fields: Observations and models. *Phys. Earth Planet. In.* **187**, 92–108.

Schubert, G., *et al.* (2001). *Mantle Convection in the Earth and Planets*. Cambridge: Cambridge University Press.

Schunk, R. W. (1977). Mathematical structure of transport equations for multispecies flows. *Rev. Geophys.* **15**, 429–445.

Schunk, R. W. (1988). The polar wind. In: *Modeling Magnetospheric Plasma*, ed. T. E. Moore, J. H. Waite, Washington, D. C.: AGU, pp. 219–228.

Schunk, R. W. and Nagy, A. (2009). *Ionospheres: Physics, Plasma Physics, and Chemistry*. New York: Cambridge University Press.

Schwarzschild, M. (1958). *Structure and Evolution of the Stars*. Princeton, NJ: Princeton University Press.

Schwieterman, E. W., *et al.* (2015). Detecting and constraining N_2 abundances in planetary atmospheres using collisional pairs. *Astrophys. J.* **810**, 57.

Scott, A. C. (2000). The Pre-Quaternary history of fire. *Palaeogeogr. Palaeoclim. Palaeoecol.* **164**, 281–329.

Scott, C., *et al.* (2008). Tracing the stepwise oxygenation of the Proterozoic ocean. *Nature* **452**, 456–459.

Scott, C., *et al.* (2014). Pyrite multiple-sulfur isotope evidence for rapid expansion and contraction of the early Paleoproterozoic seawater sulfate reservoir. *Earth Planet. Sci. Lett.* **389**, 95–104.

Seager, S. (2010). *Exoplanet Atmospheres: Physical Processes*. Princeton, N.J.: Princeton University Press.

Seager, S. (2013). Exoplanet habitability. *Science* **340**, 577–581.

Seager, S. and Bains, W. (2015). The search for signs of life on exoplanets at the interface of chemistry and planetary science. *Science Advances* **1**, http://dx.doi.org/10.1126/sciadv.1500047.

Seager, S., *et al.* (2013a). A biomass-based model to estimate the plausibility of exoplanet biosignature gases. *Astrophys. J.* **775**, 104.

Seager, S., *et al.* (2013b). Biosignature gases in H_2-dominated atmospheres on rocky exoplanets. *Astrophys. J.* **777**, 95.

Seager, S., *et al.* (2015). *Exo-S: Starshade Probe-Class Exoplanet Direct Imaging Mission Concept Final Report.* Pasadena, California.

Seager, S. and Sasselov, D. D. (2000). Theoretical transmission spectra during extrasolar giant planet transits. *Astrophys. J.* **537**, 916–921.

Seager, S., *et al.* (2012). An astrophysical view of Earth-based metabolic biosignature gases. *Astrobiology* **12**, 61–82.

Seager, S., *et al.* (2005). Vegetation's red edge: a possible spectroscopic biosignature of extraterrestrial plants. *Astrobiology* **5**, 372–390.

Sefton-Nash, E. and Catling, D. C. (2008). Hematitic concretions at Meridiani Planum, Mars: Their growth timescale and possible relationship with iron sulfates. *Earth Planet. Sci. Lett.* **269**, 365–375.

Segatz, M., *et al.* (1988). Tidal dissipation, surface heat flow, and figure of viscoelastic models of Io. *Icarus* **75**, 187–206.

Segura, A., *et al.* (2005). Biosignatures from earth-like planets around M dwarfs. *Astrobiology* **5**, 706–725.

Segura, A., *et al.* (2003). Ozone concentrations and ultraviolet fluxes on Earth-like planets around other stars. *Astrobiology* **3**, 689–708.

Segura, A., *et al.* (2007). Abiotic production of O_2 and O_3 in high-CO_2 terrestrial atmospheres. *Astrobiology* **7**, 494–495.

Segura, T. L., *et al.* (2008). Modeling the environmental effects of moderate-sized impacts on Mars. *J. Geophys. Res.* **113**.

Segura, T. L., *et al.* (2002). Environmental effects of large impacts on Mars. *Science* **298**, 1977–1980.

Segura, T. L., *et al.* (2013). The effects of impacts on the climates of terrestrial planets. In: *Comparative Climatology of Terrestrial Planets*, ed. S. J. Mackwell, *et al.*, Tucson, AZ: University of Arizona Press.

Seiff, A. and Kirk, D. B. (1977). Structure of the atmosphere of Mars in summer at mid-latitudes. *J. Geophys. Res.* **82**, 4364–4378.

Seiff, A., *et al.* (1980). Measurements of thermal structure and thermal constrasts in the atmosphere of Venus and related dynamical observations: Results from the four Pioneer Venus probes. *J. Geophys. Res.* **85**, 7903–7933.

Seinfeld, J. H. and Pandis, S. N. (1998). *Atmospheric Chemistry and Physics.* New York: Wiley.

Seinfeld, J. H. and Pandis, S. N. (2006). *Atmospheric Chemistry and Physics: From Air Pollution to Climate Change.* Hoboken, N.J.: Wiley.

Sekine, Y., *et al.* (2011a). Replacement and late formation of atmospheric N_2 on undifferentiated Titan by impacts. *Nat. Geosci.* **4**, 359–362.

Sekine, Y., *et al.* (2005). The role of Fischer–Tropsch catalysis in the origin of methane-rich Titan. *Icarus* **178**, 154–164.

Sekine, Y., *et al.* (2011b). Osmium evidence for synchronicity between a rise in atmospheric oxygen and Palaeoproterozoic deglaciation. *Nat. Commun.* **2**.

Sekine, Y., *et al.* (2011c). Manganese enrichment in the Gowganda Formation of the Huronian Supergroup: A highly oxidizing shallow-marine environment after the last Huronian glaciation. *Earth Planet. Sci. Lett.* **307**, 201–210.

Sekiya, M., *et al.* (1981). Dissipation of the primordial terrestrial atmosphere due to irradiation of the solar far-UV during T-Tauri stage. *Prog. Theor. Phys.* **66**, 1301–1316.

Sekiya, M., *et al.* (1980a). Dissipation of the primordial terrestrial atmosphere due to irradiation of the solar EUV. *Prog. Theor. Phys.* **64**, 1968–1985.

Sekiya, M., *et al.* (1980b). Dissipation of the rare-gases contained in the primordial Earth's atmosphere. *Earth Planet. Sci. Lett.* **50**, 197–201.

Sellers, W. D. (1969). A climate model based on the energy balance of the earth-atmosphere system. *J. Appl. Meteor.* **8**, 392–400.

Selsis, F., *et al.* (2002). Signature of life on exoplanets: Can Darwin produce false positive detections? *Astron. Astrophys.* **388**, 985–1003.

Senft, L. E. and Stewart, S. T. (2007). Modeling impact cratering in layered surfaces. *J. Geophys. Res.* **112**.

Senft, L. E. and Stewart, S. T. (2008). Impact crater formation in icy layered terrains on Mars. *Meteorit. Planet. Sci.* **43**, 1993–2013.

Senft, L. E. and Stewart, S. T. (2011). Modeling the morphological diversity of impact craters on icy satellites. *Icarus* **214**, 67–81.

Sessions, A. L., *et al.* (2009). The continuing puzzle of the Great Oxidation Event. *Curr. Biol.* **19**, R567–R574.

Settle, M. (1979). Formation and deposition of volcanic sulfate aerosols on Mars. *J. Geophys. Res.* **84**, 8343–8354.

Setzmann, U. and Wagner, W. (1991). A new equation of state and tables of thermodynamic properties for methane covering the range from the meling line to 625 K at pressures up to 1000 MPa. *J. Phys. Chem. Ref. Data* **20**, 1061–1155.

Sforna, M. C., *et al.* (2014). Structural characterization by Raman hyperspectral mapping of organic carbon in the 3.46 billion-year-old Apex chert, Western Australia. *Geochim. Cosmochim. Acta* **124**, 18–33.

Shaheen, R., *et al.* (2015). Carbonate formation events in ALH 84001 trace the evolution of the Martian atmosphere. *P. Natl. Acad. Sci. USA* **112**, 336–341.

Shalygin, E. V., *et al.* (2015). Active volcanism on Venus in the Ganiki Chasma rift zone. *Geophys. Res. Lett.* **42**, 4762–4769.

Shanks, W. C. and Seyfried, W. E. (1987). Stable isotope studies of vent fluids and chimney minerals, Southern Juan-De-Fuca Ridge: Sodium metasomatism and seawater sulfate reduction. *J. Geophys. Res.* **92**, 11 387–11 399.

Shapiro, R. (1995). The prebiotic role of adenine: A critical analysis. *Origins Life Evol. Biosph.* **25**, 83–98.

Shapley, H. (1953). *Climatic Change: Evidence, Causes, and Effects*. Cambridge: Harvard University Press.

Sharma, A., *et al.* (2002). Microbial activity at gigapascal pressures. *Science* **295**, 1514–1516.

Sharp, C. M. and Burrows, A. (2007). Atomic and molecular opacities for brown dwarf and giant planet atmospheres. *Astrophys. J. Suppl. S.* **168**, 140–166.

Sharp, R. P. (1973). Mars: Fretted and chaotic terrain. *J. Geophys. Res.* **78**, 4073–4083.

Shea, M. A. and Smart, D. F. (2000). Fifty years of cosmic radiation data. *Space Sci. Rev.* **93**, 229–262.

Sheldon, N. D. (2006). Precambrian paleosols and atmospheric CO_2 levels. *Precambrian Res.* **147**, 148–155.

Sheldon, N. D. and Tabor, N. J. (2009). Quantitative paleoenvironmental and paleoclimatic reconstruction using paleosols. *Earth Sci. Rev.* **95**, 1–52.

Shematovich, V. I., *et al.* (2003). Nitrogen loss from Titan. *J. Geophys. Res. – Planets* **108**.

Shen, Y. and Buick, R. (2004). The antiquity of microbial sulfate reduction. *Earth Sci. Rev.* **64**, 243–272.

Shen, Y., *et al.* (2001). Isotopic evidence for microbial sulfate reduction in the early Archean era. *Nature* **410**, 77–81.

Shen, Y., *et al.* (2003). Evidence for low sulphate and anoxia in a mid-Proterozoic marine basin. *Nature* **423**, 632–635.

Shen, Y. N., *et al.* (2002). Middle proterozoic ocean chemistry: Evidence from the McArthur Basin, northern Australia. *Am. J. Sci.* **302**, 81–109.

Shen, Y. N., *et al.* (2008). On the coevolution of Ediacaran oceans and animals. *P. Natl. Acad. Sci. USA* **105**, 7376–7381.

Shetty, S. and Marcus, P. S. (2010). Changes in Jupiter's Great Red Spot (1979-2006) and Oval BA (2000-2006). *Icarus* **210**, 182–201.

Shibuya, T., *et al.* (2012). Depth variation of carbon and oxygen isotopes of calcites in Archean altered upper oceanic crust: Implications for the CO_2 flux from ocean to oceanic crust in the Archean. *Earth Planet. Sci. Lett.* **321**, 64–73.

Shields, A. L., *et al.* (2013). The effect of host star spectral energy distribution and ice-albedo feedback on the climate of extrasolar planets. *Astrobiology* **13**, 715–739.

Shields, G. and Veizer, J. (2002). Precambrian marine carbon isotope database: version 1.1. *Geol. Geochem. Geophys.* **3**, 10.1029/2001GC000266.

Shields, G. A. (2007). A normalized seawater strontium isotope curve: Possible implications for Neoproterozoic Cambrian weathering rates and further oxygenation of the Earth. *Earth* **2**, 35–42.

Shields-Zhou, G. and Och, L. (2011). The case for a Neoproterozoic Oxygenation Event: Geochemical evidence and biological consequences. *GSA Today* **21**, 4–11.

Shine, K. P., *et al.* (2012). The water vapour continuum: Brief history and recent cevelopments. *Surv. Geophys.* **33**, 535–555.

Shirey, S. B., *et al.* (2013). Diamonds and the geology of mantle carbon. *Rev. Mineral. Geochem.* **75**, 355–421.

Shizgal, B. and Blackmore, R. (1986). A collisional kinetic theory of a plane parallel evaporating planetary atmosphere. *Planet. Space Sci.* **34**, 279–291.

Shizgal, B. and Lindenfeld, M. J. (1980). Further studies of non-Maxwellian effects associated with the thermal escape of a planetary atmosphere. *Planet. Space Sci.* **28**, 159–163.

Shizgal, B. D. and Arkos, G. G. (1996). Nonthermal escape of the atmospheres of Venus, Earth and Mars. *Rev. Geophys.* **34**, 483–505.

Shklovskii, I. S. (1951). On the possibility of explaining the difference in chemical compostion of the Earth and Sun by thermal dissipation of light gases. *Astron. Zh.* **28**, 234–243.

Shklovskii, I. S. and Sagan, C. (1966). *Intelligent Life in the Universe*. San Francisco, CA: Holden-Day.

Shock, E. L. (1990). Geochemical constraints on the origin of organic compounds in hydrothermal systems. *Origins Life Evol. Biosph.* **20**, 331–367.

Showman, A. P., *et al.* (2010). Atmospheric circulation of extrasolar planets. In: *Exoplanets*, ed. S. Seager, Tucson: University of Arizona Press, pp. 471–516.

Showman, A. P., *et al.* (2013a). Doppler signatures of the atmospheric circulation on Hot Jupiters. *Astrophys. J.* **762**, doi:10.1088/0004-637X/762/1/24.

Showman, A. P. and Guillot, T. (2002). Atmospheric circulation and tides of "51 Pegasus b-like" planets. *Astron. Astrophys.* **385**, 166–180.

Showman, A. P. and Polvani, L. M. (2011). Equatorial superrotation on tidally locked exoplanets. *Astrophys. J.* **738**, 71, doi:10.1088/0004-637X/738/1/71.

Showman, A. P., *et al.* (2013b). Atmospheric circulation of terrestrial exoplanets. In: *Comparative Climatology of Terrestrial Planets*, ed. S. J. Mackwell, *et al.*, Tucson: University of Arizona Press, pp. 277–326.

Shu, F. H. (1991). *The Physics of Astrophysics: Volume II, Gas Dynamics*. Mill Valley, Calif.: University Science Books.

Shuvalov, V. (2009). Atmospheric erosion induced by oblique impacts. *Meteorit. Planet. Sci.* **44**, 1095–1105.

Shuvalov, V., *et al.* (2014). Impact induced erosion of hot and dense atmospheres. *Planet. Space Sci.* **98**, 120–127.

Sicardy, B., *et al.* (2006). Charon's size and an upper limit on its atmosphere from a stellar occultation. *Nature* **439**, 52–54.

Sicardy, B., *et al.* (2011). A Pluto-like radius and a high albedo for the dwarf planet Eris from an occultation. *Nature* **478**, 493–496.

Sicardy, B., *et al.* (2003). Large changes in Pluto's atmosphere as revealed by recent stellar occultations. *Nature* **424**, 168–170.

Sigman, D. M., *et al.* (2008). Nitrogen isotopes in the ocean. In: *Encyclopedia of Ocean Sciences* (2nd Edn), ed. J. H. Steele, *et al.*, San Diego: Academic Press, pp. 40–54.

Silburt, A., *et al.* (2015). A statistical reconstruction of the planet population around Kepler solar-type stars. *Astrophys. J.* **799**, 180.

Sillanpaa, I., *et al.* (2011). Cassini Plasma Spectrometer and hybrid model study on Titan's interaction: Effect of oxygen ions. *J. Geophys. Res.* **116**.

Silva-Tamayo, J. C., *et al.* (2010). Global perturbation of the marine Ca isotopic composition in the aftermath of the Marinoan global glaciation. *Precambrian Res.* **182**, 373–381.

Simoncini, E., *et al.* (2013). Quantifying drivers of chemical disequilibrium: theory and application to methane in the Earth's atmosphere. *Earth Syst. Dynam.* **4**, 317–331.

Simpson, G. C. (1927). Some studies in terrestrial radiation. *Mem. R. Met. Soc.* **11**, 69–95.

Sittler, E. C., *et al.* (2006). Energetic nitrogen ions within the inner magnetosphere of Saturn. *J. Geophys. Res.* **111**.

Slack, J. F. and Cannon, W. F. (2009). Extraterrestrial demise of banded iron formations 1.85 billion years ago. *Geology* **37**, 1011–1014.

Slack, J. F., *et al.* (2007). Suboxic deep seawater in the late Paleoproterozoic: Evidence from hematitic chert and iron formation related to seafloor-hydrothermal sulfide deposits, central Arizona, USA. *Earth Planet. Sci. Lett.* **255**, 243–256.

Sleep, N. H. (1994). Martian plate tectonics. *J. Geophys. Res.* **99**, 5639–5655.

Sleep, N. H. (2005). Dioxygen over geologic time. In: *Metal Ions in Biological Systems, Vol. 43, Biogeochemical Cycles of Elements*, ed. A. Sigel, *et al.*, Boca Raton, FL: Taylor & Francis, pp. 49–73.

Sleep, N. H. (2007). Plate tectonics through time. In: *Treatise on Geophysics*, Amsterdam: Elsevier, pp. 145–169.

Sleep, N. H. and Windley, B. F. (1982). Archean plate tectonics: Constraints and inferences. *J. Geol.* **90**, 363–379.

Sleep, N. H. and Zahnle, K. (1998). Refugia from asteroid impacts on early Mars and the early Earth. *J. Geophys. Res.* **103**, 28 529–28 544.

Sleep, N. H. and Zahnle, K. (2001). Carbon dioxide cycling and implications for climate on ancient Earth. *J. Geophys. Res.* **106**, 1373–1399.

Sleep, N. H., *et al.* (1989). Annihilation of ecosystems by large asteroid impacts on the early Earth. *Nature* **342**, 139–142.

Slinn, W. G. N., *et al.* (1978). Some aspects of the transfer of atmospheric trace constituents past the air–sea interface. *Atmos. Environ.* **12**, 2055–2087.

Smith, D. E., *et al.* (2009a). Time variations of Mars' gravitational field and seasonal changes in the masses of the polar ice caps. *J. Geophys. Res.* **114**.

Smith, G. R. and Hunten, D. M. (1990). Study of planetary atmospheres by absorptive occultations. *Rev. Geophys.* **28**, 117–143.

Smith, G. S. (2005). Human color vision and the unsaturated blue color of the daytime sky. *Am. J. Phys.* **73**, 590–597.

Smith, M. D., *et al.* (2001). Thermal Emission Spectrometer results: Mars atmospheric thermal structure and aerosol distribution. *J. Geophys. Res.* **106**, 23 929–23 945.

Smith, M. L., *et al.* (2014). The formation of sulfate, nitrate and perchlorate salts in the martian atmosphere. *Icarus* **231**, 51–64.

Smith, P. H., *et al.* (2009b). H_2O at the Phoenix Landing Site. *Science* **325**, 58–61.

Smrekar, S. E., *et al.* (2010). Recent hotspot volcanism on Venus from VIRTIS emissivity data. *Science* **328**, 605–608.

Smyth, W. H. and Marconi, M. L. (2006). Europa's atmosphere, gas tori, and magnetospheric implications. *Icarus* **181**, 510–526.

Snellen, I. A. G., *et al.* (2010). The orbital motion, absolute mass and high-altitude winds of exoplanet HD 209458b. *Nature* **465**, 1049–1051.

Snellen, I. A. G., *et al.* (2013). Finding extraterrestrial life using ground-based high-dispersion spectroscopy. *Astrophys. J.* **764**, 182.

Snyder, C. W. (1979). Planet Mars as seen at the end of the Viking Mission. *J. Geophys. Res.* **84**, 8487–8519.

Sobolev, V. V. (1975). *Light Scattering in Planetary Atmospheres*. Oxford: Pergamon.

Soderblom, J. M., *et al.* (2012). Modeling specular reflections from hydrocarbon lakes on Titan. *Icarus* **220**, 744–751.

Soderblom, L. A., *et al.* (1990). Tritons geyser-like plumes: Discovery and basic characterization. *Science* **250**, 410–415.

Sohl, L. E., *et al.* (1999). Paleomagnetic polarity reversals in Marinoan (ca. 600 Ma) glacial deposits of Australia: Implications for the duration of low-latitude glaciation in neoproterozoic time. *Geol. Soc. Am. Bull.* **111**, 1120–1139.

Solanki, S. K. and Krivova, N. A. (2003). Can solar variability explain global warming since 1970? *J. Geophys. Res.* **108**, 1200, doi: 10.1029/2002JA009753.

Solanki, S. K., *et al.* (2004). Unusual activity of the Sun during recent decades compared to the previous 11,000 years. *Nature* **431**, 1084–1087.

Solomon, S. (1999). Stratospheric ozone depletion: A review of concepts and history. *Rev. Geophys.* **37**, 275–316.

Som, S. M., *et al.* (2012). Air density 2.7 billion years ago limited to less than twice modern levels by fossil raindrop imprints. *Nature* **484**, 359–362.

Som, S. M., *et al.* (2016). Air pressure limited to less than half modern levels 2.7 billion years ago. *Nat. Geosci.* **9**, 448–451.

Spang, A., *et al.* (2015). Complex archaea that bridge the gap between prokaryotes and eukaryotes. *Nature* **521**, 173–179.

Sparks, W. B., *et al.* (2016). Probing for evidence of plumes on Europa with HST/STIS. *Astrophys. J.* **829**, 121.

Spencer, J. R. (1990). Nitrogen frost migration on Triton: A historical model. *Geophys. Res. Lett.* **17**, 1769–1772.

Spencer, J. R., *et al.* (2000). Discovery of gaseous S_2 in Io's Pele plume. *Science* **288**, 1208–1210.

Spencer, J. R. and Nimmo, F. (2013). Enceladus: An active ice world in the Saturn system. *Annu. Rev. Earth Planet. Sci.* **41**, 693–717.

Spencer, J. R., *et al.* (2006). Cassini encounters Enceladus: Background and the discovery of a south polar hot spot. *Science* **311**, 1401–1405.

Spergel, D., *et al.* (2015). Wide-Field InfrarRed Survey Telescope-Astrophysics Focused Telescope Assets WFIRST-AFTA 2015 Report, *arXiv preprint arXiv:1503.03757*.

Sperling, E. A., *et al.* (2013). Oxygen, ecology, and the Cambrian radiation of animals. *P. Natl. Acad. Sci. USA* **110**, 13 446–13 451.

Sperling, E. A., *et al.* (2014). Redox heterogeneity of subsurface waters in the Mesoproterozoic ocean. *Geobiology* **12**, 373–386, doi: 10.1111/gbi.12091.

Spitzer, L. J. (1952). In: *The Terrestrial Atmosphere Above 300 km*, ed. G. P. Kuiper, Chicago: University of Chicago Press, pp. 211–247.

Sprague, A. L., *et al.* (2004). Mars' south polar Ar enhancement: A tracer for south polar seasonal meridional mixing. *Science* **306**, 1364–1367.

Sprague, A. L., *et al.* (2007). Mars' atmospheric argon: Tracer for understanding Martian atmospheric circulation and dynamics. *J. Geophys. Res.* **112**, E03S02, doi:10.1029/2005je002597.

Squyres, S. W. and Knoll, A. H. (2005). Sedimentary rocks at Meridiani Planum: Origin, diagetiesis, and implications for life on Mars. *Earth Planet. Sci. Lett.* **240**, 1–10.

Squyres, S. W., *et al.* (2009). Exploration of Victoria Crater by the Mars Rover Opportunity. *Science* **324**, 1058–1061.

Squyres, S. W., *et al.* (2006). Two years at Meridiani Planum: Results from the Opportunity Rover. *Science* **313**, 1403–1407.

Stamper, C. C., *et al.* (2014). Oxidised phase relations of a primitive basalt from Grenada, Lesser Antilles. *Contrib. Mineral. Petr.* **167**, Art. 954.

Stanley, B. D., *et al.* (2011). CO_2 solubility in Martian basalts and Martian atmospheric evolution. *Geochim. Cosmochim. Acta* **75**, 5987–6003.

Stapelfeldt, K., *et al.* (2015). *Exo-C Imaging Nearby Worlds: Exoplanet Direct Imaging: coronograph probe mission study final report*. Pasadena, California.

Steele, A., *et al.* (2012). A reduced organic carbon component in martian basalts. *Science* **337**, 212–215.

Steemans, P., *et al.* (2009). Origin and radiation of the earliest vascular land plants. *Science* **324**, 353–353.

Stephan, K., *et al.* (2010). Specular reflection on Titan: Liquids in Kraken Mare. *Geophysical Research Letters* **37**.

Stephens, G. L. and Tjemkes, S. A. (1993). Water vapour and its role in the Earth's greenhouse. *Aust. J. Phys.* **46**, 149–166.

Stephens, S. K. (1995a). *Carbonate formation on Mars: Experiments and Models*. California Institute of Technology, PhD, Pasadena, CA.

string-name>Stephens, S. K. (1995b). Carbonates on Mars: Experimental results. *Proc. Lunar Planet. Sci. Conf. XXVI*, 1355–1356.

Stepinski, T. F., *et al.* (2004). Martian geomorphology from fractal analysis of drainage networks. *J. Geophys. Res.* **109**.

Stern, J. C., *et al.* (2015a). Evidence for indigenous nitrogen in sedimentary and aeolian deposits from the Curiosity rover investigations at Gale crater, Mars. *P. Natl. Acad. Sci. USA* **112**, 4245–4250.

Stem, R. J. (2007). When and how did plate tectonics begin? Theoretical and empirical considerations. *Chinese Sci. Bull.* **52**, 578–591.

Stern, S. A. (2008). On the atmospheres of objects in the Kuiper Belt. In: *The Solar System Beyond Neptune*, ed. M. A. Barucci, *et al.*, Tucson: University of Arizona Press, pp. 365–380.

Stern, S. A., *et al.* (2015b). The Pluto system: Initial results from its exploration by New Horizons. *Science* **350**, doi:10.1126/science.aad1815.

Stevens, B. and Bony, S. (2013). Water in the atmosphere. *Phys. Today* **66**, 29–34.

Stevenson, D. J. (1975). Thermodynamics and phase separation of dense fully ionized hydrogen–helium fluid mixtures. *Phys. Rev. B* **12**, 3999–4007.

Stevenson, D. J. (1983). The nature of the Earth prior to the oldest known rock record: the Hadean Earth. In: *Earth's Earliest Biosphere: Its Origin and Evolution*, ed. J. W. Schopf, Princeton, New Jersey: Princeton University Press, pp. 32–40.

Stevenson, D. J. (1987). Origin of the Moon – the collision hypothesis. *Ann. Rev. Earth Planet. Sci.* **15**, 271–315.

Stevenson, D. J. (1999). Life-sustaining planets in interstellar space? *Nature* **400**, 32–32.

Stevenson, D. J. (2001). Mars' core and magnetism. *Nature* **412**, 214–219.

Stevenson, D. J. and Salpeter, E. E. (1977a). Dynamics and helium distribution in hydrogen-helium fluid planets. *Astrophys. J. Suppl. S.* **35**, 239–261.

Stevenson, D. J. and Salpeter, E. E. (1977b). Phase diagram and transport properties for hydrogen–helium fluid planets. *Astrophys. J. Suppl. S.* **35**, 221–237.

Stevenson, K. B., *et al.* (2014). Thermal structure of an exoplanet atmosphere from phase-resolved emission spectroscopy. *Science* **346**, 838–841.

Stiles, B. W., *et al.* (2008). Determining Titan's spin state from Cassini RADAR images. *Astron. J.* **135**, 1669–1680.

Stillman, D. E., *et al.* (2014). New observations of martian southern mid-latitude recurring slope lineae (RSL) imply formation by freshwater subsurface flows. *Icarus* **233**, 328–341.

Stocker, T. F., *et al.* (2013). *IPCC, 2013: Climate Change: The Physical Science Basis*. New York: Cambridge University Press.

Stofan, E. R., *et al.* (2007). The lakes of Titan. *Nature* **445**, 61–64.

Stone, E. C., *et al.* (2013). Voyager 1 observes low-energy galactic cosmic rays in a region depleted of heliospheric ions. *Science* **341**, 150–153.

Stoney, G. J. (1898). Of atmospheres on planets and satellites. *Astrophys. J.* **7**, 25–55.

Stoney, G. J. (1900a). Note on inquiries as to the escape of gases from atmospheres. *Astrophys. J.* **12**, 201–207.

Stoney, G. J. (1900b). On the escape of gases from planetary atmospheres according to the kinetic theory. I. *Astrophys. J.* **11**, 251–258.

Stoney, G. J. (1900c). On the escape of gases from planetary atmospheres according to the kinetic theory. II. *Astrophys. J.* **11**, 357–372.

Stoney, G. J. (1904). Escape of gases from atmospheres. *Astrophys. J.* **20**, 69–78.

Strauss, H., *et al.* (2001). The sulfur isotopic composition of Neoproterozoic to early Cambrian seawater — evidence from the cyclic Hanseran evaporites, NW India. *Chem. Geol.* **175**, 17–28.

Stribling, R. and Miller, S. L. (1987). Energy yields for hydrogen cyanide and formaldehyde syntheses: the HCN and amino acid concentrations in the primitive ocean. *Origins Life Evol. Biosph.* **17**, 261–273.

Strobel, D. F. (2002). Aeronomic systems on planets, moons, and comets. In: *Atmospheres in the Solar System:*

Comparative Aeronomy ed. M. Mendillo, *et al.*, Washington, D. C.: AGU, pp. 7–22.

Strobel, D. F. (2008a). N_2 escape rates from Pluto's atmosphere. *Icarus* **193**, 612–619.

Strobel, D. F. (2008b). Titan's hydrodynamically escaping atmosphere. *Icarus* **193**, 588–594.

Strobel, D. F. (2009). Titan's hydrodynamically escaping atmosphere: Escape rates and the structure of the exobase region. *Icarus* **202**, 632–641.

Strobel, D. F. (2010). Molecular hydrogen in Titan's atmosphere: Implications of the measured tropospheric and thermospheric mole fractions. *Icarus* **208**, 878–886.

Strobel, D. F. (2012). Hydrogen and methane in Titan's atmosphere: chemistry, diffusion, escape, and the Hunten limiting flux principle. *Can. J. Phys.* **90**, 795–805.

Strobel, D. F., *et al.* (2009). Atmospheric structure and composition. In: *Titan from Cassini-Huygens*, ed. R. H. Brown, *et al.*, New York: Springer, pp. 235–257.

Strughold, H. (1953). *The Green and Red Planet*. Albuquerque: University of New Mexico Press.

Strughold, H. (1955). The ecosphere of the Sun. *Avia. Med.* **26**, 323–328.

Stubenrauch, C. J., *et al.* (2013). Assessment of global cloud datasets from satellites: Project and database initiated by the GEWEX radiation panel. *Bull. Am. Met. Soc.* **94**, 1031–1049.

Stüeken, E. E. (2016). Nitrogen in ancient mud: A biosignature? *Astrobiology* **16**, 730–735.

Stüeken, E. E., *et al.* (2015a). Selenium isotopes support free O_2 in the latest Archean. *Geology* **43**, 259–262.

Stüeken, E. E., *et al.* (2015b). The evolution of the global selenium cycle: Secular trends in Se isotopes and abundances. *Geochim. Cosmochim. Acta* **162**, 109–125.

Stüeken, E. E., *et al.* (2015c). Isotopic evidence for biological nitrogen fixation by molybdenum-nitrogenase from 3.2 Gyr. *Nature* **520**, 666–670.

Stüeken, E. E., *et al.* (2015d). Nitrogen isotope evidence for alkaline lakes on late Archean continents. *Earth Planet. Sc. Lett.* **411**, 1–10.

Stüeken, E. E., *et al.* (2012). Contributions to late Archaean sulphur cycling by life on land. *Nat. Geosci.* **5**, 722–725.

Su, H., *et al.* (2008). Variations of tropical upper tropospheric clouds with sea surface temperature and implications for radiative effects. *J. Geophys. Res.* **113**,.

Sumi, T., *et al.* (2011). Unbound or distant planetary mass population detected by gravitational microlensing. *Nature* **473**, 349–352.

Summons, R. E., *et al.* (2006). Steroids, triterpenoids and molecular oxygen. *Phil. Trans. R. Soc. B* **361**, 951–968.

Summons, R. E. and Brocks, J. J. (2004). Sedimentary hydrocarbons, biomarkers for early life. In: *Treatise on Geochemistry – Volume 8: Biogeochemistry*, ed. W. H. Schlesinger, Amsterdam: Elsevier, pp. 63–115.

Sun, S., *et al.* (2015). Primary hematitte in Neoarchean to Paleoproterozoic oceans. *GSA Bull.* **127**, 850–861.

Sutherland, J. D. (2016). The origin of life: Out of the blue. *Angew. Chem. Int. Edit.* **55**, 104–121.

Sutter, B., *et al.* (2012). The detection of carbonate in the martian soil at the Phoenix Landing site: A laboratory investigation and comparison with the Thermal and Evolved Gas Analyzer (TEGA) data. *Icarus* **218**.

Svensen, H. and Jamtveit, B. (2010). Metamorphic fluids and global environmental changes. *Elements* **6**, 179–182.

Svensmark, H. (2007). Cosmoclimatology: a new theory emerges. *Astron. Geophys.* **48**, 18–24.

Svetsov, V. V. (2007). Atmospheric erosion and replenishment induced by impacts of cosmic bodies upon the Earth and Mars. *Solar Syst. Res.* **41**, 28–41.

Swanson-Hysell, N. L., *et al.* (2010). Cryogenian glaciation and the onset of carbon-isotope decoupling. *Science* **328**, 608–611.

Swindle, T. D., *et al.* (1986). Xenon and other noble gases in Shergottites. *Geochim. Cosmochim. Acta* **50**, 1001–1015.

Symonds, R. B., *et al.* (1994). Volcanic-gas studies: Methods, results, and applications. In: *Volatiles in Magmas*, ed. M. R. Carroll and J. R. Holloway: Mineralogical Society of America.

Szczepanieccieciak, E., *et al.* (1978). Estimation of solubility of solidified substances in liquid methane by Preston-Prausnitz Method. *Cryogenics* **18**, 591–600.

Szymanski, A., *et al.* (2010). High oxidation state during formation of Martian nakhlites. *Meteorit. Planet. Sci.* **45**, 21–31.

Tajika, E. (1998). Mantle degassing of major and minor volatile elements during the Earth's history. *Geophys. Res. Lett.* **25**, 3991–3994.

Takahashi, H., *et al.* (1963). *In: Comparative Biogeochemistry,* ed. M. Florkin and H. S. Mason, New York: Academic Press, pp. 91.

Takai, K., *et al.* (2008). Cell proliferation at 122 degrees C and isotopically heavy CH_4 production by a hyperthermophilic methanogen under high-pressure cultivation. *P. Natl. Acad. Sci. USA* **105**, 10 949–10 954.

Tam, S. W. Y., *et al.* (2007). Kinetic modeling of the polar wind. *J. Atmos. Sol-Terr. Phys.* **69**, 1984–2027.

Tanaka, K. L. (1986). The stratigraphy of Mars. *J. Geophys. Res.* **91**, E139–E158.

Tanaka, K. L., *et al.* (2014). The digital global geologic map of Mars: Chronostratigraphic ages, topographic and crater morphologic characteristics, and updated resurfacing history. *Planet. Space Sci.* **95**, 11–24.

Tappert, R., *et al.* (2013). Stable carbon isotopes of C3 plant resins and ambers record changes in atmospheric oxygen since the Triassic. *Geochim. Cosmochim. Acta* **121**, 240–262.

Taran, V. Y. A. and Giggenbach, W. F. (2003). Geochemistry of light hydrocarbons in subduction-related volcanic and hydrothermal fluids. In: *Special Publication 10: Volcanic, Geothermal, and Ore-Forming Fluids: Rulers and Witnesses of Processes within the Earth*, ed. S. F. Simmons and I. Graham, Littleton, CO: Soc. of Econ. Geol., pp. 61–74.

Taylor, F. and Grinspoon, D. (2009). Climate evolution of Venus. *J. Geophys. Res.* **114**, E00B40, doi: 10.1029/2008JE003316.

Taylor, F. W. (2010). *Planetary Atmospheres*. Oxford; New York: Oxford University Press.

Taylor, F. W., *et al.* (1997). Near-infrared sounding of the lower atmosphere of Venus. In: *Venus II–Geology, Geophysics, Atmosphere, and Solar Wind Environment*, ed. S. W. Bougher, *et al.*, Tucson, AZ: University of Arizona Press.

Taylor, F. W., *et al.* (1983). The thermal balance of the middle and upper atmosphere of Venus. In: *Venus,* ed. D. M. Hunten, *et al.*, Tucson: University of Arizona Press, pp. 650–680.

Taylor, G. J. (2013). The bulk composition of Mars. *Chemie der Erde - Geochemistry* **73**, 401–420.

Taylor, J. R. (2005). *Classical Mechanics*. Sausalito, Calif.: University Science Books.

Taylor, S. R. (2001). *Solar System Evolution*. New York: Cambridge University Press.

Teanby, N. A., *et al.* (2008). Titan's winter polar vortex structure revealed by chemical tracers. *J. Geophys. Res.* **113**, E12003, doi:10.1029/2008JE003218.

Teanby, N. A., *et al.* (2012). Active upper-atmosphere chemistry and dynamics from polar circulation reversal on Titan. *Nature* **491**, 732–735.

Tellmann, S., *et al.* (2009). Structure of the Venus neutral atmosphere as observed by the Radio Science experiment VeRa on Venus Express. *J. Geophys. Res.* **114**.

Tenishev, V., *et al.* (2010). An approach to numerical simulation of the gas distribution in the atmosphere of Enceladus. *J. Geophys. Res.* **115**.

Tenishev, V., *et al.* (2014). Effect of the Tiger Stripes on the water vapor distribution in Enceladus' exosphere. *J. Geophys. Res.* **119**, 2658–2667.

Teolis, B. D., *et al.* (2010). Cassini finds an oxygen–carbon dioxide atmosphere at Saturn's icy moon Rhea. *Science* **330**, 1813–1815.

Tera, F., *et al.* (1974). Isotopic evidence for a terminal lunar cataclysm. *Earth Planet. Sci. Lett.* **22**, 1–21.

Thiemens, M. H. (2006). History and applications of mass-independent isotope effects. *Annu. Rev. Earth Pl. Sc.* **34**, 217–262.

Thiemens, M. H. and Heidenreich, J. E. (1983). The mass-independent fractionation of oxygen: A novel isotope effect and its possible cosmochemical implications. *Science* **219**, 1073–1075.

Thomas, C., *et al.* (2007). Clathrate hydrates as a sink of noble gases in Titan's atmosphere. *Astron. Astrophys.* **474**, L17–L20.

Thomas, G. E. and Stamnes, K. (1999). *Radiative Transfer in the Atmosphere and Ocean*. Cambridge; New York: Cambridge University Press.

Thomas, P. C., *et al.* (2009). Residual south polar cap of Mars: Stratigraphy, history, and implications of recent changes. *Icarus* **203**, 352–375.

Thomas, P. C., *et al.* (2016). Enceladus's measured physical libration requires a global subsurface ocean. *Icarus* **264**, 37–47.

Thomazo, C., *et al.* (2011). Extreme ^{15}N-enrichments in 2.72-Gyr-old sediments: evidence for a turning point in the nitrogen cycle. *Geobiology* **9**, 107–120.

Thomazo, C. and Papineau, D. (2013). Biogeochemical cycling of nitrogen on the early Earth. *Elements* **9**, 345–351.

Thomazo, C., *et al.* (2009). Biological activity and the Earth's surface evolution: Insights from carbon, sulfur, nitrogen and iron stable isotopes in the rock record. *Comptes Rendus Palevol* **8**, 665–678.

Thommes, E. W., *et al.* (2002). The formation of Uranus and Neptune among Jupiter and Saturn. *Astron. J.* **123**, 2862–2883.

Thompson, A. M. and Cicerone, R. J. (1986). Possible perturbations to atmospheric CO, CH_4, and OH. *J. Geophys. Res.* **91**, 853–864.

Thompson, R. O. R. Y. (1971). Why there is an intense eastward current in the North Atlantic but not in the South Atlantic. *J. Phys. Ocean.* **1**, 235–237.

Tian, F., *et al.* (2010). Photochemical and climate consequences of sulfur outgassing on early Mars. *Earth Planet. Sci. Lett.* **295**, 412–418.

Tian, F., *et al.* (2014). High stellar FUV/NUV ratio and oxygen contents in the atmospheres of potentially habitable planets. *Earth Planet. Sci. Lett.* **385**, 22–27.

Tian, F., *et al.* (2009). Thermal escape of carbon from the early Martian atmosphere. *Geophysical Research Letters* **36**.

Tian, F., *et al.* (2011). Revisiting HCN formation in Earth's early atmosphere. *Earth Planet. Sci. Lett.* **308**, 417–423.

Tian, F. and Toon, O. B. (2005). Hydrodynamic escape of nitrogen from Pluto. *Geophys. Res. Lett.* **32**.

Tian, F., *et al.* (2005). A hydrogen-rich early Earth atmosphere. *Science* **308**, 1014–1017.

Till, C. B., *et al.* (2012). The beginnings of hydrous mantle wedge melting. *Contrib. Mineral. Petr.* **163**, 669–688.

Tinetti, G., *et al.* (2006). Detectability of planetary characteristics in disk-averaged spectra II: Synthetic spectra and light-curves of earth. *Astrobiology* **6**, 881–900.

Tinsley, B. A. (1974). Hydrogen in the upper atmosphere. *Fund. Cosmic Phy.* **1**, 201–300.

Tirard, S., *et al.* (2010). The definition of life: A brief history of an elusive scientific endeavor. *Astrobiology* **10**, 1003–1009.

Titov, D. V., *et al.* (2007). Radiation in the atmosphere of Venus. In: *Exploring Venus as a Terrestrial Planet*, ed. L. W. Esposito, *et al.*, Washington, D.C.: AGU, pp. 121–138.

Titov, D. V., *et al.* (2013). Radiative energy balance in the Venus atmosphere. In: *Towards Understanding the Climate of Venus*, ed. L. Bengtsson, *et al.*, New York: Springer, pp. 23–53.

Tobie, G., *et al.* (2006). Episodic outgassing as the origin of atmospheric methane on Titan. *Nature* **440**, 61–64.

Tokano, T. (2009). The dynamics of Titan's troposphere. *Phil. Trans. R. Soc. Lond. A* **367**, 633–648.

Tokano, T. (2010). Relevance of fast westerlies at equinox for the eastward elongation of Titan's dunes. *Aeolian Res.* **2**, 113–127.

Tokano, T. (2013). Wind-induced equatorial bulge in Venus and Titan general circulation models: Implication for the simulation of superrotation. *Geophys. Res. Lett.* **40**, 4538–4543.

Tokar, R. L., *et al.* (2012). Detection of exospheric O_2^+ at Saturn's moon Dione. *Geophys. Res. Lett.* **39**.

Tolbert, N. E., *et al.* (1995). The oxygen and carbon-dioxide compensation points of C3 plants: Possible role in regulating atmospheric oxygen. *P. Natl. Acad. Sci. USA* **92**, 11 230–11 233.

Tolstikhin, I. N. and O'Nions, R. K. (1994). The Earths missing xenon: A combination of early degassing and of rare gas loss from the atmosphere. *Chem. Geol.* **115**, 1–6.

Tomasko, M. G., *et al.* (2005). Rain, winds and haze during the Huygens probe's descent to Titan's surface. *Nature* **438**, 765–778.

Tomasko, M. G., *et al.* (2008). A model of Titan's aerosols based on measurements made inside the atmosphere. *Planet. Space Sci.* **56**, 669–707.

Tomasko, M. G., *et al.* (2009). Limits on the size of aerosols from measurements of linear polarization in Titan's atmosphere. *Icarus* **204**, 271–283.

Toner, J. D., *et al.* (2014a). The formation of supercooled brines, viscous liquids, and low-temperature perchlorate

glasses in aqueous solutions relevant to Mars. *Icarus* **233**, 36–47.

Toner, J. D., *et al.* (2014b). Soluble salts at the Phoenix Lander site, Mars: A reanalysis of the Wet Chemistry Laboratory data. *Geochim. Cosmochim. Acta* **136**, 142–168.

Toner, J. D., *et al.* (2015). A revised Pitzer model for low-temperature soluble salt assemblages at the Phoenix site, *Mars. Geochim. Cosmochim. Acta* **166**, 327–343.

Tonks, W. B. and Melosh, H. J. (1993). Magma ocean formation due to giant impacts. *J. Geophys. Res.* **98**, 5319–5333.

Toon, O. B. and Farlow, N. H. (1981). Particles above the tropopause: Measurements and models of stratospheric aerosols, meteoric debris, nacreous clouds, and noctilucent clouds. *Annu. Rev. Earth Planet. Sci.* **9**, 19–58.

Toon, O. B., *et al.* (1989). Rapid calculation of radiative heating rates and photodissociation rates in inhomogeneous multiple scattering atmospheres. *J. Geophys. Res.* **94**, 16 287–16 301.

Toon, O. B., *et al.* (2010). The formation of Martian river valleys by impacts. *Annu. Rev. Earth. Pl. Sc.* **38**, 303–322.

Toon, O. B., *et al.* (1982). The ultraviolet absorber on Venus: Amorphous sulfur. *Icarus* **51**, 358–373.

Toro, E. F. (1999). *Riemann Solvers and Numerical Methods for Fluid Dynamics: A Practical Introduction.* Berlin: Springer-Verlag.

Tosca, N. J. and Knoll, A. H. (2009). Juvenile chemical sediments and the long term persistence of water at the surface of Mars. *Earth Planet. Sc. Lett.* **286**, 379–386.

Touma, J. and Wisdom, J. (1993). The chaotic obliquity of Mars. *Science* **259**, 1294–1297.

Touma, J. and Wisdom, J. (1998). Resonances in the early evolution of the Earth-Moon system. *Astron. J.* **115**, 1653–1663.

Towe, K. M. (1981). Environmental conditions surrounding the origin and early Archean evolution of life: a hypothesis. *Precambrian Res.* **16**, 1–10.

Trafton, L. (1980). Does Pluto have a substantial atmosphere. *Icarus* **44**, 53–61.

Trafton, L. M. (1966). Pressure-induced monochromatic translational absorption coefficients for homopolar and non-polar gases and gas mixtures with particular application to H_2. *Ap. J.* **146**, 558–571.

Trafton, L. M. (1998). Planetary Atmospheres: The role of collision-induced absorption. In: *Molecular Complexes in Earth's Planetary, Cometary and Interstellar Atmospheres*, ed. A. A. Vigasin and Z. Slanina, London: World Scientific, pp. 177–193.

Trafton, L. M., *et al.* (1997). Escape processes at Pluto and Charon. In: *Pluto and Charon*, ed. S. A. Stern and D. J. Tholen, Tucson: University of Arizona Press, pp. 475–522.

Trail, D., *et al.* (2007). Constraints on Hadean zircon protoliths from oxygen isotopes, Ti-thermometry, and rare earth elements. *Geochem. Geophys. Geosys.* **8**, Q06014.

Trail, D., *et al.* (2011). The oxidation state of Hadean magmas and implications for early Earth's atmosphere. *Nature* **480**, 79–83.

Trainer, M. G., *et al.* (2006). Organic haze on Titan and the early Earth. *P. Natl. Acad. Sci. USA* **103**, 18 035–18 042.

Trauger, J. T. and Lunine, J. I. (1983). Spectroscopy of molecular oxygen in the atmospheres of Venus and Mars. *Icarus* **55**, 272–281.

Treiman, A. H. and Irving, A. J. (2008). Petrology of Martian meteorite Northwest Africa 998. *Meteorit. Planet. Sci.* **43**, 829–854.

Trenberth, K. E., *et al.* (2009). Earth's global energy budget. *Bull. Am. Met. Soc.* **90**, 311–323.

Trenberth, K. E. and Guillemot, C. J. (1994). The total mass of the atmosphere. *J. Geophys. Res.* **99**, 23 079–23 088.

Trenberth, K. E. and Stepaniak, D. P. (2003). Seamless poleward atmospheric energy transports and implications for the Hadley circulation. *J. Climate* **16**, 3706–3722.

Trieloff, M., *et al.* (2003). The distribution of mantle and atmospheric argon in oceanic basalt glasses. *Geochim. Cosmochim. Acta* **67**, 1229–1245.

Trindade, R. I. F., *et al.* (2003). Low-latitude and multiple geomagnetic reversals in the Neoproterozoic Puga cap carbonate, Amazon craton. *Terra Nova* **15**, 441–446.

Tsikos, H. and Moore, J. M. (1997). Petrography and geochemistry of the Paleoproterozoic Hotazel iron-formation, Kalahari manganese field, South Africa: Implications for Precambrian manganese metallogenesis. *Econ. Geol.* **92**, 87–97.

Tucker, O. J., *et al.* (2012). Thermally driven escape from Pluto's atmosphere: A combined fluid/kinetic model. *Icarus* **217**, 408–415.

Tucker, O. J. and Johnson, R. E. (2009). Thermally driven atmospheric escape: Monte Carlo simulations for Titan's atmosphere. *Planet. Space Sci.* **57**, 1889–1894.

Tucker, O. J., *et al.* (2013). Diffusion and thermal escape of H_2 from Titan's atmosphere: Monte Carlo simulations. *Icarus* **222**, 149–158.

Turcotte, D. L. (1993). An episodic hypothesis for Venusian tectonics. *J. Geophys. Res.* **98**, 17 061–17 068.

Turner, E. C. and Bekker, A. (2015). Thick sulfate evaporite accumulations marking a mid-Neoproterozoic oxygenation event (Ten Stone Formation, Northwest Territories, Canada). *GSA Bull.*, doi: 10.1130/B31268.1.

Turner, G. (1989). The outgassing history of the Earth's atmosphere. *J. Geol. Soc. London* **146**, 147–154.

Turner, J., *et al.* (2009). Record low surface air temperature at Vostok station, *Antarctica. J. Geophys. Res.* **114**, 1–14.

Turtle, E. P., *et al.* (2011a). Rapid and extensive surface changes near Titan's equator: Evidence of April showers. *Science* **331**, 1414–1417.

Turtle, E. P., *et al.* (2011b). Shoreline retreat at Titan's Ontario Lacus and Arrakis Planitia from Cassini Imaging Science Subsystem observations. *Icarus* **212**, 957–959.

Turtle, E. P., *et al.* (2009). Cassini imaging of Titan's high-latitude lakes, clouds, and south-polar surface changes. *Geophys. Res. Lett.* **36**, doi: 10.1029/2008GL036186.

Tyburczy, J. A., *et al.* (1986). Shock-induced volatile loss from a carbonaceous chondrite – Implications for planetary accretion. *Earth Planet. Sci. Lett.* **80**, 201–207.

Tyler, S. A. and Barghoorn, E. S. (1954). Occurrence of structurally preserved plants in Pre-Cambrian rocks of the Canadian Shield. *Science* **119**, 606–608.

Tziperman, E., *et al.* (2011). Biologically induced initiation of Neoproterozoic snowball-Earth events. *P. Natl. Acad. Sci. USA* **108**, 15091–15096.

Ueno, Y., *et al.* (2009). Geological sulfur isotopes indicate elevated OCS in the Archean atmosphere, solving faint young sun paradox. *P. Natl. Acad. Sci. USA* **106**, 14 784–14 789.

Ueno, Y., *et al.* (2008). Quadruple sulfur isotope analysis of ca. 3.5 Ga Dresser Formation: New evidence for microbial sulfate reduction in the early Archean. *Geochim. Cosmochim. Acta* **72**, 5675–5691.

Ueno, Y., *et al.* (2006). Evidence from fluid inclusions for microbial methanogenesis in the early Archaean era. *Nature* **440**, 516–519.

Unwin, S. C., *et al.* (2008). Taking the measure of the universe: Precision astrometry with SIM PlanetQuest. *Publ. Astron. Soc. Pac.* **120**, 38–88.

Usui, T., *et al.* (2012). Origin of water and mantle-crust interactions on Mars inferred from hydrogen isotopes and volatile element abundances of olivine-hosted melt inclusions of primitive shergottites. *Earth Planet. Sci. Lett.* **357**, 119–129.

Utsunomiya, S., *et al.* (2003). Iron oxidation state of a 2.45-Byr-old paleosol developed on mafic volcanics. *Geochim. Cosmochim. Acta* **67**, 213–221.

Vakoch, D. A. and Dowd, M. F. (2015). *The Drake Equation: Estimating the Prevalence of Extraterrestrial Life Through the Ages*. Cambridge: Cambridge University Press.

Valeille, A., *et al.* (2010). Water loss and evolution of the upper atmosphere and exosphere over martian history. *Icarus* **206**, 28–39.

Valentine, D. L. (2002). Biogeochemistry and microbial ecology of methane oxidation in anoxic environments: a review. *Anton. Leeuw. Int. J. G.* **81**, 271–282.

Vallis, G. K. (2006). *Atmospheric and Oceanic Fluid Dynamics: Fundamentals and Large-Scale Circulation*. New York: Cambridge University Press.

van Berk, W. and Fu, Y. J. (2011). Reproducing hydrogeochemical conditions triggering the formation of carbonate and phyllosilicate alteration mineral assemblages on Mars (Nili Fossae region). *J. Geophys. Res.* **116**.

van Berk, W., *et al.* (2012). Reproducing early Martian atmospheric carbon dioxide partial pressure by modeling the formation of Mg–Fe–Ca carbonate identified in the Comanche rock outcrops on Mars. *J. Geophys. Res.* **117**.

Van Cappellen, P. and Ingall, E. D. (1996). Redox stabilization of the atmosphere and oceans by phosphorus-limited marine production. *Science* **271**, 493–496.

Van de Kamp, P. (1969). Parallax proper motion acceleration and orbital motion of Barnard's Star. *Astron. J.* **74**, 238–240.

Van de Kamp, P. (1975). Astrometric study of Barnard's star from plates taken with Sproul 61-cm refractor. *Astron. J.* **80**, 658–661.

van de Kamp, P. (1982). The planetary system of Barnard's star. *Vistas in Astronomy* **26**, 141–157.

van den Boorn, S. H. J. M., *et al.* (2007). Dual role of seawater and hydrothermal fluids in Early Archean chert formation: Evidence from silicon isotopes. *Geology* **35**, 939–942.

Van Kranendonk, M. J. (2006). Volcanic degassing, hydrothermal circulation and the flourishing of early life on Earth: A review of the evidence from c. 3490–3240 Ma rocks of the Pilbara Supergroup, Pilbara Craton, Western Australia. *Earth Sci. Rev.* **74**, 197–240.

Van Kranendonk, M. J. (2011). Morphology as an indicator of biogenicity for 2.5-3.2 Ga fossil stromatolites from the Pilbara Craton, Western Australia. In: *Advances in Stromatolite Geobiology*, ed. J. Reitner, *et al.*, Berlin: Springer-Verlag, pp. 537–554.

Van Trump, J. E. and Miller, S. L. (1973). Carbon monoxide on the primitive Earth. *Earth Planet. Sci. Lett.* **20**, 145–150.

Van Valen, L. (1971). The history and stability of atmospheric oxygen. *Science* **171**, 439–443.

Van Zeggeren, F. and Storey, S. H. (1970). *The Computation of Chemical Equilibria*. New York: Cambridge University Press.

van Zuilen, M. A., *et al.* (2002). Reassessing the evidence for the earliest traces of life. *Nature* **418**, 627–630.

Vasavada, A. R., *et al.* (2012). Lunar equatorial surface temperatures and regolith properties from the Diviner Lunar Radiometer Experiment. *J. Geophys. Res.* **117**, E00H18, doi:10.1029/2011JE003987.

Vasavada, A. R., *et al.* (1999). Near-surface temperatures on Mercury and the Moon and the stability of polar ice deposits. *Icarus* **141**, 179–193.

Vasavada, A. R. and Showman, A. P. (2005). Jovian atmospheric dynamics: an update after Galileo and Cassini. *Rep. Prog. Phys.* **68**, 1935–1996.

Veeder, G. J., *et al.* (1994). Io's heat flow from infrared radiometry: 1983–1993. *J. Geophys. Res.* **99**, 17 095–17 162.

Veizer, J. and Prokoph, A. (2015). Temperatures and oxygen isotopic composition of Phanerozoic oceans. *Earth Sci. Rev.* **146**, 92–104.

Vickery, A. M. and Melosh, H. J. (1990). Atmospheric erosion and impactor retention in large impacts, with application to mass extinctions. In: *Global Catastrophes in Earth History. Geol. Soc. Sp. Paper 247*, ed. V. L. Sharpton, P. D. Ward, Boulder, Colo.: Geol. Soc. Amer., pp. 289–300.

Vidal-Madjar, A., *et al.* (2003). An extended upper atmosphere around the extrasolar planet HD209458b. *Nature* **422**, 143–146.

Vidal-Madjar, A., *et al.* (2004). Detection of oxygen and carbon in the hydrodynamically escaping atmosphere of the extrasolar planet HD 209458b. *Astrophys. J.* **604**, L69–L72.

Vidal-Madjar, A., *et al.* (2008). Exoplanet HD 209458b (Osiris(1)): Evaporation strengthened. *Astrophys. J. Lett.* **676**, L57–L60.

Viehmann, S., *et al.* (2014). Decoupled Hf-Nd isotopes in Neoarchean seawater reveal weathering of emerged continents. *Geology* **42**, 115–118.

Villanueva, G. L., *et al.* (2015). Strong water isotopic anomalies in the martian atmosphere: Probing current and ancient reservoirs. *Science* **348**, 218–221.

Vincendon, M. (2015). Identification of Mars gully activity types associated with ice composition. *J. Geophys. Res.* **120**, doi:10.1002/2015JE004909.

Viola, D., *et al.* (2015). Expanded secondary craters in the Arcadia Planitia region, Mars: Evidence for tens of Myr-old shallow subsurface ice. *Icarus* **248**, 190–204.

Vogel, N., *et al.* (2011). Argon, krypton, and xenon in the bulk solar wind as collected by the Genesis mission. *Geochim. Cosmochim. Acta* **75**, 3057–3071.

Voigt, A. and Abbot, D. S. (2012). Sea-ice dynamics strongly promote Snowball Earth initiation and destabilize tropical sea-ice margins. *Climate of the Past* **8**, 2079–2092.

Voigt, A., *et al.* (2011). Initiation of a Marinoan Snowball Earth in a state-of-the-art atmosphere-ocean general circulation model. *Climate of the Past* **7**, 249–263.

Volkov, A. N. and Johnson, R. E. (2013). Thermal escape in the hydrodynamic regime: Reconsideration of Parker's isentropic theory based on results of kinetic simulations. *Astrophys. J.* **765**.

Volkov, A. N., *et al.* (2013). Expansion of monatomic and diatomic gases from a spherical source into a vacuum in a gravitational field. *Fluid Dynam.* **48**, 239–250.

Volkov, A. N., *et al.* (2011a). Thermally Driven Atmospheric Escape: Transition from Hydrodynamic to Jeans Escape. *Astrophys. J. Lett.* **729**.

Volkov, A. N., *et al.* (2011b). Kinetic simulations of thermal escape from a single component atmosphere. *Phys. Fluids* **23**, 066601.

Von Bloh, W., *et al.* (2003). Biogenic enhancement of weathering and the stability of the ecosphere. *Geomicrobiol. J.* **20**, 501–511.

Von Damm, K. L. (1990). Seafloor hydrothermal activity: black smoker chemistry and chimneys. *Ann. Rev. Earth Planet. Sci.* **18**, 173–204.

Von Damm, K. L. (1995). Controls on the chemistry and temporal variability of seafloor hydrothermal fluids. In: *Seafloor Hydrothermal Systems: Physical, Chemical, Biological, and Geological Interactions*, ed. S. E. Humphris, *et al.*, Washington DC: American Geophysical Union, pp. 222–247.

Vuitton, V., *et al.* (2008). Formation and distribution of benzene on Titan. *J. Geophys. Res.* **113**, E05007, doi:10.1029/2007JE002997.

Wacey, D. (2009). *Early Life on Earth: A Practical Guide*. New York: Springer.

Wacey, D., *et al.* (2011). Microfossils of sulphur-metabolizing cells in 3.4-billion-year-old rocks of Western Australia. *Nat. Geosci.* **4**, 698–702.

Wacey, D., *et al.* (2012). Taphonomy of very ancient microfossils from the similar to 3400 Ma Strelley Pool Formation and similar to 1900 Ma Gunflint Formation: New insights using a focused ion beam. *Precambrian Res.* **220**, 234–250.

Wachtershauser, G. (1988a). Before enzymes and templates: Theory of surface metabolism. *Microbiol. Rev.* **52**, 452–484.

Wachtershauser, G. (1988b). Pyrite formation, the first energy source for life - a hypothesis. *Syst. Appl. Microbiol.* **10**, 207–210.

Wachtershauser, G. (1990). Evolution of the first metabolic cycles. *P. Natl. Acad. Sci. USA* **87**, 200–204.

Wachtershauser, G. (1992). Groundworks for an evolutionary biochemistry - the iron sulfur world. *Prog. Biophys. Mol. Biol.* **58**, 85–201.

Wade, J. and Wood, B. J. (2005). Core formation and the oxidation state of the Earth. *Earth Planet. Sc. Lett.* **236**, 78–95.

Wadhwa, M. (2001). Redox state of Mars' upper mantle and crust from Eu anomalies in Shergottite pyroxenes. *Science* **291**, 1527–1530.

Wadhwa, M. (2008). Redox conditions on small bodies, the Moon and Mars. *Rev. Mineral. Geochem.* **68**, 493–510.

Waite, J. H., *et al.* (2009a). Liquid water on Enceladus from observations of ammonia and ^{40}Ar in the plume. *Nature* **460**, 487–490.

Waite, J. H., *et al.* (2005). Ion Neutral Mass Spectrometer results from the first flyby of Titan. *Science* **308**, 982–986.

Waite, J. H., *et al.* (2007). The process of tholin formation in Titan's upper atmosphere. *Science* **316**, 870–875.

Waite, J. H., *et al.* (2009b). High-altitude production of Titan's aerosols. In: *Titan from Cassini-Huygens*, ed. R. H. Brown, *et al.*, New York: Springer, pp. 201–214.

Waldbauer, J. R., *et al.* (2009). Late Archean molecular fossils from the Transvaal Supergroup record the antiquity of microbial diversity and aerobiosis. *Precam. Res.* **169**, 28–47.

Walker, J. C. G. (1977). *Evolution of the Atmosphere.* New York: Macmillan.

Walker, J. C. G. (1978). Oxygen and hydrogen in the primitive atmosphere. *Pure Appl. Geophys.* **116**, 222–231.

Walker, J. C. G. (1980). The oxygen cycle. In: *The Natural Environment and the Biogeochemical Cycles*, ed. O. Hutzinger, Berlin: Springer-Verlag, pp. 87–104.

Walker, J. C. G. (1982). The earliest atmosphere of the Earth. *Precambrian Res.* **17**, 147–171.

Walker, J. C. G. (1984). Suboxic diagenesis in banded iron formations. *Nature* **309**, 340–342.

Walker, J. C. G. (1985). Carbon dioxide on the early Earth. *Orig. Life* **16**, 117–127.

Walker, J. C. G. (1986). Impact erosion of planetary atmospheres. *Icarus* **68**, 87–98.

Walker, J. C. G. (1987). Was the Archaean biosphere upside down? *Nature* **329**, 710–712.

Walker, J. C. G. (1990). Precambrian evolution of the climate system. *Palaeogeograph. Palaeoclimat. Palaeoecol.* **82**, 261–289.

Walker, J. C. G. and Brimblecombe, P. (1985). Iron and sulfur in the pre-biologic ocean. *Precambrian Res.* **28**, 205–222.

Walker, J. C. G., *et al.* (1981). A negative feedback mechanism for the long-term stabilization of Earth's surface temperature. *J. Geophys. Res.* **86**, 9776–9782.

Walker, J. C. G. and Kasting, J. F. (1992). Effects of fuel and forest conservation on predicted levels of atmospheric carbon dioxide. *Global Planet. Change* **97**, 151–189.

Walker, J. C. G., *et al.* (1983). Environmental evolution of the Archean-Early Proterozoic Earth. In: *Earth's Earliest Biosphere: Its Origin and Evolution*, ed. J. W. Schopf, Princeton, New Jersey: Princeton University Press, pp. 260–290.

Walker, J. C. G. and Lohmann, K. C. (1989). Why the oxygen isotopic composition of seawater changes with time. *Geophys. Res. Lett.* **16**, 323–326.

Wall, S., *et al.* (2010). Active shoreline of Ontario Lacus, Titan: A morphological study of the lake and its surroundings. *Geophys. Res. Lett.* **37**.

Wallace, J. M. and Hobbs, P. V. (2006). *Atmospheric Science: An Introductory Survey.* Burlington, MA: Elsevier Academic Press.

Wallace, M. W., *et al.* (2014). Enigmatic chambered structures in Cryogenian reefs: The oldest sponge-grade organisms? *Precambrian Res.* **255**, 109–123.

Wallmann, K. (2004). Impact of atmospheric CO_2 and galactic cosmic radiation on Phanerozoic climate change and the marine delta 18-O record. *Geol. Geochem. Geophys.* **5**.

Walsh, K. J., *et al.* (2011). A low mass for Mars from Jupiter's early gas-driven migration. *Nature* **475**, 206–209.

Walsh, M. M. (1992). Microfossils and possible microfossils from the Early Archean Onverwacht Group, Barberton Mountain Land, S. Africa. *Precambrian Res.* **52**, 271–293.

Walter, F. M. and Barry, D. C. (1991). Pre- and main-sequence evolution of solar activity. In: *The Sun in Time*, ed. C. P. Sonett, *et al.*, Tucson, AZ: University of Arizona Press, pp. 633–657.

Walter, M. (1995). Biogeochemistry – Fecal pellets in world events. *Nature* **376**, 16–17.

Wang, C. L., *et al.* (2014). Rare earth element and yttrium compositions of the Paleoproterozoic Yuanjiacun BIF in the Luliang area and their implications for the Great Oxidation Event (GOE). *Sci. China Earth Sci.* **57**, 2469–2485.

Wang, J. and Fischer, D. A. (2015). Revealing a universal planet-metallicity correlation for planets of different solar-type stars. *Astron. J.* **149**.

Wang, J. G., *et al.* (2012). Evolution from an anoxic to oxic deep ocean during the Ediacaran–Cambrian transition and implications for bioradiation. *Chem. Geol.* **306**, 129–138.

Wanke, H. (1981). Constitution of terrestrial planets. *Phil. Trans. R. Soc. Lond. A* **303**, 287–302.

Wänke, H. (2001). Geochemical evidence for a close genetic relationship of Earth and Moon. *Earth Moon and Planets* **85–6**, 445–452.

Wänke, H., *et al.* (2001). Chemical composition of rocks and soils at the Pathfinder site. *Spuce Sci. Rev.* **96**, 317–330.

Wänke, H. and Dreibus, G. (1994). Chemistry and accretion history of Mars. *Phil. Trans. R. Soc. Lond. A* **349**, 285–293.

Ward, P. D. B. D. (2000). *Rare Earth: Why Complex Life is Uncommon in the Universe*. New York: Copernicus.

Ward, W. R. (1974). Climatic variations on Mars. 1. Astronomical theory of insolation. *J. Geophys. Res.* **79**, 3375–3386.

Ward, W. R. (1992). Long-term orbital and spin dynamics of Mars. In: *Mars*, ed. H. H. Kieffer, *et al.*, Tucson: University of Arizona Press, pp. 298–320.

Warneck, P. (2000). *Chemistry of the Natural Atmosphere*. New York: Academic Press.

Warren, S. G. and Brandt, R. E. (2006). Comment on "Snowball earth: A thin-ice solution with flowing sea glaciers" by David Pollard and James F. *Kasting*. *J. Geophys. Res.* **111**, C09016, doi: 10.1029/2005JC003411.

Warren, S. G., *et al.* (2002). Snowball Earth: Ice thickness on the tropical ocean. *Journal Of Geophysical Research-Oceans* **107**.

Watanabe, Y., *et al.* (2009). Anomalous fractionations of sulfur isotopes during thermochemical sulfate reduction. *Science* **324**, 370–373.

Watson, A. J., *et al.* (1984). Temperatures in a runaway greenhouse on the evolving Venus: Implications for water loss. *Earth Planet. Sci. Lett.* **68**, 1–6.

Watson, A. J., *et al.* (1981). The dynamics of a rapidly escaping atmosphere: applications to the evolution of Earth and Venus. *Icarus* **48**, 150–166.

Wayne, R. P. (2000). *Chemistry of Atmospheres: An Introduction to the Chemistry of the Atmospheres of Earth, the Planets, and their Satellites*. Oxford: Clarendon Press.

Weaver, C. P. and Ramanathan, V. (1995). Deductions from a simple climate model: Factors governing surface temperature and atmospheric thermal structure. *J. Geophys. Res.* **100**, 11 585–11 591.

Weber, R. C., *et al.* (2011). Seismic detection of the lunar core. *Science* **331**, 309–312.

Webster, C. R., *et al.* (2015). Mars methane detection and variability at Gale crater. *Science* **347**, 415–417.

Webster, C. R., *et al.* (2013). Isotope ratios of H, C, and O in CO_2 and H_2O of the martian atmosphere. *Science* **341**, 260–263.

Wedepohl, K. H. (1995). The composition of the continental crust. *Geochim. Cosmochim. Acta* **59**, 1217–1232.

Weidenschilling, S. J. (1977). Aerodynamics of solid bodies in solar nebula. *Mon. Not. R. Astron. Soc.* **180**, 57–70.

Weisberg, M. K., *et al.* (2006). Systematics and evaluation of meteorite classification. In: *Meteorites and the Early Solar System II*, ed. D. S. Lauretta and H. Y. McSween, Tucson, AZ: University of Arizona Press, pp. 19–52.

Weiss, B. P., *et al.* (2008). Paleointensity of the ancient Martian magnetic field. *Geophys. Res. Lett.* **35**.

Weiss, J. W. (2004). Planetary Parameters. In: *Jupiter: The Planet, Satellites and Magnetosphere*, ed. F. Bagenal, *et al.*, New York: Cambridge University Press.

Weiss, L. M. and Marcy, G. W. (2014). The mass-radius relation for 65 exoplanets smaller than 4 Earth radii. *Astrophys. J. Lett.* **783**, L6.

Weiss, M. C., *et al.* (2016). The physiology and habitat of the last universal common ancestor. *Nature Microbiology* **1**, 16116 doi:10.1038/nmicrobiol.2016.116.

Weitz, C. M., *et al.* (2006). Soil grain analyses at Meridiani Planum, Mars. *J. Geophys. Res.* **111**, E12S04, doi:10.1029/2005JE002541.

Weitz, C. M., *et al.* (2015). Mixtures of clays and sulfates within deposits in western Melas Chasma, Mars. *Icarus* **251**, 291–314.

Weitz, C. M., *et al.* (2010). Mars Reconnaissance Orbiter observations of light-toned layered deposits and associated fluvial landforms on the plateaus adjacent to Valles Marineris. *Icarus* **205**, 73–102.

Welhan, J. A. (1988). Origins of methane in hydrothermal systems. *Chem. Geol.* **71**, 183–198.

Wellman, C. H., *et al.* (2003). Fragments of the earliest land plants. *Nature* **425**, 282–285.

Wen, H. J., *et al.* (2015). Reconstruction of early Cambrian ocean chemistry from Mo isotopes. *Geochim. Cosmochim. Acta* **164**, 1–16.

Werner, S. C. (2008). The early martian evolution – Constraints from basin formation ages. *Icarus* **195**, 45–60.

Werner, S. C. (2009). The global martian volcanic evolutionary history. *Icarus* **201**, 44–68.

Werner, S. C., *et al.* (2014). The source crater of martian shergottite meteorites. *Science* **343**, 1343–1346.

Werner, S. C. and Tanaka, K. L. (2011). Redefinition of the crater-density and absolute-age boundaries for the chronostratigraphic system of Mars. *Icarus* **215**, 603–607.

West, R. A., *et al.* (2004). Jovian clouds and haze. In: *Jupiter: The Planet, Satellites, and Magnetosphere*, ed. F. Bagenal, *et al.*, New York: Cambridge University Press, pp. 79–104.

West, R. A., *et al.* (2009). Clouds and aerosols in Saturn's atmosphere. In: *Saturn from Cassini-Huygens*, ed. M. Dougherty, *et al.*, New York: Springer, pp. 161–180.

West, R. A., *et al.* (1991). Clouds and aerosols in the Uranian atmosphere. In: *Uranus*, ed. J. T. Bergstrahl, *et al.*, Tucson: University of Arizona Press, pp. 296–324.

West, R. A., *et al.* (2016). Cassini imaging science subsystem observations of Titan's south polar cloud. *Icarus* **270**, 399–408.

Westall, F. (2005). Life on the early Earth: A sedimentary view. *Science* **308**, 366–367.

Westall, F., *et al.* (2001). Early Archean fossil bacteria and biofilms in hydrothermally-influenced sediments from the Barberton greenstone belt, South Africa. *Precambrian Res.* **106**, 93–116.

Whelley, P. L. and Greeley, R. (2008). The distribution of dust devil activity on Mars. *J. Geophys. Res.* **113**.

Whewell, W. (1853). *Of the Plurality of Worlds: An Essay*. London: John W. Parker and Son.

Whipple, F. L. (1972). On certain aerodynamic processes for asteroids and comets. In: *From Plasma to Planet; Proceedings*, ed. A. Elvius, New York: Wiley, pp. 211–232.

Whitehill, A. R., *et al.* (2015). SO_2 photolysis as a source for sulfur mass-independent isotope signatures in stratospheric aerosols. *Atmos. Chem. Phys.* **15**, 1843–1864.

Widdel, F., *et al.* (1993). Ferrous iron oxidation by anoxygenic phototrophic bacteria. *Nature* **362**, 834–836.

Wiens, R. C., *et al.* (1986). The case for a martian origin of the shergottites, 2. Trapped and Indigenous gas components in EETA 79001 glass. *Earth Planet. Sc. Lett.* **77**, 149–158.

Wiens, R. C. and Pepin, R. O. (1988). Laboratory shock emplacement of noble gases, nitrogen, and carbon dioxide into basalt, and implications for trapped gases in shergottite EETA-79001. *Geochim. Cosmochim. Acta* **52**, 295–307.

Wiktorowicz, S. J. and Ingersoll, A. P. (2007). Liquid water oceans in ice giants. *Icarus* **186**, 436–447.

Wilde, S. A., *et al.* (2001). Evidence from detrital zircons for the existence of continental crust and oceans on Earth 4.4 Gyr ago. *Nature* **409**, 175–178.

Wildman, R. A., *et al.* (2004). Burning of forest materials under late Paleozoic high atmospheric oxygen levels. *Geology* **32**, 457–460.

Wilhelms, D. E. and Squyres, S. W. (1984). The martian hemispheric dichotomy may be due to a giant Impact. *Nature* **309**, 138–140.

Wille, M., *et al.* (2007). Evidence for a gradual rise of oxygen between 2.6 and 2.5 Ga from Mo isotopes and Re-PGE signatures in shales. *Geochim. Cosmochim. Acta* **71**, 2417–2435.

Williams, D. A., *et al.* (2005). Erosion by flowing Martian lava: New insights for Hecates Tholus from Mars Express and MER data. *J. Geophys. Res.* **110**.

Williams, G. E. (1975). Late Precambrian glacial climate and the Earth's obliquity. *Geol. Mag.* **112**, 441–465.

Williams, G. E. (1981). Sunspot Periods in the Late Precambrian Glacial Climate and solar-planetary relations. *Nature* **291**, 624–628.

Williams, G. E. (1989). Late Precambrian tidal rhythmites in South Australia and the history of the Earth's rotation. *J. Geol. Soc.* **146**, 97–111.

Williams, G. E. (1997). Precambrian length of day and the validity of tidal rhythmite paleotidal values. *Geophys. Res. Lett* **24**, 421–424.

Williams, G. E. (2000). Geological constraints on the Precambrian history of earth's rotation and the moon's orbit. *Rev. Geophys.* **38**, 37–59.

Williams, G. E. (2005). Subglacial meltwater channels and glaciofluvial deposits in the Kimberley Basin, Western Australia: 1.8 Ga low-latitude glaciation coeval with continental assembly. *Journal of the Geological Society* **162**, 111–124.

Williams, G. P. (1979). Planetary circulations. 2. Jovian quasi-geostrophic regime. *J. Atmos. Sci.* **36**, 932–968.

Williams, G. P. (2003). Jovian dynamics. Part III: Multiple, migrating, and equatorial jets. *J. Atmos. Sci.* **60**, 1270–1296.

Williams, K. E., *et al.* (2009). Ancient melting of mid-latitude snowpacks on Mars as a water source for gullies. *Icarus* **200**, 418–425.

Williams, R. G. and Follows, M. (2011). *Ocean Dynamics and the Carbon Cycle: Principles and Mechanisms*. New York: Cambridge University Press.

Williams, R. M., *et al.* (2000). Flow rates and duration within Kasei Valles, Mars: Implications for the formation of a martian ocean. *Geophys. Res. Lett.* **27**, 1073–1076.

Wills, C. and Bada, J. (2000). *The Spark of Life : Darwin and the Primeval Soup*. Cambridge, Mass.: Perseus Pub.

Willson, L. A., *et al.* (1987). Mass loss on the main sequence. *Comments Astrophys.* **12**, 17–34.

Wilson, E. H. and Atreya, S. K. (2004). Current state of modeling the photochemistry of Titan's mutually dependent atmosphere and ionosphere. *J. Geophys. Res.* **109**, E06002, doi:10.1029/2003JE002181.

Wilson, E. H. and Atreya, S. K. (2009). Titan's Carbon Budget and the Case of the Missing Ethane. *J. Phys. Chem. A* **113**, 11 221–11 226.

Wilson, H. F. and Militzer, B. (2010). Sequestration of noble gases in giant planet interiors. *Phys. Rev. Lett.* **104**, 121101.

Wilson, R. J., *et al.* (2002). Traveling waves in the Northern Hemisphere of Mars. *Geophys. Res. Lett.* **29**, 1684, doi:10.1029/2002GL014866.

Winn, J. N. and Fabrycky, D. C. (2015). The occurrence and architecture of exoplanetary systems. *Annu. Rev. Astron. Astrophys.* **53**, 409–447.

Withers, P. and Catling, D. (2010). Observations of atmospheric tides on Mars at the season and latitude of the Phoenix atmospheric entry. *Geophys. Res. Lett* **37**, L24204.

Woese, C. R. (2005). Evolving biological organization. In: *Microbial Phylogeny and Evolution: Concepts and Controversies*, ed. J. Sapp, Oxford: Oxford University Press, pp. 99–117.

Wolery, T. J. and Sleep, N. H. (1976). Hydrothermal circulation and geochemical flux at midocean ridges. *J. Geol.* **84**, 249–275.

Wolf, E. T. and Toon, O. B. (2010). Fractal organic hazes provided an ultraviolet shield for early Earth. *Science* **328**, 1266–1268.

Wolf, E. T. and Toon, O. B. (2013). Hospitable Archean climates simulated by a General Circulation Model. *Astrobiology* **13**, 656–673.

Wolf, E. T. and Toon, O. B. (2014). Delayed onset of runaway and moist greenhouse climates for Earth. *Geophys. Res. Lett.* **41**, 167–172.

Wolf, E. T. and Toon, O. B. (2015). The evolution of habitable climates under the brightening Sun. *J. Geophys. Res.* **120**, 5775–5794.

Wones, D. R. and Gilbert, M. C. (1969). The fayalite–magnetite–quartz assemblage between 600 ° and 800 °C. *Amer. J. Sci.* **267-A**, 480–488.

Wood, B. E., *et al.* (2002). Measured mass loss rates of solar-like stars as a function of age and activity. *Ap. J.* **574**, 412–425.

Wood, B. E., *et al.* (2005). New mass-loss measurements from astrospheric Ly alpha absorption. *Astrophys. J.* **628**, L143–L146.

Wood, B. J. and Halliday, A. N. (2010). The lead isotopic age of the Earth can be explained by core formation alone. *Nature* **465**, 767–U4.

Wood, B. J., *et al.* (1996). Water and carbon in Earth's mantle. *Phil. Trans. R. Soc. Lond. A.* **354**, 1495–1511.

Wood, B. J. and Virgo, D. (1989). Upper mantle oxidation state: ferric iron contents of lherzolite spinels by 57Fe Mossbauer spectroscopy and resultant oxygen fugacities. *Geochim. Cosmochim. Acta* **53**, 1277–1291.

Wood, B. J., *et al.* (2006). Accretion of the Earth and segregation of its core. *Nature* **441**, 825–833.

Woodland, A. B. and Koch, M. (2003). Variation in oxygen fugacity with depth in the upper mantle beneath the Kaapvaal craton, *Southern Africa. Earth Planet. Sci. Lett.* **214**, 295–310.

Woolf, N. J., *et al.* (2002). The spectrum of earthshine: A pale blue dot observed from the ground. *Astrophys. J.* **574**, 430–433.

Wordsworth, R., *et al.* (2010). Infrared collision-induced and far-line absorption in dense CO_2 atmospheres. *Icarus* **210**, 992–997.

Wordsworth, R., *et al.* (2013). Global modelling of the early martian climate under a denser CO_2 atmosphere: Water cycle and ice evolution. *Icarus* **222**, 1–19.

Wordsworth, R. and Pierrehumbert, R. (2013). Hydrogen–nitrogen greenhouse warming in Earth's early atmosphere. *Science* **339**, 64–67.

Wordsworth, R. and Pierrehumbert, R. (2014). Abiotic oxygen-dominated atmospheres on terrestrial habitable zone planets. *Astrophys. J. Lett.* **785**, L20.

Wordsworth, R., *et al.* Transient reducing greenhouse warming on early Mars. *Geophys. Res. Lett.*; doi:10.1002/2016GL071766, submitted.

Wray, J. J., *et al.* (2009). Diverse aqueous environments on ancient Mars revealed in the southern highlands. *Geology* **37**, 1043–1046.

Wray, J. J., *et al.* (2016). Orbital evidence for more widespread carbonate-bearing rocks on Mars. *J. Geophys. Res.* **121**, 652–677.

Wright, J. T. and Gaudi, B. S. (2013). Exoplanet detection methods. In: *Planets, Stars and Stellar Systems*, ed. T. Oswalt, *et al.*: Springer Netherlands, pp. 489–540.

Wuebbles, D. J. and Hayhoe, K. (2002). Atmospheric methane and global change. *Earth Sci. Rev.* **57**, 177–210.

Xiao, S. (2014). Oxygen and early animal evolution. In: *Treatise on Geochemistry*, ed. H. D. Holland and K. K. Turekian, New York: Elsevier, pp. 231–250.

Xiao, S. H. and Laflamme, M. (2009). On the eve of animal radiation: phylogeny, ecology and evolution of the Ediacara biota. *Trends Ecol. Evol.* **24**, 31–40.

Yamamoto, M. and Takahashi, M. (2009). Dynamical effects of solar heating below the cloud layer in a Venus-like atmosphere. *J. Geophys. Res.* **114**.

Yan, F., *et al.* (2015). High-resolution transmission spectrum of the Earth's atmosphere-seeing Earth as an exoplanet using a lunar eclipse. *Int. J. Astrobiol.* **14**, 255–266.

Yang, J., *et al.* (2014a). Strong dependence of the inner edge of the habitable zone on planetary rotation rate. *Astrophys. J. Lett.* **787**, L2.

Yang, J., *et al.* (2013). Stabilizing cloud feedback dramatically expands the habitable zone of tidally locked planets. *Astrophys. J. Lett.* **771**, L45, doi:10.1088/2041-8205/771/2/L45.

Yang, J., *et al.* (2012a). The initiation of modern "Soft Snowball" and "Hard Snowball" climates in CCSM3. Part I: The influences of solar luminosity, CO_2 concentration, and the sea ice/snow albedo parameterization. *J. Climate* **25**, 2711–2736.

Yang, J., *et al.* (2012b). The initiation of modern "Soft Snowball" and "Hard Snowball" climates in CCSM3. Part II: Climate dynamic feedbacks. *J. Climate* **25**, 2737–2754.

Yang, W. and Holland, H. D. (2002). The redox sensitive trace elements Mo, U and Re in Precambrian carbonaceous shales: Indicators of the Great Oxidation Event (abstr.). *Geol. Soc. Am. Ann. Mtng* **34**, 382.

Yang, X., *et al.* (2014b). A relatively reduced Hadean continental crust and implications for the early atmosphere and crustal rheology. *Earth Planet. Sc. Lett.* **393**, 210–219.

Yang, X. Z., *et al.* (2014c). A relatively reduced Hadean continental crust and implications for the early atmosphere and crustal rheology. *Earth Planet. Sc. Lett.* **393**, 210–219.

Yelle, R. V., *et al.* (2008). Methane escape from Titan's atmosphere. *J. Geophys. Res.* **113**.

Yelle, R. V., *et al.* (2001). Structure of the Jovian stratosphere at the Galileo probe entry site. *Icarus* **152**, 331–346.

Yelle, R. V., *et al.* (1995). Lower atmospheric structure and surface-atmosphere interactions on Triton. In: *Neptune and Triton*, ed. D. P. Cruikshank, Tucson: University of Arizona Press, pp. 1031–1105.

Yen, A. S., *et al.* (2005). An integrated view of the chemistry and mineralogy of martian soils. *Nature* **436**, 49–54.

Yen, A. S., *et al.* (2008). Hydrothermal processes at Gusev Crater: An evaluation of Paso Robles class soils. *J. Geophys. Res.* **113**, E06S10, doi:10.1029/2007JE002978.

Yeo, G. M. (1981). The Late Proterozoic Rapitan glaciation in the northern Cordillera. In: *Proterozoic Basins of Canada*, ed. F. H. Campbell: Geological Survey of Canada Paper 81–10, pp. 25–46.

Yin, L. M., *et al.* (2007). Doushantuo embryos preserved inside diapause egg cysts. *Nature* **446**, 661–663.

Yoder, C. F. (1997). Venus spin dynamics. In: *Venus II*, ed. S. W. Bougher, *et al.*, Tucson: University of Arizona Press, pp. 1087–1124.

Young, G. M. (1991). *Stratigraphy, Sedimentology, and Tectonic Setting of the Huronian Supergroup*. Toronto: G. A. Canada, 34 pp.

Young, G. M. (2002). Stratigraphic and tectonic settings of Proterozoic glaciogenic rocks and Banded Iron-Formations: Relevance to the Snowball Earth debate. *J. African Earth Sci.* **35**, 451–466.

Young, G. M. and Gostin, V. A. (1989). An exceptionally thick Upper Proterozoic (Sturtian) glacial succession in the Mount Painter Area, South Australia. *Geol. Soc. Am. Bull.* **101**, 834–845.

Young, G. M. and Nesbitt, H. W. (1999). Paleoclimatology and provenance of the glaciogenic Gowganda Formation (Paleoproterozoic), Ontario, Canada: A chemostratigraphic approach. *Geol. Soc. Am. Bull.* **111**, 264–274.

Young, G. M., *et al.* (1998). Earth's oldest reported glaciation; physical and chemical evidence from the Archean Mozaan Group (~ 2.9 Ga) of South Africa. *J. Geol.* **106**, 523–538.

Young, L. A., *et al.* (1997). Detection of gaseous methane on Pluto. *Icarus* **127**, 258–262.

Young, L. D. G. (1971). Interpretation of high resolution spectra of Mars .2. Calculations of CO_2 abundance, rotational temperature and surface pressure. *J. Quant. Spectrosc. Ra. Trans.* **11**, 1075.

Young, R. A. and Crow, R. (2014). Paleogene Grand Canyon incompatible with Tertiary paleogeography and stratigraphy. *Geosphere* **10**, 664–679.

Yuan, X. L., *et al.* (2005). Lichen-like symbiosis 600 million years ago. *Science* **308**, 1017–1020.

Yung, Y. L., *et al.* (1984). Photochemistry of the atmosphere of Titan: comparison between model and observations. *Ap. J. Supp.* **55**, 465–506.

Yung, Y. L. and DeMore, W. B. (1982). Photochemistry of the stratosphere of Venus: Implications for atmospheric evolution. *Icarus* **51**, 199–247.

Yung, Y. L. and DeMore, W. B. (1999). *Photochemistry of Planetary Atmospheres*. New York: Oxford University Press.

Yung, Y. L. and McElroy, M. B. (1977). Stability of an oxygen atmosphere on Ganymede. *Icarus* **30**, 97–103.

Yung, Y. L., *et al.* (1989). Hydrogen and deuterium loss from the terrestrial atmosphere: A quantitative assessment of nonthermal escape fluxes. *J. Geophys. Res.* **94**, 14 971–14 989.

Yung, Y. L., *et al.* (1988). HDO in the martian atmosphere: Implications for the abundance of crustal water. *Icarus* **76**, 146–159.

Zachos, J. C., *et al.* (2008). An early Cenozoic perspective on greenhouse warming and carbon-cycle dynamics. *Nature* **451**, 279–283.

Zahnle, K. (1993a). Planetary noble gases. In: *Protostars and Protoplanets II*, ed. D. C. Black and M. S. Matthews, Tucson, AZ: University of Arizona Press, pp. 1305–1338.

Zahnle, K. (2015). Play it again, SAM. *Science* **347**, 370–371.

Zahnle, K., *et al.* (2007). Emergence of a habitable planet. *Space Sci. Rev.* **129**, 35–78.

Zahnle, K., *et al.* (2006). The loss of mass-independent fractionation in sulfur due to a Palaeoproterozoic collapse of atmospheric methane. *Geobiology* **4**, 271–283.

Zahnle, K., *et al.* (1998). Cratering rates on the Galilean satellites. *Icarus* **136**, 202–222.

Zahnle, K., *et al.* (2011). Is there methane on Mars? *Icarus* **212**, 493–503.

Zahnle, K., *et al.* (2008). Photochemical instability of the ancient Martian atmosphere. *J. Geophys. Res.* **113**, E11004.

Zahnle, K., *et al.* (1990). Mass fractionation of noble gases in diffusion-limited hydrodynamic hydrogen escape. *Icarus* **84**, 502–527.

Zahnle, K., *et al.* (1992). Impact-generated atmospheres over Titan, Ganymede, and Callisto. *Icarus* **95**, 1–23.

Zahnle, K., *et al.* (2010). Earth's earliest atmospheres. *CSH Perspect. Biol.* **2**, doi: 10.1101/cshperspect. a004895.

Zahnle, K. J. (1986). Photochemistry of methane and the formation of hydrocyanic acid (HCN) in the Earth's early atmosphere. *J. Geophys. Res.* **91**, 2819–2834.

Zahnle, K. J. (1993b). Xenological constraints on the impact erosion of the early Martian atmosphere. *J. Geophys. Res.* **98**, 10 899–10 913.

Zahnle, K. J. (1998). Origins of Atmospheres. In: *Origins*, ed. C. E. Woodward, *et al.*, San Francisco: Astron. Soc. Pacific, pp. 364–391.

Zahnle, K. J. (2000). Hydrodynamic escape of ionized xenon from ancient atmospheres. *Bull. Am. Astron. Soc.* **32**, 1044.

Zahnle, K. J., *et al.* (2013). The rise of oxygen and the hydrogen hourglass. *Chem. Geol.* **362**, 26–34.

Zahnle, K. J. and Kasting, J. F. (1986). Mass fractionation during transonic escape and implications for loss of water from Mars and Venus. *Icarus* **68**, 462–480.

Zahnle, K. J., *et al.* (1988). Evolution of a steam atmosphere during Earth's accretion. *Icarus* **74**, 62–97.

Zahnle, K. J. and Sleep, N. (1997). Impacts and the early evolution of life. In: *Comets and the Origin and Evolution of Life*, ed. P. J. Thomas, *et al.*, New York: Springer, pp. 175–208.

Zahnle, K. J. and Sleep, N. H. (2002). Carbon dioxide cycling through the mantle and its implications for the climate of the ancient Earth. *Geol. Soc. Lond. Sp. Publ.* **199**, 231–257.

Zahnle, K. J. and Walker, J. C. G. (1982). The evolution of solar ultraviolet luminosity. *Rev. Geophys. Space Phys.* **20**, 280–292.

Zalucha, A. M., *et al.* (2011). An analysis of Pluto occultation light curves using an atmospheric radiative-conductive model. *Icarus* **211**, 804–818.

Zasova, L. V., *et al.* (1981). Vertical distribution of SO_2 in upper cloud layer of Venus and origin of U.V.-absorption. *Adv. Space Res.* **1**, 13–16.

Zdunkowski, W., *et al.* (2007). *Radiation in the Atmosphere: A Course in Theoretical Meteorology.* Cambridge: Cambridge University Press.

Zebker, H., *et al.* (2014). Surface of Ligeia Mare, Titan, from Cassini altimeter and radiometer analysis. *Geophy. Res. Lett.* **41**, 308–313.

Zehr, J. P., *et al.* (1995). Diversity of heterotrophic nitrogen-fixation genes in a marine cyanobacterial mat. *Appl. Environ. Microbiol.* **61**, 2527–2532.

Zelinka, M. D. and Hartmann, D. L. (2011). The observed sensitivity of high clouds to mean surface temperature anomalies in the tropics. *J. Geophys. Res.* **116**.

Zellem, R. T., *et al.* (2014). The 4.5 μm full orbit phase curve of the hot Jupiter HD 209458b. *Astrophys. J.* **790**, doi:10.1088/0004-637x/790/1/53.

Zeng, X. (2010). What is the atmosphere's effect on Earth's surface temperature. *EOS Trans. AGU* **91**, 134–135.

Zent, A. P. and Fanale, F. P. (1986). Possible Mars brines: Equilibrium and kinetic considerations. *J. Geophys. Res.* **91**, D439–D445.

Zent, A. P. and McKay, C. P. (1994). The chemical reactivity of the Martian soil and implications for future missions. *Icarus* **108**, 146–157.

Zent, A. P. and Quinn, R. C. (1995). Simultaneous adsorption of CO_2 and H_2O under Mars-like conditions and application to the evolution of the martian climate. *J. Geophys. Res.* **100**, 5341–5349.

Zerkle, A. L., *et al.* (2012). A bistable organic-rich atmosphere on the Neoarchaean Earth. *Nat. Geosci.* **5**, 359–363.

Zhang, X., *et al.* (2012). Sulfur chemistry in the middle atmosphere of Venus. *Icarus* **217**, 714–739.

Zhang, X., *et al.* (2013). Radiative forcing of the stratosphere of Jupiter, Part I: Atmospheric cooling rates from Voyager to Cassini. *Planet. Space Sci.* **88**, 3–25.

Zhang, X. L., *et al.* (2014). Triggers for the Cambrian explosion: Hypotheses and problems. *Gondwana Res.* **25**, 896–909.

Zhu, S. and Chen, H. (1995). Megascopic multicellular organisms from the 1700-million-year-old Tuanshanzi Formation in the Jixian area, North China. *Science* **270**, 620–622.

Ziering, S. and Hu, P. N. (1967). Thermal escape from planetary atmospheres in presence of a gravitational field. *Astronaut. Acta* **13**, 327–340.

Ziering, S., *et al.* (1968). Thermal escape problem. 2. Transition domain in spherical geometry. *Phys. Fluids* **11**, 1327–1334.

Zimmer, C., *et al.* (2000). Subsurface oceans on Europa and Callisto: Constraints from Galileo magnetometer observations. *Icarus* **147**, 329–347.

Zinder, S. H. (1993). Physiological ecology of methanogens. In *Methanogenesis: Ecology, Physiology, Biochemistry and Genetics*, ed. J. G. Ferry, New York: Chapman and Hall, pp. 128–206.

Zmolek, P., *et al.* (1999). Large mass independent sulfur isotope fractionations during the photopolymerization of (CS2)-C-12 and (CS2)-C-13. *J. Phys. Chem. A* **103**, 2477–2480.

Zolotov, M. Y. and Shock, E. L. (2001). Composition and stability of salts on the surface of Europa and their oceanic origin. *J. Geophys. Res.* **106**, 32 815–32 827.

Zsom, A., *et al.* (2013). Toward the minimum inner edge distance of the habitable zone. *Astrophys. J.* **778**, 109, doi: 10.1088/0004-637X/778/2/109.

Zuber, M. T., *et al.* (2007). Density of Mars' south polar layered deposits. *Science* **317**, 1718–1719.

Zugger, M. E., *et al.* (2010). Light scattering from exoplanet oceans and atmospheres. *Astrophys. J.* **723**, 1168–1179.

Index